Tenth International Workshop DICE2022 - Spacetime - Matter - Quantum Mechanics

Journal of Physics: Conference Series Volume 2533

Castiglioncello, Italy
19 - 23 September 2022

ISBN: 978-1-7138-7532-1
ISSN: 1742-6588

Printed from e-media with permission by:

Curran Associates, Inc.
57 Morehouse Lane
Red Hook, NY 12571

Some format issues inherent in the e-media version may also appear in this print version.

Institute of Physics
Dirac House, Temple Back
Bristol BS1 6BE UK

Phone: 44 1 17 929 7481
Fax: 44 1 17 920 0979

techtracking@iop.org

Additional copies of this publication are available from:

Curran Associates, Inc.
57 Morehouse Lane
Red Hook, NY 12571 USA
Phone: 845-758-0400
Fax: 845-758-2633
Email: curran@proceedings.com
Web: www.proceedings.com

TABLE OF CONTENTS

DICE 2022 : Spacetime Matter Quantum Mechanics

Peer Review Statement

Re-Thinking Geometrogenesis: Instantaneity in Quantum Gravity Scenarios 1
Silvia De Bianchi, Luciano Gabbanelli

The Hole Argument and Beyond: Part I: The Story So Far 15
Henrique Gomes, Jeremy Butterfield

The Hole Argument and Beyond: Part II: Treating Non-Isomorphic Spacetimes 42
Henrique Gomes, Jeremy Butterfield

Towards a Field-Theory Based Relativistic Quantum Information 69
Charis Anastopoulos, Bei-Lok Hu, Konstantina Savvidou

The Case of Quantum Gravity with Spontaneous Collapse of the Wave Function 86
Lajos Diósi

Casting Rubik's Group into a Unitary Representation for Reinforcement Learning 91
Sebastiano Corli, Lorenzo Moro, Davide E Galli, Enrico Prati

Solid State Neuroscience: Spiking Neural Networks as Time Matter 99
Marcelo Rozenberg

Revisiting Quantum Contextuality in an Algebraic Framework. 112
Mathias Van Den Bossche, Philippe Grangier

Quantum Fluctuations, Particles and Entanglement: Solving the Quantum Measurement Problems 123
Kenichi Konishi

Non-Abelian Gauge Structure and Flavor Mixing in Quantum Field Theory1 146
Massimo Blasone, Luca Smaldone, Giuseppe Vitiello

Decoherent Histories Approach to the Cosmological Measure Problem 156
Seth Lloyd

Relics of the Quantum Spacetime: From Synge's World Function as the Fundamental Probe of Spacetime Architecture to the Emergent Description of Gravity 167
Dawood Kothawala

Gravitation, and Quantum Theory, as Emergent Phenomena 183
Tejinder P. Singh

Representation of the Universe as Dendrogramic Hologram Empowered with Relational Interpretation 202
Oded Shor, Felix Benninger, Andrei Khrennikov

How an Exact Discrete Symmetry Can Preserve Black Hole Information Or Turning a Black Hole Inside Out 215
Gerard't Hooft

Cellular Automaton for Spinor Gravity in Four Dimensions 225
C Wetterich

Anomalies, the Dark Universe and Matter-Antimatter Asymmetry 254
Nick E. Mavromatos

Quantum Clock Frames: Uncertainty Relations, non-Hermitian Dynamics and Nonlocality in Time 270
Eliahu Cohen

Quantum Physics and Biology: The Local Wavefunction Approach 283
George F R Ellis

Expending Quantization Procedures Can Help Quantum Field Theories and Gravity 307
John R Klauder

Shadows of New Physics on Dirac Materials, Analog GUPs and Other Amusements 312
Alfredo Iorio, Boris Ivetić, Salvatore Mignemi, Pablo Pais

Lessons and Complications from Gravitationally Induced Entanglement 322
André Großardt, M Kemal Döner

Can the Schrödinger Dynamics Explain Measurement? 330
Alexey A Kryukov

Chiral and Flavor Oscillations in Lepton-Antineutrino Spin Correlations 339
V A S V Bittencourt, M Blasone, G Zanfardino

Reparameterization Invariance in Accelerated Proton Decay 350
M Blasone, G Lambiase, G G Luciano, L Petruzziello

Aspects of Wave-Particle Complementarity in Quantum Delayed-Choice Experiments 359
P R Dieguez, H S Karthik

No-Go Results on Emergent Space and Other Structures 368
Ovidiu Cristinel Stoica

Poincaré Breaking and Gauge Invariance: A Road to Emergent Gravity and New Particles 378
Durmuş Demir

Beyond Semiclassical Time: Dynamics in Quantum Cosmology 388
Leonardo Chataignier

Classical Space from Quantum Condensates 398
A Iorio, L Smaldone

Analog Simulations of Early Universe Pre-Heating and the Back-Reaction Effect 408
Salvatore Butera, Iacopo Carusotto

Time-Loops to Spot Torsion on Bidimensional Dirac Materials with Dislocations 420
Alfredo Iorio, Pablo Pais

The Periastron Advance in Curvature Based Extended Gravity and Dark Energy 430
Antonio Capolupo, Gaetano Lambiase, Antonio Tedesco

What Quantum Matter Tells About Quantisation of Gravity in a Statistical Mechanics Context 443
Pierre A Mandrin

A Metric for Planck Stars Derived from Gravity in Asymptotic Safety 452
Fabio Scardigli, Gaetano Lambiase

Quantum-Gravitational Corrections to the Power Spectrum for a Closed Universe 462
Tatevik Vardanyan, Claus Kiefer

The Quantum Waves of Minkowski Space-Time and the Minimal Acceleration from Precanonical Quantum Gravity 469
Igor V Kanatchikov

Hints at a Relationship Between Friction and Relativistic Physics 479
E Minguzzi

Indefinite Causality in Quantum Mechanics and Its Thermodynamic Applications 493
Kyrylo Simonov

On a Reformulation of Bohmian Mechanics Including Open Dissipative Systems 501
Dieter Schuch, Moise Bonilla–Licea

Dynamical Atom-Wall Casimir-Polder Effect After a Sudden Change of the Atomic Position 511
Antonio Noto, Roberto Passante, Lucia Rizzuto, Salvatore Spagnolo

Field Observables Near a Fluctuating Boundary 521
Federico Armata, Salvatore Butera, Federico Montalbano, Roberto Passante, Lucia Rizzuto

Coherent States for Generalized Uncertainty Relations and Their Cosmological Implications 528
P Jizba, G Lambiase, G G Luciano, L Petruzziello

Axion Like Particles, Fifth Force and Neutron Interferometry 539
Antonio Capolupo, Salvatore Marco Giampaolo, Aniello Quaranta

The Universe as a Telecommunication Network 546
Alessandro Capurso

A QFT Based on Einstein-Hilbert Action and Hidden Dimensions as Consistent Theory of Measurement, Dark Energy, Dark Matter and McGaugh's Universal Law for Rotating Galaxies 557
G Doyen, D Drakova

Gravitation-Induced Temperature of Single Atoms on Atomic and Cosmological Scales 567
D Drakova, G Doyen

A Brief Tour of Gauge CPT 579
K Koltko

Complete Complementarity Relations for Three-Flavor Neutrino Oscillations 589
V A S V Bittencourt, M Blasone, S De Siena, C Matrella

Fermion Mixing in Curved Spacetime 599
Antonio Capolupo, Gaetano Lambiase, Aniello Quaranta

Author Index

Journal of Physics: Conference Series **2533** (2023) 011001 doi:10.1088/1742-6596/2533/1/011001

DICE 2022 : Spacetime Matter Quantum Mechanics

Presented in this volume are the Invited Lectures and the Contributed Papers of the *Tenth International Workshop on Decoherence, Information, Complexity and Entropy – DICE 2022 "Quantum riddles and spacetime oddities"* held at **Castello Pasquini, Castiglioncello (Tuscany), September 19-23, 2022**.

These proceedings document the lively exchange of ideas during the conference for the interested public and the scientific community alike. – For further details on all aspects of the meeting, cf. the website at `http://osiris.df.unipi.it/~elze/DICE2022.html`.

Our intention in this series of conferences has always been to unite leading researchers, advanced students, and renowned scholars from various areas, in order to stimulate new ideas and their exchange across the borders of specialization. The meetings in this series have successfully grown from DICE 2002,[1] followed by DICE 2004,[2] DICE 2006,[3] DICE 2008,[4] DICE 2010,[5] DICE 2012,[6] DICE 2014,[7] DICE 2016,[8] and DICE 2018.[9] About 80 participants from all over the world attended DICE 2022, with which the conference series resumed for the first time following the covid-19 pandemic.

It has been a great honour and inspiration having four eminent scholars presenting Keynote Lectures: Nicolas Gisin (Univ. Geneva), *"Indeterministic physics and intuitionistic*

[1] *Decoherence and Entropy in Complex Systems* ed H-T Elze *Lecture Notes in Physics* **633** (Berlin: Springer, 2004)

[2] *Proceedings of the Second International Workshop on Decoherence, Information, Complexity and Entropy – DICE 2004* ed H-T Elze *Braz. Journ. Phys.* **35**, 2A & 2B (2005) pp 205–529; free access at: **www.sbfisica.org.br/bjp**

[3] *Proceedings of the Third International Workshop on Decoherence, Information, Complexity and Entropy – DICE 2006* eds H-T Elze, L Diósi and G Vitiello *Journal of Physics: Conference Series* **67** (2007); free access at: **www.iop.org/EJ/toc/1742-6596/67/1**

[4] *Proceedings of the Fourth International Workshop on Decoherence, Information, Complexity and Entropy – DICE 2008* eds H-T Elze, L Diósi, L Fronzoni, J J Halliwell and G Vitiello *Journal of Physics: Conference Series* **174** (2009); free access at: **http://www.iop.org/EJ/toc/1742-6596/174/1**

[5] *Proceedings of the Fifth International Workshop on Decoherence, Information, Complexity and Entropy – DICE 2010* eds H-T Elze, L Diósi, L Fronzoni, J J Halliwell, E Prati, G Vitiello and J Yearsley *Journal of Physics: Conference Series* **306** (2011); free access at: **http://iopscience.iop.org/1742-6596/306/1**

[6] *Proceedings of the Sixth International Workshop on Decoherence, Information, Complexity and Entropy – DICE 2012* eds H-T Elze, L Diósi, L Fronzoni, J J Halliwell, E Prati, G Vitiello and J Yearsley *Journal of Physics: Conference Series* **442** (2013); free access at: **http://iopscience.iop.org/1742-6596/442/1**

[7] *Proceedings of the Seventh International Workshop on Decoherence, Information, Complexity and Entropy – DICE 2014* eds H-T Elze, L Diósi, L Fronzoni, J J Halliwell, C Kiefer, E Prati and G Vitiello *Journal of Physics: Conference Series* **626** (2015); free access at: **http://iopscience.iop.org/issue/1742-6596/626/1**

[8] *Proceedings of the Eighth International Workshop on Decoherence, Information, Complexity and Entropy – DICE 2016* eds H-T Elze, L Diósi, L Fronzoni, J J Halliwell, C Kiefer, E Prati and G Vitiello *Journal of Physics: Conference Series* **880** (2017); free access at: **http://iopscience.iop.org/issue/1742-6596/880/1**

[9] *Proceedings of the Ninth International Workshop on Decoherence, Information, Complexity and Entropy – DICE 2018* eds H-T Elze, M Blasone, L Diósi, L Fronzoni, J J Halliwell, C Kiefer, E Prati and G Vitiello *Journal of Physics: Conference Series* **1275** (2019); free access at: **http://iopscience.iop.org/issue/1742-6596/1275/1**

mathematics" – Seth Lloyd (MIT Boston), *"Spacetime as quantum computation"* – Gerard 't Hooft (Univ. Utrecht), *"How an exact discrete symmetry can preserve black hole information"* George Ellis (Univ. Cape Town), *"Quantum physics, emergence, and time"*.

Entitled *"Spacetime - Matter - Quantum Mechanics"*, the conference took place in the very pleasant atmosphere of Castello Pasquini – in beautiful surroundings, overlooking a piece of Tuscany's coast. The 5-day program addressed aspects of the following topics:

- Emergent spacetime, matter, and symmetries
- Quantum foundations
- Key experiments
- Quantum information and AI for physics
- Quantum gravity / cosmology and high-energy physics

A Roundtable Discussion dedicated to the public under the theme *"Intelligenza Artificiale, Big Data e l'Uomo?"* formed an integral part of the program. With the participation of L. Boi (Paris), S. Corli (Milano), A. Cossu (Pisa), and A.M. Iacono (Pisa) the event attracted a large audience involved in all-evening lively discussions.

The workshop has been organized by M Blasone (Salerno), L Diósi (Budapest), H-T Elze (Pisa, *Chair*), L Fratino (Paris), L Fronzoni (Pisa), J J Halliwell (London), C Kiefer (Cologne), E Prati (Milano) and G Vitiello (Salerno), with most essential help from our conference secretaries S Corli (Milano), R Molteni (Milano), and L Petruzziello (Salerno).

Several institutions and sponsors generously supported the workshop and their representatives as well as the citizens of Rosignano / Castiglioncello are warmly thanked for the kind hospitality:

- *Comune di Rosignano* – D Donati (Sindaco di Rosignano), together with the Giunta Comunale di Rosignano, and special thanks to vicesindaca L Montagnani.
- *Associazione Armunia* – Il Presidente e il Consiglio di Amministrazione / La direzione generale e la coordinazione artistica / Lo staff tecnico e organizzativo.
- *Solvay Italia SA, Rosignano* – A De Lorenzo (Site Communication & Total Quality Manager).

Funds made available by Dipartimento di Fisica (Università di Pisa), Dipartimento di Fisica (Universitá Degli Studi di Milano), Dipartimento di Fisica (Università di Salerno), Istituto Italiano per gli Studi Filosofici–IISF (Napoli), Institute of Physics Publishing–IOP (Bristol), Springer Verlag (Heidelberg), and journals *entropy, mathematics, and symmetry*–MDPI AG (Basel) are gratefully acknowledged.

We especially thank Laura Pesce (Vitrium Galleria – http://www.laurapesce.it) for showing her illuminated glass artworks *"Bright Lights"* during DICE 2022 at Castello Pasquini.

The proceedings have been edited by H-T Elze, M Blasone, L Diósi, L Fratino, J J Halliwell, C Kiefer, E Prati, and G Vitiello and the accepted papers follow essentially the order of presentation during the conference, separating, however, Invited Lectures and Contributed Papers (talks and posters). Invited Lectures will also be published separately as a book. – A complete list and final program is available at `http://osiris.df.unipi.it/~elze/DICE2022.html`.

In the name of all participants, we would like to thank A Ashton with collaborators at IOP Publishing (Bristol) for friendly advice and most valuable immediate help during the editing process and, especially, for their continuing efforts to make the Journal of Physics: Conference Series available to all.

Journal of Physics: Conference Series **2533** (2023) 011001 doi:10.1088/1742-6596/2533/1/011001

Pisa, Salerno, Budapest, Paris, London, Cologne, Milano — April 2023

Hans-Thomas Elze Massimo Blasone Lajos Diósi Lorenzo Fratino
Jonathan Halliwell Claus Kiefer Enrico Prati Giuseppe Vitiello

Journal of Physics: Conference Series **2533** (2023) 011002 doi:10.1088/1742-6596/2533/1/011002

Peer Review Statement

All papers published in this volume have been reviewed through processes administered by the Editors. Reviews were conducted by expert referees to the professional and scientific standards expected of a proceedings journal published by IOP Publishing.

- **Type of peer review:** Undefined
- **Conference submission management system:** Morressier
- **Number of submissions received:** 50
- **Number of submissions sent for review:** 50
- **Number of submissions accepted:** 50
- **Acceptance Rate (Submissions Accepted / Submissions Received × 100):** 100
- **Average number of reviews per paper:** 1.4
- **Total number of reviewers involved:** 6
- **Contact person for queries:**
  Name: Prof Hans-Thomas Elze
  Email: elze@df.unipi.it
  Affiliation: Dipartimento di Fisica, Universita di Pisa

Tenth International Workshop DICE2022 - Spacetime - Matter - Quantum Mechanics IOP Publishing
Journal of Physics: Conference Series **2533** (2023) 012001 doi:10.1088/1742-6596/2533/1/012001

Re-thinking geometrogenesis: Instantaneity in quantum gravity scenarios

Silvia De Bianchi[1,2] **and Luciano Gabbanelli**[2,3]

[1] Department of Philosophy, University of Milan, via Festa del Perdono 7, 20122 Milan, Italy
[2] Department of Philosophy, Universitat Autònoma de Barcelona, Campus UAB 08193 Bellaterra, Barcelona, Spain
[3] Institut de Física d'Altes Energies (IFAE), Faculty of Sciences, Campus UAB 08193 Bellaterra, Barcelona, Spain

E-mail: silvia.debianchi@unimi.it

Abstract.
Recent Quantum Gravity approaches revealed that spacetime emergence opens conceptual difficulties when the theory allows for cosmological scenarios compatible with geometrogenesis. In particular, it appears extremely difficult to think of an a-temporal transition from a non-geometric to a geometric phase and vice versa. In this paper we advance the proposal of a concept of atemporality, i.e., instantaneity that is suitable for the description of the transition occurring among fundamental phases from which spacetime emerges in some Quantum Gravity approaches, including Group Field Theory and its cosmological implications. After discussing the ontology at different levels of spacetime emergence in a theory of Quantum Gravity in Section 2, we shall focus on the definition of the notion of instantaneity to interpret the atemporal transition of geometrogenesis (Section 3.1), thereby arguing that atemporality dominates at Renormalization Group flow fixed points (Section 3.2). In Section 4, we apply for the first time our notion of instantaneity to the study of geometrogenesis in the context of tensorial Group Field Theory and we conclude by suggesting that atemporality plays a significant role for the understanding of our world at different scales.

1. Introduction: Spacetime emergence and the problem of atemporality

Throughout the long history of philosophy, we learned that the question "What is time?" is an insidious one, potentially leading to misunderstanding and confusion. An entire new field, the philosophy of time, flourished in the last decades in order to clarify the nature and the structure of the concept(s) of time, and different positions characterize the academic debate in an almost ubiquitous way.[1] However, even more problematic is to assess and develop the notions of temporal non-locality, atemporality and timelessness.[2] The difficulty, if not the non-sense,[3] of

[1] For an overview of the state of the art of the philosophy of time, see Bardon (2013) and Callender (2011). For studies relevant to the elaboration of the present study, see Meincke (2019) and Strobach (2013).

[2] In studies on quantum foundations, recent work by Adlam 2018; Adlam 2022 emphasizes the need to go beyond temporal locality and to explore a relational approach to temporal nonlocality within the Internal Quantum Reference Frame Programme. However, what we are proposing in the current paper is even more radical as shown in Sections 3 and 4.

[3] Among arguments in philosophy against atemporality, consider, for instance, Wittgenstein's take on atemporality: "If the world of data is timeless, how can we speak of it at all? The stream of life, or the

Journal of Physics: Conference Series **2533** (2023) 012001 doi:10.1088/1742-6596/2533/1/012001

talking about an atemporal world is one of the major challenges for current philosophers, but it also represents a priority for practitioners of some Quantum Gravity approaches, especially for those admitting the emergence or the disappearance of space and time, and in particular for those supporting approaches predicting models compatible with the geometrogenesis scenario. The latter is compatible with solutions letting time and/or space emerge above the Planck scale, and it has been associated to a phase transition in various approaches, including Quantum Graphity and Group Field Theory. The conceptual difficulty associated to geometrogenesis consists in the fact that it denotes a phase transition from a non-geometric to a geometric phase. In other words, geometrogenesis should be interpreted as an atemporal transition. Since we completely lack a philosophy of atemporality,[4] it seems thus that philosophers cannot provide any new conceptual ground to support research programs in Quantum Gravity admitting forms of atemporality or atemporal transitions. Indeed, current studies in the philosophy of physics attempts at accommodating what physicists identify as emergent spacetime, proto-time and other similar concepts, but a deep analysis of atemporality is still missing. Yet, we claim that one cannot fully grasp the relationship between temporality and atemporality without considering current debates on emergence and recent research in theoretical physics.[5]

Recent works by Huggett and Wüthrich (2018); Wüthrich (2019); Calamari (2021) advanced some proposals according to which we should start thinking of time as being emergent from atemporality in Loop Quantum Cosmology (LQC). Le Bihan (2018, 2019) also points to this question as a fundamental one to advance in the philosophical understanding of spacetime structure. However, these studies also suggest that it is almost impossible to avoid temporality when talking about phase transitions, such as geometrogenesis and do not offer a positive description of the special kind of a-chronicity needed in approaches, such as Quantum Graphity, Causal Dynamical Triangulations (CDT) or Tensorial Group Field Theory (TGFT) that allow solutions including geometrogenesis.[6]

In this paper, we want to explore in which sense and under which conditions one can talk about atemporality in terms of instantaneity. Second, we aim to sketch out the basis of our proposal for an ontology of phase transitions that can produce a shift in the naïve claim that an atemporal world is not epistemically accessible to us. Therefore, more than asking questions about time, what we are urged to do in the specific context of Quantum Gravity, is to address the relevant question: "What is atemporality?". Before characterizing atemporality as instantaneity, we reconstruct the salient aspects of the ontology that can be associated to temporality and atemporality in Quantum Gravity and we will do so by investigating a recent proposal by Oriti (2018) regarding levels of spacetime emergence. We shall briefly discuss them from both an ontological and epistemological standpoint. In Section 3 we shall introduce a suitable notion of

stream of the world, flows on, and our propositions are so to speak verified only at instants. Our propositions are verified only by the present". Wittgenstein, Philosophical Remarks V, sec. 48.

[4] The lack of a theory of atemporality might be observed not only among those philosophers reflecting upon the notion of spacetime emergence, e.g., Wüthrich (2019), but the impossibility of a theory of atemporality is also supported by those rejecting the idea of spacetime emergence, as in the case of Esfeld (2021).

[5] Indeed, when talking about 'emergence' there is an inevitable reference to the notion of passage or transition, i.e., something emerges from something else. This is precisely the sense of emergence that philosophers of physics see as problematic, Huggett and Wüthrich (2018), yet there is a lack of a standard definition of emergence that embraces different approaches to the appearance of spacetime. Butterfield and Isham (1999, 2001) provided a general and instrumental definition that is widely used by theoretical physicists, but not sufficient to grasp the complexity of the implications that current theories of quantum gravity generate for our conception of space, time, and atemporality. A worth-mentioning work on emergence in Quantum Gravity is Crowther (2018) exploring the inter-theory relations of Quantum Gravity, thereby offering a definition of emergence clearly differentiated from reduction.

[6] For geometrogenesis as possible implication of GFT see Oriti (2007), of quantum graphity see Konopka et al. (2006), of covariant LQG (with spin foam models) see Delcamp and Dittrich (2017) and for CDT see Mielczarek (2017); Mandrysz and Mielczarek (2019).

instantaneity (Section 3.1) and we will then apply it in the context of GFT condensate cosmology (Section 3.2). In Section 4 we investigate the atemporal transition from non-geometric phases and geometrogenesis in the specific context of Tensorial Group Field Theory (TGFT). We will conclude with future perspectives to test the notion of instantaneity, such as those involving asymptotic silence or Zeno regions in black holes, which in turn can have an impact on both philosophy of physics and philosophy of time.

2. Quantum gravity and levels of emergence

Oriti (2018) introduced a multi-layered scheme applicable to any Quantum Gravity approach describing four levels of emergence, each of them representing possible moves associated to the aim of recovering the continuum and classical notions of spacetime. In table 1, we sum up these four levels, each of which has been associated by us to notions of temporality/atemporality. We also show where phase transitions occur associated with atemporality. Before describing the levels of interest, a caveat is in order here. The four levels are not necessarily implied by each other. In other words, one can have a formalism that considers just Level 0 of emergence because it deals with continuous quantum fields as it happens in quantum general relativity, even if most of quantum gravity approaches, Oriti suggests, deal with all levels of emergence, but treat the mathematical entities pertaining to Levels 1, 2 and 3 as mere tools to reach the ontological level of interest, which is Level 0. This is a consistent generalization, but up to a certain extent. Indeed, when we are explicitly using approaches to QG implying phase transitions, as clearly happens at Level 3 of emergence, there is some physical meaning to be attributed to the implications of the transition from non-geometric to geometric phase, i.e., to geometrogenesis, and we are not just using mathematical tools to reach Level 0 of emergence. It is also worth mentioning that precisely this third level of emergence attracted our attention and triggered our attempt at characterizing the atemporality associated to this transition.

Table 1. Oriti's levels of emergence and our characterization of the objects of interest and notions of temporality/atemporality at each level.

Level	Object	Phase Transitions	Temporal or Atemporal Notions
0 (ground)	continuous fields	no	temporality
1	discrete entities	no	proto-time and/or proto-space
2	phases	yes	atemporality
3	geometrogenesis	yes	atemporality (instantaneity)

The lower level (Level 0 or ground level) encompasses more traditional ideas that attempt to present a theory of quantum gravity as some sort of a quantization of general relativity. According to Oriti, one can introduce new properties (or, equivalently, remove some features) for the fundamental entities that could provide the necessary characteristics to speak of a first level of emergence. At Level 1, we abandon the formalism of quantizing the (continuum) gravitational and matter fields and introduce new types of discrete degrees of freedom endowed with a different quantum nature. The Hilbert space is rather built with pre-geometric degrees of freedom, usually with combinatorial structures, labeled by algebraic data only, see Oriti (2014a). These structures are not merely in a quantum superposition, but their eigenstates do not allow a chronogeometric interpretation at all, see Wüthrich (2018). Therefore, the mathematical space of the theory is radically different from the one described in the previous level, being fundamentally discrete, non-spatiotemporal and lacking some important features of space, time and/or spacetime, Huggett and Wüthrich (2012).[7] Approaches encompassed under this characterization start with an underlying microscopic theory in which no straightforward reference to a spatiotemporal geometry is to be found, as for instance in Markopoulou (2009).

Level 1 of emergence implies the transition from discrete structures to the continuum, thereby raising the so-called problem of the continuum limit in discrete Quantum Gravity approaches. At this level, the emergence of spacetime is realized in a stronger sense, since it is achieved through collective properties of non-spatiotemporal entities. The novelty is that this emergence is addressed by moving along the growing numbers of fundamental building blocks and is achieved by exploring their collective properties. Therefore, some kind of multi-level ontology should be taken into account for both fundamental and collective entities/properties. For instance, in Oriti's view, it is possible that the fundamental entities, although not fully spatiotemporal, carry at least seeds of the emerging spatiotemporal notions, at least in some sectors of the same theory and under certain approximations. If concrete seeds can be identified, then a proto-spatiotemporal characterization of some of the properties of these seeds can be determined. These proto-spatiotemporal features may include notions like adjacency and succession. However, in our view, succession should not be treated on the same level of "ordering".[8] Thus, we prudently suggest that both adjacency and succession are nothing else but possible expressions or results of an internal ordering.

The further step constitutes Level 2 of emergence, according to which the same pre-geometric degrees of freedom that have been characterized at Level 1 are defined through their quantum dynamics and therefore postulated as meaningful in a physical sense. With the increasing number of them, collective effects start assuming relevance in providing continuum notions. However, similarly to any system composed by many interacting degrees of freedom, the continuum limit is generally not unique, leading to different macroscopic phases, each characterized by different effective dynamics with different emergent properties and macroscopic observables. In a sense, one can claim that the underlying microscopic quantum system combines itself into very different types of emerging macroscopic systems, and each of them is regarded as a disjoint phase.[9] This reinterpretation of fundamental constituents as interacting entities that can be collectively associated in phases, some of which are non-geometric, constitutes a novelty that affects ontology at a deeper level and introduces the need of defining atemporality. The fact that these phases are now understood as collective phenomena naturally makes the spatiotemporal or geometric attributes of the quanta completely meaningless. At this level, spacetime is understood as an approximate notion of the collective entities erasing any microscopic detail for the dynamics of the macroscopic phase. Therefore, the emergence resulting from these non-spatiotemporal degrees of freedom needs to be produced in a more radical way than in the previous level. Specific techniques or approximations such as coarse-graining are needed for the proto-spatiotemporal features to appear, and they are supposed to ground spatiotemporal observables and/or their

[7] According to de Haro and de Regt (2018), the degree of connection between the microstructure of the theory and the concepts belonging to classical spacetime varies from one theory to another. For instance, in LQG the interpretation of the constant $\ell = 8\pi\gamma\ell_{Pl}^2$ as a 'quantum of area' and indeed, since the area (and volume) operators come quantized in units of ℓ, justifies that LQG seems more tied to a classical spacetime interpretation than other theories such as causal sets or GFT. It is worthwhile to mention that areas and volumes are by no means directly interpreted as classical geometric quantities, but rather as classical limits of operators with a discrete spectrum, and in terms of states that can be in superpositions of eigenvalues.

[8] In our view, internal ordering encompasses succession and adjacency and it is more fundamental, because without ordering we could not speak of anything else but proto-spatiotemporal features and we could not determine any order of atemporality.

[9] The existence of disjoint Hilbert space sectors with stable solutions to the classical (non-linear) field equations makes one consider strict analogies with different phases of statistical mechanical systems and/or with inequivalent representations of local field algebras in quantum field theory. Each phase can be associated with a given non-linear field equation and thus one can ascribe a physical meaning to each disjoint physical world Strocchi (2008, 4-6; 28).

dynamics. These procedures are performed in some specific phase, only for some specific values of the macroscopic parameters (such as the coupling constants). Nonetheless, under the very same techniques or macroscopic approximations, not all phases rely on continuum spacetime and geometric characteristics. Therefore, concepts such as locality (or localization), geometry or even continuity are erased from the ontology of quantum entities.

Among the scenarios that arise from the interacting quanta, some Quantum Gravity approaches identify the phase transition from a non-geometric to a geometric phase. This phase transition defines Level 3 of emergence. To show the existence of such a transition, i.e., geometrogenesis, in which the system becomes ordered in such a way that it can be described under a suitable approximation in terms of fields that live on a four-dimensional spacetime manifold with a metric obeying Einstein's equation, depends on how each particular theory of quantum gravity is constructed.

In our view, among various approaches GFT is particularly interesting because in it we can identify condensed phases. The ontology of GFT thus suggests that individual spacetime quanta are a priori mathematical entities, but collectively and in analogy with condensed matter systems they embody a physical behaviour represented through phase transitions. For this reason, we suspect that a pure epistemological reading of these levels of emergence is not sufficient and must also rely on an ontology of phase transitions in order to account for what Oriti (2018, 2021) labelled as Level 3 of emergence, i.e., geometrogenesis. The relevance of studying geometrogenesis and to provide an ontology of phase transitions has several implications for the philosophy of physics and for cosmology. Indeed, among the different interpretations for the different possible phase transitions available, the natural hypothesis for geometrogenesis is associated to overcoming difficulties associated to the Big Bang singularity.[10] Let us recall one of its realizations in LGQ cosmology. In some models, see for instance Bojowald (2011), this process is understood as a continuous region from a purely spatial state with no notion of time. Brahma (2020) argues that this transition represents "the emergence of time in LQG", while Huggett and Wüthrich (2018) define it as the "(a)temporal emergence of spacetime". However, since any particular model is far from being fully developed, none of them accurately explains what is meant by time 'emergence'. The only concrete 'before' that can potentially be conceived involves the idea of extrapolating local and directed time beyond its own domain of applicability. According to Huggett and Wüthrich's (2018) suggestion, this very weak and novel sense of 'before' could be understood as a past limit relative to the arrow of time in the effective direction towards the non-temporal region. Therefore, this *locus*, which does not have any temporal extent and where all time-like curves of the effective spacetime converge, can be conceived as an initial state of a dynamics that requires a sufficiently broad and novel understanding.

In what follows, we shall propose an alternative view to this proposal. To account for what Oriti (2018) called levels of emergence of spacetime, we need a deeper analysis of the ontology of phase transitions, and we must produce a suitable conceptualization of atemporality as instantaneity, such that the conceptual difficulty of thinking of a transition from atemporality to temporality and vice versa (not only in GFT, but also in LQC and any approach to quantum gravity implying geometrogenesis) is overcome. Yet the literature presented the problem of finding a fundamental description of geometrogenesis in terms of an a-chronic transition from

[10] In these models, the idea of geometrogenesis as a replacement for the Big Bang competes with the alternative idea of a bouncing scenario, where the Big Bang would connect two macroscopic phases of the theory: the 'trans'-big-bang physics is a precise mirror image of the 'cis'-big-bang physics, except that the spatial orientation is inverted. The latter has been discovered in the simplest hydrodynamic description of the system, see for instance Gielen et al. (2013); Oriti (2017); Oriti et al. (2016); de Cesare et al. (2016). Another possibility can be found in Wüthrich (2022), according to which two universes are connected at the Big Bang and the latter is reinterpreted as the birth of two twin universes, reconciling both assumptions to some extent.

a non-geometric to a geometric phase. But what if we conjecture that our physics allows us to think of the co-presence in the world of atemporality and time and that this is instantiated in phase transitions? What if we find a suitable description of atemporality to play a role in geometrogenesis, as it plays a role in any phase transition?

3. Investigating atemporality: instantaneity as function of atemporal transitions

We will now introduce the main aspects characterizing our approach to atemporality as instantaneity in Subsection 3.1, and connect them to current research on geometrogenesis as one possible implication in both GFT condensate cosmology and Tensorial Group Field Theory in Subsection 3.2.

3.1. Beyond temporality: instantaneity

Let us start with a preliminary reasoning based on the picture sketched in the previous Sections. From a purely philosophical standpoint, our first move is 1) to clearly reject any statement that atemporality equates eternity. However, we can concede that eternity could be just one type of atemporality. Furthermore, we assume that 2) the "Now" or presentness is out of the representation of the temporal series, i.e., it cannot be defined in terms of ordering of succession. One can certainly divide operational time in parts, but still, one is performing nothing but a geometrization or a spatialization of what we call "time". Finally, we want 3) to distinguish the "Now" from the instant in order to remove the idea and any associated semantics according to which time can be divisible in instants. These three premises allow us to rephrase the problem as follows: is there a notion which designates a kind of atemporality which is identifiable or that can assume meaning in empirical processes but that is also out of the time flow and that is heterogeneous enough with respect to spacetime? Such a notion is the function of atemporal transition to be understood in terms of instantaneity and we will associate it to non-trivial critical points in phase transitions. In other words, we want to characterize a form of atemporality that can give meaning to geometrogenesis understood as a-chronic transition.

Our concept of instantaneity has nothing to do with that of being part of time. We claim, indeed, that we can consistently think of forms of atemporality that can be operationally useful to describe very special objects and processes without geometrizing atemporality. In order to spell out our definition of atemporality in terms of instantaneity consider the following argument. Let R a superset containing two complementary sets:

$$R \supseteq \{A, -A\} \tag{1}$$

In set theory this simple characterization of a set is valid for a superset R, if and only if we assume locality and the principle of contradiction. This leads to the classical example discussed by Wittgenstein (1929) according to which Bob and Alice cannot sit on the same chair at the same time t.[11] However, since we are interested in formalizing atemporality we forget about the time t to define the properties of our superset. Furthermore, atemporality must be characterized as instantaneity in which no order of succession is present (since we are not even interested in defining proto-time in scenarios such as those described at geometrogenesis). Thus, in the absence of an order of succession, what is left to us is to use a logical formula to express the co-presence of two complementary sets. By dropping temporality out of the scene, we also abandon the notion of locality.[12] Furthermore, it is perfectly possible to think of the sets A and $-A$ as

[11] In Wittgenstein (1929) it is clearly stated why and how, due to the principle of contradiction, logic forces us to think that Bob and Alice cannot sit on the same chair at the same time. Wittgenstein also highlights the limits of such a view, but does not offer any application of this insight to physics.

[12] It is worth noting that combinatorial non-locality is precisely what characterizes the structure of tensor models and tensorial GFT, thus our definition of instantaneity applies to both models even if the RG flow and their potential fixed points differ from those of tensor models.

Journal of Physics: Conference Series **2533** (2023) 012001 doi:10.1088/1742-6596/2533/1/012001

co-present in R, i.e., Bob and Alice can sit on the same chair, because even the chair is not under scrutiny here, the chair or place does not make any sense. To give a simple and concrete example of what instantaneity does, we can think of it in analogy to qubits. Consider the simple case of a qubit of two states in superposition:

$$|\Psi\rangle = \frac{1}{\sqrt{2}}\left(|\uparrow\rangle + |\downarrow\rangle\right) \tag{2}$$

To the qubit no univocal solution given by a sharp state is provided. However, by changing the basis $\frac{1}{\sqrt{2}}$, we can obtain a sharp single result:

$$|\Psi\rangle = |+\rangle \tag{3}$$

Thus, just like the superposition of states of the qubit can be "unlocked", one can expect to obtain the same for instantaneity, e.g., an ordering of succession or proto-temporal structures, or fluctuations of the topology can give rise to more coherent, non-degenerate systems or to a sharp state of the qubit. However, this would be less radical than our notion of instantaneity. Consider again our qubit, but this time as containing two entangled states:

$$|\Psi\rangle = \frac{1}{\sqrt{2}}\left(|\uparrow\uparrow\rangle + |\downarrow\downarrow\rangle\right) \tag{4}$$

No change in the basis can disentangle them, they are entangled. This corresponds to what we call a "locked-in mode" in instantaneity. Both states are co-present and prima facie we cannot decide which of them constitutes the qubit, because both do so at once, just like the superset R is constituted by both sets at once. By focusing on the first state of the qubit, the second one is implied and in no time we can pass from one to the other, simply because mathematically one immediately implies the other and vice versa. This specific function of atemporal transition according to which the co-presence of two opposite or disparate states is coherent and not contradictory is instantaneity and it can be consistently employed to describe the atemporal transition at fixed points of RG flow for geometrogenesis and to think of the co-presence of two states at any critical or tri-critical point in phase transitions (See Section 4).

From this we first infer that instantaneity is connotated by the undecidability of the two states, as well as by spatial non-locality derived from temporal non-locality (Co-presence Postulate). Indeed, just like we cannot decide whether one of the two entangled states connotates the qubit, we cannot decide whether a specific state is at a critical point in phase transitions, nor we can say the contrary, but just admit both states in an indistinguishable region.

We now introduce a second postulate, i.e., the Postulate of Exclusion in order to justify our definition of instantaneity as function of atemporal transition. The condition for not having any order of succession in the instant is the exclusion of all sub-spaces of the states of a qubit or, to in analogy with set theory, the exclusion of any sub-sets of the opposite or disparate sets contained in R. However, in order to think of an instantaneous switch un-locking any possible ordering and therefore also the proto-spatiotemporal ordering of succession and/or adjacency one needs to include the excluded subspaces under the opposite state.

Be A and $-A$ two opposite sets and a_2 a subset of A. In order to have instantaneity and therefore the co-presence of two states at the critical point in phase transitions, a_2 and any other subset must be excluded. From this Postulate, the following corollaries follow:

Corollary 1: a_2 is contained in $-A$ iff A does not contain a_2 within it. $-a_2$ is contained in A iff $-A$ does not contain $-a_2$ within it.

Corollary 2: a_2 must be an external subset A.

Corollary 3: $-a_2$ must be an external subset of $-A$.

Therefore, the instant is unique and indivisible. We cannot divide A because it has no parts of it within it, and vice versa, we cannot divide $-A$ because it has not parts of it within it (Postulate of Exclusion). Since the two states contain just parts of their opposite, it only depends on a switch of subsets, whether one transition as sudden change can occur from A to $-A$ or vice versa.

We used analogies taken from qubits and set theory in order to spell out some features that instantaneity can possess. These are just analogies because we find several difficulties in expressing with language what atemporality is. However, we now proceed in applying our notion of instantaneity to phase transitions. At the critical point we find an indistinguishable region. Thus, not only there is no spatio-temporal property associated to the critical point, but at it sub-states or sub-phases are indistinguishable. We therefore infer further properties of instantaneity, i.e., indivisibility and uniqueness. However, as we tried to spell out above, instantaneity can contain the determinable sub-states defining the phases associated to the co-present sets of A and $-A$. For instance, one can say that $-A$ contains $a_2 \dots a_n$, and vice versa that A contains $-a_2 \dots - a_n$. Moreover, one can say that instantaneity stems for the absence of any subsets at the critical point, e.g., states are indistinguishable at the critical point.

Thanks to this characterization of instantaneity, one can remark that this notion of atemporality spells out that the substates or subsets are in a "locked-in mode" and captures the co-presence at a critical point of different states. Thus, the Co-presence Postulate and the Postulate of Exclusion ground the atemporal or a-chronic representation of transitions and define the mutually excluded sub-states or sub-spaces of the two co-present states as determinable states of possible phases. Let us now deepen our study and test the applicability of instantaneity in specific GFT approaches on the ground of our suggestion to consider geometrogenesis.

3.2. Instantaneity, RG fixed points and phase transitions in GFT

Our proposal cannot be successful if it is not related to the content proposed by current approaches and problems discussed in quantum gravity, including attempts at unifying it under the idea that quantum gravity is a theory of complexity that should be described by many equations of state, Mielczarek and Trzésniewski (2018). This complexity can be captured by means of a multi-layered ontology as Oriti (2018) suggests and in our view, GFT is particularly interesting because in it we can identify condensed phases of extreme interest for the cosmological implications they can have. GFT formalism postulates a microscopic description of the fundamental atoms of spacetime, proving a starting point for studying both the statistical mechanics and the effective dynamics of a large number of them Oriti (2014b). Their collective behaviour is tentatively identified with continuum and geometric physics, being both emergent and approximate. The methods for extracting macroscopic and collective structures from the fundamental quantum dynamics are generally used in quantum many-body theory, but in a background independent context by using coarse graining techniques and renormalization group; see Finocchiaro and Oriti (2021) for recent developments. These methods allow the change from the atomic description to the hydrodynamic approximation. In Oriti (2021), the hydrodynamic approximation is presented as case of radical ontological emergence,[13] and geometrogenesis is presented as one possible implication of GFT condensate cosmology. The geometrogenesis scenario allows us to describe the emergent universe in analogy with a quantum condensate (fluid) resulting from the collective condensation of the underlying degrees of freedom. This transition cannot be a temporal process since it separates time from its disappearance in the underlying non-spatiotemporal quantum phase. One possibility is to understand it as the outcome of a possible evolution in the phase diagram of the theory, namely it can be understood

[13] This categorization lies on the fact that the basic structures are different entities at both sides of the emergence process. On the one hand, we deal with discrete entities. On the other, we are in need of an ontology of continuous fields which are defined as being collectively emergent. However, these approximations are just a matter of description of the emerged system, and the ontology of different phases only depends on changes in the theoretical framework (see Section 2 above).

as a flow in its time-independent coupling constants, as a sudden becoming or switch from a non-geometric and non-spatiotemporal phase to a geometric one. These phases cannot be causally connected or temporally distinguished, since these concepts are dissolved.[14] Therefore, this transition characterizes a sort of a-chronic emergence where time does not play any role. In Section 4 we shall identify how to portray instantaneity at critical point in TGFT, but before doing so, we would like to point out that in the current state of the art of GFT cosmology –see Gabbanelli and De Bianchi (2021); Gielen and Sindoni (2016) for reviews in this topic–, the existence of condensate phases with a tentative continuum geometric interpretation is a working hypothesis and deeper insight on renormalization group techniques for understanding such phase structures is required. The phase diagram of theory shows various phases (Level 2 of emergence) that can have or not physical meaning, they can be geometrical and non geometrical, but in what Oriti labeled as Level 3 of emergence, i.e., geometrogenesis, we can identify a specific form of atemporal transition. The notion of instantaneity allows us to think of the transition from a non-geometric to a geometric phase without contradiction, because it defines the *atemporal non-place* where the transition occurs. In other words, instantaneity represents the condition for thinking of atemporal transitions, but it does not determine in anyway whether the critical point triggers the transition to geometric phases or to non-geometric ones. In other words, since it is a condition or a function for atemporal transition, it has no directionality because it is out of time. We have emphasized that at the critical point we witness an a-chronic transition in which time does not play any role, actually no typical length scale is encountered. Thus, instantaneity is attributed to the critical point of the phase transition, and this move concretely facilitates our representation of geometrogenesis in analogy with critical phenomena. If we draw an analogy between critical phenomena (including renormalization group transformations used therein) and geometrogenesis, we can investigate its critical region.

A critical region is a fluctuation-dominated regime of the system. Exactly at the critical point, the system has an essentially different property, i.e., the absence of a typical length scale, Nishimori and Ortiz (2010), §1.4, p. 9. This means that one parameter (e.g., effective temperature) does not change by a renormalization group transformation if the system is at the critical point. Thus, when the critical point corresponds to a fixed point of the renormalization group transformation, it is fundamentally atemporal and we identify it with the *locus* of instantaneity.[15] In the case of GFT condensate cosmology, since the mean-field theory does not hold at and near the critical point, we witness the co-presence of indistinguishable phases at the critical point and even if one can better grasp the properties of critical regions in different phases, no refined RG flow technique can distinguish the phases at the fixed point. Thus, our concept of instantaneity describes through the Co-presence Postulate and the Postulate of Exclusion the atemporality and the indistinguishability of phases that characterize fixed points. Furthermore, this concept still allows us to identify the emergence of geometric phases, by including possible subspaces to be ordered or parametrized out of instantaneity, for instance, one after the other or one next to the other, thereby generating connection of fundamental degrees of freedom or succession and other proto-spatiotemporal properties.

Indeed, it is worth noting that GFT approaches to geometrogenesis appeal to renormalization techniques using the universality class concept that eliminates inessential details and allows focusing on increasingly macroscopic properties of the systems. An important consequence of universality is that quantities describing essential features of critical phenomena, e.g., critical

[14] Although non-spatiotemporal analogues can be found, e.g., causation without time Baron and Miller (2014, 2015); Tallant (2019), we are not going to deepen this issue.

[15] Of course, according to our definition of instantaneity this is a no-place. The major consequence we see in the cosmological scenario is thus the impossibility of identifying an exact moment of the transition from the non-geometric to the geometric phase and therefore any question regarding the beginning of the geometric phase does not make sense anymore.

exponents, are specified only by factors such as the symmetry of the system, range of the interactions, and its spatial dimensionality (the connectivity of its elementary degrees of freedom, such as spins in the Ising model in case of a definition of the system on a lattice). This means that two apparently different critical phenomena can share the same critical exponents, one in the Ising model, for instance, and the other in a simple liquid, as long as both are in three dimensions. These two distinct physical systems belong to the same universality class in such a way that a model of magnetism shows the same critical behaviour as one for the simple liquid. Behind this behaviour there is a physical explanation depending on the fact that many characteristics of the system gradually recede as the renormalization-group transformation proceeds and eventually only essential factors, e.g., the spatial dimensionality and the symmetry of the system, survive. Thus, to rely on universality class is fundamental in order to study the relevant symmetries of GFT and those of condensed matter physics. In the current state of the art of GFT cosmology, more work is needed to determine under which conditions phase transitions and different phases can truly exist. Whether and how macroscopic continuum phases emerge remains an open issue together with the exact representation of geometrogenesis as physical atemporal transition. Furthermore, the consequences of this phase transition are expected to be measurable, being possible imprints of the universe dynamics testable, e.g., the physics of cosmological perturbations in the very first instants of the history of the universe, in particular in the CMB spectrum, Gielen (2015, 2019); Marchetti and Oriti (2021); Pithis and Sakellariadou (2019). Nevertheless, it is worth mentioning that the quantum degrees of freedom are expected to have consequences when forming collective structures. Indeed, if no phenomenology at any scale can be associated to the quantum world, whether associated to their quantum properties per se or to collective macroscopic effects, then the quantum entities are no more than mathematical tools, and the non-geometric and non-spatiotemporal phases would have no reason to be considered as physical or philosophically interesting. In what follows we provide insight on the reasons why the atemporal character of the transition dubbed geometrogenesis should attract our attention, at least from the conceptual standpoint.

4. Omega region and Ginzburg parameter: looking for a signature of instantaneity in geometrogenesis

In recent studies based on TGFT, geometrogenesis is associated to a phase transition modeled through the Landau-Ginzburg theory, see Marchetti et al. (2022). Therefore, we have an interesting case study that can be extracted from Marchetti et al. (2022), in order to exemplify what instantaneity looks like in TGFT and in geometrogenesis scenarios.

The Landau-Ginzburg theory of second-order (continuous) phase transitions is, in fact, a *phenomenological* theory which provides a description of the fluctuations determining the critical point. Near a critical point, two or several different phases, with almost the same free energy, are competing to determine the ground-state (or low-energy states). Therefore, relatively small fluctuations in the system would lead to drastic effects.

When studying the behaviour of the Ginzburg parameter Q, for values of $Q \gg 1$ large fluctuations dominate and the critical point $\mu \to 0$ shows the signature of a divergence that dramatically leads to the impossibility for the Ginzburg-Landau mean-field theory to hold. In other words, no measurement nor observable can be associated to the critical point, no time and no space can make sense at $\mu \to 0$ and therefore no correlation can be measured therein. However, thanks to the Landau-Ginzburg theory, we can identify around the critical point a region Ω in which and of which at least we can tell that two phases are co-present (in agreement with the Co-presence postulate) and that they are indistinguishable. Since the Ginzburg-Landau theory is used in describing a physical abrupt phase transition, just like geometrogenesis is supposed to imply, it means that at critical point, when $\mu \to 0$ and $Q \gg 1$ at criticality, instantaneity dominates the transition, and Q should be understood as the triggering leading to a physical process. In other words, to think of an a-chronic transition is not only possible but necessary in this case.

Nevertheless, as highlighted in Marchetti et al. (2022), when $Q \ll 1$, correlation length makes sense again, and both local and non-local geometrical variables can be computed. Fluctuations of the order parameter Φ averaged over a region Ω are small compared to the order parameter Φ_0, averaged over that region, and the mean-field theory is applicable, and the two-point function of the fluctuations is encoded by a correlation function adapted for the case in which the Ginzburg parameter appears as $|Q| \ll 1$. The behaviour of Q is of extreme importance at the transition and also the specification of the region to be averaged over should depend on statistically relevant correlations up to distances of the order of the correlation length. Since local and non-local degrees of freedom enter in different ways in the dynamics of the models, it is preferable to differentiate a priori independent parameters that might contribute to long-range correlations that appear regardless of the physics of the phase transition and that are the result of asymptotically diverging contributions.

It is worth noticing that the behaviour of the Ginzburg parameter does not simply tell us that fluctuations are large at critical point, but Marchetti et al. (2022) have found the role played by the relevant symmetry group of GFT, namely due to the noncompactness and hyperbolic nature of the Lorentz group, we can always find a transition towards a phase with non-vanishing expectation value of the field (operator) and that this phase transition is always self-consistently described in terms of mean field theory. Since such configurations are highly populated by GFT quanta, this is evidence for the existence of an interesting continuum geometric approximation to be studied in mean-field language, in such TGFTs. Such phases had so far only been conjectured to exist for Lorentzian GFT models and had been used as a working hypothesis for the TGFT condensate cosmology program, where cosmological dynamics is also extracted from the TGFT mean-field hydrodynamics, albeit around non-uniform field configurations. Out of technicalities, all this means not only that main characteristics of instantaneity are embodied at critical point: indistinguishability, co-presence of phases, non-locality understood as impossibility to compute probability distribution and correlations, absence of time, impossibility to order successive states and so forth; it also means that depending on the properties of symmetry groups, one can find conditions under which instantaneity is "unlocked". This in turn shows that it is deeply wrong to equate atemporality with eternity or an enduring 'Now' and that physics needs the notion of instantaneity to make sense of atemporal transitions, such as geometrogenesis.

Nevertheless, what is less obvious and more problematic is the philosophical interpretation of instantaneity in this context. Should we attribute to it an ontological meaning? Or should we rather be happy with a pure epistemological interpretation of it? For the time being, we conclude that:

(1) Geometrogenesis is a phase transition showing the signature of atemporality.
(2) It not only can but must also be thought of in terms of a-chronic transition, at least on the ground of the Landau-Ginzburg theory.
(3) The critical point of an atemporal transition from a non-geometric to a geometric phase embodies the characteristic of instantaneity.
(4) Instantaneity is an epistemological tool that enables us to think of atemporal transitions, such as geometrogenesis, without contradiction.
(5) From the purely physical standpoint, instantaneity cannot be associated to observables: it rather denotes the absence of them and the maximum decoupling of correlations.

Finally, we have some other hints of possible scenarios in which atemporality dominates. For instance, in regions near black holes singularities behaving like Zeno regions, physicists identified asymptotic silence, and we have evidence of the fact that from a semi-classical regime one can pass to a region of maximally decoupled points, connotated by the disappearance of causality and the collapse of the light-cone; see for instance Eglseer, Hofmann and Schneider (2021, 2017), Carlip (2012). However, we have to further investigate whether atemporality in terms of instantaneity applies to these cases or whether we need to elaborate a different notion of another kind of atemporality. Studies in the fields of LQC and QG also highlight where and under which conditions we detect asymptotic silence; Eichhorn, Mizera and Surya (2017), Mielczarek (2013). However, and this will be the subject of our future work, Quantum information theory could help in detecting the conditions under which these Zeno regions can "speak", and it does so by showing the conditions under which states in the region are maximal entangled and cannot be disentangled. If it is possible to show that there are conditions under which these regions can start "speaking" again and thus exit the asymptotic silence, we will be entitled to infer that at various scales temporality could come into being from atemporality. We therefore envisage a huge change of perspective in future studies in the philosophy of physics and the philosophy of time if and only if the study of instantaneity is taken seriously into account, together with the latest development of quantum information theory applied to Quantum Gravity scenarios and extreme quantum and cosmological regimes.

5. Conclusion

In this paper we tried to characterize for the first time and formalize the a-chronic phase transition of geometrogenesis suggested by some Quantum Gravity models. We grounded our proposal of instantaneity on what we labeled 'Postulate of Exclusion" and "Co-presence Postulate" in Subsection 3.1. Any discussion of atemporality certainly raises many new conceptual issues that can modify current philosophy and its relationship to cosmology in the light of recent scientific debates on quantum gravity. This reflection inevitably bear implications for the philosophy of physics and further studies will test whether a promising new field of research in defining such an under-explored concept like instantaneity might prove to be necessary for a future theory of quantum gravity. What is far more interesting for the time being is that we provided a notion of atemporality, by challenging ideas that have been crystallized and consolidated throughout centuries regarding the divisibility of time in instants, the equation of atemporality with eternity or eternal presentness. We would like to stimulate scholars in different areas to explore the extraordinary multiplicity of questions that can be addressed by dropping out the idea that only temporality and spatiotemporality can be formalized. Second, the present contribution emphasized the possibility of thinking of geometrogenesis in atemporal terms. This does not imply that we provided evidence for the existence of this physical transition, rather we just tried to make it consistent with its cosmological implications and the general framework of GFT and TGFT. Furthermore, several open questions arise also for physicists in these contexts. For instance, what happens to specific hydrodynamic observables across the phase transitions? Is there any remnant of them that can be constructed out of the fundamental degrees of freedom? The notions of background independent and non-spatiotemporal evolution must be consistently developed in the 'theory space' that characterizes the chosen quantum gravity formalism. A realization of this type of evolution can be found in some specific renormalization group schemes, yet it remains obscure whether one can find some relevant notion of proto-temporal evolution, some parametrization of the flow through these non-geometric phases which may coincide with some useful observables that would provide emergent notions of time within the continuum phase. If the answer is affirmative, and since in the context of the hydrodynamic approximation the only ontology is that of fields, then the only physical observables emerging from the corresponding quantum observables are relations among the values of these fields. Therefore, in the context of the relational framework, some selected fields are the only structures that can be used to define clocks and rods, or more generally, reference frames labelling manifold points, and by doing so, spacetime and evolution of other fields. This ontology could be based on the quantum degrees of freedom behaving collectively and in an emergent manner, something that still needs to be fully investigated. However, philosophers can fruitfully proceed by introducing an ontology of phase transitions, which could accommodate descriptions of fundamental collective phenomena by framing them in a more complex idea of the world which also admits atemporality at critical points and proto-temporality at other scales depending on which phase transitions are of interest and the ontology they bear with them.

Acknowledgments
This research has received funding and support from the European Union's Horizon 2020 research and innovation programme under grant agreement n. 758145 and the project CHRONOS (PID2019-108762GB-I00) of the Spanish Ministry of Science and Innovation.

References

Bardon A 2013 *A Brief History of the Philosophy of Time* (Oxford: Oxford University Press)
Baron S and Miller K 2014 Causation in a timeless world *Synthese* **191**(12) 2867–86
Baron S and Miller K 2015 Causation sans time *American Philosophical Quarterly*, **52**(1) 27–40
Bojowald M 2011 *Quantum cosmology: A fundamental description of the universe* (New York: Springer)
Brahma S 2020 *Emergence of time in loop quantum gravity* (Beyond Spacetime: The Foundations of Quantum Gravity) ed N Huggett, K Matsubara and C Wüthrich (Cambridge: Cambridge University Press) pp 53–78
Butterfield J and Isham C J 1999 *On the emergence of time in quantum gravity* (The Arguments of Time) ed J Butterfield (Oxford: Oxford University Press) pp 111–68
Butterfield J and Isham C J 2001 *Spacetime and the philosophical challenge of quantum gravity* (Physics meets Philosophy at the Planck Scale) ed C Callender and N Huggett (Cambridge: Cambridge University Press) pp 33–89
Calamari M 2021 The metaphysical challenge of loop quantum gravity *Studies in History and Philosophy of Science Part A* **86** pp 68–83
Callender C 2011 *The Oxford handbook of philosophy of time* (Oxford: Oxford University Press)
Carlip S 2012 *The small scale structure of spacetime* (Foundations of Space and Time) ed G Ellis, J Murugan, A Weltman (Cambridge: Cambridge University Press) pp 69-84
Crowther K 2018 Inter-theory relations in quantum gravity: Correspondence, reduction, and emergence *Studies in History and Philosophy of Science Part B: Studies in History and Philosophy of Modern Physics* **63** pp 74-85
de Cesare M, Pithis A and Sakellariadou M 2016 Cosmological implications of interacting group field theory models: Cyclic universe and accelerated expansion *Physical Review D* **94** 064051
de Haro S and de Regt H 2018 Interpreting theories without a spacetime *European journal for philosophy of science* **8**(3) pp 631–70
Delcamp C and Dittrich B 2017 Towards a phase diagram for spin foams *Classical and Quantum Gravity* **34**(22) 225006
Eglseer L, Hofmann S and Schneider M 2017 Quantum populations in Zeno regions inside black holes *Preprint* arXiv-1710
Eglseer L, Hofmann S and Schneider M 2021 Quantum populations near black-hole singularities *Physical Review D* **104**(10) 105010
Eichhorn A, Mizera S and Surya S 2017 Echoes of asymptotic silence in causal set quantum gravity *Classical and Quantum Gravity* **34**](16) 16LT01
Esfeld, M 2021 Against the dissapearence of spacetime in quantum gravity *Synthese* **199** 355–69
Finocchiaro M and Oriti D 2021 Renormalization of group field theories for quantum gravity: new scaling results and some suggestions *Frontiers in Physics* **8** 552354
Gabbanelli L and De Bianchi S 2021 Cosmological implications of the hydrodynamical phase of group field theory *General Relativity and Gravitation* **53**(66)
Gielen S 2015 Identifying cosmological perturbations in group field theory condensates *Journal of High Energy Physics* **10**
Gielen S 2019 Inhomogeneous universe from group field theory condensate *Journal of Cosmology and Astroparticle Physics* **02**
Gielen S, Oriti D and Sindoni L 2013 Cosmology from group field theory formalism for quantum gravity *Physical Review Letters* **111** 031301
Gielen S and Sindoni L 2016 Quantum cosmology from group field theory condensates: a review *Symmetry, Integrability and Geometry: Methods and Applications (SIGMA)* **12** 082

Journal of Physics: Conference Series **2533** (2023) 012001 doi:10.1088/1742-6596/2533/1/012001

Huggett N and Wüthrich C 2012 Emergent spacetime and empirical (in)coherence *Studies in the History and Philosophy of Modern Physics* **44**(3) pp 276–85

Huggett N and Wüthrich C 2018 The (a)temporal emergence of spacetime *Philosophy of Science* vol 85(5), ed A Woody (Cambridge: Cambridge University Press) pp 1190–203

Konopka, T, Markopoulou, F and Smolin L 2006 Quantum graphity *Preprint* hep-th/0611197

Le Bihan, B 2018 Space emergence in contemporary physics: why we do not need fundamentality, layers of reality and emergence *Disputatio* **10**(49) 71–95

Le Bihan, B 2019 Spacetime emergence in quantum gravity: functionalism and the hard problem *Synthese* **199**(2) pp 371–93

Mandrysz M and Mielczarek J 2019 Ultralocal nature of geometrogenesis *Classical and Quantum Gravity* **36**(1) 015004

Marchetti L and Oriti D 2021 Effective relational cosmological dynamics from quantum gravity *Journal of High Energy Physics* **25**

Marchetti L, Oriti D, Pithis A and Thürigen J 2022 Phase transitions in TGFT: a Landau-Ginzburg analysis of Lorentzian quantum geometric models *Preprint* arXiv:220904297

Markopoulou F 2009 *New directions in background independent quantum gravity* (Approaches to quantum gravity: Toward a New Understanding of Space, Time and Matter) ed D Oriti (Cambridge: Cambridge University Press) pp 129–66

Meincke, A S 2019 The disappearance of change: Towards a process account of persistence *International Journal of Philosophical Studies* **27**(1) pp 12–30

Mielczarek, J 2013 Asymptotic silence in loop quantum cosmology *AIP Conference Proceedings* vol 1514, no 1 (American Institute of Physics) pp 81–4

Mielczarek, J 2017 Big bang as a critical point Advances in High Energy Physics *Article ID* 4015145

Mielczarek J and Trześniewski, T 2018 Towards the map of quantum gravity *General Relativity and Gravitation* **50**(68)

Nishimori H and Ortiz G 2010 Elements of Phase Transitions and Critical Phenomena (Oxford: Oxford University Press)

Oriti, D 2007 Group field theory as the microscopic description of the quantum spacetime fluid: a new perspective on the continuum in quantum gravity *Proceedings of science (QG-Ph)* **030**

Oriti, D 2014a Disappearance and emergence of space and time in quantum gravity *Studies in History and Philosophy of Science Part B: Studies in History and Philosophy of Modern Physics* **46** pp 186–99

Oriti D 2014b *Group field theory and loop quantum gravity* (In Loop Quantum Gravity: The First 30 Years) ed A Ashtekar, J Pullin (World Scientific: Singapore) pp 125–51

Oriti D 2017 The universe as a quantum gravity condensate *Comptes Rendus Physique* **18** 235

Oriti D 2018 *Levels of spacetime emergence in quantum gravity* (Philosophy beyond spacetime: Implications from Quantum Gravity) ed C Wüthrich, B Le Bihan and N Huggett (Oxford: Oxford University Press) pp 16–40

Oriti D 2021 The complex timeless emergence of time in quantum gravity *Time and Science* ed P Harris and R Lestienne (World Scientific)

Oriti D, Sindoni, L and Wilson-Ewing E 2016 Emergent Friedmann dynamics with a quantum bounce from quantum gravity condensates *Classical and Quantum Gravity* **33** 224001

Pithis A G A and Sakellariadou M 2019 Group field theory condensate cosmology: An appetizer *Universe* **5**(6) 147

Strobach, N 2013 *The moment of change: A systematic history in the philosophy of space and time* vol 45 (Springer Science & Business Media)

Strocchi F 2008 Symmetry breaking *Lecture Notes in Physics* 732, 2nd edition (Springer)

Tallant J 2019 Causation in a timeless world? Inquiry: An Interdisciplinary Journal of Philosophy **62**(3) pp 300–16

Wittgenstein L 1929 Some remarks on logical form *Aristotelian Society Supplementary Volume* vol 9(1) (Blackwell Publishing on behalf of The Aristotelian Society) pp 162–71

Wüthrich, C 2018 The emergence of space and time *Routledge Handbook of Emergence* ed S Gibb, R F Hendry and T Lancaster (Oxford: Routledge) pp 315–26

Wüthrich, C 2019 When the actual world is not even possible *The Foundation of Reality: Fundamentality, Space, and Time* ed G Darby, D Glick and A Marmodoro (Oxford: Oxford University Press) pp 233–53

Wüthrich, C 2022 One time, two times, or no time? *Einstein and Bergson 100 Years Later: What is Time?* ed S Gozzano, E Coccia, R Ronchi and A Campo (Berlin/Boston: de Gruyter) pp 209–30

Journal of Physics: Conference Series **2533** (2023) 012003 doi:10.1088/1742-6596/2533/1/012003

The Hole Argument and Beyond: Part II: Treating Non-isomorphic Spacetimes

Henrique Gomes[1], **Jeremy Butterfield**[2]

[1] Oriel College, Oxford OX1 4EW, UK

[2] Trinity College, Cambridge CB2 1TQ, UK

E-mail: gomes.ha@gmail.com, jb56@cam.ac.uk

Abstract. In this two-part paper we review, and then develop, the assessment of the hole argument for general relativity.

The review (in Part I) discussed how to compare points in isomorphic spacetimes, i.e. models of the theory. This second Part proposes a framework for making comparisons of *non*-isomorphic spacetimes. It combines two ideas we discussed in Part I—the philosophical idea of *counterparts*, and the idea of threading points between spacetimes other than by isomorphism—with the mathematics of fibre bundles.

We first recall the ideas from Part I (Section 1). Then in Section 2 and an Appendix, we define a fibre bundle whose fibres are isomorphic copies of a given spacetime or model, and discuss connections on this fibre bundle. This material proceeds on analogy with field-space formulations of gauge theories. Finally, in Section 3, we show how this fibre bundle gives natural expressions of the philosophical ideas of counterparts, and of threading.

1. Introduction

In Part I of this paper, we reviewed, mostly from a philosophical perspective, the hole argument in general relativity. It is about comparing isomorphic spacetimes (i.e. solutions of general relativity), and especially about how to "identify" a point in one spacetime with a point in another isomorphic one.

Here in Part II, we break new ground. We address the question of how one should compare *non*-isomorphic models: in particular, how to compare points between the two models. We will propose a framework for such comparisons, that combines the mathematics of fibre bundles with two ideas we developed in Part I:

(i): the philosophical idea of *counterparts*: which is a way of treating the identity of objects—any objects, not only points—in different possible situations; (cf. Section 2.2.3 of Part I); and :

(ii) the idea of *threading points*: which is our jargon for a point in one spacetime corresponding to (being "identified" with) a point in another spacetime *other than by* their being, respectively, the argument and value of (a diffeomorphism that is) an isomorphism between the spacetimes; (cf. Section 3.2 of Part I).

Our proposed framework proceeds by analogy with field-space formulations of gauge theories. It postulates a fibre bundle whose fibres are isomorphic copies of a given model, i.e. a given Lorentzian manifold endowed with matter fields obeying the Einstein equations. So given a model based on a manifold M, Diff(M) is to be the structure group of the bundle: any element

Journal of Physics: Conference Series **2533** (2023) 012003 doi:10.1088/1742-6596/2533/1/012003

of Diff(M) drags the given model into coincidence (and so isomorphism) with another element of the fibre. So both the base-manifold (the set of orbits) of the bundle, and its structure group, are infinite-dimensional; and our exposition will thus need to address subtleties about infinite-dimensional bundles.[1]

The paper is organised as follows. In Section 2, we expound the fibre bundle of spacetimes. This exposition will use a natural analogy with ideas and results in classical gauge theories (Yang-Mills theories, both abelian and non-abelian)—for which of course the appropriateness of fibre bundle ideas is universally accepted. Though our bundle is infinite-dimensional, we shall see that much carries over from the familiar finite-dimensional case. We will first describe the bundle (Section 2.1); and then describe connections on it (Section 2.2; and in the Appendix, Sections Appendix A.2 and Appendix A.3). For the main tool for comparing elements of distinct fibres of any bundle is a connection: so in order to compare non-isomorphic spacetimes in our bundle, we need to consider connections. Then in Section 3 , we explain how Section 2's material relates to the idea of *threading*, and also gives a natural and technically precise illustration of *counterparts*.

Two final orienting remarks. (1): Broadly speaking, the ingredients of Section 2 come from physics; and Section 3 adds ingredients from philosophy. Unsurprisingly, this makes Section 2 more technical: especially since our fibre bundle of spacetimes is infinite-dimensional, while the fibre bundles familiar from gauge theory are finite-dimensional, and have space or spacetime as the base manifold. On the other hand, the technicalities in Section 2 are already in the literature; and so we can be brief, with details in the Appendix. Besides, these technicalities will not be needed for Section 3's philosophical "punch-line": that our bundle of spacetimes illustrates counterpart theory, with neat mathematical properties. Indeed, that punch-line will use little more than the idea of a section of the bundle of spacetimes.

(2): We should note at the outset how this paper uses the word 'point' in two different ways; (though the meaning will always be clear from the context). For the fibre bundles in the familiar formulations of gauge theories, the base-manifold is spacetime or space: whose elements are thus points of exactly the kind at issue in the hole argument and in our Part I. And a point in, i.e. element of, the bundle itself is an assignment of field-values to a spacetime (or spatial) point. On the other hand, for the infinite-dimensional bundles of this paper (both those for general relativity, and those for gauge theories), a point in, i.e. element of, the bundle is a field-configuration across all of spacetime (space). And accordingly, a point in the base-manifold gives the physical information (what Part I called 'qualitative profile') shared by all the field-configurations in the fibre above the point—the "gauge redundancy" having been quotiented out.

So in short: In Part I and in the hole argument, one naturally reads 'point' as a spacetime (or spatial) point, and in that context, mention of a point in a fibre bundle suggests "at most" an assignment of field-values to a spacetime (or spatial) point (and so a familiar finite-dimensional bundle). But in this paper, the points in the fibre bundles that we will be concerned with are field-configurations across all of spacetime (space). But no matter: which meaning of 'point' is intended will always be clear from the context.

[1] Although we will throughout this paper talk of Diff(M) as the structure group, almost all of our discussion would hold good for a smaller group. And since we think of the elements of a fibre as physically equivalent, such a smaller group would be motivated by the standard practice in general relativity of regarding some isomorphisms as generating new possibilities. The main example is isomorphisms that do not preserve structure at asymptotic spatial infinity (but correspond to e.g. a time translation there). For discussion and references, cf. footnote 24 of Part I, and footnote 3 below. In any case: even if we replace Diff(M) by some subgroup whose elements also preserve some desired asymptotic structure, both the bundle and this smaller structure group, would still be infinite-dimensional.

Journal of Physics: Conference Series **2533** (2023) 012003 doi:10.1088/1742-6596/2533/1/012003

2. A fibre bundle of models

In this Section, we will proceed in two stages. In Section 2.1, we will introduce the infinite-dimensional fibre bundle of Lorentzian manifolds, with the diffeomorphism group as its structure group. In classical gauge theories, the analogue of this fibre bundle is *field-space*, with the group of gauge transformations as the structure group. Then in Section 2.2 and the Appendix, we will give some details about connections on these bundles. For a connection is the main tool for comparing elements of distinct fibres, in both finite and infinite-dimensional bundles; (though as just stressed in (2) of Section 1, the elements i.e. 'points' being compared are very different).

2.1. The bundle of spacetimes

The main idea of this Subsection is that given a spacetime (M, g, T), the set of all models isomorphic to it, that are built by drag-along (in the manner of the hole argument) by all the diffeomorphisms d on M is like the fibre of a principal fibre bundle, whose structure group is Diff(M), the group of diffeomorphisms of M. We will first: (1) state this idea more precisely, and state the analogy with gauge theory's idea of *field-space*; and then (2) give some technical qualifications, which apply to both general relativity and gauge theory.

(1): *The idea, and the analogy with gauge theory:—* First, we fix the spacetime manifold M; although general relativity has models using non-diffeomorphic manifolds, all that follows uses a single M.

For simplicity, let us also set aside T, so that we take general relativity *in vacuo*. Then the class of all Lorentzian manifolds (M, g) built on our fixed M is obviously partitioned into isometry classes by Diff(M). Our task now is to describe how this class has further structure, making it an infinite-dimensional bundle, of which each fibre is an orbit of Diff(M), i.e. an isometry class.

The situation is, in spirit, identical to how in gauge theory one can treat *field-space*, i.e. the space of field configurations, as an infinite-dimensional bundle, of which each fibre is an orbit of the group of gauge transformations. Note that here 'field configuration' means an assignment of values to the fields, including gauge fields, throughout all space, or all spacetime. So field-space is infinite-dimensional, and partitioned into gauge-equivalence classes by the gauge transformations. Since as usual we allow the gauge transformations to be local, i.e. to depend on the spatial or spacetime point, the group of gauge transformations is obviously an infinite-dimensional group. So here, the parallel task is to describe how field-space has further structure, making it an infinite-dimensional bundle, of which each fibre is an orbit of the group of gauge transformations.

Since the main idea is the same in both general relativity and gauge theories, we will adopt a common language and notation.[2] The theory gives us a space Φ of models φ. For general relativity *in vacuo*, φ is a Lorentzian manifold (M, g), while for gauge theory φ is a field configuration. A group $\mathcal{G}$ acts on Φ, and this action respects the manifold structure of Φ. In our examples, $\mathcal{G}$ is: for general relativity, the set of diffeomorphisms of M; and for gauge theory, the set of gauge transformations. We denote the action by: $\mathcal{G} \times \Phi \ni (g, \varphi) \mapsto \varphi^g \in \Phi$. So in particular, for each g, the map on Φ thereby defined, i.e. $\varphi \mapsto \varphi^g$, is a diffeomorphism of Φ.

We write an orbit of this action with the usual square-bracket notation for equivalence classes: $[\varphi]$, where $\pi : \Phi \to [\Phi]$ denotes the surjective projection onto the set of equivalence classes, which we write as $[\Phi]$. And the set of elements in Φ that are equivalent to φ is also denoted by $O_\varphi := \pi^{-1}([\varphi])$, to stress that it is an orbit. So the equivalence class gives the "physical information" shared by all the elements of the orbit. And once a bundle structure is established,

[2] The exposition in this Subsection and in the next is drawn from [1, 2, 3]. Further references are given below.

then, as mentioned in the preamble to this Section: an element of the base-manifold (which in some definitions of a bundle, is defined to be the set of equivalence classes) gives the physical information shared by all the points in the fibre above the element. In the jargon of gauge theory: an element of the base-manifold is the *physical state*: the gauge redundancy having been quotiented out. So for our example of general relativity, i.e. the bundle of Lorentzian manifolds (M, g) on a given manifold M, an element of the base-manifold is the qualitative profile of metric and matter field values, which quotients out "which spacetime point is which" within M.[3]

Besides, general relativity and gauge theories are not the only physically relevant examples of an infinite-dimensional space Φ of models φ, on which a group $\mathcal{G}$ acts. Another salient example is the Riemannian analogue of our Lorentzian bundle, i.e. the bundle of Riemannian manifolds (M, g) on a given manifold M: in other words, the configuration space of Riemannian metrics g on M. It is often written as $\text{Riem}(M)$. Similarly to the Lorentzian case, a fibre of $\text{Riem}(M)$ is built from a given Riemannian manifold ('configuration') (M, g) by drag-along, in the manner of the hole argument, by any of the diffeomorphisms d on M. So here: an equivalence class under the action of the group quotients out "which *spatial* point is which". (We will return to this example in footnote 5 and at the end of Section 4.2.)

Now we think of the set of equivalence classes, $[\Phi]$, as itself a manifold. Then we consider smooth maps σ from $[\Phi]$ to Φ that send each argument $[\varphi]$ to an element of the corresponding orbit, i.e. $\sigma([\varphi]) \in \mathcal{O}_\varphi$. In short, σ is a smooth choice of a *representative* of each orbit. When the manifold Φ has a principal fiber bundle structure, such a map is a *section* of Φ; the word is also used for the map's range, i.e. the embedded sub-manifold of Φ that crosses each orbit only once. In the context of a principal bundle, given some open subset $\mathcal{U} \subset \Phi$, a section would allow us to write a local product structure, i.e. a diffeomorphism between $\pi^{-1}([\mathcal{U}])$ and $[\mathcal{U}] \times \mathcal{G}$.

So to summarise: the main idea is that the space of models Φ is very similar to a principal fiber bundle, with $\mathcal{G}$ (respectively: diffeomorphisms and vertical automorphisms) as its structure group. But there are important differences between these infinite-dimensional sets and the finite-dimensional principal fiber bundles seen in familiar formulations of gauge theories. In the finite-dimensional case, it suffices that the action of a group G on a given manifold P be free and proper for that manifold to have a principal G-bundle structure. (This structure is usually written as $G \hookrightarrow P \to P/G$, where P/G is the base-manifold.) But in the infinite-dimensional case, these properties of the group action are not enough to guarantee the necessary fibered, or local product, structure. This is taken up in (2).

(2): *Qualifications:*— Indeed, there are three qualifications to be made here.

The first is that there are symmetric φ, i.e. φ that are fixed by some $g \in \mathcal{G}$, i.e. $\varphi^g = \varphi$ (where of course g is not the identity element). Such a φ is called *reducible*, and the g that fix it are called *stabilisers*. It is easy to show that all elements of a reducible state's orbit are reducible (with stabilisers related by conjugacy). This means that the orbits are of various "sizes", and are not all isomorphic to the structure group. So $[\Phi]$ is in fact, not a single manifold, but a union of manifolds, one for each "size" of orbit.

The manifolds Φ with a group action that we discuss here are thus decomposed into different

[3] Here, we should qualify our statement above that $\text{Diff}(M)$ is the structure group. For recall footnote 1 and footnote 24 of Part I: for spacetimes that are asymptotically flat at infinity, it is standard practice to treat some isomorphisms (for vacuum spacetimes: some isometries) as generating from a given model a new physical possibility, that differs from the given model by e.g. a time-translation at spatial infinity. (For a masterly philosophical discussion, cf. [4], especially Sections 4.3 and 4.4, pages 964-970.) To allow for this, while maintaining our interpretation that elements of a given fibre are all physically equivalent, we need to allow that the structure group of the bundle of spacetimes is smaller than $\text{Diff}(M)$, i.e. its elements fix the structure at spatial infinity. But this qualification does not affect anything in the rest of this paper. Besides, to stress the analogy with field-space formulations of gauge theory, we are anyway writing the structure group as $\mathcal{G}$.

parts, with different sizes of orbits, called *strata*. That is, Φ is stratified into—is a disjoint union of—orbits of states that possess either more or fewer stabilisers: with the orbits whose elements have more stabilisers (i.e. stabiliser groups of larger dimension) being at the boundary of the orbits of states with fewer stabilisers; (and with the 'bulk' stratum being the manifold consisting of generic fields φ that have only the identity element of $\mathcal{G}$ as a stabiliser). For each stratum, we can find a section and form a product structure as in the standard picture of the principal bundle. The notion that generalises sections to cut across such sets of orbits is called *a slice*, and there are several existence theorems for the case of gauge theories and metrics.[4]

The second qualification is that even if we restrict attention to the generic, i.e. not reducible, configurations (for non-abelian field theories)—one can have at most a local product structure: no section is global. This is known as the Gribov obstruction; see [5, 6]. (More precisely: though the name 'Gribov' is associated with the infinite-dimensional case, the point here—that a bundle has a local product structure (local trivializations) but not necessarily a global one—applies of course to finite-dimensional bundles.)

So much by way of stating the first two qualifications. The third qualification touches on the topic of sections of the bundle Φ: this will look ahead to Section 2.2's discussion of connections.

Recall that a section is essentially an embedded submanifold of Φ that intersects each orbit exactly once. Since it is natural to define a submanifold of any given manifold N as the level-surface of a real-valued function on N, we consider defining a section in this way. More precisely: we aim to define a co-dimension one surface $\Sigma \subset N$, as $\mathcal{F}^{-1}(c)$, for $c \in \mathbb{R}$, and $\mathcal{F}$ a smooth and regular function, i.e. $\mathcal{F} : N \rightarrow \mathbb{R}$ such that $\mathrm{d}\mathcal{F} \neq 0$. But in the infinite-dimensional case, both the dimension and the co-dimension of a regular value surface can be infinite, and it becomes trickier to construct such a section.

Since these are infinite-dimensional spaces, they require a more comprehensive definition of manifold. But just like the finite-dimensional manifold is modeled on $\mathbb{R}^n$, the more general notion of a Frechét manifold is modeled on a complete, metrizable, locally convex topological vector space: Φ here is always a Frechét manifold and $\mathcal{G}$ is a Frechét group. These notions allow us to import many of the ideas of differential geometry: in particular, the idea of decomposing the total tangent space at a point in a bundle into a direct sum of the tangent space to the orbit, and a complement. Thus it has been shown using different techniques and at different levels of mathematical rigour ([7, 8, 9, 10, 11, 12, 13]) that both the gauge theory configuration space Φ, and the configuration space of Riemannian metrics on a fixed manifold M, have slices and thus admit a local product structure.[5]

[4] Given $\varphi \in \Phi$, a slice at φ is a submanifold Σ of Φ containing φ such that, if we denote by $\mathsf{Aut}_\varphi \subset \mathcal{G}$ the automorphism, or stabiliser, group of φ, the following conditions are met:
1. $\tilde{g} \in \mathsf{Aut}_\varphi \Rightarrow \Sigma^{\tilde{g}} = \Sigma$;
2. $g \notin \mathsf{Aut}_\varphi \Rightarrow \Sigma^g \cap \Sigma = \emptyset$;
3. There exists a local cross section *of the group* $\tau : Q \subset \mathcal{G}/\mathsf{Aut}_\varphi \rightarrow \mathcal{G}$ where Q is an open neighborhood of the identity, such that

$$\Psi : Q \times \Sigma \rightarrow \mathcal{U}_\varphi \tag{1}$$
$$(g, \varphi') \mapsto \varphi'^{\tau(g)}, \tag{2}$$

where $\mathcal{U}_\varphi$ is an open neighborhood of $\varphi \in \Phi$, is a diffeomorphism.
Here, we retain the product structure, but it depends on some choice of embedding of (a neighborhood of the identity of) the quotient of the group by the group of automorphisms into the group.

[5] Roughly speaking, one starts by endowing Φ with some $\mathcal{G}$-invariant (super)metric, and then finds the orthogonal complement to the orbits $[\varphi]$, with respect to this supermetric. But here the intersection of the orbit with its orthogonal complement might not vanish, as it does in the finite-dimensional case; and the tangent spaces to the orthogonal section and to the orbit might not sum to the total tangent space; and they might not be closed under the Frechét space topology (based on the compact-open topology). Nonetheless, in the cases at hand, that intersection is given by the kernel of an elliptic operator; and the tangent space to the orbits is everywhere a closed

This discussion leads to the topic of *connections*, which we take up in the next Subsection.

2.2. Connections on these bundles

In this Subsection, we note some properties that a *connection* on a principal fibre bundle can have. The basic idea of a connection is of course that it is a preferred way of associating points lying in two "nearby" fibres. It defines a splitting of the tangent space at any point in the bundle into the tangent space of the fibre (the vertical subspace) and a horizontal subspace—which is an "infinitesimal bridge" to associated points in nearby fibres. (Moreover, it does this in a way that respects the group action along the orbit: the 'infinitesimal bridge' is *equivariant.*)

For our purposes, we only need to focus on two properties a connection can have: (a) being integrable, and (b) being unique, or unique subject to certain criteria. Both (a) and (b) will be relevant to our invoking connections, and related ideas like sections (and in Section 5: counterparts), in order to make comparisons between points in nearby fibres.

Here we again see the two broadly different uses of 'point' that we noted in (2) at the end of Section 1. For in the infinite-dimensional bundles of the preceding Subsection, 'point' means an entire field configuration; and so 'comparing points' means comparing global states of spacetime (or space). But Part I was about comparing *spacetime* points: i.e. elements inside the manifold M. Indeed, they were about "identifying" (in less controversial language: putting in correspondence) with each other, two points in two isomorphic spacetimes—which are now configurations $\varphi_1, \varphi_2 \in \Phi$ that are in the same fibre. So it is clear that in order to relate connections in our infinite-dimensional bundles to the debates of Part I, we will need to "descend" and "look inside" the configurations φ, at how the field values are distributed over the points (i.e. the spacetime points!) in M. "Looking inside" in this way will be the topic of Section 3.

To discuss properties (a) and (b), this Subsection will proceed in three stages. (1): We will first consider connections on the finite-dimensional bundles in familiar formulations of general relativity and gauge theories. (So in (1), it will be natural to read 'point' as a point of spacetime or space, as in Part I.) (2): Then we turn to connections on the infinite-dimensional bundles (field-spaces) Φ introduced in the preceding Subsection, though with technical details postponed to the Appendix (Sections Appendix A.2 and Appendix A.3). (3): We complement (2)'s introduction to the general ideas with an intuitive example based on point-particle mechanics.

subspace of the tangent space to Φ. Ultimately, this follows from the facts that (i) the map that embeds $\mathcal{G}$ into Φ is an 'injective' differential operator (or, more correctly, an operator with injective symbol), and (ii) one can define a canonical, invariant metric on Φ, that admits a Levi-Civita connection and thus a notion of exponential map. Thus one can use the Fredholm Alternative (see [14, Sec. 5.3 and 5.9] to show: that that intersection is at most finite-dimensional, and generically is zero; that the generic orbit has the 'splitting' property, i.e. that the total tangent space decomposes into a direct sum of the tangent space to the orbit and its orthogonal complement; and that we can extend the directions transverse to the orbit, so as to construct a small patch that intersects the neighbouring orbits only once. Unlike the field space of gauge potentials, the space of metrics is not a vector or even an affine space, and so we cannot just linearly extend the directions normal to the orbit. And so we extend the normal directions by using the Riemann normal exponential map with respect to the supermetric (cf. [15]), and thereby conclude that, for a sufficiently small radius, the resulting submanifold is transverse to the neighbouring orbits, has no caustics, and is suitably equivariant with respect to the group action (since the exponential map commutes with the diffeomorphisms). Finally, to show that this 'section' is not only transverse to the orbits, but indeed that it intersects neighbouring orbits only once, the orbits must be embedded manifolds, and not just local immersions: this is guaranteed if the group action is proper; which it is in the cases of interest. This is guaranteed when the underlying manifold M is compact. When it is not, the fact that the group action from $\mathcal{G} \times \Phi$ to Φ is jointly continuous on $\mathcal{G}$ and Φ in its domain can also be used, [16, Prop. 130].

Unfortunately, the space of Lorentzian metrics is *not* known to have such a structure. It has only been shown for the space of Einstein metrics, i.e. solutions of the Einstein field equations, that admit a constant-mean-curvature (CMC) foliation and that have a compact spatial Cauchy surface ([10]). The reason is that the CMC-foliability implies a lot of control on the diffeomorphisms orthogonal to the leaves.

Journal of Physics: Conference Series **2533** (2023) 012003 doi:10.1088/1742-6596/2533/1/012003

This is called *best-matching*. It has the advantage of illustrating connections in the space of states, or configurations, like Φ for Yang-Mills and general relativity, while being finite-dimensional. (Again, there are more details in the Appendix (Section Appendix A.1).) And finally after (3), we will end the Subsection with some general comments.

But we first make some remarks about integrability that are true of both the finite- and the infinite-dimensional cases. Recall that a connection is called *integrable* if its distribution of horizontal subspaces is integrable. That is: through any point p in the bundle there is a submanifold through p (whose co-dimension is the dimension of the structure group) whose tangent space at any point is the connection's horizontal space at that point. (This is the global definition: of course, one can also consider locally-existing submanifolds, and local sections.)

Of course, a connection may well not be integrable. For if it is, then for any point p in the bundle, and any closed loop γ in the base-manifold that starts at p's projection in the base-manifold, the horizontal lift of γ to the sub-manifold through p will also be closed. That is, it will return to p: there will be no holonomy. Or in other words: the curvature of the connection is zero. And conversely, if the curvature is non-zero, i.e. there are non-zero holonomies, then the connection is not integrable.

(1): *The finite-dimensional case:*— We recall that for the finite-dimensional bundles in familiar formulations of general relativity and gauge theories, the connection represents a *potential*, and the curvature a *field-strength*. They each have a physical dimension, occur in the theory's equations and are different in different models of the theory. In that sense, they are each 'dynamical'.

More specifically, and philosophically: there is well-nigh universal consensus that in such theories, the field-strength counts as *physically real*. After all, it determines the motion of test matter located at the point in question. On the other hand, the gauge-freedom in the potential means there is debate about whether it counts as real—a debate that nowadays centres around the Aharonov-Bohm effect, especially for gauge theories.[6]

But for our purposes, we do not need to enter this debate.[7] The reason is that our main points, here and in (2) below, do not depend on whether we say that the gauge-freedom of the potential on these finite-dimensional principal fibre bundles renders that potential 'physically unreal'.

The reason why our main points (here and in (2) below) do not thus depend arises from a general and uncontroversial aspect of how physicists, and we, think of the variety of different solutions of our theories. Namely: there being many different solutions represents *contingency*, or *happenstance*. One thinks of "Nature choosing" a solution as a matter of contingent fact (often through a specification of initial and-or boundary conditions). And there is no onus on the theory (or on us) to somehow select or motivate that solution as "preferred".

If one thinks of the variety of solutions in this way (as we do), then there is no question that in the familiar formulations of general relativity and gauge theories using a finite-dimensional principal fibre bundle (and of course, associated vector bundles):

[6] We recall also that while a connection determines all these (possibly zero) holonomies around all closed loops—which are encoded at each point in the bundle by a curvature tensor (obtained, essentially, by differentiation of the connection)—the converse fails. That is: the curvature tensor does not determine the connection. (In abelian Yang-Mills theory, this follows already from the observation that two different connections that are gauge-related determine the same curvature. But in the non-abelian theory, two such connections may have different curvatures (since the curvature transforms in the adjoint representation). Nevertheless, in all cases: the curvature need not determine the connection, since non-gauge-related connections with the same curvature (even zero curvature) may exist.

[7] Since this entire Section treats general relativity and Yang-Mills theories on a par, we should mention that general relativity and non-abelian Yang-Mills theories have analogues of the Aharonov-Bohm effect (cf. e.g. [17, Ch. 4] and references therein, and [18] for a recent experiment).

Journal of Physics: Conference Series **2533** (2023) 012003 doi:10.1088/1742-6596/2533/1/012003

(i) the curvature (field-strength) is unique in the sense of being "chosen by Nature", i.e. being part of the solution that is actually physically correct;

(ii) the connection (potential) is unique, up to a gauge choice, in the very same sense.

These points, (i) and (ii), will be enough for us. For as we will see in a moment, the corresponding points do *not* hold in the infinite-dimensional case. This will prompt the suggestion that in that case, which connection (and even which curvature) is used amounts to a theoretical choice, i.e. a human convention. This may at first seem "unphysical" or "subjective". But rest assured: we will see in Section 5 that this conclusion fits very well with philosophers' notion of counterparts.[8]

(2): *The infinite-dimensional case:*— We turn to connections on the infinite-dimensional bundles (field-spaces) Φ introduced in Section 2.1; (again focussing on the properties of integrability, and uniqueness, that were labelled (a) and (b) in this Subsection's preamble). This is a large, technical and active subject. But in this paper, we will set aside all technicalities and restrict ourselves to introducing work by one of us (HG), together with others (with references): work that sets the scene for Section 3's illustration of counterpart theory. We will leave technical details and examples to the Appendix (cf. Sections Appendix A.2 and Appendix A.3). Then in (3) below, we will illustrate these ideas with a vivid finite-dimensional example, *best-matching.* It is a connection that links the discussion to Part I's theme of comparing points of spacetime or space, rather than points of field-space. So adopting our metaphor: this connection "descends" from field-space, and "looks inside" the field-configurations at how the field-values are distributed over points of spacetime or space.

One of us (HG) and coauthors ([19, 20, 21, 22, 23]) show that on each of Section 2.1's bundles—of Lorentzian manifolds, and of field configurations of Yang-Mills theories (abelian and non-abelian)—one can define various connections; and state various criteria that one can reasonably require the connection to satisfy.

As in the finite-dimensional case, a connection-form is an appropriately smooth Lie-algebra-valued one form on Φ, i.e. $\varpi : T\Phi \mapsto \mathrm{Lie}(\mathcal{G})$. Given $\xi \in \mathrm{Lie}(\mathcal{G})$ and $g(t)$ a curve in $\mathcal{G}$ tangent to ξ at $g(0) = \mathrm{Id}$, we define the *fundamental vertical vector fields* $\xi^{\#}_{\varphi} := (\frac{d}{dt}\varphi^{g(t)})_{|t=0}$: these are the vector fields that generate at each φ the tangent space to the orbit. Then, we can briefly state the conditions on the connection-form as:

$$\varpi(\xi^{\#}) = \xi \tag{3}$$
$$\mathbb{L}_{\xi^{\#}}\varpi = [\xi, \varpi], \tag{4}$$

where $\mathbb{L}_{\xi^{\#}}$ is the Lie derivative on field-space along $\xi^{\#}$, i.e. it is the infinitesimal action of R^*_g: the pull-back of the form along the group action on Φ; and $[\bullet, \bullet]$ is the commutator of $\mathrm{Lie}(\mathcal{G})$. (We here conform to the notation of [22, 23], using double-struck notation to distinguish differential geometric objects in the infinite-dimensional setting.) The finite version of (4) is the perhaps more familiar $R^*_g\varpi = \mathrm{Ad}_g\varpi$, where Ad is the adjoint action of $\mathcal{G}$ on $\mathrm{Lie}(\mathcal{G})$.

The second condition, Eq. (4), tells us the connection must be equivariant, which guarantees that the connection meshes with gauge transformations in the appropriate manner. The first condition, (3), says that the action of the connection-form on the vertical vectors, that are generated by elements of the Lie algebra of the group, give back the corresponding Lie algebra generator. This condition is important for two reasons: (i) to relate the kernel of the connection

[8] We stress that the choice, or convention, we intend here goes beyond the familiar kind of gauge-fixing, such as choosing Coulomb gauge in electromagnetism, which the familiar fibre bundle formulations describe as choices of a section. We will see in (2) that the choice, or convention, we intend here is much more abstract; and we will see in Section 3 that there being a choice does *not* mean that "anything goes", i.e. that no reasons can be adduced for making one choice rather than another.

form to a horizontal complement of the fibers (since the action on the vertical space already 'fills up' the image of the map, and the kernel and the co-kernel sum up to the domain); and (ii) because, jointly with the second condition, Eq. (4), it yields the grain of truth in the drag-along response (cf. Part I, Section 2.2.2), as we will discuss in Section 3.4.

In general, the connections—whether finite or infinite-dimensional; i.e. ω or ϖ—are of course not integrable, and so do not define sections. They are also not unique: more than one satisfies the criteria that are reasonable to impose. But in the case of ϖ, they are also not unique in the sense discussed at the end of (1) above: namely, unique in the sense of being dynamical, and so (putatively) "chosen by Nature" in the solution, i.e. field configuration, that as a matter of happenstance is actually physically correct; as is ω.

In a bit more detail: Gomes and coauthors mostly explore what they call the *Singer-DeWitt* connection. In terms of their general criteria, it is a very convenient choice. For firstly, it is induced by the kinetic term in the theory's action, as follows. In Yang-Mills theories, in a 3+1 split, the action can be written in terms of a kinetic term and a potential. The kinetic term is positive definite and defines a gauge-invariant inner product on field space; this inner product defines horizontal sections as those orthogonal to the group orbits. For more detail, cf. the Appendix, Sections Appendix A.2 and Appendix A.3.

Were we to employ a symplectic, i.e. Hamiltonian, formalism, we would replace field velocities by their momenta, but an equivalent reason can be given for the same connection. Namely, in the alternative symplectic framework, this same connection is also determined by symplectic orthogonality. Thus in the Hamiltonian formulation of electromagnetism, the part of the electric field that is determined by the instantaneous distribution of charges is encoded in a Coulombic potential, and the Singer-DeWitt connection can be obtained as a projector onto the symplectically orthogonal complement of this potential; (see [24] for a philosophical introduction). In the abelian Yang-Mills case, this connection is integrable; and it gives rise to the Coulomb gauge. In the non-abelian Yang-Mills case, it is not integrable; and does not give rise to a gauge-fixing. And in general relativity, it is again not integrable: though it is a natural choice, in particular by giving rise to the perturbative TT-gauge.

Incidentally, it is unsurprising that these connections are in the non-abelian case not integrable: for they are everywhere defined, and had they been integrable, there would be a global section. But as is well-known, no such global sections can exist in the non-abelian theory: this is the Gribov obstruction [5] that we mentioned in (2) of Section 2.1.

So much by way of introducing general ideas and results. We end this Subsection with a vivid, and tractably simple, example of a connection on a bundle whose points, i.e. elements of the bundle, are states of the physical system (rather than field-values at a spacetime or spatial point). Indeed this example is finite-dimensional, both in the fibres and in the base-manifold.

(3): *Best-matching: an illustration with Newtonian particles:*— The system is a set of point-particles embedded in euclidean space; say N of them, with masses $m_\alpha, \alpha = 1, 2, ..., N$. So the "absolutist" or "Newtonian" configuration space of this system is $\mathbb{R}^{3N}$, since for each particle, we need three real numbers to encode its position in absolute space. But suppose we are "relationists" in the tradition of Leibniz and Mach: so that we regard two such absolutist configurations that differ by a spatial translation and-or a spatial rotation as physically the same—in physics jargon, as gauge-equivalent. Then it is natural for us to define the *relational configuration space* as the set of orbits (the quotient set) of $\mathbb{R}^{3N}$ under the action of the euclidean group. Then as in the discussion in Section 2.1, one can investigate the ensuing bundle structure. Indeed, as in (2) of Section 2.1: there are reducible configurations, i.e. configurations that are fixed by an element of the euclidean group. For example, a collinear configuration (in which all the particles lie on a line) is fixed by any rotation about that line. So the orbits are of "different sizes", and the most we can hope for is a stratified orbit structure. In fact, this holds good (see

[25, Sec. III]).

And so we have a principal fibre bundle whose elements are absolutist configurations, whose structure group is the euclidean group, and whose orbits (or elements of the base-manifold) are relational configurations—forming a stratified manifold. It is a finite-dimensional "cousin" of the stratified orbit, or quotient structure, of Riem(M) (reported in footnote 5). Here, we say "cousin", despite a state of our system being a finite set of inter-particle distances $r_{\alpha\beta}(\alpha \neq \beta)$ rather than a Riemannian metric across all of M, since in both cases "there is no time"—the discussion is wholly about space, or matter in space.

There is a very intuitive connection on this bundle. It is essentially due to [26] who call it *best-matching.* But Barbour and Bertotti did not think in terms of bundles, and so did not think of best-matching as amounting to a definition of a connection on this bundle. Here is their original formulation of the best-matching procedure. First, (i) we think of each configuration, before one best-matches, as in its own copy of $\mathbb{R}^3$; then (ii) one copy, together with its configuration, is translated and-or rotated relative to the other copy, so as to minimise the mass-weighted average squared distance for each particle, labelled α, as it occurs in one configuration and as it occurs in the other, i.e. so as to minimise $\Sigma_\alpha m_\alpha \mathbb{d}r_\alpha^2$; then (iii) one says that each point in the first copy is to be identified with that point in the second copy, that it has been translated and-or rotated into coincidence with.

Barbour and Bertotti suggested a vivid word for this identification or correspondence of points between copies of $\mathbb{R}^3$. They say such a pair of points is *equi-local.* So to summarise: though best-matching is defined in terms of embedding the two configurations in a single copy of $\mathbb{R}^3$, it also implies a notion of equi-locality between the two copies of $\mathbb{R}^3$ in which the configurations are first given. So using our jargon of 'threading' (cf. Section 3.2 of Part I), one says that Barbour and Bertotti's equi-locality is a threading of points between two copies of $\mathbb{R}^3$.

So the idea is: from a relationist perspective, an embedding of the two configurations into a single copy of $\mathbb{R}^3$ that minimises $\Sigma_\alpha m_\alpha \mathbb{d}r_\alpha^2$ is an embedding that *best-matches* the configurations. For it erases large contributions to the sum that would arise from embedding (the centre of mass of) one configuration at a great distance from the other: contributions that according to the relationist are an artefact of the mistaken postulate of an absolute space.

It can be shown that best-matching, so defined, is equivalent to a connection: (cf. [19] for the case of spatial diffeomorphisms; and [27, Ch. 5] and [28, Sec. 7.1] for the particle case). The idea is that best-matching provides for a pair of absolutist configurations that are not congruent (and so are in different fibres/orbits of the euclidean group): a decomposition of the difference between them in to (i) a vertical gauge part (an element of the euclidean group which adjusts one of their embeddings in euclidean space, so as to minimise the functional $\Sigma_\alpha m_\alpha \mathbb{d}r_\alpha^2$ of the two configurations) and (ii) a horizontal part, of which the minimum value of this functional gives a measure. For more details, cf. the Appendix especially Section Appendix A.1: it gives the explicit construction of the Singer-DeWitt connection, and its relation to best-matching as just described above.[9]

For our purposes, the best-matching connection has another merit, in addition to being intuitive. Namely: the way that best-matching compares a pair of absolutist configurations "looks inside" the spatial manifold (viz. $\mathbb{R}^3$), in the sense discussed in the preamble to this Subsection. Indeed, best-matching does more than merely "look inside", in the sense of turning our attention to the topic of Part I: i.e. the topic of "identifying" (or: putting in correspondence) spacetime or spatial points that occur in different models of our theory. For best-matching immediately defines equi-locality as such an inter-model identification or correspondence of

[9] By requiring orthogonality with respect to a fiber, best-matching ensures that horizontal (or best-matched) velocities have zero total angular and linear momentum. In [28], the link between best-matching and connections is made fully explicit. They also formulate an extension of these ideas that is able to incorporate non-zero (but conserved) momenta.

Journal of Physics: Conference Series 2533 (2023) 012003 doi:10.1088/1742-6596/2533/1/012003

points. Namely: if two configurations are best-matched, i.e. embedded into a single copy of $\mathbb{R}^3$ so that the embeddings minimise the functional $\Sigma_\alpha m_\alpha \mathrm{d}r_\alpha^2$, then: the spatial points of that copy of $\mathbb{R}^3$ are to be taken as the points of the configurations' spaces.

To sum up the relevance of this illustrative example: these ideas of best-matching and equilocality can be adapted to the infinite-dimensional, field-theoretic case. To make it explicit: we embed two field configurations on the same smooth manifold, and minimise their difference according to a group of transformations and a notion of similarity. We thus replace:

(i) positions/configurations of a set of point-particles by the distributions of fields over spacetime or space;

(ii) $\mathbb{R}^3$ by a differentiable manifold, M, in the case of metrics on spacetime or space; and

(iii) the group of translations and rotations by the group of diffeomorphisms in the case of metrics over spacetime, or vertical automorphisms in the case of gauge theories; thus the structure group (and also the base-manifold) is infinite-dimensional.

We end with three general comments about what we have done in this Section.

(1): We have seen how a bundle of spacetimes—or in gauge theories, a bundle of field configurations—is a natural arena for making comparisons of spacetimes or configurations. But while a section, which provides a "bridge" between the fibres that it intersects, is one way to make such comparisons, some bundles do not have global sections. So one naturally turns to the more general concept of a connection, which gives "infinitesimal bridges", but might not give sections, i.e. might not be integrable.

(2): In general, neither a section nor a connection in these bundles gives what Part I (especially Section 3.2) called a *threading* of the spacetime or spatial points in the copies of the manifold M on which the elements (spacetimes or field configurations) φ of the bundle are built. For although a section or a connection associates two "nearby" elements φ, φ', it might not do so by "looking inside" the elements at the details of the distribution of field-values, and then defining the association in terms of a threading of points in the two copies M.

(3): But at the end of the Section, we saw that best-matching is a connection that *does* look inside, and provides a threading of points in the different copies of its manifold M viz.$\mathbb{R}^3$. Thus best-matching leads us back to the topics of Part I; and so to counterpart theory . . .

3. Non-isomorphic models: counterpart theory

We now turn to how a section or connection on the bundle of spacetimes relates to *counterpart theory*. We recall from Section 2.2.3 of Part I that this is the philosophical doctrine, due to David Lewis ([29], [30, p. 38-43] and [31, Ch. 4]) that any two objects—in particular, spacetime points—in two different possibilities (in philosophical jargon: possible worlds) are never strictly identical. They are distinct: though of course similar to each other in various, perhaps many, respects. And what makes true a proposition that the object a could have had property F (though in fact it lacks F) is—not that in another possible world, a *itself* is F—but that in another possible world, an object appropriately similar to a is F. That object is called the *counterpart*, at this other possible world, of a.

There will be two main themes.

(1): First, a section or connection on this bundle *illustrates* both counterpart theory for spacetime points, and the threading of spacetime points (discussed in Part I's Sections 2.2.3 and 3.2 respectively).

(2): Second, a section or connection on this bundle *generalises* counterpart theory for spacetime points. This will happen in two ways: the first is broadly philosophical, and the second mathematical. Namely:

(i): A section or connection will define a counterpart relation, not just between spacetime points, but between entire spacetimes. There is a good precedent for this in the philosophy literature. Though we did not mention it in Part I, Lewis and others propose not just that ordinary objects within two possible worlds can be counterparts, but also that vast extended regions of two spacetimes (endowed with their distributions of metric and matter fields), and even entire spacetimes, can be counterparts of one another.[10]

(ii): A section or connection will define a measure of how similar any two spacetimes (in general: in different fibres, and not on the section), i.e. any two elements of the bundle, are. This measure is defined in terms of which elements of the bundle's structure group i.e. Diff(M) project the spacetimes into the section. Besides, this measure has a neat transformation property that reveals a grain of truth in what Part I called 'the drag-along response' and associated claims (cf. Sections 2.2.2 and 3.1 of Part I).

We first give, in Section 3.1, a prospectus for how we will discuss these themes. Note that throughout this Section, we will for simplicity consider only vacuum spacetimes (M, g) that are based on a single manifold M; (as we did in Section 2).

3.1. Prospectus

First, we stress that all that follows applies equally to the bundle of field configurations in a Yang-Mills theory. So where we write 'spacetimes', one could also read this as 'field configurations'; and where we write 'Diff(M) is the structure group', one could also read this as 'the group of gauge transformations is the structure group', and so on. But in view of this paper, and Part I, being about the hole argument, and for clarity of exposition, we will focus on spacetimes.

Second, we will for simplicity discuss our themes mostly in terms of choosing a section of the bundle, rather than a connection. We agree that this is a mathematical simplification, since as discussed in Section 2.1, the bundle in fact lacks global sections. We also agree that it is a philosophical simplification, since (as we will discuss below) a section can intersect two fibres representing spacetimes with very different physical contents (i.e. distributions of metric and matter fields: what Part I called 'qualitative profiles'); and one might well feel that in some such cases, no point in one spacetime's copy of the manifold M is similar enough to a point in the other spacetime's copy to be its counterpart. But we make these simplifications only to make the exposition clearer. All the proposals that follow could be stated in terms of a connection on, rather than a section of, the bundle; so that only points (or extended regions) in two sufficiently similar spacetimes $\varphi = (M, g)$, i.e. two spacetimes in sufficiently close fibres, could be counterparts of each other.

The first theme of this Section ((1) in the preamble) is that a choice of a section in the bundle of spacetimes built on a fixed manifold M can be given *in part* by a *pointwise specification*, within the spacetimes φ. That is: by an (infinitary) specification of the *threading* of points across the copies of M in the different fibres. In terms of Section 4's metaphor: it is given by "looking inside" the spacetimes φ and then threading the spacetime points in their copies of M. Furthermore, we think of a counterpart relation for points as specifying the threading. We say 'given *in part*', since the spacetimes being non-isomorphic means that the threading of points between spacetimes does *not* determine how the field values get altered, as well as re-distributed, between the spacetimes. Rather, it is the other way around: we would like the different distributions of fields to specify the threading between their copies of M.

This theme is clearest if we think of a curve γ, between two, non-isomorphic field configurations which do not lie in the same section (taken as the embedded sub-manifold). That is, we are given a smooth map of a real interval, say [0,1], whose image or range is a

[10] Lewis' own reason for this is that he accepts arbitrary *mereological fusions* of objects. As we said in Part I (cf. (3) in Section 2.1)): 'mereology' is modern logic's name for the study of the part-whole relation; and a *fusion* of objects is, intuitively, the aggregate of them. More formally, it is the whole (i) that has them as parts, and (ii) whose every part overlaps at least one of the objects. So for Lewis, a possible world is itself an object, viz. the fusion of all the objects in it; cf. e.g. [31, Section 4.3].

1-parameter family of spacetimes φ, such that $\gamma(0) = \varphi_1$, and $\gamma(1) = \varphi_2$. Since the initial and end points of the family are not isomorphic, this family is *not* defined by a 1-parameter family of diffeomorphisms of M. Along φ_2's orbit, $\mathcal{O}_{\varphi_2}$, there will be a single distribution, φ_2', such that its copy of M is threaded by the identity diffeomorphism to φ_1's copy of M. That distribution, φ_2', lies on the same section as φ_1. All other elements of $\mathcal{O}_{\varphi_2}$ will be threaded by the drag-along from (M, φ_2'). We could project γ onto the section, in which case a 1-parameter family of diffeomorphisms would partly specify the curve along the section. To sum up: we can envisage a curve lying in the section as involving (being partly specified by) a threading of spacetime points by the diffeomorphisms—but this threading is *not* by drag-along. Thus we return to Part I's third and fourth examples of threading, viz. Geroch's definition of limits of spacetimes and quantum reference frames (Sections 3.2.3 and 3.2.4).

On the other hand, the second theme of this Section ((2) in the preamble) is that, relative to a section (or connection), we can think of entire spacetimes as counterparts. As we mentioned: this accords with some of the philosophical literature, which proposes that entire spacetimes can be counterparts. Our main comment about this theme will be that the section defines a 'measure' of how similar any two (models of) spacetimes are. This 'measure' is defined in terms of which elements of the bundle's structure group, i.e. Diff(M), project the spacetimes into the given section. Besides, this 'measure' has a neat transformation property.

We stress that in the case of sections, the 'measure' of similarity is not a measure in the sense of measure theory; nor need it be some sort of minimum distance, or a functional of such a distance. But here we will see an advantage of using infinitesimal sections, i.e. connections, so as to define counterparts in neighbouring fibres. Such counterparts will minimise a well-defined measure of difference between spacetimes, and so our measure of similarity will indeed encode an intuitive sense of similarity.[11]

It will be clearest to develop these themes by focussing on the more novel of them, i.e. (2). For it will be clear how the treatment of (1) gets subsumed by (2). So we will first, in Section 3.2, state the main idea of (2). Then in Section 3.3, we will admit that for an arbitrary section, the spacetimes that it rules to be counterparts can represent very different physical states. That is, they can be on two fibres representing possibilities that are qualitatively ('no matter which point is which') very different. Finally, Section 3.4 will be more positive: we define (even for an arbitrary section) a measure of similarity of spacetimes, with a neat transformation property.

3.2. Spacetimes as counterparts

We briefly recall the notation from Section 2.1. We denote the bundle of spacetimes built on the manifold M by Φ; and we call an element of it, i.e. a Lorentzian manifold, φ. We write each gauge-equivalence class, i.e. fibre or orbit, as $[\varphi]$, and to stress its being an orbit, as O_φ. A section, σ is a choice of a state φ' in each orbit O_φ: so φ' is the representative, according to this section, of O_φ. (Again, this is as in gauge theory: where a section is a gauge-fixing.)

We will not need to distinguish in notation between two different notions of a section:

(i) the function from the bundle's base-manifold (i.e. the set of orbits/gauge-equivalence classes) into the bundle, which is usually written σ; and

(ii) the range of this function, which is the embedded sub-manifold of the bundle. This is often written $\mathcal{F}$: think of '$\mathcal{F}$' as standing for 'folium', i.e. leaf. ($\mathcal{F}$ is also used for a real-valued function on the bundle for which the folium is the level-surface with value 0. That is: iff φ is in

[11] Though this measure of counterparthood can be given an account in terms of similarity in the case of infinitesimal connections, here, for purposes of illustration, we will not dwell on the difference. Note that although this neat measure of counterparthood is for entire spacetimes (theme (2)), we envisage that for "philosophically sensible" sections or connections (cf. Section 3.3), counterparthood for spacetimes is determined by all the counterparthood relations between their points (theme (1)).

Journal of Physics: Conference Series **2533** (2023) 012003 doi:10.1088/1742-6596/2533/1/012003

the section, $\mathcal{F}(\varphi) = 0$.)
So we shall always say 'section' and write σ.

Our key proposal is to extend the familiar idea that a section is a choice of representative states, one for each physical state. Namely: to the proposal that a section is a choice, for any state, and for any fibre/orbit, of which state in that fibre corresponds to—we shall say: is the counterpart of—the given state. That is, we propose:

> A section stipulates, for each state φ (not necessarily in the section), and each orbit $O_{\varphi'}$ (in general distinct from the orbit of φ): which state, φ'' say, in $O_{\varphi'}$ corresponds to (we say: *is the counterpart of*) φ. The stipulation is the obvious one: the counterpart state φ'' is stipulated to be the element of the section within $O_{\varphi'}$.

So the intuitive idea is that in order to go from the arbitrary given state φ to the state in $O_{\varphi'}$ that is its counterpart, one projects φ vertically along its fibre into the section, and then travels horizontally across the section to $O_{\varphi'}$, and thus arrives at the counterpart state φ''.

Note that in these last two paragraphs, the word 'state' could be replaced throughout by 'spacetime'. Similarly, for the discussion to follow. But we will continue, mostly, to say 'state', not least because this signals that our proposal applies—all our considerations apply—equally to Yang-Mills theories.

We will now develop this proposal. We do so in two Subsections. First, Section 3.3 will register a note of caution. For we will admit that since the general idea of a section is formal, what a section stipulates as the counterpart state to the given one may be very dissimilar from it.[12] But we will maintain that this allowance of dissimilarity is no disadvantage.

Then Section 3.4 will be more positive. We will show how—by considering not only projecting φ vertically along its fibre into the section, but also vertical projections along the "destination fibre" $O_{\varphi'}$—we get a natural definition (relative, of course, to the choice of section) of a *measure* of "how much" one state is a counterpart of another.

3.3. The counterpart state can be dissimilar

It is clear that, since our idea invokes *any* section of the bundle, the state in a fibre $O_{\varphi'}$ that is the counterpart of a given state φ can be very dissimilar from φ.

In fact, there are two points here. We will state them, and then urge that allowing such dissimilarities is no disadvantage. That is: this allowance is no objection to our using sections of the fibre bundle of spacetimes to make comparisons of non-isomorphic spacetimes.

First, the dissimilarity can be regardless of how we might choose to thread (in the usual jargon: trans-world identify) spacetime points: regardless of which point is which, in the comparison of the manifolds. For example, the state φ might have an intuitively uniform geometry—it might be a Lorentzian manifold (M, g) with g intuitively close to the Minkowski metric—while all the states in another fibre $O_{\varphi'}$, i.e. all the mutually isometric Lorentzian manifolds in that different fibre, are very "wrinkly". So no matter where the section intersects the fibre $O_{\varphi'}$, the manifold, (M, g') say, which *is* that intersection will have a very different geometry from Minkowski. This means that there is no judicious way to thread (trans-world identify) points between the two copies of M—one in the given $\varphi \equiv (M, g)$ and the other in (M, g') i.e. the section's intersection with $O_{\varphi'}$—that will make the qualitative geometric profiles (i.e. distributions of geometric attributes) within the two copies be similar (i.e. have equal, or nearly equal, field-values).[13]

Second, even if the initially given spacetime (M, g) and its counterpart relative to the section (M, g') *do* have similar qualitative geometric profiles: nevertheless, some assumed or

[12] Besides, as discussed in Section 3.1: even if the corresponding state is similar, the section does not itself specify a threading of spacetime points.

[13] We say 'g intuitively close to the Minkowski metric', instead of 'g the Minkowski metric', so as to set aside the subtleties about extending a slice between different strata, which we discussed in (2) of Section 2.1.

Journal of Physics: Conference Series **2533** (2023) 012003 doi:10.1088/1742-6596/2533/1/012003

mathematically natural threading of points between (M, g) and (M, g')[14] might well not "follow the spirit" of genuine similarity, i.e. of pairing together equal, or nearly equal, field-values. The obvious case of such a threading—assumed or mathematically natural, but violating the spirit of similarity—is the identity map on the base-set of the manifold M. In such a case (i) there *is* a judicious way to thread points—other than by identity map—between the two copies of M—one in the given spacetime $\varphi \equiv (M, g)$ and the other in *some member* of the fibre $O_{\varphi'}$—that makes the points thus threaded to each other have similar qualitative geometric profiles; but (ii) the section does not intersect the fibre $O_{\varphi'}$ in that member.

To sum up these two points: a section is a choice, for each physical state (global distribution of geometric attributes) O_φ, of a representative spacetime $\varphi' \in O_\varphi$. But there need be no—and even a curve in the section need not suggest a—threading of spacetime points between the representative spacetimes that keeps track of equal, or nearly equal, field-values. Here is a toy-model spatial example to illustrate these two points. For the second point, it will take the identity map on the manifold M as the assumed threading.

A Spatial Example:— Take the underlying spatial manifold M to be S^1, which we take as $U(1) = \{\exp(i\theta) : \theta \in [0, 2\pi)\}$. The only field is a real scalar. We take field configurations that differ by the action of a rotation in $U(1)$ to be physically the same: so the idea is that the identity of points is gauge. So the structure group is $U(1)$. We will consider a state φ, in which this field is mostly constant, say 0, but has a sharp peak around $\exp(i\pi) = -1$, i.e. a peak with a small support in θ, centred around $-1 \in S^1$. So the other states in the orbit of φ have a congruent sharp peak centred around various $\exp(i\theta)$, with $\theta \neq \pi$.

To illustrate the first point above—that a section can stipulate a qualitatively dissimilar state as the counterpart of our given state—we need only consider a physical state (fibre, orbit) with, say, seven peaks rather than one, distributed around the circle S^1; and a minimum value in some valley between the peaks that is much greater than 0. Clearly: no matter where exactly these peaks are located around S^1—no matter where the section intersects this fibre—the two physical states are very dissimilar. No judicious threading (in the usual jargon: trans-world identification) of points between the two copies of S^1 can make them similar.

To illustrate the second point above, we choose a physical state (fibre, orbit) that is qualitatively very similar to our given state φ, e.g. with a single peak nearly congruent to the peak in φ. But then we consider a section that intersects this fibre "perversely", i.e. intersects the fibre in a state that does not "place" the peak over the same points as in our given state φ. (The phrase 'the same points' signals our invoking the identity map on the manifold $U(1)$ as the assumed threading. This will be in play in the rest of this Example.)

In a bit more detail: consider a specific physical state (fibre, orbit) $O_{\varphi'}$ that is similar—with the field-value on most of S^1 being the same constant, i.e. 0, but with a less sharp (so: not congruent) peak somewhere on the circle. Suppose our chosen section σ "happens" to intersect $O_{\varphi'}$ at a state, φ'', where the less sharp peak is centred round $\exp(i\frac{\pi}{2})$.

That is: σ happens to stipulate that the representative (orbit-element) that is the counterpart of φ—according to σ—is a state that (i) rotates φ's sharp peak anti-clockwise by $\frac{\pi}{2}$, and (ii) widens it a little to be less sharp (so as to coincide with the peak of φ'').

We thus see that the section σ violates, by a quarter-turn, the spirit of genuine similarity. For comparing these non-isomorphic models in terms of similarity would suggest "trans-world identifying" points below the two peaks. One might say: 'although the peaks are not congruent, we should still think of the two peaks as occurring in the same approximate region, e.g. with

[14] Although a threading is of course *not* defined by a section, it might be defined by a projection of (M, g) in to the section, followed by a curve in the section, i.e. a 1 parameter family of spacetimes, ending in (M, g'), if we suppose that the curve's definition includes a 1 parameter family of diffeomorphisms (as mentioned in Section 3.1)—though of course the spacetimes are *not* given as drag-alongs by the diffeomorphisms.

Journal of Physics: Conference Series **2533** (2023) 012003 doi:10.1088/1742-6596/2533/1/012003

the summit at the same place in the circle. In other words, we should identify "same place" in terms of the values of the scalar field, allowing for approximate matching.' But our section σ does *not* do so. That is of course unsurprising, since we chose it arbitrarily.

It is also worth noting the changes between states given by following curves lying in the section. A curve γ lying in the section σ is a 1-parameter family of states. (More precisely: γ's image is such a family.) We take it as given by a 1-parameter family of diffeomorphisms of S^1, together with 1-parameter family of deformations of the scalar field. (We say 'deformations' because the field's "profile" or "shape" varies along the curve, since the points are physically different—they all lie in different orbits.)

Thus in our Example: there are curves in the section that connect φ and φ''. Some curves make the required changes in states very straightforwardly. For example: they rotate the sharp peak anti-clockwise by $\pi/2$, while also smoothly broadening the sharp peak so that at the end of the curve it is exactly the peak of φ''. And other curves make the required changes in a more complicated way. For example: the sharp peak gets rotated anti-clockwise by $5\pi/2$ (i.e. a full loop around S^1), while being smoothly deformed so as to eventually give exactly the peak of φ''. And this smooth deformation could be "inefficient": the curve in the section might first smoothly make the peak yet sharper, and only then, after the sharpening, broaden it so as to eventually coincide with φ''.

So much by way of an example: we now sum up this discussion. An arbitrary section will in general provide an unintuitive standard of which pairs of states in the various pairs of orbits are counterparts of each other. Here, 'unintuitive' means the pairs of states are in general not similar in their profiles of field-values, nor in their assignments of field-values to points. But we maintain that this allowance of dissimilarity is no objection to our using sections of the fibre bundle of spacetimes for making comparisons of non-isomorphic spacetimes. It is simply an immediate consequence of our having so far put no constraints on the choice of section—and we are at liberty to add such constraints, and so make the counterpart relations (the correspondences) given by sections track similarity better.[15]

[15] Three further comments, in increasing order of mathematics. (i): For example, such constraints might make counterparthood for points and for regions mesh appropriately; cf. footnote 11. (ii): Such constraints might also make use of 'observables' by which, generically, spacetime points can be uniquely labelled (e.g. Komar observables [32], mentioned in Part I, at footnote 20). (iii): Though we have spoken for simplicity of sections, really one should make sense of these constraints on similarity in terms of infinitesimal sections, i.e. connections: (cf. again footnote 11). This is one advantage of using connections in place of sections: connections yield a measure of similarity between spacetimes that can encode information about natural notions of distance or difference between the spacetimes.

That is, assuming any equivariant complement of the tangent spaces to the orbits can be given by an orthogonality condition for some suitable inner product on Φ, we can interpret a pair of counterparts in neighbouring orbits as minimising a corresponding notion of difference between spacetimes. Moreover, we can constrain the choice of inner products by suitable requirements on how we would like to compare the spacetimes, e.g. locally, ultralocally (that is, without taking derivatives of the metric), etc. In more detail, one such constraint is that the connection be given by orthogonality with respect to a supermetric, $\mathbb{G}$ (see the Appendix), but demand that $\mathbb{G}$ be *ultra*local: i.e. expressed as an integral on the local components of the field space vectors, without derivatives. For example, for the space of spatial metrics Riem (cf. Section Appendix A.3), it is easy to see (as first shown by DeWitt) that this ultralocality condition is only satisfied by a one-parameter group:

$$\mathbb{G}(\mathbb{X},\mathbb{Y})_g := \int \sqrt{g}\,(g^{ac}g^{bd} - \lambda g^{ab}g^{cd})(x)\mathbb{X}_{ab}(x)\mathbb{Y}_{cd}(x), \tag{5}$$

which, in dimension 3, are supermetrics (i.e. positive-definite) only for $\lambda < 1/3$. (See also footnote 26.) In the Yang-Mills case, there is similarly a one-parameter family in the non-Abelian case (arising from two inner-products on the Lie algebra), and a single such metric in the Abelian case. Cf. the discussion in (2) and (3) of Section 2.2; and for details, the Appendix.

Essentially this point is a familiar one in gauge theory. In that context, it is often pointed out that the choice of a gauge (i.e. a section of field-space) is in principle arbitrary, in the sense that the physics can be accurately stated in any gauge. But of course that arbitrariness or flexibility is entirely compatible with there being good reasons (e.g. of calculational convenience, or of theoretical clarity) to choose one gauge rather than another (see footnote 8). And these reasons need not be specific to a single problem or group of problems. For example: in the Hamiltonian treatment of electromagnetism, there are theoretical reasons to favour the Coulomb gauge that are wholly general, and so apply to any problem ([24]).

3.4. A measure of similarity—relative to a section

We now set aside the concerns Section 3.3. In this Section, we describe how relative to a choice of section σ (so: no matter how unintuitive its verdicts of counterparthood might be) there is a natural 'measure' of how similar, according to the section, two states (spacetimes) $\varphi, \varphi' \in \Phi$ are. These can be *any* two spacetimes (on the fixed manifold M): neither need be on the section.

As we pointed out at the end of Section 3.1 (and in accord with the closing comments of Section 3.3): for a section, this 'measure' will not be a measure in the sense of measure theory; nor need it be some sort of minimum distance, or a functional of such a distance. But in it infinitesimal version, i.e. for a connection, it can be such a functional: see footnote 15). Nonetheless, this 'measure' is natural in that:

(i) it is determined by the section (for any section);

(ii) it encodes the similarity as, roughly speaking, the element of the diffeomorphism group that, combined with some "horizontal travel" across the section, would transform the spacetime φ into the spacetime φ';

(iii) for two states (spacetimes) that both lie in the section, the measure is what we might call 'utter similarity' i.e. maximal—represented mathematically by the identity element of the group, $Id \in \text{Diff}(M)$:

(iv) the measure has a neat property: it transforms in an appropriate way under the action of a single group element (i.e. a change of section made by translation with a single group element);

(v) finally, the measure and its properties apply to any principal fibre bundle.

Namely: we measure the similarity of two states in terms of how much they need to slide vertically along their respective orbits, in order that their images under the sliding both lie in the section. It is convenient to define this measure —now dropping the scare-quotes—as the product of one structure-group element, with the inverse of the other.

To say this exactly, we need some notation. Since as just noted in (v) above, the ideas work for any principal fibre bundle: we write the structure group as $\mathcal{G}$ with elements g; (so g is a group element, not a Lorentzian metric!). We write the (right) action of the structure group $\mathcal{G}$ on the entire bundle (the field space) Φ with elements φ, in the usual way, as a superscript. The action is:

$$(\varphi \times g) \mapsto \varphi^g \in O_\varphi. \tag{6}$$

A section σ defines what we might call a "ticket map". This map assigns to each $\varphi \in \Phi$, the element in $\mathcal{G}$ that transforms φ to its projection in the section, i.e. to O_φ's intersection with σ. We write this element, the value of the ticket map, as $g_\sigma(\varphi)$. We say "ticket" because the element g is the ticket that projects φ vertically along its orbit to the section. So we write the ticket map as:

$$\varphi \mapsto g_\sigma(\varphi) \in \mathcal{G}; \text{ with } \varphi^{g_\sigma(\varphi)} \in O_\varphi \cap \sigma. \tag{7}$$

There is a straightforward relation between the values of the ticket map for φ and for its transform φ^g, i.e. between $g_\sigma(\varphi)$ and $g_\sigma(\varphi^g)$. We can write it in two equivalent ways. Namely:[16]

$$g_\sigma(\varphi^g) = g^{-1}g_\sigma(\varphi) \;\; ; \;\; \varphi^{g_\sigma(\varphi)} = (\varphi^g)^{g^{-1}g_\sigma(\varphi^g)} . \tag{8}$$

This relation, eq. (8), will imply that the measure of counterparthood we are about to define has a neat transformation property (cf. eq. (11)).[17]

We now define the measure of how much two states, φ_1 and φ_2, need to slide vertically along their respective orbits, in order to both lie in σ, as the product of two "tickets". But for the second ticket in this product, we take the inverse. This expresses the idea of travelling from the first state φ_1 "down" into the section σ, then travelling horizontally across the section to the orbit of φ_2, and then vertically "up" out of the section to arrive at φ_2.

That is, we define: For any section σ,

$$\boxed{\text{Counter}_\sigma(\varphi_1, \varphi_2) := g_\sigma(\varphi_1)g_\sigma(\varphi_2)^{-1}} . \tag{9}$$

And note also that

$$\text{Counter}_\sigma(\varphi_1, \varphi_2)\text{Counter}_\sigma(\varphi_2, \varphi_3) = \text{Counter}_\sigma(\varphi_1, \varphi_3) \tag{10}$$

So our proposed measure, Counter_σ, of how much any two states are counterparts, relative to the section σ, is itself an element of the structure group: in our general notation $\mathcal{G}$, but in the case of interest to us, Diff(M). Of course, 'measure' here has nothing to do with measure theory. Nevertheless, the way that Counter_σ encodes degrees of similarity, as judged by σ, by an element of $\mathcal{G}$ is natural, in that:

(a) if $\varphi_1, \varphi_2 \in \sigma$ (i.e. the states are already in the section), then $\text{Counter}_\sigma(\varphi_1, \varphi_2)$ is the identity element of $\mathcal{G}$;

(b) $\mathcal{G}$ is a topological (in particular, Lie) group, so that it makes sense to talk of being near or far from the identity element.

Thus we see how the definition of Counter_σ, eq. 9, immediately implies the features listed as (i) to (iii) above. We now turn to feature (iv).

For this, we simply apply the first equation of eq. 8 (which is equivalent to the second) to the expression $\text{Counter}_\sigma(\varphi_1^g, \varphi_2^g)$, getting, for any section σ and any g:

$$\text{Counter}_\sigma(\varphi_1^g, \varphi_2^g) = g^{-1}g_\sigma(\varphi_1)(g^{-1}g_\sigma(\varphi_2)^{-1}) = g^{-1}\,\text{Counter}_\sigma(\varphi_1, \varphi_2)g . \tag{11}$$

That is: our counterpart measure, eq. 9, transforms by conjugacy under the action of each element $g \in G$.

We end this Section with two further features—we claim: advantages—of our measure of similarity, i.e. of (9), and its two consequences (10) and (11). Broadly speaking, the first feature, (1), is mathematical; the second, (2), is philosophical.

[16] As stated in Section 3.1, we have here made the simplifying assumption that the section cuts every orbit only once; that is, in the nomenclature of [3, Sec. 3], that the corresponding conditions determining the section satisfy 'universality' and 'uniqueness'.

[17] As an example, take Coulomb gauge in Hamiltonian electromagnetism. The section $\mathcal{F}$ is defined by the regular values $\nabla^i A_i = 0$. The 'ticket map' is given by $g_\sigma(A) = -\nabla^{-2}(\nabla^i A_i)$: which, as is easy to verify, satisfies, for $A_i^g = A_i + \nabla_i g$, the equation: $g_\sigma(A^g) = g_\sigma(A) - g$. When we 'dress' A_i with $g_\sigma(A)$, as in the second equation of (8), we obtain the projection to $\mathcal{F}$, given by $A_i^{g_\sigma(A)} = A_i - \nabla_i(\nabla^{-2}(\nabla^j A_j))$: which by the transformation property of the ticket map satisfies the second equation of (8). In other words, the dressed or projected gauge potential is gauge-invariant.

Journal of Physics: Conference Series **2533** (2023) 012003 doi:10.1088/1742-6596/2533/1/012003

(1): Equation (9) has a consequence that is worth noting. Two models φ_1, φ_2 that lie in the same orbit will always be related by the unique isomorphism that connects them.[18] That is: if $\varphi_2 = \varphi_1^g$, then $\text{Counter}_\sigma(\varphi_1, \varphi_2) = g$, even if neither φ_1 nor φ_2 lie in the section σ. For with the definitions above,

$$\text{Counter}_\sigma(\varphi_1, \varphi_1^g) = g_\sigma(\varphi_1)(g^{-1}g_\sigma(\varphi_1))^{-1} = g. \tag{12}$$

Furthermore, (12), and the preceding equations, have consequences for the traditional (and our Part I's) topic of the counterparthood of single points, rather than entire spacetimes. So let us now focus on the case of Lorentzian metrics and diffeomorphisms: so that the φ are now spacetime metrics g_{ab} (but we will omit indices in what follows) and the group elements $g \in \mathcal{G}$ are now diffeomorphisms $d \in \text{Diff}(M)$.

Given two models, (M, g_1) and (M, g_2), and some choice of σ, we obtain the diffeomorphism $\tilde{d} := \text{Counter}_\sigma(g_1, g_2)$. Thus given point p in model (M, g_1)—we will use a shorthand for such a doublet: (p, g_1)—its unique counterpart (in the sense, of course: 'counterpart of points, not of entire models') at (M, g_2), is $\tilde{d}(p) =: q$. Or in the shorthand notation, the counterpart is (q, g_2).

Now suppose that g_2 is isometric to a third metric, g_3, i.e. that $d^*g_2 = g_3$ for a unique $d \in \text{Diff}(M)$. Then by (12), given any point r in model (M, g_2), its counterpart in (M, g_3) is $d(r)$. Then it follows from (10) that:

$$\text{Counter}_\sigma(g_1, g_3) = d\tilde{d}, \tag{13}$$

and so the counterpart of (p, g_1) in $(M, g_3) = (M, d^*g_2) = (M, g_2^d)$ is $d(q)$, or, in shorthand notation, it is $(d(\tilde{d}(p)), g_2^d) = (d(q), g_2^d)$.[19] In other words, (p, g_1) is the counterpart of (q, g_2) iff it is the counterpart of $(d(q), g_2^d)$. Besides, this property is independent of which section we choose.

(2): Finally, we turn to the promise we made in (2) of Section 2.2 and in Part I, that this Section's framework of counterparts for spacetimes would reveal a grain of truth in what Part I called 'the drag-along response', and in the claim we there rebutted, that the drag-along response is mathematically compulsory; (cf. Part I, Sections 2.2.2 and 3.1 respectively).

What we have in mind is not just the mathematical fact that there is a unique[20] isomorphism between two elements of a fibre. After all, this mathematical fact obviously stands "on its own feet", irrespective of any mention of counterparts. Our point is, rather, that (i) thinking of the bundle interpretatively, in terms of counterparts—with the connotations of similarity and flexibility that that word now has in philosophy—and (ii) defining (degrees of) counterparthood as elements of Diff(M), provides an interpretative context in which this unique isomorphism earns the name of being a "trans-world identification". That is: by putting the unique isomorphism in a single framework together with other diffeomorphisms that are not isomorphisms, this context supports the idea that two spacetime points that the isomorphism relates have the best possible claim to "correspond". That is: the best possible claim to be the "same" as each other—to use, for once, that suspect word 'same'! Thus we come full circle, back to the basic themes that launched the hole argument, and our endeavours here and in Part 1.

Thus, although our counterparts rescue this grain of truth from the drag-along response, they are not subject to our criticisms of that response. First, by weakening the 'identification'

[18] Assuming that the orbit does not have stabilisers, the ticket map is unique, and so is the counterpart relation. If that orbit has stabilisers, the ticket map is not unique, and so there will be an ambiguity: the counterpart relation is unique up to stabilisers.

[19] Here of course, the superscript notation g_2^d just denotes the action of Diff(M) on F; cf. (6).

[20] Again, we note that uniqueness holds only generically: recall that models with automorphisms (reducible configurations) spoil uniqueness. See footnotes 5 and 18. More specifically: entire models will, under our assumptions, always have unique counterparts, even if they are reducible. But if one of the counterpart models is reducible, the spacetime points have duplicates. So one spacetime point in a model of one isomorphism class will have more than one counterpart in the (unique counterpart) model with stabilisers.

Journal of Physics: Conference Series **2533** (2023) 012003 doi:10.1088/1742-6596/2533/1/012003

endorsed by the drag-along to 'best possible correspondence', we are not subject to the problem posed by symmetric models (see (3) in Section 2.2.2 of Part I, and footnote 20). Second, we can detach diffeomorphisms from their role of inducing the 'best possible correspondence': diffeomorphisms are allowed to map between points with different qualitative profiles, so that they don't preserve physical facts pointwise. Thus diffeomorphism symmetry is 'global': it is the totality of qualitative profiles of all the points of a model that is the same as that of diffeomorphism-shifted model—and non-trivial: it can have non-trivial consequences, e.g. through Noether's second theorem (Section 3.2.2 of Part I).

Lest the reader feel there is some sleight of hand, let us be as clear as possible about the interpretive assumptions that salvage this grain of truth from Part I's criticisms of the drag-along. The point is this: independent of our endorsement of what Part I called 'Sophistication', using general relativity will often, if not always, require us to choose a single representative model from each isomorphism class. If such choice is made consistently and the choice is smooth in the space of models, then one gets a section of the fiber bundle. What we have shown above (cf. especially (1)) is that any such choice not only allows a definition of counterparts between spacetime points in non-isomorphic spacetimes. but also: those counterparts also give an equivalence relation between spacetime points (in isomorphic and non-isomorphic models) that respects the drag-along.

4. Conclusion

Let us briefly summarise what we have done in this two-part paper. In Part I we reviewed, from a philosophical perspective, the hole argument, which is about the "identification" of points between isomorphic spacetimes. Here in Part II, we proposed a fibre bundle of spacetimes, with the diffeomorphism group for a fixed spacetime manifold M as structure group, as a framework for comparing *non*-isomorphic spacetimes: using sections, and locally connections, as "bridges" between elements in different fibres of the bundle. We stressed that this proposed framework is strongly analogous to existing field-space formulations of gauge theories. In particular, the technicalities about our proposal's connections have been developed primarily for gauge theories; as expounded in the Appendix.

For this paper's endeavour, the main contributions of Part I were two ideas. Namely: the philosophical idea of points being counterparts; and the more technical idea of threading, i.e. the idea that mathematics and physics sometimes associate, or "identify", points in different manifolds by a mapping *other than* isomorphism. These ideas were combined with the idea of a fibre bundle of spacetimes, in this paper's Section 3. There, we saw how to define a measure of similarity or counterparthood (relative to a section) for entire spacetimes, in terms of the bundle's structure group; and we even showed that there was a grain of truth in doctrines that our Part I had rebutted, to the effect that isomorphism is the only good way to associate, or "identify", points. Thus we hope that in the philosophical controversies about the hole argument, some peace might break out . . .

Of course, various questions about our proposed framework remain to be explored. Two obvious ones, from a philosophical and a physical perspective respectively, are as follows.

(a): Which of the various versions of counterpart theory, that are to be found in the logico-philosophical literature (whose details we have here ignored), fits best with our proposed comparisons of spacetimes, and of spacetime points, using sections and connections on a fibre bundle?

(b): Which of the various threading schemes for non-isomorphic spacetimes that are to be found in the mathematical physics literature—Part I described only two such schemes —fits best with the known connections on our fibre bundle: in particular, with the Singer-DeWitt connections we have expounded in the Appendix.

Good questions indeed: but sufficient unto the day is the labour thereof . . .

APPENDIX

Appendix A. The Singer DeWitt connection: construction and examples

In this Appendix, we will give an introduction of the Singer DeWitt (SdW) connection, as described in Section 2.2.

We start by giving an explicit formula for both the connection and its associated curvature. Then we apply this definition in different theories: in Section Appendix A.1, to Newtonian point-particles, with translation and rotational symmetry; in Section Appendix A.2, to pure Yang-Mills theory and its standard gauge symmetry; and in Section Appendix A.3, to the space of Riemannian metrics, with diffeomorphism symmetry.

As explained in [22, Sec. 4]: given a $\mathcal{G}$-invariant metric on field space, i.e. $\mathbb{G}(\bullet,\bullet)$ such that $\mathbb{L}_{\xi^\#}\mathbb{G} = 0$ for all $\xi \in \text{Lie}(\mathcal{G})$, we will find a unique connection ϖ associated to it (see Equation 4.5 (ibid).[21]

In more detail: A configuration-space metric and a vertical direction supply enough ingredients to define a connection if and only if the directions orthogonal to the orbit remain orthogonal by pull-back along the group action. This is particularly straightforward if the orbits are Killing directions; i.e., the fundamental vector fields $\xi^\#$ are Killing fields of $\mathbb{G}$, i.e.

$$\mathbb{L}_{\xi^\#}\mathbb{G} = 0 \qquad \text{for all } \xi \in \text{Lie}(\mathcal{G})\,. \tag{A.1}$$

For such a field-space metric $\mathbb{G}$, we *define* the *Singer-DeWitt (SdW) connection* ϖ by demanding the following orthogonality relation:

$$\mathbb{G}(\xi^\#, \widehat{H}(\mathbb{X})) \equiv \mathbb{G}(\xi^\#, \mathbb{X} - \varpi(\mathbb{X})^\#) = 0, \tag{A.2}$$

for all $\xi \in \text{Lie}(\mathcal{G})$ and all $\mathbb{X} \in \mathfrak{X}^1(\Phi)$. The preceding equation holds pointwise on Φ, where the vector field $\mathbb{X}$ identifies, for each configuration $\varphi \in \Phi$, a tangent vector $\mathbb{X}_\varphi \in \mathrm{T}_\varphi\Phi$. But we will omit subscripts. Here, $\widehat{H}$ stands for the horizontal projection induced by $\mathbb{G}$, so we cold write $\varpi(\cdot)^\# = \widehat{V}(\cdot)$.

Formally, equation (A.2), defining an SdW connection, can be solved for ϖ as follows. Let $\mathbb{Q}_{ab}$ be the pullback to $\text{Lie}(\mathcal{G})$ under $\cdot^\#$ of the metric induced from $\mathbb{G}$ on the fibres as expressed in the $\{\tau_a\}$ basis of the Lie-algebra:

$$\mathbb{Q}_{ab} = \mathbb{G}(\tau_a^\#, \tau_b^\#), \tag{A.3}$$

and $\mathbb{Q}^{ab}$ its inverse. Note that $\mathbb{Q}_{ab}$ does not in general coincide with the (point-wise extensions of the) Killing form in $\mathfrak{g}$ (the natural inner product of the Lie-algebra).

Expanding $\varpi = \varpi^a\tau_a$, equation A.2 can be written as $\mathbb{G}(\tau_a^\#, \mathbb{X}) = \mathbb{Q}_{ab}\mathbb{i}_{\mathbb{X}}\varpi^b$, which is readily inverted as

$$\varpi = \mathbb{Q}^{ab}\mathbb{G}(\tau_b^\#, \cdot)\tau_a\,. \tag{A.4}$$

Note that $\mathbb{G}(\xi^\#, \cdot)$ accepts field-space vectors and hence defines a one-form in field-space.

From the last equation, we immediately obtain the first fundamental property required of a connection-form: $\varpi(\xi^\#) = \xi$. See [22, Sec. 4.1, equations 4.7 and 4.8] for a proof that an ϖ

[21] Indeed, the metric need not even be completely preserved by the group action: it suffices that it preserves the orthogonal space to the orbits (see Equation 4.6 (ibid)).

defined by the procedure above will also transform correctly under gauge transformations if $\xi^\#$ is a Killing vector of $\mathbb{G}$.

The resulting relationship between a field-space-metric $\mathbb{G}$ and the curvature $\mathbb{F}$ of the associated ϖ is:

$$\mathbb{G}(\mathbb{F}(\mathbb{X},\mathbb{Y})^\#,\xi^\#) = \mathbb{d}(\mathbb{G}(\xi^\#))(\widehat{H}(\mathbb{Y}),\widehat{H}(\mathbb{X})) \quad \text{for all } \xi \in \mathrm{Lie}(\mathcal{G}),\ \mathbb{d}\xi = 0, \tag{A.5}$$

and any $\mathbb{X},\mathbb{Y} \in \mathfrak{X}^1(\Phi)$. On the right hand side, $\mathbb{G}(\xi^\#) \equiv \mathbb{G}(\xi^\#,\cdot)$ is a one-form on field-space, so $\mathbb{d}\mathbb{G}(\xi^\#)$ is a two-form. By horizontally projecting the dummy vector fields $\mathbb{X},\mathbb{Y}$ on the right hand side, we are taking the horizontal-horizontal part of that two-form. Formally solving for $\mathbb{F}$, we get

$$\mathbb{F} = \mathbb{Q}^{ab}\big(\mathbb{d}\mathbb{G}(\tau_b^\#)\big)_{HH}\tau_a\,, \tag{A.6}$$

which is *the main result of this section.* Note that, in these formulas, $\mathbb{d}$ acts on the one-form $\mathbb{G}(\xi^\#)$. Even if ξ is taken to be configuration-independent; i.e., $\mathbb{d}\xi = 0$, the operator $\cdot^\#$ generically introduces configuration-dependence. For a proof of (A.5) (and thus the origin of (A.6)) see [22, Sec. 4.2, equation 4.12].

Appendix A.1. The SdW connection for Newtonian particles

Given a Singer-DeWitt connection, ϖ, the best-matched configurational velocities will be given by just the horizontal projection with respect to ϖ.

To explicitly obtain the Singer-DeWitt connection in the case of N Newtonian particles and compare it with best-matching, we will take the Lagrangian to be of the form

$$\mathcal{L} = K(\mathbf{r},\dot{\mathbf{r}}) - V(\mathbf{r}); \tag{A.7}$$

where, in this non-relativistic configuration space of N particles, we revert to the standard dot notation; i.e. $\dot{\mathbf{r}}_\alpha$. The total kinetic energy of the system, which will determine our configuration space metric, is

$$K(\dot{\mathbf{r}}) = \frac{1}{2}\sum_\alpha m_\alpha|\dot{\mathbf{r}}_\alpha|^2 = \frac{1}{2}\sum_\alpha m_\alpha(\dot{\mathbf{r}}_\alpha\cdot\dot{\mathbf{r}}_\alpha)\,, \tag{A.8}$$

where $\cdot$ is the Euclidean inner product in $\mathbb{R}^3$. The kinetic term K is equivalent to a choice of inner product:

$$\mathbb{G}(\dot{\mathbf{r}},\dot{\mathbf{r}}') = \frac{1}{2}\sum_\alpha m_\alpha\dot{\mathbf{r}}_\alpha\cdot\dot{\mathbf{r}}'_\alpha. \tag{A.9}$$

Equation (A.9) is clearly invariant under the time-independent transformations:

$$\mathbf{r}_\alpha \mapsto \mathbf{r}_\alpha + \mathbf{v} \tag{A.10a}$$

$$\mathbf{r}_\alpha \mapsto \mathbf{r}_\alpha R\,. \tag{A.10b}$$

Infinitesimally, the group actions of $\mathbb{R}^3$ and SO(3) correspond to, respectively (for $\mathbf{v}\in\mathbb{R}^3$ and $\xi \in \mathfrak{so}(3)$):

$$\delta_{\mathbf{v}}\mathbf{r}_\alpha = \mathbf{v} \tag{A.11a}$$

$$\delta_\xi\mathbf{r}_\alpha = \mathbf{r}_\alpha\xi\,. \tag{A.11b}$$

In terms of the Lie derivative (where $\delta_\xi\mathbf{r} =: \xi^\#$),

$$\mathbb{L}_{\xi^\#}\mathbb{G}(\mathbb{v},\mathbb{v}') = \frac{1}{t}\lim_{t\to 0}(\mathbb{G}(\mathbb{v},\mathbb{v}') - \mathbb{G}(\mathbb{v}\xi,\mathbb{v}'\xi)) = 0\,.$$

In other words: the metric has Killing directions along $\xi^{\#}$.

For translations, the generators of the algebra $\mathbb{R}^3$ can be taken to be the unit vectors, $\tau_a = e_a$, given by $\{e_x, e_y, e_z\}$. The vertical metric $\mathbb{Q}^{ab} = \delta^{ab}$, and therefore

$$\varpi_{\text{trans}}(\dot{\mathbf{r}}) = (\sum_\alpha m_\alpha \dot{\mathbf{r}}_\alpha \cdot e_a) e_a = \sum_\alpha m_\alpha \dot{\mathbf{r}}_\alpha,$$

since $\sum |e_a\rangle\langle e_a|$ is the identity operator.

An associated connection-form defines, infinitesimally along the trajectory of the system, the standard of linear translations. The translational connection therefore yields the linear momentum of the configurational velocity. Horizontal motion coincides with a choice of coordinate system for which the total *linear* momentum vanishes. In other words, for an arbitrary velocity $\dot{\mathbf{r}}_\alpha$, the horizontal, or best-matched velocity $\dot{\mathbf{r}}_\alpha - \varpi_{\text{trans}}(\dot{\mathbf{r}})$ has vanishing linear momentum. From (A.6), the curvature clearly vanishes, since neither $\mathbb{G}$ nor $\tau_a^{\#}$ depend on the configuration. This is the main difference compared with the rotational case, which we now assess.

For rotations, we get a vertical metric that is just the moment of inertia tensor:

$$\mathbb{Q}_{ab} = M_{ab} = \frac{1}{2}\sum_\alpha m_\alpha \left(|\mathbf{c}_\alpha|^2 \delta_{ab} - c_{\alpha a} c_{\alpha b}\right) . \tag{A.12}$$

Because

$$\mathbb{G}(\cdot, J_b^{\#}) = \frac{1}{2}\sum_\alpha m_\alpha \mathrm{d}c_\alpha^i c_\alpha^k \epsilon_{ibk}, \tag{A.13}$$

where $\mathrm{d}c_\alpha^i$ is a basic configuration space 1-form (like $\mathrm{d}x$ would be in space-time), we obtain the following expression for the connection-form from (A.4):

$$\varpi(\dot{\mathbf{c}}) = (M^{-1})^{ab} \left(\frac{1}{2}\sum_\alpha m_\alpha \epsilon_{bij} c_\alpha^i \dot{c}_\alpha^j \right) J_{(a)} . \tag{A.14}$$

Given an infinitesimal change of configuration, (A.14) provides the necessary rotation for that change to carry no angular momentum. By the properties of the connection-form—arising from orthogonality to the fibre with respect to a $\mathcal{G}$-invariant metric—this adjustment is gauge-covariant; i.e., it does not depend on the orientation of the configuration that we started from. The connection-form defines a standard of orientation infinitesimally along a curve.

It is also easy to write the vertical projection. From (A.4), we have $\widehat{V} = \mathbb{Q}^{ab}\mathbb{G}(J_b^{\#}, \cdot) J_a^{\#}$.[22] If we use equation (A.12) and (A.14) we obtain:

$$\widehat{V}(\dot{\mathbf{c}}) = \varpi(\dot{\mathbf{c}})^{\#} = \sum_\alpha m_\alpha \dot{\mathbf{c}}_\alpha \times \mathbf{c}_\alpha = \mathbf{L}\, , \tag{A.15}$$

which is the angular momentum. The horizontal projection is just its complement: $\widehat{H} = \mathbb{1} - \widehat{V}$. Given a generic centre-of-mass configurational velocity $\dot{\mathbf{c}}_\alpha$, the best-matched, or horizontal velocity $\dot{\mathbf{c}}_\alpha - \widehat{V}(\dot{\mathbf{c}})$ has vanishing angular momentum.

Curvature implies that, for a closed loop in the base space, the orientation may change even for motion with zero angular momentum. We can now write the curvature using (A.6). First, from (A.13):

$$\mathrm{d}\mathbb{G}(\cdot, J_b^{\#}) = \frac{1}{2}\sum_\alpha m_\alpha \epsilon_{ibk}\, \mathrm{d}c_\alpha^i \curlywedge \mathrm{d}c_\alpha^k,$$

[22] By writing the basis $J_a^{\#} = |\ell_a\rangle$ and $\mathbb{G}(J_b^{\#}, \cdot) = \langle \ell_b|$ to match [25]'s notation, we obtain their equation 5.42: $\Pi_V = |\ell_a\rangle (M^{-1})^{ab} \langle \ell_b|$.

Journal of Physics: Conference Series **2533** (2023) 012003 doi:10.1088/1742-6596/2533/1/012003

where $\curlywedge$ is the exterior differential for forms in configuration space. From this equation it is apparent that if the group action on configuration space did not depend on the configuration, the exterior derivative $\mathbb{d}$ would have nothing to act non-trivially on. We would then obtain $\mathbb{d}\mathbb{G}(\cdot, J_b^\#) = 0$, and vanishing curvature as a consequence. This is what occurs for translations. As it stands, the curvature for the rotational bundle can be written as:

$$\mathbb{F}(\dot{c}, \dot{c}') = (M^{-1})^{ab} \left(\frac{1}{2} \sum_\alpha m_\alpha \epsilon_{bij} \widehat{H}(\dot{\mathbf{c}}_\alpha)^i \widehat{H}(\dot{\mathbf{c}}'_\alpha)^j \right) J_{(a)}, \tag{A.16}$$

which only depends on the base space through the moment of inertia tensor and the horizontal projections.

Appendix A.2. The SdW connection for Yang-Mills

The prime example for an ultralocal field-space metric is the gauge supermetric for Yang-Mills (YM) theories. In the following, to emphasize our neglecting matter fields, we introduce the notation $\Phi_{\rm pYM}$ to indicate the field-space of 'pure' Yang–Mills theory. This is constituted by: $\mathfrak{g}$-valued 1-forms over the spacetime manifold[23] M,

$$A = A^a_\mu(x)\tau_a \mathrm{d}x^\mu \in \Lambda^1(M, \mathfrak{g}), \tag{A.17}$$

where $\mathfrak{g} = \mathrm{Lie}(G)$ and $\{\tau_a\}_a$ is an orthogonal basis of the latter. The fundamental vector fields are:

$$\xi^\# = \int \delta_\xi A \frac{\mathbb{d}}{\mathbb{d}A}, \tag{A.18}$$

where

$$\delta_\xi A = \mathrm{D}\xi := \mathrm{d}\xi + [A, \xi] \tag{A.19}$$

with $[\cdot,\cdot]$ the Lie bracket on $\mathfrak{g} = \mathrm{Lie}(G)$, extended pointwise on M to $\mathrm{Lie}(\mathcal{G})$.

In this field-space, our constructions used a positive-definite supermetric. In the Lorentzian case, such an assumption is hard to substantiate. Therefore, we restrict our attention to two cases: (*i*) M is spacetime, but with Euclidean signature; or (*ii*) $M = \Sigma$ represents a (portion) of a spacelike Cauchy hypersurface, in which case: spacetime admits a Lorentzian signature, field-space is understood to be the space of field configurations on Σ, and d is the dimension of Σ rather than spacetime.

As a starting point, before considering more complex situations: consider the case where Σ is a spacelike *compact* Cauchy surface, that is assumed for now to have no boundary and trivial de Rham cohomology. Here g_{ij} is a *fixed* positive-definite metric on Σ; (it is a background structure, not part of field-space). Notice that $\mathbb{G}^{\rm g}$ is independent of A.

The gauge supermetric contracts variations of the gauge field, $\mathbb{X} = \int \mathbb{X}\frac{\mathbb{d}}{\mathbb{d}A} \in \mathrm{T}_A\Phi_{\rm pYM}$, as in[24]

$$\mathbb{G}^{\rm g}(\mathbb{X}, \mathbb{Y}) = \int_\Sigma \mathrm{d}^d x \, \sqrt{g} g^{ij} \delta_{ab} \mathbb{X}^a_i \mathbb{Y}^b_j, \qquad \mathbb{X}, \mathbb{Y} \in \mathrm{T}_A\Phi_{\rm pYM}. \tag{A.20}$$

[23] One could also take the (less popular) parametrization of the degrees of freedom for YM given by ω, a connection on a principal fiber bundle P, with the base space being spacetime M and with fibers isomorphic to G. Then the relation between A and ω requires a section $\sigma : M \to P$, and is given by $A = \sigma^*\omega$. The distinction between these parametrizations is important for non-trivial bundles, where ω exists globally but A exists only locally. Although we will stick to the more popular "physicists' " parametrization, i.e. A, our formalism can be readily adapted with minimal modifications to treating ω as the fundamental variable.

[24] For $\mathbb{G}^{\rm g}$ to be dimensionless (in units of $\hbar$), it has to be multiplied by e^{-2}, where e is the Yang–Mills coupling constant.

To illustrate the general features of the SdW connections, let us solve for the connection arising for the spatial supermetric just introduced. We have

$$0 = \mathbb{G}^{\rm g}(\xi^\sharp, \mathbb{X} - \varpi(\mathbb{X})^{\#}) = \int \mathrm{d}^d x\, \sqrt{g} g^{ij} \delta_{ab} \mathrm{D}_i \xi^a \big(\mathbb{X}^b_j - \mathrm{D}_j(\varpi(\mathbb{X})^b)\big)$$
$$= -\int \mathrm{d}^d x\, \sqrt{g} \delta_{ab} \xi^a \big(\mathrm{D}^i \mathbb{X}^b_i - \mathrm{D}^i \mathrm{D}_i(\varpi(\mathbb{X})^b)\big), \qquad \text{(A.21)}$$

and, using the arbitrariness of $\xi^a(x)$ and $\mathbb{X}^a_i(x)$, we deduce

$$\mathrm{D}^2 \varpi = \mathrm{D}^i \mathbb{d} A_i, \qquad \text{(A.22)}$$

where $\mathrm{D}^2 := (\mathrm{D}^i \mathrm{D}_i)$ is the gauge-covariant Laplacian. The horizontal vector fields are the kernel of ϖ. Contraction with (A.22) shows that, in this simple case, the horizontal vector fields are those which are (covariant-)divergence free: $\mathrm{D}^i \mathbb{X}_i = 0$, since by definition $\mathbb{d} A_i(\mathbb{X}) = \mathbb{X}_i$ for any vector $\mathbb{X}$. We see that SdW connections are generically of the form: 'inverse Laplacian of divergence'.

In an Abelian pure Yang-Mills theory, $\mathrm{D} = \mathrm{d}$, and the above equation (A.22) becomes a Poisson equation for field-space one-forms. The equation therefore has a unique solution (up to a constant). For non-Abelian theories the relevant Laplacian operator is field-dependent (or background-dependent) and the defining equation for ϖ becomes more involved.[25]

To compute the curvature of ϖ, it is most convenient to use equation (A.5), rather than trying to compute it directly from $\mathbb{F} = \mathbb{d}\varpi + \frac{1}{2}[\varpi, \varpi]$. Consider a field-space constant ξ, i.e. $\mathbb{d}\xi = 0$. We have $\mathrm{D}_i \xi^a = \partial_i \xi^a + f^a{}_{bc} A^b_i \xi^c$ with $f^a{}_{bc}$ the structure constants of $\mathfrak{g}$. Then

$$\mathbb{d}\mathbb{G}^{\rm g}(\xi^{\#}) = \mathbb{d} \int \mathrm{d}^d x \sqrt{g} g^{ij} \delta_{ab} \mathrm{D}_i \xi^a \mathbb{d} A^b_j = -\int \mathrm{d}^d x \sqrt{g} g^{ij} \xi^c f_{abc} \mathbb{d} A^a_i \mathbb{d} A^b_j \qquad (\mathbb{d}\xi = 0). \qquad \text{(A.23)}$$

The horizontal projectors on the right hand side of equation (A.5) have the effect of replacing $\mathbb{d}$ with $\mathbb{d}_H$ in the last line. On its left hand side, after an integration by parts, we have

$$\mathbb{G}^{\rm g}(\mathbb{F}^{\#}, \xi^{\#}) = -\int \mathrm{d}^d x \sqrt{g} \xi^a \delta_{ab} \mathrm{D}^i \mathrm{D}_i \mathbb{F}^b. \qquad \text{(A.24)}$$

Hence, by equating the two and using the cyclicity of the structure constants (i.e. for a compact semisimple Lie algebra), as well as the arbitrariness of ξ, we obtain

$$\mathrm{D}^2 \mathbb{F} = g^{ij} [\mathbb{d}_H A_i, \mathbb{d}_H A_j], \qquad \text{(A.25)}$$

or more explicitly, $\mathrm{D}^2 \mathbb{F}^a = f^a{}_{bc} g^{ij} \mathbb{d}_H A^b_i \mathbb{d}_H A^c_j$. This result for the curvature of the Yang-Mills DeWitt connection was r first given by Singer [6], in a context where Σ is an Euclidean spacetime without boundary (rather than a time slice).

Appendix A.3. The SdW connection for Riemannian metrics

We can now be brief. First, we establish the horizontality condition, for a closed manifold, for the space of Riemannian metrics, $g \in \mathrm{Riem(M)}$. Here $\mathcal{G} = \mathrm{Diff(M)}$; and $\mathrm{Lie}(\mathcal{G}) = \mathfrak{X}(M)$, i.e. the smooth vector fields over M.

[25] For a space topology of S^3, gauge group $SU(2)$, and appropriate analytic conditions on the Yang-Mills connection, the kernel of the Laplacian is the stabilizer of A, i.e., Lie algebra elements ξ with $D\xi = 0$. Most Yang-Mills connections have a trivial stabilizer, and the Laplacian is invertible on those.

Unlike the other two cases discussed in this Section, the kinetic term of the general relativistic action in the 3+1 formalism does not give a *bona-fide*, i.e. positive-definite supermetric. The problem is well-known ([33]), and arises from the subtraction of the product of the trace of the tangent vectors on Riem (see footnote 26). If one chooses a gauge-fixing of the Hamiltonian constraint so that the volume element $\sqrt{g}d^3x$ is constant (see [34] for details), then, since the resulting lapse is positive and 'pure-trace' directions don't preserve the gauge-fixing condition, we can find a connection for the remaining spatial diffeomorphisms (see [19]) via the same procedure as we now follow.

In this Section, we will illustrate the connection with a simplified, non-dynamical, choice of supermetric:[26]

$$\mathbb{G}(\mathbb{X}, \mathbb{Y})_g := \int \sqrt{g}\, g^{ac} g^{bd} \mathbb{X}_{ab} \mathbb{Y}_{cd}. \tag{A.26}$$

A vertical element at g_{ab} is, for $X \in \mathfrak{X}(M)$, of the form $X^{\#} := \mathbb{X}_{ab} = \nabla_{(a}X_{b)}$. Thus we obtain, as in (A.21), by a simple integration by parts, that the horizontality condition is:

$$\mathbb{X} \text{ is horizontal iff } \quad \nabla^a \mathbb{X}_{ab} = 0. \tag{A.27}$$

To obtain the connection, we have to solve

$$\nabla^a(\mathbb{X}_{ab} - \nabla_{(a}\varpi(\mathbb{X})_{b)}) = 0 \tag{A.28}$$

for ϖ in terms of $\mathbb{X}_{ab}$. Writing D for the linear differential operator defined by:

$$\nabla^a \nabla_{(a} Y_{b)}) =: D(Y)_b, \tag{A.29}$$

for $Y \in C^\infty(TM)$, we have that D is elliptic. It has a non-trivial finite-dimensional kernel when the metric has Killing directions (which we are ignoring). And thus we can invert (A.28), obtaining:

$$\varpi_b(\mathbb{X}) = D^{-1}(\nabla^a \mathbb{X}_{ab}). \tag{A.30}$$

This can be seen as the generator of the vertical transformation that takes $\mathbb{X}$ to a horizontal vector, i.e. the generator of a horizontal projector.

In other words, ϖ takes an infinitesimally different metric and gives back an infinitesimal diffeomorphism. The interpretation of this infinitesimal diffeomorphism is like that of best-matching. It drags the field-values in one copy of M by whichever infinitesimal diffeomorphism makes the dragged-along values over the first copy of M be, collectively, as close as possible to the values at the points in the other copy of M to which they have been dragged. Here of course, the phrase 'as close as possible' is to be understood according to $\mathbb{G}$.

Acknowledgements
For comments on this material, we are grateful to: the audience, and especially the organizers, at the DICE conference in Castglioncello; the audience and organizers at the August 2022 QISS meeting; the participants at the LSE-Cambridge Philosophy of Physics Bootcamp discussion group; seminar audiences in Bonn, Bristol, Oxford and Warsaw; and Sam Fletcher for comments on a previous version.

[1] Gomes H 2021 *Arxiv: 2110.07203. Submitted.*
[2] Gomes H 2021 *Arxiv: 2110.07204. Submitted.*
[3] Gomes H 2022 *Unpublished*

[26] The DeWitt supermetric with unit lapse would be: $\mathbb{G}(\mathbb{X}, \mathbb{Y})_g := \int \sqrt{g}\,(g^{ac}g^{bd} - \frac{1}{2}g^{ab}g^{cd})\mathbb{X}_{ab}\mathbb{Y}_{cd}$.

[4] Belot G 2018 *Nous* **52** 946–981 (*Preprint* https://onlinelibrary.wiley.com/doi/pdf/10.1111/nous.12200) URL https://onlinelibrary.wiley.com/doi/abs/10.1111/nous.12200
[5] Gribov V N 1978 *Nucl. Phys.* **B139** 1 [,1(1977)]
[6] Singer I M 1978 *Commun. Math. Phys.* **60** 7–12
[7] Ebin D 1970 *Symp. Pure Math., AMS, 11* **11,15**
[8] Palais R 1961 *Ann. of Math.* **73** 295–322
[9] Mitter P K and Viallet C M 1981 *Commun. Math. Phys.* **79** 457
[10] Isenberg J and Marsden J E 1982 *Physics Reports* **89** 179–222
[11] Kondracki W and Rogulski J 1983 *On the Stratification of the Orbit Space for the Action of Automorphisms on Connections. On Conjugacy Classes of Closed Subgroups. On the Notion of Stratification* Preprint: Instytut Matematyczny (Inst., Acad.) URL https://books.google.co.uk/books?id=LK0JrgEACAAJ
[12] Wilkins D R 1989 *Proceedings of the Royal Irish Academy. Section A: Mathematical and Physical Sciences* **89A** 13–34 ISSN 00358975 URL http://www.jstor.org/stable/20489307
[13] Diez T and Rudolph G 2019 *Differential Geometry and its Applications* **65** 176–211 ISSN 0926-2245 URL http://dx.doi.org/10.1016/j.difgeo.2019.03.005
[14] Gilbarg D and Trudinger N 2001 *Elliptic Partial Differential Equations of Second Order* (Springer)
[15] Gil-Medrano O and Michor P W 1991 *Quarterly Journal of Mathematics, (42), 183-202*
[16] Ebin D 1964 *PhD thesis, MIT repository.*
[17] Gomes H 2022 Why gauge? Conceptual Aspects of Gauge theories URL https://arxiv.org/abs/2203.05339
[18] Overstreet C, Asenbaum P, Curti J, Kim M and Kasevich M A 2022 *Science* **375** 226–229 (*Preprint* https://www.science.org/doi/pdf/10.1126/science.abl7152) URL https://www.science.org/doi/abs/10.1126/science.abl7152
[19] Gomes H 2011 *J. Math. Phys. 52, 082501* URL http://www.citebase.org/abstract?id=oai:arXiv.org:0807.4405
[20] Gomes H and Riello A 2017 *Journal of High Energy Physics (JHEP)* **05** 017 (*Preprint* 1608.08226) URL https://link.springer.com/article/10.10072FJHEP0528201729017
[21] Gomes H and Riello A 2018 *Physical Review D* **98**(2) 025013 URL https://link.aps.org/doi/10.1103/PhysRevD.98.025013
[22] Gomes H, Hopfmüller F and Riello A 2019 *Nuclear Physics B* **941** 249 – 315 ISSN 0550-3213 URL http://www.sciencedirect.com/science/article/pii/S0550321319300483
[23] Gomes H and Riello A 2021 *SciPost Phys.* **10**(6) 130 URL https://scipost.org/10.21468/SciPostPhys.10.6.130
[24] Gomes H and Butterfield J 2022 *Erkenntnis*
[25] Littlejohn R G and Reinsch M 1997 *Reviews of Modern Physics* **69** 213–276
[26] Barbour J B and Bertotti B 1982 *Proceedings of the Royal Society of London A: Mathematical, Physical and Engineering Sciences* **382** 295–306 ISSN 0080-4630 (*Preprint* http://rspa.royalsocietypublishing.org/content/382/1783/295.full.pdf) URL http://rspa.royalsocietypublishing.org/content/382/1783/295
[27] Mercati F 2017 *Shape Dynamics: Relativity and Relationalism* (Oxford University Press) (*Preprint* 1409.0105)
[28] Gomes H and Gryb S 2021 *Studies in History and Philosophy of Science Part A* **88** 138–155 ISSN 0039-3681 URL https://www.sciencedirect.com/science/article/pii/S0039368121000704
[29] Lewis D 1968 *Journal of Philosophy* **65** 113–126
[30] Lewis D 1973 *Counterfactuals* (Cambridge, MA, USA: Blackwell)
[31] Lewis D 1986 *On the Plurality of Worlds* (Blackwell Publishers)
[32] Komar A 1958 *Physical Review* **111**(4) 1182–1187 URL https://link.aps.org/doi/10.1103/PhysRev.111.1182
[33] Kuchař K 2011 *International Journal of Modern Physics D* **20** 3–86 (*Preprint* http://www.worldscientific.com/doi/pdf/10.1142/S0218271811019347) URL http://www.worldscientific.com/doi/abs/10.1142/S0218271811019347
[34] Gielen S, de León Ardón R and Percacci R 2018 *Classical and Quantum Gravity* **35** 195009

Journal of Physics: Conference Series **2533** (2023) 012004 doi:10.1088/1742-6596/2533/1/012004

Towards a Field-Theory based Relativistic Quantum Information

Charis Anastopoulos [(a)], **Bei-Lok Hu** [(b)(c)], **and Konstantina Savvidou** [(a)]

[(a)] Department of Physics, University of Patras, 26500 Greece
[(b)] Maryland Center for Fundamental Physics and Joint Quantum Institute
[(c)] University of Maryland, College Park, Maryland 20742-4111 U.S.A.

E-mail: anastop@upatras.gr, blhu@umd.edu, ksavvidou@upatras.gr

Abstract. We present our program for the development of quantum informational concepts in relativistic systems in terms of the unequal-time correlation functions of quantum fields. We employ two formalisms that provide the basis for further developments. (i) The Quantum Temporal Probabilities (QTP) Method for quantum field measurements and (ii) the Closed-Time-Path (CTP) formalism for causal time evolutions. We present the main ideas of QTP and we show how it relates to the CTP formalism, allowing us to express concepts of measurement theory in terms of path-integrals.We also present many links of our program to non-equilibrium quantum field theories. Details can be found in a recent paper by the authors [1].

1. Introduction

Quantum information theory (QIT) is a quantum extension of classical information theory. It has identified new and powerful informational resources for quantum computing, quantum communication, quantum metrology and more. Despite the broad domain of applicability of quantum information technologies, our understanding of QIT is far lagging behind the fully developed quantum theory of nature, namely, quantum field theory (QFT). QFT which has proven its validity and worth in the full range of physical sciences from particle-nuclear physics to atomic, optical and condensed matter physics, from quarks, nucleons to black holes and the early universe. So far, quantum information theory has been largely developed in the context of non-relativistic quantum mechanics, which is a small corner of full fledged QFT. It is ostensibly inadequate when basic relativistic effects like locality, causality and spacetime covariance, need be accounted for. Recognizing the importance of these relativistic effects and seeking to understand the essential roles they play in quantum information ushered in the emergent field of relativistic quantum information (RQI) [2].

1.1. QITs not based on QFT are incomplete

QFT satisfies the principles of quantum theory, but in addition, it is constrained by axioms that govern the effects of the spacetime structure on the causal behavior of quantum systems. Such axioms are largely missing from current quantum information theories. The latters' notion of causality, based on the sequence of successive operations on a quantum system, lacks a direct spacetime representation. As a result, current QITs cannot make crucial relativistic

Journal of Physics: Conference Series **2533** (2023) 012004 doi:10.1088/1742-6596/2533/1/012004

distinctions, for example, between timelike and spacelike correlations, they do not describe real-time signal propagation, and they ignore relativistic constraints on physical operations. A genuinely relativistic QIT must overcome such limitations [1, 3].

Furthermore, experiments that study causal information transfer or gravitational interaction in multi-partite quantum systems require a QFT treatment of interactions for consistency. A non-QFT description is likely to misrepresent either the theoretical modeling of a system or the physical interpretation of the results. This point is crucial for tests of foundational issues of quantum theory invoking quantum information concepts such as entanglement and decoherence. This is especially so for quantum information experiments in space [4, 5] and for experiments designed to explore quantum effects from gravity [6–9].

The introduction of the key concepts of spacetime covariance and causality in QIT forces us to address problems that originate from the foundations of QFT. These include the following.

Quantum States. In set-ups that involve more than two quantum measurements, the standard state-update rule implies that the quantum state is genuinely different when recorded from different Lorentz frames [10]. There is no problem with the theory's physical predictions that are expressed in terms of (multi-time) probabilities [11]. However, the usual notions of quantum information (entropy, entanglement) are defined through the quantum state, and as such, they are ambiguous in relativistic measurement set-ups.

Local Operations. It is a challenging problem to formalize the notion of a localized quantum system in QFT. There are powerful theorems demonstrating that even unsharp localization in a spatial region leads to faster than light signals [12, 13]. Hence, expressing the crucial quantum informational notion of a local operation in terms of spatial localization can lead to conflicts with relativistic causality.

Projective Measurements. There are strong arguments that ideal (i.e., projective) measurements in QFT are incompatible with causality [14, 15], essentially because they change the quantum state over a full Cauchy surface. However, existing quantum information notions (including the very notion of a qubit) presuppose maximal extraction of information through ideal measurements.

We contend that the development of consistent relativistic QIT requires a measurement theory that (i) respects causality and locality, and (ii) it is expressed in terms of quantum fields. Furthermore, this measurement theory ought to be practical, i.e., it must provide non-trivial predictions for experiments that are accessible now or in the near future.

1.2. Past work on QFT measurements
The earliest discussion of measurements on quantum fields was by Landau and Peierls [17], who derived an inequality for the localization of particles. Bohr and Rosenfeld criticized their work [18], and proved that the measurement of definite field properties requires a macroscopic test particle: the particle's charge Q must be much larger than the electron charge e.

The first explicit model for QFT measurements was Glauber's photodetection theory [19, 20] that provided a foundation for the then nascent field of quantum optics. This theory involves unnormalized probabilities for photon detection in terms of the electric field operators $\mathbf{E}(x)$ and the field's quantum state $|\psi\rangle$. The joint probability density $P(x_1, x_2, \ldots, x_n)$ for the detection of a photon at each of the spacetime points $x_1, x_2, \ldots, x_n$ is given by

$$P(x_1, x_2, \ldots, x_n) = \langle\psi|E^{(-)}(x_1)E^{(-)}(x_2)\ldots E^{(-)}(x_n)E^{(+)}(x_n)\ldots E^{(+)}(x_2)E^{(+)}(x_1)|\psi\rangle, \quad (1)$$

where $E^{(+)}$ is the positive-frequency component and $E^{(-)}$ the negative-frequency component of the projected field vector field $\mathbf{n}\cdot\mathbf{E}(x)$. The probability density (1) is essential for the definition of high-order correlations of the electromagnetic field, and consequently, for the description of phenomena like the Hanbury-Brown-Twiss effect, photon bunching and anti-bunching [21].

Glauber's theory has been highly successful, but its scope is limited in that it only applies to the quantum electromagnetic field. Furthermore, it may face causality problems in set-ups that involve the propagation of photons over long distances, because the field splitting into positive and negative frequencies misrepresents retarded propagation.

Perhaps the simplest models for QFT measurements can be constructed using the notion of an *Unruh-DeWitt* (UDW) detector [22, 23]. UDW detectors first appeared in the study of the Unruh effect, where they were employed in order to demonstrate the effects of acceleration on the quantum field vacuum. In a UDW detector, the quantum field is coupled to a point-like system that moves along a pre-determined spacetime trajectory. UDW detectors have found several applications—see, for example, Ref. [24]—besides their use as models for QFT measurements [25]. However, they are limited in that the detector degrees of freedom are not described by a QFT, a problem that may lead to non-causal signals in set-ups that involve multiple detectors.

Measurement models have also been constructed in the context of algebraic QFT [26–29]. The idea is to consider a measured system and a probe / apparatus that are both described by a QFT. The two field systems start separated and interact within a bounded spacetime region. In Minkowski spacetime, this interaction is described by an S matrix, and it leads to correlations between observables on the system and records on the probe. Then, one defines probabilities for the latter in terms of operators that are well defined on the probe's Hilbert space. This approach is fully consistent with QFT, it works for curved spacetimes even in absence of asymptotic in-out regions, but it has not yet been developed into a practical tool capable of concrete physical predictions.

1.3. The Quantum Temporal Probabilities approach

Here we present an approach toward QFT measurements in terms of the Quantum Temporal Probabilities (QTP) method [30–33]. The name of this method is due to its original motivation to provide a general framework for temporally extended quantum observables [34–36].

The key idea in QTP is to distinguish between the time parameter of Schrödinger's equation from the time variable associated to particle detection [37, 38]. The latter is then treated as a macroscopic quasi-classical variable associated to the detector degrees of freedom. A quasi-classical variable is a coarse-grained quantum variable that satisfies appropriate decoherence conditions, so that its time evolution can be well approximated by classical equations [39, 40]. Hence, the detector admits a dual description: in microscopic scales it is described by quantum theory, but its macroscopic records are expressed in terms of classical spacetime coordinates.

In QTP the detector is also described in terms of quantum fields. Glauber's detection theory and Unruh-DeWitt detector models emerge from QTP as limiting cases, the former in the limit of very short detector time-scales, the second in the limit of very short detector length-scales [1]. In comparison to the algebraic QFT approaches to measurements, QTP provides the same results to leading order in perturbation theory, but allows for the definition of observables for the spacetime coordinates, and it is embedded within a nuanced analysis of the quantum-classical transition in the detector.

A key result in QTP is that probabilities for measurements are expressed in terms of specific unequal-time field correlation functions. Such correlation functions are a staple of QFT. Powerful methods have been developed for their calculation and the analysis of their properties. The specific correlation functions relevant to QTP appear in the Closed-Time-Path (CTP) (Schwinger-Keldysh or 'in-in') formalism [41–46]. The CTP formalism improves over the S-matrix (in-out formalism), in that it allows for causal equations of motion, and it has found many applications in nuclear-particle process [47–49], early universe cosmology [50, 51],

Journal of Physics: Conference Series **2533** (2023) 012004 doi:10.1088/1742-6596/2533/1/012004

and condensed matter physics [52, 53]. We demonstrate the link between the two formalisms, and this allows us to translate between the concepts of quantum measurement theory and of quantum field theory.

In QTP, unequal-time correlation functions contain all information about measured probabilities. In particular, the detection probability for N events is a linear functional of a specific $2N$-unequal time correlation function. This has the following implication. For $N = 2$, probabilities of measurement outcomes are related to bipartite entanglement. Hence, all information about bipartite entanglement is contained in the field four-point functions. An analysis at the level of the correlation functions brings us closer to the main ideas of non-equilibrium QFT. Indeed, we can establish a natural relation between QTP and non-equilibrium formalisms that are based on CTP. QTP probabilities function as a registrar of information for the quantum field, they keep track of how much information resides in which level of correlation functions, and how this information flows from one level to the other during dynamical evolution.

We believe that the scheme presented here has good potential to systemize quantum information in QFT, and to identify the parts of this information that is relevant to the field's statistical, stochastic and thermodynamic behavior. Hence, this formalism could provide a concrete method for defining quantum information in QFT via the correlation hierarchy, as has been proposed in Ref. [54]. Such a definition would be very different from definitions of information in standard QIT that is based on the properties of the single-time quantum state.

2. Probabilities for QFT measurements

We first explain the need for a QFT measurement theory, and then present the QTP approach to such measurements. The key property of the QTP probability formula is that the probability density for n measurement events is a linear functional of a specific $2n$ unequal-time correlation function.

2.1. Why we need a QFT measurement theory

Most current applications of QFT involve the S-matrix formalism. For example, S-matrix amplitudes determine scattering cross-sections; S-matrix poles determine the spectrum of composite particles and decay rates. S-matrix theory is defined for set ups with a single state preparation and to a single detection event in the asymptotic future. This gives the impression that there is no need for an elaborate measurement theory.

This impression is wrong, because there are at least two cases, where the S-matrix formulation of QFT does not suffice. First, in quantum optics, we need joint detection probabilities in order to describe phenomena like photon bunching and anti-bunching [21]. In non-relativistic physics, joint probabilities of this type involve the use of the state-update rule, i.e., quantum state reduction. However, a universal rule for reduction is missing in QFT. In practice, joint detection probabilities relevant to experiments are constructed through heuristic arguments, for example, as, for example, in Glauber's photodetection models. Planned experiments in deep space [4, 5] that involve measurement of quantum optical correlations at long distances arguably require a first-principles construction of joint probabilities.

Second, S-matrix is insufficient whenever we are interested in expectation values of physical quantities at finite moments of time, rather than scattering amplitudes [44,45]. Examples include the description of quantum transport in many-body systems [46–48], and the backreaction of quantum fields on the spacetime metric in cosmological and black hole spacetimes [50, 51]. Powerful functional techniques, like the Schwinger-Keldysh method, have been developed to deal with such problems.

2.2. The Quantum Temporal Probabilities Approach: main ideas

The key features of the QTP approach to measurements on quantum fields are the following.

Journal of Physics: Conference Series **2533** (2023) 012004 doi:10.1088/1742-6596/2533/1/012004

(i) The apparatus is fundamentally described in terms of QFT. In particular, the interaction between the measured system and the apparatus is described by a Hamiltonian that is a local functional of quantum fields.

(ii) The measurement apparatus is also assumed to exhibit classical behavior at the macroscopic level. According to the decoherent histories approach to emergent classicality [39, 40], the apparatus pointer is a highly coarse-grained observable, so that histories for measurement outcomes satisfy appropriate decoherence conditions.

(iii) All measurements events are localized in space and in time. For example, a particle detector has a fixed location in a laboratory, and it records an event at a specific moment of time. Both the locus and the time of detection can be random variables. Hence, physical predictions are expressed in terms of probability densities

$$P(x_1, q_1; x_2, q_2, \ldots, x_n, q_n), \tag{2}$$

for multiple detection events. In Eq. (2), x_i stand for spacetime points, q_i stand for any other recorded observable and P is a probability density with respect to both x_i and q_i.

2.3. Detection probability for a single detector

Consider a QFT on Minkowski spacetime M: it is described by Heisenberg-picture fields $\hat{\phi}_r(x)$ that are defined on a Hilbert space $\mathcal{F}$. The Hilbert space carries a unitary representation of the Poincaré group, and The index r runs over spacetime and internal indices.

Let $\mathcal{K}$ be the Hilbert space associated to an apparatus. We assume that the apparatus follows a world tube $\mathcal{W}$ in Minkowski spacetime, and that its size is much larger than the scale of microscopic dynamics. We introduce a field-apparatus coupling with support in a small spacetime region around a point x. The finite spacetime extent of the interaction mimics the effect of a detection record localized at x. Working in the interaction picture, we express the coupling term as

$$\hat{V}_x = \int F_x(y)\hat{C}_a(y) \otimes \hat{J}^a(y), \tag{3}$$

where $\hat{C}_a(x)$ is a composite operator on $\mathcal{F}$ that is local with respect to the field $\hat{\phi}_r(x)$ and a runs over spacetime and internal indices. The current operators $\hat{J}^a(x)$ are defined on $\mathcal{K}$. The switching functions $F_x(y)$ are dimensionless. They vanish outside the interaction region and they depend on the motion of the apparatus. The spacetime volume associated to a switching function is $v = \int dY F_x^2(y)$.

The switching function renders the interaction term (3) Poincaré non-covariant. The use of switching functions originates from von Neumann's modeling of quantum measurements [55], where it serves to localize the system-apparatus interaction in time. Certainly, in a model with a switching function time is not a random variable. A switching function is not needed in QTP, but we employ it in ths work, because it is computationally easier, and it leads to the same expressions for probabilities to leading order in perturbation theory.

The S-matrix associated to Eq. (3) is $\hat{S}_x = \mathcal{T}\exp[-i\int d^4y F_x(y)\hat{C}_a(y) \otimes \hat{J}^a(y)]$, where $\mathcal{T}$ stands for time ordering. To leading order in the interaction,

$$\hat{S}_x = \hat{I} - i\hat{V}_x. \tag{4}$$

Let the initial state of the system be $|\psi\rangle \in \mathcal{F}$ and the initial state of the apparatus be $|\Omega\rangle$. A particle record appears if the detector transitions from $|\Omega\rangle$ to its complementary subspace $\mathcal{K}'$. Once the transition occurred, we measure a property of the particle through a pointer observable

Journal of Physics: Conference Series **2533** (2023) 012004 doi:10.1088/1742-6596/2533/1/012004

q. The latter is described by a family of positive operators $\hat{\Pi}(q)$, such that $\sum_q \hat{\Pi}(q) = \hat{I} - |\Omega\rangle\langle\Omega|$. The pointer observable is coarse-grained, and we take it to stationary with respect to the the self-dynamics of the detector. This is necessary for the record to be preserved after the end of the measurement.

Then, we compute the probability $\text{Prob}(x, q)$ that the detector is excited and records a value q,

$$\text{Prob}(x, q) = \langle\psi, \Omega|\hat{S}_x^\dagger[\hat{I} \otimes \hat{\Pi}(q)]\hat{S}_x|\psi, \Omega\rangle \tag{5}$$

To leading order in perturbation theory

$$\text{Prob}(x, q) = \int d^4y_1 d^4y_2 F_x(y_1)F_x(y_2)G_{ab}(y_1, y_2)\langle\Omega|\hat{J}^a(y_1)\hat{\Pi}(q)\hat{J}^b(y_2)|\Omega\rangle, \tag{6}$$

where

$$G_{ab}(x, x') = \langle\psi|\hat{C}_a(x)\hat{C}_b(x')|\psi\rangle, \tag{7}$$

is a correlation function for the composite operator.

The probability $\text{Prob}(x, q)$ of Eq. (6) is not a density with respect to x, because x appears as a parameter of the switching function. We define an unnormalized probability density $W(x, q)$ with respect to x by dividing $\text{Prob}(x, q)$ with the effective spacetime volume v,

$$W(x, q) = v^{-1}\text{Prob}(x, q). \tag{8}$$

The definition (8) is well justified in classical probability theory, but not in quantum theory. It combines probabilities defined with respect to different experimental set-ups, i.e., different switching functions. Nonetheless, Eq. (8) can be derived as a genuine probability density through the QTP method [1, 32], to leading order in the field-apparatus coupling.

In the QTP derivation, the interaction is present at all times, as the Hamiltonian is time-translation invariant. The functions $F_x(y)$ are not interpreted as switching functions of the interaction, but they describe the *sampling* of the spacetime point. Hence, the spacetime volume $\hat{v}$ is a measure of *coarse-graining*, i.e., of the inaccuracy in the determination of the spacetime point. This point is crucial for a rigorous derivation, because probabilities can only be defined for histories that satisfy a decoherence condition, for which coarse-graining is a prerequisite. In principle, a preferred value of v that corresponds to the coarse-graining scale at which probabilities are well-defined is determined from first-principles—see [30] for explicit calculations in simple models. This means that not all sampling functions $F_x(y)$ are acceptable: their support cannot be made arbitrarily small. Such constraints cannot be seen in the derivation that we presented here.

To further proceed, we assume that the detector carries a representation of the spacetime translation group with generators $\hat{p}^\mu$. We note that the state $|\Omega\rangle$ is not the Poincaré invariant vacuum; it defines a preferred reference system at which its center of momentum has zero three-momentum.

We choose a reference point x_0 in the detector's world-tube, and we write

$$\hat{J}^a(y) = e^{-i\hat{p}\cdot(y-x_0)}\hat{J}^a(x_0)e^{i\hat{p}\cdot(y-x_0)}. \tag{9}$$

It is convenient to take $|\Omega\rangle$ to be *approximately translation invariant*, i.e., to require that

$$\int d^4x F_x(x')\hat{J}^a(x')|\Omega\rangle \simeq \int d^4x F_x(x')e^{-i\hat{p}\cdot(x'-x_0)}\hat{J}^a(x_0)|\Omega\rangle. \tag{10}$$

The idea is that the apparatus is prepared in an initial state that is homogeneous at the length scales that correspond to position sampling and approximately static at the time scales that correspond to time sampling. With this assumption, $\langle\Omega|\hat{J}^a(y_1)\hat{\Pi}(q)\hat{J}^b(y_2)|\Omega\rangle = R^{ab}(y_2 - y_1, q)$, where

$$R^{ab}(x,q) := \langle a,q|e^{i\hat{p}\cdot(x-x_0)}|b,q\rangle \tag{11}$$

is the *detector kernel*, expressed in terms of the vectors $|a,q\rangle = \sqrt{\hat{\Pi}}(q)\hat{J}^a(x_0)|\Omega\rangle$. Note that the Fourier transform of the detector kernel, $\tilde{R}^{ab}(\xi,q) = \int d^4x e^{-i\xi\cdot x}R^{ab}(x,q)$ is given by

$$\tilde{R}^{ab}(\xi,q) = (2\pi)^4 e^{i\xi\cdot x_0}\langle a,q|\hat{E}_\xi|b,q\rangle, \tag{12}$$

where $\hat{E}_\xi = \delta^4(\hat{p}-\xi)$ is the projector onto the subspace with four-momentum ξ^μ.

The simplest switching functions F_x are Gaussians, of the form $F_x(y) = f(x-y)$, where

$$f(x) = \exp[-\frac{1}{2\delta_t^2}(x^0)^2 - \frac{1}{2\delta_x^2}\mathbf{x}^2], \tag{13}$$

where δ_t is the temporal accuracy and δ_x is the special accuracy of the detector. These quantities are macroscopic, because they correspond to the sampling of the detection event.

Gaussian switching functions satisfy the identity

$$f(x)f(x') = f^2\left(\frac{x+x'}{2}\right)\sqrt{f}(x-x'). \tag{14}$$

The spacetime volume v of the interaction region is $v = \pi^2\delta_t\delta_x^3$. We note that the function $\sigma(x) := \frac{1}{v}f^2(x)$ is a normalized probability density on M. Then, we write

$$W(x,q) = \int d^4x'\sigma(x-x')P(x',q), \tag{15}$$

where

$$P(x,q) = \int d^4y\sqrt{f}(y)R^{ab}(y,q)G_{ab}(x-\frac{1}{2}y, x+\frac{1}{2}y), \tag{16}$$

The probability distribution $W(x,q)$ is the convolution of $P(x,q)$ with the probability density $\sigma(x)$ that accounts for the accuracy of our measurements. If $P(x,q)$ is non-negative and the scale of variation with respect to x is much larger than both δ_t and δ_x, we can treat $P(x,q)$ as a fine-grained version of $W(x,q)$ and use this as the probability density for detection.

The kernel $R^{ab}(x,q)$ is typically characterized by a correlation length-scale ℓ and a correlation time-scale τ, such that $R^{ab}(x,q) \simeq 0$ if $|t(\xi)| \gg \tau$ or $|\mathbf{x}(\xi)| \gg \ell$. Both scales ℓ and τ are microscopic and characterize the constituents of the apparatus and their dynamics. If $\ell \ll \delta_x$ and $\tau \ll \delta_t$, then $R^{ab}(x,q)\sqrt{f}(x) \simeq R^{ab}(x,q)$ and we obtain an expression for the probability density $P(x,q)$ that is sampling-independent

$$P(x,q) = \int d^4yR^{ab}(y,q)G_{ab}(x-\frac{1}{2}y, x+\frac{1}{2}y). \tag{17}$$

The probability densities (17) are not normalized to unity. In general, the total probability of detection $P_{det} = \sum_q \int_{\mathcal{W}} d^4xP(q,x)$ is a small number in any perturbative calculations. There is always a probability $P(\varnothing) = 1 - P_{det}$ of no detection. We normalize probabilities by conditioning the probability densities $P(q,x)$ with respect to the existence of a detection record. Hence, we use the probability densities $P(x,q)/P_{det}$.

2.4. Multiple detectors

It is straightforward to obtain the probability formula for the field interacting with n detectors. To leading order in perturbation theory, the probability density for n measurement events is

$$W_n(x_1, q_1; x_2, q_2; \dots; x_n, q_n) = \int d^4x'_1 \dots d^4x'_n \sigma^{(1)}(x_1 - x'_1) \dots \sigma^{(n)}(x_n - x'_n) \\ P(x'_1, q_1; x'_2, q_2; \dots; x'_n, q_n) \tag{18}$$

where

$$P_n(x_1, q_1; x_2, q_2; \dots; x_n, q_n) = \int d^4y_1 \dots d^4y_n \sqrt{f^{(1)}(y_1)} \dots \sqrt{f^{(n)}(y_n)} R^{a_1 b_1}_{(1)}(y_1, q_1) \dots \\ \times \dots R^{a_n b_n}_{(n)}(y_n, q_n) G_{a_1 \dots a_n, b_1 \dots b_n}(x_1 - \frac{1}{2}y_1, \dots, x_n - \frac{1}{2}y_n; x_1 + \frac{1}{2}y_1, \dots, x_n - \frac{1}{2}y_n). \tag{19}$$

Here $R^{(i)}(x, q)$ is the measurement kernel for the i-th detector. The field correlation function $G_{a_1 \dots a_n, b_1 \dots b_n}(x_1, \dots, x_n; x'_1, \dots, x'_n)$ is given by

$$G_{a_1 \dots a_n, b_1 \dots b_n}(x_1, \dots, x_n; x'_1, \dots, x'_n) = \langle \psi | \mathcal{T}^*[\hat{C}^{(1)}_{b_1}(x'_1) \dots \hat{C}^{(n)}_{b_n}(x'_n)] \\ \times \mathcal{T}[\hat{C}^{(n)}_{a_n}(x_n) \dots \hat{C}^{(1)}_{a_1}(x_1)] | \psi \rangle \tag{20}$$

where $\mathcal{T}^*$ stands for reverse time ordering.

Again, in the appropriate regime, the probability becomes independent of the switching functions, and equal to

$$P_n(x_1, q_1; x_2, q_2; \dots; x_n, q_n) = \int d^4y_1 \dots d^4y_n R^{a_1 b_1}_{(1)}(y_1, q_1) \dots R^{a_n 1 b_n}_{(n)}(y_n, q_n) \\ \times G_{a_1 \dots a_n, b_1 \dots b_n}(x_1 - \frac{1}{2}y_1, \dots, x_n - \frac{1}{2}y_n; x_1 + \frac{1}{2}y_1, \dots, x_n + \frac{1}{2}y_n) \tag{21}$$

It is convenient to express the probability densities (21) using an abstract notation. We use small Greek indices $\alpha, \beta, \gamma \dots$ for the pairs (x, a) where x is a spacetime point and a the internal index for the composite operators $\hat{C}_a$. All indices in a time-ordered product are upper, and all indices in an anti-time-ordered product are lower. Hence, we write the correlation functions (20) as

$$G^{\alpha_1 \alpha_2 \dots \alpha_n}_{\beta_1 \beta_2 \dots \beta_n}$$

Let $x \in M$ be a spacetime point and $q \in \Gamma$ any other recorded observable. We define the set of elementary events $Z := M \times \Gamma \cup \{\varnothing\}$ by z, where $\varnothing$ is the event of no detection. Then, we express the kernel

$$\sigma[x - \frac{1}{2}(y + y')] \sqrt{f}(y - y') R^{ab}(y - y', q)$$

as $R^{\beta}_{\alpha}(z)$ where $z \in Z$, α stands for (y, a), β for (y', b). We will use the same symbol for the approximate expression $\delta[x - \frac{1}{2}(y + y')] R^{ab}(y - y', q)$. We employ the Einstein summation convention over Greek indices, in order to denote sum over the discrete index a and spacetime integral.

Then, the probability formula (21) reads

$$P_n(z_1, z_2, \dots, z_n) = G^{\alpha_1 \alpha_2 \dots \alpha_n}_{\beta_1 \beta_2 \dots \beta_n} \; {}^{(1)}R^{\beta_1}_{\alpha_1}(z_1) \; {}^{(2)}R^{\beta_2}_{\alpha_2}(z_2) \dots {}^{(n)}R^{\beta_n}_{\alpha_n}(z_n). \tag{22}$$

3. Relation of the QTP approach to the Closed-Time-Path formalism

The probability density (19) for n measurement events is a linear functional of the $2n$-point unequal-time correlation function (20). This correlation function has n time-ordered arguments and n anti-time-ordered arguments. It does not appear in the usual S-matrix description of QFT; the correlation functions in the S-matrix description involve only time-ordered arguments. Rather, the correlation function (20) appears in the Schwinger-Keldysh or 'in-in' or Closed-Time-Path (CTP) formalism of QFT. Since by now the Schwinger-Keldysh formalism is quite well known and popularly used in many fields of physics we shall not belabour it but refer the reader to some source materials [56–58] where this method is used for the exploration of themes relevant to our present discussions, foremost, quantum correlations. How the CTP formalism overcomes the deficiencies of the S-matrix formulation is discussed in many original papers on CTP, e.g., [44, 45] – a short summary can be found in our recent paper [1].

In the CTP formalism, we couple the field to two different external sources $J^a(x)$ and $\bar{J}^a(x)$, and we define the CTP generating functional

$$Z_{CTP}[J, \bar{J}] = \langle\psi_0|\hat{U}^\dagger[\bar{J}]\hat{U}[J]|\psi_0\rangle, \tag{23}$$

By definition, $Z[J, J] = 1$ and $Z^*[J, \bar{J}] = Z[\bar{J}, J]$. The state $|\psi_0\rangle$ is defined in the distant past, i.e., prior to any time at which $J(x)$ has support—it is an *in* state.

The CTP generating functional describes correlation functions with n time-ordered and m anti-time-ordered entries,

$$G^{n,m}_{a_1\ldots a_m, b_1\ldots b_n}(x_1, \ldots, x_n; x'_1, \ldots, x'_n) = \langle\psi_0|\mathcal{T}^*[\hat{C}^{(1)}_{b_1}(x'_1)\ldots\hat{C}^{(n)}_{b_m}(x'_m)] \times\mathcal{T}[\hat{C}^{(i)}_{a_n}(x_n)\ldots\hat{C}^{(1)}_{a_1}(x_1)]|\psi_0\rangle \tag{24}$$

which can standardly be expressed as functional derivatives of $Z_{CTP}[J, \bar{J}]$. For a vacuum initial state, the generating functional has a path integral expression

$$Z_{CTP}[J, \bar{J}] = \int D\phi\, D\bar{\phi}\, e^{i\{S[\phi]-S[\bar{\phi}]+\int d^4x\, [J^a(x)C_a(x)-\bar{J}(x)\bar{C}_a(x)]\}}, \tag{25}$$

where $\bar{C}$ is defined a functional of $\bar{\phi}$.

The relation between the QTP description of measurements and the CTP formalism is more transparent, if we use the index notation of Sec. 2.4. For consistency, the sources J have a lower Greek index, and the sources $\bar{J}$ have an upper Greek index. Then, we write the CTP generating functional as

$$Z_{CTP}[J, \bar{J}] = \sum_{n,m=0}^{\infty} \frac{i^{m-n}}{n!m!} G^{\alpha_1\ldots\alpha_n}_{\beta_1\ldots\beta_m} J_{\alpha_1}\ldots J_{\alpha_n}\bar{J}^{\beta_1}\ldots\bar{J}^{\beta_m}, \tag{26}$$

where $G^{\alpha_1\ldots\alpha_n}_{\beta_1\ldots\beta_m}$ represents the correlation functions (24). These satisfy

$$G^{\alpha_1\ldots\alpha_n}_{\beta_1\ldots\beta_m} = i^{n-m}\left(\frac{\partial^{n+m}Z_{CTP}[J, \bar{J}]}{\partial J_{\alpha_1}\ldots\partial J_{\alpha_n}\partial\bar{J}^{\beta_1}\ldots\partial\bar{J}^{\beta_n}}\right)_{J=\bar{J}+0}. \tag{27}$$

The probability densities (22) involve *balanced* correlation functions, i.e., correlation functions with an equal number of upper and lower indices. We can construct a generating functional that contains only such functions. The key observation is that such correlations contribute to the sum only through products of the form $J_\alpha\bar{J}^\beta$. Hence, the natural source for a diagonal generating functional Z^d_{CTP} that only involves balanced correlation functions is a 'tensor' L^β_α. We define

$$Z^d_{CTP}[L] = \sum_{n=0}^{\infty} \frac{1}{n!} G^{\alpha_1 \ldots \alpha_n}_{\beta_1 \ldots \beta_m} L^{\beta_1}_{\alpha_1} \ldots L^{\beta_N}_{\alpha_N}. \tag{28}$$

Suppose now that we consider only measurements of a single type, i.e., all detector kernels $R^\beta_\alpha(z)$ are identical. Then, we can define a moment-generating functional for all probability densities (22), in terms of sources $j(z)$,

$$Z_{QTP}[j] = \sum_{n=0}^{\infty} \sum_{z_1, z_2, \ldots, z_n} \frac{1}{n!} P_n(z_1, z_2, \ldots, z_n) j(z_1) \ldots j(z_n). \tag{29}$$

It is straightforward to show that

$$Z_{QTP}[j] = Z^d_{CTP}[R \cdot j], \tag{30}$$

where $(R \cdot j)^B_A = \sum_z R^B_A(z) J(z)$.

Eq. (30) is a fundamental relation for quantum measurements in QFT, as it relates the moment generating functional for a hierarchy of measured probability densities to the generating functional of unequal-time correlation functions

It is straightforward to write a path integral expression for $Z^d_{CTP}[L]$ for a vacuum initial state

$$Z^d_{CTP}[L] = \int D\phi \, D\bar{\phi} \; e^{iS[\phi] - iS[\bar{\phi}] + \int d^4x d^4x' C_a(x) \bar{C}_b(x') L^{ab}(x,x')}. \tag{31}$$

To obtain a simple path integral expression for a broader class of states, we recall that many field initial states can be obtained from the action of an external source $\zeta(x)$ on the vacuum, i.e., they are of the form $|\psi_0\rangle = \hat{U}[\zeta]|0\rangle$, where now we write $\hat{U}[\zeta] = \mathcal{T} \exp\left[i \int d^4X \zeta^k(X) \hat{A}_k(X)\right]$ in terms of composite operators $\hat{A}_k(x)$ that differ, in general from $\hat{C}_a(x)$—see [1] for examples.

Hence, for a quantum state that is obtained from an external source ζ_k, we write the path integral expression

$$Z^d_{CTP}[f, L] = \int D\phi \, D\bar{\phi} \; e^{iS[\phi] - iS[\bar{\phi}] + i \int d^4x \; \zeta^k(x)[A_k(x) - \bar{A}_k(x)] + \int d^4x d^4x' C_a(x) \bar{C}_b(x') L^{ab}(x,x')}, \tag{32}$$

where we must assume that the spacetime support of the kernel L^{ab} is later than the support of ζ (state preparation is prior to measurement). By Eq. (30)

$$Z_{QTP}[f, j] = \int D\phi \, D\bar{\phi} \; e^{iS[\phi] - iS[\bar{\phi}] + i \int d^4x \; \zeta^k(x)[A_k(x) - \bar{A}_k(x)]} \times e^{\sum_z \int d^4x d^4x' C_a(x) \bar{C}_b(x') R^{ab}(x,x';z) j(z)} \tag{33}$$

The probability densities for n measurement events are obtained from functional variation of $Z_{QTP}[\zeta, j]$ with respect to j at $j = 0$ For example, the single-event probability density $\hat{P}_1(x, z)$ of Eq. (17) is given by the path integral

$$P_1(x, q) = \int D\phi \, D\bar{\phi} \left(\int d^4y C_a(x + \frac{1}{2}y) \bar{C}_b(x - \frac{1}{2}y) K^{ab}(y, q) \right) \times e^{iS[\phi] - iS[\bar{\phi}] + i \int d^4x \; \zeta^k(x)[A_k(x) - \bar{A}_k(x)]}.$$

Expressions such as the above provide an explicit link between concepts of quantum measurement theory like POVMs and the practical and highly successful functional language of QFT. We believe that this link is essential for a local and covariant definition of quantum informational notions in QFT.

Journal of Physics: Conference Series **2533** (2023) 012004 doi:10.1088/1742-6596/2533/1/012004

4. Links to non-equilibrium QFT

In Sec. 2, we saw that the QTP probabilities are linear functional of balanced correlation functions. The measurements do not probe unbalanced correlation functions. Since the latter include $\langle \hat{C}_a(x) \rangle$, QTP probabilities cannot access mean field information. This limitation is not fundamental. Remember that the operator $\hat{C}(x)$ appears in the interaction term with the apparatus. This restriction means that we cannot use couplings of the form $\int d^4x \hat{C}_a(x) \otimes \hat{J}^a(X)$, in order to directly measure the operator $\hat{C}_a(x)$. At least such measurements are not possible with weak field-apparatus coupling where perturbation theory is applicable.

Suppose, for example that $\hat{C}$ coincides with the field operator $\hat{\phi}$—we drop the index a for simplicity. Then, single-detector probabilities record only local information about particles. Let the field be in a state characterized by a macroscopically large number of particles; then, it can be viewed as a thermodynamic system. Then the single-detector probability essentially coincides with a particle-number density function. If we also measure the recorded particle's momentum k, the QTP probability density $P(x,k)$ is an operationally defined version of Boltzmann's distribution function. By Eq. (17), $P(x,k)$ is a linear functional of the correlation function $G(x,x') = \langle \hat{\phi}(x)\hat{\phi}(x') \rangle$, which is usually taken to satisfy the Baym-Kadanoff equations.

From the above analysis, it follows that Boltzmann's thermodynamic entropy, defined on a Cauchy surface Σ,

$$S_B(\Sigma) = -\int_\Sigma d^3x d^3k P(x,k) \ln P(x,k) \tag{34}$$

is a Shannon-type entropy for single-detection measurements. This means that quantum informational quantities, defined through measurements, have a direct application to non-equilibrium QFT. Furthermore, n-detector QTP probabilities probe higher-order correlation functions of the quantum field, thus allowing an analysis that is not accessible by traditional methods. We shall highlight some structural similarities and connections between the QTP analysis and methods of non-equilibrium QFT below.

4.1. Stochastic correlation dynamics from two-particle irreducible effective action

As shown in Sec. 3 the generating functional of QTP correlation functions is defined in terms of non-local source terms $L^{ab}(x,x')$. It is structurally similar to the two-particle irreducible effective action (2PIEA) [46, 59] that has found many applications in non-equilibrium QFT—see, for example, [47–49]. For present purposes, it suffices to show the structural framework of the 2PIEA formalism, following the presentation in [56, 57] to define evolution equations with noise from higher-order correlation functions.

For ease of notation, we use a version of DeWitt's condensed notation, where capital indices A correspond to both the spacetime dependence and the branch of the CTP field (forward or backward in time, ϕ or $\bar{\phi}$). Hence, we will be writing ϕ_A, C_A, and so on. The action in the CTP generating functional will be $S[\phi_A] = S[\phi_a] - S[\bar{\phi}_a]$. In the two-particle irreducible representation, the (two-point) correlation function stands is an independent variable, not a functional of the mean field. Thus there is a separate source K^{AB} driving $C_A C_B$ over the usual $J^A C_A$ term in the one-particle irreducible representation—see the similarity to Eq. (33).

From the generating functional

$$Z\left[K^{AB}\right] = e^{iW\left[K^{AB}\right]} = \int D\phi_A \, e^{i\left(S + \frac{1}{2}K^{AB}C_A C_B\right)} \tag{35}$$

we have

$$G^{AB} = \left\langle \hat{C}_A \hat{C}_B \right\rangle = 2 \left. \frac{\delta W}{\delta K^{AB}} \right|_{K=0} \tag{36}$$

and

$$\left.\frac{\delta^2 W}{\delta K^{AB}\delta K^{CD}}\right|_{K=0} = \frac{i}{4}\left\{\left\langle \hat{C}_A\hat{C}_B\hat{C}_C\hat{C}_D\right\rangle - \left\langle \hat{C}_A\hat{C}_B\right\rangle\left\langle \hat{C}_C\hat{C}_D\right\rangle\right\}. \tag{37}$$

Suppose that we want to express the effective dynamics of G_{AB} in a closed form, but to go beyond the Baym-Kadanoff equations, by taking into account noise from higher-order correlations. For a non-equilibrium system, we seek a formulation in terms of a new object $\mathbf{G}_{AB}$. This is a stochastic correlation function whose expectation value over the noise average gives the usual two point functions. The fluctuations of $\mathbf{G}_{AB}$ reproduce the quantum fluctuations in the binary products of field operators. The simplest assumption is to take $\mathbf{G}_{AB}$ as a Gaussian process, defined by

$$\langle \mathbf{G}_{AB}\rangle = \left\langle \hat{C}_A\hat{C}_B\right\rangle; \qquad \langle \mathbf{G}_{AB}\mathbf{G}_{CD}\rangle = \left\langle \hat{C}_A\hat{C}_B\hat{C}_C\hat{C}_D\right\rangle \tag{38}$$

The Legendre transform of W is the two-particle irreducible effective action,

$$\Gamma_{2PI}\left[G_{AB}\right] = W\left[K^{AB}\right] - \frac{1}{2}K^{AB}G_{AB}; \qquad K^{AB} = -2\frac{\delta\Gamma}{\delta G_{AB}} \tag{39}$$

The Schwinger-Dyson equation for the propagators is simply, $\frac{\delta\Gamma_{2PI}}{\delta G_{AB}} = 0$. When including the stochastic source $\mathbf{G}_{AB}$, it becomes

$$\frac{\delta\Gamma_{2PI}}{\delta\mathbf{G}_{AB}} = -\frac{1}{2}\kappa^{AB} \tag{40}$$

where κ_{ab} is a stochastic nonlocal Gaussian source defined by

$$\left\langle\kappa^{AB}\right\rangle = 0; \qquad \left\langle\kappa^{AB}\kappa^{CD}\right\rangle = 4i\left[\frac{\delta^2\Gamma_{2PI}}{\delta G_{AB}\delta G_{CD}}\right]^{\dagger} \tag{41}$$

The noiseless Eq. (40) ($\kappa = 0$) provides the basis for the derivation of transport equations in the near equilibrium limit. Indeed, for a $\lambda\phi^4$ theory, we obtain the Boltzmann equation for a distribution function f defined from the Wigner transform of G^{ab}. The full stochastic equation (40) leads, in the same limit, to a Boltzmann - Langevin equation [56].

4.2. Correlation Histories

The two-particle irreducible formalism can be extended to an n-particle irreducible formalism, for any n. There is an effective action Γ_{nPI} for each n, from which all effective actions for $n' < n$ can be derived. Taking $n \to \infty$, we obtain a master effective action. The functional variation of the master effective action yields the hierarchy of Schwinger-Dyson equations [57].

To obtain effective closed dynamics for the correlations at order n, we must truncate the Schwinger-Dyson hierarchy upon this order. Truncation renders the master effective action complex. Its imaginary part arises from correlation functions of order higher than n, the fluctuations of which Calzetta and Hu define as *correlation noises* [56] at order n. For example, the noise κ^{AB} in Eq. (40) is the correlation noise of order two.

Calzetta and Hu defined the notion of *correlation histories* [58], in analogy to the decoherent histories program. A fine-grained correlation history corresponds to the full Schwinger-Dyson hierarchy of correlation functions. When we truncate the hierarchy at finite order n, we treat only correlation functions of order n as independent. Higher -order correlations are ignored or slaved to the lowest-order ones. A truncated hierarchy defines a *coarse-grained* correlation history. For example, mean field theory studies *coarse-grained* correlation histories at order $n = 1$; the

Baym-Kadanoff equation, or Boltzmann equation and their stochastic generalizations refer to *coarse-grained* correlation histories of order $n = 2$.

The key point is that the truncation of the master effective action always leads to dissipation and noise for the coarse-grained histories. Any truncated theory is an effective field theory in the correlation hierarchy formulation. This effective field theory does not carry the full information, this loss of information being expressed as correlation noise. The higher-order correlations are analogous to an environment in the theory of open quantum systems [60]. This noise may lead to decoherence of correlation histories [58], i.e., to the classicalization of the effective description.

The QTP approach demonstrates that the different levels of correlation histories can be accessed by the measurement of n-detector joint probabilities. Eq. (22) assigns to each initial state $|\psi\rangle$ of the field a hierarchy of joint probability distributions $P_n(z_1, z_2, \ldots, z_n)$.

In classical probability theory, a hierarchy of correlation functions defines a classical stochastic process, if it satisfies the Kolmogorov additivity condition,

$$P_{n-1}(z_1, \ldots, z_{n-1}) = \int dz_n P_n(z_1, z_2, \ldots, z_n) \tag{42}$$

Quantum probability distributions for sequential measurements do not satisfy this condition [61, 62]. Hence, the violation of Eq. (42) is a genuine signature of quantum dynamics; it cannot be reproduced by classical physics, including classical stochastic processes. It is rather different from the Leggett-Garg inequalities [63] that also refer to the behavior of quantum multi-time probabilities. In contrast, if measurements on a quantum field approximately satisfy Eq. (42), then the measurement outcomes can be simulated by a stochastic process with n-time probabilities given by the probability distributions (42). Then the generating functional Z_{QTP} corresponds to a stochastic process, i.e., it is obtained as the functional Laplace transform of a classical stochastic probability measure.

Hence, the hierarchy $P_n(z_1, z_2, \ldots, z_n)$ provides a natural and unambiguous classicality criterion. Given the relation between QTP probabilities and QFT correlation functions, this criterion can be used to probe the information content of different levels for correlation histories. For example, the validity of Eq. (42) is necessary for deriving deterministic or classical stochastic dynamics for $P_1(z)$, i.e., for deriving the Boltzmann or the Boltzmann-Langevin equation. The failure of (42) means that the four-point correlation functions are too 'quantum' to allow effective classical stochastic dynamics for the two-point correlation function.

Conversely, the failure of Eq. (42) can be used to provide a measure of irreducibly quantum information at the level $n = 2$. An example of such a measure is

$$S_Q = \int dz_2 \left| \int dz_1 P_2(z_1, z_2) - P_1(z_2) \right|. \tag{43}$$

A second informational quantity is the correlation of the probability distribution, i.e., a measure of the deviation of $P_2(z_1, z_2)$ from $P_1(z_1)P_2(z_2)$. The joint probability density P_2 is then slaved to P_1. This information is typically quantified by the correlation entropy

$$S_C = \int dz_1 dz_2 P_2(z_1, z_2) \ln \frac{P_2(z_1, z_2)}{P_1(z_1)P_1(z_2)}. \tag{44}$$

In general, the correlation entropy will contain information for both quantum correlations (if $S_Q \neq 0$) and classical stochastic ones. Indeed, S_C may not have the usual properties of correlation entropy if $S_Q \neq 0$, and other measures that will distinguish will be more convenient. The third relevant informational quantity in Boltzmann's entropy (34), defined in terms of $P_1(z_1)$. These three quantities are the most important for describing the flow of information at the level of the 2PIEA.

Hence, the QTP hierarchy functions as a registrar of information of the quantum system, keeping track of how much information resides in what order, and how it flows from one order to another through the dynamics. There is a good potential for this scheme to systemize quantum information in QFT: keeping track of the contents and the flow of information and measuring the degree of coherence in a quantum system. This conceptual scheme was suggested in Refs. [54,64] to explain black hole information loss, for instance.

5. Conclusions

We presented the QTP formalism for measurements in quantum fields and its connections to the Closed-Time-Path description of QFT. These connections provide a direct translation between the operational language of measurement theory (POVMs, effects, and so on) to the manifestly covariant description of QFT through functional methods. For example, we showed how one can express POVMs for particle detection in terms of path-integrals. We believe that this is an important step towards the formulation of a general theory of relativistic quantum information.

A key aspect of our approach is the central role of the hierarchy of unequal-time correlation functions. POVMs for measurement are linear functional of such correlation functions. The very same correlation functions define the real causal dynamics in the CTP approach, which are essential for the definition of thermodynamical observables and the construction of effective irreversible dynamics in non-equilibrium QFT.

Crucially, QFT correlation functions are covariant and causal objects. For this reason, we believe that a relativistic quantum information theory (QIT) that respects both causality and spacetime symmetry must define all informational quantities in terms of such correlation functions. This contrasts the standard approach of QIT that is based on properties of single-time quantum states, like von Neumann entropy or entanglement.

A sound theoretical foundation for relativistic QIT is important for reasons that go beyond the necessity of theoretical coherence. Such a foundation is required for the description and design of quantum experiments in space [4, 5] that will explore the effects of non-inertial motion (acceleration, rotation) and gravity on quantum correlations, including entanglement. The predictions of a QFT measurement theory may be testable in experiments that involve long separations or large relative velocities between detectors.

Acknowledgements This research was supported by a grant JSF-19-07-0001 from the Julian Schwinger Foundation.

[1] C. Anastopoulos, B. L. Hu and N. Savvidou, *Quantum Field Theory based Quantum Information: Measurements and Correlations*, Ann. Phys. 450, 169239 (2023).

[2] See, e.g., https://www.isrqi.net/

[3] C. Anastopoulos and N. Savvidou, *Quantum Information in Relativity: The Challenge of QFT Measurements*, Entropy 24, 4 (2022).

[4] D. Rideout et al, *Fundamental Quantum Optics Experiments Conceivable with Satellites – Reaching Relativistic Distances and Velocities*, Class. Quantum Grav. 29, 224011 (2012).

[5] M. Mohageg et al, *The Deep Space Quantum Link: Prospective Fundamental Physics Experiments using Long-Baseline Quantum Optics*, arXiv:2111.15591.

[6] C. Anastopoulos and B. L. Hu, *Probing a Gravitational Cat State*, Class. Quant. Grav. 32, 165022 (2015).

[7] S. Bose, A. Mazumdar, G. W. Morley, H. Ulbricht, M. Toro, M. Paternostro, A. A. Geraci, P. F. Barker, M. S. Kim, and G. Milburn, *A Spin Entanglement Witness for Quantum Gravity*, Phys. Rev. Lett. 119, 240401 (2017).

[8] C. Marletto and V. Vedral, *Gravitationally Induced Entanglement between Two Massive Particles is Sufficient Evidence of Quantum Effects in Gravity*, Phys. Rev. Lett. 119, 240402 (2017).

[9] C. Anastopoulos and B. L. Hu, *Quantum Superposition of Two Gravitational Cat States*, Class. Quant. Grav. 37, 235012 (2020).

[10] A. Peres and D. Terno, *Quantum Information and Relativity Theory*, Rev. Mod. Phys. 76, 93 (2004).

[11] R. M F. Houtappel, H. van Dam, and E. P. Wigner, *The Conceptual Basis and Use of the Geometric Invariance Principles*, Rev. Mod. Phys. 37, 595 (1965).

[12] D. Malament, *In Defense of Dogma: Why There Cannot Be A Relativistic Quantum Mechanics of (Localizable) Particles*, in "Perspectives on Quantum Reality", ed. R. Clifton (Kluwer Academic, Dordrecht 1996).

[13] G. C. Hegerfeldt, *Instantaneous Spreading and Einstein Causality in Quantum Theory*, Annalen der Physik 7, 716 (1998).

[14] R. Sorkin, *Impossible Measurements on Quantum Fields*, in "Directions in General Relativity", eds. B. L. Hu and T. A. Jacobson (Cambridge University Press, 1993).

[15] L. Borsten, I. Jubb, and G. Kells, *Impossible measurements revisited*, Phys. Rev. D104, 025012 (2021).

[16] D. Beckman, D. Gottesman, M. A. Nielsen, J. Preskill, *Causal and localizable quantum operations*, Phys. Rev. A64, 052309 (2001).

[17] L. Landau and R. Peierls, *Erweiterung des Unbestimmtheitsprinzips für die Relativistische Quantentheorie*, Zeit. Phys. 69, 56 (1931).

[18] N. Bohr and L. Rosenfeld, *On the Question of the Measurability of Electromagnetic Field Quantitie*, Mat.-fys. Medd. Dan. Vid. Selsk.12, (1933).

[19] R. J. Glauber, *The Quantum Theory of Optical Coherence*, Phys. Rev. 130, 2529 (1963).

[20] R. J. Glauber,*Coherent and Incoherent States of the Radiation Field*, Phys. Rev. 131, 2766 (1963).

[21] M. O. Scully and M. S. Zubairy, *Quantum Optics* (Cambridge University Press, 2012).

[22] W. G. Unruh, *Notes on Black Hole Evaporation*, Phys. Rev. D14, 870 (1976).

[23] B. S. DeWitt, *Quantum Gravity: the New Synthesis* in General Relativity: An Einstein Centenary Survey, ed. by S. W. Hawking and W. Israel (Cambridge University Press, Cambridge, 1979), p. 680.

[24] B. L. Hu, S-Y Lin, J. Louko, *Relativistic Quantum Information in Detectors–Field Interactions*, Class. Quantum Grav. 29, 224005 (2012).

[25] J. Polo-Gómez, L. J. Garay, L. J. and E. Martín-Martínez, *A detector-based measurement theory for quantum field theory*, Phys. Rev. D 105, 065003 (2022).

[26] K. E. Hellwig and K. Kraus, *Formal description of measurements in local quantum field theory*, Phys. Rev. D 1, 566 (1970).

[27] K. Okamura and M. Ozawa, *Measurement theory in local quantum physics*, J. Math. Phys. 57, 015209 (2015).

[28] S. Doplicher, *The measurement process in local quantum physics and the EPR paradox*, Commun. Math. Phys. 357, 407 (2018).

[29] C. J. Fewster and R. Verch, *Quantum fields and local measurements*, Comm. Math. Phys. 378, 851 (2020).

[30] C. Anastopoulos and N. Savvidou, *Time-of-Arrival Probabilities for General Particle Detectors*, Phys. Rev. A86, 012111 (2012).

[31] C. Anastopoulos and N. Savvidou, *Time-of-Arrival Correlations*, Phys. Rev. A95, 032105 (2017).

[32] C. Anastopoulos and N. Savvidou, *Time of arrival and Localization of Relativistic Particles*, J. Math. Phys. 60, 0323301 (2019).

[33] C. Anastopoulos and N. Savvidou, *Coherences of Accelerated Detectors and the Local Character of the Unruh Effect*, J. Math. Phys. 53, 012107 (2012).

[34] C. Anastopoulos and N. Savvidou, *Time-of-arrival Probabilities and Quantum Measurements*, J. Math. Phys. 47, 122106 (2006).

[35] C. Anastopoulos and N. Savvidou, *Time-of-arrival probabilities and quantum measurements. II. Application to tunneling times*, J. Math. Phys. 49, 022101 (2008).

Journal of Physics: Conference Series **2533** (2023) 012004 doi:10.1088/1742-6596/2533/1/012004

[36] C. Anastopoulos, *Time-of-arrival probabilities and quantum measurements. III. Decay of unstable states*, J. Math. Phys. 49, 022103 (2008).

[37] K. Savvidou, *The Action Operator for Continuous-time Histories* J. Math. Phys. 40, 5657 (1999); *Continuous Time in Consistent Histories*, gr-qc/9912076.

[38] N. Savvidou, *Space-time Symmetries in Histories Canonical Gravity*, in "Approaches to Quantum Gravity", edited by D. Oriti (Cambridge University Press, Cambridge 2009).

[39] M. Gell-Mann and J. B. Hartle, *Classical Equations for Quantum Systems*, Phys. Rev. D47, 3345 (1993).

[40] R. Omnés, *The Interpretation of Quantum Mechanics*, (Princeton University Press, Princeton 1994).

[41] J. S. Schwinger, *Brownian Motion of a Quantum Oscillator*, J. Math. Phys. 2, 407 (1961).

[42] L. V. Keldysh, *Diagram Technique for Nonequilibrium Processes*, Zh. Eksp. Teor. Fiz. 47, 1515 (1964).

[43] G. Zhou, Z. Su, B. Hao and L. Yu, *Equilibrium and nonequilibrium formalisms made unified*, Phys. Rep. 118, 1 (1985).

[44] B. S. DeWitt, *Effective action for expectation values*, in "Quantum Concepts in Space and Time", ed. R. Penrose and C. J. Isham (Claredon Press, Oxford, 1986)

[45] R.D. Jordan, Effective field equations for expectation values. Physical Review D 33, 444 (1986)

[46] E. Calzetta and B. L. Hu, *Nonequilibrium quantum fields: Closed-time-path effective action, Wigner function, and Boltzmann equation* , Phys. Rev. D37, 2878 (1988).

[47] E. A. Calzetta and B. L. Hu, *Nonequilibrium quantum field theory* (Cambridge University Press, 2008).

[48] J. Berges, *Introduction to nonequilibrium quantum field theory*, AIP Conf. Proc. 739, 3 (2004).

[49] J. Berges, *Nonequilibrium Quantum Fields: From Cold Atoms to Cosmology*, Lecture Notes of the Les Houches Summer School: Strongly Interacting Quantum Systems out of Equilibrium (Oxford University Press, Oxford, 2016).

[50] E. Calzetta and B. L. Hu, *Closed-time-path functional formalism in curved spacetime*, Phys. Rev. D35, 495 (1987).

[51] S. Weinberg,*Quantum contributions to cosmological correlations* Phys. Rev. D 72, 043514 (2005).

[52] A. Kamenev, *Field Theory of Non-Equilibrium Systems* (Cambridge University Press, Cambridge, 2011).

[53] J. Rammer, *Quantum Field Theory of Non-equilibrium States* (Cambridge University Press, Cambridge, 2009)

[54] B. L. Hu, *Correlation Dynamics of Quantum Fields and Black Hole Information Paradox*, talk at the International School of Astro-fundamental Physics, Sept. 1995. Proceedings edited by N. Sanchez and Zichichi (Kluwer Publishers, Dordrecht, 1996).

[55] J. von Neumann, *Mathematical Foundations of Quantum Mechanics* (Princeton University Press, Princeton, 1955).

[56] E. Calzetta and B. L. Hu, *Stochastic dynamics of correlations in quantum field theory: From the Schwinger-Dyson to Boltzmann-Langevin equation*, Phys. Rev. D61, 025012 (1999).

[57] E. Calzetta and B. L. Hu, "Correlations, Decoherence, Disspation and Noise in Quantum Field Theory", in *Heat Kernel Techniques and Quantum Gravity*, ed. S. A. Fulling (Texas A& M Press, College Station 1995).

[58] E. Calzetta and B. L. Hu, "Decoherence of Correlation Histories" in *Directions in General Relativity, Vol II: Brill Festschrift*, eds B. L. Hu and T. A. Jacobson (Cambridge University Press, Cambridge, 1993).

[59] S. A. Ramsey, and B. L. Hu, B.L., *O (N) quantum fields in curved spacetime*, Phys. Rev. D56, 661 (1997).

[60] C. Anastopoulos, *Coarse grainings and irreversibility in quantum field theory*, Phys. Rev. D56, 1009 (1997).

[61] C. Anastopoulos, *Classical versus quantum probability in sequential measurements*, Found. Phys. 36, 1601 (2006).

[62] J. Kofler and Č. Brukner, *Condition for macroscopic realism beyond the Leggett-Garg inequalities*, Phys. Rev. A 87, 052115 (2013).

[63] A. J. Leggett and A. Garg, *Quantum mechanics versus macroscopic realism: Is the flux there when nobody looks?*, Phys. Rev. Lett. 54, 857 (1985).

[64] C. Anastopoulos and N. Savvidou, *Multi-time measurements in Hawking radiation: information at higher-order correlations*, Class. Quant. Grav. 37, 025015 (2019).

The Hole Argument and Beyond: Part I: The Story so Far

Henrique Gomes[1], **Jeremy Butterfield**[2]
[1] Oriel College, Oxford OX1 4EW, UK
[2] Trinity College, Cambridge CB2 1TQ, UK
E-mail: gomes.ha@gmail.com, jb56@cam.ac.uk

Abstract. In this two-part paper, we review, and then develop, the assessment of the hole argument for general relativity. This first Part reviews the literature hitherto, focussing on the philosophical aspects. It also introduces two main ideas we will need in Part II: which will propose a framework for making comparisons of *non*-isomorphic spacetimes.

In Section 1 of this paper, we recall Einstein's original argument. Section 2 recalls the argument's revival by philosophers in the 1980s and 1990s. This includes the first main idea we will need in Part II: namely, that two spacetime points in different possible situations are never strictly *identical*—they are merely *counterparts*.

In Section 3, we report—and rebut—more recent claims to "dissolve" the argument. Our rebuttal is based on the fact that in differential geometry, and its applications in physics such as general relativity, points are in some cases identified, or correspond with each other, between one context and another, by means *other than* isometry (or isomorphism). We call such a correspondence a *threading* of points. This is the second main idea we shall use in Part II.

1. Act 1: Introduction

1.1. Our two main aims

This two-part paper has two main aims. Our first aim (here in Part I) is to give a brisk review of the philosophical discussion of the hole argument for general relativity during the last thirty-five years. This discussion centres around whether and how one spacetime point in a model (solution) of general relativity corresponds to (or: is identical to; or should be identified with) a spacetime point in another such model that is isomorphic to the given one. This isomorphism is given by a diffeomorphism on the first model's base-manifold of points. For in the hole argument, the second model, whose existence poses the threat of indeterminism, is defined by dragging along the first model's metric and other fields: the field-values get "re-distributed" on the manifold's points.[1]

The second paper (Part II) addresses our second, more positive, aim. This is about the question of how one should make comparisons of *non*-isomorphic models, i.e. solutions of general relativity: in particular, how to do so, based on comparing points between the two models. We will propose a framework for such comparisons. It proceeds by analogy with field-space formulations of gauge theories. It postulates a fibre bundle whose fibres are isomorphic

[1] So the first model can lack any automorphisms, i.e. isomorphisms to itself: it can be "wrinkly". Cf. Section 1.2.

Journal of Physics: Conference Series **2533** (2023) 012002 doi:10.1088/1742-6596/2533/1/012002

copies of a given model, i.e. a given Lorentzian manifold endowed with matter fields obeying the Einstein equations. So given a model based on a manifold M, Diff(M) is the structure group of the bundle: any element of Diff(M) drags the given model into coincidence (and so isomorphism) with another element of the fibre.

This framework will build on Part I's review, and especially on two main ideas that we develop here in Part I. The first (in Section 2.2.3) is David Lewis' *counterpart theory* framework for treating the identity of objects—any objects, not only spacetime points—in different possible situations (in philosophical jargon: different *possible worlds*). Lewis says: objects in different possible worlds are never *identical*. Rather, they are *counterparts* of each other—better or worse counterparts, according to how similar they are. Although this seems puzzling at first sight, Lewis defends it convincingly—and we apply it to spacetime points: and in Part II, to points in non-isomorphic spacetimes.

The second idea (in Section 3.2) is what we will call *threading* points: (it occurs between isomorphic spacetimes, and between non-isomorphic ones). For we will argue, *pace* some recent authors, that one cannot dismiss the hole argument as a confusion (or similarly: as a non-starter, or a *canard*), by claiming that isomorphism is the mandatory standard for when points in (the manifolds of) two models of the theory correspond (or: are identical, or should be identified with each other). For, we claim, isomorphism is *not* such a mandatory standard. Various examples show that both mathematics and physics treat the correspondence or identity of points flexibly—not always identifying them by isomorphism. We will call such a flexible correspondence or "identification" a *threading*.

This paper is organised by invoking a dramatic metaphor: as three Acts, in three Sections. Thus Act 1, in the rest of this Section, will review Einstein's original version of the argument: his version is an argument against general covariance (!). Act 2 (i.e. Section 2) recalls the revival of the argument by philosophers in the 1980s and 1990s. This began with [1]'s casting the argument as against the philosophical doctrine *'substantivalism'*. Like all 'isms' in philosophy, 'substantivalism' is a bit vague in its meaning; but the broad idea is that spacetime is, or is represented by, a manifold, and exists independently of its material content, including the metric. In Section 2, we will take substantivalism to be the claim that spacetime points are objects; and we will recall some replies to Earman and Norton that were defences of substantivalism in this sense: including replies that invoked the idea of counterparts. Then Act 3 (i.e. Section 3) will report—and rebut—claims made since 2016 (first by [2], and then by others) that facts about mathematical practice rule out the hole argument. As we mentioned above, the idea of Weatherall et al. is that isomorphism is the mandatory standard for judging whether two points in (the manifolds of) two models of the theory correspond, or "are identical". Our rebuttal is based on the fact that mathematics and physics treat identity more flexibly: in differential geometry, and its applications such as general relativity, points are in some cases identified, or correspond with each other, between one context and another, by means *other than* isometry (or isomorphism): by what we will call a *threading*.

Two final orienting remarks. (1): Although the hole argument focusses on whether two isomorphic models of general relativity are, or represent, the same physical possibility ("possible world"): the recent philosophical literature discusses this claim more generally, i.e. for an arbitrary physical theory that is formulated precisely enough to have a notion of isomorphism for its models, and for which we interpret such models as representing possibilities. Indeed, this more general claim is widely endorsed (and so will be prominent in Acts 2 and 3), and there is a popular name for it. Namely, it is common to call the claim that any two isomorphic models of a theory are, or represent, the same physical possibility: *Sophistication*. So Acts 2 and 3 are in effect a review of the prospects for 'Sophistication'.

(2): Besides, this review of Sophistication connects with our second, positive, aim. For in other work, one of us [3, 4, 5] defends Sophistication for general relativity and for classical gauge theories (i.e. Yang-Mills theory, abelian or non-abelian), using parallel reasoning in these two endeavours. So this other work treats Sophistication independently of the hole argument that we focus on here. And since all agree that it is natural to use fibre bundles in gauge theories, the fact that this other work treats general relativity and gauge theories in parallel makes it natural to use fibre bundles to discuss the "trans-world identification" of points in general relativity. Indeed, Part II's fibre bundle of models of general relativity will be a close cousin of field-space in gauge theories. In short: the parallels between general relativity and gauge theories, developed in the other work, support what we will do in Part II.

1.2. Einstein 1913

Einstein's original hole argument, from the summer of 1913, was aimed *against* general covariance. At first sight, this target is of course very surprising, given Einstein's desire in those years to achieve a generally covariant theory of gravitation, and his eventual success in doing so, in November 1915. But the explanation is now well-known; as follows. Einstein and Grossmann had just published their *Entwurf* theory. It had many features of general relativity, that Einstein wanted, and that in late 1915 he successfully implemented in his field equations. But the equations of the *Entwurf* theory are not generally covariant.[2]

Pondering this defect, Einstein—with his characteristic, and brilliant, flexibility—thought: 'maybe what I cannot get, I should anyway not want ... maybe there is a good argument against general covariance'. And thus he thought of the hole argument.

As Einstein conceives the argument, it urges that general covariance implies a radical indeterminism—which is unacceptable, so that general covariance must be rejected. The idea of the argument is that if a theory of the type he sought—in which spacetime is not a "fixed canvas", or "background arena", on which material events play out, but is instead dynamically affected by such events—were to be generally covariant, then the theory would be radically indeterministic: whatever the details of its field equations. For the state throughout spacetime except for a small patch, e.g. except for some small bounded open spacetime region—called 'the hole', though it is not topologically a hole—would not determine the state in that patch. The argument proceeds by using general covariance to construct two solutions of the putative theory (again: regardless of its detailed equations), with the following features: (i) they match exactly, for both metric and material fields, throughout spacetime outside the hole; but, according to the argument, (ii) they differ about the state of affairs inside the hole.

We will present the argument briefly, and in modern terms. For it will be clear in later Sections that the differences from the historical Einstein will not matter for either of our two aims.[3]

So we take our theory to have models that are an n-tuple of a 4-dimensional manifold, a metric on it, and various fields. And for the sake of reporting Einstein, and setting up our later discussion, we shall take models very simply, as a triple (M, g, T), where M is a manifold, g is a Lorentzian metric and T is a rank-2 tensor representing stress-energy. (We shall also often drop the T, and thus consider vacuum solutions. This does not affect any of the arguments to follow, and indeed is common in the hole argument literature.) But as we said, the argument will not depend on the theory's exact equations, or even on its field content.

We will formulate general covariance, not as about the equations of the theory remaining the same under arbitrary coordinate transformations, but as about an arbitrary active transformation of the manifold—a diffeomorphism $d : M \to M$—producing, from a given model

[2] The stumbling block, in short (and in modern jargon), was that Einstein and Grossmann believed that a static spacetime must be spatially flat. The modern understanding of the history is mainly due to work in the 1980s by Stachel and Norton. Cf. e.g. [6, p. 267-284], [7], and [8, Sec. 3.1] .

[3] From the vast literature on the history, we recommend, in addition to items in the previous footnote: [9], and [10].

Journal of Physics: Conference Series **2533** (2023) 012002 doi:10.1088/1742-6596/2533/1/012002

that satisfies the equations of motion of the theory, another such model: namely, by dragging along the fields of the given model. We will think of dragging along a field as pushing it forward. For since diffeomorphisms are smooth bijections with a smooth inverse, the differences between pulling back, and pushing forward, a geometric object such as a field, will not matter in this paper. So we will just write d^* for the drag-along by the diffeomorphism d.

So we write general covariance, in terms of models of the theory, as follows:

If $\mathbf{M}_1 := (M, g, T)$ is a model, and $d : M \to M$ is a diffeomorphism, then:
$\mathbf{M}_2 := (d(M), d^*(g), d^*(T))$ is also a model.[4]

Clearly, two such models $\mathbf{M}_1$ and $\mathbf{M}_2$—with d not the identity map—will in general differ as to how the metrical and material properties and relations encoded by g and T are distributed over the points of the manifold. In this Section's exposition of the hole argument, we will take these manifold points to be spacetime points. So what we have just said implies that the models will in general differ over which spacetime points have which metrical and material properties and relations. (But later, especially from Section 2.2.1, e.g. footnotes 15 and 16, we will be more careful about the distinction between manifold points (which might be pure-mathematical entities) and the physical spacetime points they represent.) That is, the models will in general differ about such matters, when points are taken to correspond, or be identical, according to the identity map on M. For example, if $d(p) = q$ (i.e. $p, q \in M$, and $d : p \mapsto q$), and p has scalar curvature 5 in $\mathbf{M}_1$, then: in $\mathbf{M}_2, q$ has scalar curvature 5—and p has as its scalar curvature whatever is the scalar curvature, say 7, of the point $d^{-1}(p)$ in $\mathbf{M}_1$.

There is a comment here, that is important since it has caused confusion about the interpretative, or philosophical, issues; (cf. footnote 36 in Section 3.1). Namely: we said that for $d \neq Id_M$, the models $\mathbf{M}_1$ and $\mathbf{M}_2$ will *in general* differ as to how the properties and relations are distributed over the spacetime points. The qualification 'in general' signals the fact that (for almost any theory), some models $\mathbf{M}_1$ have an *automorphism*, i.e. a non-identity diffeomorphism d of their manifold such that the dragged-along model $(d(M), d^*(g), d^*(T))$ is exactly the same as $\mathbf{M}_1$. In other words: d drags g and T into (coincidence with) themselves (respectively). So although $d : p \mapsto q$, with $q \neq p$, nevertheless p's array of metrical and material properties and relations—in philosophical jargon: p's *qualitative profile*—is the very same, in this selfsame model $\mathbf{M}_1$, as q's qualitative profile.[5]

This point is also sometimes put in terms of distinguishing two senses of the word 'isometry' (or more generally, of 'isomorphism'), and using the jargon of category theory. Thus for our case of Lorentzian manifolds (now for simplicity setting aside T), one says: given a category whose objects are pairs (M, g) of smooth manifolds M and Lorentzian metrics on M, and whose arrows are isometries, where an *isometry* d from (M, g) to (N, h) is a bijection between M and N, such that d and its inverse are smooth and also $d^*g = h$. This includes cases where we assume the underlying manifolds are *identical* as sets, $M = N$. Of course, two isometric metrics, with $g \neq h$, can share the same base space. An *automorphism* in this category consists of such an isometry with d from (M, g) to itself, i.e. such that $d \in \mathrm{Diff}(M)$ and $d^*g = g$.

Setting aside this comment about automorphisms, and returning now to the generic situation in which $\mathbf{M}_1$ and $\mathbf{M}_2$ differ as to how the properties and relations encoded by g and T are

[4] This is sometimes called 'active general covariance'. If one adopts this jargon, then the requirement that a theory be formulated so as to admit arbitrary coordinate transformations is correspondingly called 'passive general covariance'.

[5] We might write p's qualitative profile, in the model $\mathbf{M}_1 := (M, g, T)$, in an obvious notation, as a doublet $(g|_p, T|_p)$. Then *in general* the image point in a dragged-along model has the same profile as the argument point in the given "domain" model. That is: if $d(p) = q$, i.e. $d : p \mapsto q$, then q's profile in the dragged-along model $(d(M), d^*(g), d^*(T))$ is the doublet $(d^*(g)|_{d(p)}, d^*(T)|_{d(p)})$. But if d is an automorphism of (M, g, T), then $(d(M), d^*(g), d^*(T)) = (d(M), g, T)$; and so if $d(p) = q$, then q's profile is the same as p's: the doublet $(d^*(g)|_{d(p)}, d^*(T)|_{d(p)}) = (g|_p, T|_p)$.

Journal of Physics: Conference Series **2533** (2023) 012002 doi:10.1088/1742-6596/2533/1/012002

distributed over the spacetime points: let us finish our rendition of the hole argument.

First, the hole argument now makes another assumption, additional to general covariance. It is about representation i.e. about how models represent physical possibilities. Namely, for what we just called the generic situation for $\mathbf{M}_1$ and $\mathbf{M}_2$, the argument assumes:

> (*Distinct*): If d is the identity map except on a small patch $H \subset M$ ('the hole'), then $\mathbf{M}_2$ is, or represents, a different physical possibility than does $\mathbf{M}_1$.

The punchline is now immediate. General covariance and (*Distinct*) imply that $\mathbf{M}_1$ and $\mathbf{M}_2$ are, or represent, different possibilities, although they exactly match each other outside H. So the theory is indeterministic. But indeterminism is unacceptable. Hence, Einstein concluded, general covariance must be given up. Of course, with the later triumph of general relativity, (*Distinct*) became the suspect, or culprit, in the argument; (as suggested by the idea of 'Sophistication' at the end of Section 1.1, and as we will see in Sections 2 and 3).[6]

Two final remarks, both limiting the scope of what follows. The first is about (*Distinct*), the second about determinism.

(1): In (*Distinct*), the words 'is, or represents' signal the philosophical debate about how models and physical possibilities are related to each other; (and the further metaphysical debate about what physical possibilities—philosophers' 'possible worlds'—really are). Although we will glimpse these debates, in Sections 2 and 3, we will not need to take a stand on them.

(2): Determinism is a large subject: (cf. e.g. [12, 13]). But we will only need the following claims, which are agreed by all hands. The intuitive idea of determinism is of course that a theory is deterministic if among its models (or more metaphysically: the possible worlds they represent) two models matching exactly on the state of some appropriate region of spacetime (in many formulations: a time-slice, or all history up to a time slice) implies that they match throughout spacetime. So if 'exact matching' must include the facts about which point has which qualitative profile (facts that philosophers call 'non-qualitative', because they concern *which* point), then a pair of models like $\mathbf{M}_1$ and $\mathbf{M}_2$ in the hole argument are undoubtedly a counterexample to this intuitive idea (i.e. to the implication just formulated). That is: they illustrate indeterminism. Besides, if we take indeterminism to require the explicitly metaphysical idea of two different possibilities, then assuming (*Distinct*), such models are, or represent, different possibilities. Furthermore, the larger the region on which the models match exactly, i.e. the smaller the "hole" $H \subset M$ on which they disagree, the stronger is the violation of determinism. (For a statement of implication is logically weaker, when its antecedent is logically stronger.)

Agreed: there is plenty more to say—both in physics and in philosophy—about the formulation and evaluation of determinism, including for general relativity. And again, we will glimpse some of these issues later (cf. especially (3), in Section 2.2.3). But this paper will not need to take a stand on them.

2. Act 2: 1987 to 2000: Philosophical debate

With the growing success of general relativity, the hole argument was mostly forgotten within physics. But within philosophy, interest in it revived in the late 1980s: very largely on account of [1].[7] In this philosophical revival, most of the attention has been on what Section 1.2 labelled (*Distinct*), rather than the argument's other ingredients, general covariance and determinism.

[6] As we admitted, we have formulated the hole argument in anachronistically modern terms. This will not affect later Sections. But we should note that for Einstein in 1913, (*Distinct*) was an implicit assumption: and when after November 1915, he returned to re-assess the argument, he of course denied (*Distinct*). More specifically, he developed what has come to be called 'the point-coincidence argument', to convey the idea that which point has which qualitative profile of metric and material properties is what physicists now call 'gauge'. Cf. [11].

[7] Although Earman and Norton's paper spawned a large literature, we say 'very largely', not 'exclusively', because some other contemporary discussions—including previous papers by Earman and by Norton, and Earman's book [13]—also influenced philosophers. A significant recent analysis of that literature, and its current importance, is

Journal of Physics: Conference Series **2533** (2023) 012002 doi:10.1088/1742-6596/2533/1/012002

That is unsurprising in so far as (*Distinct*) is about interpreting the theory—philosophers' *metier*—while general covariance and determinism seem established features of the theory.[8]

The distinctive, and perhaps surprising, feature of this philosophical revival was to articulate a new "ingredient" of the argument: a philosophical doctrine called *'substantivalism'*. Thus Earman and Norton cast it as an argument against, not just (*Distinct*), but also substantivalism (Section 2.1). And from 1987 to about 2000, other philosophers' replies to them were mostly defences of substantivalism (Section 2.2).

2.1. 1987: Earman and Norton against substantivalism

As we mentioned in Section 1.1, 'substantivalism' is vague: the broad idea being that spacetime is, or is represented by, a manifold, and exists independently of its material content. But we will take it (as we think Earman and Norton, and the ensuing literature, mostly do) as the claim that spacetime points are *objects*. (We will shortly be more careful about the distinction, signalled by our phrase 'or is represented by', between spacetime points and manifold points.) Three clarifications of this claim (in descending importance) are in order.

(1): Beware: 'object' is here meant in the "thin" or logically weak sense of modern logic, viz. as the referent of a singular term, which is therefore subject to statements of identity and non-identity. So this sense, which derives especially from the work of Frege (and was later entrenched by the large influence of Quine), abjures the "thick" or logically strong (and so controversial) connotations of the traditional philosophical (in particular, Aristotelian) word 'substance'.

(2): Beware: confronted with this definition of 'substantivalism', a physicist—indeed, anyone who is duly modest about what we can now know, or reasonably believe, about the ultimate micro-structure of spacetime—is likely to demur. They will say: 'How could anyone claim to know either way?' But there is a distinction. The debate here is about how best to interpret general relativity (and maybe other spacetime theories): not—sorry, O physicist!—about reality itself. That is: the debate aims to settle what general relativity (and maybe other theories) *say* the world is like, thus of course leaving it to scientific enquiry—ultimately experiments—to settle whether the world is really as general relativity etc., so interpreted, say it is. So really 'substantivalism'—in this literature, and in this paper—claims that *general relativity should be interpreted as committed to* spacetime points being objects.[9]

(3): Like the literature, we have so far talked always about spacetime points, not regions. But we should clarify that this is for expository convenience, and not because substantivalists are in any way reluctant to say that regions are objects, just as much as points are. (Here, 'say that' must of course be read as 'say that general relativity says that'; cf. (2) above.) The reason is that modern logic makes light work of the passage from commitment to points, and commitment to regions. For it construes regions either as sets of points, or as mereological fusions of them.[10]

[14]. Note also that Earman and Norton's philosophical writings were prompted in part by earlier historical work, by authors such as Stachel and Norton (cited in footnotes 2 and 3).

[8] Here, 'seem established' is intended to describe the attitude of most *philosophers*. For agreed: within physics, there are both rich traditions and ongoing efforts about: (i) how to better understand general covariance (also known as: diffeomorphism-invariance), and connected issues like the definition of observables, especially in connection with the hunt for a quantum theory of gravity; and (ii) definitions of, and theorems about, determinism in general relativity. Indeed, nowadays most physicists take the current interest of the hole argument to be just its directing one's attention to topics (i) and (ii). In the sequel, we will glimpse these topics.

[9] Agreed: there is controversy even about the distinction we have invoked. Thus some philosophers, notably van Fraassen (e.g. [15]), argue that even if a theory had complete experimental success, and satisfied all our other evaluative criteria (such as being conceptually unified, meshing with our other successful theories etc.), we could rationally remain agnostic concerning everything the theory says about matters lying beyond the experimental data. But throughout this paper, we set aside these general questions about exactly what beliefs about reality, a theory's complete success (experimental and otherwise) mandates. Trying to be practical and in the spirit of physics, we say: let us worry about such questions if and when we get to be so lucky!

[10] 'Mereology' is modern logic's name for the study of the part-whole relation; and a *fusion* of objects is, intuitively,

And logicians and philosophers tend to accept as unproblematic both the operation of forming the set of some given objects, and the operation of forming their fusion. So on both approaches, a theory's being committed to points is taken to imply its being committed to regions.

With substantivalism thus clarified, it is easy to guess how it enters in to Earman and Norton's version of the hole argument. The main point is that they take substantivalism to imply (*Distinct*).[11]

That is undoubtedly plausible: in effect, as a corollary of the meaning of 'object'. Thus consider two distinct objects (which need not be spacetime points), call them a and b; and two distinct arrays of properties and relations (what Section 1.2 called 'qualitative profiles') such objects can have, call them F and G. Then surely a having F and b having G is a different possibility than a having G and b having F.[12] Or recall Section 1.2's example of a diffeomorphism $d : M \to M$ where $d(p) = q$. We said that p has scalar curvature 5 in $\mathbf{M}_1$, so that in $\mathbf{M}_2, q$ has scalar curvature 5, and p has as its scalar curvature whatever is the scalar curvature, say 7, in $\mathbf{M}_1$ of $d^{-1}(p)$). If points are objects, surely these statements should be taken at face-value as regards what are the represented possibilities. In particular: the threat that p has contradictory properties (having curvature 5, and having curvature 7) is most straightforwardly answered by saying that $\mathbf{M}_1$ and $\mathbf{M}_2$ are or represent different possibilities, i.e. by endorsing (*Distinct*).

Once one assumes that substantivalism implies (*Distinct*), the argument is clear. There is a dilemma: one can maintain that general relativity is deterministic, by denying (*Distinct*) and so also denying substantivalism; or one can be a substantivalist, be committed to (*Distinct*), and say that general relativity is indeterministic.

Then Earman and Norton go on to urge the first option. Though (like Einstein!) they give no precise definition of determinism, they emphasise that their favouring the first option is *not* because of any conviction that general relativity (or any other physical theory) must at all costs be classified as deterministic. They are entirely open to the idea that a physical theory is indeterministic. But, they say, if determinism fails 'it should fail for a reason of physics' (ibid. p. 524). And this is their objection to the dilemma's second option: it makes general relativity indeterministic for a distinctively metaphysical reason that, they say, should have no weight in classifying physical theories as indeterministic.

2.2. 1987 to 2000: Defending substantivalism

For the most part, philosophers have replied to Earman and Norton's dilemma by defending substantivalism. And they took this to involve: denying that substantivalism implies (*Distinct*). We agreed at the end of Section 2.1 that this implication is undoubtedly plausible. For if spacetime points are objects, then surely swapping which point has which qualitative profile of field-values makes a genuine difference? Yet maybe this implication can be denied—albeit at the cost of what boxers call 'fancy footwork'. So this Section is about such philosophical fancy footwork. (At the end of this Subsection, we will return to the tactic of defending substantivalism by accepting its implying (*Distinct*), and just admitting that general relativity is indeterministic.)

Besides, some of these replies went further than denying (*Distinct*): which in our formulation is only about general relativity, viz. any model $\mathbf{M}_1$ and its drag-along $\mathbf{M}_2$. For recall from (1) at

the aggregate of them. More formally, it is the whole (i) that has them as parts, and (ii) whose every part overlaps at least one of the objects. As this explanation of 'fusion' suggests, mereology is in practice a lot like set theory—but without the empty set.

[11] Their jargon is different for ours. Their 'Leibniz equivalence' states that isomorphic models represent the same physical possibility (cf. our 'Sophistication'): which they call the 'acid test of substantivalism', i.e. which, they say, substantivalists must deny [1, pp. 521-522].

[12] And this is so even if the arrays F and G are exhaustive, i.e. encode all of the objects' properties and relations. This conviction, that there are two possibilities here, is compelling and recurs in the literature. For example, it is the moral of the parable of Alice and Barbara in [16, end of Section 3]'s critique of [2]: to which we will return in Section 3.1.

Journal of Physics: Conference Series 2533 (2023) 012002 doi:10.1088/1742-6596/2533/1/012002

the end of Section 1.1 the doctrine, nowadays called *Sophistication*. Namely: for any theory—not just general relativity—whose models are or represent possibilities: *any two isomorphic models are, or represent, the same physical possibility.*[13] Thus some of these replies endorsed (or came close to endorsing) *Sophistication*; (though the label 'sophistication' was only common after 2000). So in briefly reporting some of these replies (and in the next Section), Sophistication will be a theme.

For our purposes, we can group the denials of (*Distinct*) under three headings, which we call: *metrical essentialism*, *the drag-along response* and *counterparts*. The second and third of these will be (different) responses to the difficulties faced by the first; the third—which we favour—also responds to a difficulty faced by the second. The second will be tantamount to endorsing Sophistication for general relativity (but not necessarily for other theories).[14]

We should clarify at the outset that although we will here criticise the drag-along response, we are in fact sympathetic to Sophistication. So we do not intend 'fancy footwork' as a slur. This is not least because (as mentioned in (2) at the end of Section 1.1) one of us (HG) has urged parallel reasons for it, in general relativity and in gauge theories. But in this paper, we will not argue directly for Sophistication, even just for general relativity. Nor will the framework (for both general relativity and gauge theories) that we propose in our second paper, Part II, add to the *philosophical* case for Sophistication. For that framework's merits lie in treating non-isomorphic spacetimes (in gauge theory: non-isomorphic field-configurations); and thus in placing discussion of Sophistication in a wider context.

2.2.1. Metrical essentialism 'Essentialism' is jargon for the philosophical idea that some objects (maybe all objects) have one or more attributes (i.e. properties or relations) that they could not possibly lack. That is: in any possible situation in which the object exists, it has the attribute in question.

The traditional examples are of properties that are natural kinds, in particular an organism's species. (The tradition goes back to Aristotle.) Thus a human, say Socrates, could not possibly be non-human, e.g. a frog; this oak tree could not possibly be an elm tree; and so on. But one can envisage relational examples, even everyday or biological ones. For example, the parent-child relation: this child could not possibly lack their actual parentage—different parents would imply that the child is a different person. (So in the imagined possibility, 'the child' does not refer to the actual child.)

This idea immediately suggests a way to deny (*Distinct*). For suppose that a given model $\mathbf{M}_1$ is, or represents, a possibility: a 'way that a general-relativistic world could be'. And suppose that some metric or material attribute (property or relation) of a point p in the hole $H \subset M$ that is shifted by the diffeomorphism d, i.e. $d(p) \neq p$, is both:

(i) essential to p, and (ii) not possessed by $d^{-1}(p)$.[15]

It follows that $\mathbf{M}_2$—where p, as an element of the range of d, "inherits" the qualitative profile possessed in $\mathbf{M}_1$ by $d^{-1}(p)$, and has "bequeathed" to $d(p)$ the qualitative profile it possessed in $\mathbf{M}_1$—does *not* represent a possibility. For in $\mathbf{M}_2$, p's qualitative profile lacks an attribute essential to p (the attribute having been "bequeathed", in this model, to $d(p)$). In short: $\mathbf{M}_2$

[13] Section 3.1 will address subtleties about the context of such representations, that are articulated in more recent literature.

[14] As to our labels for these responses, the second is our invention; but the first and third are established jargon in philosophy. For further references, cf. below: especially Sections 2.2.2 and 3.1 for the second response, and Section 2.2.3 for the third.

[15] If one takes the points of the manifold M to be in some sense pure-mathematical points, not genuine spacetime points, then it will suffice to suppose that some such attribute, represented by a field-value of p (cf. footnote 5) is (i) essential to the spacetime point that p represents, and (ii) not possessed by the spacetime point that $d^{-1}(p)$ represents. And this supposition seems compulsory. For after all, we supposed that $\mathbf{M}_1$ represents a possibility, even if it is not identical to one; and it surely does so by its manifold points representing spacetime points.

Journal of Physics: Conference Series 2533 (2023) 012002 doi:10.1088/1742-6596/2533/1/012002

fails to be, or represent, a possibility, because it is faithless to p's essential attribute. $\mathbf{M}_2$ "paints p with the colours" of $d^{-1}(p)$ (i.e. of $d^{-1}(p)$, within $\mathbf{M}_1$), and at least one of those colours is for p an impossible colour.[16]

Note that we have given this argument in terms of p both (i) "losing an essential property to $d(p)$", and (ii) "not re-acquiring it from $d^{-1}(p)$", as our attention passes from $\mathbf{M}_1$ to $\mathbf{M}_2$. But of course, we could equally well have chosen $d(p)$ as "the victim of loss". We could have supposed that $d(p)$ has, within $\mathbf{M}_1$, an attribute that is: (i) essential to $d(p)$, but (ii) not possessed by p; and so is, within $\mathbf{M}_2$, (i') bequeathed to $d^2(p)$ but (ii') not re-acquired from p.[17]

So much for the "mechanics" of this denial of (*Distinct*). The question now is: for what attributes is it plausible to claim that they are both (i) essential to spacetime points, and yet also (ii) re-assigned among points by dragging along with a diffeomorphism?

The reply has been: metrical attributes—hence the label *metrical essentialism*. These attributes are typically not properties, but rather relations. So they are encoded, not by scalars like the scalar curvature, but by vectors and tensors, like the metric tensor and numerical quantities derived from such tensors, like the length of a line between two points. But our argument allowed for essential relations as well as properties. So the way is open to deny (*Distinct*) by holding, for example, that the metrical relations of a spacetime point p to other, maybe to all other, points (for example: the lengths of lines from p to these other points) are essential to p.[18]

The first advocate of this reply to Earman and Norton was [18]. He invokes Aristotle in defence of the general idea of essential attributes, and he invokes Newton's Scholium to the Definitions in *Principia* in defence of spacetime points (for Newton: temporal instants and spatial points) having their metrical relations to each other essentially (p. 544).

But despite this noble pedigree, this reply has a severe limitation, due to general relativity's having non-isometric models. For most advocates of essential attributes (from Aristotle onwards) hold the view for *actual* objects, and are "modestly" silent about whether merely possible objects—of course, a unicorn is a standard example—have essential attributes. Since the debate here concerns how best to interpret general relativity, assuming it is true (cf. comment (2) in Section 2.1), the modest version of this reply would confine its claims to the (putatively) actual spacetime points, i.e. the points of the actual general relativistic cosmos. But then it says nothing about the identity of points in spacetimes non-isometric to the actual cosmos, except that none of such a spacetime's points is the same point as any actual point. (For thanks to the non-isometry, any of these points lacks at least one of the intricate web of metrical relations that the reply maintains are essential to the actual points.) In particular, it says nothing about how to reply to the hole argument applied to such a spacetime. So the challenge of the hole argument remains.

We do not claim this is knock-out objection against metrical essentialism: as always in philosophy, it can give some reply. But there are good reasons to think its prospects are bleak; (as argued by e.g. [19, Section 5] and [20, Section 4]).[19]

[16] To make the wording simpler, we have stated the argument taking p etc. as spacetime points with attributes. But if one takes the points of the manifold M to be pure-mathematical and only represent spacetime points, the argument can be rephrased, though at the price of cumbersome language, along the lines of footnote 15.

[17] To be vivid, we have used metaphors of loss, inheritance, colours etc. But it is clear that the argument could be made more formal, using pullbacks, and the doublets mentioned in footnote 5.

[18] For the interplay between (a) philosophical notions of property (including intrinsic and extrinsic properties) and relation, and (b) geometry's and mechanics' notions of scalar, vector, tensor etc, cf. [17, Section 2.1.1].

[19] We will not go in to details. We just note that it does not help essentialism either (i) to relativize essential attributes to isometry classes or (ii) to advocate a different set of attributes than the metrical ones as essential to points. For even admitting (i) and (ii), essentialism is still silent about which $\mathbf{M}$ along an isometry class should be the one whose points bear the essential attributes, and a priori all such models seem equally metaphysically possible. Another worry is that metric essentialism has difficulty expressing counterfactuals such as "if the mass

2.2.2. The drag-along response To avoid the difficulties of metrical essentialism, it is natural to try and deny (*Distinct*) by saying that $\mathbf{M}_2$ is, or represents, the very same possibility as $\mathbf{M}_1$—despite its difference from $\mathbf{M}_1$ about which point has which qualitative profile. The basic idea, in philosophical jargon, is that points are *individuated* by their pattern of qualitative properties and relations, as encoded in the metric and matter fields. (The jargon is that saying a set of objects X 'is individuated by' a set of attributes Y means that there are sufficient conditions, expressed in terms of Y, for statements of identity about members of X.) But the drag-along response takes this idea further, asserting that such individuation also applies "between" isomorphic models.[20]

Thus suppose again (cf. the start of Section 2.2.1) that the given model $\mathbf{M}_1$ is, or represents, a possibility. And suppose that some point p in the hole $H \subset M$ is shifted by d to a distinct point, $q \equiv d(p) \neq p$, that *within* $\mathbf{M}_1$ has a different qualitative profile (some different metric or material attribute) than p. Then the proposed reply to the hole argument is that, *within* $\mathbf{M}_2$, we should "rebrand" $d(p)$ as "really being" p; or in other words, "replace $d(p)$ with p".

It is this proposed reply that we call *the drag-along response.* The label is meant to suggest that a diffeomorphism "drags along the identity" of each point in its domain-model to its codomain-model, so that the identity "follows" the qualitative profile.[21]

For this paper's purposes, we need to make just three comments, (1)-(3), about the drag-along response. (1) is about making it precise; (2) is about our own attitude to it; and (3) present problems for it.

(1): *Making the response precise*:— Of course, talk of 'rebranding' or 'replacing' points, or "dragging identity", is metaphorical. And to make the proposal precise, it is of course not enough to simply replace, in the mathematics, the non-identity function d by the identity function Id on M. Doing only that would simply change the subject, and ignore the non-mathematical interpretative issues raised by the hole argument, i.e. by the given function d.

But the proposal has been made precise, namely by [22, 23]; cf. also [24] and [8, Section 4]. They postulate a fibre bundle for which they can *define* a spacetime point as, in effect, a set of field-values (what we have called a 'qualitative profile'). This means that there is simply no issue about relating points between isomorphic models: the points follow their qualitative profiles—implementing the drag-along response.

But we need not go into details about these definitions, even though this approach, like our own in Part II, uses fibre bundle ideas. For the approaches are very different. As we announced in Section 1.1, ours but not Stachel's addresses how to relate points belonging to *non*-isomorphic models; and it is less abstract, with specific examples that can be worked out.

More generally, we need not try to make the drag-along response precise. For our purposes, it suffices to say that it has two components: (a) any two isomorphic models of general relativity represent the same possibility (i.e. Sophistication for general relativity); and (b) this

of the sun was doubled, spacetime would be more curved around it". One merit of our preferred response via counterparts in Part II is that it answers these worries. We also note that although we have formulated this objection by speaking of points in spacetimes, and so in effect by taking models $\mathbf{M}_1$ etc. to *be* possibilities, not just to represent them: nevertheless the objection could be stated while respecting such subtleties (albeit more cumbersomely) along the lines of footnotes 15 and 16.

[20] We say 'takes the idea further' to signal the fact that one can take the attributes Y to individuate the Xs, within a model or possibility, without being committed to their doing so "between" models, even isomorphic ones. In general relativity, this is illustrated by Komar observables (cf. [21]). The idea is that, at least in a generic *non*-symmetric spacetime, no two points have the same set of values of certain observables e.g. scalar curvature; so that the points can be labelled, i.e. uniquely referred to, by their values. But this intra-model individuation by the observables does not imply that between two models, two points must be identified by their having the same values—which is what the drag-along response will claim.

[21] So the contrast with the previous Subsection is this: whereas metrical essentialism condemns $\mathbf{M}_2$ for representing no possibility, because it is faithless to points' essential attributes, the drag-along response takes it to represent the very same possibility as $\mathbf{M}_1$—by using attributes, i.e. qualitative profiles, to "rebrand" points.

Journal of Physics: Conference Series **2533** (2023) 012002 doi:10.1088/1742-6596/2533/1/012002

is compatible with substantivalism, since points are individuated by their qualitative profiles.[22]

(2): *Our attitude*:— Our own view about the drag-along response is fourfold.

(a): It is plausible, since in many examples its verdicts about "which point is which" are right (as the quotation in footnote 22 says).

(b): But it faces problems. In (3) below, we report one: about models that have an automorphism.

(c): As we will argue in Section 3: even for models that have no automorphisms, its verdicts are not always right. So it is certainly not compulsory as a matter of mathematics, as some authors have urged.

(d): However, there is a grain of truth in the drag-along response, and even in the claim that it is mathematically compulsory. For at the end of our second paper, i.e. Part II, we will see that for a model that has no automorphisms, and an isomorphic copy of it related by a diffeomorphism d, our fibre bundle framework picks out the map d as the preferred way to relate the models.[23]

(3): *Problems*:— The drag-along response faces problems. We pick out one that is easily explained and has been discussed in the philosophical literature.[24] It is about *symmetric* models, i.e. models that have an automorphism (additional of course to the identity map on the manifold). In short, the drag-along response is led to make contradictory assertions about the identity of some points in any such model. For suppose a model $\mathbf{M} := (M, g, T)$ has an automorphism d with $d : p \mapsto q \in M$ and $q \neq p$. Then the drag-along response's idea of rebranding or replacing points leads it to say that q "is officially" p. Besides, the problem can also be stated without depending on a single manifold M. Given such a model $\mathbf{M}$ with symmetry d (with $d(p) = q \neq p$), there will be other models $\mathbf{M}'$ with a distinct manifold M', with two isomorphisms from $\mathbf{M}'$ to $\mathbf{M}$, one of them obtained by composing the other with d. That is: there are isomorphisms $d_1 : M' \to M$, $d_2 : M' \to M$, and a point $r \in M'$ with $d_1(r) = p$ and $d_2 := d \circ d_1$ so that $d_2(r) = q$. In such a case, p and q are equally good candidates to "officially be" r—and so be identical to each other. But since $q \neq p$, the drag-along response seems committed to contradiction. Furthermore, the trouble can be, so to speak, everywhere in the model. That is: if the model $\mathbf{M}$ is homogeneous, i.e. for any two points $p, q \in M$, there is an automorphism with $p \mapsto q$, then any two points in M get officially identified by the drag-along response—so that officially there is just one spacetime point.[25]

[22] Among the many formulations of the drag-along response to be found in the literature, here is a clear one expressing both these components: ' *"individuation of objects across possible worlds" is sufficiently tied to their qualitative characteristics so that if there is a unique 1-1 correspondence between the space-time of world A and the space-time of world B that preserves all geometric properties and relations (including geometric relations among the regions, and occupancy properties like being occupied by a round red object), then it makes no sense to suppose that identification of space-time regions across these worlds goes via anything other than this isomorphism.*' [25, p. 77].

[23] (i): For automorphisms, cf. the comment after Section 1.2's formulation of general covariance. (ii): Note that in Part II's discussion, the map d is written as g, since it an element of the *group* Diff(M).

[24] Another problem is more intricate and technical. It was introduced to the philosophical literature (so far as we know) by [26], especially Sections 4.3 and 4.4, pages 964-970. In short, the problem is that, for spacetimes that are asymptotically flat at infinity, it is standard practice to treat some isomorphisms (for vacuum spacetimes: some isometries) as generating from a given model (in the same manner as the hole argument) a new physical possibility, that differs from the given model by a time-translation at spatial infinity. Obviously, this standard practice threatens Sophistication about general relativity, and thus the drag-along response. The topic is intricate and far from settled: for instance, in [5], the standard practice is reconstrued so as not to threaten Sophistication. But we cannot here pursue this topic; though we will briefly return to it in Section 3.1.

[25] This problem has a previous incarnation, in the different though related debate about the philosophical doctrine called 'spacetime structuralism': which is, roughly, the view that spacetime points can be individuated by what we have called their qualitative profiles. [27] presented the problem for homogeneous models, as a refutation of

2.2.3. Counterparts We have just seen that metrical essentialism has difficulties about non-actual points, and that the drag-along response has difficulties about two or more points being equally good candidates for being identical to a given point. Both sorts of difficulty are avoided by a framework, called *counterpart theory*, for treating the identity of objects (all sorts of objects, not just points) in different possibilities; and we now review this.

We must again be brief. For counterpart theory is a large subject in the logic and metaphysics of possibility; and even within the smaller philosophy of physics literature, its application to spacetime points is a central topic in discussion of the hole argument. With an eye on what we will need in Section 3 and Part II, we shall confine ourselves to three comments, (1) to (3). (1) introduces the main ideas as they are set out in the purely philosophical literature. (2) describes the application to spacetime points and in answer to the hole argument. Finally, (3) returns us to the topic of determinism.

(1): *The main ideas:*— Counterpart theory was invented and developed by the great metaphysician David Lewis (mainly in his [29], [30, p. 38-43] and [31, Ch. 4]). Lewis is best known for advocating a framework of *possible worlds* (i.e. maximally specific possibilities) as the best way to address all sorts of philosophical problems ranging from the nature of propositions to causation; and counterpart theory is part of that framework.[26]

The key claim of counterpart theory is that objects in different possible worlds are never *identical* to each other. Rather, they are *counterparts* of each other—better or worse counterparts, according to how similar they are. Here is one of Lewis' early formulations:

> In general: something has for *counterparts* at a given world those things existing there that resemble it closely enough in important respects of intrinsic quality and extrinsic relations, and that resemble it no less closely than do other things existing there. Ordinarily something will have one counterpart or none at a given world, but ties in similarity may give it multiple counterparts. [30, p. 39].

As Lewis shows (and as this quote suggests), the great advantage of counterpart theory over saying that an object in a possible world can be literally identical with an object in another, is *flexibility*—thanks to similarity being vague and coming in degrees. In particular, since two objects can be equally similar to a given one, counterpart theory easily makes sense of statements like 'I could have been twins'. (Cf. the drag-along response's problem with symmetric models, that we saw in comment (3) of Section 2.2.2.)[27]

spacetime structuralism, calling it an 'abysmal embarrassment'. [28] is a reply, based on views about individuation and identity developed by him and Saunders.

[26] Among philosophers, Lewis is also famous for a belief he called *modal realism*: that each non-actual world—i.e. each way the cosmos could possibly be though in fact it isn't that way—is no less real than the actual world, i.e. cosmos, around us. That belief, almost all other philosophers (including us) reject as incredible—in the literal sense. We might have said 'notorious' rather than 'famous': though all agree that Lewis makes a very good case for his modal realism; initially in [30, Ch. 4.1], and at length in [31].

But fortunately, all agree that counterpart theory can be detached from modal realism, so that from now on we will set aside the latter. And we also set aside the more popular rival accounts of the ontological status of possible worlds: of which [31] gives a fair, indeed masterly, survey.

[27] Besides, Lewis rebuts the objection that when we say 'Hubert Humphrey might have won the USA's 1968 Presidential election' (an example that became famous in the philosophical literature), we mean that the actual Humphrey, "our Humphrey", might have won—not some distinct but appropriately similar person. Lewis agrees: indeed, that is what we mean. But the question is how another world can represent concerning the actual Humphrey, that he wins. And Lewis argues that counterparts do so: to use another philosophical jargon, they are suitable *truth-makers* for propositions about possibilities for actual objects, such as Humphrey [31, p. 194-197].

Even setting aside counterpart theory, similarity is an important notion for various topics in foundations of physics. For example, similarity of models (solutions) is important for inter-theoretic reduction and approximation. [32] is a fine survey, emphasising topological ideas—to which we will return in Section 3.2.3.

Journal of Physics: Conference Series 2533 (2023) 012002 doi:10.1088/1742-6596/2533/1/012002

(2): *Counterparts of points*:— So suppose we deny "trans-world identity" for spacetime points in different possible worlds: they can only be counterparts. The first thing to say about this view is that it immediately implies that (*Distinct*) is false. Thus this view affords a third way—additional to metrical essentialism and the drag-along response—to reply to Earman and Norton's dilemma by upholding substantivalism, while denying that substantivalism implies (*Distinct*). (Recall the preamble to Section 2.2.)

The reason this view denies (*Distinct*) is analogous to, but simpler than, the reason that metrical essentialism denied it.[28] For if a given model $\mathbf{M}_1$ is, or represents, a possibility (a way that a general-relativistic world could be), then the model we called $\mathbf{M}_2$, i.e. the model obtained by dragging along the fields in $\mathbf{M}_1$, is *not* (nor represents) a different possibility. For it is based on the same manifold M as $\mathbf{M}_1$; while on this view, none of the points in any such different possibility is a point in M.[29]

For the purposes of this paper and of Part II, there are two further aspects of our use of counterpart theory for spacetime points that we need to report. (Neither is contentious.) The first is to clarify our use of the jargon 'counterpart', in relation to its use in general philosophy. The second is about determinism: which we take up in (3) below.

We have said that counterparts are more or less similar, but never identical. And as the quotation from Lewis suggests: in his treatment of statements of possibility for ordinary objects, e.g. a person such as Humphrey, the relevant respects of similarity will be vague (though determined in part by the context in which the statement is made, e.g. the intentions of the speaker) and will typically involve some balancing or trade-offs between various respects. But herein lies a contrast with our application of counterpart theory to spacetime points, as we have so far stated it, i.e. as part of a reply to the hole argument.

For the respects of similarity that we invoke, in the precise setting of a spacetime theory such as general relativity, are not vague. They are—in our application as stated so far, and in the treatment of determinism in (3) below—a matter of exact matching of the points' qualitative profiles of field-values. Here, exact matching is made precise by a diffeomorphism dragging along the geometric objects at a given point in the domain-model into coincidence (respectively) with the geometric objects at the point's image in the codomain-model. In particular, there is no need to strike a balance between various attributes, some shared and some not. It is solely a matter of the geometric objects being required to be preserved by the diffeomorphism.[30]

This contrast is not a problem for us. After all, 'counterpart' is only a word, and an application of 'counterpart theory' to a specific topic, such as spacetime points or the hole argument, can perfectly well have a special feature, such as not being vague. But the contrast does prompt a clarifying philosophical remark: as follows.

For there is another philosophical jargon (also introduced by Lewis) for objects that exactly match in this sort of way. Lewis calls such objects *duplicates*. But we prefer to retain the word 'counterpart' for three reasons; (in ascending order of importance, for us).

(i): Lewis' own use of 'duplicate' is tied to contentious doctrines of his (about what he calls 'natural properties'): doctrines which, for reasons not relevant here, we do not endorse.

(ii): Being duplicates does not exclude being identical (neither in Lewis' treatment of duplicates, nor in others'); while 'counterpart' has (thanks to Lewis' large influence in

[28] That reason is at the start of Section 2.2.1. Note the contrast with the reason that the drag-along response denies (*Distinct*): *it* takes the two models to represent the same possibility.

[29] This way to defend substantivalism in reply to Earman and Norton was first advocated by one of us (JB); [33, p. 25-26], and at more length in [19, Sec. 6]. Since then, it has been elaborated. Three excellent recent papers are: [34, 35, 36]. Note also that to make the wording simpler, we here write as if spacetime points, whether actual or possible, are elements of a model's manifold; as we did in Section 2.2.1. But as noted there (cf. footnotes 15 and 16) the argument can be rephrased without this assumption.

[30] Apart from scalar fields, the geometric objects or field-values will mostly encode extrinsic attributes. For discussion, cf. the reference in footnote 18.

Journal of Physics: Conference Series 2533 (2023) 012002 doi:10.1088/1742-6596/2533/1/012002

philosophy) a firmly established connotation of excluding identity. And for our answer to the hole argument, the non-identity is important.

(iii): For our work in Part II, viz. treating *non*-isomorphic spacetimes, regardless of answering the hole argument: exact matching of points will *not* be centre-stage. Instead, thanks to the non-isomorphism: Lewis' ideas of degrees of similarity, and striking a balance between different respects of similarity, will apply naturally to spacetime points, and will be centre-stage. And since these ideas were as much part of his original conception of counterparts as was the denial of "trans-world" identity, it is clearer for us to use the word 'counterpart', rather than e.g. 'duplicate'.[31]

(3): *Determinism revisited*:— So far in this Section's review, determinism has been a minor theme. At the end of Section 2.1, we reported that Earman and Norton agree that determinism is not sacrosanct, but say that it should be given up only for reasons of physics, not for a philosophical reason such as defending substantivalism. Then this Subsection reviewed the dominant "fancy-footwork" reply to Earman and Norton, that substantivalism does not imply (*Distinct*); so that, the hole argument notwithstanding, it does not threaten determinism.

But to adequately review this controversy, we should make two further comments about determinism: the first about philosophical strategy, the second about the definition of determinism.[32]

First: one need not agree with Earman and Norton that determinism should be given up only for reasons of physics. For 'determinism' is only a word: and physicists and philosophers could perfectly well agree to use it differently. (Recall that Earman and Norton give no precise definition of it.) That is: one can be a substantivalist in Earman and Norton's sense, and agree with them that substantivalism implies (*Distinct*), and then respond to their argument by "biting the bullet". One simply concedes that general relativity is indeterministic, for the philosophical reasons about the identity of points that the hole argument reveals. But one then adds: 'No worries. For general relativity as a physical theory need have—indeed, should have—no concern about such philosophical indeterminism.'

This is undoubtedly a tenable response. But it highlights the need to formulate some precise definition (or definitions) of determinism, especially a definition that general relativity *does* satisfy. In terms of philosophical strategy, this need is especially pressing on Earman and Norton, who stress that they respect only 'reasons of physics' in evaluations of whether determinism holds. But it hardly matters where the need is most pressing. For all parties must admit the benefit of precise definition(s) of a central term such as 'determinism'. This leads to our second comment.

Second: In the endeavour to give such a definition, one of course turns to the physics literature's treatment of the existence and uniqueness of solutions to the initial-value problem for general relativity. In the philosophical literature about the hole argument, the first such effort was by one of us ([33, p. 26-31]; and [19, Sec. 3]): focussing on the treatment by [37, Ch. 7]. In short, he argued for two claims—that meshed with the counterpart-theoretic reply to Earman and Norton.

(i): There is a definition of determinism, labelled **Dm1**, that *is* violated by general relativity. In effect, the reason it is violated is that it expresses determinism's basic idea—that exact matching of two models on the past implies exact matching throughout the models—by having a single diffeomorphism, from the whole domain-model to the whole codomain-model,

[31] For more discussion of (i) and (ii), cf. [19, pp. 23-25]. As to (iii), we will also see a natural use of the term 'duplicate' at the end of Part II.

[32] We hope to be adequate, but we must be brief. We set aside details not only about our two comments, but about other aspects of determinism in the philosophical literature: about which we recommend the three recent papers cited in footnote 29, Though brief, what follows is anyway enough for this paper and for Part II: in which details about determinism will not be center-stage.

Journal of Physics: Conference Series **2533** (2023) 012002 doi:10.1088/1742-6596/2533/1/012002

drag along the geometric objects into coincidence.

(ii): But another definition of determinism, labelled **Dm2**—that is obeyed by general relativity—can be extracted from the technical theorems expounded by Hawking and Ellis (especially that of [38]). Its main difference from **Dm1** is that it expresses the regional-to-global implication (of exact matching), not with a single global diffeomorphism, but by allowing the implied global exact matching to be specified by a *different* diffeomorphism than the one that specified the assumed regional exact matching. This freedom to use a different diffeomorphism not only suggests the flexibility of counterpart theory. It also reflects the technical theorems' allowance that one *development* (intuitively: one "continuation into the future") of given initial data can be an extension of another development—and this notion of extension includes cases where a diffeomorphism defines a new model by drag-along, as in the hole argument.

But we should register here that the letter of (ii) was wrong, although the spirit was right: (*mea culpa!*). Butterfield's "transcription" from the technical theorems erred: though not in ways that affect our summary in (i) and (ii), or that matter for the hole argument and how to reply to it. The errors and how to amend them are stated very well in excellent papers by [39, p. 8-9, and footnote 22] and [35, Section 5.1].[33] In any case, these amendments are not needed in the sequel.

3. Act 3: 2016 to 2022: The hole argument dissolved? Threading spacetime points

So much by way of reviewing philosophers' replies during the 1990s to Earman and Norton's revival of the hole argument. From about 2000, there was less discussion[34], until about 2018: when [2] argued that mathematical practice rules out the hole argument. In short, he proposed that isomorphism is the mandatory standard for judging whether two points in (the manifolds of) two models of the theory correspond, or "are identical". To put it in our jargon: the 'drag-along response' (cf. Section 2.2.2) is compulsory, as a matter of *mathematics*, so that the hole argument is dissolved, or "never gets started".

This proposal revived debate. Some authors agreed, or broadly agreed, with Weatherall (e.g. [42], [43], [44] who also discuss precursors to the proposal). Others have disagreed (e.g. [45, 46, 20, 16, 47, 48]).

We ourselves disagree; and the goal of this Section is to give our reasons. Again, we must be brief: we cannot do justice to all the details of the recent literature. We confine ourselves to stating what we consider the main reasons for disagreeing. They centre around the fact that mathematics and physics treat the identity of points (indeed, of objects in general) more flexibly than Weatherall et al. allow (Section 3.1). This will lead in to our showing some examples of points in two models corresponding, or being "identified", other than by isomorphism (Section 3.2). We propose to call this the *threading* of points. Unsurprisingly, it will play a role in Part II's discussion of *non*-isomorphic spacetimes.

3.1. Against dissolution

We have already agreed that "identifying" points in two models by an isomorphism, as in the drag-along response, is plausible; (cf. (2)(a) and footnote 22 in Section 2.2.2) . But we deny that mathematical practice *dictates* this view; and that thereby, the hole argument "never gets started".

[33] The main technical points, covered by both authors, are that **Dm2** needs to be reformulated so as to be restricted to globally hyperbolic solutions to the vacuum Einstein equations, and to their maximal developments. Landsman also adds, more philosophically, that whereas both **Dm1** and **Dm2** *assume* the existence of a class of models, i.e. 4-dimensional solutions of the theory (as do all definitions of determinism in the philosophical literature), theorems solving an initial-value problem, like that of Choquet-Bruhat and Geroch (on which Landsman focusses), have the merit of *proving* that these solutions exist, given appropriate initial data.

[34] But we recommend [40] and [41].

Journal of Physics: Conference Series **2533** (2023) 012002 doi:10.1088/1742-6596/2533/1/012002

We have three main reasons for this denial. First, in (1) below, we disagree with specific claims of these authors. We hold that mathematical practice does not close off interpretative options. In particular, it does not refute (*Distinct*) in the ways they claim. Then in (2) we argue generally—independently of general relativity—that mathematical practice treats identity more flexibly than these authors allow. (1) and (2) will have a common moral: we should beware of trying to read philosophical morals directly off mathematics. Our third reason is again specific, but more positive: in some contexts, mathematics "identifies" (or: puts in correspondence) points in two models other than by isomorphism—the next Subsection will give examples.

(1): *Dissolution remains unproven:*— We claim, in short, that the arguments claiming to dissolve the hole argument fail. These arguments have of course many details and subtleties—as do the rebuttals of those arguments. But space is short, and most of these details and subtleties are not directly needed for this paper's positive project of expounding the ideas of counterparts, and threading of points, needed in Part II. So we will confine ourselves to stating some central points that have a clear connection with other Sections of this paper. So our aim here is not—cannot be—to convince all readers that the arguments fail; but just to say where we stand, and also to give newcomers a glimpse of the issues in this literature.

First, in (1A), we will articulate one main line of argument in [2], and then endorse the reply by [16]. Then in (1B) we will give three further examples from the literature of this line of reply. Finally in (1C), we suggest a reason why debate about the hole argument continues: a reason that will lead in to our (2).

(1A): Some philosophical jargon (which is widely used) will be helpful. The ways that the two models in the hole argument differ, namely over which point in the hole $H \subset M$ has which qualitative profile, are called *haecceitistic differences*.[35]

As we have said, Weatherall takes it to be standard mathematical practice to disregard such differences, since isomorphism is to be the method of comparison between models. This goes along with his rejecting the identity map on the manifold M as a method of comparison; and with our taking him to be an advocate of the drag-along response. Thus in one of the main passages expressing this view, he writes: 'mathematical models of a physical theory are only defined up to isomorphism ... One consequence of this view is that isomorphic mathematical models in physics should be taken to have the same representational capacities. By this I mean that if a particular mathematical model may be used to represent a given physical situation, then any isomorphic model may be used to represent that situation equally well' [2, p. 332; cf. also p. 338].

As we said in this Section's preamble, Weatherall's paper led to a debate that continues: which we of course cannot resolve, and we can only give our view of it. So far as we know, the first article to take a similar position to Weatherall was [42]. Thus the last assertion of Weatherall, just quoted, gets a well-nigh synonymous formulation (and endorsement) by Fletcher, who writes: 'if two models of a physical theory are mathematically equivalent, then they have the same representational capacities' [42, p. 233]. For this, Fletcher suggests a handy acronym, 'REME'; (for 'representational equivalence by mathematical equivalence').

To this, we endorse the reply by [16], who write (using our notation):

[35] The adjective 'haecceitistic' derives from *haecceitism*. It is the doctrine that some (maybe even all) objects each have a property that no other object has, even though each of these objects may match some other object in all *qualitative* respects. Such non-qualitative properties, each unique to their instance, are called *haecceities*, and are naturally expressed with proper names that encode no qualitative information, like 'is Humphrey' (cf. the example in footnote 27), or in logical notation '$= a$'. (Beware: philosophers' usage of 'property' varies. Some require a property to encode some (though perhaps logically weak) qualitative information. But such philosophers can still accept haecceitistic differences, by allowing that two objects can be distinct *solo numero*, i.e. without any property instantiated by one but not the other.)

Journal of Physics: Conference Series **2533** (2023) 012002 doi:10.1088/1742-6596/2533/1/012002

> [though] REME is a legitimate principle of mathematical representation .. . [it] fails, by itself, to block the hole argument, or to rule out the allegedly problematic use of isomorphic models deployed in standard presentations of the argument [i.e. the use of the identity map on the manifold M as a method of comparison]. All that this use requires is that if model $\mathbf{M}_1$ (say) is taken to represent a particular possible world W, then $\mathbf{M}_2$ [i.e. the model produced by drag-along, as in Section 1.2] represents (or, better, may be taken to represent, perhaps by further specifying the representational context) a distinct (but merely haecceitistically distinct) possibility W'. This is all completely compatible with $\mathbf{M}_1$ and $\mathbf{M}_2$'s having identical representational capacities. That merely requires that $\mathbf{M}_2$ might equally well have been chosen initially to represent W (in which case, some other hole-diffeomorphic model, related to $\mathbf{M}_2$ just as $\mathbf{M}_2$ is related to $\mathbf{M}_1$, could be taken to represent W'). [16, Section 5]

In short, the reply is that being precise about one's assumptions about models' representational capacities (though of course laudable) is not enough to rule out someone, viz. the advocate of the hole argument, taking the identity of the points to have representational significance. To put it in this paper's jargon: precision about representational capacities does not show the drag-along response is compulsory, or that (*Distinct*) is untenable; (recall footnote 12).

(1B): This line of reply (and the broader moral that philosophy cannot be read off mathematics) recurs in the literature. Here are some examples: earlier ones by Pooley and others, and a current one by Menon and Read.

In an earlier paper, Pooley makes essentially the same reply, and connects it with the topic of determinism: saying as we did in (3) at the end of Section 2.2.3 that a philosopher who believes that possibilities (i.e. in philosophers' jargon: possible worlds) can differ merely haecceitistically can accept that general relativity is indeterministic. Thus he writes:

> In effect, both authors [i.e. Weatherall and Fletcher] argue that, if there are pluralities of merely haecceitistically distinct possibilities, the mathematical formalism of GR, correctly interpreted, is necessarily indifferent to differences between them. But this just means that GR does not distinguish between any two elements of such a plurality; both will count as physically possible according to GR or neither will. And that, of course, is just to admit that, according to any metaphysical view committed to such pluralities, GR is indeterministic. The indeterminism cannot be avoided by remaining loftily above the metaphysical affray. [20, end of Section 4]

Other earlier papers making the same or similar replies include: (a) Roberts in [47], and [46, especially pp. 255 and 261]; Roberts argues that his condition *weak Leibniz equivalence*, which is well-nigh synonymous with Fletcher's REME, is irrelevant to, i.e. powerless to block, the hole argument; and (b) [45, Sections 3.3.2 against Weatherall, and 4.1 against Fletcher].

Recently, the same line of reply occurs in [48]'s critique of [43]'s attempt to support [2]. Most of their critique is devoted to showing (we think rightly) that Halvorson and Manchak's efforts are misdirected.[36] But once this misadventure by Halvorson and Manchak is set aside, [48]

[36] For Halvorson and Manchak's idea is that Weatherall's dissolution of the hole argument has a loophole: a loophole that they claim to close by invoking a theorem that isometries of Lorentzian manifolds are "rigid" in the same manner as analytic complex functions, i.e. such an isometry is determined by its action on an open set. Their idea is that Weatherall's requirement that points in distinct models can only be identified by invoking an isometry will run in to trouble if there is more than one such isometry; (cf. the problem with symmetric models we rehearsed in (3) at the end of Section 2.2.2). So they cite such a theorem (viz. Theorem A1 of [49, Appendix], and claim to complete Weatherall' analysis and so "close" the hole argument.

Most of [48]'s effort goes into showing that this concern about uniqueness of isometries (and so Geroch's rigidity theorem) is irrelevant to the hole argument. The issue turns on our comment in Section 1.2 (after stating general covariance): that one must distinguish two senses of the word 'isometry' (or more generally, of 'isomorphism'). (Cf.

Journal of Physics: Conference Series **2533** (2023) 012002 doi:10.1088/1742-6596/2533/1/012002

address (in their Section 3.2, 'Reopening the hole argument') the questions that remain: are Halvorson and Manchak's (and Weatherall's) tenets justified? And do these tenets (whether or not justified) block the hole argument? They argue, using the above line of reply, that for both questions, the answer is 'No'.

(1C): Agreed, these replies are not definitive. Controversy continues: the jury is still out. For example, Fletcher addresses the objection that REME does not block the hole argument [42, p. 239, start of Section 5.2], but believes he can surmount it. Again, we think he cannot, for reasons that are spelt out by both (a) [46] in his rebuttal of what he calls *strong Leibniz equivalence*, and (b) [16] in the remainder of their closing Section 5, which also discusses Roberts—and urges the relevance of appealing to counterparts, as we advocated in Section 2.2.3.[37]

But we will not further try to sum up the *status quo* in the debate. Instead, we end by reporting (uncontentiously!) a point about Weatherall's appeal to mathematical practice. (This will also lead to (2).) Namely: in the details of this debate, it turns out that Weatherall and his kindred spirits, who seek a formal or mathematical response to the hole argument, have a different view of what is the *content* (even apart from the interpretation) of general relativity, than do authors such as Pooley, Read and ourselves. These differences even involve disagreement about the sorts of logical or metalogical apparatus that a formulation of general relativity could, or should, be expressed with. For details of this disagreement, cf. [44]; and (in reply) [35, Sections 5, 6] and [36]. So these disagreements lie, in effect, in the philosophy of logic: which is an academic field where physicists, and even philosophers of physics, are bound to be cautious about taking sides. But for our purposes here, it is fortunately enough to just note that once one agrees it is a contentious matter of judgment what is the content of general relativity, then it is unsurprising that debate about how to respond to the hole argument continues.

(2): *Mathematics is indifferent and flexible in its treatment of identity*:– We agree that modern mathematics in all its branches focusses, at least for the most part, on the "structures" or "patterns" in its subject-matter; and this means it is usually *indifferent* to the identity of the objects that instantiate the structure or pattern. Here, 'indifference to identity' means, of course, indifference to the objects' intrinsic natures, and to any attributes (intrinsic or extrinsic) by means of which one might qualitatively distinguish which object is which. So it does not mean indifference as to whether the objects are distinct: which would undermine there being a definite number of them.

We also agree that this indifference makes it often natural to treat the objects that correspond by an isomorphism (as argument and value of the bijection giving the isomorphism) as corresponding, or "the same", as urged in the quote in footnote 22. (This indifference to identity, and treating objects as the same, is of course a theme in both category theory, and structuralist philosophies of mathematics; and also in the history of pure mathematics, and especially Hilbert's influence on it.) But though this is often so, it is not *always* so. Indifference makes for flexibility about identity, not for isomorphisms being the mandatory standard for it.

Menon and Read's 'isometry$_1$' and 'isometry$_2$' in [48, Section 2.2].) In short: a model (M, g, T) can be "wrinkly" i.e. have no automorphisms; yet, by construction, it is isomorphic to any model built by dragging-along g and T by a diffeomorphism d of M.

[37] We should note that Fletcher also makes an important distinction between REME and the logically stronger principle that 'if two models of a physical theory are mathematically equivalent, then there is a unique physical state of affairs that they represent equally well' [42, p. 231]: which he gives the acronym, 'RUME' (for 'representational uniqueness by mathematical equivalence'). RUME is close to Roberts' strong Leibniz equivalence; and we concur with Pooley and Read's assessment of it (in their Section 5). But here, we just note that Fletcher invokes his distinction to argue (pp. 241-246) that Belot's point (cf. footnote 24) that some isomorphisms generate new possibilities, since they change asymptotic structure, is not a problem for him (nor for Weatherall). For these examples refute RUME, not REME: indeed, they do so in much the same way as some much simpler examples about a particle whose worldline can swerve (which can accelerate) at different spacetime points.

Journal of Physics: Conference Series **2533** (2023) 012002 doi:10.1088/1742-6596/2533/1/012002

Indeed, there are two points here. The first is that we should beware of a bad argument for isomorphisms being the mandatory standard. The second point will be more positive.

The bad argument targets our admitting, in Section 2.1, that it is 'undoubtedly plausible' that substantivalism implies (*Distinct*): since for any two distinct objects, a and b, and any two distinct attributes such objects can have, F and G, surely a having F and b having G is a different possibility than a having G and b having F; (cf. also footnote 12). We similarly remarked at the start of Section 2.2 that Sophistication (viz. that two isomorphic models are, or represent, the same possibility) needs fancy footwork.

It is tempting to reply to this that, not only *can* Sophistication claim that *its* version of substantivalism denies that there are facts about which spacetime point has which qualitative profile of field-values (i.e. that its silence about "which is which" still counts as belief in the points as objects): also, it *must* claim this—thanks to the following argument.

> Isomorphisms are *defined* as bijections that are required (only) to preserve "structure", and so they can ignore the identity of the objects they map. So Sophistication must deny that which object is which can be part of what each of a pair of isomorphic models represent.

But this is a bad argument. Its error lies in its assuming that the structure that an isomorphism must preserve cannot encode the identity of the arguments of the bijection. But it can do so. For in mathematics, isomorphism and structure are both entirely formal notions. They are therefore *flexible*. So if we wish, we *can* specify that the bijection preserve for each argument a property that only it has, so that the corresponding value of the bijection (in the codomain) has it.[38]

We can put the same point in the language of formal semantics, or model theory. Agreed: an isomorphism is a bijection from the base-set of the domain model to the base-set of the codomain model that is only required to preserve the announced "structure", as encoded in predicates etc. of the language that the models interpret. And agreed: this announced structure *usually* does not incorporate anything about the identity of the base-sets' elements. The base-sets can be disjoint, or distinct but overlapping, or the same: and if they are overlapping or the same, the bijection f can either (i) map an object o that is in common to itself, or (ii) instead not do so—provided only that o and $f(o)$ share their announced "structure", i.e. patterns of instantiation of the predicates etc. in the language. But we *are* at liberty to require any features, including about the identity of base-set elements, to be part of what is to be preserved by the bijection. One just needs to introduce into the language appropriate vocabulary, e.g. predicates, to encode these features.

Our second more positive point is that in some contexts, mathematics *needs* to treat objects' identity other than by (their corresponding by an) isomorphism. And these contexts include the treatment of points in differential geometry, both pure and applied.

So to sum up: indifference to identity makes for a flexible, even opportunistic, treatment of identity: it does not make for the mandatory strait-jacket of a uniform treatment. Besides, this moral applies to the case of interest to us: points in spacetime theories like general relativity.

3.2. Examples of threading

In this Subsection, we will give four examples of differential geometry "identifying" points other than by an isomorphism; (though in each example, the two points will be related by a diffeomorphism, as argument and value). We will talk of *threading* the points, or *a threading scheme*, so as to signal that there is a variety of ways to thus thread points: a variety that is not conveyed by saying "identifying points". Note that there is no uniform meaning to this threading

[38] Recalling the jargon in footnote 35, we can put the point as follows: by enriching the notion of isomorphism to include preserving haecceities, Sophistication can be reconciled with haecceitism.

of points—and no need for such a meaning. What matters is that differential geometry in some contexts invokes a scheme of identification *other than drag-along.*

Note that the third and fourth examples will introduce the idea of threading two points that are in two *non*-isomorphic spacetimes: a topic we develop in Part II.

3.2.1. The definition of the Lie derivative To define the Lie derivative of a field, one drags the field along the integral curves of the vector field with respect to which one is taking the Lie derivative. This means assuming an "identity" or "correspondence"—we say: threading—of the points, independent of their field-values: so that sliding the field along a curve through the points makes sense. For if points' identity was determined by field-values (as in Section 2.2.2's drag-along response), the field would *not* be dragged with respect to the points. More precisely: for d_t the flow of a vector field X^a, we would obtain for e.g. a metric g:[39]

$$\mathcal{L}_X g(x) = \lim_{t \to 0} \frac{1}{t}(g(x) - d_{t*}g(d_t^{-1}(x)) \equiv 0\,! \tag{1}$$

But to do differential geometry, pure or applied, one must of course accommodate non-vanishing Lie derivatives. An obvious example is immediate from the definition in relativity theory of a rigid continuous body, as a body for which the 4-velocity field X of its material particles Lie-drags the metric: $\mathcal{L}_X g = 0$. Therefore, to describe the motion of a *non*-rigid continuous body, we need: $\mathcal{L}_X g \neq 0$. But of course, there are many more sophisticated examples: one is in the next Subsection.

Agreed, one might object to what we have just said: that there is an alternative definition of the Lie derivative as an algebraic derivation, i.e. as a set of rules about how to apply a certain operator to various quantities defined on the manifold—a definition which does not invoke pull-backs by infinitesimal diffeomorphisms. But we submit that even with such a definition, the problem of Lie derivatives trivially vanishing would remain: at least if one keeps the standard understanding of variational symmetries in a classical field theory as induced by diffeomorphisms. For as we will see in the next Subsection's example, combining this understanding with the drag-along response implies that any Lie derivative that would emerge from the action of a symmetry must trivially vanish.

3.2.2. Noether's second theorem in general relativity Noether's second theorem in general relativity gives a central example of our theories needing a non-zero Lie derivative. We suppose, to begin with, that the action takes the form

$$\int \mathrm{d}^4x\,(\mathcal{L}_g + \mathcal{L}_m)\;; \tag{2}$$

where we assume minimal coupling, in the sense that all contributions of the matter fields to the Lagrangian are confined to the second component, $\mathcal{L}_m$, of the total Lagrangian density. We also

[39] Landsman makes essentially this comment. Like us, he opposes claims (Section 3.1) that isomorphisms being a mandatory standard for identifying points vitiates the hole argument. He writes: 'Much of differential geometry would be thrown away if one is not allowed to compute the pullback of a metric under a diffeomorphism; even the very definition of an isometry rests on the ability to say whether or not $d^*g(p)$ equals $g(p)$, at each $p \in M$. The usual definition of the action of a diffeomorphism on a tensor (field) would be a "category mistake", and with it, the Lie derivative ... [notation changed]' [39, p. 4]. Cf. also [50, Sec. 2.4].

Amusingly, the *maestro* [51, p. 198] says that the Lie derivative is sometimes called 'the fisherman's derivative', because one can think of the integral curves of the vector field as the flow lines of a fluid, e.g. a river: "the flow carries all possible differential geometric objects past the fisherman, and the fisherman sits there and differentiates them'. Thus the stationary earth of the river-bed is the analogue of the "fixed"—we say: threaded—points over which the fields (Arnold's 'differential geometric objects') slide.

Journal of Physics: Conference Series **2533** (2023) 012002 doi:10.1088/1742-6596/2533/1/012002

assume that the energy-momentum tensor for the matter fields is given by (introducing indices)

$$T^{ab} = \frac{\delta \mathcal{L}_m}{\delta g_{ab}}, \tag{3}$$

and that the action is given by the Einstein-Hilbert Lagrangian, $\mathcal{L}_g = R\sqrt{g}$.

Now we would like to apply a variation along an infinitesimal symmetry, that is along a vector field X^a that is the infinitesimal generator of a flow diffeomorphism. Assuming that general flow diffeomorphisms generate symmetries of the theory, and that X has compact support so that the boundary terms vanich, we obtain:

$$0 = \delta S = \int \mathrm{d}^4x((\frac{\delta \mathcal{L}_g}{\delta g_{ab}} + \frac{\delta \mathcal{L}_m}{\delta g_{ab}})\mathcal{L}_X g) \equiv \int \mathrm{d}^4x((G^{ab} + T^{ab})\mathcal{L}_X g_{ab}). \tag{4}$$

With the drag-along response, i.e. with the fields *not* being dragged over the spacetime points, $\mathcal{L}_X g \equiv 0$. So (4) would be satisfied identically, and yield no further conditions on the quantities involved.

But with threading, i.e. the fields being dragged over the "fixed canvas" of points, we get $\mathcal{L}_X g_{ab} = \nabla_{(a}X_{b)}$ (with ∇ the Levi-Civita covariant derivative). Then, from (4), we obtain:

$$\nabla_a G^{ab} = \nabla_a T^{ab}, \tag{5}$$

which is satisfied even off-shell: that is, even independently of the Einstein equations. Therefore, using the Bianchi identity $\nabla_a G^{ab} = 0$, we also get the off-shell relation: $\nabla_a T^{ab} = 0$, i.e. the local conservation law for the energy-momentum tensor.[40]

3.2.3. Limits of spacetimes In the two preceding Subsections, we have threaded points between two isomorphic spacetimes —though, of course, not by an isomorphism. In this Subsection and the next, we turn to threading points between two non-isomorphic spacetimes: in particular, setting aside matter fields, between non-isometric Lorentzian manifolds.

Our first example is the use of threading in the rigorous construction of the limit of a (one real parameter) family of Lorentzian manifolds, by Geroch [49]: whose exposition we will follow.[41]

We note also that both this example and the next Subsection's use only relativistic spacetimes. But there are theorems stating that a non-relativistic spacetime is a limit of a suitable family of relativistic spacetimes. The best-known example is a Newton-Cartan spacetime being such a limit "as $c \to \infty$": the null cones get wider until they finally become degenerate, i.e. tangent to a spacelike hypersurface of the Newton-Cartan spacetime. (Cf. e.g. [53]: (Proposition 1, p. 194, and Proposition 2, p. 197, which treats matter); [54] for an analysis stressing topological issues.) Space forbids our going into details: but it is intuitively clear that such theorems can be thought of in terms of threading. For example, consider the idea that as the null cones widen, curves through a cone's vertex that were (in earlier members of the sequence of spacetimes, with narrower cones) spacelike enter the cone i.e. become timelike. This formulation invokes a threading across the models of a sequence of points. And as in our other examples, the sheer mathematics is happy to treat this threading as identity, and so to use a single manifold for all

[40] We stress that this argument does not show—and we do not claim—that the drag-along response implies that energy is not locally conserved. But the argument shows, and we claim, that the drag-along response implies that a proof of conservation cannot go via Noether's second theorem.

[41] Though we confine ourselves to Geroch's exposition, we stress that the construction has had many applications. The obvious ones are in proofs of the stability of a spacetime: for in such a proof one needs to consider a sequence of spacetimes converging to the given one (and show that their geometries converge to its). See for example [37, Sec. 7.6], [26, p. 968] and (stressing topological issues) [52, Section 4], and [32, Section 5.2].

Journal of Physics: Conference Series **2533** (2023) 012002 doi:10.1088/1742-6596/2533/1/012002

the models. To use the obvious metaphor: we thread the points by their occupying the same place on "the sheet of paper" in common between the models, not by their matching as regards being spacelike or timelike to the cone's vertex (a chosen origin), nor by any other matching of metrical attributes.

Geroch's main idea is that the limit of a (1-parameter) family of spacetimes (in general, *not* isometric to each other), should be defined as the boundary of a certain 5-dimensional manifold $\mathcal{M}$, of which the given spacetimes are 4-dimensional leaves. For simplicity, Geroch sets aside matter fields, so that spacetimes are Lorentzian manifolds (M, g). He writes the family as (M_λ, g_λ) with $\lambda > 0$; and the 5-dimensional manifold of which the (M_λ, g_λ) are leaves, as $\mathcal{M}$. So the limit spacetime is $\partial\mathcal{M}$ and corresponds to $\lambda = 0$. (So in this Subsection we will follow Geroch's notation, and write a spacetime as (M_λ, g_λ), or for short, as M_λ: we do not use our previous bold-font notation $\mathbf{M}$.)

Geroch's motivation for seeking a definition is that 'one cannot speak simply of the limit of [e.g.] the Schwarzschild solution as the mass $\lambda \to 0$', for [as he shows] the spacetime one obtains in the limit [e.g. Minkowski or Kasner] depends on the choices of coordinates' (p.182). He also declares at the outset (p. 181) that to give a coordinate-free definition, he will consider (i) a 1-parameter family M_λ $(\lambda > 0)$, rather than (ii) a 1-parameter family of metrics on a given manifold: for he does not wish to presuppose a scheme that identifies a point $p_\lambda \in M_\lambda$ with a point $p_{\lambda'} \in M_{\lambda'}$ $(\lambda \neq \lambda')$. (We note that even (ii) could spell trouble for Section 2.2.2's drag-along response, since the different M_λ will not be isometric.)

He also stresses (p. 182) that 'limit spacetime' will be defined without presupposing any preferred scheme for identifying (in our jargon: threading) points between different spacetimes in the family. And he points out (p. 182, footnote) that the natural way to define a scheme is by a congruence of curves, through the 5-dimensional manifold $\mathcal{M}$, such that each curve cuts each leaf, i.e. each 4-dimensional spacetime, just once. In our jargon, this of course means that two points in (the image of) one such curve are threaded together. So agreed: as regards Geroch's general definition of limit spacetimes, threading is *not* needed. And 'all the better', one might say, since the spacetimes are not isometric; (indeed, an advocate of Section 2.2.2's drag-along response presumably would say this).[42]

But threading comes into its own (and the limited scope of the drag-along response is shown) in the *application* of the definition. For as Geroch says, the general definition is not 'useful for actually writing down limits' (p. 183). To write down limits, one needs to use a family of frames (tetrads) on the M_λ, smoothly varying along the curves of a congruence of curves in $\mathcal{M}$, each curve cutting each leaf once. Thus each such congruence specifies a threading relation: $p_\lambda \in M_\lambda$ and $p_{\lambda'} \in M_{\lambda'}$ are threaded together iff they lie on the same curve of the congruence. (Of course there are many—continuously many—congruences, so the threading relation is fiducial.) Thus Geroch writes:

> By a *family of frames* in $\mathcal{M}$ we mean an orthonormal tetrad $w(\lambda)$ of vectors tangent to M_λ and attached to a single point $p_\lambda \in M_\lambda$, for each $\lambda > 0$, such that the $w(\lambda)$ vary smoothly along the smooth curve in $\mathcal{M}$ defined by the points p_λ. [In our jargon, this curve of the congruence is one set of points threaded together.]
>
> Let $\mathcal{M}'$ be a limit space of $\mathcal{M}$, and let $w(\lambda)$ be a family of frames which assumes a limit—i.e., approaches a frame $w(0)$ at some point $p_0 \in \partial\mathcal{M}'$—as $\lambda \to 0$. Let us represent points in M_λ in a neighborhood of p_λ in terms of the system of normal

[42] For completeness, here is his definition. He assumes the M_λ 'may be put together to make a smooth (Hausdorff) 5-dimensional manifold'. Then: 'we define a *limit space* of $\mathcal{M}$ as a 5-manifold $\mathcal{M}'$ with boundary $\partial\mathcal{M}'$, equipped with a tensor field $g'^{\alpha\beta}$, a scalar field λ', and a smooth, one-to-one mapping Ψ of $\mathcal{M}$ onto the interior of $\mathcal{M}'$ such that three conditions are satisfied: 1. Ψ is an isometry, i.e. Ψ takes $g^{\alpha\beta}$ into $g'^{\alpha\beta}$, and λ into λ'; 2. $\partial\mathcal{M}'$ is the region given by $\lambda' = 0$: we require, furthermore, that $\partial\mathcal{M}'$ be connected Hausdorff and non-empty; 3. $g'^{\alpha\beta}$ has signature $(0, =, -, -, -)$ on $\partial\mathcal{M}'$.' (p. 182)

> coordinates based on $w(\lambda)$. In terms of these coordinates, the components of the metric tensor in the M_λ approach a limit as $\lambda \to 0$, and the limiting components are precisely the components of $g^{ab}(0)$ in $\partial\mathcal{M}'$ in a neighborhood of p_0. Thus, the family of frames $w(\lambda)$ uniquely defines the limit space $\mathcal{M}'$, at least in a suffiiciently small neighborhood of p_0. (p. 183)

To sum up: in order to explicitly construct limits of spacetimes, one needs to thread points between non-isometric spacetimes.[43]

3.2.4. Quantum reference frames Our last example opens that Pandora's box, the interpretation of quantum theory: especially in relation to superpositions of spacetime geometries, induced by superpositions of matter distributions. Of course, we must be brief, and cannot hope to settle any controversial matters. But our main point—that this literature invokes the idea of threading: indeed, of rival threadings—holds good, irrespective of these controversies.

So we will first present the general issue; and report a proposal of Penrose. Then we report a recent critique of his proposal by Giacomini and Brukner, that uses the framework of *quantum reference frames*. But again: what matters for us is agreed by the disputants: the need for threading points between non-isometric spacetimes.

The first point to make is the obvious one. In combining quantum theory with any spacetime theory (such as general relativity) in which spacetime is dynamical, i.e. responsive to matter, the question arises whether the metric (or more generally: geometrical or gravitational) degrees of freedom are to be given a quantum treatment. We shall (like Penrose, Giacomini and Brukner) aim for a treatment in which superpositions of distributions of the masses sourcing the gravitational field imply some sort of superposition of non-isometric spacetimes. There are then two questions. First, and fundamentally: how should we understand such a superposition? Secondly, what should we say about cases where two (or more) mass distributions that are components of the superposition are macroscopically distinguishable, so that the ensuing difference in the two geometries is also macroscopically distinguishable?

Over the years, Penrose has proposed answers to both questions (as of course, other authors have): answers that vividly raise the issue of "identifying", i.e. threading, points between non-isometric spacetimes. For example, Penrose writes: 'The basic principles of general relativity—as encompassed in the term 'the principle of general covariance' (and also 'principle of equivalence')—tell us that there is no natural way to identify the points of one space-time with corresponding spacetime points of another' [56, p. 591].

Now, each (classical) metric (or gravitational) field involved in the superposition will have a different timelike Killing vector, and thus a different direction of preferred time-translation. Penrose then argues that because of the relation between timelike Killing vectors and energy conservation laws, the fact that the former are not well-defined means that the global state is not stationary; and the difference in the energies implies an energy-time uncertainty principle, which makes the superposed quantum state unstable. This is the basic idea of Penrose's gravitationally-induced state-reduction.

[43] We are not the first to invoke this construction by Geroch in connection with the hole argument. In a paper written at about the same time as [2], [55] does so. About the hole argument, he is sympathetic to Weatherall (whom he cites), saying: 'Thus the hole argument is obviated by the fact that the application of the diffeomorphisms to the manifold cum metric results only in a different presentation of the same intrinsic physical structure, and so the worry about indeterminism evaporates, doing away with the dilemma' (p. 453). But he also argues more generally that for the question of whether spacetime points exist (or exist independent of their metric relations) to be fruitful, we must place it in a specific scientific context: and limits of spacetimes *a la* Geroch provide just such a context. Hence Curiel gives criteria for a Yes answer to the question in terms of such limits (his Definitions 1 and 2, p. 463).

Journal of Physics: Conference Series **2533** (2023) 012002 doi:10.1088/1742-6596/2533/1/012002

Recently, the framework of *quantum reference frames* (see [57, 58, 59] and references therein) has been applied to these proposals; with [57], in particular, criticising the argument for gravitationally-induced "collapse of the wave packet". They point out that there is no single reference frame transformation which makes the spacetime locally Minkowskian along a geodesic in such a superposition. This, according to them, is the source of the apparent violation of the weak Equivalence Principle, i.e., the violation of the universality of free-fall.

In more detail: they say that Penrose's argument rests on the (implicit) assumption that "all coordinate systems are in a classical relation to each other. With this assumption, it is impossible to find a classical, local coordinate system in which the metric can be made locally Minkowskian for both configurations of the mass" (ibid. p. 4). They then propose, using their framework, that one should anchor the labelling of spacetime points onto the trajectories of the masses involved. This they take to disarm Penrose's argument: the Equivalence Principle holds in each component and hence in the entire superposition.

For our concerns, what matters is of course the "anchoring of the labelling". Thus for the case of a single mass that acts as a probe particle for the metric/gravitational field (the particle being, of course, present in all elements of the superposition), they take the different spacetime coordinates of the particle in different manifolds to all correspond to the same physical point, as specified by the location of the particle (ibid. p. 5). In our language, this amounts to using the values of e.g. a mass-density function for the particle (thought of as the "same in the different possible worlds"!) so as to define how to thread points.

In Part II, we shall return to this suggestion, using our counterpart-theory framework for treating non-isometric spacetimes. For now, we close by just noting that [57] implement the idea that "threading is defined by mass-density" by having some object—other than the probe particle—define a quantum reference frame, by its being placed at the origin of a coordinate system. Thus they write:

> Such relative coordinates correspond to the distance between any physical system and the initial quantum reference frame, defining [...] the origin of a coordinate system. This allows us to operationally identify points belonging to different spacetimes, and meaningfully write the superposition state by adopting a relational view: the quantum reference frame [...] is always at the origin of the coordinate system, and the quantum states of [the probed particles] are their relative [...] states, in a position basis. [...] In this view, there is no contradiction in having different Killing vectors which are in a superposition in each amplitude. [57, p. 4][44]

4. Conclusion

This concludes our review of the philosophical discussion of the hole argument. Since Section 1.1 already gave a summary of our main themes and claims, we will not do so again here. But it is worth stressing three points. For they concern the central theme, whether isomorphic models must be regarded as representing the same possibility (called 'Sophistication'); and they all look ahead to Part II.

[44] Incidentally, there is a technical issue. One glimpses it in this quotation's mention of relative coordinates and relative distances (relative to the object at the origin of the quantum reference frame). This suggests that Giacomini and Brukner's framework secures a unique Fermi coordinate system along the trajectory of the probe particle that is identified across the two (or more) spacetimes. But distances to a given object are not sufficient to determine a local coordinate system: one also needs information about directions. In other words: to define a unique Fermi coordinate system along the particle trajectory, we need more than just the metric and the trajectory—we need some kind of anisotropy. Agreed: Brukner and Giacomini seem more interested in existence, not uniqueness, of Fermi coordinates in each member of the superposition. Moreover, uniqueness can be obtained by assuming cylindrical symmetry around the particle trajectory.

(1): *The prospects for Sophistication:*— We have emphasised that although we reject the drag-along response (Sections 2.2.2 and 3), we are sympathetic to Sophistication; (cf. (1) and (2) at the end of Section 1.1, and the preamble to Section 2.2). But we do not claim to add to the *philosophical* case in favour of Sophistication: neither in this paper, nor in Part II. In particular, the main merit of Part II's fibre bundle framework, as regards Sophistication, will be to place discussion of it in a wider context.

(2): *Counterparts and threading ...:*— This paper has set the stage for Part II's endeavour, viz. comparing non-isomorphic spacetimes. In particular, the ideas of counterparts (Section 2.2.3) and of threading points (Section 3.2) will be important tools for such comparisons.

(3): *... in a fibre bundle of spacetimes:*— We will see that these two ideas can be implemented naturally in the fibre bundle of spacetimes. For example, a gauge-fixing condition in the fibre bundle, i.e. a choice of a section of the bundle that cuts across the fibres, can be specified by intuitive counterpart relations between spacetime points. (We say 'intuitive', because these relations can invoke long-established ideas that points correspond to one another by having suitably matching, or nearly matching, values of physical quantities.) Besides, the technical setting enables us to define counterparthood, i.e. degrees of similarity, for entire spacetimes. And we can then show that counterparthood for spacetimes has a natural transformation property; and even that this transformation property reveals a grain of truth in the claims (Section 3.1) that isomorphism is the mandatory standard for when points in two models should be (that dreaded word!) "identified".

Acknowledgements
For comments on this material, we are grateful to: the audience, and especially the organizers, at the DICE conference in Castglioncello; the audience and organizers at the August 2022 QISS meeting; the participants at the LSE-Cambridge Philosophy of Physics Bootcamp discussion group; to seminar audiences at Bonn, Bristol, Oxford and Warsaw; and to Caslav Brukner, Frank Cudek, Sam Fletcher, Flaminia Giacomini, Ruward Mulder, James Read and Isaac Wilkins for comments on a previous version.

[1] Earman J and Norton J 1987 *The British Journal for the Philosophy of Science* **38** 515–525 ISSN 0007-0882 (*Preprint* https://academic.oup.com/bjps/article-pdf/38/4/515/9745359/515.pdf) URL https://doi.org/10.1093/bjps/38.4.515
[2] Weatherall J 2018 *The British Journal for the Philosophy of Science* **69** 329–350 (*Preprint* https://doi.org/10.1093/bjps/axw012) URL https://doi.org/10.1093/bjps/axw012
[3] Gomes H 2021 *Arxiv: 2110.07203. Submitted.*
[4] Gomes H 2021 *Arxiv: 2110.07204. Submitted.*
[5] Gomes H 2022 *Unpublished*
[6] Norton J 1984 *Historical Studies in the Physical Sciences* **14** 253–316
[7] Stachel J 1989 *Einstein and the History of General Relativity* pp 63–100
[8] Stachel J 2014 *Living Reviews of Relativity, 17.* URL http://www.livingreviews.org/lrr-2014-1
[9] Norton J 1989 *Foundations of Physics , vol 19*
[10] Norton J 1993 *Reports of Progress in Physics* **56** 791–861
[11] Giovanelli M 2021 *European Journal for Philosophy of Science* **11**
[12] Earman J 1986 *A Primer on Determinism* (Springer, Netherlands.)
[13] Earman J 1989 *World Enough and Spacetime* (MIT press)
[14] Weatherall J 2020 *Studies in History and Philosophy of Science Part B: Studies in History and Philosophy of Modern Physics* **70** 79–87
[15] van Fraassen B 1980 *The Scientific Image* (OUP)
[16] Pooley O and Read J 2022 *The British Journal for the Philosophy of Science*
[17] Butterfield J 2006 *British Journal for the Philosophy of Science* **57** 709–753
[18] Maudlin T 1990 *Studies in History and Philosophy of Science Part A* **21** 531–561 ISSN 0039-3681 URL https://www.sciencedirect.com/science/article/pii/0039368190900324
[19] Butterfield J 1989 *The British Journal for the Philosophy of Science* **40** 1–28 ISSN 0007-0882 (*Preprint* https://academic.oup.com/bjps/article-pdf/40/1/1/9745417/1.pdf) URL https://doi.org/10.1093/bjps/40.1.1

[20] Pooley O forthcoming *The Routledge Companion to the Philosophy of Physics* ed Knox E and Wilson A (Routledge) URL http://philsci-archive.pitt.edu/18142/
[21] Komar A 1958 *Physical Review* **111**(4) 1182–1187 URL https://link.aps.org/doi/10.1103/PhysRev.111.1182
[22] Iftime M and Stachel J 2006 *General Relativity and Gravitation* **38**, *1241–1252*
[23] Stachel J and Iftime M 2005 Fibered manifolds, natural bundles, structured sets, g-sets and all that: The hole story from space time to elementary particles URL https://arxiv.org/abs/gr-qc/0505138
[24] Stachel J 1986 *In Remo Ruffini (ed.), Proceedings of the Fourth Marcel Grossmann Meeting on General Relativity. Amsterdam: Elsevier, pp. 1857–1862*
[25] Field H 1984 *PSA: Proceedings of the Biennial Meeting of the Philosophy of Science Association* **1984** 33–90 ISSN 02708647 URL http://www.jstor.org/stable/192496
[26] Belot G 2018 *Nous* **52** 946–981 (*Preprint* https://onlinelibrary.wiley.com/doi/pdf/10.1111/nous.12200) URL https://onlinelibrary.wiley.com/doi/abs/10.1111/nous.12200
[27] Wüthrich C 2009 *Philosophy of Science* **76** 1039–1051
[28] Muller F A 2011 *Philosophy of Science* **78** 1046–1057 ISSN 00318248, 1539767X URL http://www.jstor.org/stable/10.1086/662634
[29] Lewis D 1968 *Journal of Philosophy* **65** 113–126
[30] Lewis D 1973 *Counterfactuals* (Cambridge, MA, USA: Blackwell)
[31] Lewis D 1986 *On the Plurality of Worlds* (Blackwell Publishers)
[32] Fletcher S C 2018 Similarity structure on scientific theories forthcoming in the de Gruyer volume "Topological Philosophy", edited by Bartlomiej Skowron. URL http://philsci-archive.pitt.edu/17222/
[33] Butterfield J 1987 *International Studies in the Philosophy of Science* **2** 10–32
[34] Jacobs C 2023 *forthcoming in Journal of Philosophy*
[35] Cudek F 2023 *in preparation*
[36] Cudek F 2023 *in preparation*
[37] Hawking S W and Ellis G F R 1975 *The Large Scale Structure of Space-Time (Cambridge Monographs on Mathematical Physics)* (Cambridge University Press) ISBN 0521099064 URL http://www.amazon.com/Structure-Space-Time-Cambridge-Monographs-Mathematical/dp/0521099064
[38] Choquet-Bruhat Y and Geroch R 1969 *Communications in Mathematical Physics* **14** 329–335
[39] Landsman K 2022 Reopening the hole argument URL https://arxiv.org/abs/2206.04943
[40] Pooley O 1998 *Intrinsic Dynamics or Substantivalism? A Critical Review of Some Contemporary Approaches to Spacetime* Ph.D. thesis University of Oxford
[41] Pooley O 2013 *'Substantivalist and Relationalist Approaches to Spacetime'. In The Oxford Handbook of Philosophy of Physics.* (Oxford University Press) chap 15
[42] Fletcher S C 2020 *Foundations of Physics* **50** 228–249
[43] Halvorson H and Manchak J B 2023 *The British Journal for the Philosophy of Science* **22** 237–272 (*Preprint* https://doi.org/10.1086/719193) URL https://doi.org/10.1086/719193
[44] Bradley C and Weatherall J O 2022 *Philosophy of Science* **89** 1223–1232
[45] Arledge C and Rynasiewicz R 2019 URL http://philsci-archive.pitt.edu/16343/
[46] Roberts B 2020 *Foundations of Physics* **50** 250–269
[47] Roberts B 2014 Disregarding the 'hole argument' URL http://philsci-archive.pitt.edu/11687/
[48] Menon T and Read J 2023 *submitted*
[49] Geroch R 1969 *Communications in Mathematical Physics* **13** 180 – 193 URL https://doi.org/
[50] Gomes H 2022 Why gauge? Conceptual Aspects of Gauge theories URL https://arxiv.org/abs/2203.05339
[51] Arnold V 1989 *Mathematical Methods of Classical Mechanics* vol 60 (Springer, Berlin)
[52] Fletcher S C 2016 *The British Journal for the Philosophy of Science* **67** 365–389 (*Preprint* https://doi.org/10.1093/bjps/axu044) URL https://doi.org/10.1093/bjps/axu044
[53] Malament D 1986 *From Quarks to Quasars: Philosophical Problems of Modern Physics* ed Colodny R G (University of Pittsburgh Press) pp 7–181
[54] Fletcher S C 2019 *Studies in History and Philosophy of Science Part B: Studies in History and Philosophy of Modern Physics* **68** 1–15 ISSN 1355-2198 URL https://www.sciencedirect.com/science/article/pii/S1355219818301308
[55] Curiel E 2018 *The British Journal for the Philosophy of Science* **69** 447–483 (*Preprint* https://doi.org/10.1093/bjps/axw014) URL https://doi.org/10.1093/bjps/axw014
[56] Penrose R 1996 *General Relativity and Gravitation* **28** 581–600
[57] Giacomini F and Brukner C 2022 *AVS Quantum Science* **4** 015601 (*Preprint* https://doi.org/10.1116/5.0070018) URL https://doi.org/10.1116/5.0070018
[58] Giacomini F and Brukner C 2020 Einstein's equivalence principle for superpositions of gravitational fields and quantum reference frames URL https://arxiv.org/abs/2012.13754

[59] Hardy L 2020 *Progress and Visions in Quantum Theory in View of Gravity* ed Finster F, Giulini D, Kleiner J and Tolksdorf J (Cham: Springer International Publishing) pp 189–220 ISBN 978-3-030-38941-3

Journal of Physics: Conference Series **2533** (2023) 012005 doi:10.1088/1742-6596/2533/1/012005

The case of Quantum Gravity with Spontaneous Collapse of the Wave Function

Lajos Diósi
Wigner Research Centre for Physics, H-1525 Budapest 114, P.O.Box 49, Hungary
Eötvös Loránd University, H-1117 Budapest, Pázmány Péter stny. 1/A

E-mail: diosi.lajos@wigner.hu

Abstract. When about half a century ago the concept of universal spontaneous collapse of the wave function was conceived it was an attempt to alter standard non-relativistic quantum physics. As such, it was largely ignored by relativistic field theory and quantum gravity communities. A central motivation of spontaneous collapse community has been to replace the standard collapse-by-measurement that annoyed many. With few exceptions, it did not annoy the field theory and quantum gravity communities. Concept of certain general-relativity-related universal irreversibility in quantum field theory had been initiated very long ago by Wheeler, Hawking and a few others independently from the concept of spontaneous collapse. Lately the two concepts started to converge and support each other.

1. Introduction

Allow the author to introduce the topics in a self-quotation from 2015 [1]. 'The inception of a universal gravity-related irreversibility took place originally in quantum cosmology but it turned out soon that a universal non-unitary dynamics is problematic itself. Independent investigations of the quantum measurement postulate clarified that a non-unitary dynamics is of interest already in the non-relativistic context. An intricate relationship between Newton gravity and quantized bulk matter might result in universal non-relativistic violation of unitarity - also called spontaneous decoherence. The corresponding gravity-related spontaneous decoherence model is now on the verge of detectability in optomechanical experiments. It is also a toy-model of cosmic quantum gravitational non-unitarity, illuminating that the bottle-neck of quantum gravity is the quantum measurement postulate instead of quantum cosmology.'

We want to build our message around the appearance of the concept of fundamental irreversibility in two separate fields of foundational theoretical physics. One is quantum cosmology. It has no consistent theory valid down towards the Planck scale, despite various efforts using various advanced mathematical models. At the same time, semiclassical and heuristic approaches imply that fundamental irreversibilities may be part of quantum cosmology. But quantum irreversibility itself, especially when universal, happens to be non-trivial both technically and conceptually. And here another research field gains importance. Its tasks seem opaque compared to quantum cosmology's. They include, e.g., the measurement problem, the dynamics of wave function collapse, the quantum-classical transition and hybrid dynamics, but the relevant task is the quantum theory of macroscopia, coined as the Schrödinger cat problem. You should discuss the options of macroobjects' quantum behavior, and you can do it non-relativistically, before you go for quantum cosmology! The hypothesis of spontaneous universal

Journal of Physics: Conference Series **2533** (2023) 012005 doi:10.1088/1742-6596/2533/1/012005

Table 1. Comparison of the two research fields and the status of fundamental irreversibility.

QUANTUM COSMOLOGISTS	SCHRÖDINGER CAT KILLERS*
field-string-membrane theorists, for full relativistic quantum gravity of the Universe within standard quantum theory unitary (reversible)	quantum foundation experts, for non-relativistic spontaneous wave function collapse of macroscopic bodies with modified quantum theory non-unitary (irreversible)
Fundamental Irreversibility? Possible: foamy vacuum, black holes	Fundamental Irreversibility? Mandatory: wave function collapse

*Measurement Problem Solvers

collapse in massive degrees of freedom, assuming a fundamental gravity-related irreverisibility, may have perspectives for quantum cosmology as well, especially if fundamental irreversibility remains part of it.

2. Fundamental irreversibility? — Two Communities

The two fundamental issues in question are the following: unified theory of space-time with quantized matter (cf. quantum gravity, quantum cosmology) and the physics of quantum measurement (cf. wave function collapse, quantum-classical transition). These problems were considered unrelated for long time, studied by two separate research communities. We refer to the first one as quantum cosmologists and to other as measurement problem solvers or, with a spectacular touch, as Schrödinger cat killers. Quantum cosmology has always been part of main stream physics, using heavy artillery of mathematics, while Schrödinger cat killers used light weapons and sometimes whimsical identification of their problems.

2.1. Irreversible Quantum Gravity/Cosmology at the Planck Scale

Emergence of irreversibility *within standard physics* was not derived by exact derivations but by *heuristic arguments.* A selection of earliest theoretical signatures could be the following. In 1936, Bronstein [2] was the first to point out that space-time metric must have an ambiguity δg_{ab} of different nature from the dictum of Heisenberg uncertainty relationships. Much later in 1962, Wheeler [3] proposed that space-time has a certain foamy structure at the Planckian scale. After another decade, Bekenstein [4] proved that black-holes behave thermodynamically, they have entropy:

$$S_{BH} = \frac{A}{4}, \tag{1}$$

where A is the surface aera of the black-hole in Planck units (Boltzmann's constant $k_B = 1$). Then Hawking [5] showed that, yes, in accordance with the above entropy, black-holes emit thermal radiation of temperature

$$T_{BH} = \frac{1}{8\pi M}, \tag{2}$$

where M is the black-hole mass. Later, Hawking [6] went deeper and conjectured that unitarity of standard quantum theory is lost due to the instanton mechanism. The scattering processes must be non-unitary, the non-unitary super-scattering operator \$ acts on the density operator $\widehat{\rho}$:

$$\widehat{\rho} \to \$\widehat{\rho} \tag{3}$$

where the r.h.s. does not factorize into $\widehat{S}\widehat{\rho}\widehat{S}^\dagger$ with any unitary scattering operator $\widehat{S}$. Instead of Hawking's scattering process, Banks, Susskind and Peskin [7] considered a detailed dynamical equation

$$\frac{d\widehat{\rho}}{dt} = -i[\widehat{H}, \widehat{\rho}] - \iint [\widehat{Q}(x), [\widehat{Q}(y), \widehat{\rho}\,]] h(x-y) d^3x d^3y, \tag{4}$$

where $\widehat{H}$ is the Hamiltonian, $\widehat{Q}$ is a relativistic quantum field and h is a positive kernel. The authors show that the irreversible term on the r.h.s. violates the energy-momentum conservation which cannot be restored in local field theories.

2.2. Irreversible Quantum Mechanics for Massive Objects

Among diverse incentives, including a metaphysical discontent about von Neumann's quantum measurement theory, the only relevant motivation is the problematic extendibility of quantum theory for massive degrees of freedom. This led to *heuristic modifications of standard quantum physics.* According to that, the unitarity is violated in massive degrees of freedom so that macroscopically different superpositions, aka Schrödinger cat states like

$$|\Psi\rangle = \frac{|f_1\rangle + |f_2\rangle}{\sqrt{2}}, \tag{5}$$

collapse spontaneously into one of the components. Milestones of the consept could be the following.

In 1966, Károlyházy [15] proposed that, due to a conjectured spectrum of space-time metric fluctuations δg_{ab} at the Planck scale, Schrödinger cat states collapse before they get too large. He outlined a naive qualitative model of when and how the collapse happens. Independently and much later in 1986, Ghirardi, Rimini and Weber [18] constructed a simple model of universal spontaneous collapse in exact mathematical terms. They postulated extreme rare and weak spontaneous collapses for the elementary constituents' wave functions which in turn yield robust localization of the macrosopic center-of-mass, i.e., the collapse of the Schrödinger cat state. The GRW model has no reference to gravity. However, the present author [19, 20] proposed gravity-related universal spontaneous collapse of massive superpositions, like Schrödinger cats, based on a master equation:

$$\frac{d\widehat{\rho}}{dt} = -\frac{i}{\hbar}[\widehat{H}, \widehat{\rho}] - \frac{G}{2\hbar}\iint [\widehat{f}(x), [\widehat{f}(y), \widehat{\rho}\,]] \frac{1}{|x-y|} d^3x d^3y, \tag{6}$$

where $\widehat{f}$ is the non-relativistic quantized mass density field and G is Newton's constant. The master equation corresponds to a certain ambiguity $\delta\Phi = \frac{1}{2}c^2\delta g_{00}$ of the Newton potential.

2.3. Fundamental irreversibility? — Parallel pursuits

Thus, the desired quantum gravity and the unwanted Schrödinger cats led both the heavy-armed relativistic and light-horse non-relativistic studies to the same structure of heuristic master equations. And they suffer of the same problem of spontaneous creation of energy and momentum. This coincidence is a spectacular instance to illustrate that the two, apparently distant theoretical tasks, share similar proposals and problems. Another instance has been the surprising coincidence between the Schrödinger cat's collapse rate predicted by the non-relativistic master equation [19] and by the general relativistic arguments of Penrose [10]. For possible reasons of coincidence, see ref. [25].

Mention should be made of Gell-Mann and Hartle [9] who were perhaps the only ones at the time to recognize that quantum cosmology needed a more general measurement theory than von Neumann's. For the relationship of their theory of decoherent histories to spontaneous collapse theories, see [26]. Most recently, hybrid quantum-classical master equations have been tested in semiclassical cosmology by Oppenheim et al. [14], with numerous references to spontaneous collapse models which are, in fact, the alternative formalism, see [27].

Table 2. A selection of parallel concepts related to fundamental irreversibility.

	QUANTUM COSMOLOGISTS	SCHRÖDINGER CAT KILLERS
1936	Bronstein [2]: ambiguity δg_{ab}	
1962	Wheeler [3]: space-time foam	
1966		Károlyh. [15]: ambiguity δg_{ab} collapses Ψ
1973	Bekenstein [4]: black hole entropy	
1975	Hawking [5]: black holes radiate	
1976		Pearle [16]: searching collapse dynamics
1983	Hawking [6]: instantons break unitarity	
1984	BPS [7]: energy non-conservation	Gisin [17]: prototype collapse dynamics
1986		GRW [18]: universal collapse
1987		D. [19]: G-related decoherence
1989	Ellis & al. [8]: wormholes collapse Ψ	D. [20]: G-related collapse
1990	G-M&H [9]: decoherent histories	GRP [21]: G-unrelated collapse
1996	Penrose [10]: ambiguity δg_{ab} collapses Ψ	
2014		Bedingham & al. [22]: relativisation
2016	Kwon&Hogan [11]: holographic noise	Tilloy&D [23]: semiclassical G & collapse
2017		Bassi & al. [24]: review of G-collapse
2021	Sudarsky [12]: cosmology & collapse	
2022	Anastopoulos&Hu [13]: G-decoherence	
	Oppenheim & al. [14]: hybrid dynamics	

3. Summary

The failures of quantum gravity models may not be due to the inadequacy of the quantization methods tried so far, but the inadequacy of standard quantum theory for macroobjects, e.g., the spontaneous decay of massive superpositions, aka Schrödinger-cats. We recalled that the idea of fundamental irreversibility has also been raised in quantum cosmology. The roots of irreversibility may be common in quantum cosmology and in speculative models of spontaneous collapse of the wave function. This may be so despite the fact that one is predicted at the Planck scale while the other comes from non-relativistic discussion of massive superpositions [28]. Considerations of quantum cosmologists and advocates of spontaneous collapse (referred here extravagantly as Schrödinger cat killers) will have to come closer to each other. To support this suggestion, here are two quotes. The author (cat killer) claimed 'the measurement problem culminates in quantum cosmology" [29]. The cosmologist claims 'the principles of general relativity must influence, and actually change, the very formalism of quantum mechanics' [30].

Acknowledgments

This research was funded by the Foundational Questions Institute and Fetzer Franklin Fund, a donor-advised fund of the Silicon Valley Community Foundation (Grant No's. FQXi-RFPCPW-2008, FQXi-MGA-2103), the National Research, Development and Innovation Office (Hungary) "Frontline" Research Excellence Program (Grant No. KKP133827), and the John Templeton Foundation (Grant 62099).

Journal of Physics: Conference Series **2533** (2023) 012005 doi:10.1088/1742-6596/2533/1/012005

References

[1] Diósi L 2015 *Gravity related spontaneous decoherence: from Wheeler-Bekenstein-Hawking to optomechanics* (talk given in Erice) www.wigner.hu/∼diosi/slides/erice2015.pdf
[2] Bronstein M 1936 *Phys. Z. Sowjetunion* **9** 140
[3] Wheeler J A 1962 *Geometrodynamics* (New York: Academic Press)
[4] Bekenstein J D 1973 *Phys. Rev. D* **7** 2333
[5] Hawking S W 1975 *Commun. Math. Phys.* **43** 199
[6] Hawking S W 1982 *Commun. Math. Phys.* **87** 395
[7] Banks T, Susskind L and Peskin M E 1984 *Nucl. Phys. B* **244** 125
[8] Ellis J, Mohanty S and Nanopoulos D V 1989 *Phys. Lett. B* **221** 113
[9] Gell-Mann M and Hartle J B 1993 *Phys. Rev. D* **47** 3345
[10] Penrose R 1996 *Gen. Relativ. Gravit.* **28** 581
[11] Kwon O and Hogan C J 2016 *Class. Quantum Grav.* **33** 105004
[12] Sudarsky D 2021 *Do Wave Functions Jump? Perspectives of the Work of GianCarlo Ghirardi*, ed V Allori, A Bassi, D Dürr, N Zanghi (Cham: Springer Nature Switzerland) p 291
[13] Anastopoulos C and Hu B L 2021 *AVS Quantum Science* **4** 015602
[14] Oppenheim J and Weller-Davies 2022 *J. High Energ. Phys.* **2022** 80; Oppenheim J, Sparaciari C, Šoda B and Weller-Davies Z 2022 *Preprint* arXiv:2203.01982
[15] Karolyhazy F 1966 *Nuovo Cim.* **42** 390
[16] Pearl P 1976 *Phys. Rev. D* **13** 857
[17] Gisin N 1984 *Phys. Rev. Lett.* **52** 1657
[18] Ghirardi G C, Rimini A and Weber T 1986 *Phys. Rev. D* **34** 470
[19] Diósi L 1987 *Phys. Lett. A* **120** 377
[20] Diósi L 1989 *Phys. Rev. A* **40** 1165
[21] Ghirardi G C, Rimini A and Pearl P 1990 *Phys. Rev. A* **42** 78
[22] Bedingham D, Dürr D, Ghirardi G, Goldstein S, Tumulka R and Zanghi N 2014 *J. Stat. Phys.* **154** 623
[23] Tilloy A and Diósi L 2016 *Phys. Rev. D* **93** 024026
[24] Bassi A, Grossardt A and Ulbricht H 2017 *Class. Quantum Grav.* **34** 193002
[25] Diósi L 2022 *AVS Quantum Sci.* **4** 015605
[26] Diósi L, Gisin N, Halliwell J and Percival I C 1995 *Phys. Rev. Lett.* **74** 203
[27] Diósi L 2023 *Preprint* arXiv::2302.13418
[28] Diósi L 2018 *Entropy* **20** 496; 2019 *J. Phys. Conf. Ser.* **1275**
[29] Diósi L 1992 *Quantum Chaos—Quantum Measurement* (Springer) p 299
[30] Penrose R 2014 *Found. Phys.* **44** 557

Journal of Physics: Conference Series **2533** (2023) 012006 doi:10.1088/1742-6596/2533/1/012006

Casting Rubik's Group into a Unitary Representation for Reinforcement Learning

Sebastiano Corli

Istituto di Fotonica e Nanotecnologie, Consiglio Nazionale delle Ricerche, Piazza Leonardo da Vinci 32, I-20133, Milano, Italy
Politecnico di Milano, Via Colombo 81, I-20133, Milano, Italy

Lorenzo Moro

Istituto di Fotonica e Nanotecnologie, Consiglio Nazionale delle Ricerche, Piazza Leonardo da Vinci 32, I-20133, Milano, Italy
Politecnico di Milano, Via Colombo 81, I-20133, Milano, Italy

Davide E Galli

Dipartimento di Fisica Aldo Pontremoli, Università degli Studi di Milano, via Celoria 16, I-20133 Milano, Italy

Enrico Prati

Istituto di Fotonica e Nanotecnologie, Consiglio Nazionale delle Ricerche, Piazza Leonardo da Vinci 32, I-20133, Milano, Italy
Dipartimento di Fisica Aldo Pontremoli, Università degli Studi di Milano, via Celoria 16, I-20133 Milano, Italy

E-mail: `enrico.prati@unimi.it`

Abstract. Rubik's Cube is one of the most famous combinatorial puzzles involving nearly 4.3×10^{19} possible configurations. However, only a single configuration matches the solved one. Its mathematical description is expressed by the Rubik's group, whose elements define how its layers rotate. We develop a unitary representation of the Rubik's group and a quantum formalism to describe the Cube based on its geometrical constraints. Using single particle quantum states, we describe the cubies as bosons for corners and fermions for edges. By introducing a set of four Ising-like Hamiltonians, we managed to set the solved configuration of the Cube as the global ground state for all the Hamiltonians. To reach the ground state of all the Hamiltonian operators, we made use of a Deep Reinforcement Learning algorithm based on a Hamiltonian reward. The Rubik's Cube is successfully solved through four phases, each phase driven by a corresponding Hamiltonian reward based on its energy spectrum. We call our algorithm *QUBE*, as it employs quantum mechanics to tackle the combinatorial problem of solving the Rubik's Cube. Embedding combinatorial problems into the quantum mechanics formalism suggests new possible algorithms and future implementations on quantum hardware.

1. Introduction

Reinforcement Learning (RL) algorithms have been recently applied for problems of quantum control, including coherent transport of quantum states [1, 2] and digitally stimulated Raman passage [3].

Journal of Physics: Conference Series **2533** (2023) 012006 doi:10.1088/1742-6596/2533/1/012006

Furthermore, we have applied both supervised learning by quantum tensor networks [4] and an unsupervised learning algorithm on a quantum hardware, expressed as a QUBO problem [5]. Reinforcement Learning has been applied also in the field of quantum compiling, to apply a set of logic gates [6, 7] on a quantum hardware [8, 9, 10]. RL algorithms have been also employed in order to tackle problems of combinatorial optimization, such as solving the Rubik's Cube [11, 12] or finding out the native configuration for a protein from a sequence of amino acids [13, 14, 15]. Some analogies between the Rubik's Cube and the protein folding hold, as both systems share a unique native (or solved) configuration in contrast with a huge amount of other available configurations [16, 11]. The phase space of the Rubik's Cube indeed displays approximately 4.3×10^{19} available configurations [17]. The Rubik's Cube has been explored in the past for its group properties [18, 19, 17, 20, 21]. We start from Chen's formulation of the Rubik's group [17], by translating the geometrical postulates of such group into a unitary representation of it. The next step consists of introducing a set of observables, to describe the Rubik's Cube as a vector in a Hilbert space $\mathcal{H}$, and the elements from the Rubik's group as maps $\hat{U} : \mathcal{H} \to \mathcal{H}$. At last, four Ising-like Hamiltonians are introduced, based on the same observables previously defined. The method we developed involves four phases, each of them consisting of reaching the ground state of a specific Hamiltonian without moving away from the ground state of the previous one. The Hamiltonian approach provides a metric in the phase space of the Cube, i.e. a method to trace how far is a configuration from the solved one.

2. Definitions and formalism

The Cube is composed by 26 smaller cubes, called cubies [17]. Six of them show one face (the centrals), which cannot be moved from their relative positions, 12 edges show two faces and 8 corners show three faces (the corners), see Figure 1. A corner cubie cannot take the place of an edge one, and vice versa. Any configuration of the Cube can be univocally identified by two sets of features, namely the positions and orientations of the cubies. The position of a cubie marks how far a cubie is from its place in the solved configuration. The orientation of a cubie stores how, keeping the cubie in its solved position, it has been rotated around a rotation axis suitable to induce permutations. Orientation can be graphically represented as an arrow on a face of the cubie, as in Figure 1. Orientations and positions are modified by rotating the layers of the Cube. These transformations are elements from Rubik's group $(\mathfrak{R}, \circ)$, which consists of six generators, each of them corresponding to the rotation of a coloured face. In Singmaster notation [22], they are labelled as $\mathfrak{R} = \{U, D, F, B, L, R\}$, where U, D stand for up and down, F, B front and back, L, R left and right. To describe orientation and position of each cubie, one may set an orientation for the solved configuration and three cartesian axes (Figure 1). The six cubies fixed in the middle of the faces are called *centrals*. They are invariant under all the transformations and show just one face, thus they are not associated to any specific orientation. *Edge* cubies show two faces, thus two possible modes of orientation are available: the mode in the solved configuration and its flipped arrangement. *Corner* cubies exhibit three faces, but swapping two corner faces is not allowed by the Rubik's group [17]. The faces of a corner can be rotated clockwise or anti-clockwise, allowing three feasible orientation.

2.1. Single particle state

Two geometrical properties identify uniquely the state of a cubie, as clarified in Section 2. The state of position and orientation can be labelled by four numbers consisting of a triple k_i ($i = 1, ..., 3$ for each axis) and s. A description of such a state is given by a vector in a Hilbert space $\mathcal{H}$, i.e. $|cubie\rangle = |\mathbf{k}, s\rangle \in \mathcal{H}$. The focus is now to build a representation, i.e. introducing four operators such that

$$\hat{K}_i \, |\mathbf{k}, s\rangle = k_i \, |\mathbf{k}, s\rangle \, , \quad \hat{S} \, |\mathbf{k}, s\rangle = s \, |\mathbf{k}, s\rangle \tag{1}$$

$\hat{K}_i$ ($i = 1, 2, 3$) and $\hat{S}$ form a complete set of observables, i.e. their eigenvalues completely define the system without degeneracy. The statement of completeness requires $\hat{K}_i$ and $\hat{S}$ to have $|\mathbf{k}, s\rangle$ as

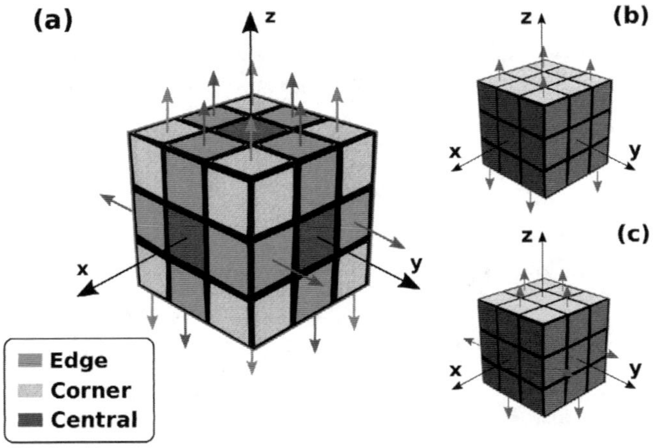

Figure 1: All the images are taken from [12]. (a) The arrows define our arbitrary choice of orientation, in the solved state, for corners and edges. (b) The faces of the Cube with their conventional colours, blue, red and white. Coordinated axes are introduced to describe the positions of the cubies, and the choice for the orientation in the solved configuration for the sole corners. (c) The previous scheme, to describe the initial edge orientation.

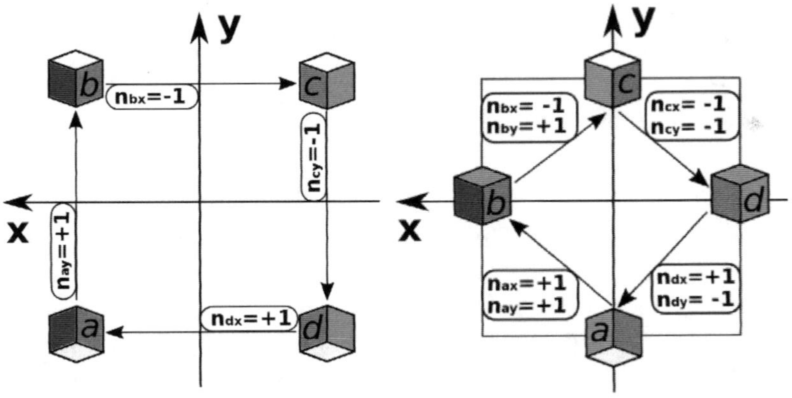

(a) (b)

Figure 2: All the images are taken from [12]. (a) The F rotation (red layer) decomposed as translations on the corners. The d cubie is moved along the x-axis of one position from its solved position, which is labelled by raising the quantum number n_{dx} (n_x is now $+1$). Conversely, c is translated in the opposite direction along the y-axis by one position, which means its quantum number n_{cy} will be lowered. (b) The F rotation decomposed as translations for each frontal edge. The edges are moved diagonally, which makes the cubies to shift along both x and y directions.

simultaneous eigenvector and thus to commute, so that $[\hat{K}_i; \hat{S}] = 0$ and $[\hat{K}_i; \hat{K}j] = 0$.

2.2. A description for the position of the cubies

Any transformation from $\mathfrak{R}$ affects the position of the involved cubies. By rotating a layer of the Rubik's Cube each cubie shifts along a particular axis. As far as any rotation is by $\pi/2$, the symmetry of translation is discrete. A translation operator $\hat{U} : \mathcal{H} \to \mathcal{H}$ over **k** should accomplish the action $\hat{U}(\mathbf{T})\,|\mathbf{k}, s\rangle = |\mathbf{k} + \mathbf{T}, s\rangle$. A good representation for both the ket state and the translation operator is given by the free particle wave function in a box, whereas the translation is given by the $U(1)$ group:

$$\psi_{\mathbf{n}}(\mathbf{x}) = \frac{1}{\sqrt{V}} e^{i\mathbf{k}\cdot\mathbf{x}}, \quad \hat{U}(\mathbf{T}) = e^{i\mathbf{T}\cdot\mathbf{x}} \tag{2}$$

Here $\mathbf{k} = 2\pi\mathbf{n}/l$ and $l^3 = V$. As **k** must be proportional to a vector **n** of integers, any translation by **T** must preserve such proportionality, with $\mathbf{k'} = \mathbf{k} + \mathbf{T} \equiv \frac{2\pi}{l}\mathbf{n'}$. Such constraint implies $\mathbf{T} = 2\pi\mathbf{t}/l$, for $\mathbf{t} = (t_1, t_2, t_3)$, $t_i \in \mathbb{Z}\,\forall i$. Thus **T** must be proportional to a vector of integers too. The operator $\hat{\mathbf{K}}$ marks the position of the cubies: when the Cube lies in its solved state, each cubie is set by $\mathbf{k} = \mathbf{n} = \mathbf{0}$.

2.3. A description for the orientation of the cubies

Considering any edge in Figure 1, its faces can be disposed in the state shown in the figure or in the reverse one, obtained by flipping them. A vectorial description for $|s\rangle$ edge cubies can be given by

$$|\uparrow\rangle = \begin{pmatrix}1\\0\end{pmatrix}, \ |\downarrow\rangle = \begin{pmatrix}0\\1\end{pmatrix}, \qquad \sigma_x = \begin{pmatrix}0 & 1\\1 & 0\end{pmatrix}, \quad \sigma_z^e = \begin{pmatrix}1 & 0\\0 & -1\end{pmatrix} \tag{3}$$

where $|\uparrow\rangle$ and $|\downarrow\rangle$ sum up the available states for $|s\rangle$, σ_x is the flip operator over such dichotomous state, while the observable σ_z measures the state of orientation. The corner orientation takes three possible states into account,

$$|+\rangle = \begin{pmatrix}1\\0\\0\end{pmatrix}, \ |0\rangle = \begin{pmatrix}0\\1\\0\end{pmatrix}, \ |-\rangle = \begin{pmatrix}0\\0\\1\end{pmatrix}, \tag{4}$$

with two available rotations (clockwise and anti-clockwise) and a corresponding observable:

$$\sigma_{xA} = \begin{pmatrix}0 & 1 & 0\\0 & 0 & 1\\1 & 0 & 0\end{pmatrix}, \ \sigma_{xC} = \begin{pmatrix}0 & 0 & 1\\1 & 0 & 0\\0 & 1 & 0\end{pmatrix}, \quad \sigma_z^c = \begin{pmatrix}1 & 0 & 0\\0 & 0 & 0\\0 & 0 & -1\end{pmatrix} \tag{5}$$

A and C indices refer to clockwise and anti-clockwise. This representation recalls the spin formalism, where corners are boson-like quantum objects with total momentum $S^2 = 1$ and edges fermion-like quantum objects with $S^2 = 1/2$.

2.4. Multiparticle state of the Cube

The Rubik's Cube is built by wedging twenty cubies together, i.e. its global state is described by 20 single particle states. The global state $|C\rangle$ can be described as a few-bodies system:

$$|C\rangle = \bigotimes_{i=1}^{12} \psi_{\mathbf{n}_i}(\mathbf{x}_i)\,|s_i^2 = \frac{1}{2}, s_i\rangle \otimes \bigotimes_{i=13}^{20} \psi_{\mathbf{n}_i}(\mathbf{x}_i)\,|s_i^2 = 1, s_i\rangle \tag{6}$$

where s_i^2 is the quantum number of $\hat{S}_i^2$ for each i-th cubie, i is an index referring to each cubie.

2.5. Action of the Rubik's group

Corners and edges cannot be transformed into each other. This constraint imposes the transformations from Rubik's group to act separately on corners and edges, so a U transformation can be built as $\hat{U} = \hat{U}_e \otimes \hat{U}_c$. After performing a translation over the momenta **k** and a possible set of spin rotations, the

Journal of Physics: Conference Series **2533** (2023) 012006 doi:10.1088/1742-6596/2533/1/012006

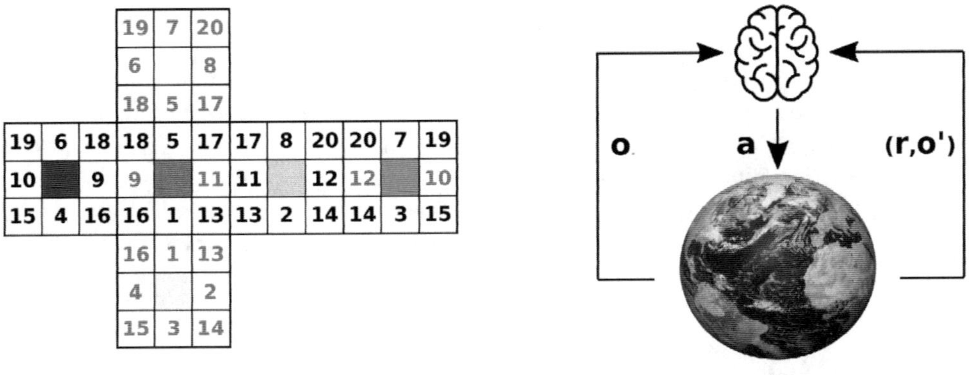

(a) (b)

Figure 3: (a) The cubies are enumerated in order to describe the transformations of the Rubik's group. The red numbers show the oriented faces of the cubies, i.e. the arrows in Figure 1. The image is taken from [12]. (b) Basic scheme for any Reinforcement Learning algorithm.

involved cubies are permuted, following the direction of the arrows as shown in Figure 2. An operator $\hat{P}_{\sigma(a,b,c,d)}$ of permutations is introduced in the overall transformation, whose action is given by

$$\hat{P}_\sigma : \bigotimes_{i=1}^{20} \mathcal{H}_i \to \bigotimes_{\sigma(i)=1}^{20} \mathcal{H}_{\sigma(i)}, \quad \hat{P}_\sigma \left|1\right\rangle \left|2\right\rangle ... \left|20\right\rangle = \left|\sigma(1)\right\rangle \left|\sigma(2)\right\rangle ... \left|\sigma(20)\right\rangle \tag{7}$$

It holds from Figure 2 that $\sigma(a, b, c, d) = (d, a, b, c)$ for both corners and edges, which means $d \to a$, $a \to b$ and so on. As an example, we bring the $\hat{F}$ operator. In order to build the rotation on the front face (the red one), one may refer to Figure 2 for the translations and to Figure 1 for the orientations, so the overall $\hat{F}$ operator can be written as

$$\hat{F}_e = \hat{P}^e_{\sigma(a,b,c,d)} e^{i\frac{2\pi}{l}(y_a+x_a+y_b-x_b-x_c-y_c-y_d+x_d)} \hat{\sigma}^a_x \hat{\sigma}^b_x \hat{\sigma}^c_x \hat{\sigma}^d_x \tag{8}$$

$$\hat{F}_c = \hat{P}^c_{\sigma(a,b,c,d)} e^{i\frac{2\pi}{l}(y_a-x_b-y_c+x_d)} \hat{\sigma}^a_{x_A} \hat{\sigma}^b_{x_C} \hat{\sigma}^c_{x_A} \hat{\sigma}^d_{x_C} \tag{9}$$

Referring to Figure $3a$, just for instance, instead of (a, b, c, d) there would be (1, 9, 5, 11) for the edges, (16, 18, 17, 13) for the corners.

2.6. Rubik's group is unitary: sketch of proof

The proof will hold only for $\hat{F}$. The adjoint of $\hat{F}$ in Equation (8) is

$$\hat{F}^\dagger_e = e^{-i\frac{2\pi}{l}(y_a+x_a+y_b-x_b-x_c-y_c-y_d+x_d)} \hat{\sigma}^a_x \hat{\sigma}^b_x \hat{\sigma}^c_x \hat{\sigma}^d_x [\hat{P}^e_{\sigma(a,b,c,d)}]^\dagger \tag{10}$$

Contracting $\hat{F}_e$ with $\hat{F}^\dagger_e$, the result is

$$\hat{F}_e \hat{F}^\dagger_e = \hat{P}^e_{\sigma(a,b,c,d)} [\hat{P}^e_{\sigma(a,b,c,d)}]^\dagger = \hat{P}^e_{\sigma(a,b,c,d)} \hat{P}^e_{\sigma^{-1}(a,b,c,d)} = \hat{\mathbb{I}}_e \tag{11}$$

An analogous result comes for $\hat{F}_c$ and $\hat{F}^\dagger_c$. The overall result is $\hat{F}\hat{F}^\dagger = \hat{\mathbb{I}}$.

Journal of Physics: Conference Series **2533** (2023) 012006 doi:10.1088/1742-6596/2533/1/012006

3. Strategy

3.1. Ising Hamiltonians

In order to pin the state corresponding to the solved configuration, it is possible to introduce a set of four Hamiltonians, each of them corresponding to a specific observable (momenta and spins) for both the corners and the edges. To shape such Hamiltonians, we took inspiration from the Ising and Potts models. The difference between these two models stands in the number of values each operator ($\hat{K}$ and $\hat{S}$) can assume: corner position and the edge orientation display two possible values (up and down for the second observable, correct position or not for each dimension of the first one), while edge position and corner orientation a range of three outcomes. Thus, the first two observables are described by the Ising model, the second ones by the Potts Hamiltonian. The Hamiltonians for the orientation $\hat{H}_o$ and the positions $\hat{H}_p$ are given by

$$\hat{H}_o = \sum_{i<j} J_{ij}[\hat{S}_i]^2[\hat{S}_j]^2 + \sum_i B_i[\hat{S}_i]^2, \; \hat{H}_p = \sum_{\substack{i<j \\ m}} J_{ij}[\hat{K}_{im}]^2[\hat{K}_{jm}]^2 + \sum_{i,m} B_{im}[\hat{K}_{im}]^2 \tag{12}$$

where i index stands for the spatial component, m points to the involved cubie. The spin operators for the edges are indeed transformed as $\hat{S} \rightarrow (\hat{S}-1)/2$, in order to yield a 0 eigenvalue when the edges are correctly oriented and -1 otherwise. The overall Hamiltonian is given by the sum of these four terms:

$$\hat{H}_{TOT} = \hat{H}_o^c + \hat{H}_o^e + \hat{H}_p^c + \hat{H}_p^e \tag{13}$$

where c and e stand for corner or edge. The ground state for $\hat{H}_{TOT}$ matches a configuration which is correctly oriented and positioned, i.e. the solution of the game is given by $E_{TOT} = 0$.

3.2. Reinforcement Learning

In a RL algorithm, an environment provides a set of possible states $\{s_i\}_{i=1}^N$ (the inputs) and possible actions $\{a_i\}_{i=1}^M$ (the outputs) [23]. When an action a is taken, the current state s of the environment is transformed into a new one s'. A visual scheme is provided in Figure 3b. The purpose of the learning consists of training an artificial intelligence agent in order to define the optimal policy, i.e. which actions lead to the solution of the problem, represented by a target state s_0. Once trained, such policy predicts the best sequence of actions from a given starting state to achieve the goal state s_0. To evaluate the effectiveness of such policy, it is used a scalar reward r returned to the agent after each couple action-state is sampled. The agent aims to maximize the total discounted return, given by the sum of all the rewards gained at the end of the problem. In Reinforcement Learning, setting a good reward $r(a, s)$ is the key feature, as it allows the agent to learn which are the best policies. In the Cube settlement, the states s are given by the vector states $|C\rangle$ as from Equation (6), while the actions are taken from the Rubik's group $\mathfrak{R}$. The reward matches the expectation value of the Hamiltonians, $r = -\langle s|\hat{H}|s\rangle$.

3.3. Phases of the game: the QUBE algorithm

The Rubik's Cube game has been therefore mapped into a problem of combinatorial optimization. In order to ease the solution, instead of searching the global ground state for $\hat{H}_{TOT}$, it is possible to divide the problem into four phases, each of them consisting of finding out the ground state for the single Hamiltonians from Equation (13). Nevertheless, once a phase i of the game is solved, i.e. the ground state for the specific Hamiltonian $\hat{H}_i$ is found, it is mandatory to block such condition. Consider $\hat{A}_{i+1}$ to be an available action from the next phase of the game: the condition described before can be stated as

$$\hat{H}_i\,|s\rangle = 0, \quad \hat{H}_i\,\hat{A}_{i+1}\,|s\rangle = \hat{H}_i\,|s'\rangle = 0 \qquad \forall \hat{A}_{i+1} \in \{\hat{A}_{i+1}\} \tag{14}$$

To achieve such result, the constraint in Equation (14) can be expressed as $\left[\hat{H}_i; \hat{A}_{i+i}\right]|s_j\rangle = 0$. Such requirement cannot be satisfied by using the generators of the Rubik's group, instead, a composition of operators are involved. Such optimization method is summed up in Figure 4b.

Journal of Physics: Conference Series **2533** (2023) 012006 doi:10.1088/1742-6596/2533/1/012006

QUBE software and parameters			
Algorithm	DDQN	**NN Architecture**	Feed-Forward
Language	Python 3.7	**Backend**	Tensorflow, numpy
CPU	4 cores/8 threads	**GPU**	3584 CUDA cores

Table 1: Table with computational parameters and settings. DDQN stands for Deep Double Q-Network, while NN for Neural Networks.

4. Results

Each phase requires a specifically designed neural networks architecture. A list with all the computational parameters and software for the QUBE algorithm is provided by Table 1. The input vector is given by the eigenvalues of the spins or momenta of the cubies, while the output vector is formed by all the possible actions the agent can take. All the NN have architectures involving two hidden layers, plus the input and output ones. As long as each episode is solved in four phases, four NNs were trained, with the task to find the ground state of a peculiar Hamiltonian from those in Equation (13). The results from the test of the game are reported in Figure 4. In Figure $4a$, it is possible to see the percentage of success from the four phases and the total percentage of success (all above 90%). In the first and fourth phases, the asymptotic percentage of success per number of episodes is 99.5% and 98.8% respectively, the complete 100% of solutions is reached in the second phase, while in the third such percentage fluctuates around 94.5%. During the third phase, the agent deals with a set of moves where a high number of cubies are involved at the end of the action: in the third phase, a total of 8 cubies per move are involved, while in the forth just 3 edges. The more cubies are involved, the more difficult will be to move them in the correct positions, and the more minima will occur in the energy landscape. A python implementation for the environment of the quantum Cube can be found at the following https://github.com/Sebastiano1994/Rubik-s-Cube-environmentgithub repository [24].

References

[1] Porotti R, Tamascelli D, Restelli M and Prati E 2019 *Communications Physics* **2** 1–9

[2] Porotti R, Tamascelli D, Restelli M and Prati E 2019 Reinforcement learning based control of coherent transport by adiabatic passage of spin qubits *Journal of Physics: Conference Series* vol 1275 (IOP Publishing) p 012019

[3] Paparelle I, Moro L and Prati E 2020 *Physics Letters A* **384** 126266

[4] Lazzarin M, Galli D E and Prati E 2022 *Physics Letters A* **434** 128056

[5] Rocutto L, Destri C and Prati E 2021 *Advanced Quantum Technologies* **4** 2000133

[6] Maronese M and Prati E 2021 *International Journal of Quantum Information* 2140002

[7] Maronese M, Destri C and Prati E 2022 *Quantum Information Processing* **21** 128

[8] Moro L, Paris M G, Restelli M and Prati E 2021 *Communications Physics* **4** 178

[9] Semola R, Moro L, Bacciu D and Prati E 2022 Deep reinforcement learning quantum control on ibmq platforms and qiskit pulse *2022 IEEE International Conference on Quantum Computing and Engineering (QCE)* (IEEE) pp 759–762

[10] Moro L, Likmeta A, Prati E, Restelli M *et al.* 2022 Goal-directed planning via hindsight experience replay *International Conference on Learning Representations* pp 1–16

[11] Agostinelli F, McAleer S, Shmakov A and Baldi P 2019 *Nature Machine Intelligence* **1** 356–363

[12] Corli S, Moro L, Galli D E and Prati E 2021 *Journal of Physics A: Mathematical and Theoretical* **54** 425302

[13] Ruff K M and Pappu R V 2021 *Journal of Molecular Biology* **433** 167208

[14] Jumper J, Evans R, Pritzel A, Green T, Figurnov M, Ronneberger O, Tunyasuvunakool K, Bates R, Žídek A, Potapenko A *et al.* 2021 *Nature* **596** 583–589

[15] AlQuraishi M 2019 *Bioinformatics* **35** 4862–4865

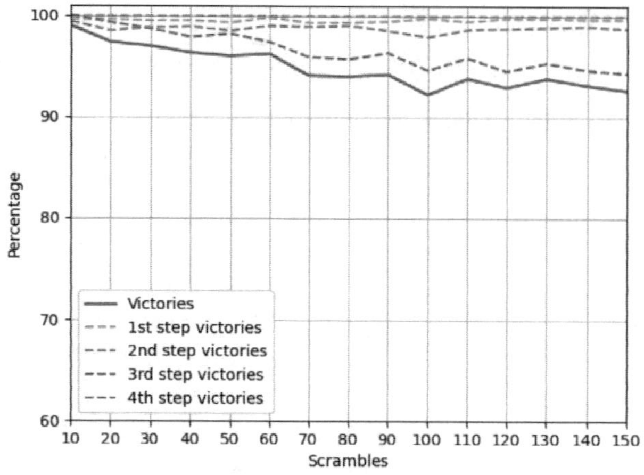

(a)

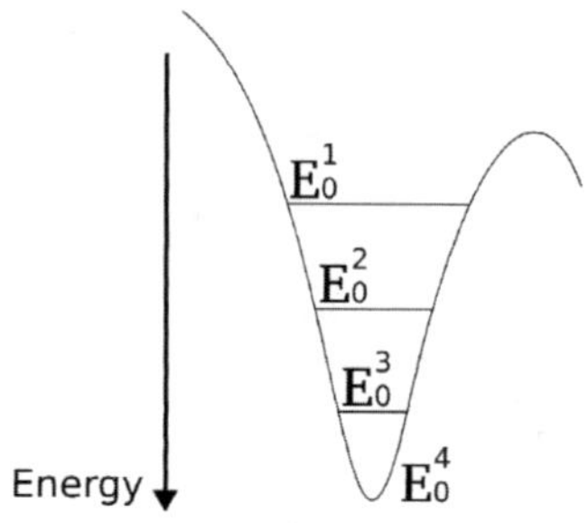

(b)

Figure 4: (a) Percentage of solutions (testing the NN models) by the number of scrambles, i.e. how many of the 12 fundamental moves are applied on the Rubik's Cube. The percentage of victories by all the four phases (solid line) are shown. In dashed lines, the percentage of victories for the first, the second, the third and the forth phases. The image is taken from [12]. (b) A graphical representation of the optimization process, split into four phases, each consisting of searching the ground state for a specific Hamiltonian.

[16] Chen Y R and Lee C L 2014 *Physical Review E* **89** 012815

[17] Chen J 2004 Group theory and the rubik's cube `http://www.math.harvard.edu/~jjchen/docs/Group\%20Theory\%20and\%20the\%20Rubik\%27s\%20Cube.pdf`

[18] Joyner D 2008 *Adventures in group theory: Rubik's Cube, Merlin's machine, and other mathematical toys* (JHU Press)

[19] Rokicki T, Kociemba H, Davidson M and Dethridge J 2014 *siam REVIEW* **56** 645–670

[20] Demaine E D, Eisenstat S and Rudoy M 2017 *arXiv preprint arXiv:1706.06708*

[21] Demaine E D, Demaine M L, Eisenstat S, Lubiw A and Winslow A 2011 Algorithms for solving rubik's cubes *European Symposium on Algorithms* (Springer) pp 689–700

[22] Singmaster D 1981 *Notes on Rubik's magic cube* (Enslow Pub Inc)

[23] Sutton R S and Barto A G 2018 *Reinforcement learning: An introduction* (MIT press)

[24] Corli S and Prati E 2021 Rubik's cube environment `https://github.com/Sebastiano1994/Rubik-s-Cube-environment`

Journal of Physics: Conference Series **2533** (2023) 012007 doi:10.1088/1742-6596/2533/1/012007

Solid State Neuroscience: Spiking Neural Networks as Time Matter

Marcelo Rozenberg
Université Paris-Saclay, CNRS Laboratoire de Physique des Solides, 91405, Orsay, France
E-mail: marcelo.rozenberg@universite-paris-saclay.fr

Abstract. We aim at building a bridge between to *a priori* disconnected fields: Neuroscience and Material Science. We construct an analogy based on identifying spikes events in time with the positions of particles of matter. We show that one may think of the dynamical states of spiking neurons and spiking neural networks as *time-matter.* Namely, a structure of spike-events in time having analogue properties to that of ordinary matter. We can define for neural systems notions equivalent to the equations of state, phase diagrams and their phase transitions. For instance, the familiar Ideal Gas Law relation (Pv = constant) emerges as analogue of the Ideal Integrate and Fire neuron model relation (I_{in}ISI = constant). We define the neural analogue of the spatial structure correlation function, that can characterize spiking states with temporal long-range order, such as regular tonic spiking. We also define the "neuro-compressibility" response function in analogy to the lattice compressibility. We show that similarly to the case of ordinary matter, the anomalous behavior of the neuro-compressibility is a precursor effect that signals the onset of changes in spiking states. We propose that the notion of neuro-compressibility may open the way to develop novel medical tools for the early diagnose of diseases. It may allow to predict impending anomalous neural states, such as Parkinson's tremors, epileptic seizures, electric cardiopathies, and perhaps may even serve as a predictor of the likelihood of regaining consciousness.

1. Introduction

The understanding of the mind is, arguably, the most mysterious scientific frontier. The ability of the mind to understand itself is puzzling. Nevertheless, it seems increasingly possible and within our reach. Neuroscience and Artificial Intelligence are making great progress in that regard, however, following very different paths and driven by very different motivations. In the first, the focus is to address fundamental questions of biology, while in the second, it is to develop brain inspired computational systems for practical applications for modern life. Evidently, there is also large overlap between them both.

The basic units that conform the physical support of the mind, namely the brain and the associated neuronal systems, are neurons. These are cells with electrical activity that interact via electric spikes, called action potentials [?]. In animals, neurons conform networks of a wide range of complexity. Ranging from a few hundred units in jelly fish, to a hundred billion in humans. A fundamental question to answer is *how and why* nature has adopted this electric signaling system. Its main functions are multifold: to sense and monitor the environment, then to produce behavior and decision making, and finally to drive the required motor actions to assure the survival of living beings. Neuroscience has already provided a good understanding of the electric behavior of individual neurons [?, ?]. A major milestone was the explanation by Hodgkin and Huxley of the physiological mechanism for the generation of the action potential [?]. At the other end, that of neural networks with large number of neurons, important contributions come from Artificial Intelligence. For instance, significant progress was made in the 80s following the pioneering work of Hopfield [?]. More recently, this area received a renewed boost of activity, enabled by the combination of new learning algorithms for Deep Convolutional Neural Networks [?] with the numerical power of modern computers [?]. However, the networks adopted in Artificial Intelligence overwhelmingly describe the neurons' activity by their firing rate and not by individual spikes. These are conceptually different, a spike is a discrete event, while the spiking rate is a continuous variable. Hence, modeling neurons in terms of the latter does not directly address the question posed above, namely, the why and how of Nature using discrete spikes.

Here we propose to look at this problem under a different light, which to my knowledge has not been discussed before. Hopefully, this may bring new insights and perhaps help to develop our intuition for the challenging problem of understanding the mechanism of spiking neural networks. We shall postulate an analogy between matter states, as in organized spatial patterns of particles (or atoms or molecules) and that of neuronal states, as organized patterns of spikes in time. Since we are attempting to build a bridge between disconnected disciplines, we shall keep our presentation pedagogical.

Ultimately, as with any definition, our analogy will be of value if it turns to be a useful besides the intrinsic academic interest. With this in mind, we shall later discuss an exciting perspective that the present approach may perhaps open. Namely, to develop novel screening tools that could allow to detect an enhanced risk to develop neural diseases that involve anomalous spiking states, such as Parkinson's, epilepsy, cardiopathies, unconsciousness.

2. The analogy

As mentioned above, here we postulate an analogy between spatial matter states, such as solid, liquid and gas, and the dynamical states of spiking neurons and neural networks. More specifically, an analogy between the organization of matter in space and that of spiking events in time, which we call *temporal-matter* states.

To motivate this, in Fig.1 we show the familiar phase diagram of water as function of pressure and temperature (p,T). Next to it we show another phase diagram, that of an electronic bursting spiking neuron, which we introduce recently [?]. In the diagram, we observe various phases that correspond to qualitatively different states: tonic spiking (TS), fast spiking (FS), and two types of bursting (IB1, IB2). The phase diagram is obtained as a function of two parameters, the excitatory current I and a circuit time constant τ_s (the circuit is shown in Fig.4).

To characterize the spiking states in the phase diagram, we need to consider the nature of the discrete spiking events in the time domain. For instance, tonic spiking is characterized by a sequence of spikes that occur at equally spaced time-intervals. In Neuroscience, the time between two consecutive spike events is called the inter-spike interval (ISI). At the transition from the TS to FS phase, one observes a sudden decrease of the ISI (i.e. a jump in the spiking frequency) as function of the parameter τ_s.

The ISI(t) characterizes the organization of the spikes in the time domain and it indicates the "time-distance" between spikes. If a sequence of spike events at times t_i are indicated by the function

$$s(t) = \sum_i \delta(t - t_i) \tag{1}$$

Then, it may be tempting to establish the analogy by simply identifying the time-position of spike events with the positions of particles in a matter state. The matter state is characterized

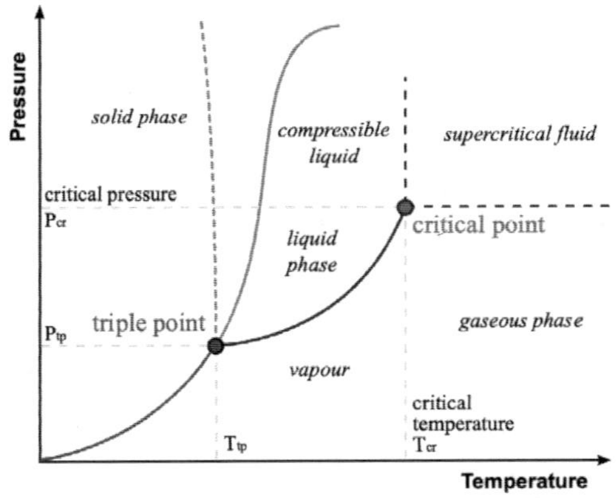

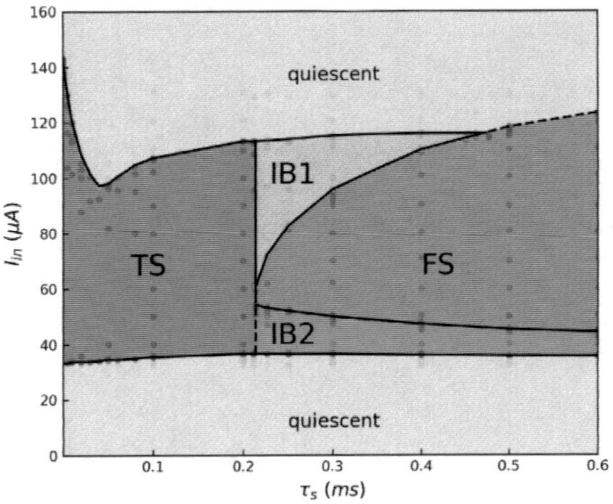

Figure 1. Left: Phase diagram of states of ordinary matter (water) showing the solid, liquid and gas phases. P and T determine the specific volume v of the state, which is the inverse of the density n. Right: Measured phase diagram of *temporal-matter* states of an electronic neuron model. The system is a two-compartment bursting neuron, showing tonic spiking (TS), fast spiking (FS) and two different intrinsic bursting states (IB1 and IB2). The input current I and the circuit time-constant τ_s determine the inter-spike interval of the state, which corresponds to the inverse of the frequency.

by its particle density function,

$$n(x) = \sum_i \delta(x - x_i) \tag{2}$$

where x_i denotes the position of the ith particle. For simplicity, we assume point particles and one dimensional space. Similarly, in Eq.1 we have assume ideal spikes represented by δ-functions, while in reality the action potentials have a typical duration of $\sim 1ms$ From the analogy, the simple tonic spiking case with equispaced spikes (i.e. constant ISI) would correspond to a perfect crystal of equally spaced particles (atoms). Thus, the familiar ISI of neuroscience would be the analogue of the familiar lattice constant a (or specific volume v) for material science.

If one applies a pressure P to matter, in general one observes the decrease of v, such as in the case of the Ideal Gas law Pv = constant. On the other hand, in neuroscience, it is well known that the ISI can be reduced by increasing the excitatory input current. Hence, one may be tempted to extend our analogy to associate the P with the input I in neural systems. We can make this more precise by introducing the simplest theoretical model of an ideal spiking neuron, the Integrate and Fire (IF) model [?].

In a very schematic view, as shown in Fig.2, a biological neuron is composed of three main parts: the dendrites, the soma and the axon. The neuron is excited through the input of electric signals arriving to the dendrites, which are called the synaptic currents. This input is integrated in the cell's body, which leads to the increase in its electric potential with respect to its resting state. By means of intense excitation, the neuron eventually reaches a threshold potential value. At that point a dramatic event takes place, an electric spike is initiated, it propagates down the axon and eventually is communicated to the dendrites of a downstream neuron. This phenomenon is called the emission of an action potential, which was first described by Hodgkin and Huxley [?]. This qualitative description can be represented by the simple (leaky) IF model

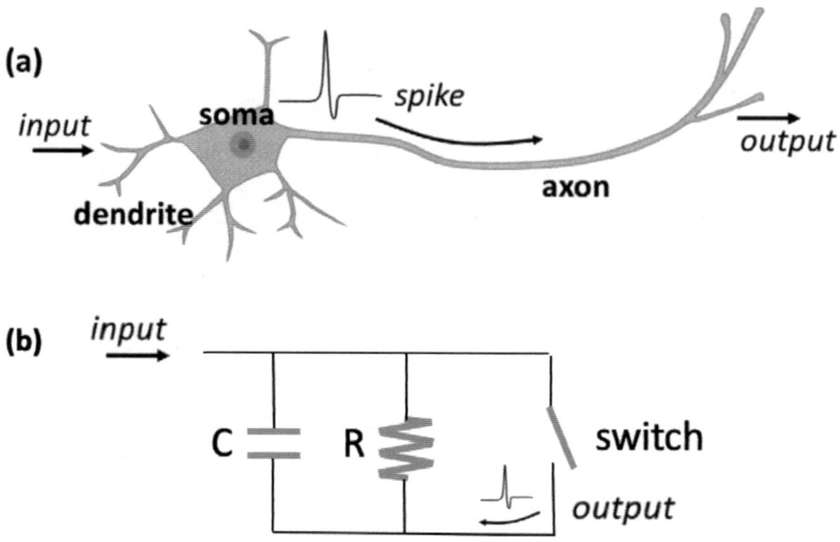

Figure 2. (a) Schematic biological neuron. (b) Electric circuit of the IF model (for the case u_{res}=0). The membrane of the cell's body (soma) is represented by the capacitor C_m, which accumulates charge of the input current. If the resistor $R \to \infty$ then the current integration is perfect, however for a finite value there is a "leaky" integration and the model is called LIF, for leaky-integrate-and-fire.

[?, ?]

$$\frac{du}{dt} = -\frac{1}{\tau}u(t) + I; \quad if\ u \geq u_{th}\ then\ spike\ and\ u = u_{rest} \tag{3}$$

where $u(t)$ represents the potential of the soma, u_{th} is the threshold potential, u_{rest} is the resting potential value and I is the input (synaptic) current. The time constant τ is a characteristic relaxation time of the neuron that represents the leakage of charge out of the soma. If the leakage is negligible, $\tau \to \infty$, and the integration is perfect, so one has an ideal IF model. The electric circuit representation of the model is straight forward. The soma is represented by a capacitor C that accumulates the charge of the input current, and the leakage by a resistor R in parallel. The threshold voltage can be represented by a switch that closes yielding the emission of the spike, which is the fast discharge of the charge accumulated in C, as a delta function of current.

The limit of zero leakage, i.e. the ideal Integrate and Fire is trivially solved. The potential u due to the integrated charge in C during an interval t is $u(t) = Q/C = (I/C)t$. Thus, the spike fire time t_f is given by the condition $u(t_f) = u_{th}$. Which leads to $It_f = u_{th}C$, or $I\text{ISI} = constant$.

Hence, we may extend our analogy by noting that the equation of state of the Ideal Gas has the same form as that of an ideal IF neuron, namely,

$$Pv = \text{constant} \longleftrightarrow I\text{ISI} = \text{constant} \tag{4}$$

where

$$P \longleftrightarrow I; \quad v \longleftrightarrow \text{ISI} \tag{5}$$

As the equation of state of a real gas departs from the ideal case, biological neurons and neuron models' equation of state, will depart from the ideal IF "neuronal equation of state" above. Interestingly, the notion of neuronal equation of state should not appear so strange.

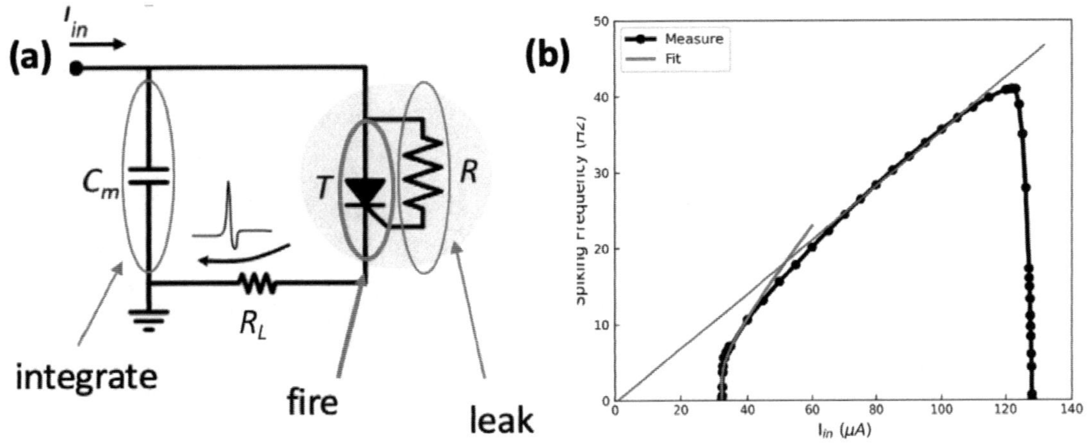

Figure 3. (a) Electric circuit that defines the physical neuron model based on the memristive properties of the thyristor. The function of the thyristor is that of a voltage controlled switch, as shown in Fig.2 above. The small "load" resistor R_L is to transform the output spike current into a voltage action potential signal for measurement convenience. (b) The "neuron equation of state" or activation function $f(I)$. Near the threshold, the equation of state follows closely the functional form of a LIF model (red line), $-1/[log(1 - I_c/I)]$ where I_c is the minimal activation current. The blue line corresponds to the Ideal IF behavior $I \propto 1/\text{ISI} = f$.

Indeed, it is nothing else than the familiar concept in neuroscience of the neuron's *activation function*, namely, the firing rate as a function of the excitatory input current, $f = f(I)$. For instance, some popular models are: the rectified linear units, or ReLU, where $f(I) = max[0, I]$; the sigmoid activation $f(I) = 1/(1 + e^{-I})$; etc. These are examples of *mathematical* neuron models, however, we may also include here *physical* neuron models, namely, models that are defined by an electronic circuit. In physical neuron models the equation of state $f(I)$ can be measured, as in real gases. We should mention that while the relation $Pv=$ constant is familiar from the Ideal Gas law, it is in general valid for any liquid or solid in the linear regime.

Recently, we introduced a *minimal* model of a physical spiking neuron, which is achieved by exploiting the memristive properties of an "old" conventional electronic component, the thyristor. The model is minimal because it provides a physical realization of the basic *leaky-integrate-and-fire* (LIF) neuron model by associating exactly one component to each one of the three functions: a capacitor to integrate, a resistor to leak and the thyristor to fire. Qualitatively, the thyristor acts as the switch in the circuit of Fig.2.

In Fig.3 we show the circuit that defines physical neuron model just described, where we identify the role of each of the three components to the functions of the LIF. We call this artificial neuron model the Memristive Spiking Neuron (MSN), which is implicitly defined by its electronic circuit. In the right panel of the figure we show the experimental neuron equation of state, which is nothing other than the activation function, as noted above. We can observe that near the excitation threshold the equation of state is well represented by the functional form of the activation function of the LIF mathematical model (red fitting line in Fig.3). At intermediate input currents, the behavior approaches the Ideal IF neuron, whose equation of state is $f \propto I$ (see 4), since $f = 1/ISI$

The same methodology allows us to consider more complex neuron models, which are also defined by their respective circuit implementation. For instance, we may consider the case of

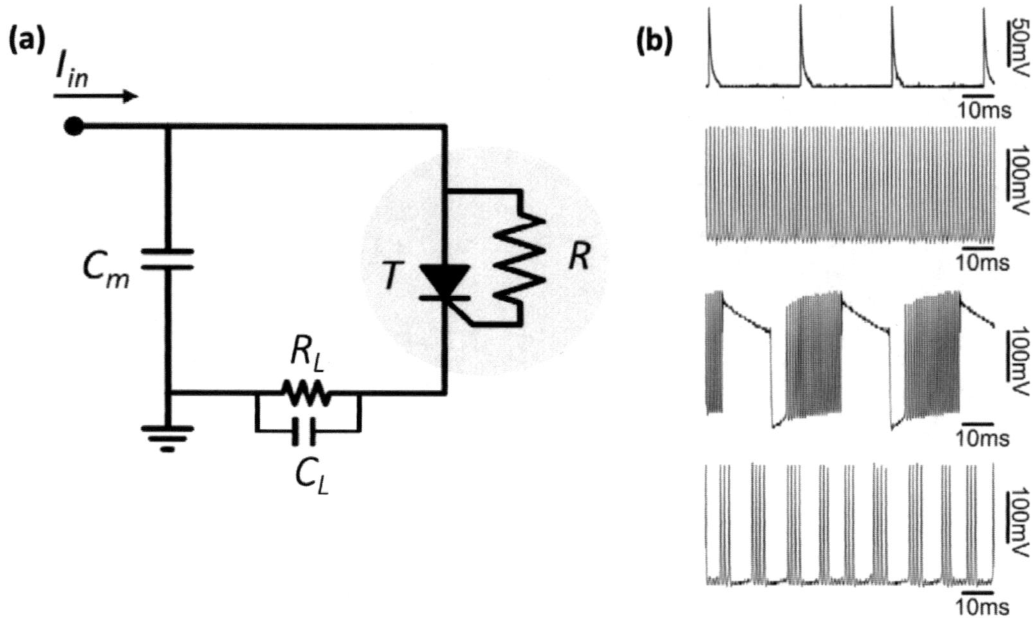

Figure 4. (a) Electric circuit that defines the physical model of a bursting neuron based on the MSN model and adding a second time constant $\tau_L = R_L C_L$. (b) The various dynamical behaviors produced by the circuit. From top to bottom: tonic spiking (TS), fast spiking (FS) and two bursting types (IB1, IB2), which correspond to the four phases of the phase diagram shown in Fig.1.

bursting neurons. From theoretical neuroscience we know that a requirement to obtain bursting behavior is by adding a second dynamical variable, besides the potential of the cell body $u(t)$, in Eq.4 and represented by the capacitor in the circuits of Fig.2 and 3. Thus, to do this we add a second RC pair to our basic MSN circuit. A simple option is to add a capacitor C_L in parallel to the (small) output resistor R_L, introducing a new time constant $\tau_L = R_L C_L$. The resulting circuit is shown in Fig.4a. In panel (b) we show the dynamical behavior that it now produces. We observe four qualitatively different spiking types: simple tonic spiking, fast spiking, and two bursting modes. These four spiking types are realized in the respective regions of the phase diagram of Fig.1 presented before.

As done before for the basic MSN, we may also obtain the equation of state of the Memristor Bursting Neuron (MBN) model. In the present case, we can consider that the time constant τ_L plays the role of a third parameter, similarly as the temperature in the case of matter systems. Hence, in Fig.5 we show the curve $f_{\tau_L}(I)$ measured at a fixed τ_L, indicated by the vertical purple line that crosses three phases. We observe the jumps in the frequency as the current drives the system from one phase to the other. This is reminiscent of changes in density when the pressure drives phase transitions at a fixed T in the phase diagram of water in Fig.1.

It is interesting to mention that a biologically realistic theoretical model of bursting neurons introduced by Pinsky and Rinzel (PR) shows a qualitatively similar behavior [?]. The PR is an example of a two-compartment model, where both the soma and the dendrites are described. We may notice that in the MBN model the $R_L C_L$ block (see Fig.4) can be considered as a second compartment, which is connected to the output of the first. In the right panel of Fig.5, we reproduce the activation function (i.e. the neuronal equation of state) of the PR model. It

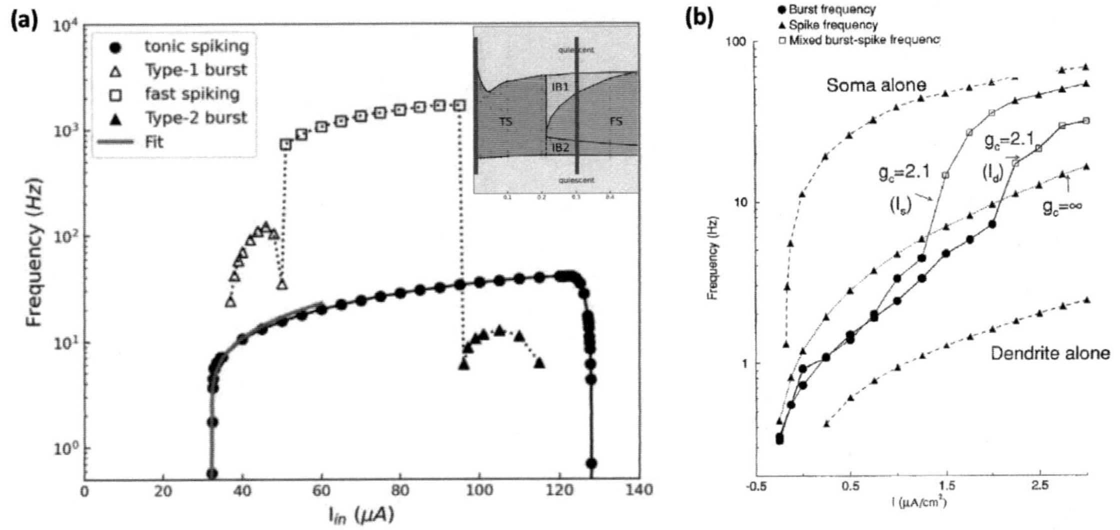

Figure 5. (a) Activation function $f(I)$ in semilog scale. The top curve corresponds to the MBN model along the right purple line at τ_L=0.3ms indicated in the phase diagram in the inset. Following the definition of Pinsky and Rinzel, the frequency is defined as the inverse of the period between spike trains (bursts). The bottom $f(I)$ curve is that of the MSN model discussed before, which corresponds to the limit $\tau_L \to 0$ of the bursting neuron model (indicated with the left purple line in the inset). (b) Activation function of the Pinsky-Rinzel model reproduced from [?].

is interesting to observe that in the simpler limits of only one compartment (soma alone and dendrite alone) the behavior of the PR is qualitatively the same as that of our basic MSN, which is also a single compartment model (see bottom curve of Fig.5a). More importantly, for the relevant case of two compartments (i.e. finite g_c) we observe a sudden changes in the firing rate as a function of excitatory input current, also in qualitative agreement with the MBN. In fact, both PR and MBN traverse the same sequence as the excitatory current is increased: initially quiescent below a critical current, then bursting, and finally jump in firing rate to the fast spiking mode. Hence, the phase transitions are abrupt, through a steep or a discontinuous change in the activation function. As we shall discuss in the next section this feature may have interesting consequences. Moreover, and we shall show that the $f(I)$ anomalies may be considered as the counterparts of certain phenomena occurring in phase transitions in matter systems.

3. Correlation and Response Functions

Correlation and response functions are useful concepts in material science, where they serve to characterize different states of matter. For instance, a regularity in the arrangement of positions of atoms is revealed by Bragg peaks in the x-ray spectra, which are maxima of the structure factor. In real space, the regularity is revealed by the pair correlation function

$$g(x) = \frac{\int n(x+x')n(x')dx'}{\int n^2(x')dx'} \tag{6}$$

where $n(x)$ indicates the particle density (such as electrons, atoms, molecules, etc) at position x, and where we consider one dimension for simplicity. In the case of a crystalline order, $g(x)$

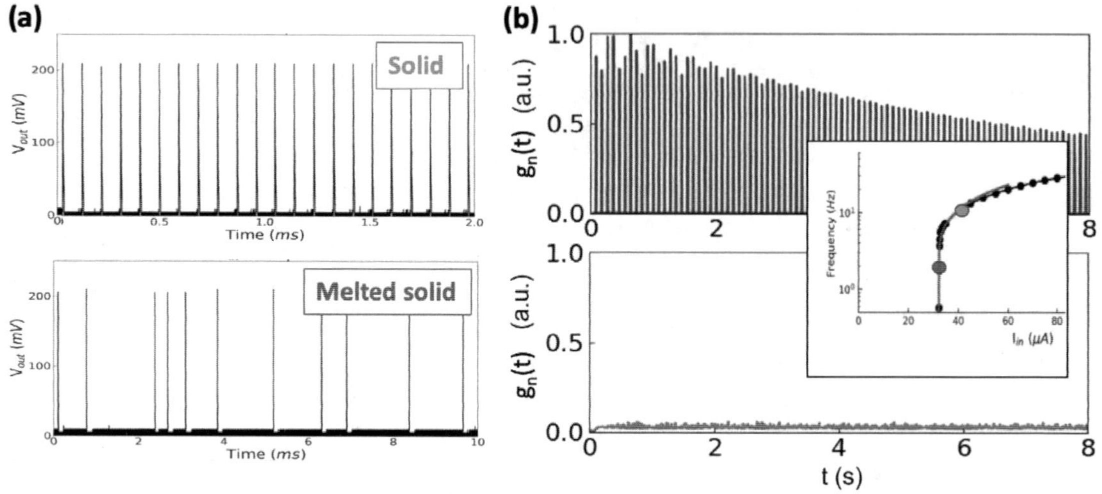

Figure 6. (a) Measured spike traces $s(t)$ of the MSN model at two values of the input current: a "solid" state measured at $I = 45.3\mu$A (top) and a "melted" state at $I = 36.4\mu$A. The states are indicated with the green and blue dots in the activation function of the neuron reproduced in the inset. (b) The neural correlation function $g_n(t)$ computed for the respective traces shown in the left side panels (only a small portion of the measured traces is shown).

shows structure with peaks. For a simple arrangement of particles along one dimension with a lattice constant a, the peaks will be at a, $2a$, $3a$, ... In contrast, for a disordered state, such as gas or liquid, the $g(x)$ is mostly featureless. The study of $g(x)$ is routinely done in condensed matter physics for the study of phase transitions (see, for example, [?]).

In our analogy, spiking systems are thought of as temporal-matter states, so it is natural to explore the behavior of the correlation-function analogue to $g(x)$. Since the position of particles correspond to position of spikes, by analogy we can define the neural correlation function $g_n(t)$ as

$$g_n(t) = \frac{\int s(t+t')s(t')dt'}{\int s^2(t')dt'} \tag{7}$$

where $s(t)$ indicates a given spike trace. The function $g_n(t)$ can characterize different spiking states of a neuron. In Fig.6 we provide a concrete example, which is realized in the basic MSN model described before (see Fig.3).

We can observe the two qualitatively different tonic spiking behaviors in the two left panels of the figure. They correspond to two constant current inputs (36.4μA and 45.3μA). In the first case, at higher input current (top panel), the spiking is perfectly regular. In contrast for a smaller current close to the threshold, the trace changes dramatically. The interval between spikes become very irregular. By analogy between spike and particles, we can think of the first case as that of a solid and the second as of the melting of the solid state. This qualitative description can be made more precise by the correlation function $g_n(t)$, shown on the right side panels of Fig.6.

The top panel shows a succession of delta functions at equally spaced at times, multiples of the constant inter-spike interval, t_k=kISI. This indicates the long range order in time. It tells that given the presence of a spike at time $t = 0$ we have a high probability (1) of finding another spike event at times t_k (k=1, 2, 3, ...). In contrast, the $g_n(t)$ shown in the bottom panel is featureless, showing a small approximate constant value. This indicates total lack of order as the presence of a spike a time $t = 0$ does not permit to predict the presence of ulterior spikes events. The emission of spikes is random as the positions of particles in a liquid or a gas.

The peaks of the $g_n(t)$ are very narrow, delta-function-like, because the spikes are very narrow with respect to the duration of the ISI. In a solid, the atoms have a size that is smaller but of the same order as the lattice spacing, so instead of narrow deltas one observes broad peaks in the $g(x)$ [?].

One may understand quite intuitively the physical origin of this dramatic changes in the time-structure. The key point is to realize that the "melted" state occurs in a regime where the activation function $f(I)$ is very steep, at the onset of neuron excitability, i.e. near the threshold. Therefore, small variations of the input current will reflect on significant variations in the ISI. This observation motivates the following important insight.

This enhanced sensitivity to current fluctuations is due to the large slope df/dI in the activation function. Then, what is this feature related to, if one follows the analogy back to the matter systems? We recall that ISI plays the role of the lattice spacing, or specific volume (see Eqs.4 and 5), then f=1/ISI corresponds the particle density $n(x) = 1/v(x)$. On the other hand, since the input current I is like the pressure, then it follows that the slope df/dI corresponds to dn/dP. This last quantity is closely related to the compressibility of matter systems

$$\beta = \frac{1}{n}\,(\frac{dn}{dP}) \tag{8}$$

which is the inverse of the bulk modulus. We can therefore follow the analogy and introduce the concept of "neuro-compressibility",

$$\beta_n = \frac{1}{f}\,(\frac{df}{dI}) \tag{9}$$

It is important to mention that this quantity may be measured using experimental methods such as dynamic clamp, where a controlled synaptic current can be injected into a neuron while its activity is monitored [?]. Moreover, this definition may turn out to have important consequences, as we discuss next.

Anomalies in the compressibility of materials are precursor signatures of structural phase transitions. A sudden increase in the compressibility of a solid indicates the "softening" of a vibrational mode (a phonon mode), which leads to a change in the structure, or possibly a phase change. Then, the question is, what would the analogue phenomenon be for a neuronal system? For a single neuron, the enhancement of β_n would indicate the proximity to a qualitative change in the spiking mode of a neuron, i.e. a "bifurcation" in its dynamics. This can in fact be seen in the panels of Fig.5. There, we observe that all the changes in the spiking modes for both, the MSN and for the MBN models, occur at current values where there are enhancements in df/dI or jumps. Most notably, this is not only a feature of our artificial neuron circuits, but also can be clearly seen in the biologically realistic Pinsky-Rinzel model activation functions that we reproduced in Fig.5 [?]. The enhancements seen in the PR data occur at the onset of change from quiescent to spiking and also in the change from burst to tonic spiking (black circles and black triangles), in very good qualitative agreement to our electronic neuron model,

We may then speculate on an important implication of our observations. It would be interesting to explore if neuro-compressibility anomalies are also found across the boundaries of qualitatively different states in *neuron networks*. If that is the case, an intriguing and exciting possibility would be to investigate if anomalies in β_n are also detected (by small current stimulation) in animal models of epilepsy and Parkinson's disease. If this were the case, then one may envision a pathway to a novel diagnose tool for early detection or a risk predictor of mental diseases associated to abnormal spike patterns in humans. In even further speculation, one may also search for anomalies at the onset of regaining, or loosing, consciousness, which is another challenging frontier of research [?].

Journal of Physics: Conference Series **2533** (2023) 012007 doi:10.1088/1742-6596/2533/1/012007

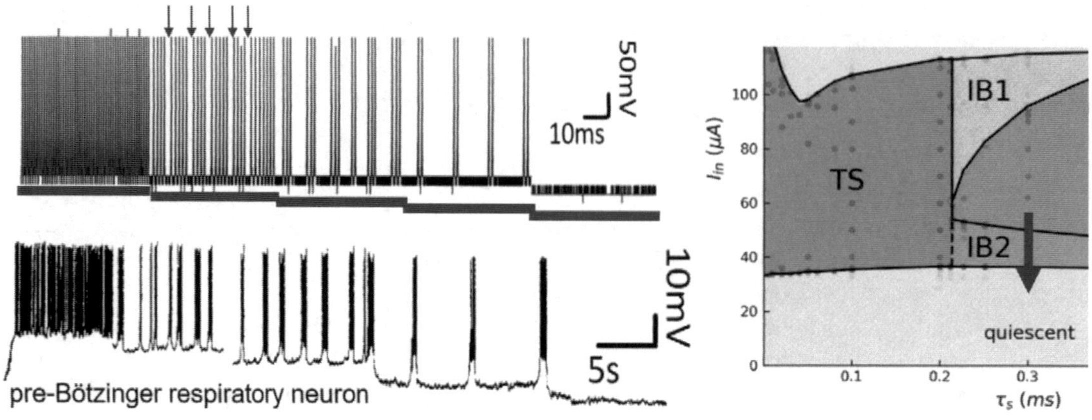

Figure 7. Top left: Measured spike traces of the MBN as a function of decreasing current in discrete steps (purple). The thin purple arrows indicate the missing spikes. Bottom left: Experimental trace of pre-Bötzinger bursting neurons from rats. The data is obtained by changing excitatory input current in discrete steps. Adapted from [?]. The circuit parameters may be easily adjusted to fit the experimental data [?]. Right: Phase diagram of the MBN where the purple arrow indicates the evolution from the fast spiking (blue) to the bursting type 1 phase (green) by decreasing current at constant τ_L.

4. Bursting spikes as an ananlogue of formation of clusters defects

Here we describe another interesting connection between common a phenomena in spiking neurons and in material science. We shall show that missing spikes in a trace of a fast spiking state, can be thought of as the analogue of missing atoms, i.e, defects, in a crystal structure.

From Fig.7 we observe that the proliferation of missing spikes in a fast spiking state is a route to generate bursting behavior. This is illustrated in the sequence of traces shown in the figure, which were obtained for a step-wise decreasing input current to the MBN. The thick purple arrow in the vertical path followed in phase diagram (from blue to green to grey). In the top trace we indicate with small purple arrows the missing arrows, showing that one may understand the onset of bursting as the result of skipping spike events, which are initially few (i.e. dilute). As the current intensity is further reduced, the missing spikes become more numerous (i.e dense) and occurring in *clusters* of inactivity, which give rise to the stuttering mode bursts [?]. In our analogy, we think of spikes as of atoms in a lattice, therefore, the initial continuous fast spiking state is like a perfect crystal. The missing spikes then play the analogue role as vacancy defects, i.e. lacking atoms. Moreover, the missing spikes are the result of decreasing current, which in the analogy represents pressure. It is then interesting to observe that in thin-film deposition, which is a topic in material science, the partial pressure of oxygen $P(O_2)$ is a relevant parameter for the quality of the growth of crystalline oxides. Moreover, it is well known that reducing the $P(O_2)$ induces the creation of oxygen vacancy defects in the crystal structure [?, ?], which often cluster together forming dislocations [?], This is in full qualitative analogy to the spiking traces in the stuttering bursting mode shown in Fig.7.

We would like to emphasize that the path of phase transformations, i.e. the evolution from fast spiking, to bursting, to quiescent does not seem to be just a peculiarity of our MBN circuit model. In the lower panel of the Fig.7 we illustrate the striking resemblance of the traces of the MBN with those measured in bursting neuron of rats [?]. Quite remarkably, the experimental traces were obtained by solely changing the intensity of the excitatory DC current.

Journal of Physics: Conference Series **2533** (2023) 012007 doi:10.1088/1742-6596/2533/1/012007

5. Neural Networks

We now consider one important final aspect in our analogy that may eventually bring new light to the issue of how to think about inter-neuron coupling. So far we have considered essentially individual neurons, but we may ask what would it be to extend the analogy to multi-neuron systems, i.e. to neural networks.

As a first glimpse into this question, we shall consider the simplest network case, namely, just two neurons that are mutually excitatory or inhibitory. We focus first in a system of two identical neurons, each excited with equal input currents. The currents are above threshold, so the neurons are active and theirs spikes are transformed via conductances into mutually injected synaptic currents that are positive for the excitatory case and negative for the inhibitory one. This is schematically shown in Fig.8.

We shall see that our analogy takes an interesting twist, as the dynamical states of the two-neuron system can be considered as an analogue of a complex crystal, i.e. a crystal with two atoms in the unit cell. Moreover, the coupling between neurons is mediated by synaptic currents, which can be excitatory or inhibitory. Since in our analogy current plays the role of pressure, then the synaptic currents should also play such a role. More precisely, an excitatory synaptic current should correspond to a repulsive inter-particle interactions (positive pressure), and an inhibitory synaptic current should be an attractive interaction (negative pressure). We shall consider the two cases, which we study using realistic electronic circuit simulations (LTspice).

We first consider two spiking neurons with mutual inhibition. When one neuron spikes it inhibits the other, and vise-versa, so they try to avoid spiking at unison since we are synaptic current are instantaneous. After a transient period, they expectedly find a stable dynamical state where they alternate to emit spikes as shown in the top panel of Fig.8. Using our analogy, this spiking pattern corresponds two a perfect molecular crystal where the unit cell has an A-B atom pair (or a basis).

In the bottom panel of the figure we consider the second case, that of mutually excitatory neurons. Again after a transient time, the system adopts a periodic spiking pattern. However, in contrast to the previous case, now both neurons fire at unison. This is also intuitive, the excitatory synaptic current of emitted by a neuron that spiked promotes the spiking of the other one, and vise versa. So, naturally, the both spike at the same time, which is nothing else that excitatory synapses promote synchrony in neural networks [?]. Following our analogy, spiking at unison corresponds to a "dimerization" of the lattice. Namely, that the distance between the A-B pair atoms is reduced as due to an attractive interaction between the A and B atomic species.

These two cases are consistent with our analogy, where current is interpreted as pressure. Indeed, the volume of the unit cell in the inhibitory case is large, as expected for positive effective pressure between A and B, while in the excitatory case the volume is fully collapsed to zero, as expected for a negative pressure within the unit cell.

It is an interesting perspective for future work to consider increasingly complex networks of several neurons (motifs). The periodic states that emerge constitute spiking *sequences*, which are of great relevance for automatic motor behavior. By virtue of the analogy that we introduced in the present work, those periodic sequences should correspond to a variety of molecular crystals. It would be exciting to explore if new intuition for Neuroscience could be brought from from those traditional areas of Condensed Matter Physics and Chemistry [?].

6. Conclusion

In this work we introduced the idea that the dynamical states of neural networks may be though of as realizations of "time matter" states.

We started from the notion that a trace of spiking events of a neuron can be analogue to a snapshot of particles or atoms arranged in space. We then went on to explore and show that

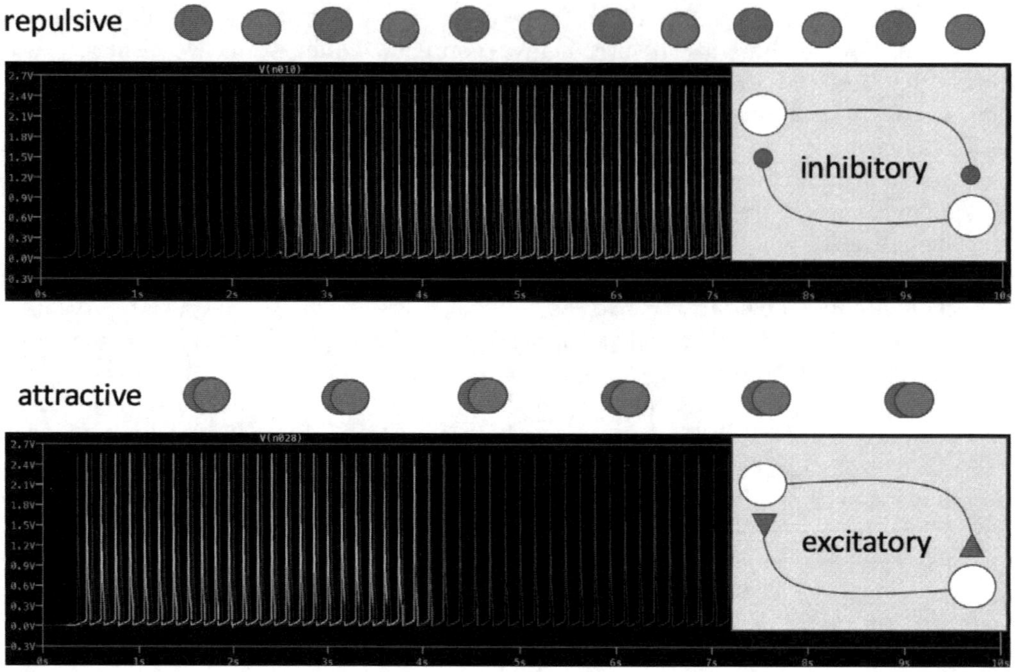

Figure 8. Top: two 1D space lattices (red and blue) with repulsive interaction and two neurons with inhibitory interactions. The neurons are prepared to start spiking synchronously and spontaneously adopt an alternating spiking pattern. Bottom: two 1D lattices with attractive interactions and two neuron network with excitatory interaction. They are prepared to alternate but spontaneously begin to spike in unison. Spike traces were produced by realistic simulations with LTSpice.

the analogy may be pushed far beyond that literary statement, and may provide new intuition in the challenging problem of understanding and designing spiking neural networks.

We identified analogue roles of basic quantities of Physics with those of Neuroscience, such as pressure and volume with input currents and inter-spike intervals. We then logically build on this assumption to show connections between correlation and response functions in both fields. Perhaps most significant was the finding that a *neuro-compressibility* can be defined, with possible far reaching consequences, including medical ones, that may be experimentally tested.

An exciting new road of discovery may be open ahead.

7. Acknowledgments
I acknowledge support from the French ANR "MoMA" project ANR-19-CE30-0020. I also thank Jiaming Wu for help with the figures and to Lorenzo Fratino and Thomas Elze for the invitation to the stimulating DICE Workshop.

References
[1] Bear M, Connors B and Paradiso M 2016 *Neuroscience: Exploring the Brain* (Wolters Kluwer, Philadelphia, 4Th ed. international ed.)

[2] Gerstner W, Kistler W, Naud R and Paninski L 2014 *Neuronal Dynamics: From Single Neurons to Networks and Models of Cognition* (Cambridge University Press) ISBN 9781107060838
[3] Ermentrout G and Terman D 2010 *Mathematical Foundations of Neuroscience* Interdisciplinary Applied Mathematics (Springer New York) ISBN 9780387877082
[4] Hodgkin A and Huxley A 1952 *The Journal of Physiology* **117** 500–544
[5] Hopfield J J 1982 *Proceedings of the national academy of sciences* **79** 2554–2558
[6] LeCun Y, Bengio Y and Hinton G 2015 *Nature* **521** 436–444
[7] Merolla P A, Arthur J V, Alvarez-Icaza R, Cassidy A S, Sawada J, Akopyan F, Jackson B L, Imam N, Guo C, Nakamura Y *et al.* 2014 *Science* **345** 668–673
[8] Wu J, Schneegans O, Wang K, Stoliar P and Rozenberg M 2023 *submitted*
[9] Pinsky P F and Rinzel J 1994 *Journal of Computational Neuroscience* **1** 39–60
[10] Debela T T, Wang X D, Cao Q P, Lu Y H, Zhang D X, Fecht H J, Tanaka H and Jiang J Z 2014 *Phys. Rev. B* **89**(10) 104205
[11] Goaillard J M and Marder E 2006 *Physiology* **21** 197–207 pMID: 16714478
[12] Storm J F, Boly M, Casali A G, Massimini M, Olcese U, Pennartz C M A and Wilke M 2017 *The Journal of Neuroscience* **37** 10882
[13] Golomb D, Donner K, Shacham L, Shlosberg D, Amitai Y and Hansel D 2007 *PLOS Computational Biology* **3** e156–
[14] Ghenzi N, Rozenberg M J, Llopis R, Levy P, Hueso L E and Stoliar P 2015 *Applied Physics Letters* **106** 123509
[15] Liu X, Zhou K, Wang L, Wang B and Li Y 2009 *Journal of the American Chemical Society* **131** 3140–3141
[16] Szot K, Speier W, Bihlmayer G and Waser R 2006 *Nature Materials* **5** 312–320
[17] Butera Jr R, Rinzel J and Smith J 1999 *J Neurophysiol* **82** 382–397
[18] Hansel D, Mato G and Meunier C 1995 *Neural Computation* **7** 307–337
[19] Corpinot M K and Bučar D K 2019 *Crystal Growth & Design* **19** 1426–1453

Journal of Physics: Conference Series **2533** (2023) 012008 doi:10.1088/1742-6596/2533/1/012008

Revisiting Quantum Contextuality in an Algebraic Framework.

Mathias Van Den Bossche[1] **and Philippe Grangier**[2]

[1] Thales Alenia Space, 26, avenue J.-F. Champollion, 31037 Toulouse, France
[2] Laboratoire Charles Fabry, IOGS, CNRS, Université Paris Saclay, F91127 Palaiseau, France.

E-mail: `philippe.grangier@institutoptique.fr`

Abstract. Within the framework of quantum contextuality, we discuss the ideas of extracontextuality and extravalence, that allow one to relate Kochen-Specker's and Gleason's theorems. We emphasize that whereas Kochen-Specker's is essentially a no-go theorem, Gleason's provides a mathematical justification of Born's rule. Our extracontextual approach requires however a way to describe the "Heisenberg cut". Following an article by John von Neumann on infinite tensor products, this can be done by noticing that the usual formalism of quantum mechanics, associated with unitary equivalence of representations, stops working when countable infinities of particles (or degrees of freedom) are encountered. This is because the dimension of the corresponding Hilbert space becomes uncountably infinite, leading to the loss of unitary equivalence, and to sectorisation. Such an intrinsically contextual approach provides a unified mathematical model including both quantum and classical physics, that appear as required incommensurable facets in the description of nature.

1. Introduction

The usual formalism of quantum physics makes use of separable Hilbert spaces[1] and operators acting in it, that form a type-I von Neumann algebra, according to the classification introduced in the 1930's by Murray and von Neumann [1, 2]. In this article we will argue that this mathematical framework is unduly restrictive, and that it is worthwhile exploring what happens when the Hilbert space is non-separable and the operators acting on its vectors form a type-II or type-III von Neumann algebra. The motivation is that this extension appears as a reasonable framework to describe macroscopic systems with such mathematical tools, as explained below. However, operator algebras are unfriendly for non-mathematically oriented physicists, and they appear more as a subfield of mathematics than of physics. It addition, it is not so obvious to find a clear-cut situation where type-I fails, and where higher-type operator algebras are required.

In this paper, we want to show how such sophisticated mathematical objects can be very physically relevant to describe large systems, and may shine some light on foundational questions of quantum mechanics. Correspondingly, it will appear that the clear-cut situation where type-I fails is nothing but the so-called quantum measurement problem.

In Section 2 we will start from a simple discussion of contextuality in standard Quantum Mechanis (QM), and show how it relates to the whole structure of QM through Gleason's theorem and Born's rule. This leads however to a picture that cannot avoid including the "Heisenberg

[1] This means that there exists a countable basis in such spaces.

Journal of Physics: Conference Series **2533** (2023) 012008 doi:10.1088/1742-6596/2533/1/012008

cut", separating (quantum) systems from (classical) contexts. In Section 3 we recall some basic results of [1], based on John von Neumann's work on infinite tensor products [2], and we show how build up a complete and consistent picture of QM.

2. Quantum contextuality revisited.

2.1. From Kochen-Specker to Gleason.

For the clarity of the presentation, we will introduce quantum contextuality in a simple way, based upon usual textbook quantum mechanics[2]. Then we will open some questions, and propose a way to answer them. Finally, we will draw some conclusions from our approach.

Within basic textbook quantum mechanics (QM), let us consider a complete quantum measurement of a quantity A, where "complete" means with no degeneracy left; usually such a measurement is associated with a complete set of commuting operators (CSCO), that we summarize in a single "observable" operator $\hat{A}$. Denoting as a_i and $|u_i\rangle$ the eigenvalues and eigenstates of $\hat{A}$, one has from the spectral theorem $\hat{A} = \sum_i a_i |u_i\rangle\langle u_i|$ with standard Dirac's notations. In the following $\hat{A}$ will be identified with a **context**, that can be understood with different meanings: on the physical side the context corresponds to a macroscopic and operational device, able to measure the quantity A, whereas on the mathematical side it corresponds to the observable $\hat{A}$, or also to the above set of rank-one orthogonal projectors $\{|u_i\rangle\langle u_i|\}$, where i goes from 1 to N, the dimension of the Hilbert space. In usual QM these physical and mathematical objects correspond to each other, and we will come back to that in the last part of the paper. We emphasize that in our approach QM is basically a non-classical probability theory, relative to events that happen to physical objects made of systems within contexts; in this framework the role of "agents" is the same as in classical probability theory.

According again to standard QM, a measurement of $\hat{A}$ will give one of the eigenvalues a_i, whereas other complete measurements $\hat{B}$, $\hat{C}$... associated with other contexts, will similarly give one of their eigenvalues b_j, c_k... In each context the corresponding observable gets one value among N possible ones, so one says that the value a_i is assigned to the observable $\hat{A}$.

Now, what is quantum contextuality, in its simplest definition ? It is the observation that in QM, **it is impossible to assign simultaneously values with certainty to all observables in all possible contexts, each one seen as the experimental background of a measurement.** This simple observation clashes with classical physics, where the possibility of such an assignment is considered essential to express the laws of nature. The formal proof of quantum contextuality is given by the Kochen-Specker (KS) theorem, showing a contradiction between QM and all models attributing non-contextual values to some (well chosen) set of observables. Such a result is "negative", or in other terms it is a no-go theorem, showing that a large class of non-contextual models (typically the models with non-contextual hidden variables) contradict QM predictions and experimental observations [3].

However, it does not tell much about what is actually happening in QM: the whole purpose of the KS discussion is to tell what QM is not, rather than to tell what QM is – which is the real issue we are interested in.

As a step in this direction, we may ask the question: if it is impossible to assign simultaneously values with certainty to all observables in all possible contexts, how to make sense of what is being observed and measured ? There is an answer from observation: given a result in a context, one cannot in general specify results in other contexts, but one can specify the **probabilities** of such results. Even more interestingly, the assignment of probabilities to the value of all

[2] Here we only consider projective measurements (PVM), where, as explained in [3], all the notions of compatibility, i.e., joint measurability, nondisturbance, outcome repeatability, etc., are equivalent to the commutativity of the observables. This is the original framework of the discussion on quantum contextuality, and though it has been extended in many other interesting directions [3], it is enough for our current purpose.

Journal of Physics: Conference Series **2533** (2023) 012008 doi:10.1088/1742-6596/2533/1/012008

observables (that is, to any given measurement result) turns out to be non-contextual! In other words, all measurement results within different contexts can be predicted simultaneously, albeit as a probability rule, and not a certainty rule.

Contrary to the previous negative result, this is major asset: as we will show in more details below, it is the basis of Gleason's theorem, mathematically establishing Born's rule. It is therefore a "positive" result, much more interesting and powerful than Kochen-Specker's theorem. On the mathematical side, we note also that Kochen-Specker's theorem can be seen as a corollary of Gleason's one, whereas the reverse is not true [3]. It is also important to note that there is a special case for probabilities: if we restrict ourselves only to certainties (probabilities $p = 0$ or $p = 1$), they are actually values (eigenvalues of projection operators), not probabilities (average values of projection operators). Certainties are therefore contextual, as it is not possible to assign truth values (0 or 1) to all measurement results, this is another useful way to see Kochen-Specker's theorem. But one can assign non-contextual probabilities as said above, with values given by Born's law or Gleason's theorem.

As a conclusion of this introduction, QM is both contextual (for the measurement results) and non contextual (for the probabilities assigned to these results). Though this sounds hardly comprehensible, everybody agrees on that, and this is the current status of quantum contextuality as presented in [3]. Our purpose here is to make sense of this situation, by introducing a few additional ideas and definitions.

2.2. What is quantum (extra)contextuality ?

In this section we will introduce some new ideas, still within the scope of the usual textbook QM. Then at some point we will diverge, as it will be seen below.

First, let us introduce a distinction between a usual pure quantum state $|\psi\rangle$, described as a vector in the relevant Hilbert space $\mathcal{H}$ for the considered system, and the same vector considered as an eigenstate of a complete measurement, therefore within a given context as introduced above. We will define a **modality** as the association of $|\psi\rangle$ and the context $\hat{A}$. By construction a modality is certain and repeatable: the same $|\psi\rangle$, and the associated eigenvalue, will be found again and again as long as the same $\hat{A}$ is measured on the same system.

From this definition one has $|\psi\rangle = |u_i\rangle$, where $|u_i\rangle$ is an eigenstate of $\hat{A}$, and the modality "$|\psi\rangle \equiv |u_i\rangle$ considered as an eigenvector of $\hat{A}$" will be denoted as $|\psi\rangle_A$. It is clear that the same vector $|\psi\rangle$ may appear as an eigenstate of many other observables (actually, an infinity), as soon as the dimension of $\mathcal{H}$ is at least 3. Since the changes of context are continuous, both physically and mathematically, there is actually an infinity of such observables. So there will be many modalities, e.g. $|\psi\rangle_A$, $|\psi\rangle_B$, $|\psi\rangle_C$, all associated with the same $|\psi\rangle$: in the langage of usual QM, they *are* the same $|\psi\rangle$; but here they are different modalities, because they belong to different sets of mutually exclusive outcomes. So now we start moving away from the usual language: let us consider that the different modalities $|\psi\rangle_A$, $|\psi\rangle_B$, $|\psi\rangle_C$ are indeed different, though they are all associated with the same vector $|\psi\rangle$, or equivalently the same projector $|\psi\rangle\langle\psi|$ (projectors are actually more suitable for reasons that will appear below). Clearly this creates an equivalence class between modalities, that we will call an extravalence class, where all modalities are associated with the same $|\psi\rangle$. This means that certainty can be transferred between contexts, because $|\psi\rangle$ is an eigenstate in all the corresponding contexts, and the associated physical phenomenon will be called extracontextuality.

So far so good, but does this help with the previous annoying statement that QM is both contextual for measurement results, and non contextual for probabilities assignments ? To see that consider the following statements

(i) when making measurements, certainty and reproducibility are warranted for a modality

Journal of Physics: Conference Series **2533** (2023) 012008 doi:10.1088/1742-6596/2533/1/012008

in a given context, and are also warranted for extravalent modalities when changing context. But they are not warranted beyond that, due to contextuality in the assignment of measurement results as established by the KS theorem.

(ii) in other cases, when changing the context from $\hat{A}$ to $\hat{B}$, the result is probabilistic and given by Born's rule $|\langle\phi|\psi\rangle|^2$ for an initial modality $|\psi\rangle_A$ and final one $|\phi\rangle_B$. This probability depends only on the extravalence classes of the initial and final modalities, in agreement with the non-contextuality in the assignment of probabilities, a basic hypothesis of Gleason's theorem.

So by attributing the vector $|\psi\rangle$ not to a "quantum state", but to an extravalence class of modalities, one gets a more transparent picture of what QM is telling us: certainty and reproducibility do exist, not only within a context, but also accross contexts within an extravalence class; when changing context and extravalence class, probabilities are needed, but they are non-contextual because they connect extravalence classes, whatever the contexts are.

This argument tells the "how", but not yet the "why": what would be needed is to explain, or at least justify, the origins of statements (i) and (ii). This can be done within the framework called CSM, for Contexts, Systems and Modalities [6, 7, 8, 9], and we will briefly summarize here some elements taken from [8]. In this framework systems within contexts are defined as basic objects, and modalities are properties associated with certainties, in agreement with the first part of statement (i). Its second part, i.e. the contextuality of the assignment of measurement results, appears as a consequence of a quantization postulate: for a given system, the maximum number of mutually exclusive modalities is bounded to M; this requires a probabilistic description when contexts and extravalence classes are being changed [8].

To get the probability law (ii), one assumes that the probabilities when changing contexts depend only on the extravalence class, and one attributes a projector $|\psi\rangle\langle\psi|$ to each extravalence class. These are strong assumptions, but they do agree with empirical evidence, and are justified in detail in [8], based on induction. Then it is easy to check that all hypotheses for Gleason's theorem are satisfied, and as a consequence, that Born's rule is justified as the only acceptable probability law. This argument can be made even more convincing by using Uhlhorn's theorem in addition to Gleason's, in order to associate unitarity transforms to context changes [10].

We see now that the consequences drawn from extracontextuality are considerably stronger than the ones drawn from contextuality alone: we not only get no-go theorems, but we reach the correct probabilistic structure of QM, essentially by providing a physical justification of Gleason's and Uhlhorn's hypotheses. This automatically ensures that Kochen-Specker's theorem is true also, as a corollary of Gleason's, without the need of examining many different scenarii in many different dimensions [3].

2.3. Confronting the Heisenberg cut.

A conclusion of the above is that extracontextuality and subsequently extravalence tell "how much non-contextual" QM can be, given that it is neither non-contextual (as it would be the case in classical physics) nor fully contextual (then there would be no connection between different contexts, and thus no objective theory at all).

Another conclusion is that the usual state vector $|\psi\rangle$ is incomplete indeed, not due to any "hidden variable", but because it gives access to an actual physical modality $|\psi\rangle_A$ **only when the context $\hat{A}$ has been specified**. Correspondingly, $|\psi\rangle$ can be ($\rightarrow$is?) turned into a non-trivial ($p \neq 0, 1$) probability distribution only within a new measurement context, that does not admit $|\psi\rangle$ as a modality, otherwise one has again $p = 0$ or 1. This feature implies that $|\psi\rangle$ is predictively incomplete [11], and allows the violation of Bell's inequalities, without requiring any nonlocality at the elementary level [12].

On the other hand, the modalities $|\psi\rangle_A$, $|\psi\rangle_B$, $|\psi\rangle_C$ are indeed complete, with deep reaching consequences on all the usual "paradoxes" of QM. For instance, in the framework of the Einstein-Bohr debate, it means that $|\psi\rangle$ is incomplete indeed, as claimed by Einstein, Podolsky and Rosen, and that it must be completed by "the very conditions that allow future predictions" (i.e. the context, either $\hat{A}$, or $\hat{B}$, or $\hat{C}$), as claimed by Bohr [13].

The above ideas emphasize that a well defined quantum property, i.e. a modality, belongs to a system within a context. This is a quite objective statement, but clearly the system and the context are not described in the same way. A system with M mutually exclusive modalities is associated with a Hilbert space $\mathcal{H}$ of dimension M, whereas the context (as a CSCO) is associated with operators acting in $\mathcal{H}$. Such operators are constructed from macroscopic data, such as orientations or positions of the apparatus, and the whole construction makes clear that **both** systems and contexts are needed, and that there is no proper way to make one "emerge" from the other, although the consistent relationship between both will be explained below.

This leads to the (in)famous question of the "Heisenberg cut" between system and context and of the universality of the quantum description. Within the CSM framework there are two ways to answer this question :

– The first and simplest one is to postulate that quantum objects are made of systems within contexts [5, 6]. This fits quite well with usual textbook QM, and allows retrieving Born's rule from Gleason's theorem [8], as explained above. The main benefit here is that the usual "quantum paradoxes" just vanish: QM is certainly not classical, but it results from contextual quantization, which fits quite well within the usual physical realism [5, 6, 9] - though not within **classical** physical realism.

– A second way is to include both systems and contexts in the same description as incommensurable objects, along the lines introduced e.g. by von Neumann in [2]. More details are given in [1, 14], but this approach requires to use algebraic tools that are not considered in the textbook QM until now. More technically, the operators used in textbook QM belong to type-I algebras in the Murray-von Neumann classification, whereas the approach quoted here requires typically higher-order algebras, that are typically type-III [1, 14]. These algebraic tools open a way to include both systems and contexts in a unified framework. The price to pay is to "manipulate infinities", which happens to be quite possible mathematically, even though not popular in physics nowadays. We develop these ideas in the next section.

3. Modelling macroscopic quantum systems

This section assumes that the usual formalism derived from the contextuality postulates to describe microscopic quantum systems should in a certain manner still hold when considering macroscopic systems built out of a large number of microscopic quantum system. But one has to take into account some changes of the mathematical tools that they involve. This change, however, matches surprisingly well the properties of macroscopic systems.

3.1. Macroscopic limit modeling

The basic assumption we make to extend the description of microscopic quantum systems to macroscopic systems is reminiscent from statistical physics. In this domain, it is assumed that a macroscopic object is well described as a system where the number N of particles it contains $N \to \infty$. This allows e.g. using the central limit theorem as exact, as well as obtaining non-analytical correlation functions in critical systems.

This assumption has key consequences in our case. As a matter of facts, it means that the composite quantum system that we will be considering has now an infinite number of elements, and thus that the Hilbert space to accomodate its states is the result of an infinite tensor product of elementary Hilbert spaces. Even in the simplest case of an infinite chain of spins 1/2 where the elementary Hilbert spaces have a finite dimension 2, the dimension of the macroscopic

Journal of Physics: Conference Series **2533** (2023) 012008 doi:10.1088/1742-6596/2533/1/012008

Hilbert state space is $2^{\aleph_0} = \aleph_1$, i.e. has the power of continuum. This means that there is no countable basis that spans this macroscopic Hilbert state space, i.e. it is non-separable. The operators acting in this space form in most cases a type-III von Neumann algebra, that contains by construction no finite non-zero projection. Type III is used in quantum field theory and statistical physics, for systems with an infinite number of particles or degrees of freedom [14].

3.2. Main properties of macroscopic system state space

The purpose of this section is a pedestrian introduction to the properties of the Hilbert spaces obtained by the approach suggested in the previous section, i.e. infinite tensor products of separable Hilbert spaces that result from the $N \to \infty$ limit. Following [2], we will first outline the meaning of the convergence of an infinite product of complex numbers, then outline the construction of a Hilbert space that results from an infinite tensor product, that will be denoted as ITP. Then we will give the most important properties of such a Hilbert space, and of the physically relevant operators acting in it. For a more detailed presentation we refer to [1, 2].

Infinite complex products – The basic predictions of quantum physics lie in probability amplitudes, i.e. scalar products between states in a Hilbert space. As scalar products of tensor products are products of elementary scalar products, $(\langle\psi_1|\otimes\langle\psi_2|)(|\phi_1\rangle\otimes|\phi_2\rangle) = \langle\psi_1|\phi_1\rangle\langle\psi_2|\phi_2\rangle$, infinite tensor product require defining properly the existence of infinite products. Let I be an infinite set of indices (countable or not), and let $\{z_\alpha\}, \alpha \in I$ be a set of complex numbers labelled in I. The infinite product of the z_α's converges to complex value Z iff one can get as close as desired to Z with a finite partial product of selected z_α's with distinct α's. One can show that this value is unique. In the case where I is a countable ordered set, this is equivalent to the convergence of the series of the logarithm of the product to $\ln(Z)$. In terms of α values, it means that all but a finite number of the z_α's accumulate in the vicinity of 1 in the complex plane. A notion of quasi-convergence can be introduced usefully, that corresponds to accumulation of z_α's in the vicinity of the unit circle, with no constraint on the argument.

Infinite tensor products (ITP) – Let us consider an infinite set of separable Hilbert spaces (i.e. with finite dimension or countably infinite dimension) $\{\mathcal{H}_\alpha\}$. One defines a (quasi-) convergent sequence $\{|\psi_\alpha\rangle\}_{\alpha\in I}$ of vectors by selecting one ψ_α in each $\mathcal{H}_\alpha$ (assuming the axiom of choice) and requiring that $\prod_{\alpha\in I}\langle\psi_\alpha|\psi_\alpha\rangle$ (resp. quasi-)converges as defined above. One has to distinguish the trivially convergent sequences (where at least one of the factors is 0) from the non-trivial convergent sequences on which the following is built. With these notions in mind, one can define linear forms on the set of convergent sequences, first by selecting another convergent sequence $\{|\phi_\alpha\rangle\}_{\alpha\in I}$ and forming $\phi[\{|\psi_\alpha\rangle\}] := \prod_{\alpha\in I}\langle\phi_\alpha|\psi_\alpha\rangle$, and then by expanding this to a linear combination of such linear forms with N convergent sequences $\{|\phi^k_\alpha\rangle\}_{\alpha\in I,k=1,N}$ and N complex coefficients Φ_k defining $\Phi[\{|\psi_\alpha\rangle\}] := \sum_{k=1,N}\Phi^*_k\prod_{\alpha\in I}\langle\phi^k_\alpha|\psi_\alpha\rangle$. One can form the scalar product of such linear forms and show that their set is a Hilbert space. This scalar product allows defining a distance (the norm of the difference of vectors) and thus, taking $N \to \infty$, proceed with the topological completion of the set (just as the rational numbers are completed to build the real numbers). Then, the infinite tensor product of the $\mathcal{H}_\alpha$'s is defined as the dual of this completion of the set of linear forms, noted $\mathcal{H} = \otimes_{\alpha\in I}\mathcal{H}_\alpha$. So, not only is it well defined but it can be shown to be a metrisable Hilbert space. Although this construction appears rather lengthy, it is rigorous and it shows that the language of quantum physics keeps existing with an infinite number of elements.

Sectorisation – With the above definition, one can explore the properties of $\mathcal{H}$. One can define a relation between two vectors $|\Psi\rangle := \otimes_{\alpha\in I}|\psi_\alpha\rangle$ and $|\Phi\rangle := \otimes_{\alpha\in I}|\phi_\alpha\rangle$ in $\mathcal{H}$ as $|\Psi\rangle \sim |\Phi\rangle \Leftrightarrow \sum_{\alpha\in I}|\langle\phi_\alpha|\psi_\alpha\rangle - 1|$ converges. One can show that this is an equivalence relation. Let us call 'sectors' its equivalence classes. It can be shown that when the above two vectors

Journal of Physics: Conference Series **2533** (2023) 012008 doi:10.1088/1742-6596/2533/1/012008

belong to the same sector, it means that $|\phi_\alpha\rangle \neq |\psi_\alpha\rangle$ for only a finite number of α's. This means that the space spanned by the vectors of one sector is of countable dimension, and that there are uncountably many sectors. It can be proven that in each sector there is at least one vector which all elements are normed, so is itself normed. The most important property is that **sectors are orthogonal** i.e. if $|\Psi\rangle$ and $|\Phi\rangle$ are in different sectors, then $\langle\Psi|\Phi\rangle = 0$ and if they are in the same sector, $\langle\Psi|\Phi\rangle = 0$ means that at least one $\langle\psi_\alpha|\phi_\alpha\rangle = 0$. Beyond being orthogonal, their direct sum is $\mathcal{H}$. Finally, one can show that the sectorisation builds up progressively, in the sense that if $|\Psi\rangle$ and $|\Phi\rangle$ are not in the same sector, for any $\varepsilon > 0$ one can find a finite set $J \subset I$ of N indices α's, all distinct, so as to build $|\Psi_N\rangle := \otimes_{\alpha\in J}|\psi_\alpha\rangle$ and $|\Phi_N\rangle := \otimes_{\alpha\in J}|\phi_\alpha\rangle$ such that $|\langle\Psi_N|\Phi_N\rangle| < \varepsilon$. The physical meaning of this results will be detailed in the next section. Before doing so, let us explain the main properties of operators that act in $\mathcal{H}$.

Composite, bounded operators – Let us consider now the operators that are built from linear operators $\hat{A}_\alpha$'s that act in the elementary $\mathcal{H}_\alpha$'s. This requires first to define the completion of $\hat{A}_\alpha$ as $\tilde{A}_\alpha := \hat{A}_\alpha \otimes (\otimes_{\beta\neq\alpha}\hat{I}_\beta)$ (where $\hat{I}_\beta$ is the identity in $\mathcal{H}_\beta$). One can show that for each α, the bounded $\hat{A}_\alpha$'s form a ring $\mathcal{B}_\alpha$ and so the $\tilde{A}_\alpha$'s form a 'completed' ring $\tilde{\mathcal{B}}_\alpha$ (i.e. stable under addition and multiplication). One can then merge all the $\mathcal{B}_\alpha$'s in a larger ring $\mathcal{B}^\#$ that is the ring of all bounded operators on $\mathcal{H}$ that can be built from elementary operators, and that is in general a sub-ring of the ring $\mathcal{B}$ of all bounded operators on $\mathcal{H}$ (if I is finite, $\mathcal{B}^\# = \mathcal{B}$). The most important property is that **the operators of $\mathcal{B}^\#$ commute with the projectors on each sector**. In other words, operators built from elementary operators do not connect different sectors; however, more general operators connecting sectors do exist, but they cannot be formed "from the bottom". Finally, as for sectorisation, one can show that there is a gradual onset of this property. Let $\hat{A}$ be a bounded operator in $\mathcal{B}^\#$. If $|\Psi\rangle$ and $|\Phi\rangle$ are not in the same sector, for any $\varepsilon > 0$ one can find a finite set $J \subset I$ of N indices α's, all distinct, so as to build $|\Psi_N\rangle := \otimes_{\alpha\in J}|\psi_\alpha\rangle$, $|\Phi_N\rangle := \otimes_{\alpha\in J}|\phi_\alpha\rangle$ and the restriction $\hat{A}_N$ of $\hat{A}$ to the $\otimes_{\alpha\in J}\mathcal{H}_\alpha$ such that $|\langle\Psi_N|\hat{A}_N|\Phi_N\rangle| < \varepsilon$. Let us come now to the meaning these surprising mathematical properties have, if used as the language of quantum physics.

3.3. Consequences on the properties of macroscopic systems

Let us consider a quantum system made of N degrees of freedom, each described by a microscopic or elementary Hilbert space $\mathcal{H}_\alpha, \alpha = 1, ..., N$. The full Hibert space of the system is then $\mathcal{H} = \otimes_{\alpha=1}^N \mathcal{H}_\alpha$. As explained in section 3.1, we will consider that a macroscopic quantum system is well modelled, as in Statistical Physics, by the macroscopic limit $N \to \infty$. This means that the Hilbert space of this system shall have all the properties outlined in section 3.2. Let us expand on these properties now.

Sectorisation is the first significant property of infinite tensor products. In this case, it means that $\mathcal{H}$ splits into uncountably many orthogonal subspaces $\{\mathcal{C}\}$ each of which is of countable dimension. Figure 1 examplifies this with two-dimensional (blue or red) $\mathcal{H}_\alpha$'s. A the top, we have a given state $|\Psi\rangle$ that defines a sector $\mathcal{C}_\Psi$. In $|\Psi'\rangle$ in the middle, a finite number of elementary states (namely, 3) have been changed from the first state. At the bottom, an infinite number of elementary states have been changed to build $|\Phi\rangle$ (namely the state of every odd element). One sees that there is a countable infinite way of changing a finite number of states in this system. It could be even finite if one decided to change only the state of a fixed finite set of elements. The subspace spanned by the states where only a finite number of elements are changed is thus separable. The second state $|\Psi'\rangle$ thus remains in sector $\mathcal{C}_\Psi$. One can say that even though this first two states are *microscopically* different, they are *macroscopically* similar (e.g. 'blue'). Now considering the third state, an infinite number of states have been changed. By the above properties, this means that the bottom state $|\Phi\rangle$ is in a different sector $\mathcal{C}_\Phi$. It is thus orthogonal to the first two states, even if each changed elementary state is not

Journal of Physics: Conference Series **2533** (2023) 012008 doi:10.1088/1742-6596/2533/1/012008

orthogonal to its initial counterpart. For instance assuming the elements are spins $\frac{1}{2}$, and that $|\psi_\alpha\rangle = |\uparrow\rangle$ while $|\phi_\alpha\rangle = |+\rangle = (|\uparrow\rangle + |\downarrow\rangle)/\sqrt{2}$, then $\langle\psi_\alpha|\phi_\alpha\rangle \neq 0$, yet $\langle\Psi|\Phi\rangle = 2^{-N/4} \to 0$ in the macroscopic limit, so indeed $\langle\Phi|\Psi\rangle = 0$. One also sees that there is uncountably many ways to change an infinite number of elementary states in this way. There are thus uncountably many macroscopically different states.

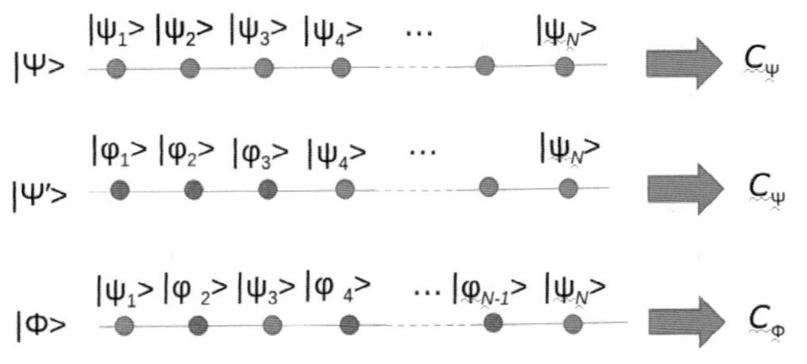

Figure 1. Illustration of sectors on a macroscopic chain.

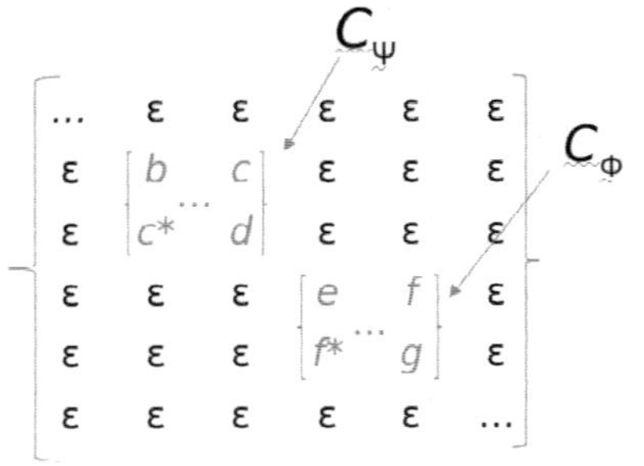

Figure 2. Typical shape of an operator built from elementary operators in an ITP.

Self-decoherence – Let us now turn to the main property of bounded operators built from microscopic operators, which is that they commute with the projectors on each sector in the macroscopic limit. This means that such a bounded operator does not connect states from different sectors. This has two consequences. First, building an operator such as a Hamiltonian as the combination of elementary, microscopic operators will yield operators that have vanishing elements between sectors, or that can be taken to arbitrarily small values by combining enough elementary subsystems, see Fig. 2. This is also true of evolution operators built this way, that will thus not take a state from one sector to another. This means in particular that evolution cannot lead a state in one sector to a superposition of states from different sectors. As a second example, the density operator of a large system will also have vanishing elements between sectors (or that can be taken to arbitrarily small values by combining enough elementary subsystems). This means that the larger the system, the smaller the correlations between sectors, so the macroscopic system behaves macroscopically as in a mixed state, with microscopic quantum correlations at most. This is reminiscent of Zeh-Zurek decoherence, apart from the fact that no assumption of information loss to the environment and partial trace to model it is needed. Decoherence appears thus to be an intrinsic property of large quantum systems.

Superposition of macroscopic system states can be considered to be produced by two different mechanisms, evolution (as in Rabi oscillations) or measurement (at least QND) on a non-orthogonal state. From the above considerations, one sees that macroscopic evolution from an initial state will not bring it to a superposed state of macroscopically different sectors. Similarly, a projective measurement on a non-orthogonal state does not yield any superposed result (or rather, has a vanishing probability amplitude to succeed) because the projection of the state of one sector on the state of another sector vanishes. This means that the usual processes to produce superposed states fail for macroscopic systems. Consistently with the block-diagonality of density operator, there is no way to have macroscopically superposed states – there are no such Schödinger cats. As superposition is the way to code the measurement indeterminism in the quantum formalism, this means that **macroscopic systems are deterministic**, while superpositions do make sense in the microscopic realm, that has a share of randomness.

Quantum measurements can be considered as involving a special case of macroscopic systems. As a matter of facts, in a typical experimental setup, we have a microscopic quantum system S and a macroscopic measurement device D which form a macroscopic system $S \cup D$. If the states of S are spanned by the M eigenstates $|u_k\rangle$ of some observable $\hat{A}$, the state of S can be written as $|\psi\rangle = \sum_{k=1}^{M} \psi_k |u_k\rangle$. The measurement device is made so that starting from an initial 'neutral' state $|D_0\rangle$, it transitions to a macroscopic state $|D_k\rangle$ that corresponds to state $|u_k\rangle$ of the system. Usually, one assumes that first, a unitary evolution takes $S \cup D$ from $|in\rangle := \sum_{k=1}^{M} \psi_k |u_k\rangle \otimes |D_0\rangle$ to a superposed state $|\text{out}\rangle := \sum_{k=1}^{M} \psi_k |u_k\rangle \otimes |D_k\rangle$ and second, that decoherence due to leak of entanglement to the device, the laboratory and the full universe, makes the density operator $|\text{out}\rangle\langle\text{out}|$ becomes diagonal (modelled by tracing over the non-system dimensions) and thus $S \cup D$ ends up in a mixed state.

The above considerations show that if modelling large systems with ITPs is correct, a new scenario has to be proposed for the quantum measurement process. A first pre-measurement step occurs where the system becomes entangled with ancilla degrees of freedom, which remain described with a separable Hilbert space. A second step then occurs where ancillae themselves become entangled with a number of secondary degrees of freedom (e.g. photons that probe the ancillae) that are selected so as to highlight their different states with a different behaviour. This involves adding energy in the $S \cup D$ (e.g to accelerate photoelectrons) and triggers a macroscopic cascade where the number of quantum degrees of freedom in interaction increases exponentially and reaches a extremely large number. If we assume that the resulting very large system is correctly modelled by infinite tensor products, it means that the various final states $|D_k\rangle$ of D (that need to be macroscopically different), lie in distinct sectors.

As a result, no unitary evolution will lead $S \cup D$ to a superposed state such as $|out\rangle$. On the contrary, as $S \cup D$ gets nearer to such a state, the off-diagonal terms of the density operator shall decrease steadily, and vanish before any such state is reached. At the end of the process, one of the $|D_k\rangle$ is probabilistically selected. The fact that this selection is random can be seen as forced by the absence of a one-to-one map between the separable state space of S and the non-separable state space of $S \cup D$, so there is no determined "way to go" for the evolution. This argument appears as an instantiation in the infinite case of the CSM reasoning, showing that quantum randomness is a necessary consequence of contextual quantization [8].

3.4. Connecting back the results with the postulates

The CSM approach allows building usual quantum mechanics thanks to a limited number of empirical, realist and objective postulates and two key theorems of linear algebra. It can be summarised the following way.

- *Quantum systems* are physical objects that can be characterised by physical quantities which value is obtained through measurements.
- *Contextuality:* Measurements are done by chosing a context, i.e. what concrete quantities are measured simultaneously. Once a context is chosen, a quantum system can provide only a limited amount ($M \leq \aleph_0$) of different issues to the measurements. These issues are the modalities of the system within the selected context.
- *Contexts* themselves are continuously infinitely many, they are self-contextual and they can provide as much information on themselves as desired.
- *Quantum randomness:* When changing the context, there is in general no certainty on what modality will be selected by a new measurement. Only probabilities can be predicted.
- *Extravalence:* When the number of mutually exclusive modalities (later on, the dimension of the Hilbert space) is at least 3, certainty and repeatability can be transferred through

modalities that belong to infinitely many different contexts, defining an equivalence relation between modalitities called extravalence. The usual state vector $|\psi\rangle$ or projector $|\psi\rangle\langle\psi|$ is associated with an extravalence class, not with a particular modality.

- *Unitary transformations* are provided by Uhlhorn's theorem applied to the above objects. Let us assume that M mutually orthogonal projectors are associated with the M mutually exclusive modalities within a given context, and that the same projector is associated to all modalities in an extravalence class. For consistency, the M projectors in a context should remain orthogonal under a map Γ corresponding to a change of context. Then Uhlhorn's theorem proves that Γ should be unitary or anti-unitary, showing that the structure of the Hilbert space is a consequence of the physical behavior required for the modalities.
- *Born's rule* is finally obtained by applying Gleason's theorem to the above objects, to provide the value of the probability to obtain a certain modality in a given context.

Let us compare the results of the previous section to these basic elements.

- Macroscopic systems that differ by an infinite number of elementary, microscopic elements can be in an uncountable (i.e. continuous) number of states. They cannot be superposed and thus are deterministic. In other words, they are classical and non-contextual. Contexts, as experimental device setups are such macroscopic systems, and what infinite tensor products tell is that they have precisely the properties postulated about them.
- Macroscopic systems that differ by a finite number of elementary, microscopic elements can be in in a finite, or at most in a countably infinite number of states. They are thus still contextual, and the finite part that differs can be seen as a quantum system. They are described by Hilbert space operators and Born's rule.

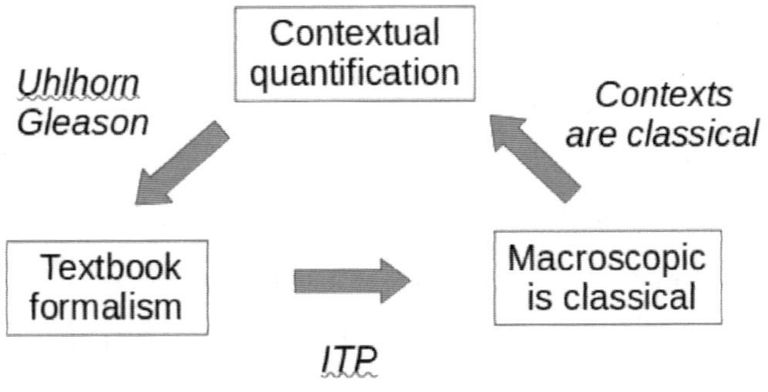

Figure 3. Closing the loop of CSM.

As a result, CSM, usual QM and ITP provide altogether a fully consistent conceptual setup to model both quantum systems and the classical systems they interact with, as illustrated in Fig. 3. CSM has the interest of being based on not-so-abstract ideas that can be used to develop an intuition on quantum systems, that is highly desirable to develop the emerging quantum technologies. This is in part thanks to its objective and realist basis. Finally, it leaves the much exotic aspect of randomness at the heart of the approach, even if it provides justifications for it. We also note that randomness seems to have a strong link with relativistic causality, and is thus all the more likely to be a fundamental aspect of our universe.

4. Conclusion

We have sketched a presentation of QM that relies on simple objective, realist and empirically understandable postulates. These postulates allow one to derive the usual textbook formalism up to dynamical equations. We push further the consequences by applying applying this formalism to the macroscopic limit that involves using type-III von Neumann algebras. The properties of these algebras allow understanding the behavior of large systems that fully support the initial assumptions. We highlight that the qualitative difference between quantum and classical system maps in this model on Cantor's hypothesis of continuum, i.e. that there is no other "infinity" between the discrete one ($\aleph_0$) and the continuous one ($\aleph_1$). The non-unitary and genuinely random aspect of measurement occurs on the verge of this limit, that occurs physically when a macroscopic cascade maps a quantum phenomenon on a classical one.

Acknowledgements. PG thanks Franck Laloë, Roger Balian, Nayla Farouki, Alexia Auffèves, and Ehtibar Dzhafarov for many discussions.

[1] Van Den Bossche M and Grangier P 2023 "Contextual unification of classical and quantum physics" *arXiv:2209.01463 [quant-ph], to appear in Found. Phys.*.
[2] von Neumann J 1939 "On infinite direct products" *Compositio Mathematica* vol 6 pp 1-77. http://www.numdam.org/item?id=CM_1939__6__1_0
[3] Budroni C, Cabello A, Gühne O, Kleinmann M, Larsson J-A "Quantum Contextuality" *arXiv:2102.13036*
[4] Grangier P 2002 "Contextual objectivity: a realistic interpretation of quantum mechanics" *European Journal of Physics* vol 23:3 p 331 (2002) [arXiv:quant-ph/0012122].
[5] Auffèves A and Grangier P 2016, "Contexts, Systems and Modalities: a new ontology for quantum mechanics", *Found. Phys.* vol 46 p 121 [arXiv:1409.2120].
[6] Auffèves A and Grangier P 2018 "Extracontextuality and extravalence in quantum mechanics", *Phil. Trans. R. Soc. A* vol 376 p 20170311 [arXiv:1801.01398].
[7] Auffèves A and Grangier P 2020 "Deriving Born's rule from an Inference to the Best Explanation", *Found. Phys.* vol 50 pp 1781-1793 [arXiv:1910.13738].
[8] Grangier P "Revisiting quantum mysteries", *http://arxiv.org/abs/2105.14448*
[9] Auffèves A and Grangier P "Revisiting Born's rule through Uhlhorn's and Gleason's theorems", *https://arxiv.org/abs/2111.10758*
[10] Jarrett J-P 1984 "On the physical significance of the locality condition in the Bell arguments", *Noûs* vol 18 p 569.
[11] Grangier P 2021 "Contextual inferences, nonlocality, and the incompleteness of quantum mechanics", *Entropy* vol 23:12 p 1660; https://doi.org/10.3390/e23121660
[12] Farouki N and Grangier P 2021 "The Einstein-Bohr debate: finding a common ground of understanding ?" *Found. Sci.* vol 26 pp 97-101 [arXiv:1907.11267]
[13] Grangier P 2021 "Completing the quantum formalism in a contextually objective framework", *Found. Phys.* vol 51 p 76 [arXiv:2003.03121]
[14] Cabello A, Estebaranz J-M and Garcia Alcaine G 1996 "Bell-Kochen-Specker theorem: a proof with 18 vectors", *Phys. Lett. A* vol 212 pp 183180.
[15] Mermin N D 1990 "Simple Unified Form for the Major No-Hidden-Variables Theorems", *Phys. Rev. Lett.* vol 65:27 p 3373.
[16] Khrennikov A 2016 "Randomness: quantum versus classical", *Int. J. Quantum Information* vol 14 p 1640009.
[17] Dzhafarov E N and Kujala J V 2016 "Context-content systems of random variables: The contextuality-by-default theory", *J. Math. Psych.* vol 74, pp 11-33.

Journal of Physics: Conference Series **2533** (2023) 012009 doi:10.1088/1742-6596/2533/1/012009

Quantum fluctuations, particles and entanglement: solving the quantum measurement problems

Kenichi Konishi

INFN, Sezione di Pisa, Largo Pontecorvo, 3, Ed. C, 56127 Pisa, Italy,
Department of Physics "E. Fermi", University of Pisa,
Largo Pontecorvo, 3, Ed. C, 56127 Pisa, Italy

E-mail: `kenichi.konishi@unipi.it`

Abstract. The so-called quantum measurement problems are solved from a new perspective. One of the main observations is that the basic entities of our world are *particles*, elementary or composite. It follows that each elementary process, hence each measurement process at its core, is a spacetime, pointlike, event. Another key idea is that, when a microsystem ψ gets into contact with the experimental device, factorization of ψ rapidly fails and entangled mixed states appear. The wave functions for the microsystem-apparatus coupled system for different measurement outcomes then lack overlapping spacetime support. It means that the aftermath of each measurement is a single term in the sum: a "wave-function collapse". Our discussion leading to a diagonal density matrix, $\rho = \mathrm{diag}(|c_1|^2, \ldots, |c_n|^2, \ldots)$ shows how the information encoded in the wave function $|\psi\rangle = \sum_n c_n|n\rangle$ gets transcribed, via entanglement with the experimental device and environment, into the relative frequencies $\mathcal{P}_n = |c_n|^2$ for various experimental outcomes $F = f_n$. Our discussion represents the first, significant steps towards filling in the logical gaps in the conventional interpretation based on Born's rule, replacing it with a clearer understanding of quantum mechanics. Accepting objective reality of quantum fluctuations, independent of any experiments, and independently of human presence, one renounces the idea that in a fundamental, complete theory of Nature the result of each single experiment must necessarily be predictable.

1. Quantum measurement problems

Quantum mechanics is described by the following few formulas,

$$[x_i, p_j] = i\hbar\,\delta_{ij}\ , \qquad \Delta x \Delta p \geq \frac{\hbar}{2}\ ;$$
$$i\hbar\frac{d\psi}{dt} = H\psi\ ;$$
$$H\psi_n = E_n\psi_n\ , \qquad H = \frac{\mathbf{p}^2}{2m} + V\ , \qquad \mathbf{p} = -i\hbar\nabla\ . \tag{1.1}$$

The rest is, essentially, just details. In spite of the fundamental remarkable simplicities, quantum mechanics (QM) has led to

(i) An amazing success and detailed confirmation in atomic physics, ushering us into a century

Journal of Physics: Conference Series **2533** (2023) 012009 doi:10.1088/1742-6596/2533/1/012009

of impressive advance in basic scientific knowledge and modern technologies, *unprecedented in human history.* We still live in the wake of this ongoing revolution.

(ii) Further pursuit into subatomic physics has eventually led to the standard model of the fundamental interactions [1]- [4], a non-Abelian gauge theory based on the gauge group,

$$SU(3)_{QCD} \times \{SU(2)_L \times U(1)\}_{GWS}, \tag{1.2}$$

(GWS stands for Glashow-Weinberg-Salam, and QCD for Quantum Chromodynamics), describing very accurately the strong (nuclear) interactions and electroweak interactions, down to the distance scales of order of $O(10^{-18})$ *cm.*

(iii) One cannot forget about the beautiful phenomena in condensed matter physics, such as superconductivity, quantum Hall effect, Bose-Einstein condensation of cold atoms, etc. etc.

We can certainly talk about

100 years of extraordinary scientific achievements.

But, at the same time, the QM predictions are, supposedly, given by the probabilistic (Born's) rule. And this has led to

100 years of uneasy feeling and doubts,

that something fundamental is missing in our understanding of QM. These are often expressed as various (apparent) puzzles or conceptual difficulties, which are all loosely called the "quantum measurement problems" [5]- [7]. They are actually a collection of different questions:

(a) Wave function collapse?

(b) Macroscopic superposition of states?

(c) Forever branching manyworlds tree?

(d) Spontaneous collapse of micro state, of the experimental device, and of the world?

(e) Wave function just a bookkeeping device?

(f) Quantum jumps vs Schrödinger equation?

(g) Born's rule from the Schrödinger equation involving everything?

(h) Quntum nonlocality, Hidden variables?

(i) How is it possible that the fundamental, complete theory of physics is incapable of unique prediction?

The following discussion, based on [8], will address all of them.

The actual measurement devices can vastly differ in their nature, size, materials used, and technologies employed, from a simple photographic plate, cloud and bubble chambers filled with liquid or vapour, spark chambers and MWPC made with metal plates, wires and gas, the neutrino detectors made of a huge tank of pure water and tens of thousands of photomultipliers, to the state-of-the-art silicon detectors and some future apparatus which uses superconducting materials for dark matter search. See Fig. 1. Such an enormous diversity of the experimental devices requires that an equally vast simplification be made, to capture the essence of quantum measurements.

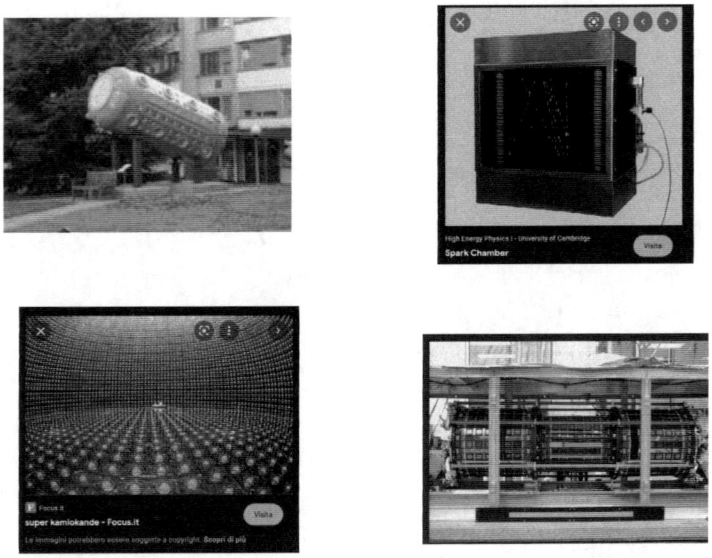

Figure 1: A few examples of quantum measurement devices.

Indeed, independently of the details, a good experimental device faithfully reflects the quantum fluctuations of the system, described by the wave function $|\psi\rangle$. Indeed, the result of each single experiment is, in general, apparently random and unpredictable [1]. Yet, the information encoded in the wave function $|\psi\rangle$ manifests itself as the expectation values for any variable [2] F (and all functions thereof),

$$\bar{F} \equiv \langle\psi|F|\psi\rangle \ , \tag{1.3}$$

that is, as a (frequency-) average of the experimental outcomes, $\bar{f}_n$,

$$\bar{f}_n = \sum_n f_n \mathcal{P}_n \ , \qquad \mathcal{P}_n = |\langle n|\psi\rangle|^2 \ , \tag{1.4}$$

where $F|n\rangle = f_n|n\rangle$. $\mathcal{P}_n$ is the relative frequency for f_n. Such a prediction of quantum mechanics is verified by countless experiments. That is, "QM works".

The prediction (1.4) follows, if one assumes that the *probability* of each single measurement

[1] An exception occurs when the state ψ is one of the eigenstates of the quantity F, with $F = f_m$, i.e., $|\psi\rangle = |m\rangle$. In such a case, a good experiment produces the same result $F = f_m$ every time.

[2] Throughout, we use simply the word "variable", for a dynamical variable, physical quantity, such as energy, momentum, angular momentum, position, etc., which is measurable (known as an observable). Also, we use the same symbol F for the physical variable itself and for the self-adjoint operator representing it. The alert reader will not have any difficulty telling which is meant, each time.

of F in the state ψ to give f_n, is given by $P_n = |\langle n|\psi\rangle|^2$, that is, by Born's rule. Born's rule is presented in most textbooks as one of the fundamental postulates of quantum mechanics [3].

So what *is* the problem? Well, "the quantum measurement problems", **(a)** $\sim$ **(i)**, above!

Throughout, we will denote the *(normalized) relative frequency* with the calligraphy font $\mathcal{P}_n$, and write the standard *probability* by using the normal font P_n. Although they express mathematically the same numbers, they differ conceptually, and lead to distinct logical conclusions.

2. Three aspects of QM

Before analyzing the measurement processes themselves in Sec. 3, we must first discuss briefly three familiar aspects of QM. Though singly well known, clarifying their precise roles in the measurement processes will be essential in the following discussions.

2.1. Expectations values versus Born's rule

To fix the idea, we consider the system described by the wave function,

$$|\psi\rangle = \sum_n c_n\,|n\rangle\ , \qquad F|n\rangle = f_n|n\rangle\ , \qquad \sum_n |c_n|^2 = 1\ , \tag{2.1}$$

where F is the variable we are interested in. Repeated, identical measurement of F will give an average, (1.4).

Now, (1.4) certainly follows from Born's rule, (1). In other words, the latter is a sufficient condition for the former. The problem is: is Born's rule also necessary?

The answer is yes, if the following equality holds

$$\bar{O(f_n)} \equiv \sum_n \mathcal{P}_n O(f_n) = \langle\psi|O(F)|\psi\rangle \equiv \sum_n |c_n|^2 O(f_n)\ , \tag{2.2}$$

for general function $O(F)$ of F. Then it can be shown that

$$\mathcal{P}_n = |c_n|^2 = P_n\ . \tag{2.3}$$

The proof and illustrations are given in [8].

2.2. Pure and mixed states

A system described by a wave function (or a state vector) ψ is a pure state. It contains the complete knowledge of the system.

Otherwise, the system is in a mixed state (or a mixture). It is described by the density matrix, ρ.

For instance, a subsystem $A \subset \Sigma$ of a closed system Σ, described by the wave function,

$$|\Psi\rangle = \sum_{n,\alpha} c_{n,\alpha}|n\rangle|\alpha\rangle\ , \qquad \sum_{n,\alpha} |^2 = 1\ , \tag{2.4}$$

[3] A rare exception is the famous book by Dirac (1958) [9], where slightly difference nuance is used.

Journal of Physics: Conference Series **2533** (2023) 012009 doi:10.1088/1742-6596/2533/1/012009

the expectation value of G (pertinent to the subsystem A) is given by

$$\bar{G} = \langle\Psi|G|\Psi\rangle = \text{Tr}\,\mathbb{G}\rho\,, \qquad \mathbb{G}_{mn} \equiv \langle m|G|n\rangle\,, \tag{2.5}$$

where the density matrix is given in this case by

$$\rho_{nm} \equiv \sum_\alpha c_{n,\alpha} c^*_{m,\alpha}\,. \tag{2.6}$$

More generally the density matrix represents any sort of ignorance about the system, and will have forms different from (2.6), but the trace formula

$$\bar{G} = \text{Tr}\,\mathbb{G}\rho\,, \tag{2.7}$$

is valid always.

2.3. Factorization vis-à-vis entanglement

The question is, symbolically, this: how can we study a single hydrogen atom, an electron, etc., when we know that the wave function involving identical fermions must be antisymmetrized (Fermi-Dirac (FD) statistics)?

The answer is the lack of the spatial support in the wrong component. For instance, for an electron in the laboratory and another in the Sun, the wave function will look like

$$\Psi = \frac{1}{\sqrt{2}}\big(\psi_{lab}(\mathbf{r}_1)\psi_{sun}(\mathbf{r}_2) - \psi_{lab}(\mathbf{r}_2)\psi_{sun}(\mathbf{r}_1)\big) \sim \psi_{lab}(\mathbf{r}_1)\,, \tag{2.8}$$

for

$$\mathbf{r}_1 \in Lab.\,, \qquad \mathbf{r}_2 \in Sun \tag{2.9}$$

as clearly there is no support for the wrong component, $\psi_{lab}(\mathbf{r}_2)$ or $\psi_{sun}(\mathbf{r}_1)$.

One sees how a pure state emerges as the result of factorization.

(i) If the anstisymmetrization is in the spin state, instead of the position states, then the wrong component is negligible, but $\psi_{lab}(\mathbf{r}_1)$ is in a spin mixed state;

(ii) If one takes two hydrogen atoms (two electrons) $\sim 2\,m$ apart, in the next room in the laboratory, instead of the second in the Sun, the factorization failure (the effect of the wrong component) is still very small,

$$\sim 10^{-18}\,, \tag{2.10}$$

as it follows by using Bohr's radius! And this gives some idea about the exactness of some of our statement below, even if some of the argument about the spacetime localization below concerns *spacetime support* of the wave functions, rather than the spatial support, considered here.

(iii) Actually, in spite of the impression this example might have given to the readers, the FD or BE statistics are not fundamental for the purpose of explaining entanglement. *Any systems* which have interacted in the past are entangled, and in general, in a mixed state.

We note that factorization and entanglement are two faces of the same medal, both characerizing quantum mechanics universally. Note in particular that

- **Factorization** works remarkably strongly; it makes QM a workable, sensible, physics theory, in spite of the fact that everything is in principle entangled in our universe. It is **factorization** that makes the pure state a significant concept, both from theoretical and experimental points of view;
- **Entanglement** leads to the fascinating, characteristic phenomena in QM, such as quantum nonlocality, violation of Bell's [7] and CHSH [10] inequalities, quantum cryptgraphy, etc. At the same time, the uncontrolled or uncontrollable entanglement with the unobserved systems, is responsible for docoherence, and emergence of classical physics [11]- [16].

A reflection

In principle, we always live in a mixed state: everything is interrelated, and interacted or interacting with each other, in our universe. It is the notion of a pure quantum state, described by a wave function ψ, which is extraordinary, and exceptional. *They arise as a result of factorization.*

These pure states are carefully prepared by experimentalists in many small bubbles in the world, called physics laboratories, equipped with a good clean room and sofiscated (and expensive) vacuum and cryogenic technologies.

But they occur also naturally, on the earth, in the form of radioactivity, which produces α (α particles), β (electrons) or γ (photons) rays. They also fall from the sky, in the form of cosmic rays (energetic protons, pions, etc.). Then there was light. Light, as it turns out, is a gas of non-interacting, thus, free, photons. They are all pure quantum states.

In a hindsight, we see why these phenomena (providing pure quantum states for free!) have led to the discovery of quantum physics by Planck, at the dawn of the 20th century.

3. The measurement processes

Now we come to our main problem: the quantum measurement processes.

3.1. Effective spacetime localization - state-vector reduction

The measurement of the variable F in the state

$$|\psi\rangle = \sum_n c_n\,|n\rangle\ , \qquad F|n\rangle = f_n|n\rangle\ , \tag{3.1}$$

is often assumed to proceed as

$$|\psi\rangle \otimes |\Phi_0\rangle \otimes |Env_0\rangle \;=\; \left(\sum_n c_n|n\rangle\right) \otimes |\Phi_0\rangle \otimes |Env_0\rangle \tag{3.2}$$

$$\longrightarrow\; \left(\sum_n c_n|n\rangle \otimes |\Phi_n\rangle\right) \otimes |Env_0\rangle \tag{3.3}$$

$$\longrightarrow\; \sum_n c_n|n\rangle \otimes |\Phi_n\rangle \otimes |Env_n\rangle\ , \tag{3.4}$$

where in the first step (3.3) the experimental device reads the measurement results, f_n, and in the final step (3.4) the "environment" comes be aware of that result (the experimentalist has seen the result on her/his computer screen). These formulas appear to suggest a coherent superposition of distinct macroscopic states leading to paradoxes, debates and endless confusion (the notorious

Journal of Physics: Conference Series **2533** (2023) 012009 doi:10.1088/1742-6596/2533/1/012009

example of this being Schrödinger's cat "paradox", see below).

Actually, the factorized form in (3.2)-(3.4), with various $\otimes$ symbols, is not correct, except for the factorization of $|\psi\rangle$ in the first line, namely, before the experiment.

We note that the factorized form $|\Phi\rangle \otimes |Env\rangle$ is incorrect, even before the measurement. The air molecules, the casing of the device, etc., all show that $|\Phi\rangle$ and $|Env\rangle$ are actually entangled. Not only, the boundaries between $|\Phi\rangle$ and $|Env\rangle$ are badly defined, and to be regarded as just a convention. Nevertheless, any experimentalist knows what the relevant part of the device is. She (or he) knows perfectly well that there is no need to push the boundary between the measuring device $|\Phi\rangle$ and the rest of the world $|Env\rangle$ up to inside the human brain, in order to ensure good, precise measurement results.

Keeping this state of matter in mind, we denote the measurement-device-environment entangled "state", as

$$|\Phi; Env\rangle \;, \tag{3.5}$$

below. The measurement process can be represented more appropriately as [4]

$$|\psi\rangle \otimes |\Phi_0; Env_0\rangle \;=\; \Big(\sum_n c_n |n\rangle\Big) \otimes |\Phi_0; Env_0\rangle \tag{3.6}$$

$$\longrightarrow \sum_n c_n \, |\tilde{\Phi}_n; Env_0\rangle \tag{3.7}$$

$$\longrightarrow \sum_n c_n \, |\tilde{\Phi}_n; Env_n\rangle \;. \tag{3.8}$$

where $\tilde{\Phi}_n$ stands for the entangled state of the microsystem-apparatus with the reading of the measurement result, f_n. The exact timing of passage from the first stage of measurement-registering of the result on the apparatus (3.7) to the second (3.8) (e.g., the moment in which the experimentalist sees the result on her or his computer screen; others read about it in Physical Review), is entirely immaterial.

Emergence of the mixed state and the state-vector reduction (the "wave function collapse") occurs as follows.

Just before the measurement, the system is described by a wave function of factorized form by assumption,

$$|\Psi\rangle_{t=0_-} = \Big(\sum_n c_n |n\rangle\Big) \otimes |\Phi_0; Env_0\rangle \;. \tag{3.9}$$

The expectation value of any generic variable G in this state (prior to the measurement) is given by

$$\bar{G} = \langle \Psi | G | \Psi \rangle = \mathrm{Tr}\, \mathbb{G} \rho \;, \qquad (\mathbb{G})_{mn} \equiv \langle m | G | n \rangle \;, \tag{3.10}$$

[4] The device-enviroment "state", $|\Phi_0; Env_0\rangle$, can never be identical, at two different measurement instants, see discussions below (3.19).

Journal of Physics: Conference Series **2533** (2023) 012009 doi:10.1088/1742-6596/2533/1/012009

with the density matrix having the form

$$\rho = \begin{pmatrix} |c_1|^2 & c_1 c_2* & c_1 c_3^* & \dots & \dots \\ c_2 c_1^* & |c_2|^2 & c_2 c_3^* & \dots & \\ c_3 c_1^* & & \ddots & & \\ \vdots & & & |c_n|^2 & \\ & & & & \ddots \end{pmatrix} , \qquad (3.11)$$

characteristic of a pure state, i.e., $\rho_{nm} = c_n c_m^*$. The normalization condition

$$\langle \Phi_0; Env_0 | \Phi_0; Env_0 \rangle = 1 \qquad (3.12)$$

has been used. In the case of the particular variable F (whose eigenstates are $|n\rangle$'s), $\mathbb{F}$ is diagonal, and its quantum average is given by the known formula

$$\bar{F} = \mathrm{Tr}(\mathbb{F}\rho) = \sum_n f_n |c_n|^2 \,. \qquad (3.13)$$

As soon as the microsystem gets into contact with the experimental device, and the measurement events (a chain ionization process, hadronic cascade, etc.) have taken place, the total wave function takes an entangled form, (3.7)

$$|\Psi\rangle_{t=0_+} = \sum_n c_n |\tilde{\Phi}_n; Env_0\rangle \,. \qquad (3.14)$$

$\tilde{\Phi}_n$ denotes the entangled microsystem-apparatus state with the reading of the measurement result, f_n.

Now, the wave function describing $\tilde{\Phi}_m$ (the aftermath of a measurement, with the result, $F = f_m$) and that for $\tilde{\Phi}_n$ (the aftermath of a measurement, with the result, $F = f_n$), corresponding to two distinct spacetime events, have no overlapping spacetime support, as illustrated in Fig. 2. (3.14) is a mixed state.

Treating the formula (3.14) as a mixed state, and not a coherent superposition of states, eliminates at once the contradiction found in a recent paper [17]. The argument of [17] may be regarded as an independent confirmation of the fact that the state after the measurement, (3.14), cannot be a coherent superposition [5].

The expectation values of a variable G are then given by

$$\bar{G} = \langle \Psi | G | \Psi \rangle = \sum_{m,n} c_m^* c_n \langle \tilde{\Phi}_m; Env_0 | \, G \, | \tilde{\Phi}_n; Env_0 \rangle \,; \qquad (3.15)$$

but the fact $\tilde{\Phi}_m$ and $\tilde{\Phi}_n$ $(m \neq n)$ have no common spacetime support, means that *for any local*

[5] A possible issue in [17] could be the fact that the information transmission from the first laboratory to the second about the result of the measurement made in the first, might render invalid the assumption that the two laboratories are isolated quantum systems.

Journal of Physics: Conference Series **2533** (2023) 012009 doi:10.1088/1742-6596/2533/1/012009

operator G, orthogonality relations,

$$\langle \tilde{\Phi}_m; Env_0 | \, G \, | \tilde{\Phi}_n; Env_0 \rangle = 0 \;, \qquad m \neq n \;, \tag{3.16}$$

hold. Consequently, $\bar{G}$ is given by the sum of the diagonal terms

$$\bar{G} = \sum_n |c_n|^2 G_{nn} \;, \qquad G_{nn} = \langle \tilde{\Phi}_n; Env_0 | \, G \, | \tilde{\Phi}_n; Env_0 \rangle \;. \tag{3.17}$$

This means that the density matrix has been reduced to a diagonal form,

$$\rho = \begin{pmatrix} |c_1|^2 & & & & \\ & |c_2|^2 & & & \\ & & \ddots & & \\ & & & |c_n|^2 & \\ & & & & \ddots \end{pmatrix} . \tag{3.18}$$

For the particular case of the variable F, we recover the standard prediction,

$$\bar{F} = \sum_n |c_n|^2 f_n \;, \qquad \therefore \;\; \mathcal{P}_n = |c_n|^2 \;, \tag{3.19}$$

where $\mathcal{P}_n$ are the normalized relative frequencies for different outcomes f_n.

The emergence of the mixed state (3.18) is often attributed to the fact that a typical experimental device is made of a macroscopic (difficult-to-specify) number of atoms and molecules, and it is not possible to keep track of the phase relations among different terms in (3.7) to any significant extent.

Also, the macroscopic device Φ, evolving in t entangled with Env, can never be in an identical quantum state $|\Phi_0; Env_0\rangle$ at two different measurement instants. This is in stark contrast with the *identical quantum state* for the microsystem, $|\psi\rangle$, which can be and is indeed produced via e.g., a repeatable experiment (state preparation - see Sec. 3.2) for each measurement.

The observations above are both certainly correct, but we need another, crucial ingredient for the decoherence in the measurement processes: the lack of the common spacetime supports in the wave functions, and the consequent orthogonality and decoherence among terms corresponding to different measurement results, (3.16). Importance of this is that *it implies that the result of each measurement event is a state-vector reduction,*

$$\Big(\sum_n c_n |n\rangle \Big) \otimes |\Phi_0; Env\rangle \Longrightarrow |\tilde{\Phi}_m; Env\rangle \;: \tag{3.20}$$

i.e., with a single term present, the instant after the measurement (e.g., with $F = f_m$*). This state of affair is perceived by us as a "wave-function collapse".*

Summarizing:

A: The spacetime eventlike nature of the triggering particle-measuring-device interactions introduces an effective spacetime localization of each measurement event;

B: At the moment the microsystem - a pure quantum state $|\psi\rangle$ - gets into contact with the

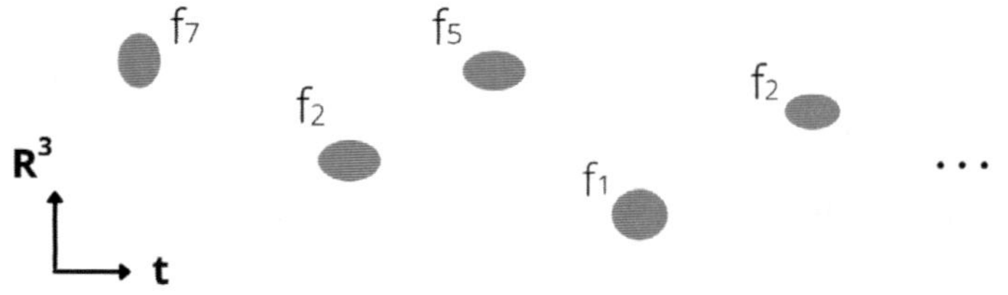

Figure 2: Measurement of a variable F. Each blob describes a single measurement event (a chain ionization reaction, hadronic cascade, etc.) occurring at a localized region in spacetime, with the experimental results, $F = f_7, f_2, f_5, f_1, f_2$, etc. The wave functions describing the different measurement events $\tilde{\Phi}$ have no overlapping spacetime supports, therefore are orthogonal.

measurement device-environment state $|\Phi_0; Env_0\rangle$ which is a mixed state, factorization of $|\psi\rangle$ gets lost rapidly, and an entangled, mixed (classical) state $|\Phi_n; Env_0\rangle$ with the unique recording, $F = f_n$, is generated.

The result is the state-vector reduction, (3.20).

The effective spacetime localization and apparent wave-function collapse, can be nicely visualized in the "double-slit" experiments by Tonomura et. al, [18], see Fig. 3.

3.2. Repeatable, nonrepeatable and intermediate types of state reductions

There are actually several types of measurements. The first is known as *repeatable, or of the first kind*: it proceeds as

$$\left(\sum_n c_n|n\rangle\right) \otimes |\Phi_0; Env_0\rangle \longrightarrow |m\rangle \otimes |\Phi_m; Env\rangle \ , \tag{3.21}$$

which corresponds to the expression often found in a QM textbook,

$$|\psi\rangle = \left(\sum_n c_n|n\rangle\right) \longrightarrow |m\rangle \ . \tag{3.22}$$

In these experiments, the measuring device has acquired and recorded the information about the microsystem under observation, but the latter remains unscathed, as a factorized, pure

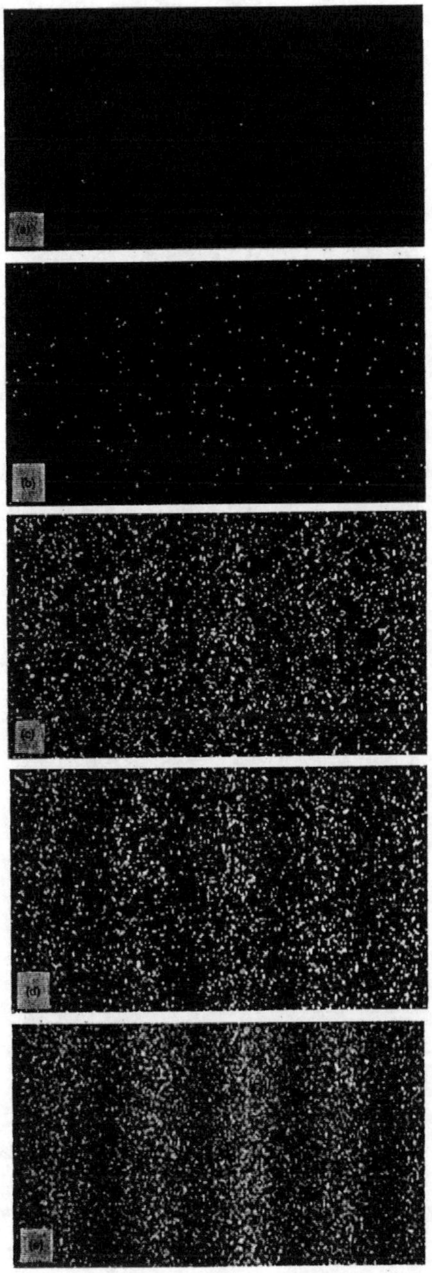

Figure 3: The pictures correspond, from the top to the bottom, to successive exposures to 10, 100, 3000, 20000 and 70000 electrons, respectively. Reproduced from [18], A . Tonomura, J . Endo, T . Matsuda, T . Kawasaki and H. Ezawa, "Dimonstration of single-electron buildup of interference pattern", American Journal of Physics 57, 117 (1989); doi: 10.1119/1.16104, with the permission of the American Association of Physics Teachers.

Journal of Physics: Conference Series **2533** (2023) 012009 doi:10.1088/1742-6596/2533/1/012009

state. These do represent special class of measurements, but they are actually neither rare nor particularly difficult to realize. A variant of the Stern-Gerlach experiment (see Fig. 4) is a good example.

The very possibility of the repeatable measurements is actually of fundamental importance for QM. They allow the preparation of a desired, *identical*, pure quantum state $|\psi\rangle$ as the initial state, whenever we wish. This is important because in QM the experimental check of the theory is done necessarily with many repeated, *identical* experiments, as QM predicts only the relative frequencies for various outcomes [6].

In contrast, a macroscopic body made of, say, 10^{25} atoms and molecules (e.g., an experimental device), can never be in an identical quantum state at different times or in different laboratories, even though they might look identical at macroscopic scales. This is in fact one of the underlying reasons for the state-vector reduction (see the discussion above, (3.20)).

The second type of the measurement may be simply termed *non repeatable.* This is a general type of the measurement - state reduction, as in (3.20). The result $F = f_m$ is recorded, but the original microsystem itself (e.g., the electron in Tonomura's experiment) simply gets lost, in $|\tilde{\Phi}_m; Env\rangle$.

There are also third, *intermediate types* of measurements. The detection of particle tracks in Wilson, or bubble chambers, or in the silicon vertex detectors, is an example of the measurement of this sort. Note that in these experiments the microscopic particle interacts with the atoms in the devise, generates a local chain ionization cloud, but its information (the position or momentum) does not get lost completely, creating successive ionization clouds (particle tracks) on its way. This kind of processes is particularly important in the momentum or energy measurement.

3.3. Unitarity and linearity

The microscopic state ψ

$$|\psi\rangle = \sum_n c_n |n\rangle \,, \tag{3.23}$$

if *left undisturbed*, evolves in time according to the Schrödinger equation,

$$i\hbar \frac{d}{dt} |\psi\rangle = H\,|\psi\rangle \,, \tag{3.24}$$

(H being the Hamiltonian) or as

$$|\psi\rangle \longrightarrow U\,|\psi\rangle = \sum_n c_n\, U|n\rangle \,, \qquad U = e^{-iHt/\hbar} \,, \quad U^\dagger U = \mathbf{1} \;: \tag{3.25}$$

it is a linear, unitary evolution. Unitarity means that

$$\langle\psi|\psi\rangle = \sum_n |c_n|^2 = 1 \,, \tag{3.26}$$

[6] Another, related fact is that the atoms of the same kind, in their ground state, are all rigorously identical, in contrast to classical particles. Such a characteristics of quantum mechanical world is at the basis of the regularity of the macroscopic world, such as crystalline structures and extremely precise biological phenomena such as the reproduction.

Journal of Physics: Conference Series **2533** (2023) 012009 doi:10.1088/1742-6596/2533/1/012009

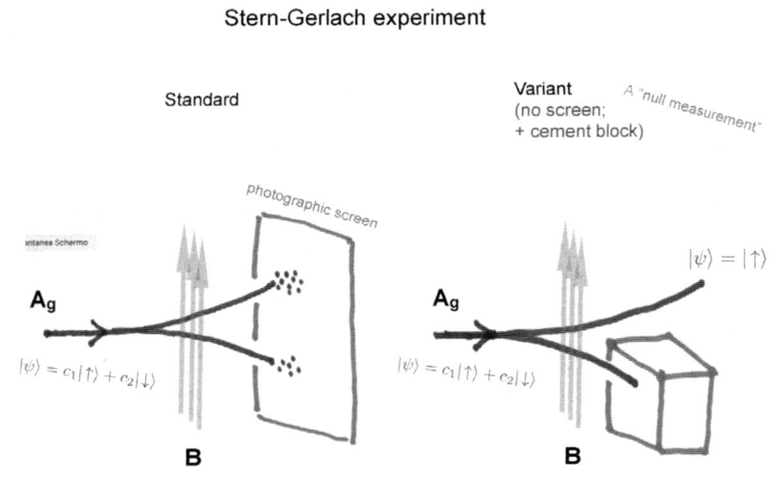

Figure 4: A variant (right) of the standard SG experiment (left). A cement block stops the lower beam containing the spin-down atoms. No measurement is done (known as a "null measurement"), but the atoms coming out of the region consist purely of spin-up state, which can be used as the initial wave function $|\psi(0)\rangle = |\uparrow\rangle$.

i.e., the state norms are conserved in time. Linearity means (3.25), i.e., superposition of different states in (3.23) continue to be coherent superposition of the corresponding states.

During the measurement process, formally written as (3.7), the time evolution might still look unitary and linear. However, as we have seen, the wave functions associated with different terms in (3.7) do not overlap in spacetime, so it represents actually a mixed state. The coherent superposition of states is no longer there, see (3.16). Most significantly, the real time evolution of the system is the state-vector reduction, (3.20), meaning that *linearity in the sense of (3.25) is lost during the measurement.*

On the other hand, unitarity is maintained [7]: the sum of all possible outcomes, occurring at different measurements at different times, adds up to unity of the total normalized frequencies, corresponding to the "norm" of the state, (3.7). In terms of the relative frequencies for different outcomes, $\mathcal{P}_n$, unitarity means

$$\sum_n \mathcal{P}_n = 1 \,, \qquad \sum_n |c_n|^2 = 1 \,, \tag{3.27}$$

even though different (sometimes the same, but repeated) results f_n refer to distinct measurements *made at different times.*

Note that this is indeed how experimentalists view the meaning of unitarity. Unitarity means

[7] Note that linearity and unitarity are two distinct concepts in quantum mechanics (e.g., the momentum operator is linear but not unitary). Dirac noted (Chap 27 of [9]) that the evolution operator must satisfy both, which are two independent requirements. They are both automatically met, once the choice $U = e^{-iHt/\hbar}$ is made with a self-adjoint Hamiltonian operator, and as long as the pure-state evolution (3.25) *before* the measurement, is concerned.

that $\mathcal{N}_n/\mathcal{N}$ [8], to be identified with the theoretical formula, $\mathcal{P}_n = |c_n|^2$, should satisfy

$$\sum_n \mathcal{N}_n/\mathcal{N} = 1 \,, \tag{3.28}$$

which might look trivial. However, this contains an implicit, important assumption that the experimental device has no systematic bias, and registers all possible results f_n with equal efficiency and with no losses. Only in such an ideal measurement setting we can expect that the experimental results $\mathcal{N}_n/\mathcal{N}$ will approach the theoretical prediction, $\mathcal{P}_n$, in the limit of *large* $\mathcal{N}$ (e.g., Tonomura's experiment, Fig. 3).

It might be of some interest to compare the concept of unitarity in the measurement processes as described here, with a more abstract one, in the conventional thinking based on Born's rule. The latter means in fact

$$\sum_n P_n = 1 \,, \tag{3.29}$$

i.e., that the total *probability* for a single experiment to give all possible results f_n, is unity, and that it is conserved in time. The request (3.29), from the mathematical, logical point of view, appears quite indispensable, and indeed has always been considered as a sacrosanct principle in the traditional approach to quantum mechanics.

However, once P_n is translated into the physically directly meaningful quantity, $\mathcal{P}_n$ $(= P_n)$, the expected relative frequencies for various results f_n which will be found in distinct measurements and at different times, unitarity as an absolute sacred principle may appear to lose part of the *aura* surrounding it.

To conclude, one must be cautious in applying the concepts such as linearity or unitarity of the time evolution, valid in the context of (isolated) pure states, to the measurement processes where entanglement between the microsystem and the measurement device as well as with the whole world, plays an essential role. Their consequences in these, complicated mixed states may well (and indeed, do) look differently from what one is accustomed to, from the study of pure quantum states kept in isolation, (3.24), (3.25).

3.4. Measurements, emergence of classical physics, macroscopic quantum states, and models of wave-function collapses

The state after a measurement, $|\tilde{\Phi}_m; Env\rangle$ (in (3.20), is an entangled, *classical* state, with the recording of the measurement result. The basic reason for that is the so-called envioronment-induced decoherence, studied in [11]- [12]. Emergence of classical physics in this context, is certainly one of the important ingredients in the understanding of the quantum measurement processes (the unique aftermath of each measurement - the wave function collapse).

In author's view, however, it is useful to consider the whole problem from a wider perspective of what may be called "great twin puzzles of physics today": the general quantum measurement problems [5]- [7] on the one hand, and emergence of the classical mechanics from QM, on the other. Even though these two classes of the problems are often discussed together [11]- [16], they are actually mostly independent issues [9]. It is better to discuss them separately, and independently.

[8] $\mathcal{N}_n$ is the number of times the experimentalist finds the result $F = f_n$, $n = 1, 2, \ldots$; $\mathcal{N}$ is the total number of the measurements made.

[9] This is so, even if there are a few key questions which link the two classes of the problems: the classical behavior of the measurement devices mentioned above is one of them. Another is the concept of the center-of-mass (CM)

Here we are concerned with the first of them, the general measurement problems; for the second (how Newton's law emerges from QM), see a recent work, [19].

For instance, we know that sufficiently close to $T = 0$, *any* matter, in any state, is quantum mechanical. At $T = 0$ any system is in the unique ground state (Planck-Nernst law, or the first law of thermodynamics).

At $T > 0$ but sufficiently close to the absolute zero temperature, a macroscopic body can be in a coherent superposition,

$$|\Psi\rangle = c_1|\Psi_1\rangle + c_2|\Psi_2\rangle \tag{3.30}$$

where $|\Psi_1\rangle$ and $|\Psi_2\rangle$ are macroscopically distinct states. For experimental efforts to realize macroscopic (or mesoscopic) quantum states at very low temperatures, see [20]- [35].

To talk about the macroscopic quantum states as (3.30) *in connection with* the measurement problems is, however, slightly out of place. Indeed, the scope of any experimental device is opposite: to acquire the information about the microscopic system ψ under consideration, and to record it in the form of a *unique, classical* state.

When $|\Psi\rangle$ represents a system with an infinite number of degrees of freedom, there is an exception to the possibility of superposition of (for instance) two, degenerate macroscopic states. The system, at sufficiently low temperatures, will choose one of them as its ground states. It could be tempting to relate such a phenomena of spontaneous symmetry breaking [36, 37] to the wave-function collapse. For instance, one might try to construct a dynamical model [38] of spin 1/2 particle, interacting with a doubly degenerate (2D) Ising model ground states, so as to mimic the state-vector reduction, (3.20).

More generally, the Ising model system may be replaced by some classical "pointer" states; the idea is to construct a model of nonunitary evolution of the microsystem ψ coupled to the pointer, to reproduce effectively (3.20), with correct relative frequencies for different outcomes [10].

Whether or not such a model of the wave-function collapse can be successfully constructed eventually, the conclusion from our discussion in this section (and in [8]) is the following:

No dynamical models of wave-function collapse are needed.

In our view, "wave-function collapse" is one of the worst misnomers in the quantum mechanics discussions. The words evoke in our mind a mysterious, nonlinear evolution which shrinks almost instantaneously whatever distribution present in ψ before the measurement. No such processes exist. Every experimentalist knows what happens in each measurement: they are chain-ionizations and amlification, hadronic cascades in a calorimeter, recording of the particle tracks, etc. They represent all nonadiabatic, irreversible processes. However complicated in detail, they are all processes we understand well in principle, in terms of the standard theory of the fundamental interactions [1]- [4].

The state reduction (3.20) is just the way we perceive the fact that the result of each measurement is necessarily one of the terms in (3.7). The wave functions of the states of the microsystem-device coupled system, corresponding to different measurements, have *no common spacetime support*, hence the wave function apparently - only apparently - "collapses".

position and momentum of a macroscopic body, which requires necessarily a measurement to define them as the initial condition for studying the time evolution of the system.

[10] We will not try to make an exhaustive list of references here: some earlier ones are discussed in [7]; many are cited in [38] and in [39]. There are also some talks presented in this conference.

Journal of Physics: Conference Series **2533** (2023) 012009 doi:10.1088/1742-6596/2533/1/012009

4. Three puzzles

Before concluding, let us go through quickly three familiar puzzles and their resolution.

4.1. Quantum jumps versus Schrödinger's equation

A question often debated is how to reconcile the smooth time evolution of the wave function described by the Schrödinger equation with sudden "quantum jumps", occurring in radioactive nuclei (α and β decays), or in atoms in excited states, and so on.

Let us consider an α decay from a metastable nucleus, X,

$$ {}^{A}_{N}(X) \Longrightarrow {}^{A-4}_{N-2}(Y) + \alpha \,, \tag{4.1} $$

where N is the atomic number and A is the mass number. The wave function for the system may be written as

$$ |\psi\rangle = |X\rangle + |Y\rangle|\alpha\rangle \,. \tag{4.2} $$

We do not bother to write the coefficients in front of the two terms: their norms are not conserved due to the decay. The first ($|X\rangle$) corresponds to a bound state; the second ($|Y\rangle|\alpha\rangle$) an unbounded system. The interference between them is absent. (4.2) is not the wave function of a pure state. It is a mixed state.

The state being metastable, its energy has a small imaginary part,

$$ E = E_R - i\,\frac{\Gamma}{2} \,, \tag{4.3} $$

where Γ represents the total decay rate per unit time (or the level width), and $\tau = \hbar/\Gamma$ is the mean lifetime. The wave function has a time dependence (for instance, see [40]),

$$ \tilde{\psi}_X(t) = e^{-iEt/\hbar}\tilde{\psi}_X(0) = e^{-iE_Rt/\hbar}e^{-\Gamma t/2\hbar}\tilde{\psi}_X(0) \,. \tag{4.4} $$

Now how can one reconcile such a smooth time dependence with sudden quantum jumps, such as α decay, a spontaneous emission of photons from an excited atom? This sort of question, probably mixed up with a philosophical confrontation between Heisenberg's view (based on the matrix mechanics, for the transition elements) and Schrödinger's one (with a smooth differential equations) of quantum mechanics, dominated the earlier debates on quantum mechanics [11].

Actually, from the very way the wave function $\tilde{\psi}$ for a metastable state is defined, where Γ represents the total decay rates of all possible decay processes of the parent particle, it is quite clear that $\tilde{\psi}$ represents an effective description of the metastable "quantum state", in which the coarse-grained time dependence (decay events) has been smoothed out. It does not have the same status as the wave function of a genuine quantum state ψ.

In conclusion, the conundrum of apparent impossibility of reconciling the smooth time-dependent Schrödinger equation with quantum jumps, appears to have been caused by the confusion between the concept of the true wave function ψ and that of an effective "wave function" for a metastable "state", $\tilde{\psi}$.

See [8] for more discussions.

[11] In fact, the two questions must be distinguished. The latter, more philosophical "puzzle" was solved via the proof of equivalence of the Hilbert spaces ℓ_2 and L^2 by Schrödinger himself. See e.g., Tomonaga [41]: it is not the subject of the discussion here.

4.2. Schrödinger's cat

In the discussion involving Schrödinger's cat, the initial "wave function" is a combination (4.2), between the undecayed nucleus $|X\rangle$ and the state after decay, $|Y\rangle|\alpha\rangle$. To maximally simplify the discussion, let us eliminate altogether the intermediate, diabolic device which upon receipt of the α particle leads to the poisoning of the cat, and treat the cat directly as the measuring device, $|\Phi\rangle$. When it detects the α particle, it dies; when it does not, it remains alive. There are two terms in the measurement process, (3.6), (3.7), which here reads

$$|\psi\rangle \otimes |\Phi_0\rangle = (|X\rangle + |Y\rangle|\alpha\rangle) \otimes |\Phi_0\rangle \tag{4.5}$$
$$\longrightarrow \quad |X\rangle \otimes |\Phi^{(alive)}\rangle + |Y\rangle|\alpha\rangle \otimes |\Phi^{(dead)}\rangle \tag{4.6}$$

(we dropped also the rest of the world, $|Env\rangle$). Such a process appears to lead to the superposition of the dead and alive cat, which is certainly an unusual notion, difficult to conceive, to say the least.

Actually, there are some abuse and/or misuse of concepts in this argument, each of which individually invalidates it, eliminating the notorious conundrum.

The first is the fact, as seen in the last section, that the "state" of the undecayed nucleus, X, is not a pure state. The "wave function" $\tilde{\psi}_X$ describing it is not a proper wave function but an effective one, in which the coarse-grained time dependence has been averaged out. Second, the linear superposition $|\psi\rangle \sim |X\rangle + |Y\rangle|\alpha\rangle$, as discussed around (4.2), does not represent a pure state, but a mixed state: $|Y\rangle|\alpha\rangle$ represents the decay product of $|X\rangle$, unbounded and incapable of interfering with the latter.

Finally, even putting aside these two issues, the process (4.5), (4.6) has clearly all the characteristics of a general measurement process discussed in Sec. 3. As explained there, the wave functions describing the two terms of (4.6) lack overlapping spacetime supports. There are no interferences between the terms involving the coupled system involving the microsystem and the classical measurement device (the cat here). The formula (4.6) represents a mixture.

Equivalently, and perhaps more intuitively, one can use the notion of the wave-function collapse. The exact timing of the spontaneous α decay cannot be predicted, as it reflects a quantum fluctuation. But the moment an α particle is emitted, and the cat gets hit by it (the instant of the measurement), the wave function collapses,

$$|\psi\rangle \otimes |\Phi_0\rangle \Longrightarrow |Y\rangle|\alpha\rangle \otimes |\Phi^{(dead)}\rangle\,. \tag{4.7}$$

Recapitulating, Schrödinger's cat paradox, was caused by improper use of concepts such as the superposition principle, unitary and linearity of evolution, and by mixing up the pure and mixed states [12].

Setting aside these arguments, *the idea of superposition of dead and alive cat states, represents in itself a self-contradictory, impossible notion.* A macroscopic superposition of states such as (3.30) or (4.6), requires the body temperature of the system close to $T = 0$, whereas a living cat needs room (body-) temperatures for its biological functions: it is necessarily a mixed state [19].

[12] What happens actually is simple: as long as the nucleus has not decayed (no α particle emission) the cat remains alive. The exact timing of the α decay cannot be predicted. But the moment the α particle is emitted, and the cat gets hit, it dies. That is all. There are no problems in describing this process appropriately in quantum mechanics, as seen above.

Journal of Physics: Conference Series **2533** (2023) 012009 doi:10.1088/1742-6596/2533/1/012009

4.3. Quantum entanglement, quantum nonlocality and EPR paradox

Let us consider now the famous Einstein-Podolsky-Rosen setting (in Bohm's version) of a total spin 0 system decaying into two spin $\frac{1}{2}$ particles flying away from each other. The (spin) wave function is given by

$$|\Psi_0\rangle = \frac{1}{\sqrt{2}}(|\uparrow\rangle|\downarrow\rangle - |\downarrow\rangle|\uparrow\rangle) \tag{4.8}$$

where $|\uparrow\rangle$ ($|\downarrow\rangle$) represents the state of the 1st and 2nd spins, $s_{1,z} = \pm\frac{1}{2}$, $s_{2,z} = \pm\frac{1}{2}$. As spin components of the single spins do not commute with the total spin $\mathbb{S}^2_{tot}$, e.g.,

$$[\mathbb{S}^2_{tot}, s_{1,i}] \neq 0\,, \quad [\mathbb{S}^2_{tot}, s_{2,j}] \neq 0\,, \qquad i,j = x,y,z\,, \tag{4.9}$$

$s_{1\,z}, s_{2\,y}$, etc., are fluctuating between $\pm\frac{1}{2}$ in the state Ψ_0. Note that Ψ_0 can be written in infinitely many different ways, reflecting possible fluctuation modes of the subsystems,

$$|\Psi_0\rangle = \frac{1}{\sqrt{2}}(|+\rangle|-\rangle - |-\rangle|+\rangle) = \frac{1}{2}[\,|\uparrow\rangle(|+\rangle - |-\rangle) - (|+\rangle - |-\rangle)|\uparrow\rangle\,] = \dots\,. \tag{4.10}$$

It is of fundamental importance to realize that such (correlated) fluctuations of the two subsystems - **quantum entanglement** - are present however distant the subsystems may be, and *even if they are relatively spacelikely separated.* Nothing in Ψ_0 tells us otherwise. Nothing in quantum mechanical law says that such entanglements are present only when the subsystems are nearby, e.g., less than 10^{-8} cm, or less than 10^{+28} cm. Simply,

Quantum mechanics does not contain any fundamental parameter with the dimension of a length [13].

In the EPR-Bohm experiment, this means that an experimental result at one arm, say, $s_{1\,z} = \frac{1}{2}$, might appear to imply "instantaneously" that the second experiment at the other arm would be in the state $s_{2\,z} = -\frac{1}{2}$, even before it is actually performed, and however distant they may be. This could sound paradoxical (sometimes the term "quantum nonlocality" is used). The second experimentalist, not having access to the first experiment, might have (should have?) expected to find the results $s_{2\,z} = \pm\frac{1}{2}$, a priori with equal probabilities.

This kind of argument has led to the introduction of the hidden-variable hypothesis (see [7] for discussions), although such a deduction was not really justified. We limit ourselves here to noting that this argument had logical flaws. First, the contemporaneity of the two experiments is not really an issue: the experimentalists must just ensure that they are studying the two spin $\frac{1}{2}$ particles from the same decay event - the coincidence check. The exact time ordering, which depends on the reference system chosen, cannot matter. In any case, as the two subsystems are spacelikely separated, it is untrue that the second experimentalist would know instantaneously the result of the first experiment, or actually, even whether or not the first measurement has been indeed performed. The information transmission is itself a dynamical process. Finally, the two "simultaneous" experiments would capture only those states of the two subsystems fluctuating according to the entangled wave function, (4.8), (4.10). In conclusion, for the second experimentalist the system presents itself as a mixture with an unknown density matrix. *He (she)*

[13] String theory, or quantum gravity, does have one, of the order of the Planck length, $\ell_P \sim 10^{-33}$ cm. We will not discuss here either possible genuine modifications of quantum mechanics at such regimes, the issues of information-loss paradox and blackhole entropy, or the consistent formulation of quantum gravity. Still, we do not take the view that either of them (or ℓ_P) might have any relevance to the measurement problems we are discussing here.

Journal of Physics: Conference Series **2533** (2023) 012009 doi:10.1088/1742-6596/2533/1/012009

simply does not know what to expect. Therefore there was no paradox, whatsoever.

It is true that the first experimental result $s_{1\,z} = \frac{1}{2}$ *does imply, instantaneously* (whatever it may mathematically mean), that the other spin is in the state $|\downarrow\rangle = \frac{1}{\sqrt{2}}(|+\rangle - |-\rangle)$, as is seen from the wave function (4.8), (4.10). This quantum nonlocality, just a name for this particular aspect of quantum entanglement, is real. *It should, however, not be confused with the dynamical concept of locality (or causality)* [14]. Quantum mechanics is perfectly consistent with causality and locality, due to the fact that all the fundamental, elementary interactions are local interactions in spacetime (see [8] and [42]). No dynamical effects, including the information transmission, propagate faster than the velocity of light. It is perhaps best to regard (4.8) simply as a particular kind of macroscopic quantum-mechanical state.

It is essential to realize that, unless the second measurement is done, any components of the second spin other than $s_{2\,z}$ are still fluctuating [15]. Thus if the two Stern-Gerlach type experiments are performed at the two ends simultaneously, by using the magnets directed to generic directions **a** and **b**, they would find, experiment by experiment, apparently random results, such as $(s_{1\,a}, s_{2\,b}) = (\frac{1}{2}, \frac{1}{2}), (\frac{1}{2}, -\frac{1}{2}), (-\frac{1}{2}, -\frac{1}{2})$, etc. But their fluctuation average is encoded in Ψ_0:

$$\langle \Psi_0 | s_a s_b | \Psi_0 \rangle = -\frac{1}{4} \cos \theta_{a,b} = -\frac{1}{4} \mathbf{a} \cdot \mathbf{b} \,. \tag{4.11}$$

A possible way to discriminate between any alternative theory with hidden variables from quantum mechanics, has been mathematically formulated, e.g., in the form of the Bell [7], or CSHS [10] inequalities for analogous, polarization correlated photon pair experiments. Beautiful experiments by Aspect et. al. (1981) [43] have subsequently demonstrated that, whenever the hidden-variable alternatives and quantum mechanics give discrepant predictions (for certain sets of polarizer axes), the experimental data confirm quantum mechanics, disproving the former.

See also Chiao et. al. (1995) [44] for a series of related experiments and discussions on quantum nonlocality. See [8,40] also for comments on and resolution to Mermin's conundrum [45] in systems with entanglement among more than two particles.

A reflection

As reviewed in Sec. 2.3, quantum correlations (entanglement) among the particles whose positions are space-like-ly separated, and hence are no longer capable to communicate with, or dynamically influence on, each other, as the two spin 1/2 particles in (4.8), (4.10), are quite a common, ordinary and ubiquitous phenomenon in QM. What is not quite ordinary, is that this "quantum nonlocality" does not represent any violation of dynamical causality or locality (sometimes called Einstein causality). Quantum mechanics is rigorously consistent with this principle, even though the proof of this fact requires working in the framework of relativistic quantum field theory, quantum theory of particles [8, 42].

Now how can these two, apparently contradicting properties of quantum mechanics coexist peacefully within the same theory? The answer lies in another fundamental aspect of QM: the wave function is not itself a measurable physical quantity (an observable). Only the relative frequencies for different outcomes (probabilities) are predicted by it.

[14] In author's view, endless debates and confusions about quantum mechanics have been caused precisely by such a confusion.

[15] This shows that the often mentioned "Bertlmann's socks" classical action-at-distance analogy, is not valid. Quantum non-locality is subtler, and is different [7].

Journal of Physics: Conference Series **2533** (2023) 012009 doi:10.1088/1742-6596/2533/1/012009

In hidden-variable models, the statistical aspect of quantum mechanics is replaced by a classical statistical (unknown) distribution of the hidden parameters $\{\lambda\}$, complementing the standard description with the wave function. But once the value(s) of $\{\lambda\}$ at $t = 0$ is (are) chosen, all possible imaginable future experimental outcomes are determined uniquely (i.e., classical evolutions).

A hidden-variable theory, with a logical structure so sharply different from that of quantum mechanics, cannot reproduce *all* of the predictions of the latter. It could, actually, if one is willing to introduce nonlocal, noncausal evolutions. This is seen clearly, e.g., in the unphysical, noncausal behavior of the trajectory $X(t)$ in Bohm's pilot-wave theory [7], [40] which is a particular type of hidden-variable model, designed by construction to reproduce all QM predictions.

If we insist, instead, that a hidden-variable model must respect local causality, *then* its predictions necessarily satisfy certain inequalities such as those in [7], [10]. Bell and Clauser et. al. have shown that *some* predictions of QM should lie outside such a bound, a fact verified, in a dramatic experimental confirmation of QM [43].

5. Conclusion

Here are the answers to the "quantum measurement problems" enlisted in the beginning:

(a) Wave function collapse?
No. See the Summary A. and B. after Eq. (3.20), and the discussion at the end of Sec. 3.4;

(b) Macroscopic superposition of states?
Yes (at temperatures close to $T = 0$ *and without environment-induced decoherence); No, otherwise;*

(c) Forever branching manyworlds tree?
No;

(d) Spontaneous collapse of micro state, of the experimental device, and of the world?
No;

(e) Wave function just a bookkeeping device?
No;

(f) Quantum jumps vs Schrödinger equation?
Explained, see Sec. 4.1;

(g) Born's rule from the Schrödinger equation involving everything?
Yes, but only effectively. No, as a result a complicated nonlinear dynamical process. See the Summary A. and B. after Eq. (3.20);

(h) Quntum nonlocality?
Yes, see the discussions in Sec. 4.3;
Hidden variables?
No. See the "reflection" at the end of Sec. 4.3.

(i) Is it possible that the fundamental, complete theory of physics is incapable of unique prediction for a single measurement?
Yes.

The key observation of this work that the basic entities (the degrees of freedom) of our world are various types of *particles*, leads to the idea that each measurement, at its core, is a spacetime, pointlike event. The crux of the absence of coherent superposition of different terms in the

Journal of Physics: Conference Series **2533** (2023) 012009 doi:10.1088/1742-6596/2533/1/012009

measuring process (3.7) is, indeed, the consequent *lack of the overlapping spacetime supports* in the associated wave functions, instant after measurement which is the process of entanglement between the microsystem and the experimental device and the environment. It is in this sense the expression (3.7) represents a mixed state. But it also implies that the aftermath of each measurement corresponds to a single term of (3.7). This (also empirical) fact, is known as the "wave-function collapse", or the state-vector reduction.

The discussions of Sec. 3 have led to the diagonal density matrix, (3.18). In particular, the information encoded in the wave function $|\psi\rangle = \sum_n c_n|n\rangle$ has been transferred via entanglement with the experimental device and environment into the relative frequencies for finding the experimental results $F = f_n$, $\mathcal{P}_n = |c_n|^2$. Combined with the equality between $\mathcal{P}_n = |c_n|^2$ (the relative frequencies) and $P_n = |c_n|^2$ (the probabilities) [8] our results represent the first steps towards filling in the logical gaps in Born's rule, by explaining the state-vector reduction, and pointing towards a more natural interpretation of quantum mechanics.

Our discussions do confirm the standard Born rule as a concise way of summarizing the predictions of quantum mechanics. But the fact that the latter is formulated in terms of the "*probability* certain result is found in an experiment, which might (or might not) be eventually performed - perhaps by an experimentalist with a Ph D [16] - has always caused confusion and conceptual difficulties [17].

In ultimate analysis, the problem of Born's rule is the fact that it presumes a human intervention.

We instead take the quantum fluctuations described by the wave function as real, and propose that they represent the fundamental laws of quantum mechanics. They are there, independently of any experiments, and in fact, whether or not we human beings are around [18].

What we propose here is a slightly unconventional way of *understanding* quantum mechanical laws. Indeed, the solid textbook materials and the standard methods of analysis in quantum mechanics, all remain unchallenged. After all, we know that quantum mechanics is a correct theory.

The new perspective, however, suggests us to dispense with, in the discussion of quantum mechanics, certain familiar concepts such as *probability, statistical ensembles, Copenhagen interpretation, etc.,* which historically have guided us, but also caused misunderstandings and introduced clouds in the discussion on quantum mechanics, for many years.

Accepting the quantum fluctuations described by the wave function as real, we simply give up pretending, or demanding, that in a fundamental, complete theory describing Nature the result of each single experiment should necessarily be predictable.

As cosmologists tell us today, all the known structures of our universe (including galaxies, stars, planets and ourselves) are believed to have grown out of uncontrollable quantum density fluctuations at some stage of the inflationary universe. In other words:

God may not play dice, but Nature apparently does.

We believe that our new, clearer understanding the quantum mechanical laws of Nature, has

[16] Borrowed from Bell's remarks, e.g., in "Quantum mechanics for cosmologists" (reprinted in [7]).
[17] Every teacher of a quantum mechanics course knows about this curious feeling of guilt he (or she) experiences, on the day of proclamation of the "fundamental postulate" (Born's rule).
[18] From this point of view, the notion of the "wave function of the universe" makes perfect sense.

far-reaching consequences and applications in all present and future science domains in physics and beyond, from biology and chemistry, quantum information and quantum computing, and possibly, to the brain science.

Acknowledgments

We thank the organizers and participants of the DICE2022 conference, Castiglioncello (Rosignano Marittimo, Tuscany), Sept. 19-23, 2022, where this talk was presented, for a stimulating atmosphere. A special gratitude is reserved to Hans-Thomas Elze, for the kind invitation and for many fruitful discussions. The author's work is supported by an INFN special initiative grant, GAST (Gauge and String Theories).

References

[1] Weinberg S 1967 *A model of Leptons*, *Phys. Rev. Lett.* **19**, 1264
[2] Salam A 1968 *Weak and electromagnetic interactions*, (Elementary Particle Theory) ed. N. Svartholm Almqvist Forlag AB, 367
[3] Glashow S L, Iliopoulos J, Maiani L 1970 *Weak Interactions with Lepton-Hadron Symmetry*, *Phys. Rev.* D **2**, 1285 (1970).
[4] Fritzsch H, Gell-Mann M, Leutwyler H 1973 *Advantages of the color octet gluon picture*, *Phys. Lett.* **47** B (1973), 365
[5] Wheeler J A, Zurek W H 1983 *Quantum Theory and Measurement* (Princeton University Press)
[6] Peres A 1995 *Quantum Theory: Concepts and Methods* (Dordrecht/Boston/London: Kluwer)
[7] Bell J S 1987 *Speakable and unspeakable in quantum mechanics* (Cambridge University Press)
[8] Konishi K 2022 *Quantum fluctuations, particles and entanglement: a discussion towards the solution of the quantum measurement problems*, *Int. Journ. Mod. Phys.* A **37**, (2022) 2250113 [arXiv:2111.14723 [quant-ph]]
[9] Dirac P A M 1958 *Principles of Quantum Mechanics* (Oxford University Press)
[10] Clauser J F, Horne M A, Shimony A, Holt R A 1969 *Proposed experiment to test local hidden-variable theories*, *Phys. Rev. Lett.*, **23** (15) 880
[11] Zurek W H 2003 *Decoherence, einselection, and the quantum origins of the classical*, *Rev. Mod. Phys.* **75**, 715-775 doi:10.1103/RevModPhys.75.715 [arXiv:quant-ph/0105127 [quant-ph]]
[12] Joos E, Zeh H D, Kiefer C, Giulini D, Kupsch J, Stamatescu I O 2002 *Decoherence and the Appearance of a Classical World in Quantum Theory* (Springer)
[13] Joos E, Zeh H D 1985 *The emergence of classical properties through interaction with the environment*, *Z. Phys.* B **59**, 223-243
[14] Zurek W H 1991 *Decoherence and the Transition from Quantum to Classical*, *Physics Today* **44**, 10, 36, https://doi.org/10.1063/1.881293
[15] Tegmark M 1993 *Apparent wave function collapse caused by scattering*, *Found. Phys. Lett.* **6**, 571 doi:10.1007/BF00662807 [arXiv:gr-qc/9310032 [gr-qc]]
[16] Zurek W H 2003 *Decoherence, einselection, and the quantum origins of the classical*, *Rev. Mod. Phys.* **75**, 715-775 doi:10.1103/RevModPhys.75.715 [arXiv:quant-ph/0105127 [quant-ph]]
[17] Frauchiger D, Renner R 2018 *Quantum theory cannot consistently describe the use of itself*, *Nature Communications* **9**, 3711
[18] Tonomura A, Endo J, Matsuda T, Kawasaki T, Ezawa H, 1989 *Dimonstration of single-electron buildup of interference pattern*, *American Journal of Physics* **57** 117
[19] Konishi K 2022 Newton's equations from quantum mechanics for a macroscopic body in the vacuum *Preprint* [arXiv:2209.07318 [quant-ph]]
[20] Leggett A J 1980 *Macroscopic Quantum Systems and the Quantum Theory of Measurement*, *Suppl. of the Progress of Theoretical Physics* **69** 80
[21] Arndt M, Nairz O, Vos-Andreae J, Keller C, van der Zouw G, Zeilinger A 1999 *Wave-particle duality of* C_{60} *molecules*, *Nature* **401** 680

[22] O'Connell A D, et. al., 2010 *Quantum ground state and single-photon control of a mechanical resonator, Nature,* **464** (7289) 697

[23] Arndt M, Hornberger K 2014 *Testing the limits of quantum mechanical superpositions, Nature Physics* **10** 271 [arXiv:1410.0270v1 [quant-ph]]

[24] Courty J M, Heidmann A, Pinard M 2001 *Quantum limits of cold damping with optomechanical coupling, Eur. Phys. J.* D **17**, 399–408

[25] Armour A D, Blencowe M P, Schwab K C 2002 *Entanglement and decoherence of a Micromechanical Resonator via Coupling to a Cooper-Pair Box, Phys. Rev. Lett.* bf 88 148301

[26] Knobel R G, Cleland A N 2003 *Nanometre-scale displacement sensing using a single electron transistor, Nature,* **424** 17

[27] LaHaye M D, Buu O, Camarota B, Schwab K C 2004 *Approaching the Quantum Limit of a Nanomechanical Resonator, Science* **304**, 74

[28] Cleland A N, Geller M R 2004 *Superconducting Qubit Storage and Entanglement with Nanomechanical Resonators, Phys. Rev. Lett.* **93** 070501

[29] Martin I, A. Shnirman I A, Tian L, Zoller P 2004 *Ground-state cooling of mechanical resonators, Phys. Rev.* B **69**, 125339

[30] Kleckner D, Bouwmeester D 2006 *Sub-kelvin optical cooling of a micromechanical resonator, Nature,* **444** 2

[31] Regal C A, Teufel J D, Lehnert K W 2008 *Measuring nanomechanical motion with a microwave cavity interferometer* (Macmillan Publishers Limited) doi:10.1038/nphys974

[32] Schliesser A, Rivière R, Anetsberger G, Arcizetandt O, Kippenberg J 2008 *Resolved-sideband cooling of a micromechanical oscillator* (Macmillan Publishers Limited) doi:10.1038/nphys939

[33] Abbott B *et al.* [LIGO Scientific] 2009 *Observation of a kilogram-scale oscillator near its quantum ground state, New J. Phys.* **11**, 073032 doi:10.1088/1367-2630/11/7/073032

[34] O'Connell A D, et. al. 2010 *Quantum ground state and single-photon control of a mechanical resonator, Nature,* **464** (7289) 697

[35] Brand C, Troyer S, Knobloch C, Cheshinovsky O, Arndt M A (2021) Single, double and triple-slit diffraction of molecular matter waves, *Preprint* arXiv:2108.06565v2 [quant-ph]

[36] Nambu Y, Jona-Lasinio G 1961 *Dynamical Model of Elementary Particles Based on an Analogy with Superconductivity. I, Physical Review* **122**: 345, *Dynamical Model of Elementary Particles Based on an Analogy with Superconductivity. II, Physical Review* **124**: 246

[37] Goldstone J 1961 *Field theories with "superconductor" solutions, Il Nuovo Cimento.* **19** (1) 154

[38] Allahverdyan A E, Balian R, Nieuwenhuizen T M 2013 *Understanding quantum measurement from the solution of dynamical models, Phys. Rep.* **525** 1-166 arXiv:1107.2138

[39] Fröhlich J 2020 *A Brief Review of the ETH-Approach to Quantum Mechanics* In: Anantharaman N., Nikeghbali A., Rassias M.T. (eds) *Frontiers in Analysis and Probability* (Springer, Cham)

[40] Konishi K, Paffuti G 2009, *Quantum Mechanics: A New Introduction* (Oxford University Press)

[41] Tomonaga S 1968 *Quantum Mechanics* Vol. II, (North Holland)

[42] Peskin M E, Schröder D V 1995 *An Introduction to Quantum Field Theory* (New York, Addison-Wesley)

[43] Aspect A, Grangier P, Roger G 1981 *Experimental realization of Einstein-Podolsky-Rosen-Bohm Gedankenexperiment: A new violation of Bell's inequalities, Phys. Rev. Lett.* **49** 91

[44] Chiao R Y, Kwiat P G, Steinberg A M 1995 Quantum Nonlocality in Two-Photon Experiments at Berkeley *Preprint* arXiv:quant-ph/9501016, and references therein

[45] D. Mermin D (1990) *Quantum mysteries revisited, American Journal of Physics,* **58**, 731

Journal of Physics: Conference Series **2533** (2023) 012010 doi:10.1088/1742-6596/2533/1/012010

Non-Abelian gauge structure and flavor mixing in quantum field theory[1]

Massimo Blasone[1], Luca Smaldone[2] and Giuseppe Vitiello[1]

[1] Physics Department "E.R. Caianiello", University of Salerno,
Via Giovanni Paolo II, 132, 84084 Fisciano (Salerno), Italy
[2] Faculty of Physics, University of Warsaw, ul. Pasteura 5, 02-093 Warsaw, Poland

E-mail: blasone@sa.infn.it, Luca.Smaldone@fuw.edu.pl, givitiello@unisa.it

Abstract.
The quantum field theory formalism describing the phenomenon of neutrino mixing and oscillations is reviewed in its essential aspects. The condensate structure of the flavor vacuum state is considered and its non-perturbative nature is discussed within the existence in the quantum field theory of the manifold of unitarily inequivalent representations of the anticommutation relations. The Poincaré structure of the theory is discussed in connection with the gauge theory features of neutrino mixing, where the possibility arises to describe flavored neutrinos as on-shell fields with definite masses. The flavor vacuum state may be thought of as a refractive medium where different refraction indexes are related to the different neutrino masses. Neglecting the condensate structure of the flavor vacuum implies neglecting second-order contributions in the gauge field coupling and leads to experimentally inconsistent results.

1. Introduction

By resorting to results of the study of neutrino mixing and oscillations [1]-[8], we review the possibility to cast such a phenomenon within the gauge theory paradigm of quantum field theory (QFT) [9, 10]. Some light is shed on the nature of flavored neutrinos fields, which being oscillating in the flavor charge, cannot be taken as asymptotic fields (in the time $t \to \infty$ limit) in the Lehmann-Symanzik-Zimmermann (LSZ) QFT formalism. Thus, our discussion leads us to consider the problem of Poincaré invariance for flavor neutrino states. In the gauge theory frame, the interaction of neutrino fields with the gauge field generates the mixing and oscillation phenomenon and the vacuum appears to have the properties of a refractive medium (see e.g. [11, 12]).

For simplicity, we consider the case of two neutrino mixing. Extension to three neutrinos is possible [3, 6, 13, 14] (and similarly to boson mixing [15, 16, 17, 18]; for a review see [19]).

Essential formal aspects are shortly summarized below. Details are in the cited references. The mixing transformations for the massive neutrino filed ν_1 and ν_2 are

$$\nu_e = \nu_1 \cos\theta + \nu_2 \sin\theta \qquad (1)$$

$$\nu_\mu = -\nu_1 \sin\theta + \nu_2 \cos\theta . \qquad (2)$$

[1] Presented by Giuseppe Vitiello

The field ν_1 and ν_2 are given by

$$\nu_i(x) = \int \frac{d^3\mathbf{k}}{(2\pi)^{3/2}} \sum_r \left[u^r_{\mathbf{k},i}(t)\alpha^r_{\mathbf{k},i} + v^r_{-\mathbf{k},i}(t)\beta^{r\dagger}_{-\mathbf{k},i}\right] e^{i\mathbf{k}\cdot\mathbf{x}}, \quad i = 1,2, \ r = 1,2, \tag{3}$$

with usual canonical anticommutation relations (CAR) and orthonormality and completeness relations for $u^r_{\mathbf{k},i}(t)$ and $v^r_{-\mathbf{k},i}(t)$. The CAR are preseved by the mixing transformations which are therefore *canonical* transformations. The vacuum state for $\alpha^r_{\mathbf{k},i}$ and $\beta^r_{-\mathbf{k},i}$ is the tensor product $|0\rangle_{1,2} = |0\rangle_1 \otimes |0\rangle_2$: $\alpha^r_{\mathbf{k},i}|0\rangle_{1,2} = 0$; $\beta^r_{-\mathbf{k},i}|0\rangle_{1,2} = 0$. $\mathcal{H}_{1,2}$ denotes the corresponding Hilbert space. The part of the Lagrangian density for the flavored neutrinos $\nu_e(\mathbf{x},t)$ and $\nu_\mu(\mathbf{x},t)$ including the mixing terms is

$$\mathcal{L} = \bar{\nu}_e\,(i\,\partial\!\!\!/ - m_e)\,\nu_e + \bar{\nu}_\mu\,(i\,\partial\!\!\!/ - m_\mu)\,\nu_\mu \;-\; m_{e\mu}\,(\bar{\nu}_e\,\nu_\mu + \bar{\nu}_\mu\,\nu_e)\,, \tag{4}$$

and is digonalized by the mixing transformations (1) and (2)

$$\mathcal{L} = \bar{\nu}_1\,(i\,\partial\!\!\!/ - m_1)\,\nu_1 + \bar{\nu}_2\,(i\,\partial\!\!\!/ - m_2)\,\nu_2. \tag{5}$$

The masses m_i, m_σ, $i = 1,2; \sigma = e,\mu$, the mixing angle θ and "coupling" $m_{e\mu}$ are related by

$$\delta m_{12}\sin 2\theta = 2m_{e\mu}, \qquad \delta m_{12}\cos 2\theta = m_\mu - m_e \equiv \delta m_{e\mu}, \qquad m_e + m_\mu = m_1 + m_2, \tag{6}$$

with $\delta m_{12} \equiv m_2 - m_1$, $m_1 \neq m_2$, $m_e = m_1\cos^2\theta + m_2\sin^2\theta$, $m_\mu = m_1\sin^2\theta + m_2\cos^2\theta$. In the Appendix A the Hamiltonian H and the momentum operator P^i, $i = 1,2,3$, are presented.

In Section 2 the flavor vacuum is obtained. The gauge theory formalism is presented in Section 3. In Section 4, the vacuum is discussed as a medium with refractive properties. Section 5 is devoted to conclusions. Some formal details are given in the Appendices.

2. The generator of the mixing transformation and the flavor vacuum state

At finite volume V, the mixing transformations (1) and (2) are induced by [3]:

$$G_\theta(t) = \exp\left[\theta \int d^3\mathbf{x} \left(\nu_1^\dagger(x)\nu_2(x) - \nu_2^\dagger(x)\nu_1(x)\right)\right] \tag{7}$$

$$\nu_\sigma(x) = G_\theta^{-1}(t)\nu_i(x)G_\theta(t)\,, \qquad (\sigma, i) = (e,1), (\mu,2), \tag{8}$$

with $G_\theta^{-1}(t) = G_{-\theta}(t) = G_\theta^\dagger(t)$. We may expand $\nu_\sigma(x)$, $\sigma = e,\mu$ in different bases [3]. When using the bases $\{u^r_{\mathbf{k},i}(t), v^r_{-\mathbf{k},i}(t)\}$ of the ν_i fields,

$$\nu_\sigma(x) = \int \frac{d^3\mathbf{k}}{(2\pi)^{3/2}} \sum_r \left[u^r_{\mathbf{k},i}(t)\alpha^r_{\mathbf{k},\sigma}(t) + v^r_{-\mathbf{k},i}(t)\beta^{r\dagger}_{-\mathbf{k},\sigma}(t)\right] e^{i\mathbf{k}\cdot\mathbf{x}}, \ (\sigma, i) = (e,1), (\mu,2). \tag{9}$$

and, from (8), $\alpha^r_{\mathbf{k},\sigma}(t) = G_\theta^{-1}(t)\alpha^r_{\mathbf{k},i}G_\theta(t)$, $(\sigma, i) = (e,1), (\mu,2)$, and h.c., and similarly for $\beta^r_{-\mathbf{k},\sigma}$ and $\beta^{r\dagger}_{-\mathbf{k},\sigma}$. In the reference frame $\mathbf{k} = (0,0,|\mathbf{k}|)$, it is, e.g. for $\alpha^r_{\mathbf{k},e}(t)$

$$\alpha^r_{\mathbf{k},e}(t) \;=\; \cos\theta\;\alpha^r_{\mathbf{k},1} \;+\; \sin\theta\;\left(U_{\mathbf{k}}(t)\;\alpha^r_{\mathbf{k},2} \;+\; \epsilon^r\;V_{\mathbf{k}}(t)\;\beta^{r\dagger}_{-\mathbf{k},2}\right), \tag{10}$$

and similarly for the other flavor operators [3]. In Eq.(10), $\epsilon^r = (-1)^r$, $U_k(t) = |U_k|e^{i(\omega_{k,2}-\omega_{k,1})t}$, $V_k(t) = |V_k|\;e^{i(\omega_{k,2}+\omega_{k,1})t}$, and $|U_k|^2 + |V_k|^2 = 1$. Their explicit expressions are in Appendix A. Eq.(10) describes a rotation nested with a Bogoliubov transformation (the U and V terms) [3, 20]. In the relativistic limit, for $|\mathbf{k}| \gg \sqrt{m_1 m_2}$, $|U_\mathbf{k}|^2 \to 1$ and

$|V_{\mathbf{k}}|^2 \to 0$, it reduces to a rotation of the ladder operators (the Pontecorvo approximation in quantum mechanics (QM)). The CAR are preserved by the transformation, e.g., from Eq.(10), $\{\alpha^r_{\mathbf{k},e}(t), \alpha^{r\dagger}_{\mathbf{k},e}(t)\} = \cos^2\theta + |U_{\mathbf{k}}|^2 \sin^2\theta + |V_{\mathbf{k}}|^2 \sin^2\theta = 1$.

The operators $\alpha^r_{\mathbf{k},\sigma}(t)$ (and $\beta^r_{\mathbf{k},\sigma}(t)$), $\sigma = e, \mu$, do not annihilate the vacuum $|0\rangle_{1,2}$. They annihilate instead the vacuum state

$$|0(\theta,t)\rangle_{e,\mu} = G^{-1}_\theta(t)\ |0\rangle_{1,2}\ . \tag{11}$$

It is ${}_{e,\mu}\langle 0(\theta,t)|0(\theta,t)\rangle_{e,\mu} = 1$. $|0(\theta,t)\rangle_{e,\mu}$ is a generalized $SU(2)$ coherent state. The generator in the exponential of Eq. (7) is indeed the $SU(2)$ operator $J(2) \equiv -(i/2)(S_+ - S_-)$ (cf. Appendix A) [3]. The associated flavor Hilbert space is denoted by $\mathcal{H}_{e,\mu}$. Explicitly,

$$|0(\theta,t)\rangle_{e,\mu} = \prod_{\mathbf{k},r} \Big[(1-\sin^2\theta\,|V_{\mathbf{k}}|^2) - \epsilon^r \sin\theta\cos\theta\, V_{\mathbf{k}}(t)\left(\alpha^{r\dagger}_{\mathbf{k},1}\beta^{r\dagger}_{-\mathbf{k},2} + \alpha^{r\dagger}_{\mathbf{k},2}\beta^{r\dagger}_{-\mathbf{k},1}\right) \tag{12}$$

$$+\, \epsilon^r \sin^2\theta V_{\mathbf{k}}(t)\left(U^*_{\mathbf{k}}(t)\alpha^{r\dagger}_{\mathbf{k},1}\beta^{r\dagger}_{-\mathbf{k},1} - U_{\mathbf{k}}(t)\alpha^{r\dagger}_{\mathbf{k},2}\beta^{r\dagger}_{-\mathbf{k},2}\right) + \sin^2\theta V^2_{\mathbf{k}}(t)\alpha^{r\dagger}_{\mathbf{k},1}\beta^{r\dagger}_{-\mathbf{k},2}\alpha^{r\dagger}_{\mathbf{k},2}\beta^{r\dagger}_{-\mathbf{k},1}\Big]\ |0\rangle_{1,2}.$$

Notice that the condensate of zero momentum pairs of massive neutrinos, $\alpha^r_{\mathbf{k},i}$ and $\beta^r_{-\mathbf{k},i}$, $i = 1, 2$, for any $\mathbf{k}$, vanishes for $V_{\mathbf{k}}(t) \to 0$. Indeed, the condensate content for the $\alpha^r_{\mathbf{k},i}$'s (and of $\beta^r_{\mathbf{k},i}$'s) is

$${}_{e,\mu}\langle 0(\theta,t)|\alpha^{r\dagger}_{\mathbf{k},i}\alpha^r_{\mathbf{k},i}|0(\theta,t)\rangle_{e,\mu} = |V_{\mathbf{k}}|^2\sin^2\theta, \quad i = 1, 2, \tag{13}$$

Flavored neutrinos are field operators intrinsically carrying infinitely many degrees of freedom and mathematical consistency requires that the QFT infinite volume limit has to be taken. In such a limit, at each time t, $|0(t)\rangle_{e,\mu}$ is orthogonal to $|0\rangle_{1,2}$ and to $|0(\theta,t')\rangle_{e,\mu}$ for $t \neq t'$, [3, 21, 22]

$$\lim_{V\to\infty}\ {}_{1,2}\langle 0|0(\theta,t)\rangle_{e,\mu} = \lim_{V\to\infty}\ e^{V\int \frac{d^3\mathbf{k}}{(2\pi)^3}\ \ln\left(1-\sin^2\theta\,|V_{\mathbf{k}}|^2\right)^2} = 0 \tag{14}$$

$$\lim_{V\to\infty}\ {}_{e,\mu}\langle 0(\theta,t')|0(\theta,t)\rangle_{e,\mu} = 0, \qquad t \neq t', \tag{15}$$

implying that, at each t, $\mathcal{H}_{e,\mu}(t)$, $\mathcal{H}_{1,2}$ and $\mathcal{H}_{e,\mu}(t')$, for $t \neq t'$, are unitarily inequivalent representations of the CAR. Eq.(14) exhibits the nonperturbative character of the flavor vacuum, i.e. $|0(\theta,t)\rangle_{e,\mu}$ cannot be expanded in terms of states of $\mathcal{H}_{1,2}$ (and of $\mathcal{H}_{e,\mu}(t')$, $t \neq t'$) in the infinite volume limit. This result is of course expected since it is well known (see e.g. [23, 24, 25]) that, under proper conditions, the spaces of the states of fields with different masses are unitarily inequivalent spaces (unitarily inequivalent representations of the CAR). Eq.(12) thus holds at finite volume, it is a formal relation. As usual in QFT, the infinite volume limit is taken after computations have been done.

The flavor vacuum $|0(t)\rangle_{e,\mu}$ is an entangled state (see Appendix B).

Neutrino oscillations are described in terms of expectation values of the flavor charges $Q_\sigma(t)$ associated to the (broken) $SU(2)$ symmetry, given by (cf. Appendix A)

$$Q_e(t) \;=\; \cos^2\theta\, Q_1 + \sin^2\theta\, Q_2 + \frac{1}{2}\sin 2\theta \int d^3\mathbf{x}\left[\nu_1^\dagger(x)\nu_2(x) + \nu_2^\dagger(x)\nu_1(x)\right], \tag{16}$$

$$Q_\mu(t) \;=\; \sin^2\theta\, Q_1 + \cos^2\theta\, Q_2 - \frac{1}{2}\sin 2\theta \int d^3\mathbf{x}\left[\nu_1^\dagger(x)\nu_2(x) + \nu_2^\dagger(x)\nu_1(x)\right]. \tag{17}$$

Denote normal ordering with respect to $|0(\theta,0)\rangle_{e,\mu}$ by $::\ A\ ::\equiv A - {}_{e,\mu}\langle 0(\theta,0)|A|0(\theta,0)\rangle_{e,\mu}$, for a generic operator A. One gets ${}_{e,\mu}\langle 0(\theta,0)|\ ::\ Q_\sigma(t)\ ::\ |0(\theta,0)\rangle_{e,\mu} = 0$, $\sigma = e, \mu$. The flavor oscillation formulas are given by [35] $\mathcal{Q}^{\mathbf{k}}_{\nu_e\to\nu_e}(t) = \langle \nu^r_{\mathbf{k},e}|\ :\,:\ Q_e(t)\ :\,:\ |\nu^r_{\mathbf{k},e}\rangle$ and

$\mathcal{Q}^{\mathbf{k}}_{\nu_e\to\nu_\mu}(t) = \langle \nu^r_{\mathbf{k},e}| :: Q_\mu(t) :: |\nu^r_{\mathbf{k},e}\rangle$, with $|\nu^r_{\mathbf{k},e}\rangle = \alpha^{r\dagger}_{\mathbf{k},e}(0)|0(\theta,0)\rangle_{e,\mu}$, at $t=0$, the one particle flavor state. We have

$$\begin{aligned}
\mathcal{Q}^{\mathbf{k}}_{\nu_e\to\nu_e}(t) &= 1-\sin^2(2\theta)\left[|U_{\mathbf{k}}|^2\sin^2\left(\frac{\omega_{k,2}-\omega_{k,1}}{2}t\right)+|V_{\mathbf{k}}|^2\sin^2\left(\frac{\omega_{k,2}+\omega_{k,1}}{2}t\right)\right], \\
\mathcal{Q}^{\mathbf{k}}_{\nu_e\to\nu_\mu}(t) &= 1-\mathcal{Q}^{\mathbf{k}}_{\nu_e\to\nu_e}(t). \qquad (18)
\end{aligned}$$

In the relativistic limit, $|\mathbf{k}| \gg \sqrt{m_1 m_2}$, $|U_{\mathbf{k}}|^2 \to 1$ and $|V_{\mathbf{k}}|^2 \to 0$, they give the Pontecorvo oscillation formulas. Note that the above corrections are of the same order of the chiral oscillation terms [36] which have been not considered here and are also vanishing in the relativistic limit. In the next Section the mixing is framed in the gauge theory framework.

3. Neutrino mixing and gauge theory

From the Lagrangian (4) the Euler–Lagrange equations are obtained

$$i\partial_0\nu_e = (-i\alpha\cdot\nabla+\beta m_e)\nu_e+\beta m_{e\mu}\nu_\mu \qquad (19)$$

$$i\partial_0\nu_\mu = (-i\alpha\cdot\nabla+\beta m_\mu)\nu_\mu+\beta m_{e\mu}\nu_e, \qquad (20)$$

where the Dirac matrices α_i, $i=1,2,3$ and β are:

$$\alpha_i = \begin{pmatrix} 0 & \sigma_i \\ \sigma_i & 0 \end{pmatrix}, \qquad \beta = \begin{pmatrix} 1\!\!1 & 0 \\ 0 & -1\!\!1 \end{pmatrix}, \qquad (21)$$

and σ_i are the Pauli matrices, $1\!\!1$ is the 2×2 identity matrix and $\gamma^0=\gamma_0=\beta$; $(\gamma^0)^2=1$; $\gamma^i = \gamma^0\alpha^i$, $\{\gamma^\mu,\gamma^\nu\} = 2g^{\mu\nu}1\!\!1$. $\nu_f = (\nu_e,\nu_\mu)^T$ is the flavor doublet and M_d the diagonal mass matrix $M_d = \mathrm{diag}(m_e, m_\mu)$. By introducing the (non-abelian) covariant derivative

$$D_0 = \partial_0 + i\,\gamma_0 m_{e\mu}\,\sigma_1, \qquad (22)$$

where $m_{e\mu} = (1/2)\,g\,\delta m_{e\mu}$, with $g\equiv\tan 2\theta$, $\delta m_{e\mu} = m_\mu - m_e = \delta m_{12}\cos 2\theta$ (cf. Eq. (6)), Eqs. (19) and (20) become [9]:

$$i\gamma^0 D_0\nu_f = (-i\gamma\cdot\nabla + M_d)\nu_f, \qquad (23)$$

The covariant derivative D_μ can be written as

$$D_\mu = \partial_\mu + i\,g\,\beta A_\mu, \qquad (24)$$

where $D_i = \partial_i$; $i=1,2,3$, and $D_0 = \partial_0 + i\,g\beta A_0 = \partial_0 + i\,g\gamma_0\delta m_{e\mu}\sigma_1/2$, with the coupling constant g of the neutrino field operators with the gauge field playing the role of the "charge". In Eqs. (23) (and (22)), the flavor mixing is thus described in terms of the coupling of the flavor field operators with the $SU(2)$ gauge field:

$$A_\mu = A^a_\mu\sigma_a\,. \qquad (25)$$

with $A^a_\mu = (A^1_0 = \delta m_{e\mu}/2, 0, 0, 0)$. The field equations acquire the manifestly covariant form:

$$(i\gamma^\mu D_\mu - M_d)\nu_f = 0, \qquad (26)$$

and are derived from the Lagrangian density (4) which can be written in the compact form

$$\mathcal{L} = \bar{\nu}_f(i\gamma^\mu D_\mu - M_d)\nu_f. \qquad (27)$$

For the local (in time) gauge transformation $\nu_f \to \nu'_f = e^{-ig\lambda_1(t)\sigma_1/2}\nu_f \equiv U(t)\nu_f$, it is:

$$U(t)D_\mu\nu_f = D'_\mu U(t)\nu_f = D'_\mu\nu'_f, \tag{28}$$

where $\lambda_1(t)$ is the time dependent phase, and, consistently with the properties of covariant derivatives [37], $D'_i = U(t)D_iU^{-1}(t) = D_i$ and $D'_0 = U(t)D_0U^{-1}(t) = \partial_t + i\,g\,(A^1_0 + \partial_t\lambda_1(t)/2)\sigma_1$.

The Lagrangian (27) is not invariant under the $SU(2)$ transformation $\nu'_f = e^{-i\theta_1\sigma_1/2}\nu_f$. The current associated to its variation $\delta\mathcal{L} = -\theta_1\,\partial_\mu j^\mu_{f,1}$ [6],

$$j^\rho_{f,1} = \frac{1}{2}\bar{\nu}_f\gamma^\rho\sigma_1\nu_f = \frac{1}{2}(\bar{\nu}_e\gamma^\rho\nu_\mu + \bar{\nu}_\mu\gamma^\rho\nu_e), \tag{29}$$

has only one component in group space. Also notice that for any μ, ν and a

$$F^a_{\mu\nu} = \epsilon^{abc}A^b_\mu A^c_\nu = 0, \quad a,b,c = 1,2,3, \tag{30}$$

since A^a_μ has only one non-zero component, the constant A^1_0. Although $F^a_{\mu\nu}$ is identically zero, however, the gauge field has physical effects.

The energy momentum tensor for the flavor neutrino fields interacting with the external gauge field is [37]

$$\widetilde{T}_{\rho\sigma} = \bar{\nu}_f i\gamma_\rho D_\sigma\nu_f - \eta_{\rho\sigma}\bar{\nu}_f(i\gamma^\lambda D_\lambda - M_d)\nu_f. \tag{31}$$

It differs from the canonical energy momentum tensor (Eq.(A.1)) just for the mixing terms in the 00 component: $T_{00} - \widetilde{T}_{00} = m_{e\mu}(\bar{\nu}_e\nu_\mu + \bar{\nu}_\mu\nu_e)$, while $T_{0i} = \widetilde{T}_{0i}$, $T_{ij} = \widetilde{T}_{ij}$. Since $[\gamma_\mu, D_0] \neq 0$, it is $\partial^\rho\widetilde{T}_{\rho i} = 0$ and $\partial^\rho\widetilde{T}_{\rho 0} \neq 0$. The conserved 3−momentum operator is $\widetilde{P}^i = \int d^3\mathbf{x}\,\widetilde{T}^{0i} = \widetilde{P}^i_e(t) + \widetilde{P}^i_\mu(t)$, $i = 1,2,3$. The Hamiltonian is $\widetilde{H}(t) \equiv \widetilde{P}^0(t) = \widetilde{H}_e(t) + \widetilde{H}_\mu(t)$. They split in the two contributions for the electron and the muon flavor, given by

$$\widetilde{\mathbf{P}}_\sigma(t) = \sum_r \int d^3\mathbf{k}\,\mathbf{k}\left(\widetilde{\alpha}^{r\dagger}_{\mathbf{k},\sigma}(t)\widetilde{\alpha}^r_{\mathbf{k},\sigma}(t) + \widetilde{\beta}^{r\dagger}_{\mathbf{k},\sigma}(t)\widetilde{\beta}^r_{\mathbf{k},\sigma}(t)\right), \tag{32}$$

$$\widetilde{H}_\sigma(t) = \sum_r \int d^3\mathbf{k}\,\omega_{\mathbf{k},\sigma}\left(\widetilde{\alpha}^{r\dagger}_{\mathbf{k},\sigma}(t)\,\widetilde{\alpha}^r_{\mathbf{k},\sigma}(t) - \widetilde{\beta}^r_{\mathbf{k},\sigma}(t)\,\widetilde{\beta}^{r\dagger}_{\mathbf{k},\sigma}(t)\right), \tag{33}$$

respectively, with $\sigma = e,\mu$, and $\widetilde{\alpha}^r_{\mathbf{k},\sigma}(t)$ and $\widetilde{\beta}^r_{\mathbf{k},\sigma}(t)$ defined below (cf. Eq.(34)). Note that $\widetilde{H}(t)$ does not generate time translations. They are generated by the Hamiltonian $H = \int d^3\mathbf{x}\,T^{00}$ including the interaction term.

Flavor states, $|\nu^r_{\mathbf{k},\sigma}\rangle = \alpha^{r\dagger}_{\mathbf{k},\sigma}|0(\theta,0)\rangle_{e,\mu}$, are eigenstates of $P^i = \int d^3\mathbf{x}\,T^{0i}$ and of the flavor charge operators Q_σ (cf. Appendix A). However, Q_σ and the Hamiltonian H do not commute, and flavor-energy uncertainty relations are obtained [6, 38].

Lorentz invariance is thus broken due to time-dependence of the flavor vacuum and flavor states, which thus do not belong to irreducible representations of the Poincaré group [39, 40]. This feature has attracted much attention, also within nonlinear realizations of the Poincaré group [41]-[45] and of its spontaneous symmetry breaking [40].

In the gauge theory mixing framework, the $\nu_\sigma(x)$ flavor field can be expanded as

$$\nu_\sigma(x) = \int \frac{d^3\mathbf{k}}{(2\pi)^{3/2}}\sum_r\left[u^r_{\mathbf{k},\sigma}(t)\widetilde{\alpha}^r_{\mathbf{k},\sigma}(t) + v^r_{-\mathbf{k},\sigma}(t)\widetilde{\beta}^{r\dagger}_{-\mathbf{k},\sigma}(t)\right]e^{i\mathbf{k}\cdot\mathbf{x}}, \quad \sigma = e,\mu, \tag{34}$$

Journal of Physics: Conference Series **2533** (2023) 012010 doi:10.1088/1742-6596/2533/1/012010

where the tilde operators are related to $\alpha^r_{\mathbf{k},\sigma}(t)$ and $\beta^{r\dagger}_{-\mathbf{k},\sigma}(t)$ through Bogoliubov transformations with angles $\xi^{\mathbf{k}}_{\sigma,j} = (\chi_\sigma - \chi_j)/2$, $\chi_\sigma = \arctan(\mu_\sigma/|\mathbf{k}|)$, $\chi_j = \arctan(m_j/|\mathbf{k}|)$ [46]. The masses for tilde operators are (m_e, m_μ); and $u^r_{\mathbf{k},\sigma}(t) = u^r_{\mathbf{k},\sigma} e^{-i\omega_{\mathbf{k},\sigma}t}$, $v^r_{-\mathbf{k},\sigma}(t) = v^r_{-\mathbf{k},\sigma} e^{i\omega_{\mathbf{k},\sigma}t}$ are solutions of

$$(-\alpha \cdot \mathbf{k} + m_\sigma \beta) u^r_{\mathbf{k},\sigma} = \omega_{\mathbf{k},\sigma} u^r_{\mathbf{k},\sigma} \tag{35}$$

$$(-\alpha \cdot \mathbf{k} + m_\sigma \beta) v^r_{-\mathbf{k},\sigma} = -\omega_{\mathbf{k},\sigma} v^r_{-\mathbf{k},\sigma}, \tag{36}$$

and $\omega_{\mathbf{k},\sigma} = \sqrt{\mathbf{k}^2 + m_\sigma^2}$. The tilde flavor vacua are $|\widetilde{0}(\theta,t)\rangle_{e\mu} = J_\mu^{-1}(\theta,t)|0(t)\rangle_{e\mu}$, where $J_\mu^{-1}(\theta,t)$ implements the Bogoliubov transformation, and

$$|\widetilde{\nu}^r_{\mathbf{k},\sigma}(\theta,t)\rangle = \widetilde{\alpha}^{r\dagger}_{\mathbf{k},\sigma}(t)|\widetilde{0}(\theta,t)\rangle_{e\mu}, \tag{37}$$

are simultaneous eigenstates of $\widetilde{H}_\sigma(t)$ and $\mathbf{P}_\sigma(t)$ (Eqs.(33) and (32)).

Flavor charges and oscillation formulas are left invariant by the $J_\mu^{-1}(\theta,t)$ Bogoliubov transformations [47], $\widetilde{Q}_\sigma = Q_\sigma$, and, since now the interaction term is absent in $\widetilde{H}$ (cf. Eq.(33)), it is $[\widetilde{Q}_\sigma(t), \widetilde{H}(t)] = 0$, $\sigma, \sigma' = e, \mu$ and a common set of eigenstates of these operators thus exists. The other Lorentz generators can be obtained: $\widetilde{M}^{\lambda\rho}(t) = \widetilde{M}^{\lambda\rho}_e(t) + \widetilde{M}^{\lambda\rho}_\mu(t)$, and the equal-time commutators are as usual. As a result, a different Poincaré structure exists as the direct sum of two Poincaré algebras and Poincaré invariance holds at each t. Flavor fields are then described as on-shell fields with masses m_e and m_μ, evolving in time through trajectories over the representations, each one with its Poincaré structure (similarly to QFT time evolution in curved background [48](for mixing in a gravity and cosmological frame see also [49]-[57]))

In this connection, β-decay could provide an experimentally observable effect. Assume that flavor neutrinos behave as fundamental fields. At the end point of the beta decay, say $A \to B + e^- + \bar{\nu}_e$, with A and B two nuclei, e.g. tritium ^{3}H and ^{3}He, the electron would have the maximal kinetic energy $K_{max} = Q - m_e$, where $Q = E_A - E_B - m \approx m_A - m_B - m$; m is the electron mass. For tritium decay, $Q = 18.6$ KeV. A different result would be obtained for massive neutrinos with m_e replaced by m_1 (or m_2). See [9] for more details.

Since in $\widetilde{H}$ there are no mixing energy contributions, it may be regarded as the free energy $F = H - TS = \widetilde{H}$, where $T \equiv g = \tan 2\theta$ and the entropy S is associated with the mixing,

$$S = \int d^3\mathbf{x}\, \bar{\nu}_f A_0 \nu_f = \frac{1}{2}\, \delta m_{e\mu} \int d^3\mathbf{x}\, (\bar{\nu}_e \nu_\mu + \bar{\nu}_\mu \nu_e), \tag{38}$$

suggesting that the gauge field A_0 may interpreted as a reservoir [58].

Finally, notice that from Eq.(A.7) one gets $|V_\mathbf{k}|^2 \sin^2\theta = (\delta m_{12})^2 \sin^2 f(k, m_1, m_2) = (A_0^1)^2 g^2 f(k, m_1, m_2)/\cos^2\theta$. Neglecting, as in the Pontecorvo approximation, the $|V_\mathbf{k}|^2$-terms amounts to neglect g^2 terms, thus spoiling the gauge structure discussed above.

4. The vacuum as a refractive medium

Neutrino flavor fields propagate one in the background field of the other one through the coupling with the gauge field as discussed in the previous Section. This suggests that the propagation medium (the flavor vacuum) behaves as a refractive medium breaking the mass degeneracy (i.e. generating the neutrino mass difference). Consider for simplicity the usual QM framework. In the (non-refractive) vacuum, two waves of same frequency f propagate with speed $\mathrm{v}_0 = \lambda f$. Denote the corresponding two degenerate states by $|1\rangle$ and $|2\rangle$, $\langle i|j\rangle = \delta_{ij}$, $i, j = 1, 2$, (not necessarily neutrino states) evolving in time as

$$\begin{pmatrix} |1(t)\rangle \\ |2(t)\rangle \end{pmatrix} = \begin{pmatrix} e^{-i\omega t} & 0 \\ 0 & e^{-i\omega t} \end{pmatrix} \begin{pmatrix} |1(0)\rangle \\ |2(0)\rangle \end{pmatrix}. \tag{39}$$

Journal of Physics: Conference Series **2533** (2023) 012010 doi:10.1088/1742-6596/2533/1/012010

Assume then that the propagation medium is a birefringent medium presenting different refraction indexes, $n_i = \mathrm{v}_0/\mathrm{v}_i$ for $|i\rangle$, $i = 1, 2$, respectively, where v_i is the propagation speed, so that the degeneracy is broken [11]. Here we limit ourselves to the assumption that such a medium might exist, the question of how and why it might be generated is outside the tasks of the present work. In such a medium, different times, t_1 and t_2 for $|1\rangle$ and $|2\rangle$, respectively, are then needed for the propagation over a path of length L, i.e.:

$$t_i = \frac{L}{\mathrm{v}_i} = n_i \frac{L}{\mathrm{v}_0} = n_i\, t\ , \qquad i = 1, 2, \tag{40}$$

with $t = L/\mathrm{v}_0$. The time evolution phase factors in Eq. (39) are now $e^{-i\omega\, t_i} = e^{-i\omega_i t}$, $i = 1, 2$, where $\omega\, t_i = 2\,\pi\, f\, n_i\, t = 2\,\pi\, f_i\, t = \omega_i\, t$, with $f_i \equiv f\, n_i$, i.e. $\lambda_i\, f = \mathrm{v}_i$ and $\lambda_i\, f_i = \mathrm{v}_0$.

The mixed states $|\phi(t)\rangle$ and $|\psi(t)\rangle$ (cf. the mixing Eqs. (1) and (2)) are now introduced

$$\begin{pmatrix} |\phi(t)\rangle \\ |\psi(t)\rangle \end{pmatrix} = e^{-i\omega_1 t} \begin{pmatrix} \cos\theta & e^{-i(\omega_2-\omega_1)t}\sin\theta \\ -\sin\theta & e^{-i(\omega_2-\omega_1)t}\cos\theta \end{pmatrix} \begin{pmatrix} |1(0)\rangle \\ |2(0)\rangle \end{pmatrix}, \tag{41}$$

with $\theta \neq 0$ and $\omega_1 \neq \omega_2$. Considering the matrix elements $\langle\varsigma(t)|i\partial_t|\varsigma(t)\rangle$, with $|\varsigma(t)\rangle$ denoting the vector $(|\phi(t)\rangle, |\psi(t)\rangle)^T$, using $\omega_{\phi\phi} = \omega_1\cos^2\theta + \omega_2\sin^2\theta$, $\omega_{\psi\psi} = \omega_1\sin^2\theta + \omega_2\cos^2\theta$, $\omega_{\phi\psi} = (1/2)(\omega_2 - \omega_1)\sin 2\theta$ and $\omega_{\psi\phi} = \omega_{\phi\psi}$, the Hamiltonian H and the evolution equations are

$$H = \omega_{\phi\phi}|\phi(t)\rangle\langle\phi(t)| + \omega_{\psi\psi}|\psi(t)\rangle\langle\psi(t)| + \omega_{\phi\psi}(|\phi(t)\rangle\langle\psi(t)| + |\psi(t)\rangle\langle\phi(t)|) \tag{42}$$

$$H|\varsigma(t)\rangle = i\partial_t|\varsigma(t)\rangle = \omega_d|\varsigma(t)\rangle + \omega_{\phi\psi}\sigma_1|\varsigma(t)\rangle, \tag{43}$$

respectively, with $\omega_d = \mathrm{diag}(\omega_{\phi\phi}, \omega_{\psi\psi})$ (cf. with Eqs.(19) and (20)). Using $g \equiv \tan 2\theta = 2\omega_{\phi\psi}/(\omega_{\psi\psi} - \omega_{\phi\phi})$, it is $\omega_{\phi\psi} = (1/2)\, g\, (\omega_{\psi\psi} - \omega_{\phi\phi}) = (1/2)\, g\, (\omega_2 - \omega_1)\cos 2\theta$, and the effect of the propagation through the refractive medium is recognized to be equivalent to the one coming from the coupling to the gauge field $A_0^1 = \frac{1}{2}(\omega_2 - \omega_1)\ \cos 2\theta = \frac{1}{2}\omega(n_2 - n_1)\ \cos 2\theta$ (which is of course zero for $n_1 = n_2$), consistently with the gauge formalism of Section 3.

5. Conclusions

In conclusion, QFT computations lead to the non-trivial condensate of the flavor vacuum and to the flavor representations of the CAR unitarily inequivalent to the representation of the massive ν_1 and ν_2. In the QM (Pontecorvo) approximation, the QFT flavor vacuum condensate is missed, i.e. $|0\rangle_{1,2}$, instead of $|0(\theta,t)\rangle_{e,\mu}$, for any t, is used, by putting $|V_{\mathbf{k}}(t)|^2 = 0$ (cf. e.g. Eq.(14)). This amounts to neglect g^2-terms in the gauge formalism (cf. Section 3) and produces physically unacceptable inconsistencies.

For example, when considering the creation of a neutrino ν_e at $t = 0$, the use of the vacuum state $|0\rangle_{1,2}$ in the Wightman function ${}_{1,2}\langle 0|\nu_e(t,\mathbf{x})\,\overline{\nu}_e(0,\mathbf{y})|0\rangle_{1,2}$, leads to the process amplitude $\mathcal{P}^{>}_{ee}(\mathbf{k},t) = \cos^2\theta + \sin^2\theta|U_{\mathbf{k}}|^2 e^{-i(\omega_{\mathbf{k},2}-\omega_{\mathbf{k},1})t}$, which is not 1 at $t = 0$, an unacceptable result [35].

On the contrary, when the Wightman function is computed in $|0(\theta,t)\rangle_{e,\mu}$, the result is $\mathcal{P}^{>}_{ee}(\mathbf{k},t) = \cos^2\theta + \sin^2\theta(|U_{\mathbf{k}}|^2 e^{-i(\omega_{\mathbf{k},2}-\omega_{\mathbf{k},1})t} + |V_{\mathbf{k}}|^2 e^{i(\omega_{\mathbf{k},1}+\omega_{\mathbf{k},2})t})$, giving the correct condition $\mathcal{P}^{>}_{ee}(\mathbf{k},0^+) = 1$ at $t = 0$. Also, it is $\sum_{\rho\sigma}|\mathcal{P}^{>}_{\rho\sigma}(\mathbf{k},t)|^2 = 1$, $\rho,\sigma = e,\mu$. For more details see [35].

Another example of wrong result coming from the use of the Pontecorvo states is the following. It is meaningful to consider flavor oscillation at a time scale larger or comparable with the oscillation time t_{osc}. Oscillation can be, however, neglected in a Δt, around the production time at the vertex, say $t = 0$, such that $\tau_W \ll \Delta t \ll t_{osc}$, with τ_W the W^+ lifetime. At tree level, in such a Δt, lepton number is conserved (lepton number violation is negligible in loop diagrams in our discussion). Then in the leading order in Δt, the computation, for example, of the flavor violating amplitude $\mathcal{A}^P_{W^+\to e^+\nu_\mu}$, in the $W^+ \to e^+ + \nu_e$ decay process, in the short time limit, turns out to be non-zero, which is clearly an experimentally unacceptable result. On the contrary, when using the QFT flavor states, the amplitudes $\mathcal{A}_{W^+\to e^+\nu_\mu}$ is found to be zero [59, 60, 61]: $\mathcal{A}_{W^+\to e^+\nu_\mu} \approx 0$, for $\tau_W \ll \Delta t \ll t_{osc}$, consistently with the experiments.

One more example comes from the use of $|0\rangle_{1,2}$ in defining the one flavor neutrino state, putting, e.g., $\alpha^{r\dagger}_{\mathbf{k},e}(t)|0\rangle_{1,2} = |\alpha^r_{\mathbf{k},e}(t)\rangle_{1,2}$. Of course, the probability amplitude ${}_{1,2}\langle 0|\alpha^r_{\mathbf{k},e}(t)\alpha^{r\dagger}_{\mathbf{k},e}(t)|0\rangle_{1,2} = {}_{1,2}\langle\alpha^r_{\mathbf{k},e}(t)|\alpha^r_{\mathbf{k},e}(t)\rangle_{1,2}$ must be 1 at $t = 0$. From Eq.(10) ${}_{1,2}\langle 0|\alpha^{r\dagger}_{\mathbf{k},e}(t)\alpha^r_{\mathbf{k},e}(t)|0\rangle_{1,2} = |V_\mathbf{k}|^2\sin^2\theta$ and from the CAR we get ${}_{1,2}\langle 0|\alpha^r_{\mathbf{k},e}(t)\alpha^{r\dagger}_{\mathbf{k},e}(t)|0\rangle_{1,2} = -{}_{1,2}\langle 0|\alpha^{r\dagger}_{\mathbf{k},e}(t)\alpha^r_{\mathbf{k},e}(t)|0\rangle_{1,2}+{}_{1,2}\langle 0|\{\alpha^r_{\mathbf{k},e}(t),\alpha^{r\dagger}_{\mathbf{k},e}(t)\}|0\rangle_{1,2} = -|V_\mathbf{k}|^2\sin^2\theta+1$, which is a wrong result since, as said, it must be 1 at $t = 0$. Such a wrong result is avoided by "adding" the condensate effects $|V_\mathbf{k}|^2\sin^2\theta$ (i.e. the g^2 gauge terms are not missed in the computation). The correct result is obtained by using, instead of $|0\rangle_{1,2}$, the flavor vacuum, ${}_{e,\mu}\langle 0(\theta,t)|\alpha^r_{\mathbf{k},e}(t)\alpha^{r\dagger}_{\mathbf{k},e}(t)|0(\theta,t)\rangle_{e,\mu} = 1$, at any time t.

Acknowledgments
L.S. is thankful to the Polish National Science Center for the supporting grant 2018/31/D/ST2/02048.

Appendix A. The Hamiltonian and the symmetry charges

The Hamiltonian H and the momentum operator P^i, $i = 1,2,3$, are derived from the canonical energy-momentum tensor $T_{\rho\sigma}$ [4, 24, 25]

$$\begin{aligned} T_{\rho\sigma} &= \bar\nu_e i\gamma_\rho\partial_\sigma\nu_e - \eta_{\rho\sigma}\bar\nu_e(i\gamma^\lambda\partial_\lambda - m_e)\nu_e + \bar\nu_\mu i\gamma_\rho\partial_\sigma\nu_\mu - \eta_{\rho\sigma}\bar\nu_\mu(i\gamma^\lambda\partial_\lambda - m_\mu)\nu_\mu \\ &\quad + \eta_{\rho\sigma}m_{e\mu}(\bar\nu_e\nu_\mu + \bar\nu_\mu\nu_e) \\ &= \bar\nu_1 i\gamma_\rho\partial_\sigma\nu_1 - \eta_{\rho\sigma}\bar\nu_1(i\gamma^\lambda\partial_\lambda - m_1)\nu_1 + \bar\nu_2 i\gamma_\rho\partial_\sigma\nu_2 - \eta_{\rho\sigma}\bar\nu_2(i\gamma^\lambda\partial_\lambda - m_2)\nu_2, \end{aligned} \tag{A.1}$$

where $\eta_{\rho\sigma} = \mathrm{diag}(+1,-1,-1,-1)$ is the Minkowskian metric tensor. One obtains

$$H = \int d^3\mathbf{x}\, T^{00} = \int d^3\mathbf{x}\, \nu_1^\dagger\left(-i\alpha\cdot\nabla + \beta m_1\right)\nu_1 + \int d^3\mathbf{x}\, \nu_2^\dagger\left(-i\alpha\cdot\nabla + \beta m_2\right)\nu_2, \tag{A.2}$$

$$P^i = \int d^3\mathbf{x}\, T^{0i} = i\int d^3\mathbf{x}\, \nu_1^\dagger\partial^i\nu_1 + i\int d^3\mathbf{x}\, \nu_2^\dagger\partial^i\nu_2, \qquad i = 1,2,3, \tag{A.3}$$

The Noether charge associated to the global $U(1)$ symmetry is $Q = Q_1 + Q_2$ [6]:

$$Q_i = \int d^3\mathbf{x}\, \nu_i^\dagger(x)\, \nu_i(x)\,, \qquad i = 1,2. \tag{A.4}$$

From the Lagrangian Eq.(4) the (non conserved) time-dependent flavor charges are derived [6]:

$$Q_\sigma(t) = \int d^3\mathbf{x}\, \nu_\sigma^\dagger(x)\, \nu_\sigma(x)\,, \quad \sigma = e,\mu. \tag{A.5}$$

It is $Q_e(t) + Q_\mu(t) = Q$. The charges $Q_\sigma(t)$ are explicitly given in Eqs.(16) and (17). The generators of the (broken) $SU(2)$ symmetry, to which $Q_\sigma(t)$ are associated, are $S_3 \equiv (1/2)(Q_1 - Q_2)$, $S_+ \equiv \int d^3\mathbf{x}\left[\nu_1^\dagger(x)\nu_2(x)\right]$, $S_- = S_+^\dagger \equiv \int d^3\mathbf{x}\left[\nu_2^\dagger(x)\nu_1(x)\right]$, with algebra: $[S_+,S_-] = 2S_3$, $[S_3,S_\pm] = \pm S_\pm$. The Casimir operator is $S_0 \equiv (1/2)(Q_1+Q_2) = (1/2)Q$.

The explicit expressions for $|U_\mathbf{k}|$ and $|V_\mathbf{k}|$ (cf. e.g. Eq (10)) are [3]:

$$|U_\mathbf{k}| = \left(\frac{\omega_{\mathbf{k},1}+m_1}{2\omega_{\mathbf{k},1}}\right)^{\frac{1}{2}}\left(\frac{\omega_{\mathbf{k},2}+m_2}{2\omega_{\mathbf{k},2}}\right)^{\frac{1}{2}}\left(1+\frac{\mathbf{k}^2}{(\omega_{\mathbf{k},1}+m_1)(\omega_{\mathbf{k},2}+m_2)}\right), \tag{A.6}$$

$$|V_\mathbf{k}| = \frac{|\mathbf{k}|}{2\sqrt{\omega_{\mathbf{k},1}\omega_{\mathbf{k},2}}}\left(\sqrt{\frac{\omega_{\mathbf{k},2}+m_2}{\omega_{\mathbf{k},1}+m_1}} - \sqrt{\frac{\omega_{\mathbf{k},1}+m_1}{\omega_{\mathbf{k},2}+m_2}}\right). \tag{A.7}$$

Appendix B. Entangled vacuum, static and dynamic neutrino entanglement

The vacuum $|0(\theta,t)\rangle_{e,\mu}$ is an entangled state of neutrino pairs $(\alpha^r_{\mathbf{k},i}, \beta^r_{-\mathbf{k},j})$ (cf. Section 2). This ca be estimated by means of the covariance $cov\,(N_a, N_b) \equiv \langle N_a N_b\rangle - \langle N_a\rangle\,\langle N_b\rangle$ [26], where N_a, N_b are number operators; a and b denote α and/or β quanta, the symbol $\langle\ldots\rangle$ denotes expectation value in $|0\rangle_{e,\mu}$. We find

$$cov\left(N_{\alpha^r_{\mathbf{k},i}}, N_{\beta^r_{-\mathbf{k},j}}\right) \;=\; \frac{1}{2}\sin^2(2\theta)\,|V_\mathbf{k}|^2\,, \qquad i,j=1,2, \quad i\neq j, \tag{B.1}$$

and

$$cov\left(N_{\alpha^r_{\mathbf{k},i}}, N_{\beta^r_{-\mathbf{k},i}}\right) \;=\; \sin^2\theta\,|V_\mathbf{k}|^2(1-\sin^2\theta\,|V_\mathbf{k}|^2), \quad i=1,2. \tag{B.2}$$

The covariance is zero is zero for the pairs $(\alpha^r_{\mathbf{k},i}, \alpha^r_{\mathbf{k},j})$ and $(\beta^r_{\mathbf{k},i}, \beta^r_{\mathbf{k},j})$, for $i\neq j$, since $\langle N_{\alpha^r_{\mathbf{k},i}} N_{\alpha^r_{\mathbf{k},j}}\rangle = \sin^4\theta\,|V_\mathbf{k}|^4 = \langle N_{\alpha^r_{\mathbf{k},i}}\rangle\,\langle N_{\alpha^r_{\mathbf{k},j}}\rangle$.

We also have $\langle(\Delta N_{\alpha^r_{\mathbf{k},i}})^2\rangle = \langle(\Delta N_{\beta^r_{\mathbf{k},i}})^2\rangle = \sin^2\theta\,|V_\mathbf{k}|^2(1-\sin^2\theta\,|V_\mathbf{k}|^2)$, where $\langle(\Delta N)^2\rangle \equiv \langle(N-\langle N\rangle)^2\rangle = \langle N^2\rangle - \langle N\rangle^2$ is the variance.

Flavor states are described as the *static* entanglement of the ν_i, $i=1,2$, neutrinos [4] when the two-qubit representation of neutrinos is adopted $|\nu_1\rangle \equiv |1\rangle_1|0\rangle_2 \equiv |10\rangle$, $|\nu_2\rangle \equiv |0\rangle_1|1\rangle_2 \equiv |01\rangle$. The variance of the Noether charges Q_i gives a measure of the static entanglement occurring in the states $|\nu^r_{\mathbf{k},\rho}\rangle$ [31]: $\sigma^2_{Q_i} \equiv \langle Q_i^2\rangle_\rho - \langle Q_i\rangle^2_\rho = (1/4)\sin^2(2\theta)$, which agrees with QM result. The *dynamical* entanglement occurs in time evolution with flavor oscillation [28]. Then the two-qubit states are, in obvious notation, $|\nu_e\rangle \equiv |1\rangle_e|0\rangle_\mu$, $|\nu_\mu\rangle \equiv |0\rangle_e|1\rangle_\mu$. A measure of the dynamic entanglement in the states $|\nu^r_{\mathbf{k},\rho}(t)\rangle$ is given by the variance of the flavor charges $\sigma^2_Q \equiv \langle Q^2_\rho(t)\rangle_\rho - \langle Q_\rho(t)\rangle^2_\rho = \mathcal{Q}_{\rho\to\rho}(t)\,(1-\mathcal{Q}_{\rho\to\rho}(t))$. In these results, the contributions from the vacuum condensate in the oscillation formulas are responsible for the differences with respect to the QM result (recovered in the relativistic limit). Remarkably, in QFT, the Leggett-Garg inequalities for temporal correlations appear to be more strongly violated than in QM [30].

References

[1] Bilenky S M and Pontecorvo B 1978 *Phys. Rept.* **41** 225
[2] Giunti C, Kim C 2007 *Fundamentals of Neutrino Physics and Astrophysics* (Oxford: Oxford Univ. Press)
[3] Blasone M and Vitiello G 1995 *Annals Phys.* **244** 283
[4] Blasone M, Jizba P, Vitiello G 2011 *Quantum Field Theory and Its Macroscopic Manifestations: Boson Condensation, Ordered Patterns, and Topological Defects* (London: Imperial College Press)
[5] Blasone M, Henning P A and Vitiello G 1999 *Phys. Lett.* B **466** 262
[6] Blasone M, Jizba P and Vitiello G 2001 *Phys. Lett.* B **517** 471
[7] Blasone M, Smaldone L and Vitiello G 2019 *J. Phys.: Conf. Series* **1275** 012023
[8] Blasone M, Jizba P, Mavromatos N E, Smaldone L 2019 *Phys. Rev.* D **100** 045027,
[9] Blasone M, Di Mauro M and Vitiello G 2011 *Phys. Lett.* B **697** 238
[10] Blasone M, Smaldone L and Vitiello G 2011 *PoS(CORFU2021)049* https://pos.sissa.it/406/
[11] Bruno A, Capolupo A, Kak S, Raimondo G, Vitiello G 2011 *Mod. Phys. Lett.* B **25**(20) 1661
[12] Weinheimer C 2010 *Prog. Part. Nucl. Phys.* **64** 205
[13] Blasone M, Capolupo A, Vitiello G 2002 *Phys. Rev.* D **66** 025033
[14] Ji C R and Mishchenko Y 2002 *Phys. Rev.* D **65** 096015
[15] Binger M and Ji C R 1999 *Phys. Rev.* D **60** 056005
[16] Ji C R and Mishchenko Y 2001 *Phys. Rev.* D **64** 076004
[17] Blasone M, Capolupo A, Romei O and Vitiello G 2001 *Phys. Rev.* D **63** 125015
[18] Blasone M, and Palmer J 2004 *Phys. Rev.* D **69** 057301
[19] Smaldone L and Vitiello G 2021 *Universe* **7** 504
[20] Blasone M, Gargiulo G, Vitiello G 2016 *Phys. Lett.* B **761** 104
[21] Hannabuss K C, Latimer D C 2000 *J. Phys.* A **33** 1369
[22] Hannabuss K C, Latimer D C 2003 *J. Phys.* A **36** L69
[23] Barton G 1963 *Introduction to Advanced Field Theory* (New York: Wiley-Interscience)
[24] Umezawa, H, Matsumoto H and Tachiki M 1982 *Thermo Field Dynamics and Condensed States* (Amsterdam: North-Holland)

[25] Bogoliubov N N, Logunov A A, Oksak A I, Todorov I T 1990 *General Principles of Quantum Field Theory* (Boston: Reidel)
[26] Gerry C, Knight P 2005 *Introductory Quantum Optics* (Cambridge: Cambridge University Press)
[27] Blasone M, Dell'Anno F, De Siena S, Di Mauro M and Illuminati F, 2008 *Phys. Rev.* D **77** 096002
[28] Blasone M, Dell'Anno F, De Siena S and Illuminati F 2009 *Europhys. Lett.* **85** 50002
[29] Dixit K, Naikoo J, Banerjee S and Alok A K 2018 *Eur. Phys. J.* C **78**(11) 914
[30] Blasone M, Illuminati F, Petruzziello L and Smaldone L Leggett-Garg inequalities in the quantum field theory of neutrino oscillations *Preprint* arXiv:2111.09979v2 [quant-ph].
[31] Blasone M, Dell'Anno F, De Siena S, Illuminati F 2014 *Europhys. Lett.* **106** 30002
[32] Blasone M, Dell'Anno F, De Siena S, Illuminati F 2015 *Europhys. Lett.* **112** 20007
[33] Alok A K, Banerjee S, Sankar S U 2016 *Nucl. Phys. B* **909** 65
[34] Blasone M, De Siena S and Matrella C 2021 *Eur. Phys. J. C* **81** 660
[35] Blasone M, Henning P A, Vitiello G 1999 *Phys. Lett.* B **451** 140.
[36] V. A. Bittencourt, A. E. Bernardini and M. Blasone, 2021 Eur. Phys. J. C **81** 1; 2022 EPL **139** 44002
[37] Leite Lopes J 1981 *Gauge Field Theories. An Introduction* (Oxford: Pergamon Press)
[38] Blasone M, Jizba P, Smaldone L 2019 *Phys. Rev.* D **99**
[39] Blasone M, Magueijo J and Pires-Pacheco P 2005 *Europhys. Lett.* **70** 600
[40] Blasone M, Jizba P, Mavromatos N E, Smaldone L 2020 *Phys. Rev.* D **102** 025021
[41] Kostelecky V A and Mewes M 2004 *Phys. Rev.* D **69** 016005
[42] Klinkhamer F R 2006 *Int. J. Mod. Phys.* **A21** 161
[43] Magueijo J and Smolin L 2002 *Phys. Rev. Lett.* **88** 190403
[44] Magueijo J and Smolin L 2003 *Phys. Rev.* D **67** 044017
[45] Exner P 1983 X *Phys. Rev.* D **28** 2621
[46] Fujii K, Habe C, Yabuki T 1999 *Phys. Rev.* D **59** 113003
[47] Blasone M, and Vitiello G 1999 *Phys. Rev.* D **60** 111302.
[48] Martellini M, Sodano P, Vitiello G 1978 Nuovo Cim. A **48** 341
[49] Blasone M, Lambiase G, Luciano G G, L. Petruzziello L 2020 *Eur. Phys. J. C* **80** 130
[50] Ellis J R, Mavromatos N E, and Nanopoulos D V 2008 *Phys. Lett.* B **665** 412
[51] Amelino-Camelia G 2007 *Nature Phys.* **3** 81
[52] Mavromatos N E and Sarkar S 2008 *New J. Phys.* **10** 073009
[53] Mavromatos N E and Sarkar S 2006 A possible quantum-gravitational origin of the neutrino mass difference? *Proc. 3rd Int. Workshop on NO-VE: Neutrino Oscillations in Venice: 50 Years after the Neutrino Experimental Discovery Preprint* hep-ph/0604081
[54] Capolupo A, Lambiase G, Vitiello G 2016 *Adv. High Energy Phys.* **2016** 3127597
[55] Blasone M, Capolupo A, Capozziello S, Carloni S, Vitiello G 2004 *Phys. Lett.* A **323** 182
[56] Capolupo A, Capozziello S, Vitiello G 2008 *Int. J. Mod. Phys.* A **23** 4979
[57] Blasone M, Lambiase G, Luciano G G, Petruzziello L, Smaldone L 2020 *Class. Quant. Grav.* **37** 155004
[58] Celeghini E, Graziano E, Nakamura K, Vitiello G 1992 *Phys. Lett.* B **285** 98
[59] Blasone M, Capolupo A, Terranova F and Vitiello G 2005 *Phys. Rev.* D **72** 013003
[60] Blasone M, Capolupo A, Ji C R, Vitiello G 2010 *Int. J. Mod. Phys. A* **25** 4179.
[61] Nishi C C 2008 *Phys. Rev.* D **78** 113007

Journal of Physics: Conference Series **2533** (2023) 012011 doi:10.1088/1742-6596/2533/1/012011

Decoherent histories approach to the cosmological measure problem

Seth Lloyd
Department of Mechanical Engineering, Massachusetts Institute of Technology
E-mail: slloyd@mit.edu

Abstract. The method of decoherent histories allows probabilities to be assigned to sequences of quantum events in systems, such as the universe as a whole, where there is no external observer to make measurements. This paper applies the method of decoherent histories to address cosmological questions. Using a series of simple examples, beginning with the harmonic oscillator, we show that systems in a stationary state such as an energy eigenstate or thermal state can exhibit decoherent histories with non-trivial dynamics. We then examine decoherent histories in a universe that undergoes eternal inflation. Decoherent histories that assign probabilities to sequences of events in the vicinity of a timelike geodesic supply a natural cosmological measure. Under reasonable conditions, such sequences of events do not suffer from the presence of unlikely statistical fluctuations that mimic reality.

1. Introduction

The consistent or decoherent histories approach to quantum mechanics is a method for assigning probabilities to sequences of events for a quantum-mechanical system [1-35]. Because it does not rely on the notion of measurement, the decoherent histories approach is useful in theories such as quantum cosmology where probabilities have to be assigned, but no external system is measuring or decohering the system – in this case, the universe itself. The cosmological measure problem addresses the question of how to assign probabilities to events in various cosmological scenarios, e.g., eternal inflation [36-54]. Some apparently reasonable cosmological measures have counter-intuitive consequences, e.g., the notion that we are all just 'Boltzmann brains' that arose from a statistical fluctuation. This paper applies the method of decoherent histories to the cosmological measure problem. First, the formalism of decoherent histories is reviewed. Second, the method is applied to simple systems to show that stationary states such as energy eigenstates or thermal states can still exhibit non-trivial decoherent histories, in contradiction to the natural intuition that systems in stationary states 'do nothing.' (This result answers a question raised by Boddy *et al.* about whether systems in stationary states can exhibit decoherent histories that fluctuate over time [54].) Decoherent histories are used to show that systems described by a stationary state of quantum jump models or environmentally induced decoherence models can still be thought of as exhibiting non-trivial temporal fluctuations. Third, decoherent histories are applied to eternal inflation [55-56]. Decoherent histories for the sequences of events that occur in the vicinity of a timelike geodesic are shown to give rise to cosmological measures that differ from conventional volume-counting measures, an approach similar to that of Nomura [57-59]. Under conventional assumptions about the potentials in the underlying physics, such histories give rise to a picture of eternal inflation in which a period of rapid inflation gives rise to a Friedman-Robertson-Walker cosmology in de Sitter space with small cosmological term. After a period long compared with the Hubble time of the de Sitter space, but short compared with the time required to generate a Boltzmann brain, quantum/thermal fluctuations give rise to another period of rapid inflation, and the cycle recommences.

2. Review of decoherent histories

The decoherent histories approach originated with Griffiths [1-3], who called this approach consistent histories, and was independently developed by Omnès [4-7] and by Gell-Mann and Hartle [8-15], who termed this approach decoherent histories. We will adopt the latter nomenclature. Consider a sequence of projective measurements that could be made on a quantum system at times $t_1 < \ldots < t_n$. (The extension to generalized measurements will be given below.) The measurement at time t_k has exhaustive and mutually exclusive outcomes α_k. This measurement corresponds to a set of projection operators $\{P^k_{\alpha_k}\}$ in the Heisenberg picture, where $P^k_{\alpha_k} P^k_{\alpha'_k} = \delta_{\alpha_k \alpha'_k} P^k_{\alpha_k}$, and $\sum_{\alpha_k} P^k_{\alpha_k} = I$. A *history* $\tilde{\alpha}$ corresponds to a sequence of outcomes $\tilde{\alpha} = \alpha_1 \ldots \alpha_n$. The decoherence functional $D(\tilde{\alpha}, \tilde{\alpha}')$ for initial state ρ_i is defined to be

$$D(\tilde{\alpha}, \tilde{\alpha}') = \mathrm{tr} P^n_{\alpha_n} \ldots P^1_{\alpha_1} \rho_i P^1_{\alpha'_1} \ldots P^n_{\alpha'_n} = \mathrm{tr} P^\dagger_{\tilde{\alpha}} \rho_i P^\dagger_{\tilde{\alpha}'}. \quad (1)$$

If one performs this sequence of measurements, then the probability for the sequence of events that constitutes the history $\tilde{\alpha}$ are given by the diagonal term of the decoherence functional $p(\tilde{\alpha}) = D(\tilde{\alpha}, \tilde{\alpha})$. Note that these probabilities are non-negative and sum to 1.

The off-diagonal terms in the decoherence function measure the degree of quantum interference between different histories. When these terms are comparable to the corresponding on-diagonal terms, it means that measurements that correspond to earlier events in the history have a strong effect on the probabilities for later events [35]. For example, in the double-slit experiment, the histories that correspond to which slit the particle goes through are coherent: consequently, a measurement that determines through which slit a particle passes has a strong effect on probability of where the particle lands on the screen. Indeed, such a measurement destroys the characteristic interference pattern of the double-slit experiment.

The presence of coherence means that the probabilities $p(\tilde{\alpha})$ fail to obey probability sum rules. For example, let P^1_1, P^1_2 project onto the states that go through slit 1 or slit 2 at time t_1, and P^2_x project onto states that land on the screen at point x. The presence of coherence means that

$$\mathrm{tr} P^2_x \rho_i = \sum_{jj'} \mathrm{tr} P^2_x P^1_j \rho P^1_{j'} \neq \sum_j \mathrm{tr} P^2_x P^1_j \rho P^1_j. \quad (2)$$

That is, the probabilities for where the particle lands on the screen in the absence of a measurement of which slit it passed through are different from the probabilities for where the particle lands on the screen given that a measurement has been made. More generally, if

$$|D(\tilde{\alpha}, \tilde{\alpha}')|^2 / D(\tilde{\alpha}, \tilde{\alpha}) D(\tilde{\alpha}', \tilde{\alpha}') \leq \epsilon << 1, \quad (3)$$

then the histories are said to be approximately decoherent: such histories obey probability sum rules to accuracy ϵ.

The decoherence functional characterizes the degree to which measurements affect the future behavior of a quantum system. If a system decoheres with respect to a particular set of measurements, then the measurements made in the past have minimal effect on the outcomes of measurements made in the future. If this is so, then we can assign probabilities to the sequence of events corresponding to the measurement outcomes *whether the measurements are actually performed or not.* That is, although they are defined in terms of the mathematical apparatus of measurements – projection operators or more generally positive operator valued measures (POVMs) – decoherent histories represent a method for assigning probabilities to sequences of events in the absence of measurement. In the words of Griffiths [1-3], decoherent histories refer to sequences of events 'that we can talk about at the breakfast table.' In the double slit experiment, we are not allowed to talk about the particle going through either one slit or the other, because to explain the interference pattern on the screen, it must go through both at once.

3. Stationary states support time-dependent decoherent histories

We now apply the method of decoherent histories to show that stationary states can exhibit non-trivial time-dependent decoherent histories.

First look at histories that trivially decohere – the histories for energy eigenstates of a closed physical system. In this case, $P_j^k = |E_j\rangle\langle E_j|$ for energy eigenstates $|E_j\rangle$. Note P_j^k is independent of the time step k. We have $P_j^k P_{j'}^k = \delta_{jj'} P_j$. and $D(\tilde{j}; \tilde{j}') \propto \delta_{\tilde{j}\tilde{j}'}$, independent of the initial state ρ_i. Here the initial projection takes ρ_i to an energy eigenstate, and subsequent projections simply confirm that the system remains in that state: histories of energy eigenstates do not exhibit time-dependent fluctuations.

Histories of other variables do exhibit fluctuations, however. A system can possess complementary sets of decoherent histories. In the harmonic oscillator, for example, even though energy does not fluctuate, phase does. Consider an harmonic oscillator with Hamiltonian $^-h\omega a^\dagger a = \hbar\omega \sum_{\ell=0}^{\infty} |\ell\rangle\langle\ell|$, where $|\ell\rangle$ is the ℓth energy eigenstate. For simplicity, restrict attention to the subspace $\mathcal{H}_N$ of states whose energy is less than $N^-h\omega$ Within this space, we can define phase states $|\phi\rangle = N^{-1/2} \sum_{\ell=0}^{N-1} e^{i\ell\phi}|\ell\rangle$. The phase states evolve in time as $|\phi\rangle \rightarrow |\phi + \omega t\rangle$. The N states $|\phi_j\rangle$ where $\phi_k = 2\pi j/N$ form an orthonormal basis for $\mathcal{H}_N$. Over time $\Delta t = 2\pi/N\omega$, we have

$$|\phi_j\rangle \rightarrow U_{\Delta t}|\phi_j\rangle = |\phi_{j+1}\rangle, \tag{4}$$

where $U_{\Delta t} = e^{-iH\Delta t/\hbar}$ and $j+1$ is defined modulo N. That is, over time Δt, the states $|\phi_j\rangle$ evolve deterministically into each other.

Suppose that the oscillator starts out in its ground state $|0\rangle$, and consider the histories defined by measurement operators $P_j = |\phi_j\rangle\langle\phi_j|$ spaced at intervals Δt. The decoherence functional is

$$D(j_1 \ldots j_n, j_1' \ldots j_n') = \mathrm{tr} P_{j_n} U_{\Delta t} \ldots U_{\Delta t} P_{j_1} |0\rangle\langle 0| P_{j_1'} U_{\Delta t}^\dagger \ldots U_{\Delta t}^\dagger P_{j_n'}. \tag{5}$$

Because of the deterministic evolution of the phase states, we have $D(j_1 \ldots j_n, j_1' \ldots j_n') = 0$ unless $j_n' = j_n$, $j_k = j_{k-1} + 1$, and $j_k' = j_{k-1}' + 1$. That is, the off-diagonal terms of $D(j_1 \ldots j_n, j_1' \ldots j_n')$ are all zero, and the on-diagonal terms reflect the deterministic nature of the time evolution. The first projection yields equal probabilities $1/N$ for all phase states $|\phi_k\rangle$, and the subsequent evolution is entirely deterministic. The phase measurement corresponds to decoherent histories, even though the initial state is the ground state.

The set of decoherent histories corresponding to phase state evolution describes a quite different type of behavior from the behavior given by the set of decoherent histories corresponding to energy eigenstates. Described in terms of energy, the system remains static. Described in terms of phase, the system fluctuates. The two types of histories, energy and phase, represent complementary ways of describing the evolution of the same physical system.

As one might expect, histories that mix phase and energy eigenstates fail to decohere. Indeed, histories that begin in an energy eigenstate, progresses through a sequence of phase states, and then end in an energy eigenstate, are fully coherent: it is straightforward to show that the off-diagonal parts of the decoherence functional are of the same size as the on-diagonal parts. In particular, if one starts in the ground state, and then ends in the ground state, phase fluctuations do not decohere. In general, histories that begin and end in a pure state decohere only if the histories are completely deterministic. When the final projector is a pure state, $P_n = |\phi\rangle\langle\phi|$, we have

$$D(\tilde{\alpha};\tilde{\alpha}') = \langle\phi|P_{\tilde{\alpha}}|\psi\rangle\langle\psi|P^{\dagger}_{\tilde{\alpha}'}|\phi\rangle, \tag{6}$$

and

$$|D(\tilde{\alpha},\tilde{\alpha}')|^2/D(\tilde{\alpha},\tilde{\alpha})D(\tilde{\alpha}',\tilde{\alpha}') = 1, \tag{7}$$

unless $D(\tilde{\alpha};\tilde{\alpha}') \propto \delta_{\tilde{\alpha}\tilde{\alpha}'}$: to be decoherent, the histories must be deterministic. So time-dependent histories of probabilistic fluctuations that begin in the vacuum and end in the vacuum are coherent.

By contrast, histories that begin in a mixed state such as a thermal state, and end in an mixed state, can be decoherent. Now consider decoherent histories of thermal states. Such states can either arise from interaction with a reservoir at temperature $T = 1/k_B\beta$, or as subsystems of a larger system that is in a pure state. The latter case arises in gravitational contexts such as Hawking radiation, Unruh radiation, and de Sitter space. The mathematical question of whether or not histories decohere depends only on the thermal form of the state and on the dynamics, not on whether the system is thermal because it is interacting with a reservoir or thermal because it is entangled. A considerable literature shows that stationary, thermal states exhibit non-trivial, temporally fluctuating, decoherent histories [12-35]. The positions of particles that begin in thermal states and that undergo Brownian motion exhibit decoherent histories, as do hydrodynamic variables – the coarse-grained values of quantum field, energies, and particle densities.

4. Decoherent histories and quantum jumps

When the system in question is an open system interacting with its environment, or equivalently a subsystem of a larger system, then the quantum jump picture yields decoherent histories for sequences of projections corresponding to the jump operators [18-22]. The decoherence of histories of quantum jumps allows one to relate the decoherent histories approach to the idea of environmentally induced decoherence. In particular, if the subsystem's dynamics can be described by a Lindblad equation, then the resulting stochastic Schrödinger equation intrinsically gives rise to decoherent histories. So, for example, a subsystem in a stationary thermal state that is a fixed point of the Lindblad equation undergoes decoherent histories described by probabilistic jumping from state to state. Such decoherent histories exist both when the open system is in a thermal state because of its interaction with a thermal environment, and when it is a subsystem of a larger system in a pure state. The automatic existence of decoherent histories corresponding to histories of the stochastic Schrödinger equation and of quantum state diffusion is particularly useful as the jump operators for systems weakly coupled to a Markovian environment represent jumps between energy eigenstates. For such systems, we are allowed to talk at the breakfast table about the system hopping thermally from energy eigenstate to energy eigenstate as described by the Bloch-Redfield equation, even though the system as a whole is a stationary thermal state.

We present here a simple derivation of why systems that evolve according to a Lindblad equation exhibit decoherent histories. In contrast to previous derivations [18-22], which focus on an Itoh calculus derivation of the relation between quantum jumps and decoherent histories, the derivation given here is based on environmentally induced decoherence. The Lindblad equation represents the most general infinitesimal completely positive (i.e., legal) time evolution for a quantum system. A general Lindblad equation takes the form

$$\frac{\partial\rho}{\partial t} = -i[H,\rho] - \gamma/2\sum_j (L_j^{\dagger}L_j\rho - 2L_j\rho L_j^{\dagger} + \rho L_j^{\dagger}L_j). \tag{8}$$

The Lindblad equation for a given system interacting with its environment can be derived by starting with system and environment in the uncorrelated state $\rho_S \otimes \rho_E$, and applying the unitary system-environment time evolution $U(\Delta t)$ over a time Δt equal to the correlation time of the environment. The system evolves to

$$\rho_S(0) \to \rho_S(\Delta t) = \mathrm{tr}_E \rho U(\Delta t)\rho_S \otimes \rho_E U^\dagger(\Delta t). \quad (9)$$

Expanding to second order in Δt yields the infinitesimal form of the Lindblad equation (8). That is, in addition to being the general infinitesimal form for a completely positive map, the Lindblad equation has a physical interpretation as an approximate infinitesimal time evolution for a system interacting unitarily with an environment whose correlations decay over a characteristic time.

To look at decoherent histories, for simplicity consider the case where there is only one Lindblad operator $L_1 = L$ and $H = 0$: $\partial\rho/\partial t = (-\gamma/2)(L^\dagger L\rho - 2L\rho L^\dagger + \rho L^\dagger L)$. Use the polar decomposition to write $L = UA$, where U is unitary, $U^\dagger = U^{-1}$, and A is Hermitian, $A = A^\dagger$. The infinitesimal dynamics generated by the Lindblad equation over time Δt is equivalent to the following measurement plus feedback procedure:

(1) Make a generalized measurement on the system with POVM operators $M_1 = A^2\gamma\Delta t$ and $M_2 = 1 - A^2\gamma\Delta t$. With probability $p_1 = \gamma\Delta t\ \mathrm{tr}A^2\rho$ the system goes to the state $\rho_1 = (1/p_1)A\rho A$, and with probability $p_0 = 1 - p_1$ the system goes to the state $\rho_0 = (1/p_0)\sqrt{1 - A^2\gamma\Delta t}\ \rho\ \sqrt{1 - A^2\gamma\Delta t} \approx \rho_0$. Because any generalized measurement can be written as a von Neumann measurement on system plus an ancilla [55], we have $\rho_1 = {p_1}^{-1}P_1\rho \otimes \sigma P_1$, $\rho_0 = {p_0}^{-1}P_0\rho \otimes \sigma P_0$, for projectors P_1, $P_0 = 1 - P_1$, and ancilla in state σ. (This technique shows how to generalize decoherent histories from projective measurements to generalized measurements [35].)

(2) Now feed back the result of the measurement. If the result of the measurement is the state 1, apply the unitary transformation U. If the result is 0, do nothing. The system is now in the state

$$\begin{aligned}\rho' &= \sqrt{1 - A^2\gamma\Delta t}\ \rho\ \sqrt{1 - A^2\gamma\Delta t} + L\rho L^\dagger\gamma\Delta t \\ &= \rho - (\gamma\Delta t/2)(L^\dagger L\rho - 2L\rho L^\dagger + \rho L^\dagger L) + O(\Delta t^2).\end{aligned} \quad (9)$$

Because the Lindblad equation is mathematically equivalent to projective measurement on system plus ancilla followed by unitary feedback, the set of histories of the system plus ancilla corresponding to the projections P_0, P_1 repeated at time intervals Δt are decoherent. (Note that in this picture the unitary time evolutions between projections depend on the previous history of projections, corresponding to the more general model of decoherent histories given by Gell-Mann and Hartle [7-15].) One can think of this demonstration of decoherence as a *derivation* of the stochastic Schrödinger equation or of a quantum jump model. Because the histories are decoherent, we can describe the time evolution of the system in terms of a stochastic process: over time Δt the system goes to the state $\rho_1 = {p_1}^{-1}L\rho L^\dagger$ with probability $p_1 = \mathrm{tr}L\rho L^\dagger\gamma\Delta t$, or remains in the state $\rho_0 = \sqrt{1 - A^2\gamma\Delta t}\ \rho\ \sqrt{1 - A^2\gamma\Delta t}$ with probability $p_0 = 1 - p_1$. The treatment of the general Lindblad equation is essentially the same, except now there are multiple types of quantum jumps that can occur, one type for each Lindblad operator, and the time evolution in between jumps includes the effect of the system Hamiltonian. Note that the histories induced by the Lindblad equation are decoherent for any initial state, including stationary states such as thermal states or energy eigenstates.

This result establishes that an open system whose time evolution is governed by a Lindblad equation can be thought of as undergoing quantum jumps even when it is in a stationary state. The histories corresponding to different sequences of quantum jumps are decoherent.

As an example, consider a harmonic oscillator with Hamiltonian $H = \hbar\omega \sum_\ell |\ell\rangle\langle\ell| = \hbar\omega a^\dagger a$ as above, interacting linearly with a bath of modes of the electromagnetic field at temperature $T = 1/\beta$. The oscillator obeys the Lindblad equation

$$\frac{\partial \rho}{\partial t} = -i[H, \rho] - \frac{\gamma_+}{2}(aa^\dagger \rho - 2a^\dagger \rho a + \rho a a^\dagger) - \frac{\gamma_-}{2}(a^\dagger a \rho - 2a\rho a^\dagger + \rho a^\dagger a), \tag{10}$$

where $\gamma_+/\gamma_- = e^{-\beta\omega}$. The thermal state $\rho_{th} = (1/Z)e^{-\beta H}$ is a stationary state of the time evolution. In the thermal state, the oscillator exhibits decoherent histories over sequences of energy eigenstates, in which the nth energy eigenstates absorbs photons at a rate $n\gamma_+$ and emits photons at a rate $n\gamma_-$.

4.1. Decoherent histories for open systems over times longer than the relaxation time

Consider an open system that relaxes over time τ to a fixed state ρ_0 of the system's Hamiltonian H, e.g., a thermal state $\rho_0 = (1/Z)e^{-\beta H}$. Suppose that the time intervals between the projectors P_{α_j} are much longer than τ. Then *any* set of histories decoheres (i.e., not merely the jump histories given by the Lindblad operators). The reason is simple: because the system relaxes to the same state independent of the input state, if one waits for $>> \tau$, the probabilities for measurement are just given by the probabilities for measurement on ρ_0, independent of whether some measurement was made long ago or not.

4.2. Time-dependent fluctuations in stationary states

The notion that a stationary quantum state does not fluctuate in time seems at first a perfectly reasonable one [54]. However, the time-independent history of a stationary state can also be decomposed as a quantum superposition of time-dependent fluctuating histories. Under a wide variety of circumstances, those histories decohere, and so we are free to describe the time evolution of such systems in terms of those histories. A recent paper [54] suggested the contrary, it is worth discussing briefly the decoherent histories in thermal states, why do Boddy *et al.* decide that decoherent histories are not possible in such states [54]? There are two reasons. First, they use a time-symmetric version of decoherent histories that includes both initial and final states. Second, they use decoherent histories with only one intermediate set of events between those initial and final states. While it is true that such histories do not decohere, it is unclear why one should to restrict one's attention so such histories.

We review their argument. The time-symmetric version of decoherent histories is appropriate when all or part of the universe possesses a final state, as in spatially and temporally compact universes whose state is computed by the Hartle-Hawking imaginary time procedure, or in the Horowitz-Maldacena model of black hole evaporation. It does not seem that such a situation holds in inflationary models, and so it is unclear why this formalism should be applied here.

The decoherence functional $D(\tilde{\alpha}, \tilde{\alpha}')$ for initial state ρ_i and final state ρ_f is defined to be

$$D(\tilde{\alpha}, \tilde{\alpha}') = Z^{-1}\mathrm{tr}\rho_f P^n_{\alpha_n} \ldots P^1_{\alpha_1} \rho_i P^1_{\alpha'_1} \ldots P^n_{\alpha'_n} = Z^{-1}\mathrm{tr}\rho_f P^\dagger_{\tilde{\alpha}} \rho_i P_{\tilde{\alpha}'}, \tag{11}$$

where $Z^{-1} = \mathrm{tr}\rho_f\rho_i$. As before, histories decohere if the off-diagonal terms in the decoherence functional are small compared with the on-diagonal ones, given that initial state is ρ_i and the final state is ρ_f. The addition of the final state in the decoherence function is equivalent to adding one additional measurement operator, whose final projection correponds to a measurement revealing that the system is in ρ_f. Such a measurement can be performed, for example, by adjoining an ancillary system in state ρ_A and performing a projection P on system and ancilla such that $\mathrm{tr}_A P I \otimes \rho_A P = \rho_f$.

Journal of Physics: Conference Series **2533** (2023) 012011 doi:10.1088/1742-6596/2533/1/012011

Boddy *et al.* consider closed quantum systems and investigate situations where the initial and final state are both stationary states of the system dynamics. They consider histories with a single projection between the initial and final state, for which the decoherence functional is

$$Z^{-1}\text{tr}\rho_f P_\alpha \rho_i P_{\alpha'}. \tag{12}$$

Boddy *et al.* note correctly that this decoherence functional is not dependent on the evolution times between the initial state, the projections, and the final state. They conclude (also correctly), that such histories do not exhibit perfect decoherence.

Two questions: first, why use the time-symmetric version of the decoherent histories formalism? Boddy *et al.*'s justification is that they are interested in histories that begin and end in a thermal state, e.g., the state of the fields de Sitter space. When looking at cosmological histories, however, there is no particular reason for making this restriction unless one desires artificially to restrict the set of possible decoherent histories. If one uses the ordinary formulation of decoherent histories, starting from initial states and evolving forward in time, thermal states such as those in de Sitter space can exhibit a wide variety of non-trivial decoherent histories. The second question is simpler: why restrict attention to histories with only one set of events? Suppose that one adds a second set of events, so that the decoherence functional is

$$Z^{-1}\text{tr}\rho_f P_{\alpha_2} P_{\alpha_1} \rho_i P_{\alpha'_1} P_{\alpha'_2}. \tag{13}$$

In this case, because the projection operators in the Heisenberg picture do depend on time, the decoherence functional depends on the time difference between the two sets of events. As noted above, decoherent histories over multiple sets of events also exhibit time dependence even when the initial and final states are stationary.

Boddy *et al.* are correct that the particular set of histories that they investigate do not decohere. It would be a mistake to conclude, however that systems that begin and end in stationary states cannot exhibit decoherent histories. Indeed, a rather trivial counterexample occurs when both ρ_i and ρ_f are the fully mixed stationary state I/d, corresponding to a thermal state with infinite temperature. In this case, the different sets of histories described above for both open and for closed systems naturally decohere. When both open and for closed systems naturally decohere. When ρ_i and ρ_f are thermal states $Z^{-1}e^{-H/kT}$ at finite temperature, then we can project those states onto a typical subspace of dimension $d \propto e^S$, where $S = H/kT - \ln Z$ is the entropy of the thermal state. Once again, these states naturally decohere as in the examples above.

5. Quantum cosmology and decoherent histories

Having established that stationary states do indeed exhibit quantum fluctuations, at least by the criterion of decoherent histories, let's turn to quantum cosmology in models of eternal inflation. In such models, vacuum energy induces an effective cosmological term Λ, causing spacetime locally to resemble de Sitter space. Inertial observers witness an event horizon at distance $\ell = \sqrt{3/\Lambda}$, and detect horizon radiation in a thermal state with temperature $T = 1/2\pi\ell$.

Look at decoherent histories corresponding to measurements made by such an inertial observer as the universe undergoes inflation and settles down into de Sitter space with a small cosmological term [57-59]; after a long time (estimated below) the observer encounters a local thermal fluctuation that gives rise to a region that possesses a high cosmological term and that is large enough to seed another inflationary epoch. Actually to make these measurements, such an inertial observer would have to be a hardy individual, capable of surviving extreme temperatures and curvatures. (Assume, however, that the observer does not fall into a black hole.) The whole point of decoherent histories, however, is to be able to assign probabilities to events whether or not the measurements corresponding to the events are actually performed. An hypothetical

Journal of Physics: Conference Series **2533** (2023) 012011 doi:10.1088/1742-6596/2533/1/012011

inertial observer suffices to assign decoherent histories. The measurements correspond to coarse grained observations of fields, energy densities, pressure, etc., in the local vicinity of the observer. As noted above in [10-35], such histories generically exhibit approximate decoherence.

Consider decoherent histories that correspond to the inertial observer making coarse-grained measurements of the fields in her vicinity together with the effective cosmological term Λ. The events along any history depend on the value of Λ_0 in the first projection in the history. Suppose for the moment that Λ_0 is large, corresponding to high energy in the false vacuum. The observer initially sees thermal fluctuations in the fields corresponding to de Sitter space at high temperature, and witnesses the surrounding spacetime inflate at a rate $\propto \sqrt{\Lambda_0}$. After a characteristic time-scale τ_0, the observer enters a region in which false vacuum decays.

That is, in such a history the universe undergoes inflation via the usual scenarios, yielding a universe more or less like our own. If at the end of inflation the cosmological term is non-zero, then the region in the vicinity of the observer will eventually settle down to de Sitter space again with effective cosmological term $\Lambda_1 < \Lambda_0$ and a horizon at distance $\ell_1 = \sqrt{3/\Lambda_1}$. The state within the horizon is thermal with temperature $T_1 = 1/2\pi\ell_1$. As above, this thermal state can exhibit decoherent histories corresponding to fluctuations in local energy density. Eventually, the region in the vicinity of the observer will exhibit a fluctuation that takes it back to a regime with high Λ and will begin inflating again.

To estimate the time τ_1 it takes for rapid inflation to recommence, we look at how long it takes for a thermal fluctuation to generate a reinflating region. By assumption, the dynamics possesses at least one quantum state for a region of radius $\ell = C\ell_0 = C\sqrt{3/\Lambda_0}$ that undergoes inflation with effective cosmological term Λ_0, where C is a positive $O(1)$ constant. The energy that has to be collected from the thermal radiation in de Sitter space to attain energy density Λ_0 over a volume ℓ^3 is

$$\Delta E = \ell^3\Lambda_0 = 3C^3\ell_0. \tag{14}$$

The de Sitter radiation has temperature $T_1 = 1/2\pi\ell_1$. To collect the energy ΔE within a region of spatial extent ℓ reduces the entropy of the surrounding de Sitter radiation by $\Delta S = \Delta E/T_1$. The probability that a thermal fluctuation at the de Sitter temperature gives rise to a region with the energy density needed to reinflate is thus $e^{-\Delta S} = e^{-6\pi C^3\ell_0\ell_1}$.

It is not enough for the energy required for reinflation to assemble itself: the fields must also be in the proper false vacuum state. If the energy is assembled in a random state, the overlap with the proper inflating state goes as e^{-S_0}, where $S_0 = \pi\ell_0^2$ is the de Sitter entropy for the fields in the region – i.e., the maximum entropy for the region. The overall thermal probability for a fluctuation that creates the inflating region then goes as

$$e^{-\Delta E/T_1 - S_0} = e^{-6\pi C^3\ell_0\ell_1 - \pi\ell_0^2} = e^{-6C^3\sqrt{S_0S_1} - S_0}. \tag{15}$$

Here $S_0 = \pi\ell_0^2$ is the entropy of the high energy density de Sitter space with cosmological term Λ_0, and S_1 is the entropy of the low energy density de Sitter space. Remarkably – given the simple and non-gravitational nature of the argument – equation (15) reproduces (up to the value of the constant C) the Farhi-Guth-Guven formula [56] for the thermal probability of exciting an inflating volume of spatial extent $\approx \ell_0$. Equation (15) shows that the decoherent histories corresponding to a hardy inertial observer reproduce the eternal inflation picture suggested by Albrecht [49].

Equation (15) gives the thermal probability for a fluctuation that can re-ignite inflation. To estimate the time it takes for such a fluctuation to arise, note that in de Sitter space with cosmological term Λ_1, the characteristic time for fluctuations to arise and decay is $\approx \pi/T = 4\pi^2\ell_1$, yielding a time

$$\tau_1 \approx 4\pi^2\ell_1 e^{(6\pi C^3\ell_0\ell_1 + \pi\ell_0^2)}, \tag{16}$$

Journal of Physics: Conference Series 2533 (2023) 012011 doi:10.1088/1742-6596/2533/1/012011

for the inertial observer to encounter another rapidly inflating region. Note that process of reinflation is much more likely to begin with a small region with large cosmological term, than with a large region with small cosmological term. This observation suggests that reinflation should typically begin at a scale close to the Planck scale.

5.1. Initial state

As in [49] this argument yields an ergodic model of eternal inflation. Inflation takes place at high energy scales; the value of the field rolls downhill, yielding an FRW universe with the usual features; eventually, the presence of a small cosmological term yields a fluctuation that causes inflation to begin again with large cosmological term. By the arguments given above, such histories are generically decoherent. The initial state of the universe as a whole can be taken to be the stationary state given by the ergodic average of the the state of the universe over time.

Note that, although ergodic, the histories seen by an inertial observer are not time-reversal invariant. Every time rapid inflation begins again at high Λ_0, it supplies a large source of free energy so that the inertial observer sees entropy increasing, consistent with the second law of thermodynamics. This time-asymmetry of the individual histories is nonetheless consistent with time-reversal invariance of the initial, stationary state of the universe *and* of its dynamics. If instead of using the theory of decoherent histories with an initial state, we use decoherent histories that end in a final state, then we can decompose the stationary state of the universe into a superposition of decoherent histories that end in a final state of high Λ_0, and that evolve backward in time. The histories that correspond to inertial observers with a fixed final state are then just the time reversed version of histories with a fixed initial state: even though they are moving in the opposite direction in time, the time-reversed inertial observers still observe entropy increasing.

6. Re-inflation versus Boltzmann brains

Now compare the probability of the inertial observer encountering another rapidly inflating region with the probability of encountering a thermal fluctuation that mimics some small piece of our universe, e.g., a 'Boltzmann brain.' The argument for the probability of recreating an inflating region via a thermal fluctuation is readily generalized to calculating the probability of recreating any system with energy ΔE and entropy $S = S_{max} - \Delta S$, where S_{max} is the maximum entropy for the system confined to the volume in which it is created. The thermal probability of such a fluctuation is $e^{-\Delta E/T_1 - \Delta S}$. Comparing with equation (14) we see that as long as the energy ΔE in the Boltzmann brain is greater than the energy $\approx \ell_0$ required to reignite inflation, then the inertial observer is more likely to encounter brains that arise by the ordinary process of evolution in an FRW universe rather than ones that arise from thermal fluctuations. If ℓ_0 is at the grand unification scale or a shorter length scale, e.g. the Planck scale, then the vast majority of brains encountered by the observer over its infinite history will be the usual kind of brains.

7. Summary

This paper investigated the question of whether stationary states can exhibit non-trivial temporal fluctuations. Viewed through the lens of decoherent histories, the answer is an unqualified Yes. Closed systems in energy eigenstates exhibit decoherent histories with non-trivial temporal fluctuations. Open systems that are subsystems of larger systems exhibit decoherent histories that correspond to the quantum jump or stochastic Schrödinger model. We then investigated decoherent histories that correspond to observations made by an inertial observer in models of eternal inflation. Such histories contain periods of inflation leading to FRW universes with small cosmological term; thermal fluctuations from the cosmological term then reignite inflation. The time scale required to reignite inflation is long compared with the

Journal of Physics: Conference Series **2533** (2023) 012011 doi:10.1088/1742-6596/2533/1/012011

horizon scale but much shorter than the time required to generate thermal fluctuations that mimic systems that evolved from initial low-entropy states.

Acknowledgments
This work was supported by ARO, DARPA, and AFOSR. The author thanks Andy Albrecht, Alan Guth, Allan Adams, Sean Carroll, Jonathan Halliwell, Jim Hartle, Don Page, and Dan Roberts for extensive discussions.

References

[1] Strite S and Morkoc H 1992 *J. Vac. Sci. Technol.* B **10** 1237
[2] Griffiths R B 1984 *J. Stat. Phys.* **36**, 219.
[3] Griffiths R B 1993 *Phys. Rev. Lett.* **70**, 2201.
[4] Griffiths R B 1996 *Phys. Rev. A* **54**, 2759; 1998 *Phys. Rev. A* **57**, 1604.
[5] Omnès R 1988 *J. Stat. Phys.* **53**, 893; 1988 *J. Stat. Phys.* **53**, 933; 1988 *J. Stat. Phys.* **53**, 957; 1989 *J. Stat. Phys.* **57**, 357.
[6] Omnès R 1990 *Ann. Phys.* **201**, 354.
[7] Omnès R 1992 *Rev. Mod. Phys.* **64**, 339.
[8] R. Omnès 1997 *J. Math. Phys.* **38**, 697.
[9] Gell-Mann M and Hartle J B 1990 in *Complexity, Entropy and the Physics of Information,* SFI Studies in the Sciences of Complexity, Vol. VIII, W.H. Zurek, ed., (Addison Wesley, Reading);
[10] Gell-Mann M and Hartle J B1990 in *Proceedings of the Third International Symposium on the Foundations of Quantum Mechanics in the Light of New Technology,* S. Kobayashi, H. Ezawa, Y. Murayama and S. Nomura, eds. (Physical Society of Japan, Tokyo).
[11] Gell-Mann M and Hartle J B 1993 *Phys. Rev. D* **47**, 3345.
[12] Gell-Mann, M. and Hartle J B 1994 in *Physical Origins of Time Asymmetry,* J. Halliwell, J. Perez-Mercader, and W. Zurek, eds., (Cambridge University Press, Cambridge), arXiv: gr-qc/9304023.
[13] Hartle J B 1994 in *Proceedings of the Cornelius Lanczos International Centenary Confererence,* J.D. Brown, M.T. Chu, D.C. Ellison and R.J.Plemmons, eds., (SIAM, Philadelphia); arXiv: gr-qc/9404017.
[14] Hartle J B 1995 in *Proceedings of the 1992 Les Houches School, Gravity and its Quantizations,* B. Julia, J. Zinn-Justin, eds. (Elsevier Science B.V.).
[15] Hartle, J B and Marolf D 1997 *Phys. Rev. D* **56**, 6247.
[16] Gell-Mann M and Hartle J B 2012 *Phys. Rev. A* **85**, 062120.
[17] Halliwell J J 1989 *Phys. Rev. D* **39**, 2912.
[18] Dowker H F and Halliwell J J 1992 *Phys. Rev. D* **46**, 1580.
[19] Diósi L, Gisin N, Halliwell J and Percival I 1995, *Phys. Rev. Lett.* **74**, 203; arXiv: gr-qc/9403047.
[20] Halliwell J J and Zoupas A 1995 *Phys. Rev. D* **52**, 7294; arXiv: quant-ph/9503008.
[21] Halliwell J J 1995 *Ann. NY Acad. Sci.* **755**, 726–740; arXiv: gr-qc/9304039.
[22] Brun T A and Halliwell J J 1996 *Phys. Rev. D* **54** 2899-2912; arXiv:quant-ph/9601004.
[23] Brun T A 1997 *Phys. Rev. Lett.* **78** 1833-1837; arXiv:quant-ph/9606025.
[24] Halliwell J J 1998 *Phys. Rev. D* **58** 105015; arXiv:quant-ph/9805062.
[25] Halliwell J J 1999 *Phys. Rev. Lett.* **83** 2481; arXiv: quant-ph/9905094.
[26] Halliwell J J and Thorwart J 2001 *Phys. Rev. D* **64**, 124018.
[27] Halliwell J J. and Thorwart J 2002 *Phys. Rev. D* **65**, 104009.
[28] Halliwell J J 2003 *Phys Rev D* **68**, 025018; arXiv: quant-ph/0305084.
[29] Halliwell J J 2005 *Contemp. Phys.* **46**, 93-104.
[30] Halliwell J J and Wallden P 2006 *Phys. Rev. D* **73**, 024011.
[31] Halliwell J J 2009 *Phys. Rev. D* **80**, 124032.
[32] Halliwell J J 2005 *Phys. Rev. A* **72**, 042109.
[33] Halliwell J J 2001 *Phys. Rev. D* **63** 085013.
[34] Halliwell J J *J. Phys.: Conf. Ser.* **306**, 012023 (2011); arXiv: 1108.5991.
[35] Dowker F and Kent A 1996 *J. Stat. Phys.* **82** 1575-1646; arXiv: gr-qc/9412067.
[36] Lloyd S 2005 Decoherent histories and generalized measurements 2005 *Preprint* arXiv: quant-ph/0504155.
[37] Boltzmann L 1895 *Nature* **51**, 413.
[38] Eddington A S 1931, reprinted in *The Book of the Cosmos: Imagining the Universe from Heraclitus to Hawking,* D. R. Danielson, ed., (Perseus, Cambridge, Mass., 2000), p. 406.
[39] Feynman R P 1965 The character of physical law (MIT Press Cambridge) p.173.
[40] Page D N 1983 *Nature* **304**, 39.

[41] Barrow J and Tipler F 1986 The Anthropic Cosmological Principle (Oxford University Press, Oxford).
[42] Coule D 2003 *Int. J. Mod. Phys. D* **12**, 963; arXiv: gr-qc/0202104.
[43] Albrecht A 2004, in *Science and Ultimate Reality: From Quantum to Cosmos, honoring John Wheeler's 90th birthday,* J.D. Barrow, P.C.W. Davies, C.L. Harper, eds. (Cambridge University Press); arXiv: astro-ph/0210527.
[44] Albrecht A and Sorbo L 2004 *Phys. Rev. D* **70**, 063528; arXiv: hep-th/0410270.
[45] Reese M 1997 *Before the Beginning*, p. 221 (Addison-Wesley).
[46] Bousso R and Freivogel B 2007 *JHEP* **0706**, 018; arXiv: hep-th/0610132.
[47] Linde A 2007 *J. Cosmol. Astropart. Phys.* **0701** 022 .
[48] Page D N 2008 *Phys. Rev. D* **78**, 063535, 063536; arXiv: hep-th/0610079, hep-th/0611158.
[49] Linde A, Vanchurin V, and Winitzki S 2009 *JCAP* **0901**, 031; arXiv: 0812.0005.
[50] Albrecht A 2009 *J. Phys.: Conf. Ser.* **174**, 012006; arXiv:0906.1047.
[51] de Simone A, Guth A H, Linde A, Noorbala M, Salem M P and Vilenkin A 2010 *Phys. Rev. D* **82**, 063520; arXiv: 0808.3778.
[52] Linde A and Noorbala M 2010 *JCAP* **1009**, 008.
[53] Albrecht A and Phillips D 1012 Origin of probabilities and their application to the multiverse *Preprint* arXiv: 1212.0953.
[54] Albrecht A 2014 Tuning, Ergodicity, Equilibrium and Cosmology *Preprint* arXiv: 1401.7309.
[55] Boddy K K, Carroll S M and Pollack J 2014 De Sitter Space Without Quantum Fluctuations *Preprint* arXiv: 1405.0298.
[56] Nielsen M A and Chuang, I L 2000 *Quantum computation and quantum information* (Cambridge University Press, Cambridge).
[57] Farhi E, Guth A H and Guven J 1990 *Nucl. Phys. B* **339**, 417–490.
[58] Nomura Y 2011 *JHEP* **11** 063; arXiv:1104.2324.
[59] Nomura Y 2013 *Found. Phys.* — **43** 978; arXiv: 1110.4630.
[60] Nomura Y 2015 *Phys. Lett.* **B749** 514; arXiv:1502.05401.

Journal of Physics: Conference Series **2533** (2023) 012012 doi:10.1088/1742-6596/2533/1/012012

Relics of the quantum spacetime:

from Synge's world function as the fundamental probe of spacetime architecture to the emergent description of gravity

Dawood Kothawala

Centre for Strings, Gravitation and Cosmology, Department of Physics, Indian Institute of Technology Madras, Chennai 600036, India

E-mail: `dawood@iitm.ac.in`

Abstract. All our observations that characterise space and time are expressed in terms of non-local, bi-tensorial objects such as geodesic intervals between events and two-point (Green) functions. In this contribution, I highlight the importance of characterising spacetime geometry in terms of such non-local objects, focusing particularly on two important bi-tensors that play a particular fundamental role – Synge's World function and the van Vleck determinant. I will first discuss how these bi-tensors help capture information about spacetime geometry, and then describe their role in characterising quantum spacetime endowed with a lower bound, say ℓ_0, on spacetime intervals. Incorporating such a length scale in a Lorentz covariant manner necessitates a description of spacetime geometry in terms of above bi-tensors, and naturally replaces the conventional description based on the metric tensor $g_{ab}(x)$ with a description in terms of a non-local bi-tensor $q_{ab}(x, y)$. The non-analytic structure of $q_{ab}(x, y)$ which renders a perturbative expansion in ℓ_0 meaningless, also generically leaves a non-trivial "relic" in the limit $\ell_0 \to 0$. I present some results where such a relic term is manifest; specifically, I will discuss how this: (i) suggests a description of gravitational dynamics different from the one based on Einstein-Hilbert lagrangian, (ii) implies dimensional reduction to 2 at small scales, (iii) connects with the notion of cosmological constant itself being a non-local vestige of the small scale structure of spacetime, (iv) helps address the issues of spacetime singularities. I will conclude by discussing the ramifications of these ideas for quantum gravity.

Keywords: Synge world function, van Vleck determinant, non-locality, minimal length, small scale structure of spacetime; geodesic structure; quantum gravity.

Journal of Physics: Conference Series **2533** (2023) 012012 doi:10.1088/1742-6596/2533/1/012012

Now it seems that the ***empirical notions on which the metrical determinations of space are founded***, the notion of a solid body and of a ray of light, ***cease to be valid for the infinitely small***. We are therefore quite at liberty to suppose that the metric relations of space in the infinitely small ***do not conform to the hypotheses of geometry***; and we ought in fact to suppose it, if we can thereby obtain a simpler explanation of phenomena.

The question of the ***validity of the hypotheses of geometry in the infinitely small*** is bound up with the question of the ground of the metric relations of space ...

The answer to these questions can only be got by starting from the conception of phenomena which has hitherto been justified by experience, and which Newton assumed as a foundation, and ***by making in this conception the successive changes required by facts which it cannot explain*** ...

This leads us into the domain of another science, of physic, into which the object of this work does not allow us to go today.

from: **On the Hypotheses which lie at the Bases of Geometry**, **Bernhard Riemann**, Gottingen lecture, 1854 (translated by W. Clifford)

1. Introduction

Riemann, in his classic lecture on the foundations of geometry, clearly highlights the importance of understanding space in the context of physical effects and observations, while alerting that our usual conceptions of space might not hold at very small scales. The statements quoted above are particularly prescient, since they it is only in recent times that studies in quantum gravity have revealed how true Riemann's insights were. All our physical measurements are characterised in terms of extended solid objects and light rays ("rods and clocks") which we employ as probes, and hence, it is but to be expected that the quantum behaviour of these probes must play an important role in characterising the structure of spacetime. In brief:

We should be able to re-construct spacetime purely in terms of quantities that characterise such measurements.

While easy to state, this basic, fundamental fact has been forgotten in the long history of gravitational physics, in which our geometric description of gravitational dynamics has been rooted in terms of local, tensorial quantities such as $g_{ab}(x), R_{abcd}(x), etc.$. While such quantities can, and do, characterise motion and measurements, they do not capture the essence of the fact highlighted above: spacetime must be re-constructed from observables tied to measurements. This shift of viewpoint, which might seem to be a matter of taste at the classical level, becomes absolutely essential at the quantum level where local, tensorial objects might not make much sense. In fact, if the quantum effects are non-analytic, then the limit in which one recovers locality might not commute with the limit $\hbar \to 0$, and hence:

Even the classical limit might carry some vestige of an inherently quantum spacetime.

In this contribution, we will describe a formalism attempts to re-construct spacetime in terms of non-local bi-tensors, $Q_{ab\ldots i'j'\ldots}(x, x')$, with the (un-)primed indices referring to an event with coordinate x (x') in some suitable chart. In particular, I will first describe how classical metric can be described in terms of such quantities, and then try to obtain an "effective metric" for

Journal of Physics: Conference Series **2533** (2023) 012012 doi:10.1088/1742-6596/2533/1/012012

the quantum spacetime by imposing the condition that there exists a lower bound on spacetime intervals. The brief plan of this contribution is:

- **Sec 2:** Small scale structure of spacetime – the zero point length
- **Sec 3:** Reconstructing spacetime geometry from Synge World function – replacing the notion of "metric" with two-point functions
- **Sec 4:** Quantum aspects of the reconstructed spacetime
  - Relics of the quantum spacetime – the limit $\ell_0 \to 0$
  - Effective dimensional reduction at small scales
  - Spacetime singularities
  - Causal structure and "silence" at small scales
- **Sec 5:** Emergent gravity as a relic of quantum spacetime
- **Sec 6:** A broader outlook – bi-tensors as tools to probe classical and quantum spacetime

2. Small scale structure of spacetime – the *zero point* length of spacetime

In almost any attempt to combine the principles of quantum mechanics and general relativity, there are atleast two results, rooted in semi-classical formalism but widely considered robust to the new physics which the domain of quantum gravity might bring in. These are:

(a) Thermal nature of acceleration horizons (Unruh effect and Hawking radiation), and
(b) Existence of a lower bound to spacetime intervals – a minimal length scale – which arises very naturally when one tries to combine the principles of quantum mechanics and general relativity.

We will largely be interested in (b) in this contribution, but in doing so, will be led to an intriguing connection with one of the implications of (a) - the emergent paradigm of gravity. We will discuss this connection towards the end. The most basic manner in which (b) can be introduced in an effective manner into the structure of spacetime - without resorting to any specific framework of quantum gravity - is to change focus from a description in terms of metric tensor, and instead work directly with the distance function $d(x, y)$ between spacetime events. Existence of a minimal length scale can then be introduced in terms of the modifications of the distance function $d(x, y)$ between spacetime events. However, in Lorentzian geometry, a better variable to use is the squared geodesic interval $\sigma(x, y)^2$, which is one half of the so called Synge world function $\Omega(x, y)$. This is related to the distance function by $d(x, y) = \sqrt{|\sigma(x, y)^2|}$. The existence of a lower bound to spacetime intervals can then be stated in a very simple manner: The presence of a lower bound to geodesic intervals manifests itself as the modification of the squared geodesic interval

$$\begin{aligned} \sigma^2 &\to \mathcal{S}(\sigma^2) \\ \mathcal{S}(0) &= \pm\ell_0^2 \neq 0 \end{aligned} \tag{1}$$

This is the most basic, Lorentz invariant, statement that in the coincidence limit $y \to x$, $\sigma^2 \to 0$, the modified geodesic distance should be non-zero. [1]

[1] As an aside, let us point out that any such modification would generically violate the axioms of metric spaces. This is not unexpected and, in fact, it is not difficult to show that even the equality part of the triangle inequality can not hold in presence of a minimal length. The key reason for this is clear - the condition $\mathcal{S}(0) \neq 0$ essentially implies that one can not localise an event to an accuracy better than ℓ_0. Therefore, given an ordered sequence of points p, q, and r on a geodesic, a statement about geodesic distances between these can not hold to an accuracy better than ℓ_0, since none of these points can be localised to that accuracy.

Given the above simple fact, how far can one go towards constructing variables analogous to the metric tensor $g_{ab}(x)$ which we conventionally use to describe spacetime? The answer to the above question lies in the perhaps not so well appreciated importance of the world function, $\Omega(x, y) = \frac{1}{2}\sigma(x, y)^2$: much of the information about $g_{ab}(x)$, $R_{abcd}(x)$, etc. can be derived from coincidence limits of derivatives of $\Omega(x, y)$. This has important consequence, which will become apparent as we proceed.

Notation: We work in D dimensions, and use the sign convention $(-, +, +, \ldots)$ for Lorentzian spaces. Latin alphabets denote spacetime indices. Also, for notational convenience, we use ℓ_0^2 throughout to denote short distance cutoff on geodesic distances; for timelike/spacelike cases, the replacement $\ell_0^2 \to \epsilon\ell_0^2$ must be made in the final results after which $\ell_0^2 > 0$. For convenience, we give below a quick list of some of the most recurring symbols/notation used in the text:

• $D_k \overset{\text{def}}{:\Longrightarrow} D - k$, where D is the dimension of spacetime.
• p_0 represents a fixed spacetime event (the *base point*), and p the variable event (the *field point*). Occasionally, coordinate labels y, x are used to represent p_0, p respectively.
• $\mathbf{Ric}(p_0)$ denotes the Ricci scalar; similarly $\widetilde{\mathbf{Ric}}(p; p_0)$ denotes the Ricci bi-scalar built from the qmetric (see text).
• $\mathcal{S}_{ab} = R_{ambn}q^m q^n$; $\mathcal{S} = g^{ab}\mathcal{E}_{ab} = R_{ab}q^a q^b$.
• $\left[\widetilde{\mathbf{Ric}}\right](p_0)$ is the coincidence limit of $\widetilde{\mathbf{Ric}}(p; p_0)$.

A note about References: Before proceeding to describe the framework, results, and implications based on the above idea, I give below a brief survey of the references *vis a vis* the various results they contain. ***No explicit citations will be given in the main body of the text unless absolutely required. Also, the*** *Bibliography* ***at the end contains references describing the formalism reviewed here.*** The reader is referred to these papers for further references to important papers directly and/or indirectly relevant to this work.

- A framework based on the above was first described in Ref. [1], where the key mathematical inputs and conceptual points were discussed in detail.
- A generalisation of this was subsequently studied in Ref. [2], which presented the important result for the Ricci bi-scalar $\widetilde{\mathbf{Ric}}(p; p_0)$ constructed from $q_{ab}(x; y)$. In particular, it was pointed out that the limit $p \to p_0$ of $\widetilde{\mathbf{Ric}}(p; p_0)$ is non-trivial in it's dependence on ℓ_0. A similar computation for the Gibbons-Hawking-York (GHY) boundary term was presented in Ref. [3]. Ref. [2] in particular presented and highlighted the conceptual importance of the results concerning minimal length and qmetric for the *Emergent Gravity* paradigm and the *Cosmological Constant* problem. It also provides a broader perspective on the entire framework, with examples from several other areas of theoretical physics that are helpful in better understanding of the result.
- Perhaps the most important missing term in the above computations was the dependence of $q_{ab}(x; y)$ on the van Vleck determinant $\Delta(x, y)$. This was studied in detail in Ref. [4], which presents all the derivations and results in a mathematical rigorous manner, and derives $q_{ab}(x; y)$ in it's final form.
- This final form was then employed in Ref. [7] to study the volumes and areas of equi-geodesic surfaces of "radius" $\sigma(x; 0)^2 = R$ using $q_{ab}(x; 0)$, leading to yet another important result: The effective dimension $D_{\text{eff}} \to 2$ in the limit $x \to 0$, regardless of the dimensionality D of the original spacetime.
- Implications of the qmetric for caustics of geodesic congruences, via Raychaudhuri equation, was discussed in Ref. [8].

Journal of Physics: Conference Series **2533** (2023) 012012 doi:10.1088/1742-6596/2533/1/012012

I will now describe and summarise these results, their implications, and future outlook.

3. Reconstructing spacetime geometry from Synge World function – replacing the notion of "metric" with two-point functions

3.1. Geodesic distances $\sigma(x, y)^2$ as more fundamental than the metric $g_{ab}(x)$

As mentioned above, our key idea is to replace the description of spacetime in terms of the metric tensor, with one that uses geodesic distances as the key variables. The possibility of doing so is based on the existence of mathematical identities relating the local tensorial quantities (metric curvature, etc.) that characterise a spacetime to the coincidence limit (denoted by "[...]") of derivatives of $\sigma(x, y)^2$. For example, in terms of Synge's world function $\Omega(x, y) = (1/2)\sigma(x, y)^2$, we have

$$\begin{aligned} g_{a'b'} = g_{ab} &= \left[\nabla_a \nabla_b \Omega(x, x')\right] = \left[\nabla_{a'} \nabla_{b'} \Omega(x, x')\right] \\ R_{a'(c'd')b'} &= (3/2)\left[\nabla_a \nabla_b \nabla_c \nabla_d \Omega(x, x')\right] \end{aligned} \tag{2}$$

As we argued above, since a Lorentz invariant lower bound is best introduced through modification of geodesic distances, it then makes sense to "reconstruct" the "metric" from the distance function. While this procedure gives the conventional metric tensor $g_{ab}(x)$ in absence of ℓ_0, we shall see that it leads to a non-trivial bi-tensor $q_{ab}(p; p_0)$ when ℓ_0 is introduced. It is this object which we shall call as the *qmetric*, and use as the key variable in our description of small scale structure of spacetime. Before proceeding to discuss the qmetric, it is worth emphasizing at this stage why the conventional Taylor expansion of the metric in Riemann normal coordinates (RNC):

$$g^{\mathrm{RNC}}_{ab}(x; y) = \underset{?}{\boldsymbol{\eta}_{ab}(y)} - \frac{1}{3}\underset{?}{\boldsymbol{R}_{acbd}(y)}\ (x-y)^c (x-y)^d + \text{higher order terms} \tag{3}$$

is not expected to hold for points x that lie within a geodesic distance of the order of ℓ_0 from y. First, it is not clear whether the leading term would be Minkowski metric tensor (due to quantum fluctuations). Moreover, such an expansion would break down near a point y if any of the basis components of the Riemann tensor diverge at y. Since a minimal length scale is expected to have a quantum gravitational origin, and since our qmetric is derived assuming such a length scale exists, we may therefore not expect that

$$q_{ab}(x; y) \not\equiv g^{\mathrm{RNC}}_{ab}(x; y) \tag{4}$$

This is illustrated in Fig. 1. Indeed, as it will turn out, $q_{ab}(x; y)$ has a much more non-trivial structure.

Journal of Physics: Conference Series **2533** (2023) 012012 doi:10.1088/1742-6596/2533/1/012012

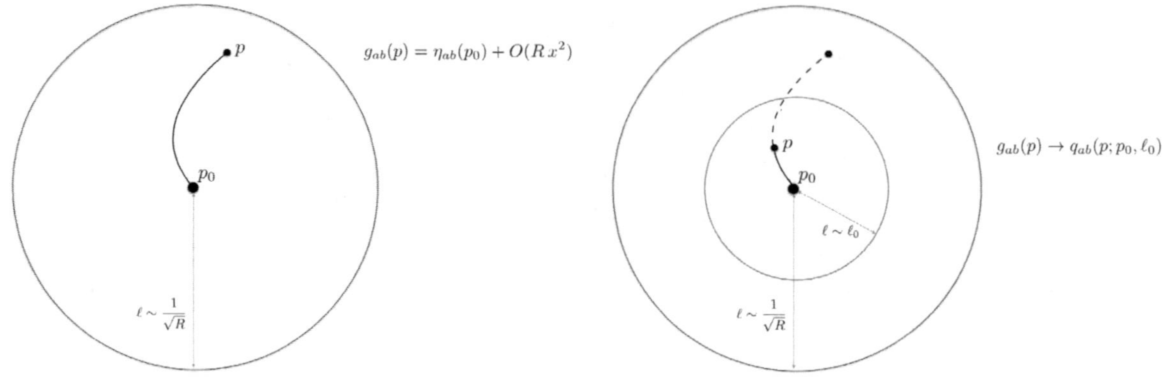

(a) Conventional expansion of the metric near an event p_0. (b) The qmetric at small scales.

Figure 1. The modification of conventional metric description by the bi-tensor q_{ab} at small scales.

3.2. Geodesic structure of spacetime

Since we are proposing to use geodesic distance as the key variable, it is natural to expect the mathematical construction of the qmetric would be based on the geodesic structure of spacetime. This is characterized by the what we shall call as the *equi-geodesic* surfaces $\Sigma_{G,p_0} \equiv \{p \in \mathcal{N} | \Omega(p, p_0) = \text{constant}\}$ in a geodesically convex neighbourhood $\mathcal{N}$ of an event p_0. We begin by discussing the intrinsic and extrinsic geometry of *equi-geodesic* surfaces Σ_{G,p_0}. This is illustrated in Fig. 2. The intrinsic and extrinsic geometry of these surfaces are characterised

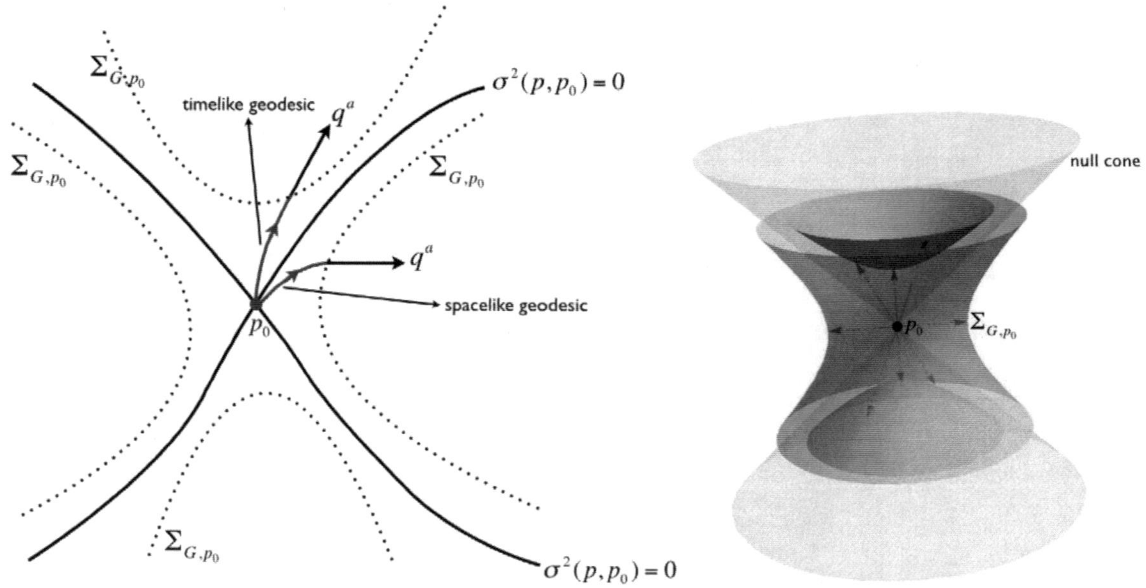

(a) Equi-geodesic surfaces σ_G attached to an event p_0 in an arbitrary curved spacetime. (b) σ_G in Minkowski spacetime.

Figure 2. The geodesic structure of spacetime.

by their intrinsic curvature scalar R_Σ and the extrinsic curvature K_{ab}. These are the quantities that will appear in the Ricci bi-scalar $\widetilde{\mathbf{Ric}}(p;p_0)$ constructed from the qmetric. Moreover, in smooth regions of the background spacetime (that is, away from curvature singularities), the following expansions will turn out to be useful:

$$K_{ab} = \frac{1}{\lambda}h_{ab} - \frac{1}{3}\lambda\mathcal{S}_{ab} + \frac{1}{12}\lambda^2\nabla_{\boldsymbol{t}}\mathcal{S}_{ab} - \frac{1}{60}\lambda^3 F_{ab} + O(\lambda^4)$$

$$K = \frac{D_1}{\lambda} - \frac{1}{3}\lambda\mathcal{S} + \frac{1}{12}\lambda^2\nabla_{\boldsymbol{t}}\mathcal{S} - \frac{1}{60}\lambda^3 F + O(\lambda^4)$$

$$R_\Sigma = \frac{\epsilon D_1 D_2}{\lambda^2} + R - \frac{2\epsilon(D+1)}{3}\mathcal{S} + O(\lambda)$$

where: $\mathcal{S}_{ab} = R_{aibj}t^i t^j$, $\mathcal{S} = R_{ab}t^a t^b$ and $F_{ab} = \nabla_{\boldsymbol{t}}^2\mathcal{S}_{ab} + (4/3)\mathcal{S}_{ak}\mathcal{S}^k{}_b$, with $F = F_{ab}g^{ab}$.

3.3. qmetric: the final form

We now put together our key inputs, collect certain mathematical identities related to the above mentioned bi-tensors, and attempt to construct the qmetric. Let us first state the inputs clearly, before implementing them mathematically:

Q1: Geodesic distances are modified so that they have a Lorentz invariant lower bound. As already stated above, this is summarised by $\sigma^2 \to \mathcal{S}\left[\sigma^2\right]$ with $\mathcal{S}\left[0\right] = \ell_0^2$.

The precise details of the function $\mathcal{S}\left[\sigma^2\right]$ must come from a complete framework of quantum gravity, and we do not impose or choose any particular form, except for demonstrating certain qualitative features of the results. [2]

Q2: The modified d'Alembartian $\widetilde{{}_{p_0}\Box_p}$ yields the following modification for the two point functions $G(p,p_0)$ of fields in *all maximally symmetric spacetimes*: $G\left[\sigma^2\right] \to \widetilde{G}\left[\sigma^2\right] = G\left[\mathcal{S}_{\ell_0}\left[\sigma^2\right]\right]$

We have not discussed this particular input before, but it comes mainly from the fact that existence of a minimal length is expected to regularise the divergences in QFT, and, in particular, modify the structure of the two-point functions. We refer the reader to the references for further discussion on this point. Here, we only point out one immediate implication of the same. Since the leading form of the two point function in any spacetime is given by

$$G(x,y) := \frac{\sqrt{\Delta(x,y)}}{\left(\sigma(x,y)^2\right)^{\frac{D-2}{2}}} \times (1 + \text{subdominant terms})$$

this second input clearly selects the bi-tensors $\Delta(x,y)$ and $\sigma(x,y)^2$ on which we can expect the qmetric to depend.

I will now sketch how the above inputs can be implemented.

(i) To implement **Q1**, we use the defining Hamilton-Jacobi eq for the world fn

$$g^{ab}\,\partial_a\Omega\,\partial_b\Omega = 2\Omega \quad\longrightarrow\quad q^{ab}\partial_a\mathcal{S}_{\ell_0}\partial_b\mathcal{S}_{\ell_0} = 4\mathcal{S}_{\ell_0}$$

[2] We must, however, state that there exist several motivations from older works that selects the form $\mathcal{S}(z) = z + \ell_0^2$.

(ii) To implement **Q2**, we use the following non-local expression for the d' Alembartian $\widetilde{\Box}$ constructed from $q_{ab}(x;y)$

$$\widetilde{\Box} = A^{-1}\left\{\Box_g + \frac{1}{2}D_3\; g^{ij}\partial_i \ln A\; \partial_j + \epsilon \not{\partial} \ln A\; \not{\partial}\right\} \;+\; \epsilon Q\left\{\left[\nabla_i q^i + \frac{1}{2}D_1 \not{\partial} \ln A\right]\not{\partial} + \not{\partial}^2\right\} + \sqrt{\epsilon\sigma^2}\alpha' \not{\partial}$$

(where $\not{\partial} = q^i\partial_i$) and following identities satisfied by the VVD

$$\Delta(p,p_0) = \frac{1}{\sqrt{|g(p)|}\sqrt{|g(p_0)|}}\det\left\{\overset{(p)}{\nabla}_a \overset{(p_0)}{\nabla}_b\; \Omega(p,p_0)\right\}$$

$$I1: \qquad \nabla_{\boldsymbol{q}} \ln \Delta = \frac{D_1}{\sqrt{\epsilon\sigma^2}} - K$$

$$I2: \qquad \nabla^2_{\boldsymbol{q}} \ln \Delta = -\frac{D_1}{\epsilon\sigma^2} + K^2_{ab} + R_{ab}q^a q^b$$

Using the above, the qmetric can now be derived, though the details of the derivation are lengthy and can be found in the references, specifically Ref. [4].

$$q_{ab}(x;y) \;= \frac{S_{\ell_0}}{\sigma^2}\left(\frac{\Delta}{\Delta_S}\right)^{+\frac{2}{D_1}} g_{ab}(x) \;+\; \epsilon\left\{\frac{\sigma^2 S'^2_{\ell_0}}{S_{\ell_0}} - \frac{S_{\ell_0}}{\sigma^2}\left(\frac{\Delta}{\Delta_S}\right)^{+\frac{2}{D_1}}\right\} t_a(x;y)\; t_b(x;y)$$

where $t_a = g_{ab}q^b$ (see Fig. 2). Note that the qmetric is:

- A non-local bi-tensor, disformally coupled to g_{ab}
- Singular in the limit $\sigma^2 \to 0$
- $\lim_{\ell_0 \to 0} q_{ab}(x;y) = g_{ab}(x)$

As an important point, we must highlight that although the qmetric above has been constructed in a timelike or spacelike neighbourhood of any event p_0, there does exist a generalisation for events separated from p_0 along null intervals, by A. Pesci; this can be found in Refs. [5]. Before we discuss some implications of this work, let me address a couple of often-asked questions:

Question 1: How generic will such a result be as far as quantum gravity is concerned?
The key insight coming from our analysis, going beyond the specifics, is the following: It seems likely that non-local but (hopefully) Lorentz in(co)variant deformation of the spacetime geometry at small scales, will be a consequence of quantum fluctuations of the (as yet unknown) microscopic degrees of freedom of quantum gravity. This leads to an inevitable $O(1)$ modification due to a minimal spacetime length. Regardless of the precise form of the deformation, the results will now essentially depend on the two limits, $\xi \to 0$ and $\xi \to \infty$ (with $\xi = \sigma^2/\ell_0^2$) being inequivalent. Any dimensionless deformation function $A[\sigma, \ell_0]$ can only depend on ξ; $A[\sigma, \ell_0] = A[\xi]$. Hence, unless $A[\xi]$ has the same limit at $\xi \to 0$ and $\xi \to \infty$, (e.g., it is symmetric under $\xi \to 1/\xi$) one will generically obtain the kind of non-trivial result that we have obtained.

Question 2: To what extent can we think of the qmetric as a metric?
This issue is somewhat irrelevant to our results and – more generally – when one treats a distance function $d(x,y)$ defined on a manifold as more fundamental than the metric, and the metric itself as a derived quantity which enables us to construct geometric invariants for a manifold. As explained in detail in Ref. [1], the non-local character of the qmetric is simply a physical

characterization of quantum fluctuations which leave their imprint (in this approach) in the form of a lower bound on the intervals: $d(x, y) \geq \ell_0$. At the smallest of the scales, we do not expect any local tensorial object to describe the quantum spacetime geometry accurately anyway, and hence it is not unexpected that more general mathematical objects must replace the conventional ones. Our work provides a first, mathematically concrete, step in this direction. In fact, non-local effective actions have been considered for quite some time in the context of quantum gravity (e.g., DeWitt proposed such an action in 1981 [6]), and $q_{ab}(p, p_0)$ might serve as an important mathematical object to build such actions. That is all we need as far as physics is concerned.

4. Quantum aspects of the reconstructed spacetime

Having described the geometrical construction, we now highlight the key implications and results that follow naturally from the reconstructed metric.

4.1. Relics of the quantum spacetime – the limit $\ell_0 \to 0$

Given the singular behaviour of $\boldsymbol{q}$, it is unclear whether local scalars constructed out of q_{ab} reduce, in the limit $\ell_0 \to 0$, to their corresponding form in $\boldsymbol{g}$. Therefore, the question of interest is whether the limit $\ell_0 \to 0$ is a singular limit of a fundamentally non-local formalism. One can expect to get interesting and non-trivial, insights into the classical theory if

$$\lim_{\ell_0 \to 0} \lim_{y \to x} \neq \lim_{y \to x} \lim_{\ell_0 \to 0}$$

Going back to the question of $\ell_0 \to 0$ limit of "local scalars" constructed out of q_{ab}, let us start by constructing the simplest geometrical quantity - the Ricci "scalar":

$$\left[\widetilde{\mathbf{Ric}}\right](p_0) \stackrel{?}{=} \mathbf{Ric}(p_0) + \text{terms of order } \ell_0$$

Is the answer to the "?" above yes? If so, one would have proved that the coincidence limit of the Ricci bi-scalar constructed from the qmetric has an expansion in ℓ_0 whos leading term is the Ricci scalar of $g_{ab}(x)$. If the limits highlighted in the title of this section were the same, this would indeed be the case. This, however, turns out not to be the case! In fact, not only do we have

$$\lim_{\ell_0 \to 0} \lim_{\sigma^2 \to 0} \widetilde{\mathbf{Ric}}(p; p_0) \neq \mathbf{Ric}(p_0)$$

but what we do get for these limits turns out to have important implications for the emergent gravity paradigm.

The **exact** form of the Ricci scalar can be written in a compact form in terms of geometric quantities associated with $\sigma^2 =$ const surface, Σ

$$\widetilde{\mathbf{Ric}}(p; p_0) = \left[\frac{\sigma^2}{\mathcal{S}_{\ell_0}} \zeta^{-2/D_1} \mathcal{R}_{\Sigma_{G,p_0}} - \frac{D_1 D_2}{\mathcal{S}_{\ell_0}} + 4(D+1)(\ln \Delta_{\mathcal{S}})^{\bullet}\right] - \frac{\mathcal{S}_{\ell_0}}{\lambda^2 \mathcal{S}_{\ell_0}'^2} \left\{K_{ab}K^{ab} - \frac{1}{D_1} K^2\right\}$$
$$+ 4\mathcal{S}_{\ell_0} \left\{-\frac{D}{D_1} [(\ln \Delta_{\mathcal{S}})^{\bullet}]^2 + 2(\ln \Delta_{\mathcal{S}})^{\bullet\bullet}\right\}$$

where $\Delta_{\mathcal{S}}$ is defined as Δ with σ^2 replaced by $\mathcal{S}$, and $\zeta = \Delta/\Delta_{\mathcal{S}}$, $(\ln \Delta_{\mathcal{S}})^{\bullet} = \mathrm{d} \ln \Delta_{\mathcal{S}} / \mathrm{d}\mathcal{S}_{\ell_0}$, $(\ln \Delta_{\mathcal{S}})^{\bullet\bullet} = \mathrm{d}(\ln \Delta_{\mathcal{S}})^{\bullet} / \mathrm{d}\mathcal{S}_{\ell_0}$.

Journal of Physics: Conference Series **2533** (2023) 012012 doi:10.1088/1742-6596/2533/1/012012

We emphasize that the above expression is exact, with no Taylor expansions anywhere. However, we now analyse its behaviour in a smooth region of spacetime, where the Taylor expansions of various quantities associated with the equi-geodesic surfaces given above will hold. After lengthy algebra, one obtains

$$\boxed{\left[\widetilde{\mathbf{Ric}}\right](p_0) = \lim_{p\to p_0} \widetilde{\mathbf{Ric}}(p;p_0) = \underbrace{\alpha\left[R_{ab}t^a t^b\right]_{p_0}}_{O(1)\text{ term}} + \ell_0^2 \underbrace{\left[\text{ curvature squared terms }\right]_{p_0}}_{O(\ell_0^2)\text{ term}}}$$

thereby giving a non-trivial limit. Although $q_{ab} \to g_{ab}$ when $\ell_0 = 0$, $\left[\widetilde{\mathbf{Ric}}\right](p_0) \neq \mathbf{Ric}(p_0)$ in the same limit!

The "zero-point length" leaves it's vestige ... much like the grin of the Cheshire cat!

Most importantly, the leading term above is precisely the entropy density which arises as Noether charge of diff-invariance, and plays a prominent role in the *Emergent gravity* paradigm! An exactly similar analysis can be done for the GHY surface term in the Einstein-Hilbert action, and can be found in Ref. [3], and we shall not discuss it here.

Instead, we briefly comment on the conceptual implications of the fact that the limit $\lim_{\ell_0\to 0}\lim_{p\to p_0}$ for certain observable $\mathcal{O}_q$, constructed using the qmetric, may not correspond to the classical expression for $\mathcal{O}_g$. Our result for the Ricci scalar demonstrates the possibility of this happening, and we would like to briefly highlight and put in context an important conceptual issue that arises from this: Our result suggests that non-local and non-analytic effects of a minimal length might leave residues which are independent of ℓ_0: *How far can we trust such relics*? In fact, quantum relics of similar nature are not unfamiliar in physics; for e.g. effects of Lorentz violating regulators at higher energies can generically get dragged to lower energies due to radiative corrections, leaving $O(1)$ residual effects [Collins et. al., Polchinski] [10]; conformal anomaly; $D \to 4$ limit in dimensional regularization. [see, for e.g., Birrell & Davies]; non-relativistic relic from c^{-1} expansion of the relativistic point particle wave fn [Padmanabhan et. al.] [11]; non-local quantum residue of discreteness of the causal set type [Sorkin] [12]. These examples show that when certain limits are taken in a theoretical model the resulting theory could contain relics of the more exact description. In all such cases no amount of study of the approximate theory will give us a clue as to where the relic came from (e.g., the study of Schrodinger equation for the helium atom can never lead us to the Fermi statistics for the electrons). We believe our result is of similar nature, and is nearly impossible to understand within the context of classical gravity itself. Our analysis throws light on this and shows that it could be a valid relic of quantum gravity.

An immediate implication of this result would then be to see whether the relic term explains some phenomenon which is difficult to explain in the conventional framework. We will discuss this in the rest of this section.

4.2. Effective dimensional reduction at small scales

The qmetric we have obtained captures the essence of existence of a minimal length scale, and it is therefore worth using it to explore geometrical quantities which could yield further information about the small scale structure of spacetime. The key idea is to compute the area of, and spacetime volume bounded by, the equi-geodesic surfaces. This exercise was done in Ref. [7],

Journal of Physics: Conference Series **2533** (2023) 012012 doi:10.1088/1742-6596/2533/1/012012

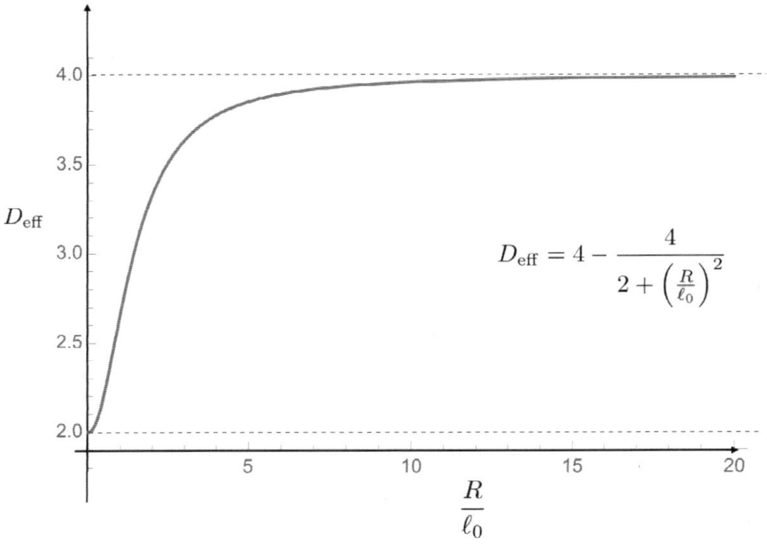

Figure 3. The effective dimension associated with the qmetric goes from D at large scales to 2 at small scales.

where the following expressions were derived (for definiteness, the results were presented for the choice $\mathcal{S}(z) = z + \ell_0^2$)

$$\sqrt{q} = \sigma \left(\sigma^2 + \ell_0^2\right)^{(D-2)/2} \left[1 - \frac{1}{6}\mathcal{S}_g \left(\sigma^2 + \ell_0^2\right)\right] \sqrt{h_\Omega}$$

$$\sqrt{h} = \left(\sigma^2 + \ell_0^2\right)^{(D-1)/2} \left[1 - \frac{1}{6}\mathcal{S}_g \left(\sigma^2 + \ell_0^2\right)\right] \sqrt{h_\Omega}$$

where $\mathcal{S}_g = R_{ab}q^a q^b$ and h_Ω depends on coordinates on the equi-geodesic surfaces. We can now compute the volume $V_D(R, \ell_0)$ of an qui-geodesic surface of size R: $\sigma^2 = R^2$ (in Euclidean), and read off the dimensionality of spacetime from the scaling of volume w.r.t. R. For example, in the standard case $V_D(R, \ell_0 = 0) \propto R^D$. With this is mind, we can define an "effective" *geometric* dimension as

$$D_{\text{eff}} = D + \frac{\mathrm{d}}{\mathrm{d}\ln R}\left\{\ln\left(\frac{V_D(R, \ell_0)}{V_D(R, \ell_0 = 0)}\right)\right\}$$

(The scaling with $V_D(R, \ell_0 = 0)$ is to mod out the effect of curvature, and does *not* affect the final result. See Ref. [7] for details.) The above result blends in very well with the fact that several approaches to quantum gravity suggest such a dimensional reduction to 2 at small scales. See Carlip [13] for a review.

4.3. Spacetime singularities

It is generally believed that the quantum theory of gravity must have something useful to say about the issue of curvature singularities that occur generically in classical general relativity. Since a consistent quantum theory of gravity is nowhere yet in sight, one can not attack the

Journal of Physics: Conference Series **2533** (2023) 012012 doi:10.1088/1742-6596/2533/1/012012

problem of singularity removal head on, but can take a cue from hints from general results in semi-classical gravity such as existence of a zero-point length. With this motivation, and as a first step, implications of the qmetric for the Raychaudhuri equation was analysed in [8], with two key results being: (i) existence of an upper bound on expansions of null and time-like geodesics, and (ii) additional terms in the Raychaudhuri equation related to the Van Vleck determinant associated with the modified geodesic interval.

A more direct approach to spacetime singularities is to construct the qmetric explicitly for a spacetime with curvature singularities, and see if the corresponding quantities are finite as one approaches the singularity. This program is complicated by two major factors: (i) Since curvature components would generically blow up near a singularity, one can no longer use Taylor expansions to analyse their local behaviour. One must instead use exact expressions for the World function and van Vleck determinant, which are not available. Nevertheless, some generic arguments can still be made, and specific results can be obtained for FLRW spacetimes. These indicate that (at least some) curvature quantities that blow up in the background spacetime are indeed rendered finite when constructed using the qmetric. These results are highly non-trivial in the sense that they involve considerable subtleties while taking the limits in presence of singularities, and can not be obtained by naive dimensional considerations (we already know ℓ_0 disappears in the coincidence limit in a regular (non-singular) region of spacetime). A more powerful set of tools can be employed to address these and related issues [9].

4.4. Causal structure and "silence" at small scales

One very straightforward and generic consequence of the non-local structure of the qmetric is to shrink the light cones at every event p of spacetime, with respect to the light cones of the original metric. One may now define a suitable, covariant averaging scheme that takes into account such a shrinking of light cones due to all points $p_0 \in I^-(p)$ - the interior of the past light cone of p - over a time scale τ. One can give some very general arguments leading to specific results concerning the outcome of such an averaging for the causal structure associated with the qmetric. Several of these can be explicitly demonstrated for Minkowski and de Sitter spacetimes. One interesting fall-out of such an analysis is the possibility that the quantum spacetime, as described by the qmetric, might have a Euclidean regime at very small scales [9].

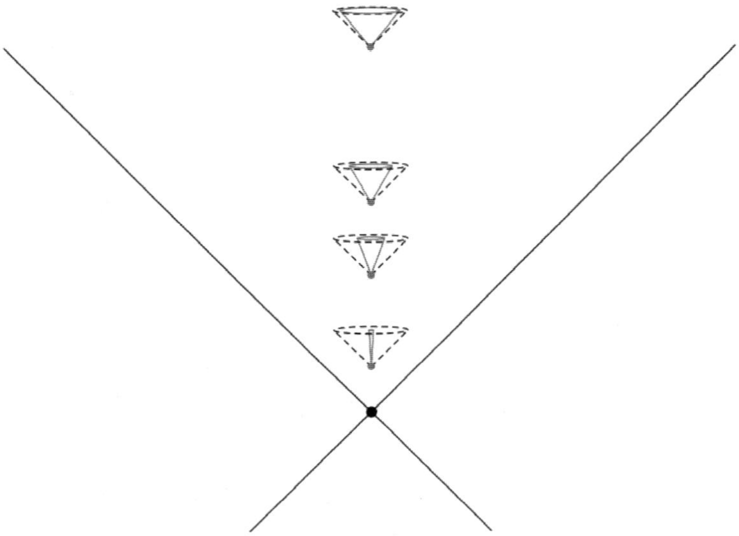

Figure 4. Shrinking of null cones associated with qmetric (red) as compared to the light cones associated with the original metric (black). The base-point p_0 here is placed at the origin.

5. Emergent gravity as a relic of quantum spacetime

We now come to perhaps the most unexpected aspect of the formalism we have described - its connection with the so called emergent gravity paradigm. The mathematical results derived here seem to strongly support this paradigm, in which gravitational dynamics is described in terms of thermodynamics of future causal horizon of an event p_0. Two of the key ideas in this context – the use of *local Rindler frames* as probes of spacetime curvature (due to Jacobson), and a variational principle based on *entropy functional* (due to Padmanabhan et. al.) – find a unified and purely geometric description in our framework in terms of equi-geodesic *surfaces* (which replace the Rindler *trajectories*) and the $\ell_0 = 0$ term of the coincidence limit of Ricci bi-scalar of the qmetric, which happens to have the same form as the entropy functional.

Our formalism amswers one of the most frequently asked questions about the emergent gravity paradigm: *Why choose it over the conventional general relativity based on the Einstein-Hilbert lagrangian*? Most of our conventional intuition about quantum gravity is built on the idea that classical gravity can be obtained as a "Taylor series expansion" in ℓ_0^2 starting from quantum gravity. This, in turn, assumes that all quantum gravitational effects are analytic in ℓ_0^2 and will lead to some sensible classical limits when $\ell_0 \to 0$ limit is taken. Some thought shows that this is a highly questionable assumption and we should take seriously the possibility that quantum gravity could have features which are non-analytic in ℓ_0. In that case, the process of taking limits might involve manipulating singular quantities leading to unexpected (but interesting) results. Our results imply that this is indeed what happens in the case of gravity, and hence

Emergent nature of gravitational dynamics, gravitational action, and space and time itself, might be an unmistakable relic of a quantum spacetime endowed with a zero point length.

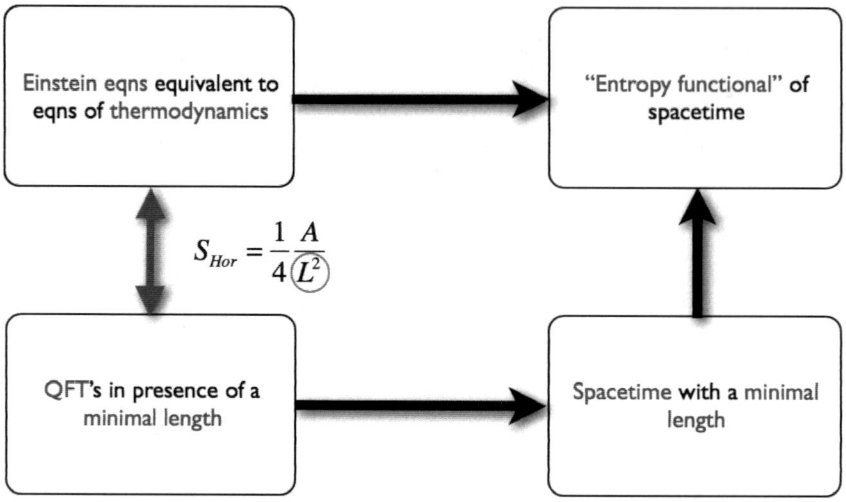

Figure 5. Implications of minimal length for Emergent gravity and spacetime thermodynamics.

A more comprehensive discussion of these ideas developing the emergent gravity paradigm based on above tools and results can be found in the recent reviews by T. Padmanabhan [14].

6. A broader outlook – *bi-tensors* as tools to probe classical and quantum spacetime

If one looks deeper into the structure of our fundamental theories – General Relativity and Quantum Field Theory – some of the most important mathematical features of these theories, such as geodesic deviation, focussing of geodesics, causal structure, singularity structure of two point functions $G(x, y)$ etc, are characterised in terms of bi-tensors. Characterising the small scale topology of spacetime is yet another aspect which would require a description directly in terms of the distance function $d(p, p_0)$. This contribution describes some very generic aspects of the formalism based on these ideas which were first presented in Ref. [1] and developed in further conceptual and technical details in Refs. [2, 3], and improved rigor basis in Ref. [4]. Although our focus has been to reconstruct spacetime metric as a non-local bi-tensor, the basic ideas go beyond this and can be applied to the matter sector as well. [3]

The key underlying idea – that curvature of spacetime might be solely expressible in terms of behaviour of it's geodesics and related bi-tensorial quantites – is powerful and more useful from an operational point of view. In absence of non-locality, such a description would eventually coincide with the standard one in terms of local tensors such as $g_{ab}(p), R_{abcd}(p)$ etc. However, if quantum gravity is inherently non-local, then the non-local character of bi-tensors, combined with the non-analytic deformation of geometry necessitated by the scale of non-locality (here brought in through a minimal length), might lead to a very different description of spacetime curvature at smallest of scales; in particular, it may leave a relic independent of the details or value of the short distance cut-off, thereby acting as a crucial guidepost towards our understanding of classical gravity itself.

[3] The idea of gravity being described fundamentally by a non-local action, with geodesic distance playing the key role, seems to be *conceptually* in tune with certain ideas presented in DeWitt and Alvarez et. al.

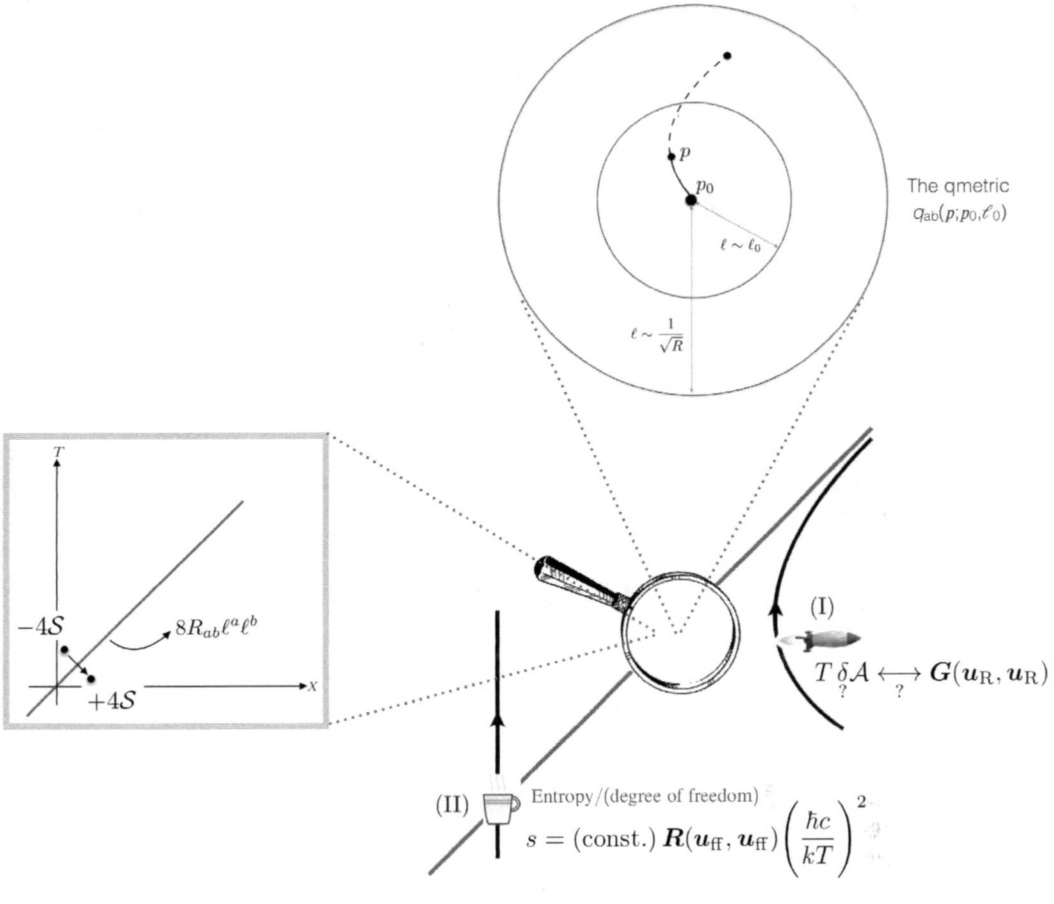

Figure 6. Small scale structure of spacetime: Implications and Future outlook

Since all the relevant comments, remarks and discussions have already been given in the respective sections, I conclude this contribution with a succinct, pictorial summary of the key features of the formalism and its inter-connectedness and relevance for many other important issues in semi-classical and quantum gravity. [4]

Acknowledgments
I thank the organisers of the Tenth International Workshop DICE2022, held at Castiglioncello, for kindly inviting me to present this work at the conference, and Prof Elze in particular for coordinating the entire effort and ensuring it runs smoothly. I would also like to thank T. Padmanabhan, Sumanta Chakraborty, and A. Pesci for discussions and collaboration on several aspects of this work.

[4] It is my hope that the serious reader who has read through this contribution and looked at the *References* given below, would understand what Fig. 6 represents!

Journal of Physics: Conference Series **2533** (2023) 012012 doi:10.1088/1742-6596/2533/1/012012

[1] Dawood Kothawala, Phys. Rev. D **88**, 104029 (2013) [arXiv:1307.5618].
[2] Dawood Kothawala, T. Padmanabhan, Phys. Rev. D **90**, 12406 (2014) [arXiv:1405.4967].
[3] Dawood Kothawala, T. Padmanabhan, Phys. Lett. B **748**, 67-69 (2015) [arXiv:1408.3963].
[4] D. Jaffino Stargen, Dawood Kothawala, Phys. Rev. D **92**, 024046 (2015) [arXiv:1503.03793].
[5] A. Pesci, Gen. Rel. Grav. **54**, 72 (2022) [arXiv:2207.12155]; A. Pesci, Phys. Rev. D **102**, 124057 (2020) [arXiv:1911.04135]; A. Pesci, Class. Quantum Grav. **36**, 075009 (2019) [arXiv:1812.01275].
[6] B. S. DeWitt, Phys. Rev. Lett. **47**, 1647 (1981).
[7] T. Padmanabhan, Sumanta Chakraborty, Dawood Kothawala, Gen. Rel. Grav. **48**, 55 (2016) [arXiv:1507.05669].
[8] A. Pesci, Particles 1 **230** (2018) [arXiv:1809.08007]; Sumanta Chakraborty, Dawood Kothawala, Alessandro Pesci, Phys. Lett. B **797**, 134877 (2019) [arXiv:1904.09053].
[9] Dawood Kothawala, *work in progress.*
[10] J. Collins, A. Perez, D. Sudarsky, L. Urrutia, H. Vucetich, Phys. Rev. Lett. **93** (2004) [arXiv:gr-qc/0403053]; J. Polchinski, Class. Quantum Grav. **29** (2012) [arXiv:1106.6346].
[11] Hamsa Padmanabhan, T. Padmanabhan, Phys.Rev., D **84**, 085018 (2011) [arXiv:1110.1314].
[12] Rafael D. Sorkin, AIP Conf. Proc. **957**, 142-153 (2007) [arXiv:0710.1675v1].
[13] S. Carlip, *Foundations of Space and Time*, Ed. by G. Ellis, J. Murugan, A. Weltman (Cambridge University Press, 2012) [arXiv:1009.1136]; AIP Conf.Proc. **1196**, 72 (2009) [arXiv:0909.3329], and references therein.
[14] T. Padmanabhan, Int. Jour. Mod. Phys. D **25**, 1630020 (2016) [arXiv:1603.08658]; T. Padmanabhan, Int. Jour. Mod. Phys. D **27**, 1846004 (2018) [arXiv:1805.07218]; T. Padmanabhan, Int. Jour. Mod. Phys. D **29**, 2030001 (2020) [arXiv:1909.02015].

Journal of Physics: Conference Series **2533** (2023) 012014 doi:10.1088/1742-6596/2533/1/012014

Representation of the universe as dendrogramic hologram empowered with relational interpretation

Oded Shor[1,2,5], Felix Benninger[1,2,3] and Andrei Khrennikov[4]

[1] Felsenstein Medical Research Center, Beilinson Hospital, Petach Tikva, Israel

[2] Sackler Faculty of Medicine, Tel Aviv University, Tel Aviv 6997801, Israel

[3] Department of Neurology, Rabin Medical Center, Petach Tikva 4941492, Israel

[4] Faculty of Technology, Department of Mathematics, Linnaeus University, 3 51 95 Växjö, Sweden

[5] Author to whom any correspondence should be addressed.

shor.oded@gmail.com

Abstract. This is a brief review on the basics of recently established *Dendrogramic Holographic theory* (DH-theory). This is the special model of the event-universe based on the clustering transformation of experimental data into dendrogram, a finite tree which branches encoding the events. These event-branches are coupled via the hierarchic interrelation determined by the dendrogram. Such relational universe differs from the universe with space-time mathematically described by the real numbers. Dendrogram is endowed with the common root ultrametric. Finite dendrograms correspond to the epistemic level of description; in the limit we obtain an infinite tree providing the ontic description. In the simplest model, the tree is homogeneous, p-adic tree. It can be endowed with the algebraic structure of the ring of p-adic integers. Hence, DH-theory is a part (but very special) of p-adic theoretical physics. In this paper we discuss the foundations of DH-theory and its applications to quantum-classical interrelation including the novel interpretation of the violations of the CHSH-inequality, to general relativity, and to emergence of quantum mechanics from the event-picture of the universe. Since both quantum theory and general relativity can be emergent from DH-theory, creation of the latter can be viewed as a step towards unification of these two fundamental physical theories.

1. Introduction

Inspired by the relational interpretation of quantum theory[1], we developed a novel model of universe based on representation of events by branches of a dendrogram. We employ the two levels of description, ontic and epistemic. The latter is generated by an observer: experimental data given by ordinary time series is transformed into a dendrogram – a finite tree – expressing the internal hierarchic relational structure within the dataset. Each detection event corresponds to a branch of the epistemic dendrogram. Ontic universe is portrayed as an infinite tree and finite epistemic trees are its subtrees. Such trees are endowed with natural ultrametric determined by common roots of event-branches. Infinite homogeneous tree with p branches (p>1) leaving each vertex can be equipped with algebraic structure of the ring of p-adic integers. Hence, our modelling can be considered as a part of p-adic theoretical physics which foundations were set by Vladimirov, Volovich, Witten, Parisi[2–9]. However, in contrast with standard exploration of p-adics in physics, we work not with separate branches of the p-adic tree, p-adic numbers, but with batches of branches unified into dendrograms. This is a kind of holographic holistic imaging, and we call our model of universe and processes in it – *dendrogramic holographic (DH) theory*[10–14].

DH-theory provides a novel mathematical model of relational event-universe. This universe has unusual geometry and consequently very special portraying of physical phenomena. Realization of the project of DH-reconsideration of the basic physical theories will lead, as we hope, to creation of new physics which is not rigidly based on the notion of space-time. The latter can be invented (non-uniquely) from hierarchic dendrograms.

In this review we briefly present the theoretical models supported by numerical simulation which were developed up to now. Introduction is a good place to mention the seeds of DH-theory:
1. *Leibniz principle* (identity of indiscernibles). If, for every property F, object x has F if and only if object y has F, then x is identical to y.

2. The relational and event interpretations of physical theories (Wheeler, Rovelli, Barbour, Smolin [1,15–21]).

3.The Copenhagen interpretation of quantum mechanics, especially Bohr's viewpoint on outcomes of measurement which cannot be treated as objective properties of systems.

4. Ontic-epistemic structuring of scientific theories (from Hertz and Boltzman to Schrödinger, Primas, Atmanspacher, Khrennikov,…[22–27]).

5. P-adic (ultrametric) physics: string theory, cosmology, complex disordered systems -- spin glasses (Volovich, Vladimirov, Witten, Parisi, Dragovich, Kozyrev, Khrennikov,…[2–9]).

It may be surprising, but DH-theory was initiated not in physical studies, but in medicine. The starting point of DH-theory was representation of EEG-data collected in medical studies on brain diseases by dendrograms -- with the aid of clustering algorithms. The branches of trees are encoded by p-adic numbers (sequences of 0/1) -- natural numbers with p-adic distance between them. In this way a time series of real EEG-data was transferred into the time series of dendrograms. From the latter, ˊwe constructed the *Bohmian quantum potential* on p-adics and it was used for the medical diagnoses of mental diseases, *depression, schizophrenia and cognitive decline* [28]. Construction of the quantum potential was done following the framework presented in [28] time series of dendrograms generate probability distribution on the set of natural numbers endowed with p-adic distance (reflecting hierarchic order structure of encoding in dendrograms). Then the probability distribution is mapped on the segment [0,1] of the real line, by using the Monna map. This line of research is interesting by itself, not only due to the exploring the hierarchic relational structure in real data expressed in dendrogram-language, but also due to the exploring of the quantum potential. The latter expresses the holistic image of hierarchy in information processing by the brain. Up to our knowledge, this is the first application of quantum(-like) modelling in medical diagnostics.

The medical studies stimulated physical research and led to the DH-theory. One of its distinguishing features is the novel viewpoint to quantum-classical interrelation and, to the problem of violation of the Bell inequalities (section 6). Another fruit of DH-reconsideration of physical theories is the DH-approach to general relativity (GR) (section 7). Combination of DH approaches to quantum theory and GR might lead to their unification. The unification is rather unexpected – merge of these fundamental physical theories in DH. Both quantum theory and GR can be emergent from DH-theory.

Journal of Physics: Conference Series **2533** (2023) 012014 doi:10.1088/1742-6596/2533/1/012014

2. Representation of real experimental data by dendrograms

An observer O starting with an ensemble of events A,B,C,…H (figure1) generates by using a clustering algorithm the dendrogramic representation of his event-universe. DH-theory operates within such dendrogramic model of the event-universe. We discuss the distinguishing features of this representation in further sections

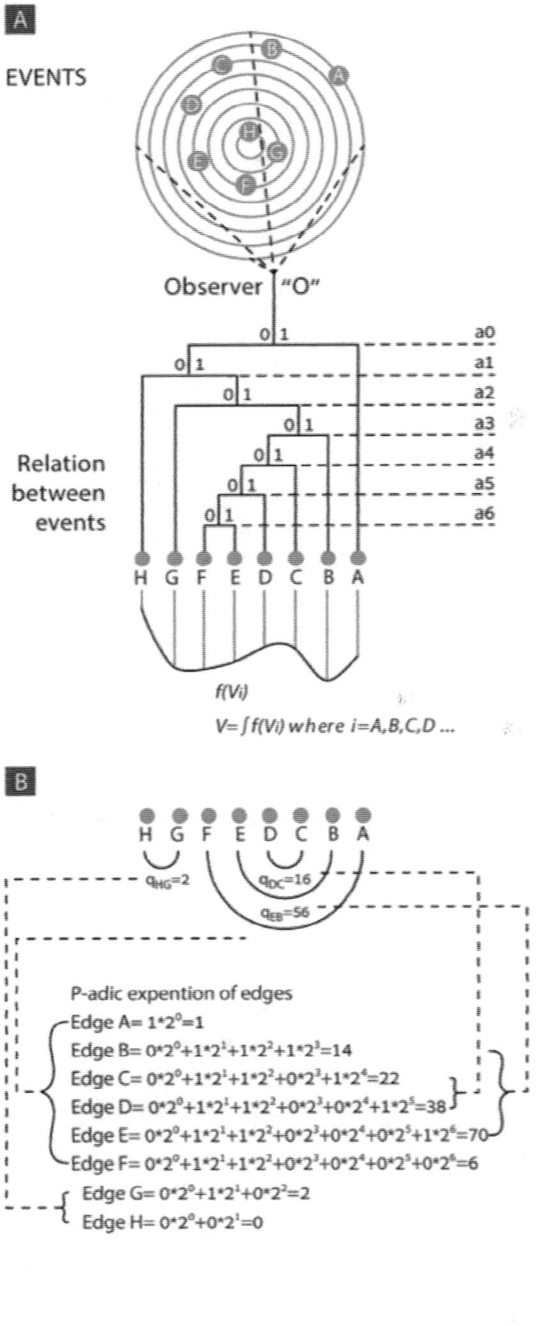

Figure 1. Relational observation of events. Observer O discriminates events A–H and constructs an object, a dendrogram, which describes the relations between these events. (**B**) Each edge of the dendrogram is a binary string of 0s and 1s which can be represented as a finite p-adic expansion. Each edge summation of its finite p-adic expansion results in a natural number. Subtracting between two edges' finite p-adic expansion results in "potential gap"—qij.

Journal of Physics: Conference Series **2533** (2023) 012014 doi:10.1088/1742-6596/2533/1/012014

3. Ontic-epistemic framework for dendrogramic representation of the universe

According to Bohr, the outcomes of measurements are not the objective properties of systems. They quantitively represent interrelation between a system S and an observer O (measurement device of O). The outcomes are generated in the process of complex interaction between a system and measuring device. This is the essence of the Copenhagen interpretation of QM.

We employ the ontic-epistemic structuring of scientific theories. Ontic description is observer independent description of reality - as it is. Epistemic description is based on knowledge which O can extract within experiments. Ontic theory is not verifiable experimentally. It is unapproachable by the observer; O constructs its approximate epistemic representation by collecting data. These approximations are mathematically described by finite dendrograms growing up with collection of new data. The infinite ontic dendrogram is stationary and encodes the future evolution of observer's knowledge.

DH-theory leads to extending of the Copenhagen (Bohr's) interpretation of outcomes of observables or by using the event-terminology, detection-events. In our theory, observer O has the free choice not only for performing or not performing some observation, but *even the free choice to decompose the collected data into blocks and to treat these blocks as the (epistemic) events*[11].

As was emphasized, the fundamental entities of the epistemic theory are not systems, but events. This line of thought is close to the ideas of Wheeler, Rovelli, Barbour, Smolin and if we go deeper into history of science – Leibnitz. Events are hierarchically interrelated with the aid of clustering algorithms. Event's place on dendrogram is like system's place in the real space-time. We point out that different clustering algorithms can generate different relational structures and lead to different epistemic universes. However, as was shown in our papers, the general theoretical properties of corresponding models are algorithm invariant.

4. P-adic representation of events on dendrograms

We recall that p-adic ultrametric, where p>1 is a natural number, is introduced on the set of infinite sequences **Zp** $=\{ x = (a_0, a_1, a_2, ..., a_n, \ldots), a_n=\mathbf{0,1,\ldots,p\text{-}1}\}$ in the following way.
Let $x = (a_0, a_1, a_2, ..., a_n, \ldots), y = (b_0, b_1, b_2, ..., b_n, \ldots) \in$ **Zp, i.e.,** $a_j, b_j=0,1,\ldots,$ p-1. We set
$r_p(x, y) = 1/p^k$, if $a_j= b_j, j = 0, 1, ..., k - 1$, and $a_k \neq b_k$.
This is not only metric on the set of sequences, but even ultrametric, i.e., it satisfies the strong triangle inequality, for any three sequences x, y, z: $r_p(x, y) <= \max [r_p(x, z), r_p(z, y)]$
Each sequence can be geometrically represented as a branch of a tree, i.e., the space of sequences **Zp** can be visualized as the p-adic tree (figure 2).

Figure 2. Treelike representation of 2-adic space

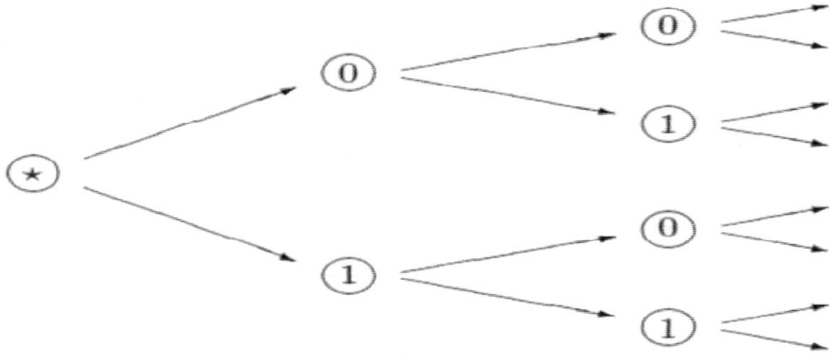

Now we list the unusual properties of p-adic geometry that is really exotic geometry:
a). All triangles are isosceles. This is a consequence of the strong triangle inequality.
b). Define "open" and "closed" balls, $B(R, a) = \{ x: r_p(a, x) <= R\}$ and $B\text{-}(R, a) = \{ x: r_p(a, x) <R\}$. Both balls are both closed and open in the metric space **Zp**, each point in a ball can be selected as its center; two balls are either disjoint or one is included in another.

c). Geometrically a ball is a batch of infinite branches having the finite common root-branch.
d). **Zp** is disordered, totally disconnected, and topologically zero dimensional.
However, such exotic geometries naturally arise from data series through application of clustering algorithms.

On Zp it is possible to introduce the algebraic structure of a ring, with the operations of multiplication, addition, subtraction. Sequences can be represented as numbers given by power series of p,

$$x = a_0 + a_1 p + a_2 p^2 + a_n p^n \quad a_j = 0,1, \dots p - 1$$

They converge in p-adic metric.

Zp, the limit of growing finite dendrograms, is the ontic representation of universe, finite dendrograms are epistemic representations of growing complexity.
The points of Zp are elementary ontic events, subsets of Zp are ontic events. End-points of branches of a finite dendrogram are epistemic elementary events, sets of these points are epistemic events. A huge batch of ontic events is identified with an epistemic point-event. By extending dendrograms size we split epistemic points of n-level to numerous events of m-level, m >>n.

5. From epistemic (explicate) to ontic (implicate): p-adic universe

By considering the data blocks of increasing size, in the limit, *O* can reconstruct the ontic description. The dendrogram-based epistemic theory leads to the p-adic geometry of the ontic universe. In the limit, dendrograms of increasing size generate the infinite p-adic tree endowed with the p-adic ultrametric. Ontic events are represented by p-adic numbers corresponding to infinite branches of the the ontic tree (figure 3).

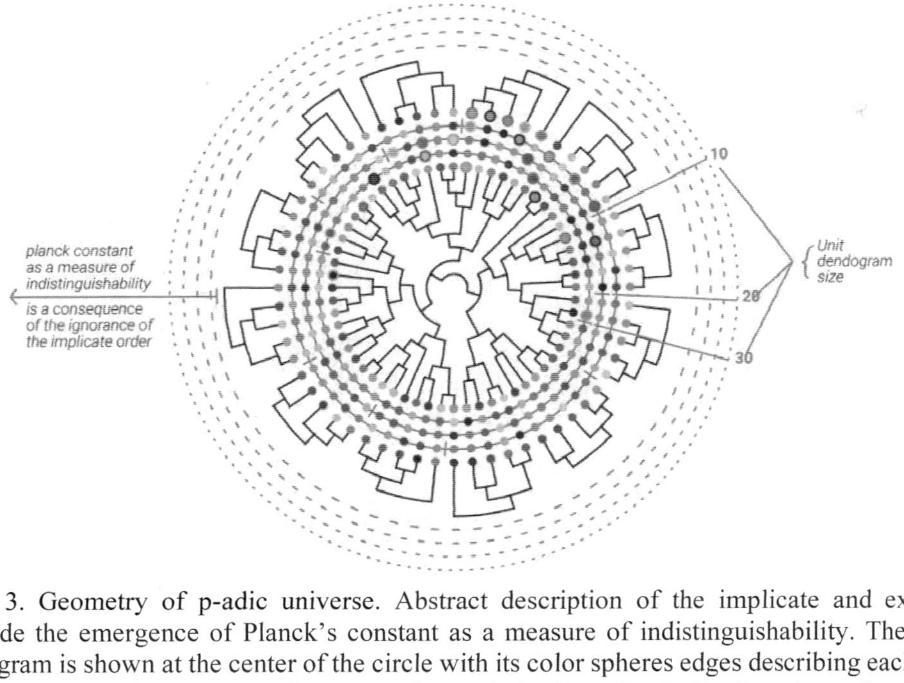

Figure 3. Geometry of p-adic universe. Abstract description of the implicate and extrinsic order alongside the emergence of Planck's constant as a measure of indistinguishability. The "Universal" dendrogram is shown at the center of the circle with its color spheres edges describing each edge 2-adic expansion values of edge route. The outer spheres circles describe the 2-adic values of each edge in the "unit" temporal dendrogram with decreasing edge numbers N = 20, 10, 5. Each unit dendrogram 2-adic values were best-matched to its corresponding edges in the smaller (higher edge number) circle compared to its non-corresponding edges (see for example 5 spheres with rings in each circle). Planck's constant is the measure of indistinguishability, which is a consequence of the ignorance of each dendrogram edge from the infinite sub dendrograms that continue from it (outer dotted circles).

Journal of Physics: Conference Series **2533** (2023) 012014 doi:10.1088/1742-6596/2533/1/012014

6. Dendrogram Viewpoint on Classical–Quantum Interrelation

In our epistemic model, the sharp classical–quantum separation disappears. The degree of classicality of experimental data is based on its complexity—the size and topological complexity of its dendrogram representation. Quantum data is characterized by a low complexity of its dendrogram. Thus, electrons and atoms can behave as quantum in some measurements because they have a very simple hierarchic structure of interrelation between their components. In our approach, even classical physical systems can exhibit quantum(-like) behavior within a hierarchic representation of experimental data.

DH-theory provides the possibility of introducing hidden variables (of the special type) beyond any experimental data and considering a class of realistic observables, those represented as functions of hidden variables. Hence, straightforwardly, it seems that Bell type inequalities cannot be violated under the assumption of locality.

However, we showed [11] the violation of the CHSH inequality for special selection of "settings" that determine the observables. The reason for this is precisely the violation of ergodicity; the measure-theoretic and frequency averages do not coincide (figure 4).

Figure 4. Correlations between two dendrogramic time-series. The possibility to violate the CHSH-inequality. The emergence of quantum correlations. Illustration of the three steps process (TSP) leading to quantum correlations. Two sets of raw data (black dots) are clustered into two sets of six dendrograms each with five edges

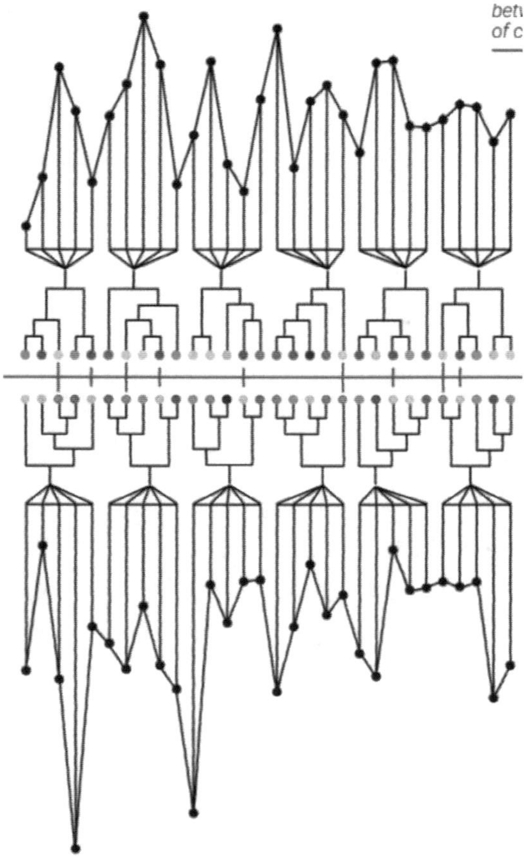

7. Dendrogram approach to general relativity

The basic idea is to consider geodesics corresponding to some metric of GR, and represent each of them by a dendrogram, as on the following figure. We studied some features of dendrogramic representation of geometry around Schwarzschild black hole (see [12]). Thus, DH-theory is a kind of information model of GR. We found a surprising informational interrelation between the fundamental

Journal of Physics: Conference Series **2533** (2023) 012014 doi:10.1088/1742-6596/2533/1/012014

constants, h, c, G, and their DH analogues, $h(D)$, $c(D)$, $G(D)$, where D is a dendrogram. Since "usual QM" can be emergent from DH-theory (section 8), GR can be represented with the quantum formalism, as a Bohmian mechanics.

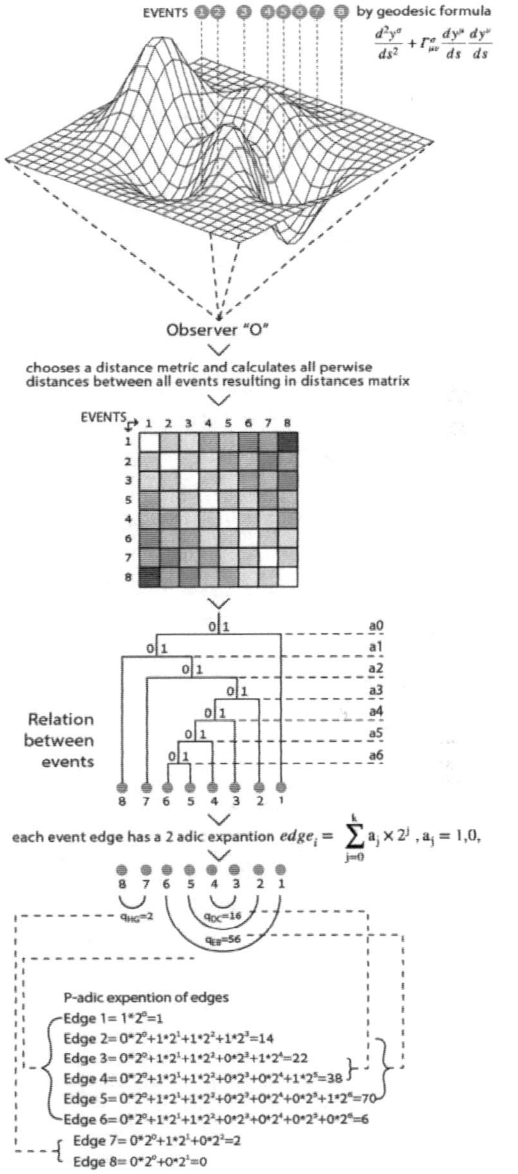

Figure 5. Illustration that describes the construction of a dendrogramic tree and its 2-adic representation from a single discrete geodesic with 8 events.
A. An observer "O" measures discreate dynamics along a spacetime geodesic.
B. Choosing a distance metric he calculates pairwise distances between the 8 discrete events along a geodesic and constructs a distance matrix. C. By applying a hierarchical clustering algorithm on the distance matrix he constructs an Agglomerative hierarchical cluster tree. D. In this heirarchical tree each event, which have a unique branch, can be represented by a binary string or p-adic expansion. E. A numerical example is given for calculating the sum of 2-adic expansions of each event 2-adic representation. More over 2-adic differences between two events, which are marked as q_{ik}, are shown as a numerical example.

Often to proceed in creation of a quantum gravitational model, one point out to Mach's principle as first assumption. We recall that heuristically Mach's principle states that the inertial forces acting on a body are consequence of the quantity and distribution of matter in the universe. DH-theory does not need to postulate Mach's principle. The latter is, in fact, a direct consequence of dendrogram representation of events. A dendrogram describes relations between observational events. In DH-theory an event has no meaning without observer and the rest of the observed universe.

We remark that Einstein attempted appeal to the Machian viewpoint in GR- constructing[29]. Although Einsteinian GR has Machian flavour, Einstein himself noted that he did not fully integrated Mach's principle in the theory. Sciama, in 1953, developed on the grounds of Mach's principle a theory that suggests that "inertia is not an intrinsic property of matter", but a consequence of matter relations in the universe [30]. Furthermore, his theory "implies that the gravitational constant at any point is determined by the total gravitational potential at that point and so by the distribution of matter in the universe"; he coupled local phenomena in the whole universe. Finally, we mention the theory developed by Dicke (1957), where he formulated a gravitational theory with a changing speed of light as a function of relations to the whole universe matter distribution [31,32].

8. Emergence of QM from DH-theory

Generalization and modification of QM-emergence scheme developed by Smolin [33]. One quantizes not events but the differences between them, *views,* and through analytic derivation one arrives at *Bohmian mechanics* [34]. For each event i, a set of views V_i^k from i to other events k=1,...N. The view V_i^k encodes the difference between the events, i and k. In Smolin's works, the Cartesian coordinates still play a crucial role. In DHT, we start directly with the event representation given by a dendrogram – events are its branches. The views are determined by the hierarchic relational structure of the event-tree.

We can informally present the view-scheme as following. The key person is an observer O who ``looks'' at the Universe. (Here ``looks'' means that he uses all needed measurement devices to collect information from the environment.) O determine all unique events which he can discriminate one from the other; O determines some epistemic level of discrimination. From the ensemble of events identified at this epistemic level of discrimination, O generate a dendrogram. For its generation, O employs some fixed clustering algorithm *A*.

This is the delicate point of DH-theory, since one can employ a huge variety of clustering algorithms and each of them can generate its own specific dendrogram D(*A)*. However, the basics of theoretical considerations are same for all algorithms. The structure of DH-theory for *A1* is the same as of DH-theory for *A2*. We tested this via numerical simulation, by playing with the variety of different algorithms.

We now turn to the dendrogram D=D(O,*A*) created by observer O. We call D the universal dendrogram of the observer. As all epistemic quantities, it is always finite – in contrast to the ontic dendrogram which is infinite. O's universe is described by relations between discernible events, see the dendrogram shown in figure 1A. Each event in O's universe is uniquely described by its p-adic expansion as in figure 1B.

9. Relational Dynamics of the Event Universe

We studied via numerical simulation the dynamics of the event universe generated by the appearance of a new event [13]. Generally, each new event can generate the complete reconstruction of the whole dendrogramic universe. The elementary events created up to the moment t = k are hierarchically ordered via the dendrogram's structure. The appearance of a new event at t = k + 1 can, in principle, destroy the previous hierarchic structure. Hence, the past is not frozen. It is in permanent motion. The dynamics that are local in the physical space become nonlocal in dendrogramic space. The DH-dynamics modifies both future and past. This is dynamics generates permanent recombination of events in dendrogram universe.

Thus, all events are permanently in dynamic motion. An event that happened a milliard years ago is not frozen at the same point in dendrogram space. Contemporary events disturb it and vice versa. The position of a newly generated event in the dendrogram space is determined by all previous events. Nevertheless, this event motion has some degree of regularity and stability. The appearance of a new event does not generate a totally random redistribution of events. Only very special dendrogramic configurations can be created with high probability (figure 6 and 7).

10. Emergence of Time Arrow from Increase of Dendrogram complexity

During evolution, the dendrogram's complexity increases and determines the arrow of time in the event universe. We used the complexity measure from particle shape dynamics, which was shown to increase in both directions away from a Janus point and thus determine the arrow of time in symmetrical manner away from a Janus point [15,16,35].

There are evidence that for some positive potentials prescribing configurations of finitely many particles at two different times have solutions. Typically they have a minimum size between the two prescribed configurations and that the size increases to infinity in both time directions. Such solutions were shown to exist for the (negative) 1/r Newton potential in [36] and are called Janus-point solutions in [35] where the point of minimum size being the Janus point.

dynamics theory is a relational theory with close ideological resemblance to DH-theory, as both rely on Mach's principle and Leibniz's relationalism and principles.

Principle 1. According to Mach, only the spatial relations between bodies matter. "When we say that a body K alters its direction and velocity solely through the influence of another body K0 , we have inserted a conception that is impossible to come at unless other bodies A, B, C... are present with reference to which the motion of the body K has been estimated." Absolute space does not exist– only the spatial relations between these bodies. This is to be the principle of spatial relationalism.

More over for Mach, the flow of time is perceived only through the changes of spatial relations. "It is utterly beyond our power to measure the changes of things by time. Quite the contrary, time is an abstraction, at which we arrive by means of the changes of things... " thus, principle 2, the flow of time is a measure of change which is the principle of temporal relationalism [37].

By using the complexity measure on dendrograms and its p-adic string representation, we demonstrate the emergence of a time arrow from the p-adic zero-dimensional field, where space and time are absent (figure 8).

Figure 6. Probability density function of natural numbers representing the p-adic expansions of added events. We generated only one base 50 random number and added 600,000 times another random number (*En+1)* to them, resulting in 51 dendrogram edges. We show the distribution of the possible value of the natural number *Vn+1* representing *En+1*. About 80 such natural number values occurred out of 51

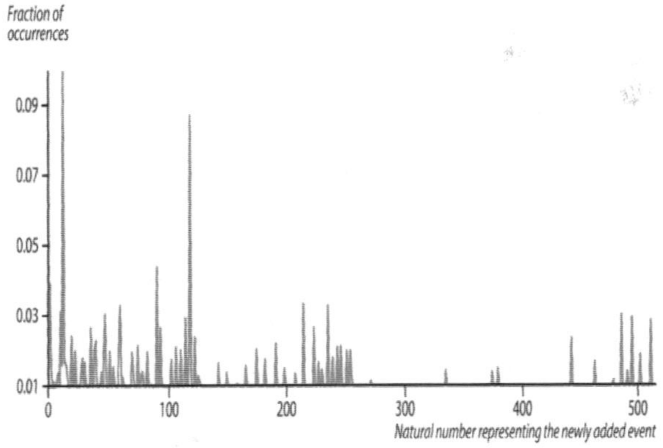

Figure 7. Possible rearrangements of dendrogram structure upon adding a new event. We generated only one base 50 random number events E1-E50 and added 600,000 times another random number (En+1) to them, resulting in 51 dendrogram edges. The extra event was added with the restriction that Dn and Dn+1 both had a number of levels equal to 9. We checked, for each iteration, how many of the base 50 events' p-adic representation changed upon adding the extra event E51.

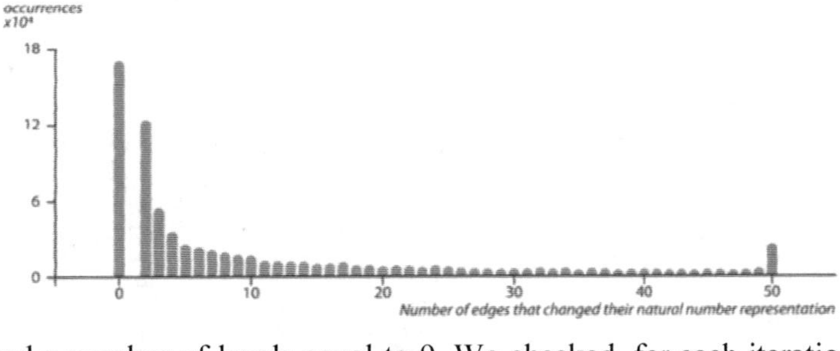

Journal of Physics: Conference Series **2533** (2023) 012014 doi:10.1088/1742-6596/2533/1/012014

The dendrogram dynamics generates not only a redistribution of events, but also increase the dendrogram's size and thus its number of levels. The number of levels N = N(D_n) for dendrogram D_n with n events increases with the increase in the number of events n. Similarly, for entropy, N(D_n) can be used to determine the arrow of time. However, in contrast to entropy (governed by the second law of thermodynamics, N(D_n) it has local minimums on the increasing phone.

11. Shape complexity measure

However, the increasing number of levels, which is an expectable feature, is not satisfactory enough to serve as a "thermodynamic" law. To capture the scale-free dynamic features of dendrogram dynamics, we use the *shape complexity measure*, which was introduced by Barbour et al. in the framework of N-body shape dynamics theory (figure 8) [15,16].

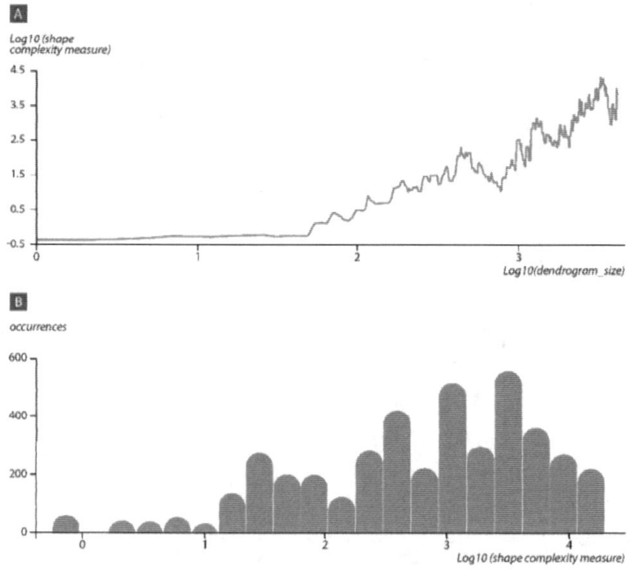

Figure 8. Shape complexity measure of dynamically growing dendrograms. (**A**). We started with only one base 50 random number events E1−E50 and added each time an extra event En, where n = 51, 52, …, 5000, resulting in a dendrogram sequence D50,D51….D5000. For each dendrogram we computed its shape complexity measure. (**B**) Histogram of the shape complexity measure values of the 4950 dendrogram structures. sequence D50,D51….D5000. For each dendrogram we computed its shape complexity measure. (**B**) Histogram of the shape complexity measure values of the 4950 dendrogram structures. sequence D50,D51….D5000. For each dendrogram we computed its shape complexity measure. (**B**) Histogram of the shape complexity measure values of the 4950 dendrogram structures.

the Newtonian N-body problem of gravitating point particles of masses m_i , i = 1, . . . N. It has an unbounded phase space since its scale variable, the root-mean-square length is

$$l_{rms} = \frac{\sqrt[2]{\sum_{i=1}^{N-1}\sum_{j=i+1}^{N} m_i m_j r_{ij}^2}}{M} \quad (1)$$

Where $M = \sum_{i=1}^{N} m_i$ and $r_{ij} = |r_i - r_j|$
is unbounded from above. the graph of $l_{rms}(t)$ is concave upwards and tends to infinity either side of a unique minimum or 'Janus point'. the mean harmonic length is defined as

$$\frac{1}{l_{mhl}} = \frac{1}{M^2}\left(\sum_{i=1}^{N-1}\sum_{j=i+1}^{N}\frac{m_i m_j}{r_{ij}}\right) \quad (2)$$

resulting in the shape complexity measure

Journal of Physics: Conference Series **2533** (2023) 012014 doi:10.1088/1742-6596/2533/1/012014

$$\textit{shape complexity measure} = \frac{l_{rms}}{l_{mhl}} \tag{3}$$

Note that the shape complexity measure depends only on the shape of the particle distribution.

12. DH-theory in progress and challenge

We start this section with recollection of basics of DH-theory.

- ▶ Basic objects are dendrograms, finite trees, which are generated by *hierarchic clustering algorithms* from real time series of experimental data.
- ▶ Thus, DH-theory studies the hierarchic relations inside experimental datasets, not simply order with respect to the position on the real space. In particular, the real numbers and space-time lose their exceptional role. But, they can be reconstructed from DH-theory.
- ▶ DH starting point was pure data analysis, analysis of hierarchic structuring in EEG-data collected for medical diagnostics. This analysis led us to theoretical and philosophical conclusions on foundations of classical and quantum physics and their interrelation.
- ▶ The DH-interrelation quantum-classical interrelation is not so sharp as in the standard theory. The degree of "quantumness" is checked with the aid of different tests, e.g., the violation of Bell type inequalities. This situation is like characterization of a random sequence as passing a batch of statistical tests, say NIST-tests.

Now we briefly list the recent steps in development of DH-theory, the dendrogram model for the event universe.

- ▶ Generally, our main tool for modelling was numerical simulation which was successfully applied for a variety of problems. It helped us to understand better the basic features of processes in the dendrogramic event-universe which differs essentially from the conventional universe based on the real space-time.
- ▶ The analytic part of DH-theory is reduced to the emergence of QM from DH-theory which is based on the calculus for views -- from one event to other events.
- ▶ Starting with geodesics of some GR-metric, we constructed the corresponding dentrogramic representation which can serve as the base for GR-quantization via emergence of DH-theory in QM
- ▶ We modelled dynamics of the relational recombination of the dendrogramic universe generated by appearance of a new event. Although such event disturbs the past event-tree, statistically recombinations are not global. We observe in data relational smoothness in evolution of the universe.
- ▶ By introducing shape complexity for dendrograms, we were able to determine the arrow of time.

References

[1] Rovelli C 1996 Relational quantum mechanics *International Journal of Theoretical Physics* **35** 1637–78

[2] Vladimirov V S, Volovich I V, Zelenov E I 1994 *P-adic Analysis and Mathematical Physics* (World Scientific, Singapore)

[3] Volovich I V 2010 Number theory as the ultimate physical theory *P-Adic Numbers, Ultrametric Analysis, and Applications* **2** 77–87
[4] Volovich I V 1987 p-adic string *Class. Quantum Gravity* **4** 83–87
[5] Dragovich B, Khrennikov A Y, Kozyrev S V and Volovich I v. 2009 On p-adic mathematical physics *P-Adic Numbers, Ultrametric Analysis, and Applications* **1** 1–36
[6] Aref'eva I Y, Dragovich B, Frampton P H and Volovich I V 1991 The wave function of the Universe andp-adic gravity *Int. J. Mod. Phys* **A6** 4341–4358
[7] Freund P G O and Witten E 1987 Adelic string amplitudes *Physics Letters B* **199** 191–4
[8] Parisi G and Sourlas N 2000 P-adic numbers and replica symmetry breaking *The European Physical Journal B - Condensed Matter and Complex Systems* **14** 535–42
[9] Parisi G 1988 On p-adic functional integrals *Mod Phys Lett A* **3** 639–43
[10] Shor O, Benninger F and Khrennikov A 2021 Representation of the universe as a dendrogramic hologram endowed with relational interpretation *Entropy* **23**
[11] Shor O, Benninger F and Khrennikov A 2021 Dendrogramic representation of data: Chsh violation vs. nonergodicity *Entropy* **23**
[12] Shor O, Benninger F and Khrennikov A 2022 Towards Unification of General Relativity and Quantum Theory: Dendrogram Representation of the Event-Universe *Entropy* **24**
[13] Shor O, Benninger F and Khrennikov A 2022 Dendrographic Hologram Theory: Predictability of Relational Dynamics of the Event Universe and the Emergence of Time Arrow *Symmetry (Basel)* **14**
[14] Shor O, Benninger F and Khrennikov A 2022 Emergent quantum mechanics of the event-universe, quantization of events via Denrographic Hologram Theory
[15] Barbour J 2000 *The End of Time* (Oxford University Press, New York)
[16] Barbour J 2012 Shape Dynamics. An Introduction *Quantum Field Theory and Gravity* 257–97
[17] Smolin L 2016 Quantum Mechanics and the Principle of Maximal Variety *Found Phys* **46** 736–58
[18] Smolin L 2018 The Dynamics of Difference *Found Phys* **48** 121–34
[19] Cortês M and Smolin L 2014 The universe as a process of unique events *Physical Review D - Particles, Fields, Gravitation and Cosmology* **90** 1–30
[20] Barbour J and Smolin L 1992 Extremal variety as the foundation of a cosmological quantum theory *arXiv preprint hep-th/9203041*
[21] Wheeler J A 1990 *Information, Physics, Quantum: The Search for Links* (Physics Department, Princeton University)
[22] Boltzmann L and Boltzmann L 1979 *Über die Frage nach der objektiven Existenz der Vorgänge in der unbelebten Natur* (Springer)
[23] Hertz H 1899 *The principles of mechanics presented in a new form* (Macmillian and Company, Limited)
[24] Atmanspacher H 2002 Determinism is ontic, determinability is epistemic *Between chance and choice: interdisciplinary perspectives on determinism* 49–74
[25] Atmanspacher H and Primas H 2003 *Epistemic and ontic quantum realities* (Springer)
[26] Khrennikov A 2019 Hertz's viewpoint on quantum theory *Act Nerv Super (Praha)* **61** 24–30
[27] Khrennikov A 2017 Quantum epistemology from subquantum ontology: Quantum mechanics from theory of classical random fields *Ann Phys (N Y)* **377** 147–63

Journal of Physics: Conference Series **2533** (2023) 012014 doi:10.1088/1742-6596/2533/1/012014

[28] Shor O, Glik A, Yaniv-Rosenfeld A, Valevski A, Weizman A, Khrennikov A and Benninger F 2021 EEG p-adic quantum potential accurately identifies depression, schizophrenia and cognitive decline *PLoS One* **16** e0255529-

[29] Einstein A 1911 On the Influence of Gravitation on the Propagation of Light *Ann Phys* **35** 906

[30] Sciama D W 1953 On the origin of inertia *Mon Not R Astron Soc* **113** 34–42

[31] Dicke R H 1957 Gravitation without a principle of equivalence *Rev Mod Phys* **29** 363

[32] Brans C and Dicke R H 1961 Mach's principle and a relativistic theory of gravitation *Physical review* **124** 925

[33] Smolin L 2016 Quantum Mechanics and the Principle of Maximal Variety *Found Phys* **46** 736–58

[34] Shor O, Benninger F and Khrennikov A 2022 Emergent quantum mechanics of the event-universe, quantization of events via Denrographic Hologram Theory *arXiv preprint arXiv:2208.01931*

[35] Barbour J 2020 *The Janus point: A new theory of time* (Random House)

[36] Barbour J, Koslowski T and Mercati F 2014 Identification of a Gravitational Arrow of Time *Phys Rev Lett* **113** 181101

[37] Mach E and McCormack T J 2013 *The Science of Mechanics* (Cambridge University Press)

Gravitation, and quantum theory, as emergent phenomena

Tejinder P. Singh[a,b]

[a]Inter-University Centre for Astronomy and Astrophysics, Post Bag 4, Ganeshkhind, Pune 411007, India
[b]Tata Institute of Fundamental Research, Homi Bhabha Road, Mumbai 400005, India

E-mail: `tejinder.singh@iucaa.in`, `tpsingh@tifr.res.in`

Abstract. There must exist a reformulation of quantum field theory, even at low energies, which does not depend on classical time. The octonionic theory proposes such a reformulation, leading to a pre-quantum pre-spacetime theory. The ingredients for constructing such a theory, which is also a unification of the standard model with gravitation, are : (i) the pre-quantum theory of trace dynamics – a matrix-valued Lagrangian dynamics, (ii) the spectral action principle of non-commutative geometry, (iii) the number system known as the octonions, for constructing a non-commutative manifold and for defining elementary particles via Clifford algebras, (iv) a Lagrangian with $E_8 \times E_8$ symmetry. The split bioctonions define a sixteen dimensional space (with left-right symmetry) whose geometry (evolving in Connes time) relates to the four known fundamental forces, while predicting two new forces, $SU(3)_{grav}$ and $U(1)_{grav}$. This latter interaction is possibly the theoretical origin of MOND. Coupling constants of the standard model result from left-right symmetry breaking, and their values are theoretically determined by the characteristic equation of the exceptional Jordan algebra of the octonions. The quantum-to-classical transition, precipitated by the entanglement of a critical number of fermions, is responsible for the emergence of classical spacetime, and also for the familiar formulation of quantum theory on a spacetime background.

1. Introduction: Quantum theory without classical time, as a route to quantum gravity and unification

In the words of Edward Witten [1],

"If one wants to summarise our knowledge of physics in the briefest possible terms, there are three really fundamental observations: (i) Space-time is a pseudo-Riemannian manifold M, endowed with a metric tensor and governed by geometrical laws. (ii) Over M is a vector bundle X with a nonabelian gauge group G. (iii) Fermions are sections of $(\hat{S}_+ \otimes V_R) \oplus (\hat{S}_- \otimes V_{\bar{R}})$. R and $\bar{R}$ are not isomorphic; their failure to be isomorphic explains why the light fermions are light and presumably has its origins in a representation difference Δ in some underlying theory. All of this must be supplemented with the understanding that the geometrical laws obeyed by the metric tensor, the gauge fields, and the fermions are to be interpreted in quantum mechanical terms" [*from the CERN preprint 'Physics and Geometry' (1987)*].

We attempt to bring the above-quoted three observations into one unified framework, in the octonionic theory [2, 3] summarised in the present article. This unification is based on Adler's theory of trace dynamics [4] and on the use of the number system of octonions as coordinate

Journal of Physics: Conference Series 2533 (2023) 012013 doi:10.1088/1742-6596/2533/1/012013

systems. The resulting theoretical framework has some commonalities with string theory (extended objects, $E_8 \times E_8$ symmetry) but also significant differences (octonions as coordinates, no compactification, trace dynamics instead of quantum dynamics). These differences help overcome the challenges to string theory as a theory of unification - we do not have multiverses, landscape or swampland; but rather the emergence of the standard model and of spacetime geometry, from an underlying higher dimensional geometric theory possessing $E_8 \times E_8$ symmetry.

The central premise for this new approach to unification is the following. The classical spacetime manifold (labelled by real numbers) exists if and only if the universe is dominated by classical bodies (planets, stars, galaxies). Therefore, the use of a classical spacetime background in quantum field theory is an approximation, because the classical objects which facilitate such a background to exist are themselves a limiting case of quantum systems. There must hence exist a reformulation of quantum field theory, which should make no reference to classical spacetime, and such a reformulation must exist at all energy scales, not just at the Planck energy scale. For there is nothing that in principle prevents a low energy universe from being entirely devoid of classical bodies, and even in such a universe we must be able to describe the dynamics of elementary particles. It turns out that a description of this kind, which uses octonions instead of real numbers as coordinates, explains why the standard model (including its 26 free parameters) is what it is, and offers a way to unify it with gravitation.

To reiterate the basic principle, consider that every elementary particle in the universe is in a quantum superposition of two or more position states. The corresponding gravitational fields will also be in a quantum superposition. Consequently, the operational distinguishability of space-time points will be lost, this latter being an implication of the Einstein hole argument, which requires the spacetime manifold to be overlaid by a classical metric. This argument holds true at every energy scale. The loss of classical spacetime hence necessitates that there be a reformulation of quantum field theory without classical spacetime; in particular needed also (and with far-reaching consequences) when we describe the standard model of particle physics.

We develop such a reformulation using only Planck length L_p, Planck time τ_p, and Planck's constant $\hbar$ as the fundamental constants of the theory. Every other dimensionful constant is expressible in terms of these (e.g. Newton's gravitational constant $G_N = L_p^5/\hbar\tau_p^3$) and every dimensionless constant (made with or without using these three constants) must be derivable from first principles. We have traded $\hbar$ for Planck energy, with the latter now assumed a derived quantity ($= \hbar/\tau_p$). When every physical subsystem in a chosen system has an action of the order $\hbar$, the point structure of spacetime is lost, irrespective of energy scale. In addition, if the length scale / time scale of interest is Planck length / Planck time, the energy scale is the Planck scale, and quantum gravitational effects become significant. Clearly, the point structure of spacetime can in principle be lost even if quantum gravity effects are not significant. The familiar formulation of quantum dynamics on a classical spacetime background is possible only when the universe is dominated by physical subsystems each of which has an action much larger than $\hbar$.

Fig. 1 lists the various ingredients which go into the construction of the proposed theory of unification. Fig. 2 highlights another aspect of this unification. Standard model gauge fields and fermions are assumed to live on a 4D spacetime curved by gravitation. When we take the square-root of the Klein-Gordon equation to write the Dirac equation, let us also take the square root of the Minkowski line element, i.e. describe the spacetime in spinorial language, as a twistor space, using complex numbers. If we were to replace complex numbers by the quaternions, the Dirac operator can be shown to be the gradient operator on the quaternionic space. Next, we replace the quaternions by the octonions (more precisely, complex split bioctonions). Standard model gauge fields and pre-gravitation are found to describe the geometry of this bioctonionic space ($E_8 \times E_8$ symmetry), and fermions in reality live in this space, not in spacetime. When very many degrees of freedom get entangled so that several physical subsystems each have action much

Journal of Physics: Conference Series **2533** (2023) 012013 doi:10.1088/1742-6596/2533/1/012013

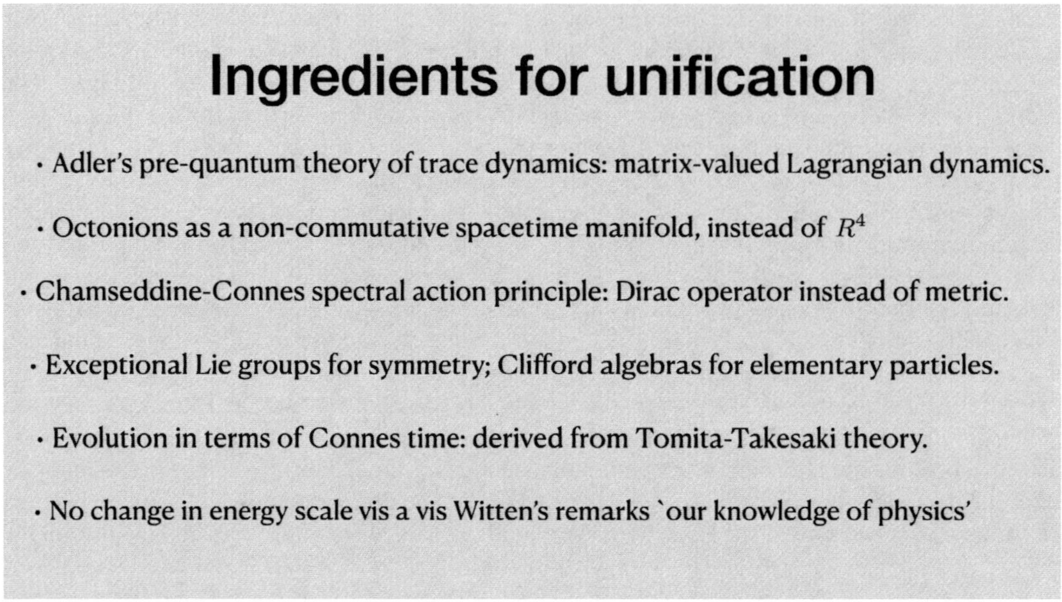

Figure 1. The ingredients for unification in the octonionic theory

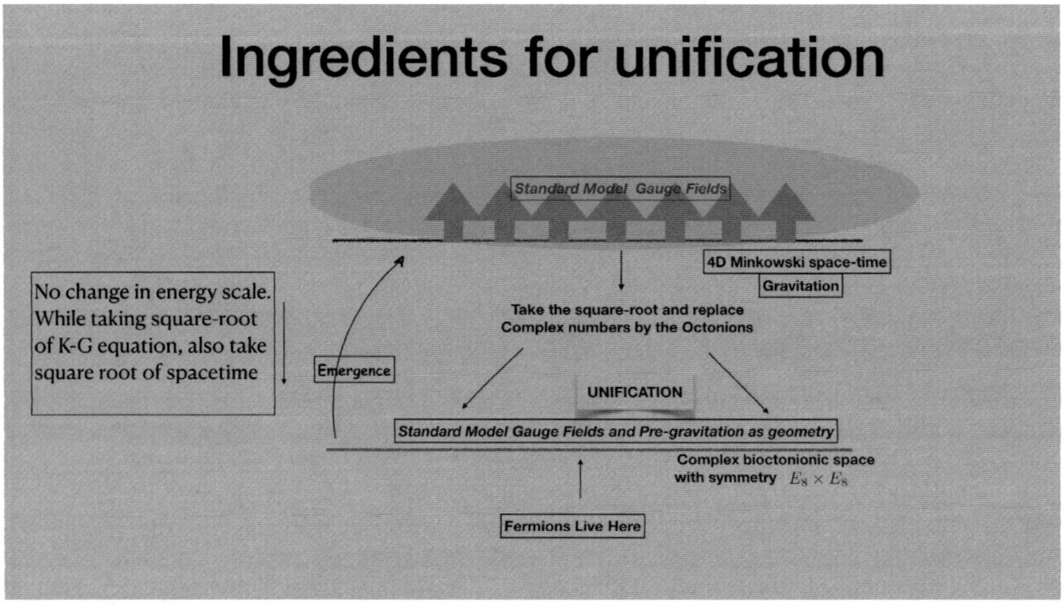

Figure 2. Ingredients for unification: taking the square root of Minkowski spacetime

larger than $\hbar$, classical spacetime emerges ('Emergence' in Fig. 2). Those subsystems which still have action order $\hbar$ are quantum in nature, and should strictly be described on bioctonionic space, but can be described, to a good approximation, on the emergent 4D spacetime background accompanied by the aforesaid vector bundle.

In the following sections we describe these ingredients, and the proposed theory, in some detail.

Journal of Physics: Conference Series **2533** (2023) 012013 doi:10.1088/1742-6596/2533/1/012013

2. From Newtonian dynamics to trace dynamics: quantum theory as an emergent phenomenon

Consider that we wish to define Newtonian dynamics using, not real numbers, but matrices. For instance, given the action

$$S = \sum_i \int d\tau \, \frac{1}{2} m_i \left(\frac{dq_i}{d\tau}\right)^2 \tag{1}$$

for a collection of point particles with configuration variables labelled by real numbers q_i, we replace real numbers by matrices, $q_i \to \mathbf{q_i}$. The Lagrangian becomes a matrix polynomial, and the new Lagrangian for this matrix dynamics is defined as the trace of the matrix polynomial:

$$S = \sum_i \int d\tau \, Tr \left[\frac{1}{2}\frac{L_p^2}{L^2}\left(\frac{d\mathbf{q}_i}{d\tau}\right)^2\right] \tag{2}$$

[Keeping in view what lies ahead, the mass parameter m has been replaced by (square of a) length parameter L (measured in units of Planck length L_p)]. Lagrange equations of motion are derived by varying the trace Lagrangian with respect to the matrix variables, and an equivalent Hamiltonian formulation and a phase space dynamics is also developed. Hamilton's equations of motion are precursors of the Heisenberg equations of motion of quantum theory; the latter are emergent.

This is the theory of trace dynamics, developed by Adler and collaborators [5, 6]. The matrices (equivalently operators) in this matrix-valued Lagrangian dynamics have the same status as in Heisenberg's matrix mechanics; their eigenvalues are the values that the corresponding classical dynamical variables take during evolution. Trace dynamics is more general than quantum theory, because it possesses an additional conserved Noether charge, because of the invariance of the trace Hamiltonian under global unitary transformations. This charge, denoted $\tilde{C}$ and known as the Adler-Millard charge, is given in terms of canonical configuration variables and their corresponding momenta, as

$$\tilde{C} = \sum_i [\mathbf{q}_B, \mathbf{p}_B] - \{\mathbf{q}_F, \mathbf{p}_F\} \tag{3}$$

The matrices have Grassmann numbers as their entries (analogous to the case in quantum field theory); the matrices q_B are made of even grade Grassmann numbers and known as bosonic matrices; whereas the q_F made of odd grade Grassmann numbers are known as fermionic matrices. This conserved charge has the dimensions of action, and is what make trace dynamics into a pre-quantum theory, from which quantum field theory is emergent. The continuum field theory generalisation of trace dynamics is achieved by considering a classical field as a collection of point particles (one per space-time point), generalising each such point particle to a matrix, and then integrating the trace Lagrangian over space-time volume so as to obtain the action. Furthermore, one can start from a Lorentz-invariant classical dynamics and construct its generalisation to a relativistic trace dynamics.

It is assumed that trace dynamics holds at some time scale resolution not accessed by current laboratory experiments, say Planck time τ_p. We then ask what is the emergent coarse-grained dynamics, if the system is observed not at Planck time resolution, but at some lower resolution $\tau \gg \tau_p$? The standard techniques of statistical thermodynamics are employed to construct a phase space density distribution of the trace dynamical system, whose emergent coarse-grained dynamics is determined by maximising the Boltzmann entropy subject to constraints representing conserved quantities. It is shown that at thermodynamic equilibrium the Adler-Millard charge is equipartitioned over all degrees of freedom so that the canonical average of each commutator $[q_B, p_B]$ and each anti-commutator $\{q_F, p_F\}$ is assumed to be equal to $i\bar{h}$. This is how the quantum commutation relations emerge from the underlying trace dynamics. Also, in this emergent thermodynamic equilibrium, the canonically averaged Hamilton's equations of motion become Heisenberg's equations of motion of quantum theory. Identification of canonical averages of functions of dynamical variables (in their ground state) with Wightman functions in relativistic quantum mechanics enables the transition from trace dynamics to quantum field theory. Quantum theory is thus shown to be an emergent (equilibrium) thermodynamic phenomenon.

At equilibrium, the Adler-Millard charge is anti-self-adjoint, and the Hamiltonian of the theory is self-adjoint. Statistical fluctuations in this charge, when significant, can drive the quantum system away from equilibrium (the charge is no longer equipartitioned). If these fluctuations are themselves dominantly self-adjoint, the Hamiltonian of the theory picks up an anti-self-adjoint component, which gets amplified if a large number of degrees of freedom are entangled with each other. This drives the system to classicality, via a Ghirardi-Rimini-Weber type of spontaneous collapse process. Thus, macroscopic classical systems are far from equilibrium emergent states in trace dynamics.

If the fluctuations in the Adler-Millard charge are dominantly anti-self-adjoint, the Hamiltonian of the theory continues to be self-adjoint and the system can be said to be in a quantum non-equilibrium, perhaps analogous to the quantum non-equilibrium talked of in Bohmian mechanics. The quantum commutation relations no longer hold. As a consequence, such trace dynamical systems violate the Tsirelson bound (obeyed by quantum systems) of the CHSH inequality. It has long been a puzzle as to why the Popescu-Rohrlich bound of 4 permitted by relativistic causality is higher than the Tsirelson bound of $2\sqrt{2}$ obeyed by quantum mechanical systems. We now know the answer: trace dynamics, being more general than quantum theory, permits supra-quantum non-local correlations when the Adler-Millard charge is not equipartitioned and is dominantly anti-self-adjoint [7]. This situation is exhibited in Fig. 3 and the experimental search for such supra-quantum correlations is of great interest: a confirmation will be a conclusive signature that quantum theory is approximate, not exact. In fact, the quantum state is an attractor, being the equilibrium state, to which classical systems as well as supra-quantum systems evolve. Since the Adler-Millard charge has dimensions of action,

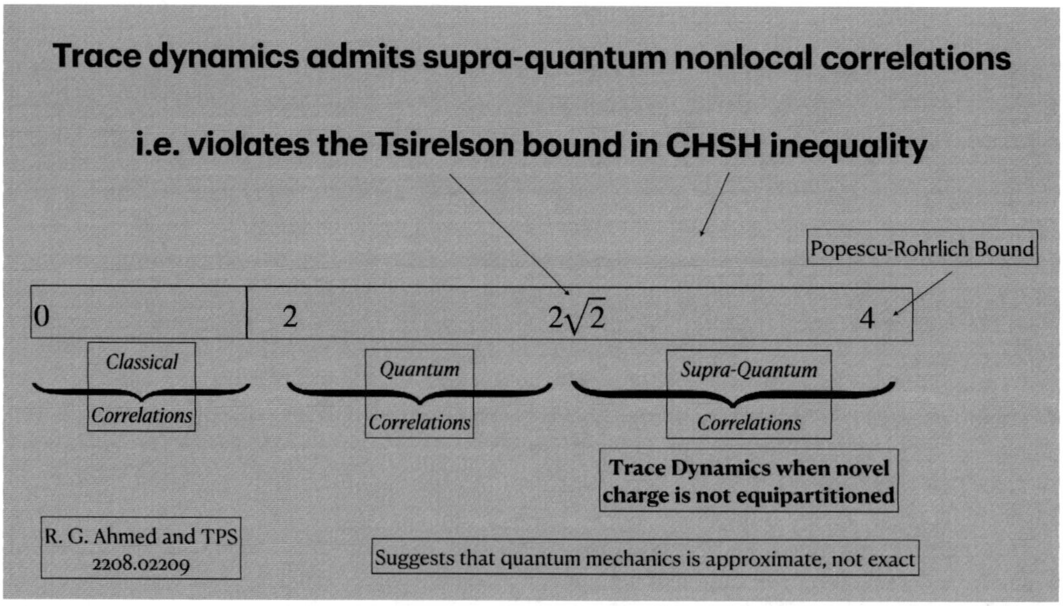

Figure 3. Trace dynamics violates the Tsirelson bound in the CHSH inequality

Journal of Physics: Conference Series **2533** (2023) 012013 doi:10.1088/1742-6596/2533/1/012013

its conjugate variable must be dimensionless. This conjugate variable is a time parameter τ measured in units of Planck time τ_p; however this is not the time coordinate (this latter is usually denoted t) of the spacetime manifold of special relativity. Rather the time τ is the Connes-Tomita-Takesaki time parameter (to be discussed in the next section) which is a unique feature of non-commutative geometry resulting from the Tomita-Takesaki theory. The conjugate of coordinate time is energy; the conjugate of Connes time is the Adler-Millard charge. When $\tau/\tau_p \gg 1$, the Adler-Millard charge is equipartitioned; and when $\tau/\tau_p \sim 1$, the charge is not equipartitioned and supra-quantum non-local quantum correlations arise. It might hence seem that such correlations can only be realised at Planck time resolution. However, in the octonionic theory presently under review, an 'octonionic' inflation in the very early universe resets the Planck energy scale to the TeV scale; hence it might be possible to experimentally detect violation of the Tsirelson bound by performing Bell type experiments at around the TeV scale. Also, it appears that trace dynamics needs non-commutative geometry for a consistent interpretation of the Adler-Millard charge. As originally formulated, trace dynamics assumes a classical spacetime manifold, which could be flat, or endowed with classical gravitation.

We generalise trace dynamics to remove the classical spacetime manifold, by using the non-commuting octonions as coordinates instead. Also, a matrix-valued description of gravitation is developed, by using the spectral action principle of Connes' non-commutative geometry. This leads to a pre-quantum, pre-spacetime theory from which classical spacetime geometry and quantum theory are emergent. Also, trace dynamics does not specify the fundamental Lagrangian of the universe. In the octonionic theory, we propose a trace dynamics Lagrangian with $E_8 \times E_8$ symmetry, which unifies pre-gravitation with the standard model.

3. Choosing the Lagrangian: the spectral action principle

In order to decide as to how gravitation should be incorporated in trace dynamics, we appeal to the celebrated spectral action principle [8], used in particular to make the transition from Riemannian geometry to non-commutative geometry. According to this principle, the Einstein-Hilbert action can be expressed in terms of the eigenvalues of the Dirac operator D on a Riemannian geometry, via a truncated heat kernel expansion in powers of L_p^{-2}:

$$Tr[L_P^2 D^2] \sim L_p^{-2} \int d^4x \, \sqrt{g}\, R + \mathcal{O}(L_P^0) = \sum_i L_p^2 \, \lambda_i^2 \qquad (4)$$

It has also been shown that these Dirac eigenvalues can be the dynamical observables of general relativity [9], in place of the metric. In the spirit of trace dynamics, every eigenvalue λ_i is raised to the status of a bosonic matrix/operator $\hat{\lambda}_i \equiv \dot{q}_{Bi}$, this being the very Dirac operator D of which it is an eigenvalue. Thus in generalised trace dynamics (a pre-quantum, pre-spacetime theory) we have a collection of 'atoms of space-time', as many atoms as there were Dirac eigenvalues, each atom being associated with a copy of the Dirac operator: $\dot{q}_{Bi} \equiv LD$, where L is a newly introduced length parameter which characterises a space-time atom. The Einstein-Hilbert action $\sum_i L_p^2 \, \lambda_i^2$ transits to $\sum_i Tr\ (L_p^2/L^2)\ [\dot{q}_{Bi}^2]$. Since the Dirac eigenvalues have been made operators, space-time is lost, and this is simultaneously a transition to trace dynamics and to non-commutative geometry, hence showing the deep connection between the new geometry and the new dynamics. But with one caveat: what was earlier the (dimensionless) action is now the dimensionless (trace) Lagrangian; the integral over time, which will make it into a trace dynamics action, is missing! Here, Connes time parameter τ, a unique feature of non-commutative geometry, comes to our rescue. The trace dynamics action for atoms of space-time matter, scaled with respect to Planck's constant $\hbar$, is given by

$$\frac{S}{\hbar} = \int \frac{d\tau}{\tau_p} \sum_i Tr\ (L_p^2/L^2)\ [\dot{q}_{Bi}^2] \qquad (5)$$

Journal of Physics: Conference Series **2533** (2023) 012013 doi:10.1088/1742-6596/2533/1/012013

The spectral action principle has been shown to hold also when Yang-Mills gauge fields are present, besides gravitation; with the gauge field A^μ introduced in the conventional manner of modifying the Dirac operator: $D \to D + A^\mu$. The eigenvalues of the gauge potential are raised to the status of matrices, and these matrices are identified with the configuration variables q_{Bi} of the corresponding Dirac operators $\dot{q}_{Bi}$. Matter (fermionic) degrees of freedom q_{Fi} and $\dot{q}_{Fi}$ are introduced such that the eigenvalues of these matrices relate to the classical relativistic action of point particles (mass term as well as currents related to gauge fields).

Henceforth, we will focus on just one STM (space-time-matter) atom and try to understand its properties; leaving for later the question of interaction between several atoms. If we keep only the dotted terms, we have the action for gravitation coupled to (as we shall see, right handed) fermions:

$$\frac{S}{\hbar} = \int \frac{d\tau}{\tau_p}\, Tr \left\{ \frac{L_p^2}{L^2} \left[\dot{q}_B^\dagger + \frac{L_p^2}{L^2}\beta_1 \dot{q}_F^\dagger \right] \times \left[\dot{q}_B + \frac{L_p^2}{L^2}\beta_2 \dot{q}_F \right] \right\} \tag{6}$$

β_1 and β_2 are constant Grassmann elements introduced to make the Lagrangian bosonic. When the undotted terms are also included, we get the action also for the Yang-Mills fields and left-handed fermions, all put together this defines an 'atom of space-time-matter', or an *aikyon*.

$$\frac{S}{\hbar} = \int \frac{d\tau}{\tau_p}\, Tr \left\{ \frac{L_p^2}{L^2} \left[\left(\dot{q}_B^\dagger + i\frac{\alpha}{L} q_B^\dagger \right) + \frac{L_p^2}{L^2}\beta_1 \left(\dot{q}_F^\dagger + i\frac{\alpha}{L} q_F^\dagger \right) \right] \times \left[\left(\dot{q}_B + i\frac{\alpha}{L} q_B \right) + \frac{L_p^2}{L^2}\beta_2 \left(\dot{q}_F + i\frac{\alpha}{L} q_F \right) \right] \right\} \tag{7}$$

An aikyon is an elementary particle (say an electron) along with all the bosonic fields it produces. (The word aikyon derives from the Sanskrit word *aikya* meaning 'oneness'.) This is the fundamental action principle for the octonionic theory. The claim is that when these matrices are defined on the space of split bioctonions, and the Lagrangian has $E_8 \times E_8$ symmetry, this action principle describes (after left-right symmetry breaking) standard model gauge fields and chiral fermions and the Higgs, as also an additional Higgs and (a generalisation of) Einstein's general relativity which now includes also an $SU(3)_{grav}$ interaction and an $U(1)_{grav}$ interaction, with the latter showing strong evidence for being the origin of (relativistic) MOND [10]. General relativity itself is inferred as the right-handed counterpart of the weak force, both being broken symmetries! The constant α is the Yang-Mills coupling constant, originating from $E_8 \times E_8$ - it arises as a result of left-right symmetry breaking which separates the unified dynamical variable $\dot{Q}_B$ into its gravitational and gauge sectors: $\dot{Q}_B \equiv \dot{q}_B + (\alpha/L)q_B$.

The Hamiltonian of the theory is not self-adjoint. Assuming that the theory holds at Planck (Connes) time resolution, and assuming that the anti-self-adjoint part of the Hamiltonian is insignificant, coarse graining to a lower time resolution gives the sought for reformulation of quantum theory which does not depend on classical spacetime. This is also a quantum theory of gravity. Evolution continues to be defined through Connes time. If in the underlying theory, sufficiently many degrees of freedom get entangled, the imaginary part of the Hamiltonian becomes important, leading to collapse of superpositions [a deterministic, non-unitary and norm-preserving evolution]. If this system is observed only under a coarse-grained approximation, the outcomes of collapse appear random, while obeying the Born probability rule, and thus offering a theoretical underpinning for models of objective collapse [11]. Collapse is the opposite of raising each eigenvalue to the status of a matrix: spontaneous collapse sends the matrix back to one of its eigenvalues. In this process one also recovers classical space-time and general relativity, as the Dirac operators collapse to (distinct) eigenvalues, and the spectral action principle ensures recovery of the Einstein-Hilbert action coupled to Yang-Mills fields and relativistic point particles, with additional corrections. Those degrees of freedom which are not sufficiently entangled continue to obey the underlying trace dynamics (no spacetime) but can be described to a good approximation by conventional quantum field theory on a classical

Journal of Physics: Conference Series **2533** (2023) 012013 doi:10.1088/1742-6596/2533/1/012013

spacetime background. In making this approximation, the origins of the standard model of particle physics are lost. We note that this spacetime background has become available only because the universe is dominated by classical objects.

Everything that is said in the previous paragraph is independent of energy scale. When we say that Connes time τ is being measured at Planck time resolution, it does not imply that the system is being probed at Planck energies. The conjugate variable for Connes time is not energy, but the Adler-Millard charge: at Planck time resolution this charge is not equipartitioned; at lower resolution it is. The variable conjugate to energy is the coordinate time of special relativity - this time does not flow and in fact, calling it time (time being that which has an arrow and distinguishes past from future) seems like a misnomer! When evolution is described through Connes time, the emergent quantum theory is analogous to the Stueckelberg-Hurwitz formulation of relativistic quantum mechanics.

So far, we have a matrix-valued Lagrangian dynamics, which is a generalisation of classical real-number valued dynamics. We have also made a transition from Riemannian geometry to Connes' non-commutative geometry. What remains is to transit from the real-number valued coordinate system which labels the 4D space-time manifold, and to instead work with the non-commuting numbers known as quaternions and octonions. The dynamical matrices (which replace vectors) have matrix-valued 'coordinate' components over the field of quaternions/octonions, instead of over the field of real numbers. We then have a pre-quantum, pre-spacetime dynamics in higher dimensions, which we employ to describe the standard model as well as gravitation, because the (broken) symmetries of bi-octonionic space coincide with the ones observed in nature.

4. Octonions as coordinate systems: a non-commutative manifold

At the beginning of this article, we quoted Witten's remarks: "(i) Space-time is a pseudo-Riemannian manifold M, endowed with a metric tensor and governed by geometrical laws. (ii) Over M is a vector bundle X with a non-abelian gauge group G." We are going to employ octonionic space to unify this vector bundle and the 4D space-time manifold into a new higher dimensional space. This is done without change of energy scale, at the very energies at which the standard model is formulated at present. The algebra automorphisms of the octonions unify space-time diffeomorphisms and standard model gauge field transformations into one common symmetry ($E_8 \times E_8$). The octonionic coordinate space is defined separately for every atom of space-time-matter, one coordinate copy per atom.

There are only four division algebras: reals, complex numbers, quaternions and octonions, denoted $\mathbf{R}, \mathbf{C}, \mathbf{H}, \mathbf{O}$. A quaternion $\mathbf{H}$

$$\mathbf{H} = a_0+a_1\hat{i}+a_2\hat{j}+a_3\hat{k}; \quad \hat{i}^2 = \hat{j}^2 = \hat{k}^2 = -1; \quad \hat{i}\hat{j} = \hat{k} = -\hat{j}\hat{i}; \ \hat{j}\hat{k} = \hat{i} = -\hat{k}\hat{j}; \ \hat{k}\hat{i} = \hat{j} = -\hat{i}\hat{k} \quad (8)$$

can be used to define a vector and its rotations in 3D space. A split biquaternion is defined as

$$\mathbf{H} \oplus \omega\tilde{\mathbf{H}} = (a_0 + a_1\hat{i} + a_2\hat{j} + a_3\hat{k}) \oplus \omega(a_0 - a_1\hat{i} - a_2\hat{j} - a_3\hat{k}) \quad (9)$$

Here ω is the split complex number (i.e. $\omega^* = -\omega, \omega^2 = 1$) made from the imaginary directions of a quaternion. Complexified split biquaternions are key to defining chiral leptons in this theory. Furthermore, the Dirac operator is nothing but the gradient operator on quaternionic space - the gamma matrices present in the Dirac operator when defined on Minkowski spacetime mimic the true nature of spacetime, which is quaternionic and non-commutative. The Lagrangian we have constructed in (7) is essentially the square of the Dirac operator (squared momentum / kinetic energy) of a free particle.

An octonion is defined as [12]

$$\mathbf{O} = a_0 + a_1\mathbf{e}_1 + a_2\mathbf{e}_2 + a_3\mathbf{e}_3 + a_4\mathbf{e}_4 + a_5\mathbf{e}_5 + a_6\mathbf{e}_6 + a_7\mathbf{e}_7 \quad (10)$$

The seven imaginary direction anti-commute, each of them squares to -1, and octonionic multiplication obeys the Fano plane rules. A split bioctonion is defined as

$$\begin{aligned}\mathbf{O} + \omega\tilde{\mathbf{O}} =& (a_0 + a_1\mathbf{e}_1 + a_2\mathbf{e}_2 + a_3\mathbf{e}_3 + a_4\mathbf{e}_4 + a_5\mathbf{e}_5 + a_6\mathbf{e}_6 + a_7\mathbf{e}_7) + \\ & \omega(a_0 - a_1\mathbf{e}_1 - a_2\mathbf{e}_2 - a_3\mathbf{e}_3 - a_4\mathbf{e}_4 - a_5\mathbf{e}_5 - a_6\mathbf{e}_6 - a_7\mathbf{e}_7)\end{aligned} \tag{11}$$

This time the split complex number ω is made from the imaginary directions of the octonion. Complexified split bioctonions are central to defining chiral quarks and leptons. Whereas split biquaternions are adequate for chiral leptons, the extension to split bioctonions is essential for bringing in chiral quarks: QCD is the geometry of extra spatial dimensions (there being four such extra dimensions).

Bosons and fermions are defined on split bioctonionic space; for instance

$$Q_B = Q_0 + Q_1\mathbf{e}_1 + Q_2\mathbf{e}_2 + Q_3\mathbf{e}_3 + Q_4\mathbf{e}_4 + Q_5\mathbf{e}_5 + Q_6\mathbf{e}_6 + Q_7\mathbf{e}_7 \tag{12}$$

shows the matrix-valued components Q_i of a bosonic matrix Q_B over octonionic space. In the action (7) the undotted matrices are defined over octonionic space and dotted matrices over the split part of the bioctonionic space. Keeping this in mind, consider the modulus square of the split bioctonion:

$$\begin{aligned}|O + \omega\tilde{O}|^2 = (\tilde{O} \ominus \omega O) \times (O \oplus \omega\tilde{O}) &= \tilde{O}O \oplus \omega\tilde{O}\tilde{O} \ominus \omega OO \ominus O\tilde{O} \\ &= (a_0^2 + a_1^2 + a_2^2 + a_3^2 + a_4^2 + a_5^2 + a_6^2 + a_7^2) \oplus \\ &\quad \omega(a_0^2 - a_1^2 - a_2^2 - a_3^2 - a_4^2 - a_5^2 - a_6^2 - a_7^2 + Im1) \ominus \\ &\quad \omega(a_0^2 - a_1^2 - a_2^2 - a_3^2 - a_4^2 - a_5^2 - a_6^2 - a_7^2 + Im2) \ominus \\ &\quad (a_0^2 + a_1^2 + a_2^2 + a_3^2 + a_4^2 + a_5^2 + a_6^2 + a_7^2)\end{aligned} \tag{13}$$

The four expressions in the four lines after the second equality demonstrate the unified presence of the vector bundle (lines one and four, Euclidean line-element) and space-time (lines two and three, Lorentzian line element, with imaginary corrections). Inspecting the bosonic part of the Lagrangian (7) we see that the two Euclidean elements are for the dotted quadratic term $\dot{q}_B^\dagger \dot{q}_B$ and the undotted term $q_B^\dagger q_B$ respectively. As has been analysed in [13], and supported by the results in [14], the undotted term represents an interaction with $SU(3)$ symmetry that is identified with $SU(3)_{color}$, whereas the dotted term is a new $SU(3)$ symmetry interpreted as $SU(3)_{grav}$. The Lorentzian elements in lines two and three in the above equation are for the mixed terms $\dot{q}_B^\dagger q_B$ and $q_B^\dagger \dot{q}_B$. They represent an $SU(2)_L$ symmetry and an $SU(2)_R$ symmetry - the former, along with a contribution from the undotted quadratic term, represent the electroweak symmetry [13, 14]. The latter, along with a contribution from the dotted quadratic term, represents a $SU(2)_R \times U(1)_{grav}$ symmetry which is the precursor of general relativity modified by a $U(1)_{grav}$. This symmetry is the right-handed counterpart of electroweak and is possibly a renormalisable theory - this might help us understand why general relativity is not renormalisable (it being a broken symmetry like the weak force), whereas the $U(1)_{grav}$ is possibly the theoretical origin of MOND.

The imaginary corrections arise from multiplying an octonion onto itself; when they are significant, they might help understand why in the macroscopic limit space-time becomes classical. Because these corrections contribute an anti-self-adjoint part to the trace Hamiltonian. Whereas the Euclidean sector has no imaginary terms, is responsible for the strong force and for the newly proposed $SU(3)_{grav}$; it remains quantum and moreover does not take part in the cosmological expansion of space-time. Also, it is evident that the weak force is a space-time symmetry, not an internal symmetry, unlike the strong force. Together, gravitation and the weak force are broken symmetries in a 6D space-time, related to $SU(2)_L \times U(1)_Y \times SU(2)_R \times U(1)_{grav}$, and stemming from the group-theoretic relation $SL(2,H) \sim SO(1,5)$. It can be argued that the two additional spatial dimensions here have a thickness of the order $(L_P/L)^{1/3} \sim 10^{-13}$ cm, where $L \sim 10^{28}$ cm is the size of the observed universe. This is not too far off from the range of the weak force, and also explains why we land up clubbing the weak force with the strong force as an internal symmetry, and not along with gravitation as a spacetime symmetry. If we were to club the four additional spatial directions (strong force) with the 6D spacetime, we have in effect a 10D space-time, motivated also by the group theory relation $SL(2,O) \sim SO(1,9)$. This 10D space-time evolves in Connes time τ.

5. Octonions, Clifford algebras, and elementary particles

Spinors can be defined as minimal left ideals of Clifford algebras. Furthermore, one can use division algebras to construct Clifford algebras. And in specific cases, the symmetry properties of the corresponding spinors coincide with those of elementary fermions of the standard model. This gives strong evidence that complex quaternions and complex octonions are the natural home for defining states of quarks and leptons of three generations. Hence, not only do the octonions serve to define the coordinate system on the non-commutative manifold, but they also serve to naturally define states of quarks, leptons and gauge bosons, including those for gravitation. The standard model group symmetries do not have to be imposed by hand on these states; rather these symmetries are already present as subgroups in the symmetries of the octonion algebra. The five exceptional Lie groups - G_2, F_4, E_6, E_7, E_8 - all associated with the octonions, play a very important role in the deduction of the standard model. The all-encompassing role of octonions in defining non-commutative space-time, internal geometry, and particle states should be contrasted with the situation in quantum field theory: complex numbers for quantum states, and real numbers for space-time.

In the present context, the most important Clifford algebras are $Cl(2), Cl(3), Cl(6)$ and $Cl(7)$. The algebra $Cl(2)$ is generated by complex quaternions, keeping one of the quaternionic imaginary directions fixed. Spinors made from $Cl(2)$ are the left handed and right handed Weyl spinors, and the associated symmetry is the Lorentz algebra $SL(2,C)$. Octonions being a non-associative algebra, do not generate a Clifford algebra. However, maps are associative; therefore octonionic maps generate a Clifford algebra. The exceptional Lie group G_2 is the automorphism group of the seven imaginary directions of the octonion. G_2 has two maximal subgroups, $SU(3)$ and $SO(4)$. The former is the element stabiliser group (i.e. the automorphism group when one of the imaginary octonionic directions is kept fixed), whereas the latter is the stabiliser group of the quaternions inside the octonions. Keeping one of the imaginaries fixed, complexified octonionic maps generate the Clifford algebra $Cl(6)$. Spinors made from this algebra (there being eight of them) obey an $SU(3)$ symmetry: two out of these eight states are singlets of $SU(3)$, three are anti-triplets and three are triplets. A number operator N made from the generators of $Cl(6)$ and having a $U(1)$ symmetry has the eigenvalues 0 and 3 for the singlet states, eigenvalue 1 for the three anti-triplet states, and 2 for the triplet states. Defining $Q = N/3$ as the electric charge operator, we conclude that the singlet states are the neutrino and the positron, the anti-triplet is the anti-down quark, and the triplet is the up quark. Anti-particles are defined simply by the complex conjugation of these states. The $SU(3)$ symmetry is identified with $SU(3)_{color}$ of QCD, and the $U(1)$ symmetry with electromagnetism, $U(1)_{em}$. This inference is fully supported by the analysis of the fundamental Lagrangian in (7), as shown in [13]. The Clifford algebra $Cl(6)$ describes one generation of standard model quarks and leptons under the unbroken symmetry $SU(3)_c \times U(1)_{em}$ [15, 16] . As noted earlier, this is an internal symmetry on Euclidean space (the vector bundle) and hence remains unbroken; whereas the weak interaction and general relativity, being space-time symmetries, are broken, as a result of the quantum-to-classical transition.

Among Clifford algebras, the algebras $Cl(3)$ and $Cl(7)$ are very special. They are the only ones, upto Bott periodicity, that have two irreducible representations (called pinors). The algebra $Cl(3)$, made from complex quaternions using all three imaginary quaternionic directions, is the algebra of complex split biquaternions. Each of the two quaternion copies corresponds to a $Cl(2)$ each, and one copy is the parity reverse of the other. This naturally enables the construction of one generation of chiral leptons and their anti-particles - left handed neutrino, right handed (sterile) neutrino, left handed electron and right handed electron, and their anti-particles [17].

Analogously, the algebra $Cl(7)$ corresponds to complex split bioctonions and two copies of $Cl(6)$; thus describing one generation of chiral quarks and leptons. However, we introduce a significant subtlety: whereas the $U(1)_{em}$ associated with left-handed quarks and leptons has electric charge as its quantum number, the quantum number associated with the right-handed quarks and leptons is the square-root of mass (in Planck units). It takes the values $0, 1/3, 2/3, 1$ for the right-handed (sterile) neutrino, electron, up quark and down quark, respectively. We have switched the position of the electron and the down quark; the former is now a triplet, and the latter is a singlet, of the symmetry $SU(3)_{grav}$ associated with the right handed electron and the right handed up quark. Since the $SU(3)$ symmetry is unbroken, being an internal symmetry, the interaction $SU(3)_{color} \times SU(3)_{grav}$ is not parity violating. This might pave the way for resolving the strong CP problem, if we assume that $SU(3)_{grav}$ is much weaker than $SU(3)_{color}$.

In this theory, there is evidence that there are three, and only three, fermion generations. This evidence comes from the triality of the group $SO(8)$, from the exceptional Jordan algebra $J_3(8)$, and from the symmetry $SU(3)_{gen}$ which arises in the branching of $E_8 \times E_8$ as we will see in the next section. The octonionic spinor states for the second and third generation can be obtained by applying $SU(3)$ rotations on the states of the first generation [18, 19]. It remains to be understood why there is only one copy of the gauge bosons, as opposed to three.

6. $E_8 \times E_8$ unification of the standard model and pre-gravitation

We propose that there are six fundamental forces (not four), and that they are described by the following symmetry

$$SU(3)_C \otimes SU(2)_L \otimes U(1)_Y \otimes SU(3)_{grav} \otimes SU(2)_R \otimes U(1)_g \tag{14}$$

The first three of these are the standard model forces, and each of them has a gravitational counterpart, shown by the last three groups, which includes $SU(2)_R$, the precursor of general relativity. $SU(3)_{grav}$ and $U(1)_g$ are newly predicted. Whereas the sources for the standard model forces are color, weak isospin and hypercharge, the corresponding gravitational sources are gravi-color (non-zero for right-handed electron and right-handed up quark), gravi-isospin and gravi-hypercharge. Gravi-hypercharge is related to square-root mass in precisely the same way in which hypercharge is related to electric charge.

The origin of these forces lies in a specific branching of $E_8 \times E_8$. Thus the rep $(248, 1) \oplus (1, 248)$ is broken into two separate E_8. Each E_8 then branches as $SU(3) \otimes E_6$. This $SU(3)$ is mapped to an 8D vector space, which is identified with an octonion in case of the first E_8, and with the split part of a bioctonion in case of the second E_8. Together, the tensor product $SU(3) \otimes SU(3)$ maps to the split bioctonion (11) which in turn gives rise to the line-element shown in (13). Thus the branching brings with it the sought for unification of the vector bundle with space-time.

E_6 is the only exceptional Lie group which has complex representations, and the two E_6 between them define three fermion generations, standard model gauge bosons, pre-gravitation, and two Higgs. Each E_6 branches as

$$E_6 \rightarrow SU(3) \otimes SU(3) \otimes SU(3) \tag{15}$$

and we have

$$248 = (8,1) \oplus (1,78) \oplus (3,27) \oplus (\bar{3},\overline{27}) \tag{16}$$

One of the $SU(3)$ branches as $SU(2) \times U(1)$ and the two E_6 are interpreted as

$$\text{First } E_6 : SU(3)_{LHgen} \otimes SU(3)_c \times SU(2)_L \times U(1)_Y \tag{17}$$

$$\text{Second } E_6 : SU(3)_{RHgen} \otimes SU(3)_{grav} \times SU(2)_R \times U(1)_g \tag{18}$$

We get three generations of chiral fermions; for the left-handed ones the $U(1)$ quantum number is electric charge, and for the right handed ones the $U(1)$ quantum number is square root mass. Both electric charge as well as square-root mass are emergent entities, arising after the left-right symmetry breaking. The standard model Higgs gives mass to the left-handed fermions, and the newly predicted Higgs gives electric charge to the right-handed fermions. For further details of this unification proposal, the reader is referred to [14].

This symmetry breaking, which is also the same as electroweak symmetry breaking, is enabled by the quantum-to-classical transition, in the very early universe as well as in the present low-energy universe. This transition happens when a very large number of fermions get entangled: for this to happen in the very early universe the universe must cool below a critical temperature (in this case the EW scale). Thus the role of the energy scale is only indirect; critical entanglement is what is actually responsible for symmetry breaking. Classical space-time emerges along with the dominance of classical objects, whose gravitation obeys the laws of general relativity in the vicinity of compact objects, and obeys the $U(1)_{grav}$ interaction in regions where gravitational acceleration is below the MOND critical acceleration. In the present context, symmetry breaking implies that the broken $SU(2)_R$ symmetry becomes classical (general relativity) whereas the broken $SU(2)_L$ symmetry, i.e. the weak force, remains quantum in nature, and is short range. Even in today's universe, it is the act of quantum measurement (wave function collapse followed by the quantum-to-classical transition) which is responsible for the breaking of left-right symmetry. A quantum system, even at low energies, lives in its own split bioctonionic space.

The left-right symmetry breaking in the very early universe is also responsible for the separation of matter (has positive sign for square-root mass: $+\sqrt{m}$) and anti-matter (has negative sign for square-root mass: $-\sqrt{m}$). Like sign square-root masses attract pre-gravitationally, whereas unlike sign square-root masses repel pre-gravitationally. This results in the separation of matter from anti-matter while preserving CPT symmetry of the matter-antimatter mirror universe, an idea which has been independently proposed earlier in [20]. Pre-gravitation described by the $SU(2)_R$ symmetry is mediated by spin one gauge bosons, just like the standard model forces. After separation of matter and anti-matter in the very early universe, gravitation in our universe (fundamental observables still being the eigenvalues of the Dirac operator) appears as an 'attractive only' interaction. This of course suggests that the quantum of gravitation is spin-2: however, a spin-2 graviton, even if it exists, cannot be fundamental but only composite.

Prior to the breaking of the $E_8 \times E_8$ symmetry, the dotted and undotted dynamical variables are unified into one, and the fundamental action (7) takes the form

$$\frac{S}{\hbar} = \frac{1}{2}\int \frac{d\tau}{\tau_p}\, Tr\left[\frac{L_p^2}{L^2}\dot{Q}^\dagger_{1bioct}\, \dot{Q}_{2bioct}\right] \tag{19}$$

where

$$\dot{Q}^\dagger_{1bioct} = \dot{Q}^\dagger_B + \frac{L_p^2}{L^2}\beta_1 \dot{Q}^\dagger_F; \qquad \dot{Q}_{2bioct} = \dot{Q}_B + \frac{L_p^2}{L^2}\beta_2 \dot{Q}_F \tag{20}$$

and

$$\dot{Q}_B = \frac{1}{L}(i\alpha q_B + L\dot{q}_B); \qquad \dot{Q}_F = \frac{1}{L}(i\alpha q_F + L\dot{q}_F); \tag{21}$$

The pre-gravitation sector (dotted variables) and the standard model sector (undotted variables) are unified, and so are the right handed fermions (dotted) and the left handed fermions (undotted). The bosonic and fermionic sectors are unified as well, and the only parameter in the action (19) is the dimensionless area L_p^2/L^2 (a scale-invariant theory). The trace Lagrangian is invariant under unitary transformations generated by the generators of $E_8 \times E_8$. The Yang-Mills coupling constant α arises after symmetry breaking, and because it appears as the relative weight between states of left handed fermions and states of right handed fermions, its value is determined by the octonion algebra. Similarly the value of L_p/L is also determined by the octonion algebra, and there are no free parameters at all in the action (19). The two dynamical variables in this action are functions only of Connes time, since the manifold (and the accompanying coordinate labels) emerge only after symmetry breaking.

This action can be thought of as describing a 2-brane with area L_p^2/L^2. The theory of interactions between 2-branes remains to be developed: interactions will be described as 'collisions and scattering' in phase space, between matrix-valued dynamical variables.

We have at hand a matrix-valued Lagrangian dynamics on split bioctonionic space, which is the analog of the Heisenberg formulation of quantum dynamics. Dynamical variables evolve in (Connes) time, whereas states defined using the split bioctonions are time-independent.

7. Applications: deriving the free parameters of the standard model

From the Lagrange equations of motion, it is possible to deduce that the three fermion generations for a given value of the electric charge obey the Dirac equation in 10D spacetime. This equation is the eigenvalue equation for the (complexified) exceptional Jordan algebra $J_3(8)_C$ which has the symmetry group E_6. Here, $J_3(8)$, known as the exceptional Jordan algebra, comprises of 3×3 Hermitean matrices with octonionic entries, and its automorphism group is F_4 [21, 22, 23]. The characteristic equation of this algebra, a cubic equation with real eigenvalues, is of great interest as its eigenvalues uniquely determine the values of several fundamental constants of the standard model.

Consider a matrix of this algebra whose off-diagonal entries are the octonionic states representing fermion states of three generations with a given electric charge [i.e. one out of $(0, 1/3, 2/3, 1)$]. Find the eigenvalues and eigenmatrices of this matrix, using the cubic characteristic equation. The eigenvalues, which are invariants (given the electric charge) and which we call Jordan eigenvalues, are shown in Fig. 4. These twelve numbers are of great importance; the observed mass-ratios, as well as the low energy fine structure constant 1/137 are derived from them. The eigenvalues for the anti-particles are opposite in sign to the values for their corresponding particles.

The octonionic states which have been employed here are charge eigenstates, which define the left-handed fermions. However, charge eigenstates are not mass eigenstates: the latter define the right-handed fermions, where after switching the position of the down quark and the electron the associated $U(1)$ quantum number takes the values $(0, 1/3, 2/3, 1)$ for square-root mass. Moreover, it takes the same values for all three generations, just as the electric charge takes the same values for all three generations. Why then do mass ratios appear so weird?! Note though that the first generation (square-root) mass ratios are very simple, being the same set as the charge ratios but with position of electron and down-quark interchanged. Only the second and third generation mass ratios are weird. If we calculate the Jordan eigenvalues for square-root mass eigenstates we get the same numbers as shown in Fig. 4 except that the position of the down quark family and that of the charged lepton family gets interchanged.

Journal of Physics: Conference Series **2533** (2023) 012013 doi:10.1088/1742-6596/2533/1/012013

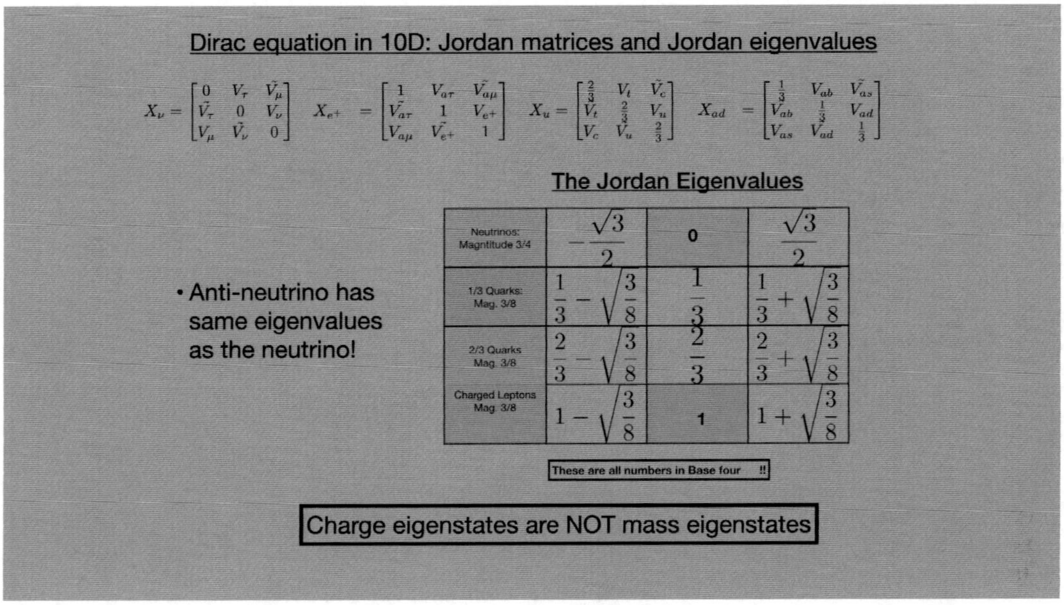

Figure 4. The Jordan eigenvalues for the neutrino family, down quark family, up quark family, and electron family.

What exactly is the interpretation of the Jordan eigenvalues and where do the weird mass ratios come from? Interpretation of these eigenvalues: prior to left-right symmetry breaking the relative weight of the left handed states and right handed states (and hence the to-be-Yang-Mills coupling constant) is simply unity, and we can express this as

$$1 = \exp[0] = \exp[Q + \sqrt{M}] \tag{22}$$

From examining the Jordan eigenmatrices corresponding to the Jordan eigenvalues we conclude that the solution of this eigenvalue problem expresses charge eigenstates (or square-root mass eigenstates) as superpositions of eigenstates of 'charge-square-root-mass', these being eigenstates of a $U(1)$ quantum number which necessarily takes the value $1/3$ (the value is $1/3$ for each generation, and $1/3 + 1/3 + 1/3 = 1$). This then is the interpretation of the Jordan eigenvalues: if we solve the eigenvalue problem for the Dirac equation in 10D spacetime for a given value of the electric charge, or for a given value of square-root mass, the three solutions are eigenmatrices of charge-square-root mass and the Jordan eigenvalues are the corresponding eigenvalues.

This permits eigenstates of square-root mass to be written as a superposition of eigenstates of electric charge. And since all our particle physics measurements are in the end based on the electromagnetic interaction (and not on the gravitational interaction) the mass ratios are determined by the relative weights of the charge eigenstates in this superposition, and the relative weights are in turn ratios of Jordan eigenvalues. This leads to a derivation of the strange observed mass ratios, which are in fact simple fractions as shown in Fig. 5. Had we been making our laboratory measurements using square-root-mass eigenstates, the mass ratios would have come out $(0, 1/3, 2/3, 1)$ for each of the three generations and the electric charge ratios would have been weird!

The theoretically calculated mass ratios match with observed values only if the neutrino is assumed to be Majorana. That is because the Jordan eigenvalues come out to be different when the neutrino is assumed to be a Dirac particle, and these values give drastically wrong ratios which do not agree with experiment. On the other hand, the so-called Koide ratio comes out

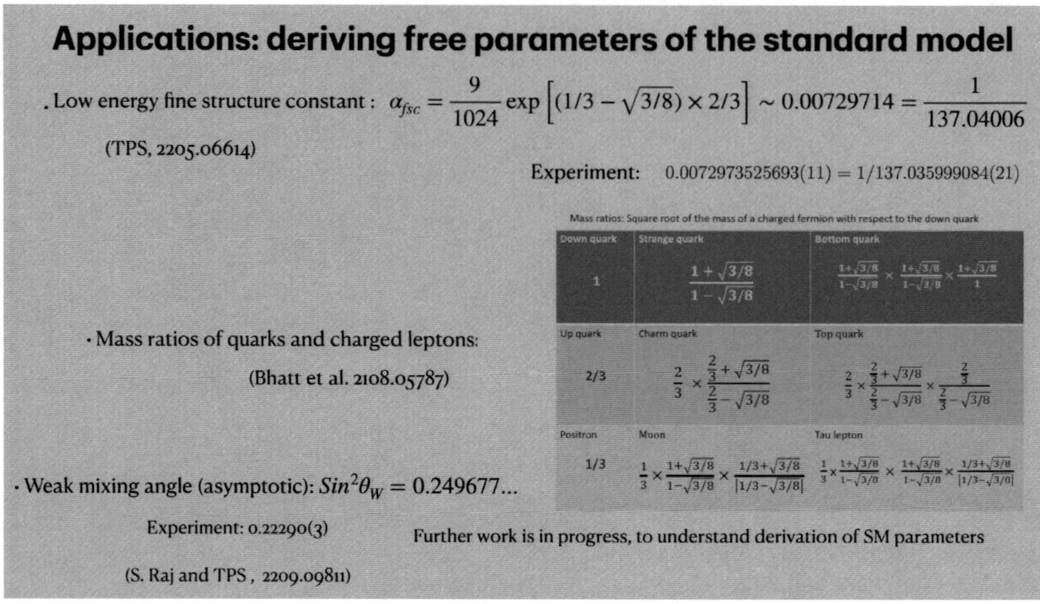

Figure 5. Standard model parameters from eigenvalues of the exceptional Jordan algebra.

to be exactly 2/3 if the neutrino is assumed Dirac, and the ratio departs slightly from 2/3, as is known from data, if the neutrino is assumed Majorana [24]. This gives support to our claim that only after wave function collapse the left-right symmetry is broken. The neutrino is fundamentally a Dirac particle, but is interpreted as a left-handed Majorana neutrino when its weak interaction is measured; it would register as a right-handed sterile Majorana neutrino if its gravitational interaction could be measured.

From the Lagrangian in (7) the expression for the low-energy fine structure constant can be read off to be $e^2/\hbar c \equiv \alpha^2 \times (L_P/L)^4$. Here, the constant α is related to the Q in (22) by $\ln \alpha = Q$. A coefficient β in front of the dotted part in $\dot{Q}_B = \frac{1}{L}(i\alpha q_B + L\dot{q}_B)$ is defined by $\ln \beta = \sqrt{M}$; with $\ln \alpha + \ln \beta = \ln \alpha\beta = Q + \sqrt{M} = 0 \implies \beta = 1/\alpha$. The constants Q and $\sqrt{M}$ arise from the partition of the charge-square-root-mass quantum number 1/3 into the weights $Q = (1/3 - \sqrt{3/8})/3$ and $\sqrt{M} = (\sqrt{3/8} - 1)/3$ determined by the Jordan eigenvalues: $(1/3 - \sqrt{3/8})$ for the left-handed down quark and $(\sqrt{3/8} - 1)$ for its right handed counterpart, the electron. This gives that $\alpha^2 = \exp[2/3 \times (1/3 - \sqrt{3/8})]$. From the construction of the octonionic states for the fermions, it can be concluded that [18] $(L_P/L)^4 = (\sqrt{1/32})^4$ and multiplying this by nine (because the square of the electric charge of the electron is nine times that of the down quark) gives the value of the low energy fine structure constant as shown in Fig. 5. Comparison with the experimental value is discussed in [18].

How do we know that this derivation is for the low energy limit? In fact this derivation is for the low interaction flat limit (13) in which the matrix-valued dynamical observables go to unity. We have then made an assumption (based on observations, and yet to be proved in this theory) that for electromagnetism, the low interaction limit is realised at low energies. The running of this constant with energy remains to be worked out, and for now we work with the running as derived using conventional methods of quantum field theory.

We note in passing that, analogous to the fine structure constant, the low energy $U(1)_{grav}$ gravitational fine structure constant $\alpha_g \equiv Gm_e^2/\hbar c$ is given by $\alpha_g^{1/2} = \beta^2(L_P/L)^4$ and has the

Journal of Physics: Conference Series **2533** (2023) 012013 doi:10.1088/1742-6596/2533/1/012013

value

$$\alpha_g^{1/2} = \frac{9}{1024}\exp[2/3 \times (\sqrt{3/8} - 1/3)] = 0.010586... = 1/94.4642... \tag{23}$$

Therefore the gravitational fine structure constant α_g is

$$\alpha_g = 0.00011206378... = 1/8923.4892... \tag{24}$$

The ratio of the strength of electromagnetic force to $U(1)_{grav}$ gravitational force is therefore

$$\alpha/\alpha_g = 65.11559... \tag{25}$$

This ratio is inevitably the primordial value, and is scaled down by an octonionic inflation which ends with the left-right symmetry breaking. This same inflation brings down the Higgs mass (as well as particle masses) from their Planck scale values to the presently observed values. These aspects are currently under investigation.

We have also derived [13] the weak mixing angle by investigating the bosonic part of the Lagrangian (7) and obtained the value shown in Fig. 5. In all, we have so far been able to derive ten standard model parameters: mass ratios, fine structure constant, and the weak mixing angle. If the octonionic theory is the correct theory of unification, it must also yield the remaining parameters from first principles: QCD coupling constant, quark mixing parameters (CKM matrix), neutrino mixing parameters (PMNS matrix) and Higgs mass. This is currently under investigation. We predict three right handed sterile neutrinos and this opens up the possibility that the observed neutrino oscillations are mediated by the sterile neutrinos, and that the neutrinos are in fact massless. Also, in the octonionic theory the cosmological constant is actually zero, and the role of dark energy is played by the uncollapsed atoms of space-ti me-matter: a preliminary attempt at showing this has been made in [25].

8. Left-right symmetry breaking: emergence of classical space-time and gravitation

While we have remarked on this topic in earlier sections, it is important enough for us to reiterate some key points, as summarised in Fig. 6. We have argued above that a line-element constructed from the split bioctonions in Eqn. (13) admits a 6D space-time, with an associated Lorentz symmetry $SO(1,5)$. Two of the spatial dimensions, whose geometry is responsible for the weak force, are much thinner than the other three, so that in the classical world we have a 4D spacetime with $SO(1,3)$ symmetry. The $SU(2)_R$ symmetry can be utilised to construct the equations of general relativity, in view of the spectral action principle. Emergence, initiated by spontaneous localisation, is the opposite of going from classical dynamics to trace dynamics. The Clifford algebra associated with the 6D space-time is $Cl(3)$ (related to complex split biquaternions), which in turn is made from two copies of $Cl(2)$. These account for chiral leptons - the neutrino and the electron, their anti-particles, and their second and third generation counterparts. When the strong interaction of quarks is included, one transits from $Cl(3)$ to $Cl(7)$ (complex split bioctonions) and to 10D spacetime. Quarks clearly cannot be confined to 6D or 4D spacetime: their strong interaction (this being the geometry of the four additional spatial directions) does not permit this, and this could be a possible explanation for quark confinement. On the other hand, all our measurements take place in 4D. Hence quarks only manifest in 4D spacetime through colorless composites such as protons and neutrons: not having color means the additional four spatial dimensions are not being probed by the composite state, but only by its constituents.

The point structure of spacetime is defined by the positions (the eigenvalues of the position operator to which collapse takes place) of the entangled fermions, these of course are the macroscopic classical objects of our universe. Since these positions commute, they impose a

Journal of Physics: Conference Series **2533** (2023) 012013 doi:10.1088/1742-6596/2533/1/012013

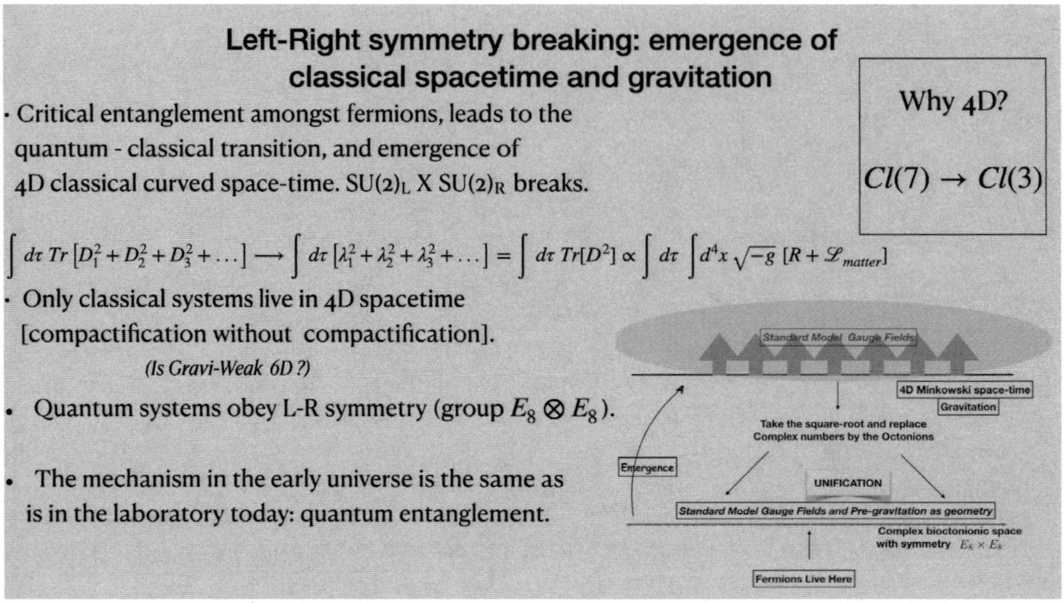

Figure 6. Critical entanglement of fermions separates the emergent classical space-time from the internal symmetry of the strong interaction.

commutative point structure on the 4D spacetime. It helps to note that in objective collapse models, which is what is recovered from the octonionic theory, classical objects are nothing but short-lived quantum superpositions. This permits replacing the non-commutative algebra of quaternions by real numbers, which commute, thus enabling the transition to the classical 4D spacetime manifold to a great accuracy.

For a discussion of spin in trace dynamics see [26] and for proposal of a ground state in quantum gravity see [27].

9. Are there any testable predictions?

Figure 7 below lists predictions of the octonionic theory which could become testable with future development in technology. There is no smoking gun prediction yet, where a measurable effect has quantitatively different predictions from quantum field theory, as compared to the prediction of the octonionic theory. There are postdictions; i.e. the theoretical derivation of the free parameters of the standard model.

We believe we have provided adequate evidence that this is the correct path to quantum gravity and unification, and to a proper understanding of the standard model. The octonionic theory is entirely motivated by addressing a foundational problem of quantum theory: the assumption of classical spacetime in quantum theory is an approximation even at low energies; an assumption which needs to be dropped. And when it is dropped, doing so opens a path to quantum gravity and to unification, and to a first principles derivation of the experimentally measured constants of the standard model.

Undoubtedly, much remains to be done still, before one can claim to have a theory that fully explains current data of the standard model. The following are the key aspects that remain to be addressed, and are currently being investigated.

- SM parameters: Higgs mass, W and Z masses, quark mixing matrix, neutrinos mixing matrix, QCD coupling constant.

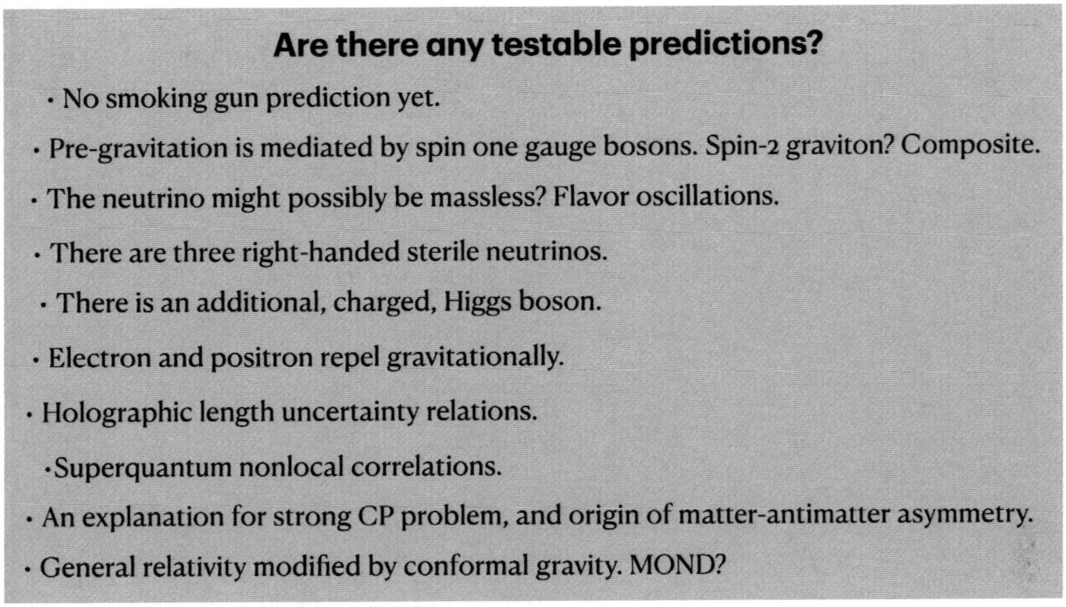

Figure 7. Predictions of the octonionic theory.

- The quantum-to-classical transition: spontaneous localisation.
- Consequences of working with a quaternionic / octonionic spinor spacetime: role in condensed matter systems?
- Develop an EFT to take account of octonionic corrections to QFT.
- Understanding the gravi-weak interaction.
- Implications for cosmology: are MOND and RMOND consequences of the octonionic theory?

Acknowledgments
The proposed octonionic theory relies very significantly on important earlier work by previous researchers. These include theories of spontaneous collapse of the wave function, the theory of trace dynamics, the spectral action principle, and the very detailed and comprehensive research on applications of the octonions to particle physics. To all these researchers, I am deeply grateful. Without the foundations laid by them, the octonionic theory would simply not exist. I would like to thank my collaborators without whose contribution the developments reported here would not have been possible. In particular I wish to thank Vatsalya Vaibhav, Priyank Kaushik, Rabsan Galib Ahmed and Sherry Raj for our collaborative work during the last two years, which has brought the octonionic theory to its present robust and believable shape. It is a pleasure to thank Felix Finster, José M. Isidro, Claudio Paganini, Cenalo Vaz, Kinjalk Lochan, Sukratu Barve, Kartik Kakade, Avnish Singh and Aditya Ankur Patel for our ongoing present collaboration. I would like to thank the participants of DICE2022 for stimulating discussions, and the organisers for providing a vibrant platform for exchanging new ideas and results.

A note on the references: the list of references below is far from being exhaustive. By and large only those works have been listed which directly impact the research reviewed in this article. In particular, the very large body of research on octonions in particle physics is not explicitly referred to; for a comprehensive listing please see [3].

Journal of Physics: Conference Series **2533** (2023) 012013 doi:10.1088/1742-6596/2533/1/012013

References

[1] Witten E 1987 Physics and geometry *Preprint CERN cer-000093203 https://cds.cern.ch/record/181783/files/cer-000093203.pdf*

[2] Singh T P 2021 Trace dynamics and division algebras: towards quantum gravity and unification *Zeitschrift fur Naturforschung* A **76** 131 *Preprint 2009.05574v4*

[3] Singh T P 2021 Quantum theory without classical time: a route to quantum gravity and unification *Preprint 2110.0206v1*

[4] Adler S L 2004 Quantum theory as an emergent phenomenon (Cambridge University Press, Cambridge)

[5] Adler S L 1994 Generalized quantum dynamics. *Nucl. Phys. B* **415** 195

[6] Adler S L and Millard A C 1996 Generalised quantum dynamics as pre-quantum mechanics *Nucl. Phys. B* **473** 199

[7] Ahmed R G and Singh T P 2022 A violation of the Tsirelson bound in the pre-quantum theory of trace dynamics *Preprint 2208.02209*

[8] Chamseddine A H and Connes A 1997 The spectral action principle *Commun. Math. Phys.* **186** 731 *Preprint hep-th/9606001*

[9] Landi G and Rovelli C 1997 General relativity in terms of Dirac eigenvalues *Phys. Rev. Lett.* **78** 3051 *Preprint gr-qc/9612034*

[10] Finster F, Isidro J M, Paganini C, Roy R and Singh T P 2023 *Work in progress*

[11] Kakade K and Singh A and Singh T P 2023 *Work in progress*

[12] Baez J 2002 The octonions *Bull. Am. Math. Soc.* **39** 145 *Preprint math/0105155*

[13] Raj S and Singh T P 2022 A Lagrangian with $E_8 \times E_8$ symmetry for the standard model and pre-gravitation I. – The bosonic Lagrangian, and a theoretical derivation of the weak mixing angle *Preprint 2208.09811*

[14] Kaushik P and Vaibhav V and Singh T P 2022 An $E_8 \times E_8$ unification of the standard model with pre-gravitation, on an octonion-valued twistor space *Preprint 2206.06911v2*

[15] Furey C 2016 Standard model physics from an algebra? *Preprint 1611.09182*

[16] Furey C 2015 Charge quantisation from a number operator *Phys. Lett.* B **742** 195 *Prreprint 1603.04078v2*

[17] Vaibhav V and Singh T P 2021 Left-Right symmetric fermions and sterile neutrinos from complex split biquaternions and bioctonions *Preprint 2108.01858*

[18] Singh T P 2022 Quantum gravity effects in the infra-red: a theoretical derivation of the low energy fine structure constant and mass ratios of elementary particles *The European Journal of Physics* **137** 664 *Preprint 2205.06614v2*

[19] Bhatt V, Mondal R, Vaibhav V and Singh T P 2022 Majorana neutrinos, exceptional Jordan algebra, and mass ratios for charged fermions 2022 *J. Phys. G: Nucl. Part. Phys.* **49** 045007 *Preprint 2108.05787v2*

[20] Boyle L, Finn K and Turok N 2018 *Phys. Rev. Lett.* **121** 251301 *Preprint 1803.08928v3*

[21] Ramond P 1976 Introduction to exceptional Lie groups and algebras *Preprint calt-68-577 https://inspirehep.net/literature/111550*

[22] Dray T and Manogue C A The exceptional Jordan eigenvalue problem *IJTP* **38** 2901

[23] Manogue C A and Dray T 2009 Octonions, E_6, and particle physics *J. Phys. Conf. Series* **254** 012005 *Preprint 0911.2253v2*

[24] Singh T P 2022 Why do elementary particles have such strange mass ratios? *Physics* **4** 948 *Preprint 2209.03205*

[25] Singh T P 2020 Dark energy as a large scale quantum gravitational phenomenon *Modern Physics Letters A* **35** 2050195 *Preprint 1911.02955v3*

[26] Singh T P 2020 Octonions, trace dynamics and non-commutative geometry: a case for unification in spontaneous quantum gravity *Zeitschrift fur Naturforschung* A **75** 1051 *Preprint 2006.16274v3*

[27] Roy A K, Sahu A and Singh T P 2021 Trace dynamics and a ground state in quantum gravity *Mod.Phys.Lett.* A **36** 2150019 *Preprint 2104.14344*

Journal of Physics: Conference Series **2533** (2023) 012015 doi:10.1088/1742-6596/2533/1/012015

How an exact discrete symmetry can preserve black hole information or Turning a black hole inside out

Gerard 't Hooft

Faculty of Science, Department of Physics, Institute for Theoretical Physics, Princetonplein 5, 3584 CC Utrecht, Netherlands

E-mail: g.thooft@uu.nl

Abstract. To apply the laws of General Relativity to quantum black holes, one first needs to remove the horizon singularity by means of Kruskal-Szekeres coordinates. This however doubles spacetime, which thereby is equipped with an exact binary symmetry. All particles near a black hole share the same symmetry, and conservation of this symmetry may completely remove the information paradox: the quantum black hole has no interior, or equivalently, the black hole interior is a quantum clone of the exterior region. These observations, totally overlooked in most of the literature on quantum black holes, resolve some issues concerning conservation of information. Some other problems do remain .

1. Introduction

A silent revolution took place in the author's understanding of the quantum mechanical equations for black holes, as first mentioned in Ref. [1]. In summary, what happens is that the space-time singularity associated to the horizon of a black hole, can easily be removed by an appropriate coordinate transformation, showing that a black hole horizon is about as regular as the apparent singularity of planet Earth at its North and South poles: it is a coordinate singularity. Removing this singularity should answer all questions about the nature of space and time there, but it leads to a new question instead. Apparently, beyond the coordinate singularity, space and time are extended into a new copy of the existing space-time. Our new insight, about which we also reported in Ref. [2], is that this is not a new space-time region at all, but merely an exact mirror image of existing space-time. The black hole carries along a clone, not only of itself, but also of the entire surrounding universe.

$$xyz \tag{1}$$

An observer[1], moving from ordinary space-time to the cloned region will notice no differences at all, but in fact space-time in this region is quite remarkable, because the clone is connected to the original *via* time reversal. Clearly, this is not an ordinary symmetry, but it does allow us to handle it as such. We can limit ourselves to states in Hilbert space that are either symmetric or anti-symmetric[2] under the switch between the black hole and its clone. As we shall show briefly, this reduction has a big effect on the expected properties of the black hole. Its entropy is now only half of what is usually found, and consequently, the temperature of the emitted radiation doubles.

But we shall come to that; let us first return to some basic physical issues concerning quantum gravity. Usually, our starting point is chosen to be Newton's force law,

$$F = \frac{G\, m_1\, m_2}{r^2}\ , \tag{2}$$

where one imposes the condition that this law needs to be made compatible with the highly successful laws of special relativity.

As is well-known, one then ends up with General Relativity. It is found that space and time are curved, and the curvature is everywhere in the vicinity of the source. Making the next step, needed to impose the laws of quantum mechanics, is now very hard, and usually comes with new assumptions and unprovable principles, such as string theory and holography. There is nothing wrong with such approaches, except that the results are generally uncertain. One has to understand Nature's (or God's[3]) way; you have to be religious to understand that, and if you are not religious, you just miss the boat.

An approach that requires (almost[4]) no religion is the path offered by black holes. If we remove the coordinate singularities by a coordinate transformation, we only encounter small amounts of local curvature. However, the effects of gravity still diverge in a remarkable way, which is the effects of the gravitational force acting between in- and out-going matter. This diverges badly, in a process that explodes exponentially in time, but in a way that triggered our interest: this divergence *only* occurs on the horizon. Elsewhere in space-time, one may assume that standard knowledge of general relativity and quantum mechanics offers sufficient clues as to what happens there, so, *unintelligible curvature* does not occur everywhere, and we may attempt to couple understandable features of regions of space and time, while reserving our attention to the tiny subset of space-time points right at the horizon. This is where a black is connected to its clone at the other side.

Instead of Newton's law, Eq. (1) , we focus on the Shapiro effect. This effect brings about gravitational lensing, as is well known in astronomy:

$$\delta u^- = -8\pi G\, p^-\, \log|\tilde{x}_1 - \tilde{x}_2|\ , \tag{3}$$

where p^- is the momentum of the source, which we take as a vector in the light cone minus-direction, while u^- is the displacement it generates in a reference mass, also in the minus-direction. $\tilde{x}_1$ and $\tilde{x}_2$ are the transverse components of the locations of source and effect. The

[1] The concept of an observer can perhaps better be replaced by the concept of observable variables, so that some problems such as finite-size effects and finiteness of memory space, are more easy to avoid, but clearly, utmost diligence would be asked for.

[2] symmetric for the real parts of wave functions, and antisymmetric for the imaginary parts, as will be explained, see section 4.

[3] I should henceforth avoid using God's name, as this apparently suggests religious feelings that I do not have.

[4] A critical reader may observe that we do make use of a belief: the fact that black holes should in no way differ from ordinary particles in the way they obey quantum mechanical laws. Yes, I admit that this is a belief.

Journal of Physics: Conference Series **2533** (2023) 012015 doi:10.1088/1742-6596/2533/1/012015

most noticeable distinction with Newton's law is the dependence on these coordinates. It goes as $1/r^2$ in the Newtonian case, while it is only logarithmic in the case of Shapiro. The logarithm reflects the fact that the associated phenomenon takes place in a two-dimensional subspace of space-time. Laplace's field equation generates logarithms in two dimensions.

The Shapiro equation (1) offers an easy way to handle the gravitational interactions between in- and out-particles. One finds that, in the reference frame of the out-particles, the momentum p^- of the in-particles increases exponentially as a function of the time separation between in and out. Remarkably, here *momenta* of the in-particles affect the *positions* of the out-particles. By applying Fourier transformations, one derives directly that (apart from a sign) the same relation holds if for both particles momenta and positions are interchanged. This is time reflection symmetry, see Section 3. Readers familiar with these issues might want to proceed directly to Section 4, where some subtle novelties enter.

2. Perturbations around the stationary black hole

Many researchers now continue with the black hole as it is being formed by some implosion event.[3] It then seems as if the details of this implosion are intimately responsible for what happens next. We claim however that this only holds for time spans shorter than the *scrambling time*. This is the time it takes for a particle to reach the horizon within a few Planck lengths, in natural units:

$$T_S = \mathcal{O}\big(M_{\rm BH} \log(M_{\rm BH}/M_{\rm Planck})\big) \ . \qquad (4)$$

Classically we have the no-hair theorem stating that the details of the implosion event are immaterial. In practice one expects that only minute quantum details will continue to depend on details of the original implosion, but this situation would be similar to what happens to an atom after its formation at the beginning of the universe. Surely such details will determine which of its quantum states it will be in right now, but as physicists we should be primarily interested in a proper categorisation of all possible quantum states, and the Schrödinger equation dictating how these evolve one into another. We here take the same attitude regarding black holes. Its history before scrambling time, or its evaporation process after scrambling time, should not be relevant for its physical equations at present.

Soon after its formation, the black hole is almost stable, absorbing and emitting particles every now and then, according to a Schrödinger equation and a Hamiltonian connecting all microstates. It is our job to use basic principles for deriving the proper classification of states and the fundamental Hamiltonian. At some decent distance from the horizon, the categorisation is clear and the Hamiltonian is known, it is called the Standard Model of the subatomic particles, which can be put in a gently curved space-time. At all space-time points where the temperature dropped to below a few TeV, we know how it generates the answers to our questions.

We now assume that the complete set of states is determined by the spectrum of particles excited thermally at all distances beyond a few Planck lengths from the horizon, where the exact numbers are still to be derived. Since the particles are thermal and the local temperature is expected to have dropped well below the Planck temperature, the effects of these particles on the black hole space-time metric must have dropped to values small enough to neglect them for the time being. We plan to postpone these details to being justified a posteriori (See Section 6). Right now we can observe that the sets of states and operators used at time scales that may vary over more than one unit of the scrambling time will definitely not commute with one another. The Schwarzschild metric, in Schwarzschild's coordinates, demonstrates clearly the time dilation

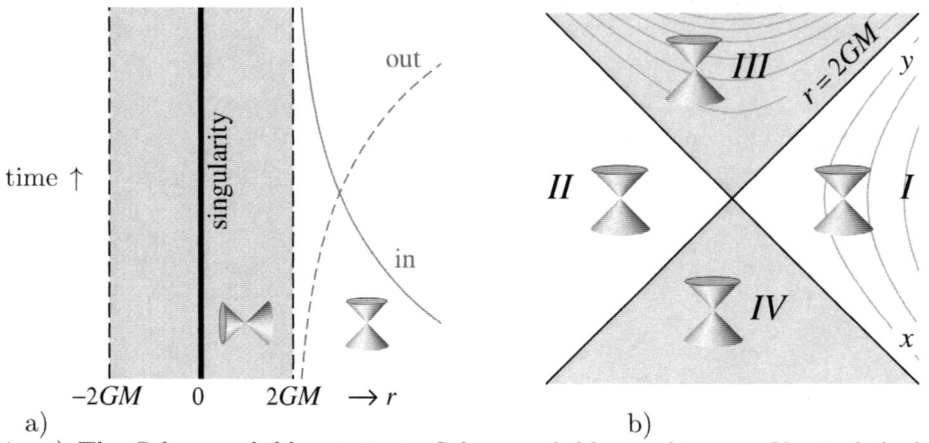

Figure 1. a) The Schwarzschild metric, in Schwarzschild coordinates. Vertical dashed lines: the event horizon. Curved lines: light like radial geodesics: y = const (smooth), and x = const (dashed). Cones: the orientation of the local light cones. Angular coordinates $(\theta,\,\varphi)$ are not shown. b) The same space-time in Tortoise coordinates $(x,\,y)$, showing regions $I - IV$ and the orientations of the local light cones. Curves show r = const. lines.

invariance:

$$\mathrm{d}s^2 \;=\; \frac{1}{1-\frac{2GM}{r}}\mathrm{d}r^2 - \left(1-\tfrac{2GM}{r}\right)\mathrm{d}t^2 + r^2\mathrm{d}\Omega^2\ , \tag{5}$$

$$\text{where}\qquad \Omega \;\equiv\; (\theta,\,\varphi)\ ,$$

$$\mathrm{d}\Omega \;\equiv\; (\mathrm{d}\theta,\,\sin\theta\,\mathrm{d}\varphi)\ . \tag{6}$$

It is sketched in Figure 1a. The regular coordinate frame is the Kruskal [4] Szekeres [5] coordinates, to be referred to as tortoise coordinates x and y, see Figure 1b. One defines

$$x\,y \;=\; \left(\frac{r}{2GM}-1\right)e^{r/2GM}\ ; \tag{7}$$

$$y/x \;=\; e^{t/2GM}\ , \tag{8}$$

in terms of which, the spacetime described by Eq. (5) is regular at the horizon:

$$\mathrm{d}s^2 \;=\; \frac{32(GM)^3}{r}\,e^{-r/2GM}\mathrm{d}x\,\mathrm{d}y \;+\; r^2\mathrm{d}\Omega^2\ . \tag{9}$$

All of space-time as we know it, is covered by the set $\{x\geq 0\,,\ y\geq 0\}$ (region I). As we mentioned in the Introduction, this new coordinate frame allows for an analytic continuation towards another black hole space-time, defined by the new region II: $\{x\leq 0\,,\ y\leq 0\}$. In contrast, the regions III where $x\leq 0$ and $y\geq 0$, and IV where $x\geq 0,\ y\leq 0$, are situated beyond the infinite future or infinite past, and these are physically less bothering.

Both coordinate frames depicted in Figure 1 are of importance, and we have to understand better how they are related. The *arrow of time* is oriented upward in all outside regions of the Schwarzschild frame Fig. 1a. In its inside regions it is pointing inwards, and this means that a local observer will be grilled at the singularity soon after having passed the horizon. But we are more interested in the way regions I and II are linked together, in Fig. 1b.

The arrow of time can be read off from Eq. (2). It implies that, as a local observer in the horizon region sees time always moving in the upwards direction inside its local light cone, an outside observer using the growing time coordinate t, will be moving downwards in region II. Wave functions in I will rotate as e^{-iEt}, but in region II as e^{+iEt}. However, as soon as the Standard Model applies, such negative energy modes are forbidden.

Either one can say that the Dirac ket states seen by the local observer relate to the bra states of the global observer, or one can say that, in the static black hole approximation, the states seen by the local observer in region II describe a universe that is *almost full* rather than almost empty. Considering the fact that, as derived by Hawking [14], quantum effects generate particles at the crossing points of the two horizons, which are seen as particles in *both* regions, the idea that we have a space full of particles in region II leads to a satisfactory image of a quantum mechanical leak at that point, producing a steady shower of Hawking particles.

At first sight one might object that there is no such leak; a time translation for the outside observer leads to a Lorentz transformation for the local observer, and it might seem that no particles should leak out. Now of course quantum effects may generate tunnelling, and there is a more calculable genuine effect that generates such tunnelling: the Shapiro effect mentioned in the Introduction.

3. The Shapiro effect

How to employ the Shapiro effect [6] in understanding the dynamics of the Hawking effect, was elaborately explained in our previous publications [1, 7]. Eq. (1) is derived by first considering the Schwarzschild metric of a particle at rest, and then subjecting that to a Lorentz transformation with a large γ factor. The particle momentum is the limit $p^- \to \gamma m$ for large γ, and small rest mass m. The result [8, 9] still describes flat space-time, but General Relativity tells us that there is curvature, delta-peaked at the location of the shock wave associated to the fast source particle. Next, one generalises the equation for the case that the source mass goes inwards through an S_2 surface, the horizon, rather than a flat shock wave. The result [10] is almost the same, but the transverse separation $\tilde{x} - \tilde{x}'$ is now described by an angular separation, the logarithm turns into a Laplace function on the sphere. We see that Eq. (1) turns into

$$\delta u^- = Gp^- f(\Omega, \Omega') \quad \text{with} \quad (1 - \Delta_\Omega) f(\Omega, \Omega') = 8\pi G \delta^2(\Omega, \Omega') \ . \tag{10}$$

Here, Ω is the solid angle at which the out-particle leaves, and Ω' is where the in-particle entered the black hole horizon.

We add normalisation constants that turn the tortoise coordinate x into the lightcone coordinate u^- and y into u^+. For large black holes, the metric in the $u^\pm$ coordinates is almost flat. Henceforth, we ignore the slight curvature that remains.

With $u^\pm$ we indicate the positions at the future or past horizon, *i.e.*, the time at which the particles enter or leave. $p^\pm$ are their momenta. These equations are linear in the position and momentum variables, and consequently, it is easy to generalise this result to handle the case of many particles going in or out. These lead to momentum and position distributions on the horizons:

$$p^\pm \to p^\pm(\Omega) , \qquad u^\pm \to u^\pm(\Omega) \ . \tag{11}$$

And this allows us to use spherical wave expansions[5]

$$u^\pm(\Omega)=\sum_{\ell,m}u^\pm_{\ell m}Y_{\ell m}(\Omega)\ ,\qquad p^\pm(\Omega)=\sum_{\ell,m}p^\pm_{\ell m}Y_{\ell m}(\Omega)\ . \tag{12}$$

Since, for each spherical harmonic, $\Delta=-\ell(\ell+1)$, we can write Eq. (3) as

$$\delta u^-_{\ell m}=\frac{8\pi G}{\ell^2+\ell+1}p^-_{\ell m}\ . \tag{13}$$

Now we can use the commutator equations for all in- and out-going particles $i,\ j$,

$$[u^\pm_i,\ p^\mp_j]=i\delta_{ij}\ , \tag{14}$$

to arrive at the algebra

$$[u^+_{\ell m},\ u^-_{\ell' m'}]=i\lambda_\ell\,\delta_{mm'}\,\delta_{\ell\ell'}\ ;\qquad \lambda_\ell=\frac{8\pi G}{\ell^2+\ell+1}\ . \tag{15}$$

A few remarks:
We dropped the symbol δ from $\delta u^\pm$, which means that it is assumed that all displacements δu added up give the positions u themselves, provided coordinates are chosen appropriately. The same holds for the momenta.

Secondly, all different $(\ell,\ m)$ modes decouple, so we obtain new, one-dimensional quantum systems that generate the total Hilbert space in a very simple manner. These equations are highly trivial.

Furthermore, the Fourier transformation from momenta to positions and back is a *unitary* transformation. Hence, the algebra (3) should act as a perfect, information preserving, mapping from in- to out-particles (and back).

4. Restoring unitarity

But then there is a more serious problem that has to be addressed. The Fourier transforms arrived at in the previous section, are only unitary if the data stretch over the entire real line both for the u^+ and the u^- variables. But the physical data in the observable part of the universe are exclusively on the half-lines $u^+>0$ and $u^->0$. If we drop the data on the negative half-lines, the Fourier transformation is not unitary at all.

The cure to this problem may well come from the observation proposed in the Introduction section: just choose region II to be a quantum clone of region I. In that case, also the Fourier variables in region II are quantum clones of those in region I. Is this an elegant solution to our problem?

Unfortunately no. If all wave functions were real, then indeed the equation $\psi(u)=\psi(-u)$ copies correctly in Fourier space as well: $\hat\psi(p)=\hat\psi(-p)$. But wave functions are not in general real. Can we impose reality as an extra condition?

In ordinary quantum mechanics, the condition that wave functions are real is not impossible. Taking the Hamiltonian to be an imaginary, antisymmetric matrix would be allowed in quantum

[5] These distribution functions are to be defined accurately: the momenta emerge as sums of δ distributions, while positions are represented as centre of mass positions. See for example Ref. [2], sections 3 and 4.

Journal of Physics: Conference Series **2533** (2023) 012015 doi:10.1088/1742-6596/2533/1/012015

mechanics, provided that we accept the corresponding symmetry in the energy spectrum: every state with energy E in the energy spectrum, is associated to a state with the opposite sign of the energy. Reality of the wave function implies a symmetry

$$E \leftrightarrow -E \tag{16}$$

in the energy spectrum. Remember that, right at the beginning of our treatment, it was imposed that for ordinary particles the effect of the particles to the total mass of the black hole may be ignored, so indeed, we may assume the condition that the entire energy spectrum of particles is shifted to obey the symmetry (4).

This line of arguments needs to be carefully checked to ascertain that it be realistic.

In previous work, it was proposed to combine the transformation $u \leftrightarrow -u$ with the *antipodal mapping*:

$$\Omega = (\theta, \varphi) \leftrightarrow -\Omega \equiv (\pi - \theta, \varphi + \pi) \ , \tag{17}$$

but this was found to lead to other difficulties: the antipodal mapping is the mapping that replaces all spherical harmonic functions $Y_{\ell m}$ by $(-1)^\ell Y_{\ell m}$. But since we already had that the mapping $I \leftrightarrow II$ corresponds to $u^\pm \leftrightarrow -u^\pm$, this would imply

$$u^\pm_{\ell,m}(\Omega) = (-1)^{\ell+1} u^\pm_{\ell,m}(\Omega) \, , \tag{18}$$

so that $u^\pm = 0$ if ℓ is even, which cannot be squared with the commutation equation saying that all u variables obey $[u^+, u^-] = i\lambda_\ell$, see Eq. (3). Therefore the transition to antipodes is herewith dismissed.

An other possibility was proposed im 1984 by the author [11]: should we regard all states in region II as the bra states $\langle\psi|$ associated to the ket states $|\psi\rangle$ in region I? The elegance of this proposal is that, together, these structures form the density matrices of the system in region I. Is it possible to accept the mathematics of the states in region I and II taken together, as representing the quantum density states? Actually, this proposal almost coincides with our attempt to impose the symmetry (4). It is not as yet ruled out, and it seems to be the only possibility left. Note that we are forced to supplant our complex wave functions by real ones. It is as if the use of complex global $U(1)$ symmetries is forbidden, which might enforce violation of exact, additive, infinite conservation laws. These are replaced by a *discrete* $\mathbb{Z}_2$ *symmetry* that switches the signs of the tortoise coordinates x and y.

And it does lead to a more general, really important observation: this proposal replaces Hawking's calculation for the temperature of a black hole by another one If we use density matrices to compute probabilities, we find that probabilities are proportional to the matrix elements of the density matrix, whereas Hawking regards these as quantum vectorstates, and consequently, he averages over the squares of the states describing region II as these can't be observed.

Thus, in the thermal Boltzmann factors, we now disagree. In our expressions, the temperature of radiation emitted by a black hole, of a given mass, must be twice the temperature found by Hawking.

5. Black hole thermodynamics in a nut shell

In statistical physics and quantum field theory it is well-known that a unitary evolution operator, e^{-iHt} can be analytically continued towards imaginary time $t \to -i\beta$ to become the operator

$e^{-\beta E}$, which is exactly the operator that produces a thermal probability distribution if the temperature T obeys $\beta = 1/kT$, with k being the Boltzmann constant. [12, 13]. One also finds that the distribution of the energy levels is given by the entropy $S(E)$ if we enumerate the energy eigenstates. In a grand canonical ensemble, one would write $\varrho(E)\mathrm{d}E$ for the number of energy levels between E and $E + \mathrm{d}E$, to define the free energy $F(\beta)$ by

$$Z(\beta) = \int_0^\infty \varrho(E)\mathrm{d}E\, e^{-\beta E} = \int_0^\infty e^{S(E) - \beta E}\mathrm{d}E = e^{-\beta F} \ , \tag{19}$$

where $S(E) = \log(\varrho(E)$ is the entropy of the system if E is handled as the expectation value $\langle E\rangle$. The usual thermodynamical equations emerge if we assume $\langle E\rangle$ to be the macroscopic energy, F is the free energy, and we use

$$\langle E\rangle = -\frac{\partial}{\partial\beta}\log Z(\beta) \ . \tag{20}$$

In black holes however, one expects the integral (5) to diverge, as the entropy increases as E^2. This leads to a negative specific heat , which destabilises the black hole, and therefore it is better to go to a micro-canonical ensemble, where we keep the total energy fixed. In a micro-canonical sample, we define the entropy by ordering all energy eigenstates, writing them as $E(n)$, and we focus on a single value for n. For large systems, such as large black holes, one can take the continuum limit. Then we write

$$e^{S(E)} = \frac{\mathrm{d}n}{\mathrm{d}E} \ , \qquad \frac{\mathrm{d}S(E)}{\mathrm{d}E} = \beta(E) \ . \tag{21}$$

Here, one defines

$$\log Z = -\beta F = E - S/\beta = E - TS \tag{22}$$

(in units where $k = 1$). While in the grand canonical ensemble, the temperature could be chosen, here in the micro-canonical ensemble, β is a fixed function of the black hole mass, being the total energy E.

The fact that β is fixed, originates from the fixed background configuration, being the Schwarzschild metric.

What is the number β for a black hole with mass M? Consider all states on a trajectory in complex space-time. Choose the trajectory in tortoise coordinates to be

$$x = A\, e^{i\varphi} \ , \quad y = A\, e^{-i\varphi} \ , \quad xy = |x|^2 = A^2 \quad \text{is fixed} \ , \quad t = -4iGM\varphi \ . \tag{23}$$

We are dealing with a quantum system where all states in the stretch at imaginary time t ranging from zero to $-i\beta$ can be excited to arbitrary values. The total number of energy eigenstates is then given by $e^{S(E)} = \mathrm{Tr}\left(U(-i\beta)\right)$, where β is the total length of the path. The length of this path is fixed if we assume the end point of the path is the first point where x and y take on the values they had at the starting point. This is at $\varphi = 2\pi$ in Eq. (5). One thus finds that

$$kT = 1/\beta = 1/8\pi GM \ , \tag{24}$$

which is $kT = \frac{\hbar c^3}{8\pi G\, M_{\mathrm{BH}}}$ when units are put back in. This is Hawking's famous result [14].

Now consider the theory where all states in region II are determined by a mapping from region I , where the exact nature of the mapping is immaterial; it just simply must be a unique mapping. Let us again follow our trajectory (5) in imaginary time. At $\varphi = \pi$ we are at $t = -4\pi GM$ and we arrive at the spot where (x, y) take the values $(-x, -y)$, which is where the clone of the starting point begins. Therefore, the trajectory closes at $\varphi = \pi$, which is at time $t = -4\pi iGM$, so that $\beta = 4\pi GM$, and the temperature turns into

$$kT = \frac{\hbar c^3}{4\pi G M_{\rm BH}} \ . \tag{25}$$

Inspecting Eq. (5), we find that not only β but also the total entropy takes on half the value of Hawking's calculation, Eq. (5). We can understand why this is so. In terms of our local theory, the entropy is an integral over Euclidean spacetime. It counts all independent quantum states in space-time, *but the states localised after the point* $(-x, -y)$ *are not independent,* they are the states in region II, which are the same states as the ones seen before in region I. They should therefore not be counted again.

6. Conclusions

Our revision of the black hole temperature, Eq. (5), is a direct consequence of the fact that we identify the states in the second region of the Penrose diagram with the ones in region I. We do not merely claim that the particles emerging there, carry all information of the in-going particles, they *are* the in-going particles. This is easiest to comprehend if we return to the Schwarzschild coordinates. There, one can argue that the cusp singularity at the horizon is somewhat more delicate than just a coordinate artefact. It asks us not to consider all states you get by filling up all of space-time there with any states we like, but we have to fill in region II exactly as region I. Here, the black hole we describe, fundamentally differs from Rindler space, Rindler space is not just a big black hole because, from the outset, it contains different regions I and II. One can't create such configurations from the Schwarzschild metric; this is where the Kruskal-Szekeres coordinates for a black hole are somewhat deceptive, suggesting a world that does not exist. It does exist in Rindler spacetime.

Note that whenever we throw something in, it is seen as in going material both in region I and in region II. For instance, when we throw in a dust shell of matter, both regions I and II undergo modification. After a large amount of time (more than the scrambling time), the dust shell should become invisible. This only happens if a dust shell is assumed to enter also in region II. This does not only hold for dust shells but for all events in the outside world.

At the beginning of our presentation, in Section 2, we mentioned some a posteriori verifications. In particular we claim that a stationary black hole is the appropriate background for an accurate description of quantum black hole dynamics, as soon as it is large enough compared to the Planck scale. To really appreciate our procedures, our paper was not enough; one should do much more test calculations to appreciate how our picture hangs together. We do hope that our general philosophy will be receiving more attention than it had so-far. The fact that the Hawking temperature is modified by a factor 2 indicates that more is going on than merely a change in semantics.

References

[1] 't Hooft G 2022 *Quantum clones inside black holes*, Universe **8**(10) 537; https://doi.org/10.3390/ universe 8100537; http://arxiv.org/abs/ 2206.04608
[2] 't Hooft G 2022 *How studying black hole theory may help us to quantise gravity*, presented at the Conference on "Eternity between Space and Time", Padova, May 19-21, 2022; http://arXiv:2211.10723v2 [gr-qc]

[3] Ashtekar A and Bojowald M 2005 *Black hole evaporation: A paradigm*, Class. Quant. Grav. **22** 3349-3362; arXiv:gr-qc/0504029v2
[4] Kruskal M D 1960 *Maximal extension of Schwarzschild metric*, Phys. Rev. **119** 1743
[5] Szekeres G 1960 *On the singularities of a Riemannian manifold*, Publ. Math. Debrecen **7** 285 (now available at General Relativity and Gravitation **34** (2002) 2001)
[6] Shapiro I I 1964 *Fourth Test of General Relativity*, Phys. Rev. Lett. **13**(26): 789-791. Bibcode: 1964 PhRvL..13..789S. doi: 10.1103/PhysRevLett.13.789.
[7] 't Hooft G 2021 *The Black Hole Firewall Transformation and Realism in Quantum Mechanics*, Universe **7** 298. https: //doi.org/ 10.3390/ universe 7080298. http: //arxiv.org/abs/ 2106.11152v2
[8] Bonnor W B 1969 *The gravitational field of light*, Commun. Math. Phys. **13** 163
[9] Aichelburg P C and Sexl R U 1971 *On the Gravitational field of a massless particle*, Gen. Rel. and Gravitation **2** 303-312
[10] Dray T and 't Hooft G 1985 *The gravitational shock wave of a massless particle*, Nucl. Phys. **B253** 173
[11] 't Hooft G 1984 *An ambiguity of the equivalence principle and Hawking's temperature*, J. of Geometry and Physics **1** 45-52
[12] Symanzik K 1966 *Euclidean quantum field theory, I, Equations for a scalar model*, J. Math. Phys. **7** 510
[13] Symanzik K 1969, *Euclidean quantum field theory*, in "Local Quantum Theory", ed. R Jost (New York: Academic Press)
[14] Hawking S W 1975 *Particle creation by black holes*, Commun. Math. Phys. **43**(3) 199

Journal of Physics: Conference Series **2533** (2023) 012016 doi:10.1088/1742-6596/2533/1/012016

Cellular automaton for spinor gravity in four dimensions

C Wetterich

Institut für Theoretische Physik, Universität Heidelberg, Philosophenweg 16, D-69120 Heidelberg

E-mail: c.wetterich@thphys.uni-heidelberg.de

Abstract. Certain fermionic quantum field theories are equivalent to probabilistic cellular automata, with fermionic occupation numbers associated to bits. We construct an automaton that represents a discrete model of spinor gravity in four dimensions. Local Lorentz symmetry is exact on the discrete level and diffeomorphism symmetry emerges in the naive continuum limit. Our setting could serve as a model for quantum gravity if diffeomorphism symmetry is realized in the true continuum limit and suitable collective fields for vierbein and metric acquire non-vanishing expectation values. The discussion of this interesting specific model reveals may key qualitative features of the continuum limit for probabilistic cellular automata. This limit obtains for a very large number of cells if the probabilistic information is sufficiently smooth. It is associated to coarse graining. The automaton property that every bit configuration is updated at every discrete time step to precisely one new bit configuration does no longer hold on the coarse grained level. A coarse grained configuration of occupation numbers can evolve into many different configurations with certain probabilities. This characteristic feature of quantum field theories can come along with the emergence of continuous space-time symmetries.

1 Introduction

Probabilistic automata are quantum systems in a discrete real formulation for the time evolution. At every given time t we consider bit configurations ρ. The probabilistic information is encoded in a real wave function $q_\rho(t)$, with $p_\rho(t) = q_\rho^2(t)$ the probability to find at t the bit configuration ρ. The updating to a new bit configuration τ at the next time step $t+\varepsilon$ is given by the evolution law

$$q_\tau(t+\varepsilon) = \left(\widehat{S}(t)\right)_{\tau\rho} q_\rho(t) \ . \tag{1}$$

The step evolution operator $\widehat{S}(t)$ is a unique jump matrix with a unique element one in each row and column, and zero elements otherwise,

$$\widehat{S}_{\tau\rho} = \delta_{\tau,\bar{\tau}(\rho)} = \delta_{\bar{\rho}(\tau),\rho} \ . \tag{2}$$

The map $\rho \to \bar{\tau}(\rho)$ defines the updating map from ρ to $\bar{\tau}(\rho)$, where we assume invertibility with $\bar{\rho}(\tau)$ the inverse map. The probability distribution is updated with the same step evolution operator, such that the probability $p_\tau(t+\varepsilon)$ coincides with the probability $p_{\bar{\rho}(\tau)}(t)$ for the configuration $\bar{\rho}(\tau)$ from which it originates at the previous time t. One can change the sign of elements of $\widehat{S}$ without affecting the updating of the probabilities.

The probabilistic automata discussed here have a deterministic time evolution, with a probabilistic nature only due to probabilistic initial conditions at some initial time t_{in}. One can define an "overall probability distribution" for bit configurations at all times as a type of generalized Ising model [1, 2], for which the probabilities vanish for all time-neighboring configurations not allowed by the updating rule. The probabilistic information is encoded in boundary terms involving only the bit configurations at initial and final time. The overall probabilities are positive for arbitrary signs of the components $q_\tau(t)$. The overall probability distribution defines a classical statistical system. One only employs the standard laws for "classical" probabilities of the overall system. The "quantum axioms" follow from a projection on the "time-local subsystem" [2].

Two simple steps have to be taken in order to realize standard continuous complex quantum mechanics. The first concerns a complex structure. We assume the presence of an involution K such that the N-component vector $q(t)$ can be divided in an $N/2$-component vector $q_R(t)$ which is even under K, and another $N/2$-component vector $q_I(t)$ which is odd under K. The equal size of the even and odd parts is guaranteed by a second map I, $I^2 = -1$, $\{K, I\} = 0$. This pair of discrete transformations allows for a map from the N-component real wave function q to the $N/2$-component complex wave function φ,

$$q \to \varphi(q) \ , \quad \varphi(t) = q_R(t) + iq_I(t) \ , \tag{3}$$

with the properties

$$\varphi(Kq) = \varphi^*(q) \ , \quad \varphi(Iq) = i\varphi(q) \ . \tag{4}$$

A rather general involution K suitable for our purpose is particle-hole conjugation [3]. This interchanges the values one and zero of the bits, or maps corresponding to occupation numbers $n_\gamma(t) \to \big(1 - n_\gamma(t)\big)$, or flips the sign of Ising spins $s_\gamma(t) = 2n_\gamma(t) - 1$. We focus on particle-hole invariant updatings for which $\widehat{S}$ is block diagonal, acting in the same way on q_R and q_I. This results in the discrete complex evolution equation

$$\varphi(t + \varepsilon) = \widehat{S}_R(t)\varphi(t) \ , \tag{5}$$

with real $N/2 \times N/2$ matrix $\widehat{S}_R$. A more general step evolution operator $\widehat{S} = \widehat{S}_R \otimes 1_2 + \widehat{S}_I I$ remains compatible with the complex structure as well, defining the complex matrix

$$U(t) = \widehat{S}_R(t) + i\widehat{S}_I(t) \ . \tag{6}$$

The second step concerns a formulation in continuous time. Unique jump matrices are orthogonal matrices,

$$\widehat{S}(t)\widehat{S}^T(t) = 1 \ . \tag{7}$$

In the complex formulation the corresponding $N/2 \times N/2$-matrices (6) are unitary matrices. Combining a certain number of time steps from t to $t + \Delta t$ results in the unitary evolution operator

$$\begin{aligned} U(t + \Delta t, t) &= U(t + \Delta t - \varepsilon) \dots U(t + \varepsilon)U(t) \ , \\ \varphi(t + \Delta t) &= U(t + \Delta t, t)\varphi(t) \ . \end{aligned} \tag{8}$$

This evolution operator can be used to define a hermitian Hamiltonian

$$U(t + \Delta t) = \exp\Big\{ -i\Delta t H(t) \Big\} \ , \quad H^\dagger(t) = H(t) \ . \tag{9}$$

As a consequence, the time evolution obeys the continuous Schrödinger equation

$$i\partial_t \varphi(t) = H(t)\varphi(t) \ , \tag{10}$$

where $H(t)$ is taken piecewise constant in the interval between t and Δt. Indeed, the solution of eq. (10) reproduces eq. (8) for discrete time points $t = t_{\text{in}} + n\Delta t$ with integer n. It can be taken as a suitable interpolation for continuous time. This closes the formal argument that probabilistic automata evolve as quantum systems. We will be interested in situations where characteristic time scales are much longer than Δt.

We consider at every t configurations of a large number of bits $n_\gamma(t) = (1, 0)$, where γ denotes both discrete positions in space $\vec{x}$ and "internal quantum numbers" α, $\gamma = (\vec{x}, \alpha)$, such that $n_\gamma(t) = n_\alpha(t, \vec{x})$. We identify the position $\vec{x}$ with the cells of a cellular automaton. The "updating" from t to $t+\varepsilon$ of the bit configuration $\{n_\alpha(t, \vec{x})\}$ within a given cell $\vec{x}$ is influenced only by the bit configurations of a few neighboring cells at t. Analogously, the bit configuration within a cell $\vec{x}$ at $t+\varepsilon$ influences only the state of a few neighboring cells at $t+2\varepsilon$. This cellular property generates backward and forward "light cones" and therefore the characteristic causal structure of local quantum field theories. Probabilistic cellular automata of this type can be interpreted as local quantum field theories.

More generally, cellular automata encode a "unitary time evolution" where no information is lost in a discrete setting with locality properties [4–20]. They are the basis for interesting attempts to obtain quantum mechanics from deterministic physics [21–25]. For a deterministic automaton a given bit configuration $\bar{\rho}$ at t corresponds to a sharp wave function with elements $q_\rho(t) = \pm\delta_{\rho\bar{\rho}}$. This is mapped at $t+\varepsilon$ to a unique configuration $\bar{\tau}(\bar{\rho})$ by $q_\tau(t+\varepsilon) = S_{\tau\rho}(t) q_\rho = S_{\tau\bar{\rho}}(t) = \pm\delta_{\tau\bar{\tau}}$, where $\bar{\tau}$ corresponds to the unique non-zero element $S_{\bar{\tau}\bar{\rho}}(t)$ associated to $\bar{\rho}$. A probabilistic automaton is defined by a probability distribution for the initial bit configurations, with the probability $p_\rho(t)$ to find at t the bit-configuration ρ. The relation $p_\rho(t) = q_\rho^2(t)$ introduces the concept of a wave function for classical probabilistic systems [26].

We are interested in the continuum limit for which characteristic time and length scales are much larger than Δt. This assumes a sufficiently smooth wave function $\varphi(t)$ both in time and space. For our cellular automata this wave function is a multi-particle wave function and the smoothness in space should apply to all particle positions. One-particle wave functions $\varphi_\sigma(t, \vec{x})$ are only a particular case. Computing the continuum limit for a very large number of cells is a challenging enterprise. In this note we only take first steps by discussing the "naive continuum limit" which will be explained later. In this context the possibility to group a sequence of step evolution operators to a combined evolution operator (8) is of great help in order to realize continuous rotation symmetry in the naive continuum limit. This can be extended to other continuous symmetries as diffeomorphism invariance. In this note we will discuss a specific model of spinor gravity. The techniques developed here should be viewed as building blocks for the construction of rather general probabilistic cellular automata with an interesting naive continuum limit.

Configurations of bits or Ising spins are equivalent to the states of a multi-fermion system in the occupation number basis. The fermionic occupation numbers $n_\alpha(\vec{x}) = (1, 0)$ for a fermion of type α situated at the position $\vec{x}$ define a basis for the quantum states of the multi-fermion system. In the occupation number basis the general wave function at time t has components $\varphi_\tau(t)$, with $\tau = \{n_\alpha(\vec{x})\}$ denoting the possible configurations of occupation numbers. This is precisely the wave function of a probabilistic cellular automaton. For a large total number M of occupation numbers $\gamma = (\vec{x}, \alpha)$ the number of configurations τ is huge, $N = 2^M$. This underlines that only a probabilistic description is meaningful. As long as M remains finite the space of functions $\varphi(t)$ is a finite-dimensional complex vector space restricted to normalized vectors $\sum_\tau \varphi_t^*(t)\varphi_\tau(t) = 1$. A continuum limit for $M \to \infty$ is possible for smooth enough wave functions belonging to a Hilbert space. Probabilistic cellular automata are fermionic quantum field theories.

In the other direction it has been established that some particular discrete fermionic quantum field theories are equivalent to probabilistic cellular automata [3, 27, 28]. This holds in the sense that all observables have the same expectation value, including all correlations, in both pictures. The evolution of appropriate wave functions is identical as well. The basic idea is simple. The equivalence is demonstrated by computing the step evolution operator for a suitable discretized fermionic quantum field theory. It is sufficient to show that in a real discrete formulation the step evolution operator $\widehat{S}(t)$ is a unique jump matrix. This means that for each discrete time step every configuration of occupation numbers evolves to a single definite new configuration.

For computing the automaton associated to a fermionic quantum field theory we reverse the construction above. Consider a fermion system with Hamiltonian H in the occupation number basis. For discrete time steps from t to $t+\varepsilon$ the unitary evolution operator is given by $U(t) = \exp[-i\varepsilon H(t)]$, where we assume that H is constant in the interval $[t, t+\varepsilon]$. The evolution of the complex wave function $\varphi(t)$ can always be written in a real formulation as

$$\begin{aligned} \varphi(t+\varepsilon) =& U(t)\varphi(t) \;, \quad q(t+\varepsilon) = \widehat{S}(t)q(t) \;, \\ q =& \begin{pmatrix} \varphi_R \\ \varphi_I \end{pmatrix} \;, \quad \widehat{S} = \begin{pmatrix} U_R & -U_I \\ U_I & U_R \end{pmatrix} \;, \end{aligned} \tag{11}$$

with subscripts R, I indicating the real and imaginary parts and $\widehat{S}$ the real step evolution operator. We actually will work here with Majorana fermions and discretized time. In this case the wave function is real from the beginning and the evolution operator $\widehat{S}$ is an orthogonal matrix. (By introducing an appropriate complex structure (inverting eq. (11)) Majorana fermions in four dimensions are equivalent to Weyl fermions in a complex setting [29, 30].)

An automaton obtains if the real step evolution operator $\widehat{S}(t)$ is a "unique jump matrix", which contains in each row and column precisely one element ± 1, with all other elements vanishing. In this case the step evolution operator maps each configuration of occupation numbers to precisely one new configuration. The unique jump property of $\widehat{S}(t)$ is a particular restriction on the dynamics of the fermion system – generic fermion systems are not described by an automaton. Nevertheless, rather rich families of two-dimensional fermionic quantum field theories with interactions have been found to admit a description as automata [3]. They include models with Lorentz symmetry in the naive continuum limit, as discrete versions of Thirring [31–34] or Gross-Neveu [35–39] type models, models with local non-abelian gauge symmetries or discrete versions of spinor gravity. The true continuum limit for these models with the associated renormalization program has not yet been investigated, however, except for the trivial cases of free fermions.

In this note we ask the question if some interesting four-dimensional fermionic quantum field theories with Lorentz symmetry are equivalent to cellular automata. Instead of a wide general discussion we concentrate for a positive answer on a discrete model of spinor gravity in four dimensions. The local Lorentz symmetry is exact in the discrete version, while the diffeomorphism symmetry of general coordinate transformations is realized only in the naive continuum limit and assumed to emerge in the true continuum limit.

The true continuum limit and its relation to the naive continuum limit are a central issue for the present paper. Establishing fully this limit for probabilistic cellular automata with interactions has to wait for the development of more sophisticated methods as functional renormalization for cellular automata or dedicated numerical simulations for the associated generalized Ising models. We develop here first steps that reveal the crucial importance of coarse graining and the new qualitative features associated to it.

Our method proceeds as follows. We start from the Grassmann functional integral for a fermionic quantum field theory formulated on a discrete space-time lattice. For this discrete fermion model we compute the step evolution operator according to the general methods of refs. [3, 27, 28]. The naive continuum limit is defined by taking the limit of a vanishing lattice distance ε in the microscopic action. This neglects the renormalization flow of couplings and the question if symmetry violating interactions correspond to "irrelevant operators" as suggested by the naive continuum limit. At a later stage functional methods for Grassmann integrals should be employed to establish the true continuum limit.

We will discuss several general building blocks for constructing discrete fermionic quantum field theories which are equivalent to probabilistic cellular automata and realize important symmetries in the naive continuum limit. This concerns a general strategy for implementing exact local gauge symmetries for cellular automata, the realization of a non-trivial evolution with the unique jump property by using an updating with shifted cells, the implementation of continuous or discrete symmetries in the naive continuum limit by sequences of different step evolution operators, and the emergence of diffeomorphism invariance in the naive continuum limit without the use of an explicit metric or vierbein. We explain all these techniques by a focus on a particular model for spinor gravity.

In sect. 2 we briefly introduce spinor gravity and discuss the properties that an associated automaton has to obey. Sect. 3 discusses simple automata with local Lorentz symmetry corresponding to fermionic actions which are quadratic in the invariants. The naive continuum limit does not yet account for propagating particles for this type of automata, however. In sect. 4 we introduce interactions, as reflected by terms with four or more Lorentz-invariants in the fermionic action. The naive continuum limit admits now a non-trivial evolution. Sect. 5 introduces alternating updatings which lead to a rotation invariant naive continuum limit. In this limit diffeomorphism invariance is realized as well. As a result, the naive continuum limit is the spinor gravity model of sect. 2. We draw conclusions in sect. 6.

2 Spinor gravity automaton

In this section we briefly describe a particular continuum model for spinor gravity. Our spinor gravity automaton will be associated to a particular discretization of this model which realizes a unique jump step evolution operator $\widehat{S}$. We establish the properties that the discrete Grassmann functional integral has to obey in order to implement the unique jump property of $\widehat{S}$.

Spinor gravity model

We investigate four-dimensional spinor gravity models [40–43] based on Majorana fermions with eight flavors. The quantum field theory is formulated as a functional integral over Grassmann variables $\psi_\beta^A(x)$, with spinor index $\beta = 1\dots 4$, and flavor index $A = 1\dots 8$, $x = (t, \vec{x})$. The totally antisymmetric product of four Majorana spinors,

$$\begin{aligned}\chi^A(x) =& \frac{1}{24}\varepsilon^{\eta\delta\alpha\beta}\psi_\eta^A(x)\psi_\delta^A(x)\psi_\alpha^A(x)\psi_\beta^A(x)\\ =& \psi_1^A(x)\psi_2^A(x)\psi_3^A(x)\psi_4^A(x)\ ,\end{aligned} \tag{12}$$

is invariant under local Lorentz transformations

$$\delta\psi_\beta^A - \frac{1}{2}\varepsilon_{np}(x)\left(\Sigma^{np}\right)_{\beta\eta}\psi_\eta^A. \tag{13}$$

Summation over repeated indices is always implied, with the exception of A in the present case. An action that is invariant under local Lorentz transformations as well as diffeomorphisms can be constructed as

$$\begin{aligned}S = &\int_x G_{ABCD,EFGH}\varepsilon^{\mu\nu\rho\sigma}\\ &\times \partial_\mu\chi^A\partial_\nu\chi^B\partial_\rho\chi^C\partial_\sigma\chi^D\chi^E\chi^F\chi^G\chi^H\ .\end{aligned} \tag{14}$$

The invariance with respect to general coordinate transformations is achieved by contracting four derivatives with the totally antisymmetric tensor $\varepsilon^{\mu\nu\rho\sigma}$. The coefficients G define the particular model, where we note the total antisymmetry in the first four flavor indices, and symmetry in the last four indices. We will find a particular choice of G and a particular discretization on a lattice for which this model is equivalent to a cellular automaton.

The action (14) shares all continuous symmetries of a fundamental theory of quantum gravity. The idea of spinor gravity is that appropriate collective fields based on products of spinor fields $\psi^A_\beta(x)$ acquire non-vanishing vacuum expectation values corresponding to a vierbein and associated metric. Indeed, a quantity with the homogeneous transformation properties of a vierbein (vector with respect to Lorentz symmetry and vector with respect to diffeomorphisms) can be formed from Grassmann variables (no sum over A in eqs. (15), (16), (19)),

$$e_\mu{}^{Am} = \widehat{\psi}^A \gamma^m \partial_\mu \psi^A = \widehat{\psi}^A_\gamma (\gamma^m)_{\gamma\delta} \partial_\mu \psi^A_\delta \ . \tag{15}$$

Here

$$\widehat{\psi}^A_\gamma = -\frac{1}{24}\varepsilon_{\gamma\delta\alpha\beta}\psi^A_\delta\psi^A_\alpha\psi^A_\beta \ , \quad \chi^A = \widehat{\psi}^A\psi^A = \widehat{\psi}^A_\gamma\psi^A_\gamma \ , \tag{16}$$

and the real Dirac matrices in a Majorana basis are given by [44, 45], in terms of the Pauli matrices τ_k,

$$\gamma^0 = \begin{pmatrix} 0 & \tau_1 \\ -\tau_1 & 0 \end{pmatrix} \ , \quad \gamma^1 = \begin{pmatrix} -\tau_1 & 0 \\ 0 & \tau_1 \end{pmatrix} \ ,$$
$$\gamma^2 = \begin{pmatrix} -\tau_3 & 0 \\ 0 & -\tau_3 \end{pmatrix} \ , \quad \gamma^3 = \begin{pmatrix} 0 & \tau_1 \\ \tau_1 & 0 \end{pmatrix} \ . \tag{17}$$

The possible inhomogeneous part, due to the derivative in eq. (15) acting on the transformation parameters $\varepsilon_{np}(x)$, cancels by virtue of the identity

$$\widehat{\psi}^A \gamma^m \Sigma^{np} \psi^A = 0 \ , \tag{18}$$

where the generators of the Lorentz group obey

$$\Sigma^{np} = -\frac{1}{4}[\gamma^n, \gamma^p] \ . \tag{19}$$

In the presence of suitable expectation values an expansion of the multi-fermion action (14) in quadratic order in ψ can induce an effective kinetic term for single fermions. If the expectation values are invariant under a global Lorentz symmetry which combines the global subgroup of the local Lorentz transformations with appropriate general coordinate transformations, this kinetic term can generate a (generalized) Dirac equation for the propagation of single fermions. The effective fermion kinetic term will become visible in a mean field approximation. Beyond this approximation it can be found in the quantum effective action which includes collective fields.

Step evolution operator for discrete fermionic quantum field theory

For a formulation of a fermionic quantum field theory on a discrete spacetime-lattice a general formulation for the extraction of the step evolution operator is available [1, 27]. This differs from other, more particular, fermionic descriptions of two-dimensional Ising models [46–51] or fermion-boson equivalence [52–55]. Our formulation employs local factors $\mathcal{K}(t)$

$$e^{-S} = e^{-\sum_t \mathcal{L}(t)} = \prod_t \mathcal{K}(t) \ , \tag{20}$$

Journal of Physics: Conference Series **2533** (2023) 012016 doi:10.1088/1742-6596/2533/1/012016

which involve only Grassmann variables $\psi(t)$ and $\psi(t+\varepsilon)$ at neighboring times. A double expansion of $\mathcal{K}(t)$ in Grassmann basis elements $g_\rho(t)$ and $g_\tau(t+\varepsilon)$,

$$\mathcal{K}(t) = g_\tau(t+\varepsilon)\widehat{S}_{\tau\rho}(t)g'_\rho(t) \ , \tag{21}$$

yields directly the elements of the step evolution operator $\widehat{S}(t)$. The Grassmann basis elements $g_\rho(t)$ are products of Grassmann variables $\psi^A_\beta(t,\vec{x})$, where g'_ρ is related to g_ρ by a sign ± 1 [27].

For an intuitive interpretation the product $g_\tau(t+\varepsilon)g'_\rho(t)$ describes at t a bit $n_\alpha(\vec{x}) = n^A_\beta(\vec{x}) = 1$ of type $\alpha = (\beta, A)$ at position $\vec{x}$ for every factor $\psi^A_\beta(t,\vec{x})$ present in $g'_\rho(t)$, while at $t+\varepsilon$ one has a bit $n^B_\delta(\vec{y}) = 1$ for every factor $\psi^B_\delta(t+\varepsilon,\vec{y})$ in $g_\tau(t+\varepsilon)$. The bits of all types for which no corresponding Grassmann variables appear in $g_\tau(t+\varepsilon)g'_\rho(t)$ take a zero value. Both ρ and τ therefore specify bit-configurations. Then the non-zero elements of the unique jump step evolution operator $\widehat{S}_{\tau\rho}(t)$ indicate for each bit configuration ρ at t a new bit configuration τ at $t+\varepsilon$. This specifies the updating of the automaton in the step from t to $t+\varepsilon$. The uniqueness of the updating requires that each element $g'_\rho(t)$ in $\mathcal{K}(t)$ is multiplied by one single element $g_\tau(t+\varepsilon)$, with coefficient ± 1. The bits $n^A_\beta(\vec{x}) = 1$ can be identified with a fermionic particle of type (β, A) present at the position $\vec{x}$, while $n^A_\beta(\vec{x}) = 0$ corresponds to a "hole" of type (β, A) or an absent fermionic particle at the position $\vec{x}$. The bits $n^A_\beta(\vec{x}) = (1, 0)$ equal the corresponding fermionic occupation numbers. The task consists in finding Grassmann expressions for $\mathcal{L}(t)$ which realize the unique jump property for $\mathcal{K}(t) = \exp l - \mathcal{L}(t)\big)$.

Due to the modulo two property of Grassmann functional integrals [30] the identification (21) holds only for odd t. For even t the role of particles and holes is interchanged, and one employs a different, but analogue, formula for the relation between the local factor $\mathcal{K}(t)$ and the step evolution operator $\widehat{S}(t)$ [28]. We will focus here on models with particle-hole symmetry for which this complication is absent.

A bit-fermion map associates generalized Ising models and Grassmann functional integrals which lead to identical step evolution operators. Besides the identical evolution operators this bit-fermion map extends to wave-functions, operators for observables and expectation values [1]. We observe that the map from a Grassmann functional integral to an evolution sequence of bit configurations is unique, since $\mathcal{K}(t)$ defines $\widehat{S}(t)$ uniquely. The opposite is not true, since a given bit sequence specifies the non-zero elements $\widehat{S}_{\tau\rho}(t)$ only up to a sign. All choices of signs are equivalent, being related by local discrete gauge transformations [56, 57]. We will choose signs such that the naive continuum limit of the discrete fermion models becomes simple. The bit-fermion map extends to arbitrary Grassmann functional integrals with time-local action (20), and to arbitrary step evolution operators. We focus here on the unique jump step evolution operators corresponding to automata.

Discrete spinor gravity model

The aim of this note is to find a discretization of the spinor gravity models (14), such that an appropriate choice of G yields the unique jump step evolution operator for an automaton. We group several microscopic cells corresponding to lattice points $\vec{x}$ into "local cells". These local cells are defined as non-overlapping cubes in a three-dimensional lattice. The updating of the bit-configuration in a local cell from t to $t+\varepsilon$ depends only on the bit configuration in the same local cell. For the next updating step from $t+\varepsilon$ to $t+2\varepsilon$ the positions of the local cells in the three-dimensional lattice are shifted, such that the bit configuration at $t+2\varepsilon$ is influenced by the bit configurations in nine neighboring local cells at t. This locality property characterizes cellular automata. It induces the causal structure characteristic for local quantum field theories.

The task of finding a discrete model of the cellular automaton type remains rather complex, since we have to guarantee that every allowed bit or fermion configuration at t is updated to precisely one allowed configuration at $t+\varepsilon$. The allowed configurations will be those for which at every site of the three-dimensional lattice the fermions can be grouped into a certain number of Lorentz invariants χ^A. The total number of fermions therefore equals 0 mod4. The action (14) contains no terms with zero, one, two or three derivatives. It is odd under an arbitrary exchange of coordinates. The corresponding discrete symmetries will provide important guiding lines for the construction of a lattice action. We will realize these symmetries of the naive continuum limit by a sequence of 24 different updatings which change the relative orientations in position and flavor space. Combining this sequence in a "large updating step" by $\Delta t = 24\varepsilon$ the large updatings are then repeated.

3 Automaton for free particles

This section discusses an automaton for "free particles", in the sense that the fermionic action is quadratic in local Lorentz invariant combinations χ^A. The particles are gauge invariant composites of fermions, somewhat similar to mesons. We first formulate a switch automaton in disconnected cells corresponding to hypercubes in a four-dimensional lattice. Propagation is introduced by shifting the cells in neighboring time layers. Comparison with the switch automaton without this shift reveals already important aspects of the naive continuum limit. At this level there is no standard propagation in the naive continuum limit.

Local updatings

The updating from t to $t+\varepsilon$ is done independently for each cell which corresponds to a cube with eight sites u at the corners. Our basic building blocks are the local Lorentz-invariants at every site,

$$\chi_u^A(t) = \psi_{u,1}^A(t)\psi_{u,2}^A(t)\psi_{u,3}^A(t)\psi_{u,4}^A(t) \ . \tag{22}$$

At every corner u of a cube and for every flavor $A = 1 \dots N_f$ they are products of the four Grassmann variables $\psi_{u,\beta}^A(t)$. The index u labels the positions at the corners of a cube as $(x,y,z) = 1$, $(x+\varepsilon,y,z) = 2$, $(x,y+\varepsilon,z) = 3$, $(x,y,z+\varepsilon) = 4$, $(x+\varepsilon,y+\varepsilon,z) = 5$, $(x+\varepsilon,y,z+\varepsilon) = 6$, $(x,y+\varepsilon,z+\varepsilon) = 7$, $(x+\varepsilon,y+\varepsilon,z+\varepsilon) = 8$, as shown in Fig. 3. The local

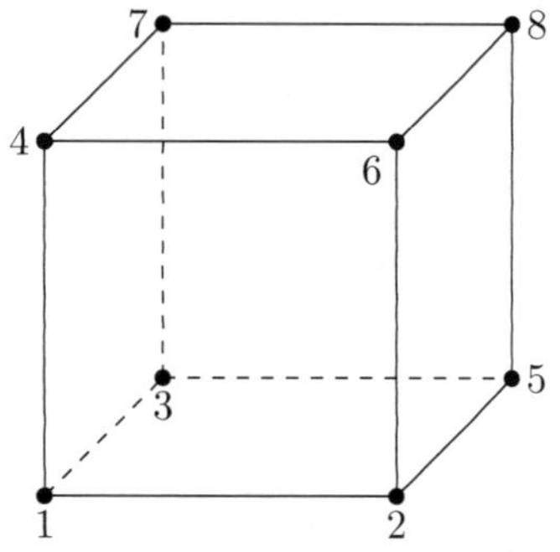

Figure 1. Labeling of corners of cube.

updating of the automaton is described by a four-dimensional hypercube which combines the cube at t with the same cube at $t+\varepsilon$. The cubes constitute the local cells of the automaton. The state of the cube at $t+\varepsilon$ is determined as a function of the state of the same cube at t. We place invariants χ^A at various positions in the hypercube. Denoting the center of the hypercube by $w^\mu = (t+\frac{\varepsilon}{2}, x+\frac{\varepsilon}{2}, y+\frac{\varepsilon}{2}, z+\frac{\varepsilon}{2})$, the positions of the corners of the hypercube are given by shifting each coordinate by $\pm\varepsilon/2$. The signs for the different corners corresponding to our labels u are listed in table 1.

Table 1. Labeling of the positions of corners of the four-dimensional hypercube. With w^μ the position of the center of the hypercube the locations are at $w^\mu + s^\mu\varepsilon/2$, with signs $s^\mu = (s^0, s^1, s^2, s^3)$ listed in the table.

$\chi_1(t)$	$----$		$\chi_1(t+\varepsilon)$	$+---$
$\chi_2(t)$	$-+--$		$\chi_2(t+\varepsilon)$	$++--$
$\chi_3(t)$	$--+-$		$\chi_3(t+\varepsilon)$	$+-+-$
$\chi_4(t)$	$---+$		$\chi_4(t+\varepsilon)$	$+--+$
$\chi_5(t)$	$-++-$		$\chi_5(t+\varepsilon)$	$+++-$
$\chi_6(t)$	$-+-+$		$\chi_6(t+\varepsilon)$	$++-+$
$\chi_7(t)$	$--++$		$\chi_7(t+\varepsilon)$	$+-++$
$\chi_8(t)$	$-+++$		$\chi_8(t+\varepsilon)$	$++++$

Time-local factors $\mathcal{K}(t)$ are products of space-time-local factors $\mathcal{M}(w)$

$$\mathcal{K}(t) = \prod_{x,y,z} \mathcal{M}(w) = \prod_{x,y,z} \mathcal{M}(t,x,y,z) \ , \tag{23}$$

with $w = w(t,x,y,z)$. Here $\mathcal{M}(w)$ is a function of the local Lorentz-invariants χ^A placed at different corners of the hypercube w. Thus $\mathcal{K}(t)$ involves Grassmann variables at t and $t+\varepsilon$. The product over x,y,z extends over a three-dimensional cubic lattice with lattice distance 2ε. Therefore two neighboring cubes at a given t do not overlap and we can discuss the updating for each hypercube separately. A connection between different cubes or cells of the automaton will be established by shifting the positions of the hypercubes at the next time step from $t+\varepsilon$ to $t+2\varepsilon$. We will express the space-time-local factors $\mathcal{M}(w)$ in an exponential form

$$\mathcal{M}(w) = \exp\big(-L(w)\big) \ . \tag{24}$$

The quantity $L(w)$ describes the elementary processes of the automaton, with $\mathcal{L}(t) = \sum_{x,y,z} L(w)$.

Transport factors

Let us introduce transport factors involving one invariant at t and one invariant at $t+\varepsilon$, for which we use the shorthands

$$\chi_u^A = \chi_u^A(t) \ , \quad \chi'^A_u = \chi_u^A(t+\varepsilon) \ . \tag{25}$$

We define

$$\begin{aligned} T^{AB}_{x\pm} &= \chi_1^A \chi'^B_2 \pm \chi_2^A \chi'^B_1 + \chi_3^A \chi'^B_5 \pm \chi_5^A \chi'^B_3 \\ &\quad + \chi_4^A \chi'^B_6 \pm \chi_6^A \chi'^B_4 + \chi_7^A \chi'^B_8 \pm \chi_8^A \chi'^B_7 \ , \end{aligned} \tag{26}$$

$$\begin{aligned} T^{AB}_{y\pm} &= \chi_1^A \chi'^B_3 \pm \chi_3^A \chi'^B_1 + \chi_2^A \chi'^B_5 \pm \chi_5^A \chi'^B_2 \\ &\quad + \chi_4^A \chi'^B_7 \pm \chi_7^A \chi'^B_4 + \chi_6^A \chi'^B_8 \pm \chi_8^A \chi'^B_6 \ , \end{aligned} \tag{27}$$

$$T_{z\pm}^{AB} = \chi_1^A \chi'^B_4 \pm \chi_4^A \chi'^B_1 + \chi_2^A \chi'^B_6 \pm \chi_6^A \chi'^B_2 \\ + \chi_3^A \chi'^B_7 \pm \chi_7^A \chi'^B_3 + \chi_5^A \chi'^B_8 \pm \chi_8^A \chi'^B_5 \, . \tag{28}$$

Without going into formal details, the interpretation of a term $\chi_u^A \chi_v'^B$ is simple. The four fermions χ_u^A are annihilated, and the four fermions forming $\chi_v'^B$ are created. Thus the transport factor T_x^{AB} moves an invariant of flavor A, present at t at position (x, y, z), $(x, y+\varepsilon, z)$, $(x, y, z+\varepsilon)$ or $(x, y+\varepsilon, z+\varepsilon)$ (positions 1,3,4,7) to new positions with x shifted to $x+\varepsilon$. At $t+\varepsilon$ the corresponding positions are $(x+\varepsilon, y, z)$, $(x+\varepsilon, y+\varepsilon, z)$, $(x+\varepsilon, y, z+\varepsilon)$ or $(x+\varepsilon, y+\varepsilon, z+\varepsilon)$ (positions 2,5,6,8) respectively. Similarly, invariants at $x+\varepsilon$ are moved to x. Simultaneously, the flavor changes from A to B. Thus T_x^{AB} describes a simultaneous switch in position and flavor. The transport factors T_y and T_z describe similar switches, with position changes now in the y- or z-directions. The position changes can be viewed as appropriate reflections within the cube. We also consider the flavor change without motion

$$T_0^{AB} = \sum_{u=1}^{8} \chi_u^A \chi'^B_u \, . \tag{29}$$

A given invariant stays at its position and only changes its flavor from A to B.

We can combine space reflections with flavor rotations by defining

$$\widetilde{T}_x^{AB} = T_{x-}^{AB} - T_{x-}^{BA} \, , \qquad \widetilde{T}_y^{AB} = T_{y-}^{AB} - T_{y-}^{BA} \, , \\ \widetilde{T}_z^{AB} = T_{z-}^{AB} - T_{z-}^{BA} \, , \qquad \widetilde{T}_0^{AB} = T_0^{AB} - T_0^{BA} \, . \tag{30}$$

The flavor A is rotated to B, while the flavor B is rotated to A with an additional minus sign. This corresponds to a $\pi/2$-rotation in flavor space.

Switch automaton

Let us consider eight flavors, $A, B = 1 \dots 8$. A simple switch automaton can be defined by

$$-L_0 = \widetilde{T}_x^{12} + \widetilde{T}_y^{34} + \widetilde{T}_z^{56} + \widetilde{T}_0^{78} \, . \tag{31}$$

The fundamental process acts separately in the flavor subspaces $(1,2)$, $(3,4)$, $(5,6)$ and $(7,8)$. We can therefore write

$$\mathcal{M}_0 = \exp\big(-L_0\big) \\ = \exp\big(\widetilde{T}_x^{12}\big) \exp\big(\widetilde{T}_y^{34}\big) \exp\big(\widetilde{T}_z^{56}\big) \exp\big(\widetilde{T}_0^{78}\big) \, . \tag{32}$$

We want to show that $\mathcal{M}_0$ generates a unique jump step evolution operator in the space of Lorentz-invariant four fermion states χ_u^A within the cube w. It describes a unique rule how an arbitrary number of invariants, situated on arbitrary corners of the cube, is mapped to a new configuration of invariants in the cube. This updating map conserves the number of invariants in the cube. We can make the proof for each of the flavor subspaces independently.

Expanding the first factor

$$\exp\big(\widetilde{T}_x^{12}\big) = 1 + \widetilde{T}_x^{12} + \frac{1}{2}\big(\widetilde{T}_x^{12}\big)^2 + \cdots + \frac{1}{16!}\big(\widetilde{T}_x^{12}\big)^{16} \, , \tag{33}$$

the series terminates since $\big(\widetilde{T}_x^{12}\big)^{17} = 0$. This is due to the property $\big(\chi_u^A\big)^2 = 0$, which is a direct consequence of the Grassmann property $\big(\psi_{u,\gamma}^A\big)^2 = 0$. In order sixteen the product $\big(\widetilde{T}_x^{12}\big)^{16}$

Journal of Physics: Conference Series **2533** (2023) 012016 doi:10.1088/1742-6596/2533/1/012016

contains one invariant $\chi_u^1(t)$ and one invariant $\chi_u^2(t)$ at every site u. The same happens at $t+\varepsilon$. The product of sixteen sums in $\left(\widetilde{T}_x^{12}\right)^{16}$ involves 16! possibilities to combine these terms from the various factors. For each term the overall factor equals one, such that the factor $1/(16!)$ from the expansion of the exponential is compensated by 16! equal contributions. We identify

$$\frac{1}{16!}\left(\widetilde{T}_x^{12}\right)^{16} = \prod_{u=1}^{8} \chi_u^1(t)\chi_u^2(t)\chi_u^1(t+\varepsilon)\chi_u^2(t+\varepsilon)$$
$$= \prod_{\gamma=1}^{4}\prod_{u=1}^{8}\prod_{A=1}^{2} \psi_{u,\gamma}^A(t)\psi_{u,\gamma}^A(t+\varepsilon) , \tag{34}$$

with the map from the totally filled (empty) state at t to the totally filled (empty) state at $t+\varepsilon$. Similarly, the first factor 1 in the expansion of the exponential (33) maps the totally empty (filled) state to itself. The association of given Grassmann elements with particles or holes, or filled and empty, changes with time due to the modulo two property of the Grassmann functional integral [27]. The corresponding two terms in the series (33) define a unit step evolution operator in the sectors of zero and sixteen invariants.

The elementary process $-L_0$ accounts for a reflection in position space accompanied by a flavor rotation for a single Lorentz-invariant. Within the restricted flavor space $(1, 2)$ the "switch factor" $\widetilde{T}_x^{12}$ is a sum of 16 terms, each relevant for the updating of a state with precisely one invariant of flavor 1 or 2, positioned at an arbitrary site u. The updating rule to a new state with only one invariant is unique, as appropriate for an automaton. For example, flavor 2 at position 5 is mapped to flavor 1 at position 3. Thus $-L_0$ describes a unique jump step evolution operator in the subsector with a single invariant.

The term $\left(\widetilde{T}_x^{12}\right)^2/2$ in the expansion (33) describes the updating for states with two invariants (8 fermions). Due to $\left(\chi_u^A\right)^2 = 0$ the two invariants either carry different flavor, or are positioned at different sites, or both. Every possible combination $\chi_u^A(t)\chi_v^B(t)$ appears in the product of the two sums. There are two possibilities with the same sign compensating the factor 1/2 in the series (33). The overall coefficient of this term is therefore ± 1. Every pair of invariants $\chi_u^A(t)\chi_v^B(t)$ is multiplied by a unique factor $\chi_{u'}^{A'}(t+\varepsilon)\chi_{v'}^{B'}(t+\varepsilon)$, corresponding to the unique updating rule for the automaton. A similar discussion applies to the term $\left(\widetilde{T}_x^{12}\right)^3/(3!)$ which accounts for the unique updating of states with three different invariants in the cube. This generalizes to the higher terms in the expansion.

In conclusion, $\exp\left(\widetilde{T}_x^{12}\right)$ describes the updating rule for an arbitrary number of different invariants of flavor 1 or 2 in the cell. Correspondingly, the factor $\exp\left(\widetilde{T}_y^{34}\right)$ accounts for the unique updating of an arbitrary number of different invariants of flavor 3 and 4, and similarly for $\exp\left(\widetilde{T}_z^{56}\right)$ for the flavors 5 and 6, and for $\exp\left(\widetilde{T}_0^{78}\right)$ for the flavors 7 and 8. Taking things together the factor $\exp\left(-L_0\right)$ describes the unique updating for an arbitrary number of different invariants in the cell. The maximal total number of invariants (for the totally filled or empty state) amounts to $8N_f = 64$. The expansion of $\mathcal{M}_0$ in the occupation number basis generates a unique jump step evolution operator for invariants in a given cube of block or local cell. The combined local factor (23) generates an invertible updating rule for an arbitrary number of invariants in the space-lattice, and a corresponding unique jump evolution operator $\widehat{S}(t)$.

Naive continuum limit

A naive continuum limit for the switch automaton can be constructed by introducing lattice derivatives which turn into partial derivatives in the limit $\varepsilon \to 0$. This limit requires a sufficiently smooth probabilistic information, i.e. smoothness of the wave function $q_\tau(t)$. The

Journal of Physics: Conference Series **2533** (2023) 012016 doi:10.1088/1742-6596/2533/1/012016

naive continuum limit simply keeps only the leading order in an expansion in ε. This does not take into account fluctuation effects which typically induce running couplings which may or may not be compatible with the naive continuum limit. This issue is in analogy to the discretization of other quantum field theories as lattice gauge theories.

We express the invariants at the different corners of the cube in terms of an average field $\bar{\chi}$ and lattice derivatives as

$$\begin{aligned}
\chi_1(t) =& \bar{\chi} - \frac{\varepsilon}{2}(\partial_t + \partial_x + \partial_y + \partial_z)\chi + \dots \\
\chi_2(t) =& \bar{\chi} - \frac{\varepsilon}{2}(\partial_t - \partial_x + \partial_y + \partial_z)\chi + \dots \\
\chi_3(t) =& \bar{\chi} - \frac{\varepsilon}{2}(\partial_t + \partial_x - \partial_y + \partial_z)\chi + \dots \\
\chi_4(t) =& \bar{\chi} - \frac{\varepsilon}{2}(\partial_t + \partial_x + \partial_y - \partial_z)\chi + \dots \\
\chi_5(t) =& \bar{\chi} - \frac{\varepsilon}{2}(\partial_t - \partial_x - \partial_y + \partial_z)\chi + \dots \\
\chi_6(t) =& \bar{\chi} - \frac{\varepsilon}{2}(\partial_t - \partial_x + \partial_y - \partial_z)\chi + \dots \\
\chi_7(t) =& \bar{\chi} - \frac{\varepsilon}{2}(\partial_t + \partial_x - \partial_y - \partial_z)\chi + \dots \\
\chi_8(t) =& \bar{\chi} - \frac{\varepsilon}{2}(\partial_t - \partial_x - \partial_y - \partial_z)\chi + \dots \; .
\end{aligned} \tag{35}$$

Similar relations hold for $\chi_w(t+\varepsilon)$, with ∂_t replaced by $-\partial_t$ in eq. (35). This yields relations of the type

$$\begin{aligned}
\chi_8(t) - \chi_1(t) =& \varepsilon(\partial_t + \partial_x + \partial_y + \partial_z)\chi \, , \\
\frac{1}{2}\big(\chi_8(t) + \chi_1(t)\big) =& \bar{\chi} + \dots \; .
\end{aligned} \tag{36}$$

The dots in eqs. (35), (36) denote corrections $\sim \varepsilon^2$ with two or more derivatives. More precisely, we define

$$\varepsilon\partial_t\chi = \frac{1}{8}\sum_{u=1}^{8}\big(\chi_u(t+\varepsilon) - \chi_u(t)\big) \, , \tag{37}$$

$$\bar{\chi} = \frac{1}{16}\sum_{u=1}^{8}\big(\chi_u(t+\varepsilon) + \chi_u(t)\big) \, , \tag{38}$$

and

$$\begin{aligned}
\varepsilon\partial_x\chi =& \frac{1}{8}\Big\{\big[\chi_2(t) + \chi_5(t) + \chi_6(t) + \chi_8(t) \\
&- \chi_1(t) - \chi_3(t) - \chi_4(t) - \chi_7(t)\big] + \big(t \leftrightarrow t+\varepsilon\big)\Big\} \\
=& \frac{1}{8}\Big\{\big[\chi(t, x+\varepsilon, y, z) + \chi(t, x+\varepsilon, y+\varepsilon, z) \\
&+ \chi(t, x+\varepsilon, y, z+\varepsilon) + \chi(t, x+\varepsilon, y+\varepsilon, z+\varepsilon) \\
&- \chi(t, x, y, z) - \chi(t, x, y+\varepsilon, z) - \chi(t, x, y, z+\varepsilon) \\
&- \chi(t, x, y+\varepsilon, z+\varepsilon)\big] + \big(t \leftrightarrow t+\varepsilon\big)\Big\} \, ,
\end{aligned} \tag{39}$$

with similar definitions for $\partial_y\chi$ and $\partial_z\chi$. This partially specifies the dots in eqs. (35), (36).

Expressing the transport factors (26)- (29) in terms of lattice derivatives one finds

$$T_{x-}^{AB} = 4\varepsilon\big(\chi^A\partial_x\chi^B - \chi^B\partial_x\chi^A\big) \\ - 2\varepsilon^2\big(\partial_t\chi^A\partial_x\chi^B + \partial_t\chi^B\partial_x\chi^A\big) + \dots \tag{40}$$

and

$$T_{x+}^{AB} = 8\chi^A\chi^B + 4\varepsilon\big(\chi^A\partial_t\chi^B - \chi^B\partial_t\chi^A\big) + \dots . \tag{41}$$

For $T_{y\pm}^{AB}$ and $T_{z\pm}^{AB}$ one obtains similar expressions, with ∂_x replaced by ∂_y or ∂_z, respectively. The first two terms in an expansion of T_0^{AB} read

$$T_0^{AB} = 8\chi^A\chi^B + 4\varepsilon\big(\chi^A\partial_t\chi^B - \chi^B\partial_t\chi^A\big) + \dots . \tag{42}$$

It differs from T_{x+}^{AB}, T_{y+}^{AB}, T_{z+}^{AB} only in the order ε^2. The combinations (30) are antisymmetric in flavor space, such that

$$\begin{aligned} \widetilde{T}_x^{12} &= 8\varepsilon\big(\chi^1\partial_x\chi^2 - \chi^2\partial_x\chi^1\big) , \\ \widetilde{T}_y^{34} &= 8\varepsilon\big(\chi^3\partial_y\chi^4 - \chi^4\partial_y\chi^3\big) , \\ \widetilde{T}_z^{56} &= 8\varepsilon\big(\chi^5\partial_z\chi^6 - \chi^6\partial_z\chi^5\big) , \\ \widetilde{T}_0^{78} &= 8\varepsilon\big(\chi^7\partial_t\chi^8 - \chi^8\partial_t\chi^7\big) . \end{aligned} \tag{43}$$

The naive continuum limit for the time-local factor of the switch automaton (31) reads

$$\widetilde{\mathcal{K}}(t) = \exp\Big\{8\varepsilon\sum_{xyz}\widetilde{T}(t,x,y,z)\Big\} , \tag{44}$$

with fermionic action

$$S = -8\varepsilon\sum_{t,x,y,z}\tilde{T}(t,x,y,z) \tag{45}$$

and

$$\widetilde{T} = \chi^1\partial_x\chi^2 - \chi^2\partial_x\chi^1 + \chi^3\partial_y\chi^4 - \chi^4\partial_y\chi^3 \\ + \chi^5\partial_z\chi^6 - \chi^6\partial_z\chi^5 + \chi^7\partial_t\chi^8 - \chi^8\partial_t\chi^7 . \tag{46}$$

Here the "continuum invariants" $\chi^A \equiv \chi^A(t,x,y,z)$ are related to the local Lorentz invariants in the discrete formulation by eqs. (37)- (39), with χ^A identified with $\bar{\chi}^A$. The action is quadratic in the invariants. This holds both for the discrete version and the naive continuum limit.

If we keep in the action (45) the discrete version of $\widetilde{T}$,

$$8\varepsilon\widetilde{T} = \widetilde{T}_x^{12} + \widetilde{T}_y^{34} + \widetilde{T}_z^{56} + \widetilde{T}_0^{78} , \tag{47}$$

the four-dimensional fermionic quantum field theory constitutes an example for a probabilistic automaton. The "quadratic action" does not describe a standard propagation of particles for increasing time, however. The only time-dependence is the oscillation between the flavors 7 and 8. Since different flavor sectors are not connected the time evolution is not transported to the other sectors. Inspection of the action (14) for spinor gravity indicates that it may be composed of four factors of the type $\widetilde{T}_j^{AB}$. This connects the different flavors. We will indeed construct a discrete spinor gravity model with this structure. For this reason we discuss a few more properties of the switch automaton.

Journal of Physics: Conference Series **2533** (2023) 012016 doi:10.1088/1742-6596/2533/1/012016

Table 2. Action of discrete reflection symmetries. For the symmetries P_x, P_y, P_z, S_{xy}, S_{yz}, S_{zx} the bilinears χ'_u transform in the same way as χ_u. Invariant bilinears are not listed.

T	$t \leftrightarrow -t$	$\chi_u \leftrightarrow \chi'_u$
P_x	$x \leftrightarrow -x$	$\chi_1 \leftrightarrow \chi_2$, $\chi_3 \leftrightarrow \chi_5$, $\chi_4 \leftrightarrow \chi_6$, $\chi_7 \leftrightarrow \chi_8$
P_y	$y \leftrightarrow -y$	$\chi_1 \leftrightarrow \chi_3$, $\chi_2 \leftrightarrow \chi_5$, $\chi_4 \leftrightarrow \chi_7$, $\chi_6 \leftrightarrow \chi_8$
P_z	$z \leftrightarrow -z$	$\chi_1 \leftrightarrow \chi_4$, $\chi_2 \leftrightarrow \chi_6$, $\chi_3 \leftrightarrow \chi_7$, $\chi_5 \leftrightarrow \chi_8$
P	$P_xP_yP_z$	$\chi_u \leftrightarrow \chi_{9-u}$
S_{xy}	$x \leftrightarrow y$	$\chi_2 \leftrightarrow \chi_3$, $\chi_6 \leftrightarrow \chi_7$
S_{yz}	$y \leftrightarrow z$	$\chi_3 \leftrightarrow \chi_4$, $\chi_5 \leftrightarrow \chi_6$
S_{zx}	$z \leftrightarrow x$	$\chi_2 \leftrightarrow \chi_4$, $\chi_5 \leftrightarrow \chi_7$
S_{tx}	$t \leftrightarrow x$	$\chi_2 \leftrightarrow \chi'_1$, $\chi_5 \leftrightarrow \chi'_3$, $\chi_6 \leftrightarrow \chi'_4$, $\chi_8 \leftrightarrow \chi'_7$
S_{ty}	$t \leftrightarrow y$	$\chi_3 \leftrightarrow \chi'_1$, $\chi_5 \leftrightarrow \chi'_2$, $\chi_7 \leftrightarrow \chi'_4$, $\chi_8 \leftrightarrow \chi'_6$
S_{tz}	$t \leftrightarrow z$	$\chi_4 \leftrightarrow \chi'_1$, $\chi_6 \leftrightarrow \chi'_2$, $\chi_7 \leftrightarrow \chi'_3$, $\chi_8 \leftrightarrow \chi'_5$

Table 3. Rotations in the xy, yz and zx planes. The transformation of χ'_w is the same as for χ_u.

R_{xy}	$\chi_1 \to \chi_3 \to \chi_5 \to \chi_2 \to \chi_1$	$\chi_4 \to \chi_7 \to \chi_8 \to \chi_6 \to \chi_4$
R_{yz}	$\chi_1 \to \chi_4 \to \chi_7 \to \chi_3 \to \chi_1$	$\chi_2 \to \chi_6 \to \chi_8 \to \chi_5 \to \chi_2$
R_{zx}	$\chi_1 \to \chi_2 \to \chi_6 \to \chi_4 \to \chi_1$	$\chi_3 \to \chi_5 \to \chi_8 \to \chi_7 \to \chi_3$

Discrete transformations and symmetries

Eq. (43) can be understood in terms of discrete transformations as reflections and lattice rotations. These discrete transformations in the hypercube are an important construction principle. Instead of changing coordinates we realize discrete transformations as discrete transformations on the local Lorentz-invariants χ_u, or the corresponding Grassmann variables $\psi_{u,\beta}$. We list various reflections in table 2, and some of the rotations in table 3. These discrete transformations are flavor blind, i.e. they act in the same way on all flavors.

The factor T^{12}_{x-} or $\tilde{T}^{12}_x$, abbreviated here by T_x, changes sign under the reflection of the x-coordinate P_x, while it is invariant under the two other coordinate reflections P_y and P_z. Similarly, T_y is odd under P_y and T_z is odd under P_z,

$$\begin{aligned} &P_x(T_x) = -T_x\ , \quad P_x(T_y) = T_y\ , \quad P_x(T_z) = T_z\ ,\\ &P_y(T_x) = T_x\ , \quad P_y(T_y) = -T_y\ , \quad P_y(T_z) = T_z\ ,\\ &P_z(T_x) = T_x\ , \quad P_z(T_y) = T_y\ , \quad P_z(T_z) = -T_z\ . \end{aligned} \tag{48}$$

One finds for the diagonal reflections

$$\begin{aligned} &S_{xy}(T_x) = T_y\ , \quad S_{xy}(T_y) = T_x\ , \quad S_{xy}(T_z) = T_z\ ,\\ &S_{yz}(T_x) = T_x\ , \quad S_{yz}(T_y) = T_z\ , \quad S_{yz}(T_z) = T_y\ ,\\ &S_{zx}(T_x) = T_z\ , \quad S_{zx}(T_y) = T_y\ , \quad S_{zx}(T_z) = T_x\ , \end{aligned} \tag{49}$$

and $\pi/2$ rotations result in

$$\begin{aligned} R_{xy}(T_x) &= -T_y\ , & R_{xy}(T_y) &= T_x\ , & R_{xy}(T_z) &= T_z\ , \\ R_{yz}(T_x) &= T_x\ , & R_{yz}(T_y) &= -T_z\ , & R_{yz}(T_z) &= T_y\ , \\ R_{zx}(T_x) &= T_z\ , & R_{zx}(T_y) &= T_y\ , & R_{zx}(T_z) &= -T_x\ . \end{aligned} \tag{50}$$

On the level of the lattice derivatives P_x inverts the sign of ∂_x while keeping the sign of ∂_y and ∂_z. This holds analogously for the other two reflections P_y and P_z. The reflection S_{xy} exchanges $\partial_x \leftrightarrow \partial_y$, keeping ∂_z fixed. Finally, the rotation R_{xy} transforms $\partial_x \to -\partial_y \to -\partial_x \to \partial_y \to \partial_x$. These transformation properties explain to a large extent eqs. (40)- (42). We observe that in these expressions the first terms $\sim \varepsilon$ are T-even, while the second terms $\sim \varepsilon^2$ are T-odd. The factor $\widetilde{T}_x^{12}$ is odd under a combination of the reflection $t \leftrightarrow t+\varepsilon$ with a flavor exchange $1 \leftrightarrow 2$. This holds analogously for $\widetilde{T}_y^{34}$ and $\widetilde{T}_z^{56}$.

Updating with shifted cells

The spacetime-local updating of a single cell associates to each configuration with an arbitrary number of invariants (up to $N_f = 8$ invariants for any given site), located at arbitrary corners of the cube, and carrying arbitrary flavors, a new such configuration. The updating involves reflections within the cube and flavor rotations. Employing the updating twice leads to an identity transformation up to signs. A sequence of updatings of the same given cell leads to a very simple, but somewhat boring automaton.

A richer structure obtains if we shift the position of the cells in consecutive updating steps. At the second updating step we will use a shifted lattice of cells. While for the updating from t to $t+\varepsilon$ the centers of the hypercubes obey, with integers n_μ,

$$\begin{aligned} w &= (w^0, w^1, w^2, w^3) \\ &= \varepsilon\big(2n_0 + \frac{1}{2}, 2n_1 + \frac{1}{2}, 2n_2 + \frac{1}{2}, 2n_3 + \frac{1}{2}\big)\ , \end{aligned} \tag{51}$$

we employ for the updating from $t+\varepsilon$ to $t+2\varepsilon$ centers of hypercubes located at

$$w = \big(2n_0 + \frac{3}{2}, 2n_1 + \frac{3}{2}, 2n_2 + \frac{3}{2}, 2n_3 + \frac{3}{2}\big)\ . \tag{52}$$

The eight corners of the cube in the first updating step belong to eight different cubes in the second updating step. We illustrate this setting in Fig. 3 for the case of two space dimensions. For example, the point $(\varepsilon, 0, \varepsilon, \varepsilon)$, which belongs to the hypercube $w = (\varepsilon/2)(1,1,1,1)$ in the first updating step, will belong to the hypercube at $w = (\varepsilon/2)(3,-1,3,3)$ in the second updating step. A different point in the same cell of the first updating, say $(\varepsilon, \varepsilon, 0, 0)$, belongs for the second step to the hypercube at $w = (\varepsilon/2)(3,3,-1,-1)$. For the updating from $t+2\varepsilon$ to $t+3\varepsilon$ we use the same cells as for the one from t to $t+\varepsilon$. The use of the same cells for t and $t+2n_0\varepsilon$ results in a sequence of alternating cell lattices.

The updating in the second step is again done for each cell separately. Combining two steps, however, the configuration of a given cell at $t+2\varepsilon$ is influenced by the configurations of eight cells at t, and of 27 cells at $t-\varepsilon$. The cells that can influence a given cell w at t form the interior of a "past light cone" of this cell. Similarly, the configuration of a given cell at t influences the configurations of eight cells at $t+2\varepsilon$. The cells that can be influenced by a given cell w form the interior of the "future light cone" of this cell. The linear extension of the light cone of w grows $\sim \varepsilon|\Delta t|$, with Δt the time distance to the cell w in the future or past. The updating with shifted cells therefore connects different parts of the lattice of cells. It exhibits a causal structure based on locality properties.

Journal of Physics: Conference Series **2533** (2023) 012016 doi:10.1088/1742-6596/2533/1/012016

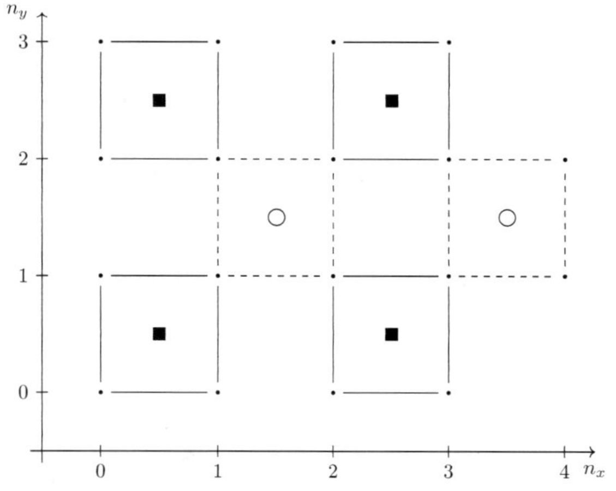

Figure 2. Sequence of updatings with shifted cells for two space dimensions. The cells in the first updating step are the squares with centers indicated by the small filled solid squares. The sites of a given cell are the corners of a square, linked by the solid lines. The cells do not overlap. At the second updating step the cells are the squares with centers at the circles. Sites of a given cell are joined by the dashed lines. Again, the different cells do not overlap. Each site of a cell for the first step is found in a different cell for the second step.

In principle, one may employ different updating rules for the individual cells at different steps in a sequence of updatings. We will indeed do so in later parts of this note. For the moment let us briefly discuss the automaton that follows if we apply the same updating rule (31) for the switch automaton for all updating steps. Consider at t a single invariant of flavor 1 located at $(x, y, z) = (0, 0, 0)$ or $u = 1$. After the first updating we have at the site $(\varepsilon, 0, 0)$ or $u = 2$ an invariant of flavor 2. In the second step this invariant at $(\varepsilon, 0, 0)$ is situated at the site $u = 7$ of the shifted cube. The updating moves it to the site $u = 8$, now corresponding to the position $(2\varepsilon, 0, 0)$. The flavor turns back to 1, and the total result of the two steps is a transport in the positive x-direction by two units, keeping the flavor fixed. This extends to further steps, such that the invariant moves on a straight line in the x-direction.

The same behavior generalizes to invariants of flavor 1 or 2 which are situated at t at the positions 1, 3, 4, 7. They move in the positive x-direction. Similarly, invariants positioned at 2, 5, 6 or 8 move in the negative x-direction. The "right movers" and "left movers" do not interfere. For example, one "right mover" at position 1 and one "left mover" at position 2 will be found for the second updating step in different cells and subsequently continue their motion, not influenced by the presence or absence of the other invariant. This generalizes to several invariants of flavor 1 or 2 in the cell at t. They move independently to the positive or negative x-direction, depending on their position at t.

The situation for invariants with flavors 3 or 4 is analogous. They move in the positive or negative y-direction, according to their position in the cube at t. For the flavors 5 and 6 the motion is in the positive or negative z-direction. Finally, the flavors 7 and 8 stay at their sites, alternating only in the type of flavor. In summary, this automaton describes a simple setting where flavors 1, 2 move on straight lines in the x-direction, flavors 3, 4 in the y-direction and flavors 5, 6 in the z-directions. The flavors 7, 8 are stationary. The flavor switches at alternating time steps. This automaton describes a type of motion of free particles, combined into Lorentz invariants, with direction of motion given by the flavor.

For the two alternating lattices of cells the translation of a given point to the location in the corresponding cell is simple. If the point (x, y, z) corresponds to the position 1 for the cells in the first lattice, it corresponds to the position 8 for the cells of the shifted lattice. Similarly, the point $(x + \varepsilon, y, z)$ has position 2 in the cells of the first lattice, and position 7 in the cells of the second lattice. Between the two lattices points in the cell are shifted according to maximal diagonals, or $u \leftrightarrow 9 - u$.

Journal of Physics: Conference Series **2533** (2023) 012016 doi:10.1088/1742-6596/2533/1/012016

Continuum limit for shifted cells

The naive continuum limit for the automaton with shifted cells is again given by eqs (45), (46). The shift of the cells is not visible in this limit. For the continuum limit there is no difference between the automaton with shifted cells and the automaton for which the same cells are used at every time step. On the other hand, we have discussed that on the discrete level the behavior of the two types of automata is rather different. This points to important properties, and also important limitations, of the naive continuum limit.

The issue can be understood if we enhance the resolution of the continuum description. For the flavors 1 and 2 we may introduce an extra label $\eta = +1$ for those invariants that are located at t at $(\overline{x} - \frac{\varepsilon}{2}, y, z)$, and $\eta = -1$ if the position at t is $(\overline{x} + \frac{\varepsilon}{2}, y, z)$. For the automaton with shifted cells the invariants with $\eta = +1$ propagate towards increasing x, while the ones with $\eta = -1$ move to decreasing x. For a single time step one has an additional flavor switch, while for an even number of time steps the invariants $\eta = +1$ propagate as "right -movers" in the x-direction without flavor change, while $\eta = -1$ denotes "left-movers". It is possible to define a refined continuum limit for right-moving and left-moving particles. For this refined continuum limit the difference to the automaton without a shift is visible. Without the shift both types $\eta = \pm 1$ are stationary, not changing their position for an even number of time steps.

The naive continuum limit employed here averages over the positions in the cell. It makes no difference between $\eta = +1$ and $\eta = -1$. For wave functions whose variation within a single cell can be neglected the probability of an invariant moving to the right or to the left is the same. The average position remains stationary, just as for the automaton without the shift. One expects that some form of dissipation should distinguish the two automata. This information is not kept, however, in the naive continuum limit.

This simple discussion reveals a general property of the continuum limit. It involves some form of "coarse graining" with the associated loss of information. On the one side, this loss of information is crucial to ensure a certain "universality" of the continuum limit. On the other side, not all possible discretizations are expected to belong to the same "universality class".

4 Automaton with interactions

In our setting interactions correspond to terms in the action for the Grassmann functional integral that involve terms with more than two invariants. This will bring us already close to the action (14) of the spinor gravity model where each term involves eight invariants. While it is straightforward to add interaction terms in L, the issue is to maintain the uniqueness of the updating which restricts the admitted properties of $\exp\big(-L\big)$.

Interaction

We use interactions in order to construct an automaton with local Lorentz invariance which leads to propagation of all invariants in the naive continuum limit. For this purpose we divide the eight flavors into four generations, each coming with two types. We use a double index notation, $A = (a, \sigma_a)$, $a = 1 \ldots 4$, $\sigma_a = \pm 1$, where a denotes the generation and σ_a the associated type. Les us consider an automaton for which an interaction occurs if four invariants belonging all to different generations meet in a cube. We modify the updating rule such that for four invariants of different generations the updating still operates a reflection within the cube, but the flavor change is different. For this particular type of configuration the generation is maintained, but the flavor switch is now between types $\sigma_a = \pm 1$. With flavor labels $A = (1, 2, 3, 4, 5, 6, 7, 8) = (1+, 2+, 3+, 4+, 1-, 2-, 3-, 4-)$ the new rule is implemented by an additional term in L

$$-\Delta L = \widetilde{T}_x^{15} \widetilde{T}_y^{26} \widetilde{T}_z^{37} \widetilde{T}_0^{48} - \widetilde{T}_x^{12} \widetilde{T}_y^{34} \widetilde{T}_z^{56} \widetilde{T}_0^{78} \ . \tag{53}$$

Journal of Physics: Conference Series **2533** (2023) 012016 doi:10.1088/1742-6596/2533/1/012016

Both therms in eq. (53) are products of four sums, with 16 terms in each sum. Each term in the total sum of $16^4 = 2^{16}$ terms contains precisely one invariant $\chi_u^{1\pm}$, one invariant $\chi_v^{2\pm}$, one invariant $\chi_w^{3\pm}$ and one invariant $\chi_z^{4\pm}$, such that all four generations are present in the cube, with either type $\sigma_a = +1$ or type $\sigma_a = -1$. The first term in eq. (53) accounts for the simultaneous reflection for all four generations, where generation 1 is reflected in the x-direction, generation 2 in the y-direction, generation 3 in the z-direction, while generation 4 is stationary. Furthermore, the type is switched from 1+ to 1− and vice versa. This is the scattering term.

The second term subtracts the part that would arise in the same sector from $\exp(-L_0)$, with L_0 corresponding to the switch automaton for free particles. Indeed, the cellular automaton property requires that each term in

$$\exp\big(-L\big) = \exp\big(-L_0 - \Delta L\big) \ , \tag{54}$$

which contains a certain product of factors $\chi_u^A(t)$, is multiplied by precisely one product of factors $\chi_u^A(t+\varepsilon)$. The term $L_0^4/24$ in the expansion of $\exp(-L_0)$ produces in the sector with four invariants of different generations at t precisely the second term in eq. (53), with a plus sign. Subtracting it in $-\Delta L$ guarantees that a given factor, say $\chi_1^{1+}(t)\chi_4^{2-}(t)\chi_8^{3-}(t)\chi_5^{4+}(t)$ is precisely multiplied by $\chi_2^{1-}(t+\varepsilon)\chi_7^{2+}(t+\varepsilon)\chi_5^{3+}(t+\varepsilon)\chi_5^{4-}(t+\varepsilon)$, such that the updating is unique.

Uniqueness of updating

The uniqueness of the updating must be maintained for every possible combination of invariants in the cube. Furthermore, the updating map should be invertible and we want to implement particle-hole symmetry. This typically requires further "correction terms" in L, which involve five or more invariants $\chi(t)$, and five or more invariants $\chi(t+\varepsilon)$. For example, on the level of five invariants one has

$$\exp\big[-\big((L_0 + \Delta L + \Delta L')\big] = -\frac{1}{5!}L_0^5 + L_0\Delta L - \Delta L' + \dots \ . \tag{55}$$

We have to specify the updating rule for configurations where five invariants are present in the cube, two of them of the same generation, and the other three of the three different complementary generations. For example, we may postulate that in this situation no scattering occurs, fixing $\Delta L' = L_0\Delta L$. For a different rule in this sector $\Delta L'$ will be different. A simple possible updating rule could involve scattering only if the number of invariants of each generation in a cell is equal. In this case the reflections in the cube are accompanied by a switch of all types σ, keeping the generation unchanged. Specifying the updating rule fixes $\Delta L'$ up to possible signs of the different terms.

In the presence of the interaction term (53) the naive continuum limit of the time-local factor becomes

$$\tilde{\mathcal{K}}(t) = \exp\left\{\sum_{xyz}\Big[8\varepsilon\tilde{T} + (8\varepsilon)^4(\tilde{I}_1 - \tilde{I}_2) + \mathcal{O}(\varepsilon^5)\Big]\right\} , \tag{56}$$

with

$$\begin{aligned}\tilde{I}_1 = &\big(\chi^1\partial_x\chi^5 - \chi^5\partial_x\chi^1\big)\big(\chi^2\partial_y\chi^6 - \chi^6\partial_y\chi^2\big)\\ &\times\big(\chi^3\partial_z\chi^7 - \chi^7\partial_z\chi^3\big)\big(\chi^4\partial_t\chi^8 - \chi^8\partial_t\chi^4\big) \ ,\end{aligned} \tag{57}$$

and

$$\begin{aligned}\tilde{I}_2 = &\big(\chi^1\partial_x\chi^2 - \chi^2\partial_x\chi^1\big)\big(\chi^3\partial_y\chi^4 - \chi^4\partial_y\chi^3\big)\\ &\times\big(\chi^5\partial_z\chi^6 - \chi^6\partial_z\chi^5\big)\big(\chi^7\partial_t\chi^8 - \chi^8\partial_t\chi^7\big) \ .\end{aligned} \tag{58}$$

Journal of Physics: Conference Series **2533** (2023) 012016 doi:10.1088/1742-6596/2533/1/012016

The corresponding term $\sim \tilde{I}_1 - \tilde{I}_2$ in the fermionic action shows an important feature of the spinor gravity action (14), namely the presence of four derivatives for the four different cartesian directions. The flavors are now all mixed. The field equations exhibit a time evolution which defines a type of propagation for all flavors.

The term $\Delta L'$ enters only in the order ε^5 and will be neglected in the naive continuum limit. This highlights the limitations of the naive continuum limit. Different automata with different updating rules for five or more invariants in a cell entail different $\Delta L'$. They have nevertheless the same naive continuum limit. It is not clear a priori which updating rules lead to the universality class corresponding to the naive continuum limit, and which ones not. The universality class is typically characterized by continuous and discrete symmetries. It is not guaranteed that such symmetries are sufficient for a complete characterization of the possible universality classes.

The fermionic action for a simple updating rule may get rather complex. While the terms with a relatively low number of Grassmann variables reflect directly the updating rule and are rather simple, higher order terms as $\Delta L'$ are often needed to maintain the unique jump property of the automaton. In the naive continuum these higher order terms vanish. The true continuum limit has to investigate the renormalization flow of these higher order terms.

5 Alternating updating and coarse graining

Sequences of different updating steps are a powerful construction principle in order to realize symmetries in the naive continuum limit. Many symmetries lead in the naive continuum limit to a situation where a given configuration of occupation numbers or bits can evolve to several different configurations – it has "different options". These options are typically related by symmetry transformations. The presence of different options is not compatible with a unique jump step evolution operator. For an automaton with a sequence of different updating steps each individual step may permit only one of the options. This preserves the automaton property, but violates the symmetry. Constructing the Hamiltonian by the evolution from t to $t + \Delta t$ as in eq. (9) amounts to some effective averaging or coarse graining. For coarse grained automata the unique jump property is typically lost. This permits us to realize the symmetry on a coarse grained level.

As an example, we may take lattice rotations by $\pi/2$ in some of the planes, $x \to y \to -x \to -y \to x$. A given single updating step may not respect the rotation symmetry as is the case for the switch automaton characterized by L_0 or the automaton with interaction specified by eq. (55). Rotating the updating for each step of the sequence by $\pi/2$ produces after four steps a rotation invariant naive continuum limit. In the naive continuum limit the contributions of the four updating steps are simply added without taking into account the ordering in time of the sequence. It remains an open issue if this approximation can be justified by the true continuum limit.

In eq. (56) the term $\tilde{I}_1 - \tilde{I}_2$ shows already some of the structures of spinor gravity. It does not yet show the characteristic contraction of four derivatives with the totally antisymmetric tensor $\varepsilon^{\mu\nu\rho\sigma}$. This contraction guarantees rotation symmetry. It is the basis of diffeomorphism invariance of the naive continuum limit. Furthermore, the local factor (56) contains the term $\sim \tilde{T}$ which is not present for the spinor gravity model. We will introduce alternating updatings in order to ensure rotation invariance of the naive continuum limit and the absence of terms quadratic in the invariants. This highlights further the coarse graining involved for the continuum limit.

Journal of Physics: Conference Series **2533** (2023) 012016 doi:10.1088/1742-6596/2533/1/012016

Reflected updatings

The automaton with interactions defined by the local factor or associated fermionic action (54), with eqs. (31), (53), is invariant under certain lattice reflections and rotations combined with flavor rotations. It is not invariant under lattice reflections or rotations alone. As a consequence, the naive continuum limit (56) is not rotation invariant. Rotation invariance of the naive continuum limit can be achieved if we change the updating rule of a given cell for consecutive steps of the updating. We will first consider a sequence of six steps. For each step the updating is given for individual cells, and we alternate the cell-lattice according to the previous discussion of shifted cells.

The first step is given by the rules (31), (53). In the second step we employ an updating that is obtained by acting with a reflection S_{yz} on the previous one, together with an additional minus sign. This yields

$$\begin{aligned} -L(t+\varepsilon) = &-\widetilde{T}_x^{12} - \widetilde{T}_y^{56} - \widetilde{T}_z^{34} - \widetilde{T}_0^{78} \\ &- \widetilde{T}_x^{15}\widetilde{T}_y^{37}\widetilde{T}_z^{26}\widetilde{T}_0^{48} + \widetilde{T}_x^{12}\widetilde{T}_y^{56}\widetilde{T}_z^{34}\widetilde{T}_0^{78} \ . \end{aligned} \tag{59}$$

The third step obtains from the second step by applying the reflection $-S_{zx}$, and the fourth step obtains from the previous one by applying the reflection $-S_{xy}$. For the fifth step we reflect the fourth step by $-S_{yz}$, and finally the sixth step is generated from the fifth step by the reflection $-S_{zx}$. Note that the first step obtains from the sixth step by $-S_{xy}$.

In the naive continuum limit the precise position of t or $t+\varepsilon$ is no longer important. One finds

$$\begin{aligned} -\big(L_0(t) + L_0(t+\varepsilon)\big) &= \widetilde{T}_y^{34} + \widetilde{T}_z^{56} - \widetilde{T}_z^{34} - \widetilde{T}_y^{56} \ , \\ -\big(L_0(t+2\varepsilon) + L_0(t+3\varepsilon)\big) &= \widetilde{T}_x^{34} + \widetilde{T}_y^{56} - \widetilde{T}_y^{34} - \widetilde{T}_x^{56} \ , \\ -\big(L_0(t+4\varepsilon) + L_0(t+5\varepsilon)\big) &= \widetilde{T}_z^{34} + \widetilde{T}_x^{56} - \widetilde{T}_x^{34} - \widetilde{T}_z^{56} \ . \end{aligned} \tag{60}$$

The sum of all six terms vanishes. What remains are the interaction terms that take the form

$$\begin{aligned} -\sum_{n_s=0}^{5} L(t+n_s\varepsilon) = \varepsilon^{ijk}\Big\{&\widetilde{T}_i^{15}\widetilde{T}_j^{26}\widetilde{T}_k^{37}\widetilde{T}_0^{48} \\ &-\widetilde{T}_i^{12}\widetilde{T}_j^{34}\widetilde{T}_k^{56}\widetilde{T}_0^{78}\Big\} \ , \end{aligned} \tag{61}$$

with $i,j,k = 1\dots3$, and ε^{ijk} the totally antisymmetric Levi-Civita tensor. (Here we employ $i=(1,2,3)=(x,y,z)$, and similar for j and k.) Insertion of eq. (43) yields

$$\begin{aligned} -\sum_{n_s=0}^{5} L(t+n_s\varepsilon) &= (8\varepsilon)^4\varepsilon^{ijk} \\ \times\Big\{&\big(\chi^1\partial_i\chi^5 - \chi^5\partial_i\chi^1\big)\big(\chi^2\partial_j\chi^6 - \chi^6\partial_j\chi^2\big) \\ \times&\big(\chi^3\partial_k\chi^7 - \chi^7\partial_k\chi^3\big)\big(\chi^4\partial_t\chi^8 - \chi^8\partial_t\chi^4\big) \\ -&\big(\chi^1\partial_i\chi^2 - \chi^2\partial_i\chi^1\big)\big(\chi^3\partial_j\chi^4 - \chi^4\partial_j\chi^3\big) \\ \times&\big(\chi^5\partial_k\chi^6 - \chi^6\partial_k\chi^5\big)\big(\chi^7\partial_t\chi^8 - \chi^8\partial_t\chi^7\big)\Big\} \ . \end{aligned} \tag{62}$$

Due to the contraction with the rotation invariant tensor ε^{ijk} the naive continuum limit is invariant under rotations in three-dimensional space, keeping the flavor fixed.

Journal of Physics: Conference Series **2533** (2023) 012016 doi:10.1088/1742-6596/2533/1/012016

Table 4. Updating for sequence of six steps for free particles. The first three columns denote the different flavors, with a flip of flavors at each step. We indicate the reflections within a cell for each step. For the lattice with shifted cells the direction of the reflection indicates the direction of motion in the corresponding positive or negative direction, depending on the starting point. The last column indicates the overall sign of $L_0(t + n_s\varepsilon)$.

	1, 2	3, 4	5, 6	sign
t	P_x	P_y	P_z	$+$
$t+\varepsilon$	P_x	P_z	P_y	$-$
$t+2\varepsilon$	P_z	P_x	P_y	$+$
$t+3\varepsilon$	P_z	P_y	P_x	$-$
$t+4\varepsilon$	P_y	P_z	P_x	$+$
$t+5\varepsilon$	P_y	P_x	P_z	$-$

Diffusion

The vanishing continuum limit for L_0,

$$-\sum_{n_s=0}^{5} L_0(t + n_s\varepsilon) = 0 \ , \tag{63}$$

does not imply that particles do not move in the absence of interactions. A single invariant of flavor 1 or 2, situated at t at position $(x, y, z) = (0, 0, 0)$, will be found at $t + 6\varepsilon$ at the position $(2\varepsilon, 2\varepsilon, 2\varepsilon)$, without changing its flavor. Similarly, an original position $(\varepsilon, 0, 0)$ is moved to $(-\varepsilon, 2\varepsilon, 2\varepsilon)$ at $t + 6\varepsilon$, while $(0, \varepsilon, 0)$ changes to $(2\varepsilon, -\varepsilon, 2\varepsilon)$ and $(0, 0, \varepsilon)$ to $(2\varepsilon, 2\varepsilon, -\varepsilon)$. The remaining four sites move as

$$\begin{aligned}
&(\varepsilon, 0, \varepsilon) \to (-\varepsilon, 2\varepsilon, -\varepsilon) \ , \quad && (\varepsilon, \varepsilon, 0) \to (-\varepsilon, -\varepsilon, 2\varepsilon) \ , \\
&(0, \varepsilon, \varepsilon) \to (2\varepsilon, -\varepsilon, -\varepsilon) \ , \quad && (\varepsilon, \varepsilon, \varepsilon) \to (-\varepsilon, -\varepsilon, -\varepsilon) \ .
\end{aligned}$$

The average position for the eight corners of the cell at t is given by $(\varepsilon/2, \varepsilon/2, \varepsilon/2)$. This average position remains the same at $t + 6\varepsilon$. We conclude that the average position does not move for the sequence of six steps. There is, however, an effective diffusion since each site in the cell at t is moved at $t + 6\varepsilon$ to a different cell. The naive continuum limit does not capture this diffusion.

The situation for the other flavors is similar. We can follow the six updating steps without interactions by the rules summarized in table 4. For the flavors 5 and 6 the positions move from t to $t + 6\varepsilon$ as

$$\begin{aligned}
&(0, 0, 0) \to (-2\varepsilon, -2\varepsilon, 2\varepsilon) \ , \quad && (\varepsilon, 0, 0) \to (3\varepsilon, -2\varepsilon, 2\varepsilon) \ , \\
&(0, \varepsilon, 0) \to (-2\varepsilon, 3\varepsilon, 2\varepsilon) \ , \quad && (0, 0, \varepsilon) \to (-2\varepsilon, -2\varepsilon, -\varepsilon) \ , \\
&(\varepsilon, \varepsilon, 0) \to (3\varepsilon, 3\varepsilon, 2\varepsilon) \ , \quad && (\varepsilon, 0, \varepsilon) \to (3\varepsilon, -2\varepsilon, -\varepsilon) \ , \\
&(0, \varepsilon, \varepsilon) \to (-2\varepsilon, 3\varepsilon, -\varepsilon) \ , \quad && (\varepsilon, \varepsilon, \varepsilon) \to (3\varepsilon, 3\varepsilon, -\varepsilon) \ .
\end{aligned}$$

Again, the average position remains invariant. Correspondingly, the position change for the flavors 3 and 4 obeys

$$\begin{aligned}
&(0,0,0) \to (2\varepsilon, 2\varepsilon, -2\varepsilon) \ , \quad && (\varepsilon,0,0) \to (-\varepsilon, 2\varepsilon, -2\varepsilon) \ , \\
&(0,\varepsilon,0) \to (2\varepsilon, -\varepsilon, -2\varepsilon) \ , \quad && (0,0,\varepsilon) \to (2\varepsilon, 2\varepsilon, 3\varepsilon) \ , \\
&(\varepsilon,\varepsilon,0) \to (-\varepsilon, -\varepsilon, -2\varepsilon) \ , \quad && (\varepsilon,0,\varepsilon) \to (-\varepsilon, 2\varepsilon, 3\varepsilon) \ , \\
&(0,\varepsilon,\varepsilon) \to (2\varepsilon, -\varepsilon, 3\varepsilon) \ , \quad && (\varepsilon,\varepsilon,\varepsilon) \to (-\varepsilon, -\varepsilon, 3\varepsilon) \ ,
\end{aligned}$$

with conserved average position. Finally, the flavors 7 and 8 remain at their positions.

We conclude that the motion of invariants on individual sites is rather complex for the combined updating from t to $t + 6\varepsilon$. Simplicity arises on the averaged level with no motion of the average position. It is this averaged property that is captured by the naive continuum limit with vanishing combined L_0. We recall that we have neglected terms with two or more derivatives acting on at least one invariant. These terms could describe part of the diffusive properties. We will discuss the justification of their neglection later.

What remains in the continuum limit are the interaction terms (62). They have no longer the unique jump property of an automaton. Indeed, if we leave the information on which particular site the invariants are located at t, and only keep the information about the number of invariants of given flavors in a cell, the average updating is no longer unique. For configurations with a fixed number of invariants with given flavors in a cell there remain several different possibilities or options for their evolution to different flavors and different cells. A coarse graining from the sites of a cell to the whole cell looses information. On the coarse grained level the automaton property is no longer present. This is a characteristic feature of any continuum limit.

The naive continuum limit can also be seen as a coarse graining over six time steps. In this coarse graining the order of the individual time steps is no longer resolved. Without the information on the precise order of the six time steps the automaton property is lost. Different orders of the updating steps lead to different possibilities for the detailed positions of the various invariants at $t + 6\varepsilon$. This feature is closely related to the fact that the different updating steps do not commute. This is directly visible in the different updating rules for different flavors on the level of invariants at resolved sites. The different updatings correspond to different orders of the individual steps. On the level of the invariant average position the order of the steps no longer matters. The sum of terms in the naive continuum limits (61), (62), (63) does not retain any information about the order of the steps.

Flavor rotations and reflections

Instead of the coordinate reflections defining the six steps we can achieve the same automaton by rotations in flavor space. The propagation term L_0 is invariant under a combination of the reflection S_{xy} and an exchange of the flavor pairs $(1, 2)$ and $(3, 4)$. The latter can be realized by the rotation in flavor space $F_{12}F_{24}$, where we define

$$F_{AB}: \ \chi^A \to \chi^B \to -\chi^A \to -\chi^B \to \chi^A \ . \tag{64}$$

Combining appropriate rotations, as $F_{13}F_{42}$ or $F_{14}F_{23}$, with flavor reflections $G_{12}G_{78}$, one can realize the additional minus sign in our choice of the updating steps. Here we define the flavor reflection as

$$G_{AB}: \ \chi^A \leftrightarrow \chi^B \ . \tag{65}$$

In other words, a change of the updating rule by the reflection S_{xy} with an additional minus sign can equivalently be achieved by an appropriate rotation in flavor space. (We use here rotations for generalized rotations including reflections.) Similarly, one can replace the transformation $-S_{yz}$ by the flavor rotations $F_{35}F_{64}$ or $F_{36}F_{45}$, combined with $G_{12}G_{78}$, and $-S_{zx}$ by $F_{15}F_{62}$

or $F_{25}F_{16}$, combined with $G_{34}G_{78}$. Thus the sequence of the different choices of L_0 for the six updating steps can equivalently be realized by relating the different steps by appropriate flavor rotations.

For the interaction part, i.e. the first term in eq. (53), we choose a different sequence of flavor rotations from one step to the next. For example, the transformation $-S_{xy}$ is now equivalently realized by the flavor rotation $F_{12}F_{65}G_{37}$. In short, we can realize the different choices of L_0 and ΔL for the six different steps by appropriate rotations in flavor space.

Symmetric continuum limit

At the present stage the naive continuum limit (62) remains still rather asymmetric in flavor space. For example, the invariant χ^8 is present only in the combinations with a time derivative. Also time and space appear in different roles. We can realize a higher degree of symmetry in space-time and in flavor space by continuing the six steps by another eighteen steps obtained by appropriate flavor rotations. The total sequence will then have 24 steps. After these 24 steps the sequence of updatings is repeated.

For L_0 step seven from $t+6\varepsilon$ to $t+7\varepsilon$ is obtained from step one by the flavor rotation $F_{17}F_{82}$ or $F_{18}F_{27}$. This exchanges the flavor pairs $(1,2)$ and $(7,8)$ with an additional minus sign, and therefore maps in eq. (31) $\tilde{T}^{12}_x+\tilde{T}^{78}_0 \to -(\tilde{T}^{78}_0+\tilde{T}^{12}_x)$. We combine this with the flavor reflection $G_{34}G_{56}$ which changes the sign of $\tilde{T}^{34}_y+\tilde{T}^{56}_z$. In the naive continuum limit (62) this subtracts a term for which ∂_t and ∂_x are exchanged. For the interaction term the map from step one to step seven involves a different flavor rotation $F_{14}F_{85}$ or $F_{15}F_{48}$, combined with a reflection G_{26} or G_{37}. Combined with the part from L_0 this subtracts from ΔL in eq. (53) a term for which ∂_t and ∂_x are exchanged. Taken together, the continuum limit of $L=L_0+\Delta L$ is antisymmetric in the exchange $t\leftrightarrow x$. The following five steps obtain from step seven by the same sequence of reflections (or equivalent flavor rotations) $-S_{yz}$, $-S_{zx}$, $-S_{xy}$, $-S_{yz}$, $-S_{zx}$. Summing up the contributions of the first twelve steps subtracts from eq. (62) a term for which ∂_t and ∂_x are exchanged. The expression $\sum_{n_s=0}^{11} L(t+n_s\varepsilon)$ is antisymmetric in ∂_t and ∂_x.

We repeat this procedure for the next six steps, performing now flavor rotations such that a further term is subtracted from L for which ∂_t and ∂_y are exchanged. Finally, the last six steps involve flavor rotations such that a term with ∂_t and ∂_z exchanged is subtracted. In summary, the naive continuum limit for the sequence of twenty four steps reads

$$\bar{L}(t)=\sum_{n_s=0}^{23} L(t+n_s\varepsilon)=(8\varepsilon)^4\varepsilon^{\mu\nu\rho\sigma}A_{\mu\nu\rho\sigma}\ , \tag{66}$$

with

$$\begin{aligned}A_{\mu\nu\rho\sigma}=\Big\{&(\chi^1\partial_\mu\chi^5-\chi^5\partial_\mu\chi^1)(\chi^2\partial_\nu\chi^6-\chi^6\partial_\nu\chi^2)\\ &\times(\chi^3\partial_\rho\chi^7-\chi^7\partial_\rho\chi^3)(\chi^4\partial_\sigma\chi^8-\chi^8\partial_\sigma\chi^4)\\ &-(\chi^1\partial_\mu\chi^2-\chi^2\partial_\mu\chi^1)(\chi^3\partial_\nu\chi^4-\chi^4\partial_\nu\chi^3)\\ &\times(\chi^5\partial_\rho\chi^6-\chi^6\partial_\rho\chi^5)(\chi^7\partial_\sigma\chi^8-\chi^8\partial_\sigma\chi^7)\Big\}\ .\end{aligned} \tag{67}$$

Here $\mu,\nu,\rho,\sigma=0,\dots 3$, $\partial_0=\partial_t$, and $\varepsilon^{\mu\nu\rho\sigma}$ is the four-index totally antisymmetric tensor, with $\varepsilon^{0123}=-\varepsilon^{1230}=1$. One recognizes the structure of the action (14).

The final step of the continuum limit replaces for the action $S=\sum_{n_s=0}^{M_t}\sum_{x,y,z}L(t+$

$n_s\varepsilon, x, y, z)$ the sums by integrals, $\int_x = \int \mathrm{d}t\,\mathrm{d}x\,\mathrm{d}y\,\mathrm{d}z$,

$$S = \sum_w L(w) = \frac{1}{8\cdot 24}\int_x \varepsilon^{-4}\bar{L}(t,x,y,z)$$
$$= \frac{64}{3}\int_x \varepsilon^{\mu\nu\rho\sigma}A_{\mu\nu\rho\sigma}\ . \tag{68}$$

This employs a lattice discretization of spacetime by the shifted lattice, $\int_x = 8\varepsilon^4\sum_w$, and a factor 1/24 for the 24 steps. The naive continuum limit yields indeed the diffeomorphism invariant action (14). Eq. (67) fixes the flavor coefficient G in eq. (14). At this stage we have realized the construction of a probabilistic cellular automaton whose naive continuum limit coincides precisely with the action of spinor gravity in sect. 2.

Higher derivatives

By construction, the naive continuum limit for the sequence of updatings realizes several discrete symmetries, as lattice rotations by $\pi/2$ and several reflections, completed by flavor rotations and reflections. These symmetries restrict the terms that can appear in the fermionic action in the naive continuum limit. In particular, it forbids terms with more than one derivative acting on an invariant χ^A that have been neglected so far. By virtue of these restrictions the naive continuum limit realizes symmetries beyond the discrete ones. This concerns continuous rotations in space and even diffeomorphism symmetry.

So far we have neglected possible terms involving two or more derivatives acting on a given invariant. They could arise from the omitted terms in eqs. (35), (36). In principle, such terms contribute to the naive continuum limit. Each derivative comes with a factor ε, such that for the naive continuum limit all terms with five or more derivatives vanish. We still need to discuss possible terms with up to four derivatives, for which at least two act on the same invariant. Since the interaction term $\Delta_L + L_0^4/24$ involves already four derivatives the possibly problematic terms can only arise from L_0. By virtue of discrete symmetries we will establish that no such terms appear in the naive continuum limit, thus confirming eq. (67).

For the naive continuum limit of L_0 we can consider different flavor pairs separately. Consider first the flavors 1 and 2. The part $L_0^{(1,2)}$ is odd under P_x and even under P_y, P_z, T. Furthermore, it is even under S_{yz} and odd under the flavor exchange $1 \leftrightarrow 2$. It contains two factors of χ such that possible higher derivative terms beyond eq. (43) could be of the form $\partial_y\chi^1\partial_y\partial_x\chi^2 - \partial_y\chi^2\partial_y\partial_x\chi^1$, and similar with ∂_y replaced by ∂_z or ∂_t. (We identify terms related by partial integration.) Another possibility is $\partial_x\chi^1\partial_x^2\chi^2 - \partial_x\chi^2\partial_x^2\chi^1$. Further terms involve five or more derivatives and vanish in the naive continuum limit.

The sum of the first six terms in the sequence of updatings is by construction antisymmetric in $x \leftrightarrow y$, $x \leftrightarrow z$, $y \leftrightarrow z$. Adding the contributions of the different steps leaves a possible term

$$L_{0,HD}^{(1,2)} \sim (\partial_z^2 - \partial_y^2)\chi^1\partial_x\chi^2 + (\partial_x^2 - \partial_z^2)\chi^1\partial_y\chi^2$$
$$+ (\partial_y^2 - \partial_x^2)\chi^1\partial_z\chi^2 - (1 \leftrightarrow 2)\ , \tag{69}$$

which is indeed odd under S_{yz}. On the other hand, $S_{yz}T_x = T_x$ implies that the higher derivative terms are even under S_{yz} and we conclude $L_{0,HD}^{(1,2)} = 0$. This holds similarly for $L_{0,HD}^{(3,4)}$ and $L_{0,HD}^{(5,6)}$.

For the flavor sector $(7,8)$ $L_0^{(7,8)}$ vanishes trivially since there are three terms with positive sign and three identical terms with negative sign. Finally, the other eighteen terms in the sequence have again a vanishing naive continuum limit of L_0 since they obtain from the first six terms by exchange of flavor pairs. In summary, for each individual term in the sequence of

Journal of Physics: Conference Series **2533** (2023) 012016 doi:10.1088/1742-6596/2533/1/012016

updatings higher derivative terms are expected for the naive continuum limit of L_0. It is the particular sequence of terms that implies a cancellation of the higher derivative terms in L_0 though the factor ε^{ijk} in eq. (61).

Signs of Grassmann basis elements and wave functions

The choice of the sequence of updatings involves particular minus signs which matter for the naive continuum limit. Replacing them by other signs would lead to a different naive continuum limit. One may wonder if this poses a problem for the meaning and the robustness of the naive continuum limit. We also admit arbitrary signs for the wave function of the cellular automaton, corresponding to arbitrary signs of the wave function for the fermionic quantum field theory. These signs do not matter for the probabilities $p_\tau = q_\tau^2$, but they play a role for other observables as the momentum observable which are represented by off-diagonal operators [2]. In our formulation the choice of signs is related to a discrete local gauge symmetry which we briefly recapitulate here.

The choice of signs for the different updating steps is related to a particular choice of relative signs for the Grassmann basis elements which are used to implement the updating rule for the automaton in the fermionic formulation. In general, the choice of signs for the Grassmann basis elements is arbitrary. For the general bit-fermion map the association of bit configurations and Grassmann basis elements is only fixed up to a sign. A similar observation can be made for the relation between the real wave function and the probability distribution. The relation $p_\tau = q_\tau^2$ fixes q_τ only up to a sign.

For our construction a change of sign of a Grassmann basis element can be associated to a change of sign of an element of the step evolution operator according to eq. (21), or to a change of sign of a component of the wave function. The independence of physics from the choice of signs is reflected by a local symmetry of discrete Z_2-transformations. [3, 27, 28]. Our particular choice of signs can be viewed as a particular choice of gauge.

Making a gauge transformation to a different choice of signs is typically not compatible with a smooth continuum limit. Starting from a smooth wave function encoding the probabilistic information about the state of the automaton, and applying a local change of sign, typically leads to a discontinuity of the wave function or its derivatives. Our choice of signs leading to the naive continuum limit should be seen as a gauge choice which permits a description in terms of a smooth wave function. For the continuum limit it is sufficient that one choice of signs exists for which the wave function is sufficiently smooth.

The local Z_2-gauge transformations corresponding to sign-changes of the elements $q_\tau(t)$ act simultaneously on Grassmann basis elements, wave functions, step evolution operator and operators for observables. For a given choice of Grassmann basis functions and step evolution operator the gauge is fixed. Changing for this choice of gauge only the signs in the wave function leads to a different physical situation. It can be shown [2] that arbitrary signs of the wave function lead to a positive overall probability distribution for all times.

Coarse graining and continuum limit

Automata with a sequence of rotated or reflected updatings demonstrate important coarse graining properties of the continuum limit. For the naive continuum limit the order of the updatings plays no role. This corresponds to a coarse graining in time. The 24 time steps of a cycle are no longer resolved in their order - only an average over these time steps is considered. On the discrete level the order of the updating steps matters since the step evolution operators for the different steps do not commute. The association of the naive continuum limit with the true continuum limit therefore relies on the basic assumption that for sufficiently smooth wave functions this non-commutativity, which is formally of higher order in ε, does not play a role. This is similar to the situation for the Feynman path integral formulation of quantum mechanics. In the continuum limit the non-commutativity of the kinetic and potential parts in the Hamiltonian is neglected there.

The naive continuum limit has a higher degree of symmetry as compared to the discrete formulation. In particular, it implements continuous rotation symmetry while only a discrete subgroup is respected by the lattice formulation. This situation is well known from lattice gauge theories or similar discrete formulations of continuum models. The question is if the renormalization flow of operators violating the enhanced symmetries drives them to zero in the infrared limit or not.

With a given set of symmetries one expects only one or a few universality classes of the renormalization flow. The microscopic information beyond a few relevant parameters characterizing these universality classes is lost in the infrared limit. This induces the high predictive power of universal critical behavior. For the particular discretization of the continuous spinor gravity model that we have constructed in the present paper one may hope that it belongs indeed to the universality class with diffeomorphism symmetry. A dedicated renormalization group investigation of the discrete model will be required in order to settle this central question.

6 Discussion

We have constructed a probabilistic cellular automaton which is equivalent to a discrete quantum field theory for fermions. Its naive continuum limit is a model of spinor gravity in four dimensions. Quantum gravity with fermions is characterized by two important properties: the symmetries of diffeomorphisms (general coordinate transformations) and local Lorentz symmetry. While our automaton realizes invariance under local Lorentz transformations directly on the discrete level, the diffeomorphism symmetry shows up only in the naive continuum limit.

The updating rule for the automaton appears to be rather complicated if we combine 24 "microscopic updatings" to a single step that is repeated. In a different view the microscopic steps are very simple, and the rule for a sequence of steps related by flavor rotations is also rather simple. It would not be difficult to implement this automaton numerically for sharp initial bit configurations. Realizing the evolution of a probability distribution over initial configurations will require more work due to the large number of possible initial configurations. We may consider our model as an example for an interesting discrete four-dimensional fermionic quantum field theory with interactions that can be mapped to a probabilistic cellular automaton.

A key question concerns the true continuum limit of this model. The naive continuum limit is suggestive by neglecting terms $O(\varepsilon)$ in the continuum action. Qualitatively, this can be justified if the wave function changes only very little for time steps of the order of a few times ε or for similar distances in space. In this respect it is not sufficient to start with a smooth initial wave function. The smoothness has to be preserved by the updating process. The conditions for this to happen are poorly known and we cannot assert at the present stage under which conditions the naive continuum limit constitutes a qualitatively correct guide.

For any continuum limit the probabilistic nature of the automaton or fermionic quantum field theory is crucial since only the continuity in the probabilistic information (wave function) permits to take $\varepsilon \to 0$. No continuum limit exists for the sharp wave function of a deterministic automaton. We have seen that taking the continuum limit corresponds to some type of coarse graining. On the coarse grained level the unique jump property of the automaton is lost. If individual time steps and their order can no longer be resolved, or if the information about the precise position in space of individual bits is no longer available, a given smoothed or averaged configuration of bits or fermions can develop into several distinct such configurations in a probabilistic way. We can still define an evolution operator on the coarse grained level. If no information is lost in the course of the time evolution on the coarse grained level, the evolution operator will be an orthogonal matrix in the real formulation, and unitary in the presence of a complex structure which is compatible with the evolution. Rows and columns do, however, contain now many non-zero elements.

The feature of coarse graining is shared by the naive and the true continuum limit. The true continuum takes the effects of fluctuations into account. This leads to a dependence of effective couplings on the scale of the coarse graining (running couplings). It is therefore not guaranteed that the neglection of couplings which are proportional to powers of ε on the microscopic level remains justified. The emergence of a continuum version of spinor gravity in the time continuum limit requires that terms violating rotation symmetry die out, similar to the continuum limit in lattice gauge theories. Furthermore, all terms violating diffeomorphism invariance of the coarse grained effective action have to die out as fluctuations on larger and larger scales are included consecutively. In the renormalization group language such symmetry violating couplings have to correspond to irrelevant parameters.

Even if our model could realize diffeomorphism invariance in the true continuum limit there remains still a long way to go before it can be accepted as a model of quantum gravity. It is necessary that composite vierbeins (15) and a composite metric, as $g_{\mu\nu} = \sum_A e_\mu{}^{Am} e_\nu{}^{An} \eta_{mn}$ or some other flavor structure, acquire a vacuum expectation value which defines a geometry. In the presence of such expectation values single fermions should propagate according to the Dirac equation in flat space, with a convenient generalization involving a composite spin connection in curved space. If this path toward gravity would reveal itself as successful, one could attempt to construct realistic grand unified models for particle physics coupled to quantum gravity. A generalization of the invariants χ^A to become also invariant under local SO(10)-gauge transformations with fermions in the fundamental spinor representation is straightforward. At present, we may only state that it seems not to be excluded that the Universe can be described by a probabilistic automaton.

References

[1] Wetterich C 2017 *Nucl. Phys. B* **917** 241–271 (*Preprint* 1612.06695)

[2] Wetterich C 2020 (*Preprint* 2011.02867)

[3] Wetterich C 2022 (*Preprint* 2203.14081)

[4] von Neumann J 1951 *The general and logical theory of automata.* (Oxford, England: Wiley) pp 1–41 Cerebral mechanisms in behavior; the Hixon Symposium.

[5] Ulam S 1950 Random processes and transformations *Proceedings of the International Congress on Mathematics* vol 2 pp 264–275

[6] Zuse K 1969 (Vieweg, Teubner Verlag)

[7] Hardy J, de Pazzis O and Pomeau Y 1976 *Phys. Rev. A* **13**(5) 1949–1961

[8] Toom A L 1978 (Springer Berlin Heidelberg)

[9] R L Dobrushin VI Kryukov A L T 1978 (Manchester University Press)

[10] Vichniac G Y 1984 *Physica D: Nonlinear Phenomena* **10** 96–116

[11] LOUIS P Y and Nardi F R 2018 *Probabilistic Cellular Automata* (Springer)

[12] Wolfram S 1983 *Rev. Mod. Phys.* **55**(3) 601–644

[13] Creutz M 1986 *Annals of Physics* **167** 62–72

[14] Toffoli T and Margolus N H 1990 *Physica D: Nonlinear Phenomena* **45** 229–253 ISSN 0167-2789

[15] Lindenmayer A and Rozenberg G 1976 Automata, languages, development (North Holland)

[16] Hedlund G A 1969 *Mathematical systems theory* **3** 320–375 ISSN 1433-0490

[17] Gardner M 1970 *Scientific American* **223** 120–123

[18] Preston K and Duff M J B 1984 (Springer US)
[19] Richardson D 1972 *Journal of Computer and System Sciences* **6** 373–388 ISSN 0022-0000
[20] Amoroso S and Patt Y 1972 *Journal of Computer and System Sciences* **6** 448–464 ISSN 0022-0000
[21] 't Hooft G 2014 (*Preprint* 1405.1548)
[22] Elze H T 2014 *EPJ Web Conf.* **78** 02005 (*Preprint* 1407.2160)
[23] 't Hooft G 2010 *Int. J. Mod. Phys. A* **25** 4385–4396
[24] 't Hooft G 2021 *Found. Phys.* **51** 63 (*Preprint* 2010.02019)
[25] Hooft G t 2021 (*Preprint* 2103.04335)
[26] Wetterich C 2010 *Annalen Phys.* **522** 807 (*Preprint* 0904.3048)
[27] Wetterich C 2021 *Nuclear Physics B* **963** 115296 ISSN 0550-3213 (*Preprint* 2007.06366)
[28] Wetterich C 2021 Fermionic quantum field theories as probabilistic cellular automata (*Preprint* 2111.06728)
[29] Wetterich C 1983 *Nuclear Physics B* **211** 177–188 ISSN 0550-3213
[30] Wetterich C 2011 *Nucl. Phys. B* **852** 174–234 (*Preprint* 1002.3556)
[31] Thirring W E 1958 *Annals of Physics* **3** 91 – 112 ISSN 0003-4916
[32] Klaiber B 1968 *Lect. Theor. Phys. A* **10** 141–176
[33] Abdalla E, Abdalla M and Rothe K 1991 *Nonperturbative methods in two-dimensional quantum field theory*
[34] Faber M and Ivanov A 2001 (*Preprint* hep-th/0112183)
[35] Gross D J and Neveu A 1974 *Phys. Rev. D* **10**(10) 3235–3253
[36] Wetzel W 1984 *Phys. Lett. B* **153** 297–299. 10 p
[37] Shankar R 1985 *Phys. Rev. Lett.* **55**(5) 453–456
[38] Rosenstein B, Warr B J and Park S H 1991 *Physics Reports* **205** 59–108
[39] Stoll J, Zorbach N, Koenigstein A, Steil M J and Rechenberger S 2021 Bosonic fluctuations in the $(1+1)$-dimensional gross-neveu(-yukawa) model at varying μ and t and finite n (*Preprint* 2108.10616)
[40] Wetterich C 2011 *Physics Letters B* **704** 612–619 ISSN 0370-2693 (*Preprint* 1108.1313)
[41] Wetterich C 2013 *Lecture Notes in Physics* 67–92 ISSN 1616-6361 (*Preprint* 1201.2871)
[42] Hebecker A and Wetterich C 2003 *Physics Letters B* **574** 269–275 ISSN 0370-2693
[43] Wetterich C 2004 *Physical Review D* **70** ISSN 1550-2368
[44] Wetterich C 2011 *Annals Phys.* **326** 2243–2293 (*Preprint* 1102.3586)
[45] Wetterich C 2011 (*Preprint* 1111.4115)
[46] Plechko V N 2005 *Phys. Part. Nucl.* **36** S203–S208 (*Preprint* hep-th/0512263)
[47] Berezin F 1966 *English translation Academic Press, New York*
[48] Berezin F A 1969 *Russian Mathematical Surveys* **24** 1–22
[49] Samuel S 1980 *J. Math. Phys.* **21** 2806–2814
[50] Itzykson C 1982 *Nuclear Physics B* **210** 448–476 ISSN 0550-3213
[51] Plechko V N 1985 *Theoretical and Mathematical Physics* **64** 748–756
[52] Furuya K, Saraví R and Schaposnik F 1982 *Nuclear Physics B* **208** 159 – 181 ISSN 0550-3213
[53] Naón C M 1985 *Phys. Rev. D* **31**(8) 2035–2044

[54] Coleman S 1975 *Phys. Rev. D* **11**(8) 2088–2097

[55] Damgaard P, Nielsen H and Sollacher R 1992 *Nuclear Physics B* **385** 227 – 250 ISSN 0550-3213

[56] Wetterich C 2018 *Nucl. Phys. B* **927** 35–96 (*Preprint* 1611.04820)

[57] Wetterich C 2018 *Annals Phys.* **393** 1–70 (*Preprint* 1706.01772)

Journal of Physics: Conference Series **2533** (2023) 012017 doi:10.1088/1742-6596/2533/1/012017

Anomalies, the Dark Universe and Matter-Antimatter asymmetry

Nick E. Mavromatos[1,2]

[1] Department of Physics, School of Applied Mathematical and Physical Sciences, National Technical University of Athens, Zografou Campus, Athens, 15780, Greece
[2] Theoretical Particle Physics and Cosmology Group, Department of Physics, King's College London, Strand, London, WC2R 2LS.

E-mail: mavroman@mail.ntua.gr

Abstract. I review a (3+1)-dimensional, string-inspired cosmological model with gravitational anomalies (of Chern-Simons (CS) type) at early epochs, and a totally-antisymmetric torsion, dual to a massless axion-like field ("gravitational axion"), which couples to the CS term. Under appropriate conditions, primordial gravitational waves can condense, leading to a condensate of the CS anomaly term. As a consequence, one obtains inflation in this theory, of running-vacuum-model (RVM) type, without the need for external inflatons. At the end of the inflationary era, chiral fermionic matter is generated, whose gravitational anomalies cancel the primordial ones. On the other hand, chiral anomalies of gauge type, which are also generated by the chiral matter, remain present during the post-inflationary epochs and become responsible for the generation of a non-perturbative mass for the torsion-related gravitational axion, which, in this way, might play the rôle of a Dark Matter component of geometrical origin. Moreover, in this model, stringy non-perturbative effects during the RVM inflationary phase generate periodic structures for the potential of axion-like particles that arise due to compactification, and co-exist with the gravitational axions. Such periodic potential modulations may lead to an enhanced production of primordial black holes during inflation, which in turn affects the profile of the generated gravitational waves during the radiation era, with potentially observable consequences. This model also entails an unconventional mechanism for Leptogenesis, due to Lorentz-violating backgrounds of the gravitational axions that are generated during inflation, as a consequence of the anomaly condensates, and remain undiluted in the radiation era.

1. Introduction

One of the currently unsolved greatest mysteries in fundamental physics is the nature of the dark sector of the Universe, which according to fits to the plethora of the available cosmological data [1] corresponds to a current-epoch energy budget for the Universe, in which $\sim 70\%$ is dark energy and about $\sim 26\%$ Dark Matter (DM), and only a $\sim 5\%$ corresponds to the known matter (mostly baryonic). The fits to the data have been made within the framework of a de-Sitter spacetime, with a a positive cosmological constant $\Lambda > 0$, in which there is a Cold-Dark-Matter (CDM) component, the so-called ΛCDM framework.

Although at large scales, such ΛCDM-based fits prove to be quite successful, nonetheless there are important issues concerning the microscopic foundations of ΛCDM, most importantly its compatibility with UltraViolet (UV) complete models of quantum gravity (QG), which low-energy cosmologies could be embedded to [2, 3, 4, 5], but also recently its small-scale

Journal of Physics: Conference Series **2533** (2023) 012017 doi:10.1088/1742-6596/2533/1/012017

phenomenology, in particular tensions in the data [6], concerning a discrepancy between the value of the current-era Hubble parameter H_0 inferred by direct local measurements of nearby galaxies and that based on ΛCDM fits (H_0 tension), but also tensions concerning the growth of structures in the Universe (σ_8 tension). Such tensions, although still susceptible to more mundane explanations, *e.g.* they could be due to statistical and other astrophysical uncertainties [7], nonetheless, due to their persisting nature, they prompted scientists to look for alternative to ΛCDM models as potential explanations [8].

An interesting framework that provides a potential alleviation of both tensions (H_0 and σ_8 [9]), but also leads to a smooth description of the evolution of the Universe from a dynamically-induced inflationary era, without the need for external inflatons, to the present era [10, 11, 12], including an explanation of the Universe thermal history [13, 14, 15], is the so-called Running Vacuum Model (RVM) [16, 17, 18, 19]. According to the latter, the equation of state (EoS) of the ground state of the evolving Universe is that of a de Sitter spacetime,

$$\rho_{\rm vac} = -p_{\rm vac}\,, \tag{1}$$

where $\rho_{\rm vac}$ ($p_{\rm vac}$) is the vacuum energy density (pressure), but there is no cosmological constant. Instead the energy density, due to reasons of general covariance, is a function of (even) powers of the Hubble parameter $H(t)$, which evolves with the cosmic time t:

$$\rho_{\rm vac}(H(t)) = \frac{3}{\kappa^2}\Big(c_0 + \nu\, H(t)^2 + \alpha_I\,\frac{H(t)^4}{H_I} + \dots\Big)\,, \tag{2}$$

with $\kappa = M_{\rm Pl}^{-1}$ the gravitational constant in (3+1)-dimensions, and $M_{\rm Pl} = 2.435 \times 10^{18}$ GeV the reduced Planck mass scale. The quantities c_0, ν, α_I are dimensionless constants, and $H_I \sim 10^{-5}\,M_{\rm Pl}$ is the inflationary scale, as measured by recent data [1]. The equation (2) was initially conjectured in [16, 17, 18], based on a renormalization-group (RG) interpretation of the cosmic evolution, in which ln(H) plays the rôle of a RG scale, and $d\rho/{\rm ln}H$ is the analogue of a RG β function. From this point of view, the constant c_0 is an arbitrary constant, which would play the rôle of a cosmological constant in the current era. It worths noticing that from a phenomenological point of view, the entire history of the Universe can be described by the terms exhibited explicitly on the right-hand side of (2). At early epochs, the non-linear H^4 terms are dominant and lead to dynamical inflation without the need for external inflatons [10, 11].

The RVM vacuum expansion (2), together with the EoS (1) have been explicitly verified within the context of a string-inspired cosmological model with gravitational anomalies and torsion at early epochs, proposed in [20, 21, 22, 23], which will be the topic of discussion in the present review. In particular, primordial gravitational waves (GW), which could themselves have been produced in a pre-inflationary era due to various mechanisms suggested in [22], can lead to condensates of gravitational anomaly terms the model possesses, which in turn lead to terms H^4 in the cosmological vacuum energy density of RVM type (2). It is important to note that in the string context, there is no cosmological constant c_0 term, though, in agreement with the swampland criteria [2, 3, 4, 5], but also perturbative scattering-matrix arguments [24, 25], which imply that a non-zero c_0 constant term would be in conflict with the UV completion (UV) of the string-inspired cosmology by the underlying microscopic string theory, or more general (a yet to be determined) quantum-gravity (QG) theory. In the model of [20, 21, 22, 23] during inflation $c_0 = 0$, while it has been argued [20] that in modern eras a *metastable* de Sitter spacetime may arise, e.g. as a result of some condensate of cosmic electromagnetic fields.

In more recent works [26, 27, 28], such an expansion, but with the H^4 terms having been replaced by H^6 (and higher order) terms, has been derived in the context of a novel renormalisation approach to the stress tensor of quantum scalar fields in curved space times. In such an approach, in agreement with the aforementioned incompatibility of string theory with de Sitter spacetimes, the constant c_0 is not a constant, but a slowly evolving function of cosmic time, which arises as a result of the specific techniques used in the absorption of UV infinities in the expression for the stress-energy tensor of the theory. Moreover, in this framework, as a consequence of the back reaction of the matter fields onto the background spacetime geometry, which is assumed to be a Robertson-Walker expanding Universe spacespacetimetime, the equation of state of the system is different from (1) [28].

Nonetheless, integrating out graviton fields in a path integral, which is essential in any consistent treatment of deriving cosmologies from string or any other UV complete QG theory, will result [29, 30, 31] in the appearance of non-polynomial terms, $H^m \ln(H)$, $m = 2, 4, \ldots$, in (2). Such terms, which are due to QG corrections, may contribute to the current-era phenomenology of such models, in particular to the alleviation of the cosmological-data tensions. A detailed analysis and comparison with the results of [9] within the traditional RVM framework, is in progress [32]. Such terms will not affect qualitatively or quantitatively the arguments of [20, 21, 22, 23] on the anomaly-condensate-induced RVM inflation. We also note that $H^2 \ln(H)$ structures also characterise the (renormalised) quantum field theories in curved space time of [26, 27, 28], but there are some differences from the QG models, which may be phenomenologically distinguishable [32].

The structure of the talk is the following: in the next section 2, we discuss the potentially important rôle of (gravitational) anomalies and torsion on inducing RVM-type inflation in the string-inspired cosmological model, without the need for external inflatons. In section 3, we discuss the rôle of periodic modulations of the potential of axion fields arising from compactification in the underlying microscopic string theory model on inducing an enhanced density of primordial black holes during inflation, which in turn will affect the profile of GW during the radiation era [33], with in principle observable consequences in future interferometers like LISA [34, 35]. In section 4, we explain how, in the above anomalous class of string-inspired models, one may generate a matter-antimatter asymmetry in the Universe as a consequence of torsion-induced axion backgrounds which violate spontaneously Lorentz and CPT symmetries. Finally, section 5 contains our conclusions and outlook.

2. Anomalies and Running-Vacuum-Model-type Inflation

As discussed in detail in [20, 21, 22, 23], the basic assumption in the model is that at early epochs, only fields that exist in the massless bosonic gravitational multiplet of the string appear in the effective action, as external (thus classical) fields. Chiral fermionic matter, as well as all the other (non-gravitational) fields of the theory, are generated at the end of the RVM inflation, as a result of the decay of the RVM vacuum.

In string theory [36], the massless gravitational multiplet consists of (spin-zero, scalar) dilatons, $\Phi(x)$, (spin-two) gravitons, $g_{\mu\nu}(x)$, and (spin-one) Kalb-Ramond (KR) antisymmetric tensor fields, $B_{\mu\nu}(x) = -B_{\nu\mu}(x)$. Due to a U(1) gauge symmetry of the closed string sector, $B_{\mu\nu} \to B_{\mu\nu} + \partial_\mu\theta_\nu - \partial_\nu\theta_\mu$, the effective low-energy target-spacetime action of the massless gravitational mutliplet depends only on the field strength of the KR field, $H_{\mu\nu\rho} = \partial_{[\mu}B_{\nu\rho]}$, where $[\ldots]$ denotes full antisymmetrisation of the respective indices. After string compactification to (3+1)-dimensions, the effective action reads:[1]

$$S = \int d^4x \Big(-\frac{1}{2\kappa^2}\, R - \frac{1}{6\,\kappa^2} e^{-4\Phi}\, H_{\mu\nu\rho}\, H^{\mu\nu\rho} + 2\partial_\mu\Phi\, \partial^\mu\Phi + \ldots \Big), \tag{3}$$

where $\ldots$ denote higher-derivative terms, which are subleading in a low-energy expansion (in terms of the effective (3+1)-dimensional reduced Planck energy scale, $M_{\rm Pl} = \kappa^{-1}$ (in units

[1] We follow the conventions $(+,-,-,-)$ for the metric signature, and the Riemann curvature tensor definition: $R^\lambda{}_{\mu\nu\sigma} = \partial_\nu\, \Gamma^\lambda{}_{\mu\sigma} + \Gamma^\rho{}_{\mu\sigma}\, \Gamma^\lambda{}_{\rho\nu} - (\nu \leftrightarrow \sigma)$, with Ricci tensor $R_{\mu\nu} = R^\lambda{}_{\mu\lambda\nu}$, and Ricci scalar $R = R_{\mu\nu}g^{\mu\nu}$.

$\hbar = c = 1$ we work throughout)). As already mentioned, according to the basic assumption of [20, 21, 22, 23], the action (3) characterises the early epoch of the string Universe. The H^2 terms in (3) can be absorbed in a generalised curvature scheme, $\overline{R}(\overline{\Gamma})$, corresponding to a geometry with (totally antisymmetric) torsion $H_{\mu\nu\rho}$, that is corresponding to a generalised spin connection:

$$\overline{\Gamma}^{\mu}{}_{\nu\rho} = \Gamma^{\mu}{}_{\nu\rho} + \frac{1}{\sqrt{3}}\, H^{\mu}{}_{\nu\rho} \neq \overline{\Gamma}^{\mu}{}_{\rho\nu} \,, \tag{4}$$

where $\Gamma^{\mu}{}_{\nu\rho} = \Gamma^{\mu}{}_{\rho\nu}$ is the torsion-free Christoffel connection, and $\frac{1}{\sqrt{3}}\, H^{\mu}{}_{\nu\rho} = -\frac{1}{\sqrt{3}}\, H^{\mu}{}_{\rho\nu}$ is the contorsion [37].

In what follows, we shall consider a constant dilaton configuration, $\Phi = \Phi_0 =$ constant, where the constant can be set to 0, without loss of generality. Such a solution is assumed to arise self consistently from the stabilisation of Φ through some appropriate string-theory-induced potential [21]. Then, on taking into account the modification of $H_{\mu\nu\rho}$ by the Chern Simons (CS) gravitational terms, which are required by the Green-Schwarz mechanism [38] for anomaly cancellation in strings, and implementing the associated Bianchi identity for the curl of $H_{\mu\nu\rho}$ via a pseudoscalar (axion-like) Lagrange multiplier field $b(x)$ (termed KR or gravitational axion) [39, 40], the effective action (3) can be expressed as:

$$\begin{aligned} S^{\rm eff}_B &= \int d^4x\sqrt{-g}\Big[-\frac{1}{2\kappa^2}\,R + \frac{1}{2}\,\partial_\mu b\,\partial^\mu b + \sqrt{\frac{2}{3}}\,\frac{\alpha'}{96\,\kappa}\, b(x)\, R_{\mu\nu\rho\sigma}\, \widetilde{R}^{\mu\nu\rho\sigma} + \dots\Big] \\ &= \int d^4x\,\sqrt{-g}\Big[-\frac{1}{2\kappa^2}\,R + \frac{1}{2}\,\partial_\mu b\,\partial^\mu b - \sqrt{\frac{2}{3}}\,\frac{\alpha'}{96\,\kappa}\,\partial_\mu b(x)\,\mathcal{K}^\mu + \dots\Big]\,, \end{aligned} \tag{5}$$

where $\widetilde{(\dots)}$ denotes the usual curved-spacetime dual tensor [20], and, in the second equality, we took into account that the CS gravitational anomaly term is a total derivative of the topological current $\mathcal{K}^\mu$, $\mu = 0, \dots 3$. Notice, as already mentioned, that there is no cosmological constant term appearing in the effective action, in agreement with the swampland criteria for embedding the low-energy theory into an UV complete string theory [2, 3, 4, 5], or for consistency with perturbative string theory, which requires a well-defined scattering matrix [24, 25]. The action (5) is a typical action of a CS gravity [41, 42], with a massless KR axion-like field $b(x)$. The coefficient of the CS-KR-axion interaction term, that is the inverse of the KR axion coupling, f_b, is fixed by the underlying string theory:

$$f_b^{-1} \equiv \sqrt{\frac{2}{3}}\,\frac{\alpha'}{96\,\kappa}\,, \tag{6}$$

where $\alpha' = M_s^{-2}$ is the Regge slope of the string, with M_s the string mass scale, that may be different from $M_{\rm Pl}$. The M_s may play the rôle of an UV cutoff of the effective low-energy gravitational theory.

For completeness we note that, classically, the duality between $b(x)$ and the torsion three-form field $H_{\mu\nu\rho}$ may be expressed via the relation [39, 40]:

$$-3\sqrt{2}\,\partial_\sigma b(x) = \kappa^{-1}\,\sqrt{-g}\,\epsilon_{\mu\nu\rho\sigma}\,H^{\mu\nu\rho}\,, \tag{7}$$

where $\epsilon_{\mu\nu\rho\sigma}$ is the Levi-Civita symbol, such that $\epsilon_{0123} = +1$ *etc.* This relation corresponds to saddle points in the path integral over the torsion field $H_{\mu\nu\rho}$ of the action (3) for $\Phi = 0$, after the insertion of the Bianchi identity constraint.

In [22, 23] we have discussed detailed ways for the production of condensates of the CS term

$$\mathcal{C}_{\rm CS} = \langle R_{\mu\nu\rho\sigma}\,\widetilde{R}^{\mu\nu\rho\sigma}\rangle\,, \tag{8}$$

which could be due to a condensation of either primordial GW or (rotating) primordial Black Holes (pBH), in the presence of both of which configurations the CS term is non vanishing [43, 44, 39, 41, 42]. Thus, the gravitational anomaly is unavoidable in our model, which involves only fields from the massless gravitational string multiplet [20, 21, 22, 23] and not gauge fields that are assumed to be generated at the end of inflation (hence, we cannot have anomaly cancellation until then).

The GW can be produced by merging of rotating pBH, or by the non-spherically-symmetric collapse of domain walls, the latter arising, e.g., in dynamically-broken supergravity (SUGRA) models that may characterise pre-inflationary epochs (the dynamical breaking of SUGRA could be achieved, for instance, through condensation of gravitino fields [29, 30]; the latter being the supersymmetric partner of gravitons still belong to the initially massles gravitational multiplet of the superstring, and hence the basic assumptions of [20, 21, 22, 23] are valid under such supersymmetric extensions. The gravitinos become very massive (even acquiring masses close to Planck mass) in such scenarios [31, 22], and eventually decouple from the spectrum). In such models, one may face a first hill-top inflation, near the origin of the double well gravitino-condensate potential [45], which provides, through instabilities of the SUGRA model [23], a natural explanation of the tunnelling process to the RVM inflationary metastable ground state of the cosmological model, and the homogeneity and isotropy conditions at the beginning of the RVM inflation [20].

The RVM phase is induced by the non-linearities of the vacuum energy of the model, which as a consequence of, e.g., the condensation of GW, assumes the form [43, 44, 46]:

$$\langle R_{\mu\nu\rho\sigma}\, \widetilde{R}^{\mu\nu\rho\sigma}\rangle = \frac{\mathcal{N}(t)}{\sqrt{-g}}\,\frac{1.1}{\pi^2}\Big(\frac{H}{M_{\rm Pl}}\Big)^3\,\mu^4\,\frac{\dot{b}(t)}{M_s^2} \equiv n_\star\,\frac{1.1}{\pi^2}\Big(\frac{H}{M_{\rm Pl}}\Big)^3\,\mu^4\,\frac{\dot{b}(t)}{M_s^2}\,, \tag{9}$$

where the overdot denotes derivative with respect to the cosmic time, and $n_\star \equiv \mathcal{N}(t)/\sqrt{-g}$ is the proper density of sources of GW, which we assume approximately constant during the RVM inflation for simplicity and concreteness. The quantity μ is the UV cutoff of the effective low-energy theory, which serves as an upper bound for the momenta of the graviton modes that are integrated over in the computation of the condensate (8) in the presence of chiral (left-right asymmetric) GW, leading to the result (9).

In the context of the low-energy string effective field theory, it is natural to identify $\mu = M_s$ in (9). One may then fine tune $n_\star$ with respect to $H(t)$, so that the anomaly equation implies an approximately constant (average) temporal component of the topological anomaly current $<\mathcal{K}^0>$ (*cf.* (5)), which is the dominant component in a (approximately) de-Sitter (inflationary) homogeneous and isotropic background spacetime:

$$\frac{d}{dt} < \mathcal{K}^0 > + 3\,H\, < \mathcal{K}^0 > \simeq n_\star\,\frac{1.1}{\pi^2}\Big(\frac{H}{M_{\rm Pl}}\Big)^3\,\mu^4\,\frac{\dot{b}(t)}{M_s^2}\,, \quad \dot{b}(t) = \sqrt{\frac{2}{3}}\frac{\alpha'}{96\,\kappa} < \mathcal{K}^0 >\,, \quad \alpha' = \mu^{-2}\,,$$
$$< \mathcal{K}^0 > \simeq \text{constant} \quad \Rightarrow \quad n_\star^{1/4} \sim 7.6 \times \Big(\frac{M_{\rm Pl}}{H}\Big)^{1/2}\,, \tag{10}$$

where the relation of $\dot{b}$ with $<\mathcal{K}^0>$ is a consequence of the KR-axion equation of motion [20, 21, 46]. If one uses the Planck cosmological data [1] to fix $H \simeq H_I$ during inflation to the range of values $H_I/M_{\rm Pl} < 10^{-5}$, then we obtain [46]

$$n_\star \gtrsim 3.3 \times 10^{13}\,, \tag{11}$$

which defines the range of values of the macroscopic number of sources (per proper volume) needed to produce an approximately constant gravitational anomaly condensate in the context of the CS cosmology of [20, 21, 22, 23].

Journal of Physics: Conference Series 2533 (2023) 012017 doi:10.1088/1742-6596/2533/1/012017

The total vacuum energy density during inflation assumes an RVM form (2) [16, 17, 18, 10, 11], as the explicit computations of [20, 22, 46] have demonstrated:

$$\rho_{\rm vac}^{\rm total} = -\frac{1}{2}\,\epsilon\, M_{\rm Pl}^2\, H^2 + 4.3 \times 10^{10}\,\sqrt{\epsilon}\,\frac{|\bar{b}(0)|}{M_{\rm Pl}}\, H^4\,, \tag{12}$$

where we have parametrised the constant $\dot{b}$ axion background during the RVM inflation as

$$b(t) = \bar{b}(t_0) + \sqrt{2\epsilon}\, H\,(t - t_0)\, M_{\rm Pl}\,, \tag{13}$$

with t_0 denoting the beginning of (the RVM) inflation. The assumption on the approximate de Sitter (positive-cosmological-constant) nature of the condensate, so as to lead to an approximately constant H_I during inflation, requires [20, 21]:

$$|\bar{b}(t_0)| \gtrsim N_e\,\sqrt{2\epsilon}\, M_{\rm Pl} = \mathcal{O}(10^2)\,\sqrt{\epsilon}\, M_{\rm Pl}\,, \tag{14}$$

with $N_e = \mathcal{O}(60-70)$ the number of e-foldings of the RVM inflation. The reader should have noticed that, as a result of the CS gravitational anomaly terms,[2] the coefficient of the H^2 term in (12) is negative, in contrast to the standard form, current-era, RVM energy density (2), in which $\nu > 0$. Such negative contributions to the energy density are characteristic of higher-curvature gravities, such as Gauss Bonnet-dilaton gravity [47].

The EoS of the string-inspired model during this inflationary phase has been explicitly computed in [22] and found to coincide with the RVM EoS (1), which justifies *a posteriori* calling this model *stringy RVM*.

An important feature of the model is the Lorentz-Violating (LV) nature of the KR axion background (13), characterised by a constant $\dot{b}$, as a consequence of the (approximately) constant anomaly condensate (10) during the RVM inflation. The LV nature stems from the duality relation (7), which implies a constant $\epsilon_{ijk}\, H^{ijk}$, $i, j, k = 1, 2, 3$, that clearly breaks spontaneously the Lorentz symmetry. Such a background survives the end of the inflationary phase, and can lead, during the radiation phase, to a LV and CPT-Violating (CPTV) Leptogenesis, according to the mechanism suggested in [48, 49, 50, 51], as we shall discuss in section 4. In the next section 3, instead, we shall study the consequences of the presence of periodic modulations of the potential of axion fields that arise from compactification in string theory, and co-exist with the KR axions [40, 52]. We shall argue that such structures enhance the densities of primordial black hole (pBH) densities during inflation, and, thus, affect the profiles of GW during the radiation era.

3. Enhanced Primordial Black Holes and modified Gravitational Wave profiles

In [33] we have considered one such toy string-inspired extension of the stringy RVM model, in which we considered the effects of one more axion field, arising from compactification, $a(x)$, with axion coupling f_a, in addition to the KR axion $b(x)$ that drove the RVM inflation. Below

[2] We note that the variation of the gravitational CS anomaly terms in (5) with respect to the graviton field is non trivial, giving rise to the so-called Cotton tensor [41, 42], which affects the Einstein equation:

$$R^{\mu\nu} - \frac{1}{2}\, g^{\mu\nu}\, R - \sqrt{\frac{2}{3}}\,\frac{\alpha'\,\kappa}{12}\, C^{\mu\nu} = \kappa^2\, T^{\mu\nu}\,.$$

The Cotton tensor is not covariantly conserved, $C^{\mu\nu}{}_{;\mu} = -\frac{1}{8}\,(\partial^\nu b) R_{\mu\nu\rho\sigma}\, \widetilde{R}^{\mu\nu\rho\sigma}$ (where ; denotes the torsion-free gravitational convariant derivative), and this implies an exchange of energy between the axion and the gravitational anomaly terms., which leads to the aforementioned negative contributions of the CS anomaly to the stress-energy tensor $T^{\mu\nu}$ of the system.

Journal of Physics: Conference Series **2533** (2023) 012017 doi:10.1088/1742-6596/2533/1/012017

we shall review briefly the situation and the consequences for the phenomenology of pBH and GW, which could in principle be observable in future interferometers [34, 35].

The presence of the anomaly condensate (8) (or (9)) leads to an approximately linear potential of the KR axion $b(x)$, but also for the axion $a(x)$, assuming a standard CS coupling of the $a(x)$ field with the gravitational anomaly [40]

$$S_{\rm aCS} = \int d^4x\sqrt{-g}\,\frac{1}{f_a}\,a(x)\,R_{\mu\nu\rho\sigma}\,\widetilde{R}^{\mu\nu\rho\sigma}\,. \tag{15}$$

To maintain the spirit of the stringy RVM model [20, 21, 22, 23], as far as the induced RVM inflation is concerned, we assume that world-sheet stringy non-perturbative effects generate dominant periodic structures for the potential of the axion field $a(x)$ only. To keep the analysis general as far as compactification effects are concerned, we also include brane compactification effects which, in the context of axion monodromy models [53], also lead to linear terms in the a-field potential, in addition to the ones due to the anomaly condensate. We assume that such effects correspond to an energy scale Λ_2. Par contrast, the gravitational anomaly effects, discussed in the previous section, are related to an energy scale Λ_0, which is defined through: $\langle R_{\mu\nu\rho\sigma}\,\widetilde{R}^{\mu\nu\rho\sigma}\rangle \equiv \widetilde{\Lambda}_0^4 \simeq \frac{1}{\pi^2}\sqrt{2\epsilon}\,\frac{\mu^4}{M_s^2\,M_{\rm Pl}^2}\,H^4$, where in the effective theory it is natural that we set $\mu = M_s$, as mentioned above. Then, the corresponding condensate-induced linear axion-b effective potential (*cf.* (5)) reads [33]

$$V(b) \simeq b\,\widetilde{\Lambda}_0^4\sqrt{\frac{2}{3}}\,\frac{M_{\rm Pl}}{96\,M_s^2} = b\,\frac{\widetilde{\Lambda}_0^4}{f_b} \equiv b\,\Lambda_0^3\,, \tag{16}$$

with the KR axion coupling f_b given in (6). Therefore, the (3+1)-dimensional spacetime effective potential for the two axion fields, $a(x)$ and $b(x)$ that we consider for the analysis in this section acquires the form:

$$V(a,\,b) = {\Lambda_1}^4\left(1 + f_a^{-1}\,\tilde{\xi}_1\,a(x)\right)\,\cos({f_a}^{-1}a(x)) + \frac{1}{f_a}\left(f_b\,{\Lambda_0}^3 + \Lambda_2^4\right)a(x) + {\Lambda_0}^3\,b(x), \tag{17}$$

where, in the study of [33], the parameters $\tilde{\xi}_1, f_a$, and the world-sheet instanton induced scales Λ_1, Λ_2 are treated as phenomenological, with the constraint though that [33]:[3]

$$\left(\frac{f_b}{f_a} + \frac{\Lambda_2^4}{f_a\,\Lambda_0^3}\right)^{1/3}\Lambda_0 \,<\, \Lambda_1 \ll \Lambda_0\,. \tag{18}$$

so that the RVM inflation is driven by the $b(x)$ field, and is only prolonged by the presence of the $a(x)$ field. The reader should notice that the non-perturbative-world-sheet-effects-generated periodic modulations of the axion-$a(x)$ potential in (17) do not lead to mass terms for the axion. The potential (17) is plotted in fig. 1 for a concrete set of parameters, given in Table 1.

[3] For practical purposes, such a hierarchy of scales (18) implies that the term Λ_2^4 can be ignored in front of $\Lambda_0^3 f_b$ in (17).

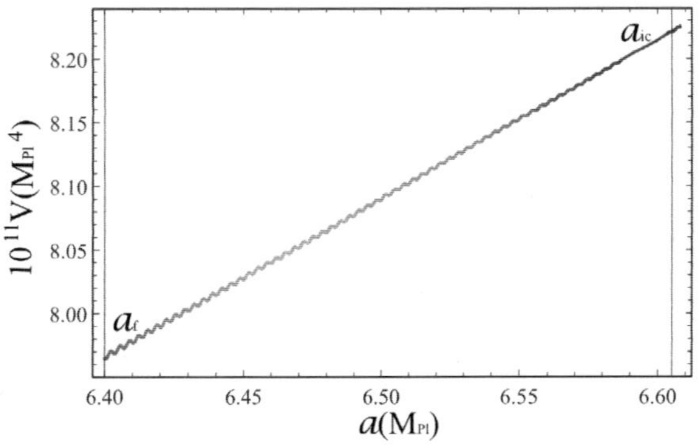

Figure 1. The potential (17) in the a-field direction for the set of parameters of Table 1. a_{ic} represents the initial condition for the a axion, before its oscillations start to become appreciable, and a_f corresponds to the value of the field when the inflation ends. Picture taken from [33].

SET	g_1	g_2	ξ	$f(M_{Pl})$	$\Lambda_0(M_{Pl})$	$\Lambda_1(M_{Pl})$	$\Lambda_3(M_{Pl})$
1	0.021	0.904	-0.15	2.5×10^{-4}	8.4×10^{-4}	8.19×10^{-4}	2.32×10^{-4}

Table 1. An example of parameters for the potential (17) [33], where, for brevity, we use the notation: $g_1 \equiv \frac{f_b}{f_a} + \frac{\Lambda_2^4}{f_a\,\Lambda_0^3}$, $\Lambda_1^4 \equiv g_2\,\Lambda_0^4$, $f \equiv f_a$, $\xi \equiv \frac{M_{\rm Pl}}{f_a}\,\tilde{\xi}_1$.

In this scenario, the KR field b dominates during the first stage of (RVM) inflation, when the oscillations of the axion-like field a are suppressed compared to its linear term. As the cosmic time passes, the value of b decreases [33], and eventually the oscillations of the field a become the leading effect. The oscillations of the a axion appear effectively as many tiny "step"-like patterns, which lead to an enhancement of the primordial spectra, a feature reflected in the background of GWs, in similar spirit to the situation governing inflationary potentials with steep-like features, including discontinuities [54, 55, 56, 57]. However, contrary to such scenarios, the evolution here is smooth, and there are no discontinuities neither in the field nor in its first derivative.

The detailed analysis of [33] has demonstrated the enhancement of the power spectra due to the oscillations of the axion a, which in turn enhances the densities of the pBH during the RVM inflation, and consequently the profile of the GW during the radiation era that succeeds the RVM inflation [10, 11, 20, 22]. This can lead to the production of a significant fraction of the DM in the form of pBHs [58, 59, 60, 61]. The situation is depicted in fig. 2, where we plot features of the power spectra P_R in the two-axion model of [33], together with the pBH abundance (as a function of pBH mass (in units of the solar mass)), as well as the energy density profile of the GW, for the set of parameters given in Table 1. Similar features persist for a large range of pBH abundances f_{PBH}, including those of order $f_{PBH} = 0.80$ [33].

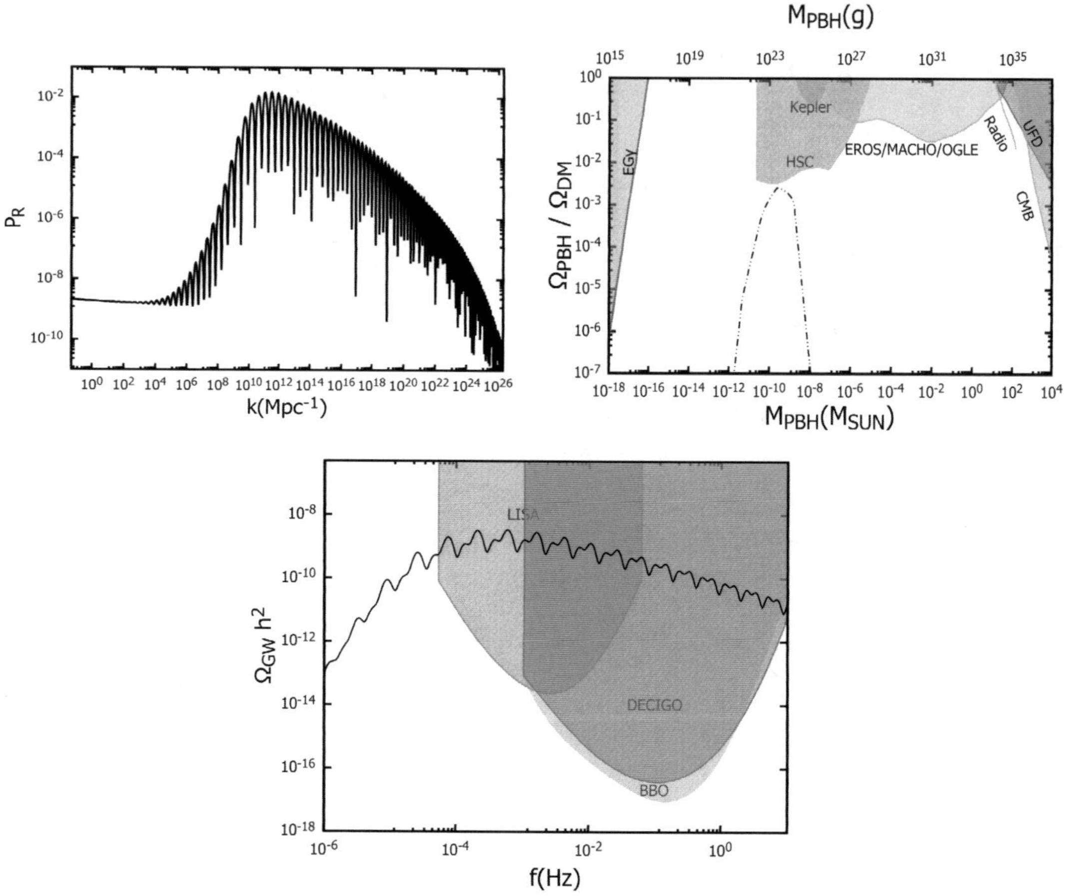

Figure 2. The results of the power spectrum, fractional abundance of PBHs and the energy density of induced GWs for the first set of parameters given in Table 1. The initial conditions for the axion fields are (in units of $M_{\rm Pl}$) $a_{ic} = 6.605$, and $b_{ic} = 11.100$. The fractional abundance of PBHs is $f_{PBH} = 0.01$. Picture taken from [33].

These results should be compared with the ones characterising other string-inspired axion-monodromy models with two axions, considered in [62], which, although leading also to enhanced power spectra and enhanced production of pBH during inflation, nonetheless are characterised by a different hierarchy of scales than (18):

$$\Lambda_0 \ll \left(\frac{f_b}{f_a} + \frac{\Lambda_2^4}{f_a\,\Lambda_0^3}\right)^{1/3} \Lambda_0 < \Lambda_1 \,, \tag{19}$$

and are such that it is the axion a that drives initially inflation, whilst the KR axion b is responsible for prolonging it. As becomes clear from fig. 3, the smooth (featureless) energy density profile of GW after inflation in that case, in contrast to the model of [33] examined above (*cf.* fig. 2), makes the models of [62] in principle distinguishable experimentally from the model of [33], e.g. in future interferometers [34, 35].

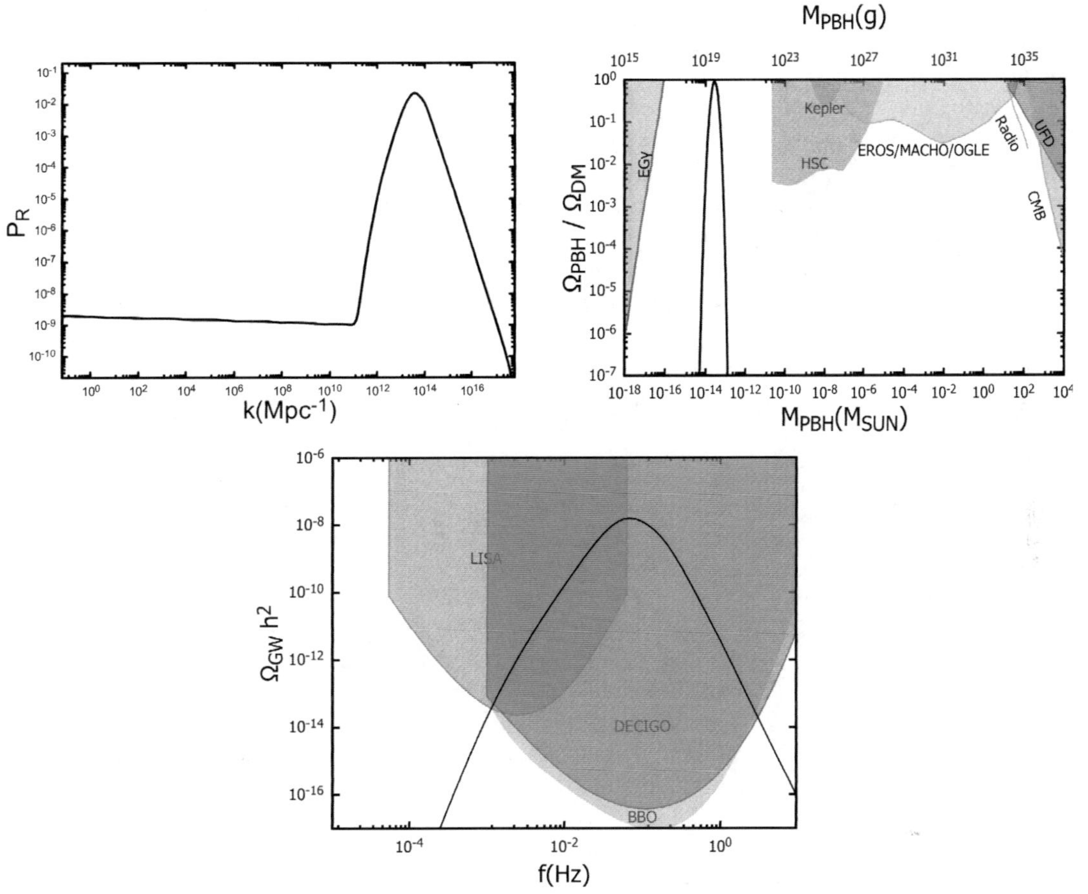

Figure 3. Upper left panel: The power spectrum for the model of [62], with the hierarchy of scales (19). Upper right panel: The abundances of PBHs. Lower panel: The energy density of GWs. The diagrams are given for the choice of parameters $\Lambda_0 = 8.4 \times 10^{-4} M_{\rm Pl}$, $\Lambda_1 = 9.7 \times 10^{-3} M_{\rm Pl}$, $g_1 = 110$, $g_2 = 1.779 \times 10^4$, $\xi = -0.09$, $f = 0.09\, M_{\rm Pl}$, which obeys the hierarchy of scales (19). The initial conditions for the axion a and b fields (in units of $M_{\rm Pl}$) are $(a_{ic}\,,\, b_{ic}) = (7.5622,\, 0.5220)$. Picture taken from [33].

Before closing this section, we would like to remark [33] that in the context of RVM inflationary models with several axions, we may expect further enhancement of the power spectra and the densities of pBH, due to a prolonged reheating period that characterises (depending on the parameters) some of the RVM models at the exit from RVM inflation [10, 11]. A complete analysis of such models and a study of the associated consequences for the profile of GW after inflation is pending.

4. Lorentz-violation-induced Matter-Antimatter Asymmetry

In this section we examine the consequences of the LV axion background (13) for inducing unconventional LV and CPTV processes for lepton asymmetry generation (Leptogenesis) during the radiation era, in models involving massive right-handed neutrinos (RHN), within the context of the mechanism suggested in [48, 49, 50, 51].

First let us review this mechanism, and then see how it can be adapted to our stringy RVM

Journal of Physics: Conference Series **2533** (2023) 012017 doi:10.1088/1742-6596/2533/1/012017

cosmological model. We consider the following fermion action in an LV and CPTV violating axial background b_μ in the presence of massive right-handed neutrinos (RHN), with standard portals coupling the RHN sector to Standard Model (SM) lepton and Higgs sectors:

$$\mathcal{L} = \mathcal{L}_{\rm SM} + i\overline{N}\,\gamma^\mu\,\partial_\mu\,N - \frac{m_N}{2}(\overline{N^c}N + \overline{N}N^c) - \overline{N}\gamma^\mu\,b_\mu\,\gamma^5 N - \sum_f y_f \overline{L}_f \tilde{\phi}^d N + \text{h.c.} \tag{20}$$

where h.c. denotes hermitian conjugate, $\mathcal{L}_{\rm SM}$ is the Standard Model (SM) Lagrangian, N is the RHN field (with N^c its charge conjugate field), of (Majorana) mass m_N, $\tilde{\phi}$ denotes the SU(2) adjoint of the Higgs field ϕ ($\tilde{\phi}^d_i \equiv \varepsilon_{ij}\phi_j$, $i, j = 1, 2$, SU(2) indices), and L_f is a lepton doublet of the SM sector, with f a generation index, $f = e, \mu, \tau$, assuming three SM generations; y_f is a Yukawa coupling, which, if non-zero, provides a non-trivial ("Higgs portal") interaction between the RHN and the SM sector, used in the seesaw mechanism for the generation of SM neutrino masses [63, 64, 65, 66, 66, 67, 68]. The LV and CPTV axial background b_μ is of Standard Model Extension (SME) type [69, 70], and it has only a constant, temporal component [48]

$$b_\mu = M_{\rm Pl}^{-1}\,\dot{b}\,\delta_{\mu 0}\,, \quad \mu = 0, \dots 3\,, \quad \dot{\overline{b}} = \text{constant}\,. \tag{21}$$

One then computes in that model, the tree-level decays of (even one) flavour of a massive RHN, N, of mass m_N, into SM particles and antiparticles, that is one computes the RHN decay rates in the channels: Channel $I: \ N \to l^- h^+\ , \ \nu\, h^0$, and Channel II : $N \to l^+ h^-\ , \ \overline{\nu}\, h^0$, where $\ell^\pm$ are charged leptons, ν ($\overline{\nu}$) are the light, "active", neutrinos (antineutrinos) in the SM sector, h^0 is the neutral Higgs field, and $h^\pm$ are the charged Higgs fields (which, at high temperatures, above the spontaneous electroweak symmetry breaking, of interest in this scenario [48, 49, 50, 51, 22], do not decouple from the physical spectrum). As a result of the non-trivial $b_0 \neq 0$ background (21), the decay rates of the Majorana RHN between the channels I and II are different, resulting in a Lepton-number asymmetry. In the limit of a weak background (21) (compared to the mass of the RHN, *i.e.* $|b_0| \ll m_N$) in the model (20), this lepton asymmetry is found proportional to the ratio b_0/m_N. Assuming a freezeout temperature T_D for the RHN $T_D \simeq m_N$, one finds (using semi-analytic techniques to compute the Lepton asymmetry ΔL) [49]:

$$\frac{\Delta L(T = T_D)}{s} \simeq 7 \times 10^{-3}\,\frac{b_0}{m_N}\,, \tag{22}$$

for Yukawa couplings $y = \mathcal{O}(10^{-5})$ and $m_N \simeq 10^5$ GeV, where s is the entropy density of the universe. The numerical coefficient in the right-hand side of (22) depends on the approximation scheme used in the computations. Requiring ΔL to be the one observed in the Universe (that is of the same order as the observed baryon asymmetry [1], $\Delta B/s \sim \mathcal{O}(10^{-10})$ for temperatures $T > 1$ GeV), we find $b_0 \simeq 10^{-8}\,m_N = \mathcal{O}(1\text{ MeV})$.

In [50], the constant b_0 analysis has been extended to the case of backgrounds that depend mildly on the temperature T:

$$b_0(T) \sim T^3\,, \tag{23}$$

which for the short period of leptogenesis can be assumed approximately constant, so that an analytic treatment is also possible in that case, yielding a lepton asymmetry at decoupling $T_D \simeq m_N$ [50]:

$$\frac{\Delta L(T = T_D)}{s} \simeq q\,\frac{\mathcal{F}}{m_N}\,, \tag{24}$$

where we have set for the T-dependent background $b_0(x) \simeq \mathcal{F}\, x^{-3}$, $x \equiv m_N/T$. The coefficient q expresses theoretical uncertainties in the approximation methods used, and in particular on the expansion point x_P of the appropriate Padè-approximant expansion used in the evaluation [50]. This yields $b_0(T_B)$ in the range:

$$b_0(T_D)/\text{keV} \in [0.36\,, 0.74]\,, \tag{25}$$

which produces phenomenologically relevant lepton asymmetries. The above lepton asymmetries are communicated to the baryon sector via appropriate sphaleron processes in the SM sector, which preserve the difference of Baryon minus Lepton (B-L) number [51].

Let us now adapt the above mechanism to our stringy RVM model [20, 21, 22, 23]. We have seen that in the presence of gravitational anomaly condensates, induced by condensation of, e.g., primordial GW, there is an induced LV axion background (13), which survives undiluted until the end of the inflationary period and the onset of radiation. In the context of the stringy RVM cosmology, Abelian and non-Abelian gauge fields of the SM sector, and chiral fermionic matter, are all generated at the end of the RVM inflation, as a result of the decay of the metastable RVM vacuum [10, 11]. In the presence of chiral fermions, the rôle of the KR field $H_{\mu\nu\rho}$ as torsion implies that its dual KR axion $b(x)$ will couple to the divergence of the axial fermion current $J^{5\mu}$, universally for all fermion species, including RHN if the model contains such fields. Specifically, the pertinent effective action will read [20]:

$$S^{\rm eff} = \int d^4x\sqrt{-g}\Big[-\frac{1}{2\kappa^2}\, R + \frac{1}{2}\,\partial_\mu b\,\partial^\mu b - \sqrt{\frac{2}{3}}\,\frac{\alpha^\prime}{96\,\kappa}\,\partial_\mu b(x)\,\mathcal{K}^\mu\Big]$$
$$+ S^{Free}_{\rm Dirac\ or\ Majorana} + \int d^4x\sqrt{-g}\,\Big[\Big(\mathcal{F}_\mu + \frac{\alpha^\prime}{2\,\kappa}\,\sqrt{\frac{3}{2}}\,\partial_\mu b\Big)\,J^{5\mu} - \frac{3\alpha^{\prime\,2}}{16\,\kappa^2}\,J^5_\mu J^{5\mu}\Big] + \dots, \tag{26}$$

where $J^{5\mu} \equiv \sum_{i=\rm fermion\ species} \overline{\psi}\gamma^5\,\gamma^\mu\,\psi$ denotes the axial fermion current , $\mathcal{F}^d = \varepsilon^{abcd}\,e_{b\lambda}\,\partial_a\,e^\lambda_{\ c}$, with $e^\mu_{\ c}$ the vielbeins (with Latin indices pertaining to the tangent space of the spacetime manifold at a given point, in a standard notation), $S^{Free}_{\rm Dirac\ or\ Majorana}$ denotes the free-fermion kinetic terms, and the $\dots$ in (26) indicate gauge field kinetic terms, as well as terms of higher order in derivatives. The action (26) is valid for both Dirac or Majorana fermions. The interaction $\partial_\mu b\, J^{5\mu}$ yields, upon partial integration, the aforementioned coupling of the axion b to the divergence of the axial current, and is due to the presence of the torsion on the covariant gravitational derivative for fermions [37, 71]. The reader should notice the repulsive four-fermion interactions $-\frac{3\alpha^{\prime\,2}}{16\,\kappa^2}\,J^5_\mu J^{5\mu}$ in the action (26), which is a characteristic feature of the Einstein-Cartan theory of fermionic torsion.

If the axial current is conserved, then the primordial anomaly terms involving the topological current $\mathcal{K}^\mu$ would remain in the radiation era, and in case there is condensation of GW, one would still obtain a LV background (13) for the KR axion well into the radiation era, which could lead to a constant b_0 leptogenesis, according to the mechanism described above (*cf.* (22)).

On the other hand, in case there are chiral anomalies, leading to the non-conservation of the axial current, $J^{5\mu}{}_{;\mu} \neq 0$, there could be a cancellation [20] of the primordial gravitational anomaly $\mathcal{K}^\mu$-terms in (26) by the gravitational anomalies generated by the chiral fermionic matter, in the sense of:

$$\partial_\mu\Big[\sqrt{-g}\,\Big(\sqrt{\frac{3}{8}}\kappa\,J^{5\mu} - \sqrt{\frac{2}{3}}\,\frac{\kappa}{96}\,\mathcal{K}^\mu\Big)\Big] = \sqrt{\frac{3}{8}}\,\frac{\alpha^\prime}{\kappa}\,\frac{e^2}{8\pi^2}\,\sqrt{-g}\,F^{\mu\nu}\,\widetilde{F}_{\mu\nu}$$
$$+ \sqrt{\frac{3}{8}}\,\frac{\alpha^\prime}{\kappa}\,\frac{\alpha_s}{8\pi}\,\sqrt{-g}\,G^a_{\mu\nu}\,\widetilde{G}^{a\mu\nu}\,, \tag{27}$$

where $F_{\mu\nu}$ denotes the electromagnetic U(1) Maxwell tensor, and $G^a_{\mu\nu}$, $a = 1, \dots 8$ is the gluon tensor associated with the strong interactions with (squared) coupling $\alpha_s = g_s^2/(4\pi)$, and the $\widetilde{(\dots)}$ denotes the corresponding duals, as usual.

Upon assuming (27), the equation of motion of the KR axion field reads:

$$\frac{1}{\sqrt{-g}}\partial_\mu\Big(\sqrt{-g}\,\partial^\mu b\Big) = -\sqrt{\frac{3}{8}}\,\frac{\alpha'}{\kappa}\,\frac{e^2}{8\pi^2}\,F^{\mu\nu}\,\widetilde{F}_{\mu\nu} - \sqrt{\frac{3}{8}}\,\frac{\alpha'}{\kappa}\,\frac{\alpha_s}{8\pi}\,G^a_{\mu\nu}\,\widetilde{G}^{a\mu\nu}\,. \tag{28}$$

If the terms on the right-hand-side of (28) are zero at the early radiation era [22], above the electroweak symmetry breaking, e.g. in the temperature range of the leptogenesis scenarios of [50], $T \sim 10^5$ GeV, then one obtains from (28),

$$\dot{b} \sim T^3\,, \tag{29}$$

as a result of the temperature dependence of the Universe scale factor $a(T) \sim 1/T$ during the radiation era, of interest to us here. Thus one faces the scenario for leptogenesis of [50], based on axial backgrounds of the form (23), with $b_0 = \dot{b}$, and the associated lepton asymmetry is given by (24), (25).

We remark at this point that for the above scenarios of Leptogenesis to be valid, it is essential that the RHN acquires a non zero mass during the leptogenesis epoch. In our approach we assumed that this is possible, without specifying the detailed mechanism by means of which such a mass is generated. One appealing scenario is the radiative mechanism for RHN Majorana mass generation of [72], which makes use of the CS gravitational anomaly itself.

Non zero chiral anomalies may appear at much lower temperatures, for instance gluon terms in (28) may appear due to QCD instanton non-perturbative effects at the QCD epoch, leading to the generation of non-perturbative periodic potentials, and thus a mass, for the KR axion, which in this way may play the rôle of a component of DM. For even lower temperatures, one may have the generation of U(1) chiral anomalies, which in turn last until the current era, affecting the temperature evolution of the $b_0(T) = \dot{b}(T)$ background [20]. Given the connection (*cf.* (7)) of $\dot{b}$ with the KR totally antisymmetric torsion, $\dot{b} \propto \epsilon^{ijk}\,H_{ijk}$, $i, j, k = 1, 2, 3$, one thus obtains a geometric origin of the DM [73]. It should also be stressed that the evolution equation of $b_0(T)$ is such that today, b_0 is well within the allowed torsion limits [20], namely, if chiral matter effects are assumed present in the late Universe, then [20]:

$$b_0^{\rm today} = \mathcal{O}(10^{-34})\ {\rm eV}\,, \tag{30}$$

otherwise it is much more suppressed. The current bounds for b_0 are [70] $b_0 < 10^{-2}$ eV, while the spatial component must satisfy $|b_i| < 10^{-22}$ eV, $i = 1, 2, 3$. The bound (30) is compatible with b_i, even if we take into account the relative velocity of an observer on Earth relative to the cosmic microwave background (CMB) frame [20].

We conclude this section by remarking that condensates of chiral U(1) anomalies [20] (or even of the combination of U(1) and non-Abelian chiral anomalies, as appears on the right-hand side of (27)) can lead to the appearance of a dark energy component of the Universe at late eras. Such condensates would necessarily be metastable if we want our low-energy model to be embeddable into string theory or perhaps into any other UV complete theory of QG. A detailed mechanism for the appearance of such a dark energy, which at late eras evolves so slowly with the cosmic time so that it approximates a cosmological constant, in agreement with the plethora of the available data, is still pending within a string theory framework. Nonetheless it seems plausible. At this point it should be remarked that when computing the dark energy of any cosmological model, the back reaction of a quantum field theoretic matter onto the space time may not be ignored, and it is possible that renormalization group effects on the stress-energy tensor of the theory play a non trivial rôle [26, 27, 28]. Thus, within our framework of generating dark energy from condensates of gauge fields in the SM sector of our string-inspired RVM cosmology, such effects should be taken properly into account.

5. Conclusions and Outlook

In this talk, we have discussed a cosmological model inspired from string theory, which however was characterised by gravitational anomalies at early epochs. The model is consistent with general covariance, but has interesting and unconventional consequences for the physics of inflation and post inflationary evolution of the Universe. In particular, condensation of primordial GW implies an RVM type inflation, which doies not require external inflaton fields. The inflation arises as a result of the non-linear terms quartic in the Hubble parameter, that characterise the energy density of the model. The anomaly condensates also result in LV and CPTV axion backgrounds, which lead to unconventional Leptogenesis during the radiation era.

In this framework, the dark sector of the Universe acquires a geometric origin, given the aforementioned association of the dark energy with condensates of anomalies, which could characterise the model beyond the RVM inflationary era, even providing (through condensates of chiral anomalies of gauge type in the modern era) a candidate for a present-era metastable de Sitter space time. Moreover, the association of the gravitational KR axion with the totally antisymmetric torsion, implies, in case the KR axion develops a non-perturbative mass through QCD instanton effects associated with chiral anomalies of the gluon field, a potential geometric origin of dark matter, a component of which could be played by such a massive KR axion.

The modern-era phenomenology of the stringy RVM model is not dissimilar to that of the conventional RVM model. The latter is known to exhibit observable in principle deviations from ΛCDM [74, 75], including the potential alleviation of the cosmological data tensions that seem to characterise the modern era of the Universe [9]. Nonetheless, the stringy RVM model includes $H^2 \ln H$ corrections in the modern-epoch energy density, as a result of integrating out graviton modes, thus a pure QG effect. Such terms, but not QG in origin, also seem to arise in the renormalisation of quantum fields in the curved spacetime of an expanding Universe as observed recently in [26, 27, 28], but there appear to be distinguishing features between those conventional quantum field theory models and the stringy RVM [32]. Although such non-polynomial-in-H terms do not affect the inflationary scenarios, nevertheless their detailed phenomenology in the modern era, especially with regards to the data tension alleviation, need yet to be explored. We hope to be able to report soon on this issue.

Acknowledgments

I would like to thank H-T. Elze and the organising committee of the 2022 edition of the DICE series of conferences for the invitation for a plenary talk and for organising such a high level and thought provoking event, which constituted another valuable addition to this excellent conference series. The work of N.E.M. is supported in part by the UK Science and Technology Facilities research Council (STFC) under the research grant ST/T000759/1. N.E.M. also acknowledges participation in the COST Association Action CA18108 "*Quantum Gravity Phenomenology in the Multimessenger Approach (QG-MM)*".

References

[1] Aghanim N *et al.* (Planck) 2020 *Astron. Astrophys.* **641** A6 [Erratum: Astron.Astrophys. 652, C4 (2021)] (*Preprint* 1807.06209)
[2] Ooguri H, Palti E, Shiu G and Vafa C 2019 *Phys. Lett. B* **788** 180–184 (*Preprint* 1810.05506)
[3] Obied G, Ooguri H, Spodyneiko L and Vafa C 2018 (*Preprint* 1806.08362)
[4] Garg S K and Krishnan C 2019 *JHEP* **11** 075 (*Preprint* 1807.05193)

[5] Palti E 2019 *Fortsch. Phys.* **67** 1900037 (*Preprint* 1903.06239)
[6] Verde L, Treu T and Riess A G 2019 *Nature Astron.* **3** 891 (*Preprint* 1907.10625)
[7] Freedman W L 2017 *Nature Astron.* **1** 0121, and references therein (*Preprint* 1706.02739)
[8] Di Valentino E, Mena O, Pan S, Visinelli L, Yang W, Melchiorri A, Mota D F, Riess A G and Silk J 2021 *Class. Quant. Grav.* **38** 153001, and references therein (*Preprint* 2103.01183)
[9] Solà Peracaula J, Gómez-Valent A, de Cruz Perez J and Moreno-Pulido C 2021 *EPL* **134** 19001 (*Preprint* 2102.12758)
[10] Lima J A S, Basilakos S and Solà J 2013 *Mon. Not. Roy. Astron. Soc.* **431** 923–929 (*Preprint* 1209.2802)
[11] Perico E L D, Lima J A S, Basilakos S and Solà J 2013 *Phys. Rev. D* **88** 063531 (*Preprint* 1306.0591)
[12] Basilakos S, Lima J A S and Solà J 2013 *Int. J. Mod. Phys. D* **22** 1342008 (*Preprint* 1307.6251)
[13] Lima J A S, Basilakos S and Solà J 2015 *Gen. Rel. Grav.* **47** 40 (*Preprint* 1412.5196)
[14] Lima J A S, Basilakos S and Solà J 2016 *Eur. Phys. J. C* **76** 228 (*Preprint* 1509.00163)
[15] Solà Peracaula J and Yu H 2020 *Gen. Rel. Grav.* **52** 17 (*Preprint* 1910.01638)
[16] Shapiro I L and Solà J 2009 *Phys. Lett. B* **682** 105–113 (*Preprint* 0910.4925)
[17] Shapiro I L and Solà J 2000 *Phys. Lett. B* **475** 236–246 (*Preprint* hep-ph/9910462)
[18] Shapiro I L and Solà J 2002 *JHEP* **02** 006 (*Preprint* hep-th/0012227)
[19] Solà J 2008 *J. Phys. A* **41** 164066 (*Preprint* 0710.4151)
[20] Basilakos S, Mavromatos N E and Solà Peracaula J 2020 *Phys. Rev. D* **101** 045001 (*Preprint* 1907.04890)
[21] Basilakos S, Mavromatos N E and Solà Peracaula J 2020 *Phys. Lett. B* **803** 135342 (*Preprint* 2001.03465)
[22] Mavromatos N E and Solà Peracaula J 2021 *Eur. Phys. J. ST* **230** 2077–2110 (*Preprint* 2012.07971)
[23] Mavromatos N E and Solà Peracaula J 2021 *Eur. Phys. J. Plus* **136** 1152 (*Preprint* 2105.02659)
[24] Hellerman S, Kaloper N and Susskind L 2001 *JHEP* **06** 003 (*Preprint* hep-th/0104180)
[25] Fischler W, Kashani-Poor A, McNees R and Paban S 2001 *JHEP* **07** 003 (*Preprint* hep-th/0104181)
[26] Moreno-Pulido C and Solà Peracaula J 2022 *Eur. Phys. J. C* **82** 551 (*Preprint* 2201.05827)
[27] Moreno-Pulido C and Solà J 2020 *Eur. Phys. J. C* **80** 692 (*Preprint* 2005.03164)
[28] Moreno-Pulido C and Solà Peracaula J 2022 *Eur. Phys. J. C* **82** 1137 (*Preprint* 2207.07111)
[29] Alexandre J, Houston N and Mavromatos N E 2013 *Phys. Rev. D* **88** 125017 (*Preprint* 1310.4122)
[30] Alexandre J, Houston N and Mavromatos N E 2015 *Int. J. Mod. Phys. D* **24** 1541004 (*Preprint* 1409.3183)
[31] Basilakos S, Mavromatos N E and Solà J 2016 *Universe* **2** 14 (*Preprint* 1505.04434)
[32] Gomez-Vàlent A, Mavromatos N E and Solà J , in progress.
[33] Mavromatos N E, Spanos V C and Stamou I D 2022 *Phys. Rev. D* **106** 063532 (*Preprint* 2206.07963)
[34] Berti E, Cardoso V and Will C M 2006 *Phys. Rev. D* **73** 064030 (*Preprint* gr-qc/0512160)
[35] Fumagalli J, Pieroni M, Renaux-Petel S and Witkowski L T 2022 *JCAP* **07** 020 (*Preprint* 2112.06903)
[36] Green M B, Schwarz J H and Witten E 2012 *Superstring Theory Vol. 2: 25th Anniversary Edition* Cambridge Monographs on Mathematical Physics (Cambridge University Press) ISBN 978-1-139-53478-9, 978-1-107-02913-2
[37] Hehl F W, Von Der Heyde P, Kerlick G D and Nester J M 1976 *Rev. Mod. Phys.* **48** 393–416
[38] Green M B and Schwarz J H 1984 *Phys. Lett. B* **149** 117–122
[39] Duncan M J, Kaloper N and Olive K A 1992 *Nucl. Phys. B* **387** 215–235
[40] Svrcek P and Witten E 2006 *JHEP* **06** 051 (*Preprint* hep-th/0605206)
[41] Jackiw R and Pi S Y 2003 *Phys. Rev. D* **68** 104012 (*Preprint* gr-qc/0308071)
[42] Alexander S and Yunes N 2009 *Phys. Rept.* **480** 1–55 (*Preprint* 0907.2562)
[43] Alexander S H S, Peskin M E and Sheikh-Jabbari M M 2006 *Phys. Rev. Lett.* **96** 081301 (*Preprint* hep-th/0403069)
[44] lyth D H, Quimbay C and Rodriguez Y 2005 *JHEP* **03** 016 (*Preprint* hep-th/0501153)
[45] Ellis J and Mavromatos N E 2013 *Phys. Rev. D* **88** 085029 (*Preprint* 1308.1906)
[46] Mavromatos N E 2022 *740. WE-Heraeus-Seminar: Experimental Tests and Signatures of Modified and Quantum Gravity Workshop* (Chapter in the Springer Book Modified and Quantum Gravity - From theory to experimental searches on all scales - WEH 740, (C. Laemmerzahl and C. Pfeifer eds) (*Preprint* 2205.07044)
[47] Kanti P, Mavromatos N E, Rizos J, Tamvakis K and Winstanley E 1996 *Phys. Rev. D* **54** 5049–5058 (*Preprint* hep-th/9511071)
[48] de Cesare M, Mavromatos N E and Sarkar S 2015 *Eur. Phys. J. C* **75** 514 (*Preprint* 1412.7077)
[49] Bossingham T, Mavromatos N E and Sarkar S 2018 *Eur. Phys. J. C* **78** 113 (*Preprint* 1712.03312)
[50] Bossingham T, Mavromatos N E and Sarkar S 2019 *Eur. Phys. J. C* **79** 50 (*Preprint* 1810.13384)
[51] Mavromatos N E and Sarkar S 2020 *Eur. Phys. J. C* **80** 558 (*Preprint* 2004.10628)
[52] Arvanitaki A, Dimopoulos S, Dubovsky S, Kaloper N and March-Russell J 2010 *Phys. Rev. D* **81** 123530 (*Preprint* 0905.4720)
[53] McAllister L, Silverstein E and Westphal A 2010 *Phys. Rev. D* **82** 046003 (*Preprint* 0808.0706)

Journal of Physics: Conference Series **2533** (2023) 012017 doi:10.1088/1742-6596/2533/1/012017

[54] Kefala K, Kodaxis G P, Stamou I D and Tetradis N 2021 *Phys. Rev. D* **104** 023506 (*Preprint* 2010.12483)
[55] Dalianis I, Kodaxis G P, Stamou I D, Tetradis N and Tsigkas-Kouvelis A 2021 *Phys. Rev. D* **104** 103510 (*Preprint* 2106.02467)
[56] Fumagalli J, Renaux-Petel S and Witkowski L T 2021 *JCAP* **08** 030 (*Preprint* 2012.02761)
[57] Fumagalli J, Renaux-Petel S e and Witkowski L T 2021 *JCAP* **08** 059 (*Preprint* 2105.06481)
[58] Carr B, Kuhnel F and Sandstad M 2016 *Phys. Rev. D* **94** 083504 (*Preprint* 1607.06077)
[59] Carr B and Kuhnel F 2020 *Ann. Rev. Nucl. Part. Sci.* **70** 355–394 (*Preprint* 2006.02838)
[60] Clesse S and García-Bellido J 2017 *Phys. Dark Univ.* **15** 142–147 (*Preprint* 1603.05234)
[61] García-Bellido J, Carr B and Clesse S 2021 *Universe* **8** 12 (*Preprint* 1904.11482)
[62] Zhou Z, Jiang J, Cai Y F, Sasaki M and Pi S 2020 *Phys. Rev. D* **102** 103527 (*Preprint* 2010.03537)
[63] Minkowski P 1977 *Phys. Lett. B* **67** 421–428
[64] Yanagida T 1979 *Conf. Proc. C* **7902131** 95–99
[65] Gell-Mann M, Ramond P and Slansky R 1979 *Conf. Proc. C* **790927** 315–321 (*Preprint* 1306.4669)
[66] Schechter J and Valle J W F 1980 *Phys. Rev. D* **22** 2227
[67] Mohapatra R N and Senjanovic G 1981 *Phys. Rev. D* **23** 165
[68] Lazarides G, Shafi Q and Wetterich C 1981 *Nucl. Phys. B* **181** 287–300
[69] Colladay D and Kostelecky V A 1998 *Phys. Rev. D* **58** 116002 (*Preprint* hep-ph/9809521)
[70] Kostelecky V A and Russell N 2011 *Rev. Mod. Phys.* **83** 11–31 (*Preprint* 0801.0287)
[71] Shapiro I L 2002 *Phys. Rept.* **357** 113 (*Preprint* hep-th/0103093)
[72] Mavromatos N E and Pilaftsis A 2012 *Phys. Rev. D* **86** 124038 (*Preprint* 1209.6387)
[73] Mavromatos N E 2022 *Phil. Trans. A. Math. Phys. Eng. Sci.* **380** 20210188 (*Preprint* 2108.02152)
[74] Solà Peracaula J, de Cruz Pérez J and Gomez-Valent A 2018 *Mon. Not. Roy. Astron. Soc.* **478** 4357–4373 (*Preprint* 1703.08218)
[75] Solà Peracaula J, Gomez-Valent A and de Cruz Pérez J 2019 *Phys. Dark Univ.* **25** 100311 (*Preprint* 1811.03505)

Journal of Physics: Conference Series **2533** (2023) 012018 doi:10.1088/1742-6596/2533/1/012018

Quantum clock frames: Uncertainty relations, non-Hermitian dynamics and nonlocality in time

Eliahu Cohen

Faculty of Engineering and the Institute of Nanotechnology and Advanced Materials, Bar-Ilan University, Ramat Gan 5290002, Israel

E-mail: eliahu.cohen@biu.ac.il

Abstract. Dynamical evolution can be reconstructed within stationary, closed quantum systems by employing the Page-Wootters "timeless approach". When conditioning upon the state of a "clock" subsystem, the rest of the system regains its time dependence. This mechanism, involving entanglement between the above subsystems has gained much attention during the last few years. After a brief introduction to the topic we will elaborate on a few recent results: The derivation of new time-energy uncertainty relations, emergence of non-Hermitian dynamics when utilizing non-inertial quantum clocks and dynamical nonlocality in quantum time.

1. Introduction

The notion of time has been perplexing since the dawn of humanity. It remains a source of mystery even nowadays, with many disciplines vigorously studying various aspects of time flow, evolution in time and time perception for different observers. Modern physics has majorly dealt with time during the last century employing relativity theory and quantum mechanics. However, time has a different role in these theories which leads to a different fundamental understanding of what it actually is. In quantum mechanics time is typically seen as a background parameter, external to the system of interest and unaffected by it, while in general relativity time is dynamical, relative and affected by mass and energy.

Moreover, quantum mechanics poses the well-known problem of time: A completely isolated quantum system, e.g. the entire universe (as we suspect), must satisfy the (timeless) Wheeler-DeWitt equation [1]. Then, how can the usual notion of evolution be recovered? To answer this question, Page and Wootters presented a framework in which time was taken to be the quantity given by a physical system, a quantum clock [2]. Time, then, becomes a dynamical quantum variable rather than a mere parameter. The Page-Wootters approach has attracted much interest in recent years leading to major development, generalization and various applications [3, 4, 5, 6, 7, 8, 9, 10, 11, 12, 13, 14, 15, 16, 17, 18, 19, 20].

This timeless framework has been shown to lead to a unitary evolution from the clock's perspective in a large class of scenarios [6]. However, when investigating a class of time-energy uncertainty relations in this setting [17], we recently found that this is not always the case. We studied von Neumann measurements of the total energy of a system that contained an internal clock. Unexpectedly, we observed that the evolution from the internal clock's perspective is, in general, non-unitary. We now believe that although our analysis was mostly nonrelativistic, we can borrow a relativistic idea to interpret the result. Because mass and energy are tightly related, measurements of a system's energy, in a sense, correspond to weighing the system, which then interferes with its dynamical time flow. Given that, we started speculating whether the non-unitarity could result from the internal clock being a non-inertial frame of reference. This seems to be indeed the case according to our more recent analysis of non-inertial quantum clocks [18], which shows that when considering an accelerated or gravitating quantum particle with an internal clock degree of freedom, the evolution given by such an internal clock is non-unitary. We did so in a non-relativistic framework with a relativistic correction due to the mass-energy equivalence. With that, the internal degrees of freedom of the particle couple to its external degrees of freedom. This, together with the system's acceleration, leads to the non-unitary evolution from the perspective of the internal clock.

Below we elaborate on these two works as well as on a different result obtained lately within the Page-Wootters framework: dynamical nonlocality in time [16]. Following our earlier studies of dynamical nonlocality in space [21, 22, 23, 24, 25], an interesting notion which seems to deserve more exploration [26], we shall study equations of motion which depend on two remote points in time.

2. Quantum mechanics in the Page-Wootters framework

The Page-Wootters framework consists of a clock system, whose state is given by a vector in a Hilbert space $\mathcal{H}_A$, and the rest of the system, represented by a state in a Hilbert space $\mathcal{H}_R$, whose evolution is studied. The joint system $|\Psi\rangle\rangle \in \mathcal{H}_A \otimes \mathcal{H}_R$ is assumed to be closed and hence

$$H_T|\Psi\rangle\rangle = 0, \tag{1}$$

where H_T is the total Hamiltonian acting on systems A and R. This equation is known as the Wheeler-DeWitt equation. Observe that the imposition of $|\Psi\rangle\rangle$ being an eigenstate with a null eigenvalue by Eq. (1) is not as restrictive as it may seem [3]. In fact, Hamiltonians that differ by constant terms are physically equivalent, which is associated with quantum states being defined up to a global phase. Then, Eq. (1) implies that the total system $|\Psi\rangle\rangle$, historically being thought of as the "Universe" [2], does not evolve with respect to an external time.

The clock system should have an observable T_A associated with its time. It is still desirable that H_A generates translations in time, i.e., $|t_0+t_A\rangle = e^{-iH_A t_A/\hbar}|t_0\rangle$, where $|t_0\rangle$ and $|t_0+t_A\rangle$ are taken to be clock states, but this does not necessarily require T_A to be a self-adjoint canonical conjugate to H_A. In fact, it is possible to construct T_A as a positive operator-valued measure (POVM), which means that the clock states are not necessarily eigenstates of T_A [27, 28, 29]. Such operators obtained from these constructions are symmetric but need not be self-adjoint. Here, however, for simplicity we assume an *ideal clock*, i.e., $[T_A, H_A] = i\hbar\mathbb{1}$, $T_A|t_A\rangle = t_A|t_A\rangle$, and $H_A = -i\hbar\partial/\partial t_A$. Although the Hamiltonians of these clocks are unbounded from below and, hence, unrealistic, they help us avoid technicalities associated with real clocks [30, 31], while providing approximations of them [32, 33]. Thus, corrections to the results presented here are expected when dealing with non-ideal clocks.

Moreover, let H_R denote the Hamiltonian of the system of interest and let $H_{int}(T_A)$ represent the time-dependent term of the evolution of system R set by clock A, which is an interaction between A and R. We have

$$H_T = H_A + H_R + H_{int}(T_A). \tag{2}$$

Observe that, H_{int} is being assumed to be independent of H_A. Inserting Eq. (2) into Eq.

Journal of Physics: Conference Series **2533** (2023) 012018 doi:10.1088/1742-6596/2533/1/012018

refeq:constraint, applying a scalar product by an eigenstate $|t_A\rangle$ of T_A on the left, and defining $|\psi(t_A)\rangle \equiv \langle t_A|\Psi\rangle\rangle$, it holds that

$$i\hbar\frac{\partial}{\partial t_A}|\psi(t_A)\rangle = [H_R + H_{int}(t_A)]\,|\psi(t_A)\rangle, \tag{3}$$

which is the time-dependent Schrödinger equation that denotes the evolution of system R with respect to the time measured by clock A. Then, the usual unitary evolution of a quantum system is recovered from the static picture introduced by Page and Wootters.

As a result, $|\Psi\rangle\rangle$ can be written as

$$|\Psi\rangle\rangle = \int dt_A\,|t_A\rangle \otimes |\psi(t_A)\rangle. \tag{4}$$

Because $|\Psi\rangle\rangle$ contains information about $|\psi(t_A)\rangle$ at every t_A, it is sometimes referred to as the *history state*.

The results in Refs. [34, 35, 11] are central to our analysis. In their approach, system R is assumed to contain a clock B, and the rest of it is simply referred to as system S. Then, while system B gives the internal time of the system $R = B + S$, system A provides time as observed by an external system. In this case, the total Hamiltonian is

$$H_T = H_A + H_B + H_S + H_{int}(T_B). \tag{5}$$

Although the time-dependent term H_{int} can be taken to be a function of both T_A and T_B in a more general scenario, we assume it does not depend on T_A for simplicity. However, nothing significant changes in our analysis if H_{int} is also a function of T_A.

With the previous H_T, the analog of the Schrödinger equation becomes

$$i\bar{h}\frac{\partial}{\partial t_A}|\psi(t_A)\rangle = (H_B + H_S + H_{int}(T_B))\,|\psi(t_A)\rangle. \tag{6}$$

This implies that the effective Hamiltonian acting on system R is

$$H_{eff} = H_B + H_S + H_{int}(T_B). \tag{7}$$

While the designation of which clock is internal or external to the system of interest seems to be arbitrary up until now, they will acquire a more concrete meaning below when measurements of energy are studied.

3. Time-energy uncertainty relations and the emergence of non-unitarity

While discussing time-energy uncertainty relations we shall demonstrate the emergence of non-Hermitian dynamics in two scenarios used for performing energy measurements: (1) Measurement carried out by an external system, and (2) Measurement carried out by an internal system. See also [17] for more details.

3.1. Measurement carried out by an external system

Here we study the von Neumann measurement of the total energy of system R in the Page and Wootters timeless framework. Specifically, we consider the case where the measurement is carried out by an external system. Then, in addition to the systems already introduced above, we also consider an external pointer system E, and the history state $|\Psi\rangle\rangle$ belongs to the space $\mathcal{H}_A \otimes \mathcal{H}_R \otimes \mathcal{H}_E$.

For simplicity, the free evolution of system E will be neglected. Thus, the measurement interaction can be represented by $H_{VN} = g(T_A)H_R P_E$, where $H_R = H_B + H_S + H_{int}(T_B)$, P_E is the conjugate momentum of the pointer E, and g is a non-negative function that differs from zero exclusively during the duration of the measurement. For notation purposes, we assume the measurement starts at $t_A = 0$ and ends at $t_A = \tau$ (both according to clock A). Moreover, we let

$$\int_{-\infty}^{\infty} g(t)dt = \int_{0}^{\tau} g(t)dt = K, \tag{8}$$

where K is a positive real constant associated with the strength (hence, the precision) of the measurement.

Moreover, we assume that P_E only takes non-negative values. As we shall see, the duration of the measurement in clock B will linearly depend on P_E. Thus, this condition will assure that the clocks considered here are "good" clocks in the sense that they do not move backward in time and, in particular, that the duration of the measurement is non-negative.

With that, the total Hamiltonian of the composed system is $H_T = H_A + H_R + g(T_A)H_R P_E$. Then, using Eq. (1) and defining $|\psi(t_A)\rangle \equiv \langle t_A|\Psi\rangle\rangle$, we obtain the Schrödinger equation

$$i\hbar\frac{\partial}{\partial t_A}|\psi(t_A)\rangle = [H_R + g(t_A)H_R P_E]\,|\psi(t_A)\rangle, \tag{9}$$

i.e., from the perspective of clock A, the evolution is generated by the effective Hamiltonian

$$H_1 = H_R + g(t_A)H_R P_E, \tag{10}$$

which is the usual Hamiltonian of a time-independent system during a measurement. This is expected since, in standard quantum mechanics, time is an external parameter, as it is in this case.

Using the Heisenberg equation of motion to study the evolution of T_B with respect to clock A, we conclude that

$$\frac{d}{dt_A}T_B = -\frac{i}{\hbar}[T_B, H_1] = I + g(t_A)P_E. \tag{11}$$

This shows that, when g vanishes, the flow of time in both clocks is the same. In fact, in this case, from the perspective of clock A, any uncertainty in clock B is associated with its initial uncertainty. In particular, if both clocks start localized and synchronized, they remain localized and synchronized.

Yet, clock B ticks faster than clock A whenever g is non-zero, i.e., during the measurement of energy. If g is a function whose integral over time grows smoothly, then the transition to a faster ticking rate also happens smoothly. However, this is not always the case. A dramatic example can be observed by assuming a highly idealized case where g is the delta function. In this case, from the perspective of clock A, there is a sudden "jump" of the pointer of clock B.

On the other hand, using again Eq. (1) and defining $|\phi(t_B)\rangle \equiv \langle t_B|\Psi\rangle\rangle$, we obtain

$$\begin{aligned} i\hbar[I + g(T_A)P_E]\frac{\partial}{\partial t_B}|\phi(t_B)\rangle &= H_A|\phi(t_B)\rangle \\ &+ [I + g(T_A)P_E]\,[H_S + H_{int}(t_B)]|\phi(t_B)\rangle. \end{aligned} \tag{12}$$

Moreover, recalling that P_E is assumed to only take non-negative values, which implies that $I + g(T_A)P_E$ is invertible, the effective Hamiltonian is

$$H_2 = [I + g(T_A)P_E]^{-1} H_A + H_S + H_{int}(t_B). \tag{13}$$

Journal of Physics: Conference Series **2533** (2023) 012018 doi:10.1088/1742-6596/2533/1/012018

Observe that the function g that controls the measurement continues to be a function of the operator T_A, while it was a function of the parameter t_A when considering A's perspective. In a scenario like this, a well-localized event in clock A has uncertainty in clock B which was previously discussed, for instance, in Ref. [11]. This means, in particular, that the start and end of the measurement, which are well-localized in A, have uncertainty in B.

Moreover, in a sense, Hamiltonian H_2 corresponds to a measurement of H_A. However, the measurement function g is controlled by the time in clock A, which is an observable from the perspective of B. Because of it, the measurement Hamiltonian would have been symmetrized in more standard treatments of the process, which is not the case here. In fact, H_2 is non-Hermitian and, as a consequence, the evolution of clock A from the perspective of clock B is non-unitary, a characteristic that will be further discussed later. For now, observe that the Heisenberg equation of motion for T_A with respect to t_B gives

$$\frac{d}{dt_B}T_A = -\frac{i}{\hbar}[T_A, H_2] = [I + g(T_A)P_E]^{-1}. \tag{14}$$

This result is expected in view of Eq. (11) since it gives the inverse of the passage of time on clock B with respect to clock A. However, once again, it should be noticed that here the argument of g is an operator, and no longer a parameter. Then, with respect to clock B, if clock A starts with some uncertainty, there will be some uncertainty about when the measurement starts and ends. Nevertheless, on average, the "flow of time" during the measurement of energy is expected to be smaller in clock A, from the perspective of both clock A and clock B. It could be if interest to further study the possible connections between this purely quantum effect and the well-known time-dilation in special relativity.

Although Eq. (14) shows consistency across different clock perspectives, it raises questions about the use of the Heisenberg equation here. While Schrödinger's and Heisenberg's representations are unitarily equivalent, since the evolution from clock B's perspective is, in general, non-unitary, it is not trivial to obtain the Heisenberg equation from the effective Hamiltonian in the Schrödinger equation. Nevertheless, this can be done with the introduction of an "indefinite metric" in the Hilbert space, a method proposed long ago by Dirac [36] (see also [37] for Pauli's perspective). The method consists of introducing a new metric to the space of states. A well-known class of systems that fit this approach is the class of PT-symmetric systems [38]. It seems, then, that these metrics are associated with particular spatial and temporal symmetries, among others. This raises many other research questions. For instance, How to transform these metrics when switching reference frames? Can modifications in informational and thermodynamic bounds be extracted from these metrics alone? To justify the latter, it is noteworthy that PT-symmetric systems can evolve faster than Hermitian systems [39, 40]. Then, the idea is to look for similar results for the type of non-Hermitian operators of interest to this research avenue.

3.2. Measurement carried out by an internal system

Here we consider once more the total energy measurement of system R, although now the measurement in question is carried out by an internal system. We assume that system R has access to an apparatus I, and the history state $|\Psi\rangle\rangle$ belongs to $\mathcal{H}_A \otimes \mathcal{H}_R \otimes \mathcal{H}_I$. Note that the apparatus may be assumed to be an internal degree of freedom of the system whose energy is left out of the measurement.

Similarly to what was done in the previous subsection, the free evolution of the pointer I will be neglected. As a result, the measurement can be represented by the following von Neumann interaction

$$H_{VN} = \frac{1}{2}[g(T_B)H_R + H_R g(T_B)]P_I, \tag{15}$$

where $H_R = H_B + H_S + H_{int}(T_B)$, P_I is the conjugate momentum of apparatus I, presupposed to only take non-negative values, and g is the same function introduced in the previous section. Observe the necessary symmetrization of the product $g(T_B)H_R$ due to the lack of commutativity between T_B and H_B, as previously considered, for instance, in Ref. [41]. Also, we assume that in clock B the measurement starts at $t_B = 0$ and ends at $t_B = \tau$.

Since the total Hamiltonian of the composed system is $H_T = H_A + H_R + H_{VN}$, using Eq. (1) and defining $|\psi(t_A)\rangle \equiv \langle t_A|\Psi\rangle\rangle$, we obtain the Schrödinger equation

$$i\hbar \frac{\partial}{\partial t_A}|\psi(t_A)\rangle = H_3|\psi(t_A)\rangle, \tag{16}$$

where

$$H_3 = H_R + \frac{1}{2}[g(T_B)H_R + H_R g(T_B)]P_I. \tag{17}$$

is the effective Hamiltonian from the perspective of clock A. With that, the Heisenberg equation of motion for T_B with respect to t_A is

$$\frac{d}{dt_A}T_B = -\frac{i}{\hbar}[T_B, H_3] = I + g(T_B)P_I. \tag{18}$$

On the other hand, using Eq. (1) and defining $|\phi(t_B)\rangle \equiv \langle t_B|\Psi\rangle\rangle$, we obtain

$$\begin{aligned} i\hbar[I + g(t_B)P_I]\frac{\partial}{\partial t_B}|\phi(t_B)\rangle = \left[H_A - \frac{i\hbar}{2}g'(t_B)P_I\right]|\phi(t_B)\rangle \\ + [I + g(t_B)P_I][H_S + H_{int}(t_B)]|\phi(t_B)\rangle, \end{aligned} \tag{19}$$

where the invertibility of $I + g(t_B)P_I$ was used. This means that, from the perspective of clock B, the evolution is generated by the effective Hamiltonian

$$H_4 = [I + g(t_B)P_I]^{-1}\left[H_A - \frac{i\hbar}{2}g'(t_B)P_I\right] + H_S + H_{int}(t_B), \tag{20}$$

which, similarly to H_2, is a non-Hermitian operator.

Then, using the Heisenberg equation of motion to study the evolution of T_A with respect to t_B, it can be concluded that

$$\frac{d}{dt_B}T_A = -\frac{i}{\hbar}[T_A, H_4] = [I + g(t_B)P_I]^{-1}, \tag{21}$$

which, as expected, up to the difference that the argument of g is now a parameter, is the inverse of the relation between the passage of time in the two clocks given by Eq. (18). Here, again, the use of the Heisenberg equation is justified with the introduction of an indefinite metric in the Hilbert space.

Could, then, the non-unitarity observed here be related to gravitational effects? After all, non-unitarity is expected to play a role in quantum gravity, see e.g. [42, 43, 44, 45]. However, gravitationally interacting clocks have been considered in the literature [46, 6, 11], and it was shown that they lead to unitary dynamics.

Then, why is the result different here? The answer to this question lies in the fact that the von Neumann interaction includes products of the time operator of a clock by the individual Hamiltonian of the same or other clocks. In fact, if such a condition is matched, the dynamics of the clocks associated with the time operator in the interaction is non-unitary with respect to the clock whose individual Hamiltonian appears multiplying it. To see that, consider a system composed of n clocks $C_1, C_2, \ldots, C_n$ and let the total Hamiltonian be

$$H_T = \sum_k H_{C_k} + f(T_{C_{r_1}}, T_{C_{r_2}}, \ldots, T_{C_{r_m}}) H_{C_s} + H_{int}, \tag{22}$$

where $m < n$, the indices $r_1, r_2, \ldots, r_m$, and s are different elements of $\{1, 2, \ldots, n\}$, and H_{int} is not a function of H_s. Then, from the perspective of C_s we can write

$$\begin{aligned} i\hbar[I + f(T_{C_{r_1}}, T_{C_{r_2}}, \ldots, T_{C_{r_m}})] \frac{\partial}{\partial t_{C_s}} |\psi(t_{C_s})\rangle = \\ = \left[\sum_{k \neq s} H_{C_k} + H_{int} \right] |\psi(t_{C_s})\rangle, \end{aligned} \tag{23}$$

i.e., the effective Hamiltonian, which is to be compared with H_2, is

$$H_{eff}^{C_s} = [I + f(T_{C_{r_1}}, T_{C_{r_2}}, \ldots, T_{C_{r_m}})]^{-1} \left[\sum_{k \neq s} H_{C_k} + H_{int} \right], \tag{24}$$

assuming that $I + f(T_{C_{r_1}}, T_{C_{r_2}}, \ldots, T_{C_{r_m}})$ is invertible. This Hamiltonian is manifestly non-Hermitian. More specifically, the parts associated with clocks $C_{r_1}, C_{r_2}, \ldots, C_{r_m}$ are non-Hermitian and, hence, have non-unitary evolution because of the lack of commutativity between their individual time and Hamiltonian operators. We find this general appearance of non-unitarity an interesting phenomenon worthy of further research because it could possibly account for an inherent irreversibility of time, even on purely quantum mechanical grounds.

4. Accelerating quantum clocks

In this section, we connect the emergence of non-unitarity just studied, with the existence of acceleration of the clock frame [18]. For that, we need to understand how acceleration affects the Hamiltonian of a system. This influence should come from the potential V that can be added to the Hamiltonian. Our analysis here will be restricted to potentials that functions of the position x only, but this should be generalized in future work. In this case, we know from Newtonian physics that $ma = -dV/dx$, where m is the mass of the system and a is its acceleration. This implies that $V(x) = -m \int_{x_0}^{x} a(x')dx'$, where it was assumed that $V(x_0) = 0$. Further assuming that the particle is moving in a single direction and that x_0 is a point very far away in the direction of motion of it, we can define the function $f(x) = \int_x^{x_0} a(x')dx'$, which is always non-negative if the particle's velocity is never lower than its initial value. In this case, $V(x) = mf(x)$. Now, with a similar argument to the one used in [47, 5, 48, 49, 6, 12], we can add a post-Newtonian correction to V by using the energy-mass equivalence, i.e., $m \mapsto m + H/c^2$, where H is the Hamiltonian of the particle (other post-Newtonian corrections could be added as well). For simplicity, we neglect the term with the static mass m and consider only H/c^2. Then, redefining f to absorb the constant $1/c^2$, we can write $V(x) = Hf(x)$.

Observe that the energy-mass equivalence implies that the potential (associated with f) also couples to internal degrees of freedom of the system [47]. Then, if the clock is an internal degree of freedom of an accelerating particle, the above discussion leads to the change

$$H_S \mapsto H_S + \frac{1}{2} [f(X)H_S + H_S f(X)], \tag{25}$$

where H_S is the full Hamiltonian of system S, which includes a particle M, before it accelerates, and a clock A, and X is its position. It can be checked that this also leads to a non-unitary evolution from the perspective of clock A. In fact, consider a system composed of two particles, M and N, each with its own internal clock, A and B, respectively. If we assume that particles A starts accelerating at some point and none of the parts interact before that, the total Hamiltonian of both systems is

$$H = H_A + H_B + H_M + H_N + \frac{1}{2}\left[f(X_M)(H_A + H_M) + (H_A + H_M)f(X_M)\right] \tag{26}$$

and the effective Hamiltonian from the perspective of clock A after the acceleration is

$$H^A_{eff} = [I + f(X_M)]^{-1}\left[H_B + H_N - \frac{1}{2}[f(X_M), H_M]\right] + H_M, \tag{27}$$

which is, generally speaking, non-Hermitian, since in most cases $[I+f(X_M)]^{-1}$ does not commute with $[f(X_M), H_M]$. Now, if the particle has a large enough momentum, it may be possible to approximate its free Hamiltonian as proportional to its momentum (this can be done in the relativistic limit or by separating the total non-relativistic momentum into a large expectation value $\langle P\rangle_0$ at an instant t_0, and a "shifted momentum operator" Π such that $P = \langle P\rangle_0 I + \Pi$ and then neglecting the quadratic term Π^2 in the free Hamiltonian). In this case, letting the proportionality constant be the unity, we first observe that $[f(X_M), H_M] = i\hbar f'(X_M)$. Moreover, the Heisenberg equation leads to $X_M/dt_A = I$. As a result, neglecting any initial uncertainty associated with X_M, we can write

$$H^A_{eff} = [I + f(t_A)]^{-1}\left[H_B + H_N - \frac{i\hbar}{2}f'(t_A)\right] + H_M. \tag{28}$$

Observe that, at this stage, we can simplify the analysis by neglecting H_M and H_N. This means that a scenario involving two clocks, A and B, with one of them, A, accelerating with an acceleration associated with $f(T_A)$, can be considered a simplification of a scenario where H_M is neglected. In this sense, the analysis of both cases where the clock is an independent particle or where it is an internal degree of freedom are equivalent.

For a concrete example, consider two inertial rocket ships, each with its own clock. Initially, both are travelling with constant speed. Assuming that there is no interaction between them, we can write

$$H = H_A + H_B, \tag{29}$$

where, without loss of generality, we simplify our discussion by not including any other system in the analysis, which is possible based on what was just discussed.

Using Eq. (1), we observe that the Hamiltonian of clock B from the perspective of clock A is simply $H^A_{eff} = H_B$. Moreover, the Heisenberg equation implies that $dT_B/dt_A = I$, i.e., the flow of time is the same in both clocks.

Now, if the rocket associated with clock A starts accelerating, as was just discussed, the effective Hamiltonian of clock B from the perspective of clock A is given by

$$H^A_{eff} = [I + f(t_A)]^{-1}\left[H_B - \frac{i\hbar}{2}f'(t_A)\right], \tag{30}$$

which implies that evolution is non-unitary. Still, from the perspective of clock B, which is still an inertial frame, clock's A evolution is unitary. In fact, the effective Hamiltonian in this case is

$$H^B_{eff} = H_A + \frac{1}{2}\left[f(T_A)H_A + H_A f(T_A)\right]. \tag{31}$$

However, there will be no perspective from which a unitary evolution is recovered if all clocks are accelerating. In fact, if g is a function associated with B's acceleration, it holds that

$$H = H_A + H_B + \frac{1}{2}\left[f(T_A)H_A + H_A f(T_A)\right] + \frac{1}{2}\left[g(T_B)H_B + H_B g(T_B)\right], \tag{32}$$

which results in non-Hermitian effective dynamics from the perspective of either clock A or B.

In future work we would like to generalize this case to motion in curvilinear coordinates and to the case of gravitationally interacting quantum systems – first within non-relativistic quantum mechanics, and then within the Klein-Gordon and Dirac equations.

5. Gravitationally interacting quantum clocks

Next is the case of gravitationally interacting clocks. We shall only present the gist of the argument, while the detailed analysis is available in [18]. In previous studies, it was shown that this type of interaction leads to a type of time dilation, although the unitarity of the evolution persists [5, 48, 49, 6, 11]. The time dilation can be even associated with loss of interferometric visibility [50].

In those analyses, a Newtonian gravitational potential of the form $V(r) = -Gm_Am_B/r$ was assumed, together with the post-Newtonian energy-mass equivalence correction, as in the above example. Then, the gravitational interaction between two clocks A and B can be added to the Hamiltonian as a term proportional to the product of their free Hamiltonians, i.e., $\lambda H_A H_B$, where $\lambda = -G/c^4 r$.

It can be noticed that the distance between the clocks is assumed to remain constant. Thus, despite the gravitational effects, both clocks work as inertial frames. In our exploration, however, we will allow the relative position of the clocks to change as a result of the gravitational interaction.

As an example, we will start by assuming that clocks A and B are internal degrees of freedom of particles M and N, respectively. Then, the gravitational potential can be written as $V(x_N - x_M) = -Gm_Mm_N/|x_N - x_M|$. Moreover, if S is much more massive than R, for some periods of time we can assume that $x_N - x_M \approx x_N - x_0$, where x_0 is the initial position of system M. Letting the latter vanish, we have $V(x_N) = -Gm_Mm_N/|x_N|$.

Now, using the mass-energy equivalence, we write $V(X_N) = -GH_AH_B|X_N|^{-1}/2c^4$. For simplicity and to make sure that any change to the ticking of either clock is due to the gravitational interaction between the systems, we assume no other interaction between them. This means that the total Hamiltonian of the composed system is

$$H = [I + f(X_N, H_B + H_N)](H_A + H_M) + H_B + H_N, \tag{33}$$

where $f(X_N, H_B) = -GH_B|X_N|^{-1}/2c^4$. As a result, the dynamics from the perspective of clock A is given by

$$H^A_{\text{eff}} = [I + f(X_N, H_B)]^{-1}(H_B + H_M + H_N), \tag{34}$$

which is non-Hermitian since $f(X_N, H_B)$ does not commute with H_N. This is so in spite of system S being assumed to be approximately inertial. This might seem surprising in view of the example with the rocket ships. However, a crucial aspect in that example is that the rockets had no interaction between them whatsoever. Here, system S interacts with the non-inertial system R.

Moreover, the effective Hamiltonian from the perspective of clock B is also non-Hermitian, as expected. More precisely,

$$H^B_{\text{eff}} = [I + f(X_N, H_A)]^{-1}(H_A + H_M + H_N). \tag{35}$$

This, however, is a simple case of Newtonian attraction. It could be interesting to study post-Newtonian and other general-relativistic corrections affecting the dynamics of gravitating quantum clocks.

It is worth highlighting that gravitationally interacting systems were also considered in a recently introduced spacetime quantum reference frame [51]. There, the weak-field limit was assumed in order to avoid the problem of ordering of operators. In this limit, the dynamics was found to be unitary. This is also consistent with our results since the non-Hermitian character of the dynamics is accentuated at higher energies. However, since the Hermiticity of the dynamics appears only as an approximation, predictions using this type of approximation might likely deviate from the ones using the non-Hermitian Hamiltonians found here in experiments that are not relatively short.

It is also noteworthy that an analysis of the dynamics of quantum systems in the presence of singularities with different clocks has revealed the manifestation of non-Hermitian dynamics and, more specifically, non-unitary evolution [52]. In fact, it was found that, in this scenario, unitarity depends on the choice of the clock — even if every clock under consideration is a counterpart of good clocks at the classical level. Moreover, it was concluded that the general covariance of general relativity turns out to be incompatible with quantum unitary dynamics.

6. Dynamical nonlocality in space and time

This section continues to employ the same timeless framework, but for a completely different purpose – introduction and preliminary analysis of dynamical nonlocality in time. However, we begin the discussion with a review of dynamical nonlocality in space. In [53, 23] we have proposed a nonlocal ontology for quantum particles based on a deterministic set of operators within the Heisenberg picture. We have shown that some operators belonging to this set may, in general, evolve in a dynamically nonlocal way. For instance, the operator $e^{iP\ell/\hbar}$, akin to the modular momentum $P_{mod} = P \bmod hI/\ell$, where ℓ has units of length, changes in time according to:

$$\frac{d}{dt}e^{iP\ell/\hbar} = -\frac{i}{\hbar}\left[e^{iP\ell/\hbar}, H\right] = -\frac{i}{\hbar}\left[V(X+\ell I) - V(X)\right]e^{iP\ell/\hbar}, \tag{36}$$

which depends on two remote spatial locations separated by a distance ℓ. This apparent nonlocality is mitigated via quantum uncertainty [23]. The operator $e^{iP\ell/\hbar}$ is important when discussing interference phenomena since its expectation value explicitly depends on the relative phase, e.g. in the case of the double slit experiment [23].

The timeless framework allows to similarly analyze the modular energy operator $e^{iH_S\tau/\hbar}$, where τ is a parameter with units of time and H_S is the Hamiltonian of the system of interest. Here, I consider the case studied in Ref. [11], where system R was assumed to be composed of the main system of interest S and a clock B, i.e.

$$H_R = H_B + H_S + H_{BS}(T_B), \tag{37}$$

where $H_{BS}(T_B)$, assumed to be such that $[H_B, H_{BS}(t)] = 0$ for a parameter t, generates the inner unitary transformation of system S controlled by the time in clock B, i.e., changes on system R when it is completely isolated.

For simplicity, the term $H_{BS}(T_B)$ is assumed to be null. This implies that the effective Hamiltonian of system R from the perspective of clock A is

$$H^A_{eff} = H_B + H_S + H_{int}(t_A). \tag{38}$$

I also restrict to cases where $[H_B, H_{int}(T_A)] = 0$, i.e., H_{int} is not a function of T_B.

Journal of Physics: Conference Series **2533** (2023) 012018 doi:10.1088/1742-6596/2533/1/012018

The goal is to obtain the dynamics of the systems from the perspective of clock B, although a similar conclusion can be drawn based on an analysis from the perspective of clock A. Then, writing $\langle t_B|\Psi\rangle = |\varphi(t_B)\rangle \in \mathcal{H}_A \otimes \mathcal{H}_S$, it holds that

$$i\hbar\frac{\partial}{\partial t_B}|\varphi(t_B)\rangle = [H_A + H_S + H_{int}(T_A)]|\varphi(t_B)\rangle, \tag{39}$$

i.e., from B's perspective, the effective Hamiltonian is

$$H^B_{eff} = H_A + H_S + H_{int}(T_A) \tag{40}$$

From the Heisenberg equation of motion, we obtain the t_B-evolution, i.e., the evolution from the perspective of the clock B of the modular energy $e^{-iH_A\tau/\hbar}$ of clock A, where τ is a parameter with units of time. Explicitly,

$$\begin{aligned} \frac{d}{dt_B}e^{iH_A\tau/\hbar} &= -\frac{i}{\hbar}\left[e^{iH_A\tau/\hbar}, H^B_{eff}\right] \\ &= -\frac{i}{\hbar}\left[e^{iH_A\tau/\hbar}H_{int}(T_A) - H_{int}(T_A)e^{iH_A\tau/\hbar}\right] \\ &= -\frac{i}{\hbar}[H_{int}(T_A + \tau I) - H_{int}(T_A)]\, e^{iH_A\tau/\hbar}, \end{aligned} \tag{41}$$

which is to be compared with the equation of motion of the modular momentum. Observe that no particular set of assumptions was made in this derivation. In fact, the result follows from the fact that $e^{iH_A\tau/\hbar}$ generates translations in the time variable of clock A and H^B_{eff} is, in general, a function of T_A. Also, although Eq. (41) seems to allow backwards-in-time signaling, this is not the case due to quantum uncertainty. Nevertheless, we find it interesting to further pursue these dynamical types of nonlocality – in space and time – and connect them with kinematic (entanglement-based) nonlocality. We have shown in [54] that kinematic nonlocality (even in theories beyond quantum mechanics) will not violate relativistic causality if generalized uncertainty relations are obeyed, which could be interesting to compare with the dynamical case. Additional analysis and specific examples are given in [16].

It may seem that the kinematic and dynamic types of nonlocality are unrelated to each other, but in some recent works [22, 55, 56, 25] connections were made between them relying, on the necessity of quantum entanglement in order to explain the latter type. We would suggest to further pursue this direction in order to unify these two types of nonlocality or alternatively, render them inherently different.

7. Discussion

Qualitative and quantitative features of time in physics remain subtle and deep but the Page-Wootters approach has shown once more to be constructive. We have utilized it above to find new time-energy uncertainty relations pertaining to the interplay between internal and external clocks during the process of energy measurement [17]. We have also employed it to introduce the phenomenon of dynamical nonlocality in quantum time, which is further exemplified in [16]. Finally, we have identified what seems to be a general feature of this approach, namely, the appearance of non-Hermiitan dynamics when non-inertial clocks are used. In particular, we showed how acceleration or (Newtonian) gravitation of massive particles lead to the emergence of non-unitarity from the perspective of quantum clocks internal to them. These results come as a consequence of the coupling between external and internal degrees of freedom of a quantum system including a clock subsystem (associated with the system's proper time).

While this latter result may seem unexpected, it is in line with the expectations of part of the community regarding the breakdown of unitarity in certain quantum gravity scenarios. In fact, using the equivalence principle, we can associate the non-unitarity to a gravitational effect, and some authors have hypothesized that deviations from unitary evolution should be an ingredient of the dynamics of quantum systems affected by gravity [42, 43, 44]. Although, in principle, it is always possible to find a locally inertial frame, in practice, every system of reference is inertial up to a certain order. Thus, further studies of these non-Hermitian systems seem to be of fundamental importance. If one wishes to approximate a clock as an inertial frame for a long period, what type of control must be applied to the clock? Analogously, if the clock is approximated as inertial, for how long can it predict the correct state of a dynamical system within some margin of error? What type of challenges does it bring to the comparison between two quantum clocks? How is the flow of information across systems affected by the non-unitary correction? These are some questions of interest for future work. There are also some possible relations to make with $\mathcal{PT}$-symmetric systems on the one hand (see some discussion in [17, 18], and gravitational approaches towards the problem of time, as well as gravitationally-related decoherence on the other hand.

Acknowledgments
This conference proceedings paper stems from collaborations with Yakir Aharonov, Marcin Nowakowski, Bar Peled, Michael Suleymanov, Amit Te'eni and above all, with Ismael L. Paiva. I wish to thank Hans-Thomas Elze for organizing DICE2022 where some of these results were presented, as well as the conference participants for helpful conversations. This work was supported by the Israel Innovation Authority under Project 73795, by the Pazy Foundation, by the Israeli Ministry of Science and Technology, and by the Quantum Science and Technology Program of the Israeli Council of Higher Education.

References

[1] DeWitt B S 1967 *Phys. Rev.* **160** 1113
[2] Page D N and Wootters W K 1983 *Phys. Rev. D* **27** 2885
[3] Giovannetti V, Lloyd S and Maccone L 2015 *Phys. Rev. D* **92** 045033
[4] Marletto C and Vedral V 2017 *Phys. Rev. D* **95** 043510
[5] Castro Ruiz E, Giacomini F and Brukner Č 2017 *Proc. Natl. Acad. Sci.* **114** E2303
[6] Smith A R H and Ahmadi M 2019 *Quantum* **3** 160
[7] Giacomini F, Castro-Ruiz E and Brukner Č 2019 *Nat. Commun.* **10** 494
[8] Diaz N L and Rossignoli R 2019 *Phys. Rev. D* **99** 045008
[9] Diaz N L, Matera J M and Rossignoli R 2019 *Phys. Rev. D* **100** 125020
[10] Martinelli T and Soares-Pinto D O 2019 *Phys. Rev. A* **99** 042124
[11] Castro-Ruiz E, Giacomini F, Belenchia A and Brukner Č 2020 *Nat. Commun.* **11** 2672
[12] Smith A R H and Ahmadi M 2020 *Nat. Commun.* **11** 5360
[13] Ballesteros A, Giacomini F and Gubitosi G 2020 *arXiv:2012.15769* URL `https://arxiv.org/abs/2012.15769`
[14] Trassinelli M 2021 *arXiv:2103.08903* URL `https://arxiv.org/abs/2103.08903`
[15] Carmo R S and Soares-Pinto D O 2021 *Phys. Rev. A* **103** 052420
[16] Paiva I L, Nowakowski M and Cohen E 2022 *Physical Review A* **105** 042207
[17] Paiva I L, Lobo A C and Cohen E 2022 *Quantum* **6** 683
[18] Paiva I L, Te'eni A, Peled B Y, Cohen E and Aharonov Y 2022 *Commun. Phys.* **5** 298
[19] Adlam E 2022 *Foundations of Physics* **52** 99
[20] Baumann V, Krumm M, Guérin P A and Brukner Č 2022 *Physical Review Research* **4** 013180
[21] Aharonov Y, Cohen E and Rohrlich D 2015 *Physical Review A* **92** 026101
[22] Aharonov Y, Cohen E and Rohrlich D 2016 *Phys. Rev. A* **93** 042110
[23] Aharonov Y, Cohen E, Colombo F, Landsberger T, Sabadini I, Struppa D C and Tollaksen J 2017 *Proc. Natl. Acad. Sci. USA* **114** 6480
[24] Paiva I L, Lenny R and Cohen E 2021 *arXiv preprint arXiv:2110.05824*
[25] Paiva I L, Dieguez P R, Angelo R M and Cohen E 2023 *Physical Review A* **107** 032213
[26] Popescu S 2010 *Nat. Phys.* **6** 151

[27] Busch P, Grabowski M and Lahti P J 1995 *Operational quantum physics* (*Lecture Notes in Physics Monographs* vol 31) (Springer)
[28] Busch P, Lahti P, Pellonpää J P and Ylinen K 2016 *Quantum measurement* (*Theoretical and Mathematical Physics* vol 23) (New York: Springer)
[29] Loveridge L and Miyadera T 2019 *Found. Phys.* **49** 549
[30] Salecker H and Wigner E P 1958 *Phys. Rev.* **109** 571
[31] Peres A 1980 *Am. J. Phys.* **48** 552
[32] Hartle J B 1988 *Phys. Rev. D* **38** 2985
[33] Singh A and Carroll S M 2018 *arXiv:1806.10134* URL `https://arxiv.org/abs/1806.10134`
[34] Höhn P A, Smith A R H and Lock M P E 2021 *Phys. Rev. D* **104** 066001
[35] Höhn P A and Vanrietvelde A 2020 *New J. Phys.* **22** 123048
[36] Dirac P A M 1942 *Proc. R. Soc. A* **180** 1
[37] Pauli W 1943 *Rev. Mod. Phys.* **15** 175
[38] Bender C M and Boettcher S 1998 *Phys. Rev. Lett.* **80** 5243
[39] Bender C M, Brody D C, Jones H F and Meister B K 2007 *Phys. Rev. Lett.* **98** 040403
[40] Zheng C, Hao L and Long G L 2013 *Philosophical Transactions of the Royal Society A: Mathematical, Physical and Engineering Sciences* **371** 20120053
[41] Massar S and Popescu S 2005 *Phys. Rev. A* **71** 042106
[42] Hawking S W 1982 *Commun. Math. Phys.* **87** 395
[43] Unruh W G and Wald R M 1995 *Phys. Rev. D* **52** 2176
[44] Penrose R 1996 *Gen. Relativ. Gravit.* **28** 581
[45] Gambini R, Porto R A and Pullin J 2007 *Gen. Relativ. Gravit.* **39** 1143
[46] Ruiz E C, Giacomini F and Brukner Č 2017 *Proc. Natl. Acad. Sci.* **114** E2303
[47] Pikovski I, Zych M, Costa F and Brukner Č 2015 *Nat. Phys.* **11** 668
[48] Sonnleitner M and Barnett S M 2018 *Phys. Rev. A* **98** 042106
[49] Zych M, Rudnicki Ł and Pikovski I 2019 *Phys. Rev. D* **99** 104029
[50] Zych M, Costa F, Pikovski I and Brukner Č 2011 *Nat. Commun.* **2** 505
[51] Giacomini F 2021 *Quantum* **5** 508
[52] Gielen S and Menendez-Pidal L 2022 *arXiv preprint arXiv:2205.15387*
[53] Aharonov Y, Landsberger T and Cohen E 2015 *arXiv preprint arXiv:1510.03084*
[54] Carmi A and Cohen E 2019 *Sci. Adv.* **5** eaav8370
[55] Vaidman L 2012 *Physical Review A* **86** 040101
[56] Marletto C and Vedral V 2020 *Physical Review Letters* **125** 040401

Journal of Physics: Conference Series **2533** (2023) 012019 doi:10.1088/1742-6596/2533/1/012019

Quantum physics and biology: the local wavefunction approach

George F R Ellis[1,2]

[1] Mathematics Department, University of Cape Town, Rondebosch 7700, Cape Town, South Africa.

[2] The New Institute, 18 Warburgstrasse, Hamburg 20354, Germany.

E-mail: george.ellis@uct.ac.za

Abstract. Is there a single linearly evolving Wave Function of the Universe that is able to lead to all the nonlinearities we see around us? This proposal seems *a priori* highly implausible. I claim that instead, in the real Universe, generically only local wave functions exist. Non-local wave functions occur for carefully engineered contexts such as Bell experiments, but there is no single wave function for a cat or macroscopic object such as a brain, let alone for the Universe as a whole. Contextual wave function collapse leads to a defensible version of the Copenhagen interpretation of quantum theory, where classical macro levels provide the context for quantum events and biological emergence. Complexity arises via multiscale adaptive modular hierarchical structures that enable logical branching to emerge from the underlying linear physics. Each emergent level is causally effective because of the meshing of upwards and downwards causation that takes place consistently with that physics. Quantum chemistry approaches in biological contexts fit this local wavefunction picture.

1. Introduction

This paper reviews three puzzles to do with the way physics relates to emergent reality. They are to do with quantum physics, with complexity, and with how these relate to each other.

1. **The puzzle of quantum linearity** How can the highly non-linear complexity we see around us arise out of a linear equation for a single wave function of the universe, as some claim? I propose that in real world applications only local wave functions exist; there is no single wave function for a living cell, or macroscopic objects such as a cat or brain. Contextual wave function collapse leads to a defensible version of the Copenhagen interpretation of quantum theory where classical macro levels provide the context for quantum events. This is discussed in Section 2.

2. **The puzzle of complexity** So how does complexity arise out of local linear dynamics? This occurs via adaptive modular hierarchical structures, where each emergent level is causally effective because of the meshing of upwards and downwards causation that takes place consistently with the underlying physics via time-dependent constraints. Such emergent structures enable logical branching to emerge from the underlying physics via macromolecular chemistry. Causal completeness is not a property of physics *per se*: it only occurs through closure of constraints involving all linked emergent levels. This is discussed in Section 3.

3. **How these relate to each other** Given the previous two understandings, in what way do linear quantum processes underlie biology? They underlie the basic features of the stability of matter, the nature of the periodic table of the elements, chemical bonding, and the folding of

Journal of Physics: Conference Series **2533** (2023) 012019 doi:10.1088/1742-6596/2533/1/012019

biomolecules. Essentially quantum effects occur in a few cases (enzyme action, photon detection, magnetic field detection), but the local wavefunction approach advocated here is applicable in all cases, and leads to the highly non-linear nature of molecular biology that cannot be described by unitary evolution of a single wave function. This is discussed in Section 4.

4. **The meta issue** is the domain of application of a theory. This is discussed in Section 5.

2. The puzzle of quantum linearity

Quantum theory [Isham (1995)] [Landshoff *et al* (1997)] is essentially a linear (Hamiltonian) theory [Silverman (2008)] [Ellis (2012)]. Nowadays it is often claimed (e.g. [Carroll (2022)]) that there is a unitarily evolving wavefunction for the entire universe from which all else follows, hence all that happens is based in linear evolution of a single wave function .

However the real world is not linear either in terms of structures or associated functions (Figure 1). How are these features compatible with the claimed linearly evolving underlying

Figure 1. A non-linear world: animals, plants, trees

dynamics of a single quantum wave function? There is a problem somewhere, even if the wave function lives in a very high dimensional space, or perhaps an infinite dimensional space: there is clearly a basic incompatibility.

I claim that the solution is that in real world applications, quantum physics applies locally everywhere, but only local wave functions exist; there is no wave function for a cat or similar macroscopic objects such as a brain, let alone for the entire universe.

2.1. Linearity and non-linearity of quantum theory

Essential linearity of Quantum theory

The time-dependent first order Schrödinger equation

$$i\hbar\frac{\partial|\Psi(t)\rangle}{\partial t} = \hat{H}\,|\Psi(t)\rangle \qquad (1)$$

for a wave function $|\Psi(t)\rangle$ ([Laughlin and Pines (2000)], see also [Isham (1995)], eq.(1.6); [Landshoff *et al* (1997)], eq.(2.8b)), is linear in $|\Psi(t)\rangle$ because it is source-free, and the Hermitian operator $\hat{H}$, which acts linearly on $|\Psi(t)\rangle$, does not itself involve $|\Psi\rangle$. Thus for arbitrary constants α, β, we have the linearity relation

$$i\hbar\frac{\partial}{\partial t}\{\alpha\,|\Psi_1(t)\rangle + \beta\,|\Psi_2(t)\rangle\} = \alpha\hat{H}\,|\Psi_1(t)\rangle + \beta\hat{H}\,|\Psi_2(t)\rangle\,. \qquad (2)$$

Key features of quantum theory result from this linearity [Ellis (2012)]:

Journal of Physics: Conference Series **2533** (2023) 012019 doi:10.1088/1742-6596/2533/1/012019

- The use of a complex Hilbert space formalism [Isham (1995)],
- Time reversible Hamiltonian evolution if $\hat{H}$ is time independent [Cappellaro (2012)],
- Interference between quantum entities, as in the 2-slit experiment [Feynman *et al* (1965)],
- Superposition of quantum states [Greenberger *et al* (1993)], [Silverman (2008)],
- The possibility of coherence [Streltsov *et al* (2015)],
- Entanglement ([Isham (1995)] §8.4), [Vedral (2014)],
- Unitarity of the S-matrix in scattering processes [Weinberg (2002)],
- The possibility of using a path integral approach [Feynman *et al* (2005)],
- Fermi-Dirac statistics ([Landshoff *et al* (1997)]:134-137), resulting from the Pauli exclusion principle [Dirac (1926)],
- Bose-Einstein statistics [Ziff *et al* (1977)].

Essential and inessential nonlinearities of quantum theory There are of course also some non-linear features of quantum theory. I classify these as inessential or essential.

Inessential nonlinearity: Non-linearities in the Hamiltonian $\hat{H}$ do not disturb the linearity relation (2). On expanding e^{iH} in a power series, these non-linearities result in the possibility of using Feynman diagrams [Feynman *et al* (2005)] as a calculational tool to solve (1). As the name implies, the 'virtual particles' represented in such diagrams are not real particles, they are rather a conceptual aid in carrying out calculations. Despite the complexity of these diagrams, the wave function is still propagated linearly.

Essential nonlinearity 1: Unless the initial state is already an eigenstate, wave function *collapse to an eigenstate* takes place ([Isham (1995)]: §8.3, §8.5), with probabilities of a specific eigenstate and associated eigenvalue being given by the Born Rule ([Landshoff *et al* (1997)]: pp.43-44, [Atkins and Friedmann (2011)]: p.23). This process is non-linear and irreversible, so it cannot be described by the linear equation (1) and information is generically lost [Ellis (2012)]. This happens all the time in the real world as photons impinge on a screen or on a CCD in a camera or on chlorophyll or rhodopsin molecules, as radioactive atoms decay, as chemical reactions take place, etc. It occurs in laboratory situations when measurements are made, but it is not generically associated with such a context: it need not be a 'measurement' in that sense.

Some presentations of quantum theory simply ignore this non-linearity, claiming that quantum theory is always unitary; but without it there is no point in knowing the wave function, as then no experimentally determinable outcomes are predicted by the theory.

There are alternatives that try to avoid this non-linearity, particularly a *many worlds* view (the Everett interpretation) where such collapse never happens ([Isham (1995)]: §8.5.4), sometimes extending to a 'many minds' view, often proclaimed in popular expositions. Why is this proposed? - it's an extended attempt to claim that only linear dynamics (1) occurs [Carroll (2022)]. However then deriving the Born rule for probabilities of outcomes, which we determine experimentally to be the case in the real universe, is problematic. A key technical problem for these theories is the non-uniqueness of relative states ([Isham (1995)]:157-159). It is also in my view inconceivable that registering a single photon on a single screen in one laboratory splits the entire universe in two, right up to the Hubble radius (and beyond?). Any 'many minds' interpretation ([Isham (1995)]:157) [Butterfield (1996)] is unviable because, apart from any other problems, there is no single wave function for a brain (see below). Tegmark's many varieties of multiverses are critiqued in [Butterfield (2014)].

There are also *hidden variable/pilot wave* theories ([Isham (1995)]:160), [Valentini (2009)] that avoid wave function collapse, with randomness of outcomes being due to the unknown values of hidden variables. They were initially crafted so that their outcomes would be identical to those of standard quantum theory ([Bohm (1957)]:110), but Bell's work led to a variety of proposed tests of those theories e.g. [Simon *et al* (2001)], [Mahler *et al* (2016)]. However if these variables do indeed exist, we have no access to their values except immediately after an experiment (when also the standard theory gives unique outcomes: the same eigenvalue and eigenvector as just measured). Their postulated existence makes no difference to the quantum mechanics impossibility of generically making unique predictions of experimental results from determinable initial data (*"One cannot even in principle know the hidden variable values"* [Genovese (2005)]:376). In any case, these theories do not feature in standard quantum chemistry texts [Buyana (1997)] [Kohn (1999)] [Atkins and Friedmann (2011)], [Karplus (2014)], which develop from the standard approach [Landshoff *et al* (1997)], so I will not consider them further.

I instead support the idea of classical outcomes arising by *Contextual Wavefunction Collapse* [Drossel and Ellis (2018)]. This is in essence a version of the Copenhagen interpretation where the measurement apparatus can be regarded as classical, as experienced in practice by physicists [Bub (2017)]. It is supported by Bohr's idea of the property of a quantum system having a meaning only within the context of a specific measurement situation, as suggested also by the Kochen-Specker Theorem ([Isham (1995)]:p.167), as elaborated recently in [Landsman (2017)]. Real quantum measurements are complex multi-stage affairs that are crucially context-dependent ([Drossel and Ellis (2018)]:§2), [Drossel (2023)].

Essential nonlinearity 2: The dual process to measurement is *state vector preparation*, which is also non-unitary. The key feature of quantum state preparation is pointed out by Isham as follows ([Isham (1995)]:74,134): selected states are drawn from some collection E_i of initial states by a suitable apparatus, for example selected to have some specific spin state, as in the Stern-Gerlach experiment; the other states are discarded. Again it is a non-linear contextual process due to the nature of the relevant apparatus ([Ellis (2012)]:§.5.2.6).

Essential nonlinearity 3: The square in Born's Rule is a basic quantum non-linearity: the wave function does not directly give probabilities, but by a square ([Isham (1995)]:5, [Landshoff *et al* (1997)]:43-44). This is a key difference between classical and quantum physics, and is essentially why complex numbers are needed in quantum theory [Renou *et al* (2021)].

Real world quantum linearities Famously, both superposition and entanglement have been demonstrated at macro scales. However to do so, one must prevent non-linear effects happening, and so need a very carefully engineered context: surfaces must be precisely machined, interactions with heat baths must be minimized, very low temperatures are usually needed at considerable cost, decoherence must be fought by isolation.

This is quite different than conditions in a biological system such as a living cell [Allen and Cowling (2011)], or a brain made of networks of billions of cells joined by synapses [Albright *et al* (2000)]. No linearity (quantum or otherwise) applies in these contexts.

2.2. Real world macroscopic non-linearities

As emphasized at the start, the real world is highly non-linear at emergent scales. A key example is ***feedback control loops*** (Figure 2). Thousands occur at all levels in biology, where they

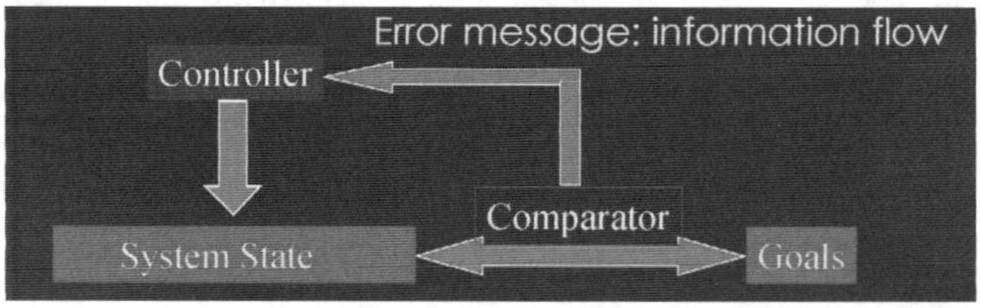

Figure 2. Feedback control loop

underlie homeostasis [Modell *et al* (2015)], for example controlling body temperature, blood

Journal of Physics: Conference Series **2533** (2023) 012019 doi:10.1088/1742-6596/2533/1/012019

pressure, electrolyte levels, and so on. These systems are key to biological function: you are ill if the relevant variable is out of range. They are central to much engineering (e.g. aircraft autopilots and chemical engineering process control) where they are labeled *cybernetic systems* [Wiener (1961)] [Ashby (1957)]. Daily life examples are thermostats for temperature control of water in a water cylinder, or of air in a room.

The general nature of feedback control is shown in Figure 2. A system goal G is set, and a comparator compares the actual system state S with that goal. If they differ, a classical error message E is sent to a controller that takes action to correct that error. Information flows through specifically constructed feedback loops, enabled by the emergent structure of the system. In the case of a thermostat for a room, the goal is a chosen comfortable temperature T_0 set on the thermostat controller. The control circuit in Figure 2 is realised by electrical wiring that either sends current to the heater, or not, depending on the sensor state; this is a carefully engineered emergent structure based in the underlying atoms. It's nature cannot be captured by coarse graining, as is the case with the kinetic theory of gases.

The outcome is not linear, rather branching dynamics of the following form occurs:

$$\{\text{IF } (T < T_o) \text{ THEN } (\textit{apply heat}) \text{ ELSE } (\textit{not})\} \Rightarrow \{T \rightarrow T_0 > 0\}. \tag{3}$$

which cannot occur for a linear source-free 1st order ODE [Dawkins (2023)]. Key points are,

- The initial data is irrelevant to the outcome: it is determined by the goal T_0;
- The branching dynamics (3) is determined physically by time-dependent constraints [Juarrero (2002)], namely the way the wires in the electric circuit control the flow of electrons, depending on the temperature T.
- If heating is needed, the topology of the circuit is closed and current flows. If it is not needed, the circuit is open and no current flows.The emergent structure thus results in discontinuous dynamics.
- Hence it is not unitary dynamics : initial states $T_1 < T_0$ and $T_1' = (1/2)T_1$ lead to the same outcome $T_0 > 0$ in a finite time.
- It therefore cannot be described by linear evolution (1) of a single wave function $|\Psi(t)\rangle$
- If the topology of the emergent structure is changed by either disconnecting the wires to the sensor, or reversing them, the outcome is different. In the first case, no homeostatic action takes place; in the second case, the system burns out. All the components are unchanged.
- The setting of the thermostat determines the average speed of motion $\langle v^2\rangle$ of molecules in the room, because they are determined by the temperature T. Thus this is top-down causation from the emergent macrostructure of the thermostat to the molecules in the room.

A second key example of emergent non-linear systems in biology is contextually-sensitive ***oscillators***, for example in the heart [Noble (2002)], with the function of pumping blood and thereby keeping cells alive; thalamocortical oscillations [Steriade *et al* (1993)]; and adaptive resonant circuits in the brain, underlying attention, learning, and recognition [Grossberg (2013)].

2.3. Atlases and local wavefunctions

How to reconcile local unitary dynamics (2) with emergent branching dynamics (3)? The key idea is *local quantum wavefunctions*. To set the scene I will first introduce the idea of local coordinates as used in in General Relativity theory.

Prolog: GR key idea: Local coordinates. The revolution of global studies in General Relativity Theory [Hawking and Ellis (1973)] arose via the concept of *local coordinates* covering part of a manifold, joined via overlap domains (Figure 3). An atlas covers the manifold globally, but often no single coordinate system exists for the whole: e.g. this is not possible even in the

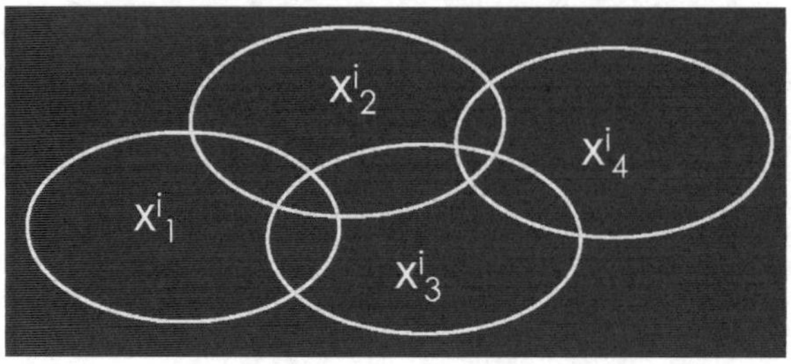

Figure 3. A coordinate atlas for local coordinates x^i_N.

case of the 2-sphere S^2. Use of atlases made global GR studies possible, in particular the study of horizons in black holes via various coordinate systems and the transformations between them.

QM Resolution: Local wave functions. Proposal: Essentially the same applies to quantum theory: Local wave functions exist everywhere, but no global wave function exists, nor even a single wave function for most macro objects, e.g. a living cell, a cat, or a brain.

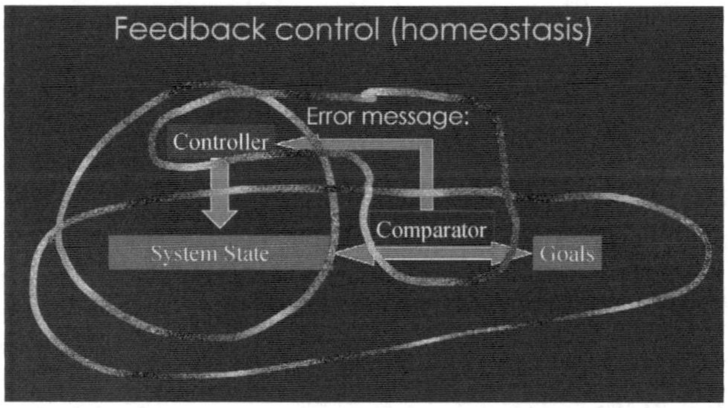

Figure 4. Local wave functions for a feedback circuit

Example: Feedback control (homeostasis) as in Figure 2. In this case, local wave functions would exist and underlie the dynamics in each of the three domains shown in Figure 4, where the dynamics is linear. The fundamental non-linearity of the feedback loop arises by combining the linear dynamics of local wave functions in each of these domains. Thus there is generically a patchwork of local wavefunctions valid in local domains, as in Figure 4. Combined, these cover the system globally, but no single wave function exists for the whole

Each cell in your body has thousands of feedback loops [Allen and Cowling (2011)]. The dynamics is highly non linear. There is no way a cell can be described by a single wave function with linear dynamics (1). But local wavefunctions as in Figure 4 will do the job.

Example: Diffraction of extended bodies A physics example of this approach is the case of diffraction of extended bodies, presented in [Schutz (2022)].

Heat Baths are fundamental to describing macroscopic systems, but cannot be described by a many particle wave function [Drossel (2017)]. From the viewpoint just put, the local wave function domains become so small and so dynamic that it no longer makes sense to separate them; one should rather use the appropriate classical concepts. [Drossel (2020)] explains:

> *"The Schrödinger equation for a macroscopic number of particles is linear in the wave function, deterministic, and invariant under time reversal. In contrast, the concepts used and calculations done in statistical physics and condensed matter physics*

involve stochasticity, nonlinearities, irreversibility, top-down effects, and elements from classical physics. This paper analyzes several methods used in condensed matter physics and statistical physics and explains how they are in fundamental ways incompatible with the above properties of the Schrödinger equation. The problems posed by reconciling these approaches to unitary quantum mechanics are of a similar type as the quantum measurement problem. This paper, therefore, argues that rather than aiming at reconciling these contrasts one should use them to identify the limits of quantum mechanics. The thermal wavelength and thermal time indicate where these limits are for (quasi-)particles that constitute the thermal degrees of freedom."

This view is strengthened by the fact that a cell is a very dynamic context: topological turbulence occurs in the membrane of a living cell [Tan *et al* (2020)], molecular machines extract order from chaos [Hoffmann (2012)], and the brain is an essentially noisy organ [Rolls and Deco (2010)].

Summary

1. Quantum physics applies locally everywhere via local wavefunctions, but does not apply globally with a single wavefunction.
2. Local wavefunctions are restricted to domains where the dynamics is linear [Ellis (2012)].
3. This split cannot be characterised simply in terms of energy or length scales: it is contextually dependent, for example sufficient isolation of a system to avoid decoherence.
4. Non-linear dynamics arises by combining local linear dynamics in non-linear ways.

2.4. Local wave functions imply macro objects are classical

This view supplies a suitable context for interpreting what Leggett states:

> "*It is quite conceivable that at the level of complex, macroscopic objects, the quantum mechanics superposition simply fails to give a correct account of the dynamics of the system*" ([Leggett (1991)]):98).

Isham confirms this possibility ([Isham (1995)]:149):

> "*A macroscopic subsystem (or ensemble of such) may not be describable by any state at all, not even a linear superposition of eigenstates*".

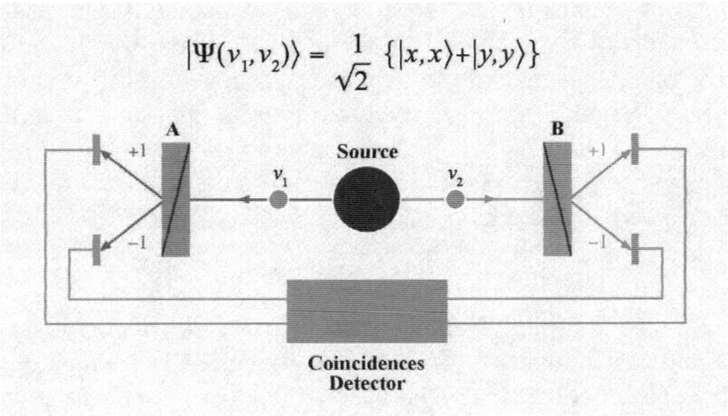

Figure 5. Bell test: Alain Aspect (adapted from [Aspect (2015)] by Mandy Darling).

This leads to a defensible version of the Copenhagen Interpretation of quantum theory. Any measuring apparatus is classical, as in the traditional view supported by Bohr: "*The measuring device is located firmly in an external classical world, thereby opening an irreducible gap between*

Journal of Physics: Conference Series **2533** (2023) 012019 doi:10.1088/1742-6596/2533/1/012019

the quantum system and the instrument of observation ([Isham (1995)]: p.132). One cannot describe it in terms of a single wave function because its dynamics are not linear, and because as it is macroscopic, in thermodynamic terms it contains effective heat baths which cannot be represented by a multiparticle wavefunction [Drossel (2017)], [Drossel (2020)].

Example: Aspect's paper on the Bell test [Aspect (2015)], see Figure 5. All is classical: the source, beam splitters, detectors, coincidence detectors), except a few interacting particles (ν_1 and ν_2). All the rest are conceived of and described in classical terms. To describe them in quantum terms would make the analysis impossible (*inter alia*, the elements of the apparatus would have no definite state).

Schrödinger's cat In particular [Silverman (2005)], there is no single wave function for a cat as a whole, made of many living cells . As stated in [Silverman (2008)] (p.26)):

> "[..] *not everything can be ascribed a wavefunction. Cats, rabbits, chickens, and other ludicrous examples of flora and fauna, which have been employed in popular accounts of quantum mysteries, do not have wave functions* ."

Thus because there is no wave function $|\Psi_{cat}\rangle$, the proverbial

$$|\Psi_{cat}\rangle = \alpha|\Psi_{alive}\rangle + \beta|\Psi_{dead}\rangle \tag{4}$$

is not a legitimate equation, as none of these wave functions exist! The only quantum feature in a typical diagram of the Schrödinger's cat setup is the set of excited atoms that trigger the detector in a random way. Everything else - the cage, the detector, the poison vial, the hammer that breaks it - is classical, including the cat.

2.5. Is there a wave function for the universe?

No, there cannot exist single wave function $|\Psi_{universe}\rangle$ for the universe as a whole, because heat-baths and cats exist in the universe, as do vast numbers of living cells, each containing thousands of feedback loops.

This statement undermines a vast amount of writing concerning alleged outcomes of the existence of such a wavefunction. Of course the idea was originally applied to minisuperspace, where only cosmological degrees of freedom were considered [Wiltshire (1996)]. This is a far cry from the way it is used today, with $|\Psi_{universe}\rangle$ supposed to apply to everything including brains, and there is much talk of minds splitting in two because of the Everett interpretation.

Astrophysical and idealised black holes Is there a single wave function for a black hole? In the case of realistic astrophysical black holes, there are many non-linear processes taking place, with heat baths involved [Begelman and Rees (2020)]. The answer will surely be no. Whether there can be a single wavefunction for an idealised black hole, as envisaged in the huge literature on the "information paradox", would seem to be a moot question. From the viewpoint of this paper, that assumption needs to be justified.

3. The puzzle of complexity

How does this all relate to complex systems such as life, digital computers, aircraft, cities? These are all associated with function or purpose, whereas there is no purpose in physics [Hartwell *et al* (1999)], ([NAS (2022)]:17,47-48). How does this arise out of physics? Via conformational properties of macromolecules in modular hierarchical structures: change of shape of macromolecules, enabled by quantum chemistry effects [Karplus (2014)], underlies biological function at higher levels [Lehn (2004)] [Lehn (2007)]. Time dependent constraints enable branching dynamics to emerge, as in the case of feedback control loops (eqn.(3)).

Journal of Physics: Conference Series **2533** (2023) 012019 doi:10.1088/1742-6596/2533/1/012019

3.1. The Hierarchy of Structure and Causation

Complex systems such as life, digital computers, aircraft, cities, are *multiscale adaptive modular hierarchical structures* [Ellis (2016)] [Martins *et al* (2010)], where each word is important.

Why structures? Because that is how desired function emerges [Brewer and Burrow (1972)]: eyes enable seeing, lungs enable breathing, and so on. Even proteins have functions enabled by their structure, see *Protein Structure and Function* [Petsko and Ringe (2004)]. One can alternatively emphasize the processes carried out by these structures, which are indeed central to function [Dupré and Nicholson (2018)] - but no processes can occur without the structures.

Why modular? The basic principle for handling a complex task is, break up the complex task into simpler tasks; create modules - semi-autonomous parts - to handle the simpler tasks; then merge the outcomes to complete the complex task. This allows abstraction, information hiding, and encapsulation [Booch (1990)]. It furthermore allows evolutionary development of modules, with inheritance: existing modules are adapted, with modifications, to fulfill new functions. Indeed this is the only way to develop truly complex systems [Simon (2019)].

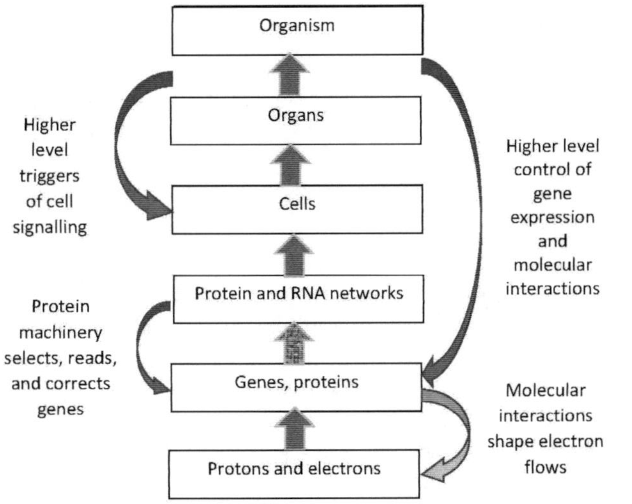

Figure 6. Branching Physics in a Biological Context *Branching biological logic chains down to cause the underlying physical dynamics to branch* From [Ellis and Kopel (2019)].

Why hierarchical? Emergent levels, each with a different nature, naturally result on combining modules to give a multiscale structure. The hierarchy of emergence for biology [Rhoades and Pflanzer (1989)] [Martins *et al* (2010)] is shown in Figure 6. There is a higher level of society, and lower levels of underlying physics (quarks, gluons; maybe strings). For practical purposes, the level of protons, neutrons, and electrons is the lowest level that counts for everyday life [Laughlin and Pines (2000)]. Each level handles different kinds of tasks, supporting other levels. Each level is needed for the whole to work, which is the true nature of causal closure [Montévil and Mossio (2015)] [Ellis (2020)]. Each level is equally significant for the functioning of the whole; none is more fundamental than any other [Noble (2012)].

The hierarchy is **multiscale** over many orders of magnitude, with huge numbers of particles constituting huge numbers of cells that together make up the whole [Martins *et al* (2010)].

Multiple realisability takes place as causation chains down from higher to lower levels: many lower level elements can give essentially the same higher level outcome, and this degeneracy is huge at the genotype level [Wagner (2014)]. Relation to function is hidden at this level.

Journal of Physics: Conference Series **2533** (2023) 012019 doi:10.1088/1742-6596/2533/1/012019

Why adaptive? The system must be able to respond effectively to the environment, which will generically be of a dynamic nature and require different responses at different times and places in order to navigate it safely, obtaining energy and information as required [NAS (2022)].

3.2. Emergence of higher effective levels

Emergence takes place over three different timescales: short term (function), medium term (development), and long term (evolution). Multiple realisability applies in all these cases.

Functionally, different descriptions, variables, and effective laws apply at each emergent level at each time, because symmetry breaking occurs between levels [Anderson (1972)]. In functional terms, the levels are linked by *upward emergence* due to coarse graining, explicit symmetry breaking, or black boxing, and by *downward causation* [Ellis (2016)] due to time dependent constraints [Juarrero (2002)], together with the ability of higher levels to create, modify, or delete lower level elements ('machresis', [Gillett (2017)]), leading to the idea of "understanding the parts in terms of the whole" [Cornish-Bowden *et al* (2014)], as for example in proteostatic regulation in neuronal compartments [Giandomenico *et al* (2022)]. Here *causation* is defined as difference making [Menzies and List (2010)].

These dynamics together enable *same level causation* to occur at every emergent level [Noble (2012)], expressed by effective laws that apply at that level [Ellis (2020a)]. These laws generically exhibit contextually determined logical branching of the form,

$$\text{IF } T(X) \text{ THEN } F1(Y) \text{ ELSE } F2(Y), \tag{5}$$

where Y is a dynamic variable at level L and X is a control variable at level L or level $L+1$. Such contextual branching is the key difference between physics and biology [Ellis and Kopel (2019)], enabling function to emerge [Hartwell *et al* (1999)]. Homeostasis such as (3) is a case of such branching due to time dependent constraints. Metabolism [Fell (1997)] is an example of machresis in physiology and cells, as are brain plasticity effects associated with memory via gene regulation [Kandel (2001)] which determine what proteins will be present.

Developmentally emergence takes place over timescales of between minutes to decades as an organism develops from a single cell to a cooperative ensemble of 10^{11} to 10^{13} cells of different types. The nature of the cells present at each place in the organism is determined by developmental processes [Wolpert *et al* (2009)] controlled by positional information conveyed by morphogens in the ecological context [Gilbert (2001)]. This is a crucial form of machresis.

In evolutionary terms, natural selection [Darwin and Wallace (1958)] [Mayr (2001)] takes place over geological or shorter timescales, depending on the reproduction timescale. Reproduction with variation, followed by natural selection of phenotypes with a relative reproductive advantage in a specific ecological context, leads to selection of genes producing proteins that result in existence of phenotypes better suited to the environment, naturalistically explaining apparent design. The selection principle is relative reproduction success.

Downward causation takes place because different environments result in different outcomes [Campbell (1974)]. Thus for example bears have adapted differently to Arctic snow (polar bears) vs Canadian forests (brown and black bears), so that they will in each case be able to hunt without being conspicuous. Thus these different contexts result in different genes (DNA) resulting in changed proteins resulting in different fur colour. The selection causal chain is

$$Environment \Rightarrow Selected\,Organisms \Rightarrow Selected\,Proteins \Rightarrow DNA\,sequence. \tag{6}$$

Other examples are given in [Wagner (2014)]: ecological advantage acts down to select preferred proteins via associated genes. This is clearly multilevel selection, with a crucial feature:

> *Selection shapes all emergent levels L_I in Figure 6 simultaneously. It has to do so, because they all work together to enable the organism to function [Noble (2012)] [Ellis (2020)]. Selection is not confined to either the gene level or the organism level.*

Journal of Physics: Conference Series **2533** (2023) 012019 doi:10.1088/1742-6596/2533/1/012019

Three closing comments follow, regarding neutral evolution, the 'Evo-Devo' viewpoint, and major evolutionary transitions.

Neutral evolution Essentially as a result of the huge multiple realisability occurring in the process (6), some population geneticists claim that non-adaptive neutral evolution, resulting from genetic drift, almost always takes place rather than adaptive selection [Kimura (1983)], [Muñoz-Gómez *et al* (2021)]. Underlying this view is a definition of "function" based in selection of DNA segments (*"Functional DNA is DNA that is currently under purifying selection - it is being maintained by natural selection"* [Moran (2023)]), rather than defining function in terms of physiological or ecological relations, as in [Hartwell *et al* (1999)], [Ellis (2020a)], [NAS (2022)].

This results in the paradox of these workers in effect denying the central achievement of evolutionary theory, namely the explanation of apparent design [Gardner (2009)]. Evolutionary theorists studying evolution in relation to physiology, ecology, or behaviour do not take this view (see e.g. [Wagner (2014)]). This paper adopts the organisational definition of function [Mossio *et al* (2009)], [Farnsworth *et al* (2017)], ([Ellis and Kopel (2019)]:§1.3).

Evo-Devo In reality, evolutionary and developmental aspects interact with each other. It is developmental systems that make it possible for an organism to exist [Gilbert (2001)], so it is they in particular that are selected by evolutionary processes [Oyama *et al* (2001)]. Their action then alters evolutionary outcomes, leading to the Evolution-Development interaction (Evo-Devo) view of the dynamics taking place [Carroll (2005)], [Carroll (2008)]. In this way gene regulatory networks are selected that enable more complex logic (AND, OR, NOT, and so on) to emerge [Jacob and Monod (1961)], [Monod *et al* (1963)] via cell signalling processes [Berridge (2007)].

Major transitions in evolution [Smith and Szathmary (1997)] correspond to new ways of transmitting information between generations, perhaps new levels of cooperation (entities that were capable of independent replication before can only replicate as part of a larger whole after), and in some key cases, adding a new level L_J to the hierarchy.

3.3. The key physics-biology link: Supramolecular chemistry
The contextual control of supramolecular shape [Lehn (2004)] is the key physics-biology link.

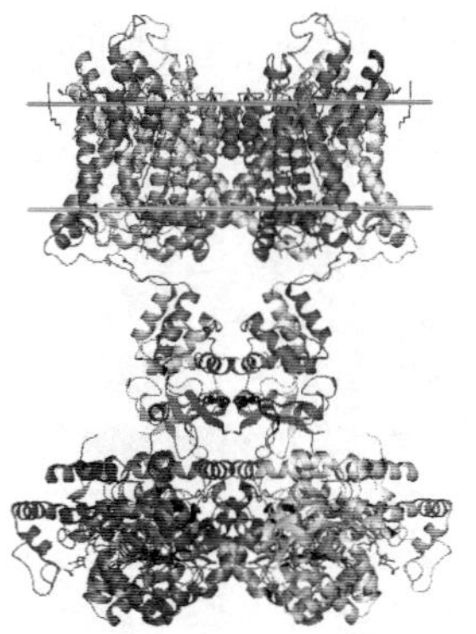

Figure 7. Membrane potassium ion channel structure when the channel is closed. *The channel shape alters according to the voltage difference across the membrane, hence allowing or (in this case) impeding ion passage.* Diagram by Andrei Lomize. From the Open Membranes Database.

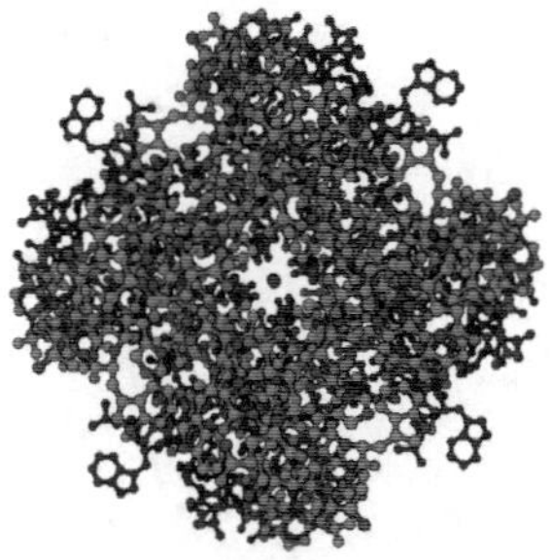

Figure 8. Membrane potassium ion channel structure when the channel is open *Top view of potassium ion (purple, at centre) moving through the potassium ion channel to the cell interior when the channel is open, because the hinged paddles have moved outwards.* (Protein Data Bank:1BL8, open access)

Voltage gated ion channels An important example is the behaviour of voltage gated ion channels in axons. This underlies the Hodgkin-Huxley equations [Hodgkin and Huxley (1952)], resulting in action potential spike chain propagation in neurons [Albright *et al* (2000)].

The key is the changing conformational shape of the protein ion channels imbedded in the cell wall, responding to electric voltage across the cell wall. The channel has four paddles that can change position, seen from the side in Figure 7 and from the top in Figure 8, which move in response to the voltage across the cell wall. They are closed when the voltage gradient is outwards (more positive in the exterior), as in Figure 7, so K^+ ions cannot enter the cell. They are open when the voltage gradient is inwards (more positive in the interior), as in Figure 8, so K^+ ions can enter the cell. This is downward causation from the scale of the ion channel to the scale of the much smaller ion, controlled by the potential gradient across the cell wall.

This context-dependent conformation change enables logic of the form (5) to emerge from the underlying physics:

$$\textbf{IF}\ (\textit{outward})\ \textbf{THEN}\ (\textit{channel closed})\ \text{IF}\ (\textit{inward})\ \textbf{THEN}\ (\textit{channel open}). \qquad (7)$$

This inflow of ions then alters the voltage across the ion channel, so circular causation occurs [Noble *et al* (2019)] and underlies the oscillations that create action potential spike chains.

When this occurs in billions of ion channels imbedded in the neural networks of the brain, it enables propagation of action potential spike chains in neurons [Albright *et al* (2000)], thereby enabling existence of adaptive resonant circuits [Grossberg (2013)] and hence underlying the emergent computational brain [Churchland and Sejnowski (1999)], able for example to calculate solutions to the Schrödinger equation (1) via conditional branching (5). The logical branching occurring in such thoughts and planning emerges in this way through the modular hierarchical structure of the brain [Churchland and Sejnowski (1999)], ([NAS (2022)]:144-152).

Such hugely complex molecules with key biological functions came into existence via evolutionary selection [Wagner (2014)]) because of the crucial role they play in enabling intelligence to emerge, and hence enhanced our ancestor's survival prospects in the past. All proteins have functions [Petsko and Ringe (2004)] which have led to their existence via adaptive selection.

Molecular recognition, logic, and information Molecular shape underlies the lock and key mechanism of molecular recognition [Cramer (1994)] [Berridge (2007)] that in turn underlies function in supramolecular chemistry [Lehn (2004)] [Lehn (2007)]. Complex chemical systems arise from components interacting via non-covalent intermolecular forces, underlying the chemistry of molecular information processing at the supramolecular level through molecular recognition processes. Cell signalling [Berridge (2007)] occurs, underlying gene regulatory networks and metabolic networks [Goelzer *et al* 2008]. Thus life is based in information use and logic [Godfrey-Smith (2007)] [Nurse (2008)] [Farnsworth *et al* (2013)].

3.4. Causal closure

Given this context, causal closure is not a property of the physics level alone, as is often claimed. Because of the confluence of upwards and downwards effects, it occurs only in the context of the interacting whole depicted in Figure 6 via causal closure of constraints [Mossio *et al* (2013)] [Montévil and Mossio (2015)] [Ellis (2020)]. In the case of digital computers, for example, it includes the reasons why an algorithm was written that is controlling flow of electrons through gates [Ellis and Drossel (2019)]. In the case of the functioning of biomolecules via the quantum physics of supramolecular chemistry [Karplus (2014)], it includes the ecological reason for that molecule existing, for example the functioning of antifreeze proteins that enable the arctic cod to survive in waters that regularly chill below 0 C ([Wagner (2014)]:107). The laws of physics by themselves, e.g. Maxwell's equations, do not determine any specific outcomes whatever. Rather they determine the possibility space within which such outcomes arise [Adlam (2022)].

Journal of Physics: Conference Series **2533** (2023) 012019 doi:10.1088/1742-6596/2533/1/012019

In summary: "*The complement of reduction is emergence. While reduction considers how the whole can be explained in terms of its parts, the concept of emergence considers the qualitatively new properties of the whole, which are not properties of the parts ... The microscopic level of a system is causally open to influences from the macroscopic environment*" [Drossel (2023)].

4. The role of quantum physics

How does all this relate to quantum physics? In many ways: it underlies the existence and stability of chemical elements; their nature, as characterised by the periodic table of the elements; the existence and nature of chemical bonds that enable polymer chains to exist and underlie supramolecular chemistry, with its logical branching underlying biological signal processing [Karplus (2014)]. Time dependent constraints occur [Juarrero (2002)], so the Hamiltonian is no longer time-independent: $\hat{H} = \hat{H}(t)$ [Ellis and Kopel (2019)] and the underlying quantum evolution based in (1) is therefore no longer unitary, and is not time reversible [Drossel (2020)].

4.1. Nature and stability of chemical elements

Stability of matter arises via the Pauli exclusion principle [Lieb (1976)] [Lieb (1991)]. The exclusion principle also underlies the existence of electron shells, and hence determines the *Periodic Table of the Elements* [Pyykko (2012)] [Scerri (2019)], which is crucial to all biology.

4.2. Quantum chemistry: bonds

Chemical Bonds are firstly due to electrostatic forces between the valence electrons of atoms ([Buyana (1997)]:186-188), resulting in two types of atomic bonds creating molecules: namely covalent and ionic. Secondly, there are two types of bonding due to molecular forces: namely van der Waals bonds and hydrogen bonds. These ultimately arise from the Schrödinger equation (1) applied in a molecular context [Laughlin and Pines (2000)] [Atkins and Friedmann (2011)].

4.3. Quantum chemistry: simple molecules

For polyatomic molecules, one can write a molecular wavefunction that is a combination of atomic wave functions. Its use to study simple molecules is discussed in ([Buyana (1997)]:§6.)

The description at the microscopic level is based on a Hamiltonian H for nuclei of mass M_i and atomic number Z_i at position $\mathbf{R}_i$ and electrons of charge e at position $\mathbf{r}_j$ [Laughlin and Pines (2000)] ([Phillips (2012)]:16). It has terms for the kinetic energies of the nuclei and of the electrons, and for the potential energies of the Coulombic interactions of the three kinds: nuclei-to-nuclei; nuclei-to-electrons; and electrons-to-electrons:

$$\begin{aligned} H \; = \; & -\sum_i \frac{\hbar^2}{2M_i}\nabla^2_{\mathbf{R}_i} - \sum_j \frac{\hbar^2}{2m_e}\nabla^2_{\mathbf{r}_j} + \sum_i \sum_{i'>i} \frac{Z_{i'} Z_i e^2}{4\pi\epsilon_0 \left|\mathbf{R}_{i'} - \mathbf{R}_{[]}\right|} \\ & - \sum_i \sum_j \frac{Z_i e^2}{4\pi\epsilon_0 \left|\mathbf{R}_i - \mathbf{r}_j\right|} + \sum_i \sum_{i'>i} \frac{e^2}{4\pi\epsilon_0 \left|\mathbf{r}_{i'} - \mathbf{r}_i\right|} \,. \end{aligned} \tag{8}$$

To derive the wave function from this Hamiltonian the actual molecular structure and electron charge distribution in the molecule, plus a series of approximations, is required [Ellis and Drossel (2019)].

The *Born-Oppenheimer* (adiabatic) approximation ([Atkins and Friedmann (2011)]:258-261) [McQuarrie (2008)] is used, which assumes that the electrons are at all times in equilibrium with the positions of the nuclei. The wave function is factorized into an electron part $\Psi_e(\mathbf{r}, \mathbf{R})$ for given positions of the nuclei, and a nucleus part $\Phi(\mathbf{R})$ (this is thus *not* an entangled state),

$$\Psi(r, \mathbf{R}) = \Phi(\mathbf{R})\Psi_e(\mathbf{r}, \mathbf{R})\,, \tag{9}$$

leading to the electron equation

$$(T_e + V_{ee} + V_{ei})\Psi_e(\mathbf{r}, \mathbf{R}) = E_e(\mathbf{R})\Psi_e(\mathbf{r},\mathbf{R}) \tag{10}$$

and the nucleus equation

$$(T_i + V_{ii} + E_{core} + E_e(\mathbf{R}))\Phi(\mathbf{R}) = E\Phi(\mathbf{R})\,. \tag{11}$$

This is in fact a mixture of classical and quantum physics ([Drossel (2020)]:222), because the nuclei are described classically [Born and Oppenheimer (1927)] [Hansen (2022)], rather than just being a solution of (1). As stated by Hendry, "*The Born-Oppenheimer wavefunction looks more like the solution to an altogether different equation: the nuclei are treated classically, and we view the electrons as constrained by the resultant field.*" [Hendry (2006)], see also [Bishop (2010)].

A series of further approximations are made to get useful results, including *Molecular orbital theory* ([Atkins and Friedmann (2011)]:262-286) and the *Hartree-Fock* approximation ([Atkins and Friedmann (2011)]:296-316). However this only works for small molecules, as Kohn stated in his Nobel lecture [Kohn (1999)], awarded for his work on density functional theory:

> *"Traditional multi-particle wave-function methods when applied to systems of many particles encounter what I call an exponential wall when the number of atoms N exceeds a critical value which currently is in the neighborhood of $N_0 \simeq 10$ (to within a factor of about 2) for a system without symmetries".*

Density Functional Theory (DFT). Because of this computational complexity of the many-body Schrödinger equation, a solution is beyond reach for larger molecules. DFT ([Atkins and Friedmann (2011)]:317-326) sidesteps this problem by making the electron density distribution $n(r)$ rather than the many-electron wave function play a central role [Hasnip *et al* (2014)]. Underlying this is the claim [Hohenberg and Kohn (1964)] that the total energy of the system is a unique functional of the electron density alone, so it is not necessary to compute the full many-body wave function. The approach is a mixture of classical and quantum methods ([Drossel (2020)]:223). However even this method will not work for macro molecules.

4.4. Quantum chemistry: macro molecules

The number of atoms in a macromolecule [Lehn (2007)] is vastly larger than 10. To handle their complexity requires *multiscale models.* Karplus states in his Nobel lecture [Karplus (2014)],

> *"To develop methods to study complex chemical systems, including biomolecules, we have to consider the two elements that govern their behavior: 1) The potential surface on which the atoms move; and 2) the laws of motion that determine the dynamics of the atoms on the potential surfaces. ...although the laws governing the motions of atoms are quantum mechanical, the key realization that made possible the simulation of the dynamics of complex systems, including biomolecules, was that a classical mechanical description of the atomic motions is adequate in most cases."*

Thus one does not solve (1) in this case. A classical background provides a basis for the quantum states. How does this method relate to biological function? Karplus states [Karplus (2014)],

> *"First, evolution determines the protein structure, which in many cases, though not all, is made up of relatively rigid units that are connected by hinges. They allow the units to move with respect to one another. Second, there is a signal, usually the binding of a ligand, that changes the equilibrium between two structures with the rigid units in different positions. ...This type of conformational change occurs in many enzymes as an essential part of their mechanism."*

This enables cell signalling [Berridge (2007)] to occur, underlying metabolic networks ([Wagner (2014)]:§2), [Goelzer *et al* 2008], gene regulatory networks [Jacob and Monod (1961)] [Monod *et al* (1963)] ([Wagner (2014)]:§5), and so developmental systems [Oyama *et al* (2001)]. Overall this is how information is important at this level in biology [Nurse (2008)].

A central issue here is, how does the classical concept of shape, key to occurrence of these processes [Lehn (2004)] [Lehn (2007)], arise out of quantum theory? This is discussed in ([Ramsey (1997)], [Bishop and Ellis (2020)]:§5.2). Molecular shape is generated by the environment [Amann (1991)] [Amann (1993)]. Biochemistry cannot be reduced to physics [Bishop (2005)] [Bishop (2010)] because microbiology interactions based in molecular shape is key in the molecular biology of the gene [Watson *et al* (2013)] and the cell [Alberts *et al* (2018)]. These emergent processes reach down to coordinate interactions of proteins, RNA, and DNA.

Overall, these studies in the quantum chemistry of biomolecules confirm the view put here: the quantum dynamics determined by the linear equation (1) is a locally valid description but does not apply even to complex biomolecules as a whole, much less to living cells or other biological structures. Their highly non-linear structure and function is enabled by local wave functions everywhere, but not by a single wave function that applies on macro scales.

4.5. Quantum circuits and qubits

A new development is simulation of molecules using superconducting quantum processors based in Kitaev-Heisenberg models and highly non-linear quantum circuits ([Tazhigulov *et al* (2022)], see eqn.(2) and Figure 2). Does this undermine what has been stated above? Is there a single wave function determining this dynamics? No, because critical to implementing a quantum computer is the ability to control the state of the qubit. A quantum algorithm is specified as a set of unitary transformations $\{U_1, U_2, U_3, ...,\}$ that are implemented by a sequence of Hamiltonians $\{H_1, H_2, H_3, ...,\}$ that are turned on and off one after the other [DiVincenzo (2000)]. This is done [McKay *et al* (2017)] by a local oscillator shaped by an Arbitrary Waveform Generator (AWG) which outputs a programmable voltage output $V(t)$ via the classical components shown in Figure 1 of that paper. This controls rotations of the Bloch sphere. Thus the quantum processes involved are controlled in a top-down way by classical components with non-linear branching dynamics, in agreement with this paper.

4.6. Essentially quantum effects (coherence, entanglement, tunneling) in biology

Essentially quantum effects in biology are considered in [Arndt *et al* (2009)] [Lambert *et al* (2013)] [Marais *et al* (2018)] [McFadden and Al-Khalili (2018)] [Cao *et al* (2020)] [Kim *et al* (2021)] [Smith *et al* 2022] and [NAS (2022)]. Claims in this regard must be treated with caution: there is some very questionable stuff out there, particularly because the brain is warm and wet, so decoherence is ubiquitous [Tegmark (2000)]. The brain is therefore probably not a quantum computer [Tegmark (2000a)].

Essentially quantum effects claimed are,

1. Magnetic field detection by birds has been claimed to be due to sustained quantum coherence and entanglement in the avian compass [Gauger *et al* (2011)], but this must be analysed in a realistic manner [Smith *et al* 2022].

2. Photon to free electron conversion by chlorophyll and rhodopsin molecules is direct evidence that the process involves quantum mechanical coherence [Engel *et al* (2007)] [Collini *et al* (2010)] ([NAS (2022)]:50-54). Electron transfer in photosynthesis at biologically relevant temperatures depends on an interplay of classical and quantum dynamics ([NAS (2022)]:52). However this claim must also be examined in the light of realistic biological contexts [Cao *et al* (2020)].

3. Proton tunneling effects apparently play an important role in enzyme functions [Masgrau *et al* (2006)] ([NAS (2022)]:53), which are key in molecular biology.

The brain Overall as regards the brain, [Litt *et al* (2006)] summarize:

> "*We argue that computation via quantum mechanical processes is irrelevant to explaining how brains produce thought, contrary to the ongoing speculations of many theorists. First, quantum effects do not have the temporal properties required for neural information processing. Second, there are substantial physical obstacles to any organic instantiation of quantum computation. Third, there is no psychological evidence that such mental phenomena as consciousness and mathematical thinking require explanation via quantum theory. We conclude that understanding brain function is unlikely to require quantum computation or similar mechanisms,*"

but see [Fisher (2017)] for a useful survey, and a claim of entanglement processes in memory. However that seems unlikely due to the effects of decoherence [Tegmark (2000)]. In any case the wave function (9) is not an entangled state, so entanglement does not play a role in standard quantum chemistry based in the Born-Oppenheimer approximation [Atkins and Friedmann (2011)].

Essentially quantum effects plausibly claimed to exist in the brain are always purely local, agreeing with the claim (§2.3) that there is no wave function for an organism or brain as a whole.

4.7. Causal closure revisited

How do the time-dependent constraints that underlie downward causation to the physics level [Ellis and Kopel (2019)] occur? In the context of macromolecular chemistry [Karplus (2014)], this occurs via the Born-Oppenheimer approximation discussed previously. The positions $\mathbf{R}_i$ of the nuclei in (8) are determined by conformational changes of biomolecules as discussed above:

$$\mathbf{R}_i = \mathbf{R}_i(t) \Rightarrow H = H(t) \tag{12}$$

and the evolution determined by (1) is no longer time reversible; it is determined by the changing biological context, where higher level dynamics shape outcomes.

For example, suppose Sue is walking down the street and sees an automobile accident at time t_1. This triggers her brain to determine what happened via computational processes at the emergent brain level [Churchland and Sejnowski (1999)], thereby shaping her immediately successive thoughts ("I must call an ambulance") based in her character (helpful) and previous knowledge (ambulances bring help). Prior knowledge of the state of every particle in her body at any time $t < t_1$ cannot predict this outcome of her brain state at times $t > t_1$, because the accident occurred at an external location: the atoms and electrons in her brain were simply not involved. Causal completeness based in the total details of her initial brain state is not possible. Incoming data is handled predictively [Clark (2013)] and shapes the pattern of action potential spike chains in her brain to reflect her choice of action for times $t > t_1$ via a process of interlevel causal closure involving such spike chains, which are shaped by rational thought that alters the positions $\mathbf{R}_i(t)$ of nuclei in the voltage gated ion channels in her brain, as well as past experiences that have shaped neural connections ([Drossel (2023)]:§6). Circular causation takes place [Noble *et al* (2019)] that involves all these levels, and indeed also the level of society in which the action is situated, which leads to the existence and functionality of the ambulance.

5. Conclusion

Consideration of the effects discussed here will confirm that the local wavefunction view proposed in Section 2 gives a correct description of what is going on.

Quantum theory underlies macromolecular chemistry (§4.4), but not by use of the Schrödinger equation (1) for a single wave function [Karplus (2014)]. Even in the case of simple molecules, and certainly macromolecules, one represents electrons as being influenced in a downward way by nuclei that are solved for separately and then used as a classical background for the electron motions. This is similar to the way electron motions are calculated in solid state physics contexts such as transistors [Ellis and Drossel (2019)], with a crystal lattice providing a background for existence of quasi-particles such as phonons that interact with the electrons [Ellis (2020a)]. Essentially quantum effects occur in very few cases (§4.5).

The local wavefunction view put here (§2.3) covers all these cases successfully, and is adequate to determine how biology arises from physics, in contrast to the proposal [Carroll (2022)] that there is a single global wave function that will suffice to determine the dynamics of everything. There are local wavefunctions everywhere, but no single wavefunction for complex systems, or even for any feedback control loops. There is therefore no wave function for cells, hearts, or brains as a whole. *Inter alia*, this appears to completely undermine any "many minds" worldview based in some version of the Everett interpretation [Zeh (1970)] - provided we believe as usual that mental functions are based in brain operations [Albright *et al* (2000)].

5.1. *Domains of validity*

The issue underlying this all is, What is the domain of validity of a theory? The meta issue is,

Each theory has a restricted domain of application.

Newtonian physics is very good within its domain of application; so is Galilean gravity, Newton's theory of gravity, Maxwell's theory of electromagnetism, Einstein's theory of gravitation, and so on. But each has its limits, as is well known; other theories take over outside that domain.

The question not asked often at present is, setting aside the issue of quantum gravity (where it is indeed often asked),

What is the domain of application of Quantum Theory in a specific context?

Nowadays it is often taken to be universally applicable at all scales, for example to cats, brains, and indeed the universe as a whole. But it is no exception to the general rule just stated: its domain of application is limited too. For example, it does not apply as such to heat baths [Drossel (2017)] or feedback control (this paper, §2.2, §2.3), cats, or brains (§2.4).

Rather quantum theory is locally applicable everywhere at all times in domains small enough that the essential dynamics is linear [Ellis (2012)]. Local wavefunctions allow this, as explained in this paper. Adaptive modular hierarchical structures then enable complex emergence out of these simple local dynamics (§3); and those structures are classical. There is no single wavefunction that underlies their existence, particularly because they are multiscale [Martins *et al* (2010)].

This is true not only in biology but also for example in engineering contexts such as digital computers [Ellis and Drossel (2019)], and so in aircraft, cellphones and so on. It also applies to apparatus used in quantum physics experiments: which can therefore be thought of as classical.

5.2. *Testing domains of validity*

The key question then is,

How does one test the domain of validity of quantum theory in a specific context?

Constructively, one does so by engineering situations where wave functions are not local, as in the experiments on entangled photons that won the 2022 Noble Prize in Physics for Aspect, Clauser, and Zeilinger. In this case violations of Bell inequalities, or quantum teleportation, established that there was a macroscopic domain of validity for a wave function. But those were not situations that occur naturally: they were carefully shaped precisely so as to give this result.

Journal of Physics: Conference Series **2533** (2023) 012019 doi:10.1088/1742-6596/2533/1/012019

The claims of this paper are not negated by a remarkable experiment [Lee *et al* (2021)] where a tardigrade - a microscopic multicellular organism known to tolerate extreme physiochemical conditions via a latent state of life known as cryptobiosis - was entangled with superconducting qubits. The experiment involved putting the tardigrade in an environment where linearity could occur, namely a temperature of sub 10 mK temperatures and pressure of 6×10^6 mba, hence preventing any significant interaction with heat baths. No metabolism or feedback loops were operational in these conditions: the organism was in a state of suspended animation, effectively being a single frozen crystal. Because of the extraordinary nature of the organism, this did not destroy its ability to function biologically when reverted to ordinary conditions for life. But it was never both entangled and functioning biologically at the same time. This situation is of course highly artificial, and cannot occur naturally, *inter alia* because the entire Universe is pervaded by primordial black body radiation at a temperature of $2.73K$ ($\gg 10mK$) at the present time [Hawking and Ellis (1973)]; the requisite temperature can only be created artificially. From the viewpoint of this paper, the experiment demonstrates how the domains of validity of wave-functions for a specific entity can vary with time, depending on context.

Negatively, unless you believe standard quantum theory to be committed to macrorealism (agreed on all sides to be an unfortunate name), one can examine cases where standard quantum theory can be shown to be invalid, as in the case of the Leggett-Garg inequalities [Leggett and Garg (1985)] [Timpson and Maroney (2013)]. Experiments based on these inequalities can be carried out to rule out quantum macrorealism, as for example in [Knee *et al* (2012)]. This establishes the principle of tests for macrorealism in larger system. One can also establish no-go theorems pertaining to the existence of long range entanglement by giving upper bounds to quantum correlations [Kuwahara and Saito (2022)].

There is a community of quantum physicists and philosophers looking at these issues, see for example [Briggs *et al* (2013)], [Ares *et al* (2021)]. I will not attempt an in-depth analysis here; many of the issues discussed there either directly relate to the proposals of the present paper, or can be adapted to do so. Perhaps the main thing missing in those papers is an interaction with the literature on quantum chemistry. This paper fills that gap to some extent.

5.3. Irreversibility and the direction of time

The Schrödinger equation (1) and Hamiltonian (8) are time symmetric, so where does the arrow of time associated with quantum physics outcomes come from? How does an essentially time symmetric quantum theory produce the arrow of time in biology?

The basic time asymmetry that occurs in the physical universe is the expansion of the Universe itself [Ellis (2020b)]. This determines a global *Direction of Time* that points from the start of the Universe to the present day. Together with special conditions in the early universe [Layzer (1975)] [Albert (2001)], this reaches down to affect all the local *arrows of time* (electrodynamic, thermodynamic, wave, diffusion, fracture, biological, mental, and quantum) [Ellis (2014)] [Ellis (2020b)] [Ellis and Drossel (2020)]. The quantum arrow of time arises firstly because realistic quantum collapse processes interact with heat baths [Drossel and Ellis (2018)], and so are time asymmetric [Ellis and Drossel (2020)]. Secondly, local microbiology functioning affected by the Second Law of Thermodynamics [Eddington (1927)] reaches down to alter positions of molecules as in (12), and hence changes the Hamiltonian (8) in a time dependent way. Hence the evolution determined by (1) together with (8), (12), as well as the wave function collapse process and associated loss of of information, are both time asymmetric [Drossel (2023)].

We can of course measure the passing of time via suitable clocks, and an interesting aspect is the feedback that occurs as regards the Second Law: any system which measures the passage of time dissipates entropy at a rate that is proportional to the accuracy of the timekeeping [Pearson *et al* (2021)]. This is an emergent classical relation with the correct direction of time.

Acknowledgements: I thank Andrew Briggs, Howard Wiseman, and Ruth Kastner for helpful comments, Martin Ciupa for a helpful reference, and particularly Jeremy Butterfield for a careful reading of the paper leading to comments resulting in many improvements.

I thank the University of Cape Town Research Committee for financial support.

References

[Acín and Masanes (2016)] Acín, A, and Masanes, L (2016) "Certified randomness in quantum physics." *Nature* **540**: 213-219.

[Adlam (2022)] Adlam, E (2022) "Laws of nature as constraints" *Foundations of Physics* **52**:28.

[Albert (2001)] Albert, D Z (2001) *Time and chance* () (Cambridge, Mass.: Harvard University Press)

[Alberts *et al* (2018)] Alberts, B, Johnson, A, Lewis, J, Morgan, D, Raff, M, and Roberts, P W (2018) *Molecular biology of the cell* (Garland Science)

[Albright *et al* (2000)] Albright, T D, Jessell, T M, Kendel, E R, and Posner, I M (2000) "Neural Science: A century of progress and the mysteries that remain" *Cell*:**100**/*Neuron* (Millenium Review):**25** S1-S55.

[Allen and Cowling (2011)] Allen, T, and Cowling, G (2011) *The cell: A very short introduction* (Oxford: Oxford University Press).

[Amann (1991)] Amann, A (1991) "Chirality: A superselection rule generated by the molecular environment?" *J Math Chemistry* **6**:1-15.

[Amann (1993)] Amann, A (1991) "The gestalt problem in quantum theory: generation of molecular shape by the environment" *Synthese* **97**:125-156.

[Anderson (1972)] Anderson, P W (1972) "More is different: broken symmetry and the nature of the hierarchical structure of science" *Science* **177**:393-396.

[Ares *et al* (2021)] Ares, N, Pearson, A N, and Briggs, G A D (2021) "Eight Oxford Questions: Quantum Mechanics under a new light" In *Do Wave Functions Jump? Perspectives of the work of Giancarlo Ghirardi*, Allori, V, Bassi, A, Durr, D, and Zanghi, N, Eds (Cham: Springer), pp.361-384.

[Arndt *et al* (2009)] Arndt, M, Juffmann, T, and Vedral, V (2009) "Quantum physics meets biology" *HFSP journal* **3**:386-400.

[Ashby (1957)] Ashby, W R (1957) *An introduction to cybernetics* (London: Chapman and Hall).

[Aspect (2015)] Aspect, A (2015) "Closing the door on Einstein and Bohr's quantum debate" *Physics* **8**:123.

[Atkins (2014)] Atkins, P (2014) *Physical Chemistry: A Very Short Introduction* (Oxford: Oxford University Press).

[Atkins and Friedmann (2011)] Atkins, P and Friedmann R (2011) *Molecular Quantum Mechanics* (Oxford: Oxford University Press).

[Bauer *et al* (2018)] Bauer, G, *et al* (2018) "The science of living matter for tomorrow" *Cell Systems* **6**:400-402.

[Begelman and Rees (2020)] Begelman, M, and Rees, M (2020) *Gravity's Fatal Attraction: Black Holes in the Universe* (Cambridge : Cambridge University Press.)

[Berridge (2007)] Berridge, M J (2007) *Cell signalling biology* (London: Portland Press). `http://www.cellsignallingbiology.co.uk/csb/`

[Bishop (2005)] Bishop, R C (2005) "Patching physics and chemistry together" *Philosophy of Science* **72**:710-722.

[Bishop (2010)] Bishop, R C (2010) "Whence chemistry?" *Studies in History and Philosophy of Science Part B: Studies in History and Philosophy of Modern Physics* **41**:171-177.

[Bishop and Ellis (2020)] Bishop, R C and Ellis, G (2020) "Contextual emergence of physical properties" *Foundations of Physics* **50**:481-510.

[Booch (1990)] Booch, G (1990) *Object oriented design with applications* (Benjamin-Cummings Publishing Co)

[Bohm (1957)] . Bohm, D (1957) "Causality and Chance in Modern Physics" (Routledge and Kegan Paul and D. Van Nostrand).

[Born and Oppenheimer (1927)] Born, M and Oppenheimer, J R (1927) "Zur Quantentheorie der Molekeln" *Annalen der Physik* **389**: 457-484.

[Brewer and Burrow (1972)] Brewer, C V, and Burrow, C D (1972) *Life: Form and Function: an Introduction to Biology* (Macmillan).

[Briggs *et al* (2013)] Briggs, G A D, Butterfield, J N, and Zeilinger, A (2013) "The Oxford Questions on the foundations of quantum physics" *Proceedings of the Royal Society A: Mathematical, Physical and Engineering Sciences* **469**:20130299.

[Bub (2017)] Bub, J (2017) "Why Bohr was (Mostly) Right" arXiv preprint https://arxiv.org/abs/1711.01604.

[Butterfield (1996)] Butterfield, J (1996) "Whither the minds?" *The British Journal for the Philosophy of Science* **47**:200-221.

[Butterfield (2014)] Butterfield, J (2014) "Our Mathematical Universe?" *Plus* Magazine 2014 (arXiv:1406.4348).

[Buyana (1997)] Buyana, T (1997) *Molecular Physics* (Singapore: World Scientific).

[Campbell (1974)] Campbell, D T (1974) "Downward causation in hierarchically organised biological systems". In *Studies in the Philosophy of Biology: Reduction and Related Problems*, eds F J Ayala and T Dobhzansky (Berkeley, CA: University of California Press), 179-186.

[Cao *et al* (2020)] Cao, J, *et al* (2020) "Quantum biology revisited" *Science Advances* B:eaaz4888.

[Cappellaro (2012)] Cappellaro, P (2012) *Quantum Theory of Raidation Interactions* MITOpenCourseWare 22.51 `https://ocw.mit.edu/courses/22-51-quantum-theory-of-radiation-interactions-fall-2012/resources/mit22_51f12_notes/`

[Carroll (2005)] Carroll, S B (2005). *Endless forms most beautiful: The new science of evo devo and the making of the animal kingdom* (WW Norton and Company).

[Carroll (2008)] Carroll, S B (2008) "Evo-devo and an expanding evolutionary synthesis: a genetic theory of morphological evolution" *Cell* **134**:2536.

[Carroll (2022)] Carroll, S M (2022) "Reality as a vector in Hilbert space". In *Quantum Mechanics and Fundamentality* (Springer, Cham), pp.211-224.

[Churchland and Sejnowski (1999)] Churchland, P S and Sejnowski, T J (1999) *The computational brain* (Cambridge, Mass: MIT press).

[Clark (2013)] Clark, A (2013) "Whatever next? Predictive brains, situated agents, and the future of cognitive science" *Behavioral and brain sciences* **36**:181-204.

[Collini *et al* (2010)] Collini, E *et al* (2010) "Coherently wired light-harvesting in photosynthetic marine algae at ambient temperature" *Nature* **463**:644647.

[Cornish-Bowden *et al* (2004)] Cornish-Bowden, A, Cardenas, M L, Letelier, J C, Soto-Andrade, J, and Abarza, F G (2004) "Understanding the parts in terms of the whole" *Biology of the Cell* **96**:713-717.

[Cramer (1994)] Cramer, F (1994) "Emil Fischer's Lock-and-Key Hypothesis after 100 years - Towards a Supracellular Chemistry" *Perspectives in supramolecular chemistry: The lock-and-key principle*, pp.1-23.

[Darwin and Wallace (1958)] Darwin, C, and Wallace, A R (1958) *Evolution by natural selection* (memorial volume) (Cambridge: Cambridge University Press).

[Dawkins (2023)] Dawkins, P (2023) "Linear Differential equations" *Paul's Online Notes* §2.1 `https://tutorial.math.lamar.edu/classes/de/linear.aspx?fbclid=IwAR37dDLbvTl9uK7tRRULedinhT5k2ALSc6rk7Sl_YehCD0jXCNpPnpLnEso`

[Dirac (1926)] Dirac, P A M (1926) "On the theory of quantum mechanics" *Proc. R. Soc. Lond.* **A112**:661677.

[DiVincenzo (2000)] DiVincenzo, D. P. (2000) "The physical implementation of quantum computation". *Fortschritte der Physik: Progress of Physics* **48**:771-783 (arXiv:quant-ph/0002077).

[Drossel (2017)] Drossel, B (2017) "Ten reasons why a thermalized system cannot be described by a many-particle wave function" *Studies in History and Philosophy of Science Part B: Studies in History and Philosophy of Modern Physics* **58**:12-21.

[Drossel (2020)] Drossel, B (2020) "What condensed matter physics and statistical physics teach us about the limits of unitary time evolution" *Quantum Studies: Mathematics and Foundations* **7**:217-231.

[Drossel (2023)] Drossel, B (2023) "The passage of time and top-down causation" *Proceedings of the MG16 Meeting on General Relativity Online; 510 July 2021* 2023: `https://www.worldscientific.com/doi/pdf/10.1142/9789811269776_0300.`

[Drossel and Ellis (2018)] Drossel, B, and Ellis, G (2018) "Contextual wavefunction collapse: An integrated theory of quantum measurement" *New Journal of Physics* **20**:113025.

[Dupré and Nicholson (2018)] Dupré, J A, and Nicholson, D J (2018) "A manifesto for a processual philosophy of biology" In *Everything Flows: Towards a Processual Philosophy of Biology* D J Nicholson and J Dupré, Eds (Oxford: Oxford University Press)

[Eddington (1927)] Eddington, A S (1927) *The Nature of the Physical World* (Cambridge Scholars Publishing).

[Ellis (2012)] Ellis, G (2012) "On the limits of quantum theory: Contextuality and the quantum-classical cut" *Annals of Physics* **327**: 1890-1932.

[Ellis (2014)] Ellis, G (2014) "The evolving block universe and the meshing together of times" *Annals of the New York Academy of Sciences* **1326**:26-41.

[Ellis (2016)] Ellis, G (2016) *How can physics underlie the mind? Top-down causation in the human context* (Springer-Verlag, Heidelberg.)

[Ellis (2020)] Ellis, G (2020) "The Causal Closure of Physics in Real World Contexts" *Foundations of Physics* **50**:1057-1097.

[Ellis (2020a)] Ellis, G (2020a) "Emergence in Solid State Physics and Biology" *Foundations of Physics* **50**:1098 1139.

[Ellis (2020b)] Ellis, G (2020b) "Physical Time and Human Time":
Preprint https://arxiv.org/abs/2210.10107.

[Ellis and Drossel (2019)] Ellis, G and Drossel, B (2019) "How downwards causation occurs in digital computers" *Foundations of Physics* **49**:1253-1277.

[Ellis and Drossel (2020)] Ellis, G and Drossel, B (2020) "Emergence of time" *Foundations of Physics* **50**:161-190.

[Ellis and Kopel (2019)] Ellis, G and Kopel, J (2019) "The dynamical emergence of biology from physics: branching causation via biomolecules" *Frontiers in Physiology* **9**:1966.

[Engel *et al* (2007)] Engel, G S *et al* (2007) "Evidence for wavelike energy transfer through quantum coherence in photosynthetic systems" *Nature* **446**:782786.

[Farnsworth *et al* (2017)] Farnsworth, K D, Albantakis, L, and Caruso, T (2017) "Unifying concepts of biological function from molecules to ecosystems" *Oikos* **126**:13671376.

[Farnsworth *et al* (2013)] Farnsworth, K D, Nelson, J, and Gershenson, C (2013) "Living is information processing: from molecules to global systems" *Acta biotheoretica* **61**:203-222.

[Fell (1997)] Fell, D (1997) *Understanding the control of metabolism* (London: Portland Press).

[Feynman *et al* (2005)] Feynman, R P, Hibbs, A R, and Styer, D F (2005) *Quantum mechanics and path integrals* (emended edition) (Minelola, NY: Dover).

[Feynman *et al* (1965)] Feynman, R P, Leighton, R B, and Sands, M (1965). *The Feynman Lectures on Physics, Vol. 3.* (Reading, Mass: Addison-Wesley).

[Fisher (2017)] Fisher, M P (2017) "Are we quantum computers, or merely clever robots" *International Journal of Modern Physics* **B31**:1743001.

[Gardner (2009)] Gardner, A (2009) "Adaptation as organism design" *Biology Letters* **5**:861-864.

[Gauger *et al* (2011)] Gauger, E M, Rieper, E, Morton, J L, Benjamin, S C, and Vedral, V (2011) "Sustained quantum coherence and entanglement in the avian compass" *Phys Rev Lett* **106**:040503.

[Genovese (2005)] Genovese, M (2005) "Research on hidden variable theories: A review of recent progresses" *Physics Reports* **413**:319-396.

[Giandomenico *et al* (2022)] Giandomenico, S L, Alvarez-Castelao, B, and Schuman, E M (2022) "Proteostatic regulation in neuronal compartments" *Trends in Neurosciences* **45**:41-52.

[Gilbert (2001)] Gilbert, S F (2001) "Ecological developmental biology: developmental biology meets the real world" *Developmental biology* **233**:1-12.

[Gillett (2017)] Gillett, C (2017) "Scientific emergentism and its move beyond (direct) downward causation". In *Philosophical and scientific perspectives on downward causation* (Routledge), pp.242-262.

[Godfrey-Smith (2007)] Godfrey-Smith, P (2007) "Information in biology"'. In *The Cambridge companion to the philosophy of biology*,ed. D L Hull and M Ruse (Cambridge: Cambridge University Press), pp. 103-119.

[Goelzer *et al* 2008] Goelzer, A, *et al* (2008) "Reconstruction and analysis of the genetic and metabolic regulatory networks of the central metabolism of Bacillus subtilis" *BMC Syst. Biol.* **2**:20.

[Greenberger *et al* (1993)] Greenberger, D M, Horne, M A, and Zeilinger, A (1993) "Multiparticle interferometry and the superposition principle" *Physics Today* **46**:22-29.

[Grossberg (2013)] Grossberg, S (2013) "Adaptive Resonance Theory: How a brain learns to consciously attend, learn, and recognize a changing world" *Neural networks* **37**:1-47.

[Hansen (2022)] Hanson, D "The Born-Oppenheimer Approximation". Chemistry Libretexts: Chemical Education Digital Library.
`https://chem.libretexts.org/Bookshelves/Physical_and_Theoretical_Chemistry_Textbook_Maps/Book%3A_Quantum_States_of_Atoms_and_Molecules_(Zielinksi_et_al)/10%3A_Theories_of_Electronic_Molecular_Structure/10.01%3A_The_Born-Oppenheimer_Approximation.`

[Hartwell *et al* (1999)] Hartwell, L, Hopfield, J, Leibler, S, and Murray, A (1999) "From molecular to modular cell biology"*Nature***402**:C47-C52.

[Hasnip *et al* (2014)] Hasnip *et al* (2014) "Density functional theory in the solid state" *Philosophical Transactions of the Royal Society* A**372**:20130270.

[Hawking and Ellis (1973)] Hawking, S W, and Ellis, G F R (1973) *The Large Scale Stucture of Spacetime* (Cambridge: Cambridge University Press).

[Hendry (2006)] Hendry, R F (2006) "Is there downward causation in chemistry?" In *Philosophy of Chemistry* (Springer, Dordrecht), 173-189.

[Hodgkin and Huxley (1952)] Hodgkin, A L and Huxley, A F (1952) "A quantitative description of membrane current and its application to conduction and excitation in nerve" *J. Physiol* **117**:500544.

[Hoffmann (2012)] Hoffmann, P (2012) *Lifes ratchet: how molecular machines extract order from chaos* (New York: Basic Books).

[Hohenberg and Kohn (1964)] Hohenberg, P, and Kohn, W (1964) "Inhomogeneous electron gas" *Phys. Rev* **136**:B864B871.

[Isham (1995)] Isham, C J (1995) *Lectures on quantum theory Mathematical and structural foundations* (London: Imperial College Press).

Journal of Physics: Conference Series **2533** (2023) 012019 doi:10.1088/1742-6596/2533/1/012019

[Jacob and Monod (1961)] Jacob, F, and Monod, J (1961) "Genetic regulatory mechanisms in the synthesis of proteins" *Journal of Molecular Biology* **3**:318356.

[Juarrero (2002)] Juarrero, A (2002) *Dynamics in Action: Intentional Behavior as a Complex System* (Cambridge, Mass: MIT Press).

[Kandel (2001)] Kandel, E R (2001) "The molecular biology of memory storage: a dialogue between genes and synapses" *Science*:**294**:1030-1038.

[Karplus (2014)] Karplus, M (2014) "Development of multiscale models for complex chemical systems: from H+ H2 to biomolecules" (Nobel lecture) *Angewandte Chemie International Edition* **53**:9992-10005.

[Kim *et al* (2021)] Kim, Y, *et al* (2021) "Quantum biology: An update and perspective" *Quantum Reports* **3**:80-126.

[Kimura (1983)] Kimura, M (1983) *The neutral theory of molecular evolution.* (Cambridge: Cambridge University Press).

[Knee *et al* (2012)] Knee, G C *et al* (2012) "Violation of a LeggettGarg inequality with ideal non-invasive measurements" *Nature communications* **3**:1-6.

[Kohn (1999)] Kohn, W (1999) "Electronic structure of matterwave functions and density functionals" (Nobel Lecture) *Reviews of Modern Physics* **71**:1253-1266.

[Kuwahara and Saito (2022)] Kuwahara, T, and Saito, K (2022) "Exponential clustering of bipartite quantum entanglement at arbitrary temperatures" *Physical Review* **X12**:021022.

[Lambert *et al* (2013)] Lambert, N, Chen, Y N, Cheng, Y C, Li, C M, Chen, G Y, and Nori, F (2013) "Quantum biology" *Nature Physics* **9**:10-18.

[Landshoff *et al* (1997)] Landshoff, P V, Metherell, A, and Rees, W G (1997) *Essential Quantum Physics* (Cambridge: Cambridge University Press).

[Landsman (2017)] Landsman, K (2017) *Foundations of quantum theory: From classical concepts to operator algebras* (Cham: Springer Nature).

[Laughlin and Pines (2000)] Laughlin, R B, and Pines, D (2000) "The theory of everything". *Proceedings of the national academy of sciences* **97**:28-31.

[Layzer (1975)] Layzer, D (1975) "The arrow of time" *Scientific American* **233**:56-69.

[Lee *et al* (2021)] Lee, K S, *et al* (2021) "Entanglement between superconducting qubits and a tardigrade" arXiv preprint arXiv:2112.07978.

[Leggett (1991)] Leggett, A J (1991). In B J Hiley, F D Peat (Eds.), *Quantum Implications: Essays in Honour of David Bohm* (Routledge, London), pp. 85-104.

[Leggett (1992)] Leggett, A J (1992) "On the nature of research in condensed-state physics" *Foundations of Physics*, **22**:221-233.

[Leggett and Garg (1985)] Leggett, A J and Garg, A (1985) "Quantum mechanics versus macroscopic realism: Is the flux there when nobody looks?" *Phys. Rev. Lett.* **54**:857860.

[Lehn (2004)] Lehn, J M (2004) "Supramolecular chemistry: from molecular information towards self-organization and complex matter" *Reports on progress in physics* **67**: 249.

[Lehn (2007)] J-M Lehn (2007) "From supramolecular chemistry towards constitutional dynamic chemistry and adaptive chemistry" *Chem. Soc. Rev.* **36**:151-160.

[Lieb (1976)] Lieb, E H (1976) "The stability of matter" *Rev Mod Phys* **48**:553-569.

[Lieb (1991)] Lieb, E H (1991) "The stability of matter". In *The Stability of Matter: From Atoms to Stars* (Berlin, Springer): pp.483-499.

[Litt *et al* (2006)] Litt, A, Eliasmith, C, Kroon, F W, Weinstein, S, and Thagard, P (2006) "Is the brain a quantum computer?" *Cognitive Science* **30**:593-603.

[Mahler *et al* (2016)] Mahler, D H, Rozema, L, Fisher, K, Vermeyden, L, Resch, K J, Wiseman, H M, and Steinberg, A (2016) "Experimental nonlocal and surreal Bohmian trajectories" *Science advances* **2**:e1501466.

[Marais *et al* (2018)] Marais, A, *et al* (2018) "The future of quantum biology" *Journal of the Royal Society Interface* **15**:20180640.

[Martins *et al* (2010)] Martins, M L, Ferreira, S C, and Vilela, M J (2010) "Multiscale models for biological systems". *Current opinion in colloid and Interface Science* **15**:18-23.

[Masgrau *et al* (2006)] Masgrau, L, *et al* (2006) "Atomic description of an enzyme reaction dominated by proton tunnelling". *Science*: **312**:237-241.

[Mayr (2001)] Mayr, E (2001) *What evolution is* (London: Basic Books).

[McKay *et al* (2017)] McKay, D C, Wood, C J, Sheldon, S, Chow, J M, and Gambetta, J M (2017) "Efficient Z gates for quantum computing" *Physical Review* **A96**: 022330.

[McFadden and Al-Khalili (2018)] McFadden, J, and Al-Khalili, J (2018) "The origins of quantum biology" *Proceedings of the Royal Society* **A474**:20180674.

[McQuarrie (2008)] McQuarrie, D A (2008) *Quantum chemistry* (Mill Valley, CA:University Science Books).

[Menzies and List (2010)] Menzies, P, and List, C (2010) "The Causal Autonomy of the Special Sciences". In *Emergence in Mind* (Oxford: Oxford University Press) (https://philpapers.org/rec/MENTCA).

[Modell *et al* (2015)] Modell, H, Cliff, W, Michael, J, McFarland, J, Wenderoth, M P, and Wright, A (2015) "A physiologist's view of homeostasis" *Advances in physiology education* **39**: 259-266.

[Monod *et al* (1963)] Monod, J, Changeux, J P, and Jacob, F (1963) "Allosteric proteins and cellular control systems" *Journal of molecular biology* **6**:306-329.

[Montévil and Mossio (2015)] Montévil, M and Mossio, M (2015) "Biological organisation as closure of constraints" *Journal of Theoretical Biology* **372**:179-191.

[Moran (2023)] Moran, L A (2023) "The function wars are over" Sandwalk blog Sunday, January 01, 2023 `https://sandwalk.blogspot.com/2023/01/the-function-wars-are-over.html`

[Mossio *et al* (2013)] Mossio, M, Bich, L, and Moreno, A (2013) "Emergence, closure and inter-level causation in biological systems" *Erkenntnis* **78**:153-178.

[Mossio *et al* (2009)] Mossio, M, Saborido, C, and Moreno, A (2009) "An organizational account of biological functions" *Br. J. Philos. Sci.* **60**:813-841.

[Muñoz-Gómez *et al* (2021)] Muñoz-Gómez, S A, Bilolikar, G, Wideman, J G, and Geiler-Samerotte, K (2021) "Constructive neutral evolution 20 years later" *Journal of molecular evolution* **89**:172-182.

[NAS (2022)] National Academies of Sciences, Engineering, and Medicine (2022) *Physics of Life* (Washington, DC: The National Academies Press) [https://doi.org/10.17226/26403].

[Noble (2002)] Noble, D (2002) "Modeling the heart – from genes to cells to the whole organ" *Science* **295**: 1678-1682

[Noble (2012)] Noble, D (2012) "A theory of biological relativity: no privileged level of causation" *Interface focus* **2**:55-64.

[Noble *et al* (2019)] Noble, R, Tasaki, K, Noble, P J, and Noble, D (2019) "Biological relativity requires circular causality but not symmetry of causation: so, where, what and when are the boundaries?" *Frontiers in physiology* **10**:827.

[Nurse (2008)] Nurse, P (2008) "Life, logic and information" *Nature* **454**:424-426.

[Oyama *et al* (2001)] Oyama, S, Griffiths, P E, and Gray, R D (2001) *Cycles of Contingency: Developmental Systems and Evolution* (Cambridge Mass: MIT Press).

[Pearson *et al* (2021)] Pearson, A N, Guryanova, Y, Erker, P, Laird, E A, Briggs, G A D, Huber, M, and Ares, N (2021) "Measuring the thermodynamic cost of timekeeping" *Physical Review* **X 11**:021029.

[Petsko and Ringe (2004)] Petsko, G A, and Ringe, D (2004) *Protein structure and function* (New Science Press).

[Phillips (2012)] Phillips, P (2012) *Advanced solid state physics* (Cambridge: Cambridge University Press).

[Pyykko (2012)] Pyykko, P (2012) "The physics behind chemistry and the periodic table" *Chemical reviews*: **112**:371-384.

[Ramsey (1997)] Ramsey, J L (1997) "Molecular shape, reduction, explanation and approximate concepts" *Synthese* **111**:233-251.

[Renou *et al* (2021)] Renou, M-O, *et al* (2021) "Quantum theory based on real numbers can be experimentally falsified" *Nature* **600**: 625-628.

[Rhoades and Pflanzer (1989)] Rhoades, R and Pflanzer, R (1989) *Human physiology* (Fort Worth: Saunders College Publishing).

[Rolls and Deco (2010)] Rolls, E T and Deco, G (2010) *The Noisy Brain: Stochastic Dynamics as a Principle of Brain Function* (Oxford: Oxford University Press).

[Scerri (2019)] Scerri, E (2019) *The periodic table: its story and its significance* (Oxford: Oxford University Press).

[Schutz (2022)] Schutz, J W (2022) "The quantum mechanics of extended bodies" *Proc Roy Soc* **A** (submitted).

[Silverman (2005)] Silverman, M P (2005) "Schroedinger's Rabbits: The Many Worlds of Quantum" *Physics in Perspective* **7**:496-498 `https://www.mpsilverman.com/reviews_commentary/schroedingers_rabbits.html`

[Silverman (2008)] Silverman, M P (2008) *Quantum Superposition: Counterintuitive Consequences of Coherence, Entanglement, and Interference.* (Berlin: Springer).

[Simon *et al* (2001)] Simon, C, Brukner, C, and Zeilinger, A (2001) "Hidden-variable theorems for real experiments" *Physical Review Letters* **86**:4427.

[Simon (2019)] Simon, H A (2019) *The Sciences of the Artificial* (Cambridge, Mass: MIT Press).

[Smith and Szathmary (1997)] Smith, J M, and Szathmary, E (1997) *The major transitions in evolution* (Oxford: Oxford University Press).

[Smith *et al* 2022] Smith, L D, Deviers, J, and Kattnig, D R (2022) "Observations about utilitarian coherence in the avian compass" *Scientific reports* **12**:1-10.

[Steriade *et al* (1993)] Steriade, M, McCormick, D A, and Sejnowski, T J (1993) "Thalamocortical oscillations in the sleeping and aroused brain" *Science* **262**:679-685.

Journal of Physics: Conference Series **2533** (2023) 012019 doi:10.1088/1742-6596/2533/1/012019

[Streltsov *et al* (2015)] Streltsov, A, Singh, U, Dhar, H S, Bera, M N, and Adesso, G (2015) "Measuring quantum coherence with entanglement" *Physical review letters* **115**:020403.

[Tan *et al* (2020)] Tan, T H, Liu, J, Miller, P W, Tekant, M, Dunkel, J, and Fakhri, N (2020) "Topological turbulence in the membrane of a living cell" *Nature Physics* **16**:657-662.

[Tanenbaum (2006)] Tanenbaum, A S (2006) *Structured Computer Organisation* (Prentice Hall, Englewood Cliffs).

[Tazhigulov *et al* (2022)] Tazhigulov, R N.*et al* (2022) "Simulating models of challenging correlated molecules and materials on the Sycamore quantum processor" *PRX Quantum* **3**:040318.

[Tegmark (2000)] Tegmark, M (2000) "Importance of quantum decoherence in brain processes" *Physical Review* **E 61**:4194.

[Tegmark (2000a)] Tegmark, M (2000) "Why the brain is probably not a quantum computer" *Information Sciences* **128**: 155-179.

[Timpson and Maroney (2013)] Timpson, C, and Maroney, O (2013) "Quantum-vs. macro-realism: what does the Leggett-Garg inequality actually test?" *The British Journal for the Philosophy of Science* https://arxiv.org/abs/1412.6139.

[Valentini (2009)] Valentini, A (2009) "Beyond the quantum" *Physics World* **22**:32.

[Vedral (2014)] Vedral, V (2014) "Quantum entanglement" *Nature Physics* **10**:256-258.

[Wagner (2014)] Wagner, A (2014) *Arrival of the fittest: solving evolution's greatest puzzle* (London: Penguin).

[Watson *et al* (2013)] Watson, J D, *et al* (2013) *Molecular Biology of the Gene* (Pearson).

[Weinberg (2002)] Weinberg, S (2002) *The Quantum Theory of Fields, vol I* (Cambridge: Cambridge University Press).

[Wiener (1961)] Wiener, N (1961) *Cybernetics or Control and Communication in the Animal and the Machine.* (Cambridge, Mass: MIT Press, Second Edition).

[Wiltshire (1996)] Wiltshire, D N (1996) "An Introduction to quantum cosmology". In B Robson, N Visvanathan and W S Woolcock (eds.), *Cosmology: The Physics of the Universe*, (World Scientific, Singapore), pp. 473-531 (arXiv:gr-qc/0101003v2).

[Wolpert *et al* (2009)] Wolpert, L *et al* (2002) *Principles of development* (Oxford: Oxford University Press).

[Zeh (1970)] Zeh, H D (1970) "On the interpretation of measurement in quantum theory" *Foundations of Physics* **1**:69-76.

[Ziff *et al* (1977)] Ziff, R M, Uhlenbeck, G E, and Kac, M (1977) "The ideal Bose-Einstein gas, revisited" *Physics Reports* **32**:169-248.

Version: 2023/03/12.

Journal of Physics: Conference Series **2533** (2023) 012020 doi:10.1088/1742-6596/2533/1/012020

Expending quantization procedures can help quantum field theories and gravity

John R Klauder

Department of Physics and Department of Mathematics, University of Florida, Gainesville, FL 32611-8440 (U.S.A.)

E-mail: `klauder@ufl.edu`

Abstract. Sometimes, canonical quantization has difficulties that can be cured using a new quantization procedure called affine quantization. After briefly introducing this new procedure, we show that its approach to simple models ensures that a valid quantization can be obtained. These valid procedures then are used to help fields and gravity.

1. Introduction

1.1. The highlights of canonical quantization

Canonical quantization (hereafter CQ) can be summarized briefly as follows. Classical variables, $-\infty < p \;\&\; q < \infty$, and a Poisson bracket $\{q, p\} = 1$, may be promoted to the basic quantum operators, $p \to P \;\&\; q \to Q$, for which $[Q, P] = i\hbar 1\!\!1$. However, a valid quantization requires that the classical variables are Cartesian, e.g., $d\sigma^2_{CQ} = \omega^{-1}\,dp^2 + \omega\,dq^2$. In that case, a classical Hamiltonian, $H(p, q) = \mathcal{H}(p, q) \Rightarrow \mathcal{H}(P, Q)$, and the latter term is the quantum Hamiltonian, which usually is a polynomial.

These important features have been developed using coherent states, e.g., $|p, q\rangle = e^{-iqP/\hbar}\, e^{ipQ/\hbar}\, |\omega\rangle$ with $(Q + iP/\omega)\,|\omega\rangle = 0$,

$$\begin{aligned} H(p,q) &= \langle p,q|\mathcal{H}(P,Q)|p,q\rangle = \langle\omega|\mathcal{H}(P+p, Q+q)|\omega\rangle \\ &= \mathcal{H}(p,q) + \mathcal{O}(\hbar; p, q)\,, \end{aligned} \tag{1}$$

which in the limit $\hbar \to 0$ leads to $H(p, q) = \mathcal{H}(p, q)$. To ensure valid results, CQ requires Cartesian phase space variables. The desired Cartesian variables can be found from the Fubini-Study metric to attain valid coordinates. In particular,

$$d\sigma^2_{CQ} = 2\hbar\,[\,\|\,d|p,q\rangle\|^2 - |\langle p,q|\,d|p,q\rangle|^2 = \omega^{-1}\,dp^2 + \omega\,dq^2\,, \tag{2}$$

an expression designed to cancel any phase factor placed on the coherent states.

A common toy model is a harmonic oscillator, now with simple parameters, in which $H = (p^2 + q^2)/2 \to \mathcal{H} = (P^2 + Q^2)/2$. This models's eigenfunctions and eigenvalues are well known, and we note that, the ground-state eigenvalue is $\hbar/2$, while the full set of eigenvalues are $E_n = \hbar(n + 1/2)$, where $n = 0, 1, 2, 3, \ldots$. A special fetcher of this model is that it has *equal spacing of the eigenvalues*, and in this case that spacing is $\hbar$.

Journal of Physics: Conference Series **2533** (2023) 012020 doi:10.1088/1742-6596/2533/1/012020

It is important to examine different classical variables for the full-harmonic oscillator, such as $q = \bar{q}^3/3$ and $p = \bar{p}/\bar{q}^2$ so that the Poisson bracket is $\{\bar{q}, \bar{p}\} = 1$. So far, such new variables may be promoted to new, and different, quantum operators, $\bar{p} \rightarrow \bar{P}$ & $\bar{q} \rightarrow \bar{Q}$. The new variables can be used for the classical Hamiltonian, and the quantum Hamiltonian would become $\bar{\mathcal{H}} = [\bar{P}(\bar{Q}^{-4})\bar{P} + \bar{Q}^6/9]/2$. Surely, this new quantum Hamiltonian does not have equally spaced eigenvalues. This example emphasizes that the full set of CQ rules are needed to get valid results. Choosing, and using, proper classical variables to promote to unique quantum operators will fully ensure valid results. Suitable coherent states lead to proper classical variables, which then can help lead to valid results [1].

Now we analyze our other quantization procedure.

1.2. The highlights of an affine quantization

Affine quantization (hereafter AQ) can be summarized briefly as follows. The primary feature of AQ is that we can remove a single coordinate point, say $q = 0$. That automatically implies that $P^\dagger \neq P$, i.e., those two operators are no longer equal. The remaining space includes $q > 0$ and $q < 0$. We discard $q < 0$ and keep $q > 0$. To restore self-adjoint operators, we introduce the dilation classical variable $d = pq \Rightarrow D = [P^\dagger Q + QP]/2 = D^\dagger$. It now follows that $[Q, D] = i^- h\, Q$ Instead of Cartesian classical variables, AQ requires that they adopt a different metric, $d\sigma^2_{AQ} = (\beta^- h)^{-1}\, q^2\, dp^2 + (\beta^- h)\, q^{-2}\, dq^2$, which is called a constant negative curvature, here with the value $-2/(\beta^- h)$. In that case, the classical primed-Hamiltonian function, $H'(pq, q) = \mathcal{H}'(pq, q) \Rightarrow \mathcal{H}'(D, Q)$, which is mow the quantum primed-Hamiltonian.

An uncommon toy model is the half-harmonic oscillator, using simple parameters again, for which the classical Hamiltonian is still the same, i.e., $H = (p^2 + q^2)/2$, but now $q > 0$. As a toy, it acts like a particle moving back and forth always bouncing off a wall at $q = 0$ where its momentum instantly reverses direction.

The quantum properties of this models's eigenfunctions and eigenvalues are recently well known, and we note that the eigenvalues are $E_n = 2\hbar(n + 1)$, where, as before, $n = 0, 1, 2, 3, ...$ [2]. Again, we find that a special fetcher of this model is that it also has *equal spacing of the eigenvalues*, and in this case, the spacing is $2\hbar$. In addition the ground state eigenvalue has become $2\hbar$. It is indisputable that the half-harmonic oscillator has a valid solution, and, thanks to the equal spacing and its values, that evidently points to a valid quantization of this model using AQ.

While CQ leads to valid results for the harmonic oscillator, it follows that models like $H = p^2/2 + V(q) \Rightarrow \mathcal{H} = P^2/2 + V(Q)$, provided that $-\infty < V(q) < \infty$, and this example will also lead to valid results. A similar comment holds for AQ, where $q > 0$, and using valid results for the half-harmonic oscillator, it follows that other models, like $H' = (d^2/q^2)/2 + V(q) \Rightarrow \mathcal{H}' = D\,(Q^{-2})\,D/2 + V(Q) = [P^2 + (3/4)\hbar^2/Q^2]/2 + V(Q)$, provided that $-\infty < V(q) < \infty$. Such models will also lead to valid results.[1]

We note that the previous paragraph shows how to create valid quantum Hamiltonians by first ensuring that their kinetic factors are well chosen.

[1] A natural variation of the half-harmonic oscillator uses $0 < q + b < \infty$, in which $0 \leq b < \infty$, only changes the ^-hterm from $(3/4)\hbar^2/Q^2$ to $(3/4)\hbar^2/(Q + b)^2$. This model also has a continuously equally spaced eigenvalues for all b, which run from $2\hbar$, for $b = 0$, down to $\hbar$, as $b \rightarrow \infty$. An excellent graph of several eigenvalues, expressed as a function of b, may be seen in [3], page 36.

2. The Quantization of Field Theories

2.1. Examining the territory

A common model, using the momentum field, $\pi(x)$, and its field, $\varphi(x)$, has the classical Hamiltonian

$$H = \int \{ \tfrac{1}{2}[\pi(x)^2 + (\overrightarrow{\nabla}\varphi(x))^2 + m^2\varphi(x)^2] + g\,\varphi(x)^p \}\; d^s\!x\,. \tag{3}$$

Let us suppose that $\varphi(x)$ represents some feature of nature, e.g., a tiny particle named $z1$ to distinguish it from particle $z2$. Note that if $\varphi(x) = 0$, that value could not distinguish the $z1$ particle from that of the $z2$ particle, or any other particle as well. Stated briefly, nothing is lost if we remove points where $\varphi(x) = 0$. In two or more spatial dimensions, paths can pass by any points where $\varphi(x) = 0$. Moreover, the integral in (3) is completely unchanged if we remove points where $\varphi(x) = 0$.

A new topic points to possibilities that $\pi(x) = \pm\infty$, and also possibly $\varphi(x) = \pm\infty$, which can lead to $H(x) = \infty$ at some points, but still find that $H = \int H(x)\, d^s\!x < \infty$. Such an example could be $H(x) = 1/(x_1^2 + ... + x_s^2)^{s/2}$ integrated over $(x_1^2 + ... + x_s^2) < 1$. Path integration involves that and similar functions. Integrable infinities are acceptable in mathematics, but any field representing nature should never reach infinity.

A way to eliminate infinity expressions is to adopt affine variables, such as $\kappa(x) = \pi(x)\,\varphi(x)$ as the dilation field. Here we remove points where $\varphi(x) = 0$, and also limit all three terms from infinity, specifically such as $0 < |\varphi(x)| < \infty$, while $0 \leq |\pi(x)|$ & $|\kappa(x)| < \infty$. Such limitations ensure correct values for all three functions.

This story leads to an infinity-absent, classical Hamiltonian, using affine variables, and which it is given by

$$H' = \int \{ \tfrac{1}{2}[\kappa(x)^2/\varphi(x)^2 + (\overrightarrow{\nabla}\varphi(x))^2 + m^2\varphi(x)^2] + g\,\varphi(x)^p \}\; d^s\!x\,. \tag{4}$$

It is noteworthy that the kinetic term already demands that $0 < |\varphi(x)| < \infty$ and $0 \leq |\kappa(x)| < \infty$ so they can represent $\pi(x)$ correctly. Moreover, it follows that $0 < |\varphi(x)|^p < \infty$.

2.2. An affine quantization of field theories

We start with traditional CQ field operators, $\hat{\pi}(x)$ and $\hat{\varphi}(x)$. Removing $\hat{\varphi}(x) = 0$ leads to the fact that now $\hat{\pi}(x)^\dagger \neq \hat{\pi}(x)$. We follow the affine procedures and introduce the dilation quantum field, $\hat{\kappa}(x) = [\hat{\pi}(x)^\dagger\,\hat{\varphi}(x) + \hat{\varphi}(x)\,\hat{\pi}(x)]/2 = \hat{\kappa}(x)^\dagger$. Like the single particles, we find that $[\hat{\varphi}(x), \hat{\kappa}(y)] = i\hbar\delta(x-y)\,\hat{\varphi}(x)$. Just like CQ, the AQ operators need scaling and regularization to get around the facts like $[\hat{\varphi}(x), \hat{\pi}(x)] = i\hbar 1\!\!1\,\infty$ and $[\hat{\varphi}(x), \hat{\kappa}(x)] = i\hbar\hat{\varphi}(x)\,\infty$. In this paper we develop expressions that are implicit and accepted as formal expressions. Using formal expressions still requires that their formulation be on the road toward valid results, and not on any other, eventually incorrect, type of road.

Using AQ rules for a formal quantum Hamiltonian of the field models, and using Scrödinger's representation, leads to

$$\mathcal{H}' = \int \{ \tfrac{1}{2}[\hat{\kappa}(x)(\varphi(x)^{-2})\hat{\kappa}(x) + (\overrightarrow{\nabla}\varphi(x))^2 + m^2\varphi(x)^2] + g\,\varphi(x)^p \}\; d^s\!x\,. \tag{5}$$

It is noteworthy that there have been several Monte Carlo (MC) versions of path integration, in which scaling and regularizing have still retained AQ validity possibility, which has to be a useful, and CQ-like, behavior in which the quantum Hamiltonian, again using Schrödinger's representation, becomes[2]

$$\mathcal{H}'' = \int \{ \tfrac{1}{2}[\hat{\pi}(x)^2 + 2\hbar^2/\varphi(x)^2 + (\overrightarrow{\nabla}\varphi(x))^2 + m^2\varphi(x)^2] + g\,\varphi(x)^p \}\; d^s\!x\,. \tag{6}$$

[2] The $2\hbar^2$ numerator term has been recently promoted from $(3/4)\hbar^2$ after further analysis. Clearly, this change boosts the contribution of the $\hbar$-term to the final result.

Observe, that while $\hat{\pi}(x)^\dagger \neq \hat{\pi}(x)$, and thanks to the $\hbar - term$, it ensures that in this equation both $\hat{\pi}(x)^\dagger$ and $(\hat{\pi}(x)^\dagger)^2$ *act* like $\hat{\pi}(x)$ and $\hat{\pi}(x)^2$. That property can aid commutation efforts.

The present MC studies for the model φ^4_4 using CQ, and with no extra $\hbar$-term, have provided 'free results', as if the interaction term was absent when in fact it was present, which is unacceptable. Other MC studies of the same model, now using AQ, which means that the $\bar{}h$-term is present, has led to 'non-free results', just as they should be, which is acceptable. That may not be the final story yet, but, at least, AQ is clearly closer to valid results than CQ is.

3. The Quantization of Einstein's Gravity

3.1. Examining the territory

Guided by introducing dilation variables and the several examples that appear, we are led to what will be called the gravity dilation field, namely $\pi^a_b(x) = \pi^{ac}(x)\, g_{bc}(x)$, with a new feature in which there is a sum over c. As required by physics, $g_{ab}(x) > 0$, meaning that $ds(x)^2 = g_{ab}(x)\, dx^a\, dx^b > 0$, provided that $\Sigma^3_{a=1}\, (dx^a)^2 > 0$.

Now, the classical ADM Hamiltonian [4], expressed in affine classical variables, uses the metric $g_{ab}(x) > 0$, $g(x) = \det[g_{ab}(x)] > 0$, and $\pi^a_b(x) = \pi^{ac}(x)\, g_{bc}(x)$, which leads to

$$H_{AQ-ADM} = \int \{ g(x)^{-1/2} [\pi^a_b(x) \pi^b_a(x) - \tfrac{1}{2}\, \pi^a_a(x) \pi^b_b(x)] \\ + g(x)^{1/2}\, {}^{(3)}R(x) \}\, d^3x\,, \tag{7}$$

where ${}^{(3)}R(x)$ is the Ricci scalar for 3 spatial dimensions.

Since the customary classical variables are the momentum field, $\pi^{ab}(x)$ and the metric field, $g_{ab}(x) > 0$, if follows that while the quantum metric is self-adjoint, i.e., $\hat{g}_{ab}(x)^\dagger = \hat{g}_{ab}(x)$, the quantum momentum is not self-adjoint, i.e., $\hat{\pi}^{ab}(x)^\dagger \neq \hat{\pi}^{ab}(x)$. Following previous procedures, we introduce the quantum dilation field, $\hat{\pi}^a_b(x) = [\hat{\pi}^{ac}(x)^\dagger\, \hat{g}_{bc}(x) + \hat{g}_{bc}(x)\, \hat{\pi}^{ac}(x)]/2 = \hat{\pi}^a_b(x)^\dagger$. It follows that the quantum Hamiltonian, in Schrödinger's representation, becomes

$$\mathcal{H}_{AQ-ADM} = \int \{ [\hat{\pi}^a_b(x)\, g(x)^{-1/2} \hat{\pi}^b_a(x) - \tfrac{1}{2}\, \hat{\pi}^a_a(x)\, g(x)^{-1/2} \hat{\pi}^b_b(x)] \\ + g(x)^{1/2}\, {}^{(3)}R(x) \}\, d^3x\,, \tag{8}$$

A special feature is that $\hat{\pi}^a_b(x)\ g(x)^{-1/2} = 0$. It is as if $\hat{\pi}^a_b(x)\, g(y)^{-1/2} = 0$, where $x \neq y$. That fact simplifies the gravity kinetic factor greatly.

The quantum gravity Hamiltonian is the center piece of a complex analysis, and it is ripe for solution. A valid quantization of the classical Hamiltonian is required to lead the constraints further toward a potentially valid quantization of gravity. A procedure to include constraints and their role in quantum gravity has been presented in [5].

4. Summary

This paper has stressed that a new quantum procedure, i.e., AQ, is prepared to find valid quantizations for problems with incomplete coordinate spaces. The results of the full- and half-harmonic oscillators separately confirmed that they performed valid quantization for CQ and AQ. Applying AQ to field theories and to gravity, exploited the use of appropriate dilation and coordinate variables. It was found that there was no longer any integrable infinities, as nature requires. While both quantum fields and quantum gravity using CQ, have been well examined, they have not as yet been able to claim either one of them as satisfactory valid quantizations. Quite possibly, these problems may be closer to validity using AQ, as this paper has implied. Several articles have already aimed at finding better field theories and gravity results; see [5 - 18].

Journal of Physics: Conference Series **2533** (2023) 012020 doi:10.1088/1742-6596/2533/1/012020

References

[1] Klauder J 2020 The Benefits of Affine Quantization *Journal of High Energy Physics, Gravitation and Cosmology* **6** 175-185; doi: 10.4236/jhepgc.2020.62014
[2] Gouba L 2021 Affine Quantization on the Half Line *Journal of High Energy Physics, Gravitation and Cosmology* **7** 352-365 *Preprint* arXiv:2005.08696
[3] Handy C 2021 Affine Quantization of the Harmonic Oscillator on the Semi-bounded Domain $(-b, \infty)$ for $b : 0 \to \infty$ *Preprint* arXiv:2111:10700
[4] Arnowitt R, Deser S and Misner C 1962 in *Gravitation: An Introduction to Current Research*, ed. L Witten (New York: Wiley & Sons, 1962) p. 227 *Preprint* arXiv:gr-qc/0405109
[5] Klauder J 2022 A Straight Forward Path to a Path Integration of Einstein's Gravity *Preprint* arXiv:2203.15141
[6] Klauder J R 2005 Essential aspects of Wiener-measure regularization for quantum mechanical path integrals *Nonlinear Analysis* **63** e1253 – e1261
[7] Klauder J R and Daubechies I 1984 Quantum Mechanical Path Integrals with Wiener Measures for all Polynomial Hamiltonians Phys. Rev. Letters **5** 1161-1164
[8] Daubechies I and Klauder J R 1985 Quantum-Mechanical Path Integrals with Wiener Measure for all Polynomial Hamiltonians II J. Nath. Phys. **26** 2239-2256
[9] Klauder J R 1997 Coherent States in Action *Preprint* arXiv:quant-ph/9710 029v1
[10] Klauder J R 2010 On the Role of Coherent States in Quantum Foundations *Preprint* arXiv:1008.4307v1
[11] Gitman D M and Tyutin I V 1990 *Quantization of Fields with Constraints* (Berlin: Springer-Verlag)
[12] Klauder J R 1997 Coherent State Quantization of Constraint Systems *Ann. Phys.* **254** 419 *Preprint* arXiv:9604033v1
[13] Kempf A and Klauder J R 2000 On the Implementation of Constraints through Projection Operators *Preprint* arXiv:quant-ph/0009072v1
[14] Klauder J R and Shabanov S V 1998 An Introduction to Coordinate-free Quantization and its Application to Constrained Systems *Preprint* arXiv:quant-ph/9804049v1
[15] Bomstad W R and Klauder J R 2006 Linearized Quantum Gravity Using the Projection Operator Formalism *Preprint* arXiv:gr-qc/0601087v2
[16] Klauder J R 1999 Noncanonical Quantization of Gravity. I. Foundations of Affine Quantum Gravity *Preprint* arXiv:gr-qc/9906013v2
[17] Klauder J R 2011 *A Modern Approach to Functional Integration* (Basel: Bitkhäuser)
[18] Klauder J R 2015 *Enhanced Quantization: Particles, Fields & Gravity* (Singapore: World Scientific)

Journal of Physics: Conference Series **2533** (2023) 012021 doi:10.1088/1742-6596/2533/1/012021

Shadows of new physics on Dirac materials, analog GUPs and other amusements

Alfredo Iorio
Faculty of Mathematics and Physics, Charles University, V Holešovičkách 2, 18000 Prague 8, Czech Republic
E-mail: iorio@ipnp.troja.mff.cuni.cz

Boris Ivetić
Wohlergasse 6, 1100 Vienna, Austria
E-mail: bivetic@yahoo.com

Salvatore Mignemi
Dipartimento di Matematica e Informatica, via Ospedale 72, 09123, Cagliari, Italy and INFN Sezione di Cagliari, Italy
E-mail: smignemi@unica.it

Pablo Pais
Faculty of Mathematics and Physics, Charles University, V Holešovičkách 2, 18000 Prague 8, Czech Republic *and*
Instituto de Ciencias Físicas y Matemáticas, Universidad Austral de Chile, Casilla 567, 5090000 Valdivia, Chile
E-mail: pais@ipnp.troja.mff.cuni.cz

Abstract. We discuss here how, when higher-order effects in the parameter $\frac{\ell}{\hbar}|\vec{p}|$, related to the lattice spacing ℓ, are considered, pristine graphene, and other Dirac materials, can be used as tabletop systems where generalized commutation relations are naturally realized. Such generalized algebras of quantization, which lead to generalized versions of the Heisenberg uncertainty principle, are under intense scrutiny these days, as they could manifest a fundamental length scale of spacetime. Despite the efforts and the many intriguing results, there are no experimental signatures of any generalized uncertainty principle (GUP). Therefore, our results here, which tell how to use tabletop physical systems to test certain GUPs in analog experiments, should be of interest to practitioners of quantum gravity. We identify three different energy regimes that we call "layers", where the physics is still of a Dirac type but within precisely described limits. The higher the energy, the more sensitive the Dirac system becomes to the effects of the lattice. Here such lattice plays the role of a discrete space where the Dirac quasi-particles live. With the goals just illustrated, we had to identify the mapping between the high-energy coordinates, X^i, and the low-energy ones, x^i, i.e., those measured in the lab. We then obtained three generalized Heisenberg algebras. For two of them we have the noticeable result that $X^i = x^i$, and for the third one we obtained an improvement with respect to an earlier work: the generalized coordinates expressed in terms of the standard phase space variables, $X^i(x,p)$, and higher order terms.

1. Introduction

In the search for new physics, it is an established tradition to look for effects of high energy phenomena at our low energy scales. One way to do it is to consider the action for the "high-energy" fields, $\hat{\Phi}_a$, as an expansion in terms of "low-energy" fields, Φ_a and a suitable parameter ϵ, so that

$$\mathcal{S}(\hat{\Phi}_a) = \mathcal{S}(\Phi_a) + S_1(\epsilon, \Phi_a, \partial) + O(\epsilon^2, \partial^2) \,. \tag{1}$$

Examples from earlier research are many. Two that come to mind are the noncommutative field theories faced with the help of the Seiberg-Witten map [1]

$$\begin{aligned} \hat{A}_\mu &= A_\mu - \frac{1}{2}\theta^{\alpha\beta} A_\alpha \left(\partial_\beta A_\mu + F_{\beta\mu}\right) + O(\theta^2) \,, \\ \hat{F}_{\mu\nu} &= F_{\mu\nu} + \theta^{\alpha\beta} \left(F_{\mu\alpha} F_{\nu\beta} - A_\alpha \partial_\beta F_{\mu\nu}\right) + O(\theta^2) \,, \end{aligned} \tag{2}$$

that for (1) gives

$$\mathcal{S}(\hat{A}_\mu) = -\frac{1}{4}\hat{F}\cdot\hat{F} = -\frac{1}{4}\left(F\cdot F - \frac{1}{2}(\theta\cdot F)(F\cdot F) + 2(F\,\theta\,F)\cdot F\right) + O(\theta^2) \,, \tag{3}$$

where $\theta \sim \ell_P^2$ and ℓ_P is a reference length, usually taken to be the Planck length. For more on this see [1, 2, 3, 4] for the theory and, e.g., [5, 6, 7] for some phenomenology.

Another, somehow more general, example is Colladay-Kostelecky's Standard Model Extension [8], that can be described by

$$\mathcal{S}(\hat{\Phi}_a) = \mathcal{S}_{SM}(\Phi_a) + \sum_{k=1}^{\infty} C^{(k)}_{\mu\ldots\nu}(\Phi_a, \partial)^{\mu\ldots\nu} \,, \tag{4}$$

with $C^{(k)} \sim \ell_P^k$. For a review of both approaches, and a comparison, see, e.g., [9].

These are the days of the generalized uncertainty principles (GUPs) [10, 11, 12]; see also the recent review [13]. Many approaches exist that consider high-energy phase space variables, (x, p), vs low-energy variables, (x_0, p_0). Here we focus on a specific one that has been introduced by Ali, Das and Vagenas (ADV) in [14, 15, 16, 17]

$$p_i = p_{0i}\left(1 - A|\vec{p}_0| + 2A^2|\vec{p}_0|^2\right) \,, \tag{5}$$

where $A = \tilde{A}\,\ell_P/\hbar$, $\tilde{A}$ is a numerical factor and

$$x_i = x_{0i} \,, \tag{6}$$

with $[x_{0i}, p_{0j}] = \imath\hbar\delta_{ij}$, but where

$$[x_i, p_j] = \imath\hbar\left[\delta_{ij} - A|\vec{p}|\left(\delta_{ij} + \frac{p_i p_j}{|\vec{p}|^2}\right) + A^2|\vec{p}|^2\left(\delta_{ij} + 3\,\frac{p_i p_j}{|\vec{p}|^2}\right)\right] \,. \tag{7}$$

There is an intense activity to find GUP-corrected physics. Besides the already cited work of [14, 15, 16, 17], other important results that use different approaches are in [18, 19, 20]. A compelling example for this paper, for instance, is in [17] where it is found that the GUP-corrected Dirac equation, in any dimension, is

$$\begin{aligned} H\,\psi &= \left(c\vec{\alpha}\cdot\vec{p} + \beta mc^2\right)\psi \\ &= \left(c\vec{\alpha}\cdot\vec{p}_0 - cA(\vec{\alpha}\cdot\vec{p}_0)^2 + \beta mc^2\right)\psi \\ &= E\,\psi \,, \end{aligned} \tag{8}$$

that in two dimensions gives

$$\left(-\imath\hbar c\sigma_1\frac{d}{dx_0} + A\,\hbar^2 c\,\frac{d^2}{dx_0^2} + \beta mc^2\right)\psi = E\,\psi \,. \tag{9}$$

This search for what we might call the "shadows of new physics" at our energy scales is fascinating. Nonetheless, not even one single experiment over the many performed

Journal of Physics: Conference Series **2533** (2023) 012021 doi:10.1088/1742-6596/2533/1/012021

over the decades has seen anything. Fortunately, due to the higher experimental control of condensed matter systems, it is now becoming increasingly popular to reproduce various aspects of fundamental physics in analog systems. Examples include the Hawking phenomenon in Bose–Einstein condensates [21], the Weyl symmetry [22, 23] and the related Hawking/Unruh phenomenon on graphene [24, 25, 26], gravitational and axial anomalies in Weyl semimetals [27], and more.

Henceforth, taking that road, we focus on the proposal of graphene as an analog of high-energy fundamental physics [22, 23, 24, 25, 26, 28, 29, 30, 31, 32], based on the fact that its low-energy excitations are massless Dirac pseudo-relativistic fermions (the matter fields ψ), propagating in a carbon two-dimensional honeycomb lattice [33]. Such behaviour is shared by a wide range of materials, called "Dirac materials" [34].

Here we shall rely heavily on [35] and, to a lesser extent, on [36, 37], and shall take the chance to explain key details of the construction that are not explicitly exposed there.

2. Variables, Supervariables, Hypervariables

In this problem we are in two-space and one-time dimensions. We have three energy "layers", each with its own appropriate phase-space variables, that we indicate as

$$(x,p)\,, \quad (X_0,P_0)\,, \quad (X,P)\,. \tag{10}$$

This notation is discussed at length in [35]. Let us say here that the *variables* (x,p) are the actual phase-space variables of the lowest energy Dirac quasiparticles of graphene that originate from the $O(\ell)$ dispersion relations, with ℓ the lattice spacing

$$E_\pm = \pm v_F|\vec{p}|\,, \tag{11}$$

where $v_F = 3/2\,|\eta_1|\,\ell/\hbar$, with $\eta_1 \simeq -2.8\text{eV}$ the nearest neighbor hopping energy; see details in [33].

Such variables have a clear physical meaning, as coordinates and momenta of the low-energy conductivity electrons moving around (quantum mechanically) on the *membrane*[1].

There we explain how to obtain the dispersion relations in the usual way (secular equation), based on the knowledge of the functions

$$\mathcal{F}_m(\vec{k}) \equiv \sum_{i=1}^{n_m} e^{i\vec{k}\cdot\vec{s}_i^{(m)}}\,, \tag{12}$$

encoding the relevant information on the geometry of the lattice relative to the mth-near neighbors. Other details on the derivation of the $\mathcal{F}_m$s are in [36], where it is explained that the n_m vectors $\vec{s}_i^{(m)}$ connect any given atom to its mth-near neighbors.

Using the fortuitous occurrence, $\mathcal{F}_2 = |\mathcal{F}_1|^2 - 3$, and various approximations explained in [35], we arrive at the dispersion relations

$$E_\pm = \eta_1(\pm|\mathcal{F}_1| - \tilde{A}|\mathcal{F}_1|^2)\,, \tag{13}$$

where $\tilde{A} \simeq 0.15$.

It is then natural to define two further sets of variables, that we call "*supervariables*", (X_0,P_0), and "*hypervariables*", (X,P), as both sets refer to functions that appear in the dispersion relations that contain *all orders*[2] in ℓ. In other words (writing things in a dimensionless fashion, i.e., not including a necessary factor of $1/\ell$)

$$|\mathcal{F}_1| \sim p + O(\ell^2)\,, \quad |\mathcal{F}_1| \sim P_0\,, \quad |\mathcal{F}_1| - \tilde{A}|\mathcal{F}_1|^2 \sim P\,. \tag{14}$$

[1] This word is used on purpose, as the hexagonal lattice appears as a continuum at that energy/wavelength.
[2] This is literally true, as the function in (13) contains all orders in ℓ. Nonetheless, one needs to decide at which order one would like to proceed, say $O(\ell^m)$, and then all quantities need be taken at order m. Furthermore, the simple expression (13) is already an approximation that holds when higher-order contributions are not considered, see [35, 36, 37].

These relations are key in our analysis and are dictated by the phenomenology of the system.

In (13), if we define the supermomenta as

$$\vec{P}_0 = -\frac{\hbar}{\ell}(\mathrm{Re}\mathcal{F}_1, \mathrm{Im}\mathcal{F}_1)\,, \tag{15}$$

then we can write

$$E_\pm = V_F(\pm|\vec{P}_0| - A\,|\vec{P}_0|^2)\,, \tag{16}$$

where $V_F \equiv |\eta_1|\,\ell/\hbar$ and $A = \ell/\hbar\,\tilde{A}$.

Notice that in the following, we shall use a slightly different definition of $\vec{P}_0$, that allows one to expand it as $\vec{P}_0 \sim \vec{p} + \cdots$, rather than $\vec{P}_0 \sim 3/2\vec{p} + \cdots$. That is $\vec{\mathcal{P}}_0 \equiv 2/3\vec{P}_0$, but, to simplify notation with an abuse, we shall write $\vec{\mathcal{P}}_0 \to \vec{P}_0$.

Let us now set $\hbar = 1$, unless otherwise stated.

Just like in the literature it is customary to use a standard Dirac Hamiltonian, in relation to the linear dispersion relations (11), we can keep going with the same logic by building-up the appropriate Hamiltonian in terms of the supermomenta $\vec{P}_0$, whose dispersion relations, descending from the secular equation

$$\det(H - E\mathbb{1}) = 0 \tag{17}$$

are indeed (16). To do that one must realize that to obtain, through (17), $|\vec{P}_0|$ in the dispersion relation, a corresponding term $\vec{\sigma}\cdot\vec{P}_0 \equiv \not{P}_0$ needs be in the Hamiltonian: $|\vec{P}_0| \to \not{P}_0$. Thus, the effective Hamiltonian we can use to describe the physics of the "superlayer" (X_0, P_0) is

$$H(P_0) = V_F \sum_k \psi_k^\dagger\,(\not{P}_0 - A\not{P}_0\not{P}_0)\,\psi_k = V_F \sum_k \psi_k^\dagger\left(\vec{\sigma}\cdot\vec{P}_0 - A|\vec{P}_0|^2\mathbb{1}\right)\psi_k\,. \tag{18}$$

Notice that we use the convention[3] $\psi_k^\dagger = (b_k^\dagger, a_k^\dagger)$, where a_k and b_k are the annihilation operators of each graphene sublattice, see [22].

Indeed, by considering the 2×2 structure of H

$$H(P_0) \sim V_F \begin{pmatrix} -A|\vec{P}_0|^2 & P_0^1 - iP_0^2 \\ P_0^1 + iP_0^2 & -A|\vec{P}_0|^2 \end{pmatrix}, \tag{19}$$

clearly the two solutions of (17)

$$\det(H(P_0) - E\mathbb{1}) = 0 \sim \det\begin{pmatrix} -A|\vec{P}_0|^2 - E & P_0^1 - iP_0^2 \\ P_0^1 + iP_0^2 & -A|\vec{P}_0|^2 - E \end{pmatrix} = 0 \sim (A|\vec{P}_0|^2 + E)^2 = |\vec{P}_0|^2\,, \tag{20}$$

are those given in (13).

We can then move one last step up, in energy and symmetry (see [35]), by defining the hypermomenta

$$\vec{P} = \vec{P}_0(1 - A\,|\vec{P}_0|)\,, \tag{21}$$

that is nothing more than a consequence of the above mentioned fortuitous occurrence, $\mathcal{F}_2 = |\mathcal{F}_1|^2 - 3$, hence a physical fact of this system; see (14) and discussion there.

With this definition we arrive at the most symmetric situation of all, that is a full linearization of the Hamiltonian

$$H(P) = V_F \sum_k \psi_k^\dagger\,\not{P}\,\psi_k\,, \tag{22}$$

[3] There are many conventions related to different choices for the pairs of inequivalent Dirac points, and different arrangements of the a and b operators to form the spinor ψ. See Appendix B of [30] for more details.

with dispersion relations

$$E_\pm = \pm V_F |\vec{P}| , \tag{23}$$

which is, in all respects but two, similar to (11). The two differences are crucial, though. First, the slope here is V_F not $v_F \simeq 1.5 V_F$. Second, the hypermomenta are generalized variables, containing, in principle, all orders in ℓ and p (see [35] and later here).

To obtain $H(P_0)$ from $H(P)$ one should first use the expansion (21), and then use the prescription $|\vec{P}_0| \to \not{P}_0$

$$\begin{aligned} H(P(P_0)) &= V_F \sum_k \psi_k^\dagger \vec{\sigma} \cdot \vec{P} \, \psi_k \\ &= V_F \sum_k \psi_k^\dagger \vec{\sigma} \cdot \left(\vec{P}_0 (1 - A|\vec{P}_0|) \right) \psi_k \\ &\to V_F \sum_k \psi_k^\dagger \left(\vec{\sigma} \cdot \vec{P}_0 - A(\vec{\sigma} \cdot \vec{P}_0)(\vec{\sigma} \cdot \vec{P}_0) \right) \psi_k \\ &= V_F \sum_k \psi_k^\dagger \left(\not{P}_0 - A|\vec{P}_0|^2 \right) \psi_k = H(P_0) . \end{aligned} \tag{24}$$

Now, let us move towards the measurable quantities of the lowest layer (x, p). We first need to expand $\mathcal{F}_1$ in powers of ℓ or p. Indeed, this means, through (15), to expand P_0 in powers of ℓ or p, and through (21), to expand P in powers of ℓ or p. The function in point is

$$\mathcal{F}_1 = e^{i\ell k_y} \left(1 + 2 \, e^{i\frac{3}{2}\ell k_x} \cos(\frac{\sqrt{3}}{2} \ell \, k_x) \right) . \tag{25}$$

This expansion, in general, needs be done around the two inequivalent Fermi points $K^D_\pm = (\pm\frac{4\pi}{3\sqrt{3}\ell}, 0)$, located where $\mathcal{F}_1 = 0$. In our case, though, we are not considering topological defects or other physical set-ups that include, in their description, both Dirac points. For us it is sufficient to do all calculations near one Dirac point, e.g., if $k_i = K^D_{+i} + p_i$.

Thus, the Hamiltonian $H(P_0)$ in (18) expressed in terms of $\vec{p}$ up to[4] $O(\ell^3)$, is

$$\begin{aligned} H(P_0(p)) = H(p) = v_F \sum_{\vec{p}} \psi_{\vec{p}}^\dagger [\sigma_1 \left(p_1 - \frac{\ell}{4}(p_1^2 - p_2^2) - \frac{\ell^2}{8} p_1 (p_1^2 + p_2^2) \right) \\ + \sigma_2 \left(p_2 + \frac{\ell}{2} p_1 p_2 - \frac{\ell^2}{8} p_2 (p_1^2 + p_2^2) \right) \\ - \frac{3}{2} A \left((p_1^2 + p_2^2) - \frac{\ell}{2} p_1^3 + \frac{3\ell}{2} p_1 p_2^2 \right)] \psi_{\vec{p}} . \end{aligned}$$

For the sake of complete control of the units, let us now reintroduce $\hbar$, and let us write the dispersion relations for this Hamiltonian as

$$E_\pm = v_F \left[\pm|\vec{p}| \mp \frac{1}{4} \frac{\ell}{\hbar} |\vec{p}|^2 \cos(3\theta) \mp \frac{1}{64} \frac{\ell^2}{\hbar^2} |\vec{p}|^3 \left(7 + \cos(6\theta)\right) - \frac{3}{2} A \, |\vec{p}|^2 + \frac{3}{4} A \, \frac{\ell}{\hbar} |\vec{p}|^3 \cos(3\theta) \right] , \tag{26}$$

where

$$A = 0.15 \, \frac{\ell}{\hbar} > 0 , \tag{27}$$

$\tan\theta = p_2/p_1$ and

$$v_F \equiv \frac{3}{2} |\eta_1| \frac{\ell}{\hbar} \simeq c/300 \simeq 10^6 m/s . \tag{28}$$

[4] One should recall that, in the Hamiltonian, there is always an overall factor $O(\ell)$, given by the Fermi velocity $v_F = \frac{3\eta_1}{2\hbar} \times \ell$, so that $O(p^m)$ and $O(\ell^m)$, in the Hamiltonian, go actually together.

Journal of Physics: Conference Series **2533** (2023) 012021 doi:10.1088/1742-6596/2533/1/012021

Let us give here another writing of the dispersion relations

$$\begin{aligned} E_\pm &= \frac{3}{2}|\eta_1| \Bigg[\pm \left(\frac{\ell}{\hbar}|\vec{p}|\right) \mp \frac{1}{4}\left(\frac{\ell}{\hbar}|\vec{p}|\right)^2 \cos(3\theta) \mp \frac{1}{64}\left(\frac{\ell}{\hbar}|\vec{p}|\right)^3 (7+\cos(6\theta)) \\ &- \frac{3}{2}0.15 \left(\frac{\ell}{\hbar}|\vec{p}|\right)^2 + \frac{3}{4}0.15\left(\frac{\ell}{\hbar}|\vec{p}|\right)^3 \cos(3\theta)\Bigg] , \end{aligned} \tag{29}$$

which makes clear that the expansion parameter is $\frac{\ell}{\hbar}|\vec{p}|$.

3. Choice of the generalized coordinates and related GUPs

From the previous discussion, it is clear that the interpretation of P_0^i and P^i as generalized momenta is suggested by the dispersion relations, hence the functional form of $P_0^i(p)$

$$\begin{aligned} P_0^1 &= p^1 + \frac{\ell}{4}\left((p^2)^2 - (p^1)^2\right) - \frac{\ell^2}{8}p^1\left((p^1)^2 + (p^2)^2\right) , \\ P_0^2 &= p^2 + \frac{\ell}{2}p^1p^2 - \frac{\ell^2}{8}p^2\left((p^1)^2 + (p^2)^2\right) . \end{aligned} \tag{30}$$

or, equivalently

$$\begin{aligned} P_0^1 &= p^1 - \frac{\ell}{4}|\vec{p}|^2\cos 2\theta - \frac{\ell^2}{8}|\vec{p}|\cos\theta , \\ P_0^2 &= p^2 + \frac{\ell}{4}|\vec{p}|^2 \sin 2\theta - \frac{\ell^2}{8}|\vec{p}|\sin\theta , \end{aligned} \tag{31}$$

Hence of $P^i(p)$, through (21), are dictated by the physics of the system. The form of the dispersion relations, in turn, tells us how the effective Hamiltonian must look like, with little or no freedom.

What is left to the interpretation is the meaning of the generalized coordinates, X_0^i and X^i. Only when this is done the picture is complete, and we have all that is necessary, (H, X, P), to describe the dynamics.

Although there is room for different interpretations for the generalized coordinates, there are two choices that are very natural, hence stand out as privileged:

- The first choice is to keep x^i as the generalized coordinates, $x^i = X_0^i = X^i$. This choice, as obvious from $[x, P(p)] \sim i\partial_p P$, leads to some form of GUP.
- The second choice is to demand the generalized coordinates to be canonical, $[X_0, P_0] = i = [X, P]$. This choice necessarily leads to both $X_0^i(x,p)$ and $X^i(x,p)$.

To summarize (see [35] for details): One can move from the standard (canonical) variables that describe the lowest energy physics of the Dirac quasi-particles of graphene, (x, p), to the next "layer", by including higher order contributions in p to P_0, that we call supervariables, (X_0, P_0). Then, considering that contributions of the same order could come by including next-to-near neighbors in the computations ($A \neq 0$), one can move one more "layer" up, to the hypervariables (X, P).

If, at every step, we only consider the two privileged choices for the generalized coordinates, the GUP and the canonical, the overall options we have are given in the

following diagram [35]

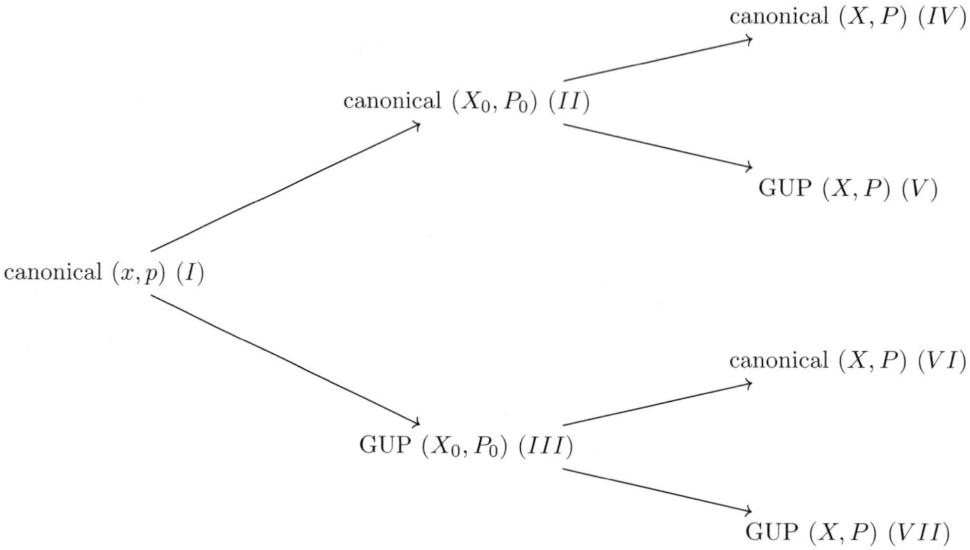

It is important to notice that:

- To include terms from $O(p^2)$ on, we have to keep $A \neq 0$, as seen in (27). Therefore, the "layer of reference" is always the "hyperworld", (X, P).
- The interesting set-ups are those leading to GUPs in the final layer, the one of reference. In the above scheme, these are (V) and (VII).
- No matter the choice, all the quantities eventually need be expressed in terms of the measurable standard variables, (x, p).

With this in mind, the GUP-corrected physics here is given by the Hamiltonian (26), and the related dispersion relations (26). We simply need to read-off all results in terms of the appropriate variables, that for the coordinates are:

(i) the case $X^i = x^i$ (that is the ladder $(I) \to (III) \to (VII)$), leading to the GUP (VII)

$$[X^i, P^j] = \mathrm{i} F^{i\,k}(\vec{P}) \left[\delta^{kj} - A|\vec{P}| \left(\delta^{kj} + \frac{P^k P^j}{|\vec{P}|^2}\right) - A^2|\vec{P}|^2 \left(\delta^{kj} + \frac{P^k P^j}{|\vec{P}|^2}\right)\right] \quad (32)$$

where

$$F^{ij}(\vec{P}) = \delta^{ij} + \frac{1}{2}\,\ell \begin{pmatrix} -P^1 & P^2 \\ P^2 & P^1 \end{pmatrix} + \frac{1}{2}\,\ell\, A\,|\vec{P}| \begin{pmatrix} -P^1 & P^2 \\ P^2 & P^1 \end{pmatrix} - \frac{1}{2}\,\ell^2 \begin{pmatrix} (P^1)^2 & P^1 P^2 \\ P^1 P^2 & (P^2)^2 \end{pmatrix},$$

Notice that $F^{ij} = F^{ji}$, and $F^i_i(\vec{P}) = 2 - \frac{\ell^2}{2}|\vec{P}|^2$.

(ii) and the case $X^i = X^i_0(x, p)$ with

$$X^1_0(x,p) = \left[1 + \frac{\ell}{2}p^1 + \frac{\ell^2}{8}\left(5(p^1)^2 + 3(p^2)^2\right)\right] x^1 + \left[-\frac{\ell}{2}p^2 + \frac{\ell^2}{4}p^1 p^2\right] x^2\,, \quad (33)$$

$$X^2_0(x,p) = \left[-\frac{\ell}{2}p^2 + \frac{\ell^2}{4}p^1 p^2\right] x^1 + \left[1 - \frac{\ell}{2}p^1 + \frac{\ell^2}{8}\left(3(p^1)^2 + 5(p^2)^2\right)\right] x^2\,.$$

That is the ladder $(I) \to (II) \to (V)$), leading to the GUP (V), an expression coinciding, at first order in A, with the ADV-like GUP (see (7))

$$[X^i, P^j] = \mathrm{i}\left[\delta^{ij} - A|\vec{P}| \left(\delta^{ij} + \frac{P^i P^j}{|\vec{P}|^2}\right) - A^2|\vec{P}|^2 \left(\delta^{ij} + \frac{P^i P^j}{|\vec{P}|^2}\right)\right]. \quad (34)$$

Journal of Physics: Conference Series **2533** (2023) 012021 doi:10.1088/1742-6596/2533/1/012021

The other commutators are trivial.

One does not need any special or exotic setup to have the GUP all-the-way ladder at work. The only thing that is necessary is to include higher order terms in the dispersion relations, and then move to the next-to-near neighbors. The coordinates are always just x^i. Since the literature is full of proposals on how GUPs affect a variety of phenomena, see, e.g., [10, 11, 12, 18, 19, 20, 32] and since we have obtained here many such GUPs, some very similar to the ones descending from quantum gravity, then we are now in the position to prove some of those theoretical conjectures in the lab.

Notice that, in going from (I) to (III), we also have a GUP

$$[X_0^i, P_0^j] = \mathrm{i}F^{ij}(\vec{P}_0), \quad [X_0^i, X_0^j] = 0 = [P_0^i, P_0^j] \,, \tag{35}$$

with $F^{ij}(\vec{P}_0)$ given in

$$F^{ij}(\vec{P}_0) = \delta^{ij} + \frac{1}{2}\ell \begin{pmatrix} -P_0^1 & P_0^2 \\ P_0^2 & P_0^1 \end{pmatrix} - \frac{1}{2}\ell^2 \begin{pmatrix} (P_0^1)^2 & P_0^1 P_0^2 \\ P_0^1 P_0^2 & (P_0^2)^2 \end{pmatrix} .$$

Since, for the reasons explained earlier, we shall always like to include the $A \neq 0$ effects, we only consider this as an intermediate mathematical step.

4. Conclusions

We have discussed here how analogs of three kinds of GUPs can be obtained when higher order terms are considered in the dispersion relations of the conductivity electrons of a generic Dirac material.

An analog system cannot behave as the target system in all respects but only in certain specific conditions, for certain types of particles and for particular regimes of the interaction. In fact, one of the aspects we have to give up in an analog is the *universality*, which should be expected in the presence of a *fundamental* GUP.

For instance, in the "three graphene worlds" we have explored here, we only have at our disposal a limited set of all the possible particles and fields, compared to those that nature gives us. Essentially, in a Dirac material we have analogs of

- massless Dirac spinors, ψ, that are the π-electrons quasiparticles;
- the U(1) gauge fields, A_μ, stemming from straining the material, that could be used to mimic electromagnetic interactions;
- the nonabelian gauge fields, A_μ^a, due to the two copies of ψ, one per Dirac point, often described as SU(2) gauge fields, that could be used to mimic internal symmetries;
- the spin connection, ω_μ^a, that always takes care of the intrinsic curvature associated to disclination defects in the lattice, and may or may not include contorsion, κ_μ^a, that takes care of the intrinsic curvature associated to dislocation defects;
- classical (and quantum) metric fields, $g_{\mu\nu}$, that emerge as a collective description of the elastic and unelastic properties of the membrane.

Some more steps could be moved when we adapt (reduce) external fields, e.g., the true electromagnetic field, to the lower dimensional, and v_F-relativistic (rather than the c-relativistic) and specific dynamics of the system. That is all.

Clearly, then, here we cannot have all types of particles and fields propagating in a GUP-compatible space, described by a suitable $g_{\mu\nu}$, but we can do that for a limited set, for instance, ψ propagating on a flat space with granular structure.

Even for this limited situation, since the recipe here consists in having control on the appropriate phase-space variables, we need to know how to recognize the "shadows of the high energy physics", so to speak, on what mimics "the low energy GUP-corrected physics".

With the results of [35], we are now able to face these issues because we now know the form of X^i and X_0^i, in terms of x^i and p^i. With those results in our hands, we may safely solve the problem of lack of universality by, e.g., limiting the set-up to only Dirac quasi-particles ψ and setting up a potential well, as many others do routinely in the same analog/effective language.

Such potential V must be a function of the correct variables, X^i, i.e. $V(X)$. But now we know $X(x,p)$, and this simply means $X_i = x_i$ in the most interesting case. Hence, the very same potential well $V(x)$ one has for the GUP-non-corrected physics (the standard one) stays for GUP-corrected physics. In an experiment like that, we would prove that the GUP-corrected dynamics can indeed be practically realized in such a simple system, and one could check many theoretical statements in a lab.

To close, a note on the generality of our results. All condensed matter systems have corrections to linear or quadratic dispersion laws. Some kind of GUP may then appear in many systems after the suitable variables change. Notice that our approach may or may not lead to sensible analogs GUPs. In other words, it is not as simple as "any suitable change of variables leads to some sensible GUP". For instance, in our case, we have seen that only by stopping at $\mathcal{F}_2$ and making other approximations, it is possible to find GUPs related to the ADV GUPs. If one keeps going in the order of the expansion, no natural ADV or ADV-like structures are recovered. Other systems may have different natural GUPs, none at all, or unknown GUPs (like some of those that we discussed here).

To close, let us stress again that, of course, we did not prove here that graphene, or other materials of that kind, are systems where *fundamental* GUPs are at work. Nor we are suggesting that, following our procedure, fundamental GUPs can be spotted in other condensed matter systems. We are saying here that those easy-to-produce tabletop real systems, by a simple redefinition of the dynamical variables, can be turned into very efficient analogs of the fundamental scenarios that are supposed to be responsible for the GUPs that the high energy physics community is eagerly looking for. This seems promising, especially in the light of how simple our recipe is and of how difficult it is to run high energy experiments.

Acknowledgements

A. I. and P. P. sincerely thank Thomas Elze and the other organizers of DICE2022, for a wonderful and very productive conference. They gladly acknowledge the financial support of Charles University Research Center (UNCE/SCI/013). P. P. is also supported by Fondo Nacional de Desarrollo Científico y Tecnológico–Chile (Fondecyt Grant No. 3200725). S. M. acknowledges support from COST action CA18108.

References

[1] Seiberg N and Witten E 1999 *JHEP* **1999** 032 (*Preprint* hep-th/9908142)
[2] Madore J, Schraml S, Schupp P and Wess J 2000 *Eur. Phys. J.* **C16** 161
[3] Iorio A and Sỳkora T 2002 *Int. J. Mod. Phys.* **A17** 2369
[4] Iorio A 2008 *Phys. Rev.* **D77** 048701
[5] Guralnik Z, Jackiw R, Pi S and Polychronakos A 2001 *Phys. Lett.* **B517** 450
[6] Jackiw R 2002 *Nucl. Phys. B Proc. Suppl.* **108** 30
[7] Castorina P, Iorio A and Zappala D 2004 *Phys. Rev.* **D69** 065008
[8] Colladay D and Kostelecky V A 1998 *Phys. Rev. D* **58** 116002 (*Preprint* hep-ph/9809521)
[9] Iorio A 2007 *J. Phys. Conf. Ser.* **67** 012008
[10] Maggiore M 1993 *Phys. Lett. B* **304** 65 (*Preprint* hep-th/9301067)
[11] Kempf A 1994 *J. Math. Phys. (NY)* **35** 4483 (*Preprint* hep-th/9311147)
[12] Scardigli F 1999 *Phys. Lett. B* **452** 39 (*Preprint* hep-th/9904025)
[13] Scardigli F 2022 Uncertainty principle and gravity (*Preprint* 2212.13776)
[14] Das S and Vagenas E C 2008 *Phys. Rev. Lett.* **101**(22) 221301
[15] Ali A F, Das S and Vagenas E C 2011 *Phys. Rev.* **D84** 044013 (*Preprint* 1107.3164)
[16] Ali A F, Das S and Vagenas E C 2009 *Phys. Lett.* **B678** 497
[17] Ali A F, Das S and Vagenas E C 2010 *Phys. Lett.* **B690** 407
[18] Buoninfante L, Luciano G G and Petruzziello L 2019 *Eur. Phys. J.* **C79** 663 (*Preprint* 1903.01382)
[19] Petruzziello L and Illuminati F 2021 *Nature Commun.* **12** 4449 (*Preprint* 2011.01255)
[20] Bosso P and Luciano G G 2021 *Eur. Phys. J. C* **81** 982 (*Preprint* 2109.15259)
[21] Muñoz de Nova J R, Golubkov K, Kolobov V I and Steinhauer J 2019 *Nature* **569** 688
[22] Iorio A 2011 *Ann. Phys. (Amsterdam)* **326** 1334
[23] Iorio A 2102 *Eur. Phys. J. Plus* **127** 156
[24] Iorio A and Lambiase G 2012 *Phys. Lett.* **B716** 334
[25] Iorio A and Lambiase G 2014 *Phys. Rev. D* **90** 025006

Journal of Physics: Conference Series **2533** (2023) 012021 doi:10.1088/1742-6596/2533/1/012021

[26] Iorio A 2015 *Int. J. Mod. Phys.* **D24** 1530013
[27] Gooth J, Niemann A C, Meng T, Grushin A G, Landsteiner K, Gotsmann B, Menges F, Schmidt M, Shekhar C, Süß V, Hühne R, Rellinghaus B, Felser C, Yan B and Nielsch K 2017 *Nature* **547** 324–327
[28] Iorio A and Pais P 2015 *Phys. Rev. D* **92** 125005
[29] Iorio A and Pais P 2022 *Phys. Rev.* **B106** 157401
[30] Iorio A and Pais P 2018 *Ann. Phys. (Amsterdam)* **398** 265
[31] Iorio A 2015 *Front. Mater.* **1** 36 (*Preprint* 1412.5160)
[32] Iorio A, Lambiase G, Pais P and Scardigli F 2020 *Phys. Rev. D* **101**(10) 105002
[33] Castro Neto A H, Guinea F, Peres N M R, Novoselov K S and Geim A K 2009 *Rev. Mod. Phys.* **81** 109
[34] Wehling T, Black-Schaffer A and Balatsky A 2014 *Adv. Phys.* **63** 1–76
[35] Iorio A, Ivetić B, Mignemi S and Pais P 2022 *Phys. Rev.* **D106** 116011
[36] Iorio A, Pais P, Elmashad I A, Ali A F, Faizal M and Abou-Salem L I 2018 *Int. J. Mod. Phys.* **D27** 1850080 (*Preprint* 1706.01332)
[37] Iorio A and Pais P 2019 *J. Phys. Conf. Ser.* **1275** 012061 (*Preprint* 1902.00116)

Journal of Physics: Conference Series **2533** (2023) 012022 doi:10.1088/1742-6596/2533/1/012022

Lessons and complications from gravitationally induced entanglement

André Großardt and M Kemal Döner

Institute for Theoretical Physics, Friedrich Schiller University Jena, Fröbelstieg 1, 07743 Jena, Germany

E-mail: andre.grossardt@uni-jena.de, kemal.doner@uni-jena.de

Abstract. We critically review arguments depicting entanglement generated between exclusively gravitationally interacting particles as evidence for the necessity of a quantized gravitational field. For proposed experiments which are supposed to witness this gravitationally induced entanglement, we discuss the implications of a positive result and defend the possibility of a semiclassical theory of quantum matter on a classical spacetime which cannot be conclusively ruled out by witnessing entanglement. For thought experiments that resolve apparent causality issues with quantum systems entangled via the Newtonian gravitational interaction by considering a quantized gravitational field, we demonstrate that quantization of gravitational degrees of freedom is not only unnecessary but may result in remaining paradoxical behavior, unless a previously disregarded decoherence channel is taken into consideration.

1. Introduction

Despite tremendous effort and progress in the detailed understanding of the various approaches to a quantum theory of the gravitational interaction, just as little is known about which way is the proper way to quantize gravity as it has been with the discovery of the principles of quantum field theory about a century ago. This lack of progress is often attributed to the absence of concrete predictions for feasible experiments from candidate theories of quantum gravity. Hopes for at least some guidance from observation have been mostly focused on generic features, such as the effects of a minimal length scale [1] or modified uncertainty relations [2, 3], which are not linked to any specific model but a common feature of a variety of different approaches.

In the same vein, driven by recent progress in laboratory quantum technology, there has been a renewed interest in experiments aimed at probing the question whether the gravitational field must be quantized at all. From early on, beginning with discussions between Feynman, Rosenfeld, and others at the 1957 Chapel Hill Conference [4], there has been a plethora of theoretical arguments for the necessity to quantize gravity, alongside similarly sophisticated refutations of these very arguments [5, 6, 7, 8, 9, 10, 11, 12, 13, 14, 15, 16, 17, 18, 19] (for a recent review see also [20, 21]). In the light of these combined arguments, the question whether gravity must be quantized remains without a definitive answer, although a consistent coupling of quantum matter to classical gravity [21, 22] faces technical difficulties that might be comparable to those faced by quantum gravity approaches. Earlier proposals [16, 23, 24] for such laboratory experiments to establish the necessity of quantizing the gravitational field focused on testing the specific model of the Schrödinger–Newton equations [25, 11, 26, 27], which results from

Journal of Physics: Conference Series **2533** (2023) 012022 doi:10.1088/1742-6596/2533/1/012022

coupling quantum matter to a classical spacetime via Einstein's equations sourced by the mean stress-energy in the quantum state of the matter fields.

Recently, there has been great interest in the idea that more generic conclusions could be drawn from the study of gravitationally induced entanglement. This idea has been sparked by the Kafri–Taylor–Milburn [28] toy model, where the Newtonian gravitational force between two particles is considered an effect of local operations with classical communication (LOCC). Interactions of this kind are known not to induce entanglement in initially separable quantum states, leading to the proposal [29, 30] that experimental evidence of entanglement between particles which only interact through the Newtonian gravitational forces between them would ascertain the 'quantumness' of the gravitational field. We critically assess the scope of such claims in section 2.

On a related note, it has been shown by Belenchia et al. [17] that an apparent paradox [31] allowing for faster-than-light signaling via the entanglement induced by the Coulombian or Newtonian forces between quantum particles is resolved by taking the quantum field character of the electromagnetic or gravitational interaction into account. However, as important as it is to understand how consistency is achieved in the electromagnetic case—and as curious as one may find the fact that similar arguments could be applied to gravity were it indeed a perturbatively defined quantum field theory at low energies—the only rigorous conclusion to be drawn from this argument is that quantization of the gravitational field would be a sufficient condition to resolve the paradox, not a necessary one. A similar case has been made by Rosenfeld [6] in response to the interpretation of his work with Bohr [32] as alleged evidence for the need to quantize the electromagnetic field—and analogies drawn for gravity. Notably, the paradox is resolved in a trivial and—as we argue in section 3—more robust way if gravity remains *unquantized.*

2. On laboratory experiments and the need to quantize gravity

A 'classical channel' interaction assumes a many-particle interaction potential $\sum_{i\neq j} V_i(\hat{x}_j)$, where each V_i is a 'classical' potential determined through measurements of the i-th particle. For such an interaction, an initially separable state $\Psi = \bigotimes_i \psi_i$ would remain separable, contrary to a typical quantum interaction potential $\sum_{i\neq j} V(|\hat{x}_i - \hat{x}_j|)$ which induces entanglement. A driving reason to consider this type of interaction as a model for the Newtonian gravitational force [28] is that it permits a consistent dynamical law for the density operator—and therefore an operational interpretation of quantum mechanics. This is in contrast to the Schrödinger–Newton model, for which finding a consistent model of the measurement process poses a challenge. Hence, despite being experimentally refuted [33] itself, the Kafri-Taylor-Milburn model has inspired different approaches how to couple quantum matter to classical gravity:

Tilloy and Diósi [34] propose a model of classical Newtonian gravity source by a mass density which is the signal of weak measurements. This model can also be interpreted in the context of collapse models [35] where the stochastic localization events become the source of the gravitational potential. This model is akin to the Schrödinger–Newton model in that it provides a joint dynamical evolution law for a quantum state Ψ for matter and a classical gravitational potential Φ. Unlike the deterministic Schrödinger–Newton model, however, this evolution is stochastic and can thereby maintain a closed, linear evolution law for the matter density operator.

Oppenheim [22] proposes a general framework for quantum-classical hybrid theories which applied to the Hamiltonian formulation of general relativity promises a consistent coupling of classical gravity and quantum matter. The model provides a deterministic evolution law for a hybrid 'classical-quantum' state $\hat{\varrho}(z)$, a subnormalized positive Hilbert space operator on each classical phase-space point z, from which both classical probability densities in phase space and quantum probabilities can be derived. Hence, like the Schrödinger evolution law in Hilbert space, a probability interpretation is necessary—not only for the quantum matter but also for the spacetime degrees of freedom. In this sense, the model can be seen as a fundamentally stochastic one, similar to quantum theory, or an effective stochastic description of a yet unknown underlying theory, comparable to classical statistical mechanics.

All of these models, including the Schrödinger–Newton model, predict no entanglement between gravitationally interacting particles, whereas perturbative quantum gravity predicts such entanglement in complete analogy to electrodynamics [36]. Hence, experiments confirming entanglement [29] would provide clear evidence against these specific semiclassical models. On the other hand, all four of these semiclassical models make further definitive predictions: the Kafri–Taylor–Milburn model has already been experimentall refuted [33], the proposals [23, 24] for testing the Schrödinger–Newton model are arguably closer to technological feasibility than witnessing entanglement, and the Tilloy–Diósi and Oppenheim models predict unaccounted decoherence or heating effects which are already being actively investigated in order to put bounds on collapse models [37, 38, 39].

What, then, is the additional benefit of testing gravitationally induced entanglement? If one believes the claims of its proponents, a huge one: namely 'that if entanglement is observed, then *all* classical models for gravity, obeying our general principles, are ruled out.' [40] This claim has been contested [41, 42, 43] as putting too tight constraints on the allowed 'classical' models, namely by a restriction to the LOCC perspective. To illustrate the loophole remaining due to this restriction, in [44] we propose a class of simple toy models that explicitly demonstrates a semiclassical model in which entanglement *is* induced through the gravitational interaction.

These proposed models are based on the de Broglie–Bohm theory where, in addition to the N-particle wave function Ψ one has the N particle coordinates $\vec{q}_i$, satisfying the guiding equation

$$\frac{\mathrm{d}\vec{q}_i(t)}{\mathrm{d}t} = \frac{\hbar}{m_i}\,\mathrm{Im}\left(\frac{\nabla_i\Psi(t;\vec{r}_1,\dots,\vec{r}_N)}{\Psi(t;\vec{r}_1,\dots,\vec{r}_N)}\right)\Bigg|_{\vec{r}_1=\vec{q}_1(t),\dots,\vec{r}_N=\vec{q}_N(t)}\,. \tag{1}$$

Now, the wave function is expected to follow a Schrödinger equation including a Newtonian gravitational potential. The distinguishing features of the various models are the different sources of said potential, i.e. the mass density ρ in the Poisson equation $\nabla^2 V = 4\pi G m\rho$ for the potential energy:

(i) In the Schrödinger–Newton model, the gravitational mass density is determined by the marginal distributions of the N particles:

$$\rho(t,\vec{r}) = \sum_i m_i \int \mathrm{d}^3r_1\cdots\mathrm{d}^3r_{i-1}\mathrm{d}^3r_{i+1}\cdots\mathrm{d}^3r_N|\Psi(t;\vec{r}_1,\dots,\vec{r}_{i-1},\vec{r},\vec{r}_{i+1},\dots,\vec{r}_N)|^2\,. \tag{2}$$

(ii) In the model by Tilloy and Diósi, the potential is sourced by the *signal* ρ, defined via

$$\rho\,\mathrm{d}t = \langle\hat{\varrho}\rangle_t\,\mathrm{d}t + \mathrm{d}W_t\,, \tag{3}$$

where $\hat{\varrho}$ is the mass density operator (smeared over some width σ in order to avoid divergences) and W_t a spatially correlated white-noise term, entering the stochastic Schrödinger equation [45] for the quantum state (with decoherence rate γ):

$$\mathrm{d}\psi_t = -\frac{\mathrm{i}}{\hbar}\hat{H}\psi_t\mathrm{d}t - \frac{\gamma}{2\hbar^2}\left(\hat{\varrho}-\langle\hat{\varrho}\rangle_t\right)^2\psi_t\mathrm{d}t + \frac{2\gamma}{\hbar}\left(\hat{\varrho}-\langle\hat{\varrho}\rangle_t\right)\psi_t\mathrm{d}W_t\,. \tag{4}$$

(iii) In the Newtonian limit of perturbative quantum gravity, the gravitational potential energy does *not* satisfy a Poisson equation but instead is an operator valued function

$$\hat{V} = -\frac{G}{2}\int \mathrm{d}^3u\,\mathrm{d}^3v\,\frac{\hat{\varrho}(\vec{u})\hat{\varrho}(\vec{v})}{|\vec{u}-\vec{v}|}\,. \tag{5}$$

Journal of Physics: Conference Series **2533** (2023) 012022 doi:10.1088/1742-6596/2533/1/012022

(iv) Finally, in de Broglie–Bohm theory the canonical choice would be to simply implement the same potential (5) in the Hamiltonian evolution of the wave function. However, the existence of the 'hidden' variables $\vec{q}_i$ leaves us with the alternative to implement a *classical* gravitational mass density sourced by the Bohmian trajectories:

$$\rho(t, \vec{r}) = \sum_i m_i \delta(\vec{r} - \vec{q}_i(t)) . \tag{6}$$

Together with the Schrödinger equation and the guiding equation (1), the mass density (6) provides a consistent nonrelativistic model for the gravitational interaction between fundamental particles which agrees with the Newtonian limit of quantum gravity (and thus the classical Newtonian theory) in the classical limit. The crucial observation is that this model has a semiclassical interpretation: the Bohmian particles can be understood as sourcing the curvature of a *classical* spacetime on which both the wave function and the particle coordinates are defined. Despite this semiclassical nature, this Bohmian trajectory sourced Newtonian interaction *will* result in entanglement between particles, providing an explicit counter-example to the claim cited above.

There are two caveats to be considered: Firstly, the model as defined above, when applied to two interacting particles, results in a relative phase that is twice that predicted by the quantum potential (5). Whereas the resulting difference in predictions for entanglement witnesses [29] could be seen as a feature making the model testable, this phase prediction contradicts the *observed* gravitational phases in the homogeneous field of the Earth [46]. It can be easily remedied by adding to the Hamiltonian a term proportional to identity operator on Hilbert space that depends solely on the trajectories $\vec{q}_i$, although at the cost of rendering the interpretation in terms of wave functions and trajectories on a classical spacetime less natural.

Second and more importantly, the interpretation in terms of a classical spacetime sourced by the trajectories assumes that the model is the nonrelativistic limit of a well-defined general relativistic theory. Considering the tremendous difficulties faced by relativistic generalizations of Bohmian mechanics, this may justifiably raise concerns. Note, however, that the same problem in principle affects *all* models listed (although possibly with different severity): The Schrödinger–Newton model can be understood as the limit of the semiclassical Einstein equations, but it cannot be defined consistent with observations unless a proper objective dynamical law for the wave-function collapse can be found [20, 21]. Similarly, the Tilloy–Diósi model is based on collapse models, whose relativistic generalization [47] is not fully understood. Finally, perturbative quantum gravity, due to its nonrenormalizability [48] is generally understood as the low energy limit of some more fundamental theory, but the existence of a consistent and predictive theory with this limit has not yet been rigorously proven.

The class of toy models we propose interpolates between the Bohmian trajectory sourced gravitational mass density (6) and the Schrödinger–Newton model (2) by taking the integrals $\int \mathrm{d}^3 r_i$ in equation (2) not over the whole configuration space $\mathbb{R}^3$ but only over a sphere of radius R around the Bohmian coordinate $\vec{q}_i(t)$. The resulting models are parameterized by this radius R, where the Schrödinger–Newton model is recovered in the limit $R \to \infty$ and the Bohmian trajectory model in the limit $R \to 0$. It can then be shown [44] that for any finite value of R one can always find an entanglement witness that confirms a nonzero entanglement between two particles due to the gravitational interaction. The amount of entanglement (as measured by the deviation of the witness function from the classically allowed values) is suppressed exponentially for large R.

In conclusion, claims that witnessing gravitationally induced entanglement would rule out *any* conceivable coupling of quantum matter to classical general relativity are evidently wrong. Nonetheless, a positive outcome of such an experiment would conclusively rule out the few explicit models in existence, as well as put even tighter constraints on the possibility of consistent semiclassical models. In this regard, experimental tests of gravitationally induced entanglement would complement direct tests of quantum gravity alternatives [16, 26, 23, 24, 33, 39], although there are good reasons [49] to believe that—comparable resources provided—they are also more technologically challenging than those direct tests.

3. On thought experiments in the Galilean sector of quantized gravity

Somewhat orthogonal to the efforts towards an experimental confirmation of the quantization of the gravitational field, outlined in the previous section, are purely theoretical consistency arguments based on thought experiments. Assume Alice and Bob, located at distant places, both possess a single quantum particle. Alice's particle at position x_A is the source of a Coulomb type potential to which Bob's particle at position x_B is sensitive, i.e. there is a force $F \sim |x_A - x_B|^{-2}$ acting on Bob's particle. The shift δx of Bob's particle position due to this force after some fixed time t_B is a strictly monotonous function of x_A, such that Bob acquires *which-way information* about Alice's particle, provided a sufficient resolution for the shift δx. Now, an essential consequence of quantum mechanics is the complementarity principle. If Alice performs an interferometry experiment with her particle, interferometric visiblity will be lost as which-way information about the particle's position can be gained. Assuming that Alice's particle is readily prepared in a superposition state at time $t = 0$, followed by the time interval $[0, t_B]$ during which Bob can choose freely to conduct or not to conduct his experiment in order to acquire which-way information about Alice's particle, Alice reunites the possible particle trajectories with the intent to create an interference pattern during the time interval $[t_B, t_A + t_B]$. As long as the total time $t_A + t_B$ is shorter than the time a light signal would take from Alice to Bob, Bob can send a binary message faster than light.

This apparent paradox has been discussed by Mari et al. [31]. Belenchia et al. [17] demonstrate how the paradoxical signaling is prohibited if the interaction between Alice's and Bob's particles is due to a quantized field. Simply put, the existence of vacuum fluctuations puts a limit on the best possible resolution which requires a minimum time t_B on Bob's side, whereas on Alice's side the experiment must be performed sufficiently slow in order to not already decohere the interference pattern by emitting radiation. Jointly, these constraints are such that both in the electromagnetic and the gravitational case they can prevent the possibility of faster-than-light signalling. Note that in the gravitational case, whereas these conditions *can* be justified from the quantization of the gravitational field, such an interpretation is by no means required; they can also be understood as a consequence of classical radiation [19, 20].

Quantum entanglement, though not explicitly mentioned in the above description, again plays a crucial role. With Alice's particle initially in a superposition state $|\psi\rangle \sim |x_A\rangle + |x'_A\rangle$, the above argument presupposes that the interaction is a quantum potential $\hat{V}(x) \sim |\hat{x}_A - x|^{-1}$ and the final state an entangled state $|\Psi\rangle \sim |x_A\rangle \otimes |x_B\rangle + |x'_A\rangle \otimes |x'_B\rangle$. If the potential instead is of the semiclassical type that results in a separable state $|\Psi\rangle \sim (|x_A\rangle + |x'_A\rangle) \otimes |x_B\rangle$, Bob cannot gain any which-way information through the position of his particle. It is, therefore, somewhat misleading to argue that the resolution of the faster-than-light signaling paradox 'provides support for the view that (linearized) gravity should have a quantum field description' [17]. Field quantization is a *sufficient* property for consistency but not a *necessary* one.

The proposal for experimental tests of gravity induced entanglement and the thought experiment outlined above have in common that they do not require many assumptions about the gravitational interaction beyond a certain analogy to the electromagnetic field, which exhibits itself in the identical form of the Newtonian gravitational interaction compared to the Coulomb interaction in the Schrödinger equation for the electromagnetic Galilean limit. As long as gravity works 'just like electromagnetism', there is no reason to doubt the positive outcome of entanglement witnessing experiments [29], and field quantization seems a perfectly valid argument to resolve the apparent paradoxes [17]. We know, however, that this analogy between

Journal of Physics: Conference Series 2533 (2023) 012022 doi:10.1088/1742-6596/2533/1/012022

gravitational and electromagnetic forces must have limitations:

(i) On the level of the classical theories, there are some fundamental differences between general relativity and Maxwell's theory. Not only are Maxwell's equations linear in contrast to the nonlinear Einstein's equations; Maxwell's theory also defines a linear connection on a line bundle, whereas the Levi–Civita connection in general relativity is the *affine* connection for the tangent bundle of spacetime. Crucially, this implies that for gravity—contrary to electrodynamics—there is no meaningful definition of a 'vacuum state' upon which a quantum field theory could be built [21].

(ii) Although a perturbative quantum field theory around a fixed background spacetime can formally be defined, quite similar to the quantization of electrodynamics, in the gravitational case this theory is known to be nonrenormalizable [48].

Hence, we propose that more promising lessons might be learned from such thought experiments which do not have a simple electromagnetic analogy.

One such effect with no immediate analogy is the universal coupling of gravity to all forms of energy. This has already been exploited for the suggestion of time-dilation related dephasing [50, 51] where it has been proposed that a nonrelativistic quantum system with internal degrees of freedom, subject to a gravitational potential $\Phi(x)$ including lowest order relativistic corrections, should be described by a Hamiltonian

$$\hat{H} = \hat{H}_{\mathrm{cm}} + \hat{H}_{\mathrm{int}} + \left(m + \hat{H}_{\mathrm{int}}\right) \Phi(\hat{x}_{\mathrm{cm}}) \,, \tag{7}$$

where $\hat{H}_{\mathrm{cm}}$ and $\hat{x}_{\mathrm{cm}}$ are the Hamiltonian and position operators acting on the center-of-mass Hilbert space and $\hat{H}_{\mathrm{int}}$ the Hamiltonian for the internal dynamics. Notably, a derivation of the Hamiltonian (7) from quantum field theory in curved spacetime is lacking with some more rigorous arguments for the conditions under which it may hold given recently [52, 53]. If correct, the term $\sim \hat{H}_{\mathrm{int}}\Phi(\hat{x}_{\mathrm{cm}})$ in equation (7) suggests entanglement between the internal and external degrees of freedom which can result in the loss of interferometric visibility.

Following the idea to focus on situations where the analogy to electromagnetism breaks down, we can combine the consequences of the Hamiltonian (7) with the thought experiment described before. Instead of a free fall experiment, Bob could sense the gravitational field sourced by Alice's particle using an interferometer himself. Instead of the displacement δx, he would then measure the COW [46] phase shift due to gravity. It has been shown by Rydving et al. [19] that the limitations from field quantization remain the same: instead of δx needing to be at least a Planck length in order to be resolvable, the interferometric fringe spacing must be at least a Planck length. In the context of equation (7) it is worth noting, however, that for a typical interferometric path with short acceleration phases splitting and recombining the trajectories with a long period of free flight, the constraint on fringe spacing depends on the dynamics during the acceleration phases, whereas the gravitational phases are determined during the free flight phase. Thus, it seems, the constraint preventing faster-than-light signaling could be overcome could one increase the mass during the free flight phase only.

The Hamiltonian (7) allows for exactly that [54]: after an initial acceleration with the rest mass m photons of energy $E \gg mc^2$ are absorbed and emitted just before entering the final acceleration phase. Does this imply that faster-than-light signaling is possible? Of course not. Firstly, the Hamiltonian (7), not being rigorously derived from a fundamental relativistic theory, could simply be inapplicable to the situation at hand. Secondly, if gravity was correctly described by a semiclassical theory and not induce entanglement, the paradoxic situation would be easily resolved.

Assuming both equation (7) and gravitationally induced entanglement to be correct models, the paradox can still be resolved. To achieve this, the comparison to electrodynamics is not helpful: a corresponding thought experiment where charge is dynamically changed is not feasible in the same way. Remembering, however, that the incoming photons carry energy and therefore gravitate, intuition suggests that the absorption and emission of these photons at the location of the particle correspond to an accelerated energy distribution which should emit gravitational waves. Rough estimates based on dimensional analysis suggest that the which-way information carried by these gravitational waves could be just enough to decohere the interference pattern in those situations that would otherwise allow faster-than-light signaling. This reasoning allows us to predict a novel type of decoherence effect: decoherence due to the emission of gravitational waves during the (otherwise coherent!) absorption or stimulated emission of photons by a massive body.

A better theoretical justification of this proposed decoherence effect is currently a work in progress. The most important lesson to be learned from this example, in our opinion, is that—unexpected experimental results such as a confirmation of the Schrödinger–Newton model or a falsification of gravitationally induced entanglement aside—the most promising thought experiments that could guide us to a proper theory of gravitating quantum matter are those without analogy in the realm of electromagnetism.

Acknowledgments
The authors gratefully acknowledge funding through the Volkswagen Foundation.

References
[1] Hossenfelder S 2013 *Living Rev. Relativ.* **16** 2
[2] Kempf A 1994 *J. Math. Phys.* **35** 4483–4496
[3] Maggiore M 1994 *Phys. Rev. D* **49** 5182–5187
[4] DeWitt C M and Rickles D (eds) 1957 *The Role of Gravitation in Physics - Report from the 1957 Chapel Hill Conference*
[5] Møller C 1962 *Colloques Internationaux CNRS* vol 91 ed Lichnerowicz A and Tonnelat M A (CNRS, Paris)
[6] Rosenfeld L 1963 *Nucl. Phys.* **40** 353–356
[7] Eppley K and Hannah E 1977 *Found. Phys.* **7** 51–68
[8] Kibble T W B 1981 *Quantum Gravity 2. A Second Oxford Symposium* ed Isham C J, Penrose R and Sciama D W (New York: Oxford University Press) pp 63–80
[9] Page D N and Geilker C D 1981 *Phys. Rev. Lett.* **47** 979–982
[10] Penrose R 1996 *Gen. Relativ. Gravit.* **28** 581–600
[11] Penrose R 1998 *Phil. Trans. R. Soc. Lond. A* **356** 1927–1939
[12] Huggett N and Callender C 2001 *Phil. Sci.* **68** S382–S394
[13] Mattingly J 2005 *Einstein Studies Volume 11. The Universe of General Relativity* Einstein Studies ed Kox A J and Eisenstaedt J (Boston: Birkhäuser) pp 327–338
[14] Mattingly J 2006 *Phys. Rev. D* **73** 064025
[15] Albers M, Kiefer C and Reginatto M 2008 *Phys. Rev. D* **78** 064051
[16] Carlip S 2008 *Class. Quantum Grav.* **25** 154010
[17] Belenchia A, Wald R, Giacomini F, Castro-Ruiz E, Brukner Č and Aspelmeyer M 2018 *Phys. Rev. D* **98** 126009
[18] Kent A 2018 *Class. Quantum Grav.* **35** 245008
[19] Rydving E, Aurell E and Pikovski I 2021 *Phys. Rev. D* **104** 086024
[20] Großardt A 2022 *AVS Quantum Sci.* **4** 010502
[21] Giulini D, Großardt A and Schwartz P K 2022 Coupling Quantum Matter and Gravity (*Preprint* 2207.05029)
[22] Oppenheim J 2021 A post-quantum theory of classical gravity? (*Preprint* 1811.03116)
[23] Yang H, Miao H, Lee D S, Helou B and Chen Y 2013 *Phys. Rev. Lett.* **110** 170401
[24] Großardt A, Bateman J, Ulbricht H and Bassi A 2016 *Phys. Rev. D* **93** 096003
[25] Diósi L 1984 *Phys. Lett. A* **105** 199–202
[26] Giulini D and Großardt A 2011 *Class. Quantum Grav.* **28** 195026
[27] Bahrami M, Großardt A, Donadi S and Bassi A 2014 *New J. Phys.* **16** 115007
[28] Kafri D, Taylor J M and Milburn G J 2014 *New J. Phys.* **16** 065020
[29] Bose S, Mazumdar A, Morley G W, Ulbricht H, Toroš M, Paternostro M, Geraci A A, Barker P F, Kim M S and Milburn G 2017 *Phys. Rev. Lett.* **119** 240401

Journal of Physics: Conference Series **2533** (2023) 012022 doi:10.1088/1742-6596/2533/1/012022

[30] Marletto C and Vedral V 2017 *Phys. Rev. Lett.* **119** 240402
[31] Mari A, De Palma G and Giovannetti V 2016 *Sci. Rep.* **6** 22777
[32] Bohr N and Rosenfeld L 1933 *K. Dan. Vidensk. Selsk. Math.-fys. Medd.* **12** 3–65
[33] Altamirano N, Corona-Ugalde P, Mann R B and Zych M 2018 *Class. Quantum Grav.* **35** 145005
[34] Tilloy A and Diósi L 2016 *Phys. Rev. D* **93** 024026
[35] Bassi A, Lochan K, Satin S, Singh T P and Ulbricht H 2013 *Rev. Mod. Phys.* **85** 471–527
[36] Christodoulou M, Di Biagio A, Aspelmeyer M, Brukner Č, Rovelli C and Howl R 2022 Locally mediated entanglement through gravity from first principles (*Preprint* 2202.03368)
[37] Vinante A, Bahrami M, Bassi A, Usenko O, Wijts G and Oosterkamp T H 2016 *Phys. Rev. Lett.* **116** 090402
[38] Bassi A, Großardt A and Ulbricht H 2017 *Class. Quantum Grav.* **34** 193002
[39] Donadi S, Piscicchia K, Curceanu C, Diósi L, Laubenstein M and Bassi A 2021 *Nat. Phys.* **17** 74–78
[40] Marletto C and Vedral V 2020 *Phys. Rev. D* **102** 086012
[41] Hall M J W and Reginatto M 2018 *J. Phys. A: Math. Theor.* **51** 085303
[42] Pal S, Batra P, Krisnanda T, Paterek T and Mahesh T S 2021 *Quantum* **5** 478 (*Preprint* 1909.11030)
[43] Fragkos V, Kopp M and Pikovski I 2022 (*Preprint* 2206.00558)
[44] Döner M K and Großardt A 2022 *Found. Phys.* **52** 101
[45] Diósi L 2015 *J. Phys.: Conf. Ser.* **626** 012008
[46] Colella R, Overhauser A W and Werner S A 1975 *Phys. Rev. Lett.* **34** 1472–1474
[47] Bedingham D J 2020 *Do Wave Functions Jump? : Perspectives of the Work of GianCarlo Ghirardi* Fundamental Theories of Physics ed Allori V, Bassi A, Dürr D and Zanghi N (Cham: Springer International Publishing) pp 191–203
[48] Feynman R P 1963 *Acta Phys. Polon.* **24** 697–722
[49] Großardt A 2020 *Phys. Rev. A* **102** 040202(R)
[50] Zych M, Costa F, Pikovski I and Brukner Č 2011 *Nat. Commun.* **2** 505
[51] Pikovski I, Zych M, Costa F and Brukner Č 2015 *Nat. Phys.* **11** 668–672
[52] Schwartz P K and Giulini D 2019 *Class. Quantum Grav.* **36** 095016
[53] Schwartz P K and Giulini D 2019 *Phys. Rev. A* **100** 052116
[54] Großardt A 2022 *Int. J. Mod. Phys. D* **31** 2242017

Can the Schrödinger dynamics explain measurement?

Alexey A Kryukov

Department of Mathematics & Natural Sciences, University of Wisconsin-Milwaukee, USA

E-mail: kryukov@uwm.edu

Abstract. The motion of a ball through an appropriate lattice of round obstacles models the behavior of a Brownian particle and can be used to describe measurement on a macro system. On another hand, such motion is chaotic and a known conjecture asserts that the Hamiltonian of the corresponding quantum system must follow the random matrix statistics of an appropriate ensemble. We use the Hamiltonian represented by a random matrix in the Gaussian unitary ensemble to study the Schrödinger evolution of non-stationary states. For Gaussian states representing a classical system, the Brownian motion that describes the behavior of the system under measurement is obtained. For general quantum states, the Born rule for the probability of transition between states is derived. It is then shown that the Schrödinger evolution with such a Hamiltonian models measurement on macroscopic and microscopic systems, provides an explanation for the classical behavior of macroscopic bodies and for irreversibility of a measurement, and identifies the boundary between micro and macro worlds.

Behavior of macroscopic bodies under a measurement can be described by Newtonian dynamics, at least in principle. For instance, suppose the position of a point-like particle is measured. The distribution of the position random variable can be explained by interaction of the particle with the surroundings and the measuring device. In principle, the effect of this interaction on the particle can be described by the Newtonian equations of motion. Because of the very large number of fluctuations in the position of the particle during the period of observation, it is more realistic to describe the measurement stochastically. For instance, the Langevin equation can be derived from the Newtonian dynamics of the particle in a medium and is well-suited for modeling a measurement of the position of the particle. It predicts the normal distribution of the position random variable during the period of observation, which is consistent with experience.

The situation in quantum theory is arguably very different. The common wisdom is that the Schrödinger equation is not sufficient to describe measurement in the micro world. The decoherence theory is seemingly the closest one can get to describing a measurement while using the Schrödinger dynamics. However, the decoherence theory does not explain how a particular measured state of the system comes to life and does not provide by itself the probability of an outcome. A modification of the Schrödinger equation or an altogether different mechanism seems to be needed to model the process of measurement. The difficulty lies in the linear property of the Schrödinger equation. Even if under the Schrödinger evolution a given state were to converge to an eigenstate of the measured observable, the sum of two such states would not do the same. This issue does not arise in Newtonian physics, which is not linear, pointing to its apparent advantage in describing measurement.

To better understand the difference between two theories in their application to measurement, let us express the relationship between Newtonian and Schrödinger dynamics in precise terms.

Journal of Physics: Conference Series **2533** (2023) 012023 doi:10.1088/1742-6596/2533/1/012023

Let M_3^σ be the set of all Gaussian states

$$g_{\mathbf{a},\sigma}(\mathbf{x}) = \left(\frac{1}{2\pi\sigma^2}\right)^{3/4} e^{-\frac{(\mathbf{x}-\mathbf{a})^2}{4\sigma^2}} \tag{1}$$

and let $M_{3,3}^\sigma$ be the set of all Gaussian wave packets

$$\varphi_{\mathbf{a},\mathbf{p},\sigma}(\mathbf{x}) = \left(\frac{1}{2\pi\sigma^2}\right)^{3/4} e^{-\frac{(\mathbf{x}-\mathbf{a})^2}{4\sigma^2}} e^{i\mathbf{px}/\hbar}. \tag{2}$$

Here, $\mathbf{x}, \mathbf{a}$ and $\mathbf{p}$ are vectors in $\mathbb{R}^3$ and the variance σ^2 is assumed to be as small as needed. In quantum mechanics, the wave packets (1) and (2) can be used to represent particles of position $\mathbf{a}$ and velocity $\mathbf{p}/m$.

The sets M_3^σ and $M_{3,3}^\sigma$ can be considered submanifolds of the projective space of states CP^{L_2} of the particle with the induced manifold structure. Moreover, the Fubini-Study metric on CP^{L_2} gives rise to a Riemannian metric on M_3^σ and $M_{3,3}^\sigma$. The map $\omega : \mathbf{a} \longrightarrow g_{\mathbf{a},\sigma}$ identifies the classical space $\mathbb{R}^3$ with the submanifold M_3^σ. Likewise, the map $\Omega : (\mathbf{a}, \mathbf{p}) \longrightarrow g_{\mathbf{a},\sigma}e^{i\mathbf{px}/\hbar}$ identifies the classical phase space $\mathbb{R}^3 \times \mathbb{R}^3$ with $M_{3,3}^\sigma$. A simple calculation demonstrates that the induced Riemannian metric on M_3^σ and $M_{3,3}^\sigma$ is the usual Euclidean metric, so that ω and Ω are isometric embeddings of the Euclidean classical space and the Euclidean phase space into CP^{L_2}.

We now claim that Newtonian dynamics is the Schrödinger dynamics of the particle whose state is constrained to the manifold $M_{3,3}^\sigma$. In that sense, it is similar to a classical dynamical system with a constraint, e.g., bead on a wire. In fact, the action functional

$$S[\varphi] = \int \overline{\varphi}(\mathbf{x}, t) \left[i\hbar\frac{\partial}{\partial t} - \widehat{h}\right] \varphi(\mathbf{x}, t) d^3\mathbf{x} dt \tag{3}$$

with the Hamiltonian $\widehat{h} = -\frac{\hbar^2}{2m}\Delta + V(\mathbf{x}, t)$ yields the Schrödinger equation. For the states φ constrained to the manifold $M_{3,3}^\sigma$, this functional reduces to the classical action

$$S = \int \left[\mathbf{p}\frac{d\mathbf{a}}{dt} - h(\mathbf{p}, \mathbf{a}, t)\right] dt, \tag{4}$$

where $h(\mathbf{p}, \mathbf{a}, t) = \frac{\mathbf{p}^2}{2m} + V(\mathbf{a}, t) + const$ is the Hamiltonian function for the particle. We used here the fact that the parameter σ can be made arbitrarily small and that the terms $g_{\mathbf{a},\sigma}^2$ form a delta sequence as σ approaches 0. It follows that the variation of the functional (3) subject to the constraint that φ belongs to $M_{3,3}^\sigma$ yields Newtonian equations of motion for the particle.

Because Gaussian states form a complete set in the Hilbert space $L_2(\mathbb{R}^3)$, one can prove the converse statement. Namely, there is a unique unitary evolution of state of the particle in $L_2(\mathbb{R}^3)$ that reduces to Newtonian motion when constrained to $M_{3,3}^\sigma$. This is exactly the Schrödinger evolution with the Hamiltonian $\widehat{h} = -\frac{\hbar^2}{2m}\Delta + V(\mathbf{x}, t)$ [1]. Furthermore, both statements can be generalized to include systems of N particles interacting via potential $V(\mathbf{x}_1, ..., \mathbf{x}_N, t)$ [2]. Thus, we have an isometric embedding of the phase space of an arbitrary classical mechanical system into the space of quantum states of the system with the property that the Schrödinger dynamics on the space of states is a unique unitary extension of the Newtonian dynamics on the phase space.

The obtained relationship between Newtonian and Schrödinger dynamics can be complemented by a relationship between the normal probability distribution for the position of a particle in $\mathbb{R}^3$ and the Born rule for the probability of transition between states in the space

Journal of Physics: Conference Series **2533** (2023) 012023 doi:10.1088/1742-6596/2533/1/012023

of states CP^{L_2}. The existence of such a relationship is clear already from the fact that for the states $g_{\mathbf{a},\sigma}$ in M_3^g the probability density $|g_{\mathbf{a},\sigma}|^2$ in the Born rule is also the normal probability density function on $\mathbb{R}^3$. So, the Born rule on CP^{L_2} implies the normal probability distribution on $\mathbb{R}^3 = M_3^g$ and is identical to it on this set.

To see this in more detail and to derive the converse statement, let us express the Born rule in terms of the probability of transition between normalized states. Let $\theta(g_{\mathbf{a},\sigma}, g_{\mathbf{b},\delta})$ be the Fubini-Study distance between the Gaussian states $g_{\mathbf{a},\sigma}$ and $g_{\mathbf{b},\delta}$. Let $(\mathbf{a}-\mathbf{b})^2$ be the square of the Euclidean distance between the corresponding points $\mathbf{a}$ and $\mathbf{b}$ in $\mathbb{R}^3$. By a direct integration, we have:

$$\left(\frac{2\sigma\delta}{\sigma^2+\delta^2}\right)^3 e^{-\frac{(\mathbf{a}-\mathbf{b})^2}{2(\sigma^2+\delta^2)}} = \cos^2\theta(g_{\mathbf{a},\sigma}, g_{\mathbf{b},\delta}). \tag{5}$$

If $\delta=\sigma$, this equation yields a relationship of the distances between $g_{\mathbf{a},\sigma}$ and $g_{\mathbf{b},\sigma}$ in the Fubini-Study metric on CP^{L_2} and between the points $\mathbf{a}$ and $\mathbf{b}$ in the Euclidean metric on $\mathbb{R}^3$:

$$e^{-\frac{(\mathbf{a}-\mathbf{b})^2}{4\sigma^2}} = \cos^2\theta(g_{\mathbf{a},\sigma}, g_{\mathbf{b},\sigma}). \tag{6}$$

On another hand, when δ approaches 0, the left side of (5) yields the normal probability density function times the volume element $(8\pi)^{\frac{3}{2}}\delta^3$. The resulting probability can be interpreted as the probability of finding the particle near point $\mathbf{b}$, given its initial position at $\mathbf{a}$. The probability on the right side is the probability of transition between the corresponding initial and end-states, calculated by the Born rule. Once again, we see that the normal probability distribution and the Born rule are identical on M_3^g. In particular, assuming the normal probability distribution of the position on $\mathbb{R}^3$, we deduce the validity of the Born rule on M_3^g.

Suppose the probability of transition between states depends only on the Fubini-Study distance between them. Then, the validity of the Born rule for the states in M_3^g signifies its validity for arbitrary quantum states. In fact, the Fubini-Study distance between states $g_{\mathbf{a},\sigma}$ and $g_{\mathbf{b},\delta}$ with $\mathbf{a}$ and $\mathbf{b}$ in $\mathbb{R}^3$ takes on all possible values in CP^{L_2} from 0 to $\pi/2$. Let then φ and ψ be any two states in CP^{L_2} and let $g_{\mathbf{a},\sigma}$ and $g_{\mathbf{b},\delta}$ be two states at the same Fubini-Study distance as the distance between φ and ψ. The probability of transition between φ and ψ is then given by

$$P_{\varphi,\psi} = P_{g_{\mathbf{a},\sigma}, g_{\mathbf{b},\delta}} = \cos^2\theta(g_{\mathbf{a},\sigma}, g_{\mathbf{b},\delta}) = \cos^2\theta(\varphi,\psi). \tag{7}$$

So, $P_{\varphi,\psi} = \cos^2\theta(\varphi,\psi)$, which is the Born rule. We conclude that under these conditions, the normal probability distribution on $\mathbb{R}^3$ implies the Born rule on the space of states.

The derived relationship between Newtonian and Schrödinger dynamics and between normal probability distribution and the Born rule indicates that there may exist a fundamental connection between measurements in the macro and micro worlds. As stated earlier, the process of measurement in classical physics does not require a separate mechanism and can be modeled within Newtonian dynamics itself. In particular, the Langevin equation for the Brownian particle can be derived from the Newtonian dynamics of the particle interacting with a harmonic oscillator heat bath [3]. Because Newtonian dynamics is a constrained Schrödinger dynamics, it is reasonable to look for a Hamiltonian that yields a Newtonian model of classical measurement of the particle under the constraint. Assuming that such a Hamiltonian exists, we shall use it to model measurements in the micro and macro worlds and study the outcomes.

To find a proper Hamiltonian, let us model the behavior of a macroscopic particle under measurement by a Brownian motion on $\mathbb{R}^3$. Under the embedding $\omega: \mathbb{R}^3 \longrightarrow CP^{L_2}$, the motion of the particle is the motion of its state, constrained to the submanifold M_3^g. We therefore need a Hamiltonian that accounts for the effect of the surroundings and the device on the particle's state and that models the Brownian motion when constrained to M_3^g.

Note that the physical Brownian motion can be identified with a regular, stochastic or a chaotic process [4]. Based on the rich experimental data and the works of Wigner [5] and BGS

[6], the general consensus currently is that Hamiltonians of all generic quantum systems whose underlying classical dynamics is chaotic follow the random matrix statistics of an appropriate ensemble. This includes complex systems such as the heavy nuclei considered in [5] as well as simple one-particle systems such as the electron in a lattice of round obstacles. The latter case is particularly relevant to the issue of measurement as the classical motion of a free particle through an appropriate lattice models the Brownian motion of the particle and yields the normal probability distribution for the position [4].

This independent argument suggests that the Hamiltonian that we are looking for exists and can be represented at any time by a random matrix. However, instead of the usual study of distribution of eigenvalues of a random matrix, our goal here is to investigate the evolution of non-stationary states of a measured particle. Under the evolution with such a Hamiltonian, the state becomes a random variable performing a random walk on the space of states CP^{L_2}.

To ensure that the Hamiltonian is Hermitian, the random matrix will be assumed to take values in the Gaussian unitary ensemble. The Hamiltonian also needs to account for the independence of action of the surroundings on the particle at different and sufficiently distant moments of time. To summarize, we conjecture that:

> **(RM)** *The Hamiltonian of a quantum system whose underlying classical dynamics describes a Brownian motion can be represented at any time by a random matrix from the Gaussian unitary ensemble. Matrices representing the Hamiltonian at two different moments of time are independent.*

This conjecture is essentially the BGS-conjecture [6] applied to a specific classical chaotic system, such as a particle in a lattice of round obstacles. The Brownian motion in the conjecture is appropriate for representing the process of measurement in the macro world. This, together with abundant experimental confirmation of the BGS-conjecture suggest that **(RM)** may describe accurately what in fact is happening in a measurement. However, for our purposes it will be sufficient to consider **(RM)** as a way to specify the model of measurement studied in the paper.

Let us prove, first of all, that we found a proper Hamiltonian. Namely, let us show that the motion of state driven by the Hamiltonian $\widehat{h}$ in **(RM)** and conditioned to stay on M_3^σ yields a random walk on $\mathbb{R}^3$ that approximates the Brownian motion and models the process of measurement on a macroscopic particle. In fact, for small time intervals $\Delta t = t_k - t_{k-1}$, the state φ_{t_N} at time t_N is approximately given by the time ordered product

$$\varphi_{t_N} = e^{-\frac{i}{\hbar}\widehat{h}(t_N)\Delta t} e^{-\frac{i}{\hbar}\widehat{h}(t_{N-1})\Delta t} ... e^{-\frac{i}{\hbar}\widehat{h}(t_1)\Delta t}\varphi_{t_0}. \tag{8}$$

For the state conditioned to stay on M_3^σ, the points $\varphi_{t_0}, \varphi_{t_1}, ...$ belong to M_3^σ and the steps can be identified with translations in the classical space. In other words, $\widehat{h}(t_k) = \xi_k \widehat{\mathbf{p}}$, where $\widehat{\mathbf{p}}$ is the momentum operator and ξ_k is a vector in $\mathbb{R}^3$. The equation (8) yields then the following expression:

$$\varphi_{t_N}(\mathbf{x}) = \varphi_{t_0}(x - \xi_1 \Delta t - \xi_2 \Delta t - ... - \xi_N \Delta t). \tag{9}$$

That is, the initial state is simply translated by the vector

$$\mathbf{d}_N = \sum_{k=1}^{N} \xi_k \Delta t \tag{10}$$

in $\mathbb{R}^3$. Now, the probability distribution of steps $-\frac{i}{\hbar}\widehat{h}(t_{k+1})\varphi_{t_k}$ must be the conditional probability distribution under the condition that the steps take place in $T_{\varphi_k} M_3^\sigma$. Because the matrix of $\widehat{h}$ is in the Gaussian unitary ensemble, the conditional probability distribution is the usual probability distribution on the subspace $T_{\varphi_k} M_3^\sigma = \mathbb{R}^3$ and the vectors ξ_k are independent and identically normally distributed random vectors. It follows that the equation (10) defines a random walk with Gaussian steps on $\mathbb{R}^3$. This is known to approximate the Brownian motion on $\mathbb{R}^3$ and can be used to model the process of measurement over the period of observation.

There is also a converse result. Namely, given a random walk $\mathbf{d}_N$ with Gaussian steps on $\mathbb{R}^3$, there is a unique Gaussian unitary ensemble such that the Scrödinger evolution with Hamiltonian represented by a random matrix in this ensemble yields the walk $\mathbf{d}_N$ when constrained to M_3^σ. In fact, the distribution of steps in the direction tangent to M_3^σ defines the distribution of all entries of the random matrix in the Gaussian unitary ensemble and therefore defines the ensemble completely.

Note also that the Gaussian orthogonal ensemble used in [5] would not result in the Brownian motion on M_3^σ in this way. In fact, the momentum operator in the derivation is Hermitian but not orthogonal. It is known that Hamiltonians with matrices in the Gaussian unitary ensemble are not invariant with respect to time reversal. So, the fact that the Brownian motion on the classical space submanifold was derived from Schrödinger evolution is tied to the use of a time-irreversible Hamiltonian.

Let us see now what kind of random walk of state is obtained for the state driven by the Hamiltonian $\widehat{h}$ and not constrained to M_3^σ. First of all, we claim that for all initial states $\{\varphi\}$ in the space of states CP^{L_2}, the vector $d\varphi = -\frac{i}{\hbar}\widehat{h}\varphi dt$ with such a Hamiltonian is a normal random vector in the tangent space $T_{\{\varphi\}}CP^{L_2}$ with a spherical distribution. In particular, the probability distribution of steps of the random walk is isotropic and homogeneous. In fact, because for all values of t, the matrix of $\widehat{h}$ is in the Gaussian unitary ensemble, the probability density function $P(\widehat{h})$ of $\widehat{h}$ is invariant with respect to conjugations by unitary transformations. That is, $P(U^{-1}\widehat{h}U) = P(\widehat{h})$ for all unitary transformations U acting in the Hilbert space of states. Also, for all unitary transformations U that leave $\{\varphi\}$ unchanged and therefore all U that act in the tangent space $T_{\{\varphi\}}CP^H$, we have

$$(U^{-1}\widehat{h}U\varphi, v) = (\widehat{h}U\varphi, Uv) = (\widehat{h}\varphi, Uv), \tag{11}$$

where v is a unit vector in $T_{\{\varphi\}}CP^H$. It follows that

$$P(\widehat{h}\varphi, v) = P(\widehat{h}\varphi, Uv), \tag{12}$$

where P is the probability density of the components of $\widehat{h}\varphi$ in the given directions. By a proper choice of U, we can make Uv to be an arbitrary unit vector in $T_{\{\varphi\}}CP^H$, proving the isotropy of the distribution. Furthermore, for all unitary operators V in H and a unit vector v in $T_{\{\varphi\}}CP^H$, we have

$$P(\widehat{h}\varphi, v) = P(V^{-1}\widehat{h}V\varphi, v) = P(\widehat{h}V\varphi, Vv). \tag{13}$$

Because $V\varphi$ is an arbitrary state and Vv is in the tangent space $T_{\{V\varphi\}}CP^H$, we conclude with the help of (11) that the probability density function is independent of the initial state of the system, proving the homogeneity of the distribution. The components of the vector $\widehat{h}\varphi dt$ are independent and normally distributed because the entries in the columns of the matrix of $\widehat{h}$ are independent and normally distributed. It follows that $-\frac{i}{\hbar}\widehat{h}\varphi dt$ is a normal random vector with a spherical distribution.

Because the steps of the obtained walk of state are independent and the distribution of steps is homogeneous and isotropic, the probability of reaching a certain state during the given period of observation can only depend on the Fubini-Study distance between the initial and final states. As we proved, under these conditions the normal probability distribution on M_3^σ implies the Born rule for the probability of transition between states. So, the Schrödinger evolution with the Hamiltonian in the Gaussian unitary ensemble yields the Brownian motion of the particle

Journal of Physics: Conference Series **2533** (2023) 012023 doi:10.1088/1742-6596/2533/1/012023

whose state is constrained to M_3^σ and the Born rule for the probability of transition between unconstrained states of the particle in CP^{L_2}.

This result combined with the obtained relationship between Newtonian and Schrödinger dynamics is telling us that in the considered model there is no fundamental difference between macroscopic and microscopic particles. The dynamics of macroscopic particles, including their behavior under a measurement is identical to the corresponding dynamics of microscopic particles whose state is constrained to the classical space submanifold of the space of states. Moreover, knowing the dynamics of a macrosystem, whether the system is measured or not, we can now predict in a unique way the dynamics of the corresponding microsystem, and vice versa.

To obtain a unified mechanism of measurement in classical and quantum mechanics, we still need to explain why in the model the state of a macroscopic particle is constrained to M_3^σ. We also need to understand the dynamics of the system consisting of a measured particle and a measuring device during measurement. Finally, the validity of the Born rule for the system driven by the Hamiltonian in (**RM**) comes with the downside that numerous final states are equally likely to occur in a measurement. We then need to explain why, for instance, a measurement of the position of a particle yields only the states of a well-defined position of the particle in $\mathbb{R}^3$.

To address the first question, note that the motion of state driven by the Hamiltonain $\widehat{h}(t)$ represented by a random matrix in the Gaussian unitary ensemble is essentially the Brownian motion on the space of states CP^{L_2}. For states constrained to M_3^σ, this motion reduces to the Brownian motion of the particle on $\mathbb{R}^3$. As is well known, the physical Brownian motion of sufficiently large particles in a medium in $\mathbb{R}^3$ is trivial. This is because the total force acting on such particles from atoms and molecules of the medium vanishes. In this case, the diffusion coefficient for the process is zero and the particle remains at rest in the lab system. Therefore, the action of the Hamiltonian $\widehat{h}(t)$ in (**RM**) that represents this situation in the direction of vectors in the tangent space $T_{\{\varphi\}}M_3^\sigma$ at initial point $\{\varphi\}$ in M_3^σ must be nearly trivial. However, because the distribution of steps for the Hamiltonian in the Gaussian unitary ensemble is isotropic and homogeneous, the distribution of all entries of $\widehat{h}(t)$ must be centered at zero and have a vanishingly small variance. In other words, if under these conditions the particle does not move in $\mathbb{R}^3$, then the state of the particle does not move in the direction of vectors in $T_{\{\varphi\}}M_3^\sigma$, and then it cannot move in any other direction in the tangent space to the space of states at $\{\varphi\}$. This reflects the fact that the state under these conditions is pushed simultaneously in all possible directions in the space of states, so that the net displacement of state is zero in the lab system.

If an external potential $V(\mathbf{x})$ is applied to such a particle, the state of the particle will evolve in the classical phase space submanifold $M_{3,3}^\sigma$ in accord with Newtonian dynamics. This can be shown by identifying components of the velocity of state $d\varphi/dt$ under the Hamiltonian $\widehat{h}+V$, as in [1]. The underlying assumption is that the operator $\widehat{h}$ is built in the usual way from the kinetic and potential energy terms of all participating particles. The effect of the external potential on the velocity appears in the form of components tangent to the classical phase space submanifold $M_{3,3}^\sigma$. The state is pushed along the submanifold and the particle moves classically in $\mathbb{R}^3$. Alternatively, note that in a frame moving relative to the lab system with linear acceleration $\mathbf{w}$, the Schrödinger equation acquires an extra term $V=-m\mathbf{w}\cdot\mathbf{x}$ [7, 8], observed in the experiment [9]. However, any differentiable potential is approximately linear within small distances. So, the action of an external potential on states in $M_{3,3}^\sigma$ with small σ is equivalent to the transformation to an accelerated reference frame, the change that preserves $M_{3,3}^\sigma$. It follows that the state initially in $M_{3,3}^\sigma$ remains confined to $M_{3,3}^\sigma$ while the particle moves classically with acceleration $\mathbf{w}=-\nabla V/m$ in $\mathbb{R}^3$. This explains why the state of a macroscopic measuring device positioned initially in the classical phase space submanifold of the space of states remains confined to the submanifold and obeys the laws of Newtonian dynamics.

Journal of Physics: Conference Series **2533** (2023) 012023 doi:10.1088/1742-6596/2533/1/012023

Furthermore, consider a microscopic particle in a medium, such as gas or liquid. A small neutral particle in the medium may be able to move freely through the medium. On another hand, the Brownian-size particles would experience a Brownian motion in the medium. Per **(RM)**, the smaller microscopic particles whose interaction with atoms and molecules of the medium cannot be neglected, will evolve by the Hamiltonian represented by a random matrix. The state of a particle will perform a random walk in CP^{L_2}. For the steps within the submanifold M_3^σ, the walk represents the Brownian motion on $\mathbb{R}^3$.

Suppose we increase the size of the particle. The Brownian motion will eventually stop, which also means that the state of the particle in CP^{L_2} in the model will remain fixed in $M_{3,3}^\sigma$. The size of the particle needed for this to happen depends on physical properties of the medium such as its dynamic viscosity and temperature. In particular, the boundary between microscopic and macroscopic bodies in the model depends on the medium the bodies are in. The cosmic background radiation is probably the ultimate medium to define such a boundary and to test for it. The state of a particle that is macroscopic for a given medium is confined to the submanifold $M_{3,3}^\sigma$ and moves in accord with Newtonian dynamics. We interpret such a motion as the classical motion of particle in $\mathbb{R}^3$. Under proper conditions, the macroscopic particle may again experience a Brownian-like motion in $\mathbb{R}^3$, this time purely classical. A large ball kicked by many feet from all sides would be a large-scale example of such a motion. In this case, the state of the particle is already constrained to $M_{3,3}^\sigma$ and its evolution is driven by the classical Hamiltonian.

Consider now a system consisting of a light particle interacting with a much heavier particle. In Newtonian physics, the effect of the light particle on the motion of the heavy particle can be neglected. In particular, the Brownian motion of the heavy particle due to its interaction with the surroundings will not be altered by its interaction with the light particle. From **(RM)**, it follows that the state of the heavier particle for the corresponding system of two microparticles in the model exists as a random variable with values in the space of states and is driven by a random Hamiltonian. In particular, the state of the system of two particles must be the product of states of the particles, i.e., the state will remain separable throughout the evolution. We know that when the heavier particle is sufficiently large, its state will be confined to the submanifold $M_{3,3}^\sigma$ and the motion of the particle will be classical.

In particular, the system consisting of a microscopic particle and a macroscopic measuring device in the model will always be in a product state with the state of the device constrained to the submanifold $M_{3,3}^\sigma$ and satisfying the classical Newtonian dynamics. The microscopic particle in this case will move in accord with the Schrödinger dynamics in the surroundings provided by the device and the environment. When the measurement occurs, the state of the particle experiences a random walk that satisfies the Schrödinger equation with Hamiltonian in **(RM)**. The walk results in the Born rule for the probability of transition to the observed state.

It remains to understand why in a measurement described by the model we can only observe the eigentstates of a measured quantity. Indeed, this seems to contradict the result that the state driven by the Hamiltonian in **(RM)** is equally likely to reach any state at a given Fubini-Study distance from the initial state. To answer, consider first the classical experiment of firing bullets into a target and measuring position of the bullets at rest. For the bullet to hit the target, the experimenter must properly aim the gun. Without this, the probability of hitting the target is small. We claim that the role of the measuring device in quantum mechanics in the proposed model is similar: it makes the state approach the target and records the outcome. The difference is that now the process takes place in the space of states.

Consider for example a setup where the position of an electron is measured by a scintillating screen, as in the double-slit experiment. What does it mean for the electron's state to approach the screen? Suppose the screen occupies region D in $\mathbb{R}^3$. Then position of the particles of the screen is well defined and the state of each particle can be identified with a point $g_{\mathbf{a},\sigma}$ in M_3^σ,

Journal of Physics: Conference Series **2533** (2023) 012023 doi:10.1088/1742-6596/2533/1/012023

with $\mathbf{a}$ in D. The state Ψ of the screen is the product of states $g_{\mathbf{a},\sigma}$, one for each of the particles of the screen. It belongs to the submanifold $M^{\sigma}_{3N} = M^{\sigma}_{3} \otimes ... \otimes M^{\sigma}_{3}$ in the N-particle space of states. Because the screen is macroscopic, as has been seen previously, the state of the particle-screen system at any time is a product $\varphi \otimes \Psi$. After the measurement, the state of the system is $g_{\mathbf{c},\sigma} \otimes \Psi$, where $g_{\mathbf{c},\sigma}$ is the state of the measured particle and the internal changes in the screen are not considered. It follows that the state $\varphi \otimes \Psi$ is close to the end-state $g_{\mathbf{c},\sigma} \otimes \Psi$ exactly when the state φ is close to $g_{\mathbf{c},\sigma}$ in the Fubini-Study metric. So, in the process of measurement, the state φ of the particle must approach the classical space submanifold M^{σ}_{3} in CP^{L_2} and the subset of states $g_{\mathbf{a},\sigma}$ with $\mathbf{a}$ in D in it.

The state $g_{\mathbf{c},\sigma}$ with a proper value of σ can be identified with the lowest energy state of the electron in the potential near the point of the screen where the electron is absorbed. The initial state φ can be decomposed into a superposition of energy eigenstates in this potential. If the coefficients of the higher-energy components of φ are not small, the state is away from M^{σ}_{3} in the Fubini-Study metric. To approach M^{σ}_{3}, the electron must loose energy so that the higher-energy components must become small or vanish completely. The process is similar to the one with the bullet hitting the target. This time, the stopping power of the screen is responsible for lowering the energy of the electron. Aiming, or providing the electron with a proper momentum directed towards the screen in $\mathbb{R}^3$ is also needed to ensure that the electron lands on the screen. However, while the motion of the bullet's state is happening in M^{σ}_{3} and is identified with a motion in $\mathbb{R}^3$, the motion of the electron's state is happening in CP^{L_2} at large. The process can be described by the Schrödinger equation with elements of the quantum field theory to account for the interaction of the electron with the electromagnetic field and for the indistinguishability of electrons. When the electron is absorbed, only the lowest energy component $g_{\mathbf{c},\sigma}$ survives, identifying the position of the electron on the screen.

Mathematically, modeling the motion of state towards the classical space M^{σ}_{3} and the screen amounts to adding a drift term to the random walk specified in **(RM)**. The drift ensures that the state arrives to the screen, while the random walk delivers the Born rule for the possible outcomes $g_{\mathbf{c},\sigma}$. A clever design of the measuring device ensures a proper drift, so that the state is guaranteed to arrive to the needed part of the space of states. As a result, the probability to find the electron at a certain point of the screen is the conditional probability under the condition that the electron hit the screen. This explains the issue of observability of just a limited subset of all possible states in a measurement in the model.

Let us summarize the lessons learnt from the proposed model of measurement. First, to answer the question in the title of the paper, the Schrödinger dynamics *is* capable of explaining the process of measurement. This is achieved by replacing the deterministic evolution of state with the motion driven by the Hamiltonian satisfying **(RM)**. In this case the distribution of end-states satisfies the Born rule for all initial states. The linear property of the evolution does not cause a problem because the distribution is homogeneous and isotropic. In particular, the random walk of a linear superposition of states satisfies the same Born rule. Second, the Newtonian dynamics of macroscopic bodies is identified with the Schrödinger dynamics constrained to the classical phase space submanifold in the space of states. Likewise, the motion of a Brownian particle in $\mathbb{R}^3$ follows from the Schrödinger evolution with Hamiltonian in **(RM)**, constrained to the classical space submanifold M^{σ}_{3}. This also points to the origin of the constraint: when the particle is sufficiently large so that its Brownian motion in the given medium trivializes, the state of the particle "freezes up" on M^{σ}_{3}, leading to its classical behavior in an external potential. Third, the irreversibility of classical and quantum measurements in the model is tied to time-irreversibility of the Hamiltonian in **(RM)**. Namely, the derivation of the Brownian motion from the Schrödinger evolution requires that the matrix that represents the Hamiltonian is in the Gaussian unitary ensemble. Such Hamiltonians do not commute with the time reversal operator, potentially providing the origin of the irreversibility. Fourth, under

Journal of Physics: Conference Series **2533** (2023) 012023 doi:10.1088/1742-6596/2533/1/012023

(RM), the state of the system consisting of a microscopic and a macroscopic particle remains separable throughout the evolution. The macroscopic measuring device behaves classically during the measurement, while the state of the microscopic particle experiences a random walk leading to the Born rule for the probability of transition to the end-states. Finally, the role of the measuring device in the model consists in creating a drift of state towards the device and in recording the result of observation.

The crux of the obtained results is the extension of the motion of particles in the classical space to the motion of states in the space of states. Going from the motion in the classical space submanifold of the space of states to the motion in the full space of states is what allowed us to connect the classical and the quantum in such a beautiful and concise way. Based on that, we hypothesize that the space of states rather than the classical space is the true arena for all physical processes.

References

[1] Kryukov A 2018 *J. Math. Phys.* **59** 052103
[2] Kryukov A 2017 *J. Math. Phys.* **58** 082103
[3] Zwanzig R 2001 *Nonequilibrium Statistical Mechanics.* (Oxford University Press, New York)
[4] Cecconi F, Cencini M, Falcioni M and Vulpiani A 2005 *Chaos* **15**(2) 26102
[5] Wigner E 1951 *Proc. Cambridge Philos. Soc.* **47** 790
[6] Bohigas O, Giannoni M and Schmit C 1984 *Physical Review Letters* **52** 1
[7] Mashhoon B 2006 *Lecture Notes in Physics* **702** 112
[8] Greenberger D 2001 *Phys. Rev. Lett.* **87** 100405
[9] Bonse U and Wroblewski T 1983 *Phys. Rev. Lett.* **51** 1401

Journal of Physics: Conference Series **2533** (2023) 012024 doi:10.1088/1742-6596/2533/1/012024

Chiral and flavor oscillations in lepton-antineutrino spin correlations

V A S V Bittencourt[1], M Blasone[2] and G Zanfardino[3]

[1]ISIS (UMR 7006), Université de Strasbourg, 67000 Strasbourg, France
[2]Dipartimento di Fisica, Università degli Studi di Salerno, Via Giovanni Paolo II, 132 84084 Fisciano, Italy & INFN Sezione di Napoli, Gruppo collegato di Salerno, Italy
[3]Dipartimento di Ingegneria Industriale, Università di Salerno, Via Giovanni Paolo II, 132 I-84084 Fisciano (SA), Italy & INFN, Sezione di Napoli, Gruppo collegato di Salerno, Italy.

Abstract. We study the evolution of quantum correlations in a lepton-antineutrino pair, as produced in weak interactions (e.g. pion decay). Assuming an inital state entangled in the spins of the two particles, we show that both chiral and flavor (neutrino) oscillations affect spin correlations. Such corrections are relevant in the non-relativistic regime. In the second part we focused our attention on the weak process $n + \nu_e \rightarrow p + e$ in which the results found in the previous sections could be observed.

1. Introduction

A remarkable property of (charged) weak interactions is their parity violating character: this is mathematically accounted for by the presence of chiral projectors in the interaction term, which allow only left handed fermion field components to interact with W bosons [1, 2]. On the other hand, weak charged currents are also characterized by the fact that only fields with definite flavors enter in the vertex. As a consequence of these facts, a fermion generated in a charged weak interaction process will have definite flavor and chirality.

States with definite chirality and finite mass undergo chiral oscillations. In fact, the Dirac equation, describing the temporal evolution of a massive fermion, includes a term which couples left and right-handed components of the bispinor, which yields chirality oscillations [3, 4], characterized by the ratio between the momentum and the mass. Chiral oscillations have been studied recently [5–7] in connection with their possible relevance for relic neutrinos of the CνB background [8] and more in general in the non-relativistic regime. In particular, the effects of chiral oscillations on the spin-spin correlations of a lepton-antineutrino pair has been investigated in Ref. [9].

In this paper we extend the analysis of Ref. [9], by taking into account also flavor oscillations for the (anti)neutrino involved in the considered process. Thus, the initial entanglement encoded only in the spins is redistributed, during time evolution, due to the chiral and flavor oscillations which occur with their characteristic time and energy scales. The result is a hyperentangled state of the system at a later time. The spin correlations display a dependence on such oscillations, thus offering a possible way for detecting them.

2. Chiral oscillations

In relativistic quantum mechanics the free evolution of a fermion is described by a wave function Ψ whose evolution is given by the Dirac equation (in natural units $\hbar = c = 1$)

$$(i\hat{\gamma}^{\mu}\partial_{\mu} - m)\Psi(x) = 0 \tag{1}$$

where the $\hat{\gamma}^{\mu}$ matrices satisfy the anticommutation relations $\{\hat{\gamma}^{\mu}, \hat{\gamma}^{\nu}\} = 2\eta^{\mu\nu}$ with $\eta^{\mu\nu} = \text{diag}\{1, -1, -1, -1\}$. The solution $\Psi(x)$ is a 4−component spinor called Dirac bispinor which has two intrinsic degrees of freedom: spin and chirality. From a group theory point of view this is due to the fact that the Dirac bispinors belong to irreducible representations of the complete Lorentz group [11].

The chiral matrix is defined by

$$\hat{\gamma}_5 \equiv i\hat{\gamma}_0\hat{\gamma}_1\hat{\gamma}_2\gamma_3 \tag{2}$$

and its expectation value $\langle\Psi|\hat{\gamma^5}|\Psi\rangle$, the chirality of a state, is not a conserved quantity due to the presence of the mass term in the Dirac equation Eq.(1), which can be recast in the form

$$\hat{H}_D\,|\Psi\rangle = (\hat{\mathbf{p}}\cdot\hat{\boldsymbol{\alpha}} + m\hat{\beta})\,|\Psi\rangle = i\partial_t\,|\Psi\rangle \tag{3}$$

with $\hat{\mathbf{p}} = (\hat{p}_1, \hat{p}_2, \hat{p}_3) \equiv -i(\partial_{x1}, \partial_{x2}, \partial_{x3})$, $\hat{\boldsymbol{\alpha}} = (\hat{\alpha}_1, \hat{\alpha}_1, \hat{\alpha}_3) \equiv (\hat{\gamma}_0\hat{\gamma}_1, \hat{\gamma}_0\hat{\gamma}_2, \hat{\gamma}_0\gamma_3)$ and $\hat{\beta} = \hat{\gamma}^0$, with $\{\hat{\alpha}_i, \hat{\alpha}_j\} = 2\delta_{ij}I_4$ and $\{\hat{\alpha}_i, \hat{\beta}\} = 0$. In terms of the $\hat{\alpha}_i$, the chiral operator becomes

$$\hat{\gamma}_5 = -i\hat{\alpha}_1\hat{\alpha}_2\hat{\alpha}_3, \tag{4}$$

which commutes only with the first term in $\hat{H}_D$. The mass term does not commute with $\hat{\gamma}^5$:

$$[\hat{\gamma}^5, \hat{H}_D] = -2im\hat{\beta}\hat{\gamma}^5 \tag{5}$$

Thus, a state with a definite chirality whose dynamics is described by the Dirac equation will undergo chiral oscillations. We can further appreciate this fact by considering chiral representation of the Dirac matrices in which [10]

$$\hat{\alpha}_i = \begin{bmatrix} \hat{\sigma}_i & 0 \\ 0 & \hat{\sigma}_i \end{bmatrix}, \hat{\beta} = \begin{bmatrix} 0 & \hat{I}_2 \\ \hat{I}_2 & 0 \end{bmatrix}, \hat{\gamma}_5 = \begin{bmatrix} \hat{I}_2, & 0 \\ 0 & -\hat{I}_2 \end{bmatrix}. \tag{6}$$

In this representation the chiral operator is diagonal with two double degenerate eigenvalues ± 1. Any Dirac bispinor can be decomposed into left and right-handed components

$$|\Psi\rangle = \begin{bmatrix} |\xi_R\rangle \\ |\xi_L\rangle \end{bmatrix} \tag{7}$$

such that the Dirac equation can be rewritten as a the set of coupled equations

$$\hat{\mathbf{p}}\cdot\hat{\boldsymbol{\sigma}}\,|\xi_R\rangle + m\,|\xi_L\rangle = i\partial_t\,|\xi_R\rangle\,, \qquad -\hat{\mathbf{p}}\cdot\hat{\boldsymbol{\sigma}}\,|\xi_L\rangle + m\,|\xi_R\rangle = i\partial_t\,|\xi_L\rangle \tag{8}$$

Note the mass term mixes the left and chiral part which allows the chiral oscillations for a state initially with a definite chirality.

In order to describe the dynamics of a given initial bispinor, we adopt the standard procedure and divide a solution of the Dirac equation into positive and negative energy solutions given by

$$|\Psi_+\rangle = e^{-i(Et-\mathbf{p}\cdot\mathbf{x})}\,|u_s(\mathbf{p}, m)\rangle\,, \qquad |\Psi_-\rangle = e^{+i(Et-\mathbf{p}\cdot\mathbf{x})}\,|v_s(\mathbf{p}, m)\rangle \tag{9}$$

where s represent the spin polarization of the particle, E and $\mathbf{p}$ are the energy and the momentum with $E^2 = p^2 + m^2$ with $p = |\mathbf{p}|$. Following Ref. [5], the solutions of the Dirac equation appearing in (9) are given in terms of the bispinors

$$|u_s(\mathbf{p}, m)\rangle = N \begin{bmatrix} f_+ \, |\eta_s(\mathbf{p})\rangle \\ f_- \, |\eta_s(\mathbf{p})\rangle \end{bmatrix}, \qquad |v_s(\mathbf{p}, m)\rangle = N \begin{bmatrix} f_+ \, |\eta_s(\mathbf{p})\rangle \\ -f_- \, |\eta_s(\mathbf{p})\rangle \end{bmatrix} \tag{10}$$

with

$$f_\pm = f_\pm(\mathbf{p}, m) = 1 \pm \frac{\hat{\mathbf{p}} \cdot \hat{\boldsymbol{\sigma}}}{E + m}, \qquad N = N(\mathbf{p}, m) = \sqrt{\frac{E + m}{4E}} \tag{11}$$

In the above equations $|\eta_s(\mathbf{p})\rangle$ is a two-component spinor encoding the polarization of the particle.

Following the scheme adopted by [7] we take $|\eta_s(\mathbf{p})\rangle$ as eigenstates of the helicity operator

$$\hat{h} = \frac{\hat{\mathbf{p}} \cdot \hat{\mathbf{\Sigma}}}{p} \tag{12}$$

where

$$\hat{\Sigma}_i = \begin{bmatrix} \hat{\sigma}_i & 0 \\ 0 & \hat{\sigma}_i \end{bmatrix} \tag{13}$$

are the generalized Pauli matrices. Since $[\hat{h}, \hat{H}_D] = 0$, helicity is conserved during the free evolution. Taking also the momentum along the z axis, $\mathbf{p} = p\hat{e}_z$, the bispinors u_s and v_s become

$$|u_\pm(\mathbf{p}, m)\rangle = N \begin{bmatrix} f_\pm \, |\pm\rangle \\ f_\mp \, |\pm\rangle \end{bmatrix}, \qquad |v_\pm(\mathbf{p}, m)\rangle = N \begin{bmatrix} f_\pm \, |\pm\rangle \\ -f_\mp \, |\pm\rangle \end{bmatrix} \tag{14}$$

where $|+\rangle$ ($|-\rangle$) is the state $|\uparrow\rangle$ ($|\downarrow\rangle$) with positive (negative) projection of the spin along z (eigenstates of s_z).

In [7] it is shown that starting with an initial state at $t = 0$ with both negative chirality and negative helicity

$$|\Psi(0)\rangle = [0, 0, 0, 1]^T \tag{15}$$

using the orthonormality relations

$$\langle u_s(\mathbf{p}, m) | v_{s'}(-\mathbf{p}, m)\rangle = 0, \qquad \langle u_s(\mathbf{p}, m) | u_{s'}(\mathbf{p}, m)\rangle = \langle v_s(\mathbf{p}, m) | v_{s'}(\mathbf{p}, m)\rangle = \delta_{s,s'} \tag{16}$$

we can write

$$|\Psi(0)\rangle = \sum_{s=+,-} \Big(\langle \Psi(0) | u_s(\mathbf{p}, m)\rangle \, |u_s(\mathbf{p}, m)\rangle + \langle \Psi(0) | v_s(-\mathbf{p}, m)\rangle \, |v_s(-\mathbf{p}, m)\rangle \Big) \tag{17}$$

and hence the time evolution is given by

$$|\Psi(t)\rangle = e^{-i\hat{H}_D t} |\Psi(0)\rangle = N\big(f_+ e^{-iEt} \, |u_-(\mathbf{p}, m)\rangle - f_- e^{iEt} \, |v_-(-\mathbf{p}, m)\rangle\big). \tag{18}$$

Since helicity is a conserved quantity, the probability that the particle does not change its chirality is given by

$$P = |\langle \Psi(0) | \Psi(t)\rangle|^2 = 1 - \frac{m^2}{E^2} \sin^2(Et) \tag{19}$$

which goes to one in the limit $\frac{p}{m} \to \infty$.

Journal of Physics: Conference Series **2533** (2023) 012024 doi:10.1088/1742-6596/2533/1/012024

3. Lepton-antineutrino entanglement and chiral oscillations

Weak interactions produce particles with a well definite chirality (left-chiral for leptons and right chiral for antileptons). If the momenta of the emitted particles are comparable with their mass, according to the time evolution of Dirac bispinors, chiral oscillations occur, as we discussed briefly in the last Section. In [9] the possibility to observe indirectly the effects of chiral oscillations has been investigated for a lepton-antineutrino pair produced by a pion decay.

The pion can decay in

$$\pi^- \to \mu^- + \bar{\nu}_\mu, \qquad \pi^- \to e^- + \bar{\nu}_e \tag{20}$$

with probability $99,9877\%$ and $0,0123\%$, respectively.

In the rest frame of the pion the two particles are emitted along opposite directions and with a well defined chirality that is right chiral for the antineutrino and left chiral for the lepton [2]. Furthermore, conservation of angular momentum imples that the total spin of the emitted particles has to vanish. Assuming that the lepton, with mass m_l, has a momentum $-p\hat{e}_z$, and the antineutrino, with mass $m_{\bar{\nu}}$, has momentum $p\hat{e}_z$, we can take the initial state in the form

$$|\Phi\rangle = \frac{|v_\uparrow(p, m_{\bar{\nu}})\rangle \otimes |u_\downarrow(-p, m_l)\rangle - |v_\downarrow(p, m_{\bar{\nu}})\rangle \otimes |u_\uparrow(-p, m_l)\rangle}{\sqrt{2}} \tag{21}$$

which is an entangled state in the spin. Since the two particles emitted in the decay have a well definite chirality, we have to project the state (21) in their respective chirality eigenstates and to normalize it, that is

$$|\Psi(0)\rangle = \frac{\hat{\Pi}^{\bar{\nu}}_R \otimes \hat{\Pi}^l_L |\Phi\rangle}{\langle\Phi|\hat{\Pi}^{\bar{\nu}}_R \otimes \hat{\Pi}^l_L|\Phi\rangle} \tag{22}$$

where the chirality projectors are given by

$$\hat{\Pi}^\alpha_{R/L} = \frac{I^\alpha \pm \hat{\gamma}^\alpha_5}{2} \tag{23}$$

with $\alpha = l, \bar{\nu}$.

After some algebraic manipulations, the state (22) can be written in the form [9]

$$|\Psi(0)\rangle = A(p, m_l, m_{\bar{\nu}})\, |\bar{\nu}_\uparrow(0)\rangle \otimes |l_\downarrow(0)\rangle - B(p, m_l, m_{\bar{\nu}})\, |\bar{\nu}_\downarrow(0)\rangle \otimes |l_\uparrow(0)\rangle \tag{24}$$

with

$$|\bar{\nu}_{\uparrow(\downarrow)}(0)\rangle = \begin{bmatrix} \uparrow (\downarrow) \\ 0 \end{bmatrix}, \qquad |l_{\uparrow(\downarrow)}(0)\rangle = \begin{bmatrix} 0 \\ \uparrow (\downarrow) \end{bmatrix} \tag{25}$$

and

$$A(p, m_l, m_{\bar{\nu}}) = N_l N_{\bar{\nu}} f^{\bar{\nu}}_+ f^l_- \left[\frac{1}{2} - \frac{p^2}{2E_l E_{\bar{\nu}}}\right]^{-\frac{1}{2}}, \qquad B(p, m_l, m_{\bar{\nu}}) = N_l N_{\bar{\nu}} f^{\bar{\nu}}_- f^l_+ \left[\frac{1}{2} - \frac{p^2}{2E_l E_{\bar{\nu}}}\right]^{-\frac{1}{2}} \tag{26}$$

with the short notation $f^\alpha_{+(-)} \equiv f_{+(-)}(p, m_\alpha)$ for $\alpha = l, \bar{\nu}$.

Note that in the relativistic limit, for the antineutrino $p \gg m_{\bar{\nu}}$ and we have $f^{\bar{\nu}}_- \to 0$: hence $A \gg B$ and only the first term in (24) contributes to the state. Time evolution of the superposition is obtained by writing explicitly $|\Psi(t)\rangle$ in terms of the bispinors (10), which are the plane wave solutions of the Dirac equation. This gives

$$|\Psi(t)\rangle = A(p, m_l, m_{\bar{\nu}})\, |\bar{\nu}_\uparrow(t)\rangle \otimes |l_\downarrow(t)\rangle - B(p, m_l, m_{\bar{\nu}})\, |\bar{\nu}_\downarrow(t)\rangle \otimes |l_\uparrow(t)\rangle \tag{27}$$

with

$$|\bar{\nu}_{\uparrow}(t)\rangle = N_{\bar{\nu}}[e^{-iE_{\bar{\nu}}t} f_{+}^{\bar{\nu}} |u_{\uparrow}(p, m_{\bar{\nu}})\rangle + e^{iE_{\bar{\nu}}t} f_{-}^{\bar{\nu}} |v_{\uparrow}(-p, m_{\bar{\nu}})\rangle] \quad (28)$$

$$|\bar{\nu}_{\downarrow}(t)\rangle = N_{\bar{\nu}}[e^{-iE_{\bar{\nu}}t} f_{-}^{\bar{\nu}} |u_{\downarrow}(p, m_{\bar{\nu}})\rangle + e^{iE_{\bar{\nu}}t} f_{+}^{\bar{\nu}} |v_{\downarrow}(-p, m_{\bar{\nu}})\rangle] \quad (29)$$

$$|l_{\uparrow}(t)\rangle = N_{l}[e^{-iE_{l}t} f_{+}^{l} |u_{\uparrow}(-p, m_{l})\rangle - e^{iE_{l}t} f_{-}^{l} |v_{\uparrow}(p, m_{l})\rangle] \quad (30)$$

$$|l_{\downarrow}(t)\rangle = N_{l}[e^{-iE_{l}t} f_{-}^{l} |u_{\downarrow}(-p, m_{l})\rangle - e^{iE_{l}t} f_{+}^{l} |v_{\downarrow}(p, m_{l})\rangle] \quad (31)$$

In [9] it has been shown that the quantum entanglement initially encoded only between the spins is redistributed by chiral oscillations to other partitions of the system. Such effects become relevant when the mass of the neutrino is comparable with its momentum. To compute the temporal evolution of the spin-spin entanglement we first obtain the spin density matrix by tracing out the chiralities

$$\rho_{S_{\bar{\nu}},S_l}(t) = \mathrm{Tr}_{C_{\bar{\nu}},C_l}\big[\rho(t)\big] = \mathrm{Tr}_{C_{\bar{\nu}},C_l}\Big[\, |\Psi(t)\rangle \langle\Psi(t)| \,\Big] \quad (32)$$

The negativity for the spin part is [9]

$$\mathcal{N}\Big[\rho_{S_{\bar{\nu}},S_l}(t)\Big] = ||\rho^{T}_{S_{\bar{\nu}},S_l}(t)|| - 1 = \mathcal{N}\Big[\rho_{S_{\bar{\nu}},S_l}(0)\Big]\Gamma(t) \quad (33)$$

with

$$\Gamma(t) = \left[1 - \frac{p^2}{m_{\bar{\nu}}^2}(\langle\gamma_5^{\bar{\nu}}\rangle(t) - 1)^2\right]^{\frac{1}{2}} \left[1 - \frac{p^2}{m_l^2}(\langle\gamma_5^{l}\rangle(t) + 1)^2\right]^{\frac{1}{2}} \equiv \Gamma_{\bar{\nu}}\Gamma_l \quad (34)$$

and

$$\langle\gamma_5^{l}\rangle(t) = -1 + \frac{m_l^2}{E_l^2}\Big[1 - \cos(2E_l t)\Big] \quad (35)$$

At $t = 0$, the negativity between the spins reads

$$\mathcal{N}[\rho_{S_l,S_{\bar{\nu}}}(0)] = 2|AB| = 2|A(p, m_l, m_{\bar{\nu}})B(p, m_l, m_{\bar{\nu}})| \quad (36)$$

Also it has been shown in [9] that the trace of the squared density operator is

$$\mathrm{Tr}[\rho^2_{S_{\bar{\nu}},S_l}(t)] = 1 - \frac{\mathcal{N}\big[\rho_{S_{\bar{\nu}},S_l}(0)\big]}{2}(1 - \Gamma(t)^2). \quad (37)$$

4. Flavor and chirality in a lepton-antineutrino entangled pair

The results in [9] revised in the previous section do not include flavor oscillations. Flavor is another intrinsic degree of freedom for the neutrino. Charged weak interactions produce neutrinos with a well defined flavor, which is not conserved during free evolution; therefore we expect that, as for the chirality, flavor oscillations will also redistribute the initial entanglement between spins. Let us consider only two flavor oscillations (electronic neutrino e and muonic neutrino μ) and suppose to start with an electronic neutrino. The free time evolution is given by [5]:

$$|\bar{\nu}_{\uparrow}(t)\rangle = (\cos^2\theta\, |\bar{\nu}_{m1\uparrow}(t)\rangle + \sin^2\theta\, |\bar{\nu}_{m2\uparrow}(t)\rangle) \otimes |\bar{\nu}_e\rangle + \sin\theta\cos\theta(|\bar{\nu}_{m1\uparrow}(t)\rangle - |\bar{\nu}_{m2\uparrow}(t)\rangle) \otimes |\bar{\nu}_\mu\rangle \quad (38)$$

$$|\bar{\nu}_{\downarrow}(t)\rangle = (\cos^2\theta\, |\bar{\nu}_{m1\downarrow}(t)\rangle + \sin^2\theta\, |\bar{\nu}_{m2\downarrow}(t)\rangle) \otimes |\bar{\nu}_e\rangle + \sin\theta\cos\theta(|\bar{\nu}_{m1\downarrow}(t)\rangle - |\bar{\nu}_{m2\downarrow}(t)\rangle) \otimes |\bar{\nu}_\mu\rangle \quad (39)$$

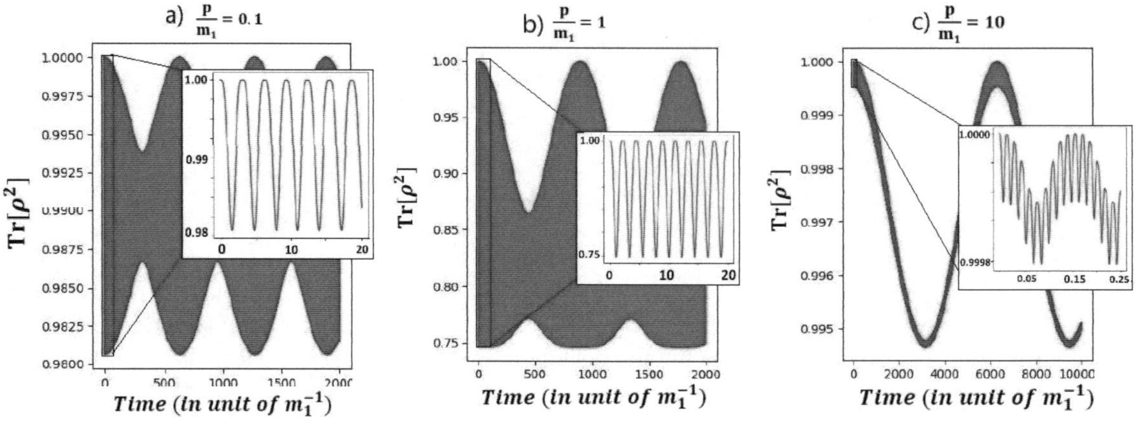

Figure 1: $\mathrm{Tr}[\rho^2_{S_l,S_\nu}]$ as a function of time for $\frac{p}{m_{\bar{\nu}_1}} = 0.1; 1; 10$. We used the values $\frac{\Delta m^2}{m_{\bar{\nu}_1}} = 0,001$ with $\Delta m^2 = m^2_{\bar{\nu}_2} - m^2_{\bar{\nu}_1}$ and $\frac{m_l}{m_{\bar{\nu}_1}} = 100$. Also we take $\sin^2\theta = 0,306$ in $a)$ for flavor mixing.

Defining the left-(right-)chiral part by $|C_\alpha^{+1(-1)}\rangle$, renaming the coefficients (11) as $f^\alpha_{+(-)} \equiv f^\alpha_{+1(-1)}$ and the bispinors (10) as $|u_\alpha\rangle \equiv |+1_\alpha\rangle$, $|v_\alpha\rangle \equiv |-1_\alpha\rangle$ for both spin up and spin down and both flavors $\alpha = \bar{\nu}_l, l$, Eqs. (28)-(31) become

$$|\bar{\nu}_\uparrow(t)\rangle = N_{\bar{\nu}} \sum_{d=\pm1} e^{-idE_{\bar{\nu}}t} f_d^{\bar{\nu}} \, |d_{\uparrow\bar{\nu}}(dp)\rangle \,, \qquad |\bar{\nu}_\downarrow(t)\rangle = N_{\bar{\nu}} \sum_{d=\pm1} e^{-idE_{\bar{\nu}}t} f_{-d}^{\bar{\nu}} \, |d_{\downarrow\bar{\nu}}(dp)\rangle \quad (40)$$

$$|l_\uparrow(t)\rangle = N_l \sum_{d=\pm1} e^{-idE_l t} f_d^l \, |d_{\uparrow l}(-dp)\rangle \,, \qquad |l_\downarrow(t)\rangle = N_l \sum_{d=\pm1} e^{-idE_l t} f_{-d}^l \, |d_{\downarrow l}(-dp)\rangle \,. \quad (41)$$

By taking into account flavor oscillations[1], the state (27) acquires the form

$$|\Psi(t)\rangle = \sum_{c,c'} \Big(\gamma_1(c,c') \, |C^c_{\bar{\nu}}\rangle \otimes |\uparrow_{\bar{\nu}}\rangle \otimes |\bar{\nu}_e\rangle \otimes |C^{c'}_l\rangle \otimes |\downarrow_l\rangle + \gamma_2(c,c') \, |C^c_{\bar{\nu}}\rangle \otimes |\uparrow_{\bar{\nu}}\rangle \otimes |\bar{\nu}_\mu\rangle \otimes |C^{c'}_l\rangle \otimes |\downarrow_l\rangle$$
$$- \gamma_3(c,c') \, |C^c_{\bar{\nu}}\rangle \otimes |\downarrow_{\bar{\nu}}\rangle \otimes |\bar{\nu}_e\rangle \otimes |C^{c'}_l\rangle \otimes |\uparrow_l\rangle - \gamma_4(c,c') \, |C^c_{\bar{\nu}}\rangle \otimes |\downarrow_{\bar{\nu}}\rangle \otimes |\bar{\nu}_\mu\rangle \otimes |C^{c'}_l\rangle \otimes |\uparrow_l\rangle \Big) \quad (42)$$

with $c, c' = \pm1$. The coefficients are given by

$$\gamma_1(c,c') = A(\omega_1^{(1}) (c,c') \cos^2\theta + \omega_2^{(1)}(c,c') \sin^2\theta), \qquad \gamma_2(c,c') = A\omega^{(2)}(c,c') \sin\theta\cos\theta,$$

$$\gamma_3(c,c') = B(\omega_1^{(3)}(c,c') \cos^2\theta + \omega_2^{(3)}(c,c') \sin^2\theta), \qquad \gamma_4(c,c') = B\omega^{(4)}(c,c') \sin\theta\cos\theta \quad (43)$$

[1] Note that the flavor oscillations here considered take into account also the non-standard oscillation term also found in QFT [12, 13].

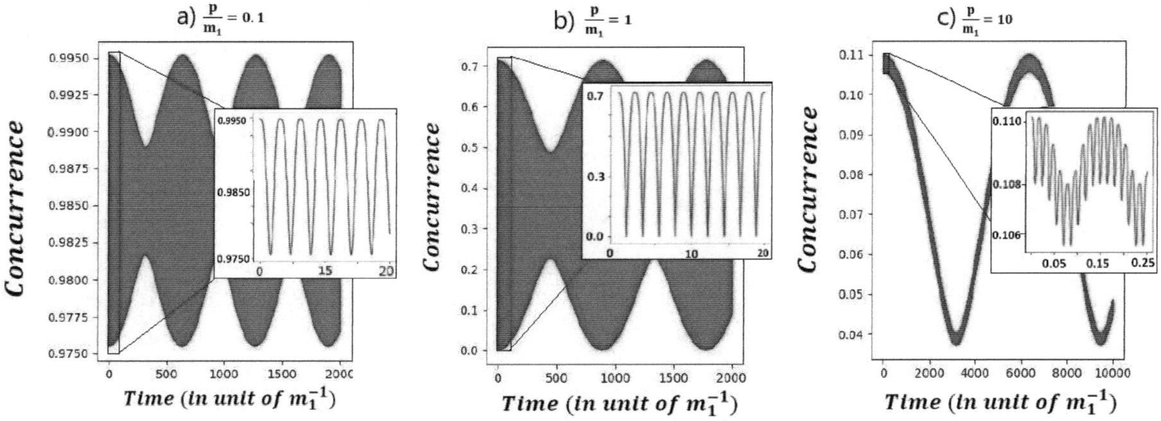

Figure 2: Concurrence $\mathcal{C}$ as a function of time for $\frac{p}{m_{\bar{\nu}_1}} = 0.1; 1; 10$. Values of the parameters are the same of Fig.1.

where A and B are given by the (26) and

$$
\begin{aligned}
\omega_1^{(1)}(c,c') &= \Big(\sum_d N_{\bar{\nu}_1}^2 c^{\frac{d-1}{2}} f_d^{\bar{\nu}_1} f_{dc}^{\bar{\nu}_1} e^{-idE_{\bar{\nu}_1}t}\Big)\Big(\sum_{d'} d' N_l^2 c'^{\frac{d'-1}{2}} f_{-d'}^l f_{d'c'}^l e^{-id'E_l t}\Big)\\
\omega_2^{(1)}(c,c') &= \Big(\sum_d N_{\bar{\nu}_2}^2 c^{\frac{d-1}{2}} f_d^{\bar{\nu}_2} f_{dc}^{\bar{\nu}_2} e^{-idE_{\bar{\nu}_2}t}\Big)\Big(\sum_{d'} d' N_l^2 c'^{\frac{d'-1}{2}} f_{-d'}^l f_{d'c'}^l e^{-id'E_l t}\Big)\\
\omega^{(2)}(c,c') &= \Big[\sum_d \Big(N_{\bar{\nu}_1}^2 c^{\frac{d-1}{2}} f_d^{\bar{\nu}_1} f_{dc}^{\bar{\nu}_1} e^{-idE_{\bar{\nu}_1}t} - N_{\bar{\nu}_2}^2 c^{\frac{d-1}{2}} f_d^{\bar{\nu}_2} f_{dc}^{\bar{\nu}_2} e^{-idE_{\bar{\nu}_2}t}\Big)\Big] \times\\
&\quad \times \Big(\sum_{d'} d' N_l^2 c'^{\frac{d'-1}{2}} f_{-d'}^l f_{d'c'}^l e^{-id'E_l t}\Big)
\end{aligned} \tag{44}
$$

and

$$
\omega_1^{(3)}(c,c') = -cc'(\omega_1^1(c,c'))^*, \quad \omega_2^{(3)}(c,c') = -cc'(\omega_2^1(c,c'))^*, \quad \omega^{(4)}(c,c') = -cc'(\omega^2(c,c'))^* \tag{45}
$$

with the notation $\bar{\nu}_i = \bar{\nu}_{m_i}$ for $i = 1, 2$.

We note that the state of the lepton-antineutrino system (42) at time t is in fact a *hyperentangled state* [14], since entanglement in the flavor degrees of freedom adds up to the initial entanglement in the spins (helicity).

In the spirit of Ref. [9], we want to investigate the effects of both chiral and flavor oscillations on the spin correlations. In order to do this, we trace out flavor and chirality from $\rho(t) = |\Psi(t)\rangle\langle\Psi(t)|$ to obtain the reduced density operator for the spin part. After some algebra, we get

$$
\rho_{S_l,S_{\bar{\nu}}} = \begin{bmatrix} A^2 & \rho_{12} \\ \rho_{12}^* & B^2 \end{bmatrix} \tag{46}
$$

with

$$
\begin{aligned}
\rho_{12} = AB\Big\{&\cos^4\theta(h_l h_{\bar{\nu}_1} + g_l g_{\bar{\nu}_1}) + \sin^4\theta(h_l h_{\bar{\nu}_2} + g_l g_{\bar{\nu}_2}) + \sin^2\theta\cos^2\theta[h_l(h_{\bar{\nu}_1} + h_{\bar{\nu}_2}) + g_l(g_{\bar{\nu}_1} + g_{\bar{\nu}_2})]\\
&+ i\cos^4\theta(g_l h_{\bar{\nu}_1} - g_{\bar{\nu}_1} h_l) + i\sin^4\theta(g_l h_{\bar{\nu}_2} - g_{\bar{\nu}_2} h_l) + i\sin^2\theta\cos^2\theta[g_l(h_{\bar{\nu}_1} + h_{\bar{\nu}_2}) - h_l(g_{\bar{\nu}_1} + g_{\bar{\nu}_2})]\Big\}
\end{aligned} \tag{47}
$$

and

$$h_i = 1 - \frac{2p^2}{E_i^2}\sin^2(E_i t), \quad g_i = \frac{p}{E_i}\sin(2E_i t) \tag{48}$$

for $i = l,\, \bar{\nu}_1,\, \bar{\nu}_2$.

Now it is possible to calculate the purity for the spin part and the result is

$$\begin{aligned} \text{Tr}[\rho^2_{S_{\bar{\nu}},S_l}(t)] &= 1 - \frac{\mathcal{N}[\rho^2_{S_{\bar{\nu}},S_l}(0)]}{2}\Big(1 - \Gamma_l^2\Big\{\cos^8\theta\Gamma^2_{\bar{\nu}_1} + \sin^8\theta\Gamma^2_{\bar{\nu}_2} + 2\sin^2\theta\cos^2\theta\Big[2\cos^4\theta\Gamma^2_{\bar{\nu}_1} \\ &\quad + 2\sin^4\theta\Gamma^2_{\bar{\nu}_1} + \sin^2\theta\cos^2\theta(\Gamma^2_{\bar{\nu}_1} + \Gamma^2_{\bar{\nu}_2}) + 2(h_{\bar{\nu}_1}h_{\bar{\nu}_2} + g_{\bar{\nu}_1}g_{\bar{\nu}_2})\Big]\Big\}\Big) \end{aligned} \tag{49}$$

with

$$\Gamma_i = h_i^2 + g_i^2$$

and $i = l, \bar{\nu}_1, \bar{\nu}_2$.

As a check, we observe that when flavor mixing is absent ($\theta = 0$), the above expression reduces to

$$\text{Tr}[\rho^2_{S_{\bar{\nu}},S_l}(t)] = 1 - \frac{\mathcal{N}[\rho^2_{S_{\bar{\nu}},S_l}(0)]}{2}\left(1 - \Gamma_l^2\Gamma^2_{\bar{\nu}_1}\right) \tag{50}$$

which coincides with Eq.(33) as we expected.

The plot of the quantity in (49) as function of time is given in Fig.1 for different values of the ratio $\frac{p}{m_{\bar{\nu}_1}}$: we see that the purity of the reduced density operator decreases in the case of $p \simeq m_{\bar{\nu}}$ and hence the redistribution of the entanglement between the spins is maximal in this case. This resonance effect has been noticed also in [9]. In the present, more general case, we can also see that the entanglement redistribution of the spins is modulated by the flavor oscillations of the antineutrino which are slower than the chiral oscillations.

It is also interesting to look how the concurrence related to the entanglement between the spins evolves in time. It is given by [15]

$$\mathcal{C} = \max\{0, \lambda_1 - \lambda_2 - \lambda_3 - \lambda_4\} \tag{51}$$

where λ_i are the squared roots of the eigenvalues (in decreasing order) of the operator $\rho\tilde{\rho}$ where $\tilde{\rho} \equiv \sigma_y \otimes \sigma_y \rho^* \sigma_y \otimes \sigma_y$ is the spin flipped density operator ρ. In our case the plots of concurrence are given in Fig.2. At $t = 0$ we have $\mathcal{C}(0) = 2|AB| = \mathcal{N}(0)$.

As expected, the concurrence decreases if the ratio $\frac{p}{m}$ increases since the coefficients A and B become more different from each other and $B \to 0$ when $\frac{p}{m} \to \infty$. Furthermore the concurrence is also modulated by both the flavor and the chiral oscillations. The effect is more relevant in the case $p = m_{\bar{\nu}_1}$ as noticed also in [9].

4.1. General case

In the previous Sections we have seen that entanglement between spins in a lepton-antineutrino pair is redistributed by both chiral and flavor oscillations during the free time evolution. Nevertheless, this scheme can be used only as a toy model because the momentum p of the two particles has been taken as a free parameter and the effects of the chiral oscillations on the entanglement between spins are relevant only in the non-relativistic limit for both particles.

Since the neutrino is much lighter than the electron ($m_{\bar{\nu}_e} \simeq 2 \cdot 10^{-6} m_e$) and the non-relativistic limit is reached in the case $\frac{p}{m_{\bar{\nu}}} \lesssim 1$, this cannot happen in the case of the pion decay, because the momentum p is fixed by the relativistic conservation laws. Indeed, in the rest frame of the pion we have

$$\begin{cases} m_\pi^2 = m_l^2 + p_l^2 + m_{\bar{\nu}_l}^2 + p_{\bar{\nu}_l}^2 \\ 0 = \mathbf{p}_l + \mathbf{p}_{\bar{\nu}_l} \end{cases}$$

which gives $p_l = p_{\bar{\nu}_l} = p$ and

$$p^2 = \frac{\Delta M_\pi^2}{2} \tag{52}$$

with $\Delta M_\pi^2 = m_\pi^2 - m_l^2 - m_{\bar{\nu}_l}^2$. In the case that the emitted particle is a muon, since $m_\mu \simeq 105.6\text{MeV}$, $m_{\bar{\nu}_\mu} \simeq 1\text{eV}$ and $m_\pi \simeq 139.6\text{MeV}$, we obtain $p^2 \simeq 17\text{MeV}$: thus the antineutrino is always relativistic and the entanglement between the spins goes to zero since the coefficient (36) also goes to zero because $B \to 0$.

To avoid this situation, let us consider a generalization of the above toy model with two initial particles a and b with masses m_a, m_b and momenta $\mathbf{p}_a$, $\mathbf{p}_b$ colliding. Suppose they emit two particles c and d with masses m_c, m_d and momenta $\mathbf{p}_c$, $\mathbf{p}_d$ and that we consider only the particles emitted on the same direction of that formed by $\mathbf{p}_a$ and $\mathbf{p}_b$. In this case the conservation laws give

$$\begin{cases} m_a^2 + p_a^2 + m_b^2 + p_b^2 = m_c^2 + p_c^2 + m_d^2 + p_d^2 \\ p_a - p_b = p_c + p_d \end{cases}$$

with $p_{a(b)} \equiv |\mathbf{p}_{a(b)}|$. Therefore the momenta of the emitted particles along the initial momentum direction become

$$p_{c,d} = \frac{p_a - p_b \pm \sqrt{\Delta M^2 + (p_a - p_b)^2}}{2} \tag{53}$$

where $\Delta M^2 \equiv m_a^2 + m_b^2 - (m_c^2 + m_d^2)$. Note that the process is possible only if $m_a^2 + m_b^2 + (p_a - p_b)^2 \geq m_c^2 + m_d^2$.

Therefore, by changing the momentum of the initial particle p_a, we can search for a regime in which both particles are non-relativistic (i.e. $p_c \simeq m_c$ and $p_d \simeq m_d$) and the effects of chiral oscillations on the redistribution of the entanglement between the spins are significant. We assume an initial entangled state for the spin part of the form

$$|\Phi\rangle = \frac{|v_\uparrow(p_c, m_c)\rangle \otimes |u_\downarrow(p_d, m_d)\rangle - |v_\downarrow(p_c, m_c)\rangle \otimes |u_\uparrow(p_d, m_d)\rangle}{\sqrt{2}} \tag{54}$$

that is the analogous of Eq.(21). The system state at $t = 0$ is obtained by the application of chirality projectors which gives

$$|\Psi(0)\rangle = A(p_c, p_d, m_c, m_d)\,|c_\uparrow(0)\rangle \otimes |d_\downarrow(0)\rangle - B(p_c, p_d, m_c, m_d)\,|c_\downarrow(0)\rangle \otimes |d_\uparrow(0)\rangle \tag{55}$$

where

$$A(p_c, p_d, m_c, m_d) = \frac{\sqrt{2} f_{+c} f_{-d}}{N_c N_d (f_{+c}^2 f_{-d}^2 + f_{-c}^2 f_{+d}^2)}; \; B(p_c, p_d, m_c, m_d) = \frac{\sqrt{2} f_{-c} f_{+d}}{N_c N_d (f_{+c}^2 f_{-d}^2 + f_{-c}^2 f_{+d}^2)} \tag{56}$$

Note that in this case, if $p_c \gg m_c$ we have $f_{-c} \to 0$ and hence $B \to 0$, while in the case $p_d \gg m_d$ we have $f_{-d} \to 0$ and hence $A \to 0$. The concurrence for the spin part is still given by Eq.(51) and at $t = 0$ (where it reaches its maximum value) it is

$$\mathcal{C}[\rho_{S_b,S_c}(0)] = 2|AB| = 2|A(p_b, p_c, m_b, m_c) B(p_b, p_c, m_b, m_c)| \tag{57}$$

which vanishes in both limit cases.

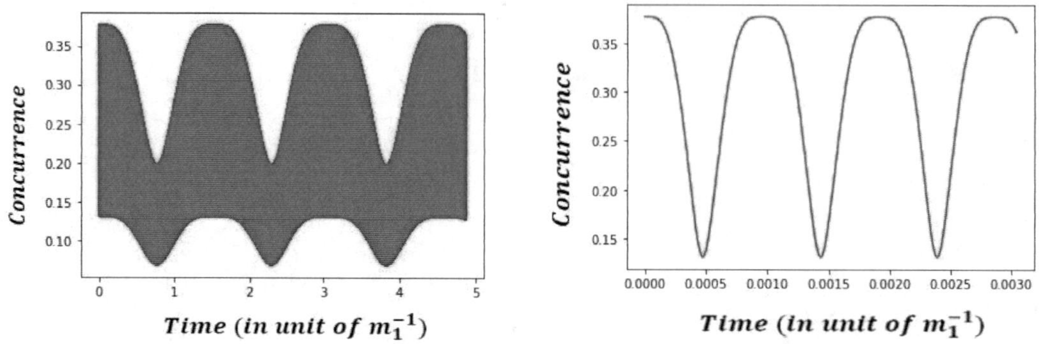

Figure 3: Concurrence shared between spins of the emitted particles as a function of time for two differents time scales for the weak process $n + \nu_e \to p + e$, supposing that the initial and final total spins of the particle is zero and that the emitted particle are entangled in the spin at $t = 0$. In this case, both particle are non-relativistic and hence the entanglement is non-negligible. The values for parameters are (in unit of m_e) $p_n = 2700$ and the real masses of the particles $m_n = 1879$, $m_p = 1876$, $m_e = 1$.

In the case of pion decay with momentum $p_a = p_\pi$ we obtain

$$p_{c,d} = \frac{p_\pi \pm \sqrt{\Delta M_\pi^2 + p_\pi^2}}{2}. \tag{58}$$

If we look for a value of the momentum p_π of the pion such that both emitted particles are non-relativistic, so that the entanglement between the spins becomes non-negligible, we see that in this case this is not allowed due to the smallness of the neutrino mass.

If we renounce to observe the effects of flavor oscillations and focus only on the chiral oscillations, we can consider another kind of decay, that is

$$n + \nu_e \to p + e \tag{59}$$

Neglecting flavor mixing, we see in Fig.3 that the concurrence is non-negligible for some values of the initial parameters and hence the effects of chiral oscillations on the entanglement between spins could in principle be observed.

5. Conclusions

In this paper, we have studied the evolution in time of spin correlations in an initially entangled state for a lepton-antineutrino system produced in a weak interaction process. In the non-relativistic regime, both chiral and flavor (neutrino) oscillations take place and affect spin correlations.

The present work represents a generalization of Ref. [9], with the addition of flavor oscillations: this makes the dynamics of oscillations much richer due to the appearance of different oscillation lengths and indeed the state of the system at time t becomes hyperentangled.

Finally, we investigated the possibility of observing the effects of chiral oscillations by looking at the entanglement between proton and electron spins produced in the process $n + \nu_e \to p + e$. We found that in this case, it is indeed possible to have both the final particles in the non-relativistic regime and thus oscillations in the spin correlations due to chiral oscillations can be relevant.

Journal of Physics: Conference Series **2533** (2023) 012024 doi:10.1088/1742-6596/2533/1/012024

References

[1] Cheng T P and Li L F 1995 *Gauge theory of elementary particle physics* (Oxford: Oxford Univ. Press)
[2] Pal P B 2014 *An introductory course of particle physics* (Boca Raton: CRC Press)
[3] Fukugita M and Yanagida T 2003 *Physics of neutrinos* (Berlin: Springer-Verlag)
[4] De Leo S and Rotelli P 1998 *Int. J. Theor. Phys.* **37** 2193
[5] Bittencourt V A, Bernardini A E and Blasone M 2021 *Eur. Phys. J. C* **81** 1
[6] Blasone M, Bittencourt V A and Bernardini A E 2022 PoS *CORFU2021* 065
[7] Bittencourt V A, Bernardini A E and Blasone M 2022 *EPL* **139** 44002
[8] Ge S F and Pasquini P 2020 *Phys. Lett. B* **811** 135961
[9] Bittencourt V A, Bernardini A E and Blasone M 2021 *Universe* **7** 293.
[10] Itzykson C and Zuber J B 2005 ed. *Quantum Field Theory* (New York: Dover Pub.)
[11] Tung W-K 1985 *Group Theory in Physics* (River Edge, Nj: World Sc.)
[12] Blasone M and Vitiello G 1995 *Annals Phys.* **244** 283
[13] Blasone M, Henning P A and Vitiello G 1999 *Phys. Lett.* B **451** 140
[14] Deng F G, Ren B C and Li X H 2017 *Science bulletin* **62** 46
[15] Wootters W K 1998 *Phys. Rev. Lett.* **80** 2245

Journal of Physics: Conference Series **2533** (2023) 012025 doi:10.1088/1742-6596/2533/1/012025

Reparameterization invariance in accelerated proton decay

M Blasone[1,2], G Lambiase[1,2], G G Luciano[3,*] and L Petruzziello[2,4,5]

[1] Dipartimento di Fisica, Università degli Studi di Salerno, Via Giovanni Paolo II, 132 I-84084 Fisciano (SA), Italy

[2] INFN Gruppo Collegato di Salerno - Sezione di Napoli c/o Dipartimento di Fisica, Università di Salerno, Italy

[3] Applied Physics Section of Environmental Science Department, Universitat de Lleida, Av. Jaume II, 69, 25001 Lleida, Spain

[4] Dipartimento di Ingegneria, Università di Salerno, Via Giovanni Paolo II, 132 I-84084 Fisciano (SA), Italy

[5] Institut für Theoretische Physik, Albert-Einstein-Allee 11, Universität Ulm, 89069 Ulm, Germany

E-mail: *giuseppegaetano.luciano@udl.cat

Abstract. The inverse β-decay of uniformly accelerated protons has proved to be a useful playground to investigate the asymptotic nature of mixed neutrino states - either flavor or mass. Based on recent results, here we explore in more depth the role of neutrino mixing in the inverse β-decay and the consistency with CP-violation in flavor oscillations. We compute the proton decay rate Γ both in the laboratory and in the comoving frame, taking into account the emergence of Unruh effect in the latter. We show that, while the flavor representation correctly leads to a CP-quantifier $A_{CP}^{(e,\mu)} \equiv \Gamma_{\nu_e,\nu_\mu} - \Gamma_{\bar{\nu}_e,\bar{\nu}_\mu}$ proportional to Jarlskog invariant, mass states give instead a vanishing value for such parameter. This implies that the mass basis cannot account for CP-asymmetry in neutrino oscillations, making its usage meaningless.

1. Introduction

In the early 1900's the study of the β-decay provided the first evidence for the existence of neutrino based on the law of energy conservation. Several decades later, the inverse process involving uniformly accelerated protons

$$p \to n + e^+ + \nu_e \,, \tag{1}$$

was first proposed as a theoretical proof of Unruh effect [1] and then used to speculate on the asymptotic behavior of mixed neutrinos [2, 3, 4, 5, 6, 7]. Toward this end, the scalar decay rate Γ of the proton was computed in both the laboratory frame (where the particle is accelerated) and comoving frame (where the proton is at rest and interacts with a thermal bath of leptons owing to the Unruh effect [8]). While the consistency between the two results is out of question in the original toy models with massless [1] or unmixed [9] neutrinos, some concerns have been raised in more realistic scenarios including flavor mixing [2] because of the possible dual nature of asymptotic neutrinos - either mass [4] or flavor [2, 3, 5, 6] states. This has led to question the thermality of Unruh effect and/or the General Covariance of Quantum Field Theory (QFT) [2].

Journal of Physics: Conference Series **2533** (2023) 012025 doi:10.1088/1742-6596/2533/1/012025

Taken for granted that General Covariance is an essential principle of QFT and that Unruh radiation preserves the characteristic thermal (Planckian-like) spectrum as far as relativistic neutrinos are considered [10, 11, 12, 13], it has been later understood that the correct treatment of the inverse β-decay must uniquely involve flavor neutrino states [3, 5]. In fact, the mass representation turns out to be inconsistent with mixing and oscillations as described in the Standard Model[1]. This has been confirmed by the recent extension to three flavors [6], which shows that mass states cannot account for CP-violation effects in neutrino oscillations processes. Although this result has been obtained specifically for the decay (1), it is expected to have more general implications, since the underlying reasoning holds true for any electroweak vertex involving neutrinos (see also [14] and [15] for an application in curved spacetime).

Starting from the above premises, in this work we revisit and further elaborate on the role of neutrino mixing in the inverse β-decay. We focus on the three-flavor analysis, showing that while the usage of flavor basis correctly results in a CP-quantifier $A_{CP}^{(e,\mu)} \equiv \Gamma_{\nu_e,\nu_\mu} - \Gamma_{\bar{\nu}_e,\bar{\nu}_\mu}$ proportional to Jarlskog invariant (as expected for any physical process that violates CP-symmetry), mass states are instead trivially consistent with the rephasing invariant parametrization of mixing matrix, as they give a vanishing (i.e. CP-violation incompatible) $A_{CP}^{(e,\mu)}$. This poses a severe contradiction with predictions from the Standard Model, thus undermining the meaning of the mass basis as asymptotic representation for neutrinos.

The manuscript is structured as follows: in the next Section we review the basics of the inverse β-decay. This framework is then studied in connection with neutrino oscillations and CP-violation effects in Sec. 3. Results and perspectives are finally summarized in Sec. 4. Throughout the paper, we use natural units $\hbar = c = k_B = 1$ and the Minkowski metric with the mostly negative signature:

$$\eta_{\mu\nu} = \mathrm{diag}(+1, -1, -1, -1). \tag{2}$$

2. Inverse β-decay and neutrino mixing

Let us consider a non-inertial proton moving with uniform acceleration a in a given laboratory frame. As first shown in [16, 17], the decay (1) is kinematically allowed because the accelerating source provides the rest energy difference between initial and final states. In order to evaluate the scalar decay rate, we treat the proton $|p\rangle$ and neutron $|n\rangle$ as unexcited and excited states of a two-level system with Hamiltonian $\hat{H}$ satisfying $\hat{H}|p\rangle = m_p|p\rangle$ and $\hat{H}|n\rangle = m_n|n\rangle$, where m_p and m_n are the rest mass of the proton and neutron, respectively. Additionally, we assume the momenta $\mathbf{k}_{e^+}, \mathbf{k}_{\nu_e}$ of the positron and neutrino to be much smaller than m_p, m_n, so that the four-velocity of the hadrons is negligibly affected by the lepton emission (*no-recoil approximation*) [1].

In the above semiclassical setting, the interaction (1) can be described by a Fermi-like current-current action in the form

$$\hat{S}_I = \int d^4x \sqrt{-g}\, \hat{J}_{h,\lambda} \hat{J}_l^{\lambda}, \tag{3}$$

where the hadronic current is

$$\hat{J}_{h,\lambda} = \hat{q}(\tau) u_\lambda \delta(x)\delta(y)\delta(u - 1/a)\,. \tag{4}$$

Here $\hat{q}(\tau) = e^{i\hat{H}\tau}\hat{q}(0)e^{-i\hat{H}\tau}$ is the monopole operator and $G_F = |\langle n|\hat{q}(0)|p\rangle|$ the Fermi constant. We have implicitly assumed that the acceleration of the proton-neutron system can be reduced to a classical (Rindler-like) trajectory of time coordinate v and proper time $\tau = v/a$. The Dirac delta picks the value $1/a$ for the spatial coordinate u, identifying a specific Rindler hyperbola

[1] We emphasize that in all the aforementioned works on the inverse β-decay, decoherence effects are neglected.

Journal of Physics: Conference Series **2533** (2023) 012025 doi:10.1088/1742-6596/2533/1/012025

in Minkowski spacetime[2]. Furthermore, the proton-neutron four-velocity is $u^\lambda = (a, 0, 0, 0)$ and $u^\lambda = (\sqrt{a^2t^2+1}, 0, 0, at)$ in Rindler and Minkowski coordinates, respectively.

On the other side, the lepton current $\hat{J}_{h,\lambda}$ reads

$$\hat{J}_l^\lambda = \sum_{\ell=e,\mu,\tau} \left(\hat{\overline{\Psi}}_{\nu_\ell} \gamma^\lambda \hat{\Psi}_\ell + \hat{\overline{\Psi}}_\ell \gamma^\lambda \hat{\Psi}_{\nu_\ell} \right), \tag{5}$$

with $\hat{\Psi}_\ell$ ($\hat{\Psi}_{\nu_\ell}$) being the charged lepton (neutrino) Dirac field (see [3] for their explicit form) and e, μ, τ the three lepton flavors.

In the Standard Model neutrinos interact with charged leptons as flavor eigenstates, which are superpositions of mass eigenstates according to

$$\begin{pmatrix} |\nu_e\rangle \\ |\nu_\mu\rangle \\ |\nu_\tau\rangle \end{pmatrix} = U \begin{pmatrix} |\nu_1\rangle \\ |\nu_2\rangle \\ |\nu_3\rangle \end{pmatrix}. \tag{6}$$

where U is the Pontecorvo-Maki-Nakagawa-Sakata (PMNS) 3×3 mixing matrix [18, 19].[3] Notice that it can be expressed in terms of four parameters, which are usually taken as three rotation angles and one phase. A number of different parametrizations have been introduced in the literature. Clearly, the adoption of a particular parametrization is arbitrary and only dictated by convenience. For a short note on the advantages and disadvantages of any parameterization, see [22]. Here, we provide two examples for later use: the first one is the standard parametrization

$$U = \begin{pmatrix} c_{12}\,c_{13} & s_{12}\,c_{13} & s_{13}\,e^{-i\delta} \\ -s_{12}\,c_{23} - c_{12}\,s_{23}\,s_{13}\,e^{i\delta} & c_{12}\,c_{23} - s_{12}\,s_{23}\,s_{13}\,e^{i\delta} & s_{23}\,c_{13} \\ s_{12}\,s_{23} - c_{12}\,c_{23}\,s_{13}\,e^{i\delta} & -c_{12}\,s_{23} - s_{12}\,c_{23}\,s_{13}\,e^{i\delta} & c_{23}\,c_{13} \end{pmatrix}, \tag{7}$$

where $c_{jk} \equiv \cos\theta_{jk}$, $s_{jk} \equiv \sin\theta_{jk}$, θ_{jk} is the ν_j-ν_k mixing angle and δ the CP-violating phase. The second parameterization is the the Kobayshi–Maskawa parameterization [23]:

$$U_{KM} = \begin{pmatrix} c_1 & s_1c_3 & s_1s_3 \\ -s_1c_2 & c_1c_2c_3 - s_2s_3e^{i\delta} & c_1c_2s_3 + s_2c_3e^{i\delta} \\ -s_1s_2 & c_1s_2c_3 + c_2s_3e^{i\delta} & c_1s_2s_3 - c_2c_3e^{i\delta} \end{pmatrix}, \tag{8}$$

with $c_i \equiv \cos\theta_i$ and $s_i \equiv \sin\theta_i$.

Using the above tools, we can now sketch the computation of the transition probability for the decay (1) (more details can be found in [3]). Calculations are performed in the S-matrix formalism for the laboratory and comoving frames separately. We work in the relativistic regime and use the flavor asymptotic representation (6) for neutrinos. Aside, we comment on the corresponding results obtained with mass asymptotic states.

[2] We are assuming the proton-neutron acceleration to be along the z-axis. In turn, Rindler coordinates (v, x', y', u) are related to Minkowski ones (t, x, y, z) by $t = u\sinh v$, $x' = x$, $y' = y$, $z = u\cosh v$.

[3] In a full-fledged QFT framework, mixing transformations are by far more complicated, since they have the structure of a Bogoliubov transformation nested into the rotation (6) [20]. However, one can prove that the states (6) well-approximate the exact field theoretical states in the relativistic regime we are considering here [21].

2.1. Laboratory frame

In the laboratory frame the transition amplitude for the inverse β-decay at tree-level takes the form [6]

$$\mathcal{A}^{(i)} \equiv \langle n| \otimes \langle e^+, \nu_e|\hat{S}_I|0\rangle \otimes |p\rangle = \frac{G_F}{2^4\pi^3}\sum_{j=1}^{3}|U_{ej}|^2\,\mathcal{I}_{\sigma_\nu\sigma_e}(\omega_{\nu_j}, \omega_e)\,, \tag{9}$$

where the superscript (i) on the left hand side labels the specific process under consideration (see Fig. 1). The function $\mathcal{I}_{\sigma_\nu\sigma_e}$ is defined by [6]

$$\mathcal{I}_{\sigma_\nu\sigma_e}(\omega_{\nu_j}, \omega_e) = \int_{-\infty}^{+\infty} d\tau\, u_\lambda \left[\bar{u}^{(+\omega_{\nu_j})}_{\sigma_\nu}\gamma^\lambda u^{(-\omega_e)}_{-\sigma_e}\right] e^{i\left[\Delta m\tau + a^{-1}(\omega_{\nu_j}+\omega_e)\sinh a\tau - a^{-1}(k^z_\nu + k^z_e)\cosh a\tau\right]}\,, \tag{10}$$

where $\Delta m \equiv m_n - m_p$ and $u^{(\pm\omega)}_\sigma$ is the Dirac spinor of polarization $\sigma = \pm$ and Minkowski frequency $\omega = \sqrt{\mathbf{k}^2 + m^2}$, respectively.

In turn, the scalar decay rate becomes

$$\Gamma^{(i)} \equiv \frac{1}{T}\sum_{\sigma_e,\sigma_\nu}\int d^3k_\nu \int d^3k_e \left|\mathcal{A}^{(i)}\right|^2, \tag{11}$$

$$= |U_{e1}|^4\,\Gamma_1 + |U_{e2}|^4\,\Gamma_2 + |U_{e3}|^4\,\Gamma_3 + \Big(|U_{e1}|^2\,|U_{e2}|^2\,\Gamma_{12} + |U_{e1}|^2\,|U_{e3}|^2\,\Gamma_{13} + |U_{e2}|^2\,|U_{e3}|^2\,\Gamma_{23} + \text{c.c.}\Big),$$

where T is the total proton-neutron proper time and we have used the simplified notation

$$\Gamma_j \equiv \frac{1}{T}\frac{G_F^2}{2^8\pi^6}\sum_{\sigma_\nu,\sigma_e}\int d^3k_\nu \int d^3k_e \left|\mathcal{I}_{\sigma_\nu\sigma_e}(\omega_{\nu_j}, \omega_e)\right|^2 \tag{12}$$

$$= \frac{G_F^2}{a\pi^6 e^{\pi\Delta m/a}}\int d^3k_\nu \int d^3k_e \left\{K_{2i\Delta m/a}\left(\frac{2(\omega_{\nu_j}+\omega_e)}{a}\right) + \frac{m_{\nu_j}m_e}{\omega_{\nu_j}\omega_e}\,\mathrm{Re}\left[K_{2i\Delta m/a+2}\left(\frac{2(\omega_{\nu_j}+\omega_e)}{a}\right)\right]\right\},$$

$$\Gamma_{jk} \equiv \frac{1}{T}\frac{G_F^2}{2^8\pi^6}\sum_{\sigma_\nu,\sigma_e}\int d^3k_\nu \int d^3k_e\, \mathcal{I}_{\sigma_\nu\sigma_e}(\omega_{\nu_j}, \omega_e)\,\mathcal{I}^*_{\sigma_\nu\sigma_e}(\omega_{\nu_k}, \omega_e)\,. \tag{13}$$

$K_{i\nu}(x)$ denotes the modified Bessel function of second kind.

On the other hand, we notice that the usage of mass asymptotic states for neutrinos would trivially lead to $\Gamma = |U_{ei}|^2\,\Gamma_i$, where the subscript $i = 1, 2, 3$ refers to the specific mass eigenstate employed in the calculation [4]. Thus, in this case there is no trace of any off-diagonal term appearing in Eq. (11). Below, we shall see how this difference is crucial for discriminating between flavor and mass representations in the context of CP-asymmetry.

Now, because of flavor oscillations, the total proton decay rate also gets non-vanishing contributions from the process $ii)\, p \to n + e^+ + \nu_\mu$ and $iii)\, p \to n + e^+ + \nu_\tau$, where the emitted neutrino is detected as muon or tau neutrino, respectively (see also Fig. 1). Following similar computations as above, one obtains

$$\Gamma^{(ii)} = |U_{\mu1}|^2|U_{e1}|^2\,\Gamma_1 + |U_{\mu2}|^2|U_{e2}|^2\,\Gamma_2 + |U_{\mu3}|^2|U_{e3}|^2\,\Gamma_3$$
$$+ \Big(U^*_{\mu1}\,U_{e1}\,U_{\mu2}\,U^*_{e2}\,\Gamma_{12} + U^*_{\mu1}\,U_{e1}\,U_{\mu3}\,U^*_{e3}\,\Gamma_{13} + U^*_{\mu2}\,U_{e2}\,U_{\mu3}\,U^*_{e3}\,\Gamma_{23} + \text{c.c.}\Big), \tag{14}$$

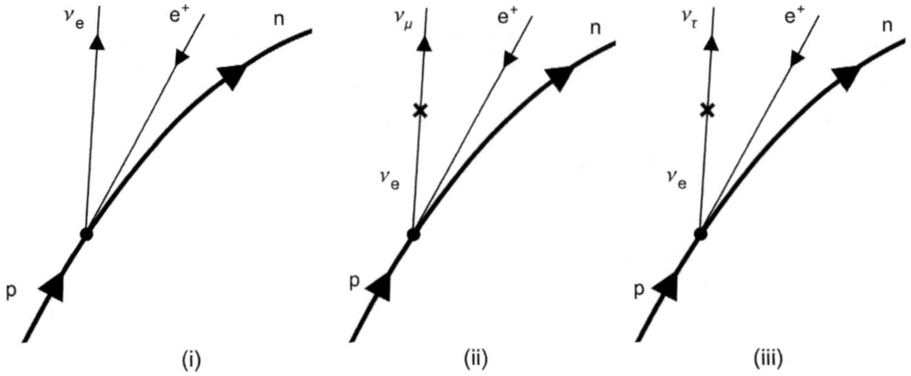

Figure 1: Inverse β-decay in the laboratory frame (time flows in the vertical direction).

$$\Gamma^{(iii)} = |U_{\tau 1}|^2|U_{e1}|^2\,\Gamma_1 + |U_{\tau 2}|^2|U_{e2}|^2\,\Gamma_2 + |U_{\tau 3}|^2|U_{e3}|^2\,\Gamma_3 + \Big(U^*_{\tau 1}U_{e1}U_{\tau 2}U^*_{e2}\,\Gamma_{12} + U^*_{\tau 1}U_{e1}U_{\tau 3}U^*_{e3}\,\Gamma_{13} + U^*_{\tau 2}U_{e2}U_{\tau 3}U^*_{e3}\,\Gamma_{23} + \text{c.c.}\Big)\,. \quad (15)$$

Overall, the decay rate in the laboratory frame is

$$\Gamma^{\rm lab} = \Gamma^{(i)} + \Gamma^{(ii)} + \Gamma^{(iii)} = |U_{e1}|^2\,\Gamma_1 + |U_{e2}|^2\,\Gamma_2 + |U_{e3}|^2\,\Gamma_3\,, \quad (16)$$

where we have taken into account the unitarity of the mixing matrix U.

2.2. Comoving frame

Above, we have argued that the inverse β-decay from the point of view of an observer co-accelerating with the proton occurs due to the Unruh effect. Indeed, in the comoving frame the proton is supposed to interact with a thermal bath of leptons that pop out from the Unruh vacuum [8] according to (see also Fig. 2) [1]

$$iv)\;\; p + e^- \to n + \nu_e\,, \qquad v)\;\; p + \overline{\nu}_e \to n + e^+\,, \qquad vi)\;\; p + e^- + \overline{\nu}_e \to n\,. \quad (17)$$

Adopting the Rindler-Fulling quantization scheme for lepton fields [1], the transition amplitude for the decay (iv) is [6]

$$\mathcal{A}^{(iv)} \equiv \langle n| \otimes \langle \nu_e|\hat{S}_I|e^-\rangle \otimes |p\rangle = \frac{G_F}{(2\pi)^2}\sum_{j=1}^{3}|U_{ej}|^2\,\mathcal{J}^{(j)}_{\sigma_\nu\sigma_e}(\omega_\nu,\omega_e)\,, \quad (18)$$

where

$$\mathcal{J}^{(j)}_{\sigma_\nu\sigma_e}(\omega_\nu,\omega_e) = \delta\big(\omega_e - \omega_\nu - \Delta m\big)\,\bar{u}^{(\omega_\nu)}_{\sigma_\nu}\gamma^0 u^{(\omega_e)}_{\sigma_e}\,, \quad (19)$$

and $\omega_{\nu(e)}$ denotes the Rindler frequency.

In order to compute the decay rate, we must now take into account that the probability for the proton to absorb (emit) a particle of frequency ω from (toward) Unruh vacuum is $n_F(\omega) = \left(e^{\omega/T_{\rm U}}+1\right)^{-1}\Big(\widetilde{n}_F(\omega) = 1 - n_F(\omega)\Big)$, with $T_{\rm U} = a/2\pi$ being the well-known Unruh temperature [8]. Bearing this in mind and evaluating the analogue of Eq. (18) for the processes (v) and (vi), we are finally led to [6]

$$\Gamma^{(iv)} + \Gamma^{(v)} + \Gamma^{(vi)} = |U_{e1}|^4\,\widetilde{\Gamma}_1 + |U_{e2}|^4\,\widetilde{\Gamma}_2 + |U_{e3}|^4\,\widetilde{\Gamma}_3 + \Big(|U_{e1}|^2\,|U_{e2}|^2\,\widetilde{\Gamma}_{12} + |U_{e1}|^2\,|U_{e3}|^2\,\widetilde{\Gamma}_{13} + |U_{e2}|^2\,|U_{e3}|^2\,\widetilde{\Gamma}_{23} + \text{c.c.}\Big)\,, \quad (20)$$

where we have defined

$$\widetilde{\Gamma}_j \equiv \frac{2\,G_F^2}{a^2\pi^7 e^{\pi\Delta m/a}}\int d\omega \Bigg\{\int d^2k_\nu\, l_{\nu_j}\left|K_{i(\omega-\Delta m)/a+1/2}\left(\frac{l_{\nu_j}}{a}\right)\right|^2 \int d^2k_e\, l_e\left|K_{i\omega/a+1/2}\left(\frac{l_e}{a}\right)\right|^2$$
$$+\, m_{\nu_j}m_e\,\mathrm{Re}\left[\int d^2k_\nu\, K^2_{i(\omega-\Delta m)/a-1/2}\left(\frac{l_{\nu_j}}{a}\right)\int d^2k_e K^2_{i\omega/a+1/2}\left(\frac{l_e}{a}\right)\right]\Bigg\}, \quad (21)$$

$$\widetilde{\Gamma}_{jk} = \frac{G_F^2}{a^2\pi^7 e^{\pi\Delta m/a}}\int d\omega \Bigg\{\int d^2k_e\, l_e\left|K_{i\omega/a+1/2}\left(\frac{l_e}{a}\right)\right|^2 \int \frac{d^2k_\nu}{\sqrt{l_{\nu_j}l_{\nu_k}}}\left((k_\nu^x)^2+(k_\nu^y)^2+m_{\nu_j}m_{\nu_k}+l_{\nu_j}l_{\nu_k}\right)$$
$$\times\left[K_{i(\omega-\Delta m)/a+1/2}\left(\frac{l_{\nu_j}}{a}\right)K_{i(\omega-\Delta m)/a-1/2}\left(\frac{l_{\nu_k}}{a}\right)\right] + \frac{m_e}{\sqrt{l_{\nu_j}l_{\nu_k}}}\int d^2k_e\int d^2k_\nu\left(l_{\nu_j}m_{\nu_k}+l_{\nu_k}m_{\nu_j}\right)$$
$$\times\left[K^2_{i\omega/a+1/2}\left(\frac{l_e}{a}\right)K_{i(\omega-\Delta m)/a-1/2}\left(\frac{l_{\nu_j}}{a}\right)K_{i(\omega-\Delta m)/a-1/2}\left(\frac{l_{\nu_k}}{a}\right)\right]\Bigg\}. \quad (22)$$

The reduced Rindler frequency is $l_{e\,(\nu)} \equiv \sqrt{(k^x)^2+(k^y)^2+m^2_{e\,(\nu)}}$.

As in the laboratory frame, non-trivial extra contributions to the total decay rate arise from the flavor-violating processes $(vii)-(xii)$ too (see Fig. 2). Adding up all the terms yields

$$\Gamma^{\text{com}} \equiv \sum_{s=iv}^{xii}\Gamma^{(s)} = |U_{e1}|^2\,\widetilde{\Gamma}_1 + |U_{e2}|^2\,\widetilde{\Gamma}_2 + |U_{e3}|^2\,\widetilde{\Gamma}_3\,. \quad (23)$$

A necessary condition for the above formalism to be consistent is that it fulfills the General Covariance of QFT. Due to the scalar nature of the definition (11) (see [1] for more details), this implies that Eqs. (16) and (23) must actually give the same result. The equivalence is straightforwardly achieved by observing that [6]

$$\Gamma_j = \widetilde{\Gamma}_j\,, \qquad \Gamma_{jk} = \widetilde{\Gamma}_{jk}\,, \quad (24)$$

where the second equality holds true at least for small neutrino mass differences [3]. By use of these relations, it follows that

$$\Gamma^{(i)} = \Gamma^{(iv)}+\Gamma^{(v)}+\Gamma^{(vi)}\,,\quad \Gamma^{(ii)} = \Gamma^{(vii)}+\Gamma^{(viii)}+\Gamma^{(ix)}\,,\quad \Gamma^{(iii)} = \Gamma^{(x)}+\Gamma^{(xi)}+\Gamma^{(xii)}\,. \quad (25)$$

In turn, Eq. (25) implies

$$\Gamma^{\text{lab}} = \Gamma^{\text{com}}\,, \quad (26)$$

as it should be. In passing, we mention that the same conclusion would be trivially reached by working with mass asymptotic states. Indeed, in that case the comoving decay rate for the three channels (17) involving, e.g. the electron neutrino, would be $\Gamma = |U_{ei}|^2\,\widetilde{\Gamma}_i$, which is still equal to the corresponding result in the laboratory frame (see below Eq. (13)).

3. CP-violation in neutrino oscillations

An important consequence of neutrino mixing and oscillations is the possibility of having CP-violation in the leptonic sector. Just as for quarks, flavor mixing between three neutrino generations opens the door to breaking CP-symmetry. Specifically, a non-vanishing value for δ in the parameterization of the mixing matrix results in a phase in the Standard Model Lagrangian

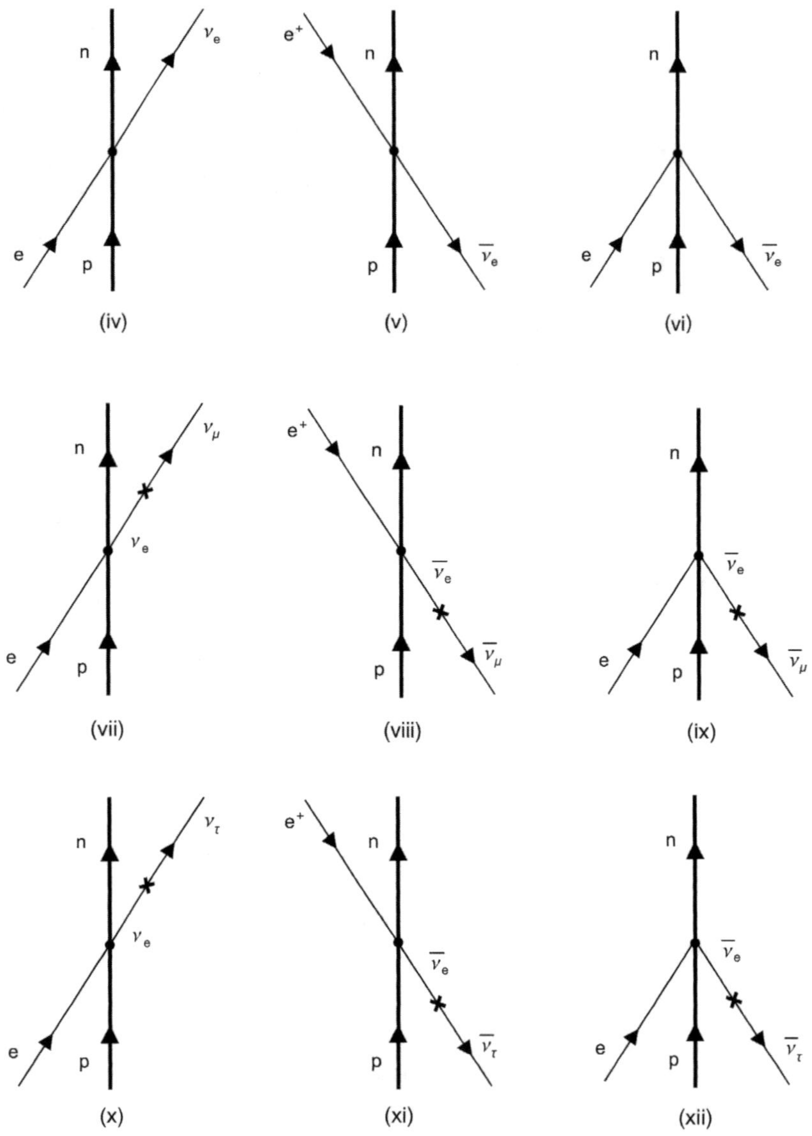

Figure 2: Inverse β-decay in the comoving frame (time flows in the vertical direction).

that violates CP [24, 25]. Beyond theoretical predictions, the statistical confidence on a non-zero δ phase is being progressively increased also by experiments (see, for instance, [25]).

CP-asymmetry is controlled by Jarlskog invariant J, which is a phase-convention-independent measure of CP violation in the Standard Model. For the case of neutrino mixing, it is defined by the relation $\mathrm{Im}\left[U_{\delta i}U^*_{\gamma i}U^*_{\delta j}U_{\gamma j}\right] \equiv J\sum_{\lambda,k}\varepsilon_{\delta\gamma\lambda}\varepsilon_{ijk}$ [26], where $\delta,\gamma,\lambda = \{e,\mu,\tau\}$ and $i,j,k = \{1,2,3\}$. It should be noticed that, since this is the sole phase-independent quantifier of CP-violation that can be built from the mixing matrix, all physical observables pertaining to phenomena that break CP-symmetry must uniquely depend on J.

CP-violation in charged-current weak interactions involving neutrinos can be taken into account by making the following general considerations, which can be promptly specialized to the inverse β-decay under examination. Let us assume that an electron neutrino ν_e emitted

Journal of Physics: Conference Series 2533 (2023) 012025 doi:10.1088/1742-6596/2533/1/012025

by a weak interaction $\hat{S}_{weak}$ is detected as a muon neutrino ν_μ after a certain time. By resorting to the flavor asymptotic representation, the transition amplitude for this process is

$$\mathcal{A}_{\nu_e,\nu_\mu} = {}_{out}\langle \nu_\mu, \ldots | S_{weak}\big(\bar{\psi}_{\nu_e} \ldots\big)| \ldots \rangle_{in}\,, \tag{27}$$

where dots replace the other fields and states involved in the process. After rotating to the mass basis, this becomes

$$\mathcal{A}_{\nu_e,\nu_\mu} = \sum_{j=1}^{3} U^*_{\mu j} U_{ej}\, \mathcal{A}_j\,, \qquad \mathcal{A}_j \equiv {}_{out}\langle \nu_j, \ldots | S_{weak}\big(\bar{\psi}_{\nu_j} \ldots\big)| \ldots \rangle_{in}\,, \tag{28}$$

which yields

$$\Gamma_{\nu_e,\nu_\mu} \sim |\mathcal{A}_{\nu_e,\nu_\mu}|^2 = |U_{\mu 1}|^2 |U_{e1}|^2 |\mathcal{A}_1|^2 + |U_{\mu 2}|^2 |U_{e2}|^2 |\mathcal{A}_2|^2 + |U_{\mu 3}|^2 |U_{e3}|^2 |\mathcal{A}_3|^2$$
$$+ \Big(U^*_{\mu 1}\, U_{e1}\, U_{\mu 2}\, U^*_{e2}\, \mathcal{A}_1\, \mathcal{A}^*_2 + U^*_{\mu 1}\, U_{e1}\, U_{\mu 3}\, U^*_{e3}\, \mathcal{A}_1\, \mathcal{A}^*_3 + U^*_{\mu 2}\, U_{e2}\, U_{\mu 3}\, U^*_{e3}\, \mathcal{A}_2\, \mathcal{A}^*_3 + \text{c.c.}\Big). \tag{29}$$

(here we have omitted the sum over σ's and integration over k's to streamline notation). By repeating the same calculations for the CP-symmetric process, the parameter $A^{(e,\mu)}_{CP} \equiv \Gamma_{\nu_e,\nu_\mu} - \Gamma_{\bar{\nu}_e,\bar{\nu}_\mu}$ takes the form

$$A^{(e,\mu)}_{CP} = 4J\Big\{ - \text{Im}\,[\mathcal{A}_1 \mathcal{A}^*_2] + \text{Im}\,[\mathcal{A}_1 \mathcal{A}^*_3] - \text{Im}\,[\mathcal{A}_2 \mathcal{A}^*_3] \Big\}. \tag{30}$$

Some comments are in order here: first, we notice that $A^{(e,\mu)}_{CP}$ is non-zero, as expected for neutrino oscillations observables in the Standard Model. Most importantly, the CP phase appears in $A^{(e,\mu)}_{CP}$ through the Jarlskog invariant J, which entails that the obtained result is independent of the choice of the parameterization of PMNS matrix. On the other hand, the approach of [4] does not account for CP-violation at all. Indeed, calculations analogous to those leading to Eq. (30) give in that case

$$A^{(e,j)}_{CP} \equiv \Gamma_{\nu_e,\nu_j} - \Gamma_{\bar{\nu}_e,\bar{\nu}_j} = |U_{ej}|^2 |\mathcal{A}_j|^2 - |U^*_{ej}|^2 |\mathcal{A}_j|^2 = 0\,, \qquad j = 1,2,3\,, \tag{31}$$

which vanishes independently of the value of δ. Clearly, this seems an unexpected result.

We also observe that both the above decay rates Γ_{ν_e,ν_μ} and Γ_{ν_e,ν_j} are invariant under re-phasing of the charged lepton fields (see for instance [27, 28]). This happens trivially for the amplitude Γ_{ν_e,ν_j}, while it occurs in a nontrivial way for each term in the second line of Eq. (29) for Γ_{ν_e,ν_μ}.

Finally, we comment on the case of Majorana neutrino, in which, as well known, additional CP violating phases are allowed in the mixing matrix. Once again, it is easy to see that Majorana phases cancel out in a trivial way for $A^{(e,j)}_{CP}$, as discussed above for the usual CP-phase. On the other hand, in the case of $A^{(e,\mu)}_{CP}$, there is still a cancellation, which is now due to the peculiar structure of the coefficients of the decay rate (29). These are indeed invariant when an entire line or column of the mixing matrix is multiplied by a phase factor.

4. Conclusions

We have reported on recent results on the role of neutrino mixing in the inverse β-decay and its consistency with CP-violation. By computing the scalar decay rate Γ of the proton in both the laboratory and comoving frames, we have shown that Pontecorvo states are consistent with the General Covariance of QFT and correctly describe CP-asymmetry in the oscillations of the

Journal of Physics: Conference Series **2533** (2023) 012025 doi:10.1088/1742-6596/2533/1/012025

emitted neutrino, at least in the relativistic approximation. On the other hand, mass states fail to account for CP-violation. This makes their usage meaningless.

Other aspects remain to be addressed. First, we plan to extend the above results to the case of exact flavor neutrino states beyond the Pontecorvo approximation [20, 29]. This is required in order to develop a full-fledged QFT analysis of the problem. Furthermore, it would be interesting to perform a similar study in curved spacetime, for instance for accelerations induced by strong gravitational fields of black holes. Research on gravity effects on neutrino mixing and oscillations appears, for instance, in [30, 31, 32, 33, 34, 35]. Work along these directions is still in progress and results will be exhibited elsewhere.

Acknowledgements

G.G.L. acknowledges the Spanish "Ministerio de Universidades" for the awarded Maria Zambrano fellowship and funding received from the European Union - NextGenerationEU. He is also grateful for participation in LISA Cosmology Working group. G.G.L. and L.P. are grateful for participation in the COST Action CA18108. L.P. acknowledges the "Angelo Della Riccia" foundation for the awarded fellowship received to support the study at Universität Ulm.

References

[1] Matsas G E A and Vanzella D A T 1999 *Phys. Rev. D* **59** 094004
[2] Ahluwalia D V, Labun L and Torrieri G 2016 *Eur. Phys. J.* A **52** 189
[3] Blasone M, Lambiase G, Luciano G G and Petruzziello L 2018 *Phys. Rev.* D **97** 105008
[4] Cozzella G, Fulling S A, Landulfo A G S, Matsas G E A and Vanzella D A T 2018 *Phys. Rev.* D **97** 105022
[5] Blasone M, Lambiase G, Luciano G G and Petruzziello L 2020 *Phys. Lett. B* **800** 135083
[6] Blasone M, Lambiase G, Luciano G G and Petruzziello L 2020 *Eur. Phys. J. C* **80** 130
[7] Salucci P *et al* 2021 *Front. in Phys.* **8** 603190
[8] Unruh W G 1976 *Phys. Rev.* D **14** 870
[9] Suzuki H and Yamada K 2003 *Phys. Rev. D* **67** 065002
[10] Blasone M, Lambiase G and Luciano G G 2017 *Phys. Rev. D* **96** 025023
[11] Luciano G G and Blasone M 2021 *Phys. Rev. D* **104** 045004
[12] Luciano G G and Blasone M 2021 *Eur. Phys. J. C* **81** 995
[13] Blasone M, Lambiase G, Luciano G G and Petruzziello L 2022 *Universe* **8** 306
[14] Luciano G G 2023 *Eur. Phys. J. Plus* **138** 83
[15] Blasone M, Lambiase G, Luciano G G, Petruzziello L Smaldone L (2020) *Class. Quant. Grav.* **37** 155004
[16] Ginzburg V L and Syrovatskii S I 1965 *Usp. Fiz. Nauk.* **87** 65
[17] Muller R 1997 *Phys. Rev. D* **56** 953
[18] Bilenky, S M and Pontecorvo B (1978) *Phys. Rept.* **41** 225
[19] Maki Z, Nakagawa M and Sakata S 1962 *Prog. Theor. Phys.* **28** 870
[20] Blasone M and Vitiello G 1995 *Annals Phys.* **244** 283
[21] Blasone M, Lambiase G, Luciano G G and Petruzziello L 2018 *PoS* **CORFU2017** 198
[22] Fritzsch H and Xing Z z 1998 *Phys. Rev. D* **57** 594
[23] Cheng T P and Li L F 1984 *Gauge theory of elementary particle physics* (Oxford University Press)
[24] Covi L, Roulet E and Vissani F 1996 *Phys. Lett. B* **384** 169
[25] Abe K *et al.* [T2K Collaboration] 2018 *Phys. Rev. Lett.* **121** 171802
[26] Jarlskog C 1985 *Phys. Rev. Lett.* **55** 1039
[27] Giunti C 2010 *Phys. Lett. B* **686** 41
[28] Johns L 2022 *P*hys. Rev. D **105** 033002
[29] Blasone M, Capolupo A and Vitiello G 2002 *Phys. Rev. D* **66** 025033
[30] Cardall C Y and Fuller G M 1997 *Phys. Rev. D* **55** 7960
[31] Fornengo N, Giunti C, Kim C W and Song J 1997 *Phys. Rev. D* **56** 1895
[32] Alexandre J, Farakos K, Mavromatos, N E and Pasipoularides P 2008 *Phys. Rev. D* **77** 105001
[33] Blasone M, Lambiase G, Luciano G G 2017 *Phys. Rev.* D **96** 025023
[34] Buoninfante L, Luciano G G, Petruzziello L and Smaldone L (2020) *Phys. Rev. D* **101** 024016
[35] Capolupo A, Lambiase G and Quaranta A (2020) *Phys. Rev. D* **101** 095022

Journal of Physics: Conference Series **2533** (2023) 012026 doi:10.1088/1742-6596/2533/1/012026

Aspects of wave-particle complementarity in quantum delayed-choice experiments

P R Dieguez, H S Karthik

International Centre for Theory of Quantum Technologies (ICTQT), University of Gdańsk, Jana Bazynskiego 8, 80-309 Gdańsk, Poland

E-mail: dieguez.pr@gmail.com; hsk1729@gmail.com

Abstract. Quantum delayed-choice experiments are designed to probe the robustness of the complementarity principle. In this proceeding, we revisit the main ideas reported recently in Ref. [Commun. Phys. 5, **82** (2022)] concerning a setup (quantum-controlled reality experiment) that has the same final visibility as the quantum version of Wheeler's delayed-choice arrangement. Besides the same final statistics, upon an operational criterion of physical realism, the authors claimed a different state of affairs for the wave-and-particle behavior in delayed-choice experiments. To this end, we put forward an analysis by employing the relative entropy of coherence to study how the local wave and particle coherences when the system is traveling inside the interferometer change accordingly to the final observed visibility.

1. Introduction

Delayed-choice experiments are a type of experimental arrangement in which the final choice of how to measure a system is made after the system has already been prepared, the experiment has already started, or even after the system has been recorded by detectors. The original proposal of Wheeler [1], who presented the experiment, considered introducing a choice of whether to complete the interferometer or not, after being guaranteed that the system has entered it thereby providing an avenue for delaying the subsequent choice of the behavior of the system. The experimenter's ability to force this delay in the choice of the final design of the interferometer to probe the system challenged the complementarity principle. Several experiments have been performed motivated by Wheeler's proposal [2, 3, 4]. The result is intriguing. The observed visibility at the output of the interferometer always revealed a statistic related to wave behavior or particle, when the arrangement is respectively settled to observe interference (complete) or not, in full agreement with the results expected when we consider that complementarity is a fundamental aspect of nature. In other words, quantum systems seem to present dichotomous properties which can not be observed in a single experimental arrangement, which also corroborates the breakdown of classical realism in which all physical properties are considered well-defined before measurements.

Quantum delayed-choice experiments, *i.e*, experiments which implement a quantum-controlled operation to delay the choice of the complete experimental arrangement until after the measurement is actually performed on the system of interest, add another layer of complexity to the concept of wave-particle duality and the role of observation in determining the behavior of physical systems. By introducing the idea of a Mach-Zehnder interferometer with a second beam splitter prepared in a spatial quantum superposition of being "present + absent" [5, 6, 7, 8, 9, 10], the complementarity principle is challenged by a nontrivial space-time ordering of events in such context.

More recently, important objections to the usual interpretations of the quantum delayed-choice experiments were reported theoretically [11] and experimentally [12] employing a two-spin nuclear magnetic resonance experiment. Adopting an operational quantifier of physical realism (introduced in Refs. [13, 14]) that explicitly depends on the quantum state, it became possible to track the behavior of the system at every stage of the interferometer [12]. To this end, the experiment showed that the visibility at the output of the quantum delayed-choice setup has no connection with the operational quantifier of wave and particle realism. Furthermore, proposing a slight variant of the quantum delayed-choice that is based on changing the causal order of the quantum-controlled operation [12], it was possible to demonstrate that by not implicitly assuming retro-action affecting the behavior of the quantum system, the relevance of quantum correlations to wave-particle duality mattered. As we discuss in this proceeding article, this new experimental arrangement has the same output visibility as the quantum delayed-choice experiment, but Bohr's original formulation of the complementarity principle was shown to have more subtle, deeper, and broader significance [12].

In this proceeding article, we review the notion of realism recently used to operationally quantify the particle and wave behavior in quantum delayed-choice experiments and discuss the ideas of the recently introduced experimental arrangement in Ref. [12] coined as a quantum-controlled reality experiment. Subsequently, we discuss its difference from the quantum version of Wheeler's delayed-choice arrangement as quantified by the operational realism measure. Moreover, we point out that the realism measure is related to another non-classicality indicator, the *quantum coherence*. Utilizing this connection, we analyze the system's coherence for the wave and particle observables in terms of the relative entropy of coherence for the system when it is traveling inside each interferometer.

2. Realism and complementarity

In this section, we discuss the notion of physical realism and revisit the operational quantifier, irrealism, introduced by Bilobran and Angelo in Ref. [13]. This is constructed under a single premise that non-selective measurements are sufficient to define elements of reality for the probed observable. The notion of irrealism for quantum states is required for explaining how this quantifier was employed in Ref. [12] to investigate, in a quantitative way, the robustness of the complementarity principle in quantum delayed-choice experiments.

EPR proposed in their celebrated work that every element of physical reality must have a correspondence in the theory as a necessary condition to consider it complete [15]. For the elements of physical reality, they define the following sufficient condition: "If without disturbing the system in any way, we can predict with certainty (that is, with a probability equal to unity) the value of a physical quantity, then there is an element of physical reality corresponding to this quantity".

Let us now discuss how this rationale could be applied to entangled spin systems described by the singlet state

$$|\phi\rangle_{\mathcal{AB}} = \frac{1}{\sqrt{2}}(|0\rangle_{\mathcal{A}}|1\rangle_{\mathcal{B}} - |1\rangle_{\mathcal{A}}|0\rangle_{\mathcal{B}}). \tag{1}$$

After the formation of the entangled pair, suppose that both systems propagated in opposite directions, so that we can classify them according to their future spatial positions in Alice's lab A and Bob's B. After this, the spin in the spatial region A is measured by Alice in some chosen direction. After the detection of the particle located in the spatial region A, Alice can conclude immediately what is the spin in the same direction for the particle in the spatial region B. Note that this conclusion is independent of the spatial distance between both particles. Upon the measurement of one of the systems, the spin of the other can be predicted with certainty. Then according to EPR, there must be an element of physical reality corresponding to this physical quantity.

The accurate prediction of the spin on a distant system requires us to accept it as the manifestation of a pre-existing value of the physical property (spin) since there is no interaction between them after the initial state was formed and separated [16]. Combined with the fact that we can measure the spin along any direction and the correlation is the same in the singlet state, this implies that even incompatible observables necessarily have elements of reality, according to EPR, defined independent of the measurement. Based on this reasoning, EPR concluded that quantum theory is an incomplete description of reality as it does not generally provide us with the values for joint measurements of non-commuting observables [15].

More recently, it has been argued that a deeper understanding of the physical realism can be attained through the concept of information, as discussed by Dieguez and Angelo in Ref. [14] through a complementarity relation between them. This connection was established by using the measure introduced by Bilobran and Angelo (BA) [13], who put forward an operational scheme to assess elements of reality of discrete-spectrum observables in quantum mechanics. The main idea relies on a single premise that a measurement establishes the reality of an observable, independently of whether we have access or not to the specific output of each measurement. This can be formally stated with the following procedure. Consider a preparation $\rho \in \mathcal{H}_{\mathcal{A}} \otimes \mathcal{H}_{\mathcal{B}}$ submitted to a protocol of unrevealed measurements (also known as non-selective measurements) of a generic observable $A = \sum_a aA_a$, with projectors $A_a = |a\rangle\langle a|$, acting on $\mathcal{H}_{\mathcal{A}}$. The resulting state is the average

$$\Phi_A(\rho) = \sum_a (A_a \otimes \mathbb{1}_B)\, \rho\, (A_a \otimes \mathbb{1}_B) = \sum_a p_a A_a \otimes \rho_{B|a}, \tag{2}$$

where $\rho_{B|a} = \langle a|\,\rho\,|a\rangle / p_a$ with $p_a = \mathrm{Tr}[(A_a \otimes \mathbb{1}_B)\rho]$. BA propose to take $\Phi_A(\rho)$ as a state of reality for A and $\rho = \Phi_A(\rho)$ as a formal criterion of reality.

Note that this premise of realism also agrees with the EPR criterion, since eigenstate preparations are elements of reality, but this also attempts to generalize EPR in the sense that it also attributes realism to a convex combination of eigenstates. With that, by employing the relative entropy $S(\rho||\sigma) \equiv \mathrm{Tr}\big[\rho(\log_2 \rho - \log_2 \sigma)\big]$, the operational quantifier of A-realism violations for a given state ρ, coined *irrealism*, can be computed as

$$\mathfrak{I}_A(\rho) \equiv \min_{\varrho} S\big(\rho||\Phi_A(\varrho)\big) = S\big(\Phi_A(\rho)\big) - S(\rho), \tag{3}$$

where $S(\rho) \equiv \mathrm{Tr}\big(\rho \log_2 \rho\big)$ is the von Neumann entropy. The irrealism of the context $\{A, \rho\}$ is bounded as $0 \leq \mathfrak{I}_A(\rho) \leq \log_2 d_{\mathcal{A}}$, with $d_{\mathcal{A}} = \dim \mathcal{H}_{\mathcal{A}}$, vanishing iff $\rho = \Phi_A(\rho)$ [13, 12].

The use of the entropic metric allows one to relate this measure with other quantities of quantum information theory [13, 14, 17]. For example, the above formula can be decomposed as

$$\mathfrak{I}_A(\rho) = \mathfrak{C}_A(\rho_{\mathcal{A}}) + D_A(\rho), \tag{4}$$

where $D_A(\rho) \equiv I_{\mathcal{A}:\mathcal{B}}(\rho) - I_{\mathcal{A}:\mathcal{B}}(\Phi_A(\rho))$ stands for the nonminimized version of the one-way quantum discord (see Refs. [13, 14] for further details), $I_{\mathcal{A}:\mathcal{B}}(\rho) \equiv S(\rho_{\mathcal{A}}) + S(\rho_{\mathcal{B}}) - S(\rho)$ is the mutual information, and $\mathfrak{C}_A(\rho_{\mathcal{A}}) \equiv S(\rho_{\mathcal{A}}||\Phi_A(\rho_{\mathcal{A}})) = S(\Phi_A(\rho_{\mathcal{A}})) - S(\rho_{\mathcal{A}})$ is the relative entropy of coherence. So, the irrealism of A is the sum of local coherence (that is, the coherence of A given the reduced state $\rho_{\mathcal{A}}$) with quantum correlations associated with measurements of A [13, 14]. Note that, for a single-partite state, ($\rho \in \mathcal{H}_{\mathcal{A}}$) or separable bipartite states, the irrealism measure reduces to the relative entropy of coherence $\mathfrak{C}_A(\rho_{\mathcal{A}})$.

To discuss the result of Ref. [12] regarding the wave and particle realism of systems during quantum delayed-choice experiments, we now introduce the quantifier presented in Ref. [14] to work with the A-realism

$$\mathfrak{R}_A(\rho) \equiv \log_2 d_{\mathcal{A}} - \mathfrak{I}_A(\rho), \tag{5}$$

which quantifies how close the scenario is to the A-realistic context. For maximally incompatible observables A and A', we can write [12, 14, 17]

$$\mathfrak{R}_A(\rho) + \mathfrak{R}_{A'}(\rho) \leq \log_2 d_{\mathcal{A}} + S(\rho_{\mathcal{A}}) - I_{\mathcal{A}:\mathcal{B}}(\rho), \tag{6}$$

which can be used to have new insights for analyzing the complementarity principle. Clearly, from the above relation one can see that correlations restrict quantum systems to have full realism, and for pure states $\rho = |\psi\rangle\langle\psi|$, the upper bound becomes $\log_2 d_{\mathcal{A}} - E(\psi)$, with $E(\psi) = S(\rho_{\mathcal{A}}) = S(\rho_{\mathcal{B}})$ being the entanglement entropy of $|\psi\rangle$ [12].

The realism measure gives a prominent role to the notion of information and can be applied in several different contexts [14, 16, 17, 18, 19, 20, 21, 22, 23, 24]. For instance, by employing a model of measurement called monitoring [14, 25], which consists of interpolating weak and strong non-selective measurements, it was possible to deduce a formal connection between information and the concept of reality for quantum states, exhibiting a dual relationship for both quantities coined as Information-reality complementarity [14]. This quantity was employed to investigate the notion of realism from the perspective of weak measurements in an experiment with photonic weak measurements [26]. Realism-based entropic uncertainty relations were derived in Ref. [18], as well as realism-based nonlocality was explored in Refs. [19, 20, 21, 22]. Realism was discussed in the context of random quantum walk [23], Hardy's paradox [24], generalized resource theory of information [27], and more recently, to analyze the Aharonov-Bohm effect [28]. Freire and Angelo presented a framework in Ref. [17], which calculates the degree of realism associated with continuous variables such as position and momentum. Employing this definition for continuous variables, the authors in Ref. [16] investigated how an initial position-momentum correlation for a Gaussian wave-packet can be employed to relate the irrealism measure with the visibility in a double-slit experiment.

On a different note, the term *interference* has long been associated with the behavior of waves through the superposition of two or more waves vectorially adding constructively or destructively. The interesting bright and dark patterns, known as interference fringes, were soon realized to be due to the coherence (constant phase difference) between the waves. The amount of coherence decided the contrast of the fringes. In the case of Young's double-slit interference, it is known that the visibility of the interference equals the degree of coherence when the slits are identical whereas in the case of a Mach-Zehnder interferometer, the visibility can be calculated using the off-diagonal terms of the density matrix representing the quantum system *using the paths as the basis* [29].

In the following sections, we review and discuss how the realism measure was employed to investigate the wave and particle behavior for the two quantum delayed-choice experiments of the Ref. [12] and realize how the degree of quantum coherence can be utilized to study the particle and wave aspects as the visibility of each interferometer is varied.

3. Realism and complementarity in quantum delayed-choice experiments

In the classical formulation of Wheeler's delayed-choice arrangement [5], a photon is considered to be in a state

$$|\wp_\theta\rangle = \frac{1}{\sqrt{2}}\left(e^{i\theta}|0\rangle + |1\rangle\right) \tag{7}$$

after passing through a balanced beam-splitter and a phase-shifter, which implements a relative phase between the paths $\{0, 1\}$ traveled by the photon in a Mach-Zehnder configuration. Clearly,

Journal of Physics: Conference Series **2533** (2023) 012026 doi:10.1088/1742-6596/2533/1/012026

the probability distribution for a path measurement of this state does not depend on the relative phase. Moreover, both detectors can click with the same probability and this is usually interpreted as the photon behaving like a particle, since, for a given output result, one can infer through which path the photon might have passed. On the other hand, if we now consider a second balanced beam-splitter, the state $|\wp_\theta\rangle$ transforms into what is regarded as a wave-like state [5]

$$|w_\theta\rangle = e^{i\theta/2}\left(\cos\frac{\theta}{2}|0\rangle + i\sin\frac{\theta}{2}|1\rangle\right), \tag{8}$$

with amplitudes clearly depending now on the phase θ. The choice to insert (or not insert) the second beam-splitter to observe a wave-like (or a particle-like) behavior then is taken after the system has entered into the interferometer. This explains the term "delay" in the delayed-choice experiments.

The quantum version of the above experiment is realized by preparing a quantum controller system ($\mathcal{C}$) in a general coherent superposition to realize the notion of a quantum-controlled Mach-Zehnder being closed (final beam-splitter in) and open (final beam-splitter out), *i.e.*, with a state described as $\cos\frac{\alpha}{2}|\text{in}\rangle + \sin\frac{\alpha}{2}|\text{out}\rangle$, with $0 \le \alpha \le \pi$. In other words, α determines the probabilities of the second beam-splitter to be found "inside" the interferometer (defining abstractly as the state $|\text{in}\rangle$) or "outside" of it ($|\text{out}\rangle$) [12]. According to this original picture, the result of such interaction at the output of the interferometer reads

$$|\psi_{\mathbf{a}}\rangle = \cos\frac{\alpha}{2}|w_\theta\rangle|\text{in}\rangle + \sin\frac{\alpha}{2}|\wp_\theta\rangle|\text{out}\rangle. \tag{9}$$

The interference pattern in the detector D_0 corresponding to the path-0 is then shown to be written as

$$\mathfrak{p}_0 = \text{Tr}(|0\rangle\langle 0|\otimes \mathbb{1}_C\,\rho_{\mathbf{a}}) = \frac{1}{2}(1+\mathcal{V}\cos\theta), \tag{10}$$

where $\rho_{\mathbf{a}} = |\psi_{\mathbf{a}}\rangle\langle\psi_{\mathbf{a}}|$ and

$$\mathcal{V} \equiv \frac{(\mathfrak{p}_0^{\max} - \mathfrak{p}_0^{\min})}{(\mathfrak{p}_0^{\max} + \mathfrak{p}_0^{\min})} = \cos^2\left(\frac{\alpha}{2}\right), \tag{11}$$

which stands for the visibility of the interference pattern computed from optimizations running over θ.

The association of $|w_\theta\rangle$ and $|\wp_\theta\rangle$ with wave- and particle-like behaviors is justified with basis on the resulting dependence of $\mathfrak{p}_0$ on the phase θ, as can be readily checked for $\alpha = 0$ and $\alpha = \pi$, respectively. Moreover, the wave-like amplitude in $|\psi_{\mathbf{a}}\rangle$ is clearly related to the visibility $\mathcal{V}$. The scenario is such that by looking at the statistics of the output of the interferometer, one infers the way the qubit traveled the interferometer.

Following Ref. [12], we revisit here the results of applying the irrealism measure to compute the wave and particle realism in a quantum-controlled interference device. The authors of Ref. [12] defined $P \equiv \sigma_z$ and $W \equiv \sigma_\perp$, with respective eigenstates $\{|\mathcal{P}_+\rangle, |\mathcal{P}_-\rangle\} \equiv \{|0\rangle, |1\rangle\}$ and $|\mathcal{W}_\pm\rangle = \frac{1}{\sqrt{2}}\left(e^{i\theta}|0\rangle \pm |1\rangle\right)$, as the relevant eigenstates for the particle and wave observables, respectively. This choice naturally connects definite paths, $|\mathcal{P}_\pm\rangle$, with particle-like elements of reality as explained in Ref. [12]. Then, it follows that

$$\mathfrak{R}_P(\wp_\theta) = 0, \tag{12}$$

and

$$\mathfrak{R}_W(\wp_\theta) = 1, \tag{13}$$

which shows that what is called a particle-like state actually corresponds to a wave reality in this framework. An important physical argument corroborating this view is that $|\wp_\theta\rangle$ encapsulates

Journal of Physics: Conference Series **2533** (2023) 012026 doi:10.1088/1742-6596/2533/1/012026

a detectable relative phase in the basis $\{|\mathcal{P}_{\pm}\rangle\}$ and can be used to create entanglement between far distant particles [11]. Therefore, in full contrast to current claims, inside the interferometer, the system always behaves as a wave, and $\mathcal{V}$ does not faithfully furnish this diagnosis [12]. Accordingly to the authors of Ref. [12], then it follows that the pattern $\mathfrak{p}_0$, for $0 < \alpha < \pi$, cannot validate the manifestation of wave and particle morphing behavior. With that, they reject retrocausal models according to which measuring the visibility at the output may determine the state of affairs inside the interferometer.

We now discuss the ideas of the experimental arrangement proposed in Ref. [12] that aims at solving the aforementioned issues and effectively superposing wave and particle elements of reality. Note that, for pedagogical reasons, we will explain the experimental setup of Ref. [12] in terms of a photon traveling a interferometer, but the actual experiment was performed employing nuclear magnetic resonance techniques with two-spin systems [12]. Consider a simple exchange in the causal order of the main operations described in the quantum version of the delayed-choice arrangement. The input state $|\psi_{in}\rangle = |0\rangle\left(\cos\frac{\alpha}{2}|\mathrm{in}\rangle + \sin\frac{\alpha}{2}|\mathrm{out}\rangle\right)$ now results in

$$|\psi_{\mathbf{b}}\rangle = \cos\frac{\alpha}{2}|w_\theta\rangle|\mathrm{in}\rangle + e^{i\theta}\sin\frac{\alpha}{2}|\wp_0\rangle|\mathrm{out}\rangle, \tag{14}$$

thus yielding precisely the same interference pattern $\mathfrak{p}_0 = \frac{1}{2}(1+\mathcal{V}\cos\theta)$ and visibility $\mathcal{V} = \cos^2\left(\frac{\alpha}{2}\right)$. There is no change in the statistics observed at the output. However, the states of the whole system when the photon is traveling inside the interferometer right after the phase-shifter are given by

$$|\phi_{\mathrm{QCRE}}\rangle = \cos\tfrac{\alpha}{2}|\mathcal{W}_+\rangle|\mathrm{in}\rangle + e^{i\theta}\sin\tfrac{\alpha}{2}|\mathcal{P}_+\rangle|\mathrm{out}\rangle. \tag{15}$$

The differences are remarkable, as described in Ref. [12]. Correlations are seen to play no role in the first setup, so there is not an effective "whole" defining the behavior of the system at this stage [12].

In this new quantum delayed-choice arrangement, when the controlled operation is deactivated, the photon just keeps traveling its original path, as a particle, this being a key difference to the original quantum delayed-choice setup. The state (15) predicts a clear superposition of two orthogonal reality scenarios: one wave-like ($|\mathcal{W}_+\rangle|\mathrm{in}\rangle$) and another particle-like ($|\mathcal{P}_+\rangle|\mathrm{out}\rangle$) [12]. In addition, it follows that

$$\mathfrak{R}_W(\phi_{\mathrm{QCRE}}) = 1 - h\left(\tfrac{1-\mathcal{V}}{2}\right), \tag{16}$$

and

$$\mathfrak{R}_P(\phi_{\mathrm{QCRE}}) = 1 - h\left(\tfrac{\mathcal{V}}{2}\right), \tag{17}$$

where $h(u) \equiv -u\log_2 u - (1-u)\log_2(1-u)$ is the binary entropy. These relations show that $\mathfrak{R}_W$ ($\mathfrak{R}_P$) is a monotonically increasing (decreasing) function of the visibility $\mathcal{V}$ (see Ref. [12] for the experimental investigation of this monotonic behavior). Then, with that result, we can see a strict equivalence between the output statistics and the wave-like behavior inside the interferometer. Note that within the domain $\alpha \in [0,\pi]$, the visibility $\mathcal{V}$ at the output and the wave and particle realism $\mathfrak{R}_{W,P}$ inside the interferometer are monotonic functions of α, meaning that these quantities can be controlled by the preparation of the quantum controller $\mathcal{C}$ [12]. Also, relation (6) here reduces to

$$\mathfrak{R}_W(\phi_{\mathrm{QCRE}}) + \mathfrak{R}_P(\phi_{\mathrm{QCRE}}) \leq 1 - h\left(\frac{1+\lambda_{\mathcal{V}}}{2}\right), \tag{18}$$

where $\lambda_{\mathcal{V}} \equiv \sqrt{2\mathcal{V}^2 - 2\mathcal{V} + 1}$. This result as a function of $\mathcal{V}$ demonstrates how quantum correlations between the qubit and the quantum controller are sufficient to deny classical realism inside the interferometer to both complementary observables at the same time, thus corroborating Bohr's original formulation of the complementarity principle [12].

Journal of Physics: Conference Series **2533** (2023) 012026 doi:10.1088/1742-6596/2533/1/012026

4. Wave and particle coherences in quantum delayed-choice experiments

It is clear from the above relations that in Wheeler's quantum delayed-choice experiment, irrealism has a direct connection with the coherence of the considered observable since the state inside the interferometer and the control are separable. In other words, since no interaction has happened yet, the breakdown with classical realism for the particle observable is due to quantum coherence on that basis. Let us now calculate the wave and particle coherences, employing the expression $\mathfrak{C}_A(\rho_\mathcal{A}) \equiv S(\rho_\mathcal{A}||\Phi_A(\rho_\mathcal{A})) = S(\Phi_A(\rho_\mathcal{A})) - S(\rho_\mathcal{A})$ and the respective eigenstates as before to define wave and particle observables $\{|\mathcal{W}_+\rangle, |\mathcal{W}_-\rangle\}$ with $|\mathcal{W}_\pm\rangle \equiv \frac{1}{\sqrt{2}}\left(e^{i\theta}|0\rangle \pm |1\rangle\right)$ and $\{|\mathcal{P}_+\rangle, |\mathcal{P}_-\rangle\} \equiv \{|0\rangle, |1\rangle\}$. Then it immediately follows for direct calculation using the state inside the quantum delayed-choice experiment that

$$\mathfrak{C}_W\left(|\mathcal{W}_+\rangle\langle\mathcal{W}_+|\right) = 0, \tag{19}$$

and

$$\mathfrak{C}_P\left(|\mathcal{W}_+\rangle\langle\mathcal{W}_+|\right) = 1. \tag{20}$$

Now, let us take a closer look at the coherence during the quantum-controlled reality experiment. Consider the state given in Eq. 15. The reduced density matrix for the system traveling the interferometer is given by $\rho_\mathcal{S} = \mathrm{Tr}_\mathcal{C}[|\phi_{\mathrm{QCRE}}\rangle\langle\phi_{\mathrm{QCRE}}|]$, one finds

$$\rho_\mathcal{S} = \frac{1}{2}\begin{pmatrix} 2-\mathcal{V} & e^{i\theta}\mathcal{V} \\ e^{-i\theta}\mathcal{V} & \mathcal{V} \end{pmatrix}. \tag{21}$$

As discussed before, the von Neumann entropy of this state is related to the amount of entanglement between the systems. By direct calculations and taking $\lambda_\mathcal{V} = \sqrt{2\mathcal{V}^2 - 2\mathcal{V} + 1}$ one can write

$$\mathfrak{C}_W\left(\rho_\mathcal{S}\right) = S\left(\Phi_\mathcal{W}(\rho_\mathcal{S})\right) - S(\rho_\mathcal{S}) = h\left(\tfrac{1-\mathcal{V}}{2}\right) - h\left(\tfrac{1+\lambda_\mathcal{V}}{2}\right), \tag{22}$$

and

$$\mathfrak{C}_P\left(\rho_\mathcal{S}\right) = S\left(\Phi_\mathcal{P}(\rho_\mathcal{S})\right) - S(\rho_\mathcal{S}) = h\left(\tfrac{\mathcal{V}}{2}\right) - h\left(\tfrac{1+\lambda_\mathcal{V}}{2}\right). \tag{23}$$

Plotting these results. one can clearly see in Fig. 1 that visibility is monotonically related to wave and particle coherences in the quantum-controlled reality experimental arrangement.

5. Discussion

Delayed-choice experiments highlight the counter-intuitive nature of quantum systems. It involves giving the experimentalist the option to delay the selection of the final design of an interferometer, which raises the question of whether this influence can alter the behavior of the quantum system. Can the system somehow anticipate a future action before it happens and acts accordingly to it? Noting that the final statistics of the classical delayed-choice arrangement preserves the interpretation of the complementarity principle, naturally, at first sight, one might jump to such a conclusion that somehow the future action of closing or opening the interferometer can influence the past behavior of a system when it is traveling inside the interferometer.

However, as we discussed in this proceeding article, this is not the end of the history regarding delayed-choice experiments. The introduction of an additional quantum-controlled system to delay the choice of how the experimental arrangement will be finally settled [5], presented us with a completely new perspective on that problem. Again, at first sight, by looking at the final statistics of this setup, one might jump to a conclusion that we are observing the quantum system exhibiting wave and particle behavior with just one experimental arrangement, thus indicating the necessity for a radical review of the complementarity principle.

After the initial proposal of the quantum-controlled interferometer more than a decade ago, several experimental investigations have supported the theoretical predictions, but, despite this,

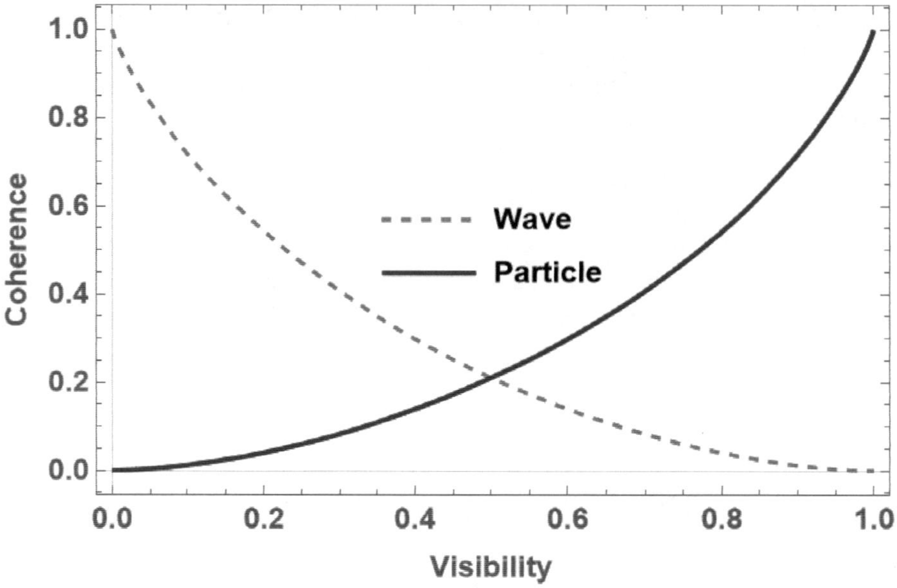

Figure 1. Wave and particle coherences in quantum-controlled reality arrangement as a function of visibility. The dashed red curve shows that wave coherence inside the interferometer decreases while visibility increases. On the other hand, the solid blue curve shows an increase in particle coherence accordingly to the interferometric visibility.

the interpretations of delayed-choice experiments are still being debated and reconsidered. In fact, by proposing and probing (using nuclear magnetic resonance techniques) a new experiment with the exact same final probability distribution and interferometric visibility as the one proposed in Ref. [5], the authors in Ref. [12] claimed contrasting conclusions about the wave and particle behavior of systems during quantum delayed-choice experiments. The central rationale of the discussion given in Ref. [12] relies on the observation that the "whole" physical context must be included and analyzed in order to investigate the robustness of the complementarity principle (and not just the final probability and interferometric visibility). This argument can be understood by comparing two experimental arrangements with the same final probability, but with a different physical realism related to the wave and particle behavior during the experiment, as discussed in section 3. For a complete discussion (including consequences for the measurement problem, and the experimental results and methods employed to probe the predictions revisited here in this proceeding article), we refer the reader to Ref. [12].

Finally, as a short contribution to that discussion, we also analyzed in section 4 how the local wave and particle coherence, quantified by the relative entropy of coherence of the reduced state which is traveling the interferometer, changes accordingly to the final visibility of the experimental setup. Higher particle (wave) coherence favors higher (lower) visibility, which contributes at the same time to the emergence of wave (particle) realism in the quantum-controlled reality setup.

Acknowledgments
P.R.D. acknowledges support by the Foundation for Polish Science (IRAP project, ICTQT, Contract No. MAB/2018/5, co-financed by EU within Smart Growth Operational Programme). H.S.K. acknowledges the support of National Science Centre, Poland through SHENG Grant No. 2018/30/Q/ST2/00625.

Journal of Physics: Conference Series **2533** (2023) 012026 doi:10.1088/1742-6596/2533/1/012026

References

[1] Wheeler J A and Zurek W H 2014 *Quantum theory and measurement* vol 15 (Princeton University Press)
[2] Jacques V, Wu E, Grosshans F, Treussart F, Grangier P, Aspect A and Roch J F 2007 *Science* **315** 966–968
[3] Manning A G, Khakimov R I, Dall R G and Truscott A G 2015 *Nature Physics* **11** 539–542
[4] Vedovato F, Agnesi C, Schiavon M, Dequal D, Calderaro L, Tomasin M, Marangon D G, Stanco A, Luceri V, Bianco G *et al.* 2017 *Science Advances* **3** e1701180
[5] Ionicioiu R and Terno D R 2011 *Physical Review Letters* **107** 230406
[6] Auccaise R, Serra R, Filgueiras J, Sarthour R, Oliveira I and Céleri L 2012 *Physical Review A* **85** 032121
[7] Roy S S, Shukla A and Mahesh T 2012 *Physical Review A* **85** 022109
[8] Peruzzo A, Shadbolt P, Brunner N, Popescu S and O'Brien J L 2012 *Science* **338** 634–637
[9] Kaiser F, Coudreau T, Milman P, Ostrowsky D B and Tanzilli S 2012 *Science* **338** 637–640
[10] Adesso G and Girolami D 2012 *Nature Photonics* **6** 579–580
[11] Angelo R and Ribeiro A 2015 *Foundations of Physics* **45** 1407–1420
[12] Dieguez P R, Guimarães J R, Peterson J P, Angelo R M and Serra R M 2022 *Communications Physics* **5** 1–8
[13] Bilobran A and Angelo R 2015 *EPL (Europhysics Letters)* **112** 40005
[14] Dieguez P and Angelo R 2018 *Physical Review A* **97** 022107
[15] Einstein A, Podolsky B and Rosen N 1935 *Physical review* **47** 777
[16] Lustosa F, Dieguez P and da Paz I 2020 *Physical Review A* **102** 052205
[17] Freire I and Angelo R 2019 *Physical Review A* **100** 022105
[18] Rudnicki Ł 2018 *Journal of Physics A: Mathematical and Theoretical* **51** 504001
[19] Gomes V and Angelo R 2018 *Phys. Rev. A* **97** 012123
[20] Gomes V and Angelo R 2019 *Physical Review A* **99** 012109
[21] Fucci D M and Angelo R 2019 *Physical Review A* **100** 062101
[22] Gomes V, Dieguez P and Vasconcelos H 2022 *Physica A: Statistical Mechanics and its Applications* 127568
[23] Orthey Jr A C and Angelo R 2019 *Physical Review A* **100** 042110
[24] Engelbert N G and Angelo R M 2020 *Foundations of Physics* **50** 105–119
[25] Dieguez P and Angelo R 2018 *Quantum Information Processing* **17** 1–13
[26] Mancino L, Sbroscia M, Roccia E, Gianani I, Cimini V, Paternostro M and Barbieri M 2018 *Physical Review A* **97** 062108
[27] Costa A C and Angelo R 2020 *Quantum Information Processing* **19** 1–14
[28] Paiva I L, Dieguez P R, Angelo R M and Cohen E 2023 *Phys. Rev. A* **107** 032213
[29] Qureshi T 2019 *Quanta* **8** 24–35

Journal of Physics: Conference Series **2533** (2023) 012027 doi:10.1088/1742-6596/2533/1/012027

No-go results on emergent space and other structures

Ovidiu Cristinel Stoica

Dept. of Theoretical Physics, NIPNE—HH, Bucharest, Romania.

E-mail: `cristi.stoica@theory.nipne.ro,holotronix@gmail.com`

Abstract. It is often claimed that, from the abstract quantum state vector and Hamiltonian, everything emerges uniquely (for example the 3D-space, a preferred factorization of the Hilbert space, and a preferred basis). This thesis, called Hilbert-Space Fundamentalism (HSF), is assumed for example in Carroll and Singh's "Mad-dog Everettianism", some proposals based on decoherence, proposals that aim to describe everything by the quantum structure alone, and proposals that spacetime emerges from a purely quantum theory of gravity.

I show that whenever the 3D-space, a preferred factorization, or any other structure emerges like this, there are infinitely many solutions. This refutes Hilbert-Space Fundamentalism.

Implications for various research programs aiming to show that certain structures emerge uniquely, and for various interpretations of Quantum Mechanics, are discussed.

1. Introduction

Definition: `MQS`. *Consider a closed quantum system with the* Hamiltonian operator $\widehat{H}$ *in the* Hilbert space $\mathcal{H}$. *We call* minimalist quantum structure (`MQS`) *the triple*

$$(\mathcal{H}, \widehat{H}, |\psi\rangle) \tag{1}$$

where the state vector $|\psi(t)\rangle \in \mathcal{H}$ *obeys the Schrödinger equation*

$$i\hbar\frac{\mathrm{d}}{\mathrm{d}\,t}|\psi(t)\rangle = \widehat{H}|\psi(t)\rangle. \tag{2}$$

In nonrelativistic Quantum Mechanics (NRQM), we normally assume the existence of a position basis $(|\mathbf{x}\rangle)_{\mathbf{x}\in\mathbb{R}^{3n}}$ of the Hilbert space and express the state vector in that basis as the *wavefunction* $\psi(\mathbf{x}) = \langle\mathbf{x}|\psi\rangle$. In Quantum Field Theory (QFT), the configuration space consists of classical fields, and the theory inherits the positions from the 3D-space on which the classical fields are defined. The position basis becomes manifest also in the Fock representation or the algebras of local observables. Similarly, the factorization into subsystems is inherited from the classical theory that is quantized.

The position basis and the factorization into subsystems seem to be artifacts of the quantization method. In NRQM one starts with a classical system expressed in terms of classical positions $\mathbf{q} = \mathbf{x}$ and momenta $\mathbf{p}$, and replaces them, and in general all classical observables $f(\mathbf{q}, \mathbf{p})$, with operators. In QFT one starts with configurations of classical fields defined on a 3D-space, and replace them with operator-valued distributions evaluated on test functions on the 3D-space.

But the world is quantum, and not the quantization of a classical world, so it makes sense to try to avoid the assumption of any preferred basis or other structure reminiscent of a classical theory. In addition, any quantum theory, including NRQM, QFT, and Quantum Gravity, is invariant under unitary transformations, so it seems that we should expect that there is no *a priori* preferred basis, and that all physically meaningful structures are basis-independent.

These remarks seem to justify the following claim:

Thesis HSF (Hilbert-space Fundamentalism Thesis [1]). *Everything about a physical system, including the 3D-space, a preferred basis, a preferred factorization of the Hilbert space (needed to represent subsystems, e.g. particles), emerge uniquely from the triple* $(\mathcal{H}, \widehat{H}, |\psi(t)\rangle)$.

This Thesis is explicit [1, 2] or implicit in various approaches to quantum gravity or to resolve the measurement problem and prove the emergence of classicality at the macro level by using decoherence. It is often believed that, in Quantum Gravity, 3D-space or 4D-spacetime are not fundamental, and have to emerge uniquely from the quantum structure, and that the factorization into subsystem has to emerge as well.

I will show that Thesis HSF is false. Whatever structures the current or imaginable various research programs based on Thesis HSF may claim to obtain as emerging from the MQS alone, they cannot emerge uniquely. This does not diminish the importance of these research programs, but only the claims of uniqueness.

Section §2 contains the proof that no structure can emerge uniquely from the MQS $(\mathcal{H}, \widehat{H}, |\psi\rangle)$.

Section §3 contains examples of research programs that are claimed or may be thought of as leading to essentially unique emergent structures of various kinds, and which the main theorem from Sec. §2 proves that they are not unique. These include generalized "preferred" bases, factorizations into subsystems, 3D-space structures, environmental decoherence *etc.*

Consequences for the foundations of Quantum Mechanics are discussed in Section §4. In particular, a "paradox" appears if we assume Thesis HSF: the present state vector supports infinitely many physically distinct worlds, in particular all the past and future states of the world, but also of alternative worlds. The interpretations of Quantum Mechanics that assume other structures in addition to the Minimal Quantum Structure $(\mathcal{H}, \widehat{H}, |\psi\rangle)$ seem to escape such problems.

2. Non-uniqueness main result

This Section contains the proof that no structure can emerge uniquely from the MQS $(\mathcal{H}, \widehat{H}, |\psi\rangle)$.

Whatever structure $\mathcal{S}$ one expects to emerge from the MQS $(\mathcal{H}, \widehat{H}, |\psi\rangle)$, it has to have physical effects. The structure $\mathcal{S}$ should be manifest for example as a set of observables $(\widehat{A}_\alpha)_{\alpha\in\mathcal{A}}$, where $\mathcal{A}$ is a set of indices or labels of the Hermitian operators $\widehat{A}_\alpha$, $\alpha \in \mathcal{A}$. Since the structure $\mathcal{S}$ is determined by the MQS, it may depend on the Hamiltonian $\widehat{H}$ and perhaps on the state vector $|\psi\rangle$, so, to express this dependency, we will denote it by $\mathcal{S}_{\widehat{H}}^{|\psi\rangle} = (\widehat{A}_\alpha^{|\psi\rangle})_{\alpha\in\mathcal{A}}$.

For more generality, we will assume that the operators $\widehat{A}_\alpha^{|\psi\rangle}$ are linear operators, $\widehat{A}_\alpha^{|\psi\rangle} \in \mathcal{L}(\mathcal{H})$. The most general proof, using any possible tensors on the Hilbert space, can be found in [3]. But since

(i) most relevant cases involve only linear operators,

(ii) the structure $\mathcal{S}$ should be characterized by observables, which consist of Hermitian operators on the Hilbert space $\mathcal{H}$,

for simplicity, here we will assume that the structure of interest $\mathcal{S}$ can be characterized by linear operators on $\mathcal{H}$, which are a particular case of tensors on $\mathcal{H}$.

For example, in NRQM for one particle, the 3D-space can be expressed by the position operators $(\widehat{\boldsymbol{x}})_{\boldsymbol{x}\in\mathbb{R}^3}$. For $\mathbf{n}$ particles, the operators $(\widehat{\mathbf{x}})_{\mathbf{x}\in\mathbb{R}^{3\mathbf{n}}}$ encode the positions in the configuration space. To recover the 3D-space from the position configuration space, we can use, in addition to the position operators $(\widehat{\mathbf{x}})_{\mathbf{x}\in\mathbb{R}^{3\mathbf{n}}}$, a factorization, which can be expressed in

Journal of Physics: Conference Series **2533** (2023) 012027 doi:10.1088/1742-6596/2533/1/012027

terms of algebras of operators acting locally on the factor Hilbert spaces [4]. A preferred basis can be as well expressed as an operator having that basis as eigenbasis, or, more generally, as a set of projection operators.

Since the operators $(\widehat{A}_\alpha^{|\psi\rangle})_{\alpha\in\mathcal{A}}$ depend on $\widehat{H}$ and $|\psi\rangle$, this dependency can be expressed as a set of *defining conditions*, which are equations or inequations of the form

$$\mathcal{C}_\beta\left(\left(\widehat{A}_\alpha^{|\psi\rangle}\right)_{\alpha\in\mathcal{A}},\widehat{H},|\psi\rangle\right)\begin{cases}=0 & \text{if } \beta\in\mathcal{B}_=,\\ \geq 0 & \text{if } \beta\in\mathcal{B}_\geq,\\ >0 & \text{if } \beta\in\mathcal{B}_>,\end{cases} \tag{3}$$

where $(\mathcal{C}_\beta)_{\beta\in\mathcal{B}}$, $\mathcal{B}=\mathcal{B}_=\sqcup\mathcal{B}_\geq\sqcup\mathcal{B}_>$, are functions constraining the operators $\left(\widehat{A}_\alpha^{|\psi\rangle}\right)_{\alpha\in\mathcal{A}}$ in function of $\widehat{H}$ and $|\psi\rangle$. They are assumed to be valued in $\mathbb{R}$ or $\mathbb{C}$, or in $\mathcal{L}(\mathcal{H})$, in which case $\geq$ and $>$ mean that the resulting operator is *positive-semidefinite*, respectively *positive-definite*. Evidently, these functions may be independent of some of these arguments.

In addition, the structure $\mathcal{S}_{\widehat{H}}^{|\psi\rangle}$, and the conditions expressing its dependency on $\widehat{H}$ and $|\psi\rangle$, have to be invariant under the unitary symmetries of $\mathcal{H}$. Violations of unitary symmetry would reflect the assumption of additional structures that break the symmetry. A symmetry breaking can appear for example by working in a particular basis and assuming it implicitly as preferred. To avoid such assumptions, $(\mathcal{C}_\beta)_{\beta\in\mathcal{B}}$ have to be *invariant functions*, which means that for any unitary operator $\widehat{S}$ on $\mathcal{H}$, any set of operators $\{\widehat{A}_\alpha^{|\psi\rangle}|\widehat{A}_\alpha^{|\psi\rangle}\in\mathcal{L}(\mathcal{H}),\alpha\in\mathcal{A}\}$, and any $\beta\in\mathcal{B}$,

$$\mathcal{C}_\beta\left(\left(\widehat{S}[\widehat{A}_\alpha]\right)_\alpha,\widehat{S}[\widehat{H}],\widehat{S}|\psi\rangle\right)=\widehat{S}\left[\mathcal{C}_\beta\left(\left(\widehat{A}_\alpha\right)_\alpha,\widehat{H},|\psi\rangle\right)\right], \tag{4}$$

where $\widehat{S}[\widehat{A}]=\widehat{S}\widehat{A}\widehat{S}^{-1}$ for operators, $\widehat{S}[|\psi\rangle]=\widehat{S}|\psi\rangle$ for vectors, and $\widehat{S}[c]=c$ for numbers.

If $\mathcal{C}_\beta$ is valued in $\mathbb{R}$ or $\mathbb{C}$, it is a *scalar* function, and eq. (4) becomes

$$\mathcal{C}_\beta\left(\left(\widehat{S}[\widehat{A}_\alpha]\right)_\alpha,\widehat{S}[\widehat{H}],\widehat{S}|\psi\rangle\right)=\mathcal{C}_\beta\left(\left(\widehat{A}_\alpha\right)_\alpha,\widehat{H},|\psi\rangle\right). \tag{5}$$

Definition 1. A *structure* $\mathcal{S}$ *of* kind $\mathcal{K}$ (or a $\mathcal{K}$*-structure*) for the Hamiltonian $\widehat{H}$ consists of

(i) a set of defining conditions $\mathcal{K}=\{\mathcal{C}_\beta\}_{\beta\in\mathcal{B}}$, called the *kind* of the structure $\mathcal{S}$, and

(ii) a set of operators $\mathcal{S}_{\widehat{H}}^{|\psi\rangle}=(\widehat{A}_\alpha^{|\psi\rangle})_{\alpha\in\mathcal{A}}$ which satisfy eq. (3) for $\{\mathcal{C}_\beta\}_{\beta\in\mathcal{B}}$.

Example 1 (Basis). A *basis* $(|\alpha\rangle)_{\alpha\in\mathcal{A}}$ of $\mathcal{H}$ can be defined by a $\mathcal{K}$-structure

$$\mathcal{S}_{\widehat{H}}^{|\psi\rangle}=\left(\widehat{A}_\alpha:=|\alpha\rangle\langle\alpha|\right)_{\alpha\in\mathcal{A}}, \tag{6}$$

where the kind $\mathcal{K}$ is given by the defining conditions, valid for all $\alpha,\alpha'\in\mathcal{A}$,

$$\begin{cases}\widehat{A}_\alpha^\dagger-\widehat{A}_\alpha=\widehat{0} & \text{(the operators } \widehat{A}_\alpha \text{ are Hermitian)},\\ \widehat{A}_\alpha\widehat{A}_{\alpha'}-\widehat{A}_\alpha\delta_{\alpha\alpha'}=\widehat{0} & \text{(the operators } \widehat{A}_\alpha \text{ are projectors)},\\ \widehat{I}_{\mathcal{H}}-\sum_{\alpha\in\mathcal{A}}\widehat{A}_\alpha=\widehat{0} & \text{(the projectors } \widehat{A}_\alpha \text{ partition the unity)},\\ \operatorname{tr}\widehat{A}_\alpha-1=0 & \text{(the projectors } \widehat{A}_\alpha \text{ have dimension} =1).\end{cases} \tag{7}$$

Since there are infinitely many bases, a preferred basis requires additional conditions. □

We should not require a structure that emerges from the `MQS` to be strictly unique, because there may be more such physically equivalent structures, for example those being related by phase or gauge transformations. If G_P is the symmetry group of all such transformations, its action on $\mathcal{H}$ preserves the rays of $\mathcal{H}$, so any element g of G_P acts as a unitary or anti-unitary operator $\widehat{g}$ on $\mathcal{H}$, as shown by Wigner and Bargmann [5, 6]. Let $\sim$ denote the *physical equivalence* relation on $\mathcal{H}$, so that $|\psi\rangle \sim |\psi'\rangle$ if and only if $|\psi'\rangle = \widehat{g}|\psi\rangle$ for some $g \in G_P$.

Definition 2. Two $\mathcal{K}$-structures $\mathcal{S}_{\widehat{H}}^{|\psi\rangle} = (\widehat{A}_\alpha^{|\psi\rangle})_{\alpha\in\mathcal{A}}$ and ${\mathcal{S}'}_{\widehat{H}}^{|\psi\rangle} = (\widehat{A'}_\alpha^{|\psi\rangle})_{\alpha\in\mathcal{A}}$ are called *physically equivalent*, $\mathcal{S}_{\widehat{H}}^{|\psi\rangle} \sim {\mathcal{S}'}_{\widehat{H}}^{|\psi\rangle}$, if there is an element $g \in G_P$,

$$\left(\widehat{A'}_\alpha^{|\psi\rangle}\right)_{\alpha\in\mathcal{A}} = \left(\widehat{g}\left[\widehat{A}_\alpha^{|\psi\rangle}\right]\right)_{\alpha\in\mathcal{A}}. \tag{8}$$

Otherwise, they are called *physically distinct*.

A $\mathcal{K}$-structure can be said to emerge uniquely or essentially uniquely from the `MQS` if it satisfies

Condition 1 (Uniqueness)**.** Any two $\mathcal{K}$-structures of the same kind $\mathcal{K}$ are physically equivalent.

Our emergent structures are also required to be *physically relevant*, *i.e.* able to distinguish some of the physically distinct states. For example, different physical states should be allowed to be in different relations with the 3D-space and with the tensor product structure, as they are in reality, because both $\psi(\mathbf{x}) = \langle\mathbf{x}|\psi\rangle$ and the reduced density matrices depend on the state.

We say that a structure $(\widehat{A}_\alpha)_{\alpha\in\mathcal{A}}$ *distinguishes* the states $|\psi\rangle$ and $|\psi'\rangle$ if there is an invariant scalar function $\mathcal{I}$ of $(\widehat{A}_\alpha)_{\alpha\in\mathcal{A}}$ and $|\psi\rangle \in \mathcal{H}$ so that

$$\mathcal{I}\left((\widehat{A}_\alpha)_{\alpha\in\mathcal{A}}, |\psi\rangle\right) \neq \mathcal{I}\left((\widehat{A}_\alpha)_{\alpha\in\mathcal{A}}, |\psi'\rangle\right). \tag{9}$$

An example of an invariant function of $\widehat{A}$ and $|\psi\rangle$ is the *mean value* $\langle\psi|\widehat{A}|\psi\rangle$. Therefore, if there is $\alpha \in \mathcal{A}$ so that $\langle\psi|\widehat{A}_\alpha|\psi\rangle \neq \langle\psi'|\widehat{A}_\alpha|\psi'\rangle$, the structure $(\widehat{A}_\alpha)_{\alpha\in\mathcal{A}}$ distinguishes $|\psi\rangle$ and $|\psi'\rangle$. In the case of the 3D-space, in NRQM, $\langle\psi|\widehat{\mathbf{x}}|\psi\rangle$ are such invariant functions. In general, they change in time, and they are also different for different states.

Proposition 1. A linear operator $\widehat{A}$ distinguishes two unit vectors $|\psi\rangle$ and $|\psi'\rangle$ if and only if there is no unitary operator $\widehat{S}$, $[\widehat{S}, \widehat{A}] = 0$, so that $|\psi'\rangle = \widehat{S}|\psi\rangle$.

Proof. Let $\widehat{S}$ be a unitary operator so that $|\psi'\rangle = \widehat{S}|\psi\rangle$. For any invariant scalar function $\mathcal{I}$,

$$\mathcal{I}\left(\widehat{A}, |\psi'\rangle\right) = \mathcal{I}\left(\widehat{A}, \widehat{S}|\psi\rangle\right) \overset{(5)}{=} \mathcal{I}\left(\widehat{S}^{-1}\widehat{A}\widehat{S}, |\psi\rangle\right). \tag{10}$$

If $[\widehat{S}, \widehat{A}] = 0$, $\widehat{S}^{-1}\widehat{A}\widehat{S} = \widehat{A}$ and eq. (9) cannot be true, because of eq. (10) becomes

$$\mathcal{I}\left(\widehat{A}, |\psi'\rangle\right) = \mathcal{I}\left(\widehat{A}, |\psi\rangle\right). \tag{11}$$

Also, if eq. (9) is true, there is no unitary operator $\widehat{S}$ so that $[\widehat{S}, \widehat{A}] = 0$ and $|\psi'\rangle = \widehat{S}|\psi\rangle$. □

Physical relevance should also take into account physical equivalence, as in Condition 1.

Condition 2 (Physical relevance)**.** For at least two physically distinct unit vectors $|\psi\rangle \nsim |\psi'\rangle \in \mathcal{H}$ not distinguished by the Hamiltonian, there is an invariant scalar function $\mathcal{I}$ of $\mathcal{S}_{\widehat{H}}^{|\psi\rangle}$ and $|\psi\rangle$ so that, for any $g, g' \in G_P$,

$$\mathcal{I}\left(\widehat{g}\left[\mathcal{S}_{\widehat{H}}^{|\psi\rangle}\right], |\psi\rangle\right) \neq \mathcal{I}\left(\widehat{g'}\left[\mathcal{S}_{\widehat{H}}^{|\psi'\rangle}\right], |\psi'\rangle\right). \tag{12}$$

Journal of Physics: Conference Series **2533** (2023) 012027 doi:10.1088/1742-6596/2533/1/012027

The role of $g, g' \in G_P$ in Condition 2 is to account for possible physical equivalences of structures. If the Hamiltonian already distinguishes the vectors $|\psi\rangle$ and $|\psi'\rangle \in \mathcal{H}$, they cannot witness if the structure $\mathcal{S}^{|\psi\rangle}_{\widehat{H}}$ is the one distinguishing them. But, most importantly, if the system changes in time from t_0 to t in a physically observable way, $|\psi(t_0)\rangle$ and $|\psi(t)\rangle = \widehat{U}_{t,t_0}|\psi(t_0)\rangle$ should be physically distinct, where

$$\widehat{U}_{t,t_0} := e^{-\frac{i}{\hbar}\widehat{H}(t-t_0)}. \tag{13}$$

The Hamiltonian does not distinguish $|\psi(t_0)\rangle$ and $|\psi(t)\rangle$, because their projections on each eigenspace of $\widehat{H}$ are equal. But interesting structures like the 3D-space or the factorization into subsystems should be able to distinguish them. They should also be able to distinguish $\widehat{g}|\psi(t_0)\rangle$ and $\widehat{g}'|\psi(t)\rangle$, for all $g, g' \in G_P$. This justifies the way Condition 2 is stated.

Now we prove the main result, that Conditions 1 and 2 cannot both be true.

Theorem 1. *Any physically relevant $\mathcal{K}$-structure is not unique.*

Proof. Figure 1 sketches the idea of the proof.

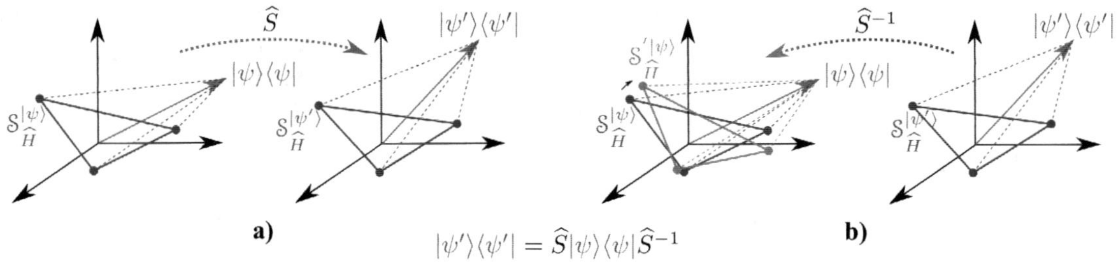

Figure 1. The idea of the proof in the vector space of linear operators (see Theorem 1).
a) Solid blue triangles represent the candidate preferred structures $\mathcal{S}^{|\psi\rangle}_{\widehat{H}}$ and $\mathcal{S}^{|\psi'\rangle}_{\widehat{H}}$. The vertices represent the linear operators $(\widehat{A}_\alpha)_{\alpha\in\mathcal{A}}$ and $(\widehat{A}'_\alpha)_{\alpha\in\mathcal{A}}$, in the vector space of linear operators on $\mathcal{H}$. The edges represent relations between these operators. The dashed blue lines represent the relations between these operators and the corresponding states, represented by $|\psi\rangle\langle\psi|$ and $|\psi'\rangle\langle\psi'|$. The Physical Relevance Condition 2 requires that there are state vectors $|\psi'\rangle \neq |\psi\rangle$ for which these relations are different.
b) The structure $\mathcal{S}'^{|\psi\rangle}_{\widehat{H}} = \widehat{S}^{-1}\left[\mathcal{S}^{\widehat{S}|\psi\rangle}_{\widehat{H}}\right]$ is of the same kind as the structure $\mathcal{S}^{|\psi\rangle}_{\widehat{H}}$. Since the relations between $\mathcal{S}'^{|\psi\rangle}_{\widehat{H}}$ and $|\psi\rangle$ are different from those between $\mathcal{S}^{|\psi\rangle}_{\widehat{H}}$ and $|\psi\rangle$, $\mathcal{S}^{|\psi\rangle}_{\widehat{H}}$ is not unique.

Let $\widehat{S}$ be a unitary operator so that $|\psi'\rangle = \widehat{S}|\psi\rangle$. Since $|\psi\rangle$ and $|\psi'\rangle$ are not distinguished by $\widehat{H}$, we can choose $\widehat{S}$ to commute with $\widehat{H}$ (Proposition 1).

Since the conditions defining the $\mathcal{K}$-structure $\mathcal{S}^{|\psi\rangle}_{\widehat{H}}$ are invariant to any unitary transformation $\widehat{S}$ that commutes with $\widehat{H}$, $\widehat{S}^{-1}$ transforms $\mathcal{S}^{\widehat{S}|\psi\rangle}_{\widehat{H}}$ into a $\mathcal{K}$-structure $\mathcal{S}'^{|\psi\rangle}_{\widehat{H}}$ for $|\psi\rangle$,

$$\mathcal{S}'^{|\psi\rangle}_{\widehat{H}} := \widehat{S}^{-1}\left[\mathcal{S}^{\widehat{S}|\psi\rangle}_{\widehat{H}}\right]. \tag{14}$$

From the uniqueness Condition 1 and from (14), there is a transformation $\widehat{g} \in G_P$ so that

$$\widehat{g}^{-1}\left[\mathcal{S}^{|\psi\rangle}_{\widehat{H}}\right] \overset{(8)}{=} \widehat{S}^{-1}\left[\mathcal{S}^{\widehat{S}|\psi\rangle}_{\widehat{H}}\right] \tag{15}$$

Journal of Physics: Conference Series **2533** (2023) 012027 doi:10.1088/1742-6596/2533/1/012027

Then, for any invariant function $\mathcal{I}$, equations (15) and (5) imply

$$\mathcal{I}\left(\widehat{g}^{-1}\left[\mathcal{S}_{\widehat{H}}^{|\psi\rangle}\right],|\psi\rangle\right) \stackrel{(15)}{=} \mathcal{I}\left(\widehat{S}^{-1}\left[\mathcal{S}_{\widehat{H}}^{\widehat{S}|\psi\rangle}\right],|\psi\rangle\right) \stackrel{(5)}{=} \mathcal{I}\left(\mathcal{S}_{\widehat{H}}^{\widehat{S}|\psi\rangle},\widehat{S}|\psi\rangle\right) = \mathcal{I}\left(\mathcal{S}_{\widehat{H}}^{\widehat{S}|\psi\rangle},|\psi'\rangle\right). \tag{16}$$

Since this contradicts Condition 2, Conditions 1 and 2 cannot both be true. □

Structures that depend on the MQS alone and are unique exist – any invariant function $f(\widehat{H},|\psi\rangle)$ leads to a unique such structure. In particular $\langle\psi|$, $|\psi\rangle\langle\psi|$, $\widehat{H}|\psi\rangle$, are structures obtained using such functions. They violate Condition 2, so Theorem 1 does not apply to them.

3. Implications of non-uniqueness for the emergence of various structures

Theorem 1, especially the fully general version from [3], shows that any research program that aims to uniquely recover any physically relevant structure from the minimal quantum structure (1) is doomed to fail. This includes the emergence of 3D-space and spacetime, the emergence of a factorization into subsystems, the emergence of a preferred basis, either for the entire universe or for a subsystem *etc.* This does not imply that such research programs are without value, only that the claim that physically relevant structures emerge uniquely is not true.

Example 2 (POVM). A *positive operator-valued measure* has the following defining conditions

$$\begin{cases} \widehat{E}_{\alpha_0}^{|\psi\rangle} \geq 0 & \text{(the operators } \widehat{E}_{\alpha_0}^{|\psi\rangle} \text{ are } \textit{positive semi-definite}\text{)}, \\ \widehat{I}_{\mathcal{H}} - \sum_\alpha \widehat{E}_\alpha^{|\psi\rangle} = \widehat{0} & \text{(the operators } (\widehat{E}_\alpha^{|\psi\rangle})_{\alpha\in\mathcal{A}} \text{ form a } \textit{resolution of the identity}\text{)}. \end{cases} \tag{17}$$

Conditions (17) are invariant. We can add more invariant conditions, as needed, for instance the conditions (7) from Example 1 can be seen as a refinement of the conditions (17). No matter what other conditions we add, the resulting $\mathcal{K}$-structure will be subject to Theorem 1. □

Example 3 (TPS). A *tensor product structure* of $\mathcal{H}$ is defined by a unitary isomorphism

$$\bigotimes_{\varepsilon\in\mathcal{E}}\mathcal{H}_\varepsilon \mapsto \mathcal{H}, \tag{18}$$

where $\mathcal{H}_\varepsilon$ are Hilbert spaces. It is unique up to local unitary transformations of each $\mathcal{H}_\varepsilon$ and permutations of the factors. A TPS can also be defined in terms of linear operators from the algebra of linear operators on $\mathcal{H}$, $\mathcal{L}(\mathcal{H})$ [4], because $\mathcal{L}(\mathcal{H})$ is spanned by the subalgebras

$$\mathcal{L}_\varepsilon := \left(\bigotimes_{\varepsilon'\neq\varepsilon}\widehat{I}_{\varepsilon'}\right) \otimes \mathcal{L}(\mathcal{H}_\varepsilon). \tag{19}$$

Therefore, $\mathcal{L}(\mathcal{H})$ also recovers the TPS (18).

So we can start with subalgebras of $\mathcal{L}(\mathcal{H})$ that satisfy the defining conditions

(i) they commute with one another, and

(ii) they span $\mathcal{L}(\mathcal{H})$, and they define a TPS [4].

Therefore, we can characterize the TPS as a $\mathcal{K}$-structure, and apply Theorem 1. □

Example 4 (Local TPS). Cotler *et al* extended the TPS with additional defining conditions of locality, in the sense that the interactions between subsystems encoded in the Hamiltonian take place only between a small number $d \in \mathbb{N}$ of subsystems [7]. If such a local TPS exists for a given value of d, it usually is unique up to local unitary transformations (as in Example 3). However, Theorem 1 shows that any method to obtain a local TPS from the Hamiltonian (or its spectrum, as in [7]) violates either Condition 1 or Condition 2. If the TPS is assumed to be physically relevant (as it is in reality, because the reduced density matrix should not be the same for all states), it follows that, for any value of d, the local TPS cannot be unique. □

Journal of Physics: Conference Series **2533** (2023) 012027 doi:10.1088/1742-6596/2533/1/012027

Example 5 ("Mad-dog Everettianism" – Emergent 3D-space). Carroll and Singh apply the result of Cotler *et al* from [7] to recover, from the Hamiltonian's spectrum, a graph $\mathcal{E}$ that encodes the topology of the 3D-space [1, 2]. At this stage distances are not defined yet. But since the reduced density matrix of $|\psi\rangle$ on a region of space $R \subset \mathcal{E}$ has to be physically relevant, so does the topology itself, so the resulting 3D-space cannot be unique, by Theorem 1.

Then, on the resulting manifold, they define a notion of distance in terms of the *mutual information* between two regions $R, R' \subset \mathcal{E}$, $I(R : R') = S_R + S_{R'} - S_{RR'}$, where $S_R = -\operatorname{tr}\widehat{\rho}_R \ln \widehat{\rho}_R$ is the von Neumann entropy of $|\psi\rangle$ in the region R. They take the distance as decreasing as the mutual information is increasing. But since when such $\mathcal{E}$ exists it cannot be unique (see Example 4), the resulting 3D-space cannot be unique.

If, on the other hand, we give up physical relevance (Condition 2) for $\mathcal{E}$ and assume it to be unique, the reduced density matrix for any region R is the same in time, because unitary evolution commutes with $\widehat{H}$ and preserves $\mathcal{E}$. This does not correspond to observations. □

Example 6 (General emergent 3D-space). Let us start from a known 3D-space, which may be $\Sigma = \mathbb{R}^3$ or a more general manifold (as in *Canonical Quantum Gravity* [8]) or even a discrete graph or hypergraph (as in other approaches to Quantum Gravity like *causal sets* [9], *Regge calculus* [10], *causal dynamical triangulations* [11], *Loop Quantum Gravity* [12, 13]). Let $\mathcal{P}$ denote the set of all types of particles, and $\mathcal{A} := \mathcal{P} \times \Sigma$. For each pair $(P, \boldsymbol{x}) \in \mathcal{A}$, define

$$\widehat{A}_{(P,\boldsymbol{x})} := \widehat{N}_P(\boldsymbol{x}) = \widehat{a}_P^\dagger(\boldsymbol{x})\widehat{a}_P(\boldsymbol{x}), \tag{20}$$

where the operator $\widehat{a}_P^\dagger(\boldsymbol{x})$ creates a particle of type P at the 3D-point $\boldsymbol{x} \in \Sigma$, $\widehat{a}_P^\dagger(\boldsymbol{x})|0\rangle = |\boldsymbol{x}\rangle_P$, $|0\rangle$ is the vacuum state, and $\widehat{N}_P(\boldsymbol{x})$ the type P *particle number operator* at $\boldsymbol{x}$. The operators $\widehat{A}_{(P,\boldsymbol{x})}$ commute with one another in both the bosonic and the fermionic case, because they consist of products of pairs of (anti)commuting operators.

Therefore, if we do not know the 3D-space Σ and want to recover it, we can try to do this by recovering commuting operators like in eq. (20). They are required to be Hermitian and to commute with one another, but we can add as many additional defining conditions as we need to describe the properties of 3D-space or of type P particle number operators. In all cases, Theorem 1 applies. Since the 3D-space has to be physically relevant, it cannot be unique. □

Example 7 (Environmental decoherence). It is said that environmental decoherence leads to the emergence of a preferred basis of a subsystem, which presumably solves the measurement problem. This of course is claimed in the context of an interpretation in which there is no collapse, like *Everett's interpretation* [14] or the *pilot-wave theory* [15]. Let us consider a decomposition

$$\mathcal{H} = \mathcal{H}_S \otimes \mathcal{H}_E, \tag{21}$$

where $\mathcal{H}_S$ is the Hilbert spaces of the subsystem and $\mathcal{H}_E$ is that of the environment. The reduced density operator of the subsystem, $\widehat{\rho}_S := \operatorname{tr}_E |\psi\rangle\langle\psi|$, evolves under continuous environmental monitoring until it becomes approximately diagonal.

From Theorem 1, there are more physically distinct factorizations as in eq. (21). If the subsystem and the environment are in separate states before decoherence diagonalizes it,

$$|\psi\rangle = |\psi_S\rangle \otimes |\psi_E\rangle, \tag{22}$$

the possible factorizations (21) are limited. But can this factorization be unique?

For a given TPS we can impose condition (21) by requiring the von Neumann entropy $\widehat{\rho}_S = \operatorname{tr}|\psi\rangle\langle\psi|$ of the partial state to vanish, $-\operatorname{tr}(\rho_S \ln \rho_s) = 0$. But if a TPS with a separable state (21) exists, it can be chosen in infinitely many physically distinct ways. Therefore, even if environmental decoherence happens and explains the measurement problem, it cannot also lead to the emergence of the preferred basis of the subsystem. □

Journal of Physics: Conference Series **2533** (2023) 012027 doi:10.1088/1742-6596/2533/1/012027

Ref. [3] contains more details and additional examples in which various structures are claimed to emerge uniquely from the MQS, a claim contradicted by Theorem 1.

These results, showing that whenever a preferred structure exists, if it is physically relevant, it is not unique, are negative and abstract. From a pedagogical point of view, it may be useful to have some positive counterexamples that can be constructed explicitly, and they exist:

Example 8 (Versatility of translations). Consider the following minimal quantum structure

$$\left(\mathcal{H}, \widehat{H} = -i\hbar\frac{\partial}{\partial\tau}, |\psi(t)\rangle\right), \tag{23}$$

where $|\psi(t)\rangle$ is an eigenstate of the Hamiltonian's conjugate operator $\widehat{\tau}$. In [16] it was shown that there are infinitely many examples of MQS of the form (23) having distinct space structures and factorizations. Three infinite lists of explicit examples are given in [16]:

1. Any deterministic time-reversible dynamical system admits a *quantum representation* (the *Koopman–von Neumann representation* [17, 18]. If it does not have time loops, its representation is of the form (23). There are infinitely many such dynamical systems with distinct choices of space and of the decomposition into subsystems, whose quantum representations are physically distinct quantum systems but have the same MQS (23).
2. Any quantum system realizing the *standard model of ideal measurements* (as for example in [19] §II.3.4 and [20] §2.2).
3. Any closed quantum system which includes an isolated clock or a sterile massless fermion in a certain eigenstate, in general a subsystem with the Hamiltonian $\widehat{H}_C = -i\hbar\frac{\partial}{\partial\tau}$.

In [21] it was shown that if a closed quantum system (*e.g.* the universe) evolves irreversibly, so that the amplitude to return to a previous state vanishes, it has to be of the form (23). This suggests that our own universe may be of this kind, which is consistent with the Page-Wootters formalism [22] for *canonical quantum gravity* [8]. □

4. Consequences for the foundations of Quantum Mechanics

Consequence 1 (Non-uniqueness of emergent structures). The main implication of Theorem 1 is the direct one, that the MQS $(\mathcal{H}, \widehat{H}, |\psi\rangle)$ is not sufficient to derive all interesting structures in physics. □

The non-uniqueness results also imply that the Thesis HSF has strange consequences:

Consequence 2 ("Travel" in time and in alternative realities). Since the factorization and the 3D-space are not determined uniquely by the MQS, Thesis HSF implies that the current state vector supports not only our present physical state, but also any state that can be obtained by a symmetry transformation as in the proof of Theorem 1. This can be the state of an alternative history, or any state from any past or future moment, since we can take $\widehat{S} = \widehat{U}_{t,t_0}$ as in eq. (13). Since there is no unique physical interpretation of the state vector, a mere passive transformation suffices to "travel" from one state to another. Even if intelligent beings populating a physical state cannot interact with those populating the state obtained by a passive transformation, this property seems at least bizarre. □

Remark 1. Consequences 1 and 2 do not imply that emergentist programs are flawed or useless. By contrast, they may be necessary, because the fundamental structures from Quantum Theory are usually imported from the classical theory that is quantized, but maybe the correct theory, particularly that includes gravity, consists of different fundamental structures. They just do not lead to unique emergent structures, but it is important to explore the relations between these structures and the Hamiltonian, and then use what we learn to extend the MQS.

Consequence 3 (Interpretations of QM). Theorem 1 implies that interpretations of QM that rely only on $|\psi\rangle$ as an abstract state vector cannot result in unique structures like 3D-space or factorization into subsystems. An example is "Mad-dog Everettianism" [2].

Journal of Physics: Conference Series **2533** (2023) 012027 doi:10.1088/1742-6596/2533/1/012027

– By contrast, Everettianism based on the wavefunction, which is the state vector expressed in the position basis, does not have this problem [23]. In general, there should preexist a fundamental factorization into subsystems and 3D-space structure, or at least other structures additional to the `MQS` from which they can be obtained.

– Bohr's interpretation, which assumes the existence of a classical measuring device, introduces by this the classical positions and momenta, and transfers them to the observed system.

– The *Pilot-Wave Theory* [15] extends the `MQS` with a preferred 3D-space and point-particles with definite positions in the 3D-space.

– In *Objective Collapse Theories* [24] the position in space plays a central role, since the wavefunction collapses spontaneously around a position in the configuration space.

– Another possibility is to break the unitary symmetry of the space of solutions, instead of the symmetry of the Hamiltonian. Some interpretations already do this, by admitting only a subset of state vectors from the Hilbert space as representing physical states. These include proposals made by Schulman [25–27], 't Hooft [28], and Stoica [29–31], with the purpose of having unitary symmetry without collapse *and* without branching into more worlds. An apparently unrelated result shows that most state vectors in the Hilbert space lead to states in which the records incorrectly represent the past history of the system [32]. These "fake records" are similar to the *Boltzmann brain* situations [26, 33]. Therefore, there are already indications that not all unit vectors from the Hilbert space represent physical states.

Which interpretation describes the real world properly is open for debate. Also, what fundamental structures are required to ensure all of the structures that are needed is an open problem. But the `MQS` $(\mathcal{H}, \widehat{H}, |\psi\rangle)$ is not sufficient to determine them uniquely.

Alternatively, we can embrace non-uniqueness and assume that there are as many 3D-space structures and factorizations as allowed by the symmetries of the `MQS`. Any interpretation that takes this step will have to admit infinitely many physically distinct worlds supported by the same state vector. The many-worlds interpretation ("mad-dog" version) will, for example, become a many-many-worlds interpretation. □

Acknowledgement. The author thanks Louis Marchildon, Tim Maudlin, Heinrich Päs, Daniel Ranard, Igor Salom, Jochen Szangolies, Iulian Toader, Lev Vaidman, Howard Wiseman, and the referees, for feedback and stimulating exchanges.

References

[1] Carroll S 2021 Reality as a vector in Hilbert space Tech. Rep. CALT-TH-2021-010 Cal-Tech
[2] Carroll S and Singh A 2019 *What is Fundamental?* ed Aguirre A, Foster B and Merali Z (Springer) pp 95–104
[3] Stoica O 2021 *Preprint arXiv:2102.08620* 1–18
[4] Zanardi P, Lidar D and Lloyd S 2004 *Phys. Rev. Lett.* **92** 060402
[5] Wigner E 1959 *Group Theory and its Application to the Quantum Mechanics of Atomic Spectra* (New York: Academic Press)
[6] Bargmann V 1954 *Ann. Math.* **59** 1–46
[7] Cotler J, Penington G and Ranard D 2019 *Comm. Math. Phys.* **368** 1267–1296
[8] DeWitt B 1967 *Phys. Rev.* **160** 1113
[9] Sorkin R 1990 *Relativity and gravitation: Classical and quantum* 150–173
[10] Regge T 1961 *Il Nuovo Cimento (1955-1965)* **19** 558–571
[11] Loll R 2019 *Class. Quant. Grav.* **37** 013002
[12] Rovelli C and Smolin L 1995 *Phy. Rev. D* **52** 5743

Journal of Physics: Conference Series **2533** (2023) 012027 doi:10.1088/1742-6596/2533/1/012027

[13] Ashtekar A and Bianchi E 2021 *Rep. Prog. Phys.* **84** 042001

[14] Everett H 1973 *The Many-Worlds Hypothesis of Quantum Mechanics* (Princeton University Press) pp 3–137

[15] Bohm D 1952 *Phys. Rev.* **85** 166–193

[16] Stoica O 2022 *Preprint arXiv:2204.01426*

[17] Koopman B 1931 *Proc. Nat. Acad. Sci. U.S.A.* **17** 315

[18] v Neumann J 1932 *Ann. Math.* 587–642

[19] Busch P, Grabowski M and Lahti P 1995 *Operational quantum physics* (Berlin, Heidelberg: Springer)

[20] Mittelstaedt P 2004 *The interpretation of quantum mechanics and the measurement process* (Cambridge, UK: Cambridge University Press)

[21] Stoica O 2022 *Preprint arXiv:2204.02270*

[22] Page D and Wootters W 1983 *Phys. Rev. D* **27** 2885

[23] Vaidman L 2016 *J. Phys. Conf. Ser.* vol 701 p 012020

[24] Ghirardi G, Rimini A and Weber T 1986 *Phys. Rev. D* **34** 470–491

[25] Schulman L 1984 *Phys. Lett. A* **102** 396–400

[26] Schulman L 1997 *Time's arrows and quantum measurement* (Cambridge University Press)

[27] Schulman L 2017 *Entropy* **19** 343

[28] 't Hooft G 2016 *The cellular automaton interpretation of quantum mechanics* vol 185 (Springer)

[29] Stoica O 2013 *EPJ Web of Conferences, TM 2012 – The Time Machine Factory [unspeakable, speakable] on Time Travel in Turin* vol 58 (EPJ Web of Conferences) p 01017 Open Access URL `http://dx.doi.org/10.1051/epjconf/20135801017`

[30] Stoica O 2017 *Quantum Stud. Math. Found.* arXiv:1607.02076

[31] Stoica O 2021 *Quantum Stud. Math. Found.* **8** arXiv:1903.07078

[32] Stoica O 2022 *Preprint arXiv:2209.13275*

[33] Albrecht A and Sorbo L 2004 *Phys. Rev. D* **70** 063528

Journal of Physics: Conference Series **2533** (2023) 012028 doi:10.1088/1742-6596/2533/1/012028

Poincaré Breaking and Gauge Invariance: A Road to Emergent Gravity and New Particles

Durmuş Demir
Sabancı University, Faculty of Engineering and Natural Sciences, 34956 İstanbul, Turkey
E-mail: durmus.demir@sabanciuniv.edu

Abstract. In this talk, we discuss how gauge symmetries broken explicitly by a Poincare-breaking UV cutoff can be restored. We show that gauge symmetries can be restored by the introduction of affine curvature in reminiscence to the Higgs field. In fact, gauge symmetries get restored and general relativity emerges at the extremum of the metric-affine action. As per this point, we show emergence of the general relativity, reveal how its parameters relate to the flat spacetime loops, elucidate the new particle spectrum it brings along, and discuss its salient signatures. We show that the resulting field-theoretic plus gravitational setup can be probed via various phenomena ranging from collider experiments to black holes.

1. Introduction
There are scales in nature that are not particle masses. The confinement scale in QCD and the fundamental scale of gravity are such scales. In general, if the UV cutoff $\Lambda_\wp$ is not the mass of an elementary particle then it breaks Poincare symmetry as it is not a Casimir invariant of the Poincare group [1]. The simplest UV cutoff as such is the Lorentz-invariant but translation-breaking cut $-\Lambda_\wp^2 \leq \ell_\mu \ell^\mu \leq \Lambda_\wp^2$ on the loop momenta ℓ_μ. Under this cut, the effective QFT remains Lorentz-invariant but breaks explicitly all the gauge symmetries since each gauge boson acquires a mass proportional to $\Lambda_\wp$ [2, 3, 4, 5]. In addition, scalar mass-squareds (vacuum energy) develop quadratic (quadratic and quartic) sensitivity to the UV cutoff $\Lambda_\wp$ [6, 7].

In this talk, we will discuss how gauge symmetries broken explicitly by a Poincare-breaking UV cutoff can be restored by a mechanism reminiscent of the usual Higgs mechanism. Indeed, gauge symmetries associated with a massive vector boson are restored (or realized) by the introduction of the Higgs field [8, 9]. In reminiscent of this, we find that gauge symmetries broken explicitly by a Poincare-breaking UV cutoff can be restored by the introduction of the affine curvature [10, 11, 12]. In the former, what is central is the Higgs potential. In the latter, however, the central object iss the metric-affine action [13, 14, 15, 16].

2. Effective QFT
For the purpose of analyzing the power-law quantum divergences in isolation, it is desirable to detach them from the logarithmic divergences in terms of the scales they involve. This scheme of regularization, the so-called detached regularization [17], extends the usual dimensional regularization [18, 19, 20] to QFTs with a UV cutoff. In detached regularization, the effective action capturing physics of the full quantum action at low energies takes the form [17]

$$S_{eff}[\eta, F; \Lambda_\wp^2, \log\mu] = S_{tree}[\eta, F] + \delta S_{log}[\eta, \log\mu, F] + \delta S_{pow}[\eta, F; \Lambda_\wp^2, \log\mu] \quad (1)$$

Journal of Physics: Conference Series **2533** (2023) 012028 doi:10.1088/1742-6596/2533/1/012028

in which $S_{tree}[\eta, F]$ stands for the tree-level QFT action and $\delta S_{log}[\eta, \log\mu, F]$ collects the logarithmic loop corrections such that both actions are gauge-invariant and independent of the UV cutoff $\Lambda_\wp$. The last piece

$$\delta S_{pow}[\eta, F; \Lambda_\wp^2, \log\mu] = \int d^4x\sqrt{-\eta}\left\{-c_O\Lambda_\wp^4 - \mathscr{M}^2\Lambda_\wp^2 - c_\phi\Lambda_\wp^2\phi^\dagger\phi + c_V\Lambda_\wp^2 \mathrm{tr}\left[V_\mu V^\mu\right]\right\} \quad (2)$$

comprises the power-law corrections. The loop factors c_ϕ and c_V depend on the QFT under concern but c_O and $\mathscr{M}^2$ are rather general at one loop [17]

$$c_O = c_O(\log\mu) \xrightarrow{\text{one loop}} \frac{(n_b - n_f)}{64\pi^2}, \quad (3)$$

$$\mathscr{M}^2 = \mathscr{M}^2(\log\mu) \xrightarrow{\text{one loop}} -\frac{1}{64\pi^2}\mathrm{str}\left[M^2\log\frac{M^2}{\mu^2}\right], \quad (4)$$

in which $n_b(n_f)$ is the total number of bosons (fermions) and M^2 is the mass-squared matrix of the QFT fields. (Details of the detached regularization and its applications can be found in [17].)

3. Effective QFT in Curved Spacetime

QFTs are specific to the flat spacetime due to the necessity of Poincare symmetry for defining the notion of particle [21, 22]. They cannot be carried into curved spacetime because curved spacetime does not allow special states like the vacuum and detectable structures like the particles [23, 24].

Effective QFTs are different. They are the QFTs of long-wavelength quantum fields. They are essent,ally the classical theory improved with quantum corrections order by order in perturbation theory. In this regard, they are like the classical field theories and have therefore a natural affinity with the classical curved spacetime. In other words, it should be natural to carry effective QFTs into curved spacetime [11, 12]. In this regard, it is necessary to carry first the flat spacetime effective QFT (with metric $\eta_{\mu\nu}$ and the partial derivative ∂_μ) to curved spacetime (with metric $g_{\mu\nu}$ and covariant derivative ∇_μ of the Levi-Civita connection ${}^g\Gamma^\lambda_{\mu\nu}$). But, for the metric $g_{\mu\nu}$ to be curved, the effective QFT in curved spacetime must involve curvature of $g_{\mu\nu}$ (like, for example, the Ricci curvature $R_{\mu\nu}({}^g\Gamma)$). It can be tempting to add the requisite curvature terms by hand (Einstein-Hilbert term $M_0^2 g^{\mu\nu}R_{\mu\nu}({}^g\Gamma)$ plus higher-curvature terms) but extension of the effective QFT by such bare terms is fundamentally inconsistent because they come to mean that the curvature sector was left unrenormalized while the QFT sector was renormalized. Similar arguments hold also for curvature terms constructed with the bare QFT parameters in $S_{tree}[\eta, F]$. In the face of these inconsistencies, one concludes that curvature must arise only in the loop-induced terms in the effective QFT [12]. To this end, definition of the Ricci curvature $[\nabla_\lambda, \nabla_\nu]V^\lambda = R_{\mu\nu}({}^g\Gamma)V^\mu$ reveals that the effective QFT in curved spacetime can develop curvature only in the gauge sector. But the gauge kinetic term $V_{\mu\nu}V^{\mu\nu}$ (with the field strength tensor $V_{\mu\nu} = \mathcal{D}_\mu V_\nu - \mathcal{D}_\nu V_\mu$ and gauge covariant derivative $\mathcal{D}_\mu$) can generate two-derivative structures like $[\nabla_\lambda, \nabla_\nu]V^\lambda$ via only by-parts integration. Indeed, $V_{\mu\nu}V^{\mu\nu}$ can be brought into the two-derivative form $V_\mu(\mathcal{D}^2)^{\mu\nu}V_\nu$ by applying by-parts integration. But by-parts integration cannot be a real source of curvature since after all $V_\mu(\mathcal{D}^2)^{\mu\nu}V_\nu$ is equivalent to $V_{\mu\nu}V^{\mu\nu}$. This observation implies that curvature must arise while passing from flat spacetime to the curved spacetime. In this regard, one notices that the kinetic construct [12]

$$I_V[\eta] = \int d^4x\sqrt{-\eta}\frac{1}{2}\mathrm{tr}[V_{\mu\nu}V^{\mu\nu}] \quad (5)$$

Journal of Physics: Conference Series **2533** (2023) 012028 doi:10.1088/1742-6596/2533/1/012028

is equivalent to

$$\bar{I}_V[\eta] \;=\; \int d^4x\sqrt{-\eta}\,\mathrm{tr}\big[V^\mu(-D^2\eta_{\mu\nu}+D_\mu D_\nu+iV_{\mu\nu})V^\nu+\partial_\mu(V_\nu V^{\mu\nu})\big] \tag{6}$$

under by-parts integration, with the flat spacetime gauge covariant derivative D_μ. Their difference vanishes identically

$$\Delta I_V[\eta] = -I_V[\eta]+\bar{I}_V[\eta] = 0 \tag{7}$$

with the general boundary term $\partial_\mu(V_\nu V^{\mu\nu})$ in $\bar{I}_V[\eta]$. But this very vanishing difference becomes non-vanishing

$$\Delta I_V[g] = -I_V[g]+\bar{I}_V[g] = -\int d^4x\sqrt{-g}\,\mathrm{tr}[V^\mu R_{\mu\nu}({}^g\Gamma)V^\nu] \tag{8}$$

when $I_V[\eta]$ and $\bar{I}_V[\eta]$ are taken to curved spacetime via the general covariance ($\eta_{\mu\nu}\to g_{\mu\nu}$ and $\partial_\mu\to\nabla_\mu$ so that $D_\mu\to\mathcal{D}_\mu$). The reason for this non-vanishing $\Delta I_V[g]$ is the aforementioned commutator $[\mathcal{D}_\lambda,\mathcal{D}_\nu]V^\lambda = (R_{\mu\nu}({}^g\Gamma)+iV_{\mu\nu})V^\mu$ in curved spacetime.

It proves useful to combine ΔI_V with the loop-induced gauge boson mass term

$$\delta S_V[\eta,\Lambda_\wp^2] = \int d^4x\sqrt{-\eta}\,c_V\Lambda_\wp^2\eta_{\mu\nu}\mathrm{tr}\,[V^\mu V^\nu] \tag{9}$$

in the power-law corrections in (2). It is clear that $\delta S_V[\eta,\Lambda_\wp^2]$ can be rewritten as

$$\delta\overline{S}_V[\eta,\Lambda_\wp^2] = \delta S_V[\eta,\Lambda_\wp^2]+c_V\Delta I_V[\eta] = S_V[\eta,\Lambda_\wp^2] \tag{10}$$

thanks to the fact that you $\Delta I_V[\eta]=0$ in flat spacetime. This identity undergoes a nontrivial change [11, 12]

$$\delta\overline{S}_V[g,\Lambda_\wp^2,R] = \delta S_V[g,\Lambda_\wp^2]+c_V\Delta I_V[g] = \int d^4x\sqrt{-g}\,c_V\mathrm{tr}\left[V^\mu\left(\Lambda_\wp^2 g_{\mu\nu}-R_{\mu\nu}({}^g\Gamma)\right)V^\nu\right] \tag{11}$$

in curved spacetime in which $\Delta I_V[g]\neq 0$ as given in the relation (8). It is worth emphasizing that $\delta\overline{S}_V[g,\Lambda_\wp^2,R]$ is the only piece that explicitly involves curvature in the entire effective QFT in curved spacetime. In other words, without $\delta\overline{S}_V[g,\Lambda_\wp^2,R]$ (more precisely $\Delta I_V[g]$), the metric $g_{\mu\nu}$ would remain flat as it would have no kinetic term at all. As a result, the effective QFT action takes the form

$$S_{eff}[g,F;\Lambda_\wp^2,\log\mu,R] = S_{tree}[g,F]+\delta S_{log}[g,\log\mu,F]+\delta\overline{S}_{pow}[\eta,F;\Lambda_\wp^2,\log\mu,R] \tag{12}$$

in the curved spacetime of the metric $g_{\mu\nu}$. Its power-law part

$$\begin{aligned}\delta\overline{S}_{pow}[g,F;\Lambda_\wp^2,\log\mu,R] = \int d^4x\sqrt{-g}\Big\{ &-\; c_O\Lambda_\wp^4-\mathscr{M}^2\Lambda_\wp^2-c_\phi\Lambda_\wp^2\phi^\dagger\phi\\ &+\; c_V\mathrm{tr}\left[V^\mu\left(\Lambda_\wp^2 g_{\mu\nu}-R_{\mu\nu}({}^g\Gamma)\right)V^\nu\right]\Big\}\end{aligned} \tag{13}$$

is obtained by taking the power-law piece in (2) to curved spacetime via the relation (8).

It is clear that the entire effective QFT in curved spacetime (12) develops a curvature-dependent term only in the gauge sector as in (13). There is no Einstein-Hilbert term and there is thus no gravitation at all. It is also clear that this curvature-dependent term breaks gauge symmetries explicitly along with the loop-induced mass $c_V\Lambda_\wp^2$ [11, 10]. In view of these features, it is necessary to first determine how the GR emerges from within the effective action (12). Besides, it is necessary to also determine if emergence of the GR can kill the gauge symmetry breaking terms. These two points will be clarified below in the framework of affine gravity with distant inspiration from the Higgs mechanism.

Journal of Physics: Conference Series **2533** (2023) 012028 doi:10.1088/1742-6596/2533/1/012028

4. Reminiscencing the Higgs Field: Affine Curvature

It is necessary to distinguish between two cases: A vector boson of mass M_V and a gauge boson of anomalous loop-induced mass (proportional to $\Lambda_\wp$). Firstly, mass M_V of a vector boson respects the Poincare symmetry simply because it is a Casimir invariant of the Poincare group [1, 9]. The loop-induced mass of a gauge boson, on the other hand, is a Poincare-breaking parameter since the cutoff $\Lambda_\wp$ itself breaks the Poincare symmetry [3]. Secondly, in the case of a massive vector boson, the goal is to restructure the vector field as a gauge field. In the case of a gauge boson with loop-induced mass, however, the goal is to kill the anomalous mass. Thirdly, being a Poincare-invariant quantity, M_V can be promoted to a suitable Higgs scalar Φ_V via

$$\Lambda_\wp^2 \mathrm{Tr}\left[V_\mu V^\mu\right] \longmapsto \Phi_V^\dagger V_\mu V^\mu \Phi_V \tag{14}$$

as in the Higgs mechanism [8, 9]. For a loop-induced gauge boson mass, however, scalar fields like the Higgs field cannot serve the purpose [8, 9] since the requisite field must be a Poincare-breaking one. In search for this Poincare-breaking field, one notes that, in a general second-quantized quantum field theory with no presumed symmetries, the Poincare (translation) invariance emerges if the Poincare-breaking terms are identified with the spacetime curvature [25]. Physically, this means that the Poincare-breaking sources in a QFT are the spots where curvature can emerge. It is all clear that the UV cutoff $\Lambda_\wp$ is the said Poincare-breaking source in an effective QFT. There is thus every reason to conclude that the sought-for Poincare-breaking field should be the spacetime curvature itself. In this regard, one notices that, in the effective action in (13), promotion of $\Lambda_\wp^2 g_{\mu\nu}$ to the Ricci curvature $R_{\mu\nu}({}^g\Gamma)$ as a Poincare-breaking spurion would completely eradicate the gauge symmetry breaking term $c_V \mathrm{tr}\left[V^\mu\left(\Lambda_\wp^2 g_{\mu\nu} - R_{\mu\nu}({}^g\Gamma)\right)V^\nu\right]$. But this promotion is inherently inconsistent simply because $R_{\mu\nu}({}^g\Gamma)$ does vanish in the flat spacetime limit while $\Lambda_\wp^2 g_{\mu\nu}$ does not. One way to resolve this inconsistency is to promote $\Lambda_\wp^2 g_{\mu\nu}$ to a curvature tensor that does not vanish in the flat spacetime limit. The simplest curvature as such is the affine Ricci curvature [13, 14, 15]

$$\mathbb{R}_{\mu\nu}(\Gamma) = \partial_\lambda \Gamma^\lambda_{\mu\nu} - \partial_\nu \Gamma^\lambda_{\mu\lambda} + \Gamma^\rho_{\lambda\rho}\Gamma^\lambda_{\mu\nu} - \Gamma^\lambda_{\rho\nu}\Gamma^\rho_{\mu\lambda} \tag{15}$$

of an affine connection $\Gamma^\lambda_{\mu\nu}$, which is a general connection independent of the metric tensor $g_{\mu\nu}$ and its Levi-Civita connection ${}^g\Gamma^\lambda_{\mu\nu}$. Needless to say, $\Gamma^\lambda_{\mu\nu}$ is not a tensor in curved spacetime but it transforms as a tensor in the flat spacetime in which coordinate transformations are the linear Lorentz transformations [26]. Now, having the affine curvature $\mathbb{R}_{\mu\nu}(\Gamma)$ at hand, the UV cutoff $\Lambda_\wp$ can be promoted to the affine curvature as

$$\Lambda_\wp^2 g_{\mu\nu} \longmapsto \mathbb{R}_{\mu\nu}(\Gamma) \tag{16}$$

in the same philosophy as the equation (14) in which $\Lambda_\wp$ is promoted to the Higgs scalar Φ_V. In essence, equation (16) is a map from UV cutoff to affine curvature, and possesses three key properties:

(i) It reduces to $\Lambda_\wp^2 \eta_{\mu\nu} \mapsto \mathbb{R}_{\mu\nu}(\Gamma)$ in the flat spacetime in which the affine connection acts as a tensor field [26],

(ii) It is meant to hold at the fundamental scale of gravity $M_{Pl} \equiv (8\pi G_N)^{-1/2}$, where G_N is Newton's gravitational constant.

(iii) It rests on the fact that both the affine curvature $\mathbb{R}_{\mu\nu}(\Gamma)$ and metric $g_{\mu\nu}$ are classical fields since effective QFTs like (12) cannot be extended with new quantum fields.

Then, the cutoff-to-curvature map (16), with these three key properties, takes the power-law correction action (13) to a metric-Palatini (metric-affine) action [15, 16]

$$\delta\overline{S}_{pow}[g, \digamma; \mathbb{R}, \log\mu, R] = \int d^4x\sqrt{-g}\Big\{-\frac{M_{Pl}^2}{2}\mathbb{R}(g,\Gamma) - \frac{\overline{c}_O}{16}\left(\mathbb{R}(g,\Gamma)\right)^2 - \frac{c_\phi}{4}\phi^\dagger\phi\,\mathbb{R}(g,\Gamma) + c_V \mathrm{tr}\left[V^\mu(\mathbb{R}_{\mu\nu}(\Gamma) - R_{\mu\nu}({}^g\Gamma))V^\nu\right]\Big\} \quad (17)$$

in which $\mathbb{R}(g,\Gamma) \equiv g^{\mu\nu}\mathbb{R}_{\mu\nu}(\Gamma)$ is the affine curvature scalar. The parameters in this action deserve a detailed discussion: First, as follows from the equation (4), the fundamental scale of gravity (the Planck scale) can be defined as

$$M_{Pl}^2 = \frac{1}{2}\mathscr{M}^2(\mu = M_{Pl}) \xrightarrow{\text{one loop}} -\frac{1}{128\pi^2}\mathrm{str}\left[\overline{M}^2\log\left(\frac{\overline{M}^2}{M_{Pl}^2}\right)\right] \quad (18)$$

in which $\overline{M}^2 = M^2(\mu = M_{Pl})$. This definition comes to mean that the gravitational scale is equal to $\mathscr{M}^2/2$ evaluated at the gravitational scale simply because the map (16) holds at the gravitational scale. It is clear that M_{Pl} remains put at its value in (18) since curvature is classical and matter loops have already been used up in forming the flat spacetime effective action in (1). It is also clear that gravity remains attractive if the bosonic sector is heavier ($\mathrm{str}[\overline{M}^2] > 0$) and if all the matter fields weigh below the gravitational scale ($\mathrm{str}[\overline{M}^2] \lesssim M_{Pl}^2$). In general, scalars, singlet fermions and vector-like fermions can weigh heavy without breaking gauge symmetries, and such heavy fields can dominate M_{Pl} through $\mathrm{str}[\overline{M}^2]$.

In line with the definition of the gravitational scale in (18), quadratic curvature coefficient $\overline{c}_O$ in (17) can be defined as

$$\overline{c}_O = c_O\,(\mu = M_{Pl}) \xrightarrow{\text{one loop}} \frac{(n_b - n_f)}{64\pi^2} \quad (19)$$

and this definition means that $n_b(n_f)$ is the number of bosons (fermions) having masses from zero way up to the gravitational scale. In other words, $n_b(n_f)$ comprises entirety of the bosons (fermions) since particles heavier than M_{Pl} are disfavored by the attractive nature of gravity. It is clear that $\overline{c}_O$ remains set at its value in (19) since curvature is classical and matter loops have already been used up in forming the flat spacetime effective action in (1).

5. Affine Dynamics: From UV Cutoff to IR Curvature

Having elucidated the parameters of the metric-Palatini action (17), it is time to go back to the total effective action (12). In fact, it takes the complete metric-Palatini form [16]

$$S_{eff}[g, \digamma; \mathbb{R}, \log\mu, R] = S_{tree}[g, \digamma] + \delta S_{log}[g, \log\mu, \digamma] + \delta\overline{S}_{pow}[g, \digamma; \mathbb{R}, \log\mu, R] \quad (20)$$

after using the action (17). This metric-Palatini theory contains both the metrical curvature $R({}^g\Gamma)$ and the affine curvature $\mathbb{R}(\Gamma)$. It remains stationary against variations in the affine connection (namely $\delta_\Gamma S_{eff}[g, \digamma; \mathbb{R}, \log\mu, R] = 0$) provided that

$${}^\Gamma\nabla_\lambda \mathbb{D}_{\mu\nu} = 0 \quad (21)$$

such that ${}^\Gamma\nabla_\lambda$ is the covariant derivative of the affine connection $\Gamma^\lambda_{\mu\nu}$, and

$$\mathbb{D}_{\mu\nu} = \left(\frac{1}{16\pi G_N} + \frac{c_\phi}{4}\phi^\dagger\phi + \frac{\overline{c}_O}{8}g^{\alpha\beta}\mathbb{R}_{\alpha\beta}(\Gamma)\right)g_{\mu\nu} - c_V\mathrm{tr}\left[V_\mu V_\nu\right] \quad (22)$$

Journal of Physics: Conference Series **2533** (2023) 012028 doi:10.1088/1742-6596/2533/1/012028

is the disformal metric of tensor fields, including the affine curvature $\mathbb{R}(\Gamma)$. The motion equation (21) implies that $\mathbb{D}_{\mu\nu}$ is covariantly-constant with respect to $\Gamma^{\lambda}_{\mu\nu}$, and this constancy leads to the exact solution

$$\begin{aligned}\Gamma^{\lambda}_{\mu\nu} &= \frac{1}{2}(\mathbb{D}^{-1})^{\lambda\rho}\left(\partial_{\mu}\mathbb{D}_{\nu\rho} + \partial_{\nu}\mathbb{D}_{\rho\mu} - \partial_{\rho}\mathbb{D}_{\mu\nu}\right)\\ &= {}^{g}\Gamma^{\lambda}_{\mu\nu} + \frac{1}{2}(\mathbb{D}^{-1})^{\lambda\rho}\left(\nabla_{\mu}\mathbb{D}_{\nu\rho} + \nabla_{\nu}\mathbb{D}_{\rho\mu} - \nabla_{\rho}\mathbb{D}_{\mu\nu}\right) \end{aligned} \tag{23}$$

in which, needless to say, ${}^{g}\Gamma^{\lambda}_{\mu\nu}$ is the Levi-Civita connection of the curved metric $g_{\mu\nu}$. The Planck scale in (18) is the largest scale and therefore it is legitimate to make the expansions

$$\Gamma^{\lambda}_{\mu\nu} = {}^{g}\Gamma^{\lambda}_{\mu\nu} + \frac{1}{M_{Pl}^{2}}\left(\nabla_{\mu}\mathbb{D}^{\lambda}_{\nu} + \nabla_{\nu}\mathbb{D}^{\lambda}_{\mu} - \nabla^{\lambda}\mathbb{D}_{\mu\nu}\right) + \mathcal{O}\left(M_{Pl}^{-4}\right) \tag{24}$$

and

$$\mathbb{R}_{\mu\nu}(\Gamma) = R_{\mu\nu}({}^{g}\Gamma) + \frac{1}{M_{Pl}^{2}}\left(\nabla^{\alpha}\nabla_{\mu}\mathbb{D}_{\alpha\nu} + \nabla^{\alpha}\nabla_{\nu}\mathbb{D}_{\alpha\mu} - \Box\mathbb{D}_{\mu\nu} - \nabla_{\mu}\nabla_{\nu}\mathbb{D}^{\alpha}_{\alpha}\right) + \mathcal{O}\left(M_{Pl}^{-4}\right) \tag{25}$$

so that both $\Gamma^{\lambda}_{\mu\nu}$ and $\mathbb{R}_{\mu\nu}(\Gamma)$ contain pure derivative terms at the next-to-leading $\mathcal{O}\left(M_{Pl}^{-2}\right)$ order [11, 12]. The expansion in (24) ensures that the affine connection $\Gamma^{\lambda}_{\mu\nu}$ is solved algebraically order by order in $1/M_{Pl}^{2}$ despite the fact that its motion equation (21) involves its own curvature $\mathbb{R}_{\mu\nu}(\Gamma)$ through $\mathbb{D}_{\mu\nu}$ [13, 14]. The expansion (25), on the other hand, ensures that the affine curvature $\mathbb{R}_{\mu\nu}(\Gamma)$ is equal to the metrical curvature $R_{\mu\nu}({}^{g}\Gamma)$ up to a doubly-Planck suppressed remainder. In essence, what happened is that the affine dynamics took the affine curvature $\mathbb{R}$ from its UV value $\Lambda_{\wp}^{2}$ in (16) to its IR value R in (25). Indeed, in the sense of holography [27, 28], the metrical curvature R sets the IR scale [29] above which QFTs hold as flat spacetime constructs [21, 22, 23].

6. Restoration of Gauge Symmetries

One consequence of the solution of the affine curvature in (25) is that the problematic loop-induced gauge boson mass term gets defused as

$$\int d^{4}x\sqrt{-g}c_{V}\mathrm{tr}\left[V^{\mu}(\mathbb{R}_{\mu\nu}(\Gamma) - R_{\mu\nu}({}^{g}\Gamma))V^{\nu}\right] \xrightarrow{\text{equation (25)}} \int d^{4}x\sqrt{-g}\left\{\text{zero} + \mathcal{O}\left(M_{Pl}^{-2}\right)\right\} \tag{26}$$

after using the solution of the affine curvature in (25) in the metric-Palatini action (17) [10, 11, 12]. The $\mathcal{O}\left(M_{Pl}^{-2}\right)$ remainder here, containing the next-to-leading order derivative term in (25), involves derivatives of the scalars ϕ and gauge fields V_{μ}, and can produce therefore no mass terms for either of them. It is worth noting that the gauge symmetries broken explicitly by a Poincare-conserving (Poincare-breaking) UV cutoff are restored via the Higgs field Φ_{V} (via the affine curvature $\mathbb{R}$) at the minimum of the Φ_{V} potential energy (at the extremum of the metric-affine action). This contrast shows that Poincare-conserving and Poincare-breaking UV cutoffs are fundamentally different and lead, respectively, to field-theoretic and gravitational completions of the effective QFT.

7. Emergence of General Relativity

One other consequence of the solution of the affine curvature in (25) is that non-gauge sector of the metric-Palatini action (17) reduces to the quadratic curvature gravity

$$\begin{aligned}\int d^{4}x\sqrt{-g}\Bigg\{ & -\frac{M_{Pl}^{2}}{2}\mathbb{R}(g,\Gamma) - \frac{\overline{c}_{O}}{16}\left(\mathbb{R}(g,\Gamma)\right)^{2} - \frac{c_{\phi}}{4}\phi^{\dagger}\phi\,\mathbb{R}(g,\Gamma)\Bigg\}\\ & \xrightarrow{\text{equation (25)}} \int d^{4}x\sqrt{-g}\left\{-\frac{M_{Pl}^{2}}{2}R - \frac{\overline{c}_{O}}{16}R^{2} - \frac{c_{\phi}}{4}\phi^{\dagger}\phi\,R + \mathcal{O}\left(M_{Pl}^{-2}\right)\right\}\end{aligned} \tag{27}$$

Journal of Physics: Conference Series **2533** (2023) 012028 doi:10.1088/1742-6596/2533/1/012028

in which $R = g^{\mu\nu} R_{\mu\nu}({}^g\Gamma)$ is the usual curvature scalar in the GR. As in (26), the $\mathcal{O}\left(M_{Pl}^{-2}\right)$ remainder here consists of the next-to-leading order and higher terms in (25). It involves derivatives of the long-wavelength fields ϕ and V_μ, produces thus no mass terms for these fields, and remains small for all practical purposes.

The reductions (26) and (27) give rise to the total QFT plus GR action

$$S_{tot}[g,F] = S_{tree}[g,F] + \delta S_{log}[g,\log\mu,F] + \int d^4x\sqrt{-g}\left\{-\frac{M_{Pl}^2}{2}R - \frac{c_O}{16}R^2 - \frac{c_\phi}{4}\phi^\dagger\phi R\right\} \quad (28)$$

in which the QFT sector

$$S_{\rm QFT}[g,F] = S_{tree}[g,F] + \delta S_{log}[g,\log\mu,F] \quad (29)$$

is the usual $\overline{MS}$-renormalized QFT resting on the matter loops in flat spacetime and evolving from scale to scale by renormalization group equations in $\log\mu$ [3]. Its gravity sector

$$S_{\rm GR}[g,\phi] = \int d^4x\sqrt{-g}\left\{-\frac{M_{Pl}^2}{2}R - \frac{c_O}{16}R^2 - \frac{c_\phi}{4}\phi^\dagger\phi R\right\} \quad (30)$$

rests on the flat spacetime loop factors and emerges from the requirement of restoring gauge symmetries. With these two unique features, it differs from all the other matter-induced gravity theories (induced [30, 31, 32], emergent [33, 34], analogue [35], broken symmetry [36, 37], and the like). It is an $R+R\phi^2+R^2$ gravity theory [38] whose each and every coupling is a flat spacetime loop factor (coefficient of $\Lambda_\wp^2$ or $\Lambda_\wp^4$ in (2)). It is the gauge symmetry-restoring emergent gravity or briefly the *symmergent gravity* [12, 11, 10], which is reformulated in a completely new setting in the present work. It should not be confused with the effective action computed in curved spacetime, which gives $M_{Pl}^2 \propto \Lambda_\wp^2$ along with $\Lambda_\wp$-sized scalar and gauge boson masses and a $\Lambda_\wp^4$-sized vacuum energy density [30, 31, 32]. Symmergent gravity, as reformulated and elucidated in the present work, stands out as a novel framework for completing effective QFTs in the UV when the UV cutoff is a Poincare-breaking one.

8. Discussions and Conclusion

In this talk, we have discussed how gauge symmetries broken explicitly by a Poincare-breaking UV cutoff can be restored, and what consequences such restoration can have. The end result is symmergent gravity as an emergent gravity plus $\overline{MS}$–renormalized QFT framework. Below, we discuss briefly some salient implications of the symmergent gravity [10, 11, 12].

First, **gravitational scale necessitates new physics beyond SM (BSM).** The gravitational scale M_{Pl} in $S_{\rm GR}(g,\phi)$ is induced by the flat spacetime matter loops as in (18). In the SM, it takes the value $M_{Pl}^2 \approx -G_F^{-1}$, where $G_F \approx (293\ \text{GeV})^{-2}$ is the Fermi scale. Its negative sign, set by the top quark contribution, is obviously unacceptable. It must be turned to positive if gravity is to be attractive, and this can be done if there exist new particles beyond the SM spectrum. These beyond-the-SM (BSM) particles are a necessity. In regard to the enormity of the gravitational scale (which can be brought to its physical shell by a conformal rescaling of the metric), the BSM sector must have

(i) either a light spectrum with numerous more bosons than fermions (for instance, $m_b \sim m_f \sim G_F^{-1/2}$ with $n_b - n_f \sim 10^{32}$),

(ii) or a heavy spectrum with few more bosons than fermions (for instance, $m_b \sim m_f \lesssim M_{Pl}$ with $n_b - n_f \gtrsim 10$),

(iii) or a sparse spectrum with net boson dominance.

Journal of Physics: Conference Series **2533** (2023) 012028 doi:10.1088/1742-6596/2533/1/012028

In general, BSM particles do not have to couple to the SM particles simply because all they are requiered to do is to saturate the super-trace in (18) at a value $M_{Pl}^2 \simeq \overline{M}_{Pl}^2$. In other words, there are no symmetries or selection rules requiring the SM particles to couple to the BSM particles. They can form therefore a fully-decoupled black sector [39, 40, 41, 11] or a feebly-coupled dark sector [11, 42, 43], with distinctive signatures at collider searches [44], dark matter searches [45], and other possible phenomena [11].

Second, **Higgs-curvature coupling can probe the BSM.** The loop factor c_ϕ in $S_{\rm GR}(g, \phi)$ couples the scalar curvature $R(g)$ to scalar fields ϕ. It is about 1.3% in the SM [10, 11]. Its deviation from this SM value indicates existence of new particles which couple to the SM Higgs boson. These BSM particles can be probed via their effects on various gravitational and astrophysical phenomena [46, 47, 48].

Third, **symmetries of the BSM sector might shed new light on the cosmological constant problem.** The vacuum energy contained in $S_{\rm QFT}(g, F)$ [17]

$$V(\mu) \xrightarrow{\text{one loop}} V\left(\langle\phi\rangle\right) + \frac{1}{32\pi^2}{\rm str}\left[\overline{M}^4\left(1 - \frac{3}{2}\log\frac{\overline{M}^2}{\mu^2}\right)\right] \tag{31}$$

gathers together field-independent $\log\mu$ corrections in (29) in the minimum $\phi = \langle\phi\rangle$ of the scalar potential $V(\phi)$. Its empirical value is $V_{emp} = \left(2.57\times 10^{-3}\ {\rm eV}\right)^4$ [49]. The cosmological constant problem is to shoot this specific value with the prediction in (31), and such a shooting is tantalizingly fine-tuned [50]. But, as a way out possible only in symmergence, it might be possible to achieve a resolution if the BSM fields enjoy appropriate symmetries and selection rules [43]. For instance, a supersymmetric BSM sector would help in eliminating $V(\mu)$ but its realization can require extra structures [51].

Fourth, **quadratic curvature term can probe the BSM.** The loop factor c_O in $S_{\rm GR}(g, \phi)$ is proportional to the boson-fermion number difference. It vanishes identically in a QFT with equal bosonic and fermionic degrees of freedom (as in the supersymmetric theories [52, 53]) and, as a result, the gravitational sector in (30) reduces to the GR. This normally is not possible since under general covariance all curvature invariants can contribute to the gravitational sector, and it simply is not possible to get the exact GR. But symmergence is able to generate the GR, the exact GR, when the SM+BSM involves equal bosonic and fermionic degrees of freedoms [11, 12].

The loop factor c_O, when nonzero, acts as probe of the BSM in strong-curvature media. It can probe the BSM sector in terms of $n_b - n_f$ via strong-curvature effects. One such effect is the Starobinsky inflation, and seems to require $n_b - n_f \approx 10^{13}$ [54]. One other effect concerns the black holes, which put limits [55, 56, 57] on c_O and the vacuum energy $V(\mu)$ through the EHT observations [58].

Fifth, **heavy BSM does not necessarily destabilize the light scalars.** Light scalars ϕ_L in $S_{\rm QFT}(g, F)$ are oversensitive to heavy fields. Indeed, their masses m_{ϕ_L} are shifted by an amount

$$\delta m^2_{\phi_L} = c_{\phi_L}\lambda_{\phi_L F_H} m^2_{F_H}\log\frac{m^2_{F_H}}{\mu^2} \tag{32}$$

if they couple with loop factor c_{ϕ_L} and coupling constant $\lambda_{\phi_L F_H}$ to heavy fields F_H of masses $m_{F_H} \gg m_{\phi_L}$. This mass correction reveals that heavier the F_H larger the shift in the ϕ_L mass and stronger the destabilization of the light scalar sector. This is the famous little hierarchy problem [59]. It is the reason the null LHC results [60, 61] have sidelined supersymmetry and other known completions.

Can symmergence have a different say on the little hierarchy problem? Can it provide a resolution? To answer these questions, one notices that the BSM sector is necessitated for generating the gravitational scale, and all that its formula in (18) involves is super-trace over mass-squareds of the SM+BSM particles. In other words, there is no symmetry principle or selection rule requiring the SM and BSM fields to interact. Indeed, the coupling $\lambda_{\phi_L \digamma_H}$ is not under any constraint since workings of symmergence do not depend on it. This means that symmergence allows $\lambda_{\phi_L \digamma_H}$ be small enough to keep $\delta m^2_{\phi_L}$ small enough. More precisely, it is possible ensure $\delta m^2_{\phi_L} \ll m^2_{\phi_L}$ if $\lambda_{\phi_L \digamma_H}$ obeys the bound

$$|\lambda_{\phi_L \digamma_H}| \lesssim \lambda_{SM} \frac{m^2_{\phi_L}}{m^2_{\digamma_H}} \tag{33}$$

in which $\lambda_{SM} \sim \mathcal{O}(1)$ is a typical SM coupling. This "small-coupling domain" is specific to symmergence. It is the domain in which the SM and BSM are sufficiently decoupled and the little hierarchy problem is naturally avoided.

Symmergence, as a framework intertwining emergent gravity and new physics sector, can have widespread effects in various phenomena. Its study in collider, astrophysical and cosmological setting can provide therefore further information.

Acknowledgements

The author is grateful to his collaborators A. Abdujabbarov, İ. Çimdiker, C. Karahan, R. Pantig, B. Puliçe, J. Rayimbaev, A. Övgün and O. Sargın for their contributions to various results presented in this talk. The author thanks to participants of DICE 2022 for enlightening conversations and discussions.

References

[1] Bogolyubov N N, Logunov A A, Oksak A I, and Todorov I T 1989 *General principles of quantum field theory* (Berlin, Germany: Springer)
[2] Umezawa H, Yukawa J, and Yamada E 1948 *Prog. Theor. Phys.* **3** 317
[3] Peskin M E and Schroeder D V 1995 *An Introduction to quantum field theory* (Reading, USA: Addison-Wesley)
[4] Bonini M, D'Attanasio M and Marchesini G 1994 *Nucl. Phys.* B **418** 81
[5] D'Attanasio M and Morris T M 1996 *Phys. Lett.* B **378**, 213
[6] Polchinski J 1984 *Nucl. Phys.* B **231**, 269
[7] Veltman M J G 1981 *Acta Phys. Polon.* B **12** 437
[8] Kibble T W B 2009 *Int. J. Mod. Phys.* A **24** 6001
[9] Vainshtein A 2006 *Surveys High Energ. Phys.* **20** 5
[10] Demir D 2016 *Adv. High Energy Phys.* **2016** 6727805
[11] Demir D 2019 *Adv. High Energy Phys.* **2019** 4652048
[12] Demir D 2021 *Gen. Rel. Grav.* **53** 22
[13] Vitagliano V, Sotiriou T P and Liberati S 20111 *Annals Phys.* **326** 1259
[14] Karahan C, Altas A and Demir D 2013 *Gen. Rel. Grav.* **45** 319
[15] Demir D and Puliçe B 2020 *JCAP* **04** 051
[16] Demir D and Puliçe B 2022 *Eur. Phys. J.* C **82** 996
[17] Demir D, Karahan C and Sargın O 2023 *Dimensional Regularization in Quantum Field Theory with Ultraviolet Cutoff*, to appear in Phys. Rev. D (2023).
[18] 't Hooft G and Veltman M J G 1972 *Nucl. Phys.* B **44** 189
[19] Bollini C and Giambiagi J 1972 *Nuovo Cim.* B **12** 20
[20] Gnendiger C *et al.* 2017 *Eur. Phys. J.* C **77** 471
[21] Hollands S and Wald R M 2008 *Gen. Rel. Grav.* **40** 2051
[22] Hollands S and Wald R M 2015 *Phys. Rept.* **574** 1
[23] Fulling S A 1973 *Phys. Rev.* D **7** 2850
[24] Bär C and Fredenhagen K (2009) *Lect. Notes Phys.* **786** 1
[25] Froggatt C D and Nielsen H B 2005 *Annalen Phys.* **517** 115
[26] Weinberg S 1972 *Gravitation and Cosmology: Principles and Applications of the General Theory of Relativity* (New York, USA: John Wiley and Sons).
[27] 't Hooft G 1993 *Conf. Proc.* C **930308** 284

[28] Cohen A G, Kaplan D B and Nelson A E 1999 *Phys. Rev. Lett.* **82** 4971
[29] Gao G, Wu F, Chen X and Shen Y G 2009 *Phys. Rev.* D **79** 043511
[30] Sakharov A D 1967 *Dokl. Akad. Nauk Ser. Fiz.* **177** 70
[31] Visser M 2002 *Mod. Phys. Lett.* A **17** 977
[32] Broda B and Szanecki M 2009 *Phys. Lett.* B **674** 64
[33] Verlinde E P 2011 *JHEP* **1104** 029
[34] Van Raamsdonk M 2010 *Gen. Rel. Grav.* **42** 2323
[35] Barcelo C, Liberati S and Visser M 2005 *Living Rev. Rel.* **8** 12
[36] Zee A 1979 *Phys. Rev. Lett.* **42** 417
[37] Adler S L 1982 *Rev. Mod. Phys.* **54** 729
[38] Buchdahl H A 1970 *Mon. Not. Roy. Astron. Soc.* **150** 1
[39] Peebles P and Vilenkin A 1999 *Phys. Rev.* D **60** 103506
[40] Tang Y and Wu Y L 2016 *Phys. Lett.* B **758** 402
[41] Ema Y, Nakayama K and Tang Y 2018 Production of Purely Gravitational Dark Matter *Prepint* arXiv:1804.07471 [hep-ph]
[42] Gori S *et al.* 2022 Dark Sector Physics at High-Intensity Experiments *Preprint* arXiv:2209.04671 [hep-ph]
[43] Demir D 2021 *Galaxies* **9** 33
[44] Cankoçak K, Demir D, Karahan C and Şen S 2020 *Eur. Phys. J.* C **80** 1188
[45] Demir D and Ün C S 2020 Scalar Dark Matter and Electroweak Stability *Preprint* arXiv:2005.03589 [hep-ph]
[46] Faraoni V 1996 *Phys. Rev.* D **53** 6813
[47] Ren J, Xianyu Z Z and He H J 2014 *JCAP* **1406**032
[48] Demir D 2014 *Phys. Lett.* B **733** 237
[49] Aghanim N *et al.* [Planck Collaboration] 2018 Planck 2018 results. VI. Cosmological parameters *Preprint* arXiv:1807.06209 [astro-ph.CO]
[50] Zeldovich Y B 1967 *JETP Lett.* **6** 316
[51] Gherghetta T and Pomarol A 2003 *Phys. Rev.* D **67** 085018
[52] Dimopoulos S and Georgi H 1981 *Nucl. Phys.* B **193** 150
[53] Anderson G W and Castano D J 1995 *Phys. Rev.* D **52** 1693
[54] Çimdiker İ. İ. 2020 *Phys. Dark Univ.* **30** 100736
[55] Çimdiker İ. İ., Demir D and Övgün A 2021 *Phys. Dark Univ.* **34** 100900
[56] Rayimbaev J, Pantig R, Övgün A, Abdujabbarov A and Demir D 2022 Quasiperiodic oscillations, weak field lensing and shadow cast around black holes in Symmergent gravity *Preprint* arXiv:2206.06599 [gr-qc]
[57] Pantig R, Övgün A and Demir D 2022 Testing Symmergent gravity through the shadow image and weak field photon deflection by a rotating black hole using the M87* and Sgr. A* results *Preprint* arXiv:2208.02969 [gr-qc]
[58] Akiyama K *et al.* [Event Horizon Telescope] 2019 *Astrophys. J. Lett.* **875** L1
[59] Birkedal A, Chacko Z and Gaillard M K 2004 *JHEP* **0410** 036
[60] ATLAS Collaboration 2022 *Nature* **607** 52
[61] CMS Collaboration 2022 *Nature* **607** 60

Journal of Physics: Conference Series **2533** (2023) 012029 doi:10.1088/1742-6596/2533/1/012029

Beyond semiclassical time: dynamics in quantum cosmology

Leonardo Chataignier[1,2]

[1]Dipartimento di Fisica e Astronomia, Università di Bologna, via Irnerio 46, 40126 Bologna, Italy

[2]I.N.F.N., Sezione di Bologna, I.S. FLAG, viale B. Pichat 6/2, 40127 Bologna, Italy

E-mail: leonardo.chataignier@unibo.it

Abstract. We review two approaches to the definition of the Hilbert space and evolution in mechanical theories with local time-reparametrization invariance, which are often used as toy models of quantum gravity. The first approach is based on the definition of invariant relational observables, whereas the second formalism consists of a perturbative construction of the Hilbert space and a weak-coupling expansion of the Hamiltonian constraint, which is frequently performed as part of the Born-Oppenheimer treatment in quantum cosmology. We discuss in which sense both approaches exhibit an inner product that is gauge-fixed via an operator version of the usual Faddeev-Popov procedure, and, in the second approach, how the unitarity of the effective Schrödinger evolution is established perturbatively. We note that a conditional probability interpretation of the physical states is possible, so that both formalisms are examples of quantum mechanics with a relational dynamics.

1. Introduction

In any approach to quantum gravity, a central issue is the combination of spacetime diffeomorphism invariance with the standard postulates of quantum mechanics. Indeed, our current accepted formulation of quantum theory refers to a fixed spacetime background and it seems to leave little room for local time reparametrization invariance, i.e., for formulating the theory [especially the wave function(al)'s probabilistic interpretation and the state update rule] without a preferred, global class of time variables. In general relativity (GR), on the contrary, spacetime diffeomorphism invariance is an important ingredient, one that is intertwined with the delicate issue of background independence (see [1, 2] for further discussion on this point) and, in particular, no preferred or global time seems to be generally available in GR. If gravity is to be quantized, one is thus faced with the choice of either modifying relativity to accommodate a preferred class of time variables, or formulating quantum theory without a fixed spacetime background. The latter choice is usually preferred and taken as a template for quantum gravity.

A background-independent quantization of gravitation faces serious challenges, as one needs to understand how to formulate the dynamics in a diffeomorphism-invariant fashion (or, at very least, explain the emergence of this diffeomorphism symmetry at the classical level). Can we maintain quantum theory's linear structure (the superposition principle) in this way? What is the Hilbert space? What are the observables? Most importantly, how can we assign a probabilistic interpretation to the wave function(al) without a preferred class of times with respect to which probabilities are conserved? With respect to which time (if any) should the

Journal of Physics: Conference Series **2533** (2023) 012029 doi:10.1088/1742-6596/2533/1/012029

state update rule be applied? Does the measurement problem become even more exacting in this case? In fact, what is the meaning of probabilities if the theory is extrapolated to the Universe, as is done in quantum cosmology?

In this conference article, we review two approaches that may be seen as templates to a version of quantum mechanics in which: no preferred time is assumed a priori; a notion of quantum diffeomorphism invariance holds; the linearity of the theory is maintained and the Hilbert space structure can be constructed, at the very least, perturbatively; a notion of unitarity and time evolution are available. In this way, the formalisms discussed here could be a starting point in our endeavour to address the above questions. Our procedures draw an analogy to well-known techniques in classical gauge systems, and we deal with mechanical models to avoid topics of field theory that are inessential to the problem at hand (quantum time diffeomorphism invariance and unitary time evolution). These models are directly applicable to homogeneous (classical and quantum) cosmology, and also to the case in which the usual Mukhanov-Sasaki variables are present [3, 4].

In Sec. 2, we briefly discuss how the classical diffeomorphism symmetry of GR can be seen as a type of gauge symmetry, and how this affects the definition of observables in the theory. In particular, in Sec. 2.2, we restrict ourselves to the mechanical case, which is then quantized in Sec. 2.3. A general, model-independent definition of gauge-invariant observables for theories with local time reparametrization invariance is proposed in Sec. 2.4. Furthermore, we discuss a perturbative construction of the classical Hamilton-Jacobi solutions, as well as the quantum Hilbert space in Sec. 3, where the perturbative unitarity of the theory is also ascertained in a weak-coupling expansion. We conclude in Sec. 4.

2. Gauge symmetry and observables

2.1. The classical theory

In classical GR, diffeomorphisms are a local symmetry of the Einstein field equations. This means that, given a solution Φ (which denotes a set of tensor fields that includes the spacetime metric) and a diffeomorphism $\phi : \mathcal{M} \to \mathcal{M}$ (where $\mathcal{M}$ is the spacetime manifold), then the transformed fields $\phi \cdot \Phi$ also constitute a solution to the same field equations [1]. It is for this reason that GR is a diffeomorphism-invariant theory.

A local symmetry of the field equations implies a type of indeterminism, which can be discerned in the initial-value problem formulation of the theory [1]. Indeed, let $\mathcal{C}$ be a Cauchy slice on which the initial data are prescribed. If the diffeomorphism ϕ coincides with the identity in a neighborhood of $\mathcal{C}$ but is nontrivial elsewhere, then Φ and $\phi \cdot \Phi$ are to two different solutions associated with the same initial conditions. To restore the one-to-one correspondence between solutions and initial data, one is led to the notion of equivalence classes of Φ under diffeomorphisms, and indeed to the proposal that observables (here understood as quantities that have a deterministic evolution) should be invariant under diffeomorphisms. An analogous situation occurs in Yang-Mills theories, where the local symmetry transformations are not diffeomorphisms but gauge transformations. In that case, one is led to the notion of gauge-invariant quantities (e.g., an electric field) in order to overcome the "gauge indeterminism". Due to this similarity, we refer to any local symmetry as a "gauge symmetry" from now on. In this terminology, GR is also a gauge theory, albeit not of the Yang-Mills type.[1]

In general, diffeomorphism invariants (such as constant spacetime scalars) lack a clear physical meaning. Which of these invariants have a reasonable physical interpretation that can be related to experiments? The so-called relational observables [2, 5–7] (sometimes also referred to as "evolving constants of motion" [8]) are invariants that have a straightforward interpretation.

[1] There are efforts to construct a gravitational theory that follows more closely the Yang-Mills paradigm. In particular, the reader is referred to Poincaré gauge theory [9].

The idea is that observers record the dynamics of fields in local regions of spacetime in a relational way: one chooses a set of reference fields χ with respect to which the values of all dynamical fields are compared. Clearly, these reference fields are physical fields that are part of a solution Φ to the field equations. Ideally, the χ fields mark the passage of time and determine points in space, and thus they serve as "generalized clocks and rods" from which "generalized referece frames" can be specified. The values of all of the fields Φ, taken relative to the clocks and rods, are the outcomes of experiments performed by the observers. These outcomes are the relational observables, which we denote by $\mathcal{O}[\Phi|\chi]$. But why are these objects invariant under diffeomorphisms?

In specifying the outcome $\mathcal{O}[\Phi|\chi]$ of an experiment (e.g., the value of an electric field relative to the values of four scalar fields that define a spacetime point [10]), one only measures physical fields. There is no need to make reference to arbitrary coordinate systems, nor to an abstract notion of a spacetime point $p \in \mathcal{M}$. For this reason, for any given value of the clocks and rods (e.g., $\chi = s$), the observable $\mathcal{O}[\Phi|\chi]$ does not depend on p, and it can be formally seen as a constant spacetime scalar, and thus a diffeomorphism invariant. More concretely, one can evidently choose χ to define an "intrinsic coordinate system" (i.e., one defined from the physical degrees of freedom instead of arbitrary constructs [11]) in the abstract spacetime manifold $\mathcal{M}$, which corresponds to gauge fixing the diffeomorphism symmetry. Then, $\mathcal{O}[\Phi|\chi]$ are simply gauge invariant extensions of gauge-fixed tensor fields (see [2] for further details on this point, and [5] for the general notion of gauge invariant extensions).

Moreover, as the outcome $\mathcal{O}[\Phi|\chi]$ only refers to physical fields, it completely encodes the dynamics (in a relational way) in the region where the experiment is performed. The observables $\mathcal{O}[\Phi|\chi]$, when computed in theory, can also be seen as conditional quantities because they yield predictions for the observed outcomes given the condition that the reference fields have a certain value (the observed value of the clocks and rods).

2.2. Worldline relational observables

Let us discuss and illustrate the construction of relational observables in a simplified, mechanical setting. This is the same as considering physics in a (0+1)-dimensional spacetime or a worldline. In this case, a diffeomorphism-invariant theory for a set of scalar fields denoted by $q(\tau)$ is given by the action

$$S = \int_a^b \mathrm{d}\tau \; \mathcal{L}(q, \dot{q}; e) = \int_a^b \mathrm{d}\tau \; \mathcal{L}(\phi^* q, \phi^* \dot{q}; \phi^* e) \; , \tag{1}$$

where τ is an arbitrary coordinate on the worldline $\mathcal{M}$, $e(\tau)$ is the 'einbein' in $\mathcal{M}$ [2], and ϕ is an arbitrary diffeomorphism on $\mathcal{M}$. Equation (1) implies that S is gauge invariant because diffeomorphisms simply relabel the arguments of the Lagrangian, which retains its form under these transformations. More precisely, the action is exactly invariant if we restrict ourselves to diffeomorphisms that coincide with the identity at the endpoints $\tau = a, \tau = b$. For more general transformations, it is possible to add a boundary term to make the action fully gauge invariant [12]. The invariance of the form of the action under (active) diffeomorphisms also implies that the Lagrangian $\mathcal{L}$ must satisfy

$$\mathcal{L}\left(q'(\tau'), \frac{\mathrm{d}q'}{\mathrm{d}\tau'}(\tau'); e'(\tau')\right) = \mathcal{L}\left(q(\tau), \frac{\mathrm{d}\tau}{\mathrm{d}\tau'}\frac{\mathrm{d}q}{\mathrm{d}\tau}(\tau); \frac{\mathrm{d}\tau}{\mathrm{d}\tau'}e(\tau)\right) = \frac{\mathrm{d}\tau}{\mathrm{d}\tau'}\mathcal{L}\left(q(\tau), \frac{\mathrm{d}q}{\mathrm{d}\tau}(\tau); e(\tau)\right) \tag{2}$$

under (passive) τ reparametrizations. This means that $\mathcal{L}$ is a homogeneous function of $\dot{q}$ and e, and therefore we must have

$$\mathcal{L}(q, \dot{q}; e) = \frac{\partial \mathcal{L}}{\partial \dot{q}^i}\dot{q}^i + \frac{\partial \mathcal{L}}{\partial e}e = p_i \dot{q}^i - eC(q, p) \; , \tag{3}$$

Journal of Physics: Conference Series **2533** (2023) 012029 doi:10.1088/1742-6596/2533/1/012029

where p are the canonical momenta conjugate to q, and $C(q,p)$ is the first-class Hamiltonian constraint. Indeed, the equation of motion of the einbein is $C = 0$. Given the Poisson bracket structure for the (q,p) pairs, one can verify that a gauge transformation (infinitesimal diffeomorphism) of the scalars reads $\delta_{\epsilon(\tau)}q(\tau) = \epsilon\dot{q} = \epsilon\{q, eC\}_{C=0} = \{q, \epsilon eC\}_{C=0} = \epsilon e\{q, C\}_{C=0}$ (and similarly for p). Thus, an invariant scalar $\mathcal{O}(q,p)$ Poisson-commutes with the constraint, $\{\mathcal{O}, C\}_{C=0} = 0$. In contrast, a phase-space function $\eta(q,p)$ that is conjugate to the constraint, $\{\eta, C\}_{C=0} = 1$, is a canonical representation of "proper time" because its equation of motion is $\dot{\eta}(\tau) = e(\tau)$, whereas "proper time" is the reparametrization invariant defined by $\eta := \int_a^b \mathrm{d}\tau\, e(\tau)$.

To find invariants that are physically meaningful, one can choose a reference field $\chi(\tau)$ to be the "generalized clock" in the worldline. This is a choice of gauge, and it is fixed by finding a diffeomorphism for which the pullback of $\chi(\tau)$ is trivial, $\phi^*\chi = \tau$. If this can be done, one can define the relational observable of any tensor field f relative to χ as the constant worldline scalar

$$\begin{aligned} \mathcal{O}[f|\chi = s] &: \mathcal{M} \to \mathbb{R} \\ p &\mapsto \mathcal{O}[f|\chi = s](p) = \phi^* f \ (\forall p) \ , \end{aligned} \tag{4}$$

for each fixed value $\chi = s$ of the reference field. Each label s corresponds to an instant. Thus, there is a family of constant scalars labelled by s, and this is why these objects are sometimes called "evolving constants of motion" [8]. Despite their conceptual simplicity, it is often difficult to build relational observables in a model-independent way (i.e., without solving the equations of motion first). One useful definition considered in [2,7,13] consists in reexpressing the pullback of f in (4) as a spacetime (worldline) integral,

$$\mathcal{O}[f|\chi = s] = \Delta_\chi \int_{\mathcal{I}} \mathrm{d}\tau\ \delta(\chi(\tau) - s) f(\tau) = \frac{\int_{\mathcal{I}} \mathrm{d}\tau\ \delta(\chi(\tau) - s) f(\tau)}{\int_{\mathcal{I}} \mathrm{d}\tau\ \delta(\chi(\tau) - s)} \ , \tag{5}$$

where $\mathcal{I}$ is an interval of $\mathcal{M}$ where $\chi(\tau)$ is monotonic, $\Delta_\chi^{-1} = \int_{\mathcal{I}} \mathrm{d}\tau\ \delta(\chi(\tau) - s)$, and $\Delta_\chi = \phi^*\{\chi, C\}$ is the 'Faddeev-Popov (FP) determinant' associated with the gauge choice $\phi^*\chi = \tau$ [2, 7, 14]. Given (5), one can explicitly compute [2] the gauge transformation $\delta_{\epsilon(\tau)}\mathcal{O}[f|\chi = s]_{C=0} = \epsilon(\tau)e\{\mathcal{O}[f|\chi = s], C\}_{C=0} = 0$; i.e., one finds that the relational observables are indeed diffeomorphism invariant.

2.3. Quantum theory. The physical Hilbert space

We now present a possible construction of the quantum theory of a diffeomorphism-invariant mechanical model. The reader is referred to [6, 13, 15] for other approaches, and to [2, 7] for more details concerning the formalism that we present here, the advantage of which is its model independence; i.e., the observables can be defined, in principle, without first solving the (classical) equations of motion.

The first step is to define an auxiliary Hilbert space $\mathcal{H}_{\mathrm{aux}}$ on which the canonical quantization of the Hamiltonian acts as a self-adjoint operator $\hat{C}$ with a complete orthonormal system of eigenstates $|E, \mathbf{k}\rangle$. The label E can be discrete or continuous and $\mathbf{k}$ stands for degeneracies. We will denote sums or integrals over E by the formal symbol $\sum_E$. From these energy eigenstates, one can construct the 'proper-time' states

$$|t, \mathbf{k}\rangle \propto \frac{1}{\sqrt{2\pi\hbar}} \sum_E \mathrm{e}^{-\frac{\mathrm{i}}{\hbar}Et} |E, \mathbf{k}\rangle \ , \tag{6}$$

which are not necessarily orthonormal with respect to the auxiliary inner product in $\mathcal{H}_{\mathrm{aux}}$ (see [15, 16]). This is, however, irrelevant for us because only physical states and invariant observables are meaningful, not the auxiliary structure of $\mathcal{H}_{\mathrm{aux}}$. Indeed, the usefulness of the states (6) will be in the definition of a physical inner product, as we will see shortly.

Journal of Physics: Conference Series **2533** (2023) 012029 doi:10.1088/1742-6596/2533/1/012029

In this theory, physical states $|\Psi\rangle$ are superpositions of $|E=0,\mathbf{k}\rangle$ so as to satisfy the quantum constraint $\hat{C}\,|\Psi\rangle = 0$, which is the celebrated Wheeler-DeWitt (WDW) equation [17]. This condition implies diffeomorphism-invariance, since $\delta_{\epsilon(\tau)}\,|\Psi\rangle = \epsilon(\tau)\mathrm{d}\,|\Psi\rangle\,/\mathrm{d}\tau = \epsilon(\tau)\hat{C}\,|\Psi\rangle = 0$. To define a Hilbert space from these physical states, it is necessary to define an inner product with respect to which $|E=0,\mathbf{k}\rangle$ are, at the very least, delta-function normalizable. The auxiliary inner product reads $\langle E=0,\mathbf{k}'|E=0,\mathbf{k}\rangle = \delta(0,0)\delta(\mathbf{k}',\mathbf{k})$. If the zero energy eigenvalue is in the continuous spectrum, then $\delta(0,0)=\infty$.

To see how the inner product can be regularized, or how the $\delta(0,0)$ factor can be discarded, let us define from (6) the operator $\hat{\mu}_t := \sum_{\mathbf{k}} |t,\mathbf{k}\rangle\,\langle t,\mathbf{k}|$, which satisfies $\int \mathrm{d}t\;\hat{\mu}_t = \hat{1}$. Then we may write

$$\begin{aligned}\langle E=0,\mathbf{k}'|E=0,\mathbf{k}\rangle &= \left(\int \mathrm{d}t\right)\left\langle E=0,\mathbf{k}'\,|\hat{\mu}_t|\,E=0,\mathbf{k}\right\rangle \\ &= \mathcal{V}\delta(\mathbf{k}',\mathbf{k}) =: \mathcal{V}(E=0,\mathbf{k}'|E=0,\mathbf{k})\ .\end{aligned} \tag{7}$$

The possibly divergent factor is clearly the "volume" of proper-time integration, $\mathcal{V} := \int \mathrm{d}t$. The physical inner product is obtained by factoring out $\mathcal{V}$, and it reads

$$(E=0,\mathbf{k}'|E=0,\mathbf{k}) := \left\langle E=0,\mathbf{k}'\,|\hat{\mu}_t|\,E=0,\mathbf{k}\right\rangle = \delta(\mathbf{k}',\mathbf{k})\ . \tag{8}$$

This regularization of the inner product is analogous to the usual FP gauge fixing that is performed in path integrals, $\mathcal{Z} = \int \mathcal{D}\Phi\;\mathrm{e}^{\frac{\mathrm{i}}{\hbar}S} = \mathcal{V}\int \mathcal{D}\Phi\;\Delta_\chi\delta(\chi-s)\mathrm{e}^{\frac{\mathrm{i}}{\hbar}S}$, where $\hat{\mu}_t$ is analogous to the gauge-fixing delta function $\delta(\chi-s)$ for the proper-time gauge (for which $\Delta_\chi = \phi^*\{\chi,C\}=1$). The inner product (8) is also obtained by the Rieffel induction procedure [18]. The physical Hilbert space is then defined to be the space of solutions of the WDW equation that are normalizable with respect to (8).

2.4. Quantum relational observables

The physical inner product (8) also motivates a relevant definition of observables. General observables should be invariant symmetric operators that have the correct Heisenberg equations of motion. Here, we simply quote the definition (see [2] for details)

$$\hat{\mathcal{O}}[f|\chi=s] := \pi\hbar\sum_E \hat{P}_E[\hat{f}(\tau),\hat{\mu}_{t=s}]_+\hat{P}_E\ , \tag{9}$$

where $\hat{P}_E$ is the projector onto the eigenspace associated with the energy E, $[\cdot,\cdot]_+$ is an anticommutator, and $\hat{f}(\tau)$ is in the Heisenberg picture. One can easily verify that (9) is an invariant in the sense that it commutes with $\hat{C}$, and we propose that (9) should be the general, model-independent definition of quantum relational observables (in the proper-time gauge) because:

(i) It can be shown that (9) reduces to an integral formula analogous to (5). For instance, if the spectrum of $\hat{C}$ spans the real line, than $\hat{\mathcal{O}}[f|\chi=s] = \frac{1}{2}\int_{-\infty}^{\infty}\mathrm{d}\tau\;\hat{f}(\tau)\hat{P}_{t=s-\tau} + \text{h.c.}$ (see [2] for further details).

(ii) By construction, $\hat{\mathcal{O}}[1|\chi=s] = \hat{1}$ (FP resolution of the identity).

(iii) In general, the observables (9) satisfy the correct Heisenberg equations of motion,

$$\mathrm{i}\hbar\frac{\mathrm{d}}{\mathrm{d}s}\hat{\mathcal{O}}[f|\chi=s] = \hat{\mathcal{O}}\left[\mathrm{i}\hbar\frac{\partial f}{\partial s} + [f,C]\middle|\,\chi=s\right]\ . \tag{10}$$

These are the correct equations because they are also satisfied by the classical observables if we replace $[\cdot,\cdot]/(\mathrm{i}\hbar) \to \{\cdot,\cdot\}$. In particular, notice that, due to (10), dynamics need not

Journal of Physics: Conference Series **2533** (2023) 012029 doi:10.1088/1742-6596/2533/1/012029

"disappear" in the quantum theory, as is often declared with regards to the WDW equation. Moreover, the evolution is with respect to a proper-time parameter that is defined directly from energy spectrum [cf. (6)], and, in this sense, it is also an intrinsic evolution, since no external or preferred time is assumed a priori.

It is also worth mentioning that, similarly to the conditional interpretation of the classical observables, the statistics of the quantum observables (9) can be seen as a conditional one. In [15], a specific class of constraint operators and another definition of relational observables was examined, and it was pointed out that the quantum dynamics of relational observables is equivalent to the one obtained in the famous Page-Wootters formalism [19], which uses conditional probabilities. Thus, it is also instructive to note that the model-independent definition (9) leads to conditional expectation values

$$\langle \hat{\mathcal{O}}[f|\chi = s]\rangle_\Psi = \frac{\left(\Psi \left| \hat{\mathcal{O}}[f|\chi = s] \right| \Psi\right)}{(\Psi|\Psi)} = \frac{\langle \Psi|\hat{f}\hat{\mu}_{t=s}|\Psi\rangle}{\langle \Psi|\hat{\mu}_{t=s}|\Psi\rangle} =: \mathrm{E}_\Psi[f|\chi = s] \ . \tag{11}$$

This is also true for more general definitions of the gauge-fixing given by $\hat{\mu}_{t=s}$ (for example, for gauges with more than one frequency sector), as was shown in [2, 7].

2.5. *A suggestion for the theory's postulates*

Equipped with a definition of the physical Hilbert space and observables, we now suggest a set of postulates for the theory, which can be seen as a conservative union of standard quantum mechanics and worldline diffeomorphism invariance. The reader is again referred to [2] for a more detailed discussion.

(i) In a quantum system with diffeomorphism invariance, the quantum state is a ray in the physical Hilbert space equipped with the inner product (8).

(ii) A measurement of a tensor field $\hat{f}(\tau)$ has an eigenvalue $f(s, \mathbf{n})$ of a self-adjoint observable $\hat{\mathcal{O}}[f|\chi = s]$ as the outcome, the probability of which is

$$p_\Psi = \frac{|(\mathbf{n}; s|\Psi)|^2}{(\Psi|\Psi)} \ , \tag{12}$$

where we denote the eigenstates of $\hat{\mathcal{O}}[f|\chi = s]$ by $|\mathbf{n}; s\rangle$.

(iii) The state of the system is updated to $|\mathbf{n}; s\rangle$ after the measurement.

3. Weak-coupling expansion

The formalism described thus far leads to a model-independent and self-contained approach. Nevertheless, it is predicated on the assumption that it is possible to find the spectrum of $\hat{C}$ and its eigenstates in an exact, analytical manner [cf. (6) and (9)]. What if we cannot solve the constraint? How do we construct the observables and the Hilbert space?

In this section, we consider an alternative, perturbative procedure that is directly applicable to toy models of the early Universe [3], and that should be useful to more realistic models, for which it may be difficult or impossible to find the spectrum of $\hat{C}$ exactly. See [2, 4] for further details. Henceforth, we set $\hbar = 1$.

3.1. *Hamilton-Jacobi theory*

We consider the important case in which the Hamilton-Jacobi (HJ) formulation of the Hamiltonian constraint C has the form

$$\frac{\kappa}{2} G^{ab}(Q) \frac{\partial \mathrm{W}}{\partial Q^a} \frac{\partial \mathrm{W}}{\partial Q^b} + \frac{1}{\kappa} V(Q) + \mathsf{h}\left(Q; \frac{\partial \mathrm{W}}{\partial q}, q\right) = 0 \ , \tag{13}$$

where a summation over repeated indices is implied, κ is a coupling constant (e.g., $\kappa = 4\pi G/3$ or $\kappa = 1/c^2$), and W is the on-shell action. We can look for a Wentzel-Kramers-Brillouin (WKB) perturbative solution of (13) by formally expanding the on-shell action as follows:

$$\mathrm{W}(Q,q) = \frac{1}{\kappa}\sum_{n=0}^{\infty} \mathrm{W}_n(Q,q)\kappa^n =: \frac{1}{\kappa}\mathrm{W}_0(Q) + \mathsf{S}(Q;q) \ . \tag{14}$$

It is straightforward to verify that, at the lowest-oder, the solution satisfies the HJ constraint solely for the Q variables, without any interaction with the q fields,

$$\frac{1}{2}G^{ab}(Q)\frac{\partial \mathrm{W}_0}{\partial Q^a}\frac{\partial \mathrm{W}_0}{\partial Q^b} + V(Q) = 0 \ . \tag{15}$$

This is the no-coupling limit. Its solution induces a foliation on the Q-configuration space, $Q \mapsto x(Q) = (t(Q), x^i(Q))$, where $t(Q)$ satisfies [4, 20]

$$G^{ab}\frac{\partial \mathrm{W}_0}{\partial Q^a}\frac{\partial t}{\partial Q^b} = 1 \ ; \tag{16}$$

i.e., it is the proper time in the no-coupling limit. Higher orders in the coupling constant κ are encoded in the $\mathsf{S}(Q;q)$ function [cf. (14)], which obeys the "corrected" time-dependent HJ equation

$$-\frac{\partial \mathsf{S}}{\partial t} = \mathsf{h} - \frac{\kappa}{4V}\mathsf{h}^2 + \frac{\kappa}{2}g^{ij}\frac{\partial \mathsf{S}}{\partial x^i}\frac{\partial \mathsf{S}}{\partial x^j} + \mathcal{O}(\kappa^2) \ . \tag{17}$$

Equation (17) implies that the no-coupling proper-time t is the time that orders the evolution of the (q, x) variables. It is a choice of gauge that is thus "preferred" by this weak-coupling expansion. The associated FP determinant is

$$|\Delta| = 1 + \kappa\frac{\mathsf{h}}{2V} + \mathcal{O}(\kappa^2) \ . \tag{18}$$

3.2. Weak-coupling expansion of the WDW equation. Unitarity

The WDW equation associated with the classical constraint (13) reads

$$\hat{C}\Psi = -\frac{\kappa}{2}\nabla^2\Psi + \frac{1}{\kappa}V(Q)\Psi + \hat{\mathsf{h}}\Psi = 0 \ , \tag{19}$$

and, in analogy to the classical expansion (14), we can solve (19) perturbatively via the ansatz

$$\Psi(Q,q) = \exp\left[\frac{\mathrm{i}}{\kappa}\sum_{n=0}^{\infty} W_n(Q,q)\kappa^n\right] =: \exp\left[\mathrm{i}\frac{1}{\kappa}\mathrm{W}_0(Q,q)\right]\psi(Q;q) \ , \tag{20}$$

which is frequently used in Born-Oppenheimer (BO) treatments of quantum cosmology [3, 4]. Just as in the classical case, we can adopt a foliation of the Q-configuration space defined from the no-coupling solution, and we find that $\psi(Q;q)$, which encodes the higher orders in κ, solves the corrected Schrödinger equation

$$\mathrm{i}\frac{\partial\tilde{\psi}}{\partial t} = \left[\hat{\mathsf{h}} - \frac{\kappa}{4V}\hat{\mathsf{h}}^2 + \frac{\kappa}{2}\hat{\Pi}_i g^{ij}\hat{\Pi}_j^\dagger + \frac{\mathcal{Q}}{M}\right]\tilde{\psi} + \mathcal{O}(\kappa^2) \ , \tag{21}$$

where t is the no-coupling proper-time, $\mathcal{Q}$ is a quantum correction to the potential, and we have defined

$$\tilde{\psi} := |2Vg|^{\frac{1}{4}}\left(1 + \frac{\kappa}{2V}\hat{\mathsf{h}}\right)\psi + \mathcal{O}(\kappa^2) \ . \tag{22}$$

Journal of Physics: Conference Series **2533** (2023) 012029 doi:10.1088/1742-6596/2533/1/012029

Equation (21) is relational in the sense that it describes the dynamics of the (q, x) variables relative to the intrinsic time t. Moreover, it clearly leads to a unitary dynamics with respect to the inner product [cf. (20), (22)]

$$\begin{aligned}(\Psi_2|\Psi_1) &:= \int \mathrm{d}x\mathrm{d}q\ \tilde{\psi}_2^*(x,q)\tilde{\psi}_1(x,q) \\ &= \int \mathrm{d}Q\mathrm{d}q\sqrt{|2Vg|}\left|\frac{\partial x}{\partial Q}\right|\ \Psi_2^*(x,q)\,\delta(t(Q)-s)\widehat{|\Delta|}\,\Psi_1(x,q) + \mathcal{O}(\kappa^2)\ ,\end{aligned} \tag{23}$$

where

$$\widehat{|\Delta|} := 1 + \frac{\kappa}{2V}\hat{\mathsf{h}} \tag{24}$$

is the quantization of the (absolute value of the) classical FP determinant [cf. (18)]. Thus, the inner product (23) is perturbatively defined and conserved, and it can be seen as a perturbative analogue of the physical inner product (8). Moreover, the inner product (23) leads to a probability density

$$p_\Psi := \left.\frac{|\tilde{\psi}(x,q)|^2}{(\Psi|\Psi)}\right|_{t=s}\ , \tag{25}$$

which refers to the measurement of the (q, x) fields based on the condition that t has (been observed to have) the value s. In this way, we can regard (25) as conditional probabilities, in analogy to (11). In fact, relational observables can be perturbatively defined from their matrix elements by inserting operators in the integrand of (23) (see [2,4]).

3.3. The early Universe

This unitary, perturbative formalism was applied to a simple model of the early Universe in the first reference of [4], which consisted of a de Sitter spacetime background [with scale factor $a = a_0\mathrm{e}^\alpha$ and cosmological constant $H_0^2/(2\kappa)$] with primordial perturbations (the Mukhanov-Sasaki variables denoted by v). The WDW constraint then has the form

$$\left[\frac{\mathrm{e}^{-3\alpha}}{a_0}\left(\frac{\kappa}{2}\frac{\partial^2}{\partial\alpha^2} + a_0^6\mathrm{e}^{6\alpha}\frac{H_0^2}{2\kappa}\right) + \frac{\mathrm{e}^{-\alpha}}{a_0}\hat{H}\right]\Psi(\alpha, v) = 0\ , \tag{26}$$

and it can be solved in perturbation theory following (20). The corrected Schrödinger equation [cf. (21)] reads

$$\mathrm{i}\frac{\partial\tilde{\psi}}{\partial\eta} = \left(\hat{H} - \kappa\frac{H_0^2\eta^4}{2}\hat{H}^2\right)\tilde{\psi} + \mathcal{O}(\kappa^2)\ , \tag{27}$$

where the usual conformal time η plays the role of the no-coupling proper time [cf. (16)]. From (27), we can compute corrections to the primordial power spectra:

$$\mathcal{P}_{S,T}(k) \simeq \mathcal{P}_{S,T;0}(k)\left\{1 + \kappa H_0^2\left(\frac{k_\star}{k}\right)^3[2.85 - 2\log(-2k\eta)]\right\}\ . \tag{28}$$

This shows that this formalism can serve as a starting point for computing phenomenological effects in canonical quantum gravity within a relational, perturbative, and unitary framework. Possible further developments are mentioned in [2,4].

4. Conclusions

In this conference article, we have briefly reviewed two relational approaches to models that exhibit a local time reparametrization invariance. Both approaches are systematic, model independent, and they can be applied either in the classical or quantum theories.

The first approach is nonperturbative, and it is predicated on the possibility of solving the WDW equation exactly. The suggested symmetric construction of relational observables leads to the correct quantum evolution via a set of diffeomorphism-invariant Heisenberg equations of motion. If the observables are also self-adjoint (or have self-adjoint extensions), then this evolution can be shown to be unitary [2, 7]. Time is intrinsic and defined from the energy spectrum of the Hamiltonian constraint.

The second approach is perturbative and consists of a weak-coupling expansion that is frequently used in the BO approach to quantum cosmology. The main result is that this formalism is, in fact, a unitary gauge theory of local time reparametrizations. The choice of gauge is induced by the weak-coupling expansion, and leads to the definition of a particular time variable: the no-coupling proper time. The inner product is perturbatively defined and it is conserved (at least up to the next-to-leading order). This settles the long-standing issue of unitarity of corrections to the Schrödinger equation in the BO approach. See [2, 4] for a comparison and comments on different BO treatments.

The two formalisms reviewed here can serve as starting points for a rigorous, relational understanding of dynamics in theories with local time reparametrization invariance, most notably quantum gravity. In particular, an adaptation of the weak-coupling expansion method to general slow-roll models could lead to unitary corrections to the usual expressions from which the power spectrum of the Cosmic Microwave Background radiation is derived. These corrections would constitute one avenue into the phenomenology of canonical quantum gravity. Moreover, the construction of quantum relational observables presented in Sec. 2.4 together with the postulates given in Sec. 2.5 could pave the way to a diffeomorphism-invariant description of the emergence of quantum mechanics. These are exciting prospects that ought to be considered in future research.

Acknowledgments

The author thanks Claus Kiefer, Manuel Krämer, David Brizuela, Branislav Nikolić, J. Brian Pitts, Alexander Y. Kamenshchik, Alessandro Tronconi, and Giovanni Venturi for discussions. The author is also grateful to the Dipartimento di Fisica e Astronomia of the Università di Bologna as well as the I.N.F.N. Sezione di Bologna for financial support.

References

[1] Smolin L 2006 *The Structural Foundations of Quantum Gravity* ed D Rickles *et al* (Oxford: Oxford University Press) pp 196-239
Giulini D 2007 Lect. Notes Phys. **721** 105
Kiefer C 2012 *Quantum Gravity* 3rd edn (International Series of Monographs on Physics) (Oxford: Oxford University Press)
Westman H and Sonego S (2008) *Found. Phys.* **38** 908
Westman H and Sonego S (2009) *Ann. Phys. (NY)* **324** 1585

[2] Chataignier L 2022 *Timeless Quantum Mechanics and the Early Universe* (Springer Theses) (Cham, Switzerland: Springer)

[3] Brizuela D, Kiefer C and Krämer M 2016 *Phys. Rev.* D **93** 104035
Brizuela D, Kiefer C and Krämer M 2016 *Phys. Rev.* D **94** 123527
Brizuela D, Kiefer C, Krämer M and Robles-Pérez S 2019 *Phys. Rev.* D **99** 104007
Steinwachs CF and van der Wild ML 2018 *Class. Quant. Grav.* **35** 135010
Steinwachs CF and van der Wild ML 2019 *Class. Quant. Grav.* **36** 245015
Bouhmadi-López M, Krämer M, Morais J and Robles-Pérez S 2019 *J. Cosmol. Astropart. Phys.* JCAP02(2019)057

Journal of Physics: Conference Series **2533** (2023) 012029 doi:10.1088/1742-6596/2533/1/012029

Kamenshchik AY, Tronconi A and Venturi G 2014 *Phys. Lett.* B **734** 72
Kamenshchik AY, Tronconi A and Venturi G 2016 *Phys. Rev.* D **94** 123524
[4] Chataignier L and Krämer M 2021 *Phys. Rev.* D **103** 066005
Chataignier L 2022 *Z. Naturforsch.* A **77** 805
[5] Henneaux M and Teitelboim C 1992 *Quantization of Gauge Systems* (Princeton, New Jersey: Princeton University Press)
[6] Tsamis NC and Woodard RP 1985 *Class. Quant. Grav.* **2** 841
Teitelboim C 1994 *Physical Origins of Time Asymmetry* ed JJ Halliwell *et al* (Cambridge: Cambridge University Press) pp 442-451
Woodard RP 1993 *Class. Quant. Grav.* **10** 483
Dittrich B 2007 *Gen. Rel. Grav.* **39** 1891
Dittrich B 2006 *Class. Quant. Grav.* **23** 6155
Tambornino J 2012 *SIGMA* **8** 017
[7] Chataignier L 2020 *Phys. Rev.* D **101** 086001
Chataignier L 2021 *Phys. Rev.* D **103** 026013
[8] Rovelli C 1990 *Phys. Rev.* D **42** 2638
Rovelli C 1991 *Class. Quant. Grav.* **8** 297
Rovelli C 1991 *Class. Quant. Grav.* **8** 317
Rovelli C 1991 *Phys. Rev.* D **43** 442
[9] 2013 *Gauge theories of gravitation - A Reader with Commentaries* ed M Blagojević and FW Hehl (Lond: Imperial College Press)
[10] Rovelli C 2002 *Phys. Rev.* D **65** 044017
[11] Pons JM, Salisbury DC, Sundermeyer KA 2009 *Phys. Rev.* D **80** 084015
Pons JM, Salisbury DC 2005 *Phys. Rev.* D **71** 124012
[12] Henneaux M, Teitelboim C and Vergara JD 1992 *Nucl. Phys.* B **387** 391
Shestakova TP and Simeone C 2004 *Grav. Cosmol.* **10** 257
[13] DeWitt B 1962 *Gravitation: An Introduction to Current Research* ed. L Witten (New York: Wiley) pp 266-381
Marolf D 1995 *Class. Quant. Grav.* **12** 1199
Marolf D 1995 *Class. Quant. Grav.* **12** 1441
Hartle JB and Marolf D 1997 *Phys. Rev.* D **56** 6247
Giddings SB, Marolf D and Hartle JB 2006 *Phys. Rev.* D **74** 064018
[14] Faddeev LD and Popov VN 1967 *Phys. Lett.* B **25** 29
Faddeev LD and Popov VN 1974 *Sov. Phys. Usp.* **16** 777 [*Usp. Fiz. Nauk* **111** 427 (1973)]
[15] Hoehn PA, Smith ARH and Lock MPE 2021 *Phys. Rev.* D **104** 06600
Hoehn PA, Smith ARH and Lock MPE 2021 *Front. in Phys.* **9** 181
[16] Busch P, Grabowski M, Lahti PJ 1995 *Operational Quantum Physics* (Lecture Notes in Physics Monographs vol 31) (Berlin: Springer)
Busch P, Lahti P, Pellonpää JP, Ylinen K 2016 *Quantum Measurement* (Theoretical and Mathematical Physics) (Switzerland: Springer International Publishing)
[17] DeWitt BS 1967 *Phys. Rev.* **160** 1113
[18] Rieffel MA 1974 *Adv. Math.* **13** 176
Landsman NP 1995 *J. Geom. Phys.* **15** 285
Marolf D 1995 *Preprint* gr-qc/9508015
Marolf D 2000 *Proc. of the 9th Marcel Grossmann Meeting on General Relativity* ed VG Gurzadian *et al* (Singapure: World Scientific Publishing Co) (*Preprint* gr-qc/0011112)
Halliwell JJ and Thorwart J 2001 *Phys. Rev.* D **64** 124018
Halliwell JJ and Thorwart J 2002 *Phys. Rev.* D **65** 104009
[19] Page DN and Wootters WK 1983 *Phys. Rev.* D **27** 2885
Wootters WK 1984 *Int. J. Theor. Phys.* **23** 701
Page DN 1994 *Physical Origins of Time Asymmetry* ed JJ Halliwell *et al* (Cambridge: Cambridge University Press) pp 287-298
Dolby CE 2004 *Preprint* gr-qc/0406034
[20] Kuchař KV 2011 *Int. J. Mod. Phys.* D **20** 3
Isham CJ 1993 *NATO Sci. Ser.* C **409** 157

Journal of Physics: Conference Series **2533** (2023) 012030 doi:10.1088/1742-6596/2533/1/012030

Classical space from quantum condensates

A Iorio [1] and L Smaldone[2]

[1] Institute of Particle and Nuclear Physics, Faculty of Mathematics and Physics, Charles University, V Holešovičkách 2, 18000 Praha 8, Czech Republic

[2] Institute of Theoretical Physics, Faculty of Physics, University of Warsaw ul. Pasteura 5, 02-093 Warsaw, Poland

E-mail: `iorio@ipnp.troja.mff.cuni.cz`

E-mail: `Luca.Smaldone@fuw.edu.pl`

Abstract. We review the boson transformation method to deal with spontaneous symmetry breaking in quantum field theory, focussing on how it describes the emergence of extended and classical objects in such quantum context. We then apply the method to the emergence of space itself, as an extended and classical object resulting from the evaporation of a quantum black hole. In particular, we show how classical torsion and curvature tensors can emerge as effects of an inhomogeneous Nambu–Goldstone boson condensation in vacuum, in $E(3)$ invariant spinor models with symmetry breaking.

1. Introduction

It has been known for a long time that classical theory of elasticity, in the presence of underlying crystal defects, can be described as an Einstein–Cartan theory in 3-dimensions with its geometry-gravity tensors, metric, g_{ij}, curvature, R^i_{jkl}, and torsion, T^i_{jk} [1, 2, 3, 4]. Similar ideas are behind the description, by means of quantum field theory (QFT) in curved spacetime[1], of the interaction between graphene quasi-particles of long-wavelength with the suitably engineered underlying carbon lattice, with disclination and dislocation topological defects [6, 7, 8, 9, 10, 11, 12, 13]. In principle, such emergent description could be derived from the underlying quantum electrodynamics, governing the elementary interactions between the electrons on the various molecular bands involved. In practice, this is not an easy task.

Within this general framework, it has been natural to put forward the proposal that both matter and space are emergent phenomena, stemming from the same fundamental quantum dynamics of basic constituents [14]. With Feynman [15] and Bekenstein [16], we call such fundamental entities "Xons" [17].

In Ref. [18] this quantum gravity (QG) scenario, the "QG quasiparticle picture", was applied to the study of black-holes (BH) information paradox. In that approach, BH evaporation [19] is viewed a consequence of Xons interaction. The emergent effects are the space outside the BH, obtained at the expenses of the horizon area, and matter, seen as "more traditional" Hawking radiation. In other words, BH evaporating here is viewed as a "factory" where both space and matter are forged. In Ref. [18] a connection between the fermionic nature of the Xons and the Bekenstein bound on entropy [20] was shown.

[1] For an approach based on deformed quantum algebras, see [5].

Despite the encouraging results just recalled, a mechanism for the emergence of (classical) space and (quatum) matter, in this QG quasiparticle picture, was only given recently in Ref. [21]. There Xons dynamics is assumed to be given by a QFT model based on a fermion field, $\Psi(x)$, on $\mathbb{R}^4$, where the latter is just a target space, with no associated geometric structure. Therefore x there is more similar to a label counting the Xon's degrees of freedom than to a coordinate. In fact, the true space and its torsion and curvature, are seen to emerge after the spontaneous symmetry breaking (SSB) of the $E(3)$ symmetry of the field equation for Ψ. In that discussion, time is just an affine parameter, which does not play any active role.

The results of [21] rely heavily on the *boson transformation method* (or simply *boson method*) of Refs. [22, 23]. The core of that idea is that classical extended objects, such as solitons, surface boundaries or topological defects, can be described in QFT as macroscopic manifestations of an inhomogeneous boson condensation in the vacuum. In particular, topological defects in a system with SSB, are due to the condensation of the Nambu–Goldstone (NG) bosons. This idea was successfully applied to many areas of physics, including topological defects in crystals [24, 25, 26], solitons [27], vortices [28, 29, 30], and strings and bags in hadron physics [31, 32].

The paper has the following structure: In Section 2 we review the boson method. In Section 3 we show that can be applied to describe emergent classical curvature and torsion. Finally, in Section 4 we propose our conclusions and perspectives.

2. The boson method

To properly introduce the boson method, we first need to recall the definition and to give some examples of what is known as the *Haag or dynamical map* [22, 23, 33, 34].

Consider a real scalar field φ satisfying

$$\left(\Box + m^2\right) \varphi(x) \ = \ j[\varphi](x) \, , \tag{1}$$

where $j[\varphi](x)$ is a functional of φ. Since in relativistic QFT the Fock space is defined in terms of asymptotic fields (e.g. *in* fields), the dynamics is completely solved once the interacting operator φ is expanded in terms of asymptotic fields[2]

$$\varphi(x) \ = \ \varphi_{in}(x) \ + \ \int \! \mathrm{d}^4 y \, \Delta_R(x-y) \, j(y) \, , \tag{2}$$

where $\Delta_R(x-y) = [\varphi(y) \, , \varphi(x)] \, \Theta(x-y)$ is the retarded propagator. This relation, known as the *Yang–Feldman equation* [35, 36], only holds in a weak-sense [36]. The iterative solution of Eq.(2) is a series involving normal ordered products of the asymptotic fields, known as Haag or dynamical map [22, 23, 33, 34]

$$\varphi(x) \ = \ v \ + \ \sum_{n=1}^{\infty} \int \! \mathrm{d}^4 x_1 \, \ldots \, \mathrm{d}^4 x_n \, F_n(x; x_1, \ldots, x_n) \ : \varphi_{in}(x_1) \, \ldots \, \varphi_{in}(x_n) : , \tag{3}$$

v being a constant. The coefficients of such expansion are truncated retarded Green's functions [22]:

$$F_n(x; x_1, \ldots, x_n) \ \equiv \ (\Box_{x_1} + m^2) \ldots \, (\Box_{x_n} + m^2) \, \langle R[\varphi(x) \, \varphi(x_1) \, \ldots \, \varphi(x_n)] \rangle \, , \tag{4}$$

where $\langle \ldots \rangle \equiv \langle 0| \ldots |0\rangle$.

In general one can consider a field Ψ satisfying the equation of motion

$$\Lambda(\partial) \, \Psi(x) \ = \ j[\Psi](x) \, , \tag{5}$$

[2] Here and in the following we do not write explicitly field renormalization constants.

where $\Lambda(\partial)$ is a differential operator, and rewrite it in the Yang–Feldman form

$$\Psi(x) \;=\; \Psi_0(x) \;+\; \left(\Lambda^{-1}\star j[\Psi]\right)(x)\,, \qquad \Lambda(\partial)\Psi_0(x) \;=\; 0\,, \tag{6}$$

where $\Lambda^{-1}\star j$ is the convolution of the interacting terms with a Green's function of the operator Λ. The appropriate Green's function depends on the choice of Ψ_0. For example, in Eq.(2), one has retarded Green's functions because $\Psi_0 = \varphi_{in}$.

Although we gave a relativistic QFT example, the present considerations are far more general. In condensed matter physics, e.g. Ψ_0 is a quasiparticle field. Solving Eq.(6), we get the dynamical map, which is indicated as

$$\Psi(x) \;=\; \Psi[x;\Psi_0]\,. \tag{7}$$

The $\lambda\phi^4$ theory, with ϕ a complex scalar field,

$$\mathcal{L}(x) \;=\; \partial_\mu\phi^\dagger(x)\partial^\mu\phi(x) - \mu^2\phi^\dagger(x)\phi(x) - \frac{\lambda}{4}|\phi^\dagger(x)\phi(x)|^2\,, \tag{8}$$

is considered in Ref. [37]. As well known, this theory is invariant under the $U(1)$ phase transformation $\phi \to e^{i\alpha}\phi$, with $\alpha \in \mathbb{R}$ and SSB occurs when $\mu^2 < 0$ and $\langle\phi\rangle = v = \sqrt{-2\mu^2/\lambda}$.

Let us write the complex field in terms of two real fields, ρ and χ

$$\phi(x) \;=\; (\rho(x) + v)e^{i\chi(x)}\,, \tag{9}$$

whose equations of motion are

$$\left[\Box - (\partial_\mu\chi(x))^2 \,+\, m^2\right]\rho(x) + \frac{3}{2}m\,g\,\rho^2(x) + \frac{1}{2}g\,\rho^3(x) \;=\; v\,(\partial_\mu\chi(x))^2\,, \tag{10}$$

$$\partial_\mu\left[(\rho(x)+v)^2\partial^\mu\chi(x)\right] \;=\; 0\,, \tag{11}$$

with $g = \sqrt{\lambda}$ and $m^2 = \lambda v^2 > 0$. The field Ψ_0 is chosen to be the *in*-field ϕ_{in}, satisfying the free field equation $\left(\Box + m^2\right)\phi_{in} \;=\; 0$ $(\Lambda(\partial) = \Box + m^2)$. Then, this is conveniently decomposed in a *radial* part ρ_{in} and in the phase χ_{in}, the latter being the NG mode, as

$$\Psi_0(x) = \phi_{in}(x) \equiv (\rho_{in}(x) + v)\,e^{i\chi_{in}(x)}\,, \tag{12}$$

satisfying the coupled equations

$$\left[\Box - (\partial_\mu\chi_{in}(x))^2 \,+\, m^2\right](\rho_{in}(x) \,+\, v) \;=\; 0\,, \tag{13}$$

$$\partial_\mu\left[(\rho_{in}(x)+v)^2\,\partial^\mu\chi_{in}(x)\right] \;=\; 0\,. \tag{14}$$

The dynamical map can be written in the compact form

$$\phi(x) \;=\; T_C\left\{(\rho_{in}(x)+v)e^{i\chi_{in}(x)}\exp\left[-i\int_C d^4y\mathcal{L}^I_{in}(y)\right]\right\}\,, \tag{15}$$

where T_C is the time-ordering along a Schwinger closed time-path from $t = -\infty$ to $t = +\infty$ and the way back, while $\mathcal{L}^I_{in} = (m^2-\mu^2)\phi^\dagger_{in}\phi_{in} - \frac{\lambda}{4}|\phi^\dagger_{in}\phi_{in}|^2 = 2m^2v\rho_{in} - gm\rho^3_{in} - \frac{g^2}{4}\rho^4_{in} + \frac{5}{4}m^2v^2$ is the (interaction picture) interacting Lagrangian [37].

In Section 3 we will follow a similar approach, taking the decomposition

$$\Psi_0(x) \;=\; \Psi_0[x;\psi(x),\varphi_1(x),\ldots,\varphi_N(x)]\,, \tag{16}$$

where ψ (the "radial mode") and φ_js (the NG modes) thus follow coupled field equations

$$\Lambda_\psi(\partial)\,\psi(x) \;=\; j_\psi[\psi,\varphi_1,\ldots,\varphi_N](x)\,, \qquad \Lambda_{\varphi_j}(\partial)\,\varphi_j(x) \;=\; j_{\varphi_j}[\psi,\varphi_1,\ldots,\varphi_N](x)\,, \tag{17}$$

$j = 1,\ldots,N$.

We have now all the needed ingredients to introduce the boson method, that is the main goal of this Section.

From Eq.(7) and the decomposition (16), we get

$$\Psi(x) \;=\; \Psi[x;\psi(x),\varphi_1(x),\ldots,\varphi_N(x)]\,. \tag{18}$$

The canonical transformation

$$\varphi_j(x) \;\rightarrow\; \varphi_j^f(x) \;\equiv\; \varphi_j(x) \;+\; f_j(x)\,, \tag{19}$$

with f_j being c-number functions, is known as *boson transformation*, which describes the condensation of φ_j quanta in the vacuum. Then, the *boson transformation theorem* [38, 22] asserts that

$$\Psi^f(x) \;=\; \Psi[x;\psi(x),\varphi_1^f(x),\ldots,\ldots,\varphi_N^f(x)]\,, \tag{20}$$

when φ_j^f satisfies the same equation as φ_j, is still a solution of the original dynamical problem, i.e.

$$\Lambda(\partial)\Psi^f(x) \;=\; j\left[\Psi^f\right](x)\,. \tag{21}$$

The solution Ψ^f describes the creation of self-sustained and extended objects, as macroscopic manifestation of the condensation in the vacuum of a large number of φ_j quanta. Physically relevant cases are those for which the f_js are not Fourier-transformable, i.e. when $\lim_{|x|\to\infty} f_j \to \infty$ or

$$[\partial_\mu\,,\partial_\nu]\,f_j(x) \;\neq\; 0\,, \tag{22}$$

at some point x.

The condition (22) can only be fulfilled when the φ_js are gapless fields, as in the case of NG bosons [22, 23]. Therefore, we turn our attention to field equations that have a symmetry group G, with Lie algebra

$$[t_a\,,\,t_b] \;=\; iC_{abc}\,t^c\,, \qquad a,b,c = 1,...,\mathrm{ord}\,G\,. \tag{23}$$

A system undergoes SSB when

$$Q_a|0\rangle \;\neq\; 0\,, \tag{24}$$

for some as, i.e., when some of the Noether charges do not annihilate the vacuum. We will then perform the boson transformation of the corresponding NG bosons

$$\varphi^a(x) \;\rightarrow\; \varphi^a(x) \;+\; f^a(x)\,, \tag{25}$$

into the dynamical map (18). Note that we now use the group index to properly count the NG fields.

When $\Psi \to \Psi^f$, one has the decomposition [22, 39]

$$\Psi^f(x) \;=\; U(x)\,\tilde{\Psi}(x)\,, \tag{26}$$

where $U(x) \equiv U[f(x)]$ is a local gauge transformation, whose rigid generator belongs to G, and where either $\tilde{\Psi}$ does not contain f or it only contains f through regular expressions (usually through regular derivatives, ∂f). It is also useful to introduce [22, 39]

$$A_\mu(x) \;=\; i\,U^{-1}(x)\partial_\mu U(x)\,, \tag{27}$$

$$F_{\mu\nu}(x) \;=\; i\,U^{-1}(x)\,[\partial_\mu\,,\partial_\nu]\,U(x)\,, \tag{28}$$

giving

$$F_{\mu\nu}(x) \;=\; \partial_\mu A_\nu(x) - \partial_\nu A_\mu(x) \;-\; i\,[A_\mu(x)\,,\,A_\nu(x)]\;. \tag{29}$$

If we write U in the form

$$U(x) \;=\; \exp\left(-i\,\alpha^a(x)\,t_a\right)\;, \tag{30}$$

where $\alpha^a(x)$ are functionals of f^a, then

$$A_\mu{}^a(x) \;=\; \partial_\mu \alpha^a(x)\,, \tag{31}$$

$$\begin{aligned} F_{\mu\nu}{}^a(x) \;&=\; \partial_\mu A_\nu{}^a(x) - \partial_\nu A_\mu{}^a(x) \;+\; C_{abc}\,A_\mu{}^b(x)\,A_\nu{}^c(x)\,,\\ &=\; [\partial_\mu\,,\,\partial_\nu]\,\alpha^a(x) \;+\; C_{abc}\,\partial_\mu\alpha^b(x)\,\partial_\nu\alpha^c(x)\,, \end{aligned} \tag{32}$$

$$A_\mu(x) \;=\; A_\mu{}^a(x)\,t_a\,, \qquad F_{\mu\nu}(x) \;=\; F_{\mu\nu}{}^a(x)\,t_a\,. \tag{33}$$

Let us remark that $F_{\mu\nu}{}^a(x)$ has a term arising from the nonabelian structure of the Lie group manifold G.

$A_\mu{}^a$ is a classical gauge field which is a physical manifestation of the inhomogeneous condensate of φ^a quanta. Since $F_{\mu\nu}{}^a$ is physically observable, one requires regularity:

$$[\partial_\mu\,,\,\partial_\nu]\,A_\lambda{}^a(x) \;=\; [\partial_\mu\,,\,\partial_\nu]\,\partial_\lambda\alpha^a(x) \;=\; 0\,, \qquad [\partial_\mu\,,\,\partial_\nu]\,F_{\lambda\rho}{}^a(x) \;=\; 0\,. \tag{34}$$

Before moving to the application of the boson method to the QG scenario of our interest here, let us go back to the example of the $\lambda\phi^4$ theory. In that case we have only one NG field χ_{in} (see Eq.(12)). Therefore, we perform the boson transformation

$$\chi_{in}^f(x) \;=\; \chi_{in}(x) \;-\; f(x)\,, \tag{35}$$

in the dynamical map (15):

$$\psi^f(x) \;=\; e^{-if(x)} T_C\left\{(\rho_{in}(x)+v)e^{i\chi_{in}(x)}\exp\left[-i\int_C d^4y\mathcal{L}_{in}^I(y)\right]\right\}\,. \tag{36}$$

Hence, in this case

$$U(x) \;=\; e^{-if(x)}\,, \qquad A_\mu(x) \;=\; \partial_\mu f(x)\,, \qquad F_{\mu\nu}(x) \;=\; [\partial_\mu,\partial_\nu]f(x)\,. \tag{37}$$

A relativistic string solution along the x_3-axis is obtained when f is the azimuthal angle in cylindrical coordinates, $f = \theta$. In that case, $\nabla\times\nabla\theta \;=\; 2\pi\hat{e}_3\,\delta(x_1)\delta(x_2)$, where $\hat{e}_3$ is the unit vector along the x_3-axis, hence $A_\mu \;=\; \partial_\mu\theta$ and $F_{12}(x) \;=\; 2\pi\delta(x_1)\delta(x_2)$, the other components being zero.

3. Classical geometry from $E(3)$ symmetry breaking

Our goal now is to apply the boson transformation method to the case of a QFT of a fermion, $\Psi(x)$, $x = (\sigma, \mathbf{x}) = (\sigma, x^1, x^2, x^3) \in \mathbb{R}\times\mathbb{R}^3$. As extensively explained in Ref.[21], this is a good description of the Xons model of Refs. [14, 17, 18, 13], for an observer whose energy scale is "low", compared to Planck's. The presentation will closely follow the one of Ref.[21].

Let us take the Xon field, $\Psi(x)$, in the irreducible spinor representation of $SO(3)$

$$\Psi(x) \;=\; \begin{pmatrix}\Psi_\uparrow(x)\\ \Psi_\downarrow(x)\end{pmatrix}\,, \tag{38}$$

Journal of Physics: Conference Series 2533 (2023) 012030 doi:10.1088/1742-6596/2533/1/012030

and let us assume that the field equation (5) for Ψ is invariant under $E(3)$. The $e(3)$ Lie algebra is

$$[P_a\,,\,P_b] \;=\; 0\,, \tag{39}$$

$$[P_a\,,\,J_b] \;=\; i\,\varepsilon_{abc}P_c\,, \tag{40}$$

$$[J_a\,,\,J_b] \;=\; i\,\varepsilon_{abc}\,J_c\,, \tag{41}$$

where $a, b, c = 1, 2, 3$.

When we perform rotations on $\Psi(x)$ we have

$$\Psi'(x) \;=\; e^{i\,\theta^a\,J_a}\,\Psi(x)\,e^{-i\,\theta^a\,J_a} \;=\; e^{-\frac{i}{2}\theta^a\sigma_a}\,\Psi(\sigma,\mathbf{x}')\,, \qquad x'^a \;\equiv\; R^a{}_b(\theta)x^b\,. \tag{42}$$

Here the θ^as are real parameters, R belongs to the fundamental representation of $SO(3)$ and σ_a are the Pauli matrices.

When we perform translations on $\Psi(x)$ we have

$$\Psi'(x) \;=\; e^{i\,u^a\,P_a}\Psi(x)e^{-i\,u^a\,P_a} \;=\; \Psi(\sigma,\mathbf{x}')\,, \qquad x'^a \;\equiv\; x^a + u^a\,, \tag{43}$$

where the u^as are real parameters.

The relevant issue here is to understand how to count NG modes [40, 41]. In fact, it is only in the direction where the symmetry is broken that one can produce a nontrivial (classical) gauge field $A_i{}^a$. A general result was presented in Ref. [41]. There, it was proved that a weak linear dependence condition of the form

$$\int\!\mathrm{d}^3x \sum_a c_a(x)\,j^0_a(x)|0\rangle \;=\; 0\,, \tag{44}$$

implies a redundancy in the number of NG bosons, which are reduced with respect to the number of broken generators (see Eq.(24)). In Eq.(44) j^0_a are the charge densities corresponding to the broken generators Q_a (see Eq.(24)), and c_a are generic functions.

Let us write $P_a = \int\!\mathrm{d}^3x p_a(x)$, $J_a = \int\!\mathrm{d}^3x j_a(x)$. The total angular momentum J_a can be decomposed into the orbital and the spin part $J_a = L_a + S_a$, with $L_a = \int\!\mathrm{d}^3x l_a(x)$, $S_a = \int\!\mathrm{d}^3x\Sigma_a(x)$.

For a scalar field, $J_a = L_a$. Since $l_a(x) = \varepsilon_{abc}x^b p_c(x)$, Eq.(44) is thus fulfilled in a strong sense and only $6 - 3 = 3$ independent NG bosons are present. Eq.(44) is also fulfilled when $S_a|0\rangle = 0$ and only three NG modes are present.

Here the key observation is that, when $S_a|0\rangle \neq 0$,

$$\int\!\mathrm{d}^3\,x\,j_a(x)|0\rangle \;=\; \int\!\mathrm{d}^3x\,\varepsilon_{abc}x^b p_c(x)|0\rangle + \int\!\mathrm{d}^3x\,\Sigma_a(x)|0\rangle\,. \tag{45}$$

Σ^a and p^a are linearly independent, thus j_a and p_a are, in general, linearly independent. In this case six NG bosons should appear in the spectrum.

In the following we will briefly review the two cases reported in the Table.

3.1. The case $P_a|0\rangle \neq 0$, $S_a|0\rangle = 0$

Let us first consider

$$P_a\,|0\rangle \;\neq\; 0\,, \qquad S_a\,|0\rangle \;=\; 0\,. \tag{46}$$

The dynamical map can be given in the compact form of Eq.(7) and we decompose $\Psi_0(x)$ as

$$\Psi_0(x) \;=\; \Psi_0[x;\psi(x), X^a(x)]\,. \tag{47}$$

Journal of Physics: Conference Series **2533** (2023) 012030 doi:10.1088/1742-6596/2533/1/012030

Table 1. Main features of the different SSB phases.

	$S\|0\rangle = 0$	$S\|0\rangle \neq 0$
NG fields	X^a	$X^a \ , \ \Theta^a$
A_i^a	$e_i{}^a \neq \delta_i^a \,, \quad \omega_i{}^a = 0$	$e_i{}^a \neq \delta_i^a \,, \quad \omega_i{}^a \neq 0$
$F_{ij}{}^a$	$T_{ij}{}^a \neq 0 \,, \quad R_{il}{}^a = 0$	$T_{ij}{}^a \neq 0 \,, \quad R_{il}{}^a \neq 0$

$X^a(x)$ $(a = 1, 2, 3)$ are the three NG modes which are the analogues of acoustic phonons in crystals [24, 25, 26, 41] (*space-phonons*), while our minimal assumption only allows for one type of fermionic matter $\psi(x)$, that is our radial mode.

Since $\Lambda(\partial)\Psi_0(x) = 0$, ψ and X^a obey coupled equations (see Eq.(17)):

$$\Lambda_\psi(\partial)\psi(x) \; = \; j_\psi[\psi, X^a](x) \,, \qquad \Lambda_X^a(\partial)X^a(x) \; = \; j_X[\psi, X^a](x) \,. \tag{48}$$

The boson transformation is then[3]

$$X^a(x) \; \to \; X_u^a(\mathbf{x}) \; \equiv \; X^a(x) \, + \, u^a(\mathbf{x}) \,, \tag{49}$$

where $X_u^a(x)$ satisfy the same equations as $X^a(x)$. By definition

$$\Psi^u(x) \; = \; \Psi[x; \psi(x), X_u^a(x)] \,, \tag{50}$$

which satisfies the field equation $\Lambda(\partial)\Psi^u = 0$, because of the boson transformation theorem. Moreover, (50) can be written as $\Psi^u = U\tilde{\Psi}$, with $[\partial_i, \partial_j]\tilde{\Psi} = 0$, and

$$U(\mathbf{x}) \; = \; \exp\left(-i \, y^a(\mathbf{x}) \, \mathcal{P}_a\right) \,. \tag{51}$$

$\alpha^a(\mathbf{x}) = y^a(\mathbf{x})$ is a functional of u so that, in general, $[\partial_i, \partial_j]y^a(\mathbf{x}) \neq 0$ and where $\mathcal{P}_a$ are a 2×2 matrix representation of the translations algebra. Note that we now indicate the indices of the coordinate with i, j, k. The reason is clear when looking at the classical gauge potential

$$A_j(\mathbf{x}) \; = \; e_j{}^a(\mathbf{x}) \, \mathcal{P}_a \,, \tag{52}$$

with

$$e_j{}^a(\mathbf{x}) \; \equiv \; \partial_j y^a(\mathbf{x}) \,. \tag{53}$$

Then y^a should be viewed as a set of flat coordinates (the metric δ_{ab} permits to contract group indices) and then $e_j{}^a(\mathbf{x})$ are identified with *Vielbeins*. The regularity condition (34) now reads

$$[\partial_i \,, \, \partial_j] \, e_k{}^a(\mathbf{x}) \; = \; [\partial_i \,, \, \partial_j] \, \partial_k y^a(\mathbf{x}) \; = \; 0 \,. \tag{54}$$

As a consequence, a metric tensor can be naturally defined on the configuration space

$$g_{ij}(\mathbf{x}) \; \equiv \; \delta_{ab} \, \partial_i y^a(\mathbf{x}) \, \partial_j y^b(\mathbf{x}) \; = \; \delta_{ab} \, e_i{}^a(\mathbf{x}) \, e_j{}^b(\mathbf{x}) \,. \tag{55}$$

Finally, one can compute the field strength Eq.(29)

$$F_{ij}(\mathbf{x}) \; = \; T_{ij}{}^a(\mathbf{x}) \, \mathcal{P}_a \,, \tag{56}$$

where

$$T_{ij}{}^a(\mathbf{x}) \; = \; \partial_i e_j{}^a(\mathbf{x}) - \partial_j e_i{}^a(\mathbf{x}) \,, \tag{57}$$

is the classical torsion tensor. We have then shown that *classical torsion emerges as a "macroscopic" manifestation of the condensation of the space-phonons.*

[3] Here we consider only static transformations, i.e. σ does not play any active role.

3.2. The case $P_a|0\rangle \neq 0$, $S_a|0\rangle \neq 0$

When

$$P_a\,|0\rangle \;\neq\; 0\,, \qquad S_a\,|0\rangle \;\neq\; 0\,, \tag{58}$$

we are spontaneously breaking the whole $E(3)$ symmetry, therefore six NG fields $X^a(x), \Theta^a(x)$ appear in the physical spectrum. Let us just stress once more that the rotational NG field is only possible because of the spin structure.

The dynamical map for the Xon field is

$$\Psi(x) \;=\; \Psi[x;\psi(x), X^a(x), \Theta^a(x)]\,. \tag{59}$$

The boson transformations

$$X^a(x) \;\to\; X^a_u \;\equiv\; X^a(x) \,+\, u^a(\mathbf{x})\,, \qquad \Theta^a(x) \;\to\; \Theta^a_\theta(\mathbf{x}) \;\equiv\; \Theta^a(x) \,+\, \theta^a(\mathbf{x})\,, \tag{60}$$

lead to

$$\Psi^{u,\theta}(x) \;=\; \Psi^{u,\theta}_0(x) \,+\, \ldots\,, \qquad \Psi^{u,\theta}_0(x) \;=\; \Psi_0[x;\psi(x), X^a_u(x), \Theta^a_\theta(x)]\,, \tag{61}$$

at the linear order of the dynamical map.

The gauge field $U(x)$ can be written as

$$U(\mathbf{x}) \;=\; \exp\left(-i\,\theta^a(\mathbf{x})\,\mathcal{J}_a - i\,y^a(\mathbf{x})\,\mathcal{P}_a\right)\,, \tag{62}$$

with $\mathcal{P}_a$ and $\mathcal{J}_a$ forming a 2×2 matrix representation of the $e(3)$ Lie algebra (39)-(41). Therefore

$$A_j(\mathbf{x}) \;=\; \omega_j{}^a(\mathbf{x})\,\mathcal{J}_a \,+\, e_j{}^a(\mathbf{x})\,\mathcal{P}_a\,. \tag{63}$$

It is clear that $\omega_j{}^a$ should be identified with the *spin connection*, hence the field strength (29) is

$$F_{ij}(\mathbf{x}) \;=\; R_{ij}{}^a(\mathbf{x})\,\mathcal{J}_a \,+\, T_{ij}{}^a(\mathbf{x})\,\mathcal{P}_a\,, \tag{64}$$

where the curvature and the torsion tensors have the usual expression

$$R^a_{ij}(\mathbf{x}) \;=\; \partial_i\omega_j{}^a(\mathbf{x}) - \partial_j\omega_i{}^a(\mathbf{x}) \,+\, \varepsilon_{abc}\,\omega_i{}^b(\mathbf{x})\omega_j{}^c(\mathbf{x})\,, \tag{65}$$

$$T_{ij}{}^a(\mathbf{x}) \;=\; \partial_i e_j{}^a(\mathbf{x}) - \partial_j e_i{}^a(\mathbf{x}) \,+\, \varepsilon_{abc}\,e_i{}^b(\mathbf{x})\,\omega_j{}^c(\mathbf{x})\,. \tag{66}$$

We have then proved that *a low-energy observer experiences non-trivial curvature and torsion, due to the non-trivial boson condensation of NG fields after the full SSB of* $E(3)$. The detailed dynamics clearly depends on the interaction with the matter ψ.

In the case of small u^a, $y^a(\mathbf{x}) \approx x^a + u^a(\mathbf{x})$ and it is reasonable to use the decomposition [21]

$$\Psi_0(x) \;=\; e^{-i[(x^a+X^a(x))\mathcal{P}_a+\Theta_a(x)\mathcal{J}_a]}\,\psi(x)\,, \tag{67}$$

that, after the boson transformation, gives

$$\Psi^{u,\theta}_0(x) \;=\; e^{-i[(y^a(\mathbf{x})+X^a(x))\mathcal{P}_a+(\Theta_a(x)+\theta^a(\mathbf{x}))\mathcal{J}_a]}\,\psi(x)\,. \tag{68}$$

If $\Lambda(\partial) \;=\; \gamma^\mu\,\partial_\mu$, γ^μ being 2×2 matrices, an average of the field equation $\Lambda(\partial)\Psi^{u,\theta}_0(x) = \gamma^\mu\,\partial_\mu\Psi^{u,\theta}_0(x) \;=\; 0$ with respect to the NG modes gives

$$i\,\gamma^\mu D_\mu\,\psi(x) \;=\; 0\,, \qquad D_\mu \;=\; \left(\partial_\sigma, \partial_j + e_j{}^a(\mathbf{x})\,\mathcal{P}_a + \omega_j{}^a(\mathbf{x})\,\mathcal{J}_a\right)\,, \tag{69}$$

which is the equation of a massless fermion field coupled with Vielbeins and spin connection [11].

Journal of Physics: Conference Series **2533** (2023) 012030 doi:10.1088/1742-6596/2533/1/012030

4. Conclusions

In this paper we reviewed the fundamental notion of dynamical map and the boson transformation method. Then we showed how this machinery can be employed to explain the emergence of classical geometric tensors as a low-energy manifestation of a non-trivial condensation in vacuum of the NG bosons in a $E(3)$-symmetric model with SSB. When translation symmetry is broken, these are the analogues of acoustic phonons in a crystal, while a non-trivial spin structure is necessary in order to have three independent NG fields related to the breaking of rotation symmetry. In the view of Ref.[21], this mechanism is realized in the QG quasiparticle picture (see Refs. [14, 18]), but it can be equally well-applied in other scenarios as condensed matter analogs of QG [6, 7, 8, 9, 10, 11, 12, 13].

An apparently unrelated issue is the the description of oscillations of neutrinos in QFT. In that context, a condensate structure of vacuum, which breaks Poincaré invariance, naturally appears [42, 43, 44, 45], and a dynamical origin of such condensate from an underlying fermionic dynamics with SSB has been extensively investigated [46, 47, 48]. It is then worth to investigate the connection with Xons.

Acknowledgements

A.I. acknowledges support from Charles University Research Center (UNCE/SCI/013). L.S. was supported by the Polish National Science Center grant 2018/31/D/ST2/02048 and thanks the Institute of Particle and Nuclear Physics of Charles University for the kind hospitality while parts of this work were completed.

References

[1] Kleinert H 1987 *Annalen der Physik* **499** 117–119
[2] Kleinert H 1989 *Gauge fields in condensed matter Vol. 2: Stresses and defects, Differential geometry, crystal melting* (World Scientific, Singapore)
[3] Katanaev M O and Volovich I V 1992 *Annals Phys.* **216** 1
[4] Zaanen J, Balm F and Beekman A J 2021 *arXiv 2109.11325 [gr-qc]*
[5] Iorio A, Lambiase G and Vitiello G 2001 *Annals Phys.* **294** 234–250
[6] Iorio A 2011 *Ann. Phys.* **326** 1334 – 1353 ISSN 0003-4916
[7] Iorio A and Lambiase G 2012 *Phys. Lett.* **B716** 334–337
[8] Iorio A and Lambiase G 2014 *Phys. Rev. D* **90** 025006
[9] Iorio A 2015 *Int. J. Mod. Phys. D* **24** 1530013
[10] Iorio A and Pais P 2015 *Phys. Rev. D* **92** 125005
[11] Iorio A and Pais P 2018 *Annals of Physics* **398** 265 – 286
[12] Ciappina M F, Iorio A, Pais P and Zampeli A 2020 *Phys. Rev. D* **101**(3) 036021
[13] Acquaviva G, Iorio A, Pais P and Smaldone L 2022 *Universe* **8**
[14] Acquaviva G, Iorio A and Scholtz M 2017 *Annals Phys.* **387** 317
[15] Feynman R, Leighton R and Sands M 2011 *The Feynman Lectures on Physics, Vol. II: The New Millennium Edition: Mainly Electromagnetism and Matter* The Feynman Lectures on Physics (Basic Books)
[16] Bekenstein J D 2003 *Scientific American* **289** 58–65
[17] Iorio A 2019 *J. Phys. Conf. Ser.* **1275** 012013
[18] Acquaviva G, Iorio A and Smaldone L 2020 *Phys. Rev. D* **102**(10) 106002
[19] Hawking S W 1975 *Commun. Math. Phys.* **43** 199–220 [Erratum: Commun.Math.Phys. 46, 206 (1976)]
[20] Bekenstein J D 1981 *Phys. Rev. D* **23** 287–298
[21] Iorio A and Smaldone L 2022 *Preprint, DOI:10.13140/RG.2.2.14442.18886* URL `https://www.researchgate.net/publication/362175062_Quantumblack_holes_as_classical_space_factories`
[22] Umezawa H, Matsumoto H and Tachiki M 1982 *Thermo-Field Dynamics and Condensed States* (North-Holland)
[23] Umezawa H 1993 *Advanced field theory: Micro, macro, and thermal physics* (AIP)
[24] Wadati M, Matsumoto H, Takahashi Y and Umezawa H 1977 *Physics Letters A* **62** 255–257
[25] Wadati M, Matsumoto H, Takahashi Y and Umezawa H 1978 *Fortschr. Phys.* **26** 357
[26] Wadati M, Matsumoto H and Umezawa H 1978 *Phys. Rev. B* **18**(8) 4077–4095
[27] Blasone M and Jizba P 2002 *Annals Phys.* **295** 230

Journal of Physics: Conference Series **2533** (2023) 012030 doi:10.1088/1742-6596/2533/1/012030

[28] Leplae L and Umezawa H 1969 *J. Math. Phys.* **10** 2038
[29] Leplae L, , Mancini F and Umezawa H 1970 *Phys. Rev. B* **2** 3594
[30] Acquaviva G, Iorio A and Smaldone L 2021 *Annals Phys.* **434** 168641
[31] Tze H and Ezawa Z 1975 *Phys. Lett. B* **55** 63
[32] Wadati M, Matsumoto H and Umezawa H 1978 *Phys. Rev. D* **18**(4) 1192–1195
[33] Haag R 1955 *Kong. Dan. Vid. Sel. Mat. Fys. Med.* **29N12** 1–37
[34] Greenberg O W 1994 Virtues of the Haag expansion in quantum field theory *1st Arctic Workshop on Future Physics and Accelerators* pp 0498–520
[35] Yang C N and Feldman D 1950 *Phys. Rev.* **79** 972–978
[36] Greiner W, Bromley D and Reinhardt J 2013 *Field Quantization* (Springer Berlin Heidelberg)
[37] Blasone M and Jizba P 2001 *Subnucl. Ser.* **37** 585–594
[38] Matsumoto H, Semenoff G, Umezawa H and Umezawa M 1980 *J. Math. Phys.* **21** 1761
[39] Matsumoto H, Umezawa H and Umezawa M 1981 *Fortsch. Phys.* **29** 441
[40] Brauner T 2010 *Symmetry* **2** 609–657
[41] Watanabe H and Murayama H 2013 *Phys. Rev. Lett.* **110** 181601
[42] Alfinito E, Blasone M, Iorio A and Vitiello G 1995 *Phys. Lett. B* **362** 91–96
[43] Blasone M, Smaldone L and Vitiello G 2019 *J. Phys. Conf. Ser.* **1275** 012023
[44] Blasone M, Jizba P, Mavromatos N E and Smaldone L 2020 *Phys. Rev. D* **102** 025021
[45] Smaldone L and Vitiello G 2021 *Universe* **7** 504
[46] Mavromatos N E, Sarkar S and Tarantino W 2009 *Phys. Rev. D* **80** 084046
[47] Mavromatos N E, Sarkar S and Tarantino W 2013 *Mod. Phys. Lett. A* **28** 1350045
[48] Blasone M, Jizba P, Mavromatos N E and Smaldone L 2019 *Phys. Rev. D* **100** 045027

Journal of Physics: Conference Series **2533** (2023) 012032 doi:10.1088/1742-6596/2533/1/012032

Time-loops to spot torsion on bidimensional Dirac materials with dislocations

Alfredo Iorio[1] **and Pablo Pais**[2,1]

[1] Faculty of Mathematics and Physics, Charles University
V Holešovičkách 2, 18000 Prague 8, Czech Republic
[2] Instituto de Ciencias Físicas y Matemáticas, Universidad Austral de Chile, Casilla 567, 5090000 Valdivia, Chile

E-mail: iorio@ipnp.troja.mff.cuni.cz, pais@ipnp.troja.mff.cuni.cz

Abstract. Assuming that, with some care, dislocations could be meaningfully described by torsion, we propose here a scenario based on an previously unexplored role of time in the low-energy Dirac field theory description of two-space-dimensional Dirac materials. Our approach is based on the realization of an exotic time-loop, that could be realized by an oscillating particle-hole pairs, to overcome well-known geometrical obstructions due to the luck of a spatial third dimension. General symmetry considerations allow to conclude that the effects we are looking for can only be seen if we move to the nonlinear response regime.

1. Introduction

Einstein's theory of General Relativity (GR) is the accepted theory of gravitation, surely for its beauty, but perhaps more solidly for its many observational successes. Recent remarkable examples of those are the discovery by the Event Horizon Telescope (EHT) of astrophysical black holes, first in the center of the galaxy M87 [1], then in the center of our Milky Way [2], and the observation of the gravitational waves by the Laser Interferometer Gravitational-wave Observatory (LIGO) [3], that opened the way to a new era in the exploration of the signals coming from the Universe surrounding us.

However, astronomical observations pointing to dark matter and dark energy, on the one hand, and theoretical challenges that mostly have to do with a quantum theory of gravity, on the other hand, call for a theory that extends GR. Since matter is characterized, at the fundamental level, by the *mass* and the *spin* of its elementary constituents[1], and since GR is a geometric theory fully based on curvature, whose source is the mass, a natural extension of GR is obtained by including torsion in the geometrical properties of the spacetime. Indeed, spin is related to torsion as mass is related to curvature. On this the literature is vast, so let us point to the seminal work of Tom Kibble [5].

Therefore, one direction to explore for extensions of GR is to include torsion as an extra feature of the spacetime geometry. This is the case of Einstein-Cartan (EC) theory of gravitation, where through a contorsion in the spin-connection, torsion appears in the physical phenomena,

[1] As fundamental particles locally obey quantum mechanics and special relativity theory, they are classified by reducible unitary representations of the Poincaré group [4], labelled by mass m and spin s.

Journal of Physics: Conference Series **2533** (2023) 012032 doi:10.1088/1742-6596/2533/1/012032

including the possibility of a minimal extension of GR. On this see the two excellent reviews [4] and [6], that have complementary points of view.

At this time, though, there are no direct experimental evidences of the effects of spacetime torsion, even though some argue that the very existence of spin itself is the only manifestation of spatiotemporal torsion [7], and that serious work is currently undertaken within certain cosmological scenarios [8]. Therefore, it comes handy that certain analog systems might furnish tabletop experimental grounds to test fundamental theories of gravity based on torsion.

Indeed, to include torsion in the geometric description of spacetime, Cartan himself was inspired by Cosserat's theory of crystal dislocations, see, e.g., the story told in [7]. On the other hand, three-dimensional gravity models have been employed to describe the (un)elastic properties of crystalline materials, where disclinations and dislocations correspond to curvature and torsion in the continuum limit, respectively, see, e.g., [9].

Given the natural occurrence of topological defects in graphene [10, 11] and other Dirac materials [12], and given that such materials have been proposed as versatile analogs of classical and quantum gravitational phenomena, as told in [13, 14, 15, 16, 17, 18, 19, 20, 21, 22, 23] and in the recent review [24], then it is very natural to seek for the role of torsion in the analog gravitational physics realized there. However, early attempts in that direction (that we shall recall later in Section 2), had negative outcomes, due to the geometrical obstruction existing in two-space-dimensional materials [11].

Here we present a possible way out from that obstruction, based on the results of [25], where time plays an unexpected role in the $(2+1)$-dimensional Dirac-like description of the low-energy quasiparticles of graphene [10].

2. The Riemann-Cartan spaces

Einstein himself [26] points out that, for a theory of gravitation as GR, more important than the metric is the parallel transport, taken into account by the affine connection. An n-dimensional (differential) manifold M, equipped with a connection Γ, is said to be *linearly connected*, and it is denoted by L_n. Given a point p on M, we say that a vector $v = v^\mu \partial_\mu$ of the tangent space at that point, T_pM, is parallel-transported along the direction dx^ν if its variation is given by

$$\delta_{||} v^\mu = v^\mu_{||}(x+dx) - v^\mu(x) = -\Gamma^\mu{}_{\rho\nu}\, v^\nu\, dx^\rho \ ,$$

where $\Gamma^\mu_{\rho\nu}$ is a *linear connection*. The covariant derivative ∇v is defined from the difference

$$v^\mu(x+dx) - v^\mu_{||}(x+dx) = v^\mu(x+dx) - v^\mu(x) - \delta_{||}\, v^\mu = (\partial_\nu v^\mu + \Gamma^\mu{}_{\nu\rho} v^\rho)\, dx^\nu \equiv (\nabla_\nu v^\mu)\, dx^\nu \ , \quad (1)$$

where in the second equality we expanded $v^\mu(x+dx)$ just up linear order in dx and, in the third equality, the order of the covariant indices of Γ must be noted. We stress here that the connection transforms under coordinate changes in an appropriate way in order for the quantity $\nabla_\nu v^\mu$ to transform as a covariant vector, and this can be generalized to an arbitrary rank tensor. Henceforth, the difference

$$T^\mu{}_{\nu\rho} \equiv (\Gamma^\mu{}_{\nu\rho} - \Gamma^\mu{}_{\rho\nu}) \ , \quad (2)$$

does transform as a third-rank tensor, and it is called *torsion tensor.*

The parallel transport is a path-dependent process and, if we parallel transport a vector around an infinitesimal loop, the components change in a way reflected by the *curvature tensor*

$$R^\mu{}_{\sigma\nu\rho} \equiv \partial_\nu \Gamma^\mu{}_{\rho\sigma} - \partial_\rho \Gamma^\mu{}_{\nu\sigma} + \Gamma^\mu{}_{\nu\tau}\, \Gamma^\tau{}_{\rho\sigma} - \Gamma^\mu{}_{\rho\tau}\, \Gamma^\tau{}_{\nu\sigma} \ , \quad (3)$$

that is a fourth-rank tensor. Importantly, both curvature and torsion can be defined in a manifold without the notion of a metric.

Journal of Physics: Conference Series **2533** (2023) 012032 doi:10.1088/1742-6596/2533/1/012032

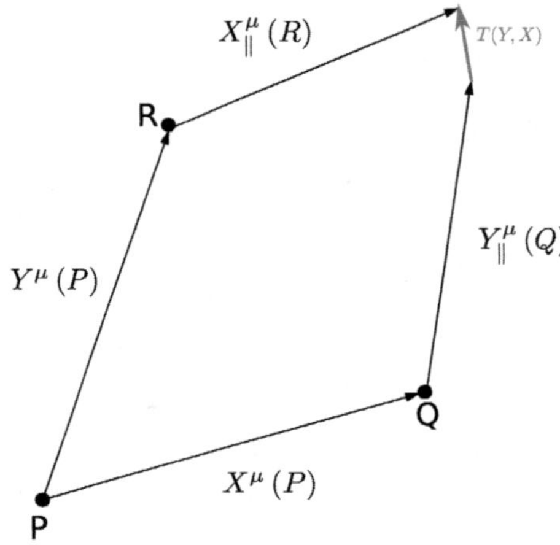

Figure 1. Geometrical interpretation of torsion in Riemann-Cartan spaces. Consider two vector fields X and Y at a point P. First parallel-transport X along Y to the infinitesimally close point R. Then, again from P, parallel-transport Y along X to reach a point Q, that in general does not coincide with R. In fact, it is precisely the failure to close the parallelogram the geometrical signal of a nonzero torsion. Its value is the difference of the two resulting vectors $T(X,Y)$. In Riemannian spaces V_n, this tensor is assumed to be zero. The picture was inspired by [7], but with the notation of [29].

From the definitions (2) and (3) it can be shown [27] that curvature and torsion are responsible for the noncommutativity of covariant derivatives of a vector in L_n,

$$\nabla_\nu \nabla_\rho v^\mu - \nabla_\rho \nabla_\nu v^\mu = R^\mu{}_{\sigma\nu\rho}\, v^\sigma - T^\sigma{}_{\nu\rho} \nabla_\sigma v^\mu \ ,$$

while for a scalar, φ, the noncommutativity is entirely due to torsion alone

$$\nabla_\nu \nabla_\rho \varphi - \nabla_\rho \nabla_\nu \varphi = T^\sigma{}_{\nu\rho} \partial_\sigma \varphi \ .$$

It is when we want to make contact with the experiments, and measure angles and distances between events in a spacetime manifold, that we introduce the *metric tensor* as a second-rank tensor defining the line element, i.e. the infinitesimal distance between two points as

$$ds^2 = g_{\mu\nu}\, dx^\mu\, dx^\nu \ , \tag{4}$$

whereas, by integration, we can define the longitude of any curve on L_n.

All the above also makes contact with the equivalence principle, as we can measure in inertial spaces *locally* diffeomorphic to Minkowski flat space. Regarding this principle, it was originally formulated by taking into account just point-like test particles and does not consider internal structures as spin. Therefore, the equivalence principle can by formulated in a U_n space. On this, see the discussion in [28].

A reasonable assumption is that the local distances do not change under parallel transportation. This can be assured if

$$\nabla_\rho g_{\mu\nu} = 0 \ . \tag{5}$$

The condition (5) for a linear connection Γ is called *metric compatibility*[2]. An n-dimensional manifold M with a linear connection preserving local distances, i.e. fulfilling condition (5), is called a *Riemann-Cartan (RC) space*, denoted by U_n. Fig. 1 gives a geometric interpretation of torsion, with details in the caption. This is the first link between the linear connection and the metric.

GR makes the further assumption that the torsion tensor vanishes, i.e., that the linear connection is symmetric. In such a case, the manifold is called (pseudo-)Riemannian and it is denoted by V_n. The unique linear connection, called *Levi-Civita connection*, can then be deduced directly from the metric [29]

$$\left\{ \begin{array}{c} \mu \\ \nu\,\rho \end{array} \right\} = \frac{1}{2} g^{\mu\sigma} \left(\partial_\nu g_{\rho\sigma} + \partial_\rho g_{\nu\sigma} - \partial_\sigma g_{\nu\rho} \right) . \tag{6}$$

The quantities (6) are called *Christoffel symbols* and the curvature associated to the Levi-Civita connection is the *Riemannian curvature tensor*, denoted by $\mathring{R}^\mu{}_{\nu\rho\sigma}$. In this way, the linear connection in a U_n space can be written as

$$\Gamma^\mu{}_{\nu\rho} = \left\{ \begin{array}{c} \mu \\ \nu\,\rho \end{array} \right\} + K^\mu{}_{\nu\rho} \,,$$

where $K^\mu{}_{\nu\rho} \equiv \frac{1}{2} (T^\mu{}_{\nu\rho} + T_\nu{}^\mu{}_\rho + T_\rho{}^\mu{}_\nu)$ is called *contorsion tensor*[3]. Notice that, contrary to the torsion tensor, K is not necessarily antisymmetric in the last two indices, unless torsion is totally antisymmetric (see below).

Apart from V_n spaces, there are other less standard "degenerated" cases of U_n spaces. One very known example is the Weitzenböck space [30], where the tensor (3) is assumed to be zero. Still, the Riemannian curvature is not necessarily zero, leading to the *teleparallel gravity* theory [31]. However, we shall not take this direction here.

It is useful to consider non-coordinate basis, e_μ^a, not only because this simplifies the notation, but also because we shall be dealing with spinors. The e_μ^a, also called *vielbeins*, are defined through the metric as

$$g_{\mu\nu} = e_\mu^a \, e_\nu^b \eta_{ab} \,,$$

where we choose the signature $\eta_{ab} = \mathrm{diag}(1, -1, \cdots)$ for the Minkowski metric. The *spin-connection*, $\omega_\mu^{ab} = e_\lambda^a (\delta_\nu^\lambda \partial_\mu + \Gamma_{\mu\nu}^\lambda) e^{b\nu}$, can be decomposed into torsion-free and contorsion contributions [29, 32], $\omega_\mu^{ab} = \mathring{\omega}_\mu^{ab} + \kappa_\mu^{ab}$, where $T_{\mu\nu}^\lambda = E_a^\lambda \kappa_\nu{}^a{}_b e_\mu^b - E_a^\lambda \kappa_\mu{}^a{}_b e_\nu^b$.

As we are interested in $(2+1)$-dimensional systems, from now on we will focus on U_3 spaces. A dynamical study of the U_3 spaces that shows how torsion is related to spin angular momentum can be found in [4] and references therein. Here we assume that the geometry is fixed and that there are no propagating gravitational degrees of freedom. Nonetheless, we shall keep in mind the natural role of torsion when spinorial fields are at stake [5]. With this, besides possible boundary terms, the graphene-Dirac action with the minimal GR extension in a U_3 space with a fixed spacetime in a RC geometry is then

$$S = \mathrm{i} \hbar v_F \int d^3x \, |e| \, \overline{\psi} \left(E_a^\mu \gamma^a \mathring{\nabla}_\mu - \frac{\mathrm{i}}{4} \gamma^5 \frac{\epsilon^{\mu\nu\rho}}{|e|} T_{\mu\nu\rho} \right) \psi \,. \tag{7}$$

[2] As $Q_{\rho\mu\nu} \equiv \nabla_\rho g_{\mu\nu}$ is a third-rank tensor, called *noonmetricity tensor*, we can wonder how far we can go by not assuming this to be zero. In fact, there are some theories where this tensor has a physical interpretation. However, in this work we always assume $Q_{\rho\mu\nu} = 0$. For more details about nomenclatures of different spaces, see [7].

[3] Some references called it *contortion tensor* [4]. However, as we are following closer the nomenclature of [29], we keep the name *contorsion*. As far as we know, there is no consensus yet about the name.

Journal of Physics: Conference Series **2533** (2023) 012032 doi:10.1088/1742-6596/2533/1/012032

The details of the deduction of each term of (7) are in [25]. Let us just mention that, being this the effective action for graphene, here we trade the speed of light, c, for the Fermi velocity, v_F. Furthermore, as for notation, $|e| = \sqrt{|g|}$, the covariant derivative, $\mathring{\nabla}_\mu$, is based on the torsion-free connection, $\mathring{\omega}_\mu^{ab}$, only, and $\gamma^5 \equiv i\gamma^0\gamma^1\gamma^2 = \begin{pmatrix} I_{2\times 2} & 0 \\ 0 & -I_{2\times 2} \end{pmatrix}$ (we used the conventions of [19] for $\gamma^0, \gamma^1, \gamma^2$ giving a γ^5 that *commutes* with the other three gamma matrices[4]).

From the action (7), the well-known fact that torsion couples to spinors through its totally antisymmetric component [6], can be read-off. Therefore, as remarked above, in this case the contorsion $K^\mu{}_{\nu\rho}$ is also antisymmetric, as can be seen from its definition. This mathematical fact is behind the obstruction pointed out some time ago leading to the conclusion that, in two-dimensional Dirac materials, torsion can have no physical role. On this see, e.g., [11, 33, 34].

3. Topological defects and their continuum limit in two-dimensional materials

In this work we focus primarily on graphene-like marerials, but many of the following considerations apply also to other two-dimensional crystals. The two most important topological defects in these materials are *disclinations* and *dislocations*[5]. These two defects are related to curvature and torsion, respectively, in a continuum limit description of graphene-like (and many other) materials [35, 36].

In a honeycomb lattice, i.e., with an hexagonal structure, the dislocation defect is a n-side polygon, with $n = 3, 4, 5$, or $n = 7, 8\,9\ldots$. The defects belonging to the first set ($n \leq 5$), carry an intrinsic *positive* curvature, while the defects belonging to the other set ($n \geq 7$) carry intrinsic *negative* curvature. Specific arrangements of such defects are then obeyed when the overall curvature is positive, see, e.g., [37], and other specific arrangements are obeyed when the overall curvature is negative, see, e.g., [38].

One way to produce dislocation defects is to have a dipole of disclinations with zero total curvature, as the pentagon-heptagon dipole of Fig. 2. This corresponds to introducing a strip in the lower half-plane of such a figure. The *Burgers vector* $\vec{b}$ characterizes this type of defect and its modulus is the width of the introduced strip. The process of cutting and sewing is called *Volterra process*, and it gives us the intuition of why these defects are topological [36]: one cannot undo them by a continuous transformation.

The torsion tensor in crystals is related to the Burgers vector through the formula [35, 9]

$$b^a = \int\int_\Sigma e^a_\lambda T^\lambda_{\mu\nu} dx^\mu \wedge dx^\nu \,, \tag{8}$$

where Σ is a surface containing the dislocation, but otherwise arbitrary, $a = 0, 1, 2$. However, the apparent simplicity of (8) should not shadow the subtleties that it involves. In fact, given a distribution of Burgers vector, it is not obvious how to assign a specific corresponding torsion tensor. See, e.g., [39], and [40] to learn on how the dislocation distributions have potential electronic device applications.

From (8) we see that, to have a nonzero Burgers vector giving rise to $\epsilon^{\mu\nu\rho}T_{\mu\nu\rho} \neq 0$ (which is necessary for a nonzero coupling in (7)), we can follow two roads: (i) a *time directed* screw dislocation (only possible if the crystal has a time direction)

$$b_t \propto \int\int T_{012} dx \wedge dy \,, \tag{9}$$

[4] This is due to the reducible, rather than irreducible, representation of the Lorentz group we use.

[5] Another important defect which could have a geometrical meaning, and has possible impacts in using graphene-like materials as analogs, is the *grain boundary* [19].

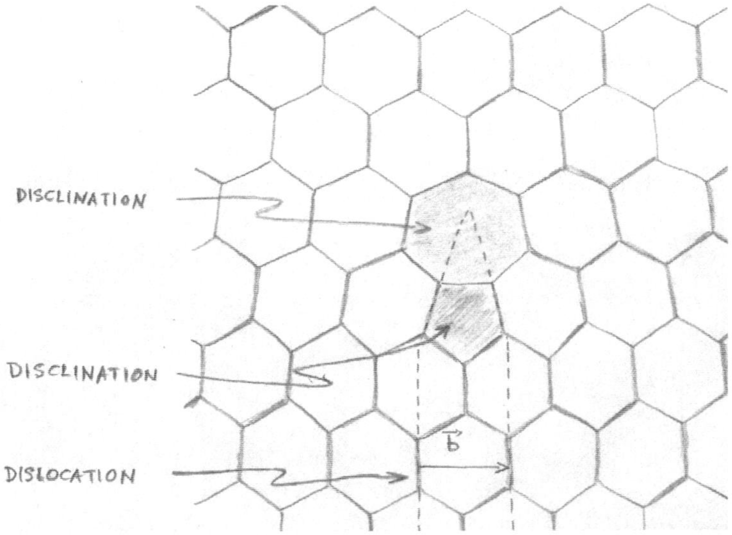

Figure 2. Edge dislocation from two disclinations. Two disclinations, an heptagon and a pentagon, adding up to zero total intrinsic curvature. This makes a dislocation with Burgers vector $\vec{b}$, as can be seen in the lower half-plane. Burgers vector in the continuum limit caries torsion. Figure taken from [25].

or (ii) an edge dislocation "felt" by an integration along a *space-time circuit* (only possible if we can actually go around a loop in time), e.g,

$$b_x \propto \int\int T_{102} dt \wedge dy \,. \tag{10}$$

4. Torsion through time crystals or through time-loops

The point we want to make here is that scenarios (i) and (ii) are, in fact, not impossible to realize in a laboratory.

The scenario (i) could be explored in the context of the fascinating time crystals introduced by Wilczek [41, 42], being nowadays the focus of intense experimental studies [43, 44]. Such lattices, discrete in all dimensions, including time, would be an interesting playground to probe ideas of quantum gravity [45]. In particular, they would be precious to explore defect-based models of classical gravity/geometry, see [46] and [35]. Despite the beauty of the scenario (i), we shall focus only on the more manageable, but still quite challenging, scenario (ii).

Let us assume the Riemann curvature to be zero, $\mathring{R}^{\mu}{}_{\nu\rho\sigma} = 0$, but torsion (or contorsion $K^{\mu}{}_{\nu\rho} \neq 0$) to be nonzero and let us choose a frame where $\mathring{\omega}^{ab}_{\mu} = 0$. The action (7) is then

$$S = \mathrm{i}\hbar v_F \int d^3x|e| \left(\overline{\psi}\gamma^\mu \partial_\mu \psi - \frac{\mathrm{i}}{4}\overline{\psi}_+ \phi \psi_+ + \frac{\mathrm{i}}{4}\overline{\psi}_- \phi \psi_- \right) , \tag{11}$$

where $\psi = (\psi_+, \psi_-)$ and $\phi \equiv \epsilon^{\mu\nu\rho} T_{\mu\nu\rho}/|e|$ is what we call *torsion field.* Even in the presence of torsion, the two irreducible spinors, ψ_+ and ψ_-, are decoupled (however, they couple to ϕ with opposite signs).

To overcome the three-dimensional geometric obstruction through the "time-loop" in the (y, t)-plane of the scenario (ii), see (10), our proposal is to make use of the particle-antiparticle description of the dynamics, encoded in the action (11). The picture goes as follows.

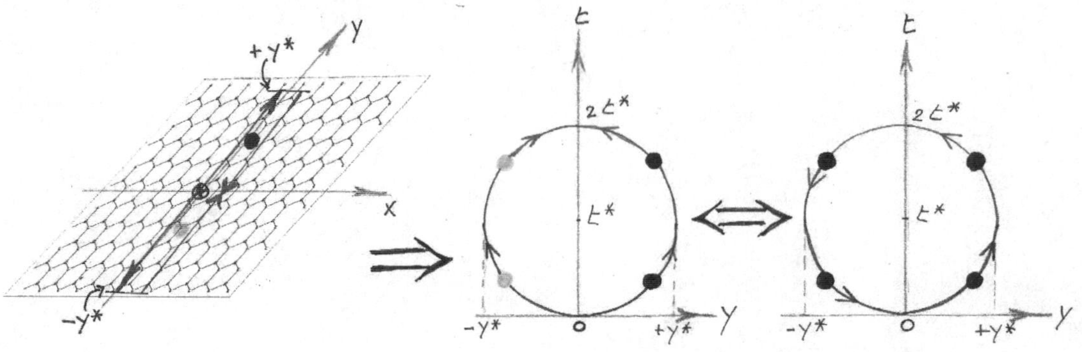

Figure 3. Idealized *time-loop*. At $t = 0$, the hole (yellow) and the particle (black) start their journey from $y = 0$, in opposite directions. Evolving forward in time, at $t = t^* > 0$, the hole reaches $-y^*$, while the particle reaches $+y^*$, (blue portion of the circuit). Then they come back to the original position, $y = 0$, at $t = 2t^*$ (red portion of the circuit). This can be repeated indefinitely. On the far right, the equivalent *time-loop*, where the hole moving forward in time is replaced by a particle moving backward in time. Figure taken from [25].

First, one realizes that the regime of the materials we are describing is the "half-filling" [10], for which the energy states of the valence band ($E < 0$) have the vacancies completely filled (being the analog of the Dirac sea), while the vacancies of the conduction band ($E > 0$) stay empty.

Think now of exciting a pair particle-hole out of this vacuum and to make them oscillate, say, along the y axis, as described in Fig. 3. This amounts to a circuit of the particle-antiparticle pair in the (y, t)-plane. What is left to do is to fully exploit the emergent relativistic-like structure of the model, and see the portion of the circuit described by the *antiparticle* moving *forward* in time, as corresponding to the same *particle* moving *backward* in time. This realizes what we may call a *time-loop*.

The pictures in Fig. 3 refer to a defect-free honeycomb graphene-like sheet. The presence of a dislocation, with Burgers vector $\vec{b}$ directed along x, would result in a failure to close the loop proportional to $\vec{b}$ [25].

Therefore, going from a first quantization to a second quantization approach, paying due attention to the subtleties involved in describing dislocation distributions by a suitable torsion tensor, the Dirac field theory emerging here can indeed include a nonzero coupling with torsion. This accounts for a field theoretical description of the effects of dislocations, only when the third dimension is taken to be time, overcoming the geometrical obstruction discussed earlier [11, 33, 34].

5. Linear and non-linear response regimes

The above is a nice idea, but the real challenge is to bring this idealized picture close to experiments. We present below the first steps in that direction, see [25] for more details.

The simplest settings we can envisage, to realize the picture above presented, need: i) an *external electromagnetic field* to excite the particle-hole pair necessary for the time-loop, and ii) that a suitable disclination/torsion provides the non-closure of the loop in the appropriate

Journal of Physics: Conference Series **2533** (2023) 012032 doi:10.1088/1742-6596/2533/1/012032

direction, something we shall refer to as *holonomy.* In other words, we are looking for *the measurable effects of a disclination/torsion-induced holonomy in a time loop.* It is only a *suitable combination* of those interactions that can produce the effect we are looking for. Notice this is more related to the *Cartan circuit* for torsion and curvature visualization [7] than the non-closing parallelogram torsion interpretation given in Fig. 1.

Therefore, the action governing the relevant microscopic dynamics is

$$S = \mathrm{i}\int d^3x\,|e|\,\left(\overline{\Psi}\gamma^\mu(\partial_\mu - \mathrm{i}g_{\rm em}\,A_\mu)\Psi - \mathrm{i}\,g_{\rm tor}\,\overline{\psi}_+\phi\psi_+ + \mathrm{i}\,g_{\rm tor}\,\overline{\psi}_-\,\phi\,\psi_-\right) \tag{12}$$

$$\to \mathrm{i}\int d^3x\,\left(\overline{\psi}\gamma^\mu\partial_\mu\psi - \mathrm{i}\,g_{\rm em}\,\hat{\jmath}^\mu_{\rm em}\,A_\mu - \mathrm{i}\,g_{\rm tor}\,\hat{\jmath}_{\rm tor}\,\phi\right) \equiv S_0[\overline{\psi},\psi] + S_I[A,\phi]\,. \tag{13}$$

where, we have set constants to one, $g_{\rm em}$ and $g_{\rm tor}$ are the electromagnetic and torsion coupling constants, respectively. In the last line, to avoid unnecessary complications, we considered only one Dirac point, say $\psi \equiv \psi_+$, and the metric is taken to be flat, $|e| = 1$, hence the indices are the flat ones, $\mu, \nu, ... \to a, b, ...$, but nonetheless, to ease the notation, we shall use Greek letters, anyway. Finally, $\hat{\jmath}^\mu_{\rm em} \equiv \overline{\psi}\gamma^\mu\psi$, while $\hat{\jmath}_{\rm tor} \equiv \overline{\psi}\psi$.

The electromagnetic field is *external*, hence a four-vector[6] $A_\mu \equiv (V, A_x, A_y, A_z)$. Nonetheless, the dynamics it induces on the electrons living on the membrane is two-dimensional, therefore, the effective vector potential may be taken to be $A_\mu \equiv (V, A_x, A_y)$, see, e.g., [47, 48]. Alternatively, the so-called *reduced QED* approach can be taken [49, 50]. In such approach, the gauge field propagates in a three-dimensional space and one direction is integrated out to obtain an effective interaction with the electrons constrained to move in a two-dimensional plane.[7]

As mentioned above, we do not consider the dynamics of defects here, hence the torsion field ϕ as well enters into the action as an external field. A different view, when ϕ is constant, is to include it into the unperturbed action, where it plays the role of a mass $S_0 \to S_m$, see, e.g., [53], where $S_m = i\int d^3x\,\overline{\psi}(\partial\!\!\!/ - m(\phi))\psi$.

We are in the situation described by the microscopic perturbation

$$S_I[F_i] = \int d^3x\,\hat{X}_i(\vec{x},t)F_i(\vec{x},t)\,, \tag{14}$$

with the system responding through $\hat{X}_i(\vec{x},t)$ to the external probes $F_i(\vec{x},t)$. The general goal is then to find

$$\hat{X}_i[F_i]\,, \tag{15}$$

to the extent of predicting a measurable effect of the combined action of the two perturbations $F_i(\vec{x},t)$: $F_1^{\rm em}(\vec{x},t) \propto A_\mu(\vec{x},t)$, that induces the response $\hat{\jmath}^\mu_{\rm em}$, and $F_2^{\rm tor}(\vec{x},t) \propto \phi(\vec{x},t)$, that induces the response $\hat{\jmath}_{\rm tor}$:

$$S_I[A,\phi] = \int d^3x\,\left(\hat{\jmath}^\mu_{\rm em}A_\mu + \hat{\jmath}_{\rm tor}\phi\right)\,, \tag{16}$$

where we have included the couplings, $g_{\rm em}$ and $g_{\rm tor}$, in the respective currents.

In fact, in our model, described microscopically by the action (13), we can indeed produce a prediction based on the charge conjugation invariance of that emergent relativistic theory. Such prediction is that

$$\chi_\mu^{\rm torem}(x,x') \sim \langle \hat{\jmath}_\mu^{\rm em}(x)\hat{\jmath}^{\rm tor}(x')\rangle \equiv 0\,. \tag{17}$$

[6] A different, if not more naturally $(2+1)-$dimensional setting would be to obtain A_μ by suitably straining the material, see, e.g., [33, 34], and [17]. In that case, a typical setting is $A_t \equiv 0$, $A_x \sim u_{xx} - u_{yy}$, $A_x \sim 2u_{xy}$, where u_{ij} is the strain tensor.

[7] This approach could shed some light on the appearance of a photon Chern-Simons term. On this see [51, 52].

Journal of Physics: Conference Series **2533** (2023) 012032 doi:10.1088/1742-6596/2533/1/012032

This is nothing more than an instance of the Furry's theorem of quantum field theory [54], that in QED reads

$$\chi^{\rm em}_{\mu_1 \cdots \mu_{2n+1}}(x_1, ..., x_{2n+1}) \sim \langle \hat{j}^{\rm em}_{\mu_1}(x_1) \cdots \hat{j}^{\rm em}_{\mu_{2n+1}}(x_{2n+1}) \rangle = 0 \,, \tag{18}$$

and for us implies

$$\chi^{\rm torem}_{\mu_1 \cdots \mu_{2n+1}}(x_1, ..., x_{2n+1}, y_1, ..., y_m) \sim \langle \hat{j}^{\rm em}_{\mu_1}(x_1) \cdots \hat{j}^{\rm em}_{\mu_{2n+1}}(x_{2n+1}) \hat{j}^{\rm tor}(y_1) \cdots \hat{j}^{\rm tor}(y_m) \rangle = 0 \,. \tag{19}$$

This result tells us that the effects we are looking for can only be seen if we move to the nonlinear response regime. We can resort to a well developed technique, the high-order harmonic generation (HHG), able to characterize structural changes both in atoms and molecules and, more recently, bulk materials (for a recent review see e.g. [55]). Therefore, in our scheme, the intra-band harmonics, governed by the intra-band (electron-hole) current, will be strongly modified, depending on whether dislocations ar there or not.

6. Conclusions

When time is duly included in the emergent relativistic-like picture of Dirac materials, the geometric obstruction to describe the effects of dislocations in terms of a suitable coupling with torsion, within the $(2+1)$-dimensional field theoretical description of the π-electrons dynamics, can be overcome. However, surely problems remain to be addressed, like a unique assignment of torsion to a given distribution of Burgers vectors. Nonetheless, when this is possible, our suggestion here opens the doors to the use of these materials as analogs of many important theoretical scenarios where torsion plays a role.

We then show the first steps toward testing the realization of these scenarios. We have focussed on one of the *Gedankenexperiments* of [25] on *time-loop*. This could spot the presence of edge dislocations, routinely produced in Dirac materials. The effect must be based on the interplay between an external electromagnetic field (necessary to excite the pair particle-hole that realizes the time-loops), and a suitable distribution of dislocations described as torsion (that will be responsible for a measurable holonomy in the time-loop).

We also can prove, on very general grounds, that a nonlinear response regime is necessary. In particular, we speculate that in a HHG technique, the structure of such response, would include manifestation of the torsion-induced holonomy of the time-loop through specific patterns of suppression and generation of higher harmonics. This sounds promising, for an experimental finding, as the laser-graphene interaction, controlling electron dynamics on an unprecedented precision scale, is the focus of intense studies, both theoretical and experimental, see, e.g., [47, 48].

Acknowledgments

We sincerely thank Thomas Elze and the other organizers for putting together an amazing conference. We are indebted to Marcelo Ciappina and Adamantia Zampeli for their collaboration on these exciting matters, and to Jorge Zanelli for many fruitful inspirational discussions. We gladly acknowledge support from Charles University Research Center (UNCE/SCI/013). P. P. is also supported by Fondo Nacional de Desarrollo Científico y Tecnológico–Chile (Fondecyt Grant No. 3200725).

References

[1] Alef W *et al.* (Event Horizon Telescope Collaboration) 2019 *Astrophys. J. Lett.* **875** L1
[2] Akiyama K *et al.* (Event Horizon Telescope Collaboration) 2022 *Astrophys. J. Lett.* **930** L12
[3] Abbott B P *et al.* (LIGO Scientific Collaboration and Virgo Collaboration) 2016 *Phys. Rev. Lett.* **116** 061102
[4] Hehl F W, Von Der Heyde P, Kerlick G D and Nester J M 1976 *Rev. Mod. Phys.* **48** 393

[5] Kibble T W B 1961 *J. Math. Phys.* **2** 212
[6] Shapiro I L 2002 *Phys. Rept.* **357** 113
[7] Hehl F W and Obukhov Y N 2007 *Ann. Fond. Broglie* **32** 157
[8] Mavromatos N E 2021 *Universe* **7** 480
[9] Katanaev M O 2005 *Phys. Usp.* **48** 675 [Usp. Fiz. Nauk175,705(2005)]
[10] Castro Neto A H, Guinea F, Peres N M R, Novoselov K S and Geim A K 2009 *Rev. Mod. Phys.* **81** 109
[11] de Juan F, Cortijo A and Vozmediano M A H 2010 *Nucl. Phys.* **B828** 625
[12] Wehling T, Black-Schaffer A and Balatsky A 2014 *Adv. Phys.* **63** 1
[13] Iorio A 2011 *Ann. Phys. (Amsterdam)* **326** 1334
[14] Iorio A and Lambiase G 2012 *Phys. Lett.* **B716** 334
[15] Iorio A and Lambiase G 2014 *Phys. Rev.* **D90** 025006
[16] Iorio A 2015 *Int. J. Mod. Phys.* **D24** 1530013
[17] Iorio A and Pais P 2015 *Phys. Rev.* **D92** 125005
[18] Iorio A and Pais P 2022 *Phys. Rev.* **B106** 157401
[19] Iorio A and Pais P 2018 *Ann. Phys. (Amsterdam)* **398** 265
[20] Iorio A 2015 *Front. Mater.* **1** 36
[21] Iorio A, Lambiase G, Pais P and Scardigli F 2020 *Phys. Rev.* **D101** 105002
[22] Iorio A, Ivetić B, Mignemi S and Pais P 2022 *Phys. Rev.* **D106** 116011
[23] Iorio A, Pais P, Elmashad I A, Ali A F, Faizal M and Abou-Salem L I 2018 *Int. J. Mod. Phys.* **D27** 1850080
[24] Acquaviva G, Iorio A, Pais P and Smaldone L 2022 *Universe* **8** 455
[25] Ciappina M F, Iorio A, Pais P and Zampeli A 2020 *Phys. Rev.* **D101** 036021
[26] Pantaleo M and Einstein A 1955 *Cinquant' anni di relatività 1905-1955* (Sansoni)
[27] Santaló L A 1973 *Vectores y Tensores con sus Aplicaciones* (Editorial Universitaria de Buenos Aires)
[28] Von Der Heyde P 1975 *Nuovo Cimento Lett. (1971-1985)* **14** 250
[29] Nakahara M 2003 *Geometry, Topology and Physics, Second Edition* Graduate student series in physics (Taylor & Francis)
[30] Weitzenböck R 1923 *Noordhoff, Groningen* 416
[31] Aldrovandi R and Pereira J 2012 *Teleparallel Gravity: An Introduction* Fundamental Theories of Physics (Springer Netherlands)
[32] Hassaine M and Zanelli J 2016 *Chern-Simons (Super)Gravity* 100 Years of General Relativity (World Scientific Publishing Company Pte Limited)
[33] Vozmediano M A H, Katsnelson M I and Guinea F 2010 *Phys. Rept.* **496** 109
[34] Amorim B *et al.* 2016 *Phys. Rept.* **617** 1
[35] Kleinert H 1989 *Gauge fields in condensed matter. Vol. 2: Stresses and defects. Differential geometry, crystal melting* (Singapore: World Scientific (1989))
[36] Ruggiero M L and Tartaglia A 2003 *Am. J. Phys.* **71** 1303
[37] Gonzalez J, Guinea F and Vozmediano M 1993 *Nucl. Phys.* **B406** 771
[38] Taioli S, Gabbrielli R, Simonucci S, Pugno N M and Iorio A *J. Phys. Condens. Matter* **28** 13LT01
[39] Lazar M 2003 *J. Phys.* **A36** 1415
[40] Banhart F, Kotakoski J and Krasheninnikov A V 2010 *ACS nano* **5** 26
[41] Wilczek F 2012 *Phys. Rev. Lett.* **109** 160401
[42] Shapere A and Wilczek F 2012 *Phys. Rev. Lett.* **109** 160402
[43] Li T, Gong Z X, Yin Z Q, Quan H T, Yin X, Zhang P, Duan L M and Zhang X 2012 *Phys. Rev. Lett.* **109** 163001
[44] Smits J, Liao L, Stoof H T C and van der Straten P 2018 *Phys. Rev. Lett.* **121** 185301
[45] Loll R 1998 *Liv. Rev. Rel.* **1** 13
[46] Katanaev M O and Volovich I V 1992 *Ann. Phys. (N.Y.)* **216** 1
[47] Heide C, Higuchi T, Weber H B and Hommelhoff P 2018 *Phys. Rev. Lett.* **121** 207401
[48] Higuchi T, Heide Christian ad Ullmann K, Weber H B and Hommelhoff P 2017 *Nature* **550** 224
[49] Marino E C 1993 *Nucl. Phys.* **B408** 551
[50] Gorbar E V, Gusynin V P and Miransky V A 2001 *Phys. Rev.* **D64** 105028
[51] Dudal D, Mizher A J and Pais P 2018 *Phys. Rev.* **D98** 065008
[52] Dudal D, Mizher A J and Pais P 2019 *Phys. Rev.* **D99** 045017
[53] Andrianopoli L, Cerchiai B L, D'Auria R, Gallerati A, Noris R, Trigiante M and Zanelli J 2020 *JHEP* **01** 084
[54] Peskin M and Schroeder D 1995 *An Introduction to Quantum Field Theory* Advanced book classics (Addison-Wesley Publishing Company)
[55] Kruchinin S Y, Krausz F and Yakovlev V S 2018 *Rev. Mod. Phys.* **90** 021002

Journal of Physics: Conference Series **2533** (2023) 012031 doi:10.1088/1742-6596/2533/1/012031

Analog simulations of early universe pre-heating and the back-reaction effect

Salvatore Butera

School of Physics and Astronomy, University of Glasgow, Glasgow G12 8QQ, UK

E-mail: `salvatore.butera@glasgow.ac.uk`

Iacopo Carusotto

Pitaevskii BEC Center, CNR-INO and Dipartimento di Fisica, Università di Trento, I-38123 Trento, Italy

E-mail: `iacopo.carusotto@unitn.it`

Abstract. We theoretically propose a ring-shaped, two-dimensional atomic Bose-Einstein condensate as analog model to investigate back-reaction effects during the pre-heating of the early universe. We study the out-of-equilibrium dynamics by which the inflaton field decays by parametrically exciting the vacuum fluctuations that initially populate the matter fields. By working at the level of the truncated Wigner approximation, our numerical simulations show how a signature of back-reaction beyond the semiclassical level is encoded in the effective friction experienced by the analog of the inflaton field, as well as in the spatial de-phasing of its oscillations and in the entanglement between the inflaton and matter fields degrees-of-freedoms.

1. Introduction

Analog models of gravity are a powerful platform where a wide range of effects of quantum fields in curved spacetime can be studied from first principles and potentially find experimental confirmation [1]. The past few decades have seen remarkable advances in the field, resulting in a surge of both theoretical and experimental works investigating different phenomena with a multitude of analog configurations (see topical reviews [2, 3], and references therein). Experimentally, exceptional results have been achieved leading to the pioneering observation of the Hawking radiation [4, 5] emanating from a sonic black hole [6, 7, 8] implemented in trans-sonically flowing Bose-Einstein condensates (BECs) of ultra-cold atoms [9, 10], as well as to the observation of cosmological particle creation both in a BEC [11] and in quantum fluids of light [12], and the detection of superradiant scattering induced by a rotating sonic black holes, implemented in a water tank exhibiting a draining vortex flow configuration [13].

The next challenge that stands in front of the analog gravity community is to extend these investigations to the so-called *back-reaction* effect [14], thus widening the objectives of the field beyond the standard test-field level of quantum fields in curved spacetime and address the role of the mutual, that is two-way, interaction between spacetime and quantum fields. Clarifying this interplay is of fundamental importance in gravity to gain a deeper understanding of those physical configurations characterized by strong gravitational interactions, in which quantum effects are important. Pioneering studies have in fact anticipated that quantum effects may have had a profound influence in driving the evolution of the early universe towards the present stage [15, 16, 17, 18, 19]. Related to this, black holes are expected to evaporate and eventually disappear due to the emitted Hawking radiation [4, 5]. Despite the amount of work carried out since the 1970's (see for example [20]), a clear picture of the physical processes involved in the back-reaction are still far from being clearly understood. The main reason is that a detailed understanding of the microscopic dynamics of spacetime is missing and only alternative top-down approaches, which attempt to include quantum fluctuations starting from the macroscopic level of the classical theory of general relativity, can be pursued [21].

To circumvent this fundamental difficulty, toy models characterized by a lower number of degrees-of-freedom have been extensively exploited in the literature, with the aim of capturing at least the main qualitative features of the back-reaction. For example, by using a zero-dimensional model for a black hole and by pursuing a fully quantum treatment of the problem, the deviation from thermal spectrum of the (analog of) Hawking radiation has been predicted [22, 23]. Similarly, in optomechanics, theoretical studies have anticipated a radiative friction experienced by an accelerated mirror in response to the dynamical Casimir emission (DCE) [24, 25, 26], and hinted at an important role of quantum fluctuations in determining the evolution of the quantum state of the mirror itself [27, 28, 29, 30].

Analog systems represent a promising platform where back-reaction effects can be investigated in the framework of quantum fields interacting with an effective background spacetime, thus retaining the full multi-mode character of the problem [31, 32]. On the one side, these systems bring the unprecedented capability to perform table-top experiments where these effects manifest in observable physical systems. On the other side, the in depth knowledge of the microscopic dynamics of these systems permits, at least in principle, the development of a self-consistent theoretical description of the problem starting from first principles. Even though the microscopic dynamics of analog systems is expected to be different from the (still unknown) physics of spacetime at the Planck scale, crucially, the mesoscopic observable effects of the back-reaction such as *fluctuation*, *dissipation* and *decoherence* are expected to be universal, since they result from a coarse-graining process, and are thus ultimately insensitive to the details of the microscopic physics. The basic mechanisms that drive the back-action of quantum fields are thus qualitatively the same, independently whether the underlying spacetime is physical (in the case of gravity) or effective (in the case of analogue systems). This assumption is well-established and is at the basis of effective field theories [33].

In this paper we study the back-reaction by considering an analog model of the pre-heating of the early Universe [34] inspired to the proposal in [35]. Our work substantially extends the short presentation recently appeared in [36]. According to the inflationary model, pre-heating is the last stage of cosmic inflation, when the inflaton field that drives the primordial exponential expansion of the universe reaches the end of its potential plateau and starts oscillating around the minimum of the potential well. Matter is thus created in the still empty universe out of the zero-point fluctuations of the matter fields, due to the parametric processes induced by the inflaton oscillations. In our analog model, we simulate this mechanism by using an elongated BEC as a platform, that we take as two-dimensional for numerical ease. Specifically, we identify the inflaton degrees-of-freedom with the relatively high energy transverse modes of the system, and the matter degrees-of-freedom with the lower energy longitudinal modes. We simulate the inflaton oscillations around its potential minimum by exciting the breathing mode of the system in the transverse direction and study the mechanism that leads to the parametric excitation of vacuum fluctuations in the longitudinal modes. By working with this quantum simulator, we aim to clarify key non-equilibrium processes by which energy is transferred from the transverse modes into the longitudinal modes, focusing on back-reaction effects. In the spirit of analog simulators and by working within the analogy described above, we expect that our results reveal the key qualitative features of the corresponding processes taking place in an actual cosmological scenario.

2. Bogoliubov theory

We perform our analog simulations by considering a dilute two-dimensional Bose gas of mass m atoms at zero temperature, whose many-body Hamiltonian reads [37]:

$$\hat{H} = \int d\mathbf{r} \left[\hat{\Psi}^\dagger(\mathbf{r}) \hat{h} \hat{\Psi}(\mathbf{r}) + \frac{U}{2} \hat{\Psi}^\dagger(\mathbf{r}) \hat{\Psi}^\dagger(\mathbf{r}) \hat{\Psi}(\mathbf{r}) \hat{\Psi}(\mathbf{r}) \right]. \tag{1}$$

Here, $\hat{h} \equiv -(\hbar^2/2m)\boldsymbol{\nabla}^2 + V_{\text{ext}}(\mathbf{r})$ is the quantum mechanical single particle Hamiltonian, where $\boldsymbol{\nabla} \equiv d/d\mathbf{r}$ the standard *nabla* operator, $V_{\text{ext}}(\mathbf{r})$ is an external potential, and $\hat{\Psi}(\mathbf{r})$ is the bosonic many-body field operator which obeys the equal time commutation rule: $[\hat{\psi}(\mathbf{r}), \hat{\psi}^\dagger(\mathbf{r}')] = \delta(\mathbf{r}-\mathbf{r}')$. The last terms in Eq. (1) accounts for a zero-range collisional interaction between atoms, whose strength we indicated by U. The Heisenberg equation for the field operator $\hat{\Psi}(\mathbf{r}, t)$ is readily obtained in the form:

$$i\hbar \frac{\partial \hat{\Psi}}{\partial t} = [\hat{\Psi}, \hat{H}] = \left(\hat{h} + g \hat{\Psi}^\dagger \hat{\Psi} \right) \hat{\Psi}. \tag{2}$$

By following the number conserving Bogoliubov formalism developed in [38, 39], we split the field operator $\hat{\Psi}(\mathbf{r}, t)$ into the *condensed* component describing atoms occupying the single particle ground state, and the *non-condensed* component that accounts for the population of the excited single-particle states:

$$\hat{\Psi}(\mathbf{r}, t) = \phi_0(\mathbf{r}, t) \hat{a}_0 + \delta \hat{\Psi}(\mathbf{r}, t). \tag{3}$$

Here the operator $\hat{a}_0$ annihilates a particle from the condensate mode $\phi_0(\mathbf{r}, t)$, while $\delta \hat{\Psi}(\mathbf{r}, t)$ annihilates a non-condensed particle at position $\mathbf{r}$. In the weakly-interacting regime, identified by the limit of large number of particles $N \to \infty$, and vanishing coupling constant $U \to 0$ (with the mean field interaction strength $U_N \equiv UN$ kept constant), a perturbative description for the system can be pursued [38] with respect to the parameter $\delta N/\sqrt{N}$, with

$$\delta N \equiv \int d\mathbf{r} \left\langle \delta \hat{\Psi}^\dagger(\mathbf{r}) \delta \hat{\Psi}(\mathbf{r}) \right\rangle \tag{4}$$

the number of particles in the excited single-particles states outside the condensate. The value of δN is set by the mean field interaction strength U_N and is of order $\delta N \sim \mathcal{O}(1)$, while N is macroscopic in the limit discussed above. Upon substitution of Eq. (3) into Eq. (2), the leading term of order $\mathcal{O}(\sqrt{N})$ gives the Gross-Pitaevskii equation (GPE), that governs the dynamics of the order parameter $\phi_0(\mathbf{r}, t)$. This reads:

$$i\hbar \frac{\partial \phi_0}{\partial t} = \left(\hat{h} + |\phi_0|^2 \right) \phi_0 \equiv \hat{H}_{GP} \phi_0. \tag{5}$$

The term of order $\mathcal{O}(1)$ in Eq. (2), that is linear in the quantum fluctuations describes the non-interacting dynamics of the elementary excitations in the system. Such excitations are described by the operator $\hat{\Lambda}(x) \equiv \frac{1}{\sqrt{N}} \hat{a}_0^\dagger \, \delta \hat{\Phi}(x)$ that transfers a particle from the non-condensed into the condensate component, and obeys the Bogoliubov-de Gennes equations [38]:

$$i\hbar \frac{d}{dt} \begin{pmatrix} \hat{\Lambda} \\ \hat{\Lambda}^\dagger \end{pmatrix} = \mathcal{L}_{\text{BdG}}[\phi_0, \phi_0^*] \begin{pmatrix} \hat{\Lambda} \\ \hat{\Lambda}^\dagger \end{pmatrix}, \tag{6}$$

where the Bogoliubov operator is defined as:

$$\mathcal{L}_{\text{BdG}}[\phi_0, \phi_0^*] = \begin{pmatrix} L_{QQ} & L_{QQ^*} \\ -L_{Q^*Q} & -L_{QQ} \end{pmatrix} \begin{pmatrix} \hat{\Lambda} \\ \hat{\Lambda}^\dagger \end{pmatrix} \tag{7}$$

with (operator-valued) components:

$$L_{QQ} = \left[\hat{H}_{\rm GP} + U_N\, Q|\phi_0(x,t)|^2 Q - \mu\right], \tag{8a}$$
$$L_{QQ^*} = N g_{1D}\, Q\phi_0^2(x,t)Q^*, \tag{8b}$$
$$L_{Q^*Q} = \left(L_{QQ^*}\right)^*. \tag{8c}$$

Here $\hat{H}_{\rm GP}$ is the Gross-Pitaevskii Hamiltonian defined in Eq. (5), while the operator $Q \equiv \mathbb{I} - |\phi_0\rangle\langle\phi_0|$ is the projector onto the non-condensed component, that is onto the Hilbert subspace spanned by all single particle excited states.

The higher order [cubic $O(1/\sqrt{N})$ and quartic $O(1/N)$] terms in the quantum fluctuations, in the Hamiltonian Eq. (1), describe the self-interaction and scattering between Bogoliubov quasi-particles, and are responsible for back-reaction and thermalization processes. We will account for these term, and thus for the full nonlinear dynamics of the problem, by using the so-called truncated Wigner approximation which we introduce in Sec. 4.

3. Physical system

We consider a dilute two-dimensional Bose gas homogeneous along the longitudinal x direction with periodic boundary conditions and trapped in the transverse y direction by an external potential $V_{\rm ext}(\mathbf{r}) = V_{\rm ext}(y)$. For numerical convenience, this is taken as harmonic of frequency ω_0 at small y with a hard-wall at $y = \pm L_y/2$ on both sides. At zero temperature, most of the atoms in the system are condensed in the ground state, whose wave function $\varphi_0(\mathbf{r})$ is given by solving the stationary GPE, obtained from Eq. (5) by posing $\phi_0(\mathbf{r},t) = \varphi_0(\mathbf{r})\exp(-i\mu t/\hbar)$ [37]:

$$\hat{H}_{GP}\varphi_0(\mathbf{r}) = \mu\,\varphi_0(\mathbf{r}), \tag{9}$$

with μ the chemical potential of the cloud.

The spectrum $\{\omega_n^r\}$ of the collective Bogoliubov modes on top of the ground-state and the corresponding eigenfunctions $\{u_n^r, v_n^r\}$ are calculated by diagonalizing the Bogoliubov operator:

$$\mathcal{L}_{\rm BdG}[\varphi_0, \varphi_0^*]\begin{pmatrix} u_n^r \\ v_n^r \end{pmatrix} = \omega_n^r \begin{pmatrix} u_n^r \\ v_n^r \end{pmatrix}. \tag{10}$$

For each mode, the integer-valued subscript n and the superscript $r = g,\, d,\, b,\, \ldots$ respectively identify the longitudinal wave vector $k = 2\pi n/L_x$ and the different excitation branches, labelled by the number of transverse nodes in the wavefunction (g = Goldstone – 0 nodes; d = dipole – 1 node; b = breathing – 2 nodes). In Fig. 1(a), we show the three lowest excitation bands for the two set of system parameters used throughout this work. The transverse profiles of the real part of the $u_0^r(y)$ functions of the $k = 0$ modes of the three $r = g, d, b$ branches are displayed in Fig. 1(b). Along the longitudinal x direction, thanks to translational invariance, each mode has a plane-wave shape.

As an explicit check of our numerical calculations, throughout this work we compare the predictions relative to a system composed by $N = 10^6$ atoms, of longitudinal(transverse) length $L_x = 140\ell_0$ ($L_y = 3.54\ell_0$) (with $\ell_0 \equiv \sqrt{\hbar/2m\omega_0}$ the transverse harmonic oscillator length) and number of grid points in the longitudinal(transverse) direction $N_x = 512$ ($N_y = 12$), with the corresponding results obtained for a system composed by $N = 750,000$ atoms, of equal transverse size but of longitudinal length $L_x = 105\ell_0$ and corresponding number of grid points $N_x = 384$. We used a different number of longitudinal grid-points in order keep constant the grid spacing δl, to the value $\delta l \approx 0.4\xi$, where $\xi = \hbar/\sqrt{2m\mu}$ is the healing length. In order to reproduce the same physical conditions and show that the results reported in the following are

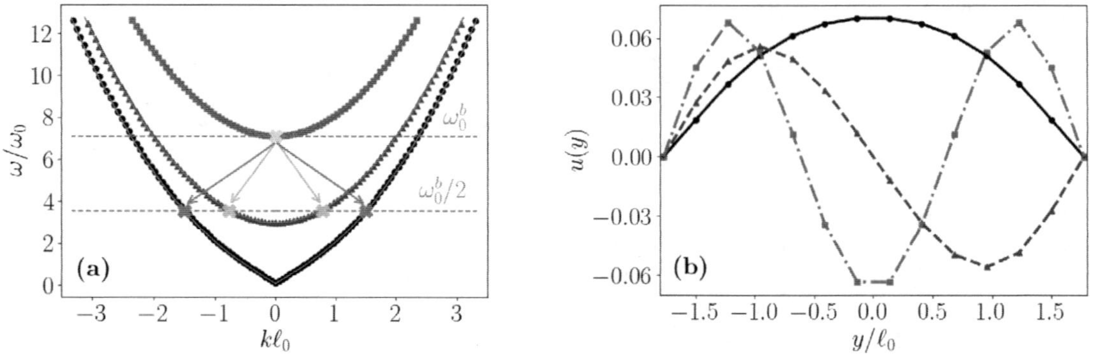

Figure 1. (a) Bogoliubov spectrum of collective excitations on top of the ground state of the system. Coloured markers provide the solution for a gas of $N = 10^6$ atoms in an integration box of size $L_{x,y}/\ell_0 = 140, 3.54$ in units of the transverse harmonic oscillator length $\ell_0 = \sqrt{\hbar/2m\omega_0}$, with $N_{x,y} = 512, 12$ grid points. The solid red lines display the solution calculated for a gas of $N = 7.5\times10^5$ atoms, of equal transverse size but different longitudinal length: $L_x/\ell_0 = 105$, with $N_x = 384$ grid points. Physical parameters have been chosen so that the equilibrium chemical potential is the same for the two configuration and equal to $\mu/\hbar\omega_0 = 2.38$. As expected, the spectra relative to these two configurations are the same. The three curves correspond to modes with zero (Goldstone, black), one (dipole, blue) and two (breathing, green) nodes in the transverse direction. The transverse profile of these modes is reported in panel (b). In panel (a), the light green cross highlights the transverse breathing mode that is excited at early times to simulate the inflaton oscillations; the light blue and grey crosses respectively highlight the dipole and Goldstone modes of opposite momenta that are resonantly excited by the parametric processes indicated by the arrows.

insensitive to the physical dimensions of the system, we adjusted the number of particles in such a way that the density of the system is the same in the two cases (the GPE Hamiltonian defined in Eq. (5) is invariant under the transformation $\psi \to \psi/\sqrt{\lambda}, \quad N \to \lambda N$). The spectrum of the system in these two configurations is the same, as expected, except for a difference in the density of eigenstates, due to the different longitudinal lengths (we do not show this feature in Fig. 1(a) as it would be difficult to visualize). Such a different density of eigenstates will result in minor deviation in the results obtained with these two configurations.

4. Numerical method

We model the nonlinear dynamics of the system beyond the Bogoliubov theory by using the truncated Wigner approximation (TWA) [40]. This technique has been used extensively in the literature for studying quantum field effects at the test-field level such as the (analog of the) Hawking radiation emanated from a sonic horizon [10], and even beyond [35, 41, 42]. Within such a semiclassical theory, quantum noise is only encoded in the initial condition of the system while the following evolution is classical and described by the GPE in Eq. (5). The quantum field operator is thus modelled as a classical field $\psi(\mathbf{r})$, whose initial configuration is sampled according to the Wigner quasi-probability distribution that describes the quantum state of the system at the initial time. Formally, the TWA results from neglecting a third-order derivative term from the Fokker-Plank-like equation for the Wigner distribution of the system, that cannot be mapped into a stochastic, classical equation of motion [40].

In our calculations, we take the Bogoliubov vacuum as the initial state of the dynamics. Within the TWA [43], such a state is constructed as the superposition of two contributions:

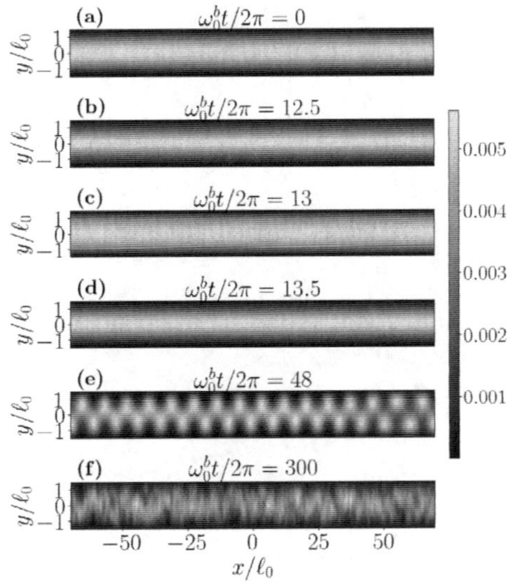

Figure 2. Time evolution of the density profile of a gas of $N = 10^6$ atoms, confined within a box of size $L_{x,y}/\ell_0 = 140, 3.54$ and $N_{x,y} = 512, 12$ grid points with periodic boundary conditions along x. We report the result obtained by propagating in time a single realization of the quantum noise that populates the initial state of the system. (a) Initial condition of the simulation. (b-d) Evolution of the system soon after the initial modulation of the transverse trapping frequency. The resulting transverse breathing oscillations of the cloud are visible. (e) Density profile of the cloud at the time instant when the exponential growth of the population in the resonant dipole and Goldstone modes saturates. Such a profile reveals the macroscopic excitation of the dominant resonant dipole mode. (f) Self-interaction and scattering processes result in a final chaotic state of the system.

one is the order parameter $\varphi_0(\mathbf{r})$ of the condensate at equilibrium, the other is a stochastic component that accounts for the zero-point fluctuations that populate the Bogoliubov vacuum:

$$\psi\left(\mathbf{r}, t=0\right) = \varphi_0(\mathbf{r}) + \sum_{\substack{r \\ n\in(+)}} \left(\beta_n^r u_n^r(\mathbf{r}) + \beta_n^{r*} v_n^{r*}(\mathbf{r})\right) \tag{11}$$

where the sum is restricted to positive norm modes only. The amplitudes β_n^r are independent, zero-mean, Gaussian random variables, that mimic the zero-point fluctuations in each mode according to the semiclassical Wigner representation for the state of a quantum system [44]. These satisfy the following correlation properties: $\langle \beta_k^2 \rangle = \left\langle \left(\beta_k^*\right)^2 \right\rangle = 0$, $\langle |\beta_k|^2 \rangle = 1/2$. Quantum expectation values of equal time correlators, symmetrically ordered respect to the field operators $\hat{\Psi}$ and $\hat{\Psi}^\dagger$, are obtained as stochastic averages over an ensemble of samples for $\psi(\mathbf{r}, t)$, each relative to a different realization of the initial noise and evolved in time according to the GPE in Eq. (5).

Specifically, we simulate the inflaton oscillations around the bottom of its potential well by exciting the transverse breathing mode of the condensate with zero longitudinal momentum (this is marked in light green in Fig. 1(a)). To this end, we impart an impulsive modulation to the trapping frequency, centred at the time t_0, having the Gaussian form:

$$\omega_0(t)/\omega_0 = 1 + Ae^{-(t-t_0)^2/2\sigma_t^2}. \tag{12}$$

Here, A is the magnitude of the modulation while $\sigma_t = 1/\omega_0^b$ is its duration. In the next section, we study the non-equilibrium dynamics of the condensate after this initial kick. We will investigate the physical processes by which the energy injected in the transverse modes is redistributed into the longitudinal modes, focusing on the early dynamics in order to elucidate the ensuing back-reaction effects.

5. Numerical results for the back-reaction effects

5.1. Parametric amplification and effective friction

We start at $t = 0$ with the system prepared in the Bogoliubov vacuum. In this state, modes are only populated by zero-point fluctuations. An example of the density profile in this configuration is shown in Fig. 2(a). We then give an impulsive modulation to the frequency

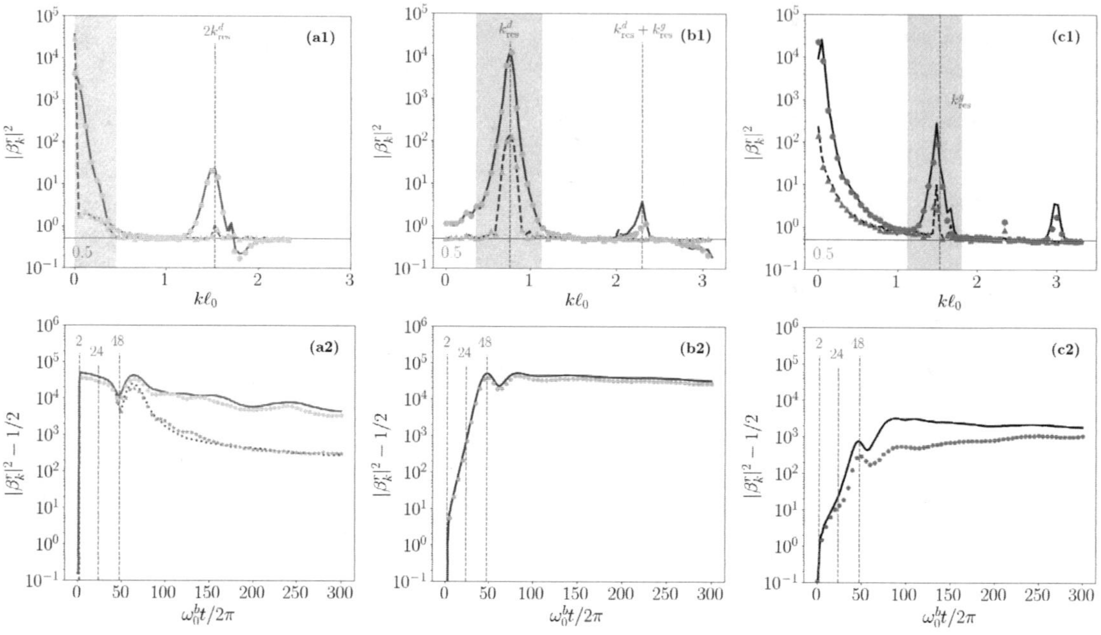

Figure 3. Numerical results of the TWA simulations. Stochastic averages are based on a sample of $\mathcal{N}_r = 1000$ independent realizations. *Panels (a1-c1):* Momentum distribution of the population in the breathing (a1), dipole (b1) and Goldstone (c1) excitation branches, at time $\omega_0^b t/2\pi = 48$ when the saturation of the parametric processes occurs (solid line), and at the intermediate time $\omega_0^b t/2\pi = 24$ (dashed line). Solid and dashed lines show results obtained for the system of longitudinal length $L_x = 140\ell_0$ and $N_x = 512$ grid points. Markers show the corresponding solution calculated for the system with $L_x = 105\ell_0$ and $N_x = 384$. *Panels (a2-c2):* Time-evolution of the integrated population in the breathing (a2), dipole (b2), and Goldstone (c2) branches, over the regions indicated by the shading in the upper panels (the same color code is used in both the upper and lower figures). In (a2), the violet dotted line and markers show the time evolution of the population in the single breathing mode at $k = 0$.

of the transverse trapping potential, of magnitude $A = 10$ and centred at the time $\omega_0 t_0/2\pi = 2$, with the result of setting the system in oscillation in the transverse breathing mode with zero longitudinal momentum. These oscillations are visible in Figs. 2(b-d), where we show the evolution of the density profile soon after the transverse kick. The nonlinear coupling between the Bogoliubov modes, engendered by the interatomic interaction, makes the vacuum fluctuations in the Goldstone and dipole modes parametrically excited by these breathing oscillations. Because of the ring configuration here considered, longitudinal momentum is conserved and parametric down-conversion process involves pairs of particles with opposite momenta as indicated by arrows in Fig. 1(a). Energy conservation makes the parametric processes to be most effective into the Goldstone and dipole modes of frequency $\omega_{\rm res}^{g,d} = \omega_0^b/2$ for which the parametric emission is resonant with the breathing mode oscillations at ω_0^b. Clear evidence of the excitation of the resonant longitudinal modes, in particular the dipole one, is visible in Fig. 2(e). Given the bosonic nature of the Bogoliubov modes, the parametric emission starts from zero-point quantum fluctuations but then gets self-stimulated as the population in the modes increases, leading to a characteristic exponential growth. This amplification process gets distorted when nonlinear effects become relevant, that is when the population in the resonant g, d modes has grown to sizable values, comparable to the population of the driving b mode. This leads to the saturation of the parametric amplification, as well as to self-interaction, scattering and thermalization processes within the Goldstone and dipole branches, that eventually drive the system into a chaotic thermal state (see Fig. 2(f)).

These dynamics are visible in Fig. 3(a1-c1), where we report a snapshot of the population in the different Bogoliubov modes, $n^r_n(t) + 1/2 \equiv \langle (\hat{b}^r_n)^\dagger \hat{b}^r_n + \hat{b}^r_n (\hat{b}^r_n)^\dagger \rangle/2 = \langle |\beta^r_n(t)|^2 \rangle_W$ at the time when the saturation occurs, and at an intermediate time. We notice the appearance of resonant peaks in both the dipole (b1) and Goldstone (c1) branches, at the values of momenta expected from the analysis of the Bogoliubov spectrum in Fig. 1(a), as well as the emergence at later times of harmonic peaks due to nonlinear processes, at values of momenta fixed by the momentum conservation constrain [35, 45, 46]. The same nonlinear effects are also responsible for the widening of the all visible peaks with time.

The initial exponential growth of the population in the resonant Goldstone and dipole modes is visible in Fig. 3(b2,c2), where we report the evolution of the population integrated within the regions indicated by the shaded areas in Fig. 3(b1,c1). This growth saturates as soon as the populations in the g, d modes have grown to a value comparable to the b mode and nonlinearities have started playing a sizeable effect. When this happens, we see a marked drop in the population in the breathing b branch, as shown in Fig. 3(a2), that is a signal of effective friction experienced by the breathing mode, due to back-reaction effects. Interestingly, such a damping is not purely monotonic, but energy gets at least partially exchanged between the transverse and longitudinal modes. This intermediate-time damped-oscillatory phenomenology is qualitatitively similar to the one predicted in [25, 29] for the back-reaction onto a moving mirror induced by the dynamical Casimir emission, where we demonstrated that a full exchange of energy between the optical field and the mirror is suppressed by purely quantum effects, but, as we are going to see in the next Sections, also displays interesting new features stemming from the many-mode nature of our system.

Finally, it is important to note how Fig. 3 shows results obtained for the two different sizes of the system given in Sec. 3. As expected, such results closely resemble each another, except for minor deviations that we ascribe to the different modes densities and thus resonances pertaining to the two configurations.

5.2. *Local observables and de-phasing*

The mode-wise analysis developed in the previous section highlighted many interesting dynamical effects: At the early time of the simulation, when the breathing mode is highly excited, we observed the parametric amplification of the vacuum fluctuations in the Goldstone and dipole modes and the simultaneous decay of the population in the transverse breathing mode with an approximately constant damping rate. During this stage of the evolution, the state of the breathing mode closely resembles a classical state and the dynamics is expected to be accurately modelled by approximating the breathing mode with its mean-field component. At later time instead, when the populations in the resonant modes have grown to a value comparable to the breathing mode, nonlinear effects result in the saturation of the parametric processes, generation of harmonic excitations, widening of peaks in the momentum spectrum and, remarkably, in a drastic modification of the qualitative features of the effective damping experienced by the breathing mode.

In this Section, we complement the momentum-space analysis by looking at spatially local quantities which have a direct connection to fluctuations in the system, thus explicitly showing the crucial role of the quantum effects in the back-reaction dynamics. Specifically, we study the spatial correlation function $C_w(X;t)$ of the transverse fluctuations of the cloud, whose transverse size we define as:

$$w(x,t) \equiv \frac{\int_0^{L_y} dy |\psi(\mathbf{r},t)|^2 y^2}{\int_0^{L_y} dy |\psi(\mathbf{r},t)|^2}. \tag{13}$$

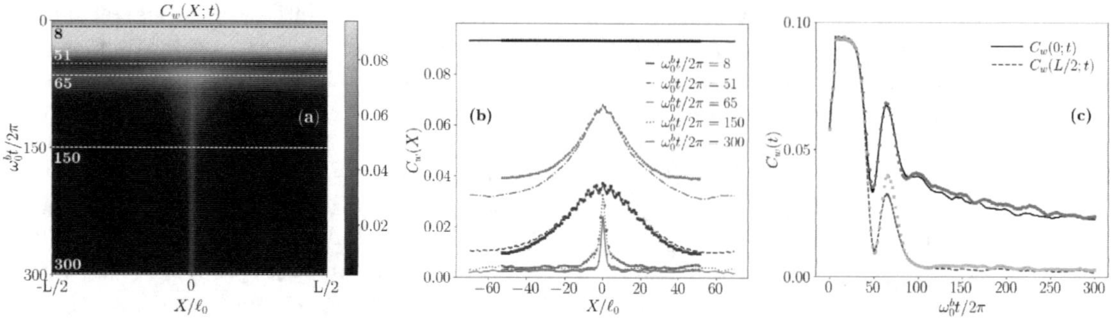

Figure 4. *Left panel:* Time-evolution of the spatial correlation function $C_w(X;t)$ of the transverse size of the cloud, calculated for the system of longitudinal length $L_x = 140\ell_0$ and $N = 10^6$ atoms. *Middle panel:* Cuts of the correlation function at the different times indicated by the horizontal dashed lines in the left panel. *Right panel:* Time evolution of the spatial correlation function for coincident $C_w(0;t)$ and opposite $C_w(L/2;t)$ points around the ring: the former gives information on the total oscillation intensity, while the latter indicates the spatial coherence of the oscillations. As in the previous Figures, lines and markers show the results obtained for the system of longitudinal length $L_x = 140\ell_0$ ($N = 10^6$) and $L_x = 105\ell_0$ ($N = 7.5 \times 10^5$), respectively.

Indicating by $\delta w(x,t) = w(x,t) - \bar{w}(0)$ the x-dependent variation of the transverse size of the cloud respect to the initial spatial average $\bar{w} \equiv L_x^{-1} \int_0^{L_x} dx\, w(x, t = 0)$, we define:

$$C_w(X;t) \equiv \left\langle \frac{\delta w(x,t)\delta w(x+X,t)}{\bar{w}^2} \right\rangle_W. \tag{14}$$

The full time evolution of $C_w(X;t)$ is illustrated in Fig. 4(a). Right after the initial kick around $\omega_0^b t = 2$, the system oscillates coherently with a uniform amplitude, so that the correlation function is large and uniform in space. At later time instead, the overall magnitude of C_w decreases as a signature of back-reaction-induced damping, and becomes sharply peaked at $X = 0$ at an even faster rate, which is a clear signal of loss of spatial coherence of the transverse oscillations. Such a de-phasing is a direct effect of the fluctuations in the system: the enhanced fluctuations of the two-mode-squeezed-like state of the emitted g, d fields directly transfer into a analogous fluctuations of the back-reaction-induced friction force. A more quantitative insight into the evolution of the spatial correlation function can be drawn from Fig. 4(b), where we report cuts of Fig. 4(a) at different times. From these plots, we clearly notice that the b mode loses spatial coherence at a much quicker rate compared to the much longer time scale of the decays of the oscillation intensity. A momentum-space signature of this effect was already visible in Fig. 3(a1), as a broadening of the population peak close to zero-momentum in the breathing branch, whose large momentum-width corresponds to the short real-space coherence length. Also for these quantities, we have checked in the Figures that the numerical results are qualitatively identical for systems of different sizes.

The results of this section explicitly demonstrates that fluctuations have dramatic consequences in determining the dynamics of a spatially-extended multi-mode system as the one here considered. In particular, our calculations confirm the outcome of previous studies carried out with simpler optomechanical configurations [29], and prove that mean-field models are expected to fail in accurately describing the physics of the back-reaction.

Journal of Physics: Conference Series **2533** (2023) 012031 doi:10.1088/1742-6596/2533/1/012031

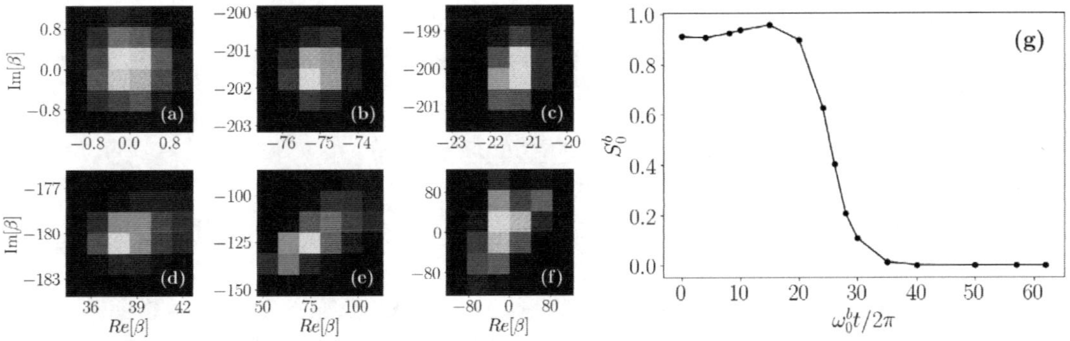

Figure 5. *Panels (a-f):* Phase space Wigner distribution for the transverse breathing mode at zero longitudinal momentum, at the time instants (a) $\omega_0^b t/2\pi = 0$, (b) $\omega_0^b t/2\pi = 10$, (c) $\omega_0^b t/2\pi = 20$, (d) $\omega_0^b t/2\pi = 30$, (e) $\omega_0^b t/2\pi = 40$,(f) $\omega_0^b t/2\pi = 50$. The Wigner function at each time has been reconstructed from the $\mathcal{N}_r = 1000$ samples of the stochastic simulations. We discretized the relevant portion of the phase space by using the number of grid points: $\mathcal{N}_b^R = \mathcal{N}_b^I = \mathcal{N}_r^{1/4} \approx 6$ along both the real and imaginary axes. *Panel (g):* Time evolution of the linear entanglement entropy S_0^b of the breathing mode as defined in the text. At the early time, when the system can be treated semiclassically, the value of the entropy is close to one. At later time instead we observe a drastic decrease of the value of the entropy, which signals entanglement of the breathing mode with the resonant Goldstone and dipole modes.

5.3. Entanglement signatures

In this last section, we further comment on the validity of the TWA that we use to solve for the dynamics of our analog system. The physical significance of the TWA has been long debated within the quantum optics community (see for example [47]), but the implications of this approximation on the dynamics of a quantum system are still not completely understood. Specifically, as detailed in Sec. 4, the TWA is based on a classical evolution of an initial quantum state and it is thus natural to wonder to what extent this formalism is able to capture quantum features of the dynamics of a physical system. While an in-depth study of the physical meaning of the TWA and of its precise limits is an important issue that needs still to be addressed in full detail, we give further evidence in this last section on the capabilities of the TWA to capture at least some features of the quantum dynamics of the system at hand.

To start with, we note that a positive initial Wigner distribution (like the one modelling the Bogoliubov vacuum) remains positive under classical evolution. This entails that quantum states characterized by negative Wigner functions cannot be modelled by the TWA. However, strictly quantum states exist that are described by positive Wigner distributions, such as squeezed states. This means that it is possible in principle that certain quantum features of the dynamics can be captured by working with the TWA. The results presented in the previous sections support this statement, but we give in this section further evidence by studying the entanglement between the zero-momentum transverse breathing mode and the resonant Goldstone and dipole modes. To this end, we use the data from our numerical simulations to construct the Wigner function $W_0^b(\beta, \beta^*)$ for the breathing mode at different time instants (see Fig. 5)(a-f). As entanglement witness, we study the linear entanglement entropy S_0^b of this mode, which is readily evaluated in terms of the corresponding Wigner distribution, as:

$$S_0^b \equiv Tr(\hat{\rho}_{b,0}^2) = \pi \int d^2\beta \, W_{b,0}^2(\beta, \beta^*).$$

In this equation, the integral is performed over the complex phase space of the breathing mode.

Journal of Physics: Conference Series **2533** (2023) 012031 doi:10.1088/1742-6596/2533/1/012031

In the case of a pure state $S_{b,0} = Tr(\hat{\rho}^2_{b,0}) = Tr(\hat{\rho}_{b,0}) = 1$, while $S_{b,0} < 1$ in case the state is mixed. This latter condition would be a signal of entanglement of the b mode with the resonant g, d modes. At the initial time, the system is in the Bogoliubov vacuum and $S_{b,0} = 1$. We expect that $S_{b,0} \approx 1$ after the transverse kick, as the state of breathing mode can be approximated as a highly excited coherent state with a Gaussian-shaped Wigner distribution (see Fig. 5(b)). At later time instead, when the effect of the back-reaction becomes sizable, the Gaussian shape gets distorted and the Wigner distribution gets broadened [(see]Fig. 5(c-f)], so the value of S^b_0 should drop below one, indicating entanglement. Indeed, this is the behaviour we observe for the linear entropy of the breathing mode, as shown in Fig. 5(g).

These results were obtained by evaluating S^b_0 from the Wigner function of the breathing mode that we reconstructed from our simulations according to Eq.(5.3). The number $\mathcal{N}_r = 1000$ of our samples limits the number $\mathcal{N}_b$ of bins we can use to discretize the complex phase space of the mode: $\mathcal{N}_b = \sqrt{\mathcal{N}_r} = \sqrt{1000} \approx 32$. As closest approximation, we used $\mathcal{N}^R_b = \mathcal{N}^I_b = 6$ to discretize $W^b_0(\beta, \beta^*)$ along the real and imaginary directions. Such a rough grid introduces an error in the result of the numerical integration, which is evident for example in the value of S^b_0 evaluated $t = 0$ in Fig. 5, which deviates from $S^b_0 = 1$ by approximately 0.1. Increasing the precision of the numerical grid by a factor n means increasing the number of numerical samples by a factor n^4, which is out of our numerical capabilities. Despite this error, we clearly see a transition from $S^b_0 \approx 1$ at early time, to $S^b_0 \ll 1$ at later time, as expected.

This result shows that entanglement between the b at $k = 0$ and the g and d resonant modes can be predicted by working within the TWA. Although a clear understanding of the limits of this approximation is still missing, our results demonstrate that the TWA allows to capture features of physics beyond the semiclassical level.

6. Conclusions

In this paper, we give an extended discussion of an analog model of the pre-heating of the early Universe first proposed in Ref.[36]. In our proposal, we make use of a two-dimensional Bose-Einstein condensate of ultra-cold atoms as simulation platform. By working at the level of the truncated Wigner approximation, our numerical results highlight the crucial role of quantum fluctuations in the back-reaction effect of particle production onto the (analog) inflaton field. We observe how the breathing oscillations of the system (which simulate the inflaton degrees-of-freedom) experience an effective damping induced by the parametric emission of excitations in the dipole and Golstone modes (that simulate the matter fields). We highlighted non-trivial dynamical effects beyond the semiclassical level: In particular, we observed a quick fluctuations induced decoherence of the initially in-phase breathing excitation, as well as signature of miltimode entanglement. The generality of the microscopic processes underlying our numerically observed results suggests the importance of going beyond semiclassical approaches [48, 49] and including quantum fluctuation features in the description of back-reaction phenomena in gravitation and cosmology.

7. Acknowledgements

Continuous support from Massimiliano Rinaldi on cosmological issues is most appreciated, as well as enlightening discussions with Stephen Barnett, Miles Blencowe, Bei-Lok Hu and Renaud Parentani. S. B. acknowledges funding from the Leverhulme Trust Grant No. ECF-2019-461, and from University of Glasgow via the Lord Kelvin/Adam Smith (LKAS) Leadership Fellowship. I.C. acknowledges support from the European Union Horizon 2020 research and innovation program under Grant Agreement No. 820392 (PhoQuS) and from the Provincia Autonoma di Trento.

Journal of Physics: Conference Series **2533** (2023) 012031 doi:10.1088/1742-6596/2533/1/012031

References

[1] Unruh W G 1981 *Phys. Rev. Lett.* **46** 1351–1353
[2] Barceló C, Liberati S and Visser M 2011 *Living Rev. Relativ.* **14**
[3] Faccio D, Belgiorno F, Cacciatori S, Gorini V, Liberati S and Moschella U 2013 *Analogue Gravity Phenomenology Analogue Spacetimes and Horizons, from Theory to Experiment* (Springer)
[4] Hawking S W 1974 *Nature* **248** 30–31
[5] Hawking S W 1975 *Commun. Math. Phys* **43** 199–220
[6] Steinhauer J 2016 *Nat. Phys.* **12** 959–965
[7] de Nova J, Golubkov K, Kolobov V and Steinhauer J 2019 *Nature* **569** 688–691
[8] Kolobov V, Golubkov K, de Nova J and Steinhauer J 2021 *Nat. Phys.* **17** 362–367
[9] Garay L J, Anglin J R, Cirac J I and Zoller P 2000 *Phys. Rev. Lett.* **85** 4643–4647
[10] Carusotto I, Fagnocchi S, Recati A, Balbinot R and Fabbri A 2008 *New J. Phys.* **10** 103001
[11] Eckel S, Kumar A, Jacobson T, Spielman I B and Campbell G K 2018 *Physical Review X* **8** 021021
[12] Steinhauer J, Abuzarli M, Aladjidi T, Bienaimè T amd Piekarski C, Liu W, Giacobino E, Bramati A and Glorieux Q 2022 *Nat. Commun.* **13** 2890
[13] Torres T, Patrick S, Coutant A, Richartz M, Tedford E and Weinfurtner S 2017 *Nat. Phys.* **13** 833–836
[14] Schander S and Thiemann T 2021 *Front. Astron. Space Sci.* **8** 692198
[15] Fischetti M V, Hartle J B and Hu B L 1979 *Phys. Rev. D* **20** 1757–1771
[16] Hartle J B and Hu B L 1979 *Phys. Rev. D* **20** 1772–1782
[17] Hartle J B and Hu B L 1980 *Phys. Rev. D* **21** 2756–2769
[18] Hartle J B 1980 *Phys. Rev. D* **22** 2091–2095
[19] Hartle J B 1981 *Phys. Rev. D* **23** 2121–2128
[20] Hu B L and Verdaguer E 2020 *Semiclassical and Stochastic Gravity: Quantum Field Effects on Curved Spacetime* Cambridge Monographs on Mathematical Physics (Cambridge University Press)
[21] Hu B 2005 *Int. J. Theor. Phys.* **44** 1785–1806
[22] Nation P D and Blencowe M P 2010 *New Journal of Physics* **12** 095013
[23] Ding S, Maslennikov G, Hablützel R and Matsukevich D 2018 *Phys. Rev. Lett.* **121** 130502
[24] Kardar M and Golestanian R 1999 *Rev. Mod. Phys.* **71** 1233–1245
[25] Butera S and Carusotto I 2019 *Phys. Rev. A* **99** 053815
[26] Macrì V, Ridolfo A, Di Stefano O, Kockum A F, Nori F and Savasta S 2018 *Phys. Rev. X* **8** 011031
[27] Dalvit D A R and Maia Neto P A 2000 *Phys. Rev. Lett.* **84** 798–801
[28] Maia Neto P A and Dalvit D A R 2000 *Phys. Rev. A* **62** 042103
[29] Butera S and Carusotto I 2019 *EPL (Europhysics Letters)* **128** 24002
[30] Butera S 2022 *Phys. Rev. D* **105** 016023
[31] Balbinot R, Fagnocchi S, Fabbri A and Procopio G P 2005 *Phys. Rev. Lett.* **94** 161302
[32] Patrick S, Goodhew H, Gooding C and Weinfurtner S 2021 *Phys. Rev. Lett.* **126** 041105
[33] Bain J 2013 *Stud. Hist. Philos. M. P.* **44** 338–345
[34] Bassett B, Tsujikawa S and Wands D 2006 *Rev. Mod. Phys.* **78** 537–589
[35] Robertson S, Michel F and Parentani R 2018 *Phys. Rev. D* **98** 056003
[36] Butera A and Carusotto I 2022 *arXiv:2207.00311*
[37] Dalfovo F, Giorgini S, Pitaevskii L and Stringari S 1999 *Rev. Mod. Phys.* **71** 463–512
[38] Castin Y and Dum R 1998 *Phys. Rev. A* **57** 3008
[39] Gardiner C W 1997 *Phys. Rev. A* **56** 1414
[40] Steel M J, Olsen M K, Plimak L I, Drummond P D, Tan S M, Collett M J, Walls D F and Graham R 1998 *Phys. Rev. A* **58** 4824–4835
[41] Jain P, Weinfurtner S, Visser M and Gardiner C W 2007 *Phys. Rev. A* **76** 033616
[42] Prokopec T and Roos T G 1997 *Phys. Rev. D* **55** 3768–3775
[43] Sinatra A, Lobo C and Castin Y 2001 *Phys. Rev. Lett.* **87** 210404
[44] Walls D F and Milburn G J 2008 *Quantum Optics* (Springer)
[45] Micha R and Tkachev I I 2003 *Phys. Rev. Lett.* **90** 121301
[46] Barroso V S, Geelmuyden A amd Fifer Z, Erne S, Avgoustidis A, Hill R J A and Weinfurtner S 2022 *arXiv:2207.02199*
[47] Reynaud S, Heidmann A, Giacobino E and Fabre C 1992 I quantum fluctuations in optical systems (*Progress in Optics* vol 30) (Elsevier) pp 1–85
[48] Anderson P R, Molina-París C and Sanders D H 2015 *Phys. Rev. D* **92** 083522
[49] Pla S, Newsome I M, Link R S, Anderson P R and Navarro-Salas J 2021 *Phys. Rev. D* **103** 105003

Journal of Physics: Conference Series **2533** (2023) 012033 doi:10.1088/1742-6596/2533/1/012033

The periastron advance in curvature based Extended Gravity and Dark Energy

Antonio Capolupo [1,2], Gaetano Lambiase [1,2] and Antonio Tedesco [1,2]

[1] Dipartimento di Fisica "E.R. Caianiello", Università degli Studi di Salerno, via G. Paolo II, I - 84084 Fisciano, Italy

[2] Istituto Nazionale di Fisica Nucleare (INFN), Gruppo collegato di Salerno, via G. Paolo II, I - 84084 Fisciano, Italy

E-mail: capolupo@sa.infn.it, lambiase@sa.infn.it, atedesco@unisa.it

Abstract. In the context of the periastron advance as a fundamental test for gravity theories, we show an approach to deduce constraints on the sizes of the new forces arising in the weak field limit of the Extended Theories of Gravity (ETG). It is also studied the force motivated by Quintessence Fields deforming the Schwarzschild geometry and associated to the Dark Energy, responsible of the acceleration of the Universe. As for the ETG, we consider the more general Scalar-Tensor Fourth Order Gravity (STFOG) and the NonCommutative Spectral Gravity (NCSG) as a special case. The solutions of the linearized field equations provide corrections to the Newtonian potential in the Yukawa-like form $V(r) = \alpha\frac{e^{-\beta r}}{r}$, where α is the parameter related to the strength of the potential and β to the range of the force. Quintessence Fields lead to a power-law correction with parameters related to the Dark Energy. By analysing the periastron advance with the use of the current data, we find improvements on the parameters of the gravitational models as well as on the bounds of the parameter β by several orders of magnitude.

1. Introduction

We report here the results presented on Ref. [82]. In the last decades, observational investigations concerning the motion of objects in galaxy clusters at extra-galactic scales have clearly highlighted how the matter present in the Universe is dominated by an invisible dark component, denominated *dark matter*. In particular we observe that the effects of dark matter act at galactic scales giving rise to rotation curves characterized by a flat behavior, differently from what we should expect from Newtonian gravity. In addition to the dark matter, the discovery that the Universe is currently accelerating allowed to realize that it is also dominated by an unknown form of energy, supposed to be responsible of this relevant phenomenon, i.e. the *dark energy* [1, 2, 3, 4, 5, 6]. However, up to now there is no result coming from experimental projects finalised to detect particles that might constitute the dark matter. If we relax the assumption that it is only a form of matter not emitting light responsible of the observed gravitational effects on galaxies and clusters, it is possible to take into consideration many other theoretical proposals. Another approach to understand the nature of dark matter is represented by the Extended Theories of Gravity (ETG), which have captured an ever increasing interest in the scientific community. The interest resides in the fact that it is possible to explain the dark matter and dark energy in a pure gravitational setting based on curvature fields. The starting idea is that there is no fundamental reason for assuming the gravitational action (from which we derive the field equations) to be just a linear function of the Ricci scalar minimally coupled to matter, namely the Hilbert Einstein action as it occurs in General Relativity [8]. In this framework, if we introduce in the action higher order terms expressed by scalar curvature invariants, we can literally *extend* the theory of gravity and General Relativity is recovered as a special case of Extended Gravity. Thanks to a more general action, the effects attributed to the presence of dark matter and dark energy are thus interpreted as results of the extra-curvature terms. Linked to this, there is the fundamental property that ETG conduct to a gravitational law which acts differently at different scales. More specifically the results of General Relativity at Solar System scales are preserved, while at galactic and extra-galactic scales the gravitational pull has an growing contribution due to the extra-curvature terms.

We emphasize that modifications of General Relativity are suggested also by the fact that Einstein's theory of gravity breaks down in the UV [9]. The deviations from General Relativity are already present in several frameworks, such as Brans-Dicke and scalar-tensor (ST) theories [10, 11, 12], braneworld theories [13, 14], and finally higher order invariants such as $f(R)$ and $f(\phi, R, R^2, R_{\mu\nu}R^{\mu\nu})$, corresponding to an Einstein's gravity plus one or multiple coupled scalar fields [15, 16, 17, 18]) [19, 20, 21, 22], Non-Commutative Spectral Geometry [23], and compactified extra dimension/Kaluza-Klein models [26]. Moreover, they can also be generated from higher-order terms in the curvature invariants, nonminimal couplings to the background geometry in the Hilbert-Einstein Lagrangian [27]. Additional terms into the action of gravity may also come from string loop effects [28], dilaton fields in string cosmology [29], and nonlocally modified gravity induced by quantum loop corrections [30].

The Newtonian limit of some models of ETG have been studied in [31], while the Minkowskian limit in [32]. Natural candidates for experimentally testing ETG are the galactic rotation curves, stellar systems and gravitational lensing [33, 34, 35] (see also [36]). In this perspective, corrections to GR were already considered by several authors [37, 38, 39, 40]. Due to the large amount of possible models, an important issue is to select viable ones by experiments and observations. Therefore, the new born *multimessenger astronomy* is giving important constraints to admit or exclude gravitational theories (see e.g. [42, 43]). However, also fine experiments can be conceived and realized in order to fix possible deviations and extensions with respect to GR. They can involve space-based setups like satellites and precise electromagnetic measurements [44]. The Yukawa-like potential potentials occur in ETG and in Non-Commutative Spectral Geometry, and Quintessence field surrounding a massive gravitational source provide a Power Law potential. We shall infer the corrections to periastron advance for Solar Planets, referring in particular to Mercury, Mars, Jupiter and Saturn, as well as to S2 star orbiting around Sagittarius A^*.

In Sec. 2 we introduce the Scalar-Tensor Fourth Order Gravity (STFOG), the particular class of the NonCommutative Spectral Geometry (NCSG) and the solutions in the weak field limit of STFOG and NCSG. Then the Quintessence Field reproducing the Dark Energy is discussed. In Sec. 3, we study the effects of these corrections to the periastron advance of Planets as well as the effects of the Quintessence field present around a Schwarzschild Black Hole, and derive a lower bound on the adiabatic index of equation of state. Finally, conclusions are drawn in the last Section.

2. Extended Gravity and Quintessence Fields

The action for the ETG is given by (see for example [7])

$$\mathcal{S} = \int d^4x\sqrt{-g}\Big[f(R, R_{\mu\nu}R^{\mu\nu}, \phi) + \omega(\phi)\phi_{;\alpha}\phi^{;\alpha} + \mathcal{X}\mathcal{L}_m\Big], \tag{1}$$

where f is a generic function of the invariant R (the Ricci scalar), the invariant $R_{\mu\nu}R^{\mu\nu} = Y$ ($R_{\mu\nu}$ is the Ricci tensor), the scalar field ϕ, g is the determinant of metric tensor $g_{\mu\nu}$ and $\mathcal{X} = 8\pi G$. The Lagrangian density $\mathcal{L}_m$ is the minimally coupled ordinary matter Lagrangian density, $\omega(\phi)$ is a generic function of the scalar field.

The field equations obtained by varying the action (1) with respect to $g_{\mu\nu}$ and ϕ, In the metric approach, are[1]:

$$f_R R_{\mu\nu} - \frac{f + \omega(\phi)\phi_{;\alpha}\phi^{;\alpha}}{2} g_{\mu\nu} - f_{R;\mu\nu} + g_{\mu\nu}\Box f_R + 2f_Y R_\mu{}^\alpha R_{\alpha\nu} + \tag{2}$$

$$-2[f_Y R^\alpha{}_{(\mu}]_{;\nu)\alpha} + \Box[f_Y R_{\mu\nu}] + [f_Y R_{\alpha\beta}]^{;\alpha\beta} g_{\mu\nu} + \omega(\phi)\phi_{;\mu}\phi_{;\nu} = \mathcal{X}\, T_{\mu\nu}\,,$$

$$2\omega(\phi)\Box\phi + \omega_\phi(\phi)\phi_{;\alpha}\phi^{;\alpha} - f_\phi = 0\,. \tag{3}$$

where:

$$f_R = \frac{\partial f}{\partial R}, \quad f_Y = \frac{\partial f}{\partial Y}, \quad \omega_\phi = \frac{d\omega}{d\phi}, \quad f_\phi = \frac{df}{d\phi},$$

and $T_{\mu\nu} = -\frac{1}{\sqrt{-g}}\frac{\delta(\sqrt{-g}\mathcal{L}_m)}{\delta g^{\mu\nu}}$ is the the energy-momentum tensor of matter. We confine ourselves to the case in which the generic function f can be expanded as follows (notice that the all other possible contributions in f are negligible [31, 40, 41])

$$\begin{aligned} f(R, R_{\alpha\beta}R^{\alpha\beta}, \phi) = \quad & f_R(0,0,\phi^{(0)})\, R + \frac{f_{RR}(0,0,\phi^{(0)})}{2} R^2 + \frac{f_{\phi\phi}(0,0,\phi^{(0)})}{2}(\phi - \phi^{(0)})^2 \\ & + f_{R\phi}(0,0,\phi^{(0)}) R\,\phi + f_Y(0,0,\phi^{(0)}) R_{\alpha\beta}R^{\alpha\beta}\,. \end{aligned} \tag{4}$$

To study the weak-field approximation, we perturb Eqs.(2) and (3) in a Minkowski background $\eta_{\mu\nu}$ [31], i.e. we look for perturbed solutions of the form

$$g_{\mu\nu} \simeq \begin{pmatrix} -1 - 2\Phi & 2\mathbf{A} \\ 2\mathbf{A} & (1 - 2\Psi)\delta_{ij} \end{pmatrix}. \tag{5}$$

and

$$\phi \sim \phi^{(0)} + \phi^{(2)} + \ldots = \phi^{(0)} + \varphi.$$

For matter described as a perfect fluid, hence $T_{00} = \rho$ and $T_{ij} = 0$, one gets that, for a ball-like source with radius $\mathcal{R}$, the gravitational potentials $\{\Phi, \Psi, A_i\}$ and the scalar field φ take the form

[1] We use, for the Ricci tensor, the convention $R_{\mu\nu} = R^\sigma{}_{\mu\sigma\nu}$, whilst for the Riemann tensor we define $R^\alpha{}_{\beta\mu\nu} = \Gamma^\alpha_{\beta\nu,\mu} + \cdots$. The affine connections are the Christoffel symbols of the metric, namely $\Gamma^\mu_{\alpha\beta} = \frac{1}{2}g^{\mu\sigma}(g_{\alpha\sigma,\beta} + g_{\beta\sigma,\alpha} - g_{\alpha\beta,\sigma})$, and we adopt the signature is $(-,+,+,+)$.

Journal of Physics: Conference Series **2533** (2023) 012033 doi:10.1088/1742-6596/2533/1/012033

$(c = 1)$ [40, 41]

$$\Phi(\mathbf{x}) = -\frac{GM}{|\mathbf{x}|}\left[1+\zeta(|\mathbf{x}|)\right], \tag{6}$$

$$\zeta(|\mathbf{x}|) \equiv g(\xi,\eta)\,F(m_+\mathcal{R})\,e^{-m_+|\mathbf{x}|} + \left[\frac{1}{3}-g(\xi,\eta)\right]F(m_-\mathcal{R})\,e^{-m_-|\mathbf{x}|} - \frac{4\,F(m_Y\mathcal{R})}{3}\,e^{-m_Y|\mathbf{x}|} \tag{7}$$

$$\Psi(\mathbf{x}) = -\frac{GM}{|\mathbf{x}|}\left[1-\psi(|\mathbf{x}|)\right], \tag{8}$$

$$\psi(|\mathbf{x}|) \equiv g(\xi,\eta)\,F(m_+\mathcal{R})\,e^{-m_+|\mathbf{x}|} + \left[\frac{1}{3}-g(\xi,\eta)\right]F(m_-\mathcal{R})\,e^{-m_-|\mathbf{x}|} + \frac{2\,F(m_Y\mathcal{R})}{3}\,e^{-m_Y|\mathbf{x}|} \tag{9}$$

$$\mathbf{A}(\mathbf{x}) = -\frac{2G\left[1-\mathcal{A}(|\mathbf{x}|)\right]}{|\mathbf{x}|^2}\,\mathbf{x}\times\mathbf{J}, \tag{10}$$

$$\mathcal{A}(|\mathbf{x}|) \equiv (1+m_Y|\mathbf{x}|)\,e^{-m_Y|\mathbf{x}|}, \tag{11}$$

$$\varphi(\mathbf{x}) = \frac{GM}{|\mathbf{x}|}\sqrt{\frac{\xi}{3}}\frac{2}{\omega_+-\omega_-}\left[F(m_+\mathcal{R})\,e^{-m_+|\mathbf{x}|} - F(m_-\mathcal{R})\,e^{-m_-|\mathbf{x}|}\right], \tag{12}$$

where $\mathbf{J} = 2M\mathcal{R}^2\mathbf{\Omega}_0/5$ is the angular momentum of the ball, $f_R(0,0,\phi^{(0)}) = 1$, $\omega(\phi^{(0)}) = 1/2$, and

$$g(\xi,\eta) = \frac{1-\eta^2+\xi+\sqrt{\eta^4+(\xi-1)^2-2\eta^2(\xi+1)}}{6\sqrt{\eta^4+(\xi-1)^2-2\eta^2(\xi+1)}}, \tag{13}$$

$$F(m\mathcal{R}) = 3\frac{m\mathcal{R}\cosh m\mathcal{R}-\sinh m\mathcal{R}}{m^3\mathcal{R}^3}, \tag{14}$$

$$\xi = 3{f_{R\phi}(0,0,\phi^{(0)})}^2, \quad \eta = \frac{m_\phi}{m_R}, \tag{15}$$

$$m_\pm^2 = m_R^2\,\omega_\pm, \tag{16}$$

$$\omega_\pm = \frac{1-\xi+\eta^2\pm\sqrt{(1-\xi+\eta^2)^2-4\eta^2}}{2}, \tag{17}$$

$${m_R}^2 \doteq -\frac{f_R(0,0,\phi^{(0)})}{3f_{RR}(0,0,\phi^{(0)})+2f_Y(0,0,\phi^{(0)})}, \tag{18}$$

$${m_Y}^2 \doteq \frac{f_R(0,0,\phi^{(0)})}{f_Y(0,0,\phi^{(0)})}, \quad {m_\phi}^2 \doteq -\frac{f_{\phi\phi}(0,0,\phi^{(0)})}{2\omega(\phi^{(0)})}. \tag{19}$$

Some ETG models studied in literature and reported in Table 1 (see [40] for further details).

Another possibility we want to discuss is related to the Quintessence Field, invoked to explain the speed-up of the present Universe [63]. Quintessence may generate a negative pressure, and, since it is diffuse everywhere in the Universe, it can be the responsible of the observed accelerated phase, as well as it is present around a massive astrophysical object deforming the spacetime around it [64]. The studies of quintessential black holes are also motivated from M-theory/superstring inspired models [65] (see [66, 67, 68, 63] for applications). The solution of Einstein's field equations for a static spherically symmetric quintessence surrounding a black hole in 4 dimension is given by [64, 66]

$$g_{\mu\nu} = \mathrm{diag}\left(-f(r), f^{-1}(r), r^2, r^2\sin^2\theta\right), \tag{20}$$

with

$$f(r) = 1-\frac{2M}{r}-\frac{c}{r^{3\omega_Q+1}}, \tag{21}$$

Journal of Physics: Conference Series **2533** (2023) 012033 doi:10.1088/1742-6596/2533/1/012033

Table 1. We report different cases of Extended Theories of Gravity including a scalar field and higher-order curvature terms. The free parameters are given as effective masses with their asymptotic behavior. Here, we assume $f_R(0,\,0,\,\phi^{(0)}) = 1$, $\omega(\phi^{(0)}) = 1/2$.

Case	**ETG**	**Parameters**				
		m_R^2	m_Y^2	m_ϕ^2	m_+^2	m_-^2
A	$f(R)$	$-\frac{f_R(0)}{3f_{RR}(0)}$	∞	0	m_R^2	∞
B	$f(R,\,R_{\alpha\beta}R^{\alpha\beta})$	$-\frac{f(0)}{3f_{RR}(0)+2f_Y(0)}$	$\frac{f_R(0)}{f_Y(0)}$	0	m_R^2	∞
C	$f(R,\,\phi) + \omega(\phi)\phi_{;\alpha}\phi^{;\alpha}$	$-\frac{f_R(0)}{3f_{RR}(0)}$	∞	$-\frac{f_{\phi\phi}(0)}{2\omega(\phi^{(0)})}$	$m_R^2 w_+$	$m_R^2 w_-$
D	$f(R,\,R_{\alpha\beta}R^{\alpha\beta},\,\phi) + \omega(\phi)\phi_{;\alpha}\phi^{;\alpha}$	$-\frac{f(0)}{3f_{RR}(0)+2f_Y(0)}$	$\frac{f_R(0)}{f_Y(0)}$	$-\frac{f_{\phi\phi}(0)}{2\omega(\phi^{(0)})}$	$m_R^2 w_+$	$m_R^2 w_-$

where ω_Q is the adiabtic index (the parameter of equation of state), $-1 \leqslant \omega_Q \leqslant -\frac{1}{3}$, and c the quintessence parameter. The cosmological constant (ΛCMD model) follows from (20) and (21) with $\omega_Q = -1$ and $c = \Lambda/3$,

$$f(r) = 1 - \frac{2M}{r} - \frac{\Lambda r^2}{3}\,. \tag{22}$$

3. The periastron advance and inference of the theoretical bounds

In this section, we proceed to evaluate the contraints on the sizes of new gravitational forces (inferred in ETG or other scenarios) by making use the data coming from the precession of Planets. For this purpose, we follow the paper by Adkins and McDonnell [45] (see also [46, 47]), where it is calculated the precession of Keplerian orbits under the influence of arbitrary central-force perturbations. In the limit of nearly circular orbits, the perturbed orbit equation takes the form ($u = 1/r$)

$$\frac{d^2u}{d\varphi^2} + u = \frac{GM}{h^2} - \frac{g(u)}{h^2} \tag{23}$$

where $g(u) = r^2\frac{F(r)}{m}|_{r=1/u}$ ($\frac{F}{m} = -\nabla V$) and $h^2 = GMa$. $g(u) = 0$ corresponds to the unperturbed solution. We refer to the corrections to the Planets precession induced by the Yukawa-like potential, $V_Y(r) = \alpha\frac{e^{-\beta r}}{r}$, and power law (PL) potentials, $V_{PL}(r) = \alpha_n r^n$.

In GR, the first post-Newtonian correction is a perturbing potential given by

$$V(r)\Big|_{\rm GR} = -\frac{GMh^2}{c^2r^3},$$

which corresponds the precession

$$\Delta\theta_p\Big|_{\rm GR} = \frac{6\pi GM}{c^2a}.$$

This gives the well known 43 arcsec per century when applied to the orbit of Mercury. The correction to the Planet precessions induced by a generic perturbing force $F(z)$ and perturbing potential $V(z)$ is [45]

$$\Delta\theta_p = -\frac{2a^2}{GM\epsilon}\int_{-1}^{1}\frac{dz\,z}{\sqrt{1-z^2}}\frac{F(z)}{(1+\epsilon z)^2} \tag{24}$$

$$= -\frac{2a}{GM\epsilon^2}\int_{-1}^{1}\frac{dz\,z}{\sqrt{1-z^2}}\frac{dV(z)}{dz}\,, \tag{25}$$

Journal of Physics: Conference Series **2533** (2023) 012033 doi:10.1088/1742-6596/2533/1/012033

where, for the sake of convenience, the correction $\Delta\theta_p$ is written in terms of the dimensionless integration variable z with a fixed range, while ϵ is the eccentricity ($\epsilon < 1$). The perturbing force $F(z)$ and $V(z)$ are evaluated at radius $r = a/(1+\epsilon z)$. In the following we refer to the Yukawa-like and power-law potentials following from different gravitational theories of gravity.

- **The Yukawa force** - The Yukawa potential (as a correction to the Newtonian potential $V_N = GM/r$) is of the form [48, 49]

$$V_Y(r) = \alpha \frac{e^{-r/\lambda}}{r} \equiv \alpha \frac{e^{-\beta r}}{r} \tag{26}$$

where α and $\lambda \equiv 1/\beta$ are the strength and the range of the interaction, respectively. As we will see, such a potential occurs in several modified theories of gravity. The precession due to a Yukawa perturbation depends on two parameters: a range parameter $\kappa = a/\lambda = \beta a$ and the eccentricity ϵ, i.e. $\Delta\theta_p(\kappa,\epsilon)$, where a is the semi-major axis. According to [45], the correction to the precession is of the integral form

$$\Delta\theta_p(\kappa,\epsilon) = -\frac{2\alpha}{GM\epsilon} I_{\epsilon,\beta}\,, \tag{27}$$

where

$$I_{\epsilon,\beta} \equiv \int_{-1}^{1} \frac{dz\, z}{\sqrt{1-z^2}} \left(1 + \frac{\kappa}{1+\epsilon z}\right) e^{-\frac{\kappa}{1+\epsilon z}}\,. \tag{28}$$

The behavior of the integral (28) is represented in Fig. 1 for several Planets.

- **Power Law potential** - The power law potential is of the form

$$V_{PL}(r) = \alpha_q\, r^q\,, \tag{29}$$

where the parameter q assume arbitrary values. The precession (24) can be exactly integrated, and leads to [45]

$$\Delta\theta_p(q) = \frac{-\pi\alpha_q}{GM} a^{q+1} \sqrt{1-\epsilon^2}\chi_q(\epsilon)\,, \tag{30}$$

where $\chi_q(\epsilon)$ is written in terms of the Hypergeometric function

$$\chi_q(\epsilon) = q(q+1)\,{}_2F_1\left(\frac{1}{2} - \frac{q}{2}, 1 - \frac{q}{2}; 2; \epsilon^2\right)\,. \tag{31}$$

3.1. Planet precession in Scalar Tensor Fourth Order Gravity

In this Section we study the periastron shift of the orbital period of objects, both astrophysical and Solar System, in Scalar Tensor Fourth Order Gravity (STFOG). As seen before, the linearised STFOG field equations lead to a gravitational potential of the Yukawa-like form ($r = |\mathbf{x}|$)

$$V(r) = \frac{GM}{r}\left(1 + \sum_{i=\pm,Y} F_i e^{-\beta_i r}\right)\,, \tag{32}$$

where F_i and β are the strength and range of the interaction corresponding to each mode $i = +, -, Y$. Comparing (32) with (26), it follows the correspondence (referring to (6) and (7))

$$\alpha \to GMF_i\,, \quad \beta \to \beta_i\,, \quad i = \pm, Y\,. \tag{33}$$

Table 2. Values of periastron advance for the first six planets of the Solar System. In the table we present the values of the eccentricity ϵ, semi-major axis a in meters, the orbital period P in years, the periastron advance predicted in General Relativity (GR).

Planet	ϵ	$a\,(10^{11}\,m)$	$P\,(yrs)$	$\Delta\phi_{GR}\,(''/century)$	$\Delta\phi_{obs}$
Mercury	0.205	0.578	0.24	43.125	42.989 ± 0.500
Venus	0.007	1.077	0.62	8.62	8.000 ± 5.000
Earth	0.017	1.496	1.00	3.87	5.000 ± 1.000
Mars	0.093	2.273	1.88	1.36	1.362 ± 0.0005
Jupiter	0.048	7.779	11.86	0.0628	0.070 ± 0.004
Saturn	0.056	14.272	29.46	0.0138	0.014 ± 0.002

Table 3. Bounds on F_i, $i = \pm, Y$ obtained from (36) using the values of periastron advance for planets of the Solar System.

Planet	$\lvert\eta\rvert$	$I_{\epsilon,\beta}^{\max}$	$\beta_i^{\max} \simeq$	$\lvert F_i\rvert \lesssim$
Mercury	0.5	0.18	$4 \times 10^{-11} m^{-1}$	0.28
Mars	5×10^{-4}	0.08	$1.1 \times 10^{-11} m^{-1}$	2.9×10^{-4}
Jupiter	4×10^{-3}	0.04	$2.5 \times 10^{-12} m^{-1}$	2.4×10^{-3}
Saturn	2×10^{-3}	0.05	$2 \times 10^{-13} m^{-1}$	1.1×10^{-3}

with

$$F_+ = g(\xi,\eta)\, F(m_+\mathcal{R})\,, \quad F_- = \left[\frac{1}{3} - g(\xi,\eta)\right] F(m_-\mathcal{R})\,, \quad F_Y = -\frac{4}{3}\, F(m_Y\mathcal{R})\,, \tag{34}$$

$$\beta_\pm = m_R\sqrt{\omega_\pm}\,, \qquad \beta_Y = m_Y\,. \tag{35}$$

We impose that the periastron shift $\Delta\theta_p(\kappa,\epsilon) = -\frac{2\alpha}{GM\epsilon} I_{\epsilon,\beta}$ given by (27), where $I_{\epsilon,\beta}$ is defined in (28), is lesser than the error η. Fixing $I_{\epsilon,\beta}$ to the maximum values, one gets the bounds on the parameters F_i:

$$|\Delta\theta_p(\kappa,\epsilon)| \lesssim \eta \quad \to \quad |F_i| \lesssim \frac{\eta\epsilon}{2I_{\epsilon,\beta_i}}\,, \; i = \pm, Y\,. \tag{36}$$

In Fig. 1 are plotted the function $I_{\epsilon,\beta}$ for the Mercury, Mars, Jupiter and Saturn planets. In Table 3 are reported the corresponding bounds on F_i. As an illustrative example, we plot $|F_\pm(\xi,\eta)|$ in Fig. 2, for $m_R = \mathcal{R}^{-1}$. The available values of the parameters $\{\xi,\eta\}$ allow to fix the masses, via Eqs. (13) , (14), (15), (18), of extra modes arising in Scalar Tensor Fourth Order Gravity. The analysis of Yukawa gravitational potential for $f(R)$ has been carried out in [50].

3.2. Non-Commutative Spectral Geometry

Following the previous section, Eqs. (27) (36), the periastron advance in NCSG for planets is given by

$$|\Delta\theta_p(\beta,\epsilon)| \lesssim \eta \quad \to \quad |I_{\epsilon,\beta}| \lesssim I_0\,, \quad I_0 \equiv \frac{3\eta\epsilon}{8}\,, \tag{37}$$

where $I_{\epsilon,\beta}$ is defined in (28). From Eq, (37) one infers the bounds on β, or equivalently an upper bound on λ. Results are reported in Table 4 (see also Fig. 3). These results show that the

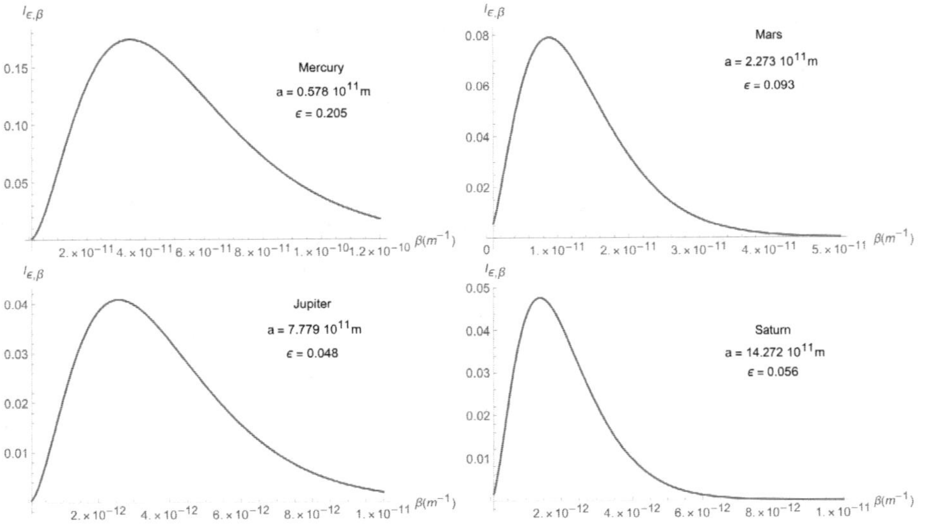

Figure 1. (a) $I_{\epsilon,\beta}$ vs β for Mercury. (b) $I_{\epsilon,\beta}$ vs β for Mars. (c) $I_{\epsilon,\beta}$ vs β for Jupiter. (d) $I_{\epsilon,\beta}$ vs β for Saturn.

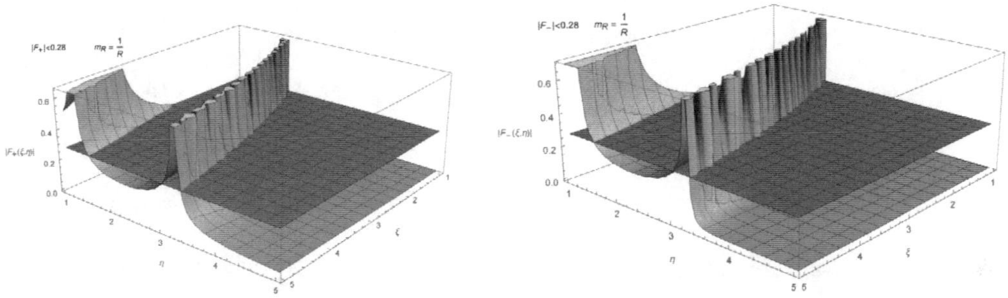

Figure 2. (a) F_+ vs $\{\xi,\eta\}$ for Mercury ($|F_+| \leq 0.28$), with $m_R = \frac{1}{\mathcal{R}}$. (b) F_- vs $\{\xi,\eta\}$ for Mercury ($|F_-| \lesssim 0.28$) with $m_R = \frac{1}{\mathcal{R}}$.

bounds on β improve several order of magnitude as compared with ones obtained using recent observations of pulsar timing, $\beta \geq 7.55 \times 10^{-13}\mathrm{m}^{-1}$ [58, 59]. Bounds on the parameter β have been obtained in different frameworks. From Gravity Probe B experiment one gets $\beta > 10^{-6}\mathrm{m}^{-1}$ [54]. A more stringent constraint on β can be obtained from laboratory experiments designed to test the fifth force, that is, by constraining λ through torsion balance experiments which implies to obtain a stronger lower bound on β (or equivalently an upper bound to the momentum f_0 in NCSG theory). The test masses have a typical size of $\sim$ 10mm and their separation is smaller than their size. As we have already mentioned above, in NCSG one has $|\alpha| \sim \mathcal{O}(1)$, so that the tightest constraint on $\lambda = \beta^{-1}$ provided by Eöt-Wash [60] and Irvine [61] experiments is [62] $\lambda \lesssim 10^{-4}$m, or equivalently $\beta \gtrsim 10^4\mathrm{m}^{-1}$.

3.3. Quintessence Fields related to the Dark Energy
The Quintessential potential reads $V_Q = -\frac{c}{r^{3\omega_Q+1}}$, so that comparing with (29) one gets

$$q \to -(3\omega_Q + 1) \qquad \alpha_q \to c\,.$$

Table 4. Lower bounds on β obtained from (37) using the values of periastron advance for planets of the Solar System.

Planet	η	$I_0 \equiv \frac{3\eta\epsilon}{8}$	$\beta(m^{-1}) >$
Mercury	0.5	0.038	1.0×10^{-10}
Mars	5×10^{-4}	1.36	7.8×10^{-11}
Jupiter	4×10^{-3}	0.0628	2.1×10^{-11}
Saturn	2×10^{-3}	0.0138	8.5×10^{-12}

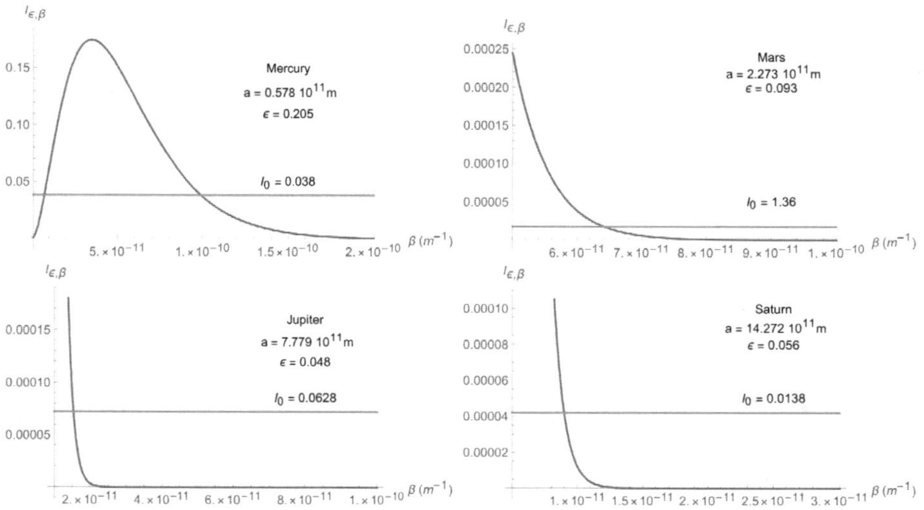

Figure 3. (a) $I_{\epsilon,\beta}$ vs β for Mercury. (b) $I_{\epsilon,\beta}$ vs β for Mars. (c) $I_{\epsilon,\beta}$ vs β for Jupiter. (d) $I_{\epsilon,\beta}$ vs β for Saturn.

The precession (30) leads to

$$|\Delta\theta_p(\omega_Q, \epsilon)| = \frac{\pi c}{GM} a^{-3\omega_Q} \sqrt{1-\epsilon^2} \chi_{\omega_Q}(\epsilon) \,, \tag{38}$$

with

$$\chi_{\omega_Q}(\epsilon) = 3\omega_Q(1+3\omega_Q)\ {}_2F_1\left(\frac{2+3\omega_Q}{2}, \frac{3+3\omega_Q}{2}; 2; \epsilon^2\right) . \tag{39}$$

By requiring $|\Delta\theta_p(\omega_Q, \epsilon)| \lesssim \eta$ one gets the bounds on the parameters $\{\omega_Q, c\}$. Results are reported in Table 5 and Fig. 4 for fixed values of c.

4. Test on S2 Star

Finally, we shortly conclude our analysis testing the modified gravity predictions for S2 Star orbiting around Sagittarius A*, the Supermassive Black Hole at the center of the Milky Way, which has got a mass equal to $M = (4.5 \pm 0.6) \times 10^6 M_\odot$ and a Schwarzschild radius $R_S = 2GM = 1.27 \times 10^{10}\, m$. The S2 Star orbit has an eccentrity $\epsilon = 0.88$ and a semi-major axis $a = 1.52917 \times 10^{14}$m. According to Ref. [75], the periastron advance is (0.2 ± 0.57) deg, hence $\eta = 0.57$ (it is expected that the interferometer GRAVITY may improve such an accuracy level). We discuss the periastron advance for the gravitational models above discussed:

Journal of Physics: Conference Series **2533** (2023) 012033 doi:10.1088/1742-6596/2533/1/012033

Table 5. Values of the parameter ω_Q obtained from (38) using the values of periastron advance for planets of the Solar System.

Planet	η	$c(m^{3\omega_Q+1}) \sim$	$\omega_Q \gtrsim$
Mercury	0.5	10^{-25}	-0.86
Mars	5×10^{-4}	10^{-30}	-0.88
Jupiter	4×10^{-3}	10^{-30}	-0.84
Saturn	2×10^{-3}	10^{-30}	-0.82

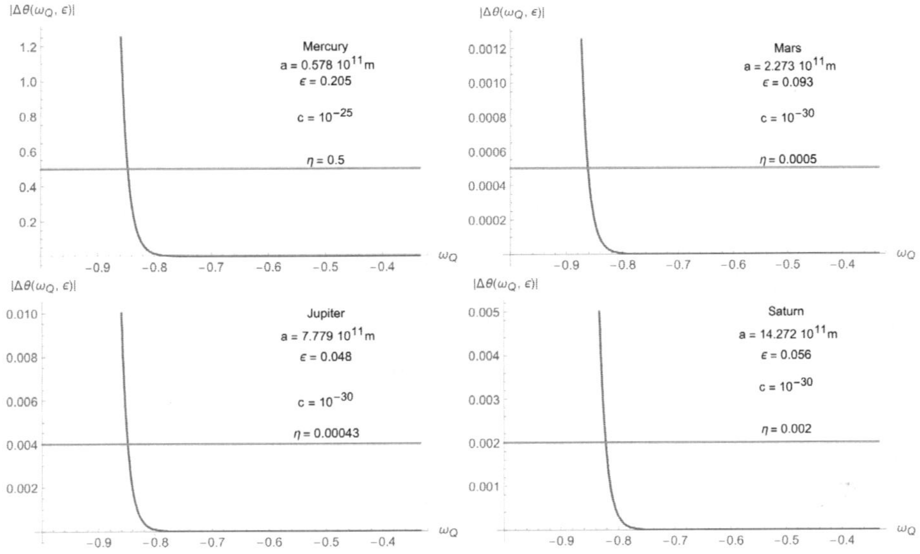

Figure 4. (a) $|\Delta\theta(\omega_Q, \epsilon)|$ vs ω_Q for Mercury. (b) $|\Delta\theta(\omega_Q, \epsilon)|$ vs ω_Q for Mars. (c) $|\Delta\theta(\omega_Q, \epsilon)|$ vs ω_Q for Jupiter. (d) $|\Delta\theta(\omega_Q, \epsilon)|$ vs ω_Q for Saturn.

- **STFOG** - Referring to Scalar-Tensor Fourth Order Gravity, from Eq. (36) one gets

$$|\Delta\theta_p(\kappa, \epsilon)| \lesssim \eta \quad \to \quad |F_i| \lesssim \frac{\eta\epsilon}{2I_{\epsilon,\beta_i}} \sim 0.36\,, \; i = \pm, Y\,. \tag{40}$$

where in Fig. 5(a) is plotted the function $I_{\epsilon,\beta}$ for the S2 star. We ave taken the maximum value of $I_{\epsilon,\beta}$ corresponding to $\beta \sim 2 \times 10^{-14} m^{-1}$ (see Fig. 5(a)). The analysis of S2 star orbit around the Galactic Centre in $f(\phi, R)$ and $f(R, \Box R)$ has been investigated in [76].

- **NCSG** - The S2 star values $\{\epsilon, \eta, a\}$ imply that, from (37),

$$|\Delta\theta_p(\beta, \epsilon)| \lesssim \eta \quad \to \quad |I_{\epsilon,\beta}| \lesssim I_0\,, \quad I_0 \equiv \frac{3\eta\epsilon}{8} \simeq 0.19\,. \tag{41}$$

Results are reported in Fig. 5(b). We can see that the lower bound on β is $\beta \gtrsim 1.1 \times 10^{-13} m^{-1}$. These bounds are compatible with the astrophysical bounds [58, 59].

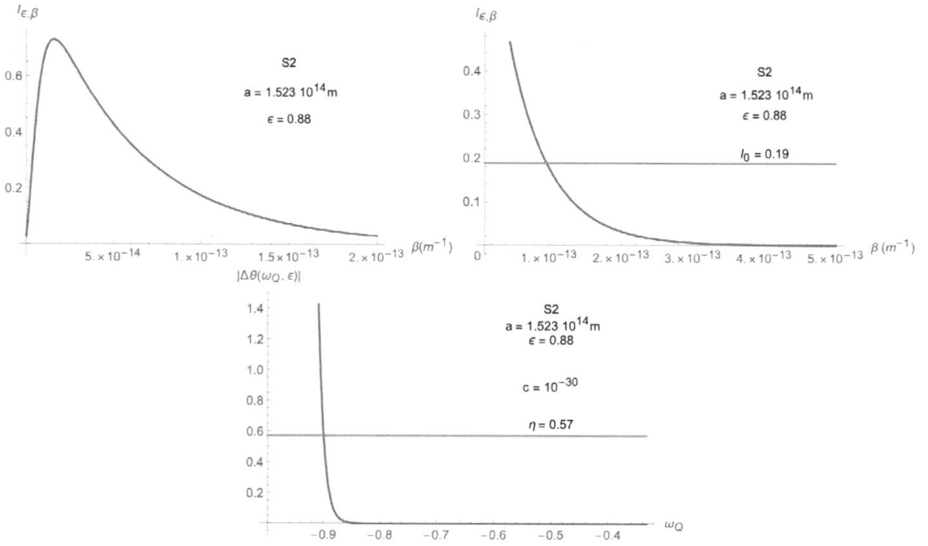

Figure 5. (a) $I_{\epsilon,\beta}$ vs β for S2 star in FOG theories. (b) $I_{\epsilon,\beta}$ vs β for S2 star. (c) $|\Delta\theta\omega_Q, \epsilon|$ vs ω_Q for S2 star. (c is in $m^{3\omega_Q+1}$ units).

- **Quintessence** - In the case of Quintessence field deforming the Schwarzschild geometry, Eq. (38) implies

$$|\Delta\theta_p(\omega_Q,\epsilon)| = \frac{\pi c}{GM} a^{-3\omega_Q}\sqrt{1-\epsilon^2}\chi_{\omega_Q}(\epsilon) \lesssim 0.57\,, \tag{42}$$

$$\chi_{\omega_Q}(\epsilon) = 3\omega_Q(1+3\omega_Q)\,{}_2F_1\left(\frac{2+3\omega_Q}{2}, \frac{3+3\omega_Q}{2}; 2; \epsilon^2\right). \tag{43}$$

Results are reported in Fig. 5(c), from which it follows that for Quintessence $|\Delta\theta_p(\omega_Q,\epsilon)| \lesssim 0.57$ provided $\omega_Q \gtrsim 0.9$. Therefore, the exact value $\omega_Q = -1$ corresponding to the cosmological constant is excluded in this range of values.

5. Conclusions

In this conference we have reported the results presented on Ref. [82]. The analysis is relative to the periastron advance of Solar system Planets in the case in which the gravitational interactions between massive bodies is described by modified theories of gravity. In these classes of theories, the corrections to the Newtonian gravitational interaction is of the Yukawa-like form, $V(r) = V_N(1+\alpha e^{-\beta r})$, where $V_N = GM/r$ is the Newtonian potential, or the power-law form, $V(r) = V_N + \alpha_q r^q$. To compute the corrections to the periastron advance, we have used results of Ref. [45] in which the general integrals are provided in terms of the central body's mass M, and the orbital parameters a and ϵ, the semi-major axis and eccentricity of the orbit, respectively. This two-body system constitutes a good model for many astrophysical scenarios, such as those at the scale of Solar System, constituted by the Sun and a planet, as well as binary system composed by a Super Massive Black Hole and an orbiting star, which are both the most suitable candidates to test a gravitational theory.

In the case of Scalar Tensor Fourth Order Gravity, we find that the parameters of the model are given by (see Eqs. (35, 34, 33) $\alpha \sim F_i$, $\beta \sim \beta_i$, with $i = \pm, Y$, $F_+ = g(\xi,\eta)\,F(m_+\mathcal{R})$, $F_- = \left[\frac{1}{3} - g(\xi,\eta)\right] F(m_-\mathcal{R})$, $F_Y = -\frac{4}{3}\,F(m_Y\mathcal{R})$, $\beta_\pm = m_R\sqrt{\omega_\pm}$, $\beta_Y = m_Y$. The greatest value of β_i is $\beta_i \sim 5 \times 10^{-11} m^{-1}$, which leads to the constraint on F_i is $F_i < 10^{-4}$. This allows to get a bound on the massive modes m_i, $i = \pm, Y$, corresponding to the extra modes presents in ETG.

For the Non-Commutative Spectral Gravity, we have found that the perihelion's shift of planets allows to constrain the parameter β at $\beta > (10^{-11} - 10^{-10}) m^{-1}$ (in this theory the parameter α is given and is of the order $\alpha \sim \mathcal{O}(1)$). Such a constraint on the parameter β improves several order of magnitude ones derived by using pulsar timing $\beta \geq 7.55 \times 10^{-13} m^{-1}$ [58, 59]. These constraints, however, are weaker compared to the ones obtained from terrestrial experimental data, Eöt-Wash [60] and Irvine [61] experiments is [62], which give $\beta \gtrsim 10^4 m^{-1}$ (a bound on β has been derived from Gravity Probe B experiment, giving $\beta > 10^{-6} m^{-1}$ [54]).

We also studied the Quintessence field surrounding a massive gravitational source. Here the parameter characterizing the gravitational field are the adiabatic index ω_Q and the quintessence parameter c. The analysis shows that c assumes tiny values, as expected, being essentially related to the cosmological constant, while $\omega_Q \gtrsim -(0.9 - 0.8)$, that is it never assumes the value -1 corresponding to the pure cosmological constant.

The case of the S2 Star around Sagittarius A*, the Super Massive Black Hole at the center of the Milky Way, has been also studied. In such a case, we have found that for STFOG and NCSG $\beta > 10^{-13} m^{-1}$, a this bound is compatible with astrophysical constraints. For the Quintessence Field we have inferred $\omega_Q \gtrsim -0.9$.

6. Acknowledgments

We acknowledge the INFN (*Istituto Nazionale di Fisica Nucleare*) for the support.

References

[1] Riess A G *et al.* 1998 *Astron. J.* **116** 1009-1038.
[2] Perlmutter S *et al.* 1999 *Astrophys. J.* **517** 565-586.
[3] Cole S *et al.* 2005 *Mon. Notices Royal Astron. Soc.* **362** 505-534.
[4] Spergel S D *et al.* 2007 *Astrophys. J., Suppl. Ser.* **170** 377-408.
[5] Carroll S M, Press W H and Turner E L 1992 *Annu. Rev. Astron. Astrophys.* **30** 499-542.
[6] Sahni V and Starobinski A A 2000 *Int. J. Mod. Phys. D* 2000, **9**, 373.
[7] Capozziello S and De Laurentis M 2011 *Phys. Rep.* **509** 167.
[8] Magnano G, Ferraris M and Francaviglia M 1987 *Gen. Rel. Grav.* **19** 465.
[9] Capozziello S and De Laurentis M 2011 *Phys. Rep.* **509** 167.
[10] Perivolaropoulos L 2010 *Phys. Rev.D* **81** 047501.
[11] Hohmann M, Jarv L, Kuusk P and Randla E 2013 *Phys. Rev. D* **88** 084054.
[12] Järv L, Kuusk P, Saal M and Vilson O 2015 *Phys. Rev. D* **91** 024041.
[13] Nojiri S and Odintsov S D 2002 *Phys. Lett. B* **548** 215.
[14] Bronnikov K A Kononogov S A and Melnikov V N 2006 *Gen. Rel. Grav* **38** 1215.
[15] Teyssandier P and Tourrenc J. 1983 *Math. Phys.* **24** 2793.
[16] Maeda K I 1989 *Phys. Rev. D* **39** 3159.
[17] Wands D 1994 *Class. Quant. Grav.* **11** 56.
[18] Schmidt H J 1987 *Astron. Nachr.* **308** 34.
[19] Berry C P L and Gair J R 2011 *Phys. Rev. D* **83** 104022.
[20] Capozziello S, Lambiase G, Sakellariadou M and Stabile A 2015 *Phys. Rev. D* **91** 044012.
[21] Schellstede G O 2016 *Gen. Rel. Grav.* **48** 118.
[22] Kaczmarek A Z and Szczesniak D 2021 *Sci. Rep.* **11** 18363.
[23] G. Lambiase, M. Sakellariadou and A. Stabile 2013 *J. Cosmol. Astropart. Phys.* 1312 020.
[24] Arkani-Hamed N, Dimopoulos S and Dvali G R 1998 *Phys. Lett. B* **429** 263.
[25] Arkani-Hamed N, Dimopoulos S and Dvali G R 1998 *Phys. Rev. D* **59** 086004.
[26] Antoniadis I, Arkani-Hamed N, Dimopoulos S and Dvali G R 1998 *Phys. Lett. B* **436** 257.
[27] Gasperini M and Veneziano G 1992 *Phys. Lett. B* **277** 256.
[28] Damour T and Polyakov A M 1994 *Gen. Rel. Grav.* **26** 1171.
[29] Gasperini M and Veneziano G 1994 *Phys. Rev. D* **50** 2519.
[30] Deser S and Woodard R, *Phys. Rev. Lett.* **99** 111301.
[31] Stabile A 2010 *Phys. Rev. D* **82** 064021.

[32] Capozziello S and Stabile A 2015 *Astrophys. Space Sci.* 358, 27.
[33] Boehmer C G, Harko T and Lobo F S N 2008 *Astropart. Phys.* **29** 386-392.
[34] Boehmer C G, Harko T and Lobo F S N 2008 *J. Cosmol. Astropart. Phys.* **0803**, 024.
[35] Stabile A and Scelza G 2012 *Phys. Rev. D* **84** 124023.
[36] Lambiase G, Mohanty S and Stabile A 2018 *Eur. Phys. J. C* **78**, 350.
[37] Weyl H 1921 *Raum zeit Materie: Vorlesungen uuber allgemeine Relativitatstheorie* Springer Berlin.
[38] Eddington A S 1924 *The Mathematical Theory of Relativity*, Cambridge University Press London.
[39] Lanczos C 1932 *Z. Phys. A: Hadrons Nucl.* **73**, 147-168.
[40] Capozziello S, Lambiase G, Sakellariadou M, Stabile A and Stabile An 2015 *Phys. Rev. D* **91** 044012.
[41] A. Stabile and S. Capozziello 2013 *Phys. Rev. D* **87** 064002.
[42] Lombriser L and Taylor A 2016 *J. Cosmol. Astropart. Phys.* **03** 031.
[43] De Laurentis M, Porth O, Bovard L, Ahmedov B and Abdujabbarov A *Phys. Rev. D* **94** 124038.
[44] Capozziello S, Lambiase G, Stabile A and Stabile A 2021 *Eur. Phys. J. Plus* **136** 144.
[45] Adkins G S and McDonnell J 2007 *Phys. Rev. D* **75** 082001.
[46] Xu F 2011 *Phys. Rev. D* **83** 084008.
[47] Chashchina O I and Silagadze Z K 2008 Phys. Rev. D **77** 107502.
[48] Yukawa H 1935 *Proc. Phys. Math Soc. Japan* **17** 48.
[49] Nieto M M and Goldman T 1991 *Phys. Rep.* **205** 221.
[50] De Laurentis M, De Martino I and Lazkoz R *Phys. Rev. D* **97** 104068.
[51] Connes A 1994 *Noncommutative Geometry* Academic Press New York.
[52] Connes and Marcolli M 2008 *Noncommutative Geometry, Quantum Fields and Motives* Hindustan Book Agency India.
[53] Sakellariadou M, Stabile A and Vitiello G 2011 *Phys. Rev. D* **84** 045026.
[54] Lambiase G, Sakellariadou M and Stabile A 2013 *J. Cosmol. Astropart. Phys.* 12 020.
[55] Stelle K S 1978 *Gen. Rel. & Grav.* **9** 353.
[56] Donoghue J F 1994 *Phys. Rev.* D **50**, 3874.
[57] Fischbach E and Talmadge C L 1999 *The search for non-newtonian gravity*, Springer Verlag.
[58] Nelson W, Ochoa J and Sakellariadou M 2010 *Phys. Rev. D* **82** 085021.
[59] Nelson W, Ochoa J and Sakellariadou M 2010 *Phys. Rev. Lett.* **105** 101602.
[60] Hoyle C D *et al.* 2001 *Phys. Rev. Lett.* **86** 1418.
[61] Hoskin J K *et al.* 1985 *Phys. Rev. D* **32**, 3084.
[62] Kapner D J *et al.* 2007 *Phys. Rev. Lett.* **98** 021101.
[63] Jamil M, Hussain S and Majeed B 2015 *Eur. Phys. J. C* **75** 24.
[64] V. Kiselev 2003 *Class. Quant. Grav.* **20** 1187.
[65] Belhaj A, El Balali A, El Hadri W, Hassouni Y and Torrente-Lujan E 2021 *Int. J. Mod. Phys. A* **36** 2150057.
[66] Chen S, Wang B and Su R 2008 *Phys. Rev. D* **77** 124011.
[67] Toshmatov B, Stuchlik Z and Ahmedov B 2017 *Eur. Phys. J. Plus* **132** 98.
[68] Ghosh S G 2016 *Eur. Phys. J. C* **76** 222.
[69] Belhaj A, Balali A E, Hadri W E, Hassouni Y and Torrente-Lujan E 2020 *Class. Quant. Grav.* **37** 215004.
[70] Israr Ali Khan S I, Amir Sultan Khan and Ali F 2020 *Int. J. Mod. Phys.* **29** 2050095.
[71] Khan S U and Ren J 2020 *Phys. Dark Univ.* **30** 100644.
[72] Abbas G, Mahmood A and Zubair M 2020 *Chin. Phys. C* **44** 095105.
[73] Javed W, Abbas J and Ovgun A 2020 *Annals Phys.* **418** 168183.
[74] Uniyal R, Chandrachani Devi N, Nandan H and Purohit K D 2015 *Gen. Rel. Grav.* **47** 16.
[75] Iorio L 2017 *Mon. Notices Royal Astron. Soc.* **472** 2249.
[76] Capozziello S, Borka D, Jovanovic P and Borka Jovanovic V 2014 *Phys. Rev. D* **90** 044052.
[77] Khoury J and Weltman A 2004 *Phys. Rev. Lett.* **93** 171104.
[78] Hinterbichler K and Khoury J 2010 *Phys. Rev. Lett.* **104** 231301.
[79] Saksťein J 2018 *Phys. Rev. D* **97** 064028.
[80] Zhang X, Zhao W, Huang H and Y. Cai 2016 *Phys. Rev. D* **93** 124003.
[81] Brax P, van de Bruck C, Davies C, Khoury J, Weltman A 2004 *Phys. Rev D* **70** 123518.
[82] Capolupo A, Lambiase G and Tedesco A 2022 *Eur. Phys. J. C* **82** 286.

Journal of Physics: Conference Series **2533** (2023) 012034 doi:10.1088/1742-6596/2533/1/012034

What quantum matter tells about quantisation of gravity in a statistical mechanics context

Pierre A Mandrin

Independent researcher, Glarnerstrasse 53a, 8854 Siebnen, Switzerland

E-mail: mandrin@hispeed.ch

Abstract. We propose a "guide" towards quantisation of gravity based on quantum matter in a statistical mechanics context. On one hand, a statistical mechanics model naturally arises from the thermodynamic interpretation of horizons in Rindler space. On the other hand, the path integral formulation of quantum field theory can be interpreted from the point of view of statistical mechanics. From these perspectives, gravity and matter are related to each other in the same way as a gas and its chemical potential are. This statistical mechanics interpretation of gravity and matter suggests that gravity should be quantised in a precise way which is determined by the quantisation of matter. Although, in a first step, quantisation of gravity applies for small perturbations of the metric with respect to the vacuum, the most central and general features of quantisation (quantisation prescription, quantum space dimensions) are supported from statistical mechanics and remain valid non-perturbatively.

1. Introduction

There is, currently, a large uncertainty on how to treat gravity in a quantum mechanics context. Although it is mainly believed that gravity is required to be quantised, it is at least unclear which type of quantisation mechanism should be used, or which variables should be the preferred ones to start with. To a large part, this situation is certainly due to a lack of available experimental data which could give insight to near-Planck scale physics.

Nevertheless, important hints should also be available from well-controlled knowledge of quantum matter. Indeed, the consequences are not obvious to investigate because we first need to understand the fundamental role of matter within the framework of gravity. By considering gravity via its statistical mechanics interpretation, matter acquires a new quality in the thermodynamic context, i.e. it acts in the same way as does the fugacity or the chemical potential in a grand canonical ensemble. The chemical potential interpretation of matter has already been proposed in [1], but an important new development step is given by identifying the grand canonical ensemble which explicitly deals with open systems both with respect to "gravitational particles" and an "energy-like" extensive quantity which comes from the Lagrangian. With the grand canonical ensemble interpretation, the quantisation of gravity can first be obtained in the perturbative regime (linearised gravity). Due to the statistical mechanical context, important features of quantisation still survive in a non-perturbative concept, namely the path integral quantisation ansatz, the preferred canonical variables for quantisation and the space dimension 3 for gravitational quantum statistical mechanics. Despite the undeniable importance of these conceptual features, their role is not to constitute the full information for presenting a complete quantum theory of gravity. Rather, the importance of the this article relies in improving the confidence for how gravity is quantised.

The concept of the statistical mechanics interpretation of the boundary term of gravity is introduced in Section 2. It is based on an argument in [2]. The newer interpretation in this article has some different features; the main one is the open system property of a compact space region, thus providing to the total action the quality of a grand canonical potential. To lowest perturbation order, a matter field has the effect to linearly shift the gravitational field away from the classical vacuum solution.

There is a subtile connection between matter and gravity: the radiated matter perceived by accelerated observers (Unruh effect) is nothing but the heat (matter particles) crossing the Rindler horizon [3, 4, 5]. An equivalent process is the Hawking radiation produced in conjunction with black hole entropy [6, 7, 8]. It is therefore natural to interpret the radiated matter as a thermodynamical manifestation of gravity, although this does not mean that gravity itself is specific to or can be identified with particular species of matter or their Hamiltonians. The open system concept and grand canonical ensemble view is introduced in Section 3.

In Section 4, we conclude from Sections 2 and 3 how quantisation of gravity follows from quantisation of matter. The same type of canonical quantisation procedure as for flat space-time Quantum Field Theory (QFT) applies for linearised gravity, with the requirement that one of the canonical variables has to be the vielbein, and path integral quantisation applies in a similar way as for QFT, appart from the evolution aspect. The path integral concept for gravity and matter has a one-to-one correspondence with the statistical mechanics formulation of gravity, i.e. it is not bound to a perturbative treatment, and the vielbeins therefore must keep their role of canonical functional variable for the non-perturbative quantisation procedure. The matter part in the path integral corresponds to the term containing the chemical potential μ in the exponent of the grand canonical partition function of a gas.

From Section 2, the general form of quantisation of gravity must have its most natural formulation in 3d-space since this is the dimension of the Rindler horizon hypersurfaces which the initial microcanonical ensemble treatment is originating from. This dimension is automatically recovered in the path integral treatment when the "external time evolution" problem is removed. The reduced dimension 3 indeed corresponds to the requirement for a unitary quantum gravity theory [9].

2. Summary of the statistical mechanical interpretation of gravity

We first shortly review the required parts from [2] and references therein and strengthen the required chain of argumentation as well as the physical picture of the statistical mechanical interpretation of space-time. As is well-known from [4, 5], space-time has a thermodynamical interpretation. Any spatial 2-surface on any 3-dimensional null-hypersurface is perceived as a Rindler horizon by a suitably located, uniformly accelerated observer, with an associated temperature $T = \kappa/(2\pi)$ (κ is the accelleration of the observer or Rindler surface gravity) and an entropy $S_{hor} = A/(4L_p^2)$ related to a surface area A, where L_p is the Planck length, and the horizon of the Rindler wedge is equivalent to a black hole horizon. It should be stressed that the unboosted original space with coordinates (t, x, y, z) does not need to be flat, it may, in general, have a non-singular metric $g_{\alpha\beta} \neq 0, \pm\infty$. By the generalised terminology "Rindler space", we therefore mean the boosted space defined by the coordinates (T, X, y, z),

Journal of Physics: Conference Series **2533** (2023) 012034 doi:10.1088/1742-6596/2533/1/012034

$$t = X \sinh T, \qquad x = X \cosh T. \tag{1}$$

According to [10, 5], a 3-dimensional null-hypersurface in Einstein gravity contributes a surface action $\mathcal{A}_{sur}$ (which corresponds in part to the boundary term), and $\mathcal{A}_{sur}$ is interpreted as the time integrated "heat" $\int dt\ Q = \int dt\ TS$ when evaluated on the Rindler horizon (taking care of the order of evaluation steps for taking the null limit). We have:

$$\mathcal{A}_{sur} = \frac{c^4}{16\pi G} \int N^c_{ab}\ f^{ab}\ d^3\Sigma_c, \qquad f^{ab} = \sqrt{-g}\ g^{ab}, \qquad N^c_{ab}\ d^3\Sigma_c = N^{ab}\ d^3x, \tag{2}$$

where, in the notation of [5], f^{ab} is the entropy density, N^c_{ab} is its conjugate momentum, and $d^3\Sigma_c$ is the 3-surface element covector. On the right hand side of (2), $N_{ab} = K_{ab} - K^c_c\gamma_{ab}$ and K_{ab} is the second fundamental form [11, 12]. The interpretation of (2) and of the elaborations of [4, 5] is that every spatial surface represents a "gas" of gravitational building blocks ("gravitational particles"). Notice that, in order to keep the entropy and thus the sample of these "particles" finite while avoiding infinitely long observer orbits, it is preferable to integrate over a finite section of 3d-hypersurface. However, this must be handled with care since cutting a finite section out of an infinite hypersurface yields an open system, we consider this in more detail in the next section.

The basic idea of our effort is that, whenever a space-time region possesses a boundary surface and we compute the physical content by varying the bulk gravitational action for this space-time region, we omit to vary the divergence term integrated over the space-time part beyond the boundary surface, and this missing information content is equivalent to the missing content that we suffer by placing a Rindler horizon exactly in that boundary surface and observing the Unruh radiation. In order to obtain a one-to-one correspondence, the observer must choose the accelleration κ so that the hidden heat contents of both frames match. In the frame of original space, the omitted part of the action is a surface contribution of the boundary term $\Delta S_{\partial V}$, and this is equal to $\mathcal{A}_{sur} = -\int dt H_{sur}$ in the language of [10], provided that a proper gauge of the metric is chosen. The boost accelleration κ must be chosen in such a way that H_{sur} in original space is equal to its value TS computed in the Rindler frame, i.e. T must have the same value in both frames since the surface areas in the yz-subspace (i.e. the entropies) are equal. There remains an issue that 3-surfaces of boundary terms are not null. However, as shown in [2], any non-null section of smooth 3d-hypersurface in the boundary term integral can be replaced to any desirable accuracy by a sequence of narrow sections of null-surfaces (strips) glued together (much like a smooth ramp would be approximated by narrow stairs), by replacing $\int_{\Delta x \Delta A} \to \int_{\Delta u \Delta A} + \int_{\Delta v \Delta A}$ (the strip widths Δu, Δv being the null decomposition of Δx). Consider now a compact space-time region V with piece-wise smooth boundary $\partial V = \sum_A B_A$, where every component B_A is non-null (time- or space-like). Then, the boundary term is the sum of the time integrated heats as perceived by all the accelerated observers related to the null-strips covering the boundary ∂V:

$$S_{\partial V} = -\int dt \sum_k T_k S_k|_{\Delta \partial V_k}, \tag{3}$$

where $\Delta\partial V_k$ are small sections covering $\partial V = \bigcup_k \Delta\partial V_k$ with the associated temperatures T_k and entropies $S_k = \ln \Omega_k$, and Ω_k is the number of distinct microstates which are compatible with the macroscopic state of $\Delta\partial V_k$.

Journal of Physics: Conference Series **2533** (2023) 012034 doi:10.1088/1742-6596/2533/1/012034

For theories of gravity beyond Einstein gravity, the definition of temperature and heat can be redefined in such a way that the interpretation of the boundary term as in (3) is preserved, while the integrands of the null-strip integrals will normally undergo a higher order modification (in terms of the metric), depending on the theory. In this sense, our statistical mechanical considerations are not restricted to Einstein gravity.

3. The grand canonical ensemble interpretation

Expression (3) is not best suited to describe finite sections of 3-surfaces. Since the bounds of a section are not bounds of an autonomous physical system, there must be an uncertainty about the number of "building blocks" of gravity that are located inside of a section. Even more, since quanta of matter are subject to the Heisenberg uncertainty relation and thereby affect the gravitational field so as to make it blurred, the "gravitational particles" may hardly be localisable, and a section may contain anything between zero and all of the "gravitational particles". A finite section therefore represents an open thermodynamic system with respect to the exchange of "gravitational particles" and also with respect to the exchange of independent extensive quantities associated with the "gravitational particles" – we will find such one (much like the "energy" of a molecular gas), and the part of the 3-surface outside the section is a particle bath and a "thermal" bath. The suitable ensemble is therefore the grand canonical ensemble for which the density of states is formally given by

$$\rho_i = \frac{\exp\left[(-\mathcal{E}_i + \mu N_i)/(k_B T)\right]}{Z} \tag{4}$$

with the grand canonical partition function

$$Z = \sum_i \exp\left[(-\mathcal{E}_i + \mu N_i)/(k_B T)\right]. \tag{5}$$

In (5), the sum is over a phase space partition with (infinitesimal) cell label i. Since (4) does not represent a classical "molecular gas", the symbol $\mathcal{E}$ may be a quantity different from the "energy". The precise interpretation of the $\mathcal{E}_i$ and "chemical potential" μ is non-trivial and cannot be derived from the 3-surface consideration of Section 2, but can be identified via general thermodynamic properties of the ensemble. A key quantity is the grand canonical potential $\Phi = k_B T \ln Z$ which is minimised in the state of thermodynamic equilibrium. Taking the logarithm of (5) yields

$$\Phi = \mathcal{U} - TS - \mu N, \tag{6}$$

where $\mathcal{U} = \langle \mathcal{E} \rangle = \sum_i \rho_i \mathcal{E}_i$ is the ensemble mean of $\mathcal{E}$, likewise $N = \langle N \rangle = \sum_i \rho_i N_i$, and $S = \langle -k_B \ln \rho \rangle = \sum_i -k_B \rho_i \ln \rho_i$. In terms of dynamics, the quantity which must be extremised is the gravitational action S_g plus boundary term $S_{g\partial V}$, supplemented by the action S_m for matter plus boundary term $S_{m\partial V}$, this altogether must be proportional to Φ in order for (3) to be satisfied under general conditions:

$$-t\Phi = S_g + S_{g\partial V} + S_m + S_{m\partial V}. \tag{7}$$

The term $S_{m\partial V}$ is required if matter contributes on the boundary, which is expected for compact V. For instance, a real scalar field has the boundary term $-\int_{\partial V} d^3x \sqrt{-g}\varphi \nabla_\perp \varphi$. As indicated by the sum of expression (3), T and μ may be allowed to vary between small 3d-volume elements. However, these variations must evolve slowly, i.e. the variation (wave) length scale e.g. of T should be much larger then the mean free path λ as given from interaction theory (such lengths are expected to be much smaller than experimentally achievable spatial resolutions), according to standard kinetic theory of gases. Accordingly, (5) should be extended to

$$Z = \sum_i \exp\left[\int_\Sigma (-\epsilon_i + \mu\eta_i)/(k_B T)\right], \tag{8}$$

where the sum over 3-volume elements has been converted to an integral, ϵ_i and η_i are 3d-densities of $\mathcal{E}_i$ and N_i, respectively, and $\Sigma \subset \partial V$ is the space-like boundary section on the past 3-surface). Then, (7) becomes:

$$\int dt \int_\Sigma \big[-\epsilon + Ts + \mu\eta\big] = S_g + S_{g\partial V} + S_m + S_{m\partial V}, \tag{9}$$

where s is the 3d entropy density. Now, whenever we vary the metric both in V and on ∂V, the effect is to change the values S_g and $S_{g\partial V}$. Notice that s and η are independent variables (since heat and "particle" density vary independently from each other), i.e. we can have a variation δs for fixed η. Minimisation of Φ only implies a change $\delta\epsilon$, i.e. we can separate, in this scenario, the gravitational part of (9) from the matter part:

$$\int dt \int_\Sigma \big[-\epsilon + Ts\big] = S_g + S_{g\partial V}, \tag{10}$$

$$\int dt \int_\Sigma \mu\eta = S_m + S_{m\partial V}. \tag{11}$$

As we see, the density of H_{sur} reappears in Ts and is also contained in $S_{g\partial V}$, therefore ϵ must contain the (densitised) gravitational bulk Lagrangian. It is also apparent that the term $S_{g\partial V}$ acts as a Legendre transformation since the variable to be varied on the boundary changes ($g_{ab} \to \Gamma^c_{ab}$) and the left-hand-side term Ts performs the same Legendre transformation in thermodynamic language ($S \to T$), i.e. ϵ is precisely the bulk Lagrangian density. From (11), μ represents the (densitised) matter Lagrangian per particle (including the divergence contribution constituting the boundary term). There is also another way to see that the $\mu\eta$-term represents matter: A variation $\delta\eta$ around $\eta = 0$ causes a change $\delta(\epsilon - Ts)$ upon minimisation of (9), which modifies the physical metric in V such that the vacuum equations of gravity are violated. In the Einstein gravity limit, we only vary the metric, and it is required to add a matter term and vary the matter field in order to restore thermal equilibrium, hence the $\mu\eta$-term corresponds to the matter contribution to the action.

In order to deepen our understanding of the role of matter in the context of gravity, it is useful to consider linearised gravity with matter. For any Lagrangian theory of gravity, the general field equations are

$$\delta(\sqrt{-g}\mathcal{L}_g) = -\frac{8\pi G}{c^4}\delta(\sqrt{-g}\mathcal{L}), \tag{12}$$

where $\mathcal{L}_g$ and $\mathcal{L}$ are the gravitational and matter Lagrangian, respectively. We linearise (12) and neglect sources for torsion (or for any other extra-variables) for consistency, thus obtaining the linearised Einstein equations with stress tensor $T_{\alpha\beta}$:

$$g_{\alpha\beta,\mu,}{}^{\mu} + g_{\mu\nu,\alpha,\beta}\eta^{\mu\nu} - g^{\mu}{}_{\alpha,\mu,\beta} - g^{\mu}{}_{\beta,\mu,\alpha} = -\frac{8\pi G}{c^4}\left[T_{\alpha\beta} - \frac{T}{2}\eta_{\alpha\beta}\right], \tag{13}$$

where the cosmological constant has been omitted. In linearised theory, it is sufficient to neglect couplings between interacting matter fields and to take into account terms up to second order only in the matter field for $T_{\alpha\beta}$. In the scope of the standard model,

$$T_{\alpha\beta} = 2\frac{\partial \mathcal{L}}{\partial g^{\alpha\beta}} - \mathcal{L} g_{\alpha\beta} \tag{14}$$

and $\mathcal{L} = K - V$ with the kinetic term K and the potential term V (K, V each admit one contribution per species). We evaluate $\partial\mathcal{L}/\partial g^{\alpha\beta}$, then eliminate all V-contributions to $T_{\alpha\beta}$ via the equations of motion of matter, e.g.

$$\mathcal{L} = \alpha\nabla_\mu(\frac{\partial K}{\partial \nabla_\mu \varphi}\varphi) - \beta\frac{\partial K}{\partial \varphi}\varphi - \gamma(\nabla_\mu \frac{\partial K}{\partial \nabla_\mu \varphi})\varphi \tag{15}$$

is the Lagrangian for every single species, where e.g. $\alpha = 1/2, \beta = \gamma = 0$ for a neutral scalar field ($\varphi = \varphi^\dagger$), and supplementing an index ($\varphi \to \varphi^\mu$) in the case of a vector field and likewise in any later expressions. For an (approximately) monochromatic field (single-k-mode idealisation for linearised theory) of a single species with Fourier transform $\tilde{\varphi}(k^\mu)$, a straightforward computation yields that it is proportional to the tetrad perturbation $h^I{}_\alpha = e^I{}_\alpha - \delta^I_\alpha$ in the neighbourhood of a space-time point and the non-oscillatory background frame is chosen to be locally Minkowski:

$$\tilde{h}^I_\alpha = \vartheta^I_\alpha \tilde{\varphi}, \qquad g_{\alpha\beta} = h^I_\alpha \,\eta_{IJ}\, h^J_\beta + g_{0\alpha\beta}. \tag{16}$$

E.g. for a neutral scalar field, we have

$$\begin{aligned} h^I_\alpha &= -2\sqrt{\frac{\pi L_P^2}{\hbar}}\,\delta^I_\alpha\,\varphi_0 \sin(k_\mu x^\mu), \\ g_{0\alpha\beta} &= \eta_{\alpha\beta} - 2\frac{\pi L_P^2}{\hbar}\,\eta_{\alpha\beta}\,\varphi_0^2 \end{aligned} \tag{17}$$

with plane wave field amplitude φ_0. (16) may be extended to admit severel species l by replacing $g_{\alpha\beta} \to \sum_l g^{(l)}_{\alpha\beta}$ and writing the proportionality for each contribution $h^{(l)I}{}_\alpha$. Despite the special limit, (16) reveals that the presence of matter acts as a shift constraint for gravity, i.e. the matter field must be compensated by a shift of the tetrad. This translates to the thermodynamics view according to which a change δN of particles causes a shift $\delta\mathcal{E}$ in order to keep the "gas" in equilibrium (minimisation of Φ). At least in the monochromatic linearised case, the particle number N corresponds to the number of building blocks of gravity to be generated in order for the matter field to emerge. Due to the fundamental nature of the thermodynamics concept, the basic interpretation of N should persist even when nonlinear corrections to (16) gradually become relevant.

4. From quantum matter to the quantisation of gravity

From the above considerations, the properties of quantum matter in the flat space-time approximation allow to infer certain properties of gravity at microscopic level. In the flat space-time approximation, the partition function (8) simplifies:

$$Z_{flat} = \sum_i \exp\left[\int_\Sigma \mu\eta_i/(k_B T)\right], \tag{18}$$

Z_{flat} corresponds to the path integral for matter for a space-time "region" with spacelike boundary contributions in the form of time slices at $t = 0$ and $t = \mathcal{T}$; we consider one species with label l:

$$Z^{(l)}_{flat}\big|_{0\to\mathcal{T}} = \langle\varphi^{(l)}(t=\mathcal{T},x^i)|e^{-iH\mathcal{T}}|\varphi^{(l)}(t=0,x^i)\rangle = \int \mathcal{D}\varphi^{(l)}\ e^{i\int_0^{\mathcal{T}} d^4x\mathcal{L}[\varphi^{(l)}]/\hbar}. \tag{19}$$

(8) together with (18) and (19) imply that the gravitational extension must be of the form

$$Z\big|_{0\to\mathcal{T}} = \int\ f(X)\mathcal{D}X\prod_l \mathcal{D}\varphi^{(l)}\ e^{i[S_{grav\Delta V}+S_{grav\Delta\partial V}]_{0\to\mathcal{T}}/L_p^2+i\sum_l[S^{(l)}_{mV}+S^{(l)}_{m\partial V}]_{0\to\mathcal{T}}/\hbar}, \tag{20}$$

where X is the (not yet determined) gravitational functional variable which is related (but does not need to be equal) to the thermodynamic quantity ϵ, the function $f(X)$ allows for an adaptation factor which has the value 1 in flat space-time approximation; f may be necessary at least to ensure invariance of the density of states under gauge transformations. The fixed time limits $0,\mathcal{T}$ are still present as a parameter to use for time evolution. However, the phase space over which the ensemble sum is taken departs from the original idea in (19), namely that the integral should have a common time interval, possibly an observable temporal distance $\langle\sqrt{\Delta x^\mu g_{\mu\nu}\Delta x^\nu}\rangle$ and not a meaningless coordinate time. Moreover, it is not clear, in the context of a covariant theory of gravity, why one should evolve paths along time rather than along space. One possible modification would be to represent all the space-time orientations in the sum over path evolutions. However, this would require to reduce the degrees of freedom afterwards in order to recover (19), and the problem of ensemble averaged evolution parameter would persist. These concerns may directly be avoided if we remove the time limit and abandon the time dimension from the beginning. We shall come back to this idea later.

From (20) and from the point of view of the present statistical mechanical interpretation, (19) does not represent a canonical partition function since it is reminiscent of the μN-part of (20). In the form (20), the gravitational path integral already reveals a basic consequence, namely that the gravitational field must be quantised in order for the matter field to be quantised, since matter plays the role of fugacity within gravity. Starting with classical statistical mechanics on the gravity side would make quantum matter (μN-part) incompatible. More precisely, when the particle number is promoted to a quantum operator $\hat{N}$, with generally mixed states in particle number representation, gravity itself must have mixed states in terms of its number of building blocks, via the relation (16), i.e. we must introduce operators $\hat{e}^I_\alpha$ and therefore $\hat{X}$. The properties of gravitational quantum mechanics is even forced to be compatible with quantum mechanics of matter since, for linearised theory, the transition functions with time slices $t=0,\mathcal{T}$ for the boundary exhibit the same formal behaviour for gravity and matter, as suggested by the following symmetric notation and admitting a single species:

$$\begin{aligned}&\langle X(0,x^i),\varphi(0,x^i)|e^{-iH\mathcal{T}}|X(\mathcal{T},x^i),\varphi(\mathcal{T},x^i)\rangle\\ &=\int\ \mathcal{D}X\mathcal{D}\varphi\ e^{i\int_0^{\mathcal{T}} d^4x\sqrt{-g}\mathcal{L}_g/L_p^2+i\int_0^{\mathcal{T}} d^4x\sqrt{-g}\mathcal{L}/\hbar}.\end{aligned} \tag{21}$$

In linearised theory, the path integral quantisation (21) is equivalent to canonical quantisation:

$$[\hat{\Pi}(x^a,t),\hat{X}(y^a,t)] \sim L_p^2\delta(x^a-y^a), \qquad [\hat{\pi}(x^a,t),\hat{\varphi}(y^a,t)] = i\hbar\delta(x^a-y^a), \tag{22}$$

where $\hat{\Pi}$ and $\hat{\pi}$ are the momenta canonically conjugate to $\hat{X}$ and $\hat{\varphi}$, respectively. Using (16) and the symmetry $X\leftrightarrow\varphi$ of (21), we see that any variation $\delta\tilde{\varphi}$ modifies in the exact same way the weights in the sum of paths as does the variation of its counter-part $\delta\tilde{e}^I_\alpha = \vartheta^I_\alpha\delta\tilde{\varphi}$, i.e. if we set $X=e^I_\alpha$, both variables will generate the exact same patterns of transition matrix elements (21),

Journal of Physics: Conference Series **2533** (2023) 012034 doi:10.1088/1742-6596/2533/1/012034

i.e. both variables have the same states ($|\varphi\rangle$ and $|e^I_\alpha\rangle$) and commute. This is also equivalent to writing the operator version of (16):

$$\hat{\hat{h}}^I_\alpha = \vartheta^I_\alpha \hat{\hat{\varphi}}. \tag{23}$$

As an example, a real scalar with plane wave amplitude $\varphi_0 = \sqrt{\hbar c^2/(2V_3\omega'_k)}$ will yield

$$\hat{h}^I_\alpha = \sqrt{\frac{\pi L_p^2}{\hbar}}\, \delta^I_\alpha\, \varphi_0 \left(-ie^{-ik_\mu x^\mu}\hat{a}_{k^i} + ie^{ik_\mu x^\mu}\hat{a}^\dagger_{k^i}\right). \tag{24}$$

The first quantisation prescription in (22) is thus satisfied if we identify $\hat{X} \to \hat{h}^I_\alpha$ or $\hat{X} \to \hat{e}^I_\alpha$. In linearised theory, the following simple commutation relations apply:

$$\left[\hat{e}^I_\alpha(x^i,t), \hat{e}^J_{\beta,0}(x'^i,t)\right] \sim L_p^2 \delta^I_\alpha \delta^J_\beta \delta(x^i - x'^i), \tag{25}$$

$$\left[\hat{e}^I_\alpha(x^i,t), \hat{e}^J_\beta(x'^i,t)\right] = \left[\hat{e}^I_{\alpha,0}(x^i,t), \hat{e}^J_{\beta,0}(x'^i,t)\right] = 0, \tag{26}$$

where the δ-function is restricted to 3 dimensions. Since we assumed vanishing torsion, (25) can be obtained directly from the right hand side of (22). For comparison, the technique equivalent to matter quantisation is to use the first derivative Einstein Lagrangian $\mathcal{L}_E \sim g^{\alpha\beta}(\Gamma^\gamma_{\alpha\delta}\Gamma^\delta_{\gamma\beta} - \Gamma^\gamma_{\alpha\beta}\Gamma^\delta_{\delta\gamma})$ [12] and then to compute the momentum $\partial(\sqrt{-g}\mathcal{L}_E)/\partial e^I_{\alpha,0}$ canonically conjugate to e^I_α, a lengthy but straight-forward calculation yields an expression of the form $\Pi^\alpha_I = F^{\alpha\beta\gamma}_{IJ}(e^K_\mu)e^J_{\beta,\gamma}$, where $F^{\alpha\beta,\gamma}_{IJ}(e^K_\mu)$ is a tensorial function of the tetrad and its inverse. One could wonder why we should not simply use the right hand side of (22) which seems to work for linearised quantisation. However, as soon as some torsion is present, the right hand side of (22) provides species-dependent and thus inconsistent commutators. Moreover, the symmetry of (21) clearly requires that we must use the calculation method by canonical conjugation using $\mathcal{L}_E$. If we extend the theory to admit some torsion (e.g. caused by the spin of a fermion species [13]), the connection with contorsion in $\mathcal{L}_E$ will naturally lead to a torsion contribution to Π^α_I. For the full gravity variable X, one may want to consider that e^I_α might have to be multiplied by a function $\epsilon(e)$ which would be unity in linearised theory, with $e = \sqrt{-g}$, yielding the operator $\hat{X} = (\epsilon(e)e^I_\alpha)\hat{}$. However, since $\epsilon(e)e^I_\alpha \sim \varphi^* = \epsilon(\varphi)\varphi$ (formally) at lowest perturbation order, the resulting commutation relation $[\hat{\pi}^*, \hat{\varphi}^*]$ for matter would be incompatible with the right hand side of (22) or not even exist; this prevents us from allowing an extra factor $\epsilon(e)$ for X. The only way to escape from this restriction would be to redefine the matter field as $\varphi^* = \epsilon(e)\varphi$, but this would require a complicated and not intuitive matter Lagrangian.

We turn back to the problem of time limits in (20) and procede with the removal of the fixed time limits from (20) and dimensional reduction. We first have to get rid of the time dynamics. This is done by choosing $\mathcal{T}$ small, i.e. $||h^I_{\alpha,0}||c\mathcal{T}/||h^I_\alpha|| \ll 1$.The integration over time can then be replaced by a constant factor $\mathcal{T}$, and only the timelike parts of the boundary terms contribute. Now that the system has lost the time dynamics, we can reformulate it explicitly in 3 dimensions using the ADM-decomposition and remove the components $g_{0\alpha}$ from our system, thus reducing at once the gauge degrees of freedom. E.g. in the case of Einstein gravity, we apply the Gauss-Codazzi equations:

$$Z\Big|_{\mathcal{T}_0} = \int f(e)\mathcal{D}e^I{}_a \prod_l \mathcal{D}\varphi^{(l)}\; e^{i\mathcal{T}_0\left(\int_{\Sigma_f} d^3x\sqrt{e}[\mathcal{R}+K_{ab}K^{ab}-K^2]/L_p^2 + \sum_l \int_{\Sigma_f} d^3x\sqrt{e}\mathcal{L}^{(l)}/\hbar + \ldots\right)}, \tag{27}$$

where $\tau_0 = \sqrt{g_{00}}\Delta\mathcal{T}$ is a fixed small timelike distance, $\mathcal{R}$ is the 3d-Riemann curvature, K_{ab} is the second fundamental form, e_a^I are triads (with 3d-indices I, a), $e = \sqrt{\det(g_{ab})}$, and the boundary terms are indicated by periods. We shall bear in mind that the sum over geometries in (27) is subject to the constraints of the initial value problem. By removing the time limits, the gravitational path integral quantisation has naturally led us to a 3-dimensional theory. This, however, is what we have expected from [9] and from the fact that we have started with 3-dimensional statistical mechanics of the Rindler horizon.

5. Conclusions

This article has shown that knowledge from horizon thermodynamics and reliable knowledge from quantum matter can be useful to enforce our confidence in the choice of the road-map towards quantisation of gravity. Although quite some open questions are not answered in a definitive way and although this procedure is not intended to yield a full theory with all its details, the findings nevertheless suggest the type of required quantisation procedure and the canonical variables required for the quantisation prescription, and new ways of interpretation of the physical situation are opened. As a key feature, the statistical mechanical interpretation of gravity within compact regions yields a grand canonical ensemble which is naturally related to the path integral concept originating from quantum field theory. In the gravitational context, matter is interpreted as the fugacity. This results in a natural gravitational extension of the path integral quantisation and provides a pair of (main) canonical variables, namely the triad and its conjugate momentum. Appart from the flat space limit, quantised gravity is preferably operating in 3d-coordinates so as to avoid the appearance of an unpleasant "external" time evolution parameter.

References

[1] Mandrin P A 2019 *J. Phys.: Conf. Ser.* **1275** 012060
[2] Mandrin P A 2017 *J. Phys.: Conf. Ser.* **880** 012028
[3] Unruh W G 1976 *Phys. Rev. D* **14** 870
[4] Jacobson T 1995 *Phys. Rev. Lett.* **75** 1260
[5] Padmanabhan T 2014 *Gen. Rel. Grav.* **46** 1673
[6] Hawking S W 1975 *Commun. math. Phys.* **43** 199
[7] Bekenstein J D 1973 *Phys. Rev. D* **7** 2333
[8] Wald R M 2001 *Liv. Rev. Rel.* **4** 2001–6wald
[9] 't Hooft G 1993 Dimensional reduction in quantum gravity *Preprint* gr-qc/9310026
[10] Majhi B R and Padmanabhan T 2013 *Eur. Phys. J.* C **73** 2651
[11] Gibbons G W and Hawking S W 1977 *Phys. Rev. D* **15** 2752
[12] York J W 1986 *Found. Phys.* **16** 249–257
[13] Hehl F W, von der Heyde P, Kerlik G D and Nester J M 1976 *Rev. Mod. Phys.* **48** 393

A metric for Planck Stars derived from Gravity in Asymptotic Safety

Fabio Scardigli[1,2] **and Gaetano Lambiase**[3]

[1] Dipartimento di Matematica, Politecnico di Milano, Piazza Leonardo da Vinci 32, 20133 Milano, Italy
[2] Institute-Lorentz for Theoretical Physics, Leiden University, P.O. Box 9506, Leiden, The Netherlands
[3] Dipartimento di Fisica "E.R. Caianiello", Universita' di Salerno, I-84084 Fisciano (Sa), Italy and INFN - Gruppo Collegato di Salerno, Italy

E-mail: fabio@phys.ntu.edu.tw

E-mail: lambiase@sa.infn.it

Abstract. The Asymptotically Safe Gravity (ASG) framework suggests a "running" Newtonian coupling constant, which depends on two free parameters $\tilde{\omega}$ and γ. The new black hole metrics inferred from such a "running" gravitational constant naturally match with a Schwarzschild metric at large radial coordinate. By further imposing the matching with the Donoghue quantum corrections to the Schwarzschild field, we find a negative value of the $\tilde{\omega}$ parameter, and this leads to a not yet explored black hole metric, which surprisingly turns out to describe the so-called Planck stars.

1. Introduction

Many models in quantum gravity (see e.g. [1, 2, 3, 4, 5] and references therein) share the property that the fundamental parameters entering the action (Newton's constant, electromagnetic coupling, cosmological constant etc), are considered scale dependent quantities. Actually, scale dependence at the level of the effective action is a generic feature of ordinary quantum field theory. In gravity theories then the scale dependence is expected to modify the horizon, the thermodynamics, the quasinormal modes spectra of classical black hole backgrounds [6, 7, 8, 9, 10]. Among the mentioned approaches based on scale-dependent gravity, which go beyond classical GR, we can find also a particular method usually known as "improved" asymptotically safe (AS) gravity [11, 12, 13]. For renormalization group (RG)-improved black hole metrics, cosmologies, and inflationary models from asymptotic safety, the reader can usefully consult e.g. Refs. [14, 15, 16, 17, 18, 19, 20, 21, 22, 23, 24, 25, 26]. In the ASG scenario usually the beta function for the gravitational coupling is integrated in order to compute the Newton's constant G as a function of some energy or momentum scale k. The "running" Newton's constant $G(k)$ so constructed is then inserted into the classical black hole solution and an "improved" lapse function is obtained, which is considered to automatically include, in this way, the quantum gravity effects. Clearly, in this approach it is essential to establish a link between the energy scale k and the radial coordinate r, so that we can write $G(k(r)) \equiv G(r)$. Only after this step is done, the improved black hole metric can be considered complete and useful. Those extended solutions, inspired by the asymptotic safety program, are expected to modify the classical black hole solutions by incorporating quantum features. However, different modified black hole metrics, in particular metrics also not affected by the central singularity, can be found, e.g, in Refs. [27, 28, 29], although those examples are not directly connected with the ASG program.

On the other hand, by reformulating General Relativity as an effective quantum field theory of gravity at low energies, John Donoghue and other authors [30, 31, 32, 33, 34, 35, 36, 37, 38] have, along the years, established a solid prediction of the quantum corrections to the Schwarzschild field, and therefore to the Newtonian potential, at least at the first order in $\hbar$. By comparing the effective Newtonian potential predicted by the ASG approach with the one computed in the framework of GR as an effective QFT, we arrive to establish, for the first time, a *negative* value for the parameter $\tilde{\omega}$, unlike previous early predictions (see Refs. [11, 22]). This in turn leads directly, without further assumptions, to a specific metric which is able to describe the principal features of the so called Planck stars. It is remakable that, while Planck stars were introduced in Ref. [39] on the grounds of plausible quite general physical considerations, here on the contrary they appear as an almost unavoidable consequence of a metric obtained from the Asymptotically Safe Gravity approach.

We work in units where $c = k_B = 1$, and ℓ_P is the Planck length defined as $G\hbar = \ell_P^2$. Then of course the Planck mass m_P satisfies $2Gm_P = \ell_P$ and $\hbar = 2m_P\ell_P$.

2. Black hole metrics from Running Newtonian coupling

So, according to the literature on the topic [11, 40, 41], the basic idea of the AS gravity approach, in order to obtain the renormalization-improved, classical Newtonian or general relativistic solutions is to replace everywhere the numerical Newton constant G with the 'running constant' $G(r)$, whose explicit form is given in Ref. [11] as

$$G(r) = \frac{Gr^3}{r^3 + \tilde{\omega}G\hbar\,(r + \gamma GM)}\,, \tag{1}$$

where, in accordance with our conventions, $c = k_B = 1$ and we retained $\hbar$. Here $\tilde{\omega}$ and γ are dimensionless numerical parameters, whose concrete value will be discussed later.

The line element for the spherically symmetric, Lorentzian metric preserves the usual form, that is

$$\mathrm{d}s^2 = F(r)\mathrm{d}t^2 - F(r)^{-1}\mathrm{d}r^2 - r^2\,\mathrm{d}\Omega^2, \tag{2}$$

where r is the radial coordinate, and $\mathrm{d}\Omega^2 = \mathrm{d}\theta^2 + \sin^2\theta\,\mathrm{d}\phi^2$ is the line element of the unit two-sphere. But now, according to the above prescriptions, the lapse function $F(r)$ of our ASG improved Schwarzschild geometry reads

$$F(r) = 1 - \frac{2MG(r)}{r} = 1 - \frac{2GMr^2}{r^3 + \tilde{\omega}G\hbar(r + \gamma GM)}\,, \tag{3}$$

with $G(r)$ given by (1) and M the mass of the black hole. Of course, we suppose $\tilde{\omega} \neq 0$, otherwise we would go back to the standard Schwarzschild metric. Two very important limiting cases should be considered.

The first corresponds to the low energy scales ($r \to \infty$, or $k \to 0$), which implies

$$F(r \to \infty) \simeq 1 - \frac{2GM}{r}, \tag{4}$$

so the standard Schwarzschild metric at large distances is recovered, and this behavior is independent from the values of $\tilde{\omega}$ and γ.

Journal of Physics: Conference Series **2533** (2023) 012035 doi:10.1088/1742-6596/2533/1/012035

The second limit corresponds to the high energy scales ($r \to 0$, or $k \to \infty$). Here we have to distinguish two subcases.
If $\gamma \neq 0$, then

$$F(r \to 0) \simeq 1 - \frac{2r^2}{\tilde{\omega}\gamma G\hbar}, \tag{5}$$

and thus the lapse function corresponds to a deSitter ($\tilde{\omega}\gamma > 0$) or an Anti-deSitter ($\tilde{\omega}\gamma < 0$) core of our metric, depending on the sign of $\tilde{\omega}\gamma$.
If $\gamma = 0$, then

$$F(r) = 1 - \frac{2GMr}{r^2 + \tilde{\omega}G\hbar}, \tag{6}$$

and therefore

$$F(r \to 0) \simeq 1 - \frac{2Mr}{\tilde{\omega}\hbar}, \tag{7}$$

so in this case we have a conic singularity at the origin. Clearly, the presence of $\hbar$ signals the quantum character of the correction that the ASG approach gives to the core of the standard Schwarzschild metric. In both cases the central Schwarzschild singularity has disappeared.

3. Values of the parameters $\tilde{\omega}$ and γ

As we said in Sec.II, the ASG-improved Newtonian potential can be obtained from the standard Newton formula

$$V(r) = -\frac{GMm}{r} \tag{8}$$

by simply replacing the experimentally observed Newton constant G with the running coupling $G(r)$ given in Eq.(1). Thus we get

$$V^{ASG}(r) = -\frac{G(r)Mm}{r} = -\frac{G\,M\,m\,r^2}{r^3 + \tilde{\omega}\,G\,\hbar\,(r + \gamma GM)}, \tag{9}$$

which can be expanded for large r as

$$V^{ASG}(r) = -\frac{GMm}{r}\left[1 - \frac{\tilde{\omega}G\hbar}{r^2} - \frac{\gamma\tilde{\omega}G^{2-}hM}{r^3} + \mathcal{O}\left(\frac{G^{2-}\hbar}{r^4}\right)\right]. \tag{10}$$

Clearly, the corrections to the standard Newtonian potential predicted by the ASG approach are all of quantum nature. This is suggested by the presence of $\hbar$ in each term of correction. In fact, there are no correction terms of classical origin, coming from some kind of post-Newtonian approximation.

On the other hand, corrections of quantum origin to the classical Newtonian potential have been elaborated by several researchers [30, 31, 32, 33, 34, 35, 36, 37, 38] in the last three decades or so. In particular, it was pointed out by Donoghue [31, 34] that the standard perturbative quantization of Einstein gravity leads to a well defined, finite prediction for the leading large distance quantum correction to Newtonian potential. The numerical coefficients of the quantum expansion have undergone a certain evolution over the years [32, 33], but the result today accepted by the community [34, 35] reads

$$V^{QGR}(r) = -\frac{GMm}{r}\left[1 + \frac{41}{10\pi}\frac{G\hbar}{r^2} + \ldots\right]. \tag{11}$$

This is an expansion at first order in $\hbar$, where the first correction term represents a genuine quantum correction proportional to $\hbar$.

Journal of Physics: Conference Series **2533** (2023) 012035 doi:10.1088/1742-6596/2533/1/012035

The comparison of the two expansions (10) and (11) allows us to fix the parameter $\tilde{\omega}$, which results to be

$$\tilde{\omega} = -\frac{41}{10\pi}\,. \tag{12}$$

The ASG parameter γ cannot be fixed by these considerations. To this aim, we refer the reader to the arguments originally developed in Ref.[11], and then taken up also by other authors (e.g.[22]). Those classical general relativistic arguments have to do with the correct identification of the infrared cutoff, and they fix $\gamma = 9/2$. Different kind of considerations, based on the generalized uncertainty principle (see Ref. [40]; see also Refs. [42]), lead to the value $\gamma = 0$. In this paper we will assume always $\gamma \geq 0$, and in some specific cases we shall comment on the special value $\gamma = 0$. However, most of the results will be qualitatively the same for all $\gamma \geq 0$.

We should here emphasize that the *sign* of the first order correction term in $\hbar$ in the expansion (11) is crucial for the physics of ASG-improved black hole. For example, Duff, in his first calculation [30] of 1974 obtained an expansion of the same kind of (11), with a *positive* coefficient of the $\hbar$ term, and therefore a negative $\tilde{\omega}$. In order to fix the $\tilde{\omega}$ parameter, many authors of ASG papers, even quite recently, (e.g.[11, 22, 23]), rely instead on the early calculations performed by Donoghue and others [31, 32] in the period 1994-1995, where the $\hbar$ coefficient obtained was negative. As a direct consequence, they get a *positive* value of $\tilde{\omega}$. This of course has the nice consequence of a black hole metric without singularity, where in particular the central singularity is wiped out, in favour of a De Sitter or an Anti de Sitter core, as can be easily inferred from Eqs.(3)(5), and is also widely discussed in the above References. However, during the years, the analytical techniques used in GR as an effective QFT have been refined, and the results now accepted by the community are those expressed, initially, in Refs.[34, 35], and then confirmed in Refs.[36], as well as in the very recent Ref.[37]. All these results coherently point to a *positive* $\bar{h}$coefficient in the expansion (11), and hence a *negative* value of $\tilde{\omega}$. This fact has deep consequences on the structure of the black hole metric (3), as we will see in the next Sections [1].

4. Study of the new ASG-improved Schwarzschild metric

The key information obtained in the previous Section is that $\tilde{\omega}$ is negative. This, as we will see, represents a major change in respect to others modified (but regular) Schwarzschild metrics present in literature [7, 10, 11, 28, 29]. Instead, some contact with our results can be found in Ref. [44], although there the authors don't deal with ASG models. So, according to the previous

[1] Before proceeding further, we should mention the ongoing lively debate among the communities working on the Effective Quantum Field Theory (EQFT) approach, and those on the ASG approach. The focus is about "if" and "to what extent" the results obtained on ASG-improved metrics by the ASG community can be compared with the results on the quantum corrected metrics obtained by the EQFT community (see e.g. Refs. [38, 43]). Here, we cannot of course enter in details. However, within the Asymptotic Safety literature the parameter $\tilde{\omega}$ is usually considered positive. In fact, considering (see e.g. Refs. [11, 40]) the running Newton's coupling $G(k) = G/(1 + \tilde{\omega}Gk^2/\hbar)$, where k is the energy or momentum scale, we see that $dG/dk < 0$ only if $\tilde{\omega} > 0$. With an $\tilde{\omega} < 0$, $G(k)$ diverges for $k \sim k_{Planck}$, and this behavior of $G(k)$ looks, in principle, outside the spirit of ASG. On the other hand, while the calculations of the EQFT community point nowadays firmly towards a positive coefficient of $\hbar$, there are still doubts about the comparability of those results with the ASG RG-improved metrics, and therefore on the possibility to infer from that comparison a negative value for $\tilde{\omega}$. In particular, it would seem that the two different frameworks effectively consider, and implement, different classes of Feynman diagrams in order to get their final metrics expressions. And we know that, for example, including or not including certain subclasses of diagrams could change the sign of the $\tilde{\omega}$ parameter. Thus, many people get the overall feeling that for the time being the available calculations are not yet able to definitely clarify the situation. This however looks to us as an even stronger motivation to explore the physical consequences of a (still well possible) *negative* $\tilde{\omega}$ parameter.

section, we consider here the case

$$\tilde{\omega} < 0 \Rightarrow \tilde{\omega} = -|\tilde{\omega}| \, ; \qquad \gamma > 0 \, . \tag{13}$$

The lapse function (3) can therefore be written as

$$F(r) = 1 - \frac{2GMr^2}{r^3 - |\tilde{\omega}|G\hbar(r + \gamma GM)} \, . \tag{14}$$

While for $\tilde{\omega} > 0$ the lapse (3) is regular everywhere when $r > 0$ (see e.g. Ref. [11]), here, with $\tilde{\omega} < 0$, the scenario is very different. First, we notice that the behavior of $F(r)$ at $r \to \infty$ remains that described by Eq.(4), namely a standard Schwarzschild metric for large r. At $r \to 0$ we have an Anti-DeSitter core, namely $F(r) \simeq 1 + 2r^2/(|\tilde{\omega}|\gamma G\hbar)$. But now the denominator $D(r)$ appearing in (14) can develop zeros. Luckly, a simple analysis is sufficient to clarify the situation (for details, see Ref. [41]).

In fact, it is possible to draw a general graph (Fig.1) of the lapse function $F(r)$ in the region $r > 0$, valid for any $\gamma > 0$ and of course $M > 0$. As we see, there is always one single positive zero $r = r_h$ of $F(r)$, which is the horizon of the ASG-improved black hole metric (14). There is also an essential singularity at $r = r_0 > 0$. The most relevant difference with the standard Schwarzschild metric is the fact that the essential (ineliminable) singularity is at $r_0 > 0$, instead of being at $r_0 = 0$. It is also clear that it is always $r_0 < r_h$, for any $M > 0$. So the singularity is always protected by the event horizon. The singularity is never naked, in full accordance with the Cosmic Censorship Conjecture.

The asymptotic behavior of the horizon r_h and of the singularity r_0 in the physical relevant limits of large M and small M can be summarized as follows (again, see Ref. [41] for details).

4.1. ***Horizon*** r_h

The behavior of the horizon for large M, $M \to \infty$, is

$$r_h \simeq 2GM + \frac{(2+\gamma)|\tilde{\omega}|\hbar}{4M} \, , \tag{15}$$

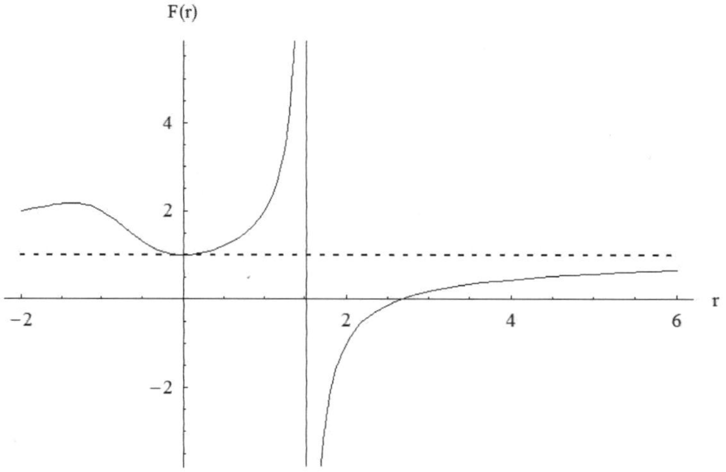

Figure 1. Lapse function $F(r)$: physical region for $r > 0$, singularity at $r = r_0$, horizon at $r = r_h$, where $F(r_h) = 0$.

Journal of Physics: Conference Series **2533** (2023) 012035 doi:10.1088/1742-6596/2533/1/012035

so the usual Schwarzschild expression for the horizon is recovered in the large M limit.

We can also investigate the behavior of the horizon in the small M limit.

We find

$$r_h \simeq \sqrt{|\tilde{\omega}|G\hbar} + \left(1+\frac{\gamma}{2}\right) GM \,. \tag{16}$$

Essentially, $r_h \geq \ell_P$, which sounds reasonable, since any length below Planck length is physically meaningless, namely unobservable.

4.2. ***Singularity*** r_0

As we have seen, the singularity of the metric (14) is located at the only positive root $r = r_0$ of the equation

$$r^3 - |\tilde{\omega}|G\hbar r - \gamma|\tilde{\omega}|G^2\hbar M = 0 \,. \tag{17}$$

The behavior of the radial coordinate $r = r_0$ of the singularity for small M is

$$r_0 \simeq \sqrt{|\tilde{\omega}|G\hbar} + \frac{\gamma}{2}GM \,, \tag{18}$$

so again $r_0 \geq \ell_P$.

Instead, for $M \to \infty$ the situation is much more interesting. The large M limit of the singularity equation (17) yields for the behavior of the root r_0

$$r_0 \simeq (\gamma|\tilde{\omega}|G^2\hbar M)^{1/3} + \frac{|\tilde{\omega}|G\hbar}{3(\gamma|\tilde{\omega}|G^2\hbar M)^{1/3}} \,. \tag{19}$$

An important physical consideration can now be stated. As already suggested by the initial analysis, the singularity results to be always protected by the horizon. Namely, by comparing Eqs.(15), (19) for large M, or instead by comparing Eqs.(16), (18) in the small M limit, we always have

$$r_0 < r_h \,, \tag{20}$$

so there are no naked singularities.

For sake of clarity, in Fig.2 the reader can find the plots of the mass function $M(r_h)$ for the horizon (red dashed line)

$$GM(r_h) = \left.\frac{r^3 - |\tilde{\omega}|\ell_P^2 r}{2r^2 + |\tilde{\omega}|\ell_P^2\gamma}\right|_{r=r_h} \,, \tag{21}$$

the mass function $M(r_0)$ for the singularity (blue dot-dashed line)

$$GM(r_0) = \left.\frac{r^3 - |\tilde{\omega}|\ell_P^2 r}{|\tilde{\omega}|\ell_P^2\gamma}\right|_{r=r_0} \,, \tag{22}$$

and the standard Schwarzschild horizon $GM = r_{SCH}/2$ (green solid line). Any horizontal line (black dashed) representing an arbitrary $M > 0$ intersects first the blue line and then the red line, namely $r_0(M) < r_h(M)$ for any $M > 0$. Notice that both the horizon and the singularity mass functions have a simple zero at $r = r_c = \sqrt{|\tilde{\omega}|}\,\ell_P$ with $r_c^2 = |\tilde{\omega}|\,\ell_P^2$.

Journal of Physics: Conference Series **2533** (2023) 012035 doi:10.1088/1742-6596/2533/1/012035

5. A metric for the Planck stars

It is of the greatest interest to examine the core of the black hole metric we obtained. For masses M larger than the Planck mass m_p, $M \gg m_p$, the central hard singularity is located at $r = r_0$, and it has, in fact, a finite positive size $r_0 \simeq (\gamma|\tilde{\omega}|G^{2-}\hbar M)^{1/3} > 0$, contrary to what happens in the standard Schwarzschild black hole, where the singularity is point-like. Observe that this finite size is completely of quantum origin: in fact, $r_0 \to 0$ if we take the classical limit $\hbar \to 0$.

But most importantly, if we presume that the whole collapsing mass M is concentrated into the central hard sphere of radius r_0, then we can compute the (non covariant) volume of this sphere, and hence the density of this matter (as seen by an observer at infinity), which will result to be finite, and precisely

$$r_0 = (\gamma|\tilde{\omega}|\ell_P^2 GM)^{1/3} = (\gamma|\tilde{\omega}|/2)^{\frac{1}{3}}\,\ell_P\left(\frac{M}{m_p}\right)^{\frac{1}{3}} \quad\Rightarrow\quad \varrho = \frac{M}{V_{core}} = \frac{3}{2\pi}\frac{m_p}{\gamma|\tilde{\omega}|\ell_P^3} \simeq \frac{\varrho_{Planck}}{2\gamma|\tilde{\omega}|}, \tag{23}$$

where we used the definitions $G\hbar = \ell_P^2$, $2Gm_p = \ell_P$, and

$$\varrho_{Planck} = \frac{m_p}{\ell_P^3}\,. \tag{24}$$

So the central hard kernel of our black hole results to have a *finite* size, and a density of the order of the Planck density. These are exactly the characteristics of the so called Planck stars, first proposed in Ref. [39], on the ground of general qualitative considerations. Many of the general properties described in [39] (see also Ref. [45]) can now be repeated for our black hole. The finite positive size of the central core, being of pure quantum origin, is presumably due to the action of the Heisenberg uncertainty principle, which prevents matter to be arbitrarily concentrated into a geometrical point of size zero. The central kernel can presumably keep trace of all the information swallowed by the black hole: we see here a possible way out of the information paradox. Of course, all the above considerations make sense only for $\gamma > 0$ strictly.

The original proposal [39] contains a certain amount of qualitative considerations, including an educated guess on the form of the metric able to describe a Planck star. Such metric was

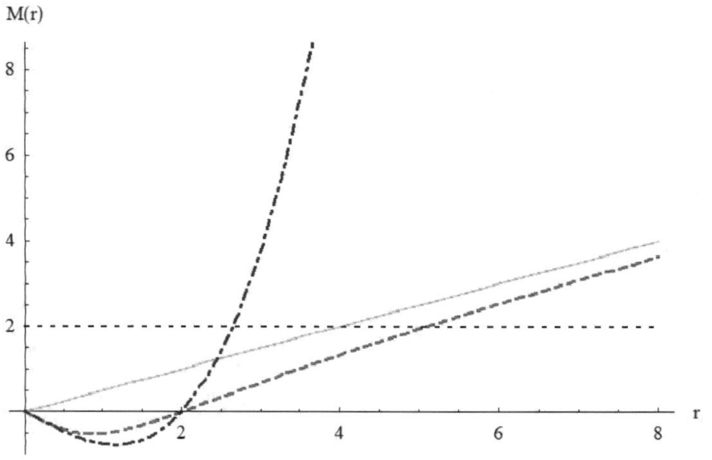

Figure 2. Horizon mass function $M(r_h)$ (red dashed line), singularity mass function $M(r_0)$ (blue dot-dashed line), and Schwarzschild mass function (green solid line). The horizontal black dashed line represents an arbitrary $M > 0$, and always intersects blue and red lines at $r_0(M) < r_h(M)$, namely there are no naked singularities.

Journal of Physics: Conference Series **2533** (2023) 012035 doi:10.1088/1742-6596/2533/1/012035

initially chosen to be the Hayward metric [28]

$$F(r) = 1 - \frac{2GMr^2}{r^3 + 2GML^2} \tag{25}$$

where L is a parameter with dimensions of a length. No particularly compelling argument, from the physical point of view, was exhibited for that choice, with the exception, perhaps, that the Hayward metric is a well known example of singularity-free metric. For large M the metric (25) develops two horizons, one inner

$$r_- \simeq L + \frac{L^2}{4GM}, \tag{26}$$

and one outer

$$r_+ \simeq 2GM - \frac{L^2}{2GM}. \tag{27}$$

However, no specific indication is contained in the metric (25) about the size of the hard kernel of a Planck star. And certainly not of a hard kernel with a size increasing with M, as instead Eq.(23) suggests (to be compared with (26)). Even worse, the Hayward metric *per se* is unable to mimic the well established quantum correction to the Newtonian potential [34] that occurs at low energies. This is due to the lack of a term $1/r^3$ in the expansion of the metric (25). The authors of Ref. [46] found a smart way to cure this shortcoming, but at the price of introducing a further metric function $H(r)$, determined through a bunch of additional constraints, so that their "modified Hayward" metric now reads

$$ds^2 = -\, H(r)F(r)dt^2 + \frac{1}{F(r)}dr^2 + r^2 d\Omega^2\,, \quad \text{with} \quad H(r) = 1 - \frac{\beta GM\alpha}{\alpha r^3 + \beta GM}, \tag{28}$$

where β is a parameter that in Ref.[46] plays the rôle of our $|\tilde{\omega}|$. The above metric [2] finally contains the $1/r^3$ term necessary to mimic the Donoghue modified Newtonian potential [34] for large r.

Although smart and working, the above solution is undeniably contorted and intricate. On the contrary, within the formalism of the Renormalization Group, the mathematical structure of the metric is dictated by the general properties of the AS Gravity, and its lapse function (14) results clearly simpler than the above product $H(r)F(r)$. The ASG metric already contains the right terms to match, at large distances, the quantum corrected Newtonian potential. Moreover, and this is quite astonishing, by simply imposing that match, the final form of the metric is uniquely fixed, and it automatically displays the size of the central hard kernel of the Planck star.

6. Conclusions

In this paper we have derived an exact value of the parameter $\tilde{\omega}$ characterizing the spherically symmetric metric suggested by the Asymptotically Safe Gravity approach. The result has been obtained by comparing the corrected Newtonian potential computed through ASG, with the analog correction suggested by the Donoghue approach to GR as a low energy effective QFT. The decisive novelty in respect to the previously computed values of $\tilde{\omega}$ is that we get a *negative* value of $\tilde{\omega}$, and this because we used the more recent results [34, 35, 36] of Donoghue, Khriplovich, and collaborators.

[2] A further requirement imposed by the authors on the function $H(r)$ is that $H(r)$ should allow for a time delay between a clock sitting at the center of the collapsed object ($r = 0$), and a clock at infinity (put a clock at $r = 0$ is in principle conceivable, just because the Hayward metric is regular at $r = 0$). To get this, authors demand that $H(r = 0) = 1 - \alpha$. They justify this further request by saying that "is a physically unmotivated restriction" to leave $H(0) = 1$. In any case, we do not have this kind of problem with the ASG metric (14), since the center $r = 0$ cannot be reached, being protected by the singularity at $r = r_0 > 0$.

Journal of Physics: Conference Series **2533** (2023) 012035 doi:10.1088/1742-6596/2533/1/012035

The fact $\tilde{\omega} < 0$ completely changes the geometry of the ASG "improved" black hole metric. Previously unexplored aspects of this metric have been studied, the most relevant one being the presence of a finite-size singularity at the core of the black hole. Surprisingly, the size of this "black kernel" results to be exactly what needed to describe the so called Planck stars. These objects were introduced years ago on the basis of semi-qualitative arguments [39], while in our context they appear as a natural mathematical consequence of the ASG metric with a negative $\tilde{\omega}$ parameter (see also e.g. [47]).

It is worth mentioning that the phenomenology of these objects could be quite rich, and presents both astrophysical and cosmological signatures, in particular in the realm of (primordial) black hole evaporation [48]. As a Planck star evaporates, with a hard kernel of finite positive size, then the final explosion may occur at "macroscopic" scale, namely at a much bigger scale than the Planck scale. So, Planck star explosions could be naturally associated with some of the measured short gamma-ray bursts (SGRBs) [49]. In Ref. [50] authors estimated that several short gamma-ray bursts per day, around 10 MeV, with isotropic distribution, can be expected coming from a region of a few hundred light years around us. On the other hand, also fast radio bursts, strong signals with millisecond duration, which are probably of extragalactic origin, have been shown in Ref. [51] to have wavelengths not far from the expected size of the exploding hole.

On the theoretical side, further investigations, aimed to better understand Penrose diagrams, energy conditions, singularity theorems, quasi-normal modes, as well as a statistical interpretation of entropy and information paradox, related to this kind of metrics are currently being carried out.

References

[1] Jacobson T 1995 *Phys. Rev. Lett.* **75** 1260
[2] Connes A 1996 *Commun. Math. Phys.* **182** 155
[3] Reuter M 1998 *Phys. Rev.* D **57** 971
[4] Rovelli C 1998 *Living Rev. Rel.* **1** 1
[5] Verlinde E P 2011 *JHEP* **04** 029
[6] Koch B, Reyes I A and Rincon Ã 2016 *Class. Quant. Grav.* **33** (22) 225010
[7] Contreras E, Rincon Ã, Koch B and Bargueno P 2017 *Int. J. Mod. Phys.* D **27** (03) 1850032
[8] Rincon Ã and Panotopoulos G 2018 *Phys. Rev.* D **97** (2) 024027
[9] Contreras E, Rincon Ã, Koch B and Bargueno P 2018 *Eur. Phys. J.* C **78** (3) 246
[10] Contreras E and Bargueno P 2018 *Mod. Phys. Lett.* A **33** (32) 1850184
[11] Bonanno A and Reuter M 2000 *Phys. Rev.* D **62** 043008
[12] Bonanno A and Reuter M 2002 *Phys. Rev.* D **65** 043508
[13] Reuter M and Weyer H 2004 *Phys. Rev.* D **69** 104022
[14] Platania A *Front. in Phys.* **8** 188
[15] Bonanno A and Reuter M 2004 *Int. J. Mod. Phys. D* **13** 107
[16] Bonanno A and Reuter M 2002 *Phys. Lett. B* **527** 9
[17] Liu L H, Prokopec T and Starobinsky A 2018 *Phys. Rev. D* **98** (4) 043505
[18] Hindmarsh M and Saltas I D 2012 *Phys. Rev. D* **86** 064029
[19] Platania A 2019 *Eur. Phys. J. C* **79** (6) 470
[20] Moti R and Shojai A 2020 *Int. J. Mod. Phys. A* **35** 2050016
[21] Koch B and Saueressig F 2014 *Class. Quant. Grav.* **31** 015006
[22] Koch B and Saueressig F 2014 Int. J. Mod. Phys. A **29** (8) 1430011
[23] Saueressig F, Alkofer N, D'Odorico G and Vidotto F 2016 PoS **FFP14** 174
[24] Bonanno A, Koch B and Platania A 2017 *Class. Quant. Grav.* **34** (9) 095012
[25] Dao-Jun Liu, Bin Yang, Yong-Jia Zhai and Xin-Zhou Li 2012 *Class. Quant. Grav.* **29** 145009
[26] Jin Li and Yuanhong Zhong 2013 *Int. J. Theor. Phys.* **52** 1583
[27] Bardeen J M 1968 *Proceedings of of International Conference GR5* (Tbilisi USSR) 174
[28] Hayward S A 2006 *Phys. Rev. Lett.* **96** 031103
[29] Frolov V P 2016 *Phys. Rev. D* **94** (10) 104056
[30] Duff M J 1974 *Phys. Rev. D* **9** 1837

Journal of Physics: Conference Series **2533** (2023) 012035 doi:10.1088/1742-6596/2533/1/012035

[31] Donoghue J F 1994 *Phys. Rev. Lett.* **72** 2996; 1994 *Phys. Rev. D* **50** 3874
[32] Hamber H W and S. Liu 1995 *Phys.Lett.B* **357** 51
Bjerrum-Bohr N E J, Donoghue J F and Holstein B R 2003 *Phys.Rev.D* **68** 084005; 2005 Erratum-ibid.D **71** 069904
[33] Khriplovich I B and Kirilin G G 2002 *J.Exp.Theor.Phys.* **95** 981
[34] Bjerrum-Bohr N E J, Donoghue J F and Holstein B R 2003 *Phys.Rev.D* **67** 084033; 2005 Erratum-ibid.D **71** 069903
[35] Khriplovich I B and Kirilin G G 2004 *J.Exp.Theor.Phys.* **98** 1063
[36] Akhundov A and Shiekh A 2008 *EJTP* **5** (17) 1;
Kiefer C 2013 *J. Phys. Conf. Ser.* **442** 012025;
Bjerrum-Bohr N E J, Donoghue J F, Holstein B R, Planté L and Vanhove P 2015 *Phys. Rev. Lett.* **114** 061301;
Donoghue J F and Holstein B R 2015 *J. Phys. G* **42** (10) 103102
[37] Fröb M B, Rein C and Verch R 2022 *JHEP* **01** 180
[38] Donoghue J F 2020 *Front. in Phys.* **8** 56
[39] Rovelli C and Vidotto F 2014 *Int. J. Mod. Phys. D* **23** (12) 1442026
[40] Lambiase G and Scardigli F 2022 *Phys. Rev. D* **105** (12) 124054
[41] Scardigli F and Lambiase G 2022 *Planck Stars from Asymptotically Safe Gravity* [arXiv:2205.07088 [gr-qc]]
[42] Scardigli F 1999 *Phys. Lett. B* **452** 39;
Adler R J and Santiago D I 1999 *Mod. Phys. Lett. A* **14** 1371;
Capozziello S, Lambiase G and Scarpetta G 2000 *Int. J. Theor. Phys.* **39** 15;
Scardigli F and Casadio R 2015 *Eur. Phys. J. C* **75** 425;
Scardigli F, Lambiase G and Vagenas E C 2017 *Phys. Lett. B* **767** 242;
Lambiase G and Scardigli F 2018 *Phys. Rev. D* **97** 075003;
Kanazawa T, Lambiase G, Vilasi G and Yoshioka A 2019 *Eur. Phys. J. C* **79** 95;
Jizba P, Kleinert H and Scardigli F 2010 *Phys. Rev. D* **81** 084030
[43] Bonanno A, Eichhorn A, Gies H, Pawlowski J M, Percacci R, Reuter M, Saueressig F and Vacca G P 2020 *Front. in Phys.* **8** 269
[44] Bargueño P, Bravo Medina S, Nowakowski M and Batic D 2017 *EPL* **117** (6) 60006
[45] Bonanno A, Casadio R and Platania A 2020 *JCAP* **01** 022
[46] De Lorenzo T, Pacilio C, Rovelli C and Speziale S 2015 *Gen. Rel. Grav.* **47** (4) 41
[47] Hajicek P and Kiefer C 2001 *Int. J. Mod. Phys. D* **10** 775;
Hossenfelder S, Modesto L, Premont-Schwarz I 2010 *Phys. Rev. D* **81** 044036;
Bambi C, Malafarina D and Modesto L 2014 *Eur. Phys. J. C* **74** 2767
[48] Goswami R, Joshi P S and Singh P 2006 *Phys. Rev. Lett.* **96** 031302;
Kavic M, Simonetti J H, Cutchin S E, Ellingson S W and Patterson C D 2008 *JCAP* **11** 017;
Griest K, Cieplak A M and Lehner M J 2013 *Phys. Rev. Lett.* **111** (18) 181302
[49] Nakar E 2007 *Phys. Rept.* **442** 166
[50] Barrau A and Rovelli C 2014 *Phys. Lett. B* **739** 405
[51] Barrau A, Rovelli C and Vidotto F 2014 *Phys. Rev. D* **90** (12) 127503

Journal of Physics: Conference Series **2533** (2023) 012036 doi:10.1088/1742-6596/2533/1/012036

Quantum-gravitational corrections to the power spectrum for a closed universe

Tatevik Vardanyan and Claus Kiefer

University of Cologne, Faculty of Mathematics and Natural Sciences, Institute for Theoretical Physics, Cologne, Germany

E-mail: `tatevik@thp.uni-koeln.de`, `kiefer@thp.uni-koeln.de`

Abstract. We study the quantum-gravitational corrections to the power spectrum of a gauge-invariant inflationary scalar perturbations in a closed model of a universe. We consider canonical quantum gravity as an approach to quantizing gravity. This leads to the Wheeler-DeWitt equation, which has been studied by applying a semiclassical Born–Oppenheimer type of approximation. At the corresponding orders of approximation, we recover both the uncorrected and quantum-gravitationally corrected Schrödinger equations for the perturbation modes from which we calculate the quantum-gravitational corrections to the power spectrum in the slow-roll regime. The results are compared to the power spectra for the flat model of the universe.

1. Introduction

Quantum-gravitational effects are expected to be strong at very high energies. Such energies have been present in the very early universe; therefore one hopes to find quantum-gravitational effects related to the anisotropies of the CMB. One assumes that these anisotropies originate from quantum fluctuations of the inflationary field and the metric.

To study the quantum-gravitational corrections to the power spectrum of inflationary perturbations, we will consider canonical quantum gravity as a conservative approach to quantize gravity. This starts with the Hamiltonian formulation of general relativity (GR) and leads to the Wheeler-DeWitt equation [1]. Applying a semiclassical approximation, we obtain a Schrödinger equation for the perturbation modes including quantum-gravitational correction terms, from which we calculate the power spectrum.

We consider a closed model of the universe because of the following arguments. The recent Planck observation [2] is consistent with a flat ΛCDM model; however, the data doesn't rule out the possibility that our universe might have slight positive curvature. In [3], the authors argue that from Planck data, the positive curvature is preferred for more than 99%. They claim that the positive curvature explains the anomalous lensing amplitude as well as it removes the tension within the Planck data set for the values of cosmological parameters obtained at different angular scales. In addition, a closed model might explain the observed low amplitude of quadrupole and octopole modes [2], which remains unexplained in the standard ΛCDM model of cosmology.

Furthermore, even though inflation would flatten any initial spatial curvature (assuming it is present before the inflation), the presence of initial spatial curvature could have a significant imprint on the power spectrum, the study of which is one of the main goals of this paper.

Journal of Physics: Conference Series **2533** (2023) 012036 doi:10.1088/1742-6596/2533/1/012036

Finally, there are debates regarding infinities in science from mathematical and philosophical perspective [4]. Thus, considering a closed model of a universe, for which the space is finite, serves to a purpose of tackling these subtle issues as well.

In what follows, in section 2 we present the general features of canonical quantum gravity and apply this quantization scheme to the inflationary universe. We derive the Wheeler-DeWitt equation for the inflationary universe and in section 3 we apply a semiclassical approximation scheme and obtain the uncorrected and the quantum-gravitationally corrected Schrödinger equations. Finally, in section 4, we present the derived uncorrected and quantum-gravitationally corrected power spectra. The results are summarized in section 5. Technical details can be found in [5], [6].

2. Derivation of the Wheeler-DeWitt equation for an inflationary closed universe

Let us briefly present the canonical quantization procedure of gravity. The first step is the construction of the Hamiltonian formulation for GR which starts with the foliation of spacetime into three-dimensional spacelike Cauchy hypersurfaces parameterized by a time function t [1]. The spacetime foliation leads to the following line element

$$ds^2 = -(N^2 - N_i N^i)dt^2 + 2N_i dx^i dt + h_{ij}dx^i dx^j, \tag{1}$$

where N is the lapse function, N_i is the shift vector, and h_{ij} is the three-metric.

Rewriting the action using (1) and taking variations of the action with respect to the lapse function and shift vector leads to the Hamiltonian and momentum constraints.

According to the procedure for the quantization of constrained systems, a classical constraint converts to quantum operators and turns into a restriction on the wave functional. The quantization of the Hamiltonian constraint leads to the Wheeler-DeWitt equation and, respectively, the momentum constraint to the quantum diffeomorphism constraint.

2.1. Hamiltonian formulation

Let us now apply the above procedure to the inflationary Friedmann universe. We impose the FLRW line element

$$ds^2 = -N^2(t)dt^2 + a^2(t)d\Omega_3^2, \tag{2}$$

where $a(t)$ is the scale factor, $d\Omega_3^2$ denotes the metric of the unit three-sphere, and due to the symmetry only the lapse function $N(t)$ enters in the line element.

Including perturbations to the metric and the scalar field ϕ with a quadratic potential $\mathcal{V}(\phi) = \frac{1}{2}m^2\phi^2$, it can be shown that the total Hamiltonian can be written as a sum of the background and perturbation parts [7],

$$H = N\Big[H_{|0} + \sum_n H_{|2}^n\Big]. \tag{3}$$

The the background Hamiltonian reads

$$H_{|0} = N\frac{1}{2}e^{-3\alpha}\Big[-\frac{1}{m_{\mathrm{P}}^2}p_\alpha{}^2 + p_\phi{}^2 + 2e^{6\alpha}\mathcal{V}(\phi) - m_{\mathrm{P}}\mathcal{K}e^{4\alpha}\Big], \tag{4}$$

where

$$p_\alpha = -m_{\mathrm{P}}^2 e^{3\alpha}\frac{\dot{\alpha}}{N} \quad \text{and} \quad p_\phi = e^{3\alpha}\frac{\dot{\phi}}{N}, \quad m_{\mathrm{P}} := \sqrt{3\pi/2G}, \quad \alpha = \ln a/a_0. \tag{5}$$

The curvature parameter $\mathcal{K}$ takes the value $\mathcal{K} = 1$ for the closed model.

Journal of Physics: Conference Series **2533** (2023) 012036 doi:10.1088/1742-6596/2533/1/012036

The perturbation Hamiltonian, denoted as $H^n_{|2}$, is composed of scalar, vector and tensor parts. Since we are interested in scalar perturbations the vector and tensor parts will be ignored. In addition, the first-order Hamiltonian constraints

$$H^n_{|1} = 0, \quad {}^S H^n_{\_1} = 0, \quad {}^V H^n_{\_1} = 0 \tag{6}$$

are the linearized energy constraint and the scalar and vector parts of the linearized momentum constraints correspondingly.

The perturbation Hamiltonian in (3) is not gauge-invariant. The gauge-invariant form for the scalar part of the perturbation Hamiltonian (which will be denoted by ${}^S H^n_{|2}$) was obtained in [8]. To bring this gauge-invariant Hamiltonian into a more compact form we perform a canonical transformation which leads to the following expression:

$${}^s H^n_{|2} = \frac{1}{2}\left[\tilde{p}_n^2 + \omega_n^2(\eta)\tilde{v}_n^2\right], \tag{7}$$

where η is the conformal time, $\tilde{v}_n$ is the gauge-invariant variable and $\tilde{p}_n$ is its conjugate momentum. The explicit expression of the function $\omega_n^2(\eta)$ can be found in [5].

In the large-wavelength limit the well-known result for the perturbation Hamiltonian in the flat model [9] is recovered.

2.2. Quantization

We consider the slow-roll regime for which the scalar field satisfies the following conditions

$$\dot{\phi} \ll \mathcal{V}(\phi), \quad \ddot{\phi} \ll 3H\phi. \tag{8}$$

Introducing the slow-roll parameters defined by $\varepsilon = 1 - \frac{\mathcal{H}'}{\mathcal{H}^2}$ and $\delta = \varepsilon - \frac{\varepsilon'}{2\mathcal{H}\varepsilon}$,[1] the Wheeler-DeWitt equation for each mode takes the following form

$$\frac{1}{2}\left\{e^{-2\alpha}\left[\frac{1}{m_{\rm P}^2}\frac{\partial^2}{\partial\alpha^2} + m_{\rm P}^2\left(e^{6\alpha}H^2\left(1 - \frac{\varepsilon}{3}\right) - \frac{\mathcal{K}}{3}e^{4\alpha}\right)\right] - \frac{\partial^2}{\partial\tilde{v}_n^2} + \omega_n^2(\eta)\tilde{v}_n^2\right\}\Psi_n(\alpha, \tilde{\phi}, \tilde{v}_n) = 0. \tag{9}$$

where we have introduced a dimensionless variable $\tilde{\phi} = m_{\rm P}^{-1}\phi$. To obtain the above equation we have made a product ansatz for the full wave functional assuming that the perturbation modes do not interact with each other.

3. Semiclassical approximation

We apply a semiclassical Born-Oppenheimer type of approximation which allows us to obtain the Schrödinger equation for the perturbations propagating on the classical background and the quantum-gravitational corrections to it following the procedure developed in [10], [11].

We make a WKB-like ansatz

$$\Psi_n(\alpha, \tilde{\phi}, \tilde{v}_n) = e^{iS(\alpha, \tilde{\phi}, \tilde{v}_n)}. \tag{10}$$

The Born-Oppenheimer approximation is implemented by expanding $S(\alpha, \phi, \tilde{v}_n)$ in terms of powers of Planck mass

$$S(\alpha, \tilde{v}_n) = m_{\rm P}^2 S_0 + m_{\rm P}^0 S_1 + m_{\rm P}^{-2} S_2 + \ldots. \tag{11}$$

[1] A dimensionless Hubble parameter $\mathcal{H}$ is defined by $\mathcal{H} = a'/a$. The derivative with respect to η is denoted by a prime.

Afterwards, we insert (10) into the Wheeler-DeWitt equation (9), collect the corresponding terms of the powers of $m_{\rm P}$ and set them equal to zero.

At order $m_{\rm P}^2$, the Hamilton–Jacobi equation for the classical background is obtained which is equivalent to the Friedmann equation. At the next order, $m_{\rm P}^0$, the obtained equation can be rewritten as a Schrödinger equation for a wave functions $\psi_n^{(0)}$, i.e.

$$i\frac{\partial}{\partial\eta}\psi_n^{(0)} = {}^sH_{|2}^n\psi_n^{(0)}, \tag{12}$$

where the conformal time parameter η has been introduced as

$$\frac{\partial}{\partial\eta} := -e^{-2\alpha}\frac{\partial S_0}{\partial\alpha}\frac{\partial}{\partial\alpha} = e^{\alpha}\sqrt{H^2\left(1-\frac{\varepsilon}{3}\right)-\frac{\mathcal{K}}{3}e^{-2\alpha}}\frac{\partial}{\partial\alpha}. \tag{13}$$

At the order of $m_{\rm P}^{-2}$, we obtain the quantum-gravitational corrections to the Schrödinger equation following the procedure in [10],

$$i\frac{\partial}{\partial\eta}\psi_n^{(1)} = {}^sH_{|2}^n\psi_n^{(1)} - \frac{\psi_n^{(1)}}{2m_{\rm P}^2\psi_n^{(0)}}\left[i\frac{\partial}{\partial\eta}\left(\frac{{}^sH_{|2}^n}{V}\right)\psi_n^{(0)} + \frac{1}{V}({}^sH_{|2}^n)^2\psi_n^{(0)}\right], \tag{14}$$

where $V\left(a,\tilde{\phi}\right) := \frac{2a^4}{m_{\rm P}^2}\mathcal{V}\left(\tilde{\phi}\right)$. Note that on the right-hand side, there is a unitarity violating term. In this paper, following the arguments in [11] we will drop the non-unitarity term because of its unphysical behavior. However, the treatment of this term will be carried on in an upcoming work with L. Chataignier following the method developed in [12], where the authors argue that this non-unitary term can be absorbed by using a suitable definition of the inner product.

To solve the Schrödinger equation (12) we make a Gaussian ansatz,

$$\psi_n^{(0)}(\eta,\tilde{v}_n) = A_n^{(0)}(\eta)e^{-\frac{1}{2}\Omega_n^{(0)}(\eta)\tilde{v}_n^2}. \tag{15}$$

Inserting this into the Schrödinger equation we get

$$\Omega_n^{(0)\prime}(\eta) = -i\Big[\left(\Omega_n^{(0)}(\eta)\right)^2 - \omega_n^2\Big]. \tag{16}$$

Similarly, using the following Gaussian ansatz for the corrected case (14),

$$\psi_n^{(1)}(\eta,\tilde{v}_n) = \left(A_n^{(0)}(\eta) + \frac{1}{m_{\rm P}^2}A_n^{(1)}(\eta)\right)\exp\Big[-\frac{1}{2}\left(\Omega_n^{(0)}(\eta) + \frac{1}{m_{\rm P}^2}\Omega_n^{(1)}(\eta)\right)\tilde{v}_n^2\Big], \tag{17}$$

leads to the equation

$$\Omega_n^{(1)\prime}(\eta) = 2\Omega_n^{(0)}(\eta)\Big[\Omega_n^{(1)}(\eta) - \frac{3}{4V}\left(\left(\Omega_n^{(0)}\right)^2(\eta) - \omega_n^2(\eta)\right)\Big]. \tag{18}$$

4. Derivation of the power spectra

4.1. Uncorrected power spectrum

Starting with the two-point correlation function one obtains the following expression for the power spectrum:

$$\mathcal{P}_{\tilde{v}} = \frac{n(n^2-\mathcal{K})}{2\pi^2}\frac{1}{2\mathcal{R}e\Omega_n^{(0)}}. \tag{19}$$

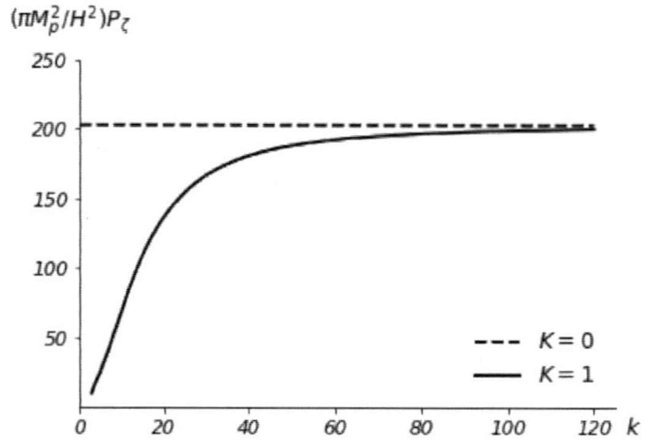

Figure 1. Plot of the power spectra of curvature perturbation at the point of horizon reentry, i.e. $\xi = 1$s . The flat case is presented with the dashed line and closed one with the thick line. The slow-roll parameters are set to $\varepsilon = 0.005$ and $\delta = -0.006$. Figure adapted from [5].

Solving (16) and substituting the real part of $\Omega_n^{(0)}(\eta)$ into (19) leads to

$$\mathcal{P}_{\tilde{v}} = \frac{n^2 - \mathcal{K}}{4\pi^2 f_\varepsilon^3(n)\xi^2}\left[\tilde{C}_{\varepsilon,\delta} - 2(\gamma - 3\lambda_{\varepsilon,\delta}(n)/2)\ln\left(\xi f_\varepsilon(n)\right)\right], \quad \text{with } \xi = \frac{n}{aH}. \tag{20}$$

For the explicit expressions of the functions $\lambda_{\varepsilon,\delta}(n)$, $f_\varepsilon(n)$ and $\tilde{C}_{\varepsilon,\delta}$ see [5].

To relate the power spectrum of scalar perturbations at the end of inflation to the temperature anisotropies of the CMB, one needs the power spectrum of curvature perturbations. Hence, we introduce a gauge-invariant variable ζ_{BST}[2] which can be related to the variable $\tilde{v}_n$. In the slow-roll regime the relation is given by

$$\zeta_{\mathrm{BST},n} = -\frac{1}{a}\frac{1}{\sqrt{2\left(\varepsilon + \frac{\mathcal{K}}{\mathcal{H}^2}\right)}M_{\mathrm{P}}}\left[1 + \frac{\mathcal{K}}{n^2 - 4\mathcal{K}}\left(\varepsilon + \frac{\mathcal{K}}{\mathcal{H}^2}\right)\right]^{1/2}\tilde{v}_n, \tag{21}$$

where we have introduced a redefined Planck mass $M_{\mathrm{P}} := 1/\sqrt{8\pi G}$. The power spectrum of ζ_{BST} reads

$$\mathcal{P}_{\zeta_{\mathrm{BST}}} = \frac{1}{2a^2\left(\varepsilon + \frac{\mathcal{K}}{\mathcal{H}^2}\right)M_{\mathrm{P}}^2}\left[1 + \frac{\mathcal{K}}{n^2 - 4\mathcal{K}}\left(\varepsilon + \frac{\mathcal{K}}{\mathcal{H}^2}\right)\right]\mathcal{P}_{\tilde{v}_n}. \tag{22}$$

In the large-wavelength limit, we recover the power spectrum for the flat model [10].

In Fig. 1, the power spectra of curvature perturbations are plotted as a function of the comoving wavelength k for the flat and the closed models. One can see that there is a significant suppression of power at large scales for the closed case compared to the flat case. Moreover, due to the finiteness of the universe, there is a cutoff at a small k.

Finally, let us point out that due to the explicit scale dependence of the power spectrum (22) we cannot apply a power-law approximation at large scales where the scale dependence is the most prominent. The power-law approximation is based on the assumption of weak scale dependence.

[2] For the flat model another variable ξ is used which cannot be used here (see the details in [5]).

4.2. Corrected power spectrum

The quantum-gravitationally corrected power spectrum takes the form [6]

$$\mathcal{P}_{\tilde{v}}^{(1)}(n) = \frac{n(n^2-\mathcal{K})}{4\pi^2}\left(\mathcal{R}e\Omega_n^{(0)} + \frac{1}{m_{\rm P}^2}\mathcal{R}e\Omega_n^{(1)}\right)^{-1}. \tag{23}$$

Solving the equation (18) and substituting $\mathcal{R}e\Omega_n^{(1)}$ into the above relation, the power spectrum takes the form [6]

$$\mathcal{P}_{\tilde{v}}^{(1)}(n) = \mathcal{P}_{\tilde{v}}^{(0)}(n)\Big(1+\Delta_{\tilde{v}}\Big), \text{ where } \Delta_{\tilde{v}} = -\frac{H^2}{m_{\rm P}^2}\frac{\left[\tilde{C}_{\varepsilon,\delta} - 2(\gamma - 3\lambda/2)\ln\left(\xi f(n,\varepsilon)\right)\right]}{n^3 f^3(n,\varepsilon)\xi^2}\mathcal{R}e\tilde{\Omega}_n^{(1)}(\xi). \tag{24}$$

The corrected power spectrum for the flat case [10] is recovered by considering the large wavelength limit.

It was estimated in [10] that the corrections $|\Delta_v| \sim 10^{-10}$. Since the corrections are so small and are within the experimental error bars, unfortunately, we don't expect to be able to observe quantum-gravitational corrections to the power spectrum in the near future.

5. Conclusion

We have calculated the quantum-gravitational corrections to the power spectrum of gauge-invariant inflationary scalar perturbations in a closed model for the universe. We have derived the Wheeler-DeWitt equation for the inflationary universe and applied a semiclassical Born–Oppenheimer type of approximation. At the corresponding orders of approximation, we obtained the uncorrected and quantum-gravitationally corrected Schrödinger equations from which we have derived the uncorrected and corrected power spectra.

We have also calculated the power spectrum of curvature perturbations which exhibits suppression of power at large scales compared to the flat model; hence, it may explain the observed lack of power for large scales.

Regarding the quantum-gravitationally corrected power spectrum, the correction leads to an additional suppression of power at large scales. Unfortunately, due to Cosmic Variance there is significant statistical uncertainty at the largest scales where this contribution is expected. Therefore, we do not expect to confirm these by observations in the near future. We emphasize, though, that this is a concrete quantitative prediction from quantum gravity.

6. Acknowledgements

We would like to thank Leonardo Chataignier for fruitful discussions. T. V. thanks the organizers of DICE2022 for providing the opportunity to present a talk.

Journal of Physics: Conference Series **2533** (2023) 012036 doi:10.1088/1742-6596/2533/1/012036

7. References

[1] Kiefer C 2012 *Quantum Gravity* 3rd edn (International Series of Monographs on Physics) (Oxford: Oxford University Press)

[2] Aghanim N *et al.* [Planck Collaboration] 2020 *Astron. Astrophys.* **641** A6

[3] Valentino E D, Melichiori A and Silk J 2020 *Nature Astronomy* **4** 196–203

[4] Ellis G F R, Meissner K A and Nicolai H 2018 *Nature Physics* **14** 770–2

[5] Kiefer C and Vardanyan T 2022 *Gen. Relativ. Gravit.* **54** 30

[6] Vardanyan T 2020 *Quantum-gravitational effects for inflationary scalar perturbations in a model for a closed universe* (unpublished master's thesis) Bonn University

[7] Halliwell J J and Hawking S W 1985 *Phys. Rev.* D **31** 1777–91

[8] Langlois D 1994 *Class. Quantum Grav.* **11** 389–407

[9] Martin J, Vennin V and Peter P 2012 *Phys. Rev.* D **86** 103524

[10] Brizuela D, Kiefer C and Krämer M 2016 *Phys. Rev.* D **94** 123527

[11] Brizuela D, Kiefer C and Krämer M 2016 *Phys. Rev.* D **93** 104035

[12] Chataignier L and Krämer M 2021 *Phys. Rev.* D **103** 066005

Journal of Physics: Conference Series **2533** (2023) 012037 doi:10.1088/1742-6596/2533/1/012037

The quantum waves of Minkowski space-time and the minimal acceleration from precanonical quantum gravity

Igor V Kanatchikov

National Quantum Information Centre in Gdansk (KCIK), 80-309 Gdańsk, Poland
IAS-Archimedes Project, Côte d'Azur, France

Abstract. We construct the simplest solutions of the previously obtained precanonical Schrödinger equation for quantum gravity, which correspond to the plane waves on the spin connection bundle and reproduce the Minkowski space-time on average. Quantum fluctuations lead to the emergence of the minimal acceleration a_0 related to the range of the Yukawa modes in the fibers of the spin connection bundle. This minimal acceleration is proportional to the square root of the cosmological constant Λ generated by the operator re-ordering in the precanonical Schrödinger equation. Thus the mysterious connection between the minimal acceleration in the dynamics of galaxies as described by Milgrom's MOND and the cosmological constant emerges as an elementary effect of precanonical quantum gravity. We also argue that the observable values of a_0 and Λ can be obtained when the scale of the parameter $\varkappa$ introduced by precanonical quantization is subnuclear, in agreement with the previously established connection between the scale of $\varkappa$ and the mass gap in quantum SU(2) Yang-Mills theory.

1. Introduction

The approaches to quantum gravity based on applying the standard methods of quantization to different versions of the gravitational Lagrangian [1, 2] almost inevitably lead to fundamental conceptual and technical problems of quantum gravity, such as the mathematical definition of the Wheeler-De Witt equation, the problem of time, the interpretation of (the measurement problem in) quantum cosmology, the problem of the correct classical limit in the loop quantum gravity, and the 122 orders of magnitude discrepancy between the theoretically plausible value of the cosmological constant and the observable one.

The approach called *precanonical quantization* [3–7] was proposed as a response to these problems. It departs from the various forms of canonical quantization, which are based on the canonical Hamiltonian formalism with a necessarily distinguished time variable, and instead uses a space-time symmetric generalization of the Hamiltonian formalism from mechanics to field theory known in the calculus of variations under the name of the De Donder-Weyl (DW) Hamiltonian formulation [8–10]. The precanonical quantization is based on the Dirac quantization of the Heisenberg-like subalgebra of Poisson-Gerstenhaber brackets of differential forms representing dynamical variables, which were found in the DW Hamiltonian formulation in [6, 11–14] and further explored and generalized e.g. in [15–18].

Quantization of brackets defined on differential forms naturally leads to a hypercomplex generalization of quantum theory where operators and wave functions are Clifford-algebra-

Journal of Physics: Conference Series **2533** (2023) 012037 doi:10.1088/1742-6596/2533/1/012037

valued [4–7]. The Clifford algebra in question is the complexified Clifford algebra of space-time. The DW Hamiltonian formulation and the quantization of Poisson-Gerstenhaber brackets of differential forms are space-time symmetric by construction. No distinction between space and time variables is required. No notion of field configurations or their initial or boundary data, i. e. the sections of the bundle whose base is space-time and whose fibers are spaces where the fields take values, is required by the procedure of precanonical quantization. The quantum dynamics of fields is described using the sections of the Clifford bundle over the bundle of field variables ϕ^a over the space-time with the coordinates x^μ. These sections are called precanonical wave functions and, in general, have the form (in $n = 1 + 3$ dimensions)

$$\Psi(\phi^a, x^\mu) = \psi + \psi_\mu \gamma^\mu + \frac{1}{2!}\psi_{\mu\nu}\gamma^{\mu\nu} + ... + \frac{1}{4!}\psi_{\mu_1\mu_2...\mu_4}\gamma^{\mu_1\mu_2...\mu_4}. \tag{1}$$

The field variables ϕ^a can be the Yang-Mills field variables A^a_μ [7, 19–21] or metric density variables $h^{\mu\nu}$ [22–25], or tetrad variables e^I_μ [26], or spin connection variables ω^{IJ}_μ [27–31]. The covariant analogue of the Schrödinger equation for the precanonical wave function has the form [22–25, 27–31]

$$i\hbar\varkappa\hat{\nabla\!\!\!\!/}\Psi - \hat{H}\Psi = 0, \tag{2}$$

where $\hat{H}$ is the operator of the covariant analogue of the Hamiltonian in the DW Hamiltonian-like formulation:

$$H := \partial_\mu\phi^a p^\mu_{\phi^a} - L, \quad p^\mu_{\phi^a} := \frac{\partial L}{\partial\partial_\mu\phi^a}, \tag{3}$$

$\hat{\nabla\!\!\!\!/}$ is the operator of the covariant Dirac operator on the space-time, and the parameter $\varkappa$ is an ultraviolet quantity of the dimension of the inverse spatial volume. It appears on purely dimensional grounds given the fact that the physical dimension of classical H is that of the mass density (in $c = 1$ units used throughout the paper). Note that $\varkappa$ is also introduced in the course of the precanonical quantization when the Poisson-Gerstenhaber brackets are replaced by commutators, and the representation of operators corresponding to differential forms is constructed in terms of Clifford-algebra-valued operators. In particular, the 3-dimensional volume element $d\mathbf{x} := dx^1 \wedge dx^2 \wedge dx^3$ is mapped to the Clifford algebra element

$$d\mathbf{x} \mapsto \frac{1}{\varkappa}\gamma_0, \tag{4}$$

where γ_I denote the flat space-time Dirac matrices, $\gamma_I\gamma_J + \gamma_J\gamma_I = 2\eta_{IJ}$, $I, J = 0, 1, 2, 3$. This map is very similar to what is known as the "quantization map" or "Chevalley map" in the Clifford algebra literature.

The relation between the description of quantum fields in terms of Clifford-algebra-valued precanonical wave functions $\Psi(\phi^a, x^\mu)$ and the standard QFT can be established if the latter is formulated in the Schrödinger functional picture [32]. In this picture, QFT is described in terms of time-dependent functionals of initial field configurations $\phi^a(\mathbf{x})$, $\mathbf{\Psi}([\phi^a(\mathbf{x})], t)$, which obey the canonical Schrödinger equation

$$i\hbar\partial_t\mathbf{\Psi} - \hat{\mathbf{H}}\mathbf{\Psi} = 0, \tag{5}$$

where $\hat{\mathbf{H}} = \int d\mathbf{x}\hat{T}^0_0$ is the operator of the canonical Hamiltonian.

Taking into account that the precanonical wave function $\Psi(\phi^a, x^\mu)$ gives the probability amplitude of detecting the field value ϕ at the space-time point x and the Schrödinger wave functional $\mathbf{\Psi}([\phi^a(\mathbf{x})], t)$ is the probability amplitude of observing the field configuration $\phi(\mathbf{x})$ on the hypersurface of constant time t, one can anticipate that the Schrödinger wave functional is a continuous product, or product integral, over the spatial points $\mathbf{x}$, of precanonical wave functions restricted to the configuration Σ given by $\phi = \phi(\mathbf{x})$, i.e. $\Psi_\Sigma(\phi = \phi(\mathbf{x}), \mathbf{x}, t)$, and

Journal of Physics: Conference Series **2533** (2023) 012037 doi:10.1088/1742-6596/2533/1/012037

transformed from the representation with a diagonal $\hat{H}$ to the representation with a diagonal $\hat{T}^0_0 = \hat{H} - \partial_i\phi^a(\mathbf{x})\hat{p}^i_{\phi_a}$. The resulting expression of the Schrödinger wave functional $\mathbf{\Psi}$ in terms of the precanonical wave functions Ψ is given by the continuous product or the Volterra product integral over $\mathbf{x}$ denoted as $\prod_{\mathbf{x}}$:

$$\mathbf{\Psi} = \mathsf{Tr}\left\{\prod_{\mathbf{x}} e^{-i\phi(\mathbf{x})\gamma^i\partial_i\phi(\mathbf{x})/\varkappa}\Psi_\Sigma(\phi(\mathbf{x}),\mathbf{x},t)\;|_{\frac{1}{\varkappa}\mapsto\gamma^0 d\mathbf{x}}\right\}, \tag{6}$$

where the inverse of the quantization map (4) is used as a natural step of transformation from the Clifford algebraic objects of precanonical quantization to the $\mathbb{C}$-valued functional of the canonical quantum field theory in the Schrödinger representation. In [34,35] we have shown, for interacting scalar fields in Minkowski space-time, that $\mathbf{\Psi}$ constructed in equation (6) satisfies the standard Schrödinger equation (5) as a consequence of (the flat space-time version of) the precanonical Schrödinger equation (2) restricted to the surface of initial data $\phi^a = \phi^a(\mathbf{x})$ at the moment of time t. A similar relation has been found also for quantum Yang-Mills theory on Minkowski space-time [19, 21] and for scalar fields on curved space-time [36–38]. The existence of this relationship shows that standard QFT based on canonical quantization is the limiting case of QFT based on precanonical quantization corresponding to the inverse quantization map $\frac{1}{\varkappa}\gamma_0 \mapsto d\mathbf{x}$ or, loosely speaking, to the limiting case of infinite $\varkappa$ corresponding to the unregularized volume of the momentum space, which equals $\delta^3(\mathbf{x}=0)$.

Let us also note that the existence of the space-time symmetric Hamilton-Jacobi (HJ) theory of fields which is associated with the DW Hamiltonian theory [8–10, 40–42] raises a question about the existence of a formulation of quantum field theories which reproduces the DW HJ theory in the classical limit. The precanonical quantization leads to such a formulation, at least in the case of scalar fields (cf. [4, 43]).

In the previous papers, the precanonical quantization has been applied to general relativity in metric variables [22–25], to the teleparallel equivalent of general relativity [26], and to general relativity in vielbein variables [27–31]. In this paper, we will briefly outline the latter in Section 2 and then, in Section 3, construct the solutions of the precanonical Schrödinger equation for quantum gravity corresponding to the quantum version of Minkowski space-time in Cartesian coordinates. We will find that quantum effects lead to the emergence of minimal acceleration related to the range of the Yukawa modes of the precanonical wave function in the spin-connection space. We will show that this minimal acceleration is related to the square root of the cosmological constant. The latter appears from the reordering of operators in the precanonical Schrödinger equation for gravity. We will also obtain realistic estimations of both quantities, albeit with an error of several orders of magnitude, when the scale of the parameter $\varkappa$ is below approximately 100 MeV, which is consistent with our previous rough estimation of the mass gap in the quantum SU(2) gauge theory [20].

2. Precanonical quantum vielbein gravity

In this section, we mainly collect together the key results from our previous considerations in [27–30]. The construction of precanonical quantum vielbein gravity starts from the Einstein-Palatini Lagrangian density

$$\mathfrak{L} = \frac{1}{8\pi G}\mathfrak{e}e^{[\alpha}_I e^{\beta]}_J(\partial_\alpha\omega^{IJ}_\beta + \omega_\alpha{}^{IK}\omega_{\beta K}{}^J) + \frac{1}{8\pi G}\Lambda\mathfrak{e}, \tag{7}$$

where the vielbein coefficients e^α_I and the spin connection coefficients $\omega_\alpha{}^{IK}$ are the independent field variables, and $\mathfrak{e} := \det(e^I_\mu)$. The DW Hamiltonian formulation leads to the constraints

$$\mathfrak{p}^\alpha_{\omega^{IJ}_\beta} := \frac{\partial\mathfrak{L}}{\partial\,\partial_\alpha\omega^{IJ}_\beta} \approx \frac{1}{8\pi G}\mathfrak{e}e^{[\alpha}_I e^{\beta]}_J, \quad \mathfrak{p}^\alpha_{e^I_\beta} := \frac{\partial\mathfrak{L}}{\partial\,\partial_\alpha e^I_\beta} \approx 0 \tag{8}$$

Journal of Physics: Conference Series **2533** (2023) 012037 doi:10.1088/1742-6596/2533/1/012037

and the DW Hamiltonian density on the surface of constraints

$$\mathfrak{H} := \mathfrak{p}_\omega \partial\omega + \mathfrak{p}_e \partial e - \mathfrak{L} \approx -\mathfrak{p}^\alpha_{\omega^{IJ}_\beta} \omega_\alpha{}^{IK} \omega_{\beta K}{}^J - \frac{1}{8\pi G} \Lambda \mathfrak{e}. \tag{9}$$

The constraints are second class according to the extension of the Dirac classification to the DW theory [39]. The calculation of the generalized Dirac brackets of forms representing the fundamental variables leads to the vanishing brackets of vielbeins and their polymomenta and very simple brackets of spin connection coefficients and their polymomenta, e.g.,

$$\{\![\mathfrak{p}^\alpha_e, e' \varpi_{\alpha'}]\!\}^D = 0,$$
$$\{\![p^\alpha_\omega, \omega' \varpi_\beta]\!\}^D = \{\![p^\alpha_\omega, \omega' \varpi_\beta]\!\} = \delta^\alpha_\beta \delta^\omega_{\omega'},$$
$$\{\![\mathfrak{p}^\alpha_e, \mathfrak{p}_\omega \varpi_{\alpha'}]\!\}^D = \{\![\mathfrak{p}^\alpha_e, \omega \varpi_{\alpha'}]\!\}^D = \{\![\mathfrak{p}^\alpha_\omega, e' \varpi_{\alpha'}]\!\}^D = 0,$$

where $\varpi_\alpha := \partial_\alpha \rfloor dx^0 \wedge dx^1 \wedge ... \wedge dx^3$ is the basis of 3-forms on 4-dimensional spacetime. Quantization of these brackets according to the following generalization of Dirac's quantization rule

$$[\hat{A}, \hat{B}] = -i\hbar \widehat{\mathfrak{e}\{\![A, B]\!\}^D} \tag{10}$$

which leads to the operator representations of the polymomenta of spin connection, the vielbeins, the DW Hamiltonian H, such that $\widehat{\mathfrak{H}} =: \widehat{\mathfrak{e}H}$, and the quantum Dirac operator:

$$\widehat{\mathfrak{p}}^\alpha_{\omega^{IJ}_\beta} = -i\hbar\varkappa\mathfrak{e}\hat{\gamma}^{[\alpha} \frac{\partial}{\partial\omega^{IJ}_{\beta]}}, \quad \text{where} \quad \hat{\gamma}^\alpha := \hat{e}^\alpha_I \underline{\gamma}^I, \tag{11}$$

$$\hat{e}^\beta_I = -8\pi i G\hbar\varkappa \underline{\gamma}^J \frac{\partial}{\partial\omega^{IJ}_\beta}, \tag{12}$$

$$\widehat{H} = 8\pi G\hbar^2\varkappa^2 \, \underline{\gamma}^{IJ} \omega_\alpha{}^{KM} \omega_{\beta M}{}^L \frac{\partial}{\partial\omega^{KL}_\beta} \frac{\partial}{\partial\omega^{IJ}_\alpha} - \frac{1}{8\pi G}\Lambda, \tag{13}$$

$$\widehat{\not\nabla} = -8\pi i G\hbar\varkappa \underline{\gamma}^{IJ} \frac{\partial}{\partial\omega^{IJ}_\mu} \left(\partial_\mu + \frac{1}{4} \omega_{\mu KL} \underline{\gamma}^{KL} \overset{\leftrightarrow}{\vee} \right), \tag{14}$$

where $\overset{\leftrightarrow}{\vee}$ denotes the commutator (antisymmetric) Clifford product $\underline{\gamma}^{IJ} \overset{\leftrightarrow}{\vee} \Psi = \frac{1}{2} [\, \underline{\gamma}^{IJ}, \Psi \,]$. Hence the precanonical Schrödinger equation for quantum gravity takes the form

$$\underline{\gamma}^{IJ} \frac{\partial}{\partial\omega^{IJ}_\mu} \left(\partial_\mu + \frac{1}{4} \omega_{\mu KL} \underline{\gamma}^{KL} \overset{\leftrightarrow}{\vee} - \frac{\partial}{\partial\omega^{KL}_\beta} \omega_\mu{}^{KM} \omega_{\beta M}{}^L \right) \Psi(\omega, x) + \lambda\Psi(\omega, x) = 0, \tag{15}$$

where $\lambda := \frac{\Lambda}{(8\pi G\hbar\varkappa)^2}$ is a dimensionless combination of the fundamental constants of the theory, which depends on the operator ordering of ω and ∂_ω.

Note that equation (15) was first obtained in [27] without explicitly specifying the action of the spin connection term on the wave function. The need for the commutator product was understood later in [36], and the coefficient $\frac{1}{2}$ in front of the commutator comes from the consideration of the Ehrenfest theorem similar to that in [7], which is still unpublished.

The scalar product of precanonical wave functions is given by

$$\langle\Phi|\Psi\rangle := \mathrm{Tr} \int \overline{\Phi} \, \widehat{[d\omega]} \, \Psi, \tag{16}$$

where $\overline{\Psi} := \underline{\gamma}^0 \Psi^\dagger \underline{\gamma}^0$ and the operator-valued invariant integration measure on the 24-dimensional space of spin connection coefficients

$$\widehat{[d\omega]} \sim \hat{\mathfrak{e}}^{-6} \prod_{\mu IJ} d\omega^{IJ}_\mu. \tag{17}$$

Journal of Physics: Conference Series **2533** (2023) 012037 doi:10.1088/1742-6596/2533/1/012037

The operator $\hat{\mathfrak{e}}^{-1}$ is constructed from the operators $\hat{e}_I^\beta$ in (12).

Thus, we arrive at the "spin connection foam" formulation of the geometry of quantum gravity in terms of the Clifford-algebra-valued wave function on the bundle of spin connection coefficients over spacetime, $\Psi(\omega, x)$, and the transition amplitudes on the total space of this bundle, $\langle\omega, x|\omega', x'\rangle$, which are the Green functions of (15). The wave function corresponds to the "quantum fuzziness" of points of the total space and the Green functions correspond to the quantum correlations between the points, i.e. a quantum analogue of the classical connection.

The normalizability of precanonical wave functions: $\langle\Psi|\Psi\rangle < \infty$, leads to the vanishing contribution of the large curvatures $R = d\omega + \omega\wedge\omega$ to the probabilistic measure defined by the norm, and that ensures the quantum-gravitational avoidance of a curvature singularity by the precanonical wave function.

In the context of quantum cosmology, $\Psi(\omega, x)$ defines the spatially homogeneous statistics of local fluctuations of the spin-connection, which is classically given by the Hubble parameter $\dot{a}/a$, not the "distribution of quantum universes according to the Hubble parameter" as suggested by the picture of the superspace of 3-geometries emerging from the canonical quantization. Hence the problem of the "external observer" of the "quantum ensemble of universes" disappears.

The evolution of matter and radiation on the background of quantum gravitational fluctuations whose statistics and correlations are predicted by (15) may lead to predictable consequences for the distribution of matter and radiation at large cosmological scales, which may be observable.

In general, analysis of solutions of precanonical Schrödinger equation for quantum gravity, equation (15), is a formidable task. In the following section, we will construct the simplest solutions which can be interpreted as a quantum wave counterpart of the Minkowski space-time.

3. Quantum wave states of Minkowski space-time

The Minkowski metric in Cartesian coordinates:

$$ds^2 = (dx^0)^2 - (dx^1)^2 - (dx^2)^2 - (dx^3)^2,$$

is characterized by the vanishing spin connection coefficients

$$\omega_\mu^{IJ} = 0. \tag{18}$$

In this case, the precanonical Schrödinger equation (15) with $\Lambda = 0$ takes the simple form

$$\underline{\gamma}^{IJ}\partial_{\omega_\mu^{IJ}}\partial_\mu\Psi = 0. \tag{19}$$

In terms of the plane waves on the total space (x^μ, ω_μ^{IJ})

$$\Psi \sim e^{ik_\mu x^\mu + i\pi^\mu_{IJ}\omega_\mu^{IJ}}\tilde{\Psi}(k_\mu, \pi^\mu_{IJ}) \tag{20}$$

we obtain

$$\underline{\gamma}^{IJ}k_\mu\pi^\mu_{IJ} = 0. \tag{21}$$

From (21), we obtain the dispersion relation

$$k_\mu\pi^\mu_{IJ}k_\nu\pi^{\nu IJ} = 0, \tag{22}$$

which reflects a strong anisotropy due to the fibred structure of the (x, ω) space.

Therefore, any solution of (19) has the form

$$\Psi(\omega, x) = \int d^4k \int d^{24}\pi\ \delta(k_\mu\pi^\mu_{IJ}k_\nu\pi^\nu_{IJ})e^{ik_\mu x^\mu + i\pi^\mu_{IJ}\omega_\mu^{IJ}}\tilde{\Psi}(k_\mu, \pi^\mu_{IJ}). \tag{23}$$

The solutions of interest should be normalizable. The normalizability on the subspace of vanishing spin connection coefficients $\omega = 0$ takes the form

$$\mathrm{Tr} \int d^{24}\omega\, \delta^{24}(\omega)\overline{\Psi}(\omega, x)\hat{\mathfrak{e}}^{-6}\Psi(\omega, x) = \mathrm{Tr}(\overline{\Psi}(0, x)\hat{\mathfrak{e}}^{-6}\Psi(0, x)) = 1, \tag{24}$$

where the short-hand notations $\overline{\Psi}(0, x)$ and $\hat{\mathfrak{e}}^{-6}\Psi(0, x)$ mean $\overline{\Psi}(\omega, x)$ and $\hat{\mathfrak{e}}^{-6}\Psi(\omega, x)$ taken at $\omega = 0$.

The states which lead to the Minkowski space-time on the classical level have to satisfy the conditions

$$\langle \hat{g}^{\mu\nu} \rangle(x) = \mathrm{Tr} \int d^{24}\omega\delta^{24}(\omega)\overline{\Psi}(\omega, x)\hat{\mathfrak{e}}^{-6}\hat{g}^{\mu\nu}\Psi(\omega, x)) = \eta^{\mu\nu},$$

where the operator of the metric derived from the representation (12) has the form

$$\hat{g}^{\mu\nu} = -(8\pi G)^2\hbar^2\varkappa^2\eta^{IK}\eta^{JL}\partial_{\omega_\mu^{IJ}}\partial_{\omega_\nu^{KL}}. \tag{25}$$

Hence the wave functions which reproduce the Minkowski space-time on average should satisfy

$$\mathrm{Tr}(\overline{\Psi}(0, x)\hat{\mathfrak{e}}^{-6}\hat{g}^{\mu\nu}(\overline{\Psi}(0, x)) = \eta^{\mu\nu}. \tag{26}$$

In terms of the Fourier components (26) implies that

$$\tilde{\Psi}(\pi, k) = \tilde{\Psi}(-\pi, k). \tag{27}$$

By comparison with the normalizability condition, we conclude that

$$-(8\pi G)^2\hbar^2\varkappa^2\eta^{IK}\eta^{JL}\partial_{\omega_\mu^{IJ}}\partial_{\omega_\nu^{KL}}\Psi(0, x) = \eta^{\mu\nu}\Psi(0, x).$$

Therefore, for the plane waves,

$$(8\pi G)^2\hbar^2\varkappa^2\pi^{\mu}_{IJ}\pi^{\nu IJ} = \eta^{\mu\nu}. \tag{28}$$

Then, from the dispersion relation, it follows

$$\eta^{\mu\nu}k_\mu k_\nu = 0. \tag{29}$$

Thus, the states corresponding to the (1+3)-dimensional Minkowski space-time in the classical limit have:

- the light-like modes (29) along the space-time dimensions (the base of the total space of the bundle of spin connection coefficients over space-time);
- 4 massive (Yukawa) modes (28) in the spin-connection spaces (the fibers of the total space of the bundle of spin connection coefficients over space-time), which propagate in 6-dimensional subspaces with the coordinates ω_μ^{IJ} for each $\mu = 0, 1, 2, 3$;
- the range of those massive modes in ω-space is $8\pi G\hbar\varkappa$, whose value we estimate below;
- the modes corresponding to the spatial $\mu = 1, 2, 3$ are tachyonic. Those tachyonic modes, however, do not violate the causality in space-time as they propagate along the fibers associated to each point of space-time rather than in the space-time itself.

Note that the spin connection has the mass dimension +1, $\varkappa$ has the mass dimension +3, and the square of the Planck length $G\hbar$ has the mass dimension −2. Hence the range of the Yukawa modes in the spin connection space, $8\pi G\hbar\varkappa$, is given in the units of mass dimension +1, which is also the mass dimension of acceleration.

If $\varkappa$ were Planckian, which is a seemingly natural first guess, then the range of Yukawa modes in the spin connection space is also Planckian, and they could be attributed to the quantum foaminess of space-time at the Planck scale, as is usually assumed. However, our study of quantum Yang-Mills theory from the perspective of precanonical quantization (see below) has produced evidence that $\varkappa$ is more likely a sub-nuclear scale quantity, which leads to a drastically different scale of the phenomena in question.

3.1. An estimation of $\varkappa$ from the mass gap in pure gauge theory

From the Lagrangian of a pure non-abelian Yang-Mills theory we can derive the corresponding DW Hamiltonian function [7, 19] and precanonically quantize it. It leads to the following expression for the DW Hamiltonian operator for the quantum pure YM field with the coupling constant g [7, 19, 20]

$$\widehat{H} = \frac{1}{2}\hbar^2\varkappa^2\frac{\partial}{\partial A^\mu_a \partial A^a_\mu} - \frac{1}{2}ig\hbar\varkappa C^a{}_{bc}A^b_\mu A^c_\nu\gamma^\nu\frac{\partial}{\partial A^a_\mu}\,. \tag{30}$$

The fact that the eigenvalues of the DW Hamiltonian operator for the pure YM field yield the spectrum of masses of the propagating modes is manifested in the precanonical Schrödinger equation in flat space-time

$$i\hbar\gamma^\mu\partial_\mu\Psi = \frac{1}{\varkappa}\widehat{H}\Psi. \tag{31}$$

In [21], we have shown that equations (30) and (31) for the wave function $\Psi(A,x)$ reproduce the functional Schrödinger equation for the wave functional $\mathbf{\Psi}([A(\mathbf{x})],t)$ after the (3+1) decomposition and the "dequantization map" $\frac{1}{\varkappa}\gamma_0 \mapsto \varpi_0 = d\mathbf{x}$ (cf. (4)).

In the temporal gauge $A^a_0 = 0$, we can limit ourselves to the operator (30) written only in terms of the spatial components A^a_i. For SU(2) theory with $a,b,c = 1,2,3$ and $i,j = 1,2,3$, we were able to estimate the gap between the ground state and the first excitation with the vanishing non-abelian charge (a "color" or rather "isospin" in the context of SU(2))

$$\langle\frac{1}{\varkappa}\hat{H}\rangle > \left(\frac{8g^2\hbar^4\varkappa}{32}\right)^{1/3}|\mathsf{ai}'_1|, \tag{32}$$

where ai'_1 is the first root of the derivative of the Airy function. This gap in the spectrum of the DW Hamiltonian operator can be identified with the mass gap

$$\Delta\mu \approx 0.86(g^2\hbar^4\varkappa)^{1/3}. \tag{33}$$

Therefore, the scale of $\varkappa$ is close to the scale of the mass gap. For SU(3) YM theory, this formula will have a different numerical coefficient in front and a different value of the coupling constant g. The SU(3) QCD mass gap lies between the pion masses at 130 MeV and the alleged glueball masses at a few GeV. The numerical factor in (33) for SU(3) may change several times, and the coupling constant g is of the order $10^{-2} - 10^{-1}$ (in the units of $\sqrt{\hbar}$). With those uncertainties, we estimate $\varkappa^{1/3}$ is below 1 GeV with an error of up to 2 orders of magnitude.

3.2. The minimal acceleration

With the GeV-scale $\varkappa$ we obtain

$$8\pi G\hbar\varkappa \sim 10^{-23\pm3\times2}\ \mathrm{cm}^{-1}. \tag{34}$$

This quantity is compatible with the scale of the Hubble radius $R_H \sim 10^{28}$ cm and the cosmological constant $|\Lambda| \sim 10^{-56}\ \mathrm{cm}^{-2}$, if the scale of $\varkappa$ is below 100 MeV, which is on the edge of our margin of error. In this case, $8\pi G\hbar\varkappa$ coincides with the scale of the minimal acceleration $a_0 \sim \sqrt{\Lambda}$ which is known from Milgrom's theory of MOND [44–48]. Note that the Yukawa modes in the spin-connection space, which set the threshold of acceleration $8\pi G\hbar\varkappa$, emerge from quantum fluctuations of spin connection around the vanishing value of the spin connection of Minkowski space-time. They establish the limit below which quantum fluctuations of space-time violate the notion of acceleration-less inertial frames which underlies classical Minkowski space-time. Note also that our value of the minimal acceleration appears here in the context of a quantum analogue of the Minkowski space-time and it may slightly change for the quantum analogues of more realistic cosmological space-times.

3.3. The cosmological constant

We have already pointed out that the expression of the precanonical Schrödinger equation for gravity (15) is defined up to the operator ordering of ω and ∂_ω. A reordering in the spin connection term will produce a constant of the order $\frac{1}{4}4^3$ added to the dimensionless $\lambda = \frac{\Lambda}{(8\pi G\hbar\varkappa)^2}$ constructed from the bare cosmological constant Λ. If the latter equals zero, then the contribution to the cosmological constant from the reordering of operators, i.e. essentially from the quantum fluctuations of the spin connection, can be estimated as $\Lambda_\omega \sim 4^2(8\pi G\hbar\varkappa)^2$. For $\varkappa \sim 10^{0\pm3\times2}$ GeV^3, in agreement with the estimation in Section 3.1, we obtain $\Lambda_\omega \sim 10^{-45\pm2\times6}$ cm^{-2}. This estimation is again consistent with the observed value of the cosmological constant if the scale of $\varkappa$ is below approximately 100 MeV. In this case, the minimal acceleration in (34) is related to Λ_ω as follows

$$a_0 \approx \frac{1}{4}\sqrt{\Lambda_\omega}, \tag{35}$$

which is close to the current observed value $a_0 \approx 1.2 \times 10^{-10}$ m·s^{-2} or $a_0 \approx 10^{-29}$ cm^{-1} in the $c = 1$ units. Thus the mysterious connection between the phenomenological minimal acceleration in the dynamics of galaxies as described by MOND (as an alternative to the dark matter) and the cosmological constant (as the simplest dark energy) emerges as an elementary effect of precanonical quantum gravity.

Acknowledgments

I gratefully acknowledge V.A. Kholodnyi for his interest and insightful comments, J. Kouneiher for his interest and encouraging support, and M. Wright for his remarks helping to improve the linguistic quality of the manuscript. I also thank Hans-Thomas Elze for his tireless organization of DICE Conferences in Castiglioncello, which provided an inspirational framework for exchanging ideas.

References

[1] Kiefer C 2012 *Quantum Gravity* (Oxford: Oxford University Press)
[2] Rovelli C 2004 *Quantum Gravity* (Cambridge: Cambridge University Press)
[3] Kanatchikov I 1998 Towards the Born-Weyl quantization of fields *Int. J. Theor. Phys.* **37** 333 (*Preprint* arXiv:quant-ph/9712058)
[4] Kanatchikov I 1999 DeDonder-Weyl theory and a hypercomplex extension of quantum mechanics to field theory *Rep. Math. Phys.* **43** 157 (*Preprint* arXiv:hep-th/9810165)
[5] Kanatchikov I 1998 On quantization of field theories in polymomentum variables *AIP Conf. Proc.* **453** 356 (*Preprint* arXiv:hep-th/9811016)
[6] Kanatchikov I 2001 Geometric (pre)quantization in the polysymplectic approach to field theory *Differential Geometry and Its Applications* eds O Kowalski D Krupka and J Slovak (Opava: Silesian University) pp 309-321 (*Preprint* arXiv:hep-th/0112263)
[7] Kanatchikov I 2015 Ehrenfest theorem in precanonical quantization *J. Geom. Symm. Phys.* **37** 43 (*Preprint* arXiv:1501.00480)
[8] Kastrup H 1983 Canonical theories of Lagrangian dynamical systems in physics *Phys. Rep.* **101** 1
[9] De Donder Th 1930 *Théorie invariantive du calcul des variations* (Paris: Gauthier-Villars)
[10] Weyl H 1935 Geodesic fields in the calculus of variation for multiple integrals *Ann. Math.* **36** 607
[11] Kanatchikov I 1993 On the canonical structure of De Donder-Weyl covariant Hamiltonian formulation of field theory 1. Graded Poisson brackets and equations of motion *Preprint* arXiv:hep-th/9312162
[12] Kanatchikov I 1998 Canonical structure of classical field theory in the polymomentum phase space *Rep. Math. Phys.* **41** 49 (*Preprint* arXiv:hep-th/9709229)
[13] Kanatchikov I 1997 On field theoretic generalizations of a Poisson algebra *Rep. Math. Phys.* **40** 225 (*Preprint* arXiv:hep-th/9710069)
[14] Kanatchikov I 1995 From the Poincaré-Cartan form to a Gerstenhaber algebra of Poisson brackets in field theory *Quantization, Coherent States, and Complex Structures* eds J-P Antoine S Twareque Ali W Lisiecki IM Mladenov and A Odzijewicz (Berlin: Springer) p 173 (*Preprint* arXiv:hep-th/9511039)

Journal of Physics: Conference Series **2533** (2023) 012037 doi:10.1088/1742-6596/2533/1/012037

[15] Hélein F and Kouneiher J 2002 Finite dimensional Hamiltonian formalism for gauge and quantum field theories *J. Math. Phys.* **43** 2306
[16] Hélein F and Kouneiher J 2004 Covariant Hamiltonian formalism for the calculus of variations with several variables: Lepage-Dedecker versus de Donder-Weyl *Adv. Theor. Math. Phys.* **8** 565
[17] Forger M Paufler C and Römer H 2003 The Poisson bracket for Poisson forms in multisymplectic field theory *Rev. Math. Phys.* **15** 705 (*Preprint* arXiv:math-ph/0202043)
[18] Kanatchikov IV 2000 On the Duffin-Kemmer-Petiau Formulation of the covariant Hamiltonian dynamics in field theory *Rep. Math. Phys.* **46** 107 (*Preprint* arXiv:hep-th/9911175)
[19] Kanatchikov I 2004 Precanonical quantization of Yang-Mills fields and the functional Schroedinger representation *Rep. Math. Phys.* **53** 181 (*Preprint* arXiv:hep-th/0301001)
[20] Kanatchikov I 2017 On the spectrum of DW Hamiltonian of quantum SU(2) gauge field *Int. J. Geom. Meth. Mod. Phys.* **14** 1750123 (*Preprint* arXiv:1706.01766)
[21] Kanatchikov I 2018 Schrödinger wave functional in quantum Yang-Mills theory from precanonical quantization *Rep. Math. Phys.* **82** 373 (*Preprint* arXiv:1805.05279)
[22] Kanatchikov I 1998 From the DeDonder-Weyl Hamiltonian formalism to quantization of gravity *Current topics in mathematical cosmology* eds M Rainer and H-J Schmidt (Singapore: World Scientific) p 472 (*Preprint* arXiv:gr-qc/9810076)
[23] Kanatchikov I 2001 Quantization of gravity: yet another way *Coherent States, Quantization and Gravity* eds M Schlichenmaier A Strasburger S T Ali and A Odzijewicz (Warsaw: Warsaw University Press) p 189 (*Preprint* arXiv:gr-qc/9912094)
[24] Kanatchikov I 2000 Precanonical perspective in quantum gravity *Nucl. Phys. Proc. Suppl.* **88** 326 (*Preprint* arXiv:gr-qc/0004066)
[25] Kanatchikov I 2001 Precanonical quantum gravity: quantization without the space-time decomposition *Int. J. Theor. Phys.* **40** 1121 (*Preprint* arXiv:gr-qc/0012074)
[26] Kanatchikov I 2023 Towards precanonical quantum teleparallel gravity *Preprint* arXiv:2302.10695
[27] Kanatchikov IV 2012 On Precanonical quantization of gravity in spin connection variables *AIP Conf. Proc.* **1514** 73 (*Preprint* arXiv:1212.6963)
[28] Kanatchikov IV 2013 De Donder-Weyl Hamiltonian formulation and precanonical quantization of vielbein gravity *J. Phys. Conf. Ser.* **442** 012041 (*Preprint* arXiv:1302.2610)
[29] Kanatchikov IV 2014 On precanonical quantization of gravity *Nonlin. Phenom. Complex Sys. (NPCS)* **17** 372 (*Preprint* arXiv:1407.3101)
[30] Kanatchikov IV 2016 Ehrenfest theorem in precanonical quantization of fields and gravity *Preprint* arXiv:1602.01083
[31] Kanatchikov IV 2015 On the "spin connection foam" picture of quantum gravity from precanonical quantization *Preprint* arXiv:1512.09137
[32] Hatfield B 1992 *Quantum field theory of point particles and strings* (Reading, MA: Perseus Books).
[33] Kanatchikov IV 2001 Precanonical quantization and the Schroedinger wave functional *Phys. Lett.* **A283** 25 (*Preprint* arXiv:hep-th/0012084)
[34] Kanatchikov IV 2014 Precanonical quantization and the Schrödinger wave functional revisited *Adv. Theor. Math. Phys.* **18** 1249 (*Preprint* arXiv:1112.5801)
[35] Kanatchikov IV 2016 On the precanonical structure of the Schrödinger wave functional *Adv. Theor. Math. Phys.* **20** 1377 (*Preprint* arXiv:1312.4518)
[36] Kanatchikov IV 2019a Schrödinger functional of a quantum scalar field in static space-times from precanonical quantization *Int. J. Geom. Meth. Mod. Phys.* **16** 1950017 (*Preprint* arXiv:1810.09968)
[37] Kanatchikov IV 2019b Precanonical structure of the Schrödinger wave functional in curved space-time *Symmetry* **11** 1413 (*Preprint* arXiv:1812.11264)
[38] Kanatchikov IV 2020 On the precanonical structure of the Schrödinger wave functional in curved space-time *Acta Phys. Polon. B Proc. Suppl.* **13** 313 (*Preprint* arXiv:1912.07401)
[39] Kanatchikov IV 2008 On a generalization of the Dirac bracket in the De Donder-Weyl Hamiltonian formalism *Differential Geometry and its Applications* eds O Kowalski D Krupka O Krupková and J Slovák (Singapore: World Scientific) p 615 (*Preprint* arXiv:0807.3127)
[40] Pietrzyk ME and Barbachoux C 2022 On the covariant Hamilton-Jacobi equation for the teleparallel equivalent of general relativity *Preprint* arXiv:2201.01295
[41] Pietrzyk ME Barbachoux C Kanatchikov I and Kouneiher J 2022 On the covariant Hamilton-Jacobi formulation of Maxwell's equations via the polysymplectic reduction *Preprint* arXiv:2212.14845
[42] Riahi N and Pietrzyk ME 2020 On the relation between the canonical Hamilton–Jacobi equation and the De Donder–Weyl Hamilton–Jacobi formulation in general relativity *Acta Phys. Polon. Suppl.* **13** 213 (*Preprint* arXiv:1912.13363)
[43] Derakhshani M Kiessling MKH and Tahvildar-Zadeh AS 2021 Covariant guiding laws for fields *Preprint*

Journal of Physics: Conference Series **2533** (2023) 012037 doi:10.1088/1742-6596/2533/1/012037

arXiv:2110.09683

[44] Milgrom M 1983a A modification of the Newtonian dynamics as a possible alternative to the hidden mass hypothesis *ApJ* **270** 365

[45] Milgrom M 1983b A modification of the Newtonian dynamics - Implications for galaxies *ApJ* **270** 371

[46] Milgrom M 1983c A modification of the Newtonian dynamics: implications for galaxy systems *ApJ* **270** 384

[47] Milgrom M 2001 MOND - a pedagogical review *Acta Phys. Pol. B* **32** 3612 (*Preprint* arXiv:astro-ph/0112069)

[48] Famaey B and McGaugh S 2012 Modified Newtonian Dynamics (MOND): observational phenomenology and relativistic extensions *Living Rev. Rel.* **15** 10 (*Preprint* arXiv:1112.3960)

Journal of Physics: Conference Series **2533** (2023) 012038 doi:10.1088/1742-6596/2533/1/012038

Hints at a relationship between friction and relativistic physics

E Minguzzi

Dipartimento di Matematica e Informatica "U. Dini",
Università degli Studi di Firenze, Via S. Marta 3, I-50139 Firenze, Italy

E-mail: ettore.minguzzi@unifi.it

Abstract. In applied mechanics Reye's law (1860) establishes, via energy arguments, that the mass of the debris produced by dry friction in the contact of rigid bodies is proportional to the work done by friction forces. This result has long been used for the determination of the distribution of pressure in the contact of rigid bodies, and hence for the design of brakes. In this work I show that when bodies losing mass due to friction are treated, as they should, as variable mass systems, a relationship analogous to the relativistic mass formula is recovered. This result suggests that mathematical structures typical of relativistic physics could have been discovered prior to 1905, without making any reference to electromagnetism, group theory or the speed of light. Also this result could point to the existence of a physical theory depending on two constants, the speed of light (Reye's constant) and a universal frictional deceleration with respect to an absolute frame. The limit of the theory for vanishing friction would give Special Relativity, as the absolute frame would become unobservable, while the limit for the speed of light going to infinity would lead to Aristotelian mechanics, i.e. a classical mechanics type theory presenting universal friction. Finally, I present a reference frame transformation that displays these features and I apply the theory to some open cosmological problems.

1. Introduction

In the last two decades I have been working on the foundations of relativity theory and causality [1], while teaching courses in Rational (Analytical) Mechanics to students in mechanical engineering. I enjoyed teaching these courses as I learned a lot particularly exploring results in mechanics that do not typically belong to the curriculum of a physicist. This work is the result of one such explorations in which my two interests in the foundations of relativity physics and in mechanics went to interact.

As a physicist I have always been puzzled by the following question: given two rigid bodies sharing a flat surface, how is the pressure between them going to distribute? If the normal force is N and the area of contact is A, it could be tempting to answer: $p = N/A$. However, this looks just as the simplest possible answer though not necessarily the correct one. For instance, should we expect this answer to change if the two bodies are in relative motion? Should the pressure depend on the relative velocity at the contact point? At first, I really did not think much about this question because I supposed that a correct answer would necessarily rely on a study of the deformation of the bodies and hence on elements of elasticity theory. Only in this way, I thought, one could really elaborate some argument leading to the justification of some pressure distribution.

Journal of Physics: Conference Series **2533** (2023) 012038 doi:10.1088/1742-6596/2533/1/012038

My arguments appeared to be correct but were in fact, as I would later realize, quite wrong. The problem of finding the distribution of pressure had been already solved in 1860 by the German mathematician Reye through an ingenious argument that involved the consideration of friction and that completely disregarded elasticity [2]. His argument is elegant and marries well with the idealized notion of rigid body that is so much used in mechanics. There is really no need to abandon the rigid body idealization just to solve the pressure distribution problem.

A historical note: Karl Theodore Reye (1838–1919) was a German mathematician who contributed significatively to projective geometry.
He studied at the polytechnic in Hanover and in Zürich, where he turned to theoretical physics under the influence of Rudolf Clausius. When he was 22, in 1860, he published the landmark paper [2] "On the theory of pivot friction" where he established a successful theory of wear under friction.

Figure 1. Karl Theodore Reye

In his study Reye considers the friction of two bodies A and B, sliding one against the other on a common surface Σ under friction. One of the bodies, say A, is considered hard so that friction wears out just the (soft) body B. It can be assumed that a given proportion of the work done by friction forces goes in the breaking of the intermolecular bonds and hence in the detachment of mass from the soft body. He then establishes [2]

Reye's wear law (1860): The wear, that is the mass of the debris produced, is proportional to the work done by friction forces:

$$m_{debris} = \frac{1}{c^2}|L_f| \tag{1}$$

where c is Reye's constant (it has dimensions of a velocity and it depends on the type of materials at contact). He also showed how to use this formula to deduce the pressure distribution on the contact surface via geometrical arguments (we shall see some examples in a moment, uniform pressure turns out to be untenable). The experimental verification of the law was sufficiently good, in fact resistance to wear was defined and measured based on its validity [3].

Opatowski [4] wrote in 1942:

> The importance of wear by friction is, in some mechanical organs, particularly brakes, so great, that the ideas which Reye published 80 years ago still form today the theoretical basis for their design.

Although there were much posterior derivations of Eq. (1), most notably by Archard [5] (1953), or by M. M. Khrushchov and M. A. Babichev (1960) for the metal-metal contact, these were actually derivations based on other laws valid in specific contact models. It is no surprise that the same law (1) can be reached starting from several different microscopical mechanisms, based on microscopic laws that apply only in specific circumstances. The strongest argument

Journal of Physics: Conference Series **2533** (2023) 012038 doi:10.1088/1742-6596/2533/1/012038

seems Reye's as it is based on energy considerations alone and as is, for this reason, universal, i.e. independent of specific microscopic models.

In Italy Reye's law, also called Reye's assumption, entered very early in the program of the university course "Meccanica applicata alle macchine" (Mechanics of machines) and has always been a traditional topic of study, particularly for its usefulness in the design of brakes.

Villaggio [6] states that Reye's work was first introduced in university courses in Italy by Modesto Panetti. Panetti wrote a paper on applications of Reye's law in 1908 [7] while first editions of his university book [8] introducing Reye's law to students in engineering were published in 1913. However, I found that this account is not completely accurate. The teaching of "Meccanica applicata alle macchine" started earlier in the "Scuola d'applicazione per gli ingegneri" established in Turin in 1860. The course was first thought by Prospero Richelmy, from the foundation of the school to 1880, then by Ferdinando Zucchetti till 1882 and then by Scipione Cappa assisted by Elia Ovazza for several years [9]. It was Cappa who introduced in his 1890 university book Reye's law [10] devoting to it considerable space (also cited in [11]). As the Italian tradition of "Meccanica applicata alle macchine" originated from this school, it appears likely that most Italian University professors, including Panetti, learned of Reye's work from Cappa's 1890 book and posterior editions. Some early references dealing with Reye's law are [12, 13], while as more recent references [14, 15] can be mentioned.

Early sources in English that mention Reye's work are certainly less numerous and include the 1891 translation 'The Constructor' by Henry Harrison Suplee of F. Reuleaux classic, or Weisbach book [16].

Reye's law admits an infinitesimal formulation. Denoting with m the mass of the soft body B, with W_f the power of the friction force (necessarily negative), with $\boldsymbol{F}_f$ the friction force and with $\boldsymbol{v}$ velocity of B relative to A,

$$\dot{m} = \frac{1}{c^2} W_f = \frac{1}{c^2} \boldsymbol{F}_f \cdot \boldsymbol{v} = -\frac{1}{c^2} F_f v,$$

where we used the fact that the friction force is oriented opposite to the relative velocity.

The power of friction forces might depend on the region of the contact surface Σ considered. Thus the body B might wear out more at some points than others. The previous formula has to be applied on a small region of Σ of area ΔA, then $F_f = \mu p \Delta A$, where μ is the friction coefficient and p is the pressure. Denoting with ρ the density, we have $\mathrm{d}m = -\rho \Delta A \mathrm{d}h$, where h is the width of the wear out part. As a result

$$\dot{h} = \frac{\mu}{\rho} \frac{1}{c^2} p v.$$

This equation shows that wear is proportional to the product of pressure and velocity at the contact point. Geometrical considerations impose specific forms for the function $h : \Sigma \to \mathbb{R}$ and hence for the product pv. This ultimately allows the determination of the pressure distribution.

For instance, Figures 2 and 3 show what happens to a block that loses mass due to wear depending on different geometrical constraints imposed on it.

An interesting design of a brake is depicted in Figure 4. Here the hard part A is the wheel, while the soft part B is the block that is pushed horizontally towards the wheel by a force $\boldsymbol{Q}$. As the wheel spins with angular velocity ω the relative velocity at each contact point might depend on time but is actually independent of the point and hence of θ. As a result the pressure is proportional to $\dot{h}$. The geometrical constraint, accurately show in the figure, shows how the block wears as it translates. The result is $\dot{h} \propto \cos\theta$ which implies $p = a \cos\theta$ for some constant a. The constant a could be determined by using the range of θ and the friction coefficient μ by imposing that the contact force between wheel and block equals Q.

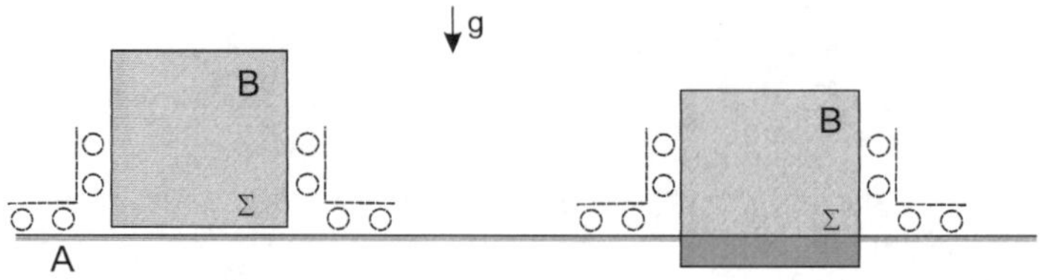

Figure 2. A block loses mass due to friction while it moves. The block is aided by an idealized device that avoids its tilting. As a result pressure is uniformly distributed as $\dot{h}$ does no depend on the contact point.

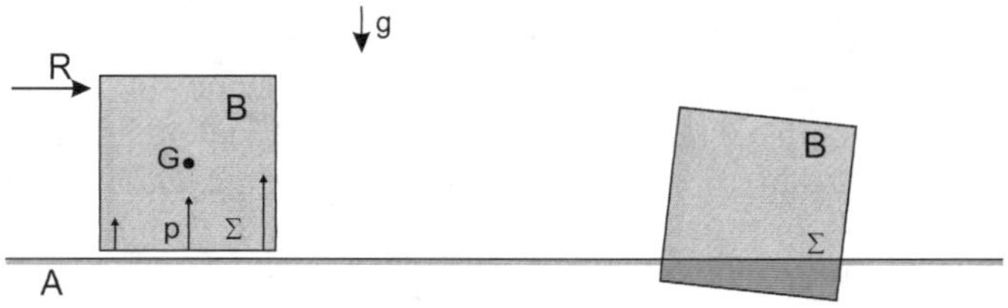

Figure 3. A block loses mass due to friction while it moves. As it is pushed by a force R, and the angular velocity is negligible, the pressure distributes so as to balance the mechanical momentum with respect to G. This results in an uneven pressure distribution and hence in the (slow) tilting of the block. In first approximation the pressure distribution must be linear, $p = a + bx$, as h varies linearly with the horizontal coordinate x.

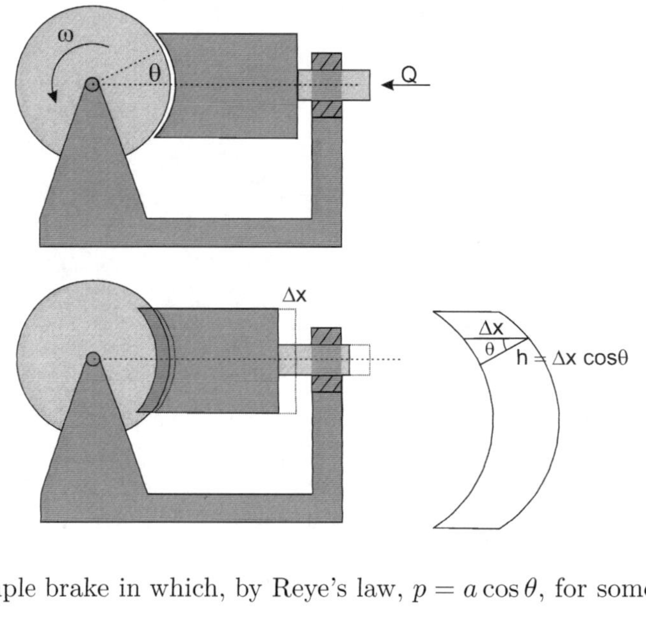

Figure 4. A simple brake in which, by Reye's law, $p = a \cos \theta$, for some constant $a > 0$.

As a further example we can consider a cylindrical annulus of mass m spinning over a flat horizontal surface (Fig. 5). The interior radius of the annulus is r while the exterior radius is R. If the axis of the cylinder does not tilt as it wears (we might assume that there is some mechanism preventing it from happening), the function $\dot{h}$ is constant (Fig. 7). Indeed, if the cylinder were to wear in an inhomogeneous way that would lead to more friction precisely in those regions that are less worn and that are therefore more exposed to contact and friction (Fig. 6). We conclude that the product of pressure and velocity is constant, and since the latter is proportional to the distance ρ from the axis, $p = a/\rho$ for some constant a. From here it is easy to calculate that $a = \frac{mg}{2\pi(R-r)}$.

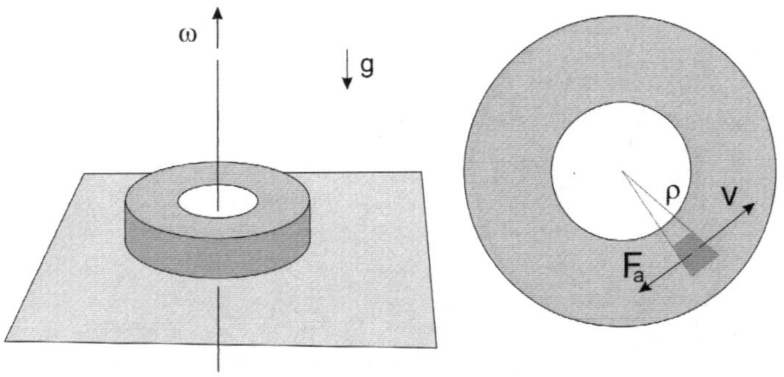

Figure 5. The surface of contact Σ. Friction and relative velocity depend on the region considered.

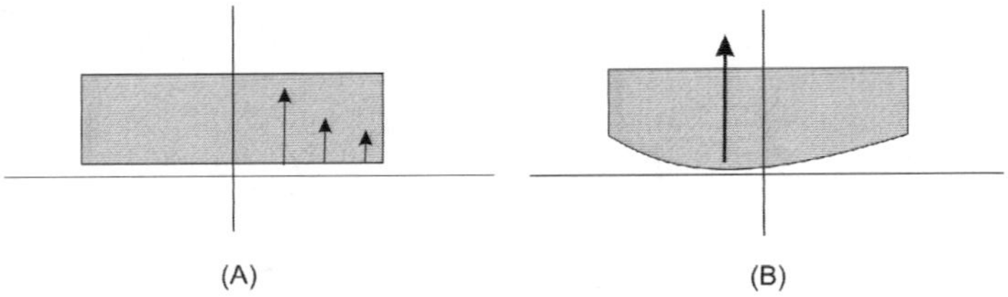

Figure 6. Inhomogeneous wearing (B) naturally leads to more load and friction in those regions that have previously lost less mass, and that are, as a result, more exposed. Ultimately, the stationary configuration is that of homogeneous wearing as illustrated in figure 7.

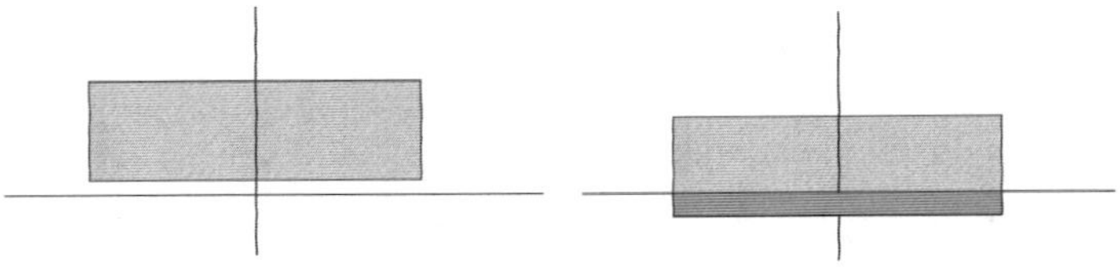

Figure 7. Homogeneous wearing follows from the request that the axis does not tilt.

2. Variable mass system

So far we were concerned with the problem of the pressure distribution and we assumed, as it is reasonable in practical problems, that the loss of mass is negligible when it comes to consider the dynamics of a body. Let us study the dynamical problem by assuming that Reye's law is

Journal of Physics: Conference Series **2533** (2023) 012038 doi:10.1088/1742-6596/2533/1/012038

exact, also well beyond the regimes of velocity in which it has been verified, and let us solve the dynamical equations with all rigor, by using the laws of classical mechanics.

Rigorously speaking, a body that loses mass should be treated as a variable mass system and hence the first cardinal equation of dynamics should be replaced by the Painlevé and Seeliger formula (1890). This formula follows from the conservation of momentum and is familiar from rocket theory [17]. In the block example we shall write (see Fig. 8)

$$m\dot{\mathbf{v}} = \mathbf{F}_f - \dot{m}(\mathbf{v} - \mathbf{v}_d) \tag{2}$$

where $\mathbf{v}_d$ is the velocity of the debris after detachment. In our case the debris do not move with respect to the hard body A (the horizontal plane) and so $\mathbf{v}_d = 0$.

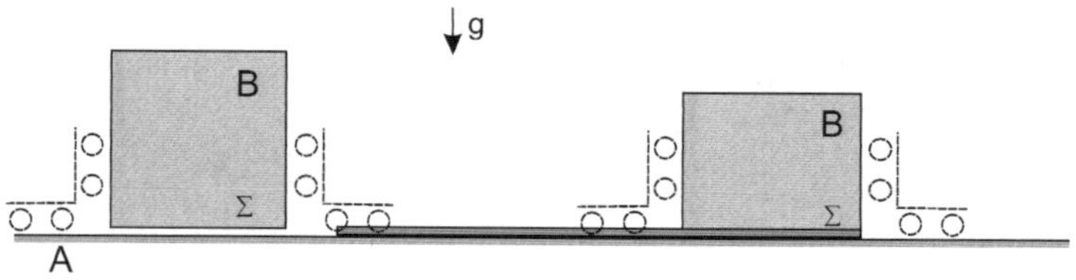

Figure 8. The block loses mass. The debris come to rest in the reference frame of the horizontal plane (hard body A).

Denoting with $\mathbf{p} := m\mathbf{v}$ the linear momentum we obtain the system of equations

$$\dot{m} = \frac{1}{c^2}\mathbf{F}_f \cdot \boldsymbol{v} = -\frac{1}{c^2}\, a_f p, \quad \text{(Reye)} \tag{3}$$

$$\dot{\mathbf{p}} = \mathbf{F}_f = -a_f m\hat{\mathbf{v}}. \quad \text{(Painlevé and Seeliger)} \tag{4}$$

where we used the formula for Coulomb's dynamical friction $\mathbf{F}_f = -\mu m g\hat{\mathbf{v}} = -m a_f\hat{\mathbf{v}}$ and where we introduced the constant $a_f := \mu g$, which is a characteristic acceleration related to friction. Actually many of the below calculations generalize to a_f dependent on velocity, and hence to Stokes friction.

2.1. A puzzling result

Multiplying Eq. (3) by m, multiplying scalarly Eq. (4) by $\frac{1}{c^2}\mathbf{p}$, and taking the difference we obtain that there is a constant $\tilde{m}$ such that

$$m^2 - \frac{1}{c^2}\mathbf{p}^2 = \tilde{m}^2, \tag{5}$$

or equivalently

$$m = \frac{\tilde{m}}{\sqrt{1-\frac{v^2}{c^2}}}, \qquad v < c \qquad if\ \tilde{m} > 0, \tag{6}$$

$$mc^2 = pc, \qquad v = c, \qquad if\ \tilde{m} = 0, \tag{7}$$

$$m = \frac{|\tilde{m}|}{\sqrt{\frac{v^2}{c^2}-1}}, \qquad v > c \qquad if\ \tilde{m} \text{ is imaginary}, \tag{8}$$

Observe that both $v(t)$ and $m(t)$ are time dependent, due to the presence of friction and wear. The just obtained result tells us that they maintain a relationship in time.

Journal of Physics: Conference Series **2533** (2023) 012038 doi:10.1088/1742-6596/2533/1/012038

As we can see, such relationship is formally coincident with the relativistic mass formula, where in the parallelism Reye's constant c corresponds to the speed of light. The three cases might be called bradyonic, luxonic and tachyonic.

The physical interpretation of $\tilde{m}$ in the bradyonic case can be read off the above formula, that is, $\tilde{m}$ is the mass for $v \to 0$, namely the mass that the block will have after wearing by friction is completed, i.e. after the block has come to rest with respect to the frame of the horizontal plane.

From the previous equations, taking into account that the mass m is always positive, we obtain

$$\dot{\mathbf{v}} = (\mathbf{F}_f - \dot{m}\mathbf{v})/m = -a_f(1 - \frac{v^2}{c^2})\hat{\mathbf{v}}. \tag{9}$$

This equation shows that if the body has velocity c it keeps moving in straight line preserving it (luxonic case). At that speed the block velocity would be basically insensitive to friction. However, mass and linear momentum decrease (recall that $p = mv$)

$$p(t) = p(0)\, e^{-\frac{1}{c}a_f t}, \qquad \text{for } v = c. \tag{10}$$

In the bradyonic casee Eq. (9) can be integrated and gives for $a_f \neq 0$

$$v(t) = c \tanh(\varphi - \frac{a_f}{c}\, t), \qquad \text{for } 0 \le v < c, \tag{11}$$

where φ is an integration constant (interpreted as the rapidity) related to the initial velocity ($c \tanh \varphi$). We observe that for the chosen Coulomb friction the time needed to reach a complete stop is finite, namely $t_f = c\varphi/a_f$.

The dependence of mass on time is

$$m(t) = \tilde{m} \cosh(\varphi - \frac{a_f}{c}\, t), \qquad \text{for bradyonic case.} \tag{12}$$

It becomes equal to the mass $\tilde{m}$ when the body comes to rest in A.

Let us integrate once more in the bradyonic case

$$x(t) = -\frac{c^2}{a_f} \ln \cosh(\varphi - \frac{a_f}{c}\, t) + x_f = -\frac{c^2}{a_f} \ln \cosh\left(\frac{a_f}{c}(t_f - t)\right) + x_f \tag{13}$$

where x_f is the coordinate of the stopping point and t_f is the stopping time.

The above 'relativistic mass formula' (6) follows just from the principles of ordinary classical mechanics and is certainly correct for small v, as Reye's law holds in that limit. Light, electromagnetism and related concepts did not enter the previous derivation. We might say that pre-1905 ordinary classical mechanics might itself suggest to investigate theories which embody mathematical results characteristic of relativistic physics.

3. Hints at new physics

The study of the previous section seems to suggest that there might be a relationship between friction and relativistic physics, in the sense that both phenomenologies might follow from a more general theory. The idea would be to develop a theory in which bodies are affected by a universal friction, and in which mass varies according to Reye's law.

The CMB radiation suggests that some privileged frame might exist after all, a frame that might exert some sort of friction on moving bodies. This friction might be extremely small but still have consequences for the formal structure of the theory.

Let us pause to recall that Newtonian Mechanics is based on the principle of Galileian invariance. As a result, the spacetime of Newtonian mechanics is a 4-dimensional affine space M endowed with a time projection $\pi : M \to T$; the space S being the model for the fibers $S_t = \pi^{-1}(t)$.

Actually, most authors develop Newtonian mechanics as Newton originally did, namely by working with a spacetime that splits as a product of space and time: $M = S \times T$. This is actually the mathematical structure of an Aristotelian spacetime with an absolute notion of 'space' and hence of 'rest' (the wordline (P, t) represents a point at 'absolute rest'). This approach to Newtonian mechanics certainly simplifies the mathematics of the theory, though the use of such an Aristotelian spacetime appears as artificial and unsatisfactory: the absolute space, which is formally present, is actually unobservable as the laws of Physics, with their Galileian invariance, do not allow for its determination.

It is interesting to observe that the original Aristotelian mechanics, with its spacetime $S \times T$, was epistemologically more satisfactory than Newtonian mechanics in the above simplified form. Indeed, Aristotle imposed that bodies not exerted upon by forces come ultimately to rest in the absolute frame. The absolute space of Aristotelian mechanics is therefore observable.

It is often claimed that Aristotle, quite naively, did not realize that the natural state of motion of bodies is that of uniform motion, as he did not recognize the role of friction forces. However, this is not a fair historical account. As Dugas [18] points out, Aristotle was very well aware that probably, in vacuum, a body would have moved in uniform motion indefinitely. However, from that he inferred that absolute vacuum is impossible. In other words, it is not because Aristotle did not recognize the role of friction that he did not arrive at the relativity principle, but rather because he regarded the principle that every body should come to absolute rest as more fundamental than the relativity principle. Indeed, some kind of deceleration is necessary in order to give physical observability to the concept of absolute space which Aristotle was not prepared to abandon. Stated in another way, Aristotle regarded as more important the spacetime structure $S \times T$, and hence the absolute space idea, rather than the relativity principle. For this reason he had to assume the ultimate deceleration of all bodies so that the absolute rest frame could become observable.

We see that while in Galileian mechanics friction has a negative connotation, as it spoils the uniform motion of bodies, in Aristotelian mechanics it has a positive connotation, has it is the main element that makes the absolute space observable and hence the very concept of absolute space epistemologically well posed.

A theory embodying a universal friction is naturally a theory that involves some absolute reference frame and conversely. Our sought theory is then going to share these features and additionally is going to interpolate between Special Relativity and Aristotelian mechanics.

So our idea is to explore a theory with the following features

(a) With respect to the previous treatment of the free block, it lives in one more space dimension, i.e. the surface of friction is now a volume. The normal force is, so to say, in the extra-direction, but we are concerned only to what happens in the observable 3-space.

(b) Friction acts over all bodies as in Eq. (4) via a coefficient a_f that is universal and unrelated to the chemical composition of the matter involved. Save for light, all bodies come ultimately to rest with respect to an absolute reference frame S_0, their mass m converging to the their absolute rest mass $\tilde{m}$.

(c) Mass is lost as in Eq. (3) and Reye's constant c is identified with the speed of light.

Mathematically it develops as a straightforward generalization of the previous theory for the moving block, where the equations there obtained are reinterpreted as applying to general bodies in the new theory. Of course, by proceeding in this way we shall miss a field theoretical description. Still this approach might give insights into a theory sharing the above features.

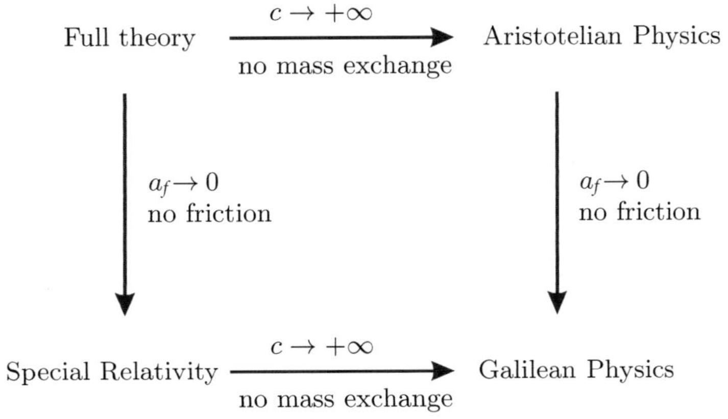

Figure 9. Limits of the theory.

The theory depends on 2-parameters a_f and c and, as mentioned, shall have two special limits, which are:

SR: For $a_f \to 0$ friction goes to zero and the absolute space ceases to exist (observe that for $a_f > 0$ we can recognize the absolute space as eventually all bodies come to rest with respect to it). In this limit the natural state of bodies is that of uniform motion (relativity principle), however, since the relativistic mass formula (6) does not depend on a_f and is preserved in the limit, one expects to recover Special Relativity rather than Galileian Relativity.

AM: For $c \to \infty$ we have by Reye's law $\dot{m} = 0$, masses are constant in time but there is still friction and the theory reduces to Aristotelian mechanics with a universal friction.

If from SR we let $c \to \infty$ we naturally obtain Galileian mechanics (GM), and similarly, from AM with $a_f \to 0$ we obtain also Galileian mechanics as bodies no more slow down to absolute rest. We arrive at the diagram of Fig. 9.

Let us show that the Lorentz transformation can be generalized so as to depend on two parameters c and a_f and so as to satisfy this scheme. Of course, we are not looking for a group of transformations as the absolute rest frame S_0 will be privileged. The group property should be recovered only for $a_f \to 0$, namely when the privileged frame becomes unobservable. We shall see in a moment that this limit group is indeed the Lorentz group.

So, let S_0 be the absolute frame that parametrizes events with coordinates (t_0, x_0, y_0, z_0), and let S be a moving frame of coordinates (t, x, y, z) whose origin moves on the abscissa of S_0.

The motion of the origin of S satisfies Eqs. (11), (13), which shows that S will come to rest with respect S_0 at time $t_f = \frac{c\varphi}{a_f}$, its velocity at $t = 0$ being $c \tanh\varphi$.

The change of frame for $t_0 < t_f$ is[1]

$$\begin{aligned} t &= -\frac{c}{a_f} \arcsin\tanh(\varphi - \frac{a_f}{c} t_0) + \frac{c}{a_f} \arcsin\tanh(\varphi), \\ x &= \cosh(\varphi - \frac{a_f}{c} t_0)\Big(x_0 + \frac{c^2}{a_f} \log \frac{\cosh(\varphi - \frac{a_f}{c} t_0)}{\cosh\varphi}\Big). \end{aligned} \quad (14)$$

It is such that $(0, 0, 0, 0) \to (0, 0, 0, 0)$, namely the frame origins meet when the respective clocks sign zero time. As mentioned, φ is an integration constant related to the velocity of

[1] For shortness, we omit in the transformation the equations $y = y_0$, $z = z_0$, that should nevertheless be included here as well as in Eqs. (15), (16), (17, (18).

the frame S with respect to S_0 at the instant when the origins meet (this velocity is $c\tanh\varphi$). Setting $x = y = z = 0$, namely studying the motion of the origin of S, one easily obtains $x_0(t) = -\frac{c^2}{a_f}\log\frac{\cosh(\varphi - \frac{a_f}{c}t_0)}{\cosh\varphi}$, which is of the form (13), with the integration constant x_f chosen so that $x_0 = 0$ for $t_0 = 0$.

For $t_0 \geq t_f$, S is at rest with respect to S_0 and the transformation is just:

$$\begin{aligned} t &= t_0, \\ x &= x_0 - \frac{c^2}{a_f}\log\cosh\varphi. \end{aligned} \tag{15}$$

Having established the compatibility of this transformation with our previous formulas, let us study its limits.

3.1. Aristotelian limit, $c \to \infty$

Let us first consider the (Aristotelian) limit $c \to \infty$, which corresponds to friction without mass loss (cf. Reye's law). Since the relative velocity when the origins meet is $v = c\tanh\varphi$, this limit should be taken while taking also the limit $\varphi \to 0$, and in such a way that $c\varphi \to v$ for some constant v to be interpreted as the initial relative velocity. In this limit from (14) we easily get for $t_0 < v/a_f$

$$\begin{aligned} t &= t_0, \\ x &= x_0 - \left(vt_0 - \frac{1}{2}a_f t_0^2\right). \end{aligned} \tag{16}$$

Observe that this is precisely the Galileian transformation from S_0 to a reference frame that is slowing down with acceleration a_f. For $t_0 \geq v/a_f$, the frame has come to rest with respect to S_0. Indeed, we get taking the limit of (15)

$$\begin{aligned} t &= t_0, \\ x &= x_0 - \frac{1}{2}\frac{v^2}{a_f}. \end{aligned} \tag{17}$$

In other words we obtain the transformation of Aristotelian mechanics in which all bodies undergo a universal deceleration till reaching absolute rest.

3.2. Special Relativity limit, $a_f \to 0$

Let us take the limit $a_f \to 0$ in Eq. (14). Observe that $t_f = \frac{c\varphi}{a_f} \to \infty$, thus we have just to consider the case $t_0 < t_f$. Let us set $v := c\tanh\varphi$, $\beta := v/c$, then we obtain from (14) in the limit

$$\begin{aligned} t &= \gamma^{-1}(\beta)\, t_0, \\ x &= \gamma(\beta)\, (x_0 - vt_0)\, . \end{aligned} \tag{18}$$

This is the *Tangherlini transformation.* It is equivalent to the Lorentz transformation of Special Relativity, the only difference is a resychronization of clocks in the moving frame S

$$t_{SR} = t - \frac{v}{c^2}x, \qquad x_{SR} = x. \tag{19}$$

The Tangherlini transformation [19, 20] has been much studied up to this day [21–23]. Tangherlini's work, popularied by Mansouri and Sexl [24], was anticipated by Scott-Iversen [25] and other authors working on non-standard synchronizations in Special Relativity.

Journal of Physics: Conference Series **2533** (2023) 012038 doi:10.1088/1742-6596/2533/1/012038

Clocks in Tangherlini's formalism are set up in such a way that the same notion of simultaneity becomes shared by all the frames. As a consequence, the speed of light is not isotropic in S (notice that the one-way speed of light depends on the synchronization adopted, as it requires two separate clocks for its measurement, while the two-way speed of light remains unaltered, see [26, 27]). Of course, in Tangherlini's formalism such a resychronization does not alter the Special Relativity physics, just its coordinate description.

4. Comparison with observations

The objective of this section is to explore the consequences of the above theory for cosmology. We will show that it predicts the Pioneer anomaly and the Hubble law, where the latter is explained through a kind of universal tired light mechanism. Couriously, the model naturally explains the observed coincidence between the Hubble constant and the Pioneer anomalous acceleration.

More generally, this model seems to be able to explain many odd phenomena that have been observed for accelerations smaller or of the order of a certain critical value a_f, although it is not able to explain other observations like the supernova light curves. Still, these results hint at the possible role of friction in the resolution of some puzzles of modern cosmology.

4.1. Pioneer anomaly

A first obvious consequence of our model is that free bodies with small velocities compared to c should show a deceleration of magnitude a_f in the absolute reference frame.

Indeed, any body should decelerate with an acceleration of magnitude a_f. The Pioneer spacecrafts do present an unmodeled acceleration approximately directed towards the Sun of magnitude [28]

$$a_P = (8.74 \pm 1.33) \times 10^{-10} m/s^2.$$

Unfortunately, according to a recent study, the anomalous acceleration seems to be non collinear with the spacecraft velocity [29], and can be accounted for by thermal radiation [30].

Also the friction model predicts a deceleration of the planets of the solar system, which decreases the radius of their orbits and hence implies an increase in the modulus of their velocity [31] (in the end $\dot{v} = a_p$, that is, the sign that one would naively expect gets inverted under Coulomb friction). However, this effect is quite small and can be cancelled by the decrease of the Sun mass due to solar wind [32].

4.2. Hubble law

Let us now consider the distribution of galaxies. Because of friction, if they are close, say at the level of the local group, they are almost at rest. Nevertheless, by a tired light effect the Hubble law still holds. Indeed, a photon sent at time t from a galaxy and received at time $t + \Delta t$ in the Milky way undergoes a redshift (see Eq. (10))

$$1 + z = \frac{E(t)}{E(t + \Delta t)} = e^{\frac{1}{c} a_f \Delta t} \simeq 1 + \frac{1}{c} a_f \Delta t.$$

If this tired light explanation for the Hubble law is correct then we must find that the observed value

$$a_H := cH = (6.9 \pm 0.9) \times 10^{-10} m/s^2$$

(if $H = (72 \pm 8)(km/s)/Mpc$) coincides with a_f, and if the above explanation for the Pioneer anomaly is correct we must expect $a_f = a_P$ and in the end

$$a_H = a_P.$$

Journal of Physics: Conference Series **2533** (2023) 012038 doi:10.1088/1742-6596/2533/1/012038

The above figures seem to confirm this prediction. This equality has been noticed by many authors [28], but this appears to be the simplest model which accounts for it. This explanation is the more striking as it appears to be compatible with Special Relativity.

We obtained a kind of tired light explanation of the Hubble law, however, contrary to usual tired light theories, it does not assume that the loss of energy is due to scattering with diffuse interstellar matter. The latter assumption would imply a modification in the direction of the photon, so that any galaxy would be seen as blurred and indefinite, contrary to observations. The universal friction mechanism preserves the direction of the photon and so avoids this problem.

Unfortunately, tired light explanations of the Hubble law cannot account for the supernova light curves [33]. Perhaps the Hubble law could be due to two mechanisms, namely tired light for sufficiently close cosmological objects (with the idea that close objects belong to the same vacuum state, the vacuum state playing the role of the hard body in the mechanical analogy) and cosmic expansion for far away objects (as they belong to different vacuums states, different states diverging from each other). In other words, in the empty region between two vacuum patches there would be no friction and so the dynamics would be completely relativistic (recall that the theory becomes relativistic for $a_f \to 0$, and by Eq. (10) there is no tired light effect in that limit). There would still be a redshift effect but due to the relative motion of the vacuum states. In general one would not expect these two redshift effects to match precisely so giving different Hubble constants depending on whether observations are local or cosmological. This tension has indeed been observed and is the subject of current debate.

If the Hubble law is at least locally due to tired light, then it should be expected to hold for close celestial objects as well. The expanding universe theory predicts the Hubble law at length scales which are well beyond the scale of homogeneity for which a Friedmann-Robertson-Walker approximation of the cosmological metric would make sense. A well known puzzle in cosmology is the "coldness of the Hubble flow" namely the observation that the Hubble law holds unreasonably well at the scale of the local group (1-10Mpc) with a velocity scatter with respect to the Hubble flow which is very small (40 km/s), although at that scale the matter distribution is very clumped [34]. A local tired light model for the Hubble law accounts for this observation quite easily, for according to this explanation the local Hubble law does not depend on an expansion dynamics.

5. Conclusions

Inspired by experimental evidence on how friction manifests itself in the wear of bodies, I introduced an absolute space theory in which mass decreases proportionally to the work done by a universal friction force. I showed that mass depends on velocity as in Special Relativity and that there is a characteristic velocity c which is insensitive to friction. Bodies with velocity smaller than c decelerate till they reach a status of absolute rest. Bodies with velocity c preserve their velocity while they lose mass. I argued that in the limit of vanishing friction, $a_f \to 0$, the theory becomes equivalent to Special Relativity (in its Tangherlini coordinate description) as the underlying absolute space becomes unobservable. This result shows that Special Relativity, often regarded as incompatible with the concept of absolute frame, could have been suggested by more detailed studies of the interaction of bodies with such a frame.

I ended the paper with the exploration of some predictions of this theory. Qualitatively, and sometimes also quantitatively, it seems to be able to account for some puzzling cosmological observations. The offered explanations are non-traditional and certainly require further details and clarifications. Still the results of this work suggest that friction might have indeed an important role in cosmology, a role that in my opinion should be explored further.

Acknowledgments
I previously posted the work arXiv:1412.0010 that presented some of the arguments of this paper.

References

[1] Minguzzi E 2019, Lorentzian causality theory, *Living Rev. Relativ.* **22** 3
[2] Reye T 1860, Zur Theorie der Zapfenreibung, *J. Der Civilingenieur* **6** 235–255
[3] Giovannini M 1935, Sulla prova di resistenza dei materiali all'usura per attrito radente, *L'Ingegnere* **13**(12) 489–496
[4] Opatowski I 1942, A theory of brakes, an example of a theoretical study of wear, *Journal of the Franklin Institute* **234**(3) 239–249
[5] Archard J 1953, Contact and rubbing of flat surfaces, *J. Appl. Phis.* **24** 981–988
[6] Villaggio P 2001, Wear of an elastic block, *Meccanica* **36** 243–249
[7] Panetti M 1908, Contributo alla teoria dei freni ad attrito, *Giornale del Genio Civile* **46** 641–658
[8] Panetti M 1947 *Meccanica Applicata* (Torino: Levrotto & Bella)
[9] Curioni G 1884 *Cenni storici e statistici sulla Scuola d'Applicazione per gl'Ingegneri fondata in Torino nell'anno 1860* (Torino: Tipografia editrice G. Candeletti)
[10] Cappa S 1890 *Lezioni di Meccanica applicata alle macchine* (Torino: Tip. e Lit. C. Giorgis)
[11] Ovazza E 1890, Sulla resistenza di attrito fra vite e madrevite, *Atti della R. Accademia delle Scienze di Torino* **26** 215–235
[12] Pomini O 1922 *Costruzione di macchine* (Milano: U. Hoepli)
[13] Rubino M 1944 *Meccanica applicata alle macchine* (Palermo: Libreria Agate, Scuola Tip. 'Boccone Del Povero')
[14] Ferrari C and Romiti A 1966 *Meccanica applicata alle macchine* (Torino: Unione Tipografico-Editrice Torinese)
[15] Funaioli E 1973 *Corso di meccanica applicata alle macchine Vol. I* 3rd ed (Bologna: Pàtron)
[16] Weisbach J 1870 *A manual of the mechanics of engineering of the construction of machines, with an introduction to the calculus* (New York: D. Van Nostrand)
[17] Irschik H and Holl H 2004, Mechanics of variable-mass systems-Part 1: Balance of mass and linear momentum, *Appl. Mech. Rev.* **57** 145–160
[18] Dugas R 1955 *A history of mechanics* (London: Routledge and Kegan Paul LTD.)
[19] Tangherlini F R 2009, The velocity of light in uniformly moving frame. A dissertation. Stanford University (1958), *The Abraham Zelmanov Journal* **2** 44–110
[20] Malykin G B and Malykin E G 2009, Tangherlini's dissertation and its significance for physics of the 21th century, *The Abraham Zelmanov Journal* **2** 121–147
[21] Guerra V and de Abreu R 2006, On the consistency between the assumption of a special system of reference and Special Relativity, *Found. Phys.* **36** 1826–1845
[22] de Abreu R and Guerra V 2008, The principle of relativity and the indeterminacy of Special Relativity, *Eur. J. Phys.* **29** 33–52
[23] Burde G 2018, Special Relativity with a preferred frame and the relativity principle, *J. Mod. Phys.* **9** 1591–1616
[24] Mansouri R and Sexl R 1977, A test theory of Special Relativity: I. Simultaneity and clock synchronization, *Gen. Relativ. Gravit.* **8** 497–513

[25] Scott-Iversen P A 1944, XIII. Introductory notes on a reformulation of the Special Theory of Relativity, *The London, Edinburgh, and Dublin Philosophical Magazine and Journal of Science* **35** 105–120

[26] Minguzzi E 2002, On the conventionality of simultaneity, *Found. Phys. Lett.* **15** 153–169

[27] Minguzzi E and Macdonald A 2003, Universal one-way light speed from a universal light speed over closed paths, *Found. Phys. Lett.* **16** 593–604

[28] Anderson J D, Laing P A, Lau E L, Liu A S, Nieto M M and Turyshev S G 2002, Study of the anomalous acceleration of Pioneer 10 and 11, *Phys. Rev. D* **65** 082004

[29] Turyshev S G, Toth V T, Ellis J and Markwardt C B 2011, Support for temporally varying behavior of the Pioneer anomaly from the extended Pioneer 10 and 11 Doppler data sets, *Phys. Rev. Lett.* **107** 081103

[30] Turyshev S G, Toth V T, Kinsella G, Lee S C, Lok S M and Ellis J 2012, Support for the thermal origin of the Pioneer anomaly, *Phys. Rev. Lett.* **108** 241101

[31] Parkyn D G 1958, The effect of friction on elliptic orbits, *The Mathematical Gazette* **42** 96–98

[32] Krasinsky G and Brumberg V 2004, Secular increase of astronomical unit from analysis of the major planet motion, and its interpretation, *Celestial Mechanics and Dynamical Astronomy* **90** 267–288

[33] Leibundgut B et al 1996, Time dilation in the light curve of the distant type Ia supernova SN 1995K, *Astrophys. J.* **466** L21–L26

[34] Baryshev Y V, Chernin A D and Teerikorpi P 2001, The cold local Hubble flow as a signature of dark energy, *Astronomy and Astrophysics* **378** 729–734

Journal of Physics: Conference Series **2533** (2023) 012039 doi:10.1088/1742-6596/2533/1/012039

Indefinite causality in quantum mechanics and its thermodynamic applications

Kyrylo Simonov

Fakultät für Mathematik, Universität Wien, Oskar-Morgenstern-Platz 1, 1090 Vienna, Austria

E-mail: `kyrylo.simonov@univie.ac.at`

Abstract. The nature of causality remains one of the key puzzles in science. In quantum theory, the causal structure is not subject to quantum uncertainty and plays rather a background role. One can ask whether the background causal structure can be dropped, for example, by respecting causality only locally. Such scenarios of local validity of quantum theory while relaxing the global definite causal order of operations can be described via the machinery of process matrices. An important example of scenarios of this kind is quantum SWITCH, a process realizing a quantum superposition of causal orders of operations. Looking for the possible applications of quantum SWITCH has been the subject of growing interest in the scientific community as it could provide communication and computational resources not realizable via standard quantum theory. In the last few years, the benefits potentially offered by quantum SWITCH for thermodynamic tasks have appeared in the spotlight. This contribution aims at reviewing the recent proposals of thermodynamic applications of quantum SWITCH and draw the perspectives.

1. Introduction

It is well known that causality plays a fundamental role in our everyday life. In particular, the causal nature of our actions is one of the first things we learn about. Nevertheless, the fundamental status of causality, being famously criticized by Bertrand Russell [1], still remains a subject of debates and one of the key puzzles in physics and philosophy. In particular, one of the difficulties with the unification of quantum theory with general relativity could arise from the very notion of time and the place of causality in quantum theory. Recently, there have appeared theoretical constructions that challenge the fixed global causal structure assumed by standard quantum theory. A significant breakthrough in this direction has been achieved by Hardy that focused on theories that borrow both indeterminism of quantum theory and dynamic treatment of space-time geometry of general relativity and formulated an operational approach to such theories known as the causaloid framework. In such theories, just as dynamic quantities manifest uncertainty in quantum theory, space-time geometry is treated as indefinite, leading thus to indefinite causal structures [2]. Later on, Chiribella et al. have shown that quantum theory allows for scenarios which respect causality only locally by constructing a process with quantum operations executed without fixed causal order, the quantum SWITCH [3]. Description of such processes without pre-assumed fixed causal order of quantum operations has been encapsulated by Oreshkov, Costa, and Brukner within the process matrix formalism [4]: its development has demonstrated that indefinite causal structures can be incorporated to quantum theory without abandoning its linear structure.

Quantum SWITCH has provided an important paradigmatic example of a process realizing indefinite causal order of operations since it has been implemented in table-top experiments [5, 6]. These have significantly increased interest in indefinite causal structures and their possible benefits, first of all, for quantum information processing. Nevertheless, despite intimate connection between information and thermodynamics, potential benefits of the quantum SWITCH for thermodynamic tasks remained in shadow and have been put under scrutiny only very recently.

Quantum SWITCH relaxes the implied causal relations between quantum operations by making their application order subject to superposition. This is achieved by coherently controlling the order of operations by another quantum system which gets entangled with the application orders of the operations. For two operations $\mathcal{A}[.]$ and $\mathcal{B}[.]$, the control system is a qubit in state ω, and the resulting evolution of both target system (prepared in a state ρ) and control qubit can be represented as a new quantum operation

$$\mathcal{S}_{\mathcal{A},\mathcal{B}}[\rho\otimes\omega]=\sum_{ij}K_{ij}(\rho\otimes\omega)K_{ij}^{\dagger}, \tag{1}$$

whose Kraus decomposition is given by the operators $K_{ij} = A_iB_j \otimes |0\rangle\langle 0| + B_jA_i \otimes |1\rangle\langle 1|$, where the sets of operators $\{A_i\}$ and $\{B_j\}$ constitute the Kraus decompositions of $\mathcal{A}$ and $\mathcal{B}$, respectively. Hence, if the control qubit is prepared in a computational basis state $|0/1\rangle$, quantum SWITCH places the channels into a specific order in which they act on the system, namely $(\mathcal{A}\circ\mathcal{B})[.]$ or $(\mathcal{B}\circ\mathcal{A})[.]$, respectively. As a consequence of that, any incoherent state of the control qubit (with respect to computational basis), i.e., $\omega = \mathrm{diag}(\phi, 1-\phi)$, realizes a convex combination of the composite channels,

$$\mathcal{S}_{\mathcal{A},\mathcal{B}}\left[\rho\otimes\begin{pmatrix}\phi & 0\\ 0 & 1-\phi\end{pmatrix}\right]=\phi\Big((\mathcal{A}\circ\mathcal{B})[\rho]\otimes|0\rangle\langle 0|\Big)+(1-\phi)\Big((\mathcal{B}\circ\mathcal{A})[\rho]\otimes|1\rangle\langle 1|\Big), \tag{2}$$

i.e., the channels are applied in a random order chosen, for example, by tossing a coin. However, any coherence in the state of the control qubit inevitably results in a combination of the $\mathcal{A}[.]$ and $\mathcal{B}[.]$ not of the form of Eq. (2) and incompatible with any well-defined causal order between them. In particular, for identical channels $\mathcal{A}[.] = \mathcal{B}[.]$, the action of quantum SWITCH can be given as

$$\mathcal{S}_{\mathcal{A},\mathcal{A}}[\rho\otimes\omega]=\frac{1}{4}\sum_{ij}\Big(\{A_i,A_j\}\rho\{A_i,A_j\}^{\dagger}\otimes\omega+[A_i,A_j]\rho[A_i,A_j]^{\dagger}\otimes Z\omega Z\Big), \tag{3}$$

where $[A_i, A_j] = A_iA_j - A_jA_i$ denotes the commutator, $\{A_i, A_j\} = A_iA_j + A_jA_i$ denotes the anti-commutator, and Z denotes the Pauli Z-operator.

2. Thermodynamic protocols with quantum-controlled thermalization

One of the most striking effects of quantum SWITCH is its ability to reduce noise in communication channels (up to its complete removal for certain channels), so that information can be transmitted even via zero capacity channels [7]. A particular example of such channels is provided by a completely depolarizing channel $\mathcal{T}_0[\rho] = \frac{I}{d}$, where I denotes a d-dimensional unity operator, that outputs a d-dimensional maximally mixed state for any d-dimensional input state ρ. From the thermodynamic point of view, the action of $\mathcal{T}_0[.]$ can be seen as a result of interchanging the states of the system and a bath at infinite temperature. Hence, it can be generalized to a family of channels $\{\mathcal{T}_\beta[.]\}_\beta$ known as thermalizing channels: they model complete thermalization of the system S with a thermal bath B at inverted temperature β swapping their states

$$\mathcal{T}_\beta[\rho]=\mathrm{Tr}_B[U_{SB}(\rho\otimes\tau_\beta)U_{SB}^{\dagger}]=\mathrm{Tr}_B[\tau_\beta\otimes\rho]=\tau_\beta, \tag{4}$$

where U_{SB} is the SWAP operator between S and B, and $\tau_\beta = \frac{e^{-\beta H}}{\mathrm{Tr}[e^{-\beta H}]}$ is a thermal state with respect to β and Hamiltonian H.

While several identical thermalizing channels $\mathcal{T}_\beta$ acting consequently on the system leave it eventually in the same thermal state τ_β, putting them into indefinite causal order via quantum SWITCH can result in an output state different from τ_β. In particular, let us consider a qubit system with Hamiltonian $H = \begin{pmatrix} 0 & 0 \\ 0 & 1 \end{pmatrix}$, which is initially prepared in a thermal state of temperature β_{in}. If the control qubit is initially prepared in the state $|+\rangle = \frac{|0\rangle+|1\rangle}{\sqrt{2}}$, quantum SWITCH outputs

$$\mathcal{S}_{\mathcal{T}_\beta,\mathcal{T}_\beta}[\rho \otimes |+\rangle\langle+|] = \frac{1}{2}\Big((\tau_\beta + \tau_\beta\tau_{\beta_{in}}\tau_\beta) \otimes |+\rangle\langle+| + (\tau_\beta - \tau_\beta\tau_{\beta_{in}}\tau_\beta) \otimes |-\rangle\langle-|\Big).$$

Therefore, if the control qubit is measured in the Fourier basis $\{|\pm\rangle\} = \{\frac{|0\rangle\pm|1\rangle}{\sqrt{2}}\}$, the system will be found in a state

$$\begin{aligned} \rho_\pm &= \frac{1}{2p_\pm}(\tau_\beta \pm \tau_\beta\tau_{\beta_{in}}\tau_\beta) \\ &= \frac{1}{2p_\pm(1+e^{-\beta})}\begin{pmatrix} 1 \pm \frac{1}{(1+e^{-\beta})(1+e^{-\beta_{in}})} & 0 \\ 0 & e^{-\beta}\Big(1 \pm \frac{e^{-(\beta+\beta_{in})}}{(1+e^{-\beta})(1+e^{-\beta_{in}})}\Big) \end{pmatrix}, \end{aligned} \tag{5}$$

with probability

$$p_\pm = \frac{1}{2}\,\mathrm{Tr}[\tau_\beta \pm \tau_\beta\tau_{\beta_{in}}\tau_\beta] = \frac{1}{2}\Big(1 \pm \frac{1+e^{-(2\beta+\beta_{in})}}{(1+e^{-\beta})^2(1+e^{-\beta_{in}})}\Big). \tag{6}$$

These output states can have temperature different from β or even be non-thermal. This important observation can be exploited to construct several novel thermodynamic protocols, which we review in the following, assuming the considered qubit system with the Hamiltonian H as working medium.

2.1. Cooling cycle from quantum-controlled thermalization

Among the first thermodynamic protocols that exploit the mentioned property of quantum-controlled thermalizing channels (and quantum control of causal orders of operations in general) is the refrigeration cycle proposed by Felce and Vedral [8] and implemented within several experimental setups [9, 10] as well as on the IBM cloud quantum computer [11]. It starts with working medium in thermal state at temperature $\beta_{in} \equiv \beta_c$ which gets thermalized in indefinite causal order with two identical cold reservoirs at the same temperature $\beta \equiv \beta_c$. Performing a measurement of the control qubit in the Fourier basis, the working medium can be found in one of the states (5) which are thermal of temperatures

$$\beta_+ = \beta_c + \ln\Big(1 + \frac{1-e^{-2\beta_c}}{1+2e^{-\beta_c}+2e^{-2\beta_c}}\Big) \geq \beta_c, \tag{7}$$

$$\beta_- = \ln\Big(1 + \frac{1-e^{-\beta_c}}{1+2e^{-\beta_c}}\Big) \leq \beta_c, \tag{8}$$

respectively. This means that after quantum-controlled thermalization the working medium is cooled or heated with the corresponding probability $p_\pm = \frac{1}{2}\Big(1 \pm \frac{1+e^{-3\beta_c}}{(1+e^{-\beta_c})^3}\Big)$. If it is heated, then heat can be consequently transferred from the working medium by thermalizing it with another

Journal of Physics: Conference Series **2533** (2023) 012039 doi:10.1088/1742-6596/2533/1/012039

(hot) reservoir at a suitable temperature $\beta_- < \beta_h < \beta_c$. The cycle can be closed by the final stroke which consists in thermalization of the working medium with a cold reservoir (regardless of outcome of the control qubit measurement) and thermalization of the cold reveservoirs with each other. The overall heat transfer from the cold reservoirs, working medium, and control qubit is given by the heat transfer to the hot reservoir

$$\begin{aligned} Q &= p_- \operatorname{Tr}[(\rho_- - \rho_h)H] \\ &= p_- \Big(\frac{e^{-\beta_-}}{1+e^{-\beta_-}} - \frac{e^{-\beta_h}}{1+e^{-\beta_h}} \Big) \\ &= p_- \Big(\frac{1+2e^{-\beta_c}}{3(1+e^{-\beta_c})} - \frac{e^{-\beta_h}}{1+e^{-\beta_h}} \Big), \end{aligned} \tag{9}$$

and can be assigned to a heat flow from the cold reservoirs to the hot one. In accordance with Landauer's principle, each run of cycle requires investing work in order to erase the register that contains the measurement outcome of control qubit. Indeed, erasure of the register by thermalization with a resetting reservoir at β_h has the cost

$$\tilde{W}_E = -\frac{p_+ \ln(p_+) + p_- \ln(p_-)}{\beta_h} = \frac{1}{\beta_h} \mathcal{H}\Big(\frac{3e^{-\beta_c}}{2(1+e^{-\beta_c})^2} \Big), \tag{10}$$

where $\mathcal{H}(x) = -x\ln(x) - (1-x)\ln(1-x)$. Therefore, performance of the obtained cycle can be estimated by

$$\eta = \frac{Q}{\tilde{W}_E} = \frac{3\beta_h e^{-\beta_c}}{2(1+e^{-\beta_c})^2 \mathcal{H}\Big(\frac{3e^{-\beta_c}}{2(1+e^{-\beta_c})^2}\Big)} \Big(\frac{1+2e^{-\beta_c}}{3(1+e^{-\beta_c})} - \frac{e^{-\beta_h}}{1+e^{-\beta_h}} \Big), \tag{11}$$

and higher efficiency can be obtained for cold reservoirs at very low temperature. However, it significantly decreases p_- and, hence, makes heat transfer from cold reservoirs to working medium problematic. Nevertheless, the heating probability p_- and the energy of the corresponding state ρ_- (hence, amount of transferred heat Q) can be increased by considering a generalization of quantum SWITCH to $N > 2$ channels and increasing the number of the coherently controlled cold reservoirs [12].

2.2. Activation of passive states via quantum-controlled thermalization

While thermalizing a system with reservoirs at its initial temperature in a quantum-controlled order still produces a thermal state, taking system in a state initially colder than the reservoirs as

$$\beta_{in} > 2\beta \tag{12}$$

and thermalizing it with them via the quantum SWITCH can produce an active state which can be used to perform work [13, 14]. Indeed, the work that can be extracted from a given state $\rho = \sum_i p_i |\psi_i\rangle\langle\psi_i|$ via unitary cycles is estimated by its ergotropy

$$W(\rho) = \operatorname{Tr}[(\rho - U_p \rho U_p^\dagger)H], \tag{13}$$

where U_p is a unitary producing the corresponding passive state $\rho_p = U_p \rho U_p^\dagger = \sum_k p_k |\epsilon_k\rangle\langle\epsilon_k|$ with decreasing populations $p_k \geq p_{k+1}$ and increasing energies $\epsilon_k \leq \epsilon_{k+1}$ (with $|\epsilon_k\rangle$ denoting corresponding eigenstates of the Hamiltonian). Obviously, any convex combination of application orders of $\mathcal{T}_\beta$ produces a state of zero ergotropy since any thermal state τ_β is

Journal of Physics: Conference Series **2533** (2023) 012039 doi:10.1088/1742-6596/2533/1/012039

a completely passive state. On the other hand, if $\mathcal{T}_\beta$ are put into indefinite causal order via quantum SWITCH, the average ergotropy

$$\langle W \rangle = p_+ W(\rho_+) + p_- W(\rho_-) = \max\Big\{0, \frac{e^{-2\beta_c} - e^{-\beta_{in}}}{2(1+e^{-2\beta_c})(1+e^{-\beta_{in}})}\Big\}. \tag{14}$$

of states (5) obtained after measurement of control qubit is non-zero if the temperature bound (12) is satisfied. It can be further improved by increasing the number of quantum-controlled reservoirs: indeed, for N reservoirs at β satisfying (12), the average ergotropy of the states obtained after measurement of the control system increases as

$$\langle W_N \rangle = 2\Big(1 - \frac{1}{N}\Big)\langle W \rangle, \tag{15}$$

hence, up to doubling the work that can be extracted in the usual quantum SWITCH scenario [14]. On the other hand, allowing for initial correlations between working medium and control qubit can change temperature bound (12) by getting it shifted towards higher temperatures of the reservoirs in the case of initial classical correlations and even completely reduced in the case of initial quantum correlations [14], hence, allowing for work extraction at the regimes beyond (12).

3. Thermodynamic cycles from quantum-controlled measurements

Recently, it has been observed that performance of a measurement on working medium can play role of a hot reservoir in a thermodynamic cycle [15, 16, 17, 18]. A particular example has been provided by a class of single-temperature cycles consisting of thermalization with a cold reservoir at inverted temperature β and subsequent generalized measurements $\mathcal{M}_a[.]$ and $\mathcal{M}_{1-b}[.]$ of tunable strengths $0 \leq a \leq 1$ and $0 \leq 1-b \leq 1$, where a measurement acts on a system's state as $\mathcal{M}_\lambda[\rho] = \begin{pmatrix} 1-\lambda & 0 \\ 0 & \lambda \end{pmatrix}$ [18]. By adjusting them, it is possible to implement via $\mathcal{M}_a$ and $\mathcal{M}_b$ exchange of energy that can have nature of heat or work. Indeed, each stroke produces a change in internal energy and entropy summarized in Table 1.

Table 1. Change of internal energy and entropy in each stroke.

Stroke 1: $\mathcal{M}_a$	Stroke 2: $\mathcal{M}_{1-b}$	Stroke 3: Thermalization
$\delta U_1 = a - \frac{e^{-\beta}}{1+e^{-\beta}}$	$\delta U_2 = 1 - b - a$	$\delta U_3 = \frac{e^{-\beta}}{1+e^{-\beta}} - 1 + b$
$\delta S_1 = \mathcal{H}(a) - \mathcal{H}\Big(\frac{e^{-\beta}}{1+e^{-\beta}}\Big)$	$\delta S_2 = \mathcal{H}(b) - \mathcal{H}(a)$	$\delta S_3 = \mathcal{H}(a) - \mathcal{H}\Big(\frac{e^{-\beta}}{1+e^{-\beta}}\Big)$

Adjusting the measurement strengths in such a way that $\delta S_i = 0$ makes the i-th stroke isentropic, and, hence, investing/extracting work to/from the medium. Otherwise, the stroke can be regarded as one transferring heat. Having this in mind, it is possible to make the cycle work in the regimes of thermal accelerator or heat engine:

- *Thermal accelerator.* In the first stroke, the measurement $\mathcal{M}_a$ realizes heat transfer to the working medium by fixing its strength in the interval $\frac{e^{-\beta}}{1+e^{-\beta}} < a < \frac{1}{2}$, so that $\delta U_1 \equiv Q_{hot} > 0$. The work is absorbed via the second measurement by fixing $b = a$, so that $\delta S_2 = 0$, and $\delta U_2 \equiv W > 0$. The last stroke implements heat transfer to the cold reservoir, so that $\delta U_3 \equiv Q_{cold} < 0$.

- *Heat engine.* Similarly to thermal accelerator, the first measurement $\mathcal{M}_a$ implements heat transfer to the working medium, , so that $\delta U_1 \equiv Q_{hot} > 0$, however, its strength is fixed in the interval $\frac{1}{2} \leq a < \frac{1}{1+e^{-\beta}}$. This choice allows one to extract work from the medium via the second measurement by fixing $b = a$, so that $\delta S_2 = 0$, and $\delta U_2 \equiv W < 0$. The last stroke implements heat transfer to the cold reservoir, so that $\delta U_3 \equiv Q_{cold} < 0$.

Dieguez, Lisboa, and Serra have proposed an extension of the considered 3-stroke cycle by increasing the number of generalizing measurements implementing heat transfer and putting them into an indefinite causal order via the quantum SWITCH [19]. Indeed, similarly to the original protocol, the new scheme has three strokes corresponding to measurement-powered heat transfer, measurement-powered work source, and thermalization with a cold reservoir. The difference lies in the heat transfer stage, which is implemented by two measurements with opposite strengths (i.e., $\mathcal{M}_\lambda$ and $\mathcal{M}_{1-\lambda}$ with $0 \leq \lambda \leq 1$) controlled via quantum SWITCH with the control qubit being initially in a certain pure state $\omega = |\theta\rangle\langle\theta|$ with $|\theta\rangle = \cos\frac{\theta}{2}|0\rangle + \sin\frac{\theta}{2}|1\rangle$. As the original cycle, it can operate in three regimes which we review below.

3.1. Thermal accelerator regime.

In the thermal accelerator regime, the cycle begins with a pair of coherently controlled measurements $\mathcal{M}_a$ and $\mathcal{M}_{1-a}$ that supports a heat flow from the working medium due to a change in its internal energy $Q^{\pm}_{hot} > 0$, which depends on the outcome of a measurement of the control qubit in the $\pm$-basis. The following measurement $\mathcal{M}_b$ has to realize the isentropic channel required to invest work: crucially, its strength depends on the outcome of the measurement of control qubit in the $|\pm\rangle$-basis. Indeed, accordingly to the obtained outcome, one picks a measurement $\mathcal{M}_{b_\pm}$ with a strength

$$b_\pm = \frac{1}{2}\Big[1 + (2\Omega_\pm - 1)\Big(1 - \frac{2e^{-\beta}}{1+e^{-\beta}}\Big)\Big], \tag{16}$$

where $p_\pm = \frac{1}{2}(1 \pm a(1-a)\sin\theta)$. Finally, the cycle is closed by thermalization with the cold reservoir delivering heat $Q^{\pm}_{cold} < 0$ from it to the working medium.

$$Q^{\pm}_{hot} \equiv \delta U_1 = \Omega_\pm\Big(1 - \frac{2e^{-\beta}}{1+e^{-\beta}}\Big), \tag{17}$$

$$W^{\pm} \equiv \delta U_2 = (1 - 2\Omega_\pm)\Big(1 - \frac{2e^{-\beta}}{1+e^{-\beta}}\Big), \tag{18}$$

$$Q^{\pm}_{cold} \equiv \delta U_3 = (\Omega_\pm - 1)\Big(1 - \frac{2e^{-\beta}}{1+e^{-\beta}}\Big). \tag{19}$$

A thermal accelerator requires heat transfer to the working medium in the first stroke (i.e., $Q^{\pm}_{hot} > 0$), which is obtained by fixing $0 < \Omega_\pm < 1$ in general. Furthermore, in the second stroke, work $W^\pm > 0$ has to be invested via $\mathcal{M}_{b_\pm}$, providing the condition $0 < \Omega_\pm < \frac{1}{2}$. Hence, the performance of the cycle depending on the obtained outcome of the control qubit measurement is given by

$$\eta^{\pm}_{acc} = -\frac{Q^{\pm}_{cold}}{W^{\pm}} = 1 - \frac{1}{2 - \frac{1}{\Omega_\pm}}. \tag{20}$$

3.2. Heat engine regime.

The heat engine regime requires a similar scheme of the cycle as the accelerator regime. However, the work supported by the measurement $\mathcal{M}_{b_\pm}$ has to be extracted from the working medium, i.e., $W^{\pm} < 0$. This is achieved by fixing $\frac{1}{2} < \Omega_{\pm} < 1$. Therefore, the performance of the cycle depending on the obtained outcome of the control qubit measurement is given by

$$\eta_{he}^{\pm} = \frac{W^{\pm}}{Q_{hot}^{\pm}} = 2 - \frac{1}{\Omega_{\pm}}. \tag{21}$$

3.3. Performance under incoherent control of the measurements.
Performance of the obtained cycle with a coherent control can be compared with its incoherent counterpart, where the corresponding pair of controlled channels is applied in an incoherent mixture of its orders obtained by tracing out the control qubit instead of measuring it. Performance of the latter for both regimes turns out to be

$$\eta_{acc}^{inc} = 1 - \frac{1}{2 - \frac{1}{\Omega_{inc}}}, \tag{22}$$

$$\eta_{he}^{inc} = 2 - \frac{1}{\Omega_{inc}}, \tag{23}$$

where $\Omega_{inc} = 2p_{\pm}\Omega_{\pm}$, taking into account that the thermal accelerator operates in the intervals of a and θ defined by the condition $0 < \Omega_{inc} < \frac{1}{2}$, whereas heat engine regime requires $\frac{1}{2} < \Omega_{inc} < 1$. Hence, for the values of a and θ that satisfy both the corresponding conditions for $\Omega_{\pm}$ and Ω_{inc}, the coherently-controlled cycle always offers a higher performance than its incoherent counterpart in one of the $|\pm\rangle$-branches. On the other hand, coherently-controlled cycle can operate in a wider interval of values of a and θ that do not necessarily satisfy the corresponding condition of Ω_{inc}, i.e., do not allow its incoherent counterpart to operate.

4. Discussion and outlook

Experimental realization of indefinite causal order of quantum operations via the quantum SWITCH has indicated an important milestone that stimulated research in its applications to information processing and communication. Recent proposals of thermal engines powered by the quantum SWITCH have supplemented this list with quantum thermodynamics. Indeed, there were proposed several protocols that exploit coherently controlled thermalization and generalized measurements in order to construct thermodynamic cycles that do not have classical counterpart or can offer regimes which outperform it. Experimental realizability of these protocols is of high importance here, and a cooling cycle with quantum-controlled thermalization has been already implemented in several table-top experiments. Beyond the reviewed protocols, quantum SWITCH allows one to outperform heat-bath algorithmic cooling technique [20] and boost the charging process of a quantum battery [21].

The development of the field of indefinite causality has just begun, and thermodynamic aspects of indefinite causality are yet to be understood. In particular, it is still debated whether indefinite causal orders can be regarded as a genuine thermodynamic resource [22, 23, 24]. A proper way to study them would be to go beyond quantum SWITCH and question usefulness of causal non-separability of a process in general. In particular, activation of thermal states via coherently controlled thermalization could be a hint for a thermodynamic analogue of the recently developed resource theories of causally non-separable processes [23, 25]: such a resource theory is expected to be richer than its communication counterpart and the existing resource theories of thermodynamics.

References

[1] Russell B 1913 *Proc. Aristot. Soc.* **13** 1
[2] Hardy L 2007 *J. Phys. A: Math. Theor.* **40** 3081

[3] Chiribella G, D'Ariano G M, Perinotti P and Valiron B 2013 *Phys. Rev. A* **88** 022318
[4] Oreshkov O, Costa F and Brukner C 2012 *Nature Commun.* **3** 1092
[5] Rubino G, Rozema L A, Feix A, Araújo M, Procopio L, Brukner Č and Walther P 2017 *Sci. Adv.* **3** e1602589
[6] Goswami K and Romero J 2020 *AVS Quantum Sci.* **2** 037101
[7] Chiribella G, Banik M, Bhattacharya S S, Guha T, Alimuddin M, Roy A, Saha S, Agrawal S and Kar G 2021 *New J. Phys.* **23** 033039
[8] Felce D and Vedral V 2020 *Phys. Rev. Lett.* **125** 070603
[9] Cao H, Wang N N, Jia Z, Zhang C, Guo Y, Liu B H, Huang Y F, Li C F and Guo G C 2022 *Phys. Rev. Research* **4** L032029
[10] Nie X, Zhu X, Huang K, Tang K, Long X, Lin Z, Tian Y, Qiu C, Xi C, Yang X, Li J, Dong Y, Xin T and Lu D 2022 *Phys. Rev. Lett.* **129** 100603
[11] Felce D, Vedral V and Tennie F 2021 Refrigeration with Indefinite Causal Orders on a Cloud Quantum Computer *Preprint* 2107.12413
[12] Nie H, Feng T, Longden S and Vedral V 2022 Quantum cooling activated by coherently-controlled thermalisation *Preprint* 2201.06954
[13] Guha D, Alimuddin M and Parashar P 2020 *Phys. Rev.* A **102** 032215
[14] Simonov K, Francica G, Guarnieri G and Paternostro M 2022 *Phys. Rev.* A **105** 032217
[15] Yi J, Talkner P and Woon Kim Y 2017, *Phys. Rev.* E **96** 022108
[16] Elouard C, Herrera-Mart D, Huard B and Auffèves A 2017 *Phys. Rev. Lett.* **118** 260603
[17] Elouard C and Jordan A N *Phys. Rev. Lett.* **120** 260601
[18] Behzadi N 2020 *J. Phys. A: Math. Theor.* **54** 015304
[19] Dieguez P R, Lisboa V F and Serra R M 2023 *Phys. Rev.* A **107** 012423
[20] Goldberg A Z and Heshami K 2023 *J. Phys. Commun.* **7** 015003
[21] Chen Y and Hasegawa Y 2021 Indefinite causal order in quantum batteries *Preprint* 2105.12466
[22] Guérin P A, Rubino G and Brukner Č 2019 *Phys. Rev.* A **99** 062317
[23] Kristjánsson H, Chiribella G, Salek S, Ebler D and Wilson M 2020 *New J. Phys.* **22** 073014
[24] Capela M, Verma H, Costa F, Céleri L C 2022 Indefinite causal order is not always a resource for thermodynamic processes *Preprint* 2208.03205
[25] Milz S, Bavaresco J and Chiribella G 2022 *Quantum* **6** 788

Journal of Physics: Conference Series **2533** (2023) 012040 doi:10.1088/1742-6596/2533/1/012040

On a Reformulation of Bohmian Mechanics Including Open Dissipative Systems

Dieter Schuch

Institute of Theoretical Physics, Goethe-University Frankfurt, Max-von-Laue-Str. 1, D-60438 Frankfurt am Main, Germany

E-mail: schuch@em.uni-frankfurt.de

Moise Bonilla–Licea

CINVESTAV-IPN, Control Automático, AP 14-740, Mexico City 07000, Mexico

E-mail: moise.bonilla@cinvestav.mx

Abstract. In Bohmian mechanics it is assumed that quantum objects are particles that follow real physical paths. These Bohmian trajectories are obtained by integrating the velocity field that occurs in the continuity equation. However, in the course of deriving these trajectories, the independent position variable x, appearing in the velocity field, is arbtirarily replaced by the time-dependent trajectory $q(t)$. We show under which restrictions such a replacement can be justified, leading to a totally different interpretation of the Bohmian trajectories. They are not real paths of physical particles but borders of regions of constant probability, so-called quantiles, and therefore must be seen in the context of descriptive statistics. Thus they provide a tool complementary to the conventional statistical formulation of quantum mechanics. A possible extension to include open dissipative systems is also discussed, showing that the inclusion of an environment does not change our interpretation of the Bohmian trajectories.

Keywords: Bohmian trajectories; descriptive statistics and quantiles; open quantum systems.

1. Introduction

Bohmian mechanics is formally identical to Madelung's hydrodynamic formulation of quantum mechanics [1] in terms of two (coupled) real equations for phase and amplitude of the complex wave function that is solution of the Schrödinger equation. Around 25 years later, Bohm used this form of quantum mechanics (related to, but independent of, an idea of de Broglie, discussed at the Solvay conference 1927, but which he abandoned due to criticisms by Pauli) for his attempt to find a deterministic version of quantum mechanics.

Although Madelung's formulation and the one by Bohm are formally identical, they differ fundamentally with respect to their ontological interpretation. While Madelung's approach is based on the same probabilistic interpretation as Schrödinger's formulation of quantum mechanics, Bohm tries to find a deterministic version of it. In this context, the so-called Bohmian trajectories play a central role as "hidden variables", although they should be in Bohm's version the variables that are actually observable. Bohm's interpretation and particularly the corresponding trajectories have some problematic aspects. One is that there is no experimental

Journal of Physics: Conference Series **2533** (2023) 012040 doi:10.1088/1742-6596/2533/1/012040

evidence for the existence of these trajectories as real geometric paths of quantum particles that really exist. To the contrary, the experiments of last year's Nobel Laureates, particularly those of Aspect *et al.* [2, 3, 4], are not in favour of this interpretation. Still one could argue, the fact that these trajectories have not yet been detected does not necessarily mean they do not exist. The other problematic aspect originates in the "derivation" or definition of these trajectories based on the velocity field that occurs in one of Madelung's equations: the continuity equation. Integration of this field would actually lead to a position field. The crucial point now is that just one value of position is chosen as the initial position for the Bohmian trajectory, *i.e.*, the independent position variable x is simply replaced by a time-dependent parameter $q(t)$ without justification. In other words, the dimension of the space of variables is reduced from (x, t) by one to $(q(t), t)$ without imposing any constraint that would justify replacing the independent variable x by the Bohmian trajectory $q(t)$.

In this paper we show how such a reduction can be achieved without *ad hoc* assumptions and without arbitrariness. As a consequence, the interpretation of $q(t)$ is no longer free but dictated by the reduction process and turns out not to be deterministic, but probabilistic; however, from a different viewpoint than that of the conventional statistical interpretation of quantum mechanics.

In our derivation the continuity equation plays a central role. We also show how our derivation can consistently be extended to include open systems with irreversible dynamics and dissipation of energy by adding a time-symmetry breaking diffusion term, leading to an effective description of the open quantum system.

The paper is organized as follows: In Section 2 the formal basis in terms of Madelung's hydrodynamic equations is provided and the usual way of obtaining the Bohmian trajectories from the above-mentioned velocity field as well as the questionable aspects of this procedure are sketched.

By comparison with a thermodynamic system it will be shown how consistently a reduction of the independent variables can be obtained via a constraint for a property of the system, in this case the internal energy. This procedure involves a Maxwell relation that, in our Bohmian context, is replaced by the continuity equation. A function $\mathrm{B}(x, t)$, we call it Bohm function, corresponding formally to the internal energy, can be defined. Imposing a constraint on this function allows consistently to reduce the number of independent variables and thus to derive the Bohmian trajectories. However, the interpretation of these trajectories is no longer arbitrary, but fixed by the definition of $\mathrm{B}(x, t)$ and has a probabilistic meaning.

As an example to demonstrate the physical meaning of the Bohmian trajectories, in Section 3 analytical Gaussian shaped solutions of the time-dependent Schrödinger equation will be considered and the Bohmian trajectories can be defined in terms of the dynamics of maximum and width, *i.e.*, of position and position uncertainty, of the Gaussian wave packet.

In Section 4 it then will be shown how this procedure can be extended to include open dissipative systems via introduction of an irreversible diffusion term into the continuity equation. It will be shown that this change does not influence the derivation obtained in Section 2, particularly the relation equivalent to the Maxwell relation is not affected essentially by this change.

Using a separation condition, a modified Schrödinger equation with complex logarithmic nonlinearity can be obtained, from which the second hydrodynamic equation corresponding to the open system can be obtained and its connection with a friction force can be shown.

Also for the open system, generalized Gaussian wave packet solutions, now for the nonlinear Schrödinger equation, are discussed. Differences and similarities compared with the case discussed in Section 3 will be shown and the explicit form of the Bohmian trajectories in terms of mean value and uncertainty of position will be provided. Section 5 summarizes and concludes the paper.

Journal of Physics: Conference Series **2533** (2023) 012040 doi:10.1088/1742-6596/2533/1/012040

2. Consistent derivation of Bohmian trajectories and correct physical interpretation

2.1. Madelung's hydrodynamic formulation and Bohmian trajectories

Shortly after Schrödinger published his communications on wave mechanics, Madelung [1] found a formulation where he replaced the complex Schrödinger equation[1]

$$\mathrm{i}\hbar\frac{\partial}{\partial t}\psi(x,t) = \Big\{ -\frac{\hbar}{2m}\frac{\partial^2}{\partial x^2} + V(x,t)\Big\}\psi(x,t) = \mathrm{H_L}\psi(x,t) \tag{1}$$

by two real equations using a polar ansatz $\psi(x,t) = \sqrt{\rho(x,t)}\exp\left\{\frac{\mathrm{i}}{\hbar}S(x,t)\right\}$ for the wave function, where $\rho = \psi^*\psi$ and $S(x,t)$, having the dimension of action, is essentially the phase of the complex function.

The two equations obtained by this ansatz resemble those known from hydrodynamics, namely a continuity equation for the (probability) density $\rho(x,t)$,

$$\frac{\partial}{\partial t}\rho + \frac{\partial}{\partial x}\left[\rho\frac{1}{m}\frac{\partial}{\partial x}S\right] = 0 \tag{2}$$

and a modified Hamilton–Jacobi-type equation for the phase $S(x,t)$,

$$\frac{\partial}{\partial t}S + \frac{1}{2m}\Big(\frac{\partial}{\partial x}S\Big)^2 + V + V_{\mathrm{qu}} = 0. \tag{3}$$

These two equations are not independent of each other, but coupled. This is because Eq. (3) for the phase S contains the density ρ via the so-called quantum potential

$$V_{\mathrm{qu}} = -\frac{\hbar}{2m}\frac{\frac{\partial^2}{\partial x^2}\sqrt{\rho}}{\sqrt{\rho}} \tag{4}$$

and the continuity equation (2) for ρ contains the phase S via the convection velocity field

$$v_{\mathrm{con}}(x,t) = \frac{1}{m}\frac{\partial}{\partial x}S(x,t). \tag{5}$$

This velocity field is the origin of the definition of the Bohmian trajectory. As v_{con} is a velocity, it is assumed the corresponding trajectory can be obtained by simply integrating this velocity. This leads to the definition of the Bohmian trajectory $q(t)$ via the so-called **guidance law**

$$\frac{\mathrm{d}}{\mathrm{d}t}q(t) = \frac{1}{m}\frac{\partial}{\partial x}S(x,t)\Big|_{x=q(t)} \tag{6}$$

which is a first order differential equation; so for given initial conditions, different trajectories will never cross each other (non-crossing rule).

This seems reasonable at first sight; but what exactly happened between Eqs. (5) and (6)? In Eq. (5), $v_{\mathrm{con}}(x,t)$ is a velocity field depending on two independent variables, (x,t). Integrating this field with respect to time should lead to a position field that again depends on the same two independent variables. In Eq. (6), however, one independent variable, position x, is (*deus ex machina*) simply replaced without justification by the time-dependent parameter $q(t)$. So the number of dimensions of the space describing the system is reduced without imposing any obvious constraints and without corresponding physical justification.

The density $\rho(x,t)$ in the continuity equation is then interpreted in terms of the number of trajectories in a given area, *i.e.*, the probability that a particle in an ensemble of trajectories lies between x and $x + \mathrm{d}x$ is given by $\rho(x,t)\mathrm{d}x = |\psi(x,t)|^2\mathrm{d}x$.

[1] The following discussion is restricted to one spatial dimension. Extension to higher dimensions is possible.

Journal of Physics: Conference Series **2533** (2023) 012040 doi:10.1088/1742-6596/2533/1/012040

Bohm originally intended for those trajectories to be real paths of quantum particles that actually exist and that they should, in principle, be observable. So the term "hidden variables" is somewhat misleading as these trajectories should in fact be the objects of experimental detection. Now there are two problems with these Bohmian trajectories. First, as mentioned above, the elimination of one independent variable without justification appears very arbitrary and needs closer analysis. This will be done subsequently. Second, there is no experimental support for the existence of these trajectories. To the contrary; not only the Stern–Gerlach-type experiments by Scully *et al.* [5, 6] indicate that these trajectories do not exist as real paths, but also the experiments by last year's Nobel Laureates, particularly those of Aspect *et al.* [2, 3, 4] are not in favour of hidden variables like the Bohmian trajectories.
So, the questions remain: what in fact is a Bohmian trajectory if it is not a particle's trajectory? How can the answer be formulated mathematically in a consistent way giving a clear physical picture?

2.2. Consistent derivation of the Bohmian trajectory and physical consequences
An example from classical thermodynamics is chosen to show a mathematically sound and consistent way of eliminating an independent variable. Via the continuity equation, this procedure can be transfered to our Bohmian situation.
For this purpose, a thermodynamic system is considered with internal energy $U(V, \mathtt{S})$, depending on the independent variables volume V and entropy[2] $\mathtt{S}$. In order to guarantee that the expression

$$\mathrm{d}U = -P\mathrm{d}V + T\mathrm{d}\mathtt{S} \tag{7}$$

is a total differential, the pressure P and the temperature T must fulfill the **Maxwell relation**

$$\frac{\partial P}{\partial \mathtt{S}} = -\frac{\partial T}{\partial V}, \tag{8}$$

leading to $\frac{\partial^2 U}{\partial \mathtt{S} \partial P} = \frac{\partial^2 U}{\partial P \partial \mathtt{S}}$.
For iso-energetic processes, $\mathrm{d}U = 0$ is valid, providing a constraint that allows the reduction of the $(\mathtt{S}, V)$ space to one that depends only on V, *i.e.*, $(\mathtt{S}, V) \xrightarrow[U=\text{const.}]{} (\sigma, V)$ with $\sigma = \sigma(V)$.
Consequently, Eq. (7) can be rewritten as

$$0 = -P\mathrm{d}V + T\mathrm{d}\sigma(V), \tag{9}$$

or

$$\frac{\mathrm{d}\sigma(V)}{\mathrm{d}V} = \frac{P}{T}, \tag{10}$$

i.e., the dependent variable $\sigma(V)$ replaces the independent variable $\mathtt{S}$. For further details, see [7]. How can this procedure be transferred to our Bohmian problem?
First, we define a function corresponding to the internal energy $U(V, \mathtt{S})$, calling it Bohm function, $\mathrm{B}(x, t)$, depending on two independent variables x and t. Next, a total differential of this function must be found, leading to

$$\mathrm{dB} = -j\mathrm{d}t + \rho\mathrm{d}x \tag{11}$$

with $j = \rho \frac{1}{m} \frac{\partial}{\partial x} S = \rho v_{\text{con}}$ and

$$\frac{\partial}{\partial x}\frac{\partial \mathrm{B}}{\partial t} = -\frac{\partial}{\partial x} j \quad , \quad \frac{\partial}{\partial t}\frac{\partial \mathrm{B}}{\partial x} = \frac{\partial}{\partial t}\rho \tag{12}$$

[2] The symbol $\mathtt{S}$ for entropy should not be mistaken for the symbol S for the phase of the wave function.

Journal of Physics: Conference Series **2533** (2023) 012040 doi:10.1088/1742-6596/2533/1/012040

which is fulfilled if the "Maxwell relation"

$$\frac{\partial}{\partial t}\rho = -\frac{\partial}{\partial x}j \tag{13}$$

is also fulfilled, *i.e.*, the continuity equation (2) in this case plays the role of the Maxwell relation. Following the procedure, as in the thermodynamic case, the constraint $\mathrm{dB} = 0$ now reduces the (x, t) space to one depending only on time t, *i.e.*, $(x, t) \xrightarrow[\mathrm{B=const.}]{} (q, t)$ with $q = q(t)$. This finally leads to

$$0 = -\rho(q, t)\frac{1}{m}\frac{\partial}{\partial q}S(q, t)\mathrm{d}t + \rho(q, t)\mathrm{d}q, \tag{14}$$

or

$$\frac{\mathrm{d}q(t)}{\mathrm{d}t} = \frac{1}{m}\frac{\partial}{\partial q}S(q, t), \tag{15}$$

i.e., the desired guidance law for the Bohmian trajectory $q(t)$. However, there is a price to be paid for this reduction of variables. The meaning of $q(t)$ is no longer arbitrary, but uniquely coupled to the definition of the Bohm function $\mathrm{B}(x, t)$. Therefore, the meaning of this function and its connection with the Bohmian trajectory is now going to be checked.
Obviously, $\mathrm{B}(x, t)$ must fulfill Eqs. (12). This is guaranteed if $\mathrm{B}(x, t)$ takes the form

$$\mathrm{B}(x, t) = \int_{-\infty}^{x} \rho(x', t)\mathrm{d}x', \tag{16}$$

leading to

$$\frac{\partial}{\partial x}\mathrm{B} = \rho(x, t) - \rho(-\infty, t) = \rho(x, t) \tag{17}$$

as $\rho(\pm\infty, t) = 0\ \forall t$ and

$$\frac{\partial}{\partial t}\mathrm{B} = \int_{-\infty}^{x} \frac{\partial}{\partial t}\rho(x', t)\mathrm{d}x' = \int_{-\infty}^{x} \frac{\partial}{\partial x'}\Big(- \rho\frac{1}{m}\frac{\partial}{\partial x'}S\Big)\mathrm{d}x' = -\rho\frac{1}{m}\frac{\partial}{\partial x}S(x, t), \tag{18}$$

providing the "Maxwell relation" (13), or the continuity equation (2), respectively, is fulfilled. Functions like (16) are known in statistical physics as cumulative probability functions (CPF), and x, in this case called **quantile**, is the border of a fixed amount of probability. For $\mathrm{dB} = 0$, x becomes a time-dependent parameter, $x \longrightarrow q(t)$. This means the limit of integration becomes time-dependent but the "amount" of probability between $-\infty$ and $q(t)$ remains unchanged; it can only be "stretched" or "squeezed" according to the movement of $q(t)$. So, even if the interval can undergo deformations (contractions or dilations) in time, the "amount" of probability between two (non-crossing) trajectories always stays constant.
This is also in agreement with the Bohmian assumption that the probability density $\rho(x, t)$ is proportional to the number of Bohmian trajectories in a given spatial interval. If there are more trajectories in this interval, the space between them is smaller; but since the "amount" of probability between the trajectories stays constant, the corresponding probability density must increase. The opposite is true for a decreasing number of trajectories.
These findings are in agreement with mathematically-motivated approaches by Oriols *et al.* [8], Brandt *et al.* [9] and Coffey *et al.* [10]. For further details see [7].

3. General Gaussian wave packets as examples

To get a better idea of the physical meaning of the Bohmian trajectories, we next consider cases where the time-dependent Schrödinger equation (1) has exact analytical solutions. This is the case when the potential $V(x,t)$ is at most quadratic in position variable, *i.e.*, for the harmonic oscillator with $V = \frac{m}{2}\omega_0^2 x^2$, in the limit $\omega_0 \longrightarrow 0$ also for the free motion, and for the parametric oscillator with $\omega = \omega(t)$. In these cases, Eq. (1) has solutions in the form of Gaussian functions that can be considered as wave packets. These Gaussian functions are uniquely determined by their maximum and width where both can be time-dependent.
A Gaussian wave packet can be written in the form

$$\psi(x,t) = N(t) \exp\left[\frac{\mathrm{i}}{\hbar}\left(\frac{m}{2}\mathcal{C}\widetilde{x}^2 + \langle p\rangle\widetilde{x} + f(t)\right)\right] \tag{19}$$

with $\widetilde{x} = x - \langle x\rangle = x - \eta(t)$, where $\langle x\rangle = \int_{-\infty}^{+\infty} \psi^* x\psi \mathrm{d}x = \eta(t)$, $\langle p\rangle = m\dot{\eta}$, $\mathcal{C}(t)$ is complex and the purely time-dependent functions $N(t)$ and $f(t)$ are not relevant for the following discussion. The equations of motion for maximum and width can be obtained by inserting (19) into (1), leading to

$$\ddot{\eta} + \omega^2\eta = 0 \tag{20}$$

for the maximum and

$$\dot{\mathcal{C}} + \mathcal{C}^2 + \omega^2 = 0 \tag{21}$$

for the width, where the connection between $\mathcal{C}$ and the width is given via the imaginary part, $\mathcal{C}_{\mathrm{I}} = \frac{\hbar}{2m\langle\widetilde{x}^2\rangle}$, with $\langle\widetilde{x}^2\rangle = \langle x^2\rangle - \langle x\rangle^2 = \sigma_x^2$. Introducing a new variable $\alpha(t)$ that is proportional to the wave packet width via $\alpha = \sqrt{\frac{2m}{\hbar}}\sigma_x$, real and imaginary parts of $\mathcal{C}(t)$ can be written as $\mathcal{C}_{\mathrm{I}} = \frac{1}{\alpha^2}$, $\mathcal{C}_{\mathrm{R}} = \frac{\dot{\alpha}}{\alpha}$, where $\mathcal{C}_{\mathrm{R}}$ is important for the definition of the Bohmian trajectory.
Using Eq. (15) and the definition of the wave packet (19) yields

$$\frac{\mathrm{d}}{\mathrm{d}t}q(t) = \frac{1}{m}\frac{\partial}{\partial q}S(q,t) = \mathcal{C}_{\mathrm{R}}\big(q(t) - \eta(t)\big) + \dot{\eta}. \tag{22}$$

In a coordinate frame moving with the maximum of the wave packet, the equation of motion for the corresponding variable $\widetilde{q} = q(t) - \eta(t)$ would be

$$\frac{\mathrm{d}}{\mathrm{d}t}\widetilde{q}(t) = \mathcal{C}_{\mathrm{R}}\widetilde{q} = \frac{\dot{\alpha}}{\alpha}\widetilde{q}, \tag{23}$$

or, expressed in terms of the Bohmian trajectory $q(t)$,

$$q(t) = \eta(t) + \big(q_0 - \eta_0\big)\frac{\alpha(t)}{\alpha_0}. \tag{24}$$

Expression (24) shows that there are two dynamical contributions to the Bohmian trajectory $q(t)$, namely, the time-dependent maximum of the wave packet and its (generally) time-dependent width. For the special case of a constant width, *i.e.*, $\alpha(t) = \alpha_0 \forall t$, $q(t)$ at initial time $t = 0$ is simply a point q_0 that is different from the initial position η_0 of the maximum. For the Bohm function (16), $\mathrm{B}(x.t)$, this means that the integration over $\rho(x,t)$ goes from $-\infty$ to q_0, providing a fixed amount of probability. As time goes on, the maximum of the Gaussian follows the classical trajectory and so does $q(t)$, *i.e.*, the distance between the maximum and $q(t)$ remains constant, as does the amount of probability defined by $\mathrm{B}(q,t)$. The situation changes when the wave packet width also becomes time-dependent. For a spreading wave packet, for instance, the value of $\rho(x,t)$ at the maximum is decreasing. Therefore, in order to keep the value of $\mathrm{B}(q,t)$

constant, an additional contribution depending on the spreading must be added to $q(t)$, which is taken into account by the second term on the rhs of (24). While $\alpha(t)$ describes the spreading at the position of the turning point of the Gaussian function, the Bohmian trajectory, however, can be considered at any point of this function; the time-dependence due to the spreading is "normalized" by dividing $\alpha(t)$ by its initial value α_0 and multiplying it by a factor that determines the position of the Bohmian trajectory with respect to the position of the maximum. Therefore, the time-dependence of the integration limit of the Bohm function, defined by Eq. (24), maintains a constant amount of probability when performing the integration in $\mathrm{B}(q(t), t)$ though maximum and width of the density $\rho(x,t)$, in this case a Gaussian distribution, can both undergo dynamical processes.
Here, only two examples, the free motion, *i.e.*, $\omega = 0$ or $V = 0$, and the harmonic oscillator, $\omega = \omega_0 = $ const., shall briefly be drawn to attention (for details, see [7]).
For the free motion, the classical solution is given by $\eta(t) = v_0 t$ and the width by $\alpha(t) = \alpha_0\sqrt{1 + \left(\frac{t}{\alpha_0^2}\right)^2} = \alpha_0\sqrt{1 + (\beta_0 t)^2}$ with frequency $\beta_0 = \frac{1}{\alpha_0^2}$, providing

$$q(t) = v_0 t + (q_0 - \eta_0)\sqrt{1 + (\beta_0 t)^2}. \tag{25}$$

For $\omega = \omega_0$ the Bohmian trajectory, *i.e.*, the limit of integration in (16), can be given by

$$q(t) = \eta(t) + (q_0 - \eta_0)\sqrt{\left(\frac{\sin\omega_0 t}{\omega_0 \alpha_0^2}\right)^2 + \cos^2\omega_0 t}, \tag{26}$$

where, for $\omega_0 = \frac{1}{\alpha_0^2}$ as in the case of the ground state and the wave packet with constant width, the square root reduces to 1, *i.e.*, there is no time-dependent contribution from the width.

4. Modifications for open systems

Key point of our derivation of the Bohmian trajectories was the application of Eq. (2) as a "Maxwell relation" in (13), equating the time-derivative of the density ρ with the spatial derivative of a (probability) current j. So far, this current was a reversible convection current, *i.e.*,

$$j = j_{\text{con}} = \rho v_{\text{con}} = \rho \frac{1}{m}\frac{\partial}{\partial x} S, \tag{27}$$

usually describing conservative systems with constant energy, but not necessarily, as it also applies, *e.g.*, to the parametric oscillator with time-dependent frequency $\omega(t)$, and thus, also time-dependent energy.
The situation becomes even more intricate when considering open systems with irreversible time-evolution and dissipation of energy due to friction forces. Can these aspects also be included in our approach?
For this purpose, it is essentially the time-reversal symmetry of the equations of motion that must be broken. In our context, this should already apply to the "Maxwell relation" (13). This can be achieved by adding a time-reversal symmetry-breaking diffusion term to the current, *i.e.*,

$$j_{\text{D}} = \rho v_{\text{D}} = -D\frac{\partial}{\partial x}\rho, \tag{28}$$

with a contribution from the amplitude to the current, where D is a diffusion coefficient.
This turns the reversible continuity equation into an irreversible Fokker–Planck-type equation, in position space called Smoluchowski equation,

$$\frac{\partial}{\partial t}\rho + \frac{\partial}{\partial x}(\rho v_{\text{con}}) - D\frac{\partial^2}{\partial x^2}\rho = 0. \tag{29}$$

Journal of Physics: Conference Series **2533** (2023) 012040 doi:10.1088/1742-6596/2533/1/012040

However, this is only one of the two hydrodynamic equations considered by Madelung. What happens to the second, the modified Hamilton–Jacobi equation? To answer this question, one can go back to a method developed by Madelung himself [11] and Mrowka [12] in the early 1950s to derive the Schrödinger equation without making use of Hamilton's formulation of classical mechanics. Omitting the details, the key point is the separation of the equation for $\rho = \psi^*\psi$ into two complex conjugate equations for ψ or ψ^*, respectively. The separation condition, apart from constant factors, turned out to be just the potential energy of the system. What changes by adding the diffusion term? Due to this diffusion term, separation in general is no longer possible as terms containing products of ψ and ψ^* appear. However, separation can be achieved if specific separation conditions are fulfilled. In our case, this condition is

$$-D\frac{\partial^2}{\partial x^2}\rho = \gamma\big(\ln\rho - \langle\ln\rho\rangle\big). \tag{30}$$

As an additional term in the equation for the density provides an imaginary term in the corresponding Schrödinger equation, thus leading to a non-Hermitian Hamiltonian, the subtraction of $\langle\ln\rho\rangle$ is necessary to guarantee normalizability of the solutions of the modified Schrödinger equation.

Separation of the Smoluchowski equation leads to a nonlinear Schrödinger equation with complex logarithmic nonlinearity (for further details, see [13, 14]),

$$\mathrm{i}\hbar\frac{\partial}{\partial t}\psi(x,t) = \left\{\mathrm{H_L} + \gamma\frac{\hbar}{\mathrm{i}}\big(\ln\rho - \langle\ln\rho\rangle\big)\right\}\psi(x,t). \tag{31}$$

With Madelung's polar ansatz, the corresponding equation for the phase of the wave function can be determined as

$$\frac{\partial}{\partial t}S + \gamma S + \frac{1}{2m}\left(\frac{\partial}{\partial x}S\right)^2 + V + V_{\mathrm{qu}} - \gamma\langle S\rangle = 0. \tag{32}$$

Taking the gradient of this equation provides an Euler-type equation for the forces according to

$$m\left(\frac{\partial}{\partial t} + v_{\mathrm{con}}\frac{\partial}{\partial x}\right)v_{\mathrm{con}} = -\frac{\partial}{\partial x}\big(V + V_{\mathrm{qu}}\big) - \gamma m v_{\mathrm{con}}, \tag{33}$$

i.e., an equation including a linear velocity-dependent friction force.

The crucial question concerning the Bohmian trajectories is how the additional diffusion term influences its derivation in our approach. The answer is clear: the convection current (27), $j = j_{\mathrm{con}}$, must simply be replaced in (13) by the total current

$$j_{\mathrm{tot}} = j_{\mathrm{con}} + j_{\mathrm{D}} = \rho\left(v_{\mathrm{con}} - D\frac{\frac{\partial}{\partial t}\rho}{\rho}\right). \tag{34}$$

Using general Gaussian wave packets as example, this can be confirmed, showing that the time-dependence of the Bohmian trajectory in terms of maximum and width of the wave packet remains unchanged.

For this purpose, we make use of analytical Gaussian wave packet solutions of the form (19) that are also given for the case of the logarithmic nonlinear Schrödinger equation. Inserting (19) into (31) provides the equations of motion for the maximum and width as

$$\ddot{\eta} + \gamma\dot{\eta} + \omega^2\eta = 0, \tag{35}$$

$$\dot{\mathcal{C}} + \gamma\mathcal{C} + \mathcal{C}^2 + \omega^2 = 0. \tag{36}$$

Journal of Physics: Conference Series **2533** (2023) 012040 doi:10.1088/1742-6596/2533/1/012040

The connection between $\mathcal{C}_{\rm I}$ and the wave packet width remains unchanged, $\mathcal{C}_{\rm I} = \frac{1}{\alpha^2}$. The additional linear term $\gamma\mathcal{C}$ changes the definition of $\mathcal{C}_{\rm R} = \frac{\dot\alpha}{\alpha} - \frac{\gamma}{2}$ and thus the expression for the convection velocity in terms of η and α according to

$$v_{\rm con} = \left(\frac{\dot\alpha}{\alpha} - \frac{\gamma}{2}\right)\widetilde{x} + \dot\eta. \tag{37}$$

However, the difference of $\mathcal{C}_{\rm R}$ in comparison with the case without environment is compensated by the diffusion velocity,

$$v_{\rm D} = -D\frac{\frac{\partial}{\partial x}\rho}{\rho} = \frac{\gamma}{2}\widetilde{x}, \tag{38}$$

leading to the total velocity $v_{\rm tot} = v_{\rm con} + v_{\rm D} = \frac{\dot\alpha}{\alpha}\widetilde{x} + \dot\eta$, or the corresponding total current $j_{\rm tot} = \rho v_{\rm tot} = \rho\left(\frac{\dot\alpha}{\alpha}\widetilde{x} + \dot\eta\right)$; *i.e.*, the dependence on (the velocity of) the maximum and the (relative) time-dependence of the width are the same in the cases with and without environment. So, finally, the Bohmian trajectory for the open system can be obtained from

$$\frac{\rm d}{{\rm d}t}q(t) = v_{\rm con}(q,t) - D\frac{\frac{\partial}{\partial q}\rho(q,t)}{\rho(q,t)} = \dot\eta + \big(q(t) - \eta(t)\big)\frac{\dot\alpha}{\alpha}, \tag{39}$$

or $\frac{\rm d}{{\rm d}t}\widetilde{q}/\widetilde{q} = \dot\alpha/\alpha$, leading to the same expression as in (24).
Therefore, our probabilistic interpretation of Bohmian trajectories, as well as their dependence on maximum and width of the wave packet solutions in cases where they exist, is unchanged even if an irreversible interaction with a dissipative environment is taken into account.

5. Conclusions

Bohmian mechanics attempts to convert quantum mechanics to a deterministic theory with the help of so-called hidden variables. In this context, Bohmian trajectories play a central role. They are obtained via integration of the velocity field that occurs in the continuity equation when quantum mechanics is written in Madelung's hydrodynamic formulation.
A major point of criticism against the Bohmian approach is the fact that these trajectories, assumed to be real geometric paths of quantum particles that actually exist, so far could never be detected experimentally. To the contrary, the experiments of last year's Nobel Laureates conspicuously reveal these paths do not exist. Nevertheless, one could argue that the fact that they have not been detected so far does not prove they cannot exist.
There is, however, a theoretical reason for questioning the interpretation of these trajectories. As mentioned above, their derivation is based on a velocity field depending on two independent variables, position and time. When integrating this field to obtain the trajectory, the independent variable position is replaced by a time-dependent parameter, the trajectory $q(t)$, without any justification. In the first part of this paper we have shown how such a reduction of the number of independent variables can be achieved consistently and mathematically sound without any *ad hoc* assumption. For this purpose we had to introduce a quantity, we call it Bohm function ${\rm B}(x,t)$ and impose a constraint on it. From the total differential of this function and with the constraint ${\rm dB} = 0$, the Bohmian trajectory can be defined without any arbitrariness, providing the continuity equation is fulfilled.
One could still argue that it is obviously possible to obtain the trajectories from the velocity field. Therefore, one would still have the freedom to choose any arbitrary initial position of the field as the initial point of a particle's trajectory without further justification. However, our derivation shows that for any given initial condition, the corresponding trajectory is uniquely defined as quantile, *i.e.*, as the limit of an integration that appears in the definition of our Bohm

function $\mathrm{B}(x,t)$ which, mathematically, is a cumulative probability function (CPF). Since the guidance law (6) or (15) is a first-order differential equation, there is only one solution for a given initial condition. In our derivation, we have clearly shown that $q(t)$ is a quantile, *i.e.*, an element of descriptive statistics. So any *ad hoc* assumption that it could also be a deterministic path of a particle must be wrong!
Our probabilistic interpretation is also in agreement with the explicit form of the trajectories obtained in the case of Gaussian wave packet solutions of the time-dependent Schrödinger equation.
Basis for our derivation was an analogy between the Maxwell relation of thermodynamics and the continuity equation that relates the temporal change of a density with a spatial change of a convection current. The simple addition of a time-symmetry-breaking diffusion current was sufficient to extend our derivation to include an effective description of open systems. By replacing the convection current with the total current, the same procedure can be applied as in the case without environment. Also the examples with the Gaussian wave packets demonstrate the consistency of our approach.
Though Bohm's original interpretation in terms of deterministic trajectories cannot be supported also from a theoretical point of view, this does not render Bohmian mechanics useless. To the contrary; the interpretation should simply be adapted to the field of descriptive statistics. Then it could play an important role as a tool complementary to the conventional statistical inference viewpoint of quantum mechanics. A potential field of application could be, for instance, the important problem of time-dependent tunnelling that we will discuss in future work.

Acknowledgements
This work was supported by CONACYT México [grant number (CVU: 633437)].

References
[1] Madelung E 1927 Quantentheorie in hydrodynamischer Form. *Z. Phys. A* **40** 322.
[2] Aspect A 1976 Proposed experiment to test the non separability of quantum mechanics. *Phys. Rev. D* **14** 1944.
[3] Aspect A, Grangier Ph and Roger G 1982 Experimental Realization of Einstein–Podolsky–Rosen–Bohm Gedankenexperiment: A New Violation of Bell's Inequalities. *Phys. Rev. Lett.* **49** 91.
[4] Aspect A, Dalibard J and Roger G 1982 Experimental Test of Bell's Inequalities Using Time-Varying Analyzers. *Phys. Rev. Lett.* **49** 1804.
[5] Englert B G, Scully M O, Süssmann G and Walther H 1992 Surrealistic Bohm trajectories. *Z. Naturforsch. A* **47** 1175.
[6] Scully M O 1998 Do Bohm trajectories always provide a trustworthy physical picture of particle motion? *Phys. Scr.-Top.* **76** 41.
[7] Bonilla–Licea M and Schuch D 2022 Bohmian Trajectories as Borders of Regions of Constant Probability. *Found. Phys.* **52** 8.
[8] Oriols X, Martín F and Suñe J 1996 Implications of the noncrossing property of Bohm trajectories in one-dimensional tunneling configurations. *Phys. Rev. A* **54** 2594.
[9] Brandt S, Dahmen H D, Gjonaj E and Stroh T 1998 Quantile motion in tunneling. *Phys. Lett. A* **249** 265.
[10] Coffey T M, Wyatt R E and Schieve W C 2008 Monte Carlo generation of Bohmian trajectories. *J. Phys. A: Math. Theor.* **41** 335304.
[11] Madelung E 1950 *Die Mathematischen Hilfsmittel des Physikers* (Berlin: Springer).
[12] Mrowka B 1951 Zur Darstellung der Quantenmechanik I. *Z. Phys.* **130** 164-173.
[13] Schuch D, Chung K–M and Hartmann H 1983 Nonlinear Schrödinger-type field equation for the description of dissipative systems. I. Derivation of the nonlinear field equation and one-dimensional example. *J. Math. Phys.* **24** 1652.
[14] Schuch D 2018 *Quantum Theory from a Nonlinear Perspective-Riccati Equations in Fundamental Physics* (Heidelberg: Springer).

Journal of Physics: Conference Series **2533** (2023) 012041 doi:10.1088/1742-6596/2533/1/012041

Dynamical atom-wall Casimir-Polder effect after a sudden change of the atomic position

Antonio Noto[1], **Roberto Passante**[1,2], **Lucia Rizzuto**[1,2], **Salvatore Spagnolo**[1]

[1]Dipartimento di Fisica e Chimica - Emilio Segr'e, Università degli Studi di Palermo, Via Archirafi 36, I-90123 Palermo, Italia

[2] INFN, Laboratori Nazionali del Sud, I-95123 Catania, Italia

E-mail: lucia.rizzuto@unipa.it

Abstract. We investigate the dynamical Casimir-Polder force between an atom and a conducting wall during the time evolution of the system from a partially dressed state. This state is obtained by a sudden change of the atomic position with respect to the plate. To evaluate the time-dependent atom-plate Casimir-Polder force we solve the Heisenberg equations for the field and atomic operators by an iterative technique. We find that the dynamical atom-plate Casimir-Polder interaction exhibits oscillation in time, and can be attractive or repulsive depending on time and the atom-wall distance. We also investigate the time dependence of global observables, such as the field and atomic Hamiltonians, and discuss some interesting features of the dynamical process bringing the interaction energy to the equilibrium configuration.

1. Introduction

Vacuum field fluctuations of the quantized electromagnetic field are a consequence of the Heisenberg uncertainty relations and are at the origin of many physical phenomena, for example the Lamb shift [1, 2], the spontaneous emission by an excited atom [3], the resonant and dispersion Casimir-Polder interactions between neutral atoms [4, 5, 6, 7, 8]. A remarkable property of these effects is their dependence on the geometry of the system, as well as on the magneto-dielectric properties of the materials [9, 10, 11, 12], and it has been shown that they can be controlled through a structured environment (cavities, waveguides, photonic crystals) [13, 14, 11, 15]. The effect of a static external field on the dispersion interactions between two atoms has been also considered [16, 17, 18, 19, 20].

These effects have been also investigated in dynamical situations, for example when atoms or macroscopic objects are set in motion or when some physical parameter of the system is modulated in time [21, 22, 23]. For example, real photons can be produced from the vacuum by a non adiabatic change of some boundary in the system (Dynamical Casimir Effect) or when an atom is subjected to an oscillatory motion in vacuum (microscopic Dynamical Casimir Effect) [24, 25]. Also, a *frictional force* may arise when two or more objects move relative to each other in vacuum at a constant velocity, an effect usually known as quantum friction [22, 26]. The effects of uniform acceleration of neutral atoms in vacuum and their relation with the Unruh effects have been also studied [27, 28]. Dynamical environments can also modify radiative properties of atoms or molecules nearby, under adiabatic conditions [29, 30, 31, 32].

A conceptually different situation also giving a dynamical (i.e. time-dependent) effect, is when some physical parameter in the system is suddenly changed, so that the system evolves starting from a non-equilibrium situation. For example, a sudden change of some atomic parameter, such as the atomic transition frequency between two levels or the electric dipole moment of one atom, gives rise to time-dependent effects, such as the dynamical Lamb shift, the dynamical Casimir-Polder force between two atoms [33] or the dynamical Casimir-Polder force between an atom and a conducting mirror [34, 35, 36]. From the physical point of view, these dynamical effects are essentially related to a change of the virtual photon cloud around the atom due to a rapid change of some physical characteristic of the atom by an external action [37, 38, 39]. It should be stressed that no real photons creation is taken into account in this case: our time-dependent effects arise from the non-equilibrium dynamics of the system, that evolves unitarily under the new Hamiltonian. We note that non-equilibrium conditions may manifest in many realistic physical situations, for example when two or more interacting bodies in a system can be at different temperatures from each other or from the temperature of the environment [40].

The main aim of this paper is to investigate the time-dependent atom-wall Casimir-Polder interaction energy between a ground-state atom and a plane mirror, when the system evolves from a nonequilibrium initial state, after a sudden change of the atomic position with respect to the mirror. Specifically, we consider an atom prepared in its dressed ground state, in the presence of a reflecting plane mirror. This state is an eigenstate of the total atom-field interacting Hamiltonian. We then suppose that at $t = 0$ the atomic position (with respect to the mirror) is abruptly changed by some external action (quench of the atomic position). This change brings the system to a non equilibrium situation since the initial (dressed) ground state of the system is no longer an eigenstate of the new Hamiltonian, and the system starts to evolve in time. We investigate the time-dependent atom-wall Casimir-Polder force during the dynamical evolution of the system, leading it to the new equilibrium configuration. We first solve the Heisenberg equations of motion of the atom-field interacting system by an iterative procedure, and then evaluate explicitly the time-dependent atom-wall Casimir–Polder interaction energy during the atomic self-dressing process. We find that, although the atom is in its ground state, the dynamical Casimir-Polder force exhibits oscillations in time, and can be attractive or repulsive depending on the atomic position and time. This is at variance with the usual (static) Casimir-Polder force between a ground-state atom and a mirror, where the force is always attractive. In the limit of long times it settles to the usual attractive Casimir-Polder interaction between a ground-state atom and a perfectly reflecting plate. We also discuss the time-dependent expectation values of global observables, such as the atomic and field Hamiltonians, on the partially dressed state and their asymptotic behavior, and discuss some aspects of the dynamical process bringing the interaction energy to the new equilibrium situation. Our results show that new effects can appear under non equilibrium situations and that the Casimir-Polder force can exhibit new features in dynamical situations.

2. Dynamical atom-wall Casimir-Polder interaction energy

Our system consists of a neutral atom, modeled as a two-level system, placed at a distance $|\mathbf{r}_0| = z_0$ from an infinite perfectly conducting plate, and interacting with the quantum electromagnetic field in the vacuum state (see Fig. (1(a))).

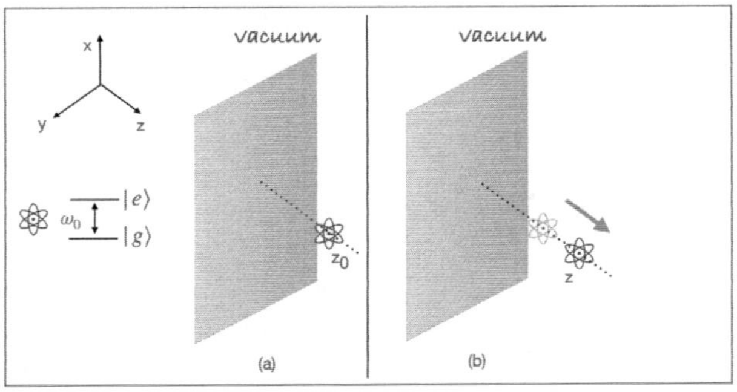

Figure 1. Pictorial description of the system. (a) The atom, modeled as two-level system, is located at a distance $|\mathbf{r}_0| = z_0$ from the mirror at $z = 0$. (b) At $t = 0$ the atom is subjected to a rapid change of its position, from $|\mathbf{r}_0| = z_0$ to the new position $|\mathbf{r}| = z$ with respect to the mirror.

We adopt the Hamiltonian in the multipolar coupling scheme within the dipole approximation, in the Coulomb gauge; the interaction Hamiltonian is then given by $H_I = -\boldsymbol{\mu}\cdot\mathbf{E}(\mathbf{x})$, where $\boldsymbol{\mu}$ is the atomic dipole moment operator and $\mathbf{E}(\mathbf{x})$ is the transverse displacement field at the atomic position $\mathbf{x}$.

The total Hamiltonian describing the atom-field interacting system, before the change of the atomic position, is given by [41]

$$\tilde{H} = \hbar\omega_0 S_z + \sum_{\mathbf{k}j} \hbar\omega_k a^\dagger_{\mathbf{k}j} a_{\mathbf{k}j} + \tilde{V}_1 + \lambda\tilde{V}_2, \tag{1}$$

$$\tilde{V}_1 = \sum_{\mathbf{k}j}\left(\epsilon_{\mathbf{k}j}(\mathbf{r}_0)a_{\mathbf{k}j}S_+ + \epsilon^\star_{\mathbf{k}j}(\mathbf{r}_0)a^\dagger_{\mathbf{k}j}S_-\right), \quad \tilde{V}_2 = \sum_{\mathbf{k}j}\left(\epsilon_{\mathbf{k}j}(\mathbf{r}_0)a^\dagger_{\mathbf{k}j}S_+ + \epsilon^\star_{\mathbf{k}j}(\mathbf{r}_0)a_{\mathbf{k}j}S_-\right),$$

where $a_{\mathbf{k}j}$ and $a^\dagger_{\mathbf{k}j}$ are the usual bosonic annihilation and creation field operators, S_+, S_- and S_z are the atomic pseudospin operators, ω_0 is the atomic transition frequency. Also, $\epsilon_{\mathbf{k}j}(\mathbf{r}_0) = -i\sqrt{\frac{2\pi\hbar c}{V}}\sum_{\mathbf{k}j}\sqrt{k}\,[\boldsymbol{\mu}\cdot\mathbf{f}_{\mathbf{k}j}(\mathbf{r}_0)]$ is the coupling constant and $\mathbf{f}_{\mathbf{k}j}(\mathbf{r})$ are the usual field mode functions satisfying the boundary conditions on the perfectly conducting walls of a cubic cavity [39]. Finally, λ is a parameter that can assume the values $\lambda = 0, -1$, so that for $\lambda = 0$ we obtain the Hamiltonian in the Rotating Wave Approximation (RWA).

We assume the atom initially prepared in its dressed ground state, that is an eigenstate of total Hamiltonian (1); at first-order in the atom-field coupling we have,

$$|\tilde{g}\rangle_d = |g, \{0_{\mathbf{k}j}\}\rangle - \lambda\sum_{\mathbf{k}j}\frac{\epsilon_{\mathbf{k}j}(\mathbf{r}_0)}{\hbar(\omega_k + \omega_0)}|e, 1_{\mathbf{k}j}\rangle\,. \tag{2}$$

We now suppose that at time $t = 0$ the atomic position is abruptly changed by some external action, from $\mathbf{r}_0$ to the final position $\mathbf{r}$ (see figure (1(b))). This is equivalent to a rapid change of the atom-field coupling that *instantaneously* brings the system to a dynamical non-equilibrium configuration. Our assumption of a sudden change of the atomic position is valid if this change occurs in a time t such that $t \ll 2r_0/c$ (that is the time taken by a virtual photon emitted by the atom to go back after reflection on the mirror) and $t \ll \omega_0^{-1}$. For an atom-plate distance of the order of 10^{-7} m, our non-adiabatic approximation requires times of the order of 10^{-16} s, which is also consistent with the requirement $t \ll \omega_0^{-1}$ for a typical atomic transition frequency $\omega_0 \sim 10^{15}\,\mathrm{s}^{-1}$.

As a consequence of this change, the ground state $|\tilde{g}\rangle_d$ in (2) it is not longer an eigenstate of the new Hamiltonian which is now

$$\begin{aligned} H &= \hbar\omega_0 S_z + \sum_{\mathbf{k}j} \hbar\omega_k a^\dagger_{\mathbf{k}j} a_{\mathbf{k}j} + \sum_{\mathbf{k}j} \left(\epsilon_{\mathbf{k}j}(\mathbf{r}) a_{\mathbf{k}j} S_+ + \epsilon^\star_{\mathbf{k}j}(\mathbf{r}) a^\dagger_{\mathbf{k}j} S_- \right) \\ &+ \lambda \sum_{\mathbf{k}j} \left(\epsilon_{\mathbf{k}j}(\mathbf{r}) a^\dagger_{\mathbf{k}j} S_+ + \epsilon^\star_{\mathbf{k}j}(\mathbf{r}) a_{\mathbf{k}j} S_- \right) , \end{aligned} \tag{3}$$

and the system will evolve in time starting from the initial partially dressed state $|\psi\rangle_p$, under the Hamiltonian (3) (note that the only difference between the new Hamiltonian (3) and the old one in (1) is the dependence on the atomic position in the coupling constant, which is now $\epsilon_{\mathbf{k}j}(\mathbf{r})$ and not $\epsilon_{\mathbf{k}j}(\mathbf{r}_0)$). It is worth noting that the bare ground-state $|g, \{0_{\mathbf{k}j}\}\rangle$ is an exact eigenstate of the total Hamiltonian in the RWA ($\lambda = 0$), and therefore it does not evolve. We stress that in our approximation, no time dependence is included in the system's Hamiltonian, the atom being supposed at rest and locally interacting with the vacuum field fluctuations at the new (static) atomic position. Therefore, no real photons creation is involved in this case, time-dependent effects only resulting from the non-equilibrium dynamics of the system that evolves unitarily under the new Hamiltonian (3). Time-dependent effects are only due to the atomic self-dressing process after the rapid perturbation of the virtual photons cloud surrounding the atom. Also, under these assumptions, we can neglect the contribution from the Röntgen term in the Hamiltonian, which is of the order of v/c, where v is the velocity of the atom. To evaluate the time-dependent Casimir-Polder interaction, we follow the same procedure adopted in [34, 35, 36, 39] where the second-order interaction energy is obtained as half of the average value of the interaction Hamiltonian $H_I^{(2)}(t)$ on the initial state of the system (in our case, the partially dressed state at $t = 0^+$ immediately after the change of the atomic position), and taking only terms depending on the atom-plate distance,

$$\Delta E^{(2)}(t) = \frac{1}{2} \; {}_p\langle\psi| \, H_I^{(2)}(t) \, |\psi\rangle_p \, , \tag{4}$$

where $|\psi\rangle_p$ is the partially dressed state of the system, and $H_I^{(2)}(t)$ is the (second-order) interaction Hamiltonian in the Heisenberg picture, obtained by solving iteratively the Heisenberg equations of motion for field and atomic operators. Equation (4) is a generalization to time-dependent situations of the well-known expression of the interaction energy obtained by second-order stationary perturbation theory [42], $\Delta E = \frac{1}{2} \; {}_d\langle g|H_I \, |g\rangle_d$, where $|\psi\rangle_d$ is the second-order dressed state of the system, and H_I is the interaction Hamiltonian in the Schrödinger representation. We note that, since the time evolution of the system is unitary, the total energy of the system is conserved during the dynamical self-dressing process. However, this energy does not coincide with the total energy of the (second-order) dressed state $|g\rangle_d$ of the Hamiltonian (3), that is ${}_p\langle\psi| \, H^{(2)}(t) \, |\psi\rangle_p \neq {}_d\langle g| \, H \, |g\rangle_d$. This is because dressed and partially dressed states have different energy (we shall discuss this relevant point below). Nevertheless, as we now show, the average value of the interaction Hamiltonian $H_I^{(2)}(t)$ on the partially dressed state $|\psi\rangle_p$, settles to its stationary value ${}_d\langle g| \, H_I \, |g\rangle_d$ in the limit of large times ($t \to \infty$), and gives the usual atom-plate Casimir-Polder interaction energy in the static case.

We first calculate the expressions of atomic and field operators in the Heisenberg representation. At the lowest significant order, we obtain

$$S_+(t) = e^{i\omega_0 t} S_+(0) + \frac{2i}{\hbar} S_z(0) e^{i\omega_0 t} \sum_{\mathbf{k}j} \epsilon_{\mathbf{k}j}(\mathbf{r}) [a^\dagger_{\mathbf{k}j}(0) G(\omega_k - \omega_0, t) + \lambda a_{\mathbf{k}j}(0) G^\star(\omega_k + \omega_0, t)] \, , \tag{5}$$

$$a_{\mathbf{k}j}(t) = e^{-i\omega_k t} a_{\mathbf{k}j}(0) + \frac{i}{\hbar} e^{-i\omega_k t} \epsilon_{\mathbf{k}j}(\mathbf{r}) \left(S_-(0) G(\omega_k - \omega_0, t) - \lambda S_+ G(\omega_k + \omega_0, t) \right) \, , \tag{6}$$

where we have introduced the function

$$G(x,t) = \int_0^t e^{ixt'} dt' = \frac{e^{ixt} - 1}{ix} . \tag{7}$$

Substituting equations (5) and (6) into the expression of the interaction Hamiltonian $V_1 + \lambda V_2$ (see Eq. (3)), and taking the average value on the partial dressed state $|\psi\rangle_p$, we obtain (up to second order in the coupling):

$$\mathcal{E}(t)_I =_p \langle\psi| H_I^{(2)}(t) |\psi\rangle_p = -\lambda^2 \frac{4\pi}{V} \sum_{\mathbf{k}j} (\boldsymbol{\mu} \cdot \mathbf{f}_{\mathbf{k}j}(\mathbf{r}))^2 \frac{\omega_k}{\omega_k + \omega_0} (1 - \cos(\omega_0 + \omega_k)t)$$
$$-\lambda^2 \frac{4\pi}{V} \sum_{\mathbf{k}j} (\boldsymbol{\mu} \cdot \mathbf{f}_{\mathbf{k}j}(\mathbf{r}))\, (\boldsymbol{\mu} \cdot \mathbf{f}_{\mathbf{k}j}(\mathbf{r}_0)) \frac{\omega_k}{\omega_k + \omega_0} \cos(\omega_0 + \omega_k)t . \tag{8}$$

The presence of the parameter λ in the equation above clearly shows that the time-dependent interaction energy is related to the virtual photons that are emitted and reabsorbed by the atom during the self-dressing process of the system. In other words, energy conserving processes arising from the rotating terms of the interaction Hamiltonian and involving real photons, do not play any role at the order considered.

In the continuum limit, using $\sum_{\mathbf{k}} \to V/(2\pi)^3 \int d^3\mathbf{k}$, after some algebra, we get

$$\mathcal{E}(t)_I = \Delta E^{(2)}_{f.s.}(t) + \Delta E^{(2)}_{b.}(t) , \tag{9}$$

where

$$\Delta E^{(2)}_{f.s.}(t) = -2\frac{\boldsymbol{\mu}_\ell \boldsymbol{\mu}_m}{\pi} \left\{ \frac{2}{3}\delta_{\ell m} \int_0^\infty dk k^3 \frac{1 - \cos\left(ct(k + k_0)\right)}{k + k_0} + F^{R'}_{\ell m} \frac{1}{R'} \int_0^\infty dk \frac{\sin kR'}{k + k_0} \cos\left(ct(k + k_0)\right) \right\} , \tag{10}$$

is the free-space contribution to the time-dependent interaction energy of the partially dressed atom ($R' = |\mathbf{r} - \mathbf{r}_0|$, and $F^x_{\ell m} = (-\delta_{\ell m}\nabla^2 + \nabla_\ell \nabla_m)^x$ is a differential operator acting on the variable x, and we used the Einstein convention for repeated indices), and

$$\Delta E^{(2)}_{b.}(t) = \frac{\boldsymbol{\mu}_\ell \boldsymbol{\mu}_m}{\pi} \left\{ \sigma_{\ell p} \left[F^{\bar{R}}_{pm} \frac{1}{\bar{R}} \int_0^\infty dk \frac{\sin k\bar{R}}{k + k_0} \left(1 - \cos\left(ct(k + k_0)\right)\right) + F^{\bar{R}'}_{pm} \frac{1}{\bar{R}'} \int_0^\infty dk \frac{\sin k\bar{R}'}{k + k_0} \cos\left(ct(k + k_0\right) \right] \right\} , \tag{11}$$

is the boundary-dependent contribution to the interaction energy. Here, $\bar{R} = |\mathbf{r} - \sigma\mathbf{r}|$ is the distance between the atomic position $\mathbf{r}$ and its image through the mirror, and $\bar{R}' = |\mathbf{r} - \sigma\mathbf{r}_0|$, where $\sigma = \mathrm{diag}(1, 1, -1)$ is the reflection matrix .

The expressions above give the time-dependent interaction energy during the self-dressing process of the atom after the abrupt change of atom's position. In particular, the free-space term (10) can be interpreted as a time-dependent Lamb-shift due to the nonadiabatic perturbation of the virtual photons cloud surrounding the atom. On the other hand, the boundary-dependent contribution (11) gives the time-dependent atom-wall Casimir-Polder interaction energy.

We now focus on this last term with the aim to discuss some features of this dynamical interaction energy. Using equation (11) in (4), for an atom with an isotropic electric dipole moment ($|\boldsymbol{\mu}_x|^2 = |\boldsymbol{\mu}_y|^2 = |\boldsymbol{\mu}_z|^2 = |\boldsymbol{\mu}|^2/3$), we obtain

$$\Delta E^{(2)}_{CP}(t) = \frac{1}{2}\Delta E^{(2)}_{b.}(t) = -\frac{2\mu^2}{3\pi}\lim_{q\to 1} D_q \left\{ \frac{1}{\bar{R}} \int_0^\infty d\chi \frac{\sin(q\chi)}{\chi+\chi_0}\left(1-\cos(a(\chi+\chi_0))\right) - \frac{1}{\bar{R}'}\int_0^\infty \frac{\sin(q\bar{\chi})}{\bar{\chi}+\bar{\chi}_0}\left(cos(\bar{a}(\bar{\chi}+\bar{\chi}_0))\right) \right\} \tag{12}$$

where we have defined the differential operator $D_q = 2 - 2\partial/\partial q + \partial^2/\partial q^2$, $a = \frac{ct}{\bar{R}}$, and $\bar{a} = \frac{ct}{\bar{R}'}$. Also, $\chi = k\bar{R}$, $\chi_0 = k_0\bar{R}$, $\bar{\chi} = k\bar{R}'$ and $\bar{\chi}_0 = k_0\bar{R}'$. The integrals in (12) can be explicitly calculated for $a < 1$ (i.e. $t < \bar{R}/c$) and $a > 1$ (i.e. $t > \bar{R}/c$) (note that the integrals diverge on the light cone $ct = \bar{R}$; the physical origin of this divergence can be ascribed to the dipole approximation and the radiation reaction field [34, 35]). We do not give here the explicit expression of the time-dependent Casimir-Polder interaction energy, because the relevant features of the atom-plate interaction can be extrapolated from figures 2 and 3. These figures show the time evolution of the Casimir-Polder interaction energy, near the light-cone, for $ct < \bar{R}$ and $ct > \bar{R}'$ (we have supposed that the atom's position is rapidly changed of a distance of the order 10^{-10} m from the initial position $z_0 = 10^{-7}$ m). Similarly to the case discussed in [35], the Casimir-Polder interaction energy oscillates in time, switching in time from an attractive to a repulsive force. In the asymptotic limit ($t \to \infty$), the interaction energy settles to its stationary value. The oscillations of the interaction energy and its strong increase around the round-trip time $t = \bar{R}/c$ (typically of the order of $10^{-16} \div 10^{-15}$ s,) suggest the possibility of observing these effects in laboratory. Also, due to the assumption of an initial partially dressed state, the value of the interaction energy at $t = 0^+$ is not zero; it represents a part of the energy stored in the system immediately after the change the atom's position. To make this point clearer, it is convenient to reconsider the interaction energy of the partially dressed state given in (8). With a simple algebraic manipulation, this expression can be written as follows

$$\mathcal{E}_I(t) = {}_p\langle\psi| H_I^{(2)}(t) |\psi\rangle_p = -\frac{4\pi}{V}\sum_{\mathbf{k}j} (\boldsymbol{\mu}\cdot\mathbf{f}_{\mathbf{k}j}(\mathbf{r}))^2 \frac{\omega_k}{\omega_k+\omega_0} + \frac{4\pi}{V}\sum_{\mathbf{k}j} \boldsymbol{\mu}\cdot\mathbf{f}_{\mathbf{k}j}(\mathbf{r}) \left(\boldsymbol{\mu}\cdot\mathbf{f}_{\mathbf{k}j}(\mathbf{r}) - \boldsymbol{\mu}\cdot\mathbf{f}_{\mathbf{k}j}(\mathbf{r}_0)\right) \frac{\omega_k}{\omega_k+\omega_0}\cos(\omega_0+\omega_k)t\,, \tag{13}$$

that clearly shows that the interaction energy at the initial time is different from zero and depends on $\mathbf{r}_0$ and $\mathbf{r}$. Thus, it keeps memory of the atom's position before and after its sudden change. On the other hand, in the limit of large times, the average value of the interaction energy settles to its static value in the new equilibrium configuration of the system, that is

$$\Delta E_I^{(static)} = \frac{1}{2}\lim_{t\to\infty} {}_p\langle\psi| H_I^{(2)}(t) |\psi\rangle_p = \frac{1}{2}\, {}_d\langle g^{(2)}| H_I |g^{(2)}\rangle_d \tag{14}$$

where $|g^{(2)}\rangle_d$ is the fully (second-order) dressed state of the interacting system (with the atom located at position $\mathbf{r}$).

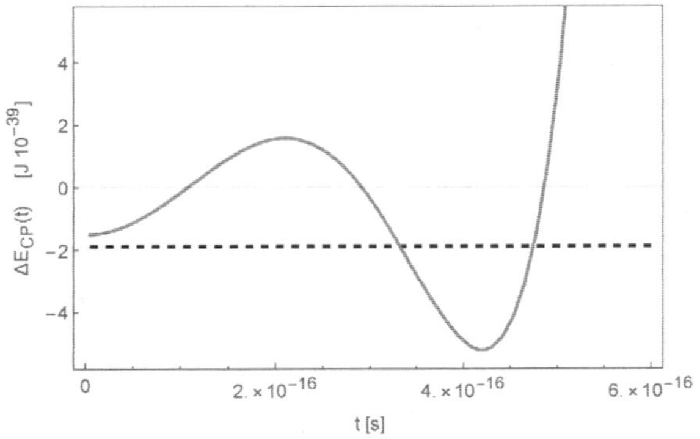

Figure 2. Dynamical Casimir-Polder interaction energy for $t < \bar{R}/c$. The orange (solid) line corresponds to the case of an initial partially dressed state, while the blue (dashed) line refers to the static CP force after the quench. The figure shows that the value of the interaction energy at $t = 0^+$ is non vanishing, as a consequence of the assumption of an initial partially dressed state. The values used are $\mu = 6.31 \times 10^{-30}$ C m, $k_0 = 5.0 \times 10^7$ m^{-1}, $r_0 = 1.001 \times 10^{-7}$ m, $r = 1.0 \times 10^{-7}$ m

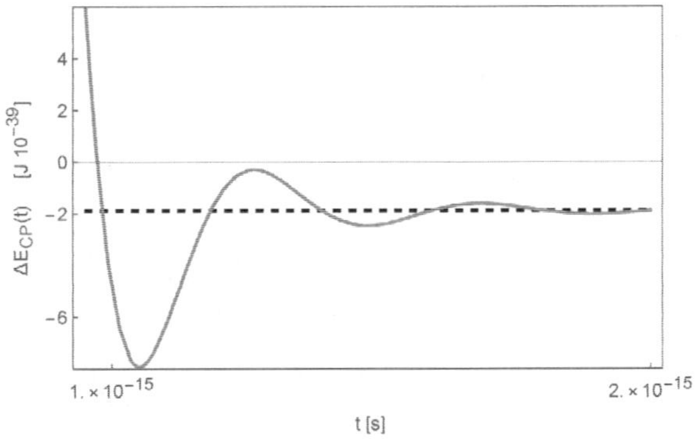

Figure 3. Dynamical Casimir-Polder interaction energy for $t > \bar{R}/c$. The orange (solid) line corresponds to the time-dependent interaction energy for $t > \bar{R}/c$, while the blue (dashed) line refers to the static interaction energy in the asymptotic limit. $\mu = 6.31 \times 10^{-30}$ C m, $k_0 = 5.0 \times 10^7$ m^{-1}, $r_0 = 1.001 \times 10^{-7}$ m, $r = 1.0 \times 10^{-7}$ m

Thus, we find that the system evolves in time and the interaction energy reaches a new equilibrium condition in the long-time limit. It is also interesting to evaluate the time-dependent average values of global observables, such as the field and atomic Hamiltonians. To do that, we first calculate the second-order dressed ground state of the total Hamiltonian (1). We get,

$$\begin{aligned} |\tilde{g}^{(2)}\rangle_d &= \left(1 - \frac{\lambda^2}{2}\sum_{\mathbf{k}j} \frac{|\epsilon_{\mathbf{k}j}(\mathbf{r}_0)|^2}{\hbar^2(\omega_k+\omega_k)^2}\right)|g,\{0_{\mathbf{k}j}\}\rangle - \lambda\sum_{\mathbf{k}j}\frac{\epsilon_{\mathbf{k}j}(\mathbf{r}_0)}{\hbar(\omega_k+\omega_0)}|e,1_{\mathbf{k}j}\rangle \\ &-\lambda\sum_{\mathbf{k}j}\sum_{\mathbf{k}'j'}\frac{\epsilon_{\mathbf{k}j}(\mathbf{r}_0)\epsilon_{\mathbf{k}'j'}(\mathbf{r}_0)}{\hbar^2(\omega_0+\omega_{k'})(\omega_0+\omega_k)}|g,1_{\mathbf{k}j},1_{\mathbf{k}'j'}\rangle\,. \end{aligned} \quad (15)$$

Journal of Physics: Conference Series **2533** (2023) 012041 doi:10.1088/1742-6596/2533/1/012041

As before, this state is a partially dressed state after the *nonadiabatic* change of the atomic position and evolves dynamically under the Hamiltonian (3). The time-dependent average values of atomic and field Hamiltonian on the partially dressed state (15) are easily evaluated, obtaining

$$\begin{aligned}\mathcal{E}_A(t) &= {}_p\langle\psi^{(2)}|H_A^{(2)}(t)|\psi^{(2)}\rangle_p = \frac{2\pi}{V}\sum_{\mathbf{k}j}(\boldsymbol{\mu}\cdot\mathbf{f}_{\mathbf{k}j}(\mathbf{r}_0))^2\frac{\omega_0\omega_k}{(\omega_0+\omega_k)^2}\\ &+\frac{4\pi}{V}\sum_{\mathbf{k}j}(\boldsymbol{\mu}\cdot\mathbf{f}_{\mathbf{k}j}(\mathbf{r}))\left(\boldsymbol{\mu}\cdot\mathbf{f}_{\mathbf{k}j}(\mathbf{r})-\boldsymbol{\mu}\cdot\mathbf{f}_{\mathbf{k}j}(\mathbf{r}_0)\right)\frac{\omega_0\omega_k}{(\omega_0+\omega_k)^2}\left(1-\cos(\omega_0+\omega_k)t\right),\end{aligned} \tag{16}$$

and

$$\begin{aligned}\mathcal{E}_F(t) &= {}_p\langle\psi^{(2)}|H_F^{(2)}(t)|\psi^{(2)}\rangle_p = \frac{2\pi}{V}\sum_{\mathbf{k}j}(\boldsymbol{\mu}\cdot\mathbf{f}_{\mathbf{k}j}(\mathbf{r}_0))^2\frac{\omega_k^2}{(\omega_0+\omega_k)^2}\\ &+\frac{4\pi}{V}\sum_{\mathbf{k}j}(\boldsymbol{\mu}\cdot\mathbf{f}_{\mathbf{k}j}(\mathbf{r}))\left(\boldsymbol{\mu}\cdot\mathbf{f}_{\mathbf{k}j}(\mathbf{r})-\boldsymbol{\mu}\cdot\mathbf{f}_{\mathbf{k}j}(\mathbf{r}_0)\right)\frac{\omega_k^2}{(\omega_0+\omega_k)^2}\left(1-\cos(\omega_0+\omega_k)t\right).\end{aligned} \tag{17}$$

In the asymptotic limit ($t\to\infty$), we have

$$\mathcal{E}_A(t)\to\frac{2\pi}{V}\sum_{\mathbf{k}j}\frac{\omega_0\omega_k\,(\boldsymbol{\mu}\cdot\mathbf{f}_{\mathbf{k}j}(\mathbf{r}_0))^2}{(\omega_0+\omega_k)^2}+\frac{4\pi}{V}\sum_{\mathbf{k}j}\frac{\omega_0\omega_k\,(\boldsymbol{\mu}\cdot\mathbf{f}_{\mathbf{k}j}(\mathbf{r}))}{(\omega_0+\omega_k)^2}\left(\boldsymbol{\mu}\cdot\mathbf{f}_{\mathbf{k}j}(\mathbf{r})-\boldsymbol{\mu}\cdot\mathbf{f}_{\mathbf{k}j}(\mathbf{r}_0)\right), \tag{18}$$

and

$$\mathcal{E}_F(t)\to\frac{2\pi}{V}\sum_{\mathbf{k}j}\frac{\omega_k^2\,(\boldsymbol{\mu}\cdot\mathbf{f}_{\mathbf{k}j}(\mathbf{r}_0))^2}{(\omega_0+\omega_k)^2}+\frac{4\pi}{V}\sum_{\mathbf{k}j}\frac{\omega_k^2\,(\boldsymbol{\mu}\cdot\mathbf{f}_{\mathbf{k}j}(\mathbf{r}))}{(\omega_0+\omega_k)^2}\left(\boldsymbol{\mu}\cdot\mathbf{f}_{\mathbf{k}j}(\mathbf{r})-\boldsymbol{\mu}\cdot\mathbf{f}_{\mathbf{k}j}(\mathbf{r}_0)\right). \tag{19}$$

A comparison with the analogous quantities obtained in the static case shows that, in the limit of large times, the time-dependent atomic and field energies does not settle to their stationary values on the fully dressed state of the system; the additional terms in (18) and (19) are related to the energy initially stored in the field and in the atom, and to the *work* done to bring the system in out-of-equilibrium configuration. In fact, we have

$$\lim_{t\to\infty}\mathcal{E}_A(t)\neq\mathcal{E}_A^{stat.}={}_d\langle g^{(2)}|\,H_A\,|g^{(2)}\rangle_d=\frac{2\pi}{V}\sum_{\mathbf{k}j}(\boldsymbol{\mu}\cdot\mathbf{f}_{\mathbf{k}j}(\mathbf{r}))^2\frac{\omega_0\omega_k}{(\omega_0+\omega_k)^2}, \tag{20}$$

$$\lim_{t\to\infty}\mathcal{E}_F(t)\neq\mathcal{E}_F^{stat.}={}_d\langle g^{(2)}|\,H_F\,|g^{(2)}\rangle_d=\frac{2\pi}{V}\sum_{\mathbf{k}j}(\boldsymbol{\mu}\cdot\mathbf{f}_{\mathbf{k}j}(\mathbf{r}))^2\frac{\omega_k^2}{(\omega_0+\omega_k)^2}. \tag{21}$$

This is different from the case where the system evolves from an initial bare state, where in the asymptotic limit both dynamical energies stored in the field and in the atom tend to twice their stationary values [39, 43]. On the other hand, at large times the energy stored in the interaction Hamiltonian coincides with its stationary value (see Equation(14)). It should be also stressed that since the time evolution of the system is unitary, the average value of the total Hamiltonian does not change during the time evolution of the system, and it is

$$\begin{aligned}\mathcal{E}_{tot}^{part.} &= {}_p\langle\psi^{(2)}|H^{(2)}(t)|\psi^{(2)}\rangle_p=\mathcal{E}_A+\mathcal{E}_F+\mathcal{E}_I=-\frac{2\pi}{V}\sum_{\mathbf{k}j}(\boldsymbol{\mu}\cdot\mathbf{f}_{\mathbf{k}j}(\mathbf{r}_0))^2\frac{\omega_k}{\omega_0+\omega_k}\\ &+\frac{4\pi}{V}\sum_{\mathbf{k}j}(\boldsymbol{\mu}\cdot\mathbf{f}_{\mathbf{k}j}(\mathbf{r}))\,(\boldsymbol{\mu}\cdot\mathbf{f}_{\mathbf{k}j}(\mathbf{r}_0))\frac{\omega_k}{\omega_0+\omega_k}-\frac{4\pi}{V}\sum_{\mathbf{k}j}(\boldsymbol{\mu}\cdot\mathbf{f}_{\mathbf{k}j}(\mathbf{r}))^2\frac{\omega_k}{\omega_0+\omega_k}.\end{aligned} \tag{22}$$

This is different from what obtained in a completely static situation, that is

$$\mathcal{E}_{tot}^{stat.} = {}_d\langle g^{(2)}| \, H_{tot} \, |g^{(2)}\rangle_d = -\frac{2\pi}{V} \sum_{\mathbf{k}j} (\boldsymbol{\mu} \cdot \mathbf{f}_{\mathbf{k}j}(\mathbf{r}))^2 \, \frac{\omega_k}{\omega_0 + \omega_k} \, . \tag{23}$$

Comparison of (22) and (23) shows that for $t \to \infty$ the total energy, similarly to the atomic and field contributions, does not settle to the value obtained in the stationary case.

3. Conclusions

In this paper, we have investigated the dynamical Casimir-Polder interaction energy between an atom and a perfect reflecting plane during the time evolution of the system after a sudden change of the atomic position, that brings the system to a nonequilibrium situation. We have discussed the dynamical processes leading the interaction energy to the new equilibrium configuration. We have shown that the time-dependent Casimir-Polder interaction energy oscillates in time, changing from attractive to repulsive, and that in the limit of large times it settles to its stationary value. Finally, we have investigated some interesting aspects of the dynamical atomic self-dressing process by investigating the time dependence of the field and atomic Hamiltonians, and discussed the different behaviour of *local* and *global* observables in the dynamical evolution of the system.

Acknowledgments

R. P. and L. R. acknowledge partial financial support from the FFR2021 grant from the University of Palermo, Italy.

References

[1] Lamb W E and Retherford R C 1947 *Phys. Rev.* **72**(3) 241–243 URL https://link.aps.org/doi/10.1103/PhysRev.72.241
[2] Maclay G J 2020 *Physics* **2** 105–149 ISSN 2624-8174 URL https://www.mdpi.com/2624-8174/2/2/8
[3] Milonni P 2019 *An Introduction to Quantum Optics and Quantum Fluctuations* (London, U.K.: Oxford University Press)
[4] Casimir H B G 1948 *Proc. Akad. Wet. Amsterdam* **51** 793–795 ISSN 0370-0348
[5] Casimir H B G and Polder D 1948 *Phys. Rev.* **73**(4) 360–372 URL https://link.aps.org/doi/10.1103/PhysRev.73.360
[6] Compagno G, Passante R and Persico F 1995 *Atom-Field Interactions and Dressed Atoms* (Cambridge, UK: Cambridge Universty Press)
[7] Salam A 2010 *Molecular Quantum Electrodynamics* (Singapore: McGraw-Hill)
[8] Passante R 2018 *Symmetry* **10** 735 URL https://www.mdpi.com/2073-8994/10/12/735
[9] Bordag M, Klimchitskaya G, Mohideen U and Mostepanenko V 2009 *Advances in the Casimir Effect* (Oxford, UK: Oxford Science Publications)
[10] Spagnolo S, Dalvit D A R and Milonni P W 2007 *Phys. Rev. A* **75**(5) 052117 URL https://link.aps.org/doi/10.1103/PhysRevA.75.052117
[11] Palacino R, Passante R, Rizzuto L, Barcellona P and Buhmann S Y 2017 *Journal of Physics B: Atomic, Molecular and Optical Physics* **50** 154001 URL https://doi.org/10.1088/1361-6455/aa75f4
[12] Armata F, Butera S, Fiscelli G, Incardone R, Notararigo V, Palacino R, Passante R, Rizzuto L and Spagnolo S 2017 *Journal of Physics: Conference Series* **880** 012064 URL https://doi.org/10.1088/1742-6596/880/1/012064
[13] Incardone R, Fukuta T, Tanaka S, Petrosky T, Rizzuto L and Passante R 2014 *Phys. Rev. A* **89**(6) 062117 URL http://link.aps.org/doi/10.1103/PhysRevA.89.062117
[14] Notararigo V, Passante R and Rizzuto L 2018 *Scientific Reports* **8**(1) 5193 URL https://doi.org/10.1038/s41598-018-23416-0
[15] John S and Quang T 1994 *Phys. Rev. A* **50**(2) 1764–1769 URL http://link.aps.org/doi/10.1103/PhysRevA.50.1764
[16] Fiscelli G, Rizzuto L and Passante R 2018 *Phys. Rev. A* **98**(1) 013849 URL https://link.aps.org/doi/10.1103/PhysRevA.98.013849

[17] Fiscelli G, Rizzuto L and Passante R 2020 *Phys. Rev. Lett.* **124**(1) 013604 URL https://link.aps.org/doi/10.1103/PhysRevLett.124.013604
[18] Abrantes P P, Pessanha V, Farina C and Souza R d M e 2021 *Phys. Rev. Lett.* **126**(10) 109301 URL https://link.aps.org/doi/10.1103/PhysRevLett.126.109301
[19] Karimpour M R, Fedorov D V and Tkatchenko A 2022 *Phys. Rev. Research* **4**(1) 013011 URL https://link.aps.org/doi/10.1103/PhysRevResearch.4.013011
[20] Karimpour M R, Fedorov D V and Tkatchenko A 2022 *The Journal of Physical Chemistry Letters* **13** 2197–2204 pMID: 35231170 (*Preprint* https://doi.org/10.1021/acs.jpclett.1c04222) URL https://doi.org/10.1021/acs.jpclett.1c04222
[21] Dodonov V 2020 *Physics* **2** 67–104 ISSN 2624-8174 URL https://www.mdpi.com/2624-8174/2/1/7
[22] Reiche D, Intravaia F and Busch K 2022 *APL Photonics* **7** 030902 (*Preprint* https://doi.org/10.1063/5.0083067) URL https://doi.org/10.1063/5.0083067
[23] Antezza M, Braggio C, Carugno G, Noto A, Passante R, Rizzuto L, Ruoso G and Spagnolo S 2014 *Phys. Rev. Lett.* **113**(2) 023601 URL https://link.aps.org/doi/10.1103/PhysRevLett.113.023601
[24] Souza R d M e, Impens F m c and Neto P A M 2018 *Phys. Rev. A* **97**(3) 032514 URL https://link.aps.org/doi/10.1103/PhysRevA.97.032514
[25] Lo L and Law C K 2018 *Phys. Rev. A* **98**(6) 063807 URL https://link.aps.org/doi/10.1103/PhysRevA.98.063807
[26] Farías M B, Fosco C D, Lombardo F C and Mazzitelli F D 2019 *Phys. Rev. D* **100**(3) 036013 URL https://link.aps.org/doi/10.1103/PhysRevD.100.036013
[27] Crispino L C B, Higuchi A and Matsas G E A 2008 *Rev. Mod. Phys.* **80**(3) 787–838 URL https://link.aps.org/doi/10.1103/RevModPhys.80.787
[28] Šoda B, Sudhir V and Kempf A 2022 *Phys. Rev. Lett.* **128**(16) 163603 URL https://link.aps.org/doi/10.1103/PhysRevLett.128.163603
[29] Glaetzle A, Hammerer K, Daley A, Blatt R and Zoller P 2010 *Optics Communications* **283** 758–765 ISSN 0030-4018 quo vadis Quantum Optics? URL https://www.sciencedirect.com/science/article/pii/S0030401809010554
[30] Ferreri A, Domina M, Rizzuto L and Passante R 2019 *Symmetry* **11** ISSN 2073-8994 URL https://www.mdpi.com/2073-8994/11/11/1384
[31] Reina M, Domina M, Ferreri A, Fiscelli G, Noto A, Passante R and Rizzuto L 2021 *Phys. Rev. A* **103**(3) 033710 URL https://link.aps.org/doi/10.1103/PhysRevA.103.033710
[32] Calajò G, Rizzuto L and Passante R 2017 *Phys. Rev. A* **96**(2) 023802 URL https://link.aps.org/doi/10.1103/PhysRevA.96.023802
[33] Passante R and Persico F 2003 *Physics Letters A* **312** 319–323 ISSN 0375-9601 URL https://www.sciencedirect.com/science/article/pii/S0375960103006789
[34] Vasile R and Passante R 2008 *Phys. Rev. A* **78**(3) 032108 URL https://link.aps.org/doi/10.1103/PhysRevA.78.032108
[35] Messina R, Vasile R and Passante R 2010 *Phys. Rev. A* **82**(6) 062501 URL https://link.aps.org/doi/10.1103/PhysRevA.82.062501
[36] Haakh H R, Henkel C, Spagnolo S, Rizzuto L and Passante R 2014 *Phys. Rev. A* **89**(2) 022509 URL https://link.aps.org/doi/10.1103/PhysRevA.89.022509
[37] Passante R and Vinci N 1996 *Physics Letters A* **213** 119–124 ISSN 0375-9601 URL https://www.sciencedirect.com/science/article/pii/0375960196001442
[38] Rizzuto L, Passante R and Persico F 2004 *Phys. Rev. A* **70**(1) 012107 URL https://link.aps.org/doi/10.1103/PhysRevA.70.012107
[39] Armata F, Vasile R, Barcellona P, Buhmann S Y, Rizzuto L and Passante R 2016 *Phys. Rev. A* **94**(4) 042511 URL https://link.aps.org/doi/10.1103/PhysRevA.94.042511
[40] Noto A, Messina R, Guizal B and Antezza M 2014 *Phys. Rev. A* **90**(2) 022120 URL https://link.aps.org/doi/10.1103/PhysRevA.90.022120
[41] Power E A and Zienau S 1959 *Philosophical Transactions of the Royal Society of London A: Mathematical, Physical and Engineering Sciences* **251** 427–454 ISSN 0080-4614 URL http://rsta.royalsocietypublishing.org/content/251/999/427
[42] Compagno G, Passante R and Persico F 1983 *Physics Letters A* **98** 253–255 ISSN 0375-9601 URL https://www.sciencedirect.com/science/article/pii/0375960183908642
[43] Armata F, Kim M S, Butera S, Rizzuto L and Passante R 2017 *Phys. Rev. D* **96**(4) 045007 URL https://link.aps.org/doi/10.1103/PhysRevD.96.045007

Field observables near a fluctuating boundary

Federico Armata[1], **Salvatore Butera**[2], **Federico Montalbano**[1], **Roberto Passante**[1,3] **and Lucia Rizzuto**[1,3]

[1] Dipartimento di Fisica e Chimica - Emilio Segrè, Università degli Studi di Palermo, Via Archirafi 36, I-90123 Palermo, Italia
[2] School of Physics and Astronomy, University of Glasgow, Glasgow G12 8QQ, United Kingdom
[3] INFN, Laboratori Nazionali del Sud, I-95123 Catania, Italia

E-mail: roberto.passante@unipa.it

Abstract. We review several aspects related to the confinement of a massless scalar field in a cavity with a movable conducting wall of finite mass, free to move around its equilibrium position to which it is bound by a harmonic potential, and whose mechanical degrees of freedom are described quantum mechanically. This system, for small displacements of the movable wall from its equilibrium position, can be described by an effective interaction Hamiltonian between the field and the mirror, quadratic in the field operators and linear in the mirror operators. In the interacting, i.e. dressed, ground state, we first consider local field observables such as the field energy density: we evaluate changes of the field energy density in the cavity with the movable wall with respect to the case of a fixed wall, and corrections to the usual Casimir forces between the two walls. We then investigate the case of two one-dimensional cavities separated by a movable wall of finite mass, with two massless scalar fields defined in the two cavities. We show that in this case correlations between the squared fields in the two cavities exist, mediated by the movable wall, at variance with the fixed-wall case.

1. Introduction

The presence of a cavity or reflecting boundaries has deep consequences on quantum fields, because they impose boundary conditions to the field operators. Well-known physical consequences are, for example, modifications of the spontaneous emission rate of atoms [1, 2, 3], the Casimir effect [4, 5] and the Casimir-Polder interactions [6, 7]. Additional effects are present in the case of moving boundaries, for example when one boundary of a cavity is put in a nonadiabatic oscillatory motion, or its magnetodielectric properties change with time: a well-known example is the dynamical Casimir effect, consisting in the emission of pairs of real photons from the vacuum [8, 9, 10, 11].

In this paper we review some recent results on observable quantities of a quantum field near a reflecting boundary with a finite mass that is free to move and whose mechanical degrees of freedom are treated quantum mechanically. Specifically, our system consists of a massless one-dimensional scalar field in a cavity with a movable boundary. The movable boundary is assumed perfectly reflecting and bound to its equilibrium position by a harmonic potential; its mechanical degrees of freedom are treated quantum mechanically, and thus it is subjected to position fluctuations. This introduces an effective field-mirror interaction, as well as an effective interaction between the field modes mediated by the movable mirror [12, 13]. Also, the effective field-mirror interaction yields virtual excitations of the field in the interacting ground state, i.e. a sort of dressing of the movable wall. Changes of local field quantities such as its energy density, and spatial correlations between field observables, are also expected due to the presence of the trembling cavity wall. We show that, indeed, local field quantities in the dressed vacuum state, such as the expectation value of the squared field or the field energy density, are modified due to the motion of the wall, as well as the existence of spatial correlations of the squared field between points at the opposite sides of the movable wall.

This paper is organized as follows. In Sec. 2 we introduce our physical system of a cavity with a movable wall, its Hamiltonian and obtain the interacting ground state at the first order; we then generalize the model to the case of two cavities separated by a movable wall, and evaluate the interacting ground state at the second order. In Sec. 3 we first evaluate the field energy density inside the cavity; then, in the two-cavity case, we evaluate the spatial correlation between field observables in the two cavities, in particular the correlation of the squared fields, showing that an anticorrelation exists. Possible observability of these phenomena is also discussed. Sec. 4 is finally devoted to our final conclusions.

2. The Hamiltonian model and the interacting ground state

We consider a massless scalar field in a one-dimensional cavity of length L, with a fixed wall at $x = 0$ and a movable wall at the average position $x = L$. We also assume that the movable mirror has a finite mass m and that it is bound to its average position by a harmonic potential of angular frequency ω_0. Both mirrors are assumed to be perfectly reflecting. The mechanical degrees of freedom of the movable mirror are described quantum mechanically, in terms of a quantum harmonic oscillator.

In the case of small displacements of the movable mirror from its equilibrium position, our system can be described by an effective Hamiltonian originally introduced by Law [12]. This effective Hamiltonian H contains an unperturbed term given by the sum of the field and mirror Hamiltonians and an interaction term that is quadratic in the field operators and linear in the mirror operators. The field annihilation and creation operators refer to the movable wall's equilibrium position. The Law Hamiltonian is [12]

$$H = \sum_k \hbar\omega_k a_k^\dagger a_k + \hbar\omega_0 b^\dagger b - (b + b^\dagger) \sum_{kj} C_{kj} N \left[(a_k + a_k^\dagger)(a_j + a_j^\dagger) \right], \tag{1}$$

where b and $b^\dagger$ are bosonic annihilation and creation operators of the movable wall, a_k and $a_k^\dagger$ are the bosonic operators of the massless 1D scalar field relative to the wall's equilibrium position, $\omega_k = ck$, N is the normal ordering operator, and $C_{kj} = (-1)^{k+j} L^{-1} \sqrt{\hbar^3 \omega_k \omega_j / (8m\omega_0)}$ is the field-mirror coupling constant. This kind of Hamiltonian has been often used to treat the mirror-field dynamics [14, 15, 16]. In the following, we will treat the field-mirror interaction term perturbatively, up to the first or second order according to the field quantities we shall consider.

The unperturbed ground state (both mirror and field in their bare vacuum state) is $|0; \{0_k\}\rangle$, where the first element is relative to the movable wall, and the second to the field. The presence of the effective mirror-field interaction yields an energy shift of the ground state given, at the second order in the coupling, by [17]

$$\Delta E_g = -\frac{\hbar^2}{4L^2 m\omega_0} \sum_{kj} \frac{\omega_k \omega_j}{\omega_0 + \omega_k + \omega_j}. \tag{2}$$

We now consider the system's interacting ground state. At first order in the coupling constant

C_{kj}, the corrected ground state is given by (apart a normalization factor)

$$|g\rangle = |0;\{0_k\}\rangle + \frac{1}{L}\left(\frac{\hbar}{8m\omega_0}\right)^{1/2}\sum_{kj}(-1)^{k+j}\frac{\sqrt{\omega_k\omega_j}}{\omega_0+\omega_k+\omega_j}|1;\{1_k1_j\}\rangle, \tag{3}$$

that contains virtual admixtures with states with one excitation in the wall's degrees of freedom and two excitations in the field. From the state (3), it is possible to evaluate the virtual photon spectrum, showing that it has a maximum when the sum of the frequencies of the two photons is equal to the oscillation frequency ω_0 of the boundary; this is analogous to the case of the dynamical Casimir effect (forced harmonic oscillation of the boundary), except that in the latter case the condition $\omega_k + \omega_j = \omega_0$ is sharp, while in our present case it is only a maximum of a quite broad frequency distribution.

In the next section, we will also consider the case of two 1D cavities of length L separated by a perfectly reflecting movable wall, and two massless quantum scalar fields defined inside each cavity. If the perfect boundary separating the two cavities were fixed in space, the fields in the two cavities cannot influence each other, of course; this means that the spatial correlation between field observables pertinent to the different cavities are vanishing. The situation changes drastically if the separating wall can move, as we shall discuss in detail in the next section.

We now introduce the Hamiltonian model for this two-cavity system that is a straightforward generalization of the previous one-cavity model. The average position of the movable wall, that separates the two cavities, is at $x = L$, while the two fixed walls are at $x = 0$ (cavity 1, at the left side of the movable wall) and at $x = 2L$ (cavity 2, at the right side of the movable wall). The Hamiltonian of our system, as mentioned, is a straightforward generalization of the Hamiltonian (1) to the present two-cavity case; it is given by

$$\begin{aligned} H &= \hbar\sum_k \omega_k a_k^\dagger a_k + \hbar\sum_k \omega_k c_k^\dagger c_k + \hbar\omega_0 b^\dagger b - \left(b+b^\dagger\right)\sum_{kj} C_{kj}^1 \mathrm{N}\left[\left(a_j+a_j^\dagger\right)\left(a_k+a_k^\dagger\right)\right] \\ &\quad - \left(b+b^\dagger\right)\sum_{kj} C_{kj}^2 \mathrm{N}\left[\left(c_j+c_j^\dagger\right)\left(c_k+c_k^\dagger\right)\right], \end{aligned} \tag{4}$$

where a_k and $a_k^\dagger$ are bosonic operators relative to the field in the left-side cavity (cavity 1), while c_k and $c_k^\dagger$ are the bosonic operators relative to the right-side cavity (cavity 2). They both refer to modes relative to the equilibrium position $x = L$ of the mobile wall. There are also two mirror-field interaction terms, with C_{kj}^1 and C_{kj}^2, respectively, the coupling constants of the mirror with the field defined in cavity 1 and in cavity 2. Also, C_{kj}^1 is equal to the coupling constant C_{kj} defined after Eq. (1) for the single-cavity case, and $C_{kj}^2 = -C_{kj}^1$, because a movement of the movable wall in one direction increases the length of one cavity and decreases the length of the other one [15, 18].

The noninteracting (bare) ground state is $|0;\{0_k\};\{0_k\}\rangle$, where the first element refers to the mirror's excitations, while the second and the third ones refer respectively to the quanta in cavity 1 and in cavity 2: in this state there are no excitations in the mirror and in any of the fields. The interaction terms in (4) give a correction yielding the interacting (dressed) state. Up to the second order in the coupling constants (we need the second-order corrected state because the correlation function we are going to calculate in Sec. 3 is nonvanishing starting from the second order in the couplings), the true ground state has the following form

$$|\tilde{g}\rangle = \left(1 - \frac{1}{2}\Lambda^2\right)|0;\{0_k\};\{0_k\}\rangle + |g^{(1)}\rangle + |g^{(2)}\rangle, \tag{5}$$

where Λ is a normalization factor, $|g^{(1)}\rangle$ is the first-order correction and $|g^{(2)}\rangle$ is the second-order correction. Using time-independent perturbation theory up to the second order, we obtain combinations of states of the form described in the following. The first order correction $|g^{(1)}\rangle$ contains states of the form $|1;\{1_j1_k\};\{0_k\}\rangle$ (one mirror excitation, two quanta in cavity 1 and zero in cavity 2) or $|1;\{0_k\};\{1_j1_k\}\rangle$ (one mirror excitation, zero quanta in cavity 1 and two quanta in cavity 2). The second-order correction $|g^{(2)}\rangle$ contains terms with zero mirror excitations and one of the following possibilities for the two fields: two quanta in one cavity and zero quanta in the other one, or two quanta in each cavity, or four quanta in one cavity and zero quanta in the other cavity (there are also states with two mirror excitations, but they do not contribute to the correlation functions we are going to calculate in Sec. 3) [18].

3. Field observables near the movable wall for the one- and two-cavity system

We first consider the one-cavity case. A relevant aspect is considering local field quantities inside the cavity such as the field energy density $\mathcal{H}(x) = \frac{1}{2}\left[\frac{1}{c^2}\phi^2(x) + \left(\frac{d\phi(x)}{dx}\right)^2\right]$. The expression of the 1D scalar field operator, with the appropriate boundary conditions at $x = 0$ and $x = L$, is

$$\phi(x) = \sqrt{\frac{\hbar c^2}{L}} \sum_j \frac{\sin(k_j x)}{\sqrt{\omega_j}} \left(a_j + a_j^\dagger\right), \tag{6}$$

where $k_j = n_j\pi/L$, with $n_j = 1, 2, ...,$ and $\omega_j = ck_j$. We now evaluate the expression for the change of the renormalized field energy density inside the cavity, evaluated on the interacting ground state (3), with respect to that for a cavity with fixed walls $\Delta\mathcal{H}(x) = \langle g|\mathcal{H}(x)|g\rangle - \langle 0|\mathcal{H}(x)|0\rangle$. The result is

$$\Delta\mathcal{H}(x) = \frac{\hbar^2}{2L^3 m\omega_0} \sum_{jk\ell} (-1)^{k+\ell} \frac{\omega_j\omega_k\omega_\ell}{(\omega_0+\omega_j+\omega_k)(\omega_0+\omega_j+\omega_\ell)} \cos\left[\frac{(\omega_k-\omega_\ell)x}{c}\right] f_{kj\ell}(\omega_M), \tag{7}$$

where x is the distance from the movable wall and $f_{kj\ell}(\omega_M)$ is a regularization function necessary to cure ultraviolet divergences and to simulate the effect of a real metal, ω_M being an upper cutoff frequency. In Ref. [17] $\Delta\mathcal{H}(x)$ was evaluated numerically using a sharp cutoff function; this, for a finite cavity size L, is equivalent to considering a finite number of field modes. The results from the numerical evaluation show that, by including the motion of the finite-mass wall, the field energy density inside the cavity changes, and that this change is particularly relevant in the proximity of the movable wall (the effect increases with increasing cutoff frequency), while it is negligible at large distances form the movable wall. This result is consistent with the following physical picture. Due to the field-mirror interaction, pairs of virtual quanta are emitted and reabsorbed by the wall (see Eq. (3)). These quanta, however, remain confined near the wall, according to the energy-time uncertainty relation: the higher their frequency is, the more are they confined near the mobile wall. It is also worth to stress that the field energy density we are considering can be probed through the atom-wall dispersion interaction energy with a polarizable body such as a ground-state atom [19]: thus, the change in the field energy density inside the cavity due to the position fluctuations of the movable wall can be in principle measured. Eq. (7) also shows that the effect found becomes larger as the mass and oscillation frequency of the movable wall are decreased.

We also wish to mention that expressions like Eq. (7) could be also evaluated with an exponential cutoff function (more realistic than a sharp one), in particular in continuum limit, $L \to \infty$, $\sum_k \to L/(2\pi)\int dk$, thus recovering the case of a single movable wall, as we will explicitly do in the next part of this section for the spatial correlations in the two-cavity case.

These results can be extended to the case of a 1D electromagnetic field or to a 3D scalar field [20]. For example, in the case of the 1D electromagnetic case, the correction to the electric

Journal of Physics: Conference Series **2533** (2023) 012042 doi:10.1088/1742-6596/2533/1/012042

and magnetic field fluctuations (or, equivalently, to the electric and magnetic energy densities) in the cavity due to the wall's position fluctuations at the first order are found to be [20]

$$\langle E_z^2(x)\rangle = \frac{\hbar^2}{m\omega_0 L^3}\sum_{j\ell n}(-1)^{\ell+n}\frac{\omega_j\omega_\ell\omega_n f_{j\ell n}(\omega_M)}{(\omega_0+\omega_j+\omega_\ell)(\omega_0+\omega_j+\omega_n)}\sin(k_\ell x)\sin(k_n x), \tag{8}$$

$$\langle B_y^2(x)\rangle = \frac{\hbar^2}{m\omega_0 L^3}\sum_{j\ell n}(-1)^{\ell+n}\frac{\omega_j\omega_\ell\omega_n f_{j\ell n}(\omega_M)}{(\omega_0+\omega_j+\omega_\ell)(\omega_0+\omega_j+\omega_n)}\cos(k_\ell x)\cos(k_n x), \tag{9}$$

where the subscripts in the electric and magnetic field operators refer to their cartesian components and $f_{j\ell n}(\omega_M)$ is an appropriate cutoff function for $\omega_j, \omega_\ell, \omega_n$. The main qualitative features of the results obtained for the field energy densities of the 1D electric and magnetic fields, as well as of the 3D scalar field, are similar to the case considered before for the 1D scalar field. We refer the reader to Ref. [20] for more details. For example, explicit evaluation of (8,9) shows significant changes of the electric and magnetic energy density in the cavity, this change rapidly increasing when approaching the movable wall.

We now consider the case of two cavities separated by a perfectly reflecting movable mirror, introduced at the end of Sec. 2. Specifically, we calculate the spatial correlation functions between field observables defined in the two cavities on the interacting ground state (5). The field operator in cavity 1 ($x_1 \in (0, L)$) is

$$\phi(x_1) = \sqrt{\frac{\hbar c^2}{L}}\sum_j \frac{\sin(k_j x_1)}{\sqrt{\omega_j}}\left(a_j + a_j^\dagger\right), \tag{10}$$

and the field operator in cavity 2 ($x_2 \in (L, 2L)$) is

$$\phi(x_2) = -\sqrt{\frac{\hbar c^2}{L}}\sum_j \frac{\sin(k_j x_2)}{\sqrt{\omega_j}}\left(c_j + c_j^\dagger\right), \tag{11}$$

with $k_j = n_j\pi/L$, $n_j = 1, 2, ...$ due to the boundary conditions of the two field operators and $\omega_j = ck_j$. It is immediate to see that the spatial correlation function on the dressed ground state (5) between these two fields (i.e. between a point in cavity 1 and a point in cavity 2) vanishes, $\langle\tilde{g}|\phi(x_1)\phi(x_2)|\tilde{g}\rangle - \langle\tilde{g}|\phi(x_1)|\tilde{g}\rangle\langle\tilde{g}|\phi(x_2)|\tilde{g}\rangle = 0$ (this is indeed true at any order in perturbation theory, within the Hamiltonian model (4) used).

The spatial correlation of the *squared* fields between the two cavities, with $x_1 \in (0, L)$ and $x_2 \in (L, 2L)$, evaluated on the state (5), after some lengthy algebraic calculation is obtained as [18]

$$\begin{aligned} C(x_1,x_2) &= \langle\tilde{g}|\phi^2(x_1)\phi^2(x_2)|\tilde{g}\rangle - \langle\tilde{g}|\phi^2(x_1)|\tilde{g}\rangle\langle\tilde{g}|\phi^2(x_2)|\tilde{g}\rangle \\ &= -\frac{\hbar^3 c^4}{L^4 m\omega_0}\sum_{pqrs}(-1)^{p+q+r+s}\Big\{\frac{\sin(k_p x_1)\sin(k_q x_1)\sin(k_r x_2)\sin(k_s x_2)}{(\omega_0+\omega_p+\omega_q)(\omega_0+\omega_r+\omega_s)} \\ &+\Big[\frac{\sin(k_p x_1)\sin(k_q x_1)\sin(k_r x_2)\sin(k_s x_2)}{(\omega_0+\omega_p+\omega_q)(\omega_p+\omega_q+\omega_r+\omega_s)} + (x_1 \leftrightarrow x_2)\Big]\Big\} f_{pqrs}(\omega_M), \end{aligned} \tag{12}$$

where

$$f_{pqrs}(\omega_M) = \exp\left[-(\omega_p+\omega_q+\omega_r+\omega_s)/\omega_M\right] \tag{13}$$

is a regularization function and ω_M is the cutoff frequency; in the case of boundaries made of a real metal, the cutoff frequency ω_M can be also identified with the metal plasma frequency.

In the continuum limit, $L \to \infty$, $\sum_j \to (L/2\pi)\int_0^\infty dk_j$, and defining in both cavities the distance from the movable wall as $\tilde{x}_1 = L - x_1$ and $\tilde{x}_2 = x_2 - L$, we recover the case of a single movable boundary, to which we are mainly interested. In the continuum limit and for generic values of the distances $\tilde{x}_1 = L - x_1$ and $\tilde{x}_2 = x_2 - L$, the spatial correlation function (12) contains four frequency integrals, three of them can be obtained analytically, and we must resort to a numerical integration for the last one. For large distances from the wall, i.e. for $\tilde{x}_{1,2} \gg c/\omega_0$, and using $\omega_M \gg \omega_0$ (condition certainly met in any realistic setup), we can however obtain an approximated analytical expression for the spatial correlation function

$$C(\tilde{x}_1, \tilde{x}_2) \simeq -\frac{\hbar^3 c^4}{2^9 \pi^4} \frac{1}{m\omega_0^3} \frac{1}{\tilde{x}_1^2 \tilde{x}_2^2}. \tag{14}$$

Equation (14) proves the existence of a nonvanishing correlation between the squared fields in the two cavities, even if the cavities are separated by a perfectly reflecting mirror. $C(\tilde{x}_1, \tilde{x}_2)$ is negative, and thus the squared fields at the opposite sides of the movable wall are anticorrelated, even if there is not any direct interaction between them. The physical origin of this result is indeed in the mutual interaction of both fields with the movable mirror. For points at the same distance $\tilde{x} = \tilde{x}_1 = \tilde{x}_2$ from the wall, the anticorrelation scales with the distance as $\tilde{x}^{-4}$, and, respectively, as m^{-1} and ω_0^{-3} from the mirror's mass and oscillation frequency. Even if Eq. (14) holds only at large distances from the mirror, the numerical evaluation of (12) shows that a nonvanishing correlation exists also at shorter distances from the movable mirror [18]. Since, according to (12), the (anti)correlation between the squared field scales as $1/m$ and $1/\omega_0^3$, the effects we have described are larger the smaller the mass and oscillation frequency of the movable wall are. In optomechanical experiments, a typical oscillation frequency is of the order of $10^4 - 10^6\,\mathrm{s}^{-1}$ and masses as low as 10^{-15} kg to 10^{-21} kg can be experimentally achieved [21, 22]. These very low values of the mass could make experimentally detectable the effects we have found for the field energy density and squared field spatial correlations, for example exploiting the connection between field energy densities and two- and many-body dispersion interactions [18, 19]. Finally, we wish to mention that the results here presented have some similarity with the emission of radiation by a fuzzy black-hole event horizon [23, 24], and we guess it could be worth to pursue more deeply this analogy.

4. Conclusion

In this paper, we have reviewed some relevant aspects related to quantum fields confined in one or two perfectly reflecting one-dimensional cavities with a wall of finite mass and free to move, bound to its equilibrium position by a harmonic potential. The mechanical degrees of freedom of the movable wall are treated quantum mechanical, and this yield a mirror-field interaction and an effective interaction between the field modes, mediated by the movable boundary. Using the Law Hamiltonian, in the case of a single cavity with a movable boundary, we have investigated the effect of the motion and position fluctuations of the wall on some local field observables (energy densities and squared field) in the interacting ground state, which contains virtual field excitations; we find significant changes of the field energy density inside the cavity. In the case of two ideal cavities separated by a movable wall, we have investigated the spatial correlations between the squared field in the two cavities, showing that an anticorrelation exists, notwithstanding they are separated by a perfectly reflecting mirror. We have discussed the dependence of such spatial anticorrelation from the relevant parameters, in particular from the distance from the movable wall and the wall's mass and oscillation frequency. Possible observability of these effects has been also discussed.

Journal of Physics: Conference Series **2533** (2023) 012042 doi:10.1088/1742-6596/2533/1/012042

Acknowledgments
R. P. and L. R. acknowledge financial support from the Julian Schwinger Foundation. R. P. and L. R. also acknowledge partial financial support from the FFR2021 grant from the University of Palermo, Italy. S. B. acknowledges funding from the Leverhulme Trust Grant No. ECF-2019-461, and from University of Glasgow via the Lord Kelvin/Adam Smith (LKAS) Leadership Fellowship

References
[1] Purcell E 1946 *Phys. Rev.* **69**(11-12) 681 URL `https://link.aps.org/doi/10.1103/PhysRev.69.674`
[2] Meschede D 1992 *Physics Reports* **211** 201–250 ISSN 0370-1573 URL `https://www.sciencedirect.com/science/article/pii/037015739290110L`
[3] Milonni P 1994 *The Quantum Vacuum: An Introduction to Quantum Electrodynamics* (San Diego, CA: Academic Press)
[4] Casimir H B G 1948 *Proc. Kon. Ned. Akad. Wet.* **B51** 793
[5] Bordag M, Mohideen U and Mostepanenko V 2001 *Physics Reports* **353** 1–205 ISSN 0370-1573 URL `https://www.sciencedirect.com/science/article/pii/S0370157301000151`
[6] Casimir H B G and Polder D 1948 *Phys. Rev.* **73**(4) 360–372 URL `https://link.aps.org/doi/10.1103/PhysRev.73.360`
[7] Messina R, Passante R, Rizzuto L, Spagnolo S and Vasile R 2008 *Journal of Physics A: Mathematical and Theoretical* **41** 164031 URL `http://stacks.iop.org/1751-8121/41/i=16/a=164031`
[8] Moore G T 1970 *Journal of Mathematical Physics* **11** 2679–2691 URL `https://doi.org/10.1063/1.1665432`
[9] Dodonov V V 2010 *Physica Scripta* **82** 038105 URL `https://doi.org/10.1088/0031-8949/82/03/038105`
[10] Dodonov V 2020 *Physics* **2** 67–104 ISSN 2624-8174 URL `https://www.mdpi.com/2624-8174/2/1/7`
[11] Mantiñan M, Mazzitelli F D and Trombetta L G 2023 *Entropy* **25** ISSN 1099-4300 URL `https://www.mdpi.com/1099-4300/25/1/151`
[12] Law C K 1995 *Phys. Rev. A* **51**(3) 2537–2541 URL `https://link.aps.org/doi/10.1103/PhysRevA.51.2537`
[13] Law C K 1994 *Phys. Rev. A* **49**(1) 433–437 URL `https://link.aps.org/doi/10.1103/PhysRevA.49.433`
[14] Cheung H K and Law C K 2011 *Phys. Rev. A* **84**(2) 023812 URL `https://link.aps.org/doi/10.1103/PhysRevA.84.023812`
[15] Armata F, Kim M S, Butera S, Rizzuto L and Passante R 2017 *Phys. Rev. D* **96**(4) 045007 URL `https://link.aps.org/doi/10.1103/PhysRevD.96.045007`
[16] Butera S 2022 *Phys. Rev. D* **105**(1) 016023 URL `https://link.aps.org/doi/10.1103/PhysRevD.105.016023`
[17] Butera S and Passante R 2013 *Phys. Rev. Lett.* **111**(6) 060403 URL `https://link.aps.org/doi/10.1103/PhysRevLett.111.060403`
[18] Montalbano F, Armata F, Rizzuto L and Passante R 2023 *Phys. Rev. D* **107**(5) 056007 URL `https://link.aps.org/doi/10.1103/PhysRevD.107.056007`
[19] Passante R 2018 *Symmetry* **10** 735 ISSN 2073-8994 URL `https://www.mdpi.com/2073-8994/10/12/735`
[20] Armata F and Passante R 2015 *Phys. Rev. D* **91**(2) 025012 URL `https://link.aps.org/doi/10.1103/PhysRevD.91.025012`
[21] Meystre P 2013 *Annalen der Physik* **525** 215–233 URL `https://onlinelibrary.wiley.com/doi/abs/10.1002/andp.201200226`
[22] Aspelmeyer M, Kippenberg T J and Marquardt F 2014 *Rev. Mod. Phys.* **86**(4) 1391–1452 URL `https://link.aps.org/doi/10.1103/RevModPhys.86.1391`
[23] Arias E, Krein G, Menezes G and Svaiter N F 2012 *International Journal of Modern Physics A* **27** 1250129
[24] Takahashi T and Soda J 2010 *Classical and Quantum Gravity* **27** 175008 URL `https://doi.org/10.1088/0264-9381/27/17/175008`

Journal of Physics: Conference Series **2533** (2023) 012043 doi:10.1088/1742-6596/2533/1/012043

Coherent states for generalized uncertainty relations and their cosmological implications

P Jizba[1,*], G Lambiase[2,3,†], G G Luciano[4,‡] and L Petruzziello[§,3,5,6]

[1]FNSPE, Czech Technical University in Prague, Břehová 7, 115 19 Praha 1, Czech Republic
[2]Dipartimento di Fisica, Università di Salerno, Via Giovanni Paolo II, 132 I-84084 Fisciano (SA), Italy
[3]INFN, Sezione di Napoli, Gruppo collegato di Salerno, Italy
[4]Applied Physics Section of Environmental Science Department, Universitat de Lleida, Av. Jaume II, 69, 25001 Lleida, Spain
[5]Dipartimento di Ingegneria, Università di Salerno, Via Giovanni Paolo II, 132 I-84084 Fisciano (SA), Italy
[6]Institut für Theoretische Physik, Albert-Einstein-Allee 11, Universität Ulm, 89069 Ulm, Germany

E-mail: [*]p.jizba@fjfi.cvut.cz, [†]lambiase@sa.infn.it, [§]lupetruzziello@unisa.it, [‡]giuseppegaetano.luciano@udl.cat

Abstract. We study coherent states associated with a generalized uncertainty principle (GUP). Our particular focus is on the negative deformation parameter β. We show that the ensuing coherent state can be identified with Tsallis' probability amplitude with the non-extensivity parameter q being a monotonically increasing function of β. Furthermore, for $\beta < 0$, we reformulate the GUP in terms of a one-parameter class of Tsallis entropy based uncertainty relations, which are again saturated by the GUP coherent states. We argue that this combination of coherent states with Tsallis entropy offers a natural conceptual framework allowing to study the quasi-classical regime of GUP in terms of non-extensive thermostatistics. We bolster this claim by discussing a generalization of Verlinde's entropic force and the ensuing implications in the late-inflation epoch. The corresponding dependence of the β parameter on the cosmological time is derived for the reheating epoch. The obtained β is consistent with both values predicted by string-theory models and the naturalness principle.

1. Introduction

The Heisenberg Uncertainty Principle (HUP) – the cornerstone of quantum mechanics – provides an intrinsic limitation on the simultaneous knowledge of position and momentum of any quantum system. While working successfully in low-energy regime, it is anticipated that modifications will occur as we approach the Planck scale, likely due to quantum gravitational effects. Several models of quantum gravity, such as String Theory, Loop Quantum Gravity, Quantum Geometry and Doubly Special Relativity, have converged to the idea that the HUP should be generalized so as to account for the emergence of a minimal length at the Planck scale. The ensuing uncertainty relations are typically referred to as Generalized Uncertainty Principles (GUPs).

The simplest version of GUP can be obtained by adding a term quadratic in the momentum

Journal of Physics: Conference Series **2533** (2023) 012043 doi:10.1088/1742-6596/2533/1/012043

uncertainty to the right-hand side of HUP [1, 2, 3, 4, 5, 6], namely

$$\delta x\, \delta p \,\geq\, \frac{\hbar}{2}\left(1 \,+\, \beta\,\frac{\delta p^2}{m_p^2}\right), \tag{1}$$

where m_p stands for the Planck mass and the β parameter quantifies the departure from HUP. Here and throughout we set $c = 1$. Note that such term is not fixed by the theory, albeit it is generally assumed to be of order unity [1, 2, 3, 4]. Clearly, the traditional HUP form is recovered for $\beta \to 0$ and/or $\delta p \ll m_p$.

The symbol δ in Eq. (1) represents the uncertainty of a given observable, and it does not necessarily have to be associated with the standard deviation. In fact, in the original HUP δ can represent, e.g. Heisenberg's "ungenauigkeiten" (i.e., error-disturbance uncertainties caused by the back-reaction in simultaneous measurement) or $\delta p = \langle\psi||p||\psi\rangle \equiv \langle|p|\rangle_\psi$. Nevertheless, in cases where δ is identified with the standard deviation (henceforth denoted by Δ), the generalized uncertainty principle (1) can be derived from the deformed commutation relation (DCR)

$$[\hat{x}, \hat{p}] \,=\, i\hbar\left(1 + \beta\,\frac{\hat{p}^2}{m_p^2}\right), \tag{2}$$

via the Cauchy–Schwarz inequality [7, 8], provided one restricts the attention to *mirror symmetric* states, i.e. states satisfying $\langle\hat{p}\rangle_\psi = 0$. Commutator (2) is typically supplemented with the commutators

$$[\hat{x}, \hat{x}] \,=\, [\hat{p}, \hat{p}] \,=\, 0\,, \tag{3}$$

which, together with (2), satisfy Jacobi identities and determine the whole symplectic structure of the model. So far, the quadratic GUP (1) has been mostly used to study phenomenology of quantum gravity in many sectors, ranging from quantum mechanics [9, 10, 11] to particle physics [12, 13, 14, 15] and cosmology [16, 17]. In contrast, comparatively less research has been conducted on the quasi-classical domain of the GUP. Nevertheless, this domain has relevant and possibly observable implications for the early Universe cosmology and astrophysics [18, 19]. To explore physics in the quasi-classical realm, it is common to use coherent states (CSs). CSs are, in a sense, privileged quantum states in the description of the quantum-to-classical transition, as they are the only states that remain pure in the decoherence process [20, 21]. Since CSs are pure, they allow for maximal resolution in phase-space, thus appearing as the closest quantum counterparts of classical points. Additionally, the CS formalism offers a convenient description which can draw upon developments in quantum optics [22].

Our principal aim here is to address the issue of CSs for GUP and, in particular, to point out the role that Tsallis' entropy plays in this context. Being equipped with the ensuing CSs, we would like to discuss some simple consequences of GUP systems in their decoherence regime. To this end, we first introduce the Schrödinger–Nieto type of minimum-uncertainty CSs [7, 23] associated with GUP. Subsequently, we show that these states coincide with Tsallis' probability amplitudes. Furthermore, by using Bekner–Babenko inequality, we recast the GUP for $\beta < 0$ in terms of a one-parameter class of Tsallis entropy-power based uncertainty relations (EPUR), which are again saturated by the GUP CSs. Finally, we invoke the Maximum Entropy Principle (MEP), i.e. the principle which posits that the thermodynamic entropy is the statistical entropy evaluated at the maximum entropy distribution, to understand the quasi-classical domain of GUP in terms of non-extensive Tsallis thermostatistics. To further illustrate our point, we will examine two relevant examples from cosmology: a) the GUP generalization of Verlinde's entropic gravity force [24] and b) its connection with conformal gravity (CG) [25, 26, 27].

The remainder of the work is organized as follows: in Sec. 2 we derive coherent states for the quadratic GUP and reformulate the GUP in terms of Tsallis entropy-power based uncertainty relations. Within this setting, we discuss in Sec. 3 some illustrative examples from the early Universe cosmology by employing Verlinde's entropic gravity together with Tsallis' non-extensive thermostatistics. Conclusions and outlook are finally summarized in Sec. 4.

2. Coherent states for GUP

2.1. GUP (1) — closer look

Let us first briefly review the steps leading to (1) from the DCR (2). To this end, we shall quantify the uncertainty of an observable $\hat{\mathcal{O}}$ with respect to a state $|\psi\rangle$ via its standard deviation. In particular, for the variance (i.e., square of the standard deviation) we have

$$(\Delta\hat{\mathcal{O}})^2 \;=\; \langle\psi|\hat{\mathcal{O}}^2|\psi\rangle - \langle\psi|\hat{\mathcal{O}}|\psi\rangle^2 \;\equiv\; \langle\hat{\mathcal{O}}^2\rangle_\psi - \langle\hat{\mathcal{O}}\rangle_\psi^2\,. \tag{4}$$

In the upcoming considerations, the subscript ψ in $\langle\cdots\rangle$ will be omitted. By specializing the analysis to the observables $\hat{x}$ and $\hat{p}$ and mirror states, it is easy to check that the DCR (2) holds. Indeed, by employing the Cauchy–Schwarz inequality, one can write

$$(\Delta x)^2(\Delta p)^2 \;=\; \langle\hat{x}^2\rangle\langle\hat{p}^2\rangle \;\geq\; |\langle\hat{x}\,\hat{p}\rangle|^2\,. \tag{5}$$

At this point, we rewrite $\hat{x}\,\hat{p}$ as the sum of Hermitian and anti-Hermitian operators, so that

$$\hat{x}\,\hat{p} \;=\; \tfrac{1}{2}[\hat{x},\hat{p}]_+ \;+\; \tfrac{1}{2}[\hat{x},\hat{p}] \;=\; \tfrac{1}{2}[\hat{x},\hat{p}]_+ \;+\; \frac{i\hbar}{2}\left(1+\beta\,\frac{\hat{p}^2}{m_p^2}\right). \tag{6}$$

Here $[,]_+$ denotes the anticommutator. By inserting this expression back into (5), we arrive at

$$(\Delta x)^2(\Delta p)^2 \;\geq\; \tfrac{1}{4}\,\langle[\hat{x},\hat{p}]_+\rangle^2 \;+\; \frac{\hbar^2}{4}\left[1+\beta\,\frac{(\Delta p)^2}{m_p^2}\right]^2 \;\geq\; \frac{\hbar^2}{4}\left[1+\beta\,\frac{(\Delta p)^2}{m_p^2}\right]^2. \tag{7}$$

To obtain the last step, we have used the Robertson trick [8] and neglected the anticommutator part. Equation (7) clearly coincides with the GUP (1), with variances in place of error disturbances. In passing, we should mention that the inequality (7) based on (2) is valid not only for pure states, but holds also for mixed states [29].

2.2. Coherent states and GUP

Coherent states in quantum mechanics are quantum states that saturate Heisenberg uncertainty relations (URs). In fact, there are more defining properties that CSs could/should satisfy and there are many names used in the literature to denote states which have part but not all properties of the CSs, like the aforementioned minimal uncertainty states, or maximal localization states, or quasi-classical states, or weak coherent states. Here, we will concentrate on the so-called Schrödinger–Nieto type CSs where the only requirement is the saturation of URs [7, 23]. Such CSs are the most classical-like states in the sense that positions and momenta are in these states as well defined/localized as QM allows. We can mention yet another important property associated with the Schrödinger–Nieto CSs, namely that they most closely approximate the classical motion of a particle as they entangle least with the environment (and so they are least perturbed by it). It is this robustness with respect to decoherence that makes them an ideal tool for quantum optics applications.

It is thus natural to ask how such CSs look like in a GUP-driven Universe. In order to find the states $|\psi\rangle$ that saturate the GUP inequality (7), we might observe from Eqs. (5) and (7) that two requirements must hold simultaneously [28], namely

$$\hat{p}|\psi\rangle \;=\; c\,\hat{x}|\psi\rangle\,, \qquad \langle[\hat{x},\hat{p}]_+\rangle \;=\; 0\,, \tag{8}$$

Journal of Physics: Conference Series **2533** (2023) 012043 doi:10.1088/1742-6596/2533/1/012043

with $c \in \mathbf{C}$. The first condition saturates the Cauchy–Schwarz inequality, whereas the second one saturates the Robertson trick. Note that from the second requirement it automatically follows that $c = -c^*$, so we can write $c = i\gamma$, with $\gamma \in \mathbf{R}$. Consequently, the state $|\psi\rangle$ that saturates the GUP inequality (7) must satisfy the equation

$$(\hat{p} - i\gamma\hat{x})\,|\psi\rangle \;=\; 0\,. \tag{9}$$

We now seek the solution to (9) in the momentum representation, i.e $|\psi\rangle \mapsto \psi(p) = \langle p|\psi\rangle$. It is not difficult to see that in momentum space the operators $\hat{x}$ and $\hat{p}$ have the form [29]

$$\hat{p}\,\psi(p) \;=\; p\,\psi(p)\,, \qquad \hat{x}\,\psi(p) \;=\; i\hbar\left(\frac{d}{dp} + \frac{\beta}{2\,m_p^2}\left[p^2, \frac{d}{dp}\right]_+\right)\psi(p)\,. \tag{10}$$

This form is chosen so that both $\hat{x}$ and $\hat{p}$ are manifestly symmetric. With this, we can cast Eq. (9) into a differential equation

$$\frac{d}{dp}\,\psi(p) \;=\; -\frac{\left(1 + \frac{\beta\gamma\hbar}{m_p^2}\right)}{\gamma\hbar\left(1 + \beta\frac{p^2}{m_p^2}\right)}\,p\,\psi(p)\,, \tag{11}$$

which has a generic solution of the form

$$\psi(p) \;=\; N\left[1 + (\beta\,p^2)/m_p^2\right]_+^{-\frac{m_p^2}{2\beta\gamma\hbar} - \frac{1}{2}}\,. \tag{12}$$

Here, $[z]_+ = \max\{z, 0\}$ guarantees that the wave functions (12) are single-valued. The normalization coefficient N ensures that $\int |\psi(p)|^2 dp = 1$ and for $\beta < 0$ it is

$$N_< \;=\; \sqrt{\sqrt{\frac{|\beta|}{m_p^2\pi}}\;\frac{\Gamma\left(\frac{1}{2} + \frac{m_p^2}{|\beta|\gamma\hbar}\right)}{\Gamma\left(\frac{m_p^2}{|\beta|\gamma\hbar}\right)}}\,, \tag{13}$$

with $\Gamma(x)$ being the Euler gamma function. In passing, we observe that as $\beta \to 0$ then Eq. (12) reduces to the usual minimum uncertainty Gaussian wave-packet (Glauber coherent state) associated with the Heisenberg uncertainty relation.

In order to determine the significance of the γ parameter, we note that (9) implies

$$0 \;=\; \langle\psi|\,(\hat{p} + i\gamma\hat{x})\,(\hat{p} - i\gamma\hat{x})\,|\psi\rangle \;=\; (\Delta p)^2 - \gamma\,|\langle\psi|\,[\hat{x},\hat{p}]\,|\psi\rangle| + \gamma^2\frac{|\langle\psi|\,[\hat{x},\hat{p}]\,|\psi\rangle|^2}{4(\Delta p)^2}\,, \tag{14}$$

where we utilized that (7) is saturated. This has a single solution for γ, which with the help of (2) can be written as

$$\gamma \;=\; \frac{2(\Delta p)^2}{\hbar\left[1 \;+\; \beta\,(\Delta p)^2/m_p^2\right]}\,. \tag{15}$$

We note that γ can be defined also in the limit $\Delta p \to \infty$, even though GUP (7) is in such a case meaningless. In addition, states (12) are clearly mirror symmetric.

Journal of Physics: Conference Series **2533** (2023) 012043 doi:10.1088/1742-6596/2533/1/012043

2.3. Tsallis' transition amplitude and Tsallis' entropies

Let us now take in (12) the following substitutions (valid for $\beta \lesseqgtr 0$):

$$q \;=\; \frac{\beta\gamma\hbar}{m_p^2 + \beta\gamma\hbar} + 1\,, \qquad b \;=\; \frac{2m_p}{\gamma\hbar} + \frac{2\beta}{m_p}\,. \tag{16}$$

With this, we can rewrite the two-parameter class of CSs (12) as

$$\psi(p) \;=\; \tilde{N}\left[1 - b\,(1-q)\frac{p^2}{2m_p}\right]_+^{\frac{1}{2(1-q)}}\,, \tag{17}$$

where $\tilde{N}$ is appropriately rescaled N. The above expression is nothing but the probability amplitude for the Tsallis distribution of a free, non-relativistic particle

$$q_{\scriptscriptstyle T}(p|q,b) \;=\; |\psi(p)|^2 \;=\; \frac{1}{Z}\left[1 - b\,(1-q)\frac{p^2}{2m_p}\right]_+^{\frac{1}{1-q}}\,, \tag{18}$$

with Z being the "partition function".

Two-parameter class of probability distributions (18) decays asymptotically following power law rather than exponential law. If we keep variance and mean as the only statistical observables, power-law type distributions are incompatible with the conventional maximum entropy prescription (MaxEnt) applied to Shannon–Gibbs entropy. It is, however, the Shannon–Gibbs' MaxEnt that provides Gaussian probability distributions associated with canonical (or Glauber) CSs. In this respect, it is interesting to observe that distributions (18) are maximizers for Tsallis (differential) entropy [30, 31, 32], i.e., entropy of the form

$$\mathcal{S}_q^T(\mathcal{F}) \;=\; \frac{1}{(1-q)}\left(\int_{\mathbb{R}} dp\,\mathcal{F}^q(p) - 1\right)\,, \tag{19}$$

where $\mathcal{F}$ is a probability density function. It can easily be seen (by applying L'Hopital's rule) that in the limit $q \to 1$ the Tsallis entropy tends to the Shannon entropy

$$\mathcal{S}^S(\mathcal{F}) \;=\; -\int_{\mathbb{R}} dp\,\mathcal{F}(p)\log_2 \mathcal{F}(p)\,, \tag{20}$$

and the ensuing maximizer, i.e. distribution (18), tends to a Gaussian distribution.

2.4. Connection with entropic uncertainty relations

Now, having the entropy that is maximized by CSs, we can rephrase the original GUP in terms of entropies. The strategy is basically the same as in conventional quantum mechanics.

To illustrate this point we start with the canonical commutation relation $[\hat{x},\hat{p}] = i\hbar$. This induces ordinary representations of $\hat{x}$ and $\hat{p}$. For instance, in momentum representation we have

$$(\hat{x})^{\mathcal{P}} \;=\; i\hbar\frac{d}{dp} \quad \text{and} \quad (\hat{p})^{\mathcal{P}} \;=\; p\,. \tag{21}$$

Since eigenstates $\langle p|\psi_x\rangle \equiv \psi_x(p)$ and $\langle x|\psi_p\rangle \equiv \psi_p(x)$ are plane waves, generic states in momentum and position representation are connected via the Fourier transform

$$\psi(x) \;=\; \int_{\mathbb{R}} e^{ip\cdot x/\hbar}\,\hat{\psi}(p)\,\frac{dp}{(2\pi\hbar)^{1/2}}\,.$$

Journal of Physics: Conference Series **2533** (2023) 012043 doi:10.1088/1742-6596/2533/1/012043

By using Beckner–Babenko's inequality for the Fourier transform duals (basically inequality between their norms) [34, 35], one can derive the Shannon entropy-based UR [37]

$$N(|\psi|^2)N(|\hat{\psi}|^2) \;\geq\; \frac{\hbar^2}{4}\,, \tag{22}$$

which for white Gaussian distributions (i.e. Gaussian distributions whose covariance matrix is proportional to the identity matrix) reduces to $\sigma_x\sigma_p \;=\; \frac{\hbar}{2}$. The function $N(\ldots)$ is the so-called Shannon entropy power, which is defined as [36]

$$\mathcal{S}^S\left(\mathcal{X}\right) \;=\; -\int d^D x\, \mathcal{F}(x) \log \mathcal{F}(x) \;=\; \mathcal{S}^S\left(\sqrt{N(\mathcal{X})}\cdot \mathcal{Z}^G\right). \tag{23}$$

Here, $\{\mathcal{Z}_i^G\}$ is a Gaussian random vector with zero mean and unit covariance matrix and D is the dimension of the random vector. So, $N(\ldots)$ is the variance of a would-be Gaussian distribution that has the same (Shannon's) information content as the random variable $\mathcal{X}$ in question. In (23), we have employed the short-hand notation $\mathcal{S}^S(A) \equiv \mathcal{S}^S(\mathcal{F}_A)$, with $\mathcal{F}_A$ being a probability density function associated with the random variable describing the system A. Equation (23) has the unique solution [36]

$$N(\mathcal{X}) \;=\; \frac{1}{2\pi e} \exp\left(\frac{2}{D}\mathcal{S}(\mathcal{X})\right). \tag{24}$$

A few observation are in order: a) entropy-power UR represented by (22) is saturated only for $\tilde{\psi}$ and ψ that are Gaussian (and white for higher dimensional random vectors), b) Shannon's entropy power is defined only for distributions with continuous spectrum (integral calculus is essential in the proof) and c) the above EPUR indicates that Shannon entropy is a pertinent entropy in the semi-classical regime of conventional quantum mechanics.

Let us now repeat the above steps for our GUP; in particular, we concentrate on $\beta < 0$. In this case, the canonical commutation relation reads $[\hat{x}, \hat{p}] \;=\; i\hbar\left(1 - |\beta|\frac{\hat{p}^2}{m_p^2}\right)$, which induces in momentum space the representation

$$(\hat{x})^{\mathcal{P}} \;=\; i\hbar\left(\frac{d}{dp} - \frac{|\beta|}{2m_p^2}\left[p^2, \frac{d}{dp}\right]_+\right) \quad \text{and} \quad (\hat{p})^{\mathcal{P}} \;=\; p\,. \tag{25}$$

Eigenstates $\langle p|\psi_x\rangle \equiv \psi_x(p)$ are not plane waves but rather they have the form

$$\psi_x(p) \;=\; B_x\, \frac{e^{-ixm_p \operatorname{arctanh}\left(p\sqrt{|\beta|}/m_p\right)/\hbar\sqrt{|\beta|}}}{\sqrt{m_p^2 \;-\; p^2|\beta|}}\,, \quad B_x = \sqrt{m_p^2/2\pi\hbar}\,. \tag{26}$$

Note that $\psi_x(p)$ are not quadratically integrable, namely $\psi_x(p) \notin L^2((-m_p/\sqrt{|\beta|}, m_p/\sqrt{|\beta|}))$, but instead $\psi_x(p) \in \mathcal{S}^{1'}((-m_p/\sqrt{|\beta|}, m_p/\sqrt{|\beta|}))$, i.e. they belong to the space of complex-valued tempered distributions (like plane waves). It can be further shown [29] that for $\beta < 0$ the operator $(\hat{x})^{\mathcal{P}}$ is self-adjoint and its spectrum is continuous.

Generic states in momentum and position representation are thus not connected via the Fourier transform, but via the *Abel transform*

$$\psi(x) \;=\; \int_{-m_p/\sqrt{|\beta|}}^{m_p/\sqrt{|\beta|}} dp\; \psi_p(x)\tilde{\psi}(p) \;=\; \int_{-m_p/\sqrt{|\beta|}}^{m_p/\sqrt{|\beta|}} \frac{dp}{\sqrt{2\pi\hbar}}\, \frac{e^{ixm_p \operatorname{arctanh}\left(p\sqrt{|\beta|}/m_p\right)/\hbar\sqrt{|\beta|}}}{\sqrt{1 - p^2|\beta|/m_p^2}}\, \tilde{\psi}(p)\,, \tag{27}$$

Journal of Physics: Conference Series **2533** (2023) 012043 doi:10.1088/1742-6596/2533/1/012043

where $\tilde{\psi}(p) = \langle p|\psi\rangle$. Analogous relation holds for the inverse transformation.

At this stage, we define the entropy power based on Tsallis entropy as

$$\mathcal{S}_q^T(\mathcal{X}) \;=\; \mathcal{S}_q^T\left(\sqrt{M_q^T(\mathcal{X})}\cdot \mathcal{Z}_{2-q}^T\right). \qquad (28)$$

Here $\{\mathcal{Z}_i^T\}$ is a Tsallisian random vector (i.e. random vector distributed according to Tsallis distribution) with zero mean and unit covariance matrix. Eq. (28) has a unique solution [29, 37]

$$M_q^T(\mathcal{X}) \;=\; A_q\left[\exp_q(\mathcal{S}_q^T(\mathcal{X}))\right]^{2/D} \;=\; A_q \exp_{1-(1-q)D/2}\left(\frac{2}{D}\mathcal{S}_q^T(\mathcal{X})\right), \qquad (29)$$

with A_q being a q-dependent constant and D dimension of the random vector. In (29) we have employed the q-exponential $e_q^x = [1+(1-q)x]^{1/(1-q)}$. As a consistency check, we might notice that in the $q \to 1$ limit we recover Shannon's EP, namely

$$\lim_{q\to 1} M_q^T(\mathcal{X}) \;=\; \frac{1}{2\pi e}\exp\left(\frac{2}{D}\mathcal{S}^S(\mathcal{X})\right) \;=\; N(\mathcal{X}). \qquad (30)$$

At this stage one can employ the fact that the Abel transform (27) can be equivalently rewritten as the Fourier transform of the function

$$\bar{\psi}(z) \;=\; \frac{\tilde{\psi}(m_p \tanh(z\sqrt{|\beta|}/m_p)/\sqrt{|\beta|})}{\cosh(z\sqrt{|\beta|}/m_p)}, \qquad (31)$$

and use again Beckner–Babenko's inequality for the Fourier transform duals. After some algebra, one obtains the Tsallis entropy-based URs [29]

$$M_{q/2}^T(|\psi|^2)M_{q'/2}^T(|\tilde{\psi}|^2) \;\geq\; \frac{\hbar^2}{4}, \qquad (32)$$

where q' and q are Hölder conjugates, i.e. $1/q + 1/q' = 1$ with $q \in \mathbb{R}^+$. This one-parameter class of entropy power URs clearly resembles the form of Shannon's entropy power UR (22) by having the unique irreducible lower bound.

It is now worth noting a few points; a) the inequality (32) is saturated if and only if ψ and $\tilde{\psi}$ are CSs of GUP (1), b) the above entropy power URs can be formulated only for $\beta < 0$ (as in such cases the spectrum is continuous) and c) entropic URs (32) indicate that Tsallis entropy is a pertinent entropy functional in the GUP framework. CSs saturating UR belong to the class of so-called pointer states, i.e. those states that are least affected by the interaction with the environment [38, 39]. Such states are particularly pertinent to the quasi-classical domain of quantum theory, as they are maximally predictable despite decoherence [40, 41]. Among all pointer states in the would-be GUP-driven Universe, only CSs (18) saturate both the "x-p" GUPs and ensuing Tsallis entropy power URs. Moreover, the existence of Tsallis entropy power UR indicates that Tsallis' entropy should be a relevant entropy in the GUP context. If we couple this observation with the fact that CSs (18) extremize Tsallis entropy, we might invoke (similarly as in conventional statistical physics) MEP, but this time with Tsallis entropy (in place of Shannon–Gibbs entropy) to discuss a statistical physics of an ensemble of non-interacting GUP-governed particles in their quasi-classical regime. In this context, non-extensive thermostatistics of Tsallis (NTT) [30] should provide the necessary mathematical framework that ought to be employed to describe the borderline between classical and quantum physics in the GUP Universe.

3. Applications

3.1. Verlinde's entropic gravity

Particularly interesting applications of the aforementioned observation can be expected in the early-Universe gravity. To do this, we must extend our one-dimensional analysis to three dimensions. This can be done, e.g., by extending the algebra (2)-(3) as, cf. [6, 42]

$$[\hat{\mathbf{x}}_i, \hat{\mathbf{p}}_j] \;=\; i\hbar\delta_{ij}\left(1 \,+\, \beta\frac{\mathbf{p}^2}{m_p^2}\right), \quad [\hat{\mathbf{x}}_i, \hat{\mathbf{x}}_j] \;=\; 2i\hbar\frac{\beta}{m_p^2}\left(\hat{\mathbf{x}}_i\hat{\mathbf{p}}_j - \hat{\mathbf{x}}_j\hat{\mathbf{p}}_i\right), \quad [\hat{\mathbf{p}}_i, \hat{\mathbf{p}}_j] \;=\; 0\,, \tag{33}$$

which in the momentum space representation (in terms of symmetric operators) yields

$$\hat{\mathbf{p}}_i\,\psi(\mathbf{p}) \;=\; \mathbf{p}_i\,\psi(\mathbf{p})\,, \qquad \hat{\mathbf{x}}_i\,\psi(\mathbf{p}) \;=\; i\hbar\left(\frac{d}{d\mathbf{p}_i} + \frac{\beta}{2\,m_p^2}\left[\mathbf{p}^2, \frac{d}{d\mathbf{p}_i}\right]_+\right)\psi(\mathbf{p})\,. \tag{34}$$

By following a similar route as in 1 dimension, the system of commutators admits only one (normalized) mirror-symmetry solution for $\psi \in L^2(\mathbb{R}^3)$, namely

$$\psi(\mathbf{p}) \;=\; N\left[1 \,+\, (\beta\,\mathbf{p}^2)/m_p^2\right]_+^{-\frac{m_p^2}{2\beta\gamma\hbar}-\frac{1}{2}} \qquad \text{with} \quad N_<^2 \;=\; \frac{\beta^{3/2}}{2\pi m_p^3\, B(5/2, m_p^2/|\beta|\gamma\hbar)}\,. \tag{35}$$

Here, $B(x, y)$ is the beta function. At this stage, we can ask what modifications to Newton's law should be expected in the quasi-classical limit of the GUP-based Universe. To answer this question, we can combine Verlinde's idea that gravity is an entropy-driven phenomenon — entropic gravity (EG), with the non-extensive thermostatistics of Tsallis.

Following Verlinde [24], we suppose that the true (unknown) microscopic degrees of freedom in any given part of space are stored in discrete bits on the holographic screen that surrounds them. A holographic screen can be considered to be spherically symmetric of area $A = 4\pi R^2$. Outside of the screen is the *emergent world*, so the screen acts as an interface between known and unknown physics. When a test particle moves away from the screen, it feels an effective force F satisfying $F\delta x = T\delta S$, where T and δS are the temperature and the entropy change on the holographic surface, respectively, and δx is the distance of the particle from the screen. In NTT the heat one-form $T\delta S$ must be replaced with [43]: $T\delta\mathcal{S}_q^T/[1 + ((1-q)/k_B)\mathcal{S}_q^T]$ (in our context $\mathcal{S}_q \mapsto \mathcal{S}_{2-q}$ as it is $\mathcal{S}_{2-q}$ that is extremized by $q_T(\mathbf{p}|q, b)$). If L is a (dimensionless) characteristic length scale (e.g. radius R/ℓ_p) then the Bekenstein–Hawking entropy $S_{BH} = \ln W(L) \propto L^2$, which implies that the total number of internal configurations W behaves as $W(L) = \phi(L)\nu^{L^2}$ for $L \gg 1$ (ϕ is a positive function satisfying $\lim_{L\to\infty} \ln\phi/L^2 = 0$ and $\nu > 1$ constant). Hence, from the outside, the holographic screen has Tsallis'entropy

$$\mathcal{S}_{2-q}^T = k_B \ln_{2-q} W(L) = \frac{k_B}{q-1}\left[\left(\phi(L)\nu^{L^2}\right)^{q-1} - 1\right]. \tag{36}$$

Consequently, the entropic force follows from the relation

$$F\delta x \;=\; \frac{T\delta S_{2-q}^T}{1 + (q-1)\left(\omega_3 L^3 + \omega_2 L^2 + \cdots\right) + \cdots}\,, \tag{37}$$

where $\omega_2, \omega_3 > 0$ are intensive coefficients — so-called Hills' coefficients, known from entropy expansion in (conventional) thermodynamics of small and mesoscopic systems [44]. To comply with Hills' expansion, we have formally included the term $\omega_3 L^3$, even if it is not supported by the EG prescription. It will be seen that such a term is cosmologically unfeasible in the quasi-classical regime, so that $\omega_3 \approx 0$.

By holographic scaling, the energy residing inside the holographic screen is related with the on-screen degrees of freedom via the equipartition theorem $E = Nk_BT/2$, with $E = M$ being the total mass enclosed by the surface and $N = A/(G\hbar)$ the number of bits connected with the area by the holographic principle [24]. EG paradigm posits that the minimum possible increase in the screen entropy (equivalent to one bit of Shannon's information) happens if a particle of radius of Compton wave length λ_C is added to a holographic sphere [24, 45]. This happens when a point-like quantum particle appears at the distance λ_C from the screen (note that λ_C is the minimal distance at which a particle stil retains its single-particle picture). By setting $\delta x = \lambda_C = \hbar/m$ and using the non-extensive version of Landauer principle [46, 47], which states that the erasure of information leads to an entropy increase $\delta S_q^T = 2\pi k_B/(3-2q)$ per erased bit, we derive the following modified Newton's law

$$F(R) = \frac{GMm}{wR^2}\frac{1}{1-\kappa_3\varepsilon_q R^3 - \kappa_2\varepsilon_q R^2}, \tag{38}$$

with $\varepsilon_q = 1-q$, $w = 1+2\varepsilon_q$ and $\kappa_n = \omega_n/\ell_p^n$ ($\ell_p = \hbar/m_p \approx 1.6\times 10^{-35}$m is the Planck length), $n = 2, 3$. Since $2\varepsilon_q$ is small (see below), we can set $w = 1$. The ensuing gravitational potential up to the first-order in ε_q then reads

$$V(R) = \frac{r_s}{2}\left[-\frac{1}{R} + \varepsilon_q\kappa_2 R + \frac{\varepsilon_q\kappa_3}{2}R^2\right], \tag{39}$$

where $r_s = 2GM$ is the Schwarzschild radius. The gravitational potential (39) coincides with the Mannheim–Kazanas gravitational potential of a static, spherically symmetric source of mass M in conformal Weyl gravity (CWG) [26, 27]. Strictly speaking, in CWG a given local gravitational source generates only a gravitational potential

$$V_{MK}(R) = -\frac{r_s}{2R} + \frac{\chi}{2}R = \frac{r_s}{2}\left[-\frac{1}{R} + (1-q)\kappa_2 R\right], \tag{40}$$

where $\kappa_2 = \omega_2/\ell_p^2$ ($\omega_2 = \pi$ is the second Hill's coefficient [29, 44]). The would-be term $\propto R^2$ corresponds to a trivial vacuum solution of CG and hence does not couple to matter sources [25, 26, 27]. Fitting with CWG thus implies $\omega_3 \approx 0$.

3.2. Some cosmological consequences

In CWG, the parameter in front of the linear term is identified with the inverse Hubble length R_H (more precisely with $1/(2R_H)$) [48]. It s quite intriguing that, for present macroscopic scales (i.e., $R_H \sim 10^{26}$m), the Mannheim–Kazanas solution has been successful in fitting more than two hundred galactic rotation curves with no adjustable parameters (other than the galactic mass-to-light ratios) with no need for dark matter or other exotic modifications of gravity [26, 27]. Despite the fact that macroscopic-scale gravity does not fall within the assumed quasi-classical regime, the idea that the coefficient in front of a linear term in (40) should be associated with the inverse Hubble length is valid even in the early Universe cosmology. This is because the argument of CWG leading to this result is independent of an actual Universe epoch [49].

In conventional cosmology, it is expected that a quasi-classical description becomes pertinent at the late-inflation epoch (after the first Hubble radius crossing) and perhaps even after its end during reheating [51, 50]. So, in this period the NTT should be a suitable framework for the description of an "inflaton gas". For instance, by viewing the "inflaton gas" as an ideal gas, the NTT predicts that the inflaton pressure for $0 < q < 1$ should satisfy a polytrope relation [52, 53]

$$p = \frac{2\pi\hbar^2}{m_i\, e^{5/3}}\rho^{5/3}, \tag{41}$$

where $\rho = N/V$ is the particle density. In this connection, it should be stressed that the relation (41) holds for $0 < q < 1$ but not in the limit $q \to 1$, see, e.g. [53]. In fact, at $q = 1$ one has the familiar pressure relation $p \propto \rho_E$. So, the NTT and extensive limits are not interchangeable. The polytrope relation of the type (41) often appears in studies on late inflation, see, e.g. [54, 55].

In order to gain information about β, we employ the CWG observation that the cosmologically viable linear term in (40) should have its parameter associated with $1/R_H$. According to CWG, the Newtonian potential (40) should dominate on short scales, while the linear one becomes prominent at large scales. Both potentials get equal at R_H, which in our case implies that $q = 1 - \ell_p^2/(\pi R_H^2)$. Note that this is compatible with the condition that $r_s = R_H$. By combining the latter expression for q with (15) and (16), we obtain $|\beta| \simeq m_p^2\,\ell_p^2/(2\pi\,(\Delta p)_\psi^2\,R_H^2)$.

To see how such β explicitly depends on a cosmological time t, we first write $R_H(t) = H^{-1}(t) = a(t)/\dot{a}(t)$, where H is the Hubble parameter and $a(t)$ is the scale factor. The latter can be evaluated, e.g., from the Vilenkin–Ford inflationary model [56], where $a(t) = A\sqrt{\sinh{(B\,t)}}$, with $B = 2\sqrt{\Lambda/3}$ (Λ is the cosmological constant). We then use the result from the relativistic equipartition theorem $(\Delta p)_\psi^2 \simeq 12\,(k_B T)^2$ and after simple algebraic manipulations, we obtain

$$|\beta| \ \equiv\ |\beta(t)| \ =\ \frac{m_p^2\,\ell_p^2\,\Lambda}{72\pi\,(k_B T)^2\tanh^2\left(2t\sqrt{\Lambda/3}\right)}\,. \tag{42}$$

For concreteness' sake, let us consider the reheating epoch, i.e. time scale $t \simeq 10^{-33}$s. By taking the mass of the inflaton $m_i = 10^{12} \div 10^{13}$GeV, T of the order of the reheating temperature $T_R \simeq 10^7 \div 10^8$GeV and the presently known value of the cosmological constant $\Lambda \simeq 10^{-52}\text{m}^{-2}$, we obtain $|\beta| \sim 10^{-2} \div 1$, which is in agreement with the values predicted by string-theory models, cf. e.g. [1, 3, 4]. This result is also consistent with the naturalness principle that dictates that not so far from the Planck scale the β should not be too large nor too small.

4. Conclusions

In this paper, we have unified two seemingly unrelated concepts, namely generalized uncertainty principle and Tsallis thermostatistics. On the one hand, the GUP strives to explore consequences of the existence of a minimal length scale. On the other hand, Tsallis statistics is a theoretical concept that accounts for systems with long-range correlations or long-time memory, for which the conventional central limit theorem does not apply. A merger of these two concepts presented here is intriguing from both a conceptual perspective and from a phenomenological point of view.

In order to substantiate our point, we have employed the NTT to generalize Verlinde's entropic force. Apart from obtaining a modified Newtonian (basically Mannheim–Kazanas-like) potential, we have argued that such a generalization should be phenomenologically pertinent at the late-inflation epoch. The corresponding dependence of the GUP β parameter on the cosmological time t has also been derived for the reheating epoch. The β parameter obtained is consistent both with values predicted by string-theory models and with the naturalness principle.

Acknowledgments

L.P. is grateful to the "Angelo Della Riccia" foundation for the awarded fellowship received to support the study at Universität Ulm. L. P. and G. G. L. acknowledge networking support by the COST Action CA18108. G. G. L. acknowledges the Spanish "Ministerio de Universidades" for the awarded Maria Zambrano fellowship and funding received from the European Union - NextGenerationEU. P. J. was in part supported by the FNSPE CTU grant RVO14000.

[1] Amati D, Ciafaloni M and Veneziano G 1987 *Phys. Lett. B* **197** 81
[2] Gross D J and Mende P F 1987 *Phys. Lett.* B **197** 129
[3] Amati D, Ciafaloni M and Veneziano G 1989 *Phys. Lett.* B **216** 41

[4] Konishi K, Paffuti G and Provero P 1990 *Phys. Lett.* B **234** 276
[5] Maggiore M 1993 *Phys. Lett.* B **319** 83
[6] Kempf A, Mangano G and Mann R B 1995 *Phys. Rev.* D **52** 1108
[7] Schrödinger E 1928 *Collected Papers on Wave Mechanics*, (Blackie & Son, London), p.41
[8] Robertson H P 1929 *Phys. Rev.* **34** 163
[9] Bernardo R C S and Esguerra P H (2018) Annals Phys. **391** 293
[10] Frassino A M and Panella O 2012 *Phys. Rev.* D **85** 045030
[11] Bosso P, Petruzziello L and Wagner F 2022 *Phys. Lett.* B **834** 137415
[12] Das S and Vagenas E C 2008 *Phys. Rev. Lett.* **101** 221301
[13] Husain V, Kothawala D and Seahra S S 2013 *Phys. Rev.* D **87** 025014
[14] Bosso P and Luciano G G 2021 *Eur. Phys. J.* C **81** 982
[15] Luciano G G and Blasone M 2021 *Eur. Phys. J.* C **81** 995
[16] Jizba P and Scardigli F 2012 *Phys. Rev.* D **86** 025029
[17] Giardino S and Salzano V 2021 *Eur. Phys. J.* C **81** 110
[18] Kiefer C 2012 *Quantum Gravity*, (Oxford Science Publications, Oxford).
[19] Kiefer C 1994 *The semiclassical approximation to quantum gravity.* In: Ehlers J and Friedrich H (eds) *Canonical Gravity: From Classical to Quantum*, Lecture Notes in Physics, vol 434. (Springer, Berlin)
[20] Dutra S M 1998 *J. Mod. Optic.* **45** 759
[21] Matacz A L 1994 *Phys. Rev.* D **49** 788
[22] Mandel L and Wolf E 1995 *Optical Coherence and Quantum Optics* (Cambridge University Press, Cambridge)
[23] Nieto M M and L.M. Simmons L M 1978 *Phys. Rev. Lett.* **41** 207
[24] Verlinde E P 2011 *JHEP* **04** 029
[25] Mannheim P D and O'Brien J G 2011 *Phys. Rev. Lett.* **106** 121101
[26] Mannheim P D and O'Brien J G 2012 *Phys. Rev.* D **85** 124020
[27] O'Brien J G, Chiarelli T L and Mannheim P D 2018 *Phys. Lett.* B **782** 433
[28] Messiah A 1961 *Quantum Mechanics* (North Holland Publishing Company, Amsterdam)
[29] Jizba P, Lambiase G, Luciano G G and Petruzziello L 2022 *Phys. Rev.* D **105** L121501
[30] Tsallis C 1988 *J. Statist. Phys.* **52** 479
[31] Jizba P, Korbel J and Zatloukal V 2017 *Phys. Rev.* E **95** 022103
[32] Jizba P and Korbel J 2019 *Phys. Rev. Lett.* **122** 120601
[33] Jizba P, Dunningham J A and Joo J 2015 *Annals of Physics* **355** 87
[34] Beckner W 1975 *Ann. of Math.* **102** 159
[35] Babenko K I 1962 *Amer. Math. Soc. Transl. Ser.* 2 **44**, 115
[36] Shannon C E 1948 *Bell Syst. Tech. J.* **27** 623
[37] Jizba P, Dunningham J A and Prokš M 2021 *Entropy* **23** 334
[38] Paz J P and Zurek W H 1999 *Phys. Rev. Lett.* **82** 5181
[39] Venugopalan A 1999 *Phys. Rev. A* **61** 012102
[40] Dalvit D A R, Dziarmaga J and Zurek W H 2005 *Phys. Rev. A* **72** 062101
[41] Paz J P and Zurek W H 2001 *Coherent Atomic Matter Waves.* In: Kaiser R, Westbrook C and David F (eds) *Les Houches-Ecole d'Ete de Physique Theorique*, vol 72. (Springer, Berlin)
[42] Tawfik A and Diab A 2014 *Int. J. Mod. Phys.* D **23** 1430025
[43] Abe S, Martínez S, Pennini F and Plastino A 2001 *Phys. Lett.* A **281** 126
[44] Hill T L 1994 *Thermodynamics of Small Systems*, (Dover, New York)
[45] Bekenstein J D 1973 *Phys. Rev.* D **7** 2333
[46] Curilef S, Plastino A R, Wedemann R S and Daffertshofer A 2008 *Phys. Lett.* A **372** 2341
[47] Curilef S, Plastino A R, Wedemann R S and Daffertshofer A 2017 *Phys. Lett.* B **767** 242
[48] Sultana J, Kazanas D and Said J L 2012 *Phys. Rev.* D **86** 084008
[49] Mannheim P D and Kazanas D 1989 *Astrophys. J.* **342** 635
[50] Burgess C P, Holman R and Hoover D 2008 *Phys. Rev.* D **77** 063534
[51] Kiefer C, Lohmar I, Polarski D and Starobinsky A A 2007 *Classical Quantum Gravity* **24** 1699
[52] Du J 2004 *Physica (Amsterdam)* **335A** 107
[53] Abe S 2000 *Phys. Lett.* A **263** 424; **267** 456(E)
[54] Barrow J D and Paliathanasis A 2016 *Phys. Rev.* D **94** 083518
[55] Barrow J D and Paliathanasis A 2018 *Gen. Relativ. Gravit.* **50** 82
[56] Vilenkin A and Ford L H 1982 *Phys. Rev.* D **26** 1231

Journal of Physics: Conference Series **2533** (2023) 012044 doi:10.1088/1742-6596/2533/1/012044

Axion like particles, fifth force and neutron interferometry

Antonio Capolupo
Dipartimento di Fisica "E.R. Caianiello" Università di Salerno, and INFN — Gruppo Collegato di Salerno, Via Giovanni Paolo II, 132, 84084 Fisciano (SA), Italy

Salvatore Marco Giampaolo
Institut Ruder Boskovic, Bijenicka cesta 54, 10000 Zagreb, Croatia

Aniello Quaranta
Dipartimento di Fisica "E.R. Caianiello" Università di Salerno, and INFN — Gruppo Collegato di Salerno, Via Giovanni Paolo II, 132, 84084 Fisciano (SA), Italy

Abstract.
We report on recent result according to which the fermion-fermion interaction mediated by axions and axion–like particles can be revealed by means of neutron interferometry. We assume that the initial neutron beam is split in two beams which are affected by differently oriented magnetic fields, in order that the phase difference depends only by the axion–induced interaction. In this way, the phase difference is directly related to the presence of axions.

Phenomena, ranging from neutrino mixing [1, 2, 3, 4, 5, 6, 7] to the dark matter and energy [8, 9], to the muon g-2 anomaly [10, 11], together with the strong CP problem (i.e. the absence of CP symmetry violation in strong interaction, solved by Peccei and Quinn [12, 13] by introducing pseudo–scalar particles known as axions [12, 13, 14, 15, 16]) show the necessity of physics beyond the standard model of particles [17]. Axions and axion–like particles (ALPs) [18, 19, 20, 21, 22], with masses from the ultralight [23, 24] $m \simeq 10^{-22}$eV to the heavy axions [25] $m \simeq 1$ TeV, could represent a possible dark matter component [26]. Many experiments searched for ALPs [27, 28, 29, 30, 31, 32, 33, 34, 35], however, up to now no evidence for ALPs has been found. ALPs interact with the electromagnetic field, moreover, they are expected to play the role of a mediating boson in the fermion–fermion interaction [36, 37]. Therefore, different experiments were designed to probe such interactions [38, 40, 39, 41].

Another extremely thriving field of physics is the neutron interferometry [42]. It has allowed to verify many theoretical effects, such as the Sagnac effect [43], the geometric phase [44, 45, 46] and the wave–particle duality.

Here, we report on a new approach to detect ALPs, which suggests the use of a neutron interferometer in which two sub–beams are subject to external magnetic fields of equal strength but different direction, as a device to reveal fermion-fermion interaction mediated by axions. [47]. Indeed, we show that a detectable neutron phase difference, depending only by the axion–induced interaction between neutrons, can be achieved by setting the magnetic fields in the arms of the interferometer, one in the direction of propagation of the relative sub-beam and the other

Journal of Physics: Conference Series **2533** (2023) 012044 doi:10.1088/1742-6596/2533/1/012044

one orthogonally to the propagation. We fix the experimental parameters in order that the phase difference depends only on the axion–mediated interaction and the contributions given by the other interactions are removed. Then we show how a neutron interferometer is sensible to the presence of ALPs in a significant portion of parameter space.

The Lagrangian describing the neutron-neutron interaction due to axion exchange is given by [36, 37]

$$\mathcal{L}_{INT} = -\sum_{j=1,2} i g_{aj} \phi \bar{\psi}_j \gamma_5 \psi_j \tag{1}$$

with ϕ the axion field, ψ_1, ψ_2 fermion fields and g_{aj} the (dimensionless) effective axion–fermion coupling constants, $g_{aj} = g_{aNN}$ in the case of axion-neutron coupling. Such constant is depending on the axion (or ALP) model and on the fermions interacting. We shall assume that the neutron velocities are non-relativistic and analyze this interaction potential within the context of ordinary quantum mechanics. In the non–relativistic limit $\mathcal{L}_{INT}$ yields a two–body potential for the neutrons [36, 37]. This axion–induced interaction is not the only one in play. However, the gravitational interaction is easily seen to be irrelevant, due to the smallness of the masses involved. In addition, we shall always assume a relative distance $r > 10^{-12}m$ among the neutrons, so that the short–range nuclear interactions can also be ignored. With this assumption the only other relevant interaction is the magnetic one between the neutron dipoles. The neutron–neutron interaction Hamiltonian comprising the magnetic and the axion–mediated interaction, generalized to an arbitrary number of neutrons can be written as [47]

$$H_{ij} = -\frac{\mathcal{A}}{r_{ij}^3}\left[\left(3 - \mathcal{B}e^{-mr_{ij}}K^{(a)}(r_{ij})\right)\sigma_i^{r_{ij}}\sigma_j^{r_{ij}} - \left(1 - \mathcal{B}e^{-mr_{ij}}K^{(b)}(r_{ij})\right)\boldsymbol{\sigma}_i \cdot \boldsymbol{\sigma}_j\right]. \tag{2}$$

In eq. (2) the two contributions are signaled by the parameters $\mathcal{A} = \frac{g^2\alpha}{16M^2}$, denoting the strength of the magnetic interaction (g is the neutron g–factor, M is the neutron mass and α is the fine-structure constant) and $\mathcal{B} = \frac{g_{aNN}^2}{\pi\alpha g^2}$, representing the relative weight of the axion interaction, which vanishes in absence of the ALP (whose mass is denoted m). The dimensionless (in natural units $c = 1 = \hbar$) functions $K(r)$ are defined as $K^{(a)}(r) = m^2r^2 + 3mr + 3$, and $K^{(b)}(r) = mr + 1$. The vector $\boldsymbol{r_{ij}} = \boldsymbol{r_i} - \boldsymbol{r_j}$ denotes the relative position of the neutrons i and j, $r_{ij} = |\boldsymbol{r_{ij}}|$ is their relative distance and the operators $\sigma_l^{r_{ij}} = \boldsymbol{\sigma}_l \cdot \hat{\boldsymbol{r}}_{\boldsymbol{ij}}$ are defined by the projection of the pauli operators $\boldsymbol{\sigma_l}$ of neutron l on $\boldsymbol{r_{ij}}$. The notation $\sigma_i^{r_{ij}}$ is used to remark that while these operators act only upon the space of the i-th particle, their form depends on the specific pair i, j considered. Eq.(2) can be recast in a more compact form by defining the functions $C(r) = \frac{\mathcal{A}}{r^3}\left(1 - \mathcal{B}e^{-mr}K^{(b)}(r)\right)$, $D(r) = \frac{\mathcal{A}}{r^3}\left(3 - \mathcal{B}e^{-mr}K^{(a)}(r)\right)$ and the symmetric matrix $K^{uv}(\boldsymbol{r}) = C(r)\delta^{uv} - D(r)R^u(\boldsymbol{r})R^v(\boldsymbol{r})$ for $u, v = x, y, z$. The symbol $R^u(\boldsymbol{r}) = \hat{\boldsymbol{r}} \cdot \hat{\boldsymbol{u}}$ denotes the projection of the vector $\boldsymbol{r}$ on the u axis. Thus $H_{ij} = \sum_{u,v} K^{uv}(\boldsymbol{r}_{ij})\sigma_i^u\sigma_j^v$. The two–neutron interaction can easily be generalized to an arbitrary number of neutrons. The total interaction Hamiltonian is simply the sum over all pairs i, j $H = \frac{1}{2}\sum_{i,j} H_{ij}$, where the factor $\frac{1}{2}$ accounts for double counting.

The study of the evolution of the n–neutron states interacting via H represents a complicated many–body problem. However, we want to consider only the evolution of the single neutron state and we are not interested to correlations and collective effects. Therefore, by using a mean field approach, we describe the interaction of a neutron with all the other nucleons by means of an effective one–particle potential. Let us consider a neutron at position $\boldsymbol{r}_i$ and the Pauli operator $\boldsymbol{\sigma}_i$, the instantaneous interaction hamiltonian due to the other neutrons is $H_i = \sum_{u,v}\sum_{j\neq i} K^{uv}(\boldsymbol{r}_{ij})\sigma_j^u\sigma_i^v$, where the sum is relative to all the other neutrons j. Since we are interested to an effective local potential for the single neutron, then we substitute the above equation with its expectation value on the state of the other nucleons. Notice that the term of a spin interacting with a magnetic field $H_i = -\mu\,(\boldsymbol{B_i}(\boldsymbol{r_i}))\cdot\boldsymbol{\sigma}$, is obtained by setting

$$\mu\boldsymbol{B_i}(\boldsymbol{r_i}) = -\sum_u\sum_j K^{uv}(\boldsymbol{r_{ij}})\langle\sigma_j^u\rangle\,. \tag{3}$$

The evolution of the single neutron state is analyzed by determining the effective magnetic field for a particular spatial spin configuration, and then by inserting the one particle operator into the Schroedinger equation.

For our purposes, we shall consider cold neutron beams with specific requirements. First of all, we deal with collimated neutron beams and a small beam width of the order of 10 μm. Experimentally, such kind of beams can be produced in different ways as shown in [48]. Neglecting the angular spread, we can assume that the beam is distributed with cylindrical symmetry around the beam axis $\hat{\boldsymbol{y}}$, and, considering it sufficiently thin, we can represent the beam as a monodimensional system. We will also assume that, by using a monochromator, one can select only neutrons around a given value of the kinetic energy K. The beam intensity is expected to decay as the neutron beam propagates. However, as a first instance, we consider the beam intensity constant and neglect the losses due to the propagation.

We now present, in Fig.(1), the scheme of the neutron interferometer eligible for the detection of the axion–mediated interaction among fermions. In such a scheme, a source, as a reactor, **SRC** generates cold or ultra–cold neutrons, which are directed to monochromator and a collimator **EXT**, making the beam (approximately) linear and monochromatic. Then a beam splitter **BS** separates the beam into two sub–beams I and II which enter in a device that selects two different spin polarizations $\boldsymbol{P}_I$ and $\boldsymbol{P}_{II}$. The sub–beams pass through regions permeated by magnetic fields of the same strength but with distinct direction, i.e. $\boldsymbol{B_J^0} = B_0\boldsymbol{P_J}$.

Making as close as possible the fractional intensities of the sub-beams χ_J with respect to the initial beam I_0: $\chi_J = I_J/I_0$, $J = I, II$, $\chi_I \simeq \chi_{II}$, one has that the average distances between successive neutrons in the two sub–beams are similar $d_I \simeq d_{II}$. Indeed, $d_J = \frac{\bar{v}_J}{I_J}$, with $\bar{v}_J$ average neutron velocity of propagation in the J sub–beam. Moreover, considering constant I_J and $\bar{v}_J$, the distances d_J are constant, and the one particle Hamiltonian H is time–independent. We assume that $d_J > 10^{-12}m$ in order that the effects of nuclear interactions are negligible.

The interference is detected in a plane **IP** where the polarized sub–beams hit, after crossing two optical paths of the same length

We choose $\boldsymbol{P}_I$ to be orthogonal to $\hat{\boldsymbol{y}}_I$ and $\boldsymbol{P}_{II}$ parallel to $\hat{\boldsymbol{y}}_{II}$. Starting from the total interaction Hamiltonian, exploiting the assumption that all the distances between two subsequent neutrons in sub–beam J are equal to d_J, with simple algebraic steps, one arrives at the effective magnetic fields [47]. It is easy to show that, within these assumptions, each neutron in sub–beam J is subject to the same effective magnetic field $\mu\boldsymbol{B_J} = \mu B_J\boldsymbol{P_J}$ given by

$$\begin{aligned}
\mu B_I &= -\frac{2\mathcal{A}\zeta(3)}{d_I^3}+\frac{2\mathcal{AB}}{d_I^3}\mathrm{Li}_3(e^{-md_I})+\frac{2\mathcal{AB}m}{d_I^2}\mathrm{Li}_2(e^{-md_I})\\
\mu B_{II} &= \frac{4\mathcal{A}\zeta(3)}{d_{II}^3}-\frac{4\mathcal{AB}}{d_{II}^3}\mathrm{Li}_3(e^{-md_{II}})-\frac{4\mathcal{AB}m}{d_{II}^2}\mathrm{Li}_2(e^{-md_{II}})\\
&\quad+\frac{2\mathcal{AB}m^2}{d_{II}}\log(1-e^{-md_{II}})
\end{aligned} \tag{4}$$

where $\zeta(s)$ stands for the Riemann zeta function while $\mathrm{Li}_s(z) = \sum_{n=1}^{\infty}\frac{z^n}{n^s}$ is the Polylogarythm function [49]. The Schroedinger equation ruling the evolution of the single neutron state, in both sub–beams, is

$$i\partial_t\psi_J = \left(-\frac{\nabla^2}{2M}+M\right)\psi_J - \boldsymbol{\sigma}\cdot\left[\mu(\boldsymbol{B_J}+\boldsymbol{B_J^0})\right]\psi_J \tag{5}$$

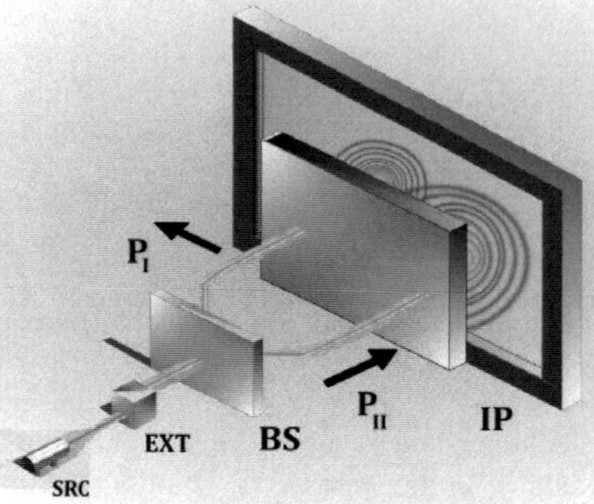

Figure 1. (color online) Diagram of the interferometric apparatus. The beam from the neutron source **SRC** is transmitted to a collimation device and a monochromator **EXT**. Then the beam is conveyed to a beam splitter **BS** which splits the beam into two subbeams with different average spin polarizations $\boldsymbol{P}_I$ and $\boldsymbol{P}_{II}$. In the end, the two subbeams interfere at the interference plane **IP**.

with ψ_J the product of a spatial wave-function and a spin function. We express the spinor for each sub–beam in the basis defined by the corresponding polarization, i.e. $\boldsymbol{\sigma}\cdot\boldsymbol{P_J}|\uparrow_J\rangle = |\uparrow_J\rangle$ and $\boldsymbol{\sigma}\cdot\boldsymbol{P_J}|\downarrow_J\rangle = -|\downarrow_J\rangle$. Moreover, we assume that the neutron, after passing the beam splitter, is in the up state for the corresponding sub–beam. We consider $t=0$ at this instant. Denoting with y the coordinate along the propagation axis, with $y=0$ and $t=0$ at the beginning of the optical path, one has $\psi_J(t) = f(t)e^{iky}|\uparrow_J\rangle$ with $f(t)$ a function which can be assumed as $f(t) = e^{-i\omega_J t}$. Then, from Eq.(5) we have $\omega_J = \frac{k^2}{2M} + M - \mu B_J - \mu B_0$, and the total phase at time t is given by $\phi_J(t) = \arg\left(\langle\psi_J(0)|\psi_J(t)\rangle\right) = -\left(\frac{k^2}{2M} + M - \mu(B_J + \mu B_0)\right)t$. The phase difference between the two beams at time t is $\Delta\phi(t) = \phi_{II}(t) - \phi_I(t) = \mu\left(B_{II} - B_I\right)t$, where μB_I and μB_{II} are given in Eq. (4). By setting $d_I = d_{II} = d$, $\Delta\phi(t)$ can be written as $\Delta\phi(t) = \left[G_m(d) + G_a(d)\right]t$, where $G_m(d) = \frac{6\mathcal{A}}{d^3}\zeta(3)$ is due to the dipole–dipole interaction and

$$G_a(d) = -\frac{6\mathcal{AB}}{d^3}\mathrm{Li}_3(e^{-md}) - \frac{6\mathcal{AB}m}{d^2}\mathrm{Li}_2(e^{-md}) + \frac{2\mathcal{AB}m^2}{d}\log(1 - e^{-md}),$$

is due to the axion–mediated interaction. Here the $n \to \infty$ limit is intended.

Neglecting a possible phase shift due to the beam splitter, the phase difference induced by the dipole-dipole interactions can be removed by setting the beam path in such a way that $G_m(d)$ is an integer multiple of 2π. This is obtained for time intervals T_k, which for any integer k are $T_k = \frac{2k\pi}{G_m(d)} = \frac{k\pi d^3}{3\mathcal{A}\zeta(3)}$. The phase difference, evaluated at T_k, is then

$$\Delta\phi(T_k) \quad = \quad \left\{\frac{k\pi\mathcal{B}}{3\zeta(3)}\left[2m^2d^2\log(1 - e^{-md}) - 6md\mathrm{Li}_2(e^{-md}) - 6\mathrm{Li}_3(e^{-md})\right]\right\}_{\mathrm{mod}\,2\pi}, \tag{6}$$

and it is different from zero only in presence of ALPs. Indeed, $\Delta\phi(T_k) = 0$ for $\mathcal{B} = 0$. Eq.(6) shows that $\Delta\phi(T_k)$ is proportional to the parameter $\mathcal{B} \propto g_p^2$. For ALP masses $m \in [10^{-6} - 1]$eV, and distances $d \in [10^{-11} - 10^{-6}]$m, one has that $\Delta\phi(T_k)$ depends only weakly on m and d, while it depends strongly on the coupling. The phase difference, modulo 2π, is plotted in Fig.(2), at the minimum recurrence time T_1, for several values of the coupling constant g_p in the mass range $[10^{-3}, 1]$ eV. For axion masses $m < 0.1$eV, one has that $\Delta\phi$ is almost independent on the distance for $d = 10^{-8}$m and $d = 10^{-6}$m. The dependence on the distance is relevant only when

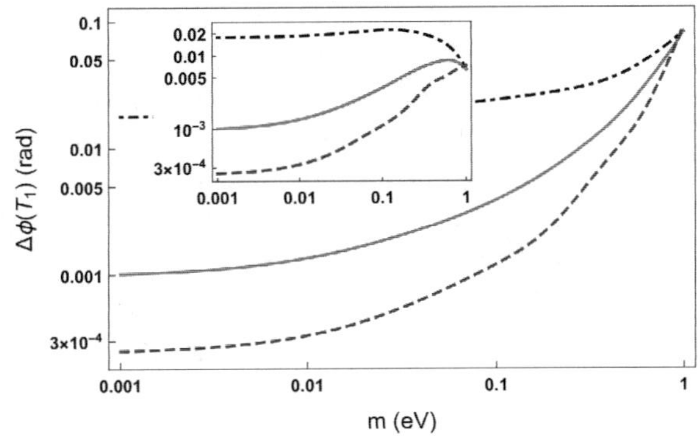

Figure 2. (color online) Logarithmic scale plot of the phase difference $|\Delta\phi(T_1)|$ modulo 2π for several values of the coupling constant in the ALP mass range $[10^{-3}, 1]$ eV for an inter–neutron distance $d = 10^{-8}$m ($d = 10^{-6}$m in the inset). In particular: the black dot–dashed line is obtained by considering the threshold from effective Casimir pressure measurements [38], $g_p = g_{CP}$, and using the following values $g_{CP} = 0.0327$ for $m = 10^{-3}$eV, $g_{CP} = 0.0348$ for $m = 0.05$eV, $g_{CP} = 0.0674$ for $m = 1$eV. For the red solid line, we used $g_p = g_{CF}$, where g_{CF} is the threshold from measurements of the difference of Casimir forces [39] and the sample values are $g_{CF} = 0.007$ for $m = 10^{-3}$eV, $g_{CF} = 0.012$ for $m = 0.05$eV, $g_{CF} = 0.066$ for $m = 1$eV. In the blue dashed line we assume $g_p = g_{IE}$, where g_{IE} is the threshold from isoelectronic experiments [40], and sample values are $g_{IE} = 0.0036$ for $m = 10^{-3}$eV, $g_{IE} = 0.006$ for $m = 0.05$eV, $g_{IE} = 0.07$ for $m = 1$eV.

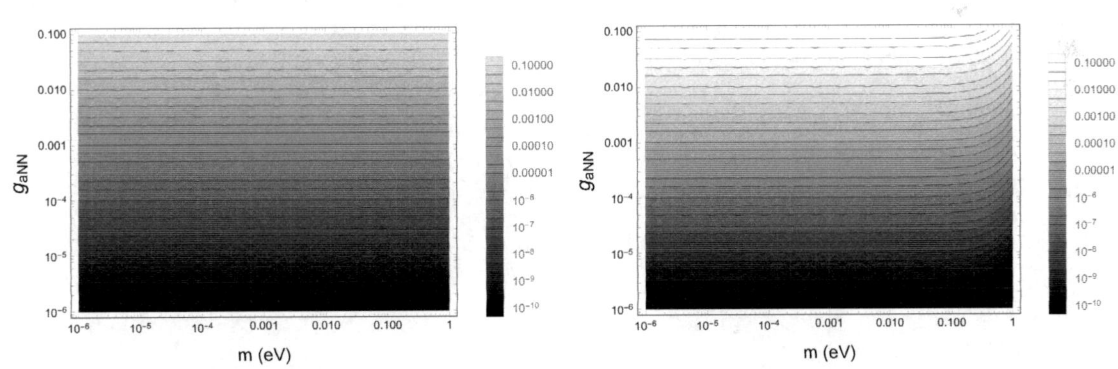

Figure 3. (color online) Contour plots of $|\Delta\phi(T_1)|$ in the mass–coupling plane for $d = 10^{-9}m$ (left panel) and $d = 10^{-6}m$ (right panel). The range of axion masses and coupling constant adopted is $(m, g_p) \in [10^{-6}, 1]$eV $\times$ $[10^{-6}, 10^{-1}]$.

the product md is quite high. On the other hand, for $md \ll 1$, $\Delta\phi$ essentially depends only on the coupling g_p.

The plots of Fig.(3) show the g_p^2 dependence of $\Delta\phi(T_1)$ and an extremely weak dependence on the ALP mass in the ranges analyzed.

Notice that in order to detect the phase difference, one has to take into account the limitation coming from the need for a recurrence time T required to isolate the axion contribution to $\Delta\phi$, and the limitation due to the total neutron flux available, which is fundamental for a sufficient small inter-neutron distance. The first limitation can be overcame by using ultra cold neutrons with velocities as small as 5 m/s [50]. These neutrons allow to have minimal recurrence times of order of the second (less than the neutron lifetime and the coherence time) and length of the interferometer of order of meters. The second limitation can be passed by using neutron beams of intensities of the order $I \simeq (10^8 - 10^{10})n/s$ [51]. Indeed, for neutron velocity of the order of $v \simeq 1$ m/s, a distance $d \simeq 10^{-8}$m is obtained in correspondence with an intensity $I \simeq 10^8$n/s.

In conclusion, we have shown that, in the time evolution of neutrons, the axion mediated fermion-fermion interaction generates a contribution to the total phase. Such a contribution can be in principle detectable, by using ultra cold neutrons in an interferometer in which the two beams are affected by differently oriented magnetic fields, in order that the phase difference depends only by the axion–induced interaction. Significant phase difference is obtained for a wide range of ALP parameters.

Acknowledgements

A.C. and A.Q thank partial financial support from MIUR and INFN. A.C. also thanks the COST Action CA1511 Cosmology and Astrophysics Network for Theoretical Advances and Training Actions (CANTATA). SMG acknowledge support from the European Regional Development Fund for the Competitiveness and Cohesion Operational Programme (KK.01.1.1.06–RBI TWIN SIN) and from the Croatian Science Fund Project No. IP-2016–6–3347 and IP-2019–4–3321. SMG also acknowledge the QuantiXLie Center of Excellence, a project co–financed by the Croatian Government and European Union through the European Regional Development Fund– the Competitiveness and Cohesion Operational Programme (Grant KK.01.1.1.01.0004).

References

[1] Bilenky S M and Pontecorvo B, *Phys. Lett. B* **61**, 248 (1976).
[2] Bilenky S M and Pontecorvo B, *Yad. Fiz.* **3**, 603 (1976).
[3] Nachtmann O, *"Elementary Particle Physics: Concepts and Phenomena"*, Springer, Berlin (1990).
[4] Capolupo A, Lambiase G, and Quaranta A, *Phys. Rev. D*, **101**, 095022 (2020).
[5] Capolupo A, Carloni S and Quaranta A, *Phys. Rev.* D **105**, 105013 (2022).
[6] Capolupo A, Giampaolo S M and Quaranta A, *Phys. Lett.* B **820**, 136489 (2021).
[7] Capolupo A, Giampaolo S M, Hiesmayr B C, Lambiase G and Quaranta A, *J. Phys. G: Nucl. Part. Phys.* **50**, 025001 (2023).
[8] Rubin V, Thonnard W K Jr, Ford N, *The Astrophysical Journal*, **238**: 471, (1980).
[9] Aghanim N et al., Planck Collaboration, *Astron.Astrophys.* **641**, A6, (2020).
[10] Keshavarzi A, Khaw K S and Yoshioka T, *Nucl. Phys.* B **975**, 115675 (2022).
[11] Capolupo A, Lambiase G and Quaranta A, *Phys. Lett.* B **829**, 137128 (2022).
[12] Peccei R D and Quinn H, *Phys. Rev. Lett.* **38**, 1440 (1977).
[13] Peccei R D and Quinn H, *Phys. Rev. D* **16** 1791 (1977).
[14] Weinberg S, *Phys. Rev. Lett.* **40**, 223 (1978).
[15] Wilczek F, *Phys. Rev. Lett.* **40**, 279 (1978).
[16] Raffelt G G, *J. Phys. A* **40**, 6607 (2007).
[17] Ellis J, *Nucl. Phys. A* 827, 187-198 (2009).
[18] Dine M, Fischler W, and Srednicki M, Phys. Lett. B **104**, 199 (1981).
[19] Zhitnitsky A, Sov. J. Nucl. Phys. **31**, 260 (1980).
[20] Particle Data Group Zyla P A et al., *Progress of Theoretical and Experimental Physics*, vol. **2020**, Issue 8, August 2020, 083C01 (2020).
[21] Kim J E, Phys. Rev. Lett. **43**, 103 (1979).
[22] Shifman M A, Vainshtein A, and Zakharov V I, Nucl. Phys. B 1**66**, 493 (1980).
[23] Kim J E and Marsch D J E, *Phys. Rev. D* **93**, 025027 (2016).
[24] De Martino I, Broadhurst T, Henry-Tye S-H, Chiueh T, Schive H-Y and Lazkoz R, *Phys. Rev. Lett.* **119**, 221103 (2017).
[25] Rubakov V A, *JETP Lett.* **65**, 621-624 (1997).
[26] Marsch D J E, *Phys. Rep.*, **643**, 1 (2016).
[27] Zavattini E et al. (PVLAS Collaboration), *Phys. Rev. D* **77**, 032006 (2008).
[28] Pugnat P et al. (OSQAR Collaboration), *Phys. Rev. D* **78**, 092003 (2008).
[29] Ballou R et al. (OSQAR Collaboration), *Phys. Rev. D* **92**, 092002 (2015).

Journal of Physics: Conference Series **2533** (2023) 012044 doi:10.1088/1742-6596/2533/1/012044

[30] Ehret K et al. (ALPS Collaboration), *Phys. Lett. B* **689**, issues 4-5, pages 149-155 (2010).
[31] Aune S et al. (CAST Collaboration), *Phys. Rev. Lett.* **107**, 261302 (2011).
[32] Du N et al. (ADMX Collaboration), *Phys. Rev. Lett.* **120**, 151301 (2018).
[33] Barbieri R et al., *Phys. Dark Univ.* **15**, 135-141 (2017).
[34] Capolupo A, Lambiase G, Vitiello G, *Adv. In High En. Phys.* 826051 (2015).
[35] Capolupo A, De Martino I, Lambiase G, Stabile A, *Phys. Lett. B* **790**, 427-435 (2019).
[36] Moody J E, Wilczek F, *Phys. Rev. D*, **30**, 130 (1984).
[37] Daido R, Takahashi F, *Phys. Lett. B*, **772**, 127 (2017).
[38] Bezerra V B, Klimchitskaya G L and Mostepanenko V M and Romero C, *Eur. Phys. J. C* **74**, 2859 (2014).
[39] Klimchitskaya G L and Mostepanenko V M, *Phys. Rev. D* **95** 123013 (2017).
[40] Klimchitskaya G L and Mostepanenko V M, *Eur. Phys. J. C* **75**, 164 (2015).
[41] Capolupo A, Giampaolo S M, Lambiase G, Quaranta A, *Phys. Lett. B* **804**, 135407 (2020).
[42] Rauch H, Werner S A, *Neutron Interferometry–Lessons in Experimental Quantum Mechanics, Wave–Particle Duality, and Entanglement*, 2nd Edition, Oxford University Press, Oxford (2015).
[43] Werner S A, Staudenmann J–L, Colella R, *Phys. Rev. Lett.*, **42**, 1103 (1979).
[44] Allman B E, Kaiser H, Werner S A, Wagh A G, Rakhecha V C, Summhammer J, *Phys. Rev. A*, **56**, 4420 (1997).
[45] Wagh A G, Rakhecha V C, *Phys. Lett. A*, **148**, Issues 1–2, pp. 17-19 (1990).
[46] Wagh A G, Rakhecha V C, Fischer P, Ioffe A, *Phys. Rev. Lett.*, **81**, 1992 (1998).
[47] Capolupo A, Giampaolo S M, Quaranta A, Neutron interferometry, fifth force and axion like particles, European Physical Journal C, **81**, 1116 (2021)
[48] Ott F, Kozhevnikov S, Thiaville A, Torrejon J, Vazquez M, *Nucl. Inst. Meth. in Phys. Res. A*, **788** (2015), pp. 29-34.
[49] Gradshteyn I S, Ryzhik I M, *Table of Integrals, Series and Products*, 7th edition, Academic Press (2007), pp. 110-111.
[50] Steyerl A, *Phys. Lett. B*, **29**, Issue 1, pp. 33-35 (1969).
[51] INTERNATIONAL ATOMIC ENERGY AGENCY, *Compendium of Neutron Beam Facilities for High Precision Nuclear Data Measurements*, IAEA-TECDOC-1743, IAEA, Vienna (2014).

Journal of Physics: Conference Series **2533** (2023) 012047 doi:10.1088/1742-6596/2533/1/012047

Gravitation-induced temperature of single atoms on atomic and cosmological scales

D Drakova[1] and G Doyen[2]

[1] Sofia University, 1 James Bourchier Blvd, 1164 Sofia , Bulgaria
[2] Ludwig-Maximilians Universität, Butenandstr. 5-13, 81377 Munich, Germany

E-mail: `nhdd@chem.uni-sofia.bg; gerold@gerold-doyen.de`

Abstract. The experimental findings on the rate of transfer of a single Xe atom from the sample surface towards the tip in the low temperature STM, induced by voltage pulses, are understood by assuming a vibrational temperature for a single Xe atom. But temperature is introduced in standard quantum physics as a parameter in the Boltzmann distribution of a statistical ensemble consisting of large number of non-interacting exact copies of the dynamical system under investigation. This represents a problem, if the system under investigation is to be described by a single pure quantum state being a solution of Schrödinger's equation. In the framework of a quantum field theory based on Einstein-Hilbert action in high dimensions (EHHD) we demonstrate that temperature for a single atom emerges if the atom is beabled in a warp resonance due to entanglement with local gravitons (gravonons), living in high spacetime dimensions (11D). Desorbing Xe atom in a limbo state in the tunnel gap, due to destruction of the beable via inelastic tunnelling, is in Boltzmann distribution over plane waves $\{\mid k\rangle\}$ since the squared overlap of the warp resonance with free plane wave states varies like $\exp(-E_k/k_BT_{vib})$ where E_k is the energy of the free particle state and T_{vib} is the vibrational temperature of Xe calculated from the shape and the extension of the warp resonance. The nature of the warp resonance varies with the tip-sample voltage (becoming more contracted at higher tunnelling currents, as the tip approaches the sample) and raises its energy to higher values. More contracted warps at higher tunnelling currents imply stronger overlap with free states of higher energy, described by Boltzmann factors of higher temperature. The mechanism generating single atom temperature generalizes to other dynamical processes where temperature is completely determined by the nature of the warp resonance in the beables. For homogeneous and isotropic systems this implies a universal value of the temperature everywhere in the system. With these considerations the mechanism is also applied to intergalactic hydrogen-atom clusters, where it explains the observed temperature of the CMB radiation and the absence of the horizon problem.

1. Introduction

Boltzmann distribution is measured experimentally for a single system and not for an ensemble of identical independent systems [1] (cf. $Xe/Ni(110)$ in the low temperature STM by Eigler et al. [2]). The question is where and how does temperature of a single atom or single universe emerge.

A single universe has no temperature since it is described by one wave function, but a single Xe atom has temperature as the experiment of Eigler [2] and the approach by Walkup et al. [3, 4] show. Temperature exists for objects within the universe when they are measured. Galaxy surveys show that the observable universe is nearly isotropic and homogeneous. The problem of

Journal of Physics: Conference Series **2533** (2023) 012047 doi:10.1088/1742-6596/2533/1/012047

thermalization of the cosmic microwave background radiation, with temperature differences as small as 10^{-5} K, is addressed in inflation theory by postulating that in an exponential expansion of nearly 10^{22} times in 10^{-32} seconds the CMB retains the equilibrium of the causally connected preinflation state.

If quantum theory is valid for the whole universe the whole story about the horizon problem cannot be relevant. The assumptions, on which the horizon problem is based, have to be thought over if a single wave function describes the whole universe. Zeilinger et al. demonstrated the existence of correlations in a quantum system of two photons on a very long distance range, which cannot be provided by messenger particles moving with light velocity [5].

The horizon problem, which is a classical problem, is discussed in terms of temperature measured here and temperature measured there. In inflation theory and in Copenhagen Quantum Mechanics (CQM) this is interpreted with the collapse of the wave function of the total universe on a statistical mixture of states, which is described by the statistical density matrix. But what is the mechanism of the collapse?

We suggest a solution of the horizon problem and Cosmic microwave background isotropy in the hypothesis of high dimensions (HDH). Recent experiments point out exactly in the direction of the HDH theory. Arguments of the kind used by inflation theory are not needed in HDH. The recent experiments show that temperature can be assigned to a single atom and it is different from the temperature of its environment. Hence temperature is not what it is believed to be. *To understand temperature we need quantum mechanics.*

2. Emerging temperature of a single atom

2.1. The experimental situation

It is a low temperature STM experiment in ultra-high vacuum with the tip positioned 5 Å aside from the Xe atom, adsorbed on a $Ni(110)$ surface. The height of the tip for this operation is adjusted in the imaging mode, such that a constant resistance of the junction is retained, $R = \frac{U}{I} = 906\ \text{k}\Omega \pm 2\%$, while the Xe atom is on the surface for the full coverage of voltages used. Pulses of varying voltage are applied between the tip and the sample, which is negatively polarized at $T_s = 4$ K. The delay time between the onset of the voltage pulse and the change of the junction conductance, owing to the jump of a Xe atom from the $Ni(110)$ terrace to the STM tip, is measured. The dependence of the transfer rate, which is the inverse of the delay time, on the applied voltage is reproduced only if the Xe atom is assigned a vibrational temperature T_{vib}, which is different from that of the environment T_s and depends upon the applied voltage. For the tunnel junction with $R = \frac{U}{I} = 906\ \text{k}\Omega \pm 2\%$ a power-law dependence of the transfer rate on the voltage of the pulse U, and consequently on the tunnel current I, according to $I^{4.9\pm0.2}$ is observed. The measurement, performed several times, however on a single Xe atom at selected values of the voltage U, does not mean statistics over 10^{23} Xe atoms, as it is implied in the theory of Walkup, Avouris and Newns based on the ergodic theorem and assuming Boltzmann statistical distribution of the Xe atom on vibrational states in the potential wells on the sample and the tip. [3, 4]. The transfer rate is calculated in a statistical approach using Boltzmann factors with the Xe atom vibrational temperatures for each value of the STM voltage as a fit parameter. We realize that the vibrational temperature of Xe is the necessary condition for reproducing the experimental results. Our aim is to give an answer to the question how does vibrational temperature for a single atom emerge.

Journal of Physics: Conference Series **2533** (2023) 012047 doi:10.1088/1742-6596/2533/1/012047

2.2. EHHD in high hidden spacial dimensions

The theory has been described and applied to many physical phenomena [6]. Soft gravonons, local gravitons, living in high hidden dimensions (HHD) with high density of states, entangle with matter fields. Even though the gravitational interaction in 11 spacetime dimensions at short distances is higher than Newton gravity in three dimensional space, $V^{(11)}_{grav}(r) = -\frac{G^{(11)}m}{\pi^7 r^8}$, it is still very weak and of short range. Therefore it involves gravonons, which are nearly degenerate with the initial particle state. These are the decisive features of the dimensionality effect, the backbone of the present theory: weak degeneracy coupling with a continuum of high density of states.

Beables, solution of Schrödinger's equation, are generated if a beabling condition is fulfilled: the ratio of the weak gravitational interaction with the gravonons and the gravonon frequency $\frac{V_{grav}}{\omega_{grav}}$ should be of the order of ≈ 10. *Localization of a matter field as a beable is the solution of the measurement problem of standard quantum mechanics.* A beabled Xe atom is experimentally observed as a localized particle on its adsorption site for weeks. After the lifetime of the gravonons in the HHD expires they recur back in three dimensional space. The effect is a fast increase of the weight of configurations in the total wave function describing a free Xe atom in a limbo state in the tunnel gap moving from the sample to the tip. This happens at the expense of the beable configurations, which lose weight during a tiny burst out of the warped space where entanglement with the gravonons is effective (fig. 1, left plot).

In our theory, in addition to the potential for a single Xe atom in a double well between the sample and the tip, a term is involved describing the interaction with the gravonons in the warped space. The pure effect of entanglement with the gravonons is studied by prediagonalizing all other quantum interactions with the exception of the gravitational interaction of the Xe atom core with gravonons in high spatial dimensions. This is a very small energy change, however, it is decisive for the localization of the Xe atom in its equilibrium adsorption site on the nickel surface. Standard quantum mechanics would not lead to atom localization on the flat nickel surface, its result is the wave function of the limbo state, which cannot be measured experimentally. In contrast, in Eigler's STM experiments the Xe atom is initially localized and does not move parallel to the surface. An atom in the limbo state, delocalized over the sample surface, cannot be manipulated and transferred towards the STM tip!

Schrödinger's time dependent equation $\mathrm{i}\frac{\partial}{\partial t}\Psi(\psi,\zeta) = H\Psi(\psi,\zeta)$ is solved with a Hamiltonian, which depends on the matter field ψ and the gravonon field ζ. The interaction Lagrangian density of EHHD for the matter field of the Xe atom and the gravitational interaction h_{00} with the gravonon field is:

$$\mathcal{L}_{int}(r) = \frac{1}{2}m^2\psi^+\psi\Big[-h_{00} - \zeta^+_{00}\zeta_{00} - h_{00}\zeta^+_{00}\zeta_{00} - \frac{3}{8}\zeta^+_{00}\zeta_{00}\zeta^+_{00}\zeta_{00}\Big]. \quad (1)$$

The solution is expanded in a configuration-interaction basis of many-particle configurations: $\{|\, g_0, 0\rangle\}$,$\{|\, w_{sample}, \kappa\rangle\}$ and $\{|\, k, \kappa\rangle\}$, where the first symbol denotes a basis state for the Xe atom (g_0 initial state, w_{sample} state in the warp resonance on the sample, k plane waves in the tunnel gap), the second simbol κ denotes a state of the gravonon field: 0 for the ground state and κ for gravonon excitations.

3. Temperature emerges in the beable-limbo transition

The variation of the weights of the beable and limbo configurations with time can be seen in fig. 1. As long as the Xe atom is not vibrationally excited and is entangled with gravonons living in the high hidden spacial dimensions the xenon atom is localized as a beable on the nickel surface

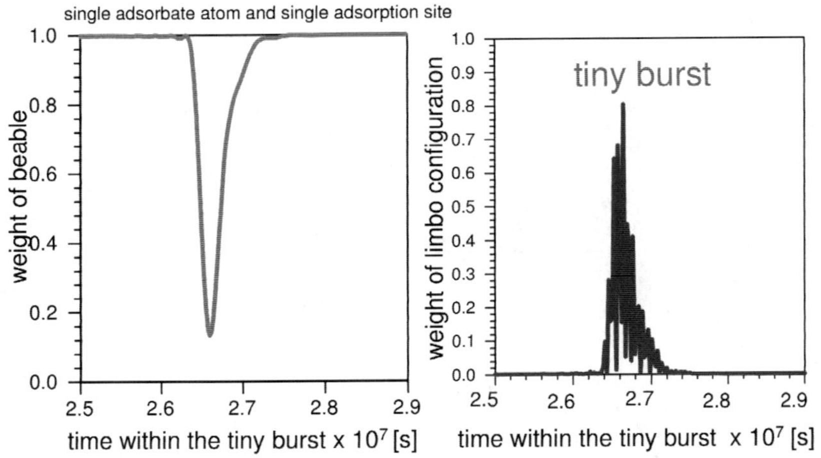

Figure 1. A Xe atom on a $Ni(110)$ surface entangled with gravonons is localized as a beable on a chosen site for days, its parallel motion being frozen (left hand side). In the presence of the STM tip in a process of inelastic electron tunnelling the beabling condition is violated and the Xe atom bursts out of the warp resonance in a limbo state during a tiny burst. The weight of configurations involving gravonons drops. The weight of configurations with plane waves, describing the atom in the limbo state, increases (right hand side).

for days, its parallel motion being frozen, despite the nearly flat potential parallel to the surface. In inelastic tunnelling a tunnelling electron in the sample can lose energy, which is gained as vibrational energy by the adsorbed Xe atom. But then the beabling condition for the atom is no more fulfilled, the atom bursts out of the warp resonance in a limbo state in the tunnel gap (fig. 2). During the tiny burst the atom is disentangled from the gravonons. Plane wave components of the limbo state with different momenta, delocalized in the tunnel gap, gain weight (fig. 1).

The weight of the plane-wave involving components of the limbo configuration is proportional to the squared overlap between the Gaussian warp resonance on the sample surface $\mid w_{sample}\rangle$ and the plane wave $\mid k\rangle$, which is the Boltzmann factor at constant temperature T (fig. 3):

$$\langle w_{sample} \mid Xe\ limbo\rangle = \sum_k \langle w_{sample} \mid k\rangle\langle k \mid Xe\ limbo\rangle. \tag{2}$$

$$\begin{aligned} \mid Xe\ limbo\rangle &= \sum_k \mid k\rangle\langle k \mid w_{sample}\rangle \\ &= Z^{-1/2}\sum_k e^{-\frac{E_k}{2k_BT}} \mid k\rangle \end{aligned} \tag{3}$$

because the overlap of a plane wave $\exp(ikx)$ with energy E_k with the warp resonance is:

$$\langle w_{sample} \mid Xe\ limbo(k)\rangle = Z^{-1/2}e^{-\frac{E_k}{2k_BT}}. \tag{4}$$

In this way the Xe atom in the limbo state is described by a coherent superposion of states, with the square root of the Boltzmann factors as coefficients, which will be measured. The final

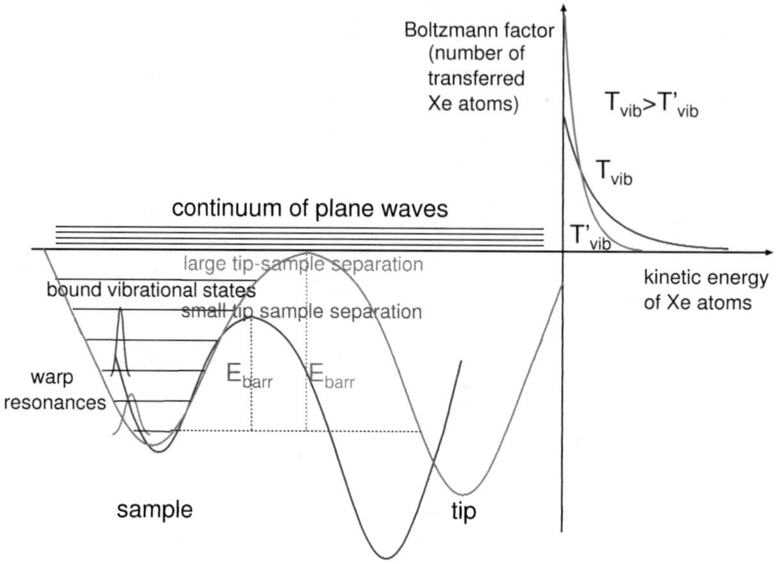

Figure 2. The Xe atom in a double well potential between the $Ni(110)$ surface and the tip at two tip-sample separations. The Xe atom is initially in a warp resonance localized as a beable on the sample surface. In inelastic electron tunnelling, when the electron loses energy, energy conservation requires that the Xe atom is vibrationally excited. In this case the degeneracy with the gravonons is violated and the Xe atom leaves the warp resonance, the beable is destroyed. The atom is in a limbo state, a coherent superposition of plane wave states above the tunnel barrier on the way to the tip. The current (i.e. number of tunnelling electrons) is measured on the tip side for each choice of STM parameters (tip-sample distance, voltage) at definite energy. This means that the Xe atom in limbo is also in a state of definite energy, yielding Boltzmann distribution over plane waves at a constant temperature. In EHHD the Boltzmann distribution arises from the overlap between the Xe warp resonance and the limbo plane wave state according to eq. (5), when the atom bursts out of the beable into limbo. Different experimental conditions (tip-sample distance, voltage) require different warp resonances on the sample which means different Boltzmann distributions depending on the temperature. This is how temperature emerges in EHHD.

measured state with the electron at energy $-\varepsilon_\gamma$, relative to its initial energy at the Fermi level in the sample, is then:

$$\begin{aligned} \mid \Psi\rangle &= Z^{-1/2} C \sum_k \mathrm{e}^{-\frac{E_k}{2k_B T}} \mid k\rangle \mathrm{e}^{\mathrm{i}\gamma x} \\ &\rightarrow Z^{-1/2} \mathrm{e}^{-\frac{\varepsilon_\gamma}{2k_B T}} \mid k(\varepsilon_\gamma)\rangle \mathrm{e}^{\mathrm{i}\gamma x} \end{aligned} \qquad (5)$$

because $E_k = \mid \varepsilon_\gamma \mid$.

The temperature is defined by the warp resonance of Xe on the nickel side where it is beabled by entanglement with gravonons. As soon as the entanglement is truncated due to vibrational excitation of Xe in the inelastic tunnelling process, the xenon atom bursts out of the warp resonance in a distribution over plane wave configurations. This is called a Boltzmann limbo state

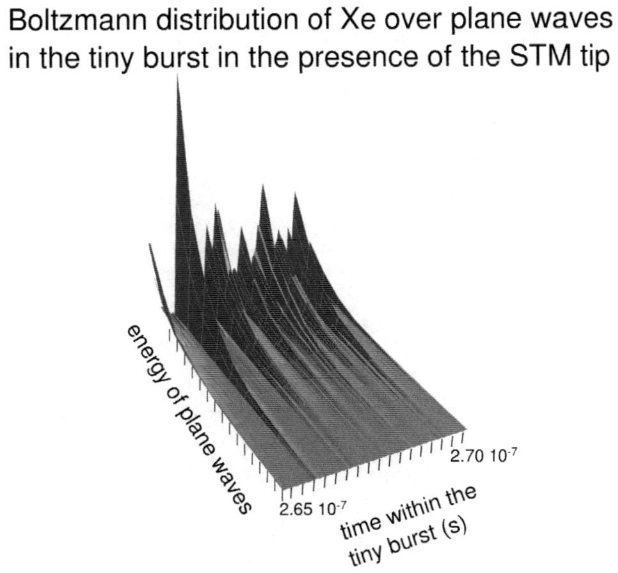

Figure 3. The weights of plane-wave-involving configurations in the limbo state of the Xe atom in the tunnel gap during the tiny burst are the Boltzmann factors with definite temperature.

with a sharp energy, it is a state describing the Xe field in the same way as in the case of the warp resonance on the sample. $\mid Xe\ limbo\rangle$ is expanded in plane waves as eq. (3) shows. The Boltzmann distribution is what is measured in the STM by making thousands of experiments. The transition is: $\mid Xe\ w_{sample}, \kappa\rangle \rightarrow \mid Xe\ limbo\ plane\ waves, 0\rangle$ each plane wave component with weight equal to the Boltzmann factor: $\mid \exp(\frac{-E_k}{2k_B T_{vib}})\mid^2$ with $E_k = \mid \varepsilon_\gamma \mid$ the energy gained as vibrational energy of Xe in inelastic electron tunnelling. The weight of each plane wave component is the squared overlap of a Gaussian warp resonance with a plane wave with energy ε_γ, which represents the Boltzmann overlap distribution. This provides the Boltzmann distribution by the measured signal with a definite vibrational temperature.

One or another plane-wave-involving component dominates in the limbo configuration with time during the tiny burst (fig. 3). In different experiments one or another component can choose a site on the tip, where the beabling condition is fulfilled. In the beable on the tip the temperature of the respective plane wave limbo component can be measured experimentally because the kinetic energy of the plane waves is transferred to free current-carrying electrons in the tip, which are measured in the experimental equipment. Measured is the current at a given energy of the tunnelling electrons, i.e. measured is the Boltzmann distribution of the electrons over energy states between the two Fermi levels, which is equivalent to Boltzmann distribution of the Xe atom over plane waves in the tunnel gap at a temperature T_{vib}, assigned to the vibrational motion of the Xe atom, which is excited by inelastically tunnelling electrons.

The vibrational temperature of a single atom, which can be measured somewhere, somehow, emerges in the transition from the beabled to the limbo state. Temperature of a single atom emerges as a quantum phenomenon.

4. Vibrational temperature of a single atom in the STM at varying voltages

We have to conclude that we also need different vibrational temperatures for different tunnelling currents. For each value of the voltage we have to make a different calculation with entanglement

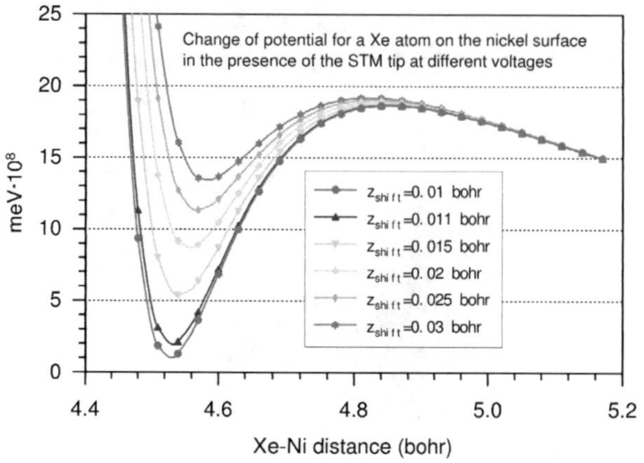

Figure 4. The changes of the potential for a Xe atom on the $Ni(110)$ surface *in the presence of the STM tip* refer to shifts of the potential well depths on the nickel surface to higher energy of the order of several 10^{-8} meV and depend upon tiny shifts z_{shift} of the equilibrium position of the Xe atom away from the nickel surface. The effects due to the STM tip are stronger for increasing voltages (negative with respect to the sample) up to $U = -0.2$ V.

to gravonons which will result in a Boltzmann factor $\mid \exp(-\frac{\varepsilon_\gamma}{2k_B T_{vib}}) \mid^2$ with a different temperature T_{vib}. The justification is that different voltages in the STM lead to change of the potential for the Xe atom, as it is schematically displayed fig. 2 and in fig. 4, where the values of z_{shift} for different voltages in the STM are taken from fig. 3 in ref. [4] and fig. 1 in [3]. Negative voltage applied to the sample, when the Xe atom can be transferred to the tip, results in reduced attractive potential well on the sample site and increased attractive well depth on the tip site. A shift of the potential well on the sample site to higher energy and larger equilibrium distance and increase of the repulsive energy are the effects of negative sample voltage in the presence of the tip.

For higher voltages in the STM, negative with respect to the sample, the shift of the potential for Xe to larger distance from the Ni surface and the attenuation of the attractive potential well depth on the sample site are more pronounced. The effect would be attenuation of the entanglement strength with the gravonons, hence the xenon atom would burst out of the warp resonance in a limbo state because of violation of the beabling condition. However, in experiment the Xe atom remains beabled, i.e. the beabling condition is still satisfied. Beabling can be warranted if the warp resonance shifts closer to the nickel surface, the cost is climbing up the repulsive branches of the potential curves in fig. 4, as it is schematically displayed in fig. 2. The kinetic energy of Xe in the new warp resonance raises, which means w_{sample} is changing its expansion, becoming more contracted. If not, the average gravitational potential within the warp resonance will be reduced because large part of it will penetrate in the repulsive part of the potential.

The scheme in fig . 5 can be used to discuss the nature of the Xe warp resonances in the presence of the STM tip. A field configuration with the Xe atom in a warp resonance in the non-shifted potential determines the energy shell and warrants optimal entanglement with the gravonons, with the beabling condition satisfied. The shifted gravitational potential is less attractive. If the Xe atom is beabled in the same warp resonance (black diffuse Gaussian in fig. 5) it needs to be shifted closer to the repulsive wall to achieve an average gravitational potential as in the absence of the STM tip (black dashed horizontal line). This, however, costs energy because large part of

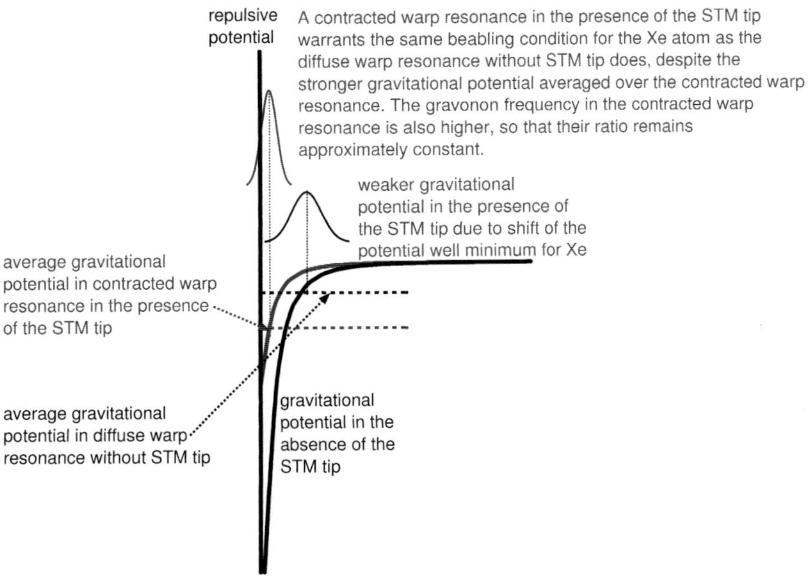

Figure 5. Non-shifted (black curve) at zero voltage and shifted (blue curve) gravitational potential for a Xe atom due to the applied voltage in the STM. The black and blue Gaussians (full curves) represent the warp resonances for Xe without and with the STM voltage applied. The field configuration with Xe in the initial warp resonance (diffuse black Gaussian) is on the energy shell, whereas the field configuration with Xe in the contracted warp resonance (contracted blue Gaussian) is no more on the energy shell. However, in the contraced warp resonance the same beabling condition for the Xe atom exists as in the initial more diffuse warp resonance. This is possible despite that the gravitational potential averaged over the contracted warp resonance has a higher value than the potential averaged over the diffuse warp resonance since the gravonon frequency in the contracted warp resonance is also higher.

the diffuse warp resonance will penetrate in the repulsive wall and the energy of Xe must shift to higher value. If the warp resonance at the higher energy becomes more contracted (the blue Gaussian in fig. 5) the average gravitational potential $V_{grav\ average} = V_{grav} \times \alpha^{3/2}$ for Xe will be stronger because the Gaussian exponent α increases (blue dashed horizontal line). This means that the kinetic energy of Xe in the warp resonance will increase and the field configuration with Xe in the contracted warp resonance will no more be on the energy shell. The situation is repaired because at the same time the frequency of the entangled gravonons increases, keeping the beabling condition $V_{grav\ average}/\omega_{grav}$ constant. The values of the frequencies of the soft gravonon modes ω_{grav} in different warp resonances are evaluated as a function of the coupling strength γ within the cluster of nickel atoms in the presence of the Xe atom in the $\gamma - \eta$ model [7]. The extreme sensitivity of γ to the shift of the equilibrium position of the Xe atom is described in ref. [8]. Thus both a diffuse and a shifted contracted warp resonance can provide the optimal beabling condition for entanglement with the gravonons, which is needed for the localization of the Xe atom on the nickel surface at both tip-sample voltages. Therefore we may scale the gravitational potential as we wish, no matter whether we have diffuse or shifted contracted warp resonances. The beabling condition is retained constant because the exponents α of the Gaussian warp resonances $| w_{sample}\rangle = N\exp(-\alpha r^2)$ enter in the averaging of the gravitational potential. Enhancement of the average value of the gravitational potential $V_{grav\ average} = V_{grav} \times \alpha^{3/2}$ in a shifted contracted warp resonance is compensated by the increase of ω_{grav} so that $V_{grav\ average}/\omega_{grav}$ remains constant.

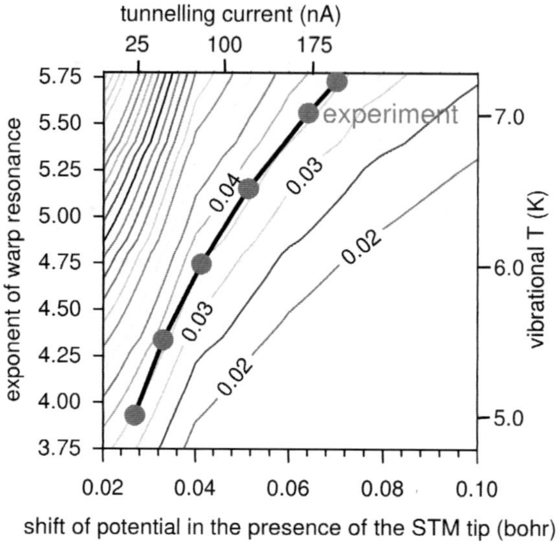

Figure 6. Lines of constant beabling condition for the Xe atom on a nickel surface $const \times \left(\frac{V_{grav} \times \alpha^{3/2}}{\omega_{grav}}\right)^2$ in Gaussian warp resonances of different expansion $\mid w\rangle = \exp(-\alpha r^2)$, centered at different positions from the nickel surface, which in experiment depend upon the applied voltage in the STM at constant resistivity. The experimental data of the rate of transfer of the Xe atom as a function of the tunnelling current is reproduced with a black curve. The experimental values of the transfer rate have been associated with vibrational temperatures of the Xe atom in ref. [4], varying with the tunnelling current.

The degree of warp resonance contraction provides the vibrational temperature at each value of the applied voltage. This is so because the exponent of the warp resonance defines the vibrational energy of Xe in the beable $\alpha = \frac{mE_o}{\hbar} = \frac{mk_B T_{vib}}{\hbar}$. In the beable-limbo transition the Xe atom bursts out of different warp resonances with different vibrational temperature in different plane wave configurations and Boltzmann factors, corresponding to different temperatures, as it is described in the previous section. The contours of constant beabling condition as function of the exponent of the warp resonance and the shift of potential in the presence of the tip, i.e. the voltage in the STM, are the result of EHHD plotted in fig. 6. The experimental curve by Eigler et al. [2] coincides with one curve of constant beabling condition, reproducing the variations of the vibrational temperature of a single Xe atom measured in the STM.

5. Temperature of the CMB radiation: light emission from a light-atom cluster

A model with intergalactic hydrogen-atom lattices has been used to study the cosmic microwave background (CMB) radiation in ref. [9]. It was concluded that the beabling condition is fulfilled for hydrogen-atom lattices with 10^6 hydrogen-atom clusters at the lattice sites and lattice constant of the order of 10 meters entangled with very soft gravonons, called Newton modes, with frequency $\omega_{Newton\ mode} = 7.6 \times 10^{-11}$ year^{-1}. The time dependence of the beabling process via entanglement with the Newton mode is calculated by solving Schrödinger's equation, yielding Hubble time of the order of $(H_o)^{-1} = \omega^{-1}_{Newton\ mode} = 1.31 \times 10^{10}$ years. The Hubble time derived within this picture is in agreement with the value measured today.

Expansion of the hydrogen-atom lattice occurs as quantum diffusion of the clusters in the limbo

[h]

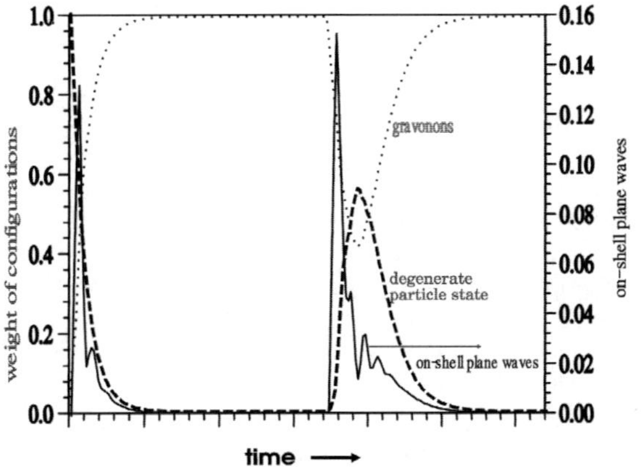

Figure 7. Partial sums of weights (i.e. squared coefficient of the field configurations) of configurations vs. time as a solution of Schrödinger's equation. The dotted curve comprises all configurations entangled to gravonons (Newton mode). The dashed curve is the sum over all configurations with the light-atom cluster in the ground state. The full curve represents all configurations where a H-atom is in a plane wave limbo state (from ref. [9]).

state. Rebeabling of the hydrogen-atom clusters on new lattice sites at the edges of the lattice far away from the initial beabling sites in the lattice gives the impression that space is expanding. But space in our theory does not expand. Just the distances in space between beabled objects expand.

A beable of the light-atom cluster exists and each hydrogen atom in the cluster is localized in a warp resonance. Beabling of the hydrogen atoms relative to each other in the cluster is possible only if the $H-H$ distances are of the order of 0.33 cm. At this lattice constant the beabling effect of entanglement with the Newton mode vibration of the light-atom lattice is more or less equal to the effect of Van der Waals interactions, which counteract the stabilization of the hydrogen atom beable due to gravity in high hidden spatial dimensions. The Newton gravonon mode is slightly perturbed by a non-adiabatic coupling to vibration of a hydrogen atom on a lattice site of the cluster, which may be resonantly excited by an external cosmic particle or phonon incident on the cluster. The vibrational excitation of the hydrogen atom in the cluster due to an incident particle or a photon destroys the beable. The hydrogen atom is in the limbo state, plane waves being used for the description of the matter field (fig. 7).

5.1. How is the CMB radiation created in EHHD

The CMB radiation is emitted within the hydrogen-atom clusters on the lattice sites of the hydrogen-atom lattice. A cosmic particle hitting a hydrogen atom in the cluster polarizes the hydrogen atom and excites it vibrationally, the hydrogen atom goes in limbo because the beabling condition is no more satisfied. The competition between Van der Waals interaction of polarized atoms and the gravitational interactions according to Newton's law shows that beabling of the hydrogen atoms relative to each other in the cluster is possible only if the $H-H$ distances are of the order of 0.33 cm when the two interactions are comparable [9]. This conditon sets limits to the lattice constant of the hydrogen-atom clusters for which the beabling condition via entanglement to the Newton mode is fulfilled and the hydrogen atoms in the cluster are beabled relative to each other.

Journal of Physics: Conference Series **2533** (2023) 012047 doi:10.1088/1742-6596/2533/1/012047

A hydrogen atom in this hydrogen-atom cluster may be resonatly excited by an incident cosmic particle with maximal energy of the order of 10^{-4} eV because the vibration of the hydrogen atom cannot excede 0.33 cm. A quantum of energy of 1.9×10^{-4} eV, corresponding to a wavelength twice the lattice constant of 0.66 cm, is the minimal energy, which can be resonantly absorbed by a single hydrogen atom. In this opto-mechanical model the hydrogen atom bursts out of the warp resonance in the limbo state with the vibrationally excited hydrogen atom in plane waves. A photon is emitted in the light-atom lattice, when the hydrogen atom is rebeabled in its ground state in a warp resonance, with the energy released in the transition of the hydrogen atom from the excited vibrational state. Since the hydrogen atom has bursted out of the warp resonance during the tiny burst in a Boltzmann distribution over plane waves the distribution of the weight, with which the photon is emitted with different energy, is also the Boltzmann distribution. This is how the temperature of the CMB radiation emerges. A hydrogen-atom cluster with lattice constant 0.33 cm emits photons with maximal wavelength $\lambda = 0.66$ cm, the corresponding CMB temperature would be $T_{CMB} = 2.2$ K. Very tiny changes in the light-atom cluster are responsible for tiny differences with respect to the exact CMB temperature measured today. For instance, CMB radiation with $T_{CMB} = 2.7$ K would be emitted from a hydrogen-atom cluster with $R_{H-H} = 0.27$ cm.

The opto-mechanical model of emerging temperature of the cosmic microwave background within the EHHD theory yields the correct order of magnitude. Obviously the hydrogen-atom cluster operates as a black body radiator. This is a mechanism for gravitation-induced temperature of single atoms. In a homogeneous and isotropic universe the conditions for atoms bursting out of beables in limbo states have identical initial conditions: identical warp resonances release the atoms in the same Boltzmann distributions over plane wave components of the limbo states with the same temperature. Upon rebeabling the atoms emit photons with the same temperature, which is the temperature of the cosmic microwave background radiation. The model provides the solution of the horizon problem in CMB.

6. Conclusion

In the EHHD theory temperature and the Boltzmann factor are the result of the *coherent time development* according to Schrödinger's equation of the wave packet of an adsorbed Xe atom localized on a nickel surface as a beable via entanglement with local gravonons. The coherent time evolution of the core motion of a single adsorbed atom entangled with the continuum of delocalized plane waves and the environmental field provided by the surface atoms (gravonon field) results in a distribution of the particle in eigenstates of the continuum of plane waves, which describe its motion in the gas phase. The approach leads to the Boltzmann distribution factor and to emerging temperature in quantum mechanics. The Boltzmann factors are the squared coefficients in the expansion of the time dependent wave packet over the plane wave components. With the mechanism of choosing one and only one warp resonance, where entanglement with gravonons is effective, and bursting out of the beable in limbo configurations, the Boltzmann factors emerge without introducing statistics.

The major implications of our theory referring to the emergence of temperature on atomic and cosmological scale can be summarized:

- In the present quantum field theory temperature emerges as a quantum mechanical phenomenon.
- The prerequsite for emerging temperature of a single atom in the STM is its localization as a beable due to entanglement with gravonons in different warp resonances at different voltages. Temperature emerges during the beable-limbo transition.

Journal of Physics: Conference Series **2533** (2023) 012047 doi:10.1088/1742-6596/2533/1/012047

- According to this theory thermalization is completely determined by the local environment. If the local environment is homogeneous and isotropic then this solves the horizon problem of CMB.
- The dimensionality effect, large density of graviton states and enhanced gravitational strength on the atomic scale, is the backbone of EHHD based on Einstein-Hilbert action in high spacetime dimensions.

References

[1] Bogoliubov N N 1968 *Lectures on Quantum Mechanics* vol. 1, (London: Macdonald & Co Ltd)
[2] Eigler D M, Lutz C P and Rudge W E 1991 *Nature* **352** 600
[3] Walkup R W, Newns D M and Avouris Ph 1993 *Phys. Rev. B* **48** 1858
[4] Walkup R W, Newns D M and Avouris Ph 1993 in *Atomic and Nanometer-Scale Modification of Materials: Fundamentals and Applications* ed. Ph Avouris NATO ASI Series E vol. 239 p 97 (Dordrecht: Kluwer Acad Publ)
[5] Ursin R et al 2007 *Nature Physics* **3** 481
[6] Doyen G and Drakova D 2015 *Found. Phys.* **45** 959
[7] Doyen G 1982 in *Vibrations at surfaces* eds. R Caudano, J-M Gilles and A A Lucas (New York: Plenum)
[8] Drakova D and Doyen G 2013 *J. Phys.: Conf. Series* **442** 012049
[9] Doyen G and Drakova D 2019 *J. Phys.: Conf. Series* **1275** 012044

The Universe as a Telecommunication Network

Alessandro Capurso
Via Vicenza 26, 00185, Rome, Italy
E-mail: ale.capurso@gmail.com

Abstract. We suggest a framework that models the universe as a Telecommunication Network, with several layers of abstraction, characteristic protocols and encoding schemes. We first define a *fundamental network* of atoms of space (*Present foliation*) as a toy model of the most elementary abstraction on spacetime. We describe a protocol to encode the information of distance, non-locality, and entanglement among these nodes. In this framework, we then conjecture the emergence of fermions from gradients of entanglement in the foliation, encoded as momenta in the fundamental network. We propose an interpretation of the asymmetry and of the half spin of matter in the model, towards a description of fermions as independent Network Entities, adjacent possible complexities from the most fundamental abstraction. We conclude our contribution considering several parallels between Nature and nodes in a layered network. The proposed framework seems a promising path to describe the emergence of a universe from information through the language of networks. We believe this perspective can connect several areas of research and deserves further investigation.

1. Introduction

Many research paths towards a theory able to describe spacetime and the evolution of the universe in a quantum formalism start from the concept of information, as a lens through which both Quantum Mechanics (QM) and General Relativity (GR) should be interpreted.

The language of information has proven to be a powerful tool to understand Nature and the relation between information and communication should be a guiding principle in the understanding of Nature through information. In this contribution, inspired by the *OSI model* reported in the ISO standard of Ref. [1], we will suggest a speculative toy model of the universe as a Telecommunication Network (TN) described through several layers.

Our work follows the recent efforts to restore the importance of time in Physics promoted by Smolin in Ref. [2] and [3]. In the context of a *Presentism* approach to evolution, extending a previous paper on time and entanglement reported in Ref. [4], we will design a *fundamental network* evolving in discrete steps and encoding the information of distance and entanglement among far atoms of space. This elementary network is not given as spacetime itself, nor is a "stage" for particles. As it will be better clarified, it is a thick foliation and a space-lattice. It is described with a holographic perspective (in the context of ER=EPR, as introduced by Susskind and Maldacena in Ref. [5] and [6]), and it is intended as a memory-network of quantum information from which spacetime and particles emerge.

Given the interdisciplinary spirit of *DICE2022*, we will conclude extending our considerations towards concepts inspired by Biology and Neuroscience. We will look for common patterns in natural evolution and TNs and suggest an interpretation of evolution in terms of the emergence of network functions, as in an *OSI-alike* TN, structured in several layers.

Our work spans multiple areas of research, and the introduced toy model is missing some details, often left to the referenced literature due to the concise nature of this contribution.

We hope the reader will find interesting stimuli in our proposal and that the introduced concepts of *network*, *encoding* and *memory* (intended in this work as an information storage) might allow a better understanding of Nature through the language of *Information and Communication Technologies* (ICT).

2. A Toy Model of the most Fundamental Network

To design a possible *fundamental network*, intended as an elementary level of abstraction from which spacetime and particles emerge, we start from a set of *virtual atoms of space* as nodes. The ensemble of these nodes represents a thick Present and a discrete space-foliation of spacetime.

The discreteness of spacetime is proposed in the context of a physically realizable universe, for which the information needed to describe it must be finite, given the *Bekenstein bound* of Ref. [7] and as discussed in Ref. [8] and [9]. Discrete and continuous band-limited signals are anyway equivalent (Nyquist-Shannon theorem), and the consequences of this has also been related to physical phenomena, as in Ref. [10].

The idea of a present instant as the only element of reality has ancient roots in eastern and western philosophies, back to Heraclit of Efes and Plato. Among the several works in the recent literature, we base our research on the physics interpretations of Presentism given in Ref. [3], [4], [11], [12], [13], [14], [15], [16], [17], [18], [19], [20], [21], [22] and the philosophical perspectives of Ref. [23], [24], [25]. In this chapter, we will report the main ideas useful to our model, leaving the details to these references.

The Present instant T_k is related to an atomic evolution cycle of the kth space of possibilities in the foliation. It is time-symmetric within its thickness, of extension $2T$, with T Planckian unit of time. It is a reference for time intervals and a maximum rate of change in spacetime (*events' resolution*).

The Present is not a "common now", as there is no "time" yet at this fundamental abstraction. An extended arrow of time emerges in the succession of the instants. This arrow comes from causality and incoming photons, in a thermal bath that allows *events' counting*. It records a global increase in entropy (info required to describe the Present), as the cycles occur, and the space-foliation in the Present expands.

An extended *causal time* is then, according to Presentism, an emerging dimension giving order to a Causal Set of events and a classical domain of *Res Extensa* behind the past boundary of the Present. Being the Present the only element of reality in the causal time, events are given as irreversible as they do not exist anymore. Only their consequences persist in the Present, through their causal cones.

From any atom of space X (*e.g. event location*), we describe the other nodes in the present foliation through an imaginary time of motion at the speed of causality c needed by virtual photons to reach the other atoms, and 2 angular degrees of freedom $\theta_{1,2}$. In this mapping of space, each node X sits at the origin of a discrete axis of space-distance developing in 2 directions, $i\,|\uparrow\rangle\, cT$ and $i\,|\downarrow\rangle\, cT$. The other nodes are identified at $n_X \in 2\mathbb{N}$ steps from X, in a relational description of the network. We define this space within the Present as the *imaginary space* through the axis icn_XT (orthogonal to the causal time) and $\theta_{1,2}$, in a *space-lattice* $\{n, \theta_{1,2}\}_X$ extending in 3D from any node. The 4D discrete lattice $\{n, \theta_{1,2}\}_{X,k}$ described from X in the succession of the instants along k is proposed as a model of an empty relational spacetime.

To define this network beyond the nodes' relative location, we consider a protocol based on 3 global references: the evolution $2T$ (for a causal time in steps, given all atoms of space synced in each cycle as quantum oscillators), the speed of causality c (to define an orthogonal space icn_XT along an imaginary time of motion from each node X) and the reference quantum of action $\hbar$ (to scale the info of entanglement among the nodes over a thermal vacuum $\hbar/T$).

Journal of Physics: Conference Series **2533** (2023) 012045 doi:10.1088/1742-6596/2533/1/012045

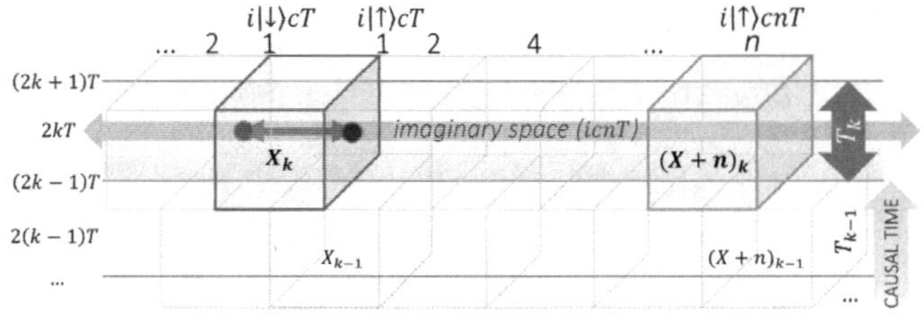

Figure 1. Each node X is at the origin of icn_XT. The others are mapped at $n_X \in 2\mathbb{N}$ from X. With 2 degrees of freedom $\theta_{1,2}$ we construct an *imaginary space* as a 3D *lattice* $\{n, \theta_{1,2}\}_X$. A 4D lattice $\{n, \theta_{1,2}\}_{X,k}$ emerges in the succession of cycles.

These universal *keys* are the basis to describe the elementary information in the proposed network. As it will be better clarified, they represent a protocol to encode the fundamental degrees of freedom of distance and entanglement in the foliation from different nodes. They are given as characteristic of all the atoms of space and measured equal in every reference frame in the emerging spacetime.

In the choice of the references to encode the most elementary information, we intend a design of the proposed fundamental network from first principles: the *invariance of the speed of causality* (to allow relativity between a causal time and an imaginary space distance), and the *invariance and continuity of information* in the measurement of $\hbar$ (leading to a *relativity of data partition* and QM, as clarified in Ref. [26] and [27]). Eventually, only with common references it is possible to describe coherently and consistently any relation between different perspectives. With no common units of measure there is no confrontation on any information and the emerging spacetime would be incoherent and disconnected.

From this protocol, the resulting lattice can be interpreted as a coherent state of oscillators, with all the nodes in the foliation evolving synched on a *common beat* $\hbar/T$. The ensemble of all these synced nodes defines the vacuum like a coherent Bose-Einstein condensate and reminds of Hawking's proposal elaborated in Ref. [28] and [29], where each atom of space is interpreted as a *virtual Planck-mass black hole.*

Defined the lattice, we want to describe the QM in this fundamental network. We start from the time symmetric Present of Ref. [12]. We map the kth cycle of a system at X to 2 Hilbert spaces, $\mathcal{H}_k$ and $\mathcal{H}_k^\dagger$, in a dual cover of icn_XT from the boundaries of the Present. We consider $\mathcal{H}_k$ of *ket* vectors $|\psi\rangle$ from the past boundary $(2k-1)T$, and $\mathcal{H}_k^\dagger$ of bra vectors $\langle\psi|$ from the future boundary $(2k+1)T$, with a max correlation $|\psi\rangle\langle\psi|$ between instants and disentanglement in case of an irreversible event at a given instant.

In this symmetric description, the state of a system at X in the Present is determined by 2 waves: the first "evolves towards the future", from $(2k-1)T$ to $2kT$ (ket $|\psi\rangle$ as *pre-selected, causal* or *initial*), the second "towards the past", from $(2k+1)T$ to $2kT$ (bra $\langle\psi|$ as *post-selected, retro-causal* or *final*). Even if called *initial* and *final*, as in the TSVF of Ref. [16] and [17], these waves occur in the same atomic $2T$, as in Ref. [12], and non-locally entangle the state at X to other atoms on the boundary at $2kT$, where the waves meet. The Hilbert space after k instants is of the form $(\mathcal{H}_k^\dagger \otimes \mathcal{H}_k) \otimes \cdots \otimes (\mathcal{H}_1^\dagger \otimes \mathcal{H}_1)$.

Beyond the time-symmetric description of Ref. [12], according to the model of the Present given in Ref. [21], non-locality in the foliation is fundamental. In Ref. [4] it is considered as holographically equivalent to an entanglement between nodes (*ER=EPR*, also inspired by Ref. [5] and [6]). This undefined but logically consistent information is encoded through the superposition of paths from $(2k-1)T$ and $(2k+1)T$ as in a path integral. These paths meet at $2kT$ as Closed Time-like Curves in the thickness of the Present (*ER=EPR=CTC*).

CTCs seems the ideal *circuit* to encode the *Res Potentia* of Ref. [22], given that no events can occur on a closed path, as shown in Ref. [30], [31], [32], [33].

Journal of Physics: Conference Series **2533** (2023) 012045 doi:10.1088/1742-6596/2533/1/012045

In our model, these CTCs are the *links* from a node X encoding a quantum information beyond a flat imaginary distance icn_XT. We call them *memory loops*, given that loops are also the simplest circuits to implement a memory. The fundamental network of possible CTCs among the nodes is equivalent to a distributed quantum memory encoding the information of distance and entanglement in the present foliation. In this memory, beyond what is real or imaginary, what matters is what has potential or consequences.

3. Fermions from Networks in Space and Time

Following the rich literature on Presentism, we have introduced a toy model in which spacetime emerges from a sequence of space-like thick foliations. Each foliation has been described as a discrete lattice of atoms of space that evolve synched on a common present cycle, as in a fundamental network.

Distance among the nodes is described through an imaginary time of motion in_XT at the speed of causality c, with steps $n_X \in 2\mathbb{N}$. Entanglement among the atoms of space, equivalent to non-locality, is given as the links of this network. These links are described as CTCs in the thickness of the Present that pierce the imaginary space in the entangled nodes, in a resulting non-local potential.

In the context of Ref. [12] and [4], we model fermions from their non-locality. Given a fermion in empty space with Center of Momentum (CoM) at a node X, we describe its forward and backward waves as a probabilistic bundle of CTCs in the Present. This bundle encodes the non-local potential from the CoM and pierce icn_XT where the particle might be in space. It represents a path integral from $\pm T$ that may resolve in a tunnelling of the CoM from X to $X+n$ (given icn_XT as the lattice of Ref. [34] and [35]). This bundle represents $|\psi\rangle\langle\psi|$ and it is shaped in a thermal distribution by the particle's energy E on the floor $\hbar/T$.

Mathematically, we define a *momentum* p_{ST} of a fermion of $E = c\sqrt{p_{ST}p^*_{ST}}$) with CoM in X at an instant k as:

$$p_{ST} = p_T + ip_S = m_Tv_q + imv_s = m_Tc + i(\gamma m_T)(\beta c) = p_T(1 + i\gamma\beta) \tag{1}$$

The values m_T and $m = \gamma m_T$ are the static and the relativistic mass respectively, $v_s = \beta c$ is the average variation of the CoM in a given imaginary direction along the evolution cycles k, while v_q depends on the particle's internal degrees of freedom from the CoM reference frame, as defined in Ref. [36] and [37].

We picture p_T as a *momentum in time* from $\pm T$ connected to the *zitterbewegung* from the CoM, orthogonal to the imaginary space and ip_S. The momentum p_{ST} is related with the *ket* $|\psi\rangle$ in $\mathcal{H}$, while p^*_{ST} is related to the *bra* $\langle\psi|$ in $\mathcal{H}^\dagger$. They shape where the particle might manifest in an event at the kth evolution cycle (random tunnel from X to n_X atoms of space from X in the evolution $\mathcal{H}^\dagger_k \otimes \mathcal{H}_k$).

To relate these momenta with the non-locality of the particle in the imaginary space, we define the wave function from the CoM at a node X as: $\psi_{E@k} = \frac{R_E(n_X)}{cT}e^{\frac{iS_E(n_X)}{\hbar}}$ (from now on we will give the indexes X and k as implicit). We define the gradient in $icnT = ir$ as $\nabla = i\frac{\partial}{\partial ir} = \frac{\partial}{cT}$(with ∂ intended as a variation in the foliation from X along n). In line with the model of *zitterbewegung* given in Ref. [36] and [37], we consider $|v_q| = |\hbar\frac{\nabla R_E}{m_TR_E}| = c$, with a thermal $R_E = \frac{p_Tc}{\hbar/T}e^{-\frac{p_Tc}{\hbar/nT}} = \frac{1}{n_c}e^{-\frac{n}{n_c}} = e^{-\left(\frac{n}{n_c}+\varphi_c\right)}$ driven by $icn_cT = i\hbar/p_T$ (Compton radius from the CoM), and derive a description of p_{ST} from ψ_E:

$$ip_S = i\nabla S_E; \quad p_T = -\hbar\nabla \ln R_E = -(\hbar/cT)\partial \ln R_E \tag{2}$$

As a synthesis able to relate the momentum of a free fermion (with CoM at X in the Present) with the "scale of non-local information" encoded in ψ_E, we derive the following equations:

$$(-p^*_{ST}c)\psi_E = (\hbar/T)\partial\psi_E; \quad (-p^*_{ST}c) = (\hbar/T)\partial \ln \psi_E \tag{3}$$

The *tempo* $\frac{1}{n_c} = \left|\frac{p_T c}{\hbar/T}\right|$ represents the absolute persisting information of a signal from $+T$ and $-T$ over the floor $\hbar/T$ and shapes ψ_E as a *Boltzmann mode* encoded in the thickness of the Present. The phase of R_E, $\varphi_c = \ln n_c$, is a summary of a local gradient of non-locality and entanglement in the imaginary space. The limits $\varphi_c \to \pm\infty$ (equivalent to *null-infinity* in $icnT = ie^{\varphi}cT$ from X) are mapped in this framework to the boundaries $-T$ (*null*) and $+T$ (*infinity*) of each instant, where the "scale of the information of entanglement" in the foliation from X is holographically summarized.

Eventually, in the model of ψ_E as an ontological information encoded in a bundle of CTCs shaped by p_T, we promote a description of massive particles as *networks of atoms of space* entangled in T_k on a *common tempo*. These local gradients of entanglement in the imaginary space embroider the Present with information densities, like *defects* in the space-lattice and *phonons in time*. Fig. 2 offers a graphical summary of the proposed toy model.

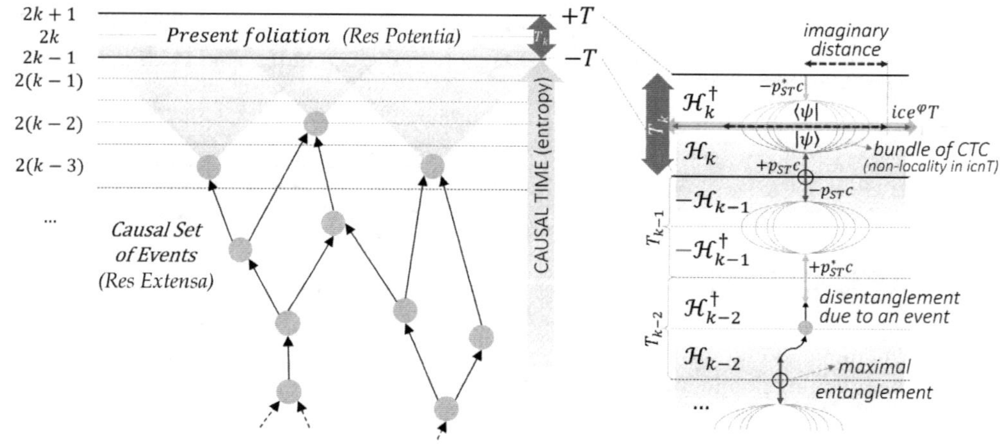

Figure 2. Model of a universe evolving in instants k. On the left, the causal set of events (*grey dots*) ordered along an emerging causal time. The Present T_k is the kth evolution cycle, bounded between $(2k-1)T$ and $(2k+1)T$. On the right, we highlight the imaginary distance axis in the present foliation T_k (blue line as $ice^{\varphi}T$). The vectors p_{ST} and p^*_{ST} (*red* in $\mathcal{H}$, *cyan* in $\mathcal{H}^\dagger$) are at the root of the initial/final wave (represented as possible paths closing in a bundle of CTCs). Maximal entanglement among successive instants (*black circle*) is broken in case of an event (*grey dot* at T_{k-2} on the right side of the Figure).

Beyond this speculative model of fermions, we would like to elaborate on a possible interpretation of the asymmetry of their wave function in the context of a Presentism approach to evolution and interpret their half spin in comparison to the *spin 1* of the photons tracing the imaginary space at c.

We interpret the asymmetry of the wave function and the half spin considering, on a given dimension of imaginary distance from the CoM, the paths from $\pm T$ in each cycle along a single direction of the 2 possible ones at a node X, namely $i\left|\uparrow\right\rangle cT$ and $i\left|\downarrow\right\rangle cT$. The indeterminacy of order along the 2 sides of the imaginary space in 2 successive cycles can be expressed with a XOR logic $\left|\uparrow\downarrow\right\rangle \oplus \left|\downarrow\uparrow\right\rangle$ and represents a logically consistent information (no over or under-determined solutions) and an entanglement in time of 2 instants, as introduced in Ref. [4].

The *spin 1/2* is connected in our toy model to the fermions' evolution over 2 cycles, in a $4T$ revolution in which they explore the possibilities in $ic\left(\left|\uparrow\downarrow\right\rangle \oplus \left|\downarrow\uparrow\right\rangle\right)T$ in both directions with undefined order. *Spin 1/2* is equivalent, in our interpretation of fermions, to a *longer period*, like an extension of perspectives across the boundary of adjacent instants.

In this model, fermions are beyond the perspective of single nodes, as it is in their entanglement in space AND in time that their full identity is defined. Even the fundamental network, in its perspective on entanglement in all nodes of space over single instants of $2T$, only captures a projection of a more complex entity at its level of abstraction: energy densities in the imaginary space and collapse events. Fermions emerged as independent entities, with an additional perspective across successive instants. Entanglement in time is as an exploration of an *adjacent possible* dimension of complexity: *memory.*

In the proposed toy model, the asymmetry in each cycle identifies the CoM of the particle in the global network as at the center of these 2 sides of imaginary space, entangled with an undefined order. The identification of the node X as at the center of this variation could be considered also from the perspective of a particle. In a holographic description, we could say that a fermion identifies its CoM in an unreferenced imaginary space though its holographic image "projected from infinity", following its momentum in time from $\pm T$. This description resembles the metaphor of the *belt in the book* from Dirac, also used by Penrose in Ref. [38]. With a belt anchored on a *book-particle* at one end and in our hand (at infinity in the imaginary time of motion) on the other side, we need not a 2π but a 4π rotation for a revolution, as for a fermion's full cycle in respect to the Present duration $2T$.

Fig. 3 offers a graphical summary of matter particles in the proposed toy model.

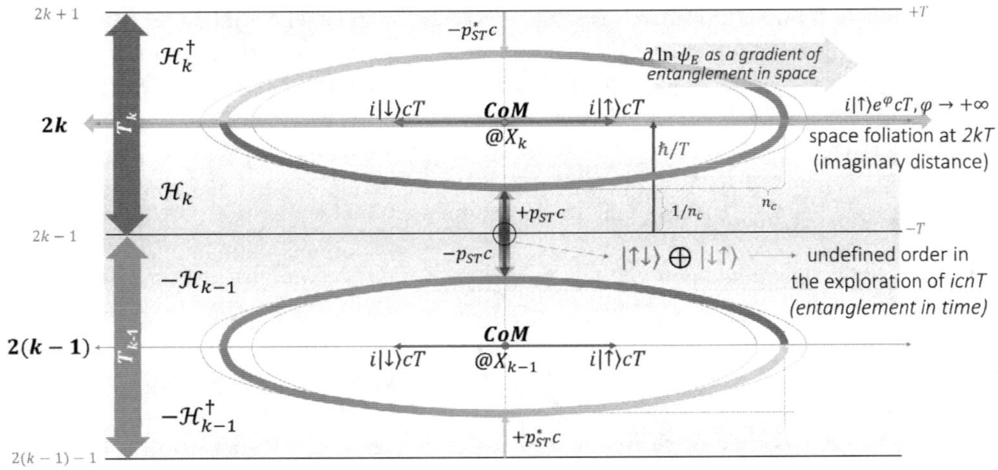

Figure 3. Fermion at X as a bundle of CTCs from the boundaries of each instant. These CTCs encode the entanglement in the foliation (*networks of atoms of space*) as a time-symmetric superposition of $|\psi\rangle$ in $\mathcal{H}$ and $\langle\psi|$ in $\mathcal{H}^\dagger$. The bold CTCs are the most probable (cross icn_XT at a distance equal to the Compton radius from the CoM at X). The momentum p_{ST} is a summary of the scale of the non-local information from the boundaries of the Present (like the *belt in our hand*), where entanglement in time assures consistency in the undefined order of exploration of $ic\,(|\uparrow\downarrow\rangle \oplus |\downarrow\uparrow\rangle)\,T$. The undefined order in 2 instants is pictured in the figure with alternated colors of the 4 CTCs segments along the $4T$ period (*red/green* in $\mathcal{H}$, *cyan/magenta* in $\mathcal{H}^\dagger$).

In a time-symmetric cycle, we can ideally extend the emerging arrow of causality (from past to future cycles) inside the Present. In this ideal logical extension, we relate the difference between matter and antimatter in our model to a phasing with inverted hemicycles in $\mathcal{H}$ and $\mathcal{H}^\dagger$ in each cycle. More specifically, given a momentum in time of opposite sign in respect to matter from the 2 signs of $|v_q|$, we consider the initial and the final wave of antimatter as "occurring in opposite direction", with $|\psi_{anti-m}\rangle$ as *initial* wave in $\mathcal{H}^\dagger$ always collapsing with the *final* wave of matter $\langle\psi_m|$ in $\mathcal{H}^\dagger$.

We can metaphorically say that antimatter is not in a *Dirac Sea* but roots its p_{ST} at $+T$, dropping like a *Dirac Rain* and as energy "borrowed from the future". In the ideal extension of the causal time within the time-symmetric Present, antimatter is evolving "backwards in time". This description may offer a possible physical interpretation of the similar proposal from Feynman and Stueckelberg.

In the search for a better understanding of QG, it is worth to note here that fermions are, in this toy model, characterized by a *PT-symmetry* (Parity-Time), as inverting $i\,|{\uparrow}\rangle\,cT$ with $i\,|{\downarrow}\rangle\,cT$ and T_k with T_{k-1} would not change the description. This kind of symmetry is weaker than the Hermitian condition of self-adjointness and, as discussed in Ref. [39], allows infinite classes of complex Hamiltonians with real and positive spectra, in an analytic continuation from a real to a complex phase space.

In our proposal, fermions show non-Hermicity, with an undefined asymmetric tunnelling probability in each cycle (exploration of the space $ic\,(|{\uparrow\downarrow}\rangle \oplus |{\downarrow\uparrow}\rangle)\,T$ in 2 instants). It seems worth to investigate the relation between the description of fermions in our model and the connection between curvature, time dilation and non-Hermicity highlighted in Ref. [35] and [40].

We also note here, leaving further studies to following contributions, an insight from Information Theory: a continuous signal with max rate of change $1/2T$ (such as our toy model, being $2T$ the atomic Present) is equivalent to a discrete signal from the samples every T (Shannon-Nyquist). If the Present is related to a max bandwidth in spacetime, the Planckian nature of T may be connected to a discrete representation of the metric and a possible QG theory. Moreover, in our model, the $2T$ present foliation is defined at the speed of causality of *spin 1* photons, while matter is represented by *spin 1/2* particles, with an extended memory and a period $2T/(1/2) = 4T$. If spin (as a *period* $2T/spin$) is connected to a *time-resolution* and a sort of particle's *memory*, the gravitational info might be related to a *spin 2* particle. In this sense, curvature comes from a max info of $2T/2$, equivalent to a sampling of $1/T$, able to represent the signal $1/2T$ of the Present in accordance with the mentioned theorem.

4. An Interconnected Perspective on Evolution

The concepts of information, network, encoding, and memory seem relevant in the study of natural phenomena and of the universe evolution, as in the toy model of spacetime and matter particles suggested in this contribution. Here we would like to further elaborate with some high-level parallels.

To describe the emergence of different entities along the evolution of the universe through the framework of TNs, we start from the insights of the *Theory of Adjacent Possible* (TAP), a combinatorial model of evolution recently extended in a wider theory of *Biocosmology* (presented in Ref. [41], [42], [43], [44], [45]).

According to TAP, we can see evolution as a combinatorial exploration of the *adjacent possible*, with the emergence over time of several new entities (that we call here *Network-Entities* or NEs) thanks to the already existing elements. This process is similar in a TN, where new NEs, functions and protocols are possible thanks to the already existing layers.

In this high-level perspective, a molecule can be described as a network of atoms, which are the result of the interconnection of several more elementary particles. Each atom can be seen as an extended NE itself of several more elementary nodes, connected through the "physical protocol" of the elementary forces.

From simpler molecules, going up the ladder of *atomic-networks*, we can consider amino-acids and proteins, capable of expressing in even more complex structures and shapes. Molecules served as a basis for cells to emerge, and cells developed their own *molecular protocols* to interconnect in greater structures and implement new encodings.

Biology, at the level of both single species and larger ecosystems, can be described with NEs when we identify the encoding schemes and the communications between the components.

Journal of Physics: Conference Series **2533** (2023) 012045 doi:10.1088/1742-6596/2533/1/012045

Eventually, human groups as well can be seen as NEs, given the agreed *imaginary codes* as protocols to interconnect, as also proposed in Ref. [46].

All these NEs are different levels of abstraction on the surrounding information across the cycles of the Present (*distances, correlations, events, ...*). This information, citing Landauer in Ref. [47], *is not a disembodied abstract entity; it is always tied to a physical representation.* Every NE has a proper description of the surrounding universe that depends on the implemented abstraction protocols and interconnection functions. Each layer can be investigated on its own, deriving a description at the equivalent level of abstraction, or "scale of the phenomena".

In the decomposition of our universe in several layers, we shall remember that information is defined in different ways in each NE, depending on the encoding and communication schemes used. In all these parallels with NEs, it is crucial to identify the shared language that allows encoding and interconnection, as in general "entropy" is not an observable and depends on the interactions considered. Each NE encodes the information content with its own protocols, with its own abstraction on the system, and with a different definition of entropy. In this sense, all the natural entities seem to encode the information of interest at the proper level of abstraction like "memory-networks" distributed in the constituent lower-level nodes and NEs.

Beyond the concepts of *network* and *information*, the relevance of the concept of *memory* seems evident in the study of evolution in several fields. In Biology, for example, DNA has often been described as a memory encoding the information needed for cellular life. History started with writing as a new form of improved trace and memory that allowed bigger networks of *homo sapiens*. In recent studies of cognitive neurology (see Ref. [48], [49]), consciousness as well has been connected to the access to a "memory", and the possible connection between consciousness, memory and information has been investigated in Ref. [50] and [51]. The relevance of a memory for an interconnected description of the *Self* has also been proposed as a philosophical perspective on existence in Ref. [52], through the synthesis *memoro ergo sum.*

The creation of more elaborated memories and strategies to encode the surrounding information might be a *leitmotiv* of evolution, and seems a key element to understand Nature, from the most fundamental physical layer of our universe to massive particles, from Biology to the most abstract entity we can scientifically discuss about, as our consciousness.

5. Synthesis

We list here the key messages at the basis of our contribution.

We suggest a toy model of spacetime and fermions as emerging from the information in NEs. We consider, as the most fundamental abstraction, a network of atoms of space separated by an imaginary time of motion at the speed of causality and synched on a common cycle. This model represents a foliation of spacetime and a discrete space-lattice evolving in atomic instants.

The description of spacetime from a discrete foliation is illustrated in the rich literature supporting *Presentism.* This approach is compatible with both QM and SR, it is strictly related with Information Theory and seems a promising path for a possible more general theory of QG.

The indeterminacy in the foliation, represented as entanglement (EPR) and non-locality (ER) among the atoms of space, is proposed as the elementary links in the fundamental network. These correlations are encoded with a time-symmetric description from the boundaries of the Present in closed paths (CTCs), *memory-loops* in the thickness of an instant. The thick Present, as a network of entangled nodes, is like a distributed memory of the *Res Potentia* in the universe.

We define an elementary protocol based on 3 references, shared by all nodes in the fundamental network and considered with no preferred reference frame in the foliation. These references are the Present temporal extension (evolution cycle $2T$ of the foliation), the speed of causality c (scaling the info of distance $icnT$ in an imaginary space), and $\hbar$ (reference of action to scale entanglement over a thermal vacuum $\hbar/T$). The resulting lattice is a coherent state of atoms of space synched on a *common beat* and reminds of a Bose-Einstein condensate.

Journal of Physics: Conference Series **2533** (2023) 012045 doi:10.1088/1742-6596/2533/1/012045

We describe fermions from their non-locality. We model a free massive particle as a bundle of CTCs encoding the superposition of the forward and the backward waves in each instant. These CTCs pierce the imaginary space where the particle could manifest in its *zitterbewegung* and are summarized as momenta at the CoM, measure of the local entanglement in the foliation. In this model, fermions are interpreted as networks of atoms of space entangled on a common momentum in time, absolute persisting information of the particle in the memory of the Present.

The asymmetry of the wave-function in fermions is connected to an exploration of the 2 sides of the imaginary space from the CoM in 2 cycles, with undefined order. This revolution over a period of $4T$ depends on the spin $1/2$, model of *extended memory* from the entanglement in time of 2 present instants. The non-Hermicity over single instants may be at the origin of curvature.

Given the Shannon-Nyquist sampling theorem, if the Present duration is $2T$, with a maximum bandwidth of $1/2T$ (events' sampling rate), we consider the equivalence with a discrete series coming from the sampling at $1/T$. If the space foliation is defined at the speed of causality of *spin 1* photons, and matter particles entangle 2 consecutive instants with a *spin 1/2* revolution, we suggest the existence of *spin 2* particles, connected to a gravitational sampling every T.

In the proposed interpretation, our universe evolved from a fundamental abstraction in new NEs of higher complexity. This evolution, as in a TAP process, allowed the emergence of fermions, atoms and so on, up to a full biological tree and several human cultures. The resulting universe appears as a global TN at different layers of abstraction. The exploration of complexity seems related to the evolution of the encoding schemes and of the memories to account for more surrounding information, in the Present and in the causal set of events seen along the cycles.

To conclude and give a more explicit representation of the proposed parallels between different forms of entities that emerged along the universe evolution and different NEs and protocol layers, we follow the OSI model of TNs reported in Ref. [1].

Given this layered description, Table 1 illustrates a synthesis of the interpretation of *the universe as a TN*, with the 3 bottom layers introduced in this contribution. At the foundations, we consider the gravitational vacuum as a *clock* and *common beat* $\hbar/T$. Then the Present, as a *memory* with an elementary protocol to encode the quantum information of non-locality in a foliation. From this *Res Potentia*, in the succession of the instants, an imaginary space of possibilities evolves towards an open future, leaving behind a *Res Extensa* and a classical spacetime that we describe through a causal set of irreversible events, experienced in a thermal bath of incoming causal photons.

Eventually, in our description, at the most elementary perspective, everything in the universe seems interconnected though the *memory of a common beat.*

We hope this speculative proposal could be of inspiration.

Abstraction	Information layer and Network functions
societies	networks of *homo sapiens* with shared classical memories and codes
homo sapiens	networks of several *cell types* (quantum and classical memories and codes)
...	...
cells	networks of *molecules* (memory-DNA and protein-codes for biological cycle)
molecules	networks of *atoms* exchanging simpler particles
atoms	networks of *particles* exchanging fermions and bosons
fermions	networks of *atoms of space* entangled on a common momentum in time, asymmetric tunnelling from the CoM in each present instant (*period* $4T$)
thick Present	*fundamental network*: thick foliation and space-lattice, causal events' resolution, $icnT$ space from *spin 1* photons, *memory* of local momenta, SR, QM (*period* $2T$)
common beat	vacuum $\hbar/T$ and energy at max sampling rate, *spin 2 particles*, QG (*period* T)

Table 1. *OSI-alike* model of the universe as a TN, with different emergent network layers.

References

[1] ISO/IEC - Information technology — Open Systems Interconnection 1994 Basic reference model — ISO/IEC 7498-1:1994 Standard Geneva URL `iso.org/standard/20269.html`
[2] Smolin L 2013 *Time Reborn: From the Crisis in Physics to the Future of the Universe* (Boston: Houghton Mifflin Harcourt)
[3] Smolin L 2018 *Foundations of Physics* **48** 121–134
[4] Capurso A 2022 *Entropy* **24** 410
[5] Susskind L 1995 *Journal of Mathematical Physics* **36** 6377–6396
[6] Maldacena J and Susskind L 2013 *Fortschritte der Physik* **61** 781–811
[7] Bekenstein J D 1981 *Phys. Rev.* D **23**(2) 287–298
[8] Gisin N 2021 *Erkenntnis* **86** 1469–1481
[9] Del Santo F and Gisin N 2021 *Entropy* **23** 1326
[10] Kempf A 2014 *Foundations of Physics* **44** 472–482
[11] Elitzur A and Dolev S 2005 *Endophysics, Time, Quantum and the Subjective* (Singapore: World Scientific Publishing) pp 589–606
[12] Aharonov Y, Popescu S and Tollaksen J 2014 *Quantum Theory: A Two-Time Success Story* ed Struppa D C and Tollaksen J M (Milano: Springer) pp 21–36
[13] Gisin N 2020 *Nature Physics* **16** 114–116
[14] Hardy L 2009 Foliable operational structures for general probabilistic theories (*Preprint* `quant-ph/0912.4740`)
[15] Hardy L 2021 Time symmetry in operational theories (*Preprint* `quant-ph/2104.00071`)
[16] Cohen E, Cortês M, Elitzur A and Smolin L 2020 *Phys. Rev. D* **102**(12) 124027
[17] Cohen E, Cortês M, Elitzur A and Smolin L 2020 *Phys. Rev. D* **102**(12) 124028
[18] Schlatter A 2018 *Entropy* **20** 741
[19] Schlatter A 2019 *Entropy* **21** 323
[20] Schlatter A 2021 *Foundations of Physics* **52** 7
[21] Kauffman S A 2022 *Entropy* **24** 554
[22] Kastner R E, Kauffman S and Epperson M 2018 *Int. Journal Quantum Found.* **4** 158–172
[23] Mariani C and Torrengo G 2021 *Synthese* **199** 3923–3944
[24] Tallant J and Ingram D 2021 *Synthese* **199** 3969–3991
[25] De Bianchi S 2022 *Time and Cosmology in Plato and the Platonic Tradition* (Leiden: BRILL) pp 156–178
[26] Brukner Č and Zeilinger A 2009 *Found. Phys.* **39** 677–689
[27] Dakić B and Brukner Č 2016 *Fundamental Theories of Physics* (Dordrecht: Springer) pp 249–282
[28] Hawking S W 1978 *Phys. Rev. D Part. Fields* **18** 1747–1753
[29] Hawking S W 2001 *The Universe in a Nutshell* (New York: Bantam Spectra)
[30] Paganini C F 2020 No events on closed causal curves (*Preprint* `gr-qc/2005.05748`)
[31] Rovelli C 2019 Can we travel to the past? irreversible physics along closed timelike curves (*Preprint* `gr-qc/1912.04702v2`)
[32] Baumeler Ä, Gilani A S and Rashid J 2022 *Quantum* **6** 673
[33] Barrett J, Lorenz R and Oreshkov O 2021 *Nat. Commun.* **12** 885
[34] Amir A, Hatano N and Nelson D R 2016 *Phys. Rev.* E **93** 4
[35] Lv C, Zhang R, Zhai Z and Zhou Q 2022 *Nat. Commun.* **13** 2184
[36] Esposito S 1999 *Found Phys Lett* **12** 165–177
[37] Salesi G 1996 *Modern Physics Letters* A **11** 1815–1823
[38] Penrose R 2004 *The Road to Reality* (London: Jonathan Cape)
[39] Bender C M and Boettcher S 1998 *Phys. Rev. Lett.* **80** 5243–5246
[40] Paiva I L, Te'eni A, Peled B Y, Cohen E and Aharonov Y 2022 *Commun. Phys.* **5** 1
[41] Björneborn L 2020 *The Palgrave Encyclopedia of the Possible* (Cham: Springer) pp 1–12
[42] Cortês M, Kauffman S A, Liddle A R and Smolin L 2022 Biocosmology: Towards the birth of a new science (*Preprint* `astro-ph.CO/2204.09378`)
[43] Cortês M, Kauffman S A, Liddle A R and Smolin L 2022 Biocosmology: Biology from a cosmological perspective (*Preprint* `physics.hist-ph/2204.09378`)
[44] Cortês M, Kauffman S A, Liddle A R and Smolin L 2022 The tap equation: evaluating combinatorial innovation in biocosmology (*Preprint* `q-bio.PE/2204.14115`)
[45] Kauffman S 2022 *Entropy* **24** 1383
[46] Harari Y N 2014 *Sapiens: A Brief History of Humankind* (London: Harvill Secker)
[47] Landauer R 1996 *Physics Letters* A **217** 188–193
[48] Buonomano D 2017 *Your Brain Is a Time Machine: The Neuroscience and Physics of Time* (New York: W W Norton and Co Inc)

[49] Budson A E, Richman K A and Kensinger E A 2022 *Cogn. Behav. Neurol.* **35** 263–297
[50] D'Ariano G M and Faggin F 2022 *Artificial Intelligence Versus Natural Intelligence* (Cham: Springer) pp 145–192
[51] Kauffman S 2019 *Act Nerv Super* **61** 61–64
[52] Capurso A 2021 Notes on a common beat (*Preprint* 10.31219/osf.io/6agcz)

Journal of Physics: Conference Series **2533** (2023) 012046 doi:10.1088/1742-6596/2533/1/012046

A QFT based on Einstein-Hilbert action and hidden dimensions as consistent theory of measurement, dark energy, dark matter and McGaugh's universal law for rotating galaxies

G Doyen[1] **and D Drakova**[2]

[1] Ludwig-Maximilians Universität, Butenandstr. 5-13, 81377 Munich, Germany
[2] Sofia University, 1 James Bourchier Blvd, 1164 Sofia , Bulgaria

E-mail: gerold@gerold-doyen.de; nhdd@chem.uni-sofia.bg

Abstract. We present a quantum field theory of gravity in 11 dimensional spacetime, based on Einstein-Hilbert action and expanded up to 4th order in the metric. In the theory the deviation of the local metric from Minkowski metric is a quantum field, called gravonon field. It is responsible for the localization of a matter particle on a definite position in four dimensional spacetime (warp resonance, beable). This is *the solution of the measurament problem of Quantum Mechanics coming out as result of the deterministic Schrödinger equation.* The quantum gravonon field is also decisive for calculating the expansion of the universe. In lowest order of the metric the interaction of the matter field with the gravonon field is the cause for the expansion of the universe and is the quantum analogue of the often mentioned dark energy. In higher orders of the metric the interaction of ordinary matter and gravonons becomes attractive. The gravonon field then assumes the role of dark matter. McGaugh's universal radial acceleration relation in rotationally supported galaxies is obtained using this quantum gravitational appoach. The concept of dark matter emerges as do the results of Milgrom's MOND analysis.

1. Introduction

During the last decades fundamental theories have been devised in order to unite gravitation (general relavity, GR) and quantum mechanics (QM). Quite a few researchers doubt that this is necessary.

1.1. Why unite QM and GR?

It is often said that GR has its own realm of validity and QM has its own. A possible connection is mainly discussed for two cases:

- A classical description fails for black holes and for the big bang.
- Gedankenexperiments about particle collisions at unreachable high energies need QM and GR to be discussed.

But these mighty demands are missing the support of experimental facts (see e.g. [1]). The leading models of quantum gravitation tend towards lower dimensional structures containing the information of the universe (e.g. loop quantum gravitation, causal sets, causal dynamical triangulation, etc.) and the universe then appears as a kind of holography, i.e., a geometrical incorporation of quantum information. But the holographic universes can only be constructed as Anti-de-Sitter (AdS) spaces, however we live in a de Sitter space. Verlinde [2] speculates that the special properties of the de Sitter geometry create the illusion of dark matter. Verlinde is in a dilemma between what can be done mathematiclly (AdS) and what is given in reality (de Sitter) and the validity of his approach is difficult to estimate (see e.g. [3]).

1.2. Gravitation and the measurement problem
If the dilemma of gravitation is its resistance towards quantization then the dilemma of quantum mechanics is its resistance towards handling the measurement problem. Sir Roger Penrose and Lajos Diósi [4, 5] suggest to deal one for the other, i.e., use the smooth fundamentally classical nature of spacetime to let the wave function collapse. The rigidity of spacetime prevents then a superposition of different curvatures and because matter is gravitationally coupled this lets the wave function collapse. If this would be correct then it would have two undesirable consequences:

- All cosmological phenomena (including big bang, black holes, baryogenesis and gravitational waves e.g.) cannot rely on quantum mechanical support.
- The measurement process stays outside any mathematical and physical theory.

This is definitely not what quantum loop gravity and string theory are aiming at.

1.3. Beauty and naturalness
The missing experimental support of the claimed symmetries in the highly elaborated theories together with their powerful and beautiful mathematics has led to an attitude which is sometimes denoted as postempirism [6, 7]: If nature does not like our beautiful theories why care about nature then?

Supersymmetry did not yet bear out its existence in experiment. None of the superparticles it predicts have been detected at the world's largest particle accelerator: the Large Hadron Collider (LHC) at CERN. What the LHC discovered instead was the Higgs particle with an annoyingly low mass of 125 MeV. For the Standard Model of Elementary Particles (SMEP) this confirms that not only does it need twenty parameters or more to fit experiment, but in addition it has to choose some values with an accuracy of ten million billion significant digits. Otherwise "particles would all be black holes, and particle physics as we know it would not exist" [8]. This is commonly denoted as 'fine tuning' implying that quantities of vastly deviating orders of magnitude have to compensate each other and this is considered to be a violation of naturalness.

In addition there is the problem of dark energy which in connection with the SMEP leads to a desaster which is sometimes called 'the mother of all discrepancies' [9] meaning that the content of dark energy in the universe as determined experimentally and estimated by quantum field theory (QFT) differs by 122 orders of magnitude.

1.4. Infinities
Both GR (at the center of a black hole) and QFT produce infinities if their mathematical apparatus is applied in order to obtain numbers to be compared to experiments. In QFT the standard perturbation theory combined with the renormalization group fails if applied to a QFT of gravitation although the method of asymptotic safety claims that the dfficulties can be circumvented [10] and in fact this method is considered by its inventors as nonperturbative.

1.5. What should be done and what can be achieved?
An important test for all quantum gravitational theories of dark matter is the reproduction of McGaugh's universal radial acceleration relation in rotationally supported galaxies [11]. The MOND calculational method [12] reproduces the experimental galaxy rotation curves without relying on dark matter or on quantum gravitation, however the dark matter hypothesis is required to understand further cosmological observations. McGaugh and colleagues [11] analyzed thousands of measurements for more than 150 galaxies and found a strong correlation between the gravitational force which is to be expected from the visible matter alone and the total observed force. The experimentally found correlation can be summarized in a simple mathematical equation. This ubiquitous cosmological observation deserves an explanation based on the present state of our theoretical insight.

This established correlation, however, is hard to reproduce using computer simulations which treat visible matter and dark matter as independent components. Many paramters have to be carefully tuned for a successful fit.

Plausible models addressing quantum gravitation should be solved non-perturbativly to produce numbers which can be compared to experiment. Questions of beauty, symmetry, naturalness and other aesthetic requirements should not be of predominant importance on this way, neither the demands formulated in subsection 1.1. This publication tries to contribute to this schedule, concentrating in particular on the problem of the rotating galaxies. In the present communication we describe a quantum field theory, based on Einstein-Hilbert action and harmonic gauge invariance, in which gravity in high dimensional spacetime (11D) in the weak interaction limit is quantized (section 2). Terms analogous to dark energy and dark matter emerge. In the presence of baryonic matter local gravitons (called 'gravonons') are created and their entanglement leads to localization of the matter field as beable in three dimensional space. This represents the solution of the measurement problem of Copenhagen Quantum Mechanics (CQM), beables being the only experimentally accessible objects [13]. The gravonon density is attenuated at small distance from the center bulge of a galaxy, whereas it is retained at large distances (section 3). The two limiting situations in the observed curves $g_{obs} = f(g_{bar})$ are discussed in section 4.

2. Basic features of EHHD, emerging dark energy and dark matter
The Einstein-Hilbert High Dimensionality theory (EHHD) has been described and applied to numerous physical phenomena [13]-[17]. Soft gravonons, local gravitons, living also in hidden dimensions with high density of states, entangle with matter fields. The dimensionality effect, the decisive feature of EHHD, which represents weak degeneracy coupling between the gravonon continuum of high density of states and matter fields, is the physics behind the localization of matter fields as beables.

2.1. Solving Schrödinger's equation
Due to the high dimensionality the density of the graviton modes is large and the still relatively weak gravitational potential couples to nearly degenerate states which are many. If a matter particle is in a local state where it couples gravitationally to other masses and is entangled to gravitons, we say that the paticle is in a warp resonance. We then have the situation that a particle localized in three dimensional space couples to a quasi continuum of states.

Schrödinger's time dependent equation $\mathrm{i}\frac{\partial}{\partial t}\Psi(\psi,\zeta) = H\Psi(\psi,\zeta)$ then predicts that the localized state decays into the continuum of graviton states (ψ: matter field, ζ: gravonon field). This means that the localized matter particle remains entangled to the gravitons and hence will

Journal of Physics: Conference Series **2533** (2023) 012046 doi:10.1088/1742-6596/2533/1/012046

remain localized as long as no other force is exerted on it. In reality the local graviton states - the gravonons - near the massive particle have a high but finite density of states. This implies that there is a recurrence time after which the system returns to a disentangled situation where the massive particle can move away even without an external force. This is the situation with quantum diffusion where the theory reproduces the experimental results [14].

2.2. The concept of gravonons and beables

Beables are generated if a beabling condition is fulfilled: the ratio of the weak gravitational potential and the gravonon frequency $\frac{V_{grav}}{\omega_{grav}}$ should be of the order of ≈ 10. Localization of a matter field as a beable is the solution of the measurement problem in Copenhagen quantum mechanics [13]. After the lifetime of the gravonons in the hidden dimensions expires they recur back in three dimensional space. The effect is a fast increase of the weight of configurations in the total wave functional describing the matter field in limbo. This happens at the expense of the beabled configurations, which lose weight during a tiny burst out of the warped space where entanglement with the gravonons is effective (cf. [15] in this volume).

2.3. The model Lagrangian

In a new application of the EHHD theory we address the problem of the universality of the law for flat rotating galaxies. A rotating star is regarded as a beabled spectator mass, the center bulge of the galaxy is denoted as M_{center} and the gravonon field as ζ_{00} indicating that only one component of the energy-momentum tensor is considered, the other components being smaller by a factor $(v/c)^2$. The interaction Lagrangian density of EHHD for the matter field ψ and the gravonon field ζ_{00} is:

$$\begin{aligned}\mathcal{L}_{int}(r) &= \frac{1}{2}m^2\psi^+\psi\Big[-h_{00}-\zeta_{00}^+\zeta_{00}-h_{00}\zeta_{00}^+\zeta_{00}-\frac{3}{8}\zeta_{00}^+\zeta_{00}\zeta_{00}^+\zeta_{00}\Big]\\ &= \frac{G^{(D)}mM_{center}}{r}\left(1+\zeta_{00}^+\zeta_{00}\right)\psi^+\psi-\frac{m}{2}\zeta_{00}^+\zeta_{00}\left(1+\frac{3}{8}\zeta_{00}^+\zeta_{00}\right)\psi^+\psi. \qquad (1)\end{aligned}$$

ψ indicates the hydrogen quantum field beabled by the considered gravonons and m the hydrogen mass. The terms in the first line of eq. (1) appear due to expansion of the metric where higher order terms compared to ref. [13] are included. Terms, involving the gravonon number operators $\zeta_{00}^+\zeta_{00}$, or the expectation value $\langle\zeta_{00}^+\zeta_{00}\rangle$ in the notation of second quantization, are associated with effects due to dark energy (last term in the second line of eq. (1)) and dark matter, responsible for the expansion and accelerated expansion of the universe [16, 17]. (notice: there is a confusion of $\pm$-signs in [13].)

The first term in the second line of eq. (1) is just Newton' gravity due to the field of the center bulge mass M_{center}, plus a term due to the gravonon field density interacting with the Newton term. What is the meaning of this term? This term is found to reproduce Milgrom's classical correction to Newton's classical gravity if $\zeta_{00}^+\zeta_{00} \rightarrow \langle\zeta_{00}^+\zeta_{00}\rangle$ is introduced. Within our quantum field theory a physical interpretation is asigned to this term: the gravonons, a quantum field, act as an additional quantum gravitational force on the rotating star. The tiny gravonons of mass one tenth of the mass of an electron [13] affect the behaviour of enormous cosmic bodies! The expectation value of the gravonon number operator must be large, i.e. $\langle\zeta_{00}^+\zeta_{00}\rangle \gg 1$ and of the order of 10, to explain 10 times stronger force than Newton's force, acting on stars far away from the central mass, as it is evidenced by McGaugh's observation [11].

2.4. The nature of dark matter in EHHD

Dark matter is commonly assumed to exist, although neutrino, axons, WIMPs etc. suggestions have not been shown as feasible or even as existing. The assumption is that dark matter is

Journal of Physics: Conference Series 2533 (2023) 012046 doi:10.1088/1742-6596/2533/1/012046

mass, distributed over space in galaxies, exercising Newton force on stars. As it is shown by eq. (1) dark matter does emerge in the EHHD Lagrangian. So what is the difference to these suggestions?

In Newton's classical theory of gravitation the gravitational force acts instantanously on stars at large distances. But a term like Newoton's potential h_{00} emerges in the EHHD Lagrangian. This is due to the harmonic gauge, used in deriving the formalism in ref. [13]. In this approach the classical Newton term h_{00} appears due to gauge invariance, even without taking into account retarded interaction potentials.

The physics behind the term additional to Newton's gravity due to dark matter, which is $\propto \frac{G^{(D)} m M_{center}}{r} \zeta_{00}^{+} \zeta_{00} \psi^{+} \psi$ can be interpreted in the following way: dark matter, i.e. the gravonon field density $\zeta_{00}^{+} \zeta_{00}$, gets the information that a gravitational force center exists somewhere in the galaxy, depending on place and time. This information is transmitted and amplified via the gravonon quantum field, modifying the classical Newton potential. This is an exact result of the EHHD formalism. The theoretical formalism yields the experimental observations of McGaugh et al. [11], as it is described in the next sections.

3. Gravonons in the $\gamma - \eta$ model

What is the effect of the gravonon density on the additional force, i.e. on the observed rotational acceleration of galaxies? One effect was discussed in the previous section. Another effect is implied by the last term in eq. (1). This is our dark energy term, which leads to the expansion of the universe. The mathematical formalism leads to this term.

It is obvious that the gravonon continuum plays the decisive role not only for expansion of the universe [16, 17]. The amplification of the gravitational interaction with matter fields beyond the classical Newton gravity is due to the effect of the gravonon field, mentioned in the previous section.

There exists another effect as well, which we might call "vacuum cleaner effect". It is the depletion of gravonon-plus-hydrogen-atom clusters by the center bulge of a galaxy. This is the same cluster of gravonons and hydrogen atoms as in the theory of the expansion of the universe [16, 17]. The expectation value of $\zeta_{00}^{+} \zeta_{00}$, which is the occupation number of the gravonon field, is needed in eq. (1). At this place a theory of the variation of $\zeta_{00}^{+} \zeta_{00}$ with the variation of the entanglement strength γ in the gravonon cluster is proposed.

3.1. Dependence of the gravonon frequency on γ and on the center bulge in a galaxy

The external Newton potential due to the center bulge changes the interaction strength in the cluster. For further progress we have to find out how the strength of gravitational interaction γ within a cluster of gravonons varies with the force exerted by the center bulge M_{center}.

If a star is observed it must be beabled, i.e. it is surrounded and entangled with clouds of gravonons. In a previous study the frequency of the soft gravonon mode of the gravonon cluster, in which hydrogen-atom clusters are bound as beables, has been determined and equals the Hubble parameter. In EHHD the Hubble time, the inverse of the gravonon frequency, is of the order of 10^{10} years reproducing the measured value today [16]. A matter field, e.g. a star, and a gravonon cluster in free space can be regarded at this place as the starting situation in the absence of any external Newton force on the matter field. The star and its gravonon environment is a dynamic system undergoing beable-limbo transitions, with the period of 10^{10} years as the

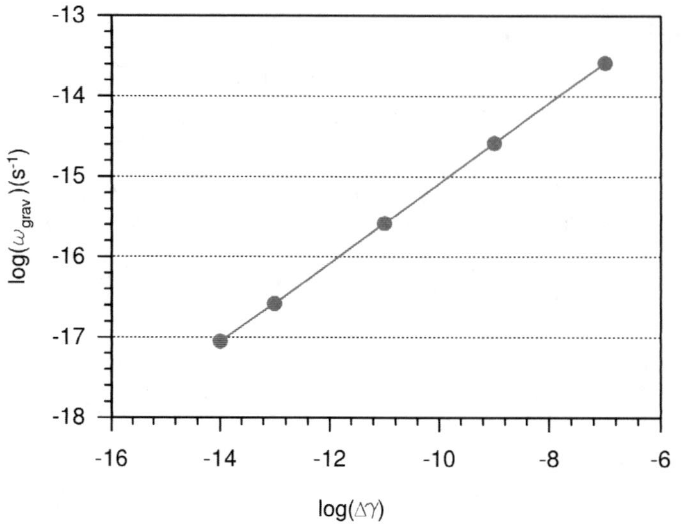

Figure 1. Vibrational frequency of a soft gravonon mode ω_{grav} (s^{-1}) of a gravonon cluster, subjected to an external gravitational potential due to the center bulge M_{center} in a galaxy, as a function of the changed entanglement strength $\Delta\gamma = 1 - \gamma$ in the cluster.

Hubble time defines.

Imagine now that the star passes close to a galaxy with a bulge in the center. It experiences an additionl weak Newton force, such that the acceleration g_{bar} is very small but not zero. It is probably 10^2 times smaller than the acceleration $g_\dagger$ in the absence of the additional Newton force. What happens in the gravonon cluster? We need to understand the changes because they will lead us to the experimental curve of McGaugh et al. The internal forces, which bind the gravonon cloud and the hydrogen atoms, are short range gravitational forces in high dimensional space. The external classical Newton force is of different nature compared to the forces within the cluster. Therefore the weak Newton force is not just a perturbation of the internal forces inside the gravonon continuum.

In the discussion of the effect of the external Newton force on the internal structure of the gravonon cluster we use the $\gamma - \eta$ model in ref. [18]. Compared to the gravonon mode of the unperturbed gravonon cluster the gravonon frequency ω_{grav} in the presence of the Newton force due to the bulge changes as $\omega_{grav} \propto \sqrt{1-\gamma}$, i.e. the gravonon frequency increases with the decrease of the entanglement strength γ in the gravonon cluster. This is the result of solving Schrödinger's equation and is displayed in fig. 1.

3.2. Dependence of γ on g_{bar}

In the stationary state of the hydrogen cluster in the gravonon environment the minimum of the potential is obtained if the attraction by the center bulge is exactly counterbalanced by the retracting force of the springs connecting the gravonon clusters:

$$GM_{center}m_{grav}/r = \frac{1}{2}K(\Delta l)^2. \tag{2}$$

r is the distance of the rotating star (assumed to be a spectator mass at the position of the gravonon cluster) from the bulge (approximated here by a point mass M_{center}). m_{grav} is the

Journal of Physics: Conference Series **2533** (2023) 012046 doi:10.1088/1742-6596/2533/1/012046

mass of the gravonon cluster, which is beabled by gravitational interaction with other gravonon clusters of the same kind (cf. ref. [16]). K is the force constant

$$K = \gamma m_{grav}\omega_{grav}^2, \tag{3}$$

where ω_{grav} is the vibrational frequency of the gravonon clusters against each other (see eq. (13) in ref. [18]). γ is the mass ratio of the considered gravonon cluster and the mean mass of environmental gravonon clusters. The potential of the center bulge is assumed not to vary significantly over the distance of the vibration amplitude of the gravonon cluster. Then we have the situation of a harmonic oscillator in a constant force field and the shift Δl of the gravonon cluster counterbalancing the attraction by the center bulge can be calculated as force divided by force constant:

$$\Delta l = GM_{center}/(r^2\omega_{grav}^2) = g_{bar}/\omega_{grav}^2. \tag{4}$$

Inserting the last two equations into eq. (2) yields

$$\gamma = 2r^3\omega_{grav}^2/GM_{center} = 2r/g_{bar}. \tag{5}$$

The result of this section, needed for the further development, is the modification of the soft gravonon mode as $\omega_{grav} \propto \sqrt{1-\gamma}$, i.e. $\omega_{grav} \propto \sqrt{g_{bar}}$, because $\Delta\gamma \propto g_{bar}$ is the change by the Newton potential due to the center bulge as it is seen in fig. 1.

Since γ is reduced from the starting value of 1, typical for the isolated gravonon cluster, ω_{grav} is increased from its starting value close to zero. The soft gravonon mode becomes harder through the effect of the center bulge in the galaxy. Therefore the time for beable-limbo transition ω_{grav}^{-1} of the gravonon cluster is reduced, beable-limbo transition of the gravonon cluster becomes faster in the presence of the bulge. It takes shorter time for the gravonon cluster to go in the limbo state. The number of gravonons entangled with matter fields in three dimensional space may be reduced in the presence of the central bulge, gravonons in limbo in three dimensional space may be swallowed by the bulge. This is the "vacuum cleaner effect" of the center bulge of a galaxy. The increased rate of beable-limbo transition is just an opportunity for the gravonons to go in the limbo state more frequently under the effect of the center bulge, reducing the value of $\langle\zeta^+\zeta\rangle$. In the next section we discuss how often the opportunity is utilized in two limiting cases.

4. Number of gravonons in two limiting situations. Contact with McGaugh's curve

The beable-limbo transition and the reverse limbo-beable transition of gravonons reminds of the chemical equilibrium achieved in a chemical reaction when the rate of the forward reaction equates with the rate of the backward reaction. However, if a beabled gravonon cluster comes in a situation where it can go in the limbo state, this does not mean necesseraly that it will make the transition. There exists an attempt probability to transit in the limbo state, which will compete with an attempt probability for the reverse limbo-beable transition, until the two counterbalance each other, achieving a steady state. For two limiting cases we use simplified kinetics of the gravonon cluster after adapting to a new center bulge mass:
(i) the steady state of beabled gravonon cluster and gravonons in limbo, when $g_{bar} \ll g_\dagger$;
(ii) the stedy state when $g_{bar} \geq g_\dagger$.
The first situation corresponds to points on the left end of McGaugh's curve, the second is found at values of g_{bar} where the observed acceleration g_{obs} merges with the Newton curve (fig. 2, right end). A detailed discussion will be presented in a future publication.

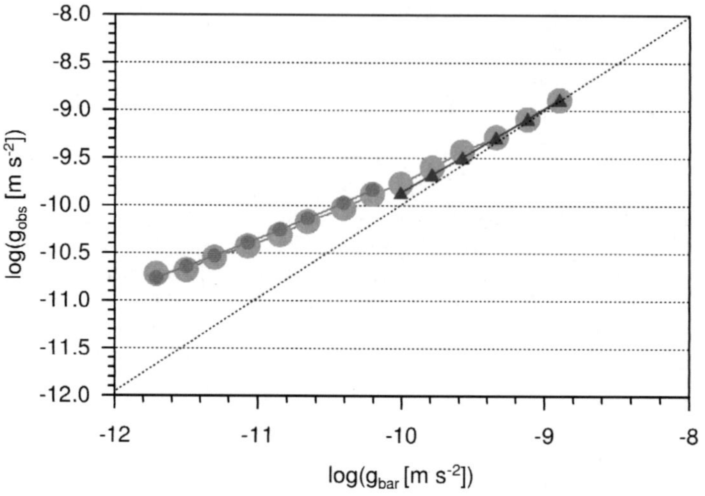

Figure 2. Result of EHHD theory: red points using eq. (10), blue symbols using eq. (15) for $\langle\zeta^+\zeta\rangle$ with $g_\dagger = 1.2E - 10$ m s^{-2}. Observed acceleration g_{obs} from McGaugh et al. [11] (gray symbols) measured for 2693 points in 153 galaxies vs. Newton acceleration g_{bar} (dashed curve).

4.1. The gravonon cluster experiences a very small acceleration $g_{bar} \ll g_\dagger$
A steady state is achieved in the competition of the two reactions:

$$\langle\zeta^+\zeta\rangle \underset{k_-}{\overset{k}{\rightleftarrows}} \langle limbo\rangle \tag{6}$$

$\langle limbo\rangle$ is the density of free isolated gravonon clusters moving in plane waves as trivial solutions of Schrödinger's equation. k_- is the rate constant for beabling a free gravonon cluster.

The kinetic equation yielding the steady state is then

$$\frac{\mathrm{d}}{\mathrm{d}t}\langle\zeta^+\zeta\rangle = 0 = -k\langle\zeta^+\zeta\rangle + k_-\langle limbo\rangle, \tag{7}$$

$$\langle\zeta^+\zeta\rangle = \frac{k_-}{k}\langle limbo\rangle. \tag{8}$$

$k_-\langle limbo\rangle$ is assumed to be constant because we have an infinite reservoir of free gravonon clusters in limbo. The desorption rate constant k for the beable-limbo transition is given by the following expression well known from chemical kinetics:

$$k = \omega_{grav} e^{-\Delta E/k_B T_{warp}}. \tag{9}$$

ω_{grav} in the denominator of eq. (8) is known to be $\propto \sqrt{g_{bar}}$ so that we get the final result:

$$\langle\zeta^+\zeta\rangle = \sqrt{g_\dagger/g_{bar}} \tag{10}$$

Journal of Physics: Conference Series **2533** (2023) 012046 doi:10.1088/1742-6596/2533/1/012046

with $\sqrt{g_\dagger} = k_-\langle limbo\rangle\sqrt{g_{bar}}/(\omega_{grav}e^{-\Delta E/k_BT_{warp}})$. This is a constant independent of g_{bar} because ω_{grav} is $\propto \sqrt{g_{bar}}$. Multiplying eq. (10) with g_{bar}, the EHHD expression for the acceleration g_{DM} due to dark matter,

$$g_{DM} = g_{obs} - g_{bar} = g_{bar}\langle\zeta^+\zeta\rangle \tag{11}$$

is the same as Milgrom's MOND result, leading to constant rotational velocities of the galaxies far from the central bulge. Then it is required that $\langle\zeta^+\zeta\rangle \propto \frac{1}{\sqrt{g_{bar}}}$ as eq. (10) due to EHHD shows.

For large values of ω_{grav} the value of $\langle\zeta^+\zeta\rangle$ becomes small, more and more gravonons go in the limbo state. This explains why in McGaugh's curve the difference $g_{obs} - g_{bar}$ reduces with increasing g_{bar} starting at very small values of $g_{bar} \approx 0$. For the steady state of eq. (7) we need gravonons in the limbo state. If g_{bar} is very small, i.e. the Newton potential is very weak, gravonons are supplied to the gravonon cluster from the infinite intergalactic reservoir to be beabled with the hydrogen atoms. The rate of the backward reaction in eq. (7) can be considered constant $k_-\langle limbo\rangle$ =const. Therefore the only influence on the gravonon number $\langle\zeta^+\zeta\rangle$ in the gravonon cluster comes from the soft gravonon mode ω_{grav} which, being in the denominator of eq. (8) and being $\propto \sqrt{g_{bar}}$, determines the dependence of the gravonon number as $\langle\zeta^+\zeta\rangle \propto \frac{1}{\sqrt{g_{bar}}}$.

4.2. The gravonon cluster experiences an acceleration $g_{bar} \geq g_\dagger$

For this situation we assume that the "vacuum cleaner effect" due to the center bulge has already diluted the concentration of limbo gravonon clusters so much that no significant new beabling of gravonon clusters out of the limbo occurs. The reverse process of beabling is ignored because the Newton force becomes strong and the gravonons in limbo state in the gravonon cluster are no more in equilibrium with the gravonons in the intergalactic space:

$$\langle\zeta^+\zeta\rangle \quad \overset{k}{\rightarrow} \quad \langle limbo\rangle. \tag{12}$$

The kinetic equation is

$$\frac{\mathrm{d}}{\mathrm{d}t}\langle\zeta^+\zeta\rangle = -k\langle\zeta^+\zeta\rangle \tag{13}$$

and the solution using k from eq. (9) is

$$\langle\zeta^+\zeta\rangle \quad = \quad \langle\zeta^+\zeta\rangle_0 e^{-\sqrt{g_{bar}}[\frac{t\cdot\omega_{grav}}{\sqrt{g_{bar}}}exp(-\frac{\Delta E}{k_BT_{warp}})]}. \tag{14}$$

If for this kinetics we take $g_{bar} = g_\dagger$ as our starting point we have $\langle\zeta^+\zeta\rangle_0 = 1$ and we get *exactly* the limit $g_{bar} \gg g_\dagger$ of eq. (5) of McGaugh. If for the time t we define $t = \frac{1}{k_-<limbo>}$ then

$$\langle\zeta^+\zeta\rangle = e^{-\sqrt{g_{bar}/g_\dagger}}. \tag{15}$$

This is the situation where g_{obs} merges with g_{bar} (fig. 2, right end). No gravonon clusters are provided from the intergalactic reservoir. The gravonon clusters are no more in equilibrium with the reservoir. Small number of gravonon clusters implies little dark matter. Effects due to Newton gravitation dominate because dark matter effects are attenuated. Why? The answer is the "vacuum cleaner effect". The displacement of the gravonon clusters in the direction of the

Journal of Physics: Conference Series **2533** (2023) 012046 doi:10.1088/1742-6596/2533/1/012046

bulge is $\Delta l = \frac{Force}{m_{grav}\omega_{grav}^2}$ as eq. (4) shows. If there is no external force $\Delta l = 0$. If the external force due to the bulge is large then the displacement of the gravonon clusters is large, they move away from the rotating star in the direction of the bulge, which exercises the Newton force. The gravonon number reduces, hence dark matter effects are attenuated as well. The Milky Way is an example of this situation. Fewer satellite galaxies are found in its vicinity compared to what is expected on the basis of the calculated dark matter, needed for the stability of our galaxy.

The agreement between the data observed g_{obs} vs. g_{bar} by McGaugh et al. with the result of our EHHD quantum field theory is the evidence that beable-limbo quantum transitions and quantum gravitation are the physics behind the universal flat rotational curves of disk galaxies. In experiment the data measured is directly associated with the occupation of gravonons in different environments.

5. Conclusion

The EHHD theory is a quantum field theory of gravity in 11 dimensional spacetime, based on Einstein-Hilbert action and expansion of the metric up to 4th order. The theory accounts for the entanglement of the local gravonon field and matter fields, which is responsible for the localization of a matter particle on a definite position in four dimensional spacetime This is *the solution of the measurament problem of Quantum Mechanics coming out as result of the deterministic Schrödinger equation.* The quantum gravonon field is the quantum analogue of the dark energy. In higher orders of the metric the interaction of ordinary matter and gravonons becomes attractive. The gravonon field then assumes the role of dark matter.

References

[1] Natalie W 2019 *Physicists find a way to see the 'Grin' of quantum gravity* in *Quanta Magazine*
[2] Verlinde E P 2016 *Preprint* arXiv: 1611.02269
[3] Brouwer M M et al. 2017 *Monthly Notices of the Royal Astronomical Society* **466** 2547
[4] Diósi L 1987 *Phys. Lett.* A **120**(8) 377; 1989 *Phys. Rev.* A **40**(3) 1165
[5] Penrose R 1996 *Gen. Rel. Gravit.* **28**(5) 581; 2014 *Found. Phys.* **44**(5) 557
[6] Hossenfelder S 2018 *Lost in Math: How Beauty Leads Physics Astray* (Basic Books)
[7] Wilczek F 2015 *A Beautiful Question: Finding Nature's Deep Design* (Penguin)
[8] Randall L 2005 *Warped Passages: Unraveling the Mysteries of the Universe's Hidden Dimensions* (Harper Collins Publishers)
[9] Zee A 2010 *Quantum field theory in a nutshell* (Princeton and Oxford: Princeton University Press 2nd ed.)
[10] Reuter M 1998 *Phys. Rev.* D **57** 971
[11] McGaugh S S, Lelli F and Schombert M J 2016 *Phys. Rev. Letters* **117** 201101; 2016 *Preprint* arXiv: 1609.05917
[12] Milgrom M 1983 *Astrophys. J.* **270** 365; 1983 *Astrophys. J.* **270** 371
[13] Doyen G and Drakova D 2015 *Found. Phys.* **45** 959
[14] Drakova D and Doyen G 2013 *J. Phys.: Conf. Series* **442** 012049
[15] Drakova D and Doyen G in the present volume
[16] Drakova D and Doyen G 2019 *J. Phys.: Conference Ser.* **1275** 012055
[17] Doyen G and Drakova D 2019 *J. Phys.: Conference Ser.* **1275** 012044
[18] Doyen G 1982 in *Vibration at Surfaces* eds. R Caudano, J-M Gilles and A A Lucas (New York: Plenum)

Journal of Physics: Conference Series **2533** (2023) 012048 doi:10.1088/1742-6596/2533/1/012048

A brief tour of gauge CPT

K Koltko
Independent Theoretical Physicist, Denver, Colorado, USA
E-mail: g8cpt@protonmail.com

Abstract. We outline the gauge theory of CPT transformations. Gauge CPT provides an alternative to the dark matter hypothesis by unveiling a natural extension of general relativity. Additionally, the new dynamical degrees of freedom revealed by gauge CPT should be essential for any approach to quantum gravity. Even though it may seem absurd to entertain the notion of gauging a discrete symmetry because there are no continuous parameters, we show otherwise via a "back door".

1. Introduction

The outstanding success of the gauge paradigm provides hope that it can again be used to solve some of the current issues in physics such as an alternative to dark matter and supplying missing ingredients required for quantum gravity. Actually, we do not have much of a choice other than the CPT symmetry if we are to stick with experimentally verified symmetries. The use of vierbein and the arbitrary choices of *where* (on the continuous spacetime manifold) one wants to apply the local CPT symmetry transformations allow for the gauging procedure.

The actual physics consists of four parts: motivation, fundamental local CPT transformations, choice of Lagrangian, and experimental/observational explanations and predictions. A good introduction to gauge CPT can be found in "The Baryonic Tully-Fisher Law and the Gauge Theory of CPT Transformations" [1]. A simplified mathematical approach is in "A free-field Lagrangian for a gauge theory of the CPT symmetry" [2].

2. Motivation

There are two components of the motivation to gauge CPT - quantum gravity and astrophysical issues. Briefly, we argue that gauging the CPT symmetry is necessary for any approach to quantum gravity for two reasons: First, PT (together) is a proper Lorentz transformation just like the continuous Lorentz rotations (boosts and spatial rotations), λ, used to obtain the metric spin connection formulation of general relativity (GR). Therefore, it would seem logical to include local PT transformations with local Lorentz rotations in order to complete the gauging of the entire group of proper Lorentz transformations. Inclusion of C is required to preserve the universal symmetry status introduced by the Lorentz rotations. Second, the CPT symmetry is born from the eminently successful *union* of special relativity with quantum theory - a "bridge" between the two theories. Because GR can be obtained by gauging global Lorentz rotations, should not CPT be brought along for the ride, too? Perhaps the new physical dynamical degrees

Journal of Physics: Conference Series **2533** (2023) 012048 doi:10.1088/1742-6596/2533/1/012048

of freedom unveiled by gauge CPT can be used to resolve at least some of the renormalization problems encountered when trying to make GR a quantum field theory? This last point is analogous to expanding the early nonrenormalizable weak interaction theories by the renormalizable $SU(2) \otimes U(1)$ electroweak theory.

We now turn to the astrophysical motivation for gauge CPT. The starting point is the flat rotation curves of spiral galaxies and the excessive gravitational lensing produced by galaxies and galactic groups. That particles (matter and photons) are not moving the way they should be *are the observational facts.* Dark matter and modified gravity theories *are not.* Even though dark matter sounds too much like the 19th century aether, it is certainly worth examining. However, the absence of its detection and the well-known theoretical issues such as the core-cusp problem should add to the motivation to examine the logical alternative of modified or extended gravity theories.

Because accelerations are caused by forces, and all known forces can be obtained by gauging select global symmetries, we search for a candidate symmetry to be gauged. That CPT is an experimentally verified symmetry is sufficient reason to attempt gauging it and examine the consequences. Also, because CPT is a universal symmetry containing Lorentz transformations, there is the intuition of obtaining the required mass independent accelerations of galactic rotation curves.

3. Fundamental local CPT transformations

The two fundamental local CPT transformations of spacetime (vierbein) and Dirac wavefunction are perhaps the most difficult part of gauge CPT; different descriptions of the derivation are found in [1], [2], [3], [4]. In the usual gauging procedure we first introduce "arbitrariness," then introduce locality in a closely intertwined procedure. For example, in the $U(1)$ gauging process, we first pick an arbitrary phase ϕ from the continuum of choices in $0 \leq \phi < 2\pi$ and apply it to a wavefunction $\psi \rightarrow e^{i\phi}\psi$. Then, we make it local by replacing ϕ with $\phi(x)$: $\psi \rightarrow e^{i\phi(x)}\psi$. In gauging CPT, the two steps are more distinct and reversed.

We make note that in Minkowski spacetime, the coordinate axes are flipped, and the origin is mapped to itself under the global PT transformation - these two properties are of crucial importance in defining local CPT transformations in curved spacetime! We assume that there is no nontrivial C operation acting on spacetime. In other words, we assume that there is no such thing as antispacetime. A discussion of why this is an important concept to consider and an attempt to find a nontrivial C spacetime operation can be found in [3]. We also note that the Dirac spinor wavefunction, ψ, transforms under Minkowski global CPT as $\psi(x_\mu) \rightarrow i\gamma^5\psi(x_\mu)$.[1] These two points are the essence of defining local CPT transformations.

The first crucially important step in the CPT gauging procedure is to make the CPT transformation local. We start by introducing a vierbein field, $e_a{}^\mu(x)$, where μ are the manifold coordinates and a are local inertial coordinates at the point x. As described in [1], we view $e_a{}^\mu(x)$ as tiny Minkowski coordinate axes centered at x_μ. When the free falling observer at x_μ applies a CPT transformation, she sees that her origin stays put ($x_\mu \rightarrow x_\mu$), her coordinate

[1] We use Bjorken-Drell conventions except for σ^{ab}. In this paper $\sigma^{ab} = \frac{1}{4}\left[\gamma^a, \gamma^b\right]$.

Journal of Physics: Conference Series **2533** (2023) 012048 doi:10.1088/1742-6596/2533/1/012048

axes flip ($e_a^{\ \mu}(x) \to -e_a^{\ \mu}(x)$), and a Dirac wavefunction at her origin picks up a factor of $i\gamma^5$ ($\psi(x_\mu) \to i\gamma^5\psi(x_\mu)$). This is the definition of local CPT transformations applied at the point x_μ in curved spacetime.

To complete the gauging procedure, we have to introduce arbitrariness. This may seem difficult because there are only two choices at a given point: do nothing, or do the above local CPT transformations. To complete gauging our discrete transformations, we introduce a real, differentiable scalar function $f(x)$ defined everywhere on our manifold. The *choice* of $f(x)$ is arbitrary. Wherever $f(x) \le 0$, we do nothing to the vierbein and Dirac fields. Wherever $f(x) > 0$, we apply the local CPT transformation to the vierbein and Dirac fields. *We emphasize that $f(x)$ is NOT a physical field!* The use of $f(x)$ is the "back door" to introduce continuous parameters - the manifold coordinates - into the gauging procedure for the discrete CPT symmetry. Just as $\phi(x)$ must disappear in the final field equations of the $U(1)$ gauge theory, $f(x)$ must also disappear from the field equations of gauge CPT. Finally, we also introduce local Lorentz rotations[2] $\lambda(x)$, with the local CPT transformation for two reasons. First, as stated before, we are gauging the entire proper Lorentz group of spacetime transformations (induced by gauging $CPT\lambda$ on the Dirac field). Second, this forces us to introduce the metric spin connection, $\omega_{\mu ab}$. If the transformation of $\omega_{\mu ab}$ can compensate for the variations induced by local $CPT\lambda$, then nothing new is happening, i.e., there is no need to introduce another gauge field.

Putting everything together, we get the two fundamental local $CPT\lambda$ transformations acting at a point x_μ [4]:

$$e_a^{\ \mu} \;\to\; \Theta[-f]\, e_a^{\ \mu} + \Theta[f]\left(-e_b^{\ \mu}\right)\Lambda_a^{\ b}, \tag{1}$$

$$\psi \to \Theta[-f]\,\psi + \Theta[f]\, i\gamma^5\Lambda_\psi\psi, \tag{2}$$

where $\Lambda_a^{\ b}$ and Λ_ψ are the spacetime vector and Dirac spinor representations of the proper Lorentz rotations, $\lambda(x)$, respectively. We define the Heaviside step function, $\Theta[y]$, as $\Theta[y] = 0$, if $y \le 0$; and $\Theta[y] = 1$, if $y > 0$. From these two transformations, one easily obtains the transformations of $\overline{\psi}$ and $\omega_{\mu ab}$:

$$\begin{aligned}\overline{\psi} \;&\to\; \Theta[-f]\,\overline{\psi} + \Theta[f]\, i\overline{\psi}\gamma^5\Lambda_{\overline{\psi}},\\ \omega_{\mu ab} \;&\to\; \Theta[-f]\,\omega_{\mu ab} + \Theta[f]\,\widetilde{\omega}_{\mu ab} + \delta[f]\,\Theta[-f]\,\varsigma_{\mu ab} + \delta[f]\,\Theta[f]\,\widetilde{\varsigma}_{\mu ab},\end{aligned}$$

where $\delta[\cdots]$ is the Dirac delta functional. Explicit expressions for $\widetilde{\omega}_{\mu ab}$, $\varsigma_{\mu ab}$, and $\widetilde{\varsigma}_{\mu ab}$ are found in the appendix from [4]. We are using the metric spin connection:

$$\omega_{\mu ab} = \frac{1}{2}\left[e_a^{\ \nu}\left(\partial_\mu e_{b\nu} - \partial_\nu e_{b\mu}\right) - e_b^{\ \nu}\left(\partial_\mu e_{a\nu} - \partial_\nu e_{a\mu}\right) - e_a^{\ \rho} e_b^{\ \sigma}\left(\partial_\rho e_{r\sigma} - \partial_\sigma e_{r\rho}\right) e^r_{\ \mu}\right].$$

Also, the volume element transforms as $ed^4x \to (\Theta[f] + \Theta[-f])\, ed^4x$, where $e = \det\left(e^a_{\ \mu}\right)$. Clearly, these transformations are well defined in curved spacetime.

The appearance of the delta functionals, arising from the differentiation of the transformed vierbein and spinors, need to be in a definite integral in order to be defined. The obvious choice is the action integral because of its fundamental role in physics. *Therefore, we use the action integral as the fundamental arena for deriving the necessary mathematics used to gauge the CPT symmetry and introduce the new gauge field.* The derivation of the mathematics used to handle the various ensuing products of the step functions and delta functionals and their differentiation is found in the appendix from [4].

[2] In most of the author's previous work, Λ, denotes Lorentz rotations. To avoid confusion with the hypothetical cosmological constant, we no longer use Λ.

4. Consequences of the transformations

Do these transformations make sense? In other words, can we recover the global CPT transformations in Minkowski spacetime from these local transformations? Yes, our starting point is the Hermitian Dirac action in Minkowski spacetime, S_D, and the extension of the Dirac action to curved spacetime by use of vierbein and the spin connection ($\partial_\mu\psi \rightarrow \partial_\mu\psi + \frac{1}{2}\omega_{\mu ab}\sigma^{ab}\psi$), S_ω [4]:

$$S_D = \int \left\{ \frac{i}{2} \left[\overline{\psi}\gamma^a e_a^{\ \mu} \partial_\mu \psi - e_a^{\ \mu} \partial_\mu \overline{\psi} \gamma^a \psi \right] - m\overline{\psi}\psi \right\} e d^4x, \tag{3}$$

$$S_\omega = \frac{i}{4} \int e_a^{\ \mu} \omega_{\mu bc} \overline{\psi} \left\{ \gamma^a, \sigma^{bc} \right\} \psi e d^4x. \tag{4}$$

We now apply the transformations to S_D noting that the transformation of the volume element, ed^4x, only introduces removable singularities where $f = 0$ which we can ignore in a definite integral:

$$\begin{aligned} S_D \rightarrow S'_D = & \int \Theta[-f] \left\{ \frac{i}{2} \left[\overline{\psi}\gamma^a e_a^{\mu} \partial_\mu \psi - e_a^{\mu} \partial_\mu \overline{\psi} \gamma^a \psi \right] - m\overline{\psi}\psi \right\} e d^4x \\ & + \int \Theta[f] \left\{ \frac{i}{2} \left(\partial_\mu \overline{\psi} \Lambda_{\overline{\psi}} + \overline{\psi} \partial_\mu \Lambda_{\overline{\psi}} \right) e_b^{\mu} \Lambda_a^{\ b} \gamma^a \Lambda_\psi \psi \right\} e d^4x \\ & - \int \Theta[f] \left\{ \frac{i}{2} \overline{\psi} \Lambda_{\overline{\psi}} \gamma^a e_b^{\mu} \Lambda_a^{\ b} \left(\partial_\mu \Lambda_\psi \psi + \Lambda_\psi \partial_\mu \psi \right) \right\} e d^4x \\ & + \int \Theta[f] \, m\overline{\psi}\psi e d^4x \\ & + \frac{i}{2} \int \delta[f] \, \partial_\mu f \left\{ \Theta[-f] \, i e_a^{\mu} \overline{\psi} \left(\gamma^a \Lambda_\psi + \Lambda_{\overline{\psi}} \gamma^a \right) \gamma^5 \psi \right\} e d^4x \\ & - \frac{i}{2} \int \delta[f] \, \partial_\mu f \left\{ \Theta[f] \, i e_b^{\mu} \Lambda_a^{\ b} \overline{\psi} \left(\Lambda_{\overline{\psi}} \gamma^a + \gamma^a \Lambda_\psi \right) \gamma^5 \psi \right\} e d^4x. \end{aligned} \tag{5}$$

By setting $f < 0$ everywhere, we recover our original S_D. Setting $f > 0$ and λ constant everywhere gives us the Minkowski Dirac action transformed by global $CPT\lambda$. So, we assert that the local $CPT\lambda$ transformations make sense. We note that S_D is not invariant under global CPT transformations - it changes sign. This has no effect on the Dirac equation but raises some subtle points when using variational calculus with gauge CPT [4].

To complete the introduction of the local $CPT\lambda$ transformations, we need δS_D and δS_ω. This is almost the same as $\delta S_D = S'_D - S_D$, $\delta S_\omega = S'_\omega - S_\omega$ but for the subtle issues regarding the sign change. However, the use of the simplified mathematical approach in [2] avoids the issues and leads to the same δS_D as in [4]. Upon substitution of the local $CPT\lambda$ transformations into S_D and S_ω, along with various manipulations, one obtains the variation of the Dirac action in curved spacetime, $\delta S = \delta S_D + \delta S_\omega$ [4]:

$$\delta S_D = -\frac{1}{4} \int \delta[f] \, \partial_\mu f \left\{ e_a^{\mu} \overline{\psi} \gamma^5 \left(\left[\gamma^a, \Lambda_{\overline{\psi}} \right] + \left[\Lambda_\psi, \gamma^a \right] \right) \psi \right\} e d^4x, \tag{6}$$

and

$$
\begin{aligned}
\delta S_\omega = & \frac{i}{16}\int \delta[f]\,\partial_\mu f\overline{\psi}\left\{\gamma^a,\sigma^{bc}\right\}\left[\eta_{ad}e_b{}^\mu\left(\Lambda_c{}^d+\Lambda^d{}_c\right)-\eta_{bd}e_c{}^\mu\left(\Lambda^d{}_a+\Lambda_a{}^d\right)\right. \\
& \left.+\eta_{bd}e_a{}^\mu\left(\Lambda^d{}_c-\Lambda_c{}^d\right)\right]\psi e d^4x \\
& +\frac{i}{16}\int \delta[f]\,\partial_\mu f\overline{\psi}\Lambda_{\overline{\psi}}\left\{\gamma^a,\sigma^{bc}\right\}\Lambda_\psi\eta_{bd}\left[e_i{}^\mu\Lambda_a{}^i\left(\Lambda_c{}^d-\Lambda^d{}_c\right)\right. \\
& \left.-e_i{}^\mu\Lambda_c{}^i\left(\delta_a^d+\Lambda_a{}^d+\Lambda^d{}_a\right)+\Lambda^d{}_i e^{i\mu}\left(\eta_{cj}\Lambda_a{}^j+\eta_{aj}\Lambda_c{}^j\right)\right]\psi e d^4x.
\end{aligned}
\tag{7}
$$

We now show that the introduction of the local CPT transformations unveils new physical phenomena distinct, yet coupled, to general relativity. This discussion is slightly modified and abridged from [4]. If $\delta S_D = 0$ identically, then nothing new is going on other than defining the CPT symmetry locally on a curved manifold. If $\delta S_D \neq 0$ but $\delta S_D + \delta S_\omega = 0$, then the local CPT transformations are just a part of general relativity without any new physics. If $\delta S_D \neq 0$ and $\delta S_D + \delta S_\omega \neq 0$, then general relativity cannot accommodate the local CPT symmetry. We are then *forced* to introduce a new gauge field X_μ in order to arrive at an expanded action invariant under local $CPT\lambda$ transformations. The proof that there exists at least one transformation such that $\delta S_D \neq 0$ and $\delta S_D + \delta S_\omega \neq 0$ is done by explicit construction using a simple choice for local Lorentz rotations corresponding to a velocity boost along the x-axis of Minkowski spacetime [4]. It is straightforward to show that such a λ applied in regions where $f > 0$ satisfies the above criteria that new physical phenomena is unveiled by gauge $CPT\lambda$. Under this transformation one obtains:

$$
\delta S_D = -\sinh\frac{\omega}{2}\int \delta[f]\,\partial_\mu f\overline{\psi}\left(e_0{}^\mu\gamma^5\gamma^1 + e_1{}^\mu\gamma^5\gamma^0\right)\psi e d^4x, \tag{8}
$$

and

$$
\delta S_\omega = \frac{3}{64}\sinh\omega\left(\cosh\omega - 1\right)\int \delta[f]\,\partial_\mu f\overline{\psi}\left(e_2{}^\mu\gamma^5\gamma^3 - e_3{}^\mu\gamma^5\gamma^2\right)\psi e d^4x, \tag{9}
$$

where $\omega = \tanh^{-1}\left(\frac{v}{c}\right)$, v and c being the velocities of the boost and light respectively. Because the $\gamma^5\gamma^a$ are linearly independent and ψ is arbitrary, we see that $\delta S_D \neq 0$, $\delta S_\omega \neq 0$, and $\delta S_D + \delta S_\omega \neq 0$ for this choice of transformation. Also see [2] for a simplified proof. The fact that $\delta S_\omega \neq 0$ means that general relativity is not invariant under local $CPT\lambda$ transformations; however, the Einstein-Hilbert action remains invariant but allows for an extension as shown in [2]. Therefore, the new gauge field X_μ must also compensate for the inhomogeneous (i.e. $\delta[f]$) terms arising from the transformation of $\omega_{\mu ab}$ under local $CPT\lambda$ transformations. Thus, we see from variational arguments that a new gauge field coupled to GR is required in order to accommodate local $CPT\lambda$ transformations.

5. Introduction of the new gauge field

We now introduce the new gauge field, X_μ, subject to only one assumption - minimal coupling. There are three reasons for minimal coupling: simplicity, the author's preference of viewing gauge fields as "compensating" fields, and a quick way to get rid of the $\Theta[\pm f]\,\delta[f]$ terms in the transformation of $\omega_{\mu ab}$.

This last reason motivates the fundamental necessary requirement for any Lagrangian allowed under gauge $CPT\lambda$ transformations: absence of $\delta[f]$ in the transformed Lagrangian. Because CPT is a discrete symmetry, singularities must occur when CPT is made local. The arena of the action integral determines which singularities are allowed and which are fatal. For example, terms with the coefficient $(\Theta[-f] + \Theta[f])$ have removable singularities at $f = 0$, and the coefficient can be simply replaced by 1. Terms containing $\delta[f]$ produce surface integrals at $f = 0$. These are finite but arbitrary because the choice of f is arbitrary; therefore, one cannot uniquely extremize the action in order to obtain field equations. So, any potential Lagrangian densities cannot contain $\delta[f]$ after local $CPT\lambda$ transformations are applied. The following derivation of the structure and transformation of X_μ is taken directly from [4] abridged with slight changes.

The first step in determining the transformation of X_μ is to notice that both δS_D and δS_ω are boundary integrals, i.e. they contain the terms $\delta[f]\,\partial_\mu f$. Also, the transformation of $\omega_{\mu ab}$ contains the terms $\Theta[-f]\,\delta[f]\,\varsigma_{\mu ab}$ and $\Theta[f]\,\delta[f]\,\widetilde{\varsigma}_{\mu ab}$ which need to be cancelled out by the transformation of X_μ under local $CPT\lambda$ transformations. Hence, the transformation of X_μ under local $CPT\lambda$ transformations will be of the form:

$$X_\mu \to \Theta[-f]\,X_\mu + \Theta[f]\,\widetilde{X}_\mu + \Theta[-f]\,\delta[f]\,Y_\mu + \Theta[f]\,\delta[f]\,\widetilde{Y}_\mu. \tag{10}$$

We introduce the covariant derivatives, $D_\mu\psi$, $D_\mu\overline{\psi}$:

$$D_\mu\psi = \partial_\mu\psi + \frac{1}{2}\omega_{\mu ab}\sigma^{ab}\psi + \beta X_\mu\psi, \tag{11}$$

and

$$D_\mu\overline{\psi} = \partial_\mu\overline{\psi} - \frac{1}{2}\omega_{\mu ab}\overline{\psi}\sigma^{ab} + \beta^*\overline{\psi}\gamma^0 X_\mu^\dagger\gamma^0, \tag{12}$$

where β is the coupling constant. The first order theory [3], [5] unveiled only the γ^5 components of X_μ. However, from δS_D and δS_ω, we see that other components are also needed. So, we treat X_μ as a matrix: $X_\mu = x_{\mu n}\Gamma^n$, where the $x_{\mu n}$ are the dynamical components of X_μ; and the Γ^n are the 16 linearly independent matrices I, γ^5, γ^a, $\gamma^5\gamma^a$, and σ^{ab}.

The replacement of $\partial_\mu\psi$ and $\partial_\mu\overline{\psi}$ in S_D by $D_\mu\psi$, $D_\mu\overline{\psi}$ results in an expanded action, $S_{D\omega X}$, which will determine Y_μ and $\widetilde{Y}_\mu$ upon requiring $\delta S_{D\omega X} = 0$. The form of $\widetilde{X}_\mu$, the transformation of X_μ under global $CPT\lambda$ transformations, is determined by requiring the $\Theta[f]$ term of the transformed $S_{D\omega X}$ to change sign, just as in S'_D.

We determine Y_μ and $\widetilde{Y}_\mu$ by requiring the transformation of the expanded action to have no terms containing $\delta[f]$. By simply substituting the transformation equations into $S_{D\omega X}$ and setting the sums of all terms containing $\Theta[-f]\,\delta[f]$ and $\Theta[f]\,\delta[f]$ separately to zero, we can straightforwardly solve for Y_μ and $\widetilde{Y}_\mu$. We obtain:

$$Y_\mu = \beta^{-1}\left[\partial_\mu f\left(I - i\gamma^5\Lambda_\psi\right) - \frac{1}{2}\varsigma_{\mu ab}\sigma^{ab}\right], \tag{13}$$

and

$$\widetilde{Y}_\mu = \beta^{-1}\left[\partial_\mu f\left(-I - i\gamma^5\Lambda_{\overline{\psi}}\right) - \frac{1}{2}\widetilde{\varsigma}_{\mu ab}\sigma^{ab}\right]. \tag{14}$$

We note that $\gamma^5\Lambda_\psi$ and $\gamma^5\Lambda_{\overline{\psi}}$ are linear combinations of I, γ^5, and σ^{ab}; so we see that only eight of the possible 16 Γ^n are needed. We assume that X_μ has only these eight $x_{\mu n}$ dynamical components: $X_\mu = x_{\mu I}I + x_{\mu 5}\gamma^5 + x_{\mu ab}\sigma^{ab}$.

We now turn our attention to finding $\widetilde{X}_\mu$ by again substituting the transformation equations into the expanded action. As mentioned above, we require that the $\Theta[f]$ terms in the expanded action change sign and cancel out the corresponding unvaried terms of the expanded action in

Journal of Physics: Conference Series **2533** (2023) 012048 doi:10.1088/1742-6596/2533/1/012048

the regions where $f > 0$. We obtain from the $\Theta[f]$ terms:

$$\begin{aligned}
& e_b^{\ \mu}\Lambda_a^{\ b}\overline{\psi}\Lambda_{\overline{\psi}}\gamma^a\left(\left(\partial_\mu\Lambda_\psi\right)\psi+\Lambda_\psi\partial_\mu\psi+\frac{1}{2}\widetilde{\omega}_{\mu cd}\sigma^{cd}\Lambda_\psi\psi+\beta\gamma^5\widetilde{X}_\mu\gamma^5\Lambda_\psi\psi\right) \\
& -e_b^{\ \mu}\Lambda_a^{\ b}\left(\partial_\mu\overline{\psi}\right)\Lambda_{\overline{\psi}}\gamma^a\Lambda_\psi\psi-e_b^{\ \mu}\Lambda_a^{\ b}\left(\overline{\psi}\partial_\mu\Lambda_{\overline{\psi}}-\frac{1}{2}\widetilde{\omega}_{\mu cd}\overline{\psi}\Lambda_{\overline{\psi}}\sigma^{cd}\right)\gamma^a\Lambda_\psi\psi \\
& -e_b^{\ \mu}\Lambda_a^{\ b}\beta^*\overline{\psi}\gamma^5\Lambda_{\overline{\psi}}\gamma^0\widetilde{X}_\mu^\dagger\gamma^0\gamma^5\gamma^a\Lambda_\psi\psi \\
= \; & e_a^{\ \mu}\overline{\psi}\gamma^a\left(\partial_\mu\psi+\frac{1}{2}\omega_{\mu cd}\sigma^{cd}\psi+\beta X_\mu\psi\right) \\
& -e_a^{\ \mu}\left(\partial_\mu\overline{\psi}-\frac{1}{2}\omega_{\mu cd}\overline{\psi}\sigma^{cd}+\beta^*\overline{\psi}\gamma^0X_\mu^\dagger\gamma^0\right)\gamma^a\psi.
\end{aligned} \tag{15}$$

The discussion on use of this equation to find $\widetilde{X}_\mu$ is found in [4]; the result is:

$$\widetilde{X}_\mu=\gamma^5\Lambda_\psi X_\mu\Lambda_{\overline{\psi}}\gamma^5=\Lambda_\psi X_\mu\Lambda_{\overline{\psi}}. \tag{16}$$

We temporarily retain the γ^5 to emphasize that the transformations are $CPT\lambda$ and not merely λ.

6. The free-field Lagrangian

Now that the transformation of X_μ has been determined, we can construct a Lagrangian density comprised of the expanded Dirac Lagrangian, free-field Lagrangian for X_μ, and a density containing the Einstein-Hilbert term of GR, κR, where $\kappa=(-16\pi G_N)^{-1}$ (this discussion is slightly modified from [1]). The total Lagrangian must satisfy the following requirements [2], [3], [4], [5]:

1) gauge covariance under local CPTλ transformations,

2) absence of any Dirac delta functionals, $\delta[...]$, in the Lagrangian under local CPTλ transformations,

3) terms containing $x_{\mu ab}$ which are not solely constrained to appear within the combination $\left(\frac{1}{2}\omega_{\mu ab}+\beta x_{\mu ab}\right)\sigma^{ab}$, and

4) some components of X_μ appearing in both types of free-field Lagrangians used in GR and the standard model (SM).

The first requirement is obvious. The second prevents pathological variations of the expanded action under local CPTλ transformations as mentioned earlier. The third ensures that $x_{\mu ab}$ has physical significance and is not just a fancy way to ignore the delta functionals appearing in the transformation of $\omega_{\mu ab}$. The fourth reflects that CPT arises from both GR and quantum theory (SM). To this end, we note that the minimal coupling term acting on matter, $\left(\frac{1}{2}\omega_{\mu ab}+\beta x_{\mu ab}\right)\sigma^{ab}\psi$, implies replacing κR by $\kappa R_{X\omega}$, where $\kappa R_{X\omega}$ is the modified Einstein-Hilbert curvature term formed by replacing the metric spin connection $\omega_{\mu ab}$ in R by $\omega_{\mu ab}+2\beta x_{\mu ab}$. We obtain [2] for the Hermitian action, S:

$$\begin{aligned} S &= \int \left\{ \frac{\mathrm{i}}{2} e_a{}^{\mu} \overline{\psi} \gamma^a \left(\partial_\mu \psi + \frac{1}{2} \omega_{\mu bc} \sigma^{bc} \psi + \beta X_\mu \psi \right) - m \overline{\psi} \psi \right\} |e| \, d^4x \\ &- \int \left\{ \frac{\mathrm{i}}{2} e_a{}^{\mu} \left(\partial_\mu \overline{\psi} - \frac{1}{2} \omega_{\mu bc} \overline{\psi} \sigma^{bc} + \beta \overline{\psi} \gamma^0 X_\mu^\dagger \gamma^0 \right) \gamma^a \psi \right\} |e| \, d^4x \\ &+ \int \left\{ \frac{1}{4} Tr \left(H_{\mu\nu} H^{\mu\nu\dagger} \right) + \frac{\kappa}{2} \left(R_{X\omega} + R_{X\omega}^\dagger \right) \right\} |e| \, d^4x, \end{aligned} \tag{17}$$

where

$$H_{\mu\nu} = \frac{\beta}{2} \left(\omega_{\mu ab} \left[\sigma^{ab}, X_\nu \right] - \omega_{\nu ab} \left[\sigma^{ab}, X_\mu \right] \right) + \beta^2 \left[X_\mu, X_\nu \right] + \beta \left(\partial_\mu X_\nu - \partial_\nu X_\mu \right). \tag{18}$$

The term $\frac{1}{4} Tr \left(H_{\mu\nu} H^{\mu\nu\dagger} \right)$ contains both a Yang-Mills term for X_μ - reflecting the quantum (SM) contribution to the origin of the CPT symmetry - and part of the coupling of GR with gauge CPT. We note that a mass term for X_μ, $mTr \left(X_\mu X^{\mu\dagger} \right)$, is not gauge covariant and not allowed [2], [3], [4], [5]. The proof that the Lagrangian density is gauge covariant is found in [2].

The Euler-Lagrange variation of S with respect to $x_{\mu ab}$, $e_a{}^{\mu}$, $\omega_{\mu ab}$, ψ, $x_{\mu I}$, and $x_{\mu 5}$ respectively gives the following field equations [1], [4], [6]:

$$\begin{aligned} & 4\beta D_\nu \left(\partial^\nu x^\mu_{cd} - \partial^\mu x^\nu_{cd} \right) Tr \left[\left(\sigma^{ab} \right)^\dagger \sigma^{cd} \right] + 2\beta \left\{ \left(\omega_{\nu rs} + 2\beta x^*_{\nu rs} \right) \right. \\ & \left. \times \left(\partial^\nu x^\mu_{cd} - \partial^\mu x^\nu_{cd} \right) Tr \left[\left[\sigma^{ab}, \sigma^{rs} \right]^\dagger \sigma^{cd} \right] \right\} \\ & + 2\beta D_\nu \left(2\beta x^\nu_{cd} x^\mu_{rs} + \omega^\nu_{cd} x^\mu_{rs} + \omega^\mu_{rs} x^\nu_{cd} \right) Tr \left[\left(\sigma^{ab} \right)^\dagger \left[\sigma^{cd}, \sigma^{rs} \right] \right] \\ & - \left\{ \beta \left(\omega_{\nu cd} + 2\beta x^*_{\nu cd} \right) \left(2\beta x^\mu_{mk} x^\nu_{rs} + \omega^\mu_{mk} x^\nu_{rs} + \omega^\nu_{rs} x^\mu_{mk} \right) \right. \\ & \left. \times Tr \left[\left[\sigma^{ab}, \sigma^{cd} \right]^\dagger \left[\sigma^{mk}, \sigma^{rs} \right] \right] \right\} \\ & + 8\kappa \eta^{bc} \left(e^{a\mu} e^{n\rho} - e^{n\mu} e^{a\rho} \right) \left(\omega_{\rho cn} + 2\beta x^*_{\rho cn} \right) \\ & + 8\kappa \eta^{ac} \left(e^{b\mu} e^{n\rho} - e^{n\mu} e^{b\rho} \right) \left(\omega_{\rho nc} + 2\beta x^*_{\rho nc} \right) \\ = \; & 4\mathrm{i} \overline{\psi} \sigma^{ab} \gamma^\mu \psi + 8\kappa D_\nu \left(e^{a\nu} e^{b\mu} - e^{a\mu} e^{b\nu} \right). \end{aligned} \tag{19}$$

The modified GR equation,

$$\frac{1}{2} \left\{ \left(R^{\mu\nu}_{X\omega} + R^{\mu\nu\dagger}_{X\omega} \right) - \frac{1}{2} g^{\mu\nu} \left(R_{X\omega} + R^\dagger_{X\omega} \right) \right\} = \frac{1}{2\kappa} T^{\mu\nu}, \tag{20}$$

where

$$\begin{aligned} T^{\mu\nu} &= \frac{g^{\mu\nu}}{4} Tr \left(H_{\rho\sigma} H^{\rho\sigma\dagger} \right) + \frac{\mathrm{i}}{2} \left[\left(D^\mu \overline{\psi} \right) \gamma^\nu \psi - \overline{\psi} \gamma^\nu D^\mu \psi \right] \\ &+ g^{\mu\nu} \left[\frac{\mathrm{i}}{2} \left(\overline{\psi} \gamma^\rho D_\rho \psi - \left(D_\rho \overline{\psi} \right) \gamma^\rho \psi \right) - m \overline{\psi} \psi \right]. \end{aligned} \tag{21}$$

The Palatini variation with respect to $\omega_{\mu ab}$,

Journal of Physics: Conference Series **2533** (2023) 012048 doi:10.1088/1742-6596/2533/1/012048

$$
\begin{aligned}
&\kappa\eta^{bc}\left(e^{n\mu}e^{a\rho}-e^{a\mu}e^{n\rho}\right)\left(\omega_{\rho cn}+\beta x^{*}_{\rho cn}+\beta x_{\rho cn}\right)\\
&+\kappa\eta^{ac}\left(e^{n\mu}e^{b\rho}-e^{b\mu}e^{n\rho}\right)\left(\omega_{\rho nc}+\beta x^{*}_{\rho nc}+\beta x_{\rho nc}\right)\\
&+\frac{i}{4}\overline{\psi}\left\{\sigma^{ab},\gamma^{\mu}\right\}\psi+\frac{\beta^2}{4}\left\{x^{*}_{\nu rs}\left(-\partial^{\nu}x^{\mu}_{cd}+\partial^{\mu}x^{\nu}_{cd}\right)\right.\\
&\times Tr\left[\left[\sigma^{ab},\sigma^{rs}\right]^{\dagger}\sigma^{cd}\right]+x_{\nu rs}\left(-\partial^{\nu}x^{*\mu}_{cd}+\partial^{\mu}x^{*\nu}_{cd}\right)\\
&\times Tr\left[\left[\sigma^{ab},\sigma^{rs}\right]\left(\sigma^{cd}\right)^{\dagger}\right]+\frac{1}{2}x_{\nu rs}\left(2\beta x^{*\mu}_{mk}x^{*\nu}_{cd}\right.\\
&\left.+\omega^{\mu}_{mk}x^{*\nu}_{cd}+x^{*\mu}_{mk}\omega^{\nu}_{cd}\right)Tr\left[\left[\sigma^{ab},\sigma^{rs}\right]\cdot\left[\sigma^{mk},\sigma^{cd}\right]^{\dagger}\right]\\
&+\frac{1}{2}x^{*}_{\nu rs}\left(2\beta x^{\mu}_{mk}x^{\nu}_{cd}+\omega^{\mu}_{mk}x^{\nu}_{cd}+x^{\mu}_{mk}\omega^{\nu}_{cd}\right)\\
&\left.\times Tr\left[\left[\sigma^{ab},\sigma^{rs}\right]^{\dagger}\cdot\left[\sigma^{mk},\sigma^{cd}\right]\right]\right\}\\
=\;&\kappa D_{\nu}\left(e^{a\mu}e^{b\nu}-e^{a\nu}e^{b\mu}\right),
\end{aligned} \tag{22}
$$

the Dirac equation,

$$
ie_{a}^{\ \mu}\gamma^{a}\left(\partial_{\mu}\psi+\frac{1}{2}\omega_{\mu bc}\sigma^{bc}\psi+\beta X_{\mu}\psi\right)-m\psi=0, \tag{23}
$$

and finally

$$
i\overline{\psi}e_{a}^{\ \mu}\gamma^{a}\psi = 4\beta\left(\partial^{\mu}x^{\nu}_{I}-\partial^{\nu}x^{\mu}_{I}\right)_{;\nu} \text{ and} \tag{24}
$$

$$
i\overline{\psi}e_{a}^{\ \mu}\gamma^{a}\gamma^{5}\psi = 4\beta\left(\partial^{\mu}x^{\nu}_{5}-\partial^{\nu}x^{\mu}_{5}\right)_{;\nu} \text{ (the chiral terms of } X_{\mu}\text{)}. \tag{25}
$$

We note that chiral and gravitational anomalies have not been taken into account.

7. Weak field gauge CPT extension of gravity

The ultimate judge to see if gauge CPT has anything to do with reality is experimental and astrophysical predictions. The arguments leading to the point neutrino emitting source, weak field gauge CPT force law are found in [1], [6]. However, early discussions of experimental predictions appear in [3], [4]. The weak field law is used to explain the Tully-Fisher law as well as four experimental predictions [1], [6]. Due to space limitations, we simply state the law:

$$
a_X = k\frac{(I_{\nu})^{\frac{1}{2}}}{r}, \tag{26}
$$

where a_X is the acceleration a point mass feels due to the (anti)neutrino source, I_{ν} is the total (anti)neutrino luminosity (power) of the point source, r is the distance of the point mass from the (anti)neutrino source, and k is the physical constant ($k = 4.25\times10^{-8}\frac{m}{(kg.s)^{\frac{1}{2}}}$) associated with gauge CPT analogous to Newton's gravitational constant, G_N.

Journal of Physics: Conference Series **2533** (2023) 012048 doi:10.1088/1742-6596/2533/1/012048

8. Conclusion

We have gauged the CPT symmetry and used it to derive a new gauge field with some components being a natural extension of GR. Further work remains, for example, the physical meaning of the $x_{\mu I} \pm x_{\mu 5}$ terms and the relationship to anomalies. Rigorous point and plane wave neutrino source solutions to the field equations are required to determine the attractive or repulsive nature of the forces as well as strictly proving the inverse square behavior. Of course, another step is to incorporate Gauge CPT with GR to see if a renormalizable quantum field theory can be obtained.

References

[1] Koltko K 2020 *Proc. Intl. Conf. Cosmology on Small Scales 2020 Excessive Extrapolations and Selected Controversies in Cosmology (Prague)* p 84

[2] Koltko K 2017 A free-field Lagrangian for a gauge theory of the CPT symmetry *Preprint* arXiv 1703.10904v3

[3] Koltko K 2000 Attempts to find additional dynamical degrees of freedom in spacetime using topological and geometrical methods *PhD Thesis* University of South Carolina, Columbia

[4] Koltko K 2013 Gauge CPT as a possible alternative to the dark matter hypothesis *Preprint* arXiv 1308.6341v1

[5] Koltko K 2002 *Found. Phys. Lett.* **15** No. 3 299

[6] Koltko K 2022 *Proc. Intl. Conf. Cosmology on Small Scales 2022 Dark Energy and the Local Hubble Expansion Problem (Prague)* p 114

Journal of Physics: Conference Series **2533** (2023) 012049 doi:10.1088/1742-6596/2533/1/012049

Complete complementarity relations for three-flavor neutrino oscillations

V A S V Bittencourt [1], M Blasone[2,3], S De Siena[4], C Matrella[2,3]

[1] ISIS (UMR 7006), Université de Strasbourg, 67000 Strasbourg, France.
[2] Dipartimento di Fisica, Università di Salerno, Via Giovanni Paolo II, 132 I-84084 Fisciano (SA), Italy.
[3] INFN, Sezione di Napoli, Gruppo collegato di Salerno, Italy.
[4] Università degli Studi di Salerno (Ret.), Fisciano, Italy.

E-mail: sant@unistra.fr,mblasone@unisa.it, silvio.desiena@gmail.com, cmatrella@unisa.it

Abstract. We study quantum correlations encoded in a three-flavor neutrino system by using complete complementarity relations (CCR). Due to the presence of local coherence in two-flavor subsystems, the CCR has an additional contribution not present in the two flavor mixing case. We investigate such coherence for the three possible bipartite subsystems of the global state both for an electron and a muon neutrino system.

1. Introduction

In the last few years, elementary particles as neutrinos have been investigated in the context of quantum information [1]. In fact, the property of neutrinos to interact very weakly and to deeply penetrate into matter makes these particles interesting candidates for applications of quantum information beyond photons[1]. The characterization of quantum correlations in such systems is therefore important for the development of algorithms and protocols that can harness not only quantum entanglement but also other resources, such as quantum discord [3] and coherence [4]. Such quantum correlations have been studied and probed in quantum optics and condensed matter physics, but only recently they have been investigated in relativistic systems of neutrinos and mesons [5]-[19]. In particular, the phenomenon of neutrino oscillations offers a rare example of quantum correlation on macroscopic scale.

The quantum nature of neutrino oscillations has been studied in terms of entanglement [5]–[8], Bell and Leggett-Garg inequalities [11]–[14] and various aspects of quantum coherence such as steering [16, 17] and Non-local Advantage of Quantum Coherence [18]–[21]. Neutrino oscillations have also been considered in the context of entropic uncertainty relations [22, 23].

Complete complementarity relations provide a way to characterize quantum correlations in bi- and multi-partite systems [24] and can be applied to the description of quantum correlations intrinsic to neutrino systems [25]. The concept of complementarity is often associated with wave-particle duality [26]: it is summarized in the statement that a quantum system may possess properties which are equally real but mutually exclusive, in the sense that the more information one has about one aspect of the system, the less information can be obtained about

[1] Classical communication using a neutrino beam was demonstrated in [2].

Journal of Physics: Conference Series **2533** (2023) 012049 doi:10.1088/1742-6596/2533/1/012049

the other. In the context of the two-slit experiment, CCRs can be formalized [27, 28] by defining a predictability P, associated to the knowledge of the path of the particle, and a visibility V, connected with the capacity to distinguish the interference fringes:

$$P^2 + V^2 \leq 1. \tag{1}$$

Complementarity relation as in Eq.(1) are saturated only for pure single-partite quantum states. In [29] it is shown that for a bipartite state we have to consider a third entry C – representing the correlation between the subsystems – in order to obtain a complete complementarity relation:

$$P_k^2 + V_k^2 + C^2 = 1, \qquad k = 1, 2. \tag{2}$$

The quantities associated with the wave-particle duality generate local, single-partite realities, while C generates an exclusive bipartite nonlocal reality. In [24] it has been shown that the CCR can be efficiently expressed in terms of the elements of the density matrix representing the system.

In this work, we briefly review the formalism of CCR for bipartite states and its application to two-flavor neutrino system in the plane-wave approximation. Then, we consider the extension to the case of three flavors (tripartite system).

The interesting new feature in the three-flavor case with respect to the results of Ref.[25] is the presence of non-vanishing local coherences for the elements of the possible bipartitions of the system. We investigate in detail such coherences for an electron and a muon neutrino system.

2. Formalism of CCR

2.1. CCR for bipartite states

Let us consider [24] a bipartite state represented as a vector in the Hilbert state $\mathcal{H}_A \otimes \mathcal{H}_B$ of dimension $d = d_A d_B$, where d_A and d_B are the dimension of the subsystem A and B, respectively. $\{|i\rangle_A \otimes |j\rangle_B = |i,j\rangle_{AB}\}_{i,j=0}^{d_A-1,d_B-1}$ represents an orthonormal basis for $\mathcal{H}_A \otimes \mathcal{H}_B$, where $\{|i\rangle_A\}_{i=0}^{d_A-1}$ and $\{|j\rangle_B\}_{j=0}^{d_B-1}$ are the local basis for the spaces $\mathcal{H}_A$ and $\mathcal{H}_B$, respectively. In this basis, the density matrix of any bipartite state is:

$$\rho_{A,B} = \sum_{i,k=0}^{d_A-1} \sum_{j,l=0}^{d_B-1} \rho_{ij,kl} |i,j\rangle_{AB} \langle k,l| . \tag{3}$$

The state of subsystem A(B) is obtained by tracing over B(A). For example, for subsystem A, we have:

$$\rho_A = \sum_{i,k=0}^{d_A-1} \left(\sum_{j=0}^{d_B-1} \rho_{ij,kj} \right) |i\rangle_A \langle k| \equiv \sum_{i,k=0}^{d_A-1} \rho_{ik}^A |i\rangle_A \langle k| , \tag{4}$$

with a similar form for the subsystem B.

In general, even if the joint state $\rho_{A,B}$ is pure, the states of the subsystems A and B are not pure, which implies that some information is missing when the state of a single subsystem is considered. The missing information is being shared via correlations with the subsystem B [30]. Hence, the complete complementarity relation to consider is:

$$P_{\text{hs}}(\rho_A) + C_{\text{hs}}(\rho_A) + C_{\text{hs}}^{nl}(\rho_{A|B}) = \frac{d_A - 1}{d_A} \tag{5}$$

where $P_{\text{hs}}(\rho_A) = \sum_{i=0}^{d_A-1} (\rho_{ii}^A)^2 - \frac{1}{d_A}$ is the predictability measure and $C_{\text{hs}}(\rho_A) = \sum_{i\neq k}^{d_A-1} |\rho_{ik}^A|^2$ is the Hilbert-Schmidt quantum coherence, a measure of visibility and $C_{\text{hs}}^{nl}(\rho_{A|B}) =$

Journal of Physics: Conference Series **2533** (2023) 012049 doi:10.1088/1742-6596/2533/1/012049

$\sum_{i\neq k,j\neq l}|\rho_{ij,kl}|^2 - 2\sum_{i\neq k,j<l}\mathrm{Re}(\rho_{ij,kj}\rho^*_{il,kl})$ is called non local quantum coherence, that is the coherence shared between A and B.

Another form of CCR can be obtained by defining the predictability and the coherence measures in terms of the von Neumann entropy:

$$C_{\mathrm{re}}(\rho_A) + P_{\mathrm{vn}}(\rho_A) + S_{\mathrm{vn}}(\rho_A) = \log_2 d_A, \tag{6}$$

where $C_{\mathrm{re}}(\rho_A) = S_{\mathrm{vn}}(\rho_{A,\,\mathrm{diag}}) - S_{\mathrm{vn}}(\rho_A)$ is the relative entropy of coherence, with $S_{\mathrm{vn}}(\rho)$ denoting the von Neumann entropy of ρ, and $\rho_{A,\,\mathrm{diag}} = \sum_{i=1}^{d_A}\rho^A_{ii}\,|i\rangle\langle i|$. $P_{\mathrm{vn}}(\rho_A) \equiv \log_2 d_A - S_{\mathrm{vn}}(\rho_{A,\,\mathrm{diag}})$, is a measure of predictability.

2.2. CCR for tripartite states

In [24] the generalization of the CCR for tri-partite pure states is obtained. Let us consider a tri-partite state represented as a vector in the Hilbert state $\mathcal{H}_A \otimes \mathcal{H}_B \otimes \mathcal{H}_C$ of dimension $d = d_A d_B d_C$, where d_A, d_B, d_C are the dimension of the subsystem A, B and C, respectively. $\{|i\rangle_A \otimes |j\rangle_B \otimes |k\rangle_C = |i,j,k\rangle_{ABC}\}_{i,j,k=0}^{d_A-1,d_B-1,d_C-1}$ represents an orthonormal basis for $\mathcal{H}_A \otimes \mathcal{H}_B \otimes \mathcal{H}_C$, where $\{|i\rangle_A\}_{i=0}^{d_A-1}$, $\{|j\rangle_B\}_{j=0}^{d_B-1}$ and $\{|k\rangle_C\}_{k=0}^{d_C-1}$ are the local basis for the spaces $\mathcal{H}_A$, $\mathcal{H}_B$ and $\mathcal{H}_C$, respectively. In this basis, the density matrix of any tri-partite state is:

$$\rho_{A,B,C} = \sum_{i,l=0}^{d_A-1}\sum_{j,m=0}^{d_B-1}\sum_{k,n=0}^{d_C-1}\rho_{ijk,lmn}\,|i,j,k\rangle_{ABC}\langle l,m,n|\,. \tag{7}$$

The state of subsystem A is obtained by tracing over B and C:

$$\rho_A = \sum_{i,l=0}^{d_A-1}\left(\sum_{j=0}^{d_B-1}\sum_{k=0}^{d_C-1}\rho_{ijk,ljk}\right)|i\rangle_A\langle l| \equiv \sum_{i,l=0}^{d_A-1}\rho^A_{il}\,|i\rangle_A\langle l|\,, \tag{8}$$

with a similar form for the subsystems B and C.

The complete complementarity relation to consider for subsystem A is:

$$P_{\mathrm{hs}}(\rho_A) + C_{\mathrm{hs}}(\rho_A) + C^{nl}_{\mathrm{hs}}(\rho_{A|BC}) = \frac{d_A-1}{d_A} \tag{9}$$

where, in this case, the non local coherence is given by:

$$\begin{aligned} C^{nl}_{\mathrm{hs}}(\rho_{A|BC}) =& \sum_{i\neq l}\left(\sum_{j\neq m,k\neq n} + \sum_{j=m,k\neq n} + \sum_{j\neq m,k=n}\right)|\rho_{ijk,lmn}|^2 \\ &- 2\sum_{i\neq l}\left(\sum_{j=m,k<n} + \sum_{j<m,k=n} + \sum_{j<m,k\neq n}\right)\mathrm{Re}(\rho_{ijk,ljk}\rho^*_{imn,lmn}). \end{aligned} \tag{10}$$

The other form of the CCR, Eq.(6), is still valid for the single-partite subsystems A, B and C. But the interesting behaviour comes out when we consider the three possible bipartite subsystems AB, AC and BC. Indeed, as it is shown in the next sections, in which we apply this formalism to a neutrino system, the local coherences for bipartite subsystems are non-zero, in contrast to the local coherence of a single-partite subsystem. For the subsystem AB, for example, is valid the following CCR:

$$C_{\mathrm{re}}(\rho_{AB}) + P_{\mathrm{vn}}(\rho_{AB}) + S_{\mathrm{vn}}(\rho_{AB}) = \log_2(d_A d_B). \tag{11}$$

3. The three-flavor neutrino model

Let us consider a three-flavor neutrino oscillation model, in which the flavor states are written in terms of the mass eigenstates:

$$|\nu_\alpha\rangle = \sum_k U_{\alpha k} |\nu_k\rangle \,, \tag{12}$$

where $k = 1, 2, 3$ and $\alpha = e, \mu, \tau$. $U_{\alpha k}$ are the elements of a 3×3 unitary PMNS matrix, characterized by the three mixing angles and a CP violating phase:

$$U = \begin{pmatrix} c_{12}c_{13} & s_{12}c_{13} & s_{13}e^{-i\delta_{CP}} \\ -s_{12}c_{23} - c_{12}s_{13}s_{23}e^{i\delta_{CP}} & c_{12}c_{23} - s_{12}s_{13}s_{23}e^{i\delta_{CP}} & c_{13}s_{23} \\ s_{12}s_{23} - c_{12}s_{13}c_{23}e^{i\delta_{CP}} & -c_{12}s_{23} - s_{12}s_{13}c_{23}e^{i\delta_{CP}} & c_{13}c_{23} \end{pmatrix}, \tag{13}$$

where $c_{ij} = \cos\theta_{ij}$ and $s_{ij} = \sin\theta_{ij}$, $(i, j = 1, 2, 3)$.

The time evolution of the flavor neutrino state $|\nu_\alpha(t)\rangle$ is given by:

$$|\nu_\alpha(t)\rangle = a_{\alpha e}(t)\,|\nu_e\rangle + a_{\alpha\mu}(t)\,|\nu_\mu\rangle + a_{\alpha\tau}(t)\,|\nu_\tau\rangle \,, \tag{14}$$

where $a_{\alpha\beta}(t) = \sum_k U_{\alpha k} e^{-iE_k t/\hbar} U^*_{\beta k}$ and E_k is the energy of the k-th mass eigenstate.

The transition flavor probability $P_{\alpha\beta} = |\langle\nu_\beta|\nu_\alpha(t)\rangle|^2$ is given by:

$$P_{\alpha\beta} = \sum_{kl} U^*_{\alpha k} U_{\beta k} U_{\alpha l} U^*_{\beta l} e^{-i(E_k - E_l)t}. \tag{15}$$

For ultra-relativistic neutrinos, we can use the approximation $E_k \simeq E + \frac{m_k^2}{2E}$, leading to $E_k - E_l \simeq \frac{\Delta m^2_{kl}}{2E}$ with $\Delta m^2_{kl} = m_k^2 - m_l^2$. E and $L \approx ct$ are the energy and the baseline of the neutrino experiment, respectively. Hence, Eq.(15) becomes:

$$P_{\alpha\beta} = \sum_{kl} U^*_{\alpha k} U_{\beta k} U_{\alpha l} U^*_{\beta l} e^{-i\frac{\Delta m^2_{kl}}{2E} L}. \tag{16}$$

It is also possible to write the oscillation probability in a more convenient way that permits us to separate a constant term to the oscillating one by exploiting the unitary relation $UU^\dagger = \mathbf{1}$, i.e. $\sum_\alpha U_{\alpha k} U^*_{\beta k} = \delta_{\alpha\beta}$. From the square of the unitary relation, by separating the real and imaginary parts of $U^*_{\alpha k} U_{\beta k} U_{\alpha l} U^*_{\beta l}$ we obtain:

$$P_{\alpha\beta} = \delta_{\alpha\beta} - 4\,\mathrm{Re}(U^*_{\alpha k} U_{\beta k} U_{\alpha l} U^*_{\beta l}) \sin^2\left(\Delta m^2_{kl} \frac{Lc^3}{4\hbar E}\right) + 2\sum_{k>l} Im\,(U^*_{\alpha k} U_{\beta k} U_{\alpha l} U^*_{\beta l}) \sin\left(\Delta m^2_{kl} \frac{Lc^3}{2\hbar E}\right). \tag{17}$$

In what follows, we will use the following oscillation parameters [31]–[34]:

$$\begin{aligned} \Delta m^2_{21} &= 7.50 \times 10^{-5} eV^2, \\ \Delta m^2_{31} &= 2.46 \times 10^{-3} eV^2, \\ \Delta m^2_{32} &= 2.38 \times 10^{-3} eV^2, \\ \theta_{12} = 33.48^\circ, \theta_{23} &= 42.3^\circ, \theta_{13} = 8.50^\circ. \end{aligned} \tag{18}$$

For simplicity, here we consider $\delta_{CP} = 0$. We can then use the following correspondence [5]:

$$\begin{aligned} |\nu_e\rangle &= |1\rangle_e \otimes |0\rangle_\mu \otimes |0\rangle_\tau = |100\rangle \,, \\ |\nu_\mu\rangle &= |0\rangle_e \otimes |1\rangle_\mu \otimes |0\rangle_\tau = |010\rangle \,, \\ |\nu_\tau\rangle &= |0\rangle_e \otimes |0\rangle_\mu \otimes |1\rangle_\tau = |010\rangle \,, \end{aligned} \tag{19}$$

where it is highlighted the composite nature of neutrino flavor states.

4. CCR for neutrino states

In [25] we have analyzed the CCR for a bipartite neutrino state. We briefly recall the principal results. For example, if we consider an initial electronic neutrino Eq.(14) becomes:

$$|\nu_e(t)\rangle = a_{ee}(t)\,|10\rangle + a_{e\mu}(t)\,|01\rangle\,. \tag{20}$$

By constructing the density matrix for the state $\rho_{A,B}$ and by tracing to obtain the density matrices for the subsystems ρ_A and ρ_B, it is simple to check that Eq.(5) is verified. In fact, $P_{\rm hs}(\rho_A) = P_{ee}^2 + P_{e\mu}^2 - \frac{1}{2}$, $C_{\rm hs}(\rho_A) = 0$ and $C_{\rm hs}^{nl}(\rho_{AB}) = 2P_{ee}P_{e\mu}$, where we use $|a_{ee}(t)|^2 = P_{ee}$, $|a_{e\mu}(t)|^2 = P_{e\mu}$ and $P_{ee} + P_{e\mu} = 1$. Furthermore, it is simple to see that $\rho_A = \rho_{A,\rm diag}$ and, consequently, $S_{\rm vn}(\rho_A) = S_{\rm vn}(\rho_{A,\rm diag})$. As result, $C_{\rm re}(\rho_A) = 0$, $P_{\rm vn}(\rho_A) = |a_{ee}|^2 \log_2 |a_{ee}|^2 + |a_{e\mu}|^2 \log_2 |a_{e\mu}|^2$ and $S_{\rm vn}(\rho_A) = -|a_{ee}|^2 \log_2 |a_{ee}|^2 - |a_{e\mu}|^2 \log_2 |a_{e\mu}|^2$. Since the dimension of subsystem A is $d_A = 2$, $\log_2 d_A = 1$ and Eq.(6) is satisfied.

It is worth to notice that in the case of a bipartite pure neutrino state, for both Eqs.(5) and (6), the local coherence term is zero. It is natural to ask what happens in the case of a tripartite neutrino state, in which there are bipartite subsystems with their own specific internal structure. We will see indeed that in this case the local coherence terms are non vanishing and depend on the chosen bipartition.

Let us suppose to have a neutrino state in a flavor $\alpha = e, \mu, \tau$ at $t = 0$. The time evolved state is given by:

$$|\nu_\alpha(t)\rangle = a_{\alpha e}(t)\,|100\rangle + a_{\alpha\mu}(t)\,|010\rangle + a_{\alpha\tau}(t)\,|001\rangle\,, \tag{21}$$

The density matrix associated to this state is given by:

$$\rho^\alpha_{e\mu\tau} = \begin{pmatrix} 0 & 0 & 0 & 0 & 0 & 0 & 0 & 0 \\ 0 & \rho^\alpha_{22} & \rho^\alpha_{23} & 0 & \rho^\alpha_{25} & 0 & 0 & 0 \\ 0 & \rho^\alpha_{32} & \rho^\alpha_{33} & 0 & \rho^\alpha_{35} & 0 & 0 & 0 \\ 0 & 0 & 0 & 0 & 0 & 0 & 0 & 0 \\ 0 & \rho^\alpha_{52} & \rho^\alpha_{53} & 0 & \rho^\alpha_{55} & 0 & 0 & 0 \\ 0 & 0 & 0 & 0 & 0 & 0 & 0 & 0 \\ 0 & 0 & 0 & 0 & 0 & 0 & 0 & 0 \\ 0 & 0 & 0 & 0 & 0 & 0 & 0 & 0 \end{pmatrix} \tag{22}$$

where the matrix elements are written as:

$$\rho^\alpha_{22} = |a_{\alpha\tau}(t)|^2; \qquad \rho^\alpha_{23} = \rho^{\alpha *}_{32} = a_{\alpha\tau}(t)a^*_{\alpha\mu}(t); \qquad \rho^\alpha_{25} = \rho^{\alpha *}_{52} = a_{\alpha\tau}(t)a^*_{\alpha e}(t); \tag{23}$$

$$\rho^\alpha_{33} = |a_{\alpha\mu}(t)|^2; \qquad \rho^\alpha_{35} = \rho^{\alpha *}_{53} = a_{\alpha\mu}(t)a^*_{\alpha e}(t); \qquad \rho^\alpha_{55} = |a_{\alpha e}(t)|^2. \tag{24}$$

The corresponding oscillation probabilities are $P_{\alpha e}(t) = |a_{\alpha e}(t)|^2$, $P_{\alpha\mu}(t) = |a_{\alpha\mu}(t)|^2$, $P_{\alpha\tau}(t) = |a_{\alpha\tau}(t)|^2$. By tracing with respect one of the subsystems we can obtain the reduced density matrix for bipartite subsystems $e\mu$, $e\tau$, $\mu\tau$, which are, respectively:

$$\rho^\alpha_{e\mu} = \begin{pmatrix} \rho^\alpha_{22} & 0 & 0 & 0 \\ 0 & \rho^\alpha_{33} & \rho^\alpha_{35} & 0 \\ 0 & \rho^\alpha_{53} & \rho^\alpha_{55} & 0 \\ 0 & 0 & 0 & 0 \end{pmatrix}, \quad \rho^\alpha_{e\tau} = \begin{pmatrix} \rho^\alpha_{33} & 0 & 0 & 0 \\ 0 & \rho^\alpha_{22} & \rho^\alpha_{25} & 0 \\ 0 & \rho^\alpha_{52} & \rho^\alpha_{55} & 0 \\ 0 & 0 & 0 & 0 \end{pmatrix}, \quad \rho^\alpha_{\mu\tau} = \begin{pmatrix} \rho^\alpha_{55} & 0 & 0 & 0 \\ 0 & \rho^\alpha_{22} & \rho^\alpha_{23} & 0 \\ 0 & \rho^\alpha_{32} & \rho^\alpha_{33} & 0 \\ 0 & 0 & 0 & 0 \end{pmatrix}. \tag{25}$$

By tracing again we can obtain the reduced density matrices of the single-partite subsystems:

$$\rho^\alpha_e = \begin{pmatrix} \rho^\alpha_{22} + \rho^\alpha_{33} & 0 \\ 0 & \rho^\alpha_{55} \end{pmatrix}, \quad \rho^\alpha_\mu = \begin{pmatrix} \rho^\alpha_{22} + \rho^\alpha_{55} & 0 \\ 0 & \rho^\alpha_{33} \end{pmatrix}, \quad \rho^\alpha_\tau = \begin{pmatrix} \rho^\alpha_{55} + \rho^\alpha_{33} & 0 \\ 0 & \rho^\alpha_{22} \end{pmatrix}. \tag{26}$$

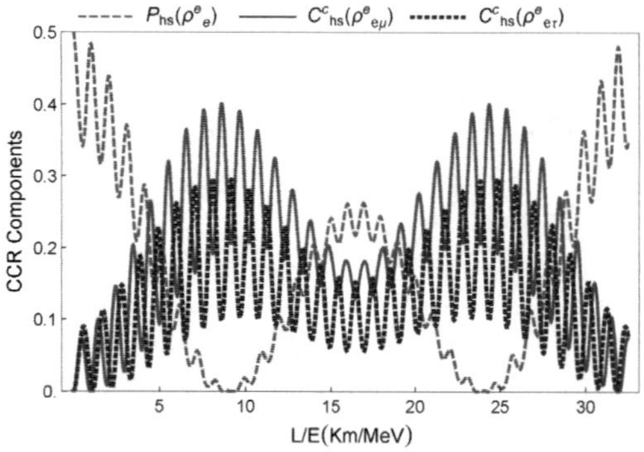

Figure 1: CCR terms, for an initial electronic neutrino, of Eq.(30) as function of L/E.

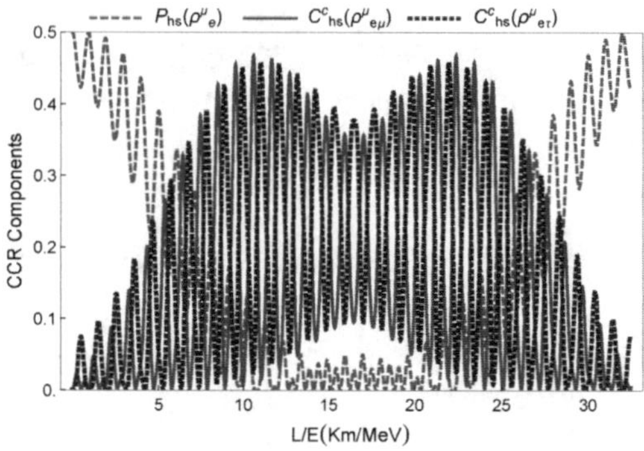

Figure 2: CCR terms, for an initial muonic neutrino, of Eq.(30) as function of L/E.

By following the above prescription, it is simple to evaluate the CCR terms of Eq.(9):

$$P_{\rm hs}(\rho_e^\alpha) = (|a_{\alpha\mu}(t)|^2 + |a_{\alpha\tau}(t)|^2)^2 + |a_{\alpha e}(t)|^2 - \frac{1}{2}, \tag{27}$$
$$C_{\rm hs}(\rho_e^\alpha) = 0, \tag{28}$$
$$C_{\rm hs}^{nl}(\rho_{e|\mu\tau}^\alpha) = 1 - |a_{\alpha e}(t)|^2 - (|a_{\alpha\mu}(t)|^2 + |a_{\alpha\tau}(t)|^2)^2. \tag{29}$$

By summing up all these terms we verify that Eq.(9) is satisfied.

For a state such as in Eq.(21), $C_{\rm hs}^{nl}(\rho_{e|\mu\tau}^\alpha) = C_{hs}(\rho_{e\mu}^\alpha) + C_{hs}(\rho_{e\tau}^\alpha)$, i.e. the non-local coherence that the subsystem e shares with $\mu\tau$ is equal to the sum of the bipartite correlations that e shares with μ and τ separately. So, Eq.(9) can be written as:

$$P_{\rm hs}(\rho_e^\alpha) + C_{hs}(\rho_{e\mu}^\alpha) + C_{hs}(\rho_{e\tau}^\alpha) = \frac{1}{2}, \tag{30}$$

with $C_{hs}(\rho_{e\mu}^\alpha) = (a_{\alpha e}(t) a_{\alpha\mu}(t)^*)^2 + (a_{\alpha\mu}(t) a_{\alpha e}^*(t))^2$ and $C_{hs}(\rho_{e\tau}^\alpha) = (a_{\alpha e}(t) a_{\alpha\tau}^*(t))^2 + (a_{\alpha\tau}(t) a_{\alpha e}^*(t))^2$.

Let us now evaluate the terms of Eq.(11) for subsystem $e\mu$. By evaluating the eigenvalues of the reduced density matrices in Eq.(25) we obtain:

$$S_{vn}(\rho_{e\mu}^\alpha) = -(P_{\alpha e} + P_{\alpha\mu}) \log_2(P_{\alpha e} + P_{\alpha\mu}) - P_{\alpha\tau} \log_2 P_{\alpha\tau}, \tag{31}$$
$$P_{vn}(\rho_{e\mu}^\alpha) = 2 + P_{\alpha e} \log_2 P_{\alpha e} + P_{\alpha\mu} \log_2 P_{\alpha\mu} + P_{\alpha\tau} \log_2 P_{\alpha\tau}, \tag{32}$$
$$C_{re}(\rho_{e\mu}^\alpha) = -P_{\alpha e} \log_2 P_{\alpha e} - P_{\alpha\mu} \log_2 P_{\alpha\mu} + (P_{\alpha e} + P_{\alpha\mu}) \log_2(P_{\alpha e} + P_{\alpha\mu}). \tag{33}$$

For completeness we also evaluate the CCR terms of Eq.(11) for subsystems $e\tau$ and $\mu\tau$. For $e\tau$ bipartition we have:

$$S_{vn}(\rho_{e\tau}^\alpha) = -(P_{\alpha e} + P_{\alpha\tau}) \log_2(P_{\alpha e} + P_{\alpha\tau}) - P_{\alpha\mu} \log_2 P_{\alpha\mu}), \tag{34}$$
$$P_{vn}(\rho_{e\tau}^\alpha) = 2 + P_{\alpha e} \log_2 P_{\alpha e} + P_{\alpha\mu} \log_2 P_{\alpha\mu} + P_{\alpha\tau} \log_2 P_{\alpha\tau}, \tag{35}$$
$$C_{re}(\rho_{e\tau}^\alpha) = -P_{\alpha e} \log_2 P_{\alpha e} - P_{\alpha\tau} \log_2 P_{\alpha\tau} + (P_{\alpha e} + P_{\alpha\tau}) \log_2(P_{\alpha e} + P_{\alpha\tau}), \tag{36}$$

and for $\mu\tau$ bipartition:

$$S_{vn}(\rho_{\mu\tau}^\alpha) = -(P_{\alpha\mu} + P_{\alpha\tau}) \log_2(P_{\alpha\mu} + P_{\alpha\tau}) - P_{\alpha e} \log_2 P_{\alpha e}), \tag{37}$$
$$P_{vn}(\rho_{\mu\tau}^\alpha) = 2 + P_{\alpha e} \log_2 P_{\alpha e} + P_{\alpha\mu} \log_2 P_{\alpha\mu} + P_{\alpha\tau} \log_2 P_{\alpha\tau}, \tag{38}$$
$$C_{re}(\rho_{\mu\tau}^\alpha) = -P_{\alpha\mu} \log_2 P_{\alpha\mu} - P_{\alpha\tau} \log_2 P_{\alpha\tau} + (P_{\alpha\mu} + P_{\alpha\tau}) \log_2(P_{\alpha\mu} + P_{\alpha\tau}). \tag{39}$$

Journal of Physics: Conference Series **2533** (2023) 012049 doi:10.1088/1742-6596/2533/1/012049

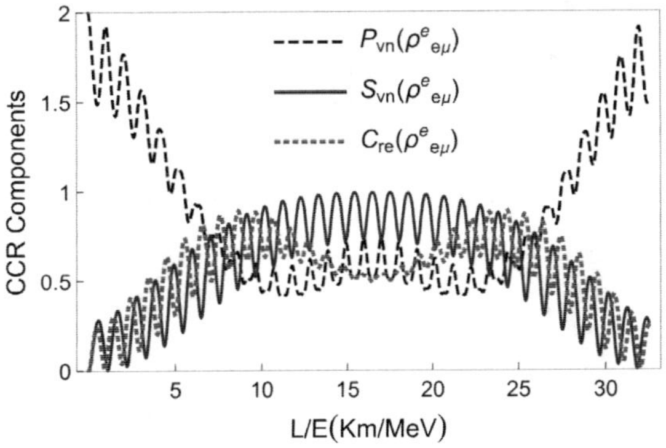

(a) $e\mu$ *subsystem*

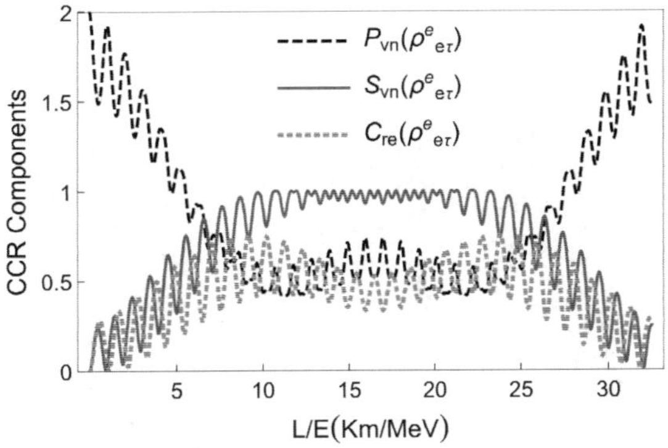

(b) $e\tau$ *subsystem*

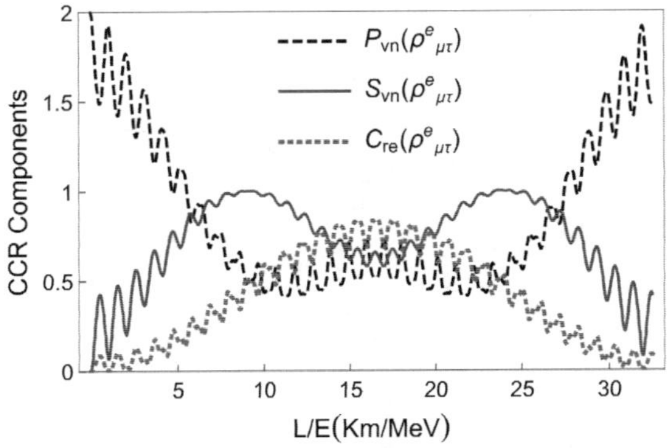

(c) $\mu\tau$ *subsystem*

Figure 3: CCR terms for bipartite subsystems $e\mu$, $e\tau$ and $\mu\tau$ as function of L/E in the case of an initial electronic neutrino.

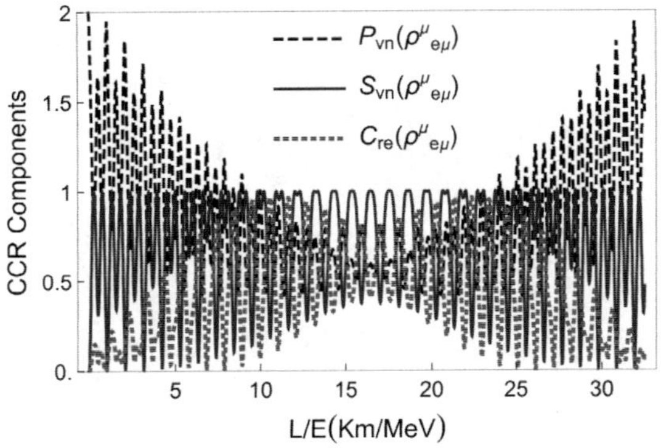

(a) $e\mu$ *subsystem*

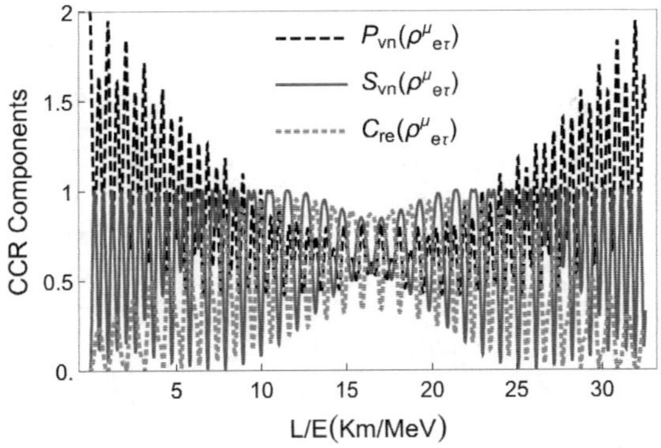

(b) $e\tau$ *subsystem*

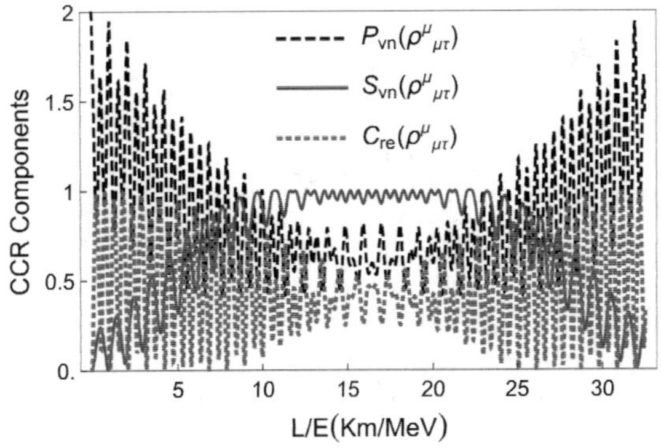

(c) $\mu\tau$ *subsystem*

Figure 4: CCR terms for bipartite subsystems $e\mu$, $e\tau$ and $\mu\tau$ as function of L/E in the case of an initial muonic neutrino.

4.1. Results for electron neutrino oscillations

Here we show the results for the case of an initial electron neutrino state, i.e. $\alpha = e$. In Fig.1 are plotted the terms of Eq.(30) as function of L/E. We can observe as the bipartite correlations between e and μ is greater than the bipartite correlations between e and τ. So, the term $C_{hs}(\rho^e_{e\mu})$ gives a greater contribution in completing the complementary relation with respect to $C_{hs}(\rho^e_{e\tau})$.

Journal of Physics: Conference Series **2533** (2023) 012049 doi:10.1088/1742-6596/2533/1/012049

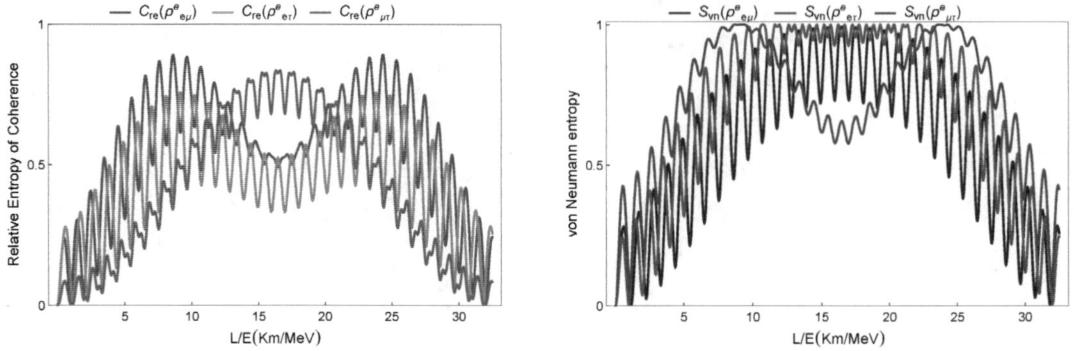

Figure 5: Comparison among $C_{re}(\rho^e_{e\mu})$, $C_{re}(\rho^e_{e\tau})$ and $C_{re}(\rho^e_{\mu\tau})$ (left panel) and $S_{vn}(\rho^e_{e\mu})$, $S_{vn}(\rho^e_{e\tau})$ and $S_{vn}(\rho^e_{\mu\tau})$ (right panel), for an electronic neutrino.

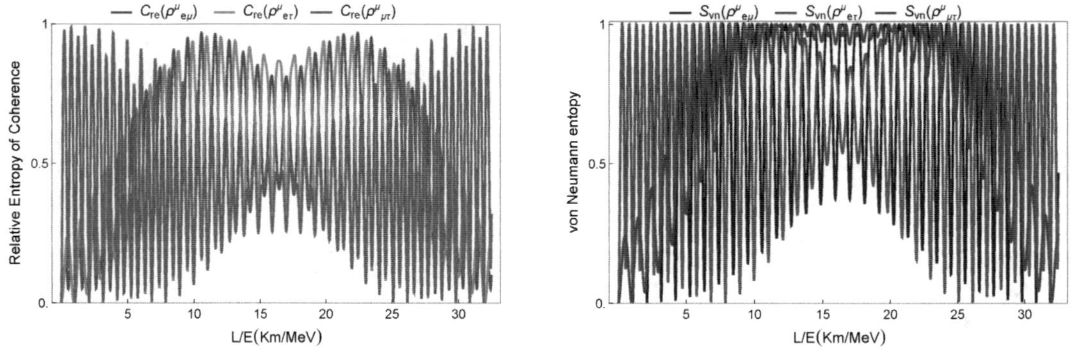

Figure 6: Comparison among $C_{re}(\rho^\mu_{e\mu})$, $C_{re}(\rho^\mu_{e\tau})$ and $C_{re}(\rho^\mu_{\mu\tau})$ (left panel) and $S_{vn}(\rho^\mu_{e\mu})$, $S_{vn}(\rho^\mu_{e\tau})$ and $S_{vn}(\rho^\mu_{\mu\tau})$ (right panel), for a muonic neutrino.

Similar considerations apply to subsystems μ and τ.

In Fig.3, the CCR terms of Eq.(11), written in terms of oscillation probabilities, for subsystems $e\mu$, $e\tau$ and $\mu\tau$ are shown as function of L/E. It is interesting to note the plateau exhibited by the von Neumann entropy in correspondence of its maximum value for $e\tau$ subsystem Fig.3(a), that persists for a relatively large range of L/E. It would be interesting to analyze this aspect for quantum information tasks. On the left panels of Fig.5 we can observe a comparison among the three bipartite local coherences $C_{re}(\rho^e_{e\mu})$, $C_{re}(\rho^e_{e\tau})$ and $C_{re}(\rho^e_{\mu\tau})$. On the right panel of Fig.5 it is shown a comparison among $S_{vn}(\rho^e_{e\mu})$, $S_{vn}(\rho^e_{e\tau})$ and $S_{vn}(\rho^e_{\mu\tau})$, representing the entanglement between subsystems $e\mu - \tau$, $e\tau - \mu$ and $\mu\tau - e$, respectively. It is worth noting how the behaviour of these terms is different depending on the bipartite subsystem considered.

4.2. Results for muon neutrino oscillations

Here we consider the case of an initial muon neutrino state, i.e. $\alpha = \mu$. In Fig.2 are shown the terms of Eq.(30) as function of L/E. We can observe that, differently to the electron case, it is difficult to recognize a dominant contribution of one of the two bipartite correlations, $C_{hs}(\rho^\mu_{e\mu})$ and $C_{hs}(\rho^\mu_{e\tau})$. It is possible to observe that the bigger the one the smaller the other, showing a sort of anti-correlation between them. However, overall, they show the same trend.

In Fig.4 the CCR terms, Eq.(11), written in terms of oscillation probabilities, for subsystems $e\mu$, $e\tau$ and $\mu\tau$ are shown as function of L/E. In this case, the plateau of the von Neumann entropy is exhibited for $\mu\tau$ subsystem (Fig.4(c)), differently to the electron case.

Journal of Physics: Conference Series **2533** (2023) 012049 doi:10.1088/1742-6596/2533/1/012049

On the left panels of Fig.6 we can observe a comparison among the three bipartite local coherences $C_{re}(\rho^{\mu}_{e\mu})$, $C_{re}(\rho^{\mu}_{e\tau})$ and $C_{re}(\rho^{\mu}_{\mu\tau})$. On the right panel of Fig.6 it is shown a comparison among $S_{vn}(\rho_{e\mu})$, $S_{vn}(\rho_{e\tau})$ and $S_{vn}(\rho_{\mu\tau})$, representing the entanglement between subsystems $e\mu - \tau$, $e\tau - \mu$ and $\mu\tau - e$, respectively. It is worth noting how the behaviour of these terms is different depending on the bipartite subsystem considered.

5. Conclusions

In this paper, we have analyzed the quantumness of a three-flavor pure neutrino state by means of the complete complementarity relations.

In particular, we focused on the quantum coherence, which can be contained either locally or in the correlations. We highlight how, in contrast to the case of a pure bipartite neutrino state, where the local coherences of subsystems are zero, for a tri-partite neutrino system the local coherences are non-vanishing and we investigate them for the three possible bipartite subsystems of the global state both for an electron and a muon neutrino system. We find a dependence of these terms on the particular bipartition chosen.

We plan to extend these consideration by using a wave packet approach for neutrino oscillation. In fact, in this case we expect that at great distances, one of the three local coherences associated with the three bipartitions will dominate the other two.

Our analysis has been performed for the case without CP violating phase, which we plan to include in future work.

References

[1] Horodecki R 2021 *Acta Phys. Pol. A* **139** 197
[2] Stancil D D et al. 2012 *Mod. Phys. Lett. A* **27** 1250077
[3] Ollivier H, Zurek W H 2001 *Phys. Rev. Lett.* **88** 017901
[4] Braumgratz T, Cramer M, Plenio M B 2014 *Phys. Rev. Lett.* **113** 140401
[5] Blasone M, Dell'Anno F, De Siena S, Illuminati F 2009 *EPL* **85** 50002
[6] Blasone M, Dell'Anno F, Di Mauro M, De Siena S, Illuminati F 2008 *Phys. Rev. D* **77** 096002
[7] Alok A K, Banarjee S, Sankar S U 2016 *Nucl. Phys. B* **909** 65
[8] Banarjee S, Alok A K, Srikanth R, Hiesmayr B C 2015 *Eur. Phys. J. C* **75**(10) 487
[9] Blasone M, Dell'Anno F, De Siena S, Illuminati F 2015 *Europhys. Lett.* **112**(2) 20007
[10] Naikoo J et al. 2020 *Nucl. Phys. B* **951** 114872
[11] Gangopadhyay D, Roy A S 2017 *Eur. Phys. J. C* **77**(4) 260
[12] Formaggio J A, Kaiser D I, Murskyj M M, Weiss T E 2016 *Phys. Rev. Lett.* **117** 050402
[13] Naikoo J, Alok A K, Banarjee S 2019 *Phys. Rev. D.* **99** 095001
[14] Wang X-Z, Ma B-Q 2022 *Eur. Phys. J. C* **82** 133
[15] Dixit K, Naikoo J, Banarjee S, Alok A K 2019 *Eur. Phys. Rev. D* **79** 96
[16] Song X-K, Huang Y, Ling J, Yung M-H 2018 *Phys. Rev. A* **98** 050302
[17] Uola R, Costa A C S, Chau Nguyen H, Uḧhne O 2020 *Rev. Mod. Phys.* **92** 15001
[18] Ming F, Song X-K, Ling L, Ye L, Wang D 2020 *Eur. Phys. J. C* **80** 275
[19] Li Y-W, Li L-J, Song X-K, Wang D 2022 *Eur. Phys. J. Plus* **137** 1267
[20] Blasone M, De Siena S, Matrella C 2022 *PoS(DISCRETE2020-2021)003*
[21] Blasone M, De Siena S, Matrella C 2021 *Eur. Phys. J. C* **81** 660
[22] Wang D, Ming F, Song X-K, Ye L, Chen J-L 2020 *Eur. Phys. J. C* **80** 800
[23] Blasone M, De Siena S, Matrella C 2022 *Eur. Phys. J. Plus* **137** 1272
[24] Basso M L W, Maziero J 2020 *J. Phys. A: Math. Theor.* **53** 465301
[25] Bittencourt V A S V, Blasone M, De Siena S, Matrella C 2022 *Eur. Phys. J. C* **82** 566
[26] Bohr N 1928 *Nature* **121** 580
[27] Englert B G 1996 *Phys. Rev. Lett.* **77** 2154
[28] Wootters W K, Zurek W H 1979 *Phys. Rev. D.* **19** 473
[29] Jakob M, Bergou J A 2010 *Opt. Commun.* **283** 827
[30] Qian X-F 2020 *Phys. Rev. Research* **2** 012016
[31] Acero M A et al. (NOvA Collaboration) 2018 *Phys. Rev. D* **98** 032012
[32] Adey D et al. (Daya Bay Collaboration) 2018 *Phys. Rev. Lett.* **121** 241805
[33] Adamson P et al. (MINOS+ Collaboration) 2020 *Phys. Rev. Lett.* **125** 131802

[34] Abe K et al. (The T2K Collaboration) 2021 *Phys. Rev. D.* **103** 112008

Journal of Physics: Conference Series **2533** (2023) 012050 doi:10.1088/1742-6596/2533/1/012050

Fermion mixing in curved spacetime

Antonio Capolupo, Gaetano Lambiase and Aniello Quaranta

Dipartimento di Fisica "E.R. Caianiello" Universita di Salerno, and INFN – Gruppo Collegato di Salerno, Via Giovanni Paolo II, 132, 84084 Fisciano (SA), Italy

E-mail: capolupo@sa.infn.it, lambiase@sa.infn.it, anquaranta@unisa.it

Abstract. We develop the quantum field theory of fermion mixing in curved spacetime and discuss the role of unitarily inequivalent representations in the particle interpretation of the theory. We derive general oscillation formulae and apply them to specific spcetimes of interest, such as spatially flat FRW metrics and the Schwarzschild spacetime. We exhibit the main deviations from the usual quantum mechanical approach.

1. Introduction

It is today accepted that neutrinos have a mass and oscillate among three flavors [1, 2, 3, 4, 5, 6, 7, 8, 9, 10, 11, 12, 13, 14]. They are the only known elementary particles to experience field mixing. These peculiarities place neutrinos beyond the standard model of particles. Many of the issues related to neutrino physics, including the origin of their mass [15], their fundamental nature [16, 17, 18, 19, 20] and the overall number of flavors [21] are still open. It is nevertheless clear that neutrinos play a fundamental role in the universe at all the scales. As they are abundantly produced in nuclear reactions, they carry important informations on astrophysical sources and they represent a valuable resource in multi-messenger astronomy [22]. Relic neutrinos, as probed by experiments like PTOLEMY [23], may be used to test cosmological theories. Several theories envisage a crucial role of neutrinos during the first phases of the universe in producing the original baryon asymmetry [24]. Furthermore neutrinos may be connected to the dark sector of the universe, contributing to dark matter [25, 26, 27, 28, 29] along with hypothetical particles such as axions [30, 31, 32], and even to dark energy [33, 34, 35, 36], for which they can also function as a probe of the underlying model [37, 38, 39]. Due to these reasons, a deep understanding of neutrino physics in gravitational backgrounds is required. The topic has been analyzed primarily in quantum mechanical approaches [40, 41, 42] both in vacuum and in matter [42, 43, 44, 45]. Here we wish to go beyond the quantum mechanical treatment and present a quantum field theoretical approach to fermion mixing in curved space. We limit ourselves to two flavors, but the formalism can be easily generalized to more families. The theory merges the features of mixing with the field quantization on curved space, bringing along several interesting aspects, such as a non-trivial structure of the vacuum. We proceed as follows: in Section 2 we introduce the flavor fields and pursue their canonical quantization; in Section 3 we derive the transition probabilities for a generic spacetime; in Section 4 we apply the formalism to some specific metrics and compute the corresponding oscillation formulae, showing their departure from their quantum mechanical counterparts; Section 5 is devoted to the conlcusions.

2. Flavor fields in curved spacetime

We start with the Dirac equations for the fields with definite masses

$$(i\gamma^{\mu}(x)D_{\mu} - m_j)\,\nu_j = 0\ , \tag{1}$$

where $j = 1, 2$, $D_{\mu} = \partial_{\mu} - \frac{i}{4}\omega_{\mu}^{AB}\sigma_{AB}$ is the spinorial covariant derivative and $\gamma^{\mu}(x) = e^{\mu}_{A}(x)\gamma^{A}$ are the curved space gamma matrices. It is understood that a tetrad $e^{\mu}_{A}(x)$, with Lorentz index $A = 0, 1, 2, 3$, is picked on the manifold, which is therefore assumed parallelizable. We will also require that the underlying manifold be globally hyperbolic, thus admitting a foliation by Cauchy surfaces Σ. The spin connection is accordingly given by $\omega_{\mu}^{AB} = e^{A}_{\nu}\Gamma^{\nu}_{\rho\mu}e^{\rho B} + e^{A}_{\nu}\partial_{\mu}e^{\nu B}$, with $\Gamma^{\nu}_{\rho\mu}$ the Christoffel symbols, and coupling to the (spacetime constant) coefficients $\sigma_{AB} = \frac{i}{2}\left[\gamma_A, \gamma_B\right]$. The Lagrangian density corresponding to Equations (1) is

$$\mathcal{L} = \sqrt{-g}\sum_{j=1,2}\bar{\nu}_j\,(i\gamma^{\mu}(x)D_{\mu} - m_j)\,\nu_j\ . \tag{2}$$

Notice that the above Lagrangian is evidently invariant under global $U(1)$ gauge transformations $\nu_j \to e^{i\beta}\nu_j$. For a given manifold we can solve the Equations (1) for "positive energy" $\phi_{k,s,j}(x)$ and "negative energy" $\psi_{k,s,j}(x)$ modes. Here k, s are generalized momentum and spin indices, while $j = 1, 2$ distinguishes the two massive fields. A crucial point is that on a generic spacetime there is, in general, no global or natural definition of "positive energy modes". One has to pick a timelike vector field $\frac{\partial}{\partial T}$ and define the energy with respect to it. While in Minkowski space this field is essentially unique, the same does not hold in a generic spacetime. Such an ambiguity is the source of the well-known particle creation phenomena in curved space (Parker effect [46], Hawking radiation [47]). Given a set of solutions $\phi_{k,s,j}, \psi_{k,s,j}$, which is complete with respect to the inner product

$$(f, g)_{\tau} = \int_{\Sigma(\tau)}\sqrt{-g}\,d\Sigma_{\mu}(\tau)\bar{f}\gamma^{\mu}(x)g \tag{3}$$

we can expand the fields ν_j as usual. Here $\tau \in \mathbb{R}$ labels the foliation by Cauchy surfaces, $g = \det g_{\mu\nu}$ and the integral is performed over the surface $\Sigma(\tau)$ with (3-)volume element $d\Sigma_{\mu}(\tau)$. We anticipate that the inner product between solutions of the same Dirac equation (that is, with the same mass) does not depend on τ, as a consequence of the $U(1)$ symmetry of the Lagrangian (2). On the other hand the product between solutions of distinct Dirac equations generally depends on τ. The fields are then expanded as

$$\nu_j(x) = \sum_{s}\int d^3k\left(a_{k,s;j}\phi_{k,s,j}(x) + b^{\dagger}_{k,s;j}\psi_{k,s,j}(x)\right) \tag{4}$$

with the (spacetime constant) coefficients a, b satisfying the canonical anticommutation relations $\left\{a_{k,s;j}, a^{\dagger}_{q,r;l}\right\} = \delta_{jl}\delta_{sr}\delta_{kq} = \left\{b_{k,s;j}, b^{\dagger}_{q,r;l}\right\}$. The next step is to introduce the flavor fields by means of an $SU(2)$ rotation through the mixing angle θ

$$\begin{aligned}\nu_e(x) &= \cos\theta\ \nu_1(x) + \sin\theta\ \nu_2(x)\\ \nu_{\mu}(x) &= -\sin\theta\ \nu_1(x) + \cos\theta\ \nu_2(x)\ ,\end{aligned} \tag{5}$$

or, which is equivalent, by means of the *generator of mixing transformations* $\mathcal{I}_{\theta}(\tau)$

$$\nu_e(x) = \mathcal{I}_{\theta}^{-1}(\tau)\nu_1(x)\mathcal{I}_{\theta}(\tau)\ ; \quad \nu_{\mu}(x) = \mathcal{I}_{\theta}^{-1}(\tau)\nu_2(x)\mathcal{I}_{\theta}(\tau)\ . \tag{6}$$

Inspection of the above equations reveals that the generator can be written as

$$\mathcal{I}_{\theta}(\tau) = \exp\left\{\theta\left[(\nu_1, \nu_2)_{\tau} - (\nu_2, \nu_1)_{\tau}\right]\right\}\ . \tag{7}$$

Journal of Physics: Conference Series **2533** (2023) 012050 doi:10.1088/1742-6596/2533/1/012050

The application of the generator on the mass fields defines the (surface-wise) expansion of the flavor fields, as

$$\begin{aligned}\nu_e(x) &= \sum_s \int d^3k \left(a_{k,s;e}(\tau)\phi_{k,s;1}(x) + b^\dagger_{k,s;e}(\tau)\psi_{k,s;1}(x)\right)\\ \nu_\mu(x) &= \sum_s \int d^3k \left(a_{k,s;\mu}(\tau)\phi_{k,s;2}(x) + b^\dagger_{k,s;\mu}(\tau)\psi_{k,s;2}(x)\right) ,\end{aligned} \tag{8}$$

where the flavor operators are by definition

$$a_{k,s;e}(\tau) = \mathcal{I}_\theta^{-1}(\tau)a_{k,s;1}\mathcal{I}_\theta(\tau) = \cos\theta a_{k,s;1} + \sin\theta \sum_r \int d^3q \left(\Lambda^*_{q,r;k,s}(\tau)a_{q,r;2} + \Xi_{q,r;k,s}(\tau)b^\dagger_{q,r;2}\right) \tag{9}$$

and similar for a_μ, b_e, b_μ. The *Bogoliubov coefficients of the mixing transformations* are given by the inner products

$$\Lambda_{q,r;k,s}(\tau) = (\phi_{q,r,2}, \phi_{k,s,1})_\tau \ ; \qquad \Xi_{q,r;k,s}(\tau) = (\psi_{k,s,1}, \phi_{q,r,2})_\tau \ . \tag{10}$$

That these coefficients define a fermionic Bogoliubov (linear canonical) transformation is ensured by the property

$$\sum_r \int d^3q \left(\Lambda^*_{k,s;q,r}(\tau)\Lambda_{k',s',q,r}(\tau) + \Xi^*_{k,s;q,r}(\tau)\Xi_{k',s',q,r}(\tau)\right) = \delta_{kk'}\delta_{ss'} \ . \tag{11}$$

The expansions of the free fields (4) and of the flavor fields (8) define two distinct Fock space representations, with the mass vacuum $|0_M\rangle$ and the flavor vacuum $|0_F(\tau)\rangle$ satisfying respectively

$$a_{k,s;i}\,|0_M\rangle = 0 = b_{k,s;i}\,|0_M\rangle \ ; \qquad a_{k,s;\alpha}(\tau)\,|0_F(\tau)\rangle = 0 = b_{k,s;\alpha}(\tau)\,|0_F(\tau)\rangle \tag{12}$$

for all $i = 1,2$ and $\alpha = e,\mu$. The appearance of a Bogoliubov transformation in the flavor operators (9) shows that the mass and the flavor representations are unitarily inequivalent. In particular, the particle content of the two vacua is different and it can be shown [10] that the flavor vacuum has the structure of a condensate of particle-antiparticle pairs with definite masses. Notice that both the flavor operators and the flavor vacuum carry an intrinsic τ dependence, regardless of the set of solutions chosen for the expansion. The one particle states $|\nu_{k,s;\alpha}(\tau)\rangle = a^\dagger_{k,s;\alpha}(\tau)\,|0_F(\tau)\rangle$ are interpreted as the states for a single neutrino of momentum k, spin s and flavor $\alpha = e,\mu$. Clearly the τ dependence of these states is due to the non-conservation of flavor by the Lagrangian (2).

3. Oscillation Formulae

In addition to the unitarily inequivalence between mass and flavor representations, it should be stressed that the flavor representations themselves are unitarily inequivalent to each other at distinct times $\tau \neq \tau'$ [10, 13]. The flavor transition amplitudes cannot be introduced by simple scalar products of the form $\langle\nu_\alpha(\tau)|\,\nu_\beta(\tau')\rangle$ because, strictly speaking, they all vanish for $\tau' \neq \tau$, as the states belong to mutually orthogonal Hilbert spaces. Rather we have to identify a suitable operator whose matrix elements give the flavor transition probability. The crucial observation is that due to the $U(1)$ invariance of the Lagrangian (2) there exists a conserved (i.e. τ independent) charge

$$\begin{aligned}Q &= \sum_{j=1,2} Q_j = \sum_{\alpha=e,\mu} Q_\alpha(\tau) ,\\ \text{with } Q_\lambda(\tau) &= \sum_s \int d^3k \left(a^\dagger_{k,s;\lambda}(\tau)a_{k,s;\lambda}(\tau) - b^\dagger_{k,s;\lambda}(\tau)b_{k,s;\lambda}(\tau)\right)\end{aligned} \tag{13}$$

and $\lambda = j, \alpha$ runs over all possible particle indices $\lambda = 1, 2, e, \mu$. Equation (13) takes into account the fact that the massive charges Q_j are separately conserved, while the flavor charges $Q_\alpha(\tau)$ are not, although their sum is. We *define* the transition probabilities as

$$P^{\alpha\to\beta}_{k,s}(\tau,\tau_0) = \langle \nu_{k,s;\alpha}(\tau_0)| \, Q_\beta(\tau) \, |\nu_{k,s;\alpha}(\tau_0)\rangle - \langle 0_F(\tau_0)| \, Q_\beta(\tau) \, |0_F(\tau_0)\rangle \ , \tag{14}$$

where τ_0 is a reference time and the subtraction of the second term on the right hand side amounts to the normal ordering with respect to $|0_F(\tau_0)\rangle$. Such a subtraction is required in order that the probabilities be well-defined. In particular, by construction, one has $P^{\alpha\to\beta}(\tau,\tau_0) \leq 1$, $\sum_{\beta=e,\mu} P^{\alpha\to\beta}(\tau,\tau_0) = 1$ and $P^{\alpha\to\beta}(\tau_0,\tau_0) = \delta^{\alpha\beta}$. It is also easy to show that Equation (14) defines a proper generalization of the standard oscillation probabilities, to which they reduce when the suitable limits are considered [13]. By trivial manipulations we arrive at the explicit expression

$$P^{e\to\mu}_{k,s}(\tau,\tau_0) = \frac{\sin^2 2\theta}{2}\left[1 - \sum_r \int d^3q \, \Re\left(\Lambda^*_{k,s;q,r}(\tau_0)\Lambda_{k,s;q,r}(\tau) + \Xi^*_{k,s;q,r}(\tau_0)\Xi_{k,s;q,r}(\tau)\right)\right] . \tag{15}$$

In addition one has $P^{e\to e}_{k,s}(\tau,\tau_0) = 1 - P^{e\to\mu}_{k,s}(\tau,\tau_0) = P^{\mu\to\mu}_{k,s}(\tau,\tau_0)$ and $P^{e\to\mu}_{k,s}(\tau,\tau_0) = P^{\mu\to e}_{k,s}(\tau,\tau_0)$. We conclude the section with some remarks about the validity of equations (14). It should be noted that the whole construction assumes a given expansion of the mass fields (4) and thus a specific choice of the set of solutions for the Dirac equation. Clearly one can consider several distinct choices for the mass field representation, and perform the same construction arriving at the transition probabilities of eqs. (14) and (15). In general both the mass vacuum $|0'_M\rangle$ and the corresponding flavor vacua $|0'_F(\tau)\rangle$, as well as the related Fock spaces, shall have different interpretations for different choices of the solutions set. Nonetheless covariance demands that the *local* observables be the same for whatever representation of the mass fields is chosen. This, in turn, implies a specific relation among the flavor representations built out of distinct mass representations [13]. On the other hand the transition probabilities are not, strictly speaking, local observables, and they generally vary according to the mass representation chosen. The conditions under which the probabilities are left invariant by changes of mass representation are more restrictive than those required for covariance. For a complete discussion we refer to [13].

4. Applications

We can now move on to apply the formulae (15) to some spacetimes of interest. We skip the computation of the Minkowskian and quantum mechanical limits, which are trivial, and can be found in [13]. For a specific metric, the computation essentially amounts to the determination of the Bogoliubov coefficients (10), which are then combined as in eq. (15) to extract the transition probabilities. Then we are only limited by the knowledge of the solutions to the Dirac equations. There is a class of metrics (spatially flat Friedmann-Robertson-Walker (FRW) spacetimes) which are of interest in cosmology and for which there is often an exact analytical solution of the Dirac equation. The general form of the spatially flat FRW metric is, in a rectangular coordinate system,

$$ds^2 = dt^2 - a^2(t)\left(dx^2 + dy^2 + dz^2\right) . \tag{16}$$

The function $a(t)$ is known as the scale factor. According to the shape of $a(t)$, the metric of eq. (16) describes various phases in the evolution of a isotropic, homogeneous and spatially flat universe. The spatial dependence of the solutions to the Dirac equations corresponding to the metric (16) is trivial. Due to the spatial translation symmetry we can put

$$\phi_{\boldsymbol{k},s;j}(x) = u_{\boldsymbol{k},s;j}(t)e^{i\boldsymbol{k}\cdot\boldsymbol{x}} \ ; \qquad \psi_{\boldsymbol{k},s;j}(x) = v_{\boldsymbol{k},s;j}(t)e^{i\boldsymbol{k}\cdot\boldsymbol{x}} \ . \tag{17}$$

Here the generalized momentum index has been explicitly substituted with the 3-vector $\boldsymbol{k}$ and $\boldsymbol{x}$ denotes the spatial 3-vector $\boldsymbol{x} \equiv (x, y, z)$. The 4-component spinors $u_{\boldsymbol{k},s;j}(t), v_{\boldsymbol{k},s;j}(t)$ depend only on time. We can anticipate from eqs. (17) that the Bogoliubov coefficients will have the form

$$\Lambda_{\boldsymbol{k},s;\boldsymbol{q},r}(t) = \delta_{rs}\delta^3(\boldsymbol{k}-\boldsymbol{q})\Lambda_{k,s}(t) \; ; \qquad \Xi_{k,s;\boldsymbol{q},r}(t) = \delta_{rs}\delta^3(\boldsymbol{k}+\boldsymbol{q})\Xi_{k,s}(t) \, . \tag{18}$$

The natural choice for the foliation is given by the constant t surfaces and we take the energy to be defined with respect to $\frac{\partial}{\partial t}$.

- $a(t) = e^{Ht}$. The exponential evolution is suited to describe inflation and epochs dominated by the cosmological constant. This form can also be considered as an approximate solution, for a suitable value of H, in time ranges where the Hubble rate $\frac{\dot{a}}{a}$ does not vary sensibly in time. The Hubble parameter H has dimensions of an energy and its value depends on the epoch considered. The solution of the Dirac equation for this scale factor can be found in [48]. The corresponding Bogoliubov coefficients are

$$\begin{aligned}\Lambda_{k,s}(t) &= \frac{\pi k e^{-Ht}}{2H\sqrt{\cos(\frac{i\pi m_2}{H})\cos(\frac{i\pi m_1}{H})}}\left[J^*_{v_2}\left(\frac{k}{H}e^{-Ht}\right)J_{v_1}\left(\frac{k}{H}e^{-Ht}\right)+J^*_{v_2-1}\left(\frac{k}{H}e^{-Ht}\right)J_{v_1-1}\left(\frac{k}{H}e^{-Ht}\right)\right]\\ \Xi_{k,s}(t) &= \frac{\pi k e^{-Ht}}{2H\sqrt{\cos(\frac{i\pi m_2}{H})\cos(\frac{i\pi m_1}{H})}}\left[J^*_{v_1}\left(\frac{k}{H}e^{-Ht}\right)J_{-v_2}\left(\frac{k}{H}e^{-Ht}\right)-J^*_{v_1-1}\left(\frac{k}{H}e^{-Ht}\right)J_{1-v_2}\left(\frac{k}{H}e^{-Ht}\right)\right]\end{aligned} \tag{19}$$

where $v_j = \frac{1}{2}\left(1+\frac{2im_j}{H}\right)$ and $J_v(x)$ denotes the Bessel function of the first kind of order v. Insertion in Eq. (15) yields the result

$$\begin{aligned}P^{e\to\mu}_{k,s}(t,t_0) &= 2\cos^2\theta\sin^2\theta\Bigg\{1-\frac{\pi^2k^2e^{-H(t+t_0)}}{4H^2\cos(\frac{i\pi m_2}{H})\cos(\frac{i\pi m_1}{H})}\\ &\times\Re\Bigg[\left[J_{v_2}(\eta_0)J^*_{v_1}(\eta_0)+J_{v_2-1}(\eta_0)J^*_{v_1-1}(\eta_0)\right]\left[J^*_{v_2}(\eta)J_{v_1}(\eta)+J^*_{v_2-1}(\eta)J_{v_1-1}(\eta)\right]\\ &+\left[J_{v_1}(\eta_0)J^*_{-v_2}(\eta_0)-J_{v_1-1}(\eta_0)J^*_{1-v_2}(\eta_0)\right]\left[J^*_{v_1}(\eta)J_{-v_2}(\eta)-J^*_{v_1-1}(\eta)J_{1-v_2}(\eta)\right]\Bigg]\Bigg\}\,,\end{aligned} \tag{20}$$

with $\eta = \frac{k}{H}e^{-Ht}$ and $\eta_0 = \frac{k}{H}e^{-Ht_0}$. To understand the qualitative behaviour of Eq. (20) we have plotted it for sample values of masses and momenta in the left panel of (Fig. 1). We notice that for small times the oscillations display an interference pattern similar to that of coupled harmonic oscillators. As time is increased, the transition probabilities gradually converge to a flat space–like oscillation and the interference pattern eventually disappears. In the right panel of (Fig. 1) we show a qualitative comparison with the Pontecorvo formulae, for different sample values of masses and momenta.

- $a(t) = a_0 t^{\frac{1}{2}}$. This evolution of the scale factor corresponds to a universe dominated by radiation. The constant a_0 has dimensions of $[\text{mass}]^{\frac{1}{2}}$. The solution for the corresponding equation may be found in [48] The mixing coefficients read

$$\begin{aligned}\Lambda_{k,s}(t) &= \frac{1}{\sqrt[4]{4m_1m_2t^2}}e^{-\frac{\pi k^2(m_1+m_2)}{4m_1m_2a_0^2}}\Bigg\{W^*_{\kappa_2,\frac{1}{4}}(-2im_2t)W_{\kappa_1,\frac{1}{4}}(-2im_1t)\\ &+ \frac{4k^2}{m_1m_2a_0^2t}\left[\frac{1}{4}W^*_{\kappa_2,\frac{1}{4}}(-2im_2t)-\frac{1}{8}\left(1+\frac{ik^2}{m_2a_0^2}\right)W^*_{\kappa_2-1,\frac{1}{4}}(-2im_2t)\right]\\ &\times \left[\frac{1}{4}W_{\kappa_1,\frac{1}{4}}(-2im_1t)-\frac{1}{8}\left(1-\frac{ik^2}{m_1a_0^2}\right)W_{\kappa_1-1,\frac{1}{4}}(-2im_1t)\right]\Bigg\}\end{aligned} \tag{21}$$

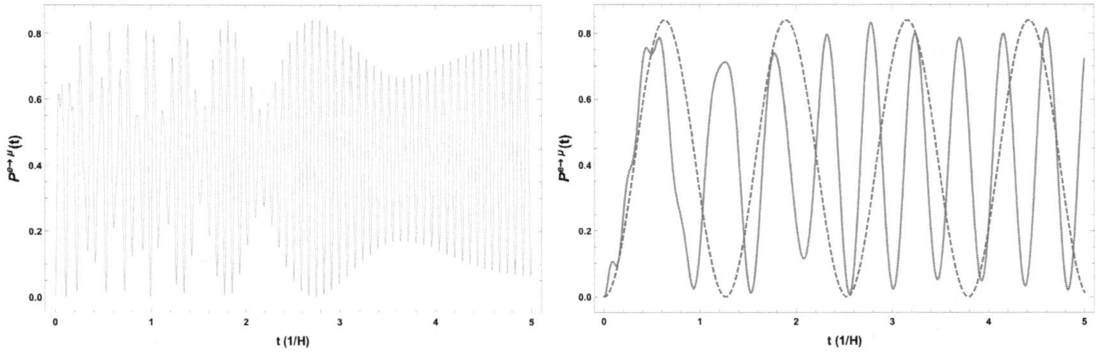

Figure 1. Plots of the oscillation formulae from Eq. (20) for different sample values of masses and momenta, chosen in order to highlight the qualitative behaviour of Eq. (20). Masses and momenta are expressed in units of H, time is expressed in units of H^{-1}. (Left panel) Plot of the $\nu_e - \nu_\mu$ transition probability Eq. (20) as a function of time. Here we have used the sample values $\sin^2(\theta) = 0.3, k = 30, m_1 = 1, m_2 = 80$ and $t_0 = 0$. (Right panel) Plot of the oscillation formulae from Eq. (20) (blue solid line) and from the Pontecorvo formulae (orange dashed line) for sample values of masses and momenta. Here we have used $\sin^2(\theta) = 0.3, k = 20, m_1 = 1, m_2 = 15$ and $t_0 = 0$.

$$\begin{aligned}\Xi_{k,s}(t) &= (-1)^s \frac{k}{\sqrt[4]{2m_1(2m_2)^3 a_0^4 t^2}} e^{-\frac{\pi k^2(m_1+m_2)}{4m_1 m_2 a_0^2}} \Big\{ W^*_{\kappa_1,\frac{1}{4}}(-2im_1t) W_{-\kappa_2,\frac{1}{4}}(2im_2t) \\ &+ \frac{k^2}{m_1 m_2 a_0^2 t}\left[\frac{1}{4} W^*_{\kappa_1,\frac{1}{4}}(-2im_1t) - \frac{1}{8}\left(1+\frac{ik^2}{m_1 a_0^2}\right) W^*_{\kappa_1-1,\frac{1}{4}}(-2im_1t)\right] \\ &\times \left[W_{-\kappa_2,\frac{1}{4}}(2im_2t) + \frac{2im_2a_0^2}{k^2} W_{-\kappa_2+1,\frac{1}{4}}(2im_2t)\right]\Big\} \end{aligned} \quad (22)$$

where $W_{\kappa,\mu}(z)$ are the Whittaker functions [49] and $\kappa_j = \frac{1}{4}\left(1+\frac{2ik^2}{a_0^2 m_j}\right)$ for $j = 1, 2$. The transition probabilities are

$$\begin{aligned}&P^{e\to\mu}_{k,s}(t) = 2\cos^2\theta \sin^2\theta \Big\{ 1 - \Re\Big[\frac{1}{\sqrt[2]{4m_1 m_2 t_0 t}} e^{-\frac{\pi k^2(m_1+m_2)}{2m_1 m_2 a_0^2}} \Big\{ W_{\kappa_2,\frac{1}{4}}(-2im_2t_0) W^*_{\kappa_1,\frac{1}{4}}(-2im_1t_0) \\ &+ \frac{4k^2}{m_1 m_2 a_0^2 t_0}\left(\frac{1}{4}W_{\kappa_2,\frac{1}{4}}(-2im_2t_0) - \frac{1}{8}\left(1-\frac{ik^2}{m_2a_0^2}\right)W_{\kappa_2-1,\frac{1}{4}}(-2im_2t_0)\right) \\ &\times \left(\frac{1}{4}W^*_{\kappa_1,\frac{1}{4}}(-2im_1t_0) - \frac{1}{8}\left(1+\frac{ik^2}{m_1a_0^2}\right)W^*_{\kappa_1-1,\frac{1}{4}}(-2im_1t_0)\right)\Big\}\Big\{ W^*_{\kappa_2,\frac{1}{4}}(-2im_2t) W_{\kappa_1,\frac{1}{4}}(-2im_1t) \\ &+ \frac{4k^2}{m_1 m_2 a_0^2 t}\left(\frac{1}{4}W^*_{\kappa_2,\frac{1}{4}}(-2im_2t) - \frac{1}{8}\left(1+\frac{ik^2}{m_2a_0^2}\right)W^*_{\kappa_2-1,\frac{1}{4}}(-2im_2t)\right) \\ &\times \left(\frac{1}{4}W_{\kappa_1,\frac{1}{4}}(-2im_1t) - \frac{1}{8}\left(1-\frac{ik^2}{m_1a_0^2}\right)W_{\kappa_1-1,\frac{1}{4}}(-2im_1t)\right)\Big\} \\ &+ \frac{k^2}{\sqrt[2]{2m_1(2m_2)^3 a_0^4 t_0 t}} e^{-\frac{\pi k^2(m_1+m_2)}{2m_1 m_2 a_0^2}} \Big\{ W_{\kappa_1,\frac{1}{4}}(-2im_1t_0) W^*_{-\kappa_2,\frac{1}{4}}(2im_2t_0) \end{aligned}$$

Journal of Physics: Conference Series **2533** (2023) 012050 doi:10.1088/1742-6596/2533/1/012050

$$
\begin{aligned}
&+\frac{k^2}{m_1 m_2 a_0^2 t_0}\left(\frac{1}{4}W_{\kappa_1,\frac{1}{4}}(-2im_1t_0)-\frac{1}{8}\left(1-\frac{ik^2}{m_1a_0^2}\right)W_{\kappa_1-1,\frac{1}{4}}(-2im_1t_0)\right)\\
&\times\left(W^*_{-\kappa_2,\frac{1}{4}}(2im_2t_0)-\frac{2im_2a_0^2}{k^2}W^*_{-\kappa_2+1,\frac{1}{4}}(2im_2t_0)\right)\Bigg\}\Bigg\{W^*_{\kappa_1,\frac{1}{4}}(-2im_1t)W_{-\kappa_2,\frac{1}{4}}(2im_2t)\\
&+\frac{k^2}{m_1 m_2 a_0^2 t}\left(\frac{1}{4}W^*_{\kappa_1,\frac{1}{4}}(-2im_1t)-\frac{1}{8}\left(1+\frac{ik^2}{m_1a_0^2}\right)W^*_{\kappa_1-1,\frac{1}{4}}(-2im_1t)\right)\\
&\times\left(W_{-\kappa_2,\frac{1}{4}}(2im_2t)+\frac{2im_2a_0^2}{k^2}W_{-\kappa_2+1,\frac{1}{4}}(2im_2t)\right)\Bigg\}\Bigg]\Bigg\}\,. \qquad (23)
\end{aligned}
$$

The expression for the transition probabilities, eq. (23) is quite involved. In order to get some insight we plot the eq. (23) for sample values of masses and momenta in (Fig. 2). Despite the analytical complexity, the probabilities of eq. (23) share some common features

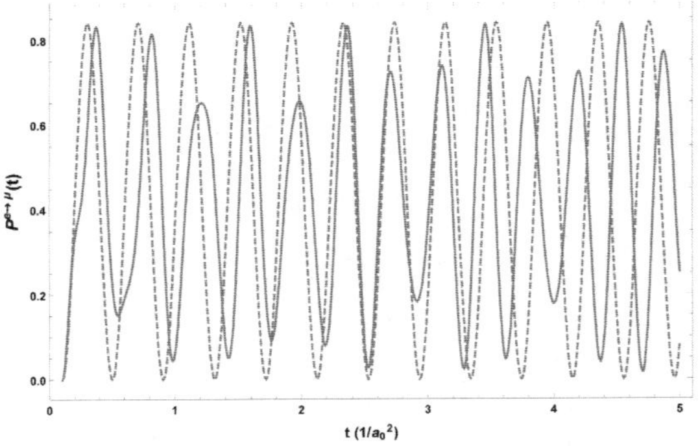

Figure 2. Plots of the $\nu_e - \nu_\mu$ transition probability as a function of time as from Eq. (23) (blue solid line) and from the Pontecorvo formulae (orange dashed line) for sample values of massed and momenta, chosen in order to highlight the qualitative behaviour of Eq. (23). Masses and momenta are expressed in units of a_0^2, time is expressed in units of a_0^{-2}. We have used the sample values $\sin^2(\theta) = 0.3, k = 5, m_1 = 1, m_2 = 20$ and $t_0 = 0.1$.

with those of eq. (20). Both show amplitude and phase variations with respect to the (flat) Pontecorvo oscillation formulae. In particular the amplitude variations present in eqs. (23) and (20) are a distinctive feature of quantum field theory in curved spacetime. The latter cannot be obtained in the quantum mechanical limit, which in curved space modifies the Pontecorvo formulae only in the phase of the oscillations [42].

As a further application we consider the (static) Schwarzschild metric

$$
ds^2 = \left(1-\frac{2GM}{r}\right)dt^2 - \left(1-\frac{2GM}{r}\right)^{-1}dr^2 - r^2 d\Omega\,. \qquad (24)
$$

Unfortunately the corresponding Dirac equation has no exact analytical solution. Nonetheless we can work out an approximate form of the transition probabilities for the propagation of neutrinos from the asymptotic past to the asymptotic future. We shall not give the details of the calculation here, and we refer to [13] for a complete treatment. The trick is to consider the asymptotic solutions of the Dirac equation at the future and past infinities and the relations among them. One is then able to write down the Bogoliubov coefficients in terms of the asymptotic (flat) solutions. The final result is

$$
\begin{aligned}
&P^{e\to\mu}_{\omega}(m,n) \approx\\
&2\cos^2\theta\sin^2\theta\,\Bigg(1-\sqrt{[1-F_H(\omega)]\,[1-F_H(\omega')]}\,\Big[|U_{\omega;\omega'}|^2\cos(\Delta^-_{\omega;m,n})+|V_{\omega;\omega'}|^2\cos(\Phi^-_{\omega;m,n})\Big]\\
&+\sqrt{F_H(\omega)\,[1-F_H(\omega')]}\,|U_{\omega;\omega'}||V_{\omega;\omega'}|\,\Big[\cos(\Theta^-_{\omega;m,n})-\cos(\Psi^-_{\omega;m,n})\Big]\\
&+\sqrt{F_H(\omega')\,[1-F_H(\omega)]}\,|U_{\omega;\omega'}||V_{\omega;\omega'}|\,\Big[\cos(\Psi^+_{\omega;m,n})-\cos(\Theta^+_{\omega;m,n})\Big]\\
&-\sqrt{F_H(\omega)F_H(\omega')}\,\Big[|U_{\omega;\omega'}|^2\cos(\Delta^+_{\omega;m,n})+|V_{\omega;\omega'}|^2\cos(\Phi^+_{\omega;m,n})\Big]\Bigg)\,. \qquad (25)
\end{aligned}
$$

Here m, n are integers labeling two families of surfaces approaching respectively the past and future timelike infinity (the limit $m, n \to \infty$ is understood). The phase factors depend on the details of the surfaces considered, as well as the energy. The quantities $U_{\omega\omega'}$ and $V_{\omega\omega'}$ denote the mixing Bogoliubov coefficients in flat space in terms of the energies $\omega, \omega' = \sqrt{\omega^2 + m_2^2 - m_1^2}$. Finally $F_H(\omega) = (1 + e^{\frac{\omega}{k_B T_H}})^{-1}$ is the Fermi-Dirac distribution at the Hawking temperature $T_H = \frac{1}{8\pi GM}$. The most remarkable feature of eq. (25) is the appearance of the Hawking temperature, which shows how Hawking radiation directly affects the propagation of mixed fermions on a black hole spacetime. The typical behavior of eq. (25) is shown in (Fig. 3).

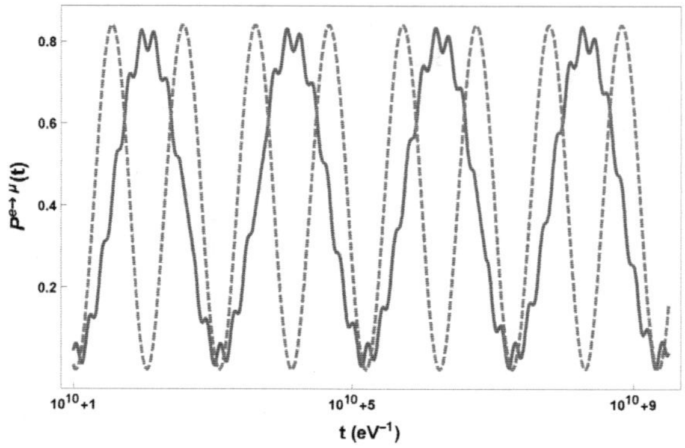

Figure 3. Plot of the $\nu_e - \nu_\mu$ transition probability from Eq. (25) (blue solid line) and from the Pontecorvo oscillation formulae (red dashed line) for late times and sample values of masses, momenta and Hawking Temperature. The phases in (25) have been chosen so as to match the flat space phases for simplicity, $\Delta^{\pm}_{\omega;m,n} \to \frac{\omega_2 - \omega_1}{2}(t \pm t_0)$, $\Phi^{\pm}_{\omega;m,n} \to \frac{\omega_2 + \omega_1}{2}(t \pm t_0)$, $\Psi^{\pm}_{\omega;m,n} \to \frac{\omega_2}{2}(t \pm t_0) + \frac{\omega_1}{2}(t \mp t_0)$, $\Theta^{\pm}_{\omega;m,n} \to \frac{\omega_2}{2}(t \pm t_0) - \frac{\omega_1}{2}(t \mp t_0)$, where t and t_0 denote respectively the future and past hypersurfaces. We have used the sample values $\sin^2(\theta) = 0.3$, $k = 30$ eV, $m_1 = 1$ eV, $m_2 = 20$ eV, $t_0 = 0$, $k_B T_H = 10^{-10}$ eV and t in the range $[10^{10} + 1, 10^{10} + 9.5]$ eV^{-1}.

5. Conclusions

We have constructed the quantum field theory of (two flavor) fermion mixing in curved spacetime. We have derived general oscillation formulae and applied them to several spacetimes of interest. The theory has a considerably richer structure when compared to its flat spacetime and its quantum mechanical counterparts, and joins the peculiarities of field mixing with the inherent ambiguity of field quantization in curved spacetime. Such a complexity is mirrored in the oscillation formulae, which involve a non-trivial evolution in both phase and amplitude. A similar analysis can also be set up for boson mixing [50, 51] The formalism developed here is suited for the analysis of neutrino oscillations in situations where gravity plays an important role. That includes extreme astrophysical environments, such as the primordial universe and black holes. Moreover the theory allows one to study the flavor vacuum in an arbitrary spacetime. Due to its condensate structure, the flavor vacuum yields a non-vanishing contribution to the energy momentum tensor of matter, inducing a new source term on the right hand side of the Einstein field equations. Recent developments [28] have shown that in cosmological spacetimes the flavor vacuum behaves as a perfect fluid with the dust (or cold dark matter) equation of state $w = 0$. This gives rise to the intriguing possibility that a pure field theoretical effect may contribute to dark matter. Future studies shall explore the link between field mixing and dark matter in a wider class of metrics.

Acknowledgements
Partial financial support from MUR and INFN is acknowledged. A.C. and G.L. also acknowledge the COST Action CA1511 Cosmology and Astrophysics Network for Theoretical Advances and Training Actions (CANTATA).

References

[1] (SNO Collaboration) Ahmad Q R et al. 2001 *Phys. Rev. Lett.* **87** 071301
[2] (Super-Kamiokande Collaboration) Fukuda Y et al. 1998 *Phys. Rev. Lett.* **81** 1562
[3] Bilenky S M and Pontecorvo B 1978 *Phys. Rep.* **41** 225
[4] Bilenky S M and Petcov S T 1987 *Rev. Mod. Phys.* **59** 671
[5] Fujii K, Habe C and Yabuki T 1999 *Phys. Rev.* D **59** 113003
[6] Fujii K, Habe C and Yabuki T 2001 *Phys. Rev.* D **64** 013011
[7] Hannabuss K C and Latimer D C 2000 *J. Phys.* A **33** 1369
[8] Hannabuss K C and Latimer D C 2003 *J. Phys.* A **36** L69
[9] Ji C-R and Mishchenko Y 2002 *Phys. Rev.* D **65** 096015
[10] Blasone M, Capolupo A and Vitiello G 2002 *Phys. Rev.* D **66** 025033
[11] Capolupo A, Ji C-R, Mishchenko Y and Vitiello G 2004 *Phys. Lett.* B **594** 1-2, 135-140
[12] Blasone M, Capolupo A, Terranova F and Vitiello G 2005 *Phys. Rev.* D **72** 013003
[13] Capolupo A, Lambiase G and Quaranta A 2020 *Phys. Rev.* D **101** 095022
[14] Blasone M and Vitiello G 1995 *Annals of Physics* **244** 2, 283-311
[15] Cai Y, Garcia J H, Schmidt M A, Vicente A and Volkas R R 2017 *Front. Phys.* **5**
[16] Bilenky S M, Hošek J and Petcov S T 1980 *Phys. Lett.* B **94** 495-498
[17] Capolupo A, Giampaolo S M and Lambiase G 2019 *Phys. Lett.* B **792** 298
[18] Buoninfante L, Capolupo A, Giampaolo S M and Lambiase G 2020 *EPJ* C **80** 1009
[19] Capolupo A, Giampaolo S M, Lambiase G and Quaranta A 2020 *Universe* **2020** 6(11), 207
[20] Capolupo A, Giampaolo S M, Hiesmayr B C, Lamnbiase G and Quaranta A 2022 On the geometric phase for Majorana and Dirac neutrinos *J. Phys.* G
[21] (Particle Data Group) Tanabashi M et al. 2018 *Phys. Rev.* D **98** 030001
[22] Franckowiak A 2017 *J. Phys.: Conf. Ser.* **888** 012009
[23] (PTOLEMY Collaboration) Betti M G et al. 2019 *JCAP* **07** 047
[24] Buchmüller W 2002 Grand Unification and Leptogenesis *Preprint* hep-ph/0204288
[25] Capolupo A 2016 *Adv. High En. Phys.* **2016** 8089142
[26] Capolupo A 2018 *Adv. High En. Phys.* **2018** 9840351
[27] Capolupo A, Giampaolo S M, Lambiase G and Quaranta A 2020 *EPJ* C **80** 423
[28] Capolupo A, Carloni S and Quaranta A 2022 *Phys. Rev.* D **105** 105013
[29] Capolupo A and Quaranta A 2022 Neutrino capture on tritium as a probe of flavor vacuum condensate and dark matter *Preprint* hep-ph/2205.09640
[30] Marsh D J E 2016 *Phys. Rep.* **643**, 1-79
[31] Capolupo A, Lambiase G, Quaranta A and Giampaolo S M 2020 *Phys. Lett.* B **804** 135407
[32] Capolupo A, Giampaolo S M and Quaranta A 2021 *EPJ* C **81** 1116
[33] Fardon R, Nelson A E and Weiner N 2004 *JCAP* **2004 (10)** 005
[34] Capolupo A, Capozziello S and Vitiello G 2009 *Phys. Lett.* A **373** 6, 601-610
[35] Capolupo A, Capozziello S and Vitiello G 2007 *Phys. Lett.* A **363** 1-2, 53-56
[36] Capolupo A, Capozziello S and Vitiello G 2008 *Int. J. Mod. Phys.* A **23** 31, 4979-4990
[37] Kaplan D B, Nelson A E and Weiner N *Phys. Rev. Lett.* **93** 091801
[38] Khalifeh A R and Jimenez R 2022 *Phys. Dark Univ.* **37** 101063
[39] Khalifeh A R and Jimenez R 2021 *Phys. Dark Univ.* **34** 100897
[40] Grossman Y and Lipkin J H 1997 *Phys. Rev.* D **55** 2760
[41] Piriz D, Roy M and Wudka J 1996 *Phys. Rev.* D **54** 1587
[42] Cardall C Y and Fuller G M 1997 *Phys. Rev.* D **55** 7960
[43] Simonov K, Capolupo A and Giampaolo S M 2019 *EPJ* C **79** 902
[44] Capolupo A, Giampaolo S M and Quaranta A 2021 *EPJ* C **81** 410
[45] Capolupo A, Giampaolo S M and Quaranta A 2021 *Phys. Lett.* B **820** 136489
[46] Parker L 1969 *Phys. Rev.* **183** 1057
[47] Hawking S W 1975 *Comm. Math. Phys.* **43** 3, 199-220
[48] Barut A O and Duru H 1987 *Phys. Rev.* D **36** 3705
[49] Abramowitz M and Stegun I A 1965 *Handbook of Mathematical Functions* (New York: Dover Publications Inc.)

Journal of Physics: Conference Series **2533** (2023) 012050 doi:10.1088/1742-6596/2533/1/012050

[50] Blasone M, Capolupo A, Romei O and Vitiello G 2001 *Phys. Rev.* D **63** 125015
[51] Capolupo A, Quaranta A and Setaro P A 2022 *Phys. Rev.* D **106** 043013

AUTHOR INDEX

Anastopoulos, Charis ... 69
Armata, Federico ... 521
Benninger, Felix ... 202
Bittencourt, V A S V ... 339, 589
Blasone, M ... 339, 350, 589
Blasone, Massimo ... 146
Bonilla–Licea, Moise ... 501
Butera, Salvatore ... 408, 521
Butterfield, Jeremy ... 15, 42
Capolupo, Antonio ... 430, 539, 599
Capurso, Alessandro ... 546
Carusotto, Iacopo ... 408
Chataignier, Leonardo ... 388
Cohen, Eliahu ... 270
Corli, Sebastiano ... 91
De Bianchi, Silvia ... 1
De Siena, S ... 589
Demir, Durmuş ... 378
Dieguez, P R ... 359
Diósi, Lajos ... 86
Döner, M Kemal ... 322
Doyen, G. ... 557, 567
Drakova, D. ... 557, 567
Ellis, George F R ... 283
Gabbanelli, Luciano ... 1
Galli, Davide E ... 91
Giampaolo, Salvatore Marco ... 539
Gomes, Henrique ... 15, 42
Grangier, Philippe ... 112
Großardt, André ... 322
Hooft, Gerard't ... 215
Hu, Bei-Lok ... 69
Iorio, A ... 398
Iorio, Alfredo ... 312, 420
Ivetić, Boris ... 312
Jizba, P. ... 528
Kanatchikov, Igor V ... 469
Karthik, H S ... 359
Khrennikov, Andrei ... 202
Kiefer, Claus ... 462
Klauder, John R ... 307
Koltko, K ... 579
Konishi, Kenichi ... 123
Kothawala, Dawood ... 167
Kryukov, Alexey A ... 330
Lambiase, G. ... 350, 528
Lambiase, Gaetano ... 430, 452, 599
Lloyd, Seth ... 156
Luciano, G G ... 350, 528
Mandrin, Pierre A ... 443
Matrella, C ... 589
Mavromatos, Nick E. ... 254
Mignemi, Salvatore ... 312
Minguzzi, E ... 479
Montalbano, Federico ... 521
Moro, Lorenzo ... 91
Noto, Antonio ... 511
Pais, Pablo ... 312, 420
Passante, Roberto ... 511, 521
Petruzziello, L ... 350, 528
Prati, Enrico ... 91
Quaranta, Aniello ... 539, 599
Rizzuto, Lucia ... 511, 521
Rozenberg, Marcelo ... 99
Savvidou, Konstantina ... 69
Scardigli, Fabio ... 452
Schuch, Dieter ... 501
Shor, Oded ... 202
Simonov, Kyrylo ... 493
Singh, Tejinder P. ... 183
Smaldone, L ... 398
Smaldone, Luca ... 146
Spagnolo, Salvatore ... 511
Stoica, Ovidiu Cristinel ... 368
Tedesco, Antonio ... 430
Van Den Bossche, Mathias ... 112
Vardanyan, Tatevik ... 462
Vitiello, Giuseppe ... 146
Wetterich, C ... 225
Zanfardino, G ... 339